Thermodynamics
Concepts and Applications

SECOND EDITION

Fully revised to match the more traditional sequence of course materials, this full-color second edition presents the basic principles and methods of thermodynamics using a clear and engaging style and a wealth of end-of-chapter problems.

It includes five new chapters on topics such as mixtures, psychrometry, chemical equilibrium, and combustion; and the discussion of the second law of thermodynamics has been expanded and divided into two chapters, allowing instructors to introduce the topic using either the cycle analysis in Chapter 6 or the definition of entropy in Chapter 7.

Online ancillaries, including a password-protected solutions manual, figures in electronic format, prepared PowerPoint lecture slides, and instructional videos, are available at www.cambridge.org/thermo.

STEPHEN R. TURNS is Professor Emeritus of Mechanical Engineering at The Pennsylvania State University. His research interests include combustion-generated air pollution, other combustion-related topics, and engineering education pedagogy. Turns was the recipient of the 2009 ASEE Mechanical Engineering Division Ralph Coats Roe Award and has been the recipient of several teaching awards at Pennsylvania State. He is the author of *An Introduction to Combustion: Concepts and Applications*, 3rd edn (also appearing in Chinese, Korean, and Portuguese editions), and of *Thermal-Fluid Sciences: An Integrated Approach*. Turns is a fellow of the American Society of Mechanical Engineering (ASME).

LAURA L. PAULEY is Professor of Mechanical Engineering at The Pennsylvania State University. She teaches courses in the thermal sciences and conducts research in engineering education and computational fluid dynamics. Pauley has worked on grants focusing on aspects of engineering education including classroom experiments, presentation skills, and the climate for women in engineering. She has also initiated efforts in curriculum reform and experimental case studies for required courses in mechanical engineering. She is the recipient of the Undergraduate Program Leadership Award (2003) and was selected as the first Arthur L. Glenn Professor of Engineering Education. Pauley is a fellow of the American Society of Mechanical Engineering (ASME).

Thermodynamics

Concepts and Applications

SECOND EDITION

Stephen R. Turns
The Pennsylvania State University

Laura L. Pauley
The Pennsylvania State University

CAMBRIDGE
UNIVERSITY PRESS

University Printing House, Cambridge CB2 8BS, United Kingdom

One Liberty Plaza, 20th Floor, New York, NY 10006, USA

477 Williamstown Road, Port Melbourne, VIC 3207, Australia

314–321, 3rd Floor, Plot 3, Splendor Forum, Jasola District Centre, New Delhi – 110025, India

79 Anson Road, #06–04/06, Singapore 079906

Cambridge University Press is part of the University of Cambridge.

It furthers the University's mission by disseminating knowledge in the pursuit of education, learning, and research at the highest international levels of excellence.

www.cambridge.org
Information on this title: www.cambridge.org/9781107179714
DOI: 10.1017/9781316840979

First published 2020

Printed in Singapore by Markono Print Media Pte Ltd

A catalogue record for this publication is available from the British Library.

Library of Congress Cataloging-in-Publication Data
Names: Turns, Stephen R., author. | Pauley, Laura L., author.
Title: Thermodynamics : concepts and applications / Stephen R. Turns (Pennsylvania State University), Laura L. Pauley (Pennsylvania State University).
Description: Second edition. | Cambridge ; New York, NY : Cambridge University Press, 2018. | Includes index.
Identifiers: LCCN 2018030834 | ISBN 9781107179714 (hardback : alk. paper)
Subjects: LCSH: Thermodynamics. | Fluid mechanics.
Classification: LCC TJ265 .T86 2018 | DDC 621.402/1–dc23
LC record available at https://lccn.loc.gov/2018030834

ISBN 978-1-107-17971-4 Hardback

This book is dedicated to Peter Gordon (SRT)
and to
Anne, Mark, and Eric Pauley (LLP).

Brief Contents

Contents

Preface to the Second Edition

This second edition of *Thermodynamics: Concepts and Applications* results from a significant reorganization and expansion of the first edition. The first edition was first published in 2006 by Steve Turns only and was well received by students and instructors alike. We received encouraging comments and helpful feedback from readers and this has been a great motivation to prepare a new and improved version with updated text and illustrations. Laura Pauley was invited as the second author in this edition to contribute an additional approach to the introduction of second-law analysis and to provide additional examples, applications, and problems.

As in the first edition, the purpose of the new edition is to introduce undergraduate students to the basic principles, aspects, and methods of thermodynamics. However, we have:

- reorganized the text to follow the order in which material tends to be presented in a thermodynamics course. For instance, entropy properties are not defined until they are used in Chapter 7.
- expanded the second-law material, dividing it into two chapters (Chapters 6 and 7). This is now presented in such a way that instructors can decide whether to introduce this topic in Chapter 6, using the cycle analysis, or in Chapter 7, starting from the definition of entropy.
- added additional cycles to Chapter 9 (which was previously Chapter 8), including: closed feedwater heaters, cogeneration cycles, combined cycles, Otto and diesel cycles.
- moved mixtures to Chapter 10, humidity to Chapter 11, and combustion to Chapters 12 and 13.
- increased the number of end-of-chapter problems from 899 in the first edition to 1363 in this new edition.
- reorganized and labeled the chapter problems by section in each chapter.
- added FE problems to the chapters covering topics on the FE exam.
- added EES or computer problems to Chapters 2 and 4–9. These problems frequently include a parameter study. The EES problems are often presented as a series of problems in which one case is solved by hand, that same problem is solved in EES, and then the EES program is adapted for a parameter study. This gives the student a structure for good analysis methods that includes checking with hand calculations.
- included a hierarchical chart of topics at the beginning of each chapter to orient the student.

Features

As in the first edition, this new edition includes a set of pedagogical features to enhance students' motivation and learning, as well as to make teaching easier and more effective for the instructor. Feedback from instructors who use this book is most welcome.

Chapter Overviews

These appear at the beginning of each chapter and include: a list of learning objectives, a chapter introduction, and a brief historical perspective (where appropriate); they conclude with a brief summary of the topics covered in each chapter. These sections will guide students as they are introduced to and study the thermodynamics concepts in each chapter.

Figures and Photos

Figures have been designed to convey the key concepts in a clear and self-explanatory way. An abundance of color photographs and images illustrates important concepts and emphasizes practical applications.

In-Text Examples

Each chapter contains many examples that follow a standard problem-solving format. The examples apply the concepts of the section and also demonstrate a methodical solution method that should be followed by the student.

Self-Tests

Self-tests follow most of the in-text examples. These self-tests allow students to solve a problem that is similar to the in-text example. This gives students an opportunity to work through the same solution steps as in the example, to gain confidence in problem solving.

Problems

A large number of hierarchically arranged problems, expanded from the previous edition, are included at the end of each chapter. The problems have been organized by section titles to assist instructors and students. The National Institute of Science and Technology (NIST) WebBook and *miniREFPROP* software for thermodynamic properties are used extensively.

Review Questions

New thought-provoking conceptual questions, instructional videos, problems requiring EES/MATLAB or other software, and FE exam practice problems are included.

Key Equations

These are highlighted by a colored background to allow students to easily find these equations and structure their learning. Instructors might specify that all homework solutions must start from a key equation that is highlighted. This can clarify which equations can be used directly and which equations must be derived in a solution.

Appendices for Additional Coverage

Appendices at the end of the book provide additional information on the topics explored in the main text.

Chapter Summaries

A section at the end of each chapter summarizes the key points developed in the chapter so that students can, at a glance, confirm that they have understood the key take-away messages. These key concepts and definitions are linked to specific end-of-chapter questions and problems.

SI Units

SI units have been consistently used throughout the book.

Organization

Chapter 1 provides motivation for studying thermodynamics by presenting many applications and introduces key definitions and concepts to provide a framework for further study. Chapter 2 develops the concept of state relations and presents various equations of state (P, $\mathcal{V}$, T relations) and calorific equations of state (h, u, etc., T, P relations). Ideal gases, real gases, and multi-phase substances (with emphasis on liquid–vapor systems) are all treated here. Second-law properties are dealt with in later chapters. Chapter 3 focuses entirely on the conservation of mass for open and closed systems, with examples reinforcing the property relations from Chapter 2. Chapter 4 provides a detailed discussion of energy, heat interactions, and work transfers, setting the stage for the study in Chapter 5 of the first law of thermodynamics.

The second-law content has been expanded and split into two chapters. The two chapters are written to allow the second law to be introduced in two different ways. Chapter 6 introduces the second-law analysis for cycles, whereas Chapter 7 introduces second-law analysis starting with the definition of entropy for a reversible isothermal process. The earliest sections of Chapter 7 are written without reference to Chapter 6 so that an instructor can start with Chapter 7 and then consider cycle analysis in Chapter 6.

Chapter 8 presents steady-flow devices, and Chapter 9 deals with how these devices can be combined to provide systems for power production, propulsion, and heating and cooling. Chapter 9 now contains traditional analyses of gas-power systems, e.g., the Otto and diesel cycles.

Topics that are likely to be covered in a second course in thermodynamics have been unembedded, expanded, and moved to separate chapters: mixtures (Chapter 10), psychrometry (Chapter 11), combustion (Chapter 12), and chemical equilibrium (Chapter 13).

Online Resources

A companion website at www.cambridge.org/thermo contains the following online ancillaries:

Accessible to All:

- Instructional Videos to assist students in developing a deep conceptual understanding of the first law and ideal-gas processes
- Instructions
- Work Sheet Links
- NIST software

Password-Protected:

- Solutions Manual
- Image Gallery
- Lectures Slides in two formats. One format includes all lecture content. The second format has open sections that can be filled using a tablet PC during a class lecture.
- Test Bank

Preface to the First Edition

WHY ANOTHER THERMODYNAMICS BOOK? With a number of excellent thermodynamics textbooks available, what purpose is served by publishing another? The answer to these important questions lies in the origins of this book. Although the subject matter focuses almost entirely on traditional thermodynamics topics, the structure of the book puts it in the broader context of the thermal-fluid sciences. The following specific features provide this integrated feel:

- Chapter 3 is devoted entirely to the conservation of mass principle and deals with some concepts traditionally treated in fluid mechanics textbooks.
- Rate laws for conduction, convection, and radiation heat transfer are introduced in Chapter 4—Energy and Energy Transfer. These rate laws are used subsequently in examples in this and later chapters.
- The concept of head loss, a usual topic in fluid mechanics, is introduced in Chapter 5 and utilized in examples.

A second feature also sets this book apart from other engineering thermodynamics textbooks: a hierarchical structure. The following arrangements illustrate the hierarchical arrangement of subject matter:

- In Chapter 2, essentially all material related to thermodynamic properties is grouped together. In this way, we are able to show clearly the hierarchy of thermodynamic state relationships: starting with the basic equation of state involving P, $\mathcal{V}$, and T; adding first-law-based calorific equations of state involving u, h, P, T, and $\mathcal{V}$; and ending with the second-law-based state relationships involving s, T, P, and $\mathcal{V}$, for example. Chapter 2 also treats properties of ideal gas mixtures and introduces standardized properties for reacting mixtures. Such an arrangement requires that Chapter 2 be revisited at appropriate places in the study of later chapters. In this sense, Chapter 2 is a resource that is to be returned to many times.
- Element conservation, a topic central to reacting systems, is considered in Chapter 3 as just one of many ways of expressing conservation of mass.
- Constant-pressure and constant-volume combustion are considered as topics within Chapter 5, Energy Conservation.
- Chemical and phase equilibria are treated as a consequence of the second law of thermodynamics and developed within Chapter 6; therefore, all topics related to the second law are found hierarchically in a single chapter.

What purpose is served by such an arrangement? First, it provides an important structure for a beginning learner. Experts who have mastered and work within a discipline organize material this way in their minds, while novices tend to treat concepts in an undifferentiated way as a collection of seemingly unrelated topics.[1]

[1] Larkin, J., McDermott, J., Simon, D. P., Simon, H. A., "Expert and Novice Performance in Solving Physics Problems," *Science*, 208: 1335–1342 (1980).

Even through all topics will not be covered sequentially in a course, when students revisit chapters they will see the connections between new and previously covered material. It is hoped that providing a useful hierarchy from the start may speed learning and aid in retention. A second reason for a hierarchical arrangement is flexibility. In general, the book has been designed to permit an instructor to select topics from within a chapter and combine them with material from other chapters in a relatively seamless manner. This flexibility allows the book to be used in many ways, depending upon the educational goals of a particular course, or a sequence of courses. To assist in selecting topics, the text distinguishes three levels; level 1 (basic) material is unmarked, level 2 (intermediate) material appears with a blue background and a blue edge stripe, and level 3 (advanced) material is denoted with a light red background and a red edge stripe. Instructors can therefore choose from the numerous topics presented to create courses that meet their specific educational objectives. The syllabus following the table of contents illustrates how the book might be used in a traditional, one-semester, engineering thermodynamics class.

In addition to structure, many other pedagogical devices are employed in this book. These include the following:

- An abundance of color photographs and images illustrate important concepts and emphasize practical applications;
- Each chapter begins with a list of learning objectives, a chapter overview, and a brief historical perspective, where appropriate, and concludes with a brief summary;
- Each chapter contains many examples that follow a standard problem-solving format;
- Self-tests follow most examples;
- Key equations are denoted by colored backgrounds;
- Each chapter concludes with a checklist of key concepts and definitions linked to specific end-of-chapter questions and problems;
- The National Institute of Science and Technology (NIST) database for thermodynamic and transport properties (included in the NIST12 v.5.2 software provided with the book) is used extensively.

All of these features are intended to enhance student motivation and learning and to make teaching easier for the instructor. For example, the many color photographs make connections to real-world devices, a strong motivator for undergraduate students. Also, the learning objectives and checklists are particularly useful. For the instructor, they aid in the selection of homework problems and the creation of quizzes and exams, or other instructional tools. For students, they can be used as self-tests of comprehension and can monitor progress. The checklists also cite topic-specific questions and problems. In his use of the book, the author utilizes the learning objectives to guide reviews of the material prior to examinations. Having well-defined learning objectives is also useful in meeting engineering accreditation requirements. Many questions and problems are included at the end of each chapter. The purpose of the questions is to reinforce conceptual understanding of the material and to provide an outlet for students to articulate such understanding. Throughout the book, students are encouraged to use the National Institute of Science and Technology (NIST) databases to obtain thermodynamic properties. The online NIST property database is easily accessible and is a powerful resource. It is a tool that will always be up to date.

The NIST12 v.5.2 software included with this book has features not available online. This user-friendly software provides extensive property data for 18 fluids and has an easy-to-use plotting capability. This invaluable resource makes dealing with properties easy and can be used to enhance student understanding.

Feedback from instructors who use this book is most welcome.

Acknowledgements to the Second Edition

Many thanks go to the people who helped create both the first and second editions of this book. Thanks goes to Chris Mordaunt for his creation of the self-tests and his meticulous reading of the manuscript and insightful comments and to the members of the solutions manual team: Jacob Stenzler and Dave Kraige, leaders of the effort, and Justin Sabourin, Yoni Malchi, and Shankar Narayanan. From the outset, Mary Newby deciphered pencil scrawls to create a word-processed manuscript. We thank Mary for her invaluable efforts.

We thank Eric Lemmon at NIST for revising *miniREFPROP* for our needs, and we acknowledge fellow textbook authors Dwight Look, Jr., Harry Sauer, Jr, and Glen Myers for permission to use selected problems from their works. We acknowledge the support of the National Science Foundation for the development of the video exercises and the efforts of the following in their creation: the first author's colleagues, Peggy Van Meter, Carla Firetto, and Tom Litzinger; undergraduate students Leigh Lesnick, Jordan Chaklos-Sracic, and Herschel Pangborn; and graduate students Chelsea Cameron and Charlyn Shaw.

Three people were critical to the creation of this book. The first is Peter Gordon, who came to the rescue in trying times and breathed life into this project. Second is Dick Benson, without whose enthusiastic support this book would not have been possible. Third, but hardly last, is Joan, the first author's wife. She cannot be thanked enough for her help, patience, and support.

Conversion Factors

Energy	1 J	$= 9.478\ 17 \times 10^{-4}$ Btu
		$= 2.388\ 5 \times 10^{-4}$ kcal
	1 Btu	$= 778.17$ ft·lb_f
Energy rate	1 W	$= 3.412\ 14$ Btu/hr
		$= 1.341 \times 10^{-3}$ horsepower
	1 ton (refrigeration)	= 3.517 kW
		= 12,000 Btu/hr
	1 hp	$= 550$ ft·lb_f/s
Force	1 N	$= 0.224\ 809\ lb_f$
Heat flux	1 W/m^2	$= 0.317\ 1$ Btu/(hr·ft^2)
Kinematic viscosity and diffusivities	1 m^2/s	$= 3.875 \times 10^4\ ft^2/hr$
Length	1 m	= 39.370 in
		= 3.280 8 ft
Mass	1 kg	$= 2.204\ 6\ lb_m$
	1 slug	$= 32.174\ lb_m$
Mass density	1 kg/m^3	$= 0.062\ 428\ lb_m/ft^3$
Mass flow rate	1 kg/s	$= 7936.6\ lb_m/hr$
Pressure	1 Pa	$= 1\ N/m^2$
		$= 0.020\ 885\ 4\ lb_f/ft^2$
		$= 1.450\ 4 \times 10^{-4}\ lb_f/in^2$
		$= 4.015 \times 10^{-3}$ in water
	0.1 MPa	= 1 bar
	1 atm	= 101.325 kPa
		= 760 mm Hg
		= 29.92 in Hg
		= 14.70 psia
Specific heat	1 J/kg·K	$= 2.388\ 6 \times 10^{-4}$ Btu/(lb_m·R)
Temperature	K	= (5/9) R
		= (5/9) (F + 459.67)
		= °C + 273.15
Time	3600 s	= 1 hr
Velocity	1 m/s	= 2.237 miles/hr
Viscosity	1 N · s/m^2	= 1 Pa·s

		$= 1$ kg/s·m
		$= 2419.1\ lb_m/ft \cdot hr$
		$= 5.8016 \times 10^{-6}\ lb_f \cdot hr/ft^2$
		$= 1000$ centipoise (CP)
Volume	$1\ m^3$	$= 1000$ liters
		$= 264.2$ gallons (US liq.)
		$= 35.31\ ft^3$

Nomenclature

$\mathbf{a}$	Acceleration vector (m/s^2)
a	Specific availability or specific energy (J/kg)
a	Specific Helmholtz free energy (J/kg) or van der Waals constant ($Pa{\cdot}m^6/kmol^2$)
a	Speed of sound (m/s)
A	Area (m^2) or Helmholtz free energy (J)
$\mathscr{A}$	Availability or energy
A/F	Air–fuel ratio (kg_{air}/kg_{fuel})
b	Van der Waals constant ($m^3/kmol$)
c	Specific heat (J/kg·K)
c_p	Constant-pressure specific heat (J/kg·K)
$\bar{c}_p$	Molar constant-pressure specific heat (J/kmol·K)
c_v	Constant-volume specific heat (J/kg·K)
$\bar{c}_v$	Molar constant-volume specific heat (J/kmol·K)
C	Heat-capacity rate (J/K·s)
C	Number of components (dimensionless)
C_p	Diffuser pressure-recovery coefficient (dimensionless)
COP	Coefficient of performance
D	Diameter (m)
e	Specific energy (J/kg)
$\bar{e}$	Molar-specific energy (J/kmol)
E	Energy (J)
$\dot{E}$	Energy rate (W)
f	Arbitrary function
F	Force (N)
F	Degrees of freedom (dimensionless)
$\mathbf{F}$	Force vector (N)
$\mathbf{F}_t$	Thrust force (N)
F/A	Fuel–air ratio (kg_{fuel}/kg_{air})
g	Gravitational acceleration (m/s^2) 9.80665 m/s^2
g	Specific Gibbs function (J/kg)
$\bar{g}$	Molar-specific Gibbs function (J/kmol)
G	Gibbs function (J)
h	Specific enthalpy (J/kg)
$\bar{h}$	Molar-specific enthalpy (J/kmol)
$\bar{h}_{conv}$	Average heat-transfer coefficient ($W/m^2{\cdot}K$ or °C)
H	Enthalpy (J)
HHV	Higher heating value (J/kg_{fuel})
$\overline{HHV}$	Higher heating value ($J/kmol_{fuel}$)
HV	Heating value (J/kg)

HX	Heat exchanger
i	Electrical current (A)
I	Rotational moment of inertia (kg/m^2)
k	Thermal conductivity (W/m·K)
k_B	Boltzmann constant, 1.380649×10^{-23} (J/K)
ke	Specific kinetic energy (J/kg)
K_p	Equilibrium constant (dimensionless)
KE	Kinetic energy (J)
$\dot{KE}$	Kinetic energy rate (W)
ℓ_{mf}	Mean free path (m)
L	Length (m)
LHV	Lower heating value (J/kg$_{fuel}$)
$\overline{LHV}$	Lower heating value (J/kmol$_{fuel}$)
$\dot{m}$	Mass flow rate (kg/s)
$\dot{m}''$	Mass flux (kg/s·m^2)
m_u	Unified atomic mass unit (kg)
M	Mass (kg)
$\mathcal{M}$	Molecular weight (kg/kmol)
M	Number of tube-bundle passes
n	Number of particles
n	Polytropic exponent (dimensionless)
N	Number of moles (kmol)
N	Number of tubes in a bundle
$\mathcal{N}_{AV}$	Avagodro's number (see Eq. 2.8)
pe	Specific potential energy (J/kg)
P	Pressure (N/m^2 or Pa) or chart parameter
P_i	Partial pressure (Pa)
$\mathcal{P}$	Power (J/s or W)
$\mathcal{P}$	Number of phases (dimensionless)
PE	Potential energy (J)
q	Heat transfer per unit mass (J/kg)
Q	Heat (or heat transfer, or heat interaction) (J)
$\dot{Q}$	Heat transfer rate (J/s or W)
$\dot{Q}''$	Heat flux (heat transfer rate per unit area) (W/m^2)
r	Radius (m)
R, R_i	Particular gas constant (J/kg·K)
R_P	Pressure ratio (dimensionless)
R_e	Electrical resistance (ohm)
R_u	Universal gas constant, 8314.472 (J/kmol·K)
s	Specific entropy (J/kg·K)
$\mathbf{s}$	Distance vector (m)
$\bar{s}$	Molar-specific entropy (J/kmol·K)
S	Entropy (J/K)
$\mathcal{S}_{gen}$	Entropy generated by irreversibilities (J/K)
$\dot{\mathcal{S}}_{gen}$	Rate of entropy generation by irreversibilities (J/K·s)
sfc	Specific fuel consumption for power-producing engines (kg/J)
SFC	Specific fuel consumption for thrust-producing engines [(kg/s)/N]
t	Time (s)

T	Temperature (K)
$\mathcal{T}$	Torque (N·m)
u	Specific internal energy (J/kg)
$\bar{u}$	Molar-specific internal energy (J/kmol)
U	Internal energy (J)
$\bar{v}_{\text{molec}}$	Mean molecular speed (m/s)
v	Specific volume (m^3/kg)
$\bar{v}$	Molar-specific volume (m^3/kmol)
V	Velocity or speed (scalar) (m/s)
$\mathbf{V}$	Velocity (vector) (m/s)
V, v	Volume (m^3)
$\dot{V}$	Volumetric flow rate (m^3/s)
$\mathcal{V}$	Voltage (V)
w	Work per mass (J/kg)
W	Work (N·m or J)
$\dot{W}$	Rate of working or power (J/s or W)
x	Quality (dimensionless)
x	Axial coordinate or distance (m)
x	Moles of carbon per mole of fuel (kmol/kmol)
X	Mole fraction (dimensionless)
y	Spatial coordinate (m)
y	Perpendicular distance from surface (m)
y	Extracted fraction (dimensionless)
Y	Mass fraction (dimensionless)
z	Elevation (m)
z	Spatial coordinate (m)
z	Vertical coordinate or distance (m)
Z	Compressibility factor, $P\bar{V}/RT$ (dimensionless)
Z	Ratio of element atoms or moles (dimensionless)

Greek

α	Fraction dissociated
β	Arbitrary property
β	Coefficient of performance
γ	Specific-heat ratio c_p/c_v (dimensionless)
δ	Increment along specific path (e.g., δW or δQ)
Δ	Difference or increment (e.g., $\Delta E = E_2 - E_1$)
ΔG_T^o	Standard-state Gibbs function change at temperature T (J/kmol)
ΔH_c	Enthalpy of combustion (J)
ΔH_R	Enthalpy of reaction (J)
Δh_R	Enthalpy of reaction (or of combustion) per mass of fuel (J/kg_{fuel})
Δh_c	Heat of combustion or heating value (J/kg_{fuel})
ε	Emissivity (dimensionless)
η	Efficiency
η_s	Isentropic efficiency of a turbine, pump, or compressor
η_{th}	Thermal efficiency
κ	Proportionality constant

μ	Chemical potential (J/kmol)
v_j	Stoichiometric coefficient of species j
ρ	Density (kg/m^3)
σ	Stefan–Boltzmann constant (5.670400×10^{-8} $W/m^2{\cdot}K^4$)
τ	Shear stress (N/m^2)
ϕ	Relative humidity
Φ	Equivalence ratio (dimensionless)
ω	Humidity ratio or specific humidity
ω	Angular velocity (rad/s)
Ω	Angle (rad)

Subscripts

a	actual
A	air
abs	absolute
act	actual
ad	adiabatic
atm	atmospheric
avg	average
b	downstream receiver
B	boundary (control surface)
bulk	bulk, or associated with the system or control volume as a whole
c	critical
c	characteristic
c	compressor
C	cold fluid
CF	counterflow
cond	conduction
conv	convection
cv	open system (control volume)
DB	dry bulb
e	exit plane
e	exhaust
elec	electrical
f	final
f	fluid (liquid)
f	formation
F	fuel
fg	difference between saturated vapor and saturated liquid states
flow	associated with a flow
fric	friction
g	gas (vapor)
g	saturated vapor
gage	gage
gen	generated within system
gen	generator
H	high-temperature reservoir

H	hot fluid
HX	heat exchanger
i	species i
i	initial
i	inner
in	inlet, into system
init	initial
internal	internal or associated with the constituent molecules, etc.
irrev	irreversible
isen	isentropic
j	index for inlets
k	index for outlets
L	low-temperature reservoir
ℓ	liquid
liq	liquid
max	maximum
max	fluid with maximum heat capacity rate
mech	mechanical
min	fluid with minimum heat capacity rate
mix	mixture
molec	molecule
n	nozzle
o	outer
OA	overall
out	outlet, out of system
Ox	oxidizer
P	pressure
P, prod	products
P	products of combustion
p	pump or pumping device
PF	parallel flow
rad	radiation
reac	reactants
ref	reference state or value
rev	reversible
rot	rotational
s	isentropic
s	sensible
s	surface
sat	saturated state
SF	steady flow
shaft	shaft
S-T	shell-and-tube
stoich	stoichiometric
stored	stored within system
sublim	sublimation
surr	surroundings
sys	system

t	turbine or throat
th	thermal
tot	total
trans	translational
v	vapor
vac	vacuum
vap	vapor
vib	vibrational
visc	viscous
wall	wall, at the wall
WB	wet bulb
x-sec	cross-sectional
1	state or station 1
2	state or station 2
∞	ambient or freestream value

Superscripts

$^{\circ}$	Denotes standard-state, e.g. pressure (P° = 1 atm)
$'$	Product

Other

$\overline{(\ )}$	Average value

CHAPTER 1
Beginnings

LEARNING OBJECTIVES

After studying Chapter 1, you should:

- Have a basic idea of what thermodynamics is and the kinds of engineering problems to which it applies.
- Be able to distinguish between a closed system and an open system (control volume).
- Have an understanding of and be able to state the formal definitions of thermodynamic properties, states, processes, and cycles.
- Understand the concept of thermodynamic equilibrium and its requirement of simultaneously satisfying thermal, mechanical, phase, and chemical equilibria.
- Be able to explain the meaning of a quasi-equilibrium process.
- Understand the distinction between primary dimensions and derived dimensions, and the distinction between dimensions and units.
- Be able to convert SI units of force, mass, energy, and power to US customary units, and vice versa.

CHAPTER 1 OVERVIEW

IN THIS CHAPTER, we introduce and define the subject of thermodynamics. We also introduce three complex practical applications of our study of thermodynamics: the fossil-fueled steam power plant, jet engines, and the spark-ignition reciprocating engine. To set the stage for more detailed developments later in the book, several of the most important concepts and definitions are presented here. These include the concepts of: open and closed thermodynamic systems; thermodynamic properties, states, and cycles; and equilibrium and quasi-equilibrium processes. The chapter concludes with an organizational overview of engineering thermodynamics and presents some ideas of how you might optimize the use of this textbook based on your particular educational objectives.

1.1 What Is Thermodynamics?

Thermodynamics is one of three disciplines known collectively as the thermal-fluid sciences, sometimes as just the thermal sciences: thermodynamics, heat transfer, and fluid mechanics. We begin with a dictionary definition [1] of thermodynamics:

***Thermodynamics* is the science that deals with the relationship of heat and mechanical energy and conversion of one into the other.**

The Greek roots, *therme*, meaning heat, and *dynamis*, meaning power or strength, suggest a more elegant definition: *the power of heat*. In its common usage in engineering, thermodynamics has come to mean the broad study of energy and its various interconversions from one form to another. Figure 1.1 illustrates a few examples that motivate our study of thermodynamics.

FIGURE 1.1 Examples of energy conversion systems: Hybrid engine converts energy stored in chemical bonds of fuel molecules or from batteries to shaft power (left) (GreenPimp / E+ / Getty images); photovoltaic solar panels and wind turbines convert solar radiation and wind energy to electricity, respectively (right) (GIPhotoStock / Cultura / Getty Images).

1.2 Some Applications

Practical applications of thermodynamics abound. Biological systems provide many examples. Consider yourself as a thermodynamic system. All of your physical activities require energy transformation. Energy stored in the chemical bonds of foodstuffs is transformed to power temperature regulation, respiration, blood circulation, muscle movements, and other body functions. Electronic devices are ubiquitous. Consider your smart phone. Charging the phone involves converting electrical energy from a wall outlet to chemical energy stored in the phone's battery. The energy in the battery then powers the electronic circuits and is ultimately transported to the surroundings as thermal energy. Another natural example of the application of thermodynamics is provided by the physical processes associated with the weather. Radiant energy from the sun heats the ground, which in turn heats the air and results in thermals and downdrafts. The evaporation and condensation of water to form clouds involve thermodynamic processes. Some of the problems at the end of this chapter focus on everyday applications of thermodynamics.

We introduce the following three practical applications, which we use as recurring themes throughout the book:

- Steam power plants,
- Spark-ignition engines, and
- Jet engines.

These applications, and others, provide a practical context for our study of thermodynamics. Many of the examples presented in subsequent chapters revisit these specific applications, as do many of the end-of-chapter problems. Where these particular examples appear, a note reminds the reader that the example relates to one of these three themes.

1.2a Steam Power Plants

There are many reasons to choose the steam power plant as an application of thermodynamics. First, and foremost, is the overwhelming importance of such power plants to our daily existence. Imagine how your life would be changed if you did not have access to electrical power (Fig. 1.2) or, less severely, if electricity had to be rationed so that it would be available to you only a few hours each day! It is easy to forget the blessings of essentially limitless electrical power available to residents of the United States. From Table 1.1, we see that the combustion of fossil fuels is the dominant source of our electricity; approximately 63.4% of the electricity produced in the United States in 2018 had its origin in the combustion of a fossil fuel, that is, coal, gas, or oil. Figure 1.3 shows a German coal-fired power plant. Nuclear power is the second largest source with approximately 19.3% of the total generation. We also note that, with 8.2%, the combined amount of electricity generated by solar and wind power has increased by almost a factor of six from 2008 to 2018. A second reason for our choice of steam power plants as an integrating application is the historical significance of steam power. The science of thermodynamics was born, in part, from a desire to understand and improve the earliest steam engines. John Newcomen's first coal-fired steam engine in 1712 (Fig. 1.4) predates the discovery of the fundamental principles of thermodynamics by more than a hundred years! The idea later to be

TABLE 1.1 Electricity Generation in the United States for 2018 [2]

Source		Billion kW·hr	%
Fossil fuels			
Coal		1146.4	27.4
Petroleum		24.6	0.6
Natural gas		1468.0	35.1
Other gases		12.2	0.3
	Subtotal	2651.2	63.4
Nuclear		807.1	19.3
Hydro pumped storage		–5.9	–0.1
Renewables			
Hydro		291.7	7.0
Wood		41.4	1.0
Waste		21.4	0.5
Geothermal		16.7	0.4
Solar		66.6	1.6
Wind		275.0	6.6
	Subtotal	712.8	17.1
Other		12.6	0.3
TOTAL		4177.8	100.0

FIGURE 1.2 A complex electrical transmission grid transmits electricity from power plants to users throughout the USA. How to deal with the intermittent power production from solar and wind energy sources is a major concern associated with integrating these renewable sources into the electrical grid (Jeff_Hu / iStock / Getty Images Plus).

known as the second law of thermodynamics was published by Sadi Carnot in 1824; and Julius Mayer first presented the conservation of energy principle, or the first law of thermodynamics, in 1842.

A timeline of important people and events in the history of the thermal sciences is presented in Appendix A.

FIGURE 1.3 Modern coal-fired power plant (Germany) (acilo / E+ / Getty Images).

In this chapter, we present the basic steam power plant cycle and illustrate some of the hardware used to accomplish this cycle. In subsequent chapters, we will add devices and complexity to the basic cycle. Figure 1.5 shows the basic steam power cycle, or Rankine cycle. (William Rankine (1820–1872), a Scottish engineer, was the author of the *Manual of the Steam Engine and Other Prime Movers* (1859) and made significant contributions to the fields of civil and mechanical engineering.) Water

FIGURE 1.4 Newcomen's first coal-fired steam engine (PRISMA ARCHIVO / Alamy Stock Photo).

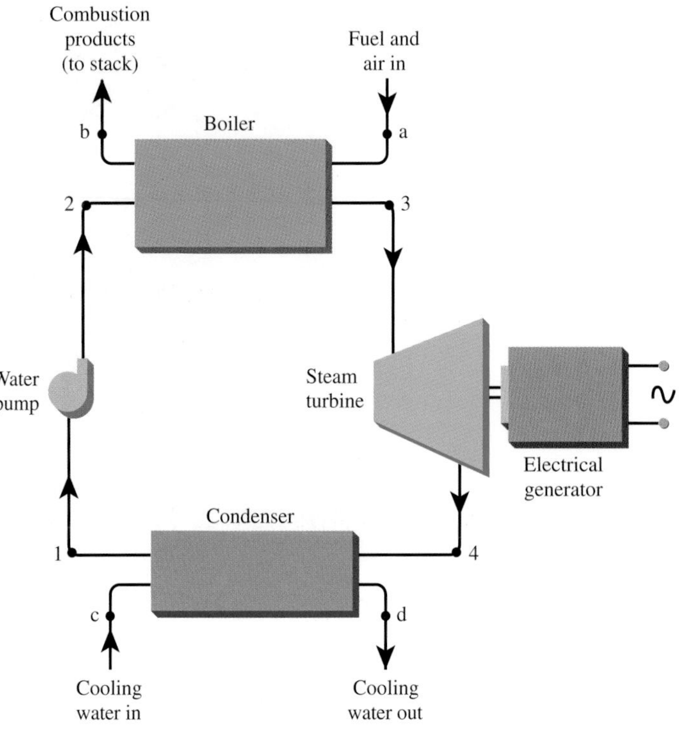

FIGURE 1.5 This basic steam power cycle is also known as the Rankine cycle.

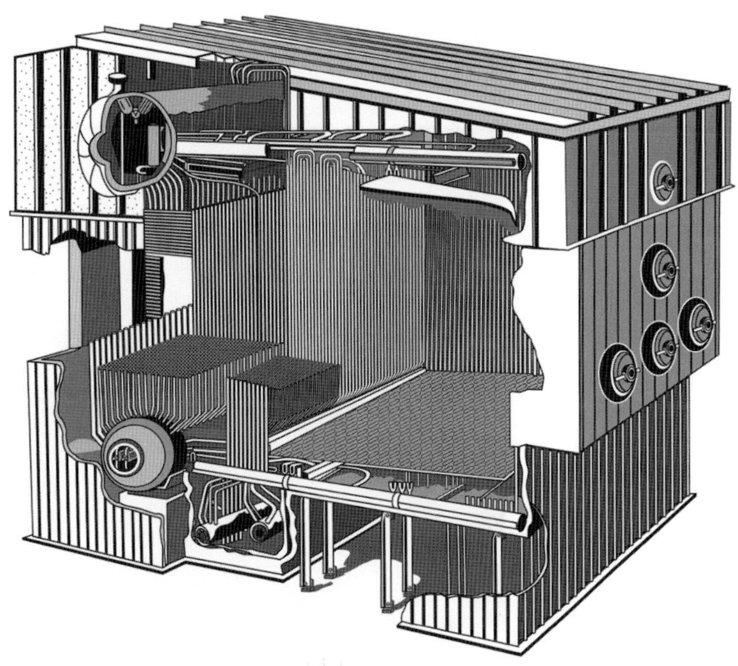

FIGURE 1.6 Cutaway view of boiler showing gas- or oil-fired burners on the right wall. Hot combustion products heat liquid water flowing through the tubes. The steam produced resides in the steam drum (tank) at the top left. This boiler has a nominal 8-m width, 12-m height, and 10-m depth. Adapted from Ref. [3]. (Courtesy of the Babcock & Wilcox Company.)

(liquid and vapor) is the working fluid in the closed loop 1–2–3–4–1. The water undergoes four processes:

Process 1–2 A pump boosts the pressure of the liquid water prior to entering the boiler. To operate the pump, an input of energy is required.

Process 2–3 Energy is added to the water in the boiler, resulting, first, in an increase in the water temperature and, second, in a phase change. The hot products of combustion provide this energy. The working fluid is liquid at state 2 and all vapor (steam) at state 3.

Process 3–4 Energy is removed from the high-temperature, high-pressure steam as it expands through a steam turbine. The output shaft of the turbine is connected to an electrical generator for the production of electricity.

Process 4–1 The low-pressure steam is returned to the liquid state as it flows through the condenser. The energy from the condensing steam is transferred to the cooling water.

Figures 1.6–1.10 illustrate the various generic components used in the Rankine cycle. Figure 1.6 shows a cutaway view of an industrial boiler; a much larger central power station utility boiler is shown in Figure 1.7. Although the design of boilers [3–6] is well beyond the scope of this book, the text offers much about the fundamental principles of their operation. For example, you will learn about the properties of water and steam in Chapter 2, whereas the necessary aspects of mass and energy conservation needed to deal with the components are treated in Chapters 3–5. Chapter 8 considers the components of a power plant (see Figs. 1.8–1.10); and Chapter 9 considers the system as whole.

As we begin our study, we emphasize the importance of safety in both the design and operation of power generation equipment. Fluids at high pressure contain enormous quantities of energy, as do spinning turbine rotors. Figure 1.11 shows the results

FIGURE 1.7 Boilers for public utility central power generation can be quite large, as are these natural gas-fired units (Ron_Thomas / E+ / Getty Images).

FIGURE 1.8 Steam turbine for power generation. Photograph and original caption reproduced with permission of the Smithsonian Institution.

The "Heart" of the Huge Westinghouse Turbine – An unusual detailed picture showing the maze of minutely fashioned blades – approximately five thousand – of the Westinghouse turbine rotor, or "spindle". Though only twenty-five feet in length this piece of machinery weighs one hundred and fifteen thousand pounds. At full speed the outside diameter of the spindle, on the left, is running nearly ten miles per minute, or a little less than 600 miles per hour. The problem of excessive heat resulting from such tremendous speed has been overcome by working the bearings under forced lubrication, about two barrels of oil being circulated through the bearings every sixty seconds to lubricate and carry away the heat generated by the rotation. The motor is that of the 45000 H.P. generating unit built by the South Philadelphia Works, Westinghouse Electric & Mfg. Co. for the Los Angeles Gas and Electric Company (Getty).

of a catastrophic boiler explosion. Similarly, environmental concerns are extremely important in power generation. Examples here are the emission of potential air pollutants from the combustion process (Fig. 1.12) and thermal interactions with the environment associated with steam condensation. Control of sulfur dioxide emissions

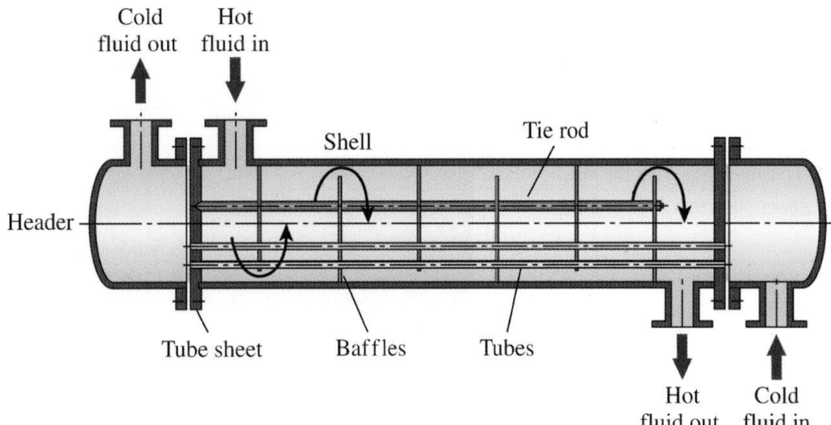

FIGURE 1.9 Cutaway view of shell-and-tube heat exchanger. Energy is transferred from the hot fluid passing through the shell to the cold fluid flowing through the tubes.

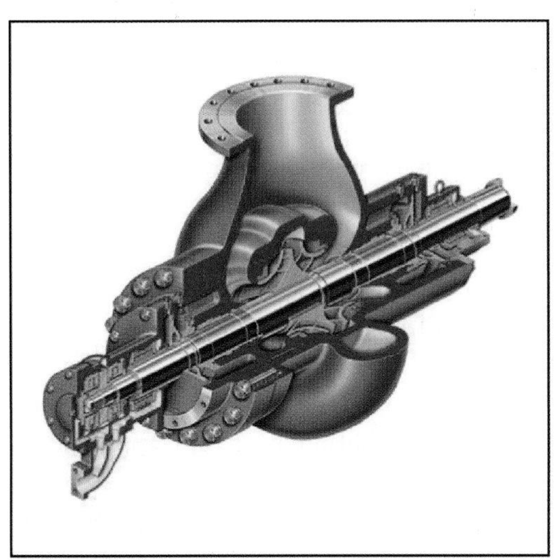

FIGURE 1.10 This pump was designed for nuclear reactor and steam generator feed applications. Feedwater pumps may be driven by electric motors or from auxiliary steam turbines. Courtesy of Flowserve Corporation.

FIGURE 1.11 A policeman inspects the site of an explosion near Bangkok, Thailand, August 19, 2014. A large boiler in a cloth dyeing factory exploded, injuring 22 people (Xinhua / Alamy Stock Photo).

FIGURE 1.12 Smog in Shanghai, China (left) (Wenjie Dong / E+ / Getty Images). Pollution controls are important components of fossil-fueled power plants. Shown here (right) is an electrostatic precipitator, which removes particulate matter from the flue gases of a power plant (nsf / Alamy Stock Photo).

from coal combustion generates large quantities of sludge requiring disposal or storage. You can find entire textbooks devoted to these topics [7, 8].

1.2b Spark-Ignition Engines

We choose the spark-ignition engine as one of our applications to revisit because there are so many of them (Fig. 1.13) – approximately 200 million are installed in automobiles and light-duty trucks in the United States alone – and because many students are particularly interested in engines. Owing to these factors, and others, many schools offer entire courses dealing with internal combustion engines, and many books are devoted to this subject, among them Refs. [9–12].

FIGURE 1.13 Spark-ignition engines power hundreds of millions of vehicles in the USA and around the world (fotog / Getty Images).

Although you may be familiar with the four-stroke engine cycle, we present it here to make sure that all readers have the same understanding. Figure 1.14 illustrates the following sequence of events:

Intake Stroke The inlet valve is open and the downward motion of the piston pulls a fresh fuel-air mixture into the cylinder. At some point near the bottom of the stroke, the intake valve closes.

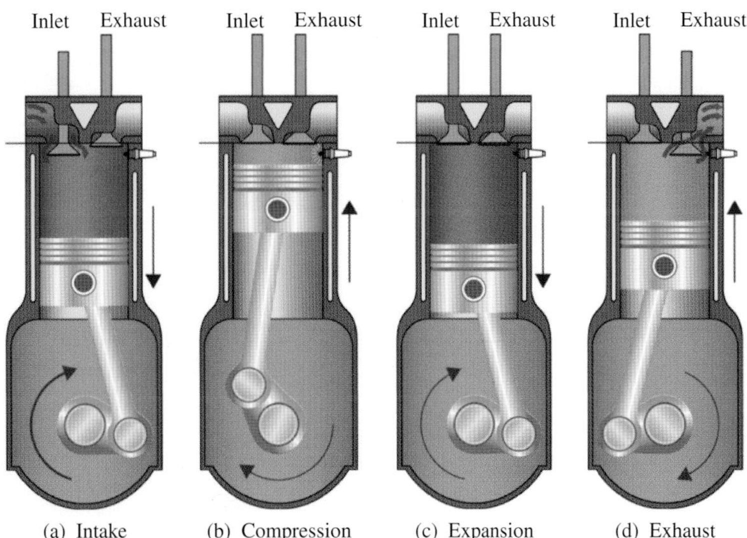

FIGURE 1.14 The mechanical cycle of the four-stroke spark-ignition engine consists of the intake stroke **(a)**, the compression stroke **(b)**, the expansion stroke **(c)**, and the exhaust stroke **(d)**. The sequence of events, however, does not execute a thermodynamic cycle. Adapted from Ref. [9] with permission. (Credit: Internal Combustion Engine Fundamentals, John Heywood. © McGraw-Hill Education.)

Compression Stroke The piston moves upward, compressing the mixture. The temperature and pressure increase. Prior to the piston reaching the top of its travel (i.e., the top center position), the spark plug ignites the mixture and a flame begins to propagate across the combustion chamber. The pressure rises above that due to compression alone.

Expansion Stroke The flame continues its travel across the combustion chamber, ideally burning all of the mixture before the piston has descended much from top center. The high pressure in the cylinder pushes the piston downward. Energy is extracted from the burned gases in the process.

Exhaust Stroke When the piston is near the bottom of its travel (bottom center), the exhaust valve opens. The hot combustion products flow rapidly out of the cylinder because of the relatively high pressure within the cylinder compared to that in the exhaust port. The piston ascends, pushing most of the remaining combustion products out of the cylinder. When the piston is somewhere near top center, the exhaust valve closes and the intake valve opens. The mechanical cycle now repeats.

In Chapters 5 and 12, we will analyze the processes that occur during the times in the cycle when both valves are closed and the gas contained within the cylinder can be treated as a thermodynamic system. With this analysis, we can model the compression, combustion, and expansion processes. A section of Chapter 9 also focuses on the Otto and diesel gas power cycles.

1.2c Jet Engines

Air travel is a common mode of transportation, with 571 billion passenger miles flown in the United States in 2012. Since you are likely to entrust your life from time to time to the successful performance of jet engines, you may find learning about these engines interesting. Figure 1.15 schematically illustrates the two general types of aircraft engines.

The schematic at the top shows a pure turbojet engine in which the jet of combustion products passing through the exhaust nozzle generates all of the thrust. This type

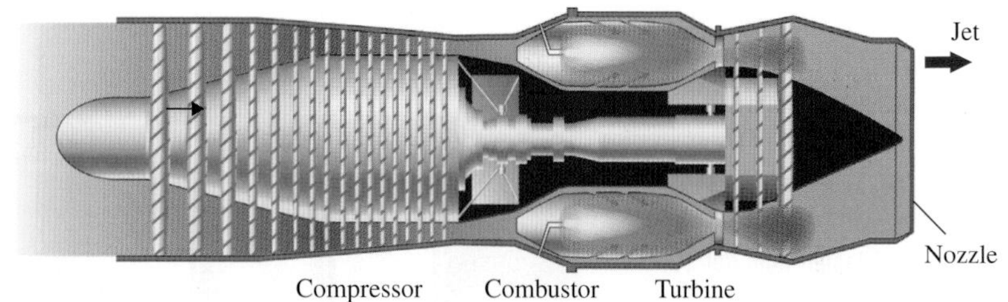

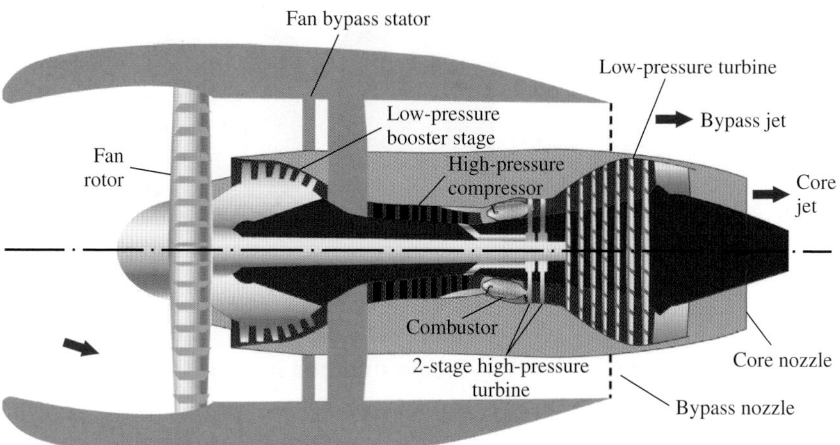

FIGURE 1.15 Schematic drawings of a single-shaft turbojet engine (top) and a two-shaft high-bypass turbofan engine (bottom). Adapted from Ref. [13].

of engine powered the supersonic J4 Phantom military jet fighter and the retired Concorde supersonic transport aircraft. In the turbojet, a multistage compressor boosts the pressure of the entering air. A portion of the high-pressure air enters the combustor, where fuel is added and burned, while the remaining air cools the combustion chamber. The hot products of combustion then mix with the cooling air, and these gases expand through a multistage turbine. In the final process, the gases accelerate through a nozzle and exit to the atmosphere to produce a high-velocity propulsive jet.[1] The compressor and turbine are rotary machines with spinning wheels of blades. Rotational speeds vary over a wide range but are of the order of 10,000–20,000 rpm. Other than that needed to drive accessories, all of the power delivered by the turbine is used to drive the compressor in the pure turbojet engine.

The second major type of jet engine is the turbofan engine (Fig. 1.15 bottom). This is the engine of choice for commercial aircraft. (See Fig. 1.16.) In the turbofan, a bypass air jet generates a significant proportion of the engine thrust. The large fan shown at the front of the engine creates this jet. A portion of the total air entering the engine bypasses the core of the engine containing the compressor and turbine, while the remainder passes through the core. The turbines drive the fan and the core compressors, generally using separate shafts for each. In the turbofan configuration, the exiting jets from both the bypass flow and the core flow provide the propulsive force.

To appreciate the physical size and performance of a typical turbojet engine, consider the GE F103 engine. These engines power the Airbus A300B, the DC-10–30, and the Boeing 747. The F103 engine has a nominal diameter of 2.7 m (9 ft) and a length of 4.8 m (16 ft), produces a maximum thrust of 125 kN (28,000 lb_f), and

[1] The basic principle here is similar to that of a toy balloon that is propelled by a jet of escaping air.

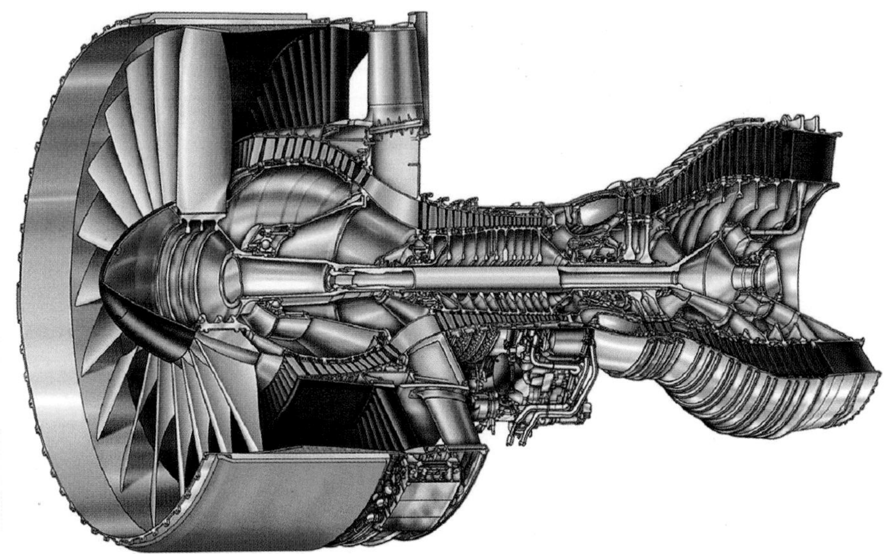

FIGURE 1.16 Cutaway view of PW4000-series turbofan engine. The PW4084 powers the Airbus A310–300 and A300–600 aircraft, and Boeing 747–400, 767–200/300, and MD-11 aircraft. Cutaway drawing courtesy of Pratt & Whitney.

weighs 37 kN (8,325 lb_f). Typical core and fan speeds are 14,500 and 8000 rpm, respectively [14].

Jet engines afford many opportunities to apply the theoretical concepts developed throughout this book. Analyzing a turbojet cycle is a major topic in Chapter 9, and individual components are treated in Chapter 8.

1.3 Learning Thermodynamics

Many students find their study of thermodynamics to be challenging. Part of the challenge results from the need to deal with many new concepts and relationships in an effective way [15]. Research has shown that in learning a new subject, beginning learners or novices may treat their newfound knowledge as unrelated bits or pieces; however, experts in the subject have an integrated knowledge, bringing together the many bits in ways that make sense and allow their knowledge to be useful [16]. Having an integrated knowledge of thermodynamics will support your problem solving ability and will lead to a deeper conceptual understanding. It can move you beyond a plug-and-chug approach. Therefore, developing integrated knowledge should be a goal of your learning. Achieving this integration is not an easy task. As a first step to assist you in this process, we present Fig. 1.17. This figure shows the typical organization of a first course in engineering thermodynamics. Here we see how the important bits and pieces of thermodynamics knowledge go together to form a cogent whole. In exploring this figure, we liken the structure of engineering thermodynamics to that of a house.

In the top levels, or top floors of our thermodynamics house, reside the exciting aspects of thermodynamics – the analysis and design of practical devices and systems. (For example, see all the previous figures.) Being able to analyze and design practical devices and systems may be an important motivation for your studying engineering. The top floors present an exhilarating view.

The lowest level, our sub-basement, contains a number of definitions, frameworks, and concepts that provide a basis for building the house above – concepts needed even before we can build a basement. Mastering the material in this lowest level is very

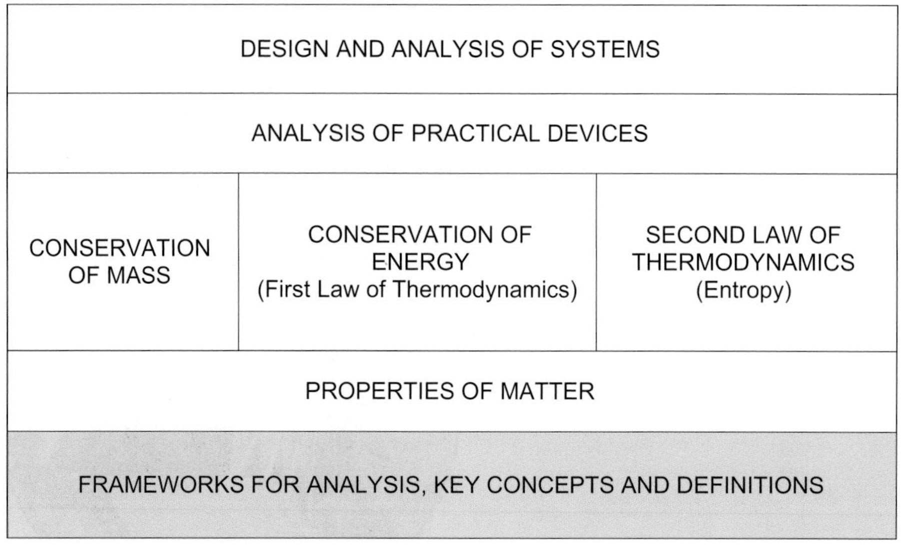

FIGURE 1.17 Hierarchical arrangements of the topics in our study of engineering thermodynamics. Chapter 1 introduces frameworks for analysis along with key concepts and definitions.

important to achieving a deep understanding of thermodynamics; unfortunately, this material provides a lot less excitement than do the top levels for most engineering students. Some people may say that it is boring. Nevertheless, it is where our study of thermodynamics must begin (here in Chapter 1).

The next level up, our basement, contains the study of the properties of matter. To create and operate practical devices and systems requires stuff, or matter; so clearly, a study of the properties of matter is necessary. For our study, the matter that matters the most is air, which we will treat as an ideal gas, and H_2O in its liquid (water) and vapor (steam) phases. Of course, we will not restrict our study to these fluids. At the end of your course of study, you should be comfortable working with a large number of fluids (liquids/gases). Although Chapter 2 contains most of the material related to thermodynamic properties, we delay a discussion of entropy and other related properties until Chapter 7. Other chapters also contain property-related topics.

Sitting on top of the properties of matter are three fundamental principles: two conservation principles – those of mass and energy – and the second law of thermodynamics. These are the big ideas associated with our study – they reside in the main living area of our thermodynamics house.

In the context of a first course in thermodynamics, the conservation-of-mass principle is straightforward and causes little difficulty for students. Chapter 3, one of the shortest, covers this principle.

The conservation-of-energy principle is at the heart of our study of thermodynamics and offers more complexity than mass conservation. The idea here is that energy is never created or destroyed, but only converted from one form to another. Although not wishing to trivialize this idea, one can think of applying energy conservation (the first law of thermodynamics) as an exercise in accounting – all energy debits and credits must always balance. Throughout the book, we stress the importance of this concept and seek ways to help the reader develop a deep understanding of it. Chapter 4 sets the stage for dealing with energy conservation, and Chapter 5 provides a detailed treatment. Subsequent chapters all link back to or use this important concept.

The second law of thermodynamics (Chapter 6), the third of our big ideas, places limits on what is possible – limits over and above those associated with conservation of energy. For example, the second law establishes the maximum possible thermal efficiency one can obtain from a Rankine cycle (Fig. 1.5). You may be surprised to learn that the maximum thermal efficiency is much less than 100%, even if friction and similar performance impediments could be eliminated. The second law also provides the basis for the thermodynamic property entropy, which we treat in Chapter 7. From the viewpoint of Fig. 1.17, the discussion of entropy could be part of Chapter 2, as entropy is just another thermodynamic property. However, the second law needs to precede the introduction of entropy to provide the context for understanding this property.

We will repeat Fig. 1.17 as we move through the book, highlighting the material treated at each juncture. Again, the purpose of this is to help you organize your knowledge of thermodynamics in useful ways. We now enter the sub-basement, the highlighted portion of Fig. 1.17.

1.4 Physical Frameworks for Analysis

In this section, we define **closed systems** and **open systems**. The latter are also known as **control volumes.** These concepts are central to almost any analysis of a thermal-fluid problem.

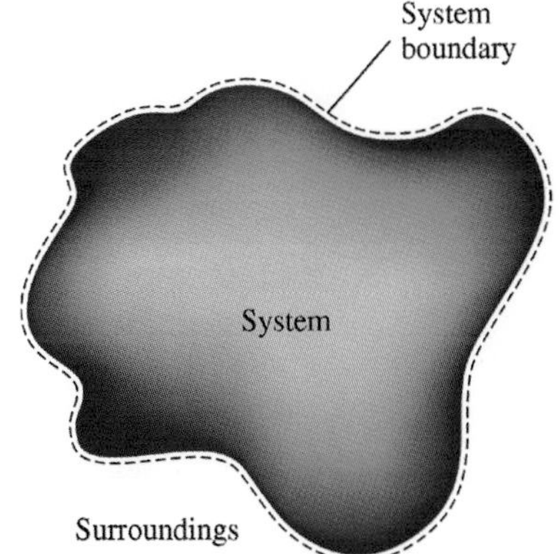

FIGURE 1.18 The system boundary separates a fixed mass, the closed system, from its surroundings.

1.4a Closed Systems

In a generic sense, a **system** is anything that we wish to analyze and distinguish from its **surroundings** or **environment.** To denote a system, all one needs to do is to create a **boundary** between the system of interest and everything else, that is, the surroundings. The boundary may be a real surface or an imaginary construct indicated by a dashed line on a sketch. Figure 1.18 illustrates the separation of a system from its surroundings by a boundary.

We will deal with two kinds of systems. The first is a **closed (or fixed-mass) system**. The specific definition of a closed system is the following:

A *closed system* is a specifically identified fixed mass of material separated from its surroundings by a real or imaginary boundary.

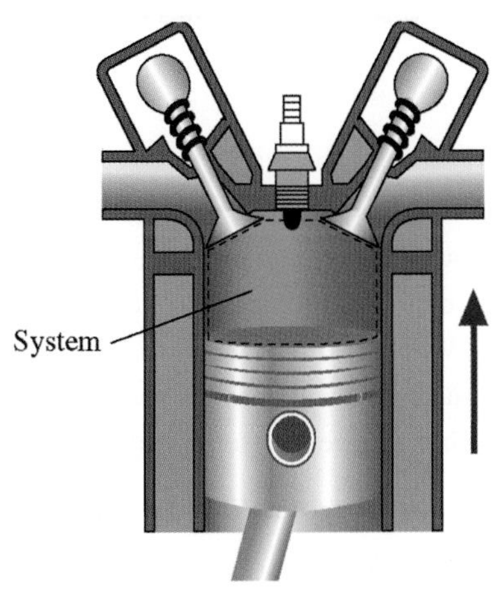

FIGURE 1.19 The gas within the cylinder of a spark-ignition engine constitutes a closed system, provided there is no leakage past the valves or the piston rings. The boundaries of this closed system deform as the piston moves.

The boundaries of a closed system need not be fixed in space but can, out of necessity, move. For example, consider the gas as the closed system of interest in the piston–cylinder arrangement shown in Fig.1.19. As the piston ascends and the gas is compressed, the closed system boundary shrinks to always enclose the same mass. Conversely, the closed system boundary expands when the piston travels downward.

Depending upon one's objectives, a closed system may be simple or complex, homogeneous or nonhomogenous. Our example of the gas enclosed in an engine cylinder (Fig. 1.19) is a relatively simple closed system. The system consists of only one substance: the fuel–air mixture. To simplify further an analysis of this particular closed system, we might assume that the fuel–air mixture has a uniform temperature, although in an operating engine the temperature will vary throughout the system. The matter within the closed system need not be a gas. Liquids and solids, of course, can be the whole system or a part of it. Again, the key distinguishing feature of a closed system is that it contains a fixed quantity of matter. No mass can cross the closed-system boundary.

To further illustrate the thermodynamic concept of a closed system, consider the computer chip module schematically illustrated in Fig. 1.20. A thermal analysis of this device can be performed to ensure that the chip stays sufficiently cool. Considering the complexity of this module, a host of possible closed systems exist. For example, we might choose a system boundary surrounding the entire device and cutting through the connecting wires, indicated as system 1. Alternatively, we might choose the chip itself (system 2) to be the closed system of interest.

Choosing system boundaries is critical to any thermodynamic analysis. One of the goals of this book is to help you develop the skills required to define and analyze thermodynamic systems.

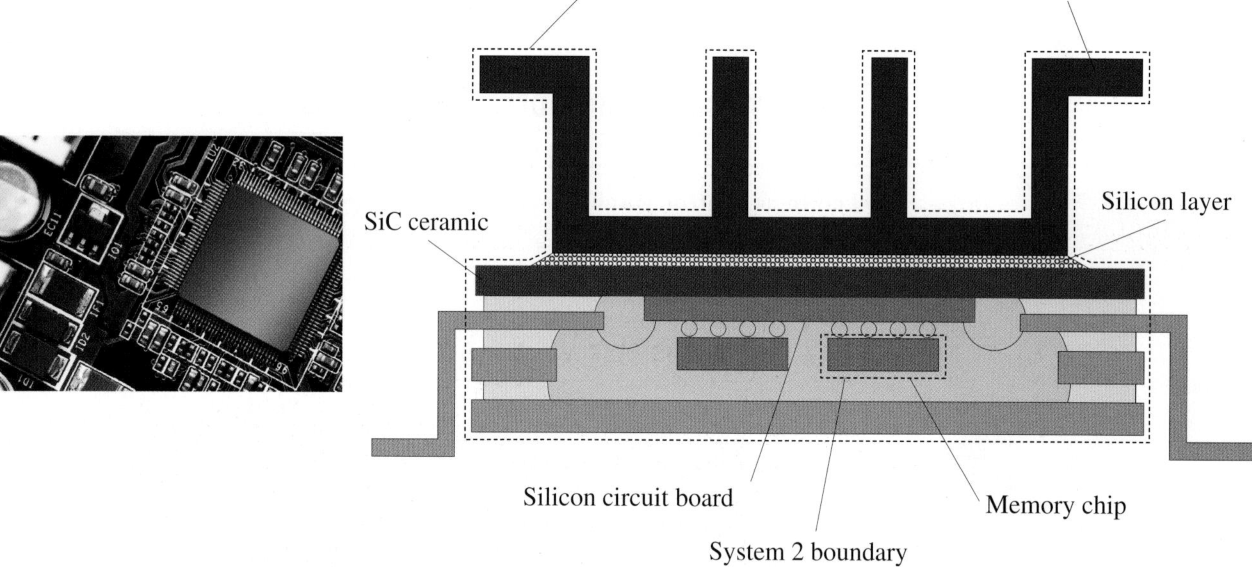

FIGURE 1.20 A computer chip is housed in a module designed to keep the chip sufficiently cool (WLADIMIR BULGAR / Science Photo Library / Getty). Various closed systems can be defined for thermal analysis of this relatively complex device. Two such choices are shown (dotted lines). Basic module sketch courtesy of *Mechanical Engineering*, Vol. 108, No. 3, March 1986, page 41; © Mechanical Engineering (The American Society of Mechanical Engineers International) (ASME Mag).

1.4b Open Systems (or Control Volumes)

In contrast to a closed system, mass may cross the boundary of, and enter and/or exit, an **open system**. Open systems are also known as control volumes. The term open system is more frequently used in thermodynamics, whereas the term control volume is more frequently used in the domain of fluid mechanics; however, there is no distinction between the two terms. We formally define an open system as follows:

An open system is a region in space separated from its surroundings by a real or imaginary boundary across which mass may pass.

Figure 1.21 illustrates an open system and its attendant boundary. Here, mass in the form of water vapor or water droplets crosses the upper part of the boundary as a result of evaporation from the hot liquid coffee. For this example, we chose a fixed boundary near the top of the cup; however, we could have chosen the regressing liquid surface to be the upper boundary. The choice of moving boundary may or may not simplify an analysis, depending on the particular situation.

Open systems may be simple or complex. Fixing the boundary in space yields the simplest open system, whereas moving open systems with deforming boundaries are the most complex. Figure 1.22 illustrates the latter, where the exiting jet of air propels an inflated balloon. In this example, the control volume both moves with respect to a fixed observer and shrinks with time. Figure 1.23 illustrates a simple open system with a fixed boundary associated with the analysis of a water pump. Note that the boundary cuts through flanged connections at the inlet and outlet of the pump. Frequently, the particular choice of an open system and its boundary is of overwhelming importance to an analysis. A wise choice can make an analysis simple, whereas a poor choice can make the analysis more difficult, or perhaps, impossible. Examples presented throughout this book provide guidance in selecting boundaries for open systems.

FIGURE 1.21 The boundary surrounds an open system containing hot liquid coffee, air, and moisture. Water vapor or droplets exit through the upper portion of the boundary (photo: Zoonar RF / Zoonar / Getty Images Plus).

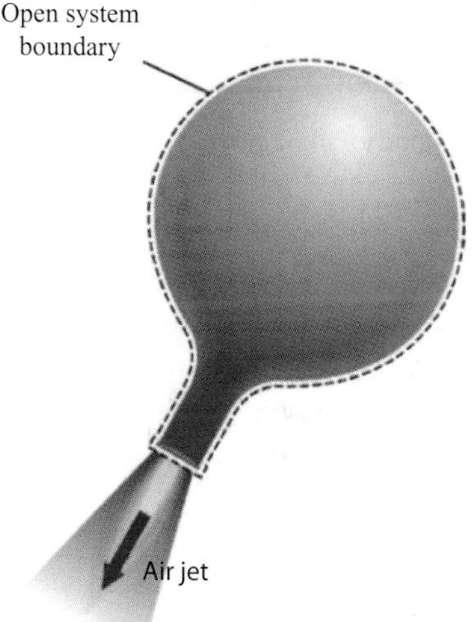

FIGURE 1.22 A rubber balloon and the air it contains constitute an open system (control volume). The open system moves through space and shrinks as the air escapes.

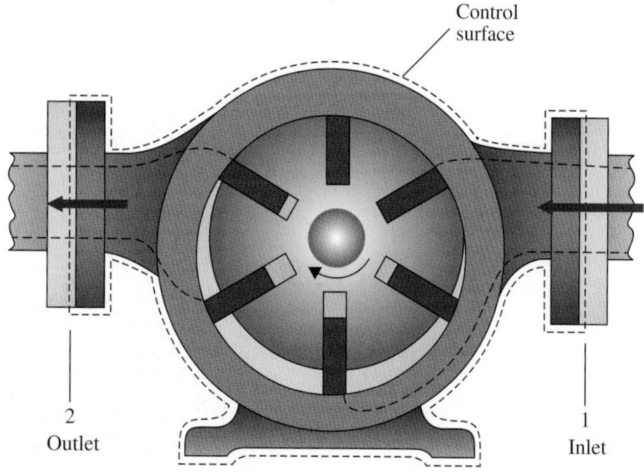

FIGURE 1.23 A sliding-vane pump and the fluid it contains constitute a simple open system. Mass crosses the boundary at both the pump inlet and outlet. Adapted from Ref. [17] with permission of McGraw-Hill.

1.5 Key Concepts and Definitions

In addition to frameworks for analysis (closed/open systems), several other basic concepts permeate our study of thermodynamics and are listed in Table 1.2. We introduce these concepts in this section, recognizing that they will be revisited again, perhaps several times, in later chapters.

TABLE 1.2 Some Fundamental Thermodynamic Concepts

Closed system
Open system (control volume)
Surroundings
Property
State
Process
Flow process
Cycle
Equilibrium
Quasi-equilibrium

1.5a Properties

Before we can begin a study of thermodynamic properties understood in their most restricted sense, we define what is meant by a **property** in general:

A *property* is a quantifiable macroscopic characteristic of a system.

Examples of system properties include mass, volume, density, pressure, temperature, height, width, and color, among others. Not all system properties, however, are thermodynamic properties. Thermodynamic properties all relate in some way to the energy of a system. In our list here, height, width, and color, for example, do not qualify as thermodynamic properties, although the others do. A precise definition of thermodynamic properties often depends on restricting the system to which they apply in subtle ways. Chapter 2 is devoted to thermodynamic properties and their interrelationships.

Properties are frequently combined to create new ones. For example, a spinning baseball in flight not only has the properties of mass and velocity but also kinetic energy and angular momentum. Thus, any closed system may possess numerous properties.

1.5b States

Another fundamental concept in thermodynamics is that of a **state**:

A thermodynamic* state *of a closed system is defined by the values of all the closed-system thermodynamic properties.

When the value of any one of a closed system's properties changes, the system undergoes a change in state. For example, hot coffee in a thermos bottle undergoes a continual change of state as it slowly cools. Not all thermodynamic properties necessarily change when the state of a system changes; the mass of the coffee in a sealed thermos remains constant while the temperature falls. An important skill in solving problems in thermodynamics is to identify which properties remain fixed and which properties change during a change in state.

1.5c Processes

One goal of studying thermal-fluid sciences is to develop an understanding of how various devices convert one form of energy to another. For example, how does the burning coal in a power plant result in the electricity supplied to your home? In analyzing such energy transformations, we formally introduce the idea of a thermodynamic **process**:

A process occurs whenever a closed system changes from one state to another state.

Figure 1.24a illustrates a process in which the gas contained in a cylinder–piston arrangement at state 1 is compressed to state 2. As is obvious from the sketch, one property, the volume, changes in going from state 1 to state 2. Without knowing more details of the process, we cannot know what other properties may have changed as well. In many thermodynamic analyses, a single property – for example, temperature, pressure, or entropy – is held constant during the process. In these particular cases, we refer to the processes as a constant-temperature (isothermal) process, constant-pressure (isobaric) process, or a constant-entropy (isentropic) process, respectively.

Although formal definitions of a process refer to systems, the terminology is also applied to open systems, in particular, steady flows in which no properties change with time. We thus define a **flow process** as follows:

A *flow process* occurs whenever the state of the fluid entering an open system is different from the state of the fluid exiting the open system.

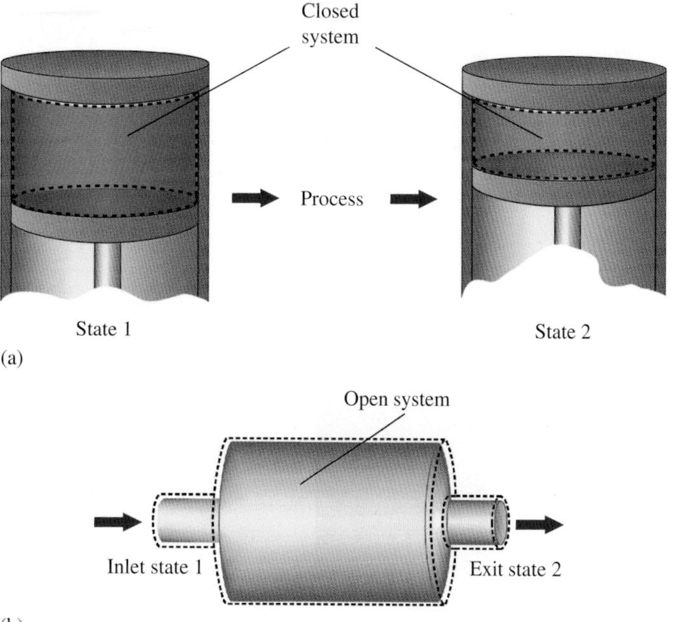

FIGURE 1.24 (a) A closed system undergoes a process that results in a change of the system state from state 1 to state 2. **(b)** In a steady-flow process, fluid enters the open system in state 1 and exits in state 2.

FIGURE 1.25 A series of steady-flow processes are associated with an operating jet engine. We explore these in Chapter 9. Photo courtesy of US Air Force.

Figure 1.24b schematically illustrates a flow process. A jet engine employs a series of steady-flow processes (Fig. 1.25). Air enters the engine and is compressed; fuel is injected and burned; the combustion products expand and exhaust into the surroundings. Various flow processes underlie the operation of myriad practical devices, many of which we will study in subsequent chapters.

1.5d Cycles

In many energy-conversion devices, the **working fluid** undergoes a thermodynamic **cycle**. Since the word *cycle* is used in many ways, we need a precise definition for our study of thermal-fluid sciences. Our definition is the following:

A thermodynamic* cycle *consists of a sequence of processes in which the working fluid returns to its original thermodynamic state.

An example of a cycle applied to a closed (fixed-mass) system is presented in Fig. 1.26. Here a gas trapped in a piston–cylinder assembly undergoes four processes. A cycle can be repeated any number of times, following the same sequence of processes. Although it is tempting to consider reciprocating internal combustion engines as operating in a cycle, the products of combustion never undergo a transformation back to fuel and air, as would be required for our definition of a cycle. There exist, however, types of reciprocating engines that do operate on thermodynamic cycles. A prime example of this is the Stirling engine [18]. The working fluid in Stirling engines is typically hydrogen or helium. All combustion takes place outside the cylinder. Figure 1.27 shows a model Stirling engine.

Thermodynamic cycles are most frequently executed by a series of flow processes, as illustrated in Fig. 1.28. As the working fluid flows through the loop, it experiences many changes in state as it passes through various devices such as pumps, boilers, and heat exchangers, but ultimately it returns to its initial state, arbitrarily chosen here to

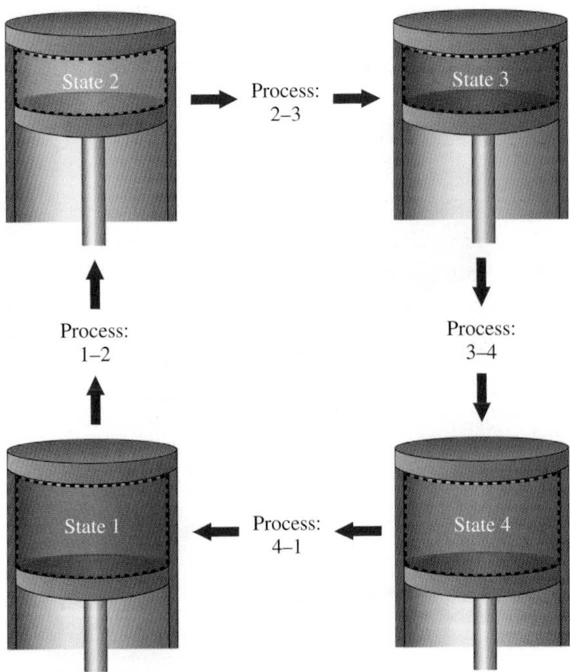

FIGURE 1.26 A closed system undergoes a cycle when a series of processes returns the system to its original state. In this sketch, the cycle consists of the state sequence 1–2–3–4–1.

FIGURE 1.27 Model hot-air Stirling engine (rico ploeg / Alamy Stock Photo).

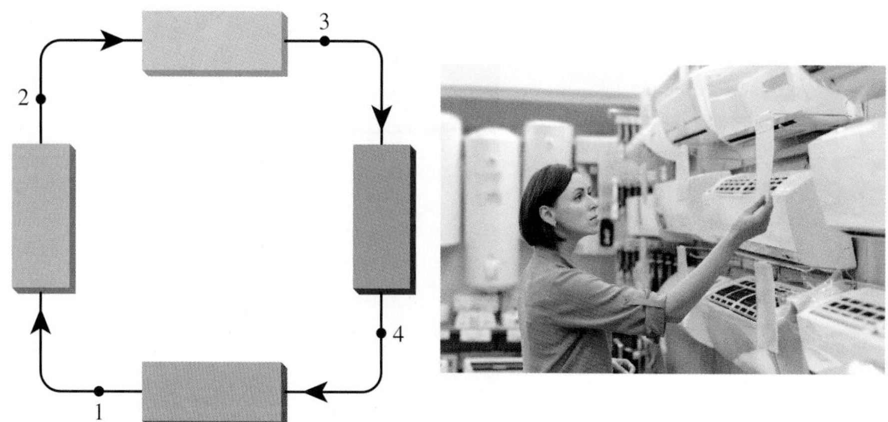

FIGURE 1.28 A cycle can also consist of a sequence of flow processes in which the flowing fluid is returned to its original state: 1–2–3–4–1. Air conditioners operate in a cycle of flow processes (97 / E+ / Getty Images).

be state 1. As we have already seen, the heart of a fossil-fueled steam power plant operates on a thermodynamic cycle of the type illustrated in Fig. 1.28 (cf. Fig. 1.5). Refrigerators and air conditioners are everyday examples of devices operating on thermodynamic cycles. In Chapter 9, we analyze various cycles for power production, propulsion, and heating and cooling.

The thermodynamic cycle is central to many statements of the second law of thermodynamics and concepts of thermal efficiency (Chapter 6).

1.5e Equilibrium and the Quasi-Equilibrium Process

The science of thermodynamics builds upon the concept of equilibrium states. For example, the common thermodynamic property **temperature** has meaning only for a system in equilibrium. In a thermodynamic sense, then, what do we mean when we say that a system is in equilibrium? In general, that no unbalanced potentials or drivers exist to promote a change of state. The state of a system in equilibrium remains unchanged for all time. For a system to be in equilibrium, we require that the system be simultaneously in thermal equilibrium, mechanical equilibrium, phase equilibrium, and chemical equilibrium. Other considerations may exist, but they are beyond our scope. We now consider each of these components of thermodynamic equilibrium.

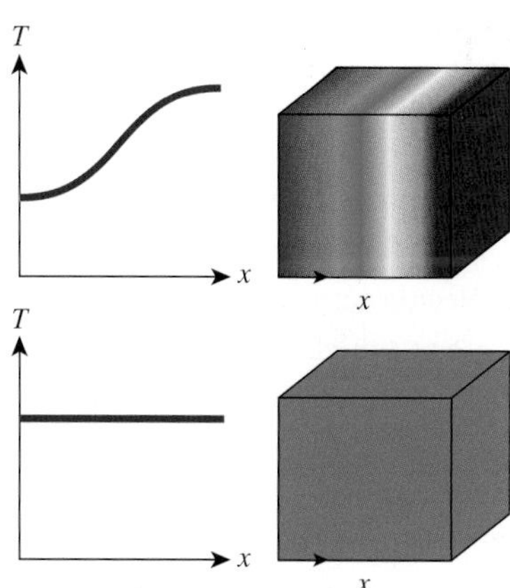

FIGURE 1.29 The flow of energy from a region of higher temperature to a region of lower temperature drives a system toward thermal equilibrium. The temperature is uniform in a system at equilibrium.

Thermal equilibrium exists when a system both has a uniform temperature and is at the same temperature as its surroundings. If, for example, the surroundings are hotter than the system under consideration, energy may flow across the system boundary from the surroundings to the system. Such an exchange would result in an increase in the temperature of the system. Equilibrium thus does not prevail. Similarly, if temperature gradients exist within the system, the initially hotter regions become cooler, and the initially cooler regions become hotter, as time passes (Fig. 1.29 top). Given sufficient time, the temperature within the system becomes uniform, provided the ultimate temperature of the system is identical to that of the surroundings (Fig. 1.29 bottom). Prior to achieving the final uniform temperature, the system is not in thermal equilibrium.

FIGURE 1.30 A familiar example of a substance that can exist in three phases is H_2O. The vapor phase (steam) is invisible, but condenses into small droplets that can be seen above the ice cube (craigratcliffe / iStock / Getty Images Plus).

Mechanical equilibrium occurs when the pressure throughout the system is uniform and there are no unbalanced forces at the system boundaries. An exception to the condition of uniform pressure exists when a system is under the influence of a gravitational field. For example, pressure increases with depth in a fluid such that pressure forces balance the weight of the fluid above. In many systems, however, we can neglect the effects of gravity and the assumption of uniform pressure is reasonable.

Phase equilibrium relates to conditions in which a substance can exist in more than one physical state; that is, any combination of vapor, liquid, and solid. For example, you are quite familiar with the three states of H_2O: water vapor (steam), liquid water, and ice (Fig. 1.30). Phase equilibrium requires that the amount of a substance in any one phase does not change with time; for example, liquid–vapor phase equilibrium implies that the rate at which molecules escape the liquid phase to enter the gas phase is exactly balanced by the rate at which molecules from the gas phase enter the liquid phase. We will discuss these ideas further in Chapters 2 and 13.

Our final condition for thermodynamic equilibrium requires the system to be in **chemical equilibrium.** For systems incapable of chemical reaction, this condition is trivial; however, for reacting systems this constraint is quite important. Chapter 13 is devoted to this topic.

From the foregoing discussion, we see that an equilibrium state is a boring proposition. Nothing happens. The system just sits there. Nevertheless, considering a system to be in an equilibrium state at the beginning and at the end of a process is indeed useful. In fact, it is this idea that motivates our discussion. Our development of the conservation of energy principle relies on the assumption that equilibrium states

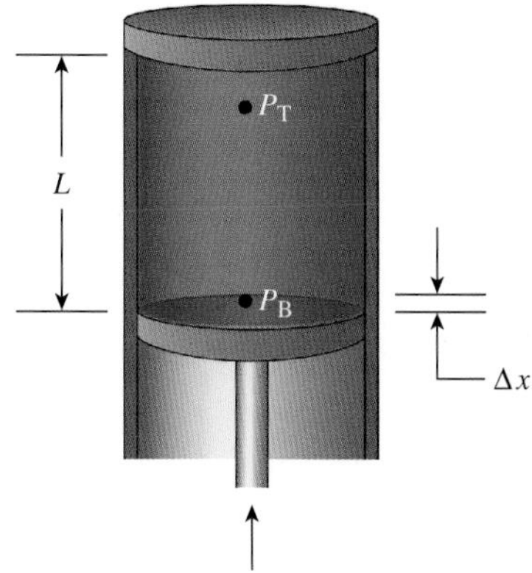

FIGURE 1.31 In the quasi-equilibrium compression of a gas, the pressure is essentially uniform throughout the gas system, that is $P_B = P_T$.

exist at the beginning and end of a process, even though during the process the system may be far from equilibrium.

This discussion suggests that it is not possible to describe the details of a process because of departures from equilibrium. In some sense, this is indeed true. However, the often-invoked assumption of a quasi-static or quasi-equilibrium process does permit an idealized description of a process as it occurs. We define a **quasi-static** or **quasi-equilibrium process** to be a process that happens sufficiently slowly such that departures from thermodynamic equilibrium are always so small that they can be neglected. For example, the compression of a gas in a perfectly insulated piston–cylinder system (Fig. 1.31) results in the simultaneous increase in temperature and pressure of the gas. If the compression is performed slowly, the pressure and temperature at each instant will be uniform throughout the gas system for all intents and purposes [i.e., $P_B(t) = P_T(t)$], and the system can be considered to be in an equilibrium state.[2] In contrast, if the piston moves rapidly, the pressure in the gas at the piston face (P_B) would be greater than at the far end of the cylinder (P_T) and equilibrium would not be achieved. The formal requirement for a quasi-equilibrium process is that the time for a system to reach equilibrium after some change is small compared to the time scale of the process. For our example of the gas compressed by the piston, the time for the pressure to equilibrate is determined by the speed of sound, that is, the propagation speed of a pressure disturbance. For room-temperature air, the sound speed is approximately 340 m/s. If we assume that at a particular instant the piston is 100 mm (= L) from the closed end of the cylinder, the change in the pressure due to the piston motion will be communicated to the closed end in a time equal to 2.9×10^{-4} s [= (0.100 m)/(340 m/s)] or 0.29 ms. If during 0.29 ms the piston moves very little (Δx), the pressure can be assumed to be uniform throughout the system for all practical purposes, and the compression can be considered as a quasi-equilibrium process. Alternatively, we can say that the speed of the pressure waves that communicate changes in pressure is much faster than the speed of the piston that creates the increasing pressure.[3] Similar arguments involving time scales can be used to ascertain whether thermal and chemical equilibrium are approximated for any real processes. Our purposes here, however, are not to define the exact conditions for which a quasi-equilibrium process might be assumed, but rather to acquaint the reader with this commonly invoked assumption. Later in the book, we use the quasi-equilibrium process as a standard to which real processes are compared.

1.6 Dimensions and Units

The primary dimensions (or base units) used in this book are *mass, length, time, temperature, electric current,* and *amount of substance*. All other dimensions, such as force, energy, and power, are derived from these primary dimensions [19]. Furthermore, we employ almost exclusively the International System of Units, or le Système International (SI) d'Unités. The primary dimensions thus have the following associated units:

Mass [=] kilogram (kg),

Length [=] meter (m),

Time [=] second (s),

[2] For the pressure to be truly uniform requires that we neglect the effect of gravity. As discussed in Chapter 2, we restrict our study to simple compressible substances, which requires neglecting gravity.

[3] Pressure waves travel at the speed of sound.

NIST physicists Steve Jefferts (foreground) and Tom Heavner with the NIST-F2 cesium fountain atomic clock, a new civilian time standard for the United States. Atomic clocks such as these have an accuracy of better than 1 second in 20 million years [20]. Photograph courtesy of NIST. Reprinted courtesy of the National Institute of Standards and Technology, US Department of Commerce.

Temperature [=] kelvin (K),

Electric current [=] ampere (A), and

Amount of substance [=] mole (mol).

The symbol [=] is used to express *has units of* and is used throughout this book. The derived dimensions most frequently used in this book are defined as follows:

$$\text{Force} = 1\ \text{kg·m/s}^2 = 1\ \text{newton (N)}. \tag{1.1}$$

$$\text{Energy} = 1\ (\text{kg·m/s}^2)\text{·m} = 1\ \text{N·m or 1 joule (J), and} \tag{1.2}$$

$$\text{Power} = 1\frac{(\text{kg·m/s}^2)\text{·m}}{\text{s}} = 1\ \text{J/s or 1 watt (W)}. \tag{1.3}$$

Unfortunately, a wide variety of non-SI units are used customarily in the United States, many of which are industry-specific. You can find conversion factors from SI units to the most common non-SI units on the inside covers of this book. We refer to certain non-SI units in several examples to provide some familiarity with these important non-SI units. To have a single location for their definition, we present the customary units most important to our study of the thermal-fluid sciences here. In terms of the four primary dimensions, mass ([=] pound-mass or lb_m), length ([=] foot), time ([=] second), and temperature ([=] degree Rankine or R), common derived dimensions and units are as follows:

$$\text{Force} = 32.174\frac{\text{lb}_\text{m}\text{·ft}}{\text{s}^2} = 1\ \text{pound-force or lb}_\text{f}, \tag{1.4}$$

$$\text{Energy} = 1\ \text{ft·lb}_\text{f} = \frac{1}{778.17}\ \text{British thermal unit or Btu, and} \tag{1.5}$$

$$\text{Power} = 1\frac{\text{ft·lb}_\text{f}}{\text{s}} = \frac{1}{550}\ \text{horsepower or hp}. \tag{1.6}$$

In some applications, the unit associated with mass is the slug. Using this unit to define force yields

$$\text{Force} = 1\frac{\text{slug·ft}}{\text{s}^2} = 1\ \text{pound-force or lb}_\text{f}, \tag{1.7}$$

Further elaboration of units is found in the discussion of thermodynamic properties in Chapter 2.

The use of the units pound-force (lb_f) and pound-mass (lb_m) can cause some confusion to the casual user. A safe rule of thumb to avoid any confusion with units is to convert all given quantities in a problem to SI units, solve the problem using SI units, and then convert the final answer to any desired customary units. The following example clarifies the usage of pound-force and pound-mass.

Example 1.1 Mass and Weight on Earth

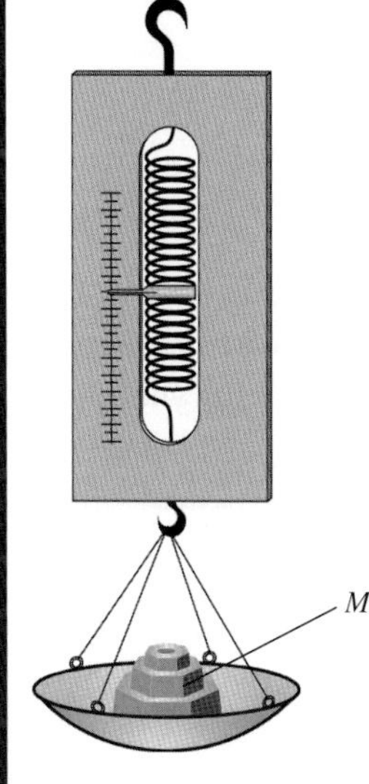

A mass of 1 lb_m is placed on a spring scale calibrated to read in pounds-force. Assuming an earth-standard gravitational acceleration of 32.174 ft/s^2, what is the scale reading? What is the equivalent reading in newtons? How do these results change if the measurement is conducted on the surface of the moon where the gravitational acceleration is 5.32 ft/s^2?

Solution

Known $M = 1\ lb_m$, $g_{earth} = 32.174\ ft/s^2$

Find Force exerted on scale

Analysis The force exerted on the scale is the weight, $F = W$, which equals the product of the mass and the gravitational acceleration, that is,

$$F = Mg_{earth}.$$

Thus,

$$F = (1\,lb_m)(32.174\,ft/s^2).$$

Using the definition of pounds-force (Eq. 1.4), we generate the identity

$$1 = \left[\frac{1\dfrac{lb_f}{ft/s^2}}{32.174\,lb_m}\right].$$

Using this identity in the previous expression for the force yields

$$F = 1\ lb_m\left[\frac{1\dfrac{lb_f}{ft/s^2}}{32.174\,lb_m}\right]32.174\,ft/s^2$$

or

$$F = 1\ lb_f.$$

To convert this result to SI units, we use the conversion factor from the front of the book:

$$F = 1\,lb_f\left[\frac{1\,N}{0.224809\,lb_f}\right] = 4.45\,N.$$

On the moon, the force is

$$\begin{aligned} F &= Mg_{moon} \\ &= (1\,lb_m)(5.32\,ft/s^2) \\ &= 1\,lb_m\left[\frac{1\dfrac{lb_f}{ft/s^2}}{32.174\,lb_m}\right]5.32\,ft/s^2 \\ F &= 0.165\,lb_f, \end{aligned}$$

or

$$F = 0.736\,\text{N}.$$

Comments From this example, we see that the definition of the US customary unit for force was chosen so that a one-pound mass produces a force of exactly one pound-force under conditions of standard earth gravity. In the SI system, numerical values for mass and force are *not* identical for conditions of standard gravity. For example, consider the weight associated with a 1-kg mass:

$$\begin{aligned} F &= Mg \\ &= (1\,\text{kg})(9.807\,\text{m/s}^2) \\ &= 1\text{kg}\left(\frac{1\,\text{N}}{\text{kg}\cdot\text{m/s}^2}\right)(9.807\,\text{m/s}^2) \\ &= 9.807\,\text{N}. \end{aligned}$$

Regardless of the system of units, care must be exercised in any unit conversions. Checking to see that the units associated with any calculated result are correct should be part of every problem solution. Such practice also helps you to spot gross errors and can save time.

Self-Test 1.1

Repeat Example 1.1 for a given mass of 1 slug.

(Answer: 32.174 lb_f, 143.12 N, 5.32 lb_f, 23.66 N)

1.7 Problem-Solving Method

For you to consistently solve engineering problems successfully (both textbook and real problems) depends on your developing a procedure that (i) aids thought processes, (ii) facilitates identification of errors along the way, (iii) allows others to easily check your work, and (iv) provides a reality check at completion. The following general procedure has these attributes. We recommend that you follow this procedure for most problems. Nearly all the examples throughout the text illustrate its use.

1. *KNOWN.* State what is known in a simple manner without rewriting the problem statement.
2. *FIND.* Indicate what quantities you want to find.
3. *SKETCH.* Draw and label useful sketches whenever possible. (What is useful generally depends on the context of the problem. Suggestions are provided at appropriate locations in the text.)
4. *MODELING PREMISES AND ASSUMPTIONS.* List the modeling premises associated with your analysis; list your initial assumptions and add others to the list as you proceed with your solution.

5. *ANALYSIS*. Analyze the problem and identify the important definitions and principles that apply to your solution.
6. *SOLUTION*. Develop a symbolic or algebraic solution to your problem, delaying the substitution of numerical values as late as possible in the process.
7. Substitute numerical values as appropriate and indicate the source of all physical data as you proceed. (The appendices of this book contain much useful data.)
8. Check the units associated with each calculation. The factor-label method is an efficient way to do this.
9. Examine your answer critically. Does it appear to be reasonable and consistent with your expectations and/or experience?
10. *COMMENT(S)*. Write out one or more comments using step 9 as your guide. What did you learn from solving the problem? Were your assumptions justified?

1.8 Mathematical Skills to Review

The mathematics required for a first course in thermodynamics is quite simple: basic algebra and calculus suffice. The authors have found that some students will benefit from a review of the following three topics:

- Linear interpolation of tabular data – a skill needed to deal with property tables in a proficient manner;
- Differentiation of simple polynomials – a skill needed to calculate thermodynamic properties given a curve-fit or other polynomial representation;
- Integration of simple polynomials – a skill needed to calculate reversible, moving-boundary, or steady-flow work and to calculate certain thermodynamic properties given a curve-fit or other polynomial representation.

A review of linear interpolation is provided in the Chapter 2 *Tutorial 1 – How to Interpolate* on page 76. You may wish to read this tutorial now and work some of the end-of-chapter problems (Chapter 1) dedicated to interpolation to hone your skill. Similarly, we include end-of-chapter problems to review the calculus skills important in your study of thermodynamics.

SUMMARY

In this chapter, we introduced the subject of thermodynamics, a foundational engineering science within the thermal-fluid science domain. We also presented the following three practical applications: fossil-fuel steam power plants, spark-ignition engines, and jet engines. From these discussions, you should have a reasonably clear idea of what this book is about. Also presented in this chapter are many concepts and definitions upon which we will build in subsequent chapters. To make later study easier, you should acquire a firm understanding of these concepts and definitions. A review of the learning objectives presented at the start of this chapter may be helpful in that regard. Finally, it is the authors' hope that your interest has been piqued in the subject matter of this book.

CHAPTER 1 KEY CONCEPTS AND DEFINITIONS CHECKLIST

Solve the Problems following the arrows to demonstrate mastery of the listed concepts and definitions.

1.1 What is thermodynamics? ➔ Problem 1.1

1.2 Some applications ➔ Problems 1.1, 1.2

1.3 Learning thermodynamics ➔ Redraw Fig. 1.17

1.4 Physical frameworks for analysis

- ☐ Closed systems ➔ Problems 1.3, 1.4, 1.6
- ☐ Open systems (or control volumes) ➔ Problems 1.3, 1.4
- ☐ Boundaries ➔ Problem 1.11
- ☐ Surroundings ➔ Problem 1.13

1.5 Key concepts and definitions

- ☐ Property ➔ Problem 1.15
- ☐ State ➔ Problem 1.15
- ☐ Process ➔ Problem 1.15
- ☐ Flow process ➔ Problems 1.17, 1.18
- ☐ Cycle ➔ Problems 1.16, 1.23
- ☐ Equilibrium ➔ Problems 1.20, 1.24
- ☐ Quasi-equilibrium ➔ Problems 1.26, 1.27, 1.28

1.6 Dimensions and units ➔ Problems 1.36, 1.39, 1.44, 1.46

1.7 Problem-solving method ➔ List key steps

1.8 Mathematical skills to review ➔ Problems 1.56, 1.59, 1.62, 1.63, 1.66, 1.67

REFERENCES

1. *Webster's New Twentieth Century Dictionary*, J. L. McKechnie (Ed.), Collins World, Cleveland, 1978.
2. Energy Information Agency, U.S. Department of Energy, "Monthly Energy Review April 2019," http://www.eia.gov/totalenergy/data/monthly/index.cfm#electricity, Release Date: April 25, 2019.
3. *Steam: Its Generation and Use*, 39th edn, Babcock & Wilcox, New York, 1978.
4. Singer, J. G. (Ed.), *Combustion Fossil Power: A Reference Book on Fuel Burning and Steam Generation*, 4th edn, Combustion Engineering, Windsor, CT, 1991.
5. Basu, P., Kefa, C, and Jestin, L., *Boilers and Burners: Design and Theory*, Springer, New York, 2000.
6. Goodall, P. M., *The Efficient Use of Steam*, IPC Science and Technology Press, Surrey, England, 1980.
7. Flagan, R. C, and Seinfeld, J. H., *Fundamentals of Air Pollution Engineering*, Prentice Hall, Englewood Cliffs, NJ, 1988.
8. Schetz, J. A. (Ed.), *Thermal Pollution Analysis*, Progress in Astronautics and Aeronautics, Vol. **36**, AIAA, New York, 1975.
9. Heywood, J. B., *Internal Combustion Engine Fundamentals*, McGraw-Hill, New York, 1988.

10. Obert, E. E, *Internal Combustion Engines and Air Pollution*, Harper & Row, New York, 1973.
11. Ferguson, C. E, *Internal Combustion Engines: Applied Thermosciences*, Wiley, New York, 1986.
12. Campbell, A. S., *Thermodynamic Analysis of Combustion Engines*, Wiley, New York, 1979.
13. Cumpsty, N, *Jet Propulsion*, Cambridge University Press, New York, 1997.
14. St. Peter, J., *History of Aircraft Gas Turbine Engine Development in the United States: A Tradition of Excellence*, International Gas Turbine Institute of the American Society of Mechanical Engineers, Atlanta, 1999.
15. Turns, S. R., and Van Meter, P. N., "Applying Knowledge from Educational Psychology and Cognitive Science to a First Course in Thermodynamics," *Proceedings of the ASEE Annual Conference & Exposition*, June 26–29, 2011.
16. Brandsford, J. D., Brown, A. L., and Cocking, R. R. (Eds.), Ch. 2, How Experts Differ from Novices, in *How People Learn*, Expanded Edition, National Academy Press, Washington, D.C., 2000.
17. White, F. M., *Fluid Mechanics*, 3rd edn, McGraw-Hill, New York, 1994.
18. White, M. A., Colendbrander, K., Olan, R. W., and Penswick, L. B., "Generators That Won't Wear Out," *Mechanical Engineering*, **118**:92–96 (1996).
19. Lide, D. R. (Ed.),*Handbook of Chemistry and Physics*, 77th edn, CRC Press, Boca Raton, FL, 1996.
20. Lombardi, M. A., Heavner, T. P., and Jefferts, S. R., "NIST Primary Frequency Standards and the Realization of the SI Second," *NCSL International Measure: The Journal of Measurement Science*, Vol. 2, No. 4, December 2007.

Some end-of-chapter problems were adapted with permission from the following:

Look, D. C. Jr., and Sauer, H. J. Jr., *Engineering Thermodynamics*, PWS, Boston, 1986.

Myers, G. E., *Engineering Thermodynamics*, Prentice Hall, Englewood Cliffs, NJ, 1989.

Pnueli, D., and Gutfinger, C, *Fluid Mechanics*, Cambridge University Press, Cambridge, England, 1992.

Chapter 1 Problem Subject Areas

1.1, 1.2	Applications of thermodynamics
1.3–1.14	Closed and open systems
1.15–1.31	Key concepts and definitions
1.32–1.55	Dimensions and units
1.56–1.69	Mathematics review

PROBLEMS

1.1, 1.2 Applications of thermodynamics

1.1 ***Conceptual problem***. Write out the definition of thermodynamics. Using this definition as a guide, list three practical situations or devices that closely relate to thermodynamics. Explain the relationships of each situation/device to thermodynamics. Do not repeat any of the examples given in the book.

1.2 ***Conceptual problem***. Make a table that has the following three headings: *Device, Form of Energy Input, Form(s) of Energy Output*. Under the heading "device" list 10 or more practical devices for which their design and operation is strongly linked to thermodynamics. Recall that thermodynamics involves the study of energy and its transformation from one form to another. Indicate the energy transformations associated with each device by filling in the columns under the other two headings. Do not necessarily restrict your device choices to those discussed in the text. Look around your indoor and outdoor environments for ideas. For example, a hair dryer converts electricity (input energy) into a flow of hot air (output of kinetic and thermal energies).

HINT: Here is a list of typical forms of energy: Chemical energy (as in a fuel), thermal energy (associated with the heat capacity and temperature of a substance), kinetic energy, mechanical work or power (a spinning shaft or moving boundary), electrical work or power, potential energy, nuclear energy, radiant energy (from the sun or a hot object), and perhaps others.

1.3–1.14 Closed and open systems

1.3 ***Conceptual problem.*** Write two or three sentences that explain the differences and similarities between a **closed system** and an **open system (control volume)**. Give practical examples of each, avoiding examples discussed in the text.

1.4 ***Conceptual problem.*** Consider a hand pump inflating a bicycle tire. Define three closed systems and three open systems that relate to the pump and/or the tire. Feel free to subdivide parts of the pump/tire to create your systems. Write a sentence describing each of your choices and draw a sketch for each, indicating your boundaries with a dashed line. Draw arrows indicating where mass crosses the boundaries of your open systems. Make sure one or more of your systems is associated in some way with air.

1.5 ***Conceptual problem.*** Consider the toy balloon shown in Fig. 1.22. (a) Define an open system that is different than the one illustrated in the figure. Write a sentence describing your choice and draw a sketch for each indicating your boundary with a dashed line. (b) Define two closed systems associated with the balloon/air system. Write a sentence describing each of your choices and draw a sketch for each, indicating your boundaries with a dashed line. Make sure one of your systems is associated with air. **HINT:** Assume there is no mixing within the air contained in the balloon.

1.6 ***Conceptual problem.*** Turn on a faucet and observe the stream of water flowing from it. Define a closed system related to the water in this situation. How do the boundaries change and move with time? What would you have to do to change your closed system to an open system?

1.7 ***Conceptual problem.*** Consider a household gas-fired hot-water heater. Perform an internet search if you are not familiar with this device. Note that there are five flows to consider: cold water in, hot water out, natural gas in, air in, and products of combustion out. Sketch an open system for this device for the following situations. Use labeled arrows to denote mass flows across your boundaries.

A. Hot water is being drawn for a shower and the heater is trying to meet the demand.

B. No hot water is being drawn, but a lot of hot water has just been used.

C. There is no demand for hot water and the water in the tank is hot, i.e., there is no need for it to be heated further.

Is case C an open or a closed system? Discuss.

1.8 ***Conceptual problem***. Can energy cross the boundary of a thermodynamic system? Discuss.

1.9 ***Conceptual problem***. From your list in Problem 1.2, select two devices, or parts of those devices, that you can represent as closed systems. Sketch each device, show the system boundary as a dotted line, and label as needed to make your sketch intelligible. Discuss why your boundary choice encloses a closed system.

1.10 ***Conceptual problem***. From your list in Problem 1.2, select two devices, or parts of those devices, that you can represent as open systems (control volumes). Sketch the devices, show the boundaries as a dotted line, and label as needed to make your sketch intelligible. Discuss why your boundary choice defines an open system.

1.11 ***Conceptual problem***. Consider a conventional toaster – the kind used to toast bread. Examine a real toaster, and then perform the following tasks: (a) Sketch two different boundaries associated with an operating toaster: one in which you define a closed system, and a second one in which you define an open system (control volume). Be sophisticated in your analyses. For example, you might want to consider whether air flows through the toaster. If so, how? What happens to the bread during the toasting process? Can the bread by itself be considered an open or a closed system? Discuss your selections justifying why each is either closed or open. (b) Repeat the problem choosing two different closed and open systems.

1.12 ***Conceptual problem***. Discuss what would have to be done to transform the open system (control volume) shown in Fig. 1.22 to a closed system. Ignore the boundary shown in Fig. 1.22. What closed systems can be defined for the situation illustrated?

1.13 ***Conceptual problem***. Consider a conventional house in the Northeastern region of the United States. Isolate the house from its surroundings by drawing a boundary to define an open system. Identify all the locations where mass enters or exits your open system. Assume the house has a natural-gas furnace, has running water, and connects to a sanitary sewer system. Make and list other assumptions as you see fit.

1.14 ***Conceptual problem***. Consider an automobile, containing a driver and several passengers, traveling along a country road. Isolate the automobile and its contents from the surroundings by drawing an appropriate boundary. Does your boundary enclose a thermodynamic closed system or an open system? Justify your choice by writing a sentence or two. How would you have to modify your boundary to convert from one to the other? What assumptions, if any, would you have to make to perform this conversion?

1.15–1.31 Key concepts and definitions

1.15 ***Conceptual problem.*** Write out the formal definitions of the following terms: property, state, and process. What is the relationship of a property to a state? What is the relationship of a state to a process? What is the relationship of a property to a process?

1.16 ***Conceptual problem.*** Can you identify any devices that operate in a thermodynamic cycle? If so, list them. Also, identify the working fluids, if known.

1.17 ***Conceptual problem.*** Distinguish between a closed-system process and a flow process. List three of each type.

1.18 ***Conceptual problem.*** Consider the diagram of a simple steam power plant shown in Fig. 1.5. Here the components act to produce a thermodynamic cycle (see Fig. 1.28). Which of the various fluids involved is the working fluid of the cycle? Explain your choice. For the other fluids involved, what flow processes do they undergo?

1.19 ***Conceptual problem.*** Consider the mechanical cycle associated with the four-stroke spark-ignition engine illustrated in Fig. 1.14. The combustion process begins near the end of the compression stroke and ends shortly after the expansion stroke begins. A typical plot of the cylinder pressure as a function of the cylinder volume is shown in the sketch below. Redraw the sketch and label the beginning and end points of each of the four mechanical strokes. Also, indicate the beginning and end of the combustion process. For each stroke, write a sentence describing why you selected the specific beginning and end points.

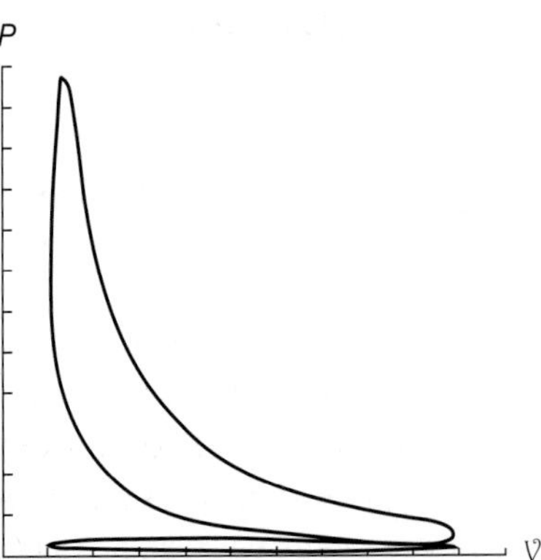

Typical pressure-volume diagram for a 4-stroke-cycle, spark-ignition engine.

1.20 ***Conceptual problem.*** List the conditions that must be established for thermodynamic equilibrium to prevail (i.e., list the subtypes of equilibrium).

1.21 ***Conceptual problem.*** Consider the following statement: For a closed system to be in thermodynamic equilibrium, the system must be all in one phase, i.e., all solid, all liquid, or all vapor. Is this statement true or false? Explain your choice.

1.22 ***Conceptual problem***. Consider the following statement: In a thermodynamic cycle, the temperature of the working fluid must be the same at the start and the end of the cycle, although the pressure does not have to be the same at the start and the end of the cycle. Is this statement true or false? Explain your choice.

1.23 ***Conceptual problem***. Consider the following statement: In a thermodynamic cycle, many, but not all, of the thermodynamics properties of the working fluid will be the same at the start and the end of the cycle. Is this statement true or false? Explain your choice.

1.24 ***Conceptual problem***. Consider the following statement: For a closed system in thermodynamic equilibrium, thermodynamic properties do not change with time. Is this statement true or false? Explain your choice.

1.25 ***Conceptual problem***. Consider the following statement: For a closed system in thermodynamic equilibrium, the temperature can vary with position within the system. Is this statement true or false? Explain your choice.

1.26 ***Conceptual problem***. Consider the following statement: During a quasi-equilibrium process, the state of the system does not change. Is this statement true or false? Explain your choice.

1.27 ***Conceptual problem***. Consider the following statement: For a quasi-equilibrium process, the property changes can be large, e.g., the final pressure can be much larger (or greater) then the initial pressure. Is this statement true or false? Explain your choice.

1.28 ***Conceptual problem***. Consider the following statement: For a quasi-equilibrium process, the property changes during the process proceed as infinitesimal departures from thermodynamic equilibrium. Is this statement true or false? Explain your choice.

1.29 ***Conceptual problem***. Explain how a very rapid expansion of a gas is an example of a non-quasi-equilibrium process.

1.30 ***Conceptual problem***. Consider a closed system consisting of a gas. The temperature in the gas varies with position along the system boundary. Is the system in equilibrium? Is it possible that the system could be undergoing a quasi-equilibrium process? Explain your answers.

1.31 ***Conceptual problem***. Consider a closed system consisting of ice and liquid water. The ice is melting at constant temperature. Is the ice–water system in equilibrium? Explain your answer. Is the ice–water system undergoing a process? Explain your answer.

1.32–1.55 Dimensions and units

1.32 ***Open-ended problem.*** Do an internet search to find examples of disasters, or near disasters, created by errors associated with units or units conversions. List six and describe the errors.

1.33 What is your approximate weight in pounds-force? In newtons? What is your mass in pounds-mass? In kilograms? In slugs?

1.34 Repeat Problem 1.33 assuming you are now on the surface of Mars, where the gravitational acceleration is 3.71 m/s^2.

1.35 A residential natural gas-fired furnace has an output of 86,000 Btu/hr. What is the output in kilowatts? How many 100-W incandescent light bulbs would be required to provide the same output as the furnace?

(Credit: nycshooter / iStock / Getty Images Plus.)

1.36 The 1964 Pontiac GTO was one of the first so-called muscle cars produced in the United States in the 1960s and 1970s. The 389-in^3 displacement V-8 engine in the GTO delivered a maximum power of 348 hp and a maximum torque of 428 $lb_f \cdot ft$. From a standing start, the GTO traveled 1/4 mile in 14.8 s, accelerating to 95 mph. Convert all the US customary units in these specifications to SI units.

(Credit: schlol / E+ / Getty Images.)

1.37 Coal-burning power plants convert the sulfur in the coal to sulfur dioxide (SO_2), a regulated air pollutant. The maximum allowable SO_2 emission for new power plants (i.e., the maximum allowed ratio of the mass of SO_2 emitted to the input of fuel energy) is 0.80 lb_m/(million Btu). Convert this emission factor to units of grams per joule.

1.38 The National Ambient Air Quality Standard (NAAQS) for lead (Pb) in the air is 1.5 μg/m^3. Convert this standard to US customary units of lb_m/ft^3.

1.39 The astronauts of *Apollo 17*, the final US manned mission to the moon, collected 741 lunar rock and soil samples, having a total mass of 111 kg. Determine the weight of samples in SI and US customary units (a) on the lunar surface (g_{moon} = 1.62 m/s^2) and (b) on the earth (g_{earth} = 9.807 m/s^2). Also determine (c) the mass of the samples in US customary units.

(Credit: Education Images / Contributor / Universal Images Group / Getty Images.)

1.40 The gravitational acceleration on the earth varies with latitude and altitude as follows [19]:

$$g(\mathrm{m/s^2}) = 9.780356\,[1 + 0.0052885\,\sin^2\theta - 0.0000059\,\sin^2(2\theta)] - 0.003086z,$$

where θ is the latitude and z is the altitude in kilometers. Use this information to determine the weight (in newtons and pounds-force) of a 54-kg mountain climber at the following locations:

A. The summit of Kilimanjaro in Tanzania (3.07° S and 5895 m)

B. The summit of Cerro Aconcagua in Argentina (32° S and 6962 m)

C. The summit of Denali in Alaska (63° N and 6194 m)

How do these values compare with the mountain climber's weight at 45° N latitude at sea level? Use spreadsheet software to perform all calculations.

1.41 The values of gravitational acceleration at the surfaces of Jupiter, Pluto, and the sun are 23.12 m/s^2, 0.72 m/s^2, and 273.98 m/s^2, respectively. Determine your weight at each of these locations in both SI and US customary units. Assume no loss of mass results from the extreme conditions.

1.42 A coal-burning power plant consumes fuel energy at a rate of 4845 million Btu/hr and produces 500 MW of net electrical power. Determine the overall efficiency of the power plant (i.e., the dimensionless ratio of the net output and input energy rates).

1.43 Derive a conversion factor relating pressure in pascals (N/m^2) and in psi (lb_f/in^2). Compare your result with the conversion factor provided at the front of this book.

1.44 The US Space Shuttle fleet was operational from 1981 to 2011. Each shuttle had three main engines. A single main engine had the following specifications, *ca.* 1987. Convert these values to SI units.

(Credit: Comstock Images / Stockbyte / Getty Images.)

Maximum Thrust	
At sea level	408,750 lb_f
In vacuum	512,300 lb_f
Pressures	
Hydrogen pump discharge	6872 psi
Oxygen pump discharge	7936 psi
Combustion chamber	3277 psi
Flow Rates	
Hydrogen	160 lb_m/s
Oxygen	970 lb_m/s
Power	
High-pressure H_2 turbopump	74,928 hp
High-pressure O_2 turbopump	28,229 hp
Weight	7000 lb_f
Length	14 ft
Diameter	7.5 ft

1.45 Determine the acceleration of gravity for which the weight of an object will be numerically equal to its mass in the SI unit system.

1.46 Use the conversion factors found at the front of the book to convert the flowing quantities to SI units:
Density, $\rho = 120\ \text{lb}_\text{m}/\text{ft}^3$
Thermal conductivity, $k = 170$ Btu/(hr·ft·F)
Convective heat-transfer coefficient, $h_{conv} = 211$ Btu/(hr·ft^2·F)
Specific heat, $c_p = 175$ Btu/(lb$_\text{m}$·F)
Viscosity, $\mu = 20$ centipoise
Viscosity, $\mu = 77$ lb$_\text{f}$·s/ft^2
Kinematic viscosity, $v = 3.0$ ft^2/s
Stefan–Boltzmann constant, $\sigma = 0.1713 \times 10^{\ 8}$ Btu/(ft^2·hr·R^4)
Acceleration, $a = 12.0$ ft/s^2

1.47 Calculate the stored chemical energy per unit mass of the following foods in units of Btu per pound-mass. Then compare your result to the energy density of gasoline, expressed as a ratio (gasoline energy/food energy). An approximate value for the energy density (heating value) of gasoline is 20,000 Btu per pound-mass. One food calorie (or large calorie) is equivalent to 4184 joules.

A. A 12-fluid ounce can of soft drink containing 120 food calories per can. One fluid ounce is 1.805 cubic inches. The density of the soft drink is 1.1 gram/milliliter.

B. A 3.8-oz bagel containing 290 food calories. One ounce (mass) is 1/16 of a pound-mass.

C. A 21.3-g serving of honey containing 64 food calories. This serving size is approximately equivalent to one tablespoon.

1.48 Calculate the stored chemical energy per unit mass of the following foods in units of Btu per pound-mass. Then compare your result to the energy density of gasoline, expressed as a ratio (gasoline energy/food energy). An approximate value for the energy density of gasoline is 20,000 Btu per pound-mass. One food calorie (or large calorie) is equivalent to 4184 joules. One ounce (mass) is 1/16 of a pound-mass.

A. A 2-ounce serving of sourdough bread containing 140 food calories.

B. One Hostess Twinkie with a mass of 1.5 ounces and containing 290 food calories.

C. An 8-ounce (mass) broiled sirloin steak containing 414 food calories.

D. A one-cup serving cup of green string beans with a mass of 4.4 ounces and containing 44 food calories.

1.49 Calculate the stored chemical energy per unit mass of the following foods in units of Btu per pound-mass. Then compare your result to the energy density of gasoline, expressed as a ratio (gasoline energy/food energy). An approximate value for the energy density of gasoline is 20,000 Btu per pound-mass. One food calorie (or large calorie) is equivalent to 4184 joules. One ounce (mass) is 1/16 of a pound-mass.

A. A one-cup serving of vanilla ice cream with a mass of 4.7 ounces and containing 273 food calories.

B. A large 9-oz pear when peeled and cored with a mass of 8.1 ounces and containing 133 food calories.

C. A 1-ounce reduced-fat (2% milk fat) cottage cheese containing 26 food calories.

D. An 8-fluid-ounce bottle of low-calorie cranberry juice cocktail containing 45 food calories. One fluid ounce is 1.805 cubic inches. The density of the cranberry juice cocktail is 1.05 gram/millilitre.

1.50 A window air conditioner has a cooling capacity of 1 ton (energy rate). Determine the cost to operate the air conditioner continuously for 2 hours if the cost of electricity is 12.2 cents per kilowatt-hour.

1.51 Estimate the number of tons of coal required to supply electricity to a small city of 50,000 households for one week. The medium-volatile bituminous coal used has an approximate energy density (heating value) of 13,000 Btu/lb$_m$. Assume the energy in the coal is transformed to electricity with an efficiency of 32%. The average monthly household electricity usage is 750 kilowatt-hours.

1.52 Vigorously riding a bicycle results in the "burning" of 10 food calories per minute. How many 100-watt light bulbs can be powered by this rate of food energy use? One food calorie (or large calorie) is equivalent to 4184 joules.

1.53 Vigorous swimming results in the "burning" of 11 food calories per minute. What is amount of energy associated with 30 minutes of swimming expressed in Btu? How many equivalent ounces (mass) of gasoline are associated with this energy expenditure? An approximate value for the energy density of gasoline is 20,000 Btu per pound-mass. One food calorie (or large calorie) is equivalent to 4184 joules. One ounce (mass) is 1/16 of a pound-mass.

1.54 Studying thermodynamics results in the "burning" of 120 food calories per hour. How many 100-watt light bulbs can be powered by this rate of food energy use? One food calorie (or large calorie) is equivalent to 4184 joules.

1.55 Playing singles tennis for an hour results in the "burning" of 428 food calories. How many minutes of operation of a 500-watt blow dryer are equivalent to the food energy used in playing tennis for an hour? One food calorie (or large calorie) is equivalent to 4184 joules.

1.56–1.69 Mathematics review

1.56 For the following tabular data, use linear interpolation to determine the value of y when x has a value of 0.14.

x	y
0.10	5.81
0.20	2.96

1.57 For the following tabular data, apply linear interpolation to determine the value of h when T has a value of 357.

T	h
350	476.4
360	486.5

1.58 For the following tabular data, apply linear interpolation to determine the value of T when P has a value of 4.13.

P	T
4.10	524.97
4.20	526.41

1.59 For the following tabular data, apply linear interpolation to determine the value of v when P has a value of 0.56.

P	v
0.50	0.3748
0.60	0.3156

1.60 For the following tabular data, apply linear interpolation to determine the value of ρ when T has a value of 333.

T	ρ
320	0.1524
340	0.1434

1.61 For the following tabular data, apply linear interpolation to determine the value of s when u has a value of 2462.0.

u	s
2461.2	7.8442
2463.5	7.8167

1.62 Determine the derivative dy/dx of the function $y = a + bx + cx^3$.

1.63 Determine the derivative dP/dT of the function $P = a \ln T + bT^{-1}$.

1.64 Determine the derivative dz/dT of the function

$$z = -0.29T + 26.3T^2 - 10.61T^3 + 1.56T^4 - 0.16/T - 18.3.$$

1.65 Determine the derivative dT/dP of the function $T = 32.4P^{-0.5} - 12.8P^{0.5}$.

1.66 Determine the indefinite integral of $y(x)$ where

$$y(x) = 19.86 - 597x^{-0.5} + 7500/x.$$

1.67 Determine $\int z dy$ where $y = 11.6z^2$.

1.68 Determine $\int P d\mathcal{V}$ where $P\mathcal{V} = 1000$.

1.69 Determine the function $h(T)$ where $h(T) = \int c dT$ and

$$c = 11.515 - \frac{172}{\sqrt{T}} + \frac{1530}{T} + \frac{0.05}{1000}(T - 4000).$$

CHAPTER 2

Thermodynamic Properties, Property Relationships, and Processes

LEARNING OBJECTIVES

After studying Chapter 2, you should:

- Be familiar with the following basic thermodynamic properties: pressure, temperature, specific volume and density, specific internal energy and enthalpy, constant-pressure and constant-volume specific heats, and the ratio of specific heats.
- Be able to explain the concepts of absolute pressure and absolute temperature and be proficient in their use.
- Be able to explain the concept of state relations and how they are used.
- Be able to write one or more forms of the ideal gas equation of state and from this derive all other (mass, molar, mass-specific, and molar-specific) forms.
- Be proficient at obtaining thermodynamic properties for liquids and gases from the NIST online WebBook (or from software) and from printed tables.
- Be able to draw T–v and P–v diagrams for two-phase substances and on these diagrams identify iso-property lines, the various phase regions, the phase boundary lines, and the critical point.
- Be able to explain the principle of corresponding states and the use of generalized compressibility charts.
- Be able to apply the concepts and skills developed in this chapter throughout this book.

CHAPTER 2 OVERVIEW

THIS IS ONE OF THE LONGER chapters in this book and should be revisited many times at various levels. To cover in detail the subject of the properties of gases and liquids would take an entire book. Reference [1] is a classic example of such a book. We begin our study of properties by defining a few basic terms and concepts. We follow this by examining ideal-gas properties that originate from the equation of state and calorific equations of state. Various approaches for obtaining properties of nonideal gases, liquids, and solids follow. We emphasize the properties of substances that have coexisting liquid and vapor phases. The concept of illustrating processes graphically, using thermodynamic property coordinates (i.e., T–v and P–v coordinates), is developed.

2.1 Why Properties Are Important in Our Study

To apply fundamental principles to solve engineering problems requires a good understanding of thermodynamic properties, their interrelationships, and their relation to various processes. Figure 2.1 illustrates how these fundamental principles sit on top of our study of properties. Although mass conservation may require only a few

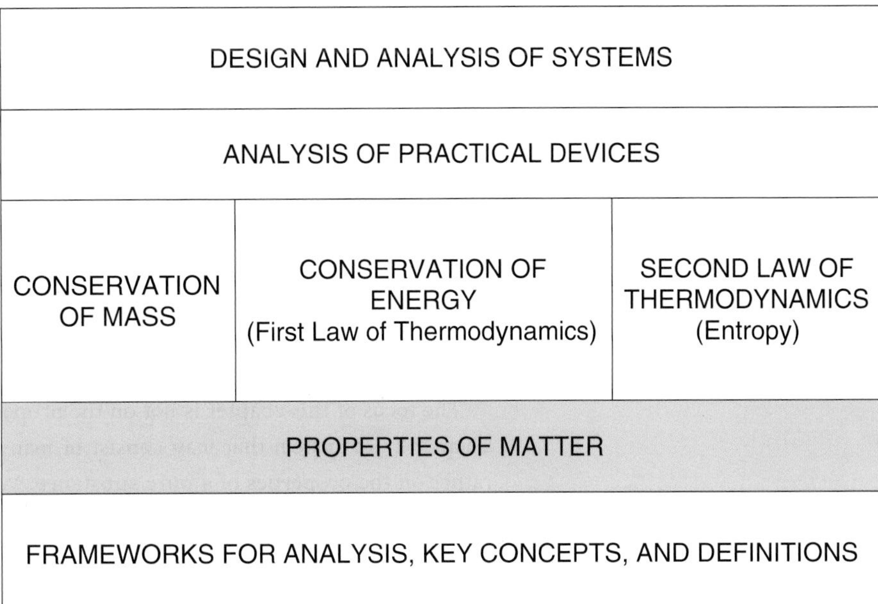

FIGURE 2.1 Chapter 2 deals with many thermodynamic properties of matter. These are key quantities needed to express the fundamental principles of conservation of mass, conservation of energy, and the second law of thermodynamics.

thermodynamic properties – usually, the mass density ρ and its relationship to pressure P and temperature T suffice – the successful application of the energy conservation principle relies heavily on thermodynamic properties. To illustrate this, we preview Chapter 5 by examining a particular conservation of energy statement as follows:

$$Q_{\text{net in}} - W_{\text{net out}} = U_2 - U_1. \tag{2.1}$$

Here, $Q_{\text{net in}}$ is the net energy that entered the system as a result of a heat-transfer process, $W_{\text{net out}}$ is the net energy that left the system as a result of work being performed, and U_1 and U_2 are the internal energies of the system at the initial state 1 and at the final state 2, respectively.

This first-law statement (Eq. 2.1) immediately presents us with an important thermodynamic property, the internal energy U. Although you are likely to have encountered this property in a study of physics, a much deeper understanding is required in our study of engineering thermodynamics. Defining internal energy and relating it to other, more common, system properties, such as temperature T and pressure P, are important topics considered in Chapter 2. We also need to know how to deal with this property when dealing with two-phase mixtures such as water and steam.

To further illustrate the importance of Chapter 2, we note that the work term in our conservation of energy statement (Eq. 2.1) also relates to system thermodynamic properties. For the special case of a reversible expansion, the work can be related to the system pressure P and volume $\mathbb{V}$, both thermodynamic properties. Specifically, to obtain the work requires the evaluation of the integral of $Pd\mathbb{V}$ from state 1 to state 2. To perform this integration, we need to know how P and $\mathbb{V}$ are related for the particular process that takes the system from state 1 to state 2. Much of Chapter 2 deals with, first, defining the relationships among various thermodynamic properties and, second, showing how the properties relate during various specific processes. For example, the ideal-gas law, $P\mathbb{V} = NR_uT$ expresses a familiar relationship among the thermodynamic properties P, $\mathbb{V}$, and T. (Here N is the number of moles contained in the system, and R_u is the universal gas constant.) From this relationship among properties, we see that the product of the system pressure and volume is constant for a constant-temperature process. In addition to exploring the ideal-gas equation of state, Chapter 2 shows how to deal with the many substances that do not behave as ideal gases. Common examples here include water, water vapor, and various refrigerants.

2.2 Key Definitions

To start our study of thermodynamic properties, you should review the definitions of **properties**, **states**, and **processes** presented in Chapter 1, as these concepts are at the heart of the present chapter.

The focus of this chapter is not on the properties of the generic system discussed in Chapter 1 – a system that may consist of many clearly identifiable subsystems – but rather on the properties of a pure substance. We formally define a **pure substance** as follows:

A pure substance is a substance that has a homogeneous and unchanging chemical composition.

Each element of the periodic table is a pure substance. Compounds, such as CO_2 and H_2O, are also pure substances. Note that pure substances may exist in various physical phases: **vapor**, **liquid**, and **solid**. You are well acquainted with these phases of H_2O (i.e., steam, water, and ice). Carbon dioxide also readily exhibits all three phases. In a compressed gas cylinder at room temperature, liquid CO_2 is present with a vapor phase above it at a pressure of 56.5 atmospheres. Solid CO_2 or "dry ice" exists at temperatures below −78.5 °C at a pressure of 1 atmosphere.

In most of our discussions of properties, we further restrict our attention to those pure substances that may be classified as **simple compressible substances**:

A* simple compressible substance *is one in which the effects of the following are negligible: motion, fluid shear, surface tension, gravity, and magnetic and electrical fields.

The adjectives ***simple*** and ***compressible*** greatly simplify the description of the state of a substance. Many systems of interest to engineering closely approximate simple compressible substances. Although we consider motion, fluid shear, and gravity as present in many engineering systems, their effects on local thermodynamic properties are quite small; thus, fluid properties are very well approximated as those of a simple compressible substance. Free surface (see Fig. 2.2), magnetic, and electrical effects are not considered in this book.

FIGURE 2.2 Surface tension complicates specifying the thermodynamic state of very small droplets. For example, the pressure inside the droplet depends on the size of the droplet (András Csapó / EyeEm / Getty images).

A more rigorous definition of a simple compressible substance is that, in a system comprising such a substance, the only reversible work mode is that associated with compression or expansion work, that is, P–$d\mathcal{V}$ work. To appreciate this definition, however, requires an understanding of what is meant by *reversible* and by P–$d\mathcal{V}$ *work*. These concepts are developed at length in Chapters 4 and 6.

We conclude this section by distinguishing between **extensive properties** and **intensive properties:**

An* extensive property *depends on how much of the substance is present or the "extent" of the system under consideration.

Examples of extensive properties are volume $\mathcal{V}$ and energy E. Clearly, numerical values for $\mathcal{V}$ and E depend on the size, or mass, of the system. In contrast, intensive properties do not depend on the extent of the system:

An* intensive property *is independent of the mass of the substance or system under consideration.

Two familiar thermodynamic properties, temperature T and pressure P, are intensive properties. Numerical values for T and P are independent of the mass or the amount of substance in the system.

Intensive properties are generally designated using lowercase symbols, although both temperature and pressure violate the rule in this text. Any extensive property can be converted to an intensive one simply by dividing by the mass M. For example, the **specific volume** v and the **specific energy** e are defined, respectively, by

$$v \equiv V/M\,[=]\ \mathrm{m^3/kg} \tag{2.2a}$$

and

$$e \equiv E/M\,[=]\ \mathrm{J/kg}. \tag{2.2b}$$

Equations with yellow backgrounds express key concepts and are the most important relationships in the chapter.

Intensive properties can also be based on the number of moles present rather than the mass. Thus, the properties defined in Eq. 2.2 would be more accurately designated as **mass-specific** properties, whereas the corresponding **molar-specific** properties would be defined by

$$\bar{v} \equiv V/N\,[=]\ \mathrm{m^3/kmol} \tag{2.3a}$$

and

$$\bar{e} \equiv E/N\,[=]\ \mathrm{J/kmol}, \tag{2.3b}$$

where N is the number of moles under consideration. Molar-specific properties are designated using lowercase symbols with an overbar, as shown in Eq. 2.3. Conversions between mass-specific properties and molar-specific properties are accomplished using the following simple relationships:

$$\bar{z} = z\mathcal{M} \tag{2.4}$$

and

$$z = \bar{z}/\mathcal{M} \tag{2.5}$$

where z (or $\bar{z}$) is any intensive property and $\mathcal{M}$ ([=] kg/kmol) is the molecular weight of the substance (see also Eq. 2.7 below).

2.3 Frequently Used Thermodynamic Properties

One of the principal objectives of this chapter is to see how various thermodynamic properties relate to one another, expressed (1) by an **equation of state**, and (2) by **calorific equations of state**. We delay our discussion of properties related to the second law and their relationships until Chapter 7. Before looking at how properties are related, we list in Table 2.1 the most common thermodynamic properties so that you might have an overview of the scope of our study. You may be familiar with many of these properties, although others will be new and may seem strange. As we proceed in our study, this strangeness should disappear as you work with these new properties. We begin with a discussion of three extensive properties: mass, number of moles, and volume.

TABLE 2.1 Common Thermodynamic Properties of Single-Phase Pure Substances

Property	Symbolic Designation		Units		Relation to Other Properties	Classification
	Extensive	Intensive*	Extensive	Intensive*		
Mass	M	—	kg	—	—	Fundamental property appearing in equation of state
Number of moles	N	—	kmol	—	—	Fundamental property appearing in equation of state
Volume and specific volume	$\mathcal{V}$	v	m^3	m^3/kg	—	Fundamental property appearing in equation of state
Pressure	—	P	—	Pa *or* N/m^2	—	Fundamental property appearing in equation of state
Temperature	—	T	—	K	—	Fundamental property appearing in equation of state
Density	—	ρ	—	kg/m^3	—	Fundamental property appearing in equation of state
Internal energy	U	u	J	J/kg	—	Based on first law and calculated from calorific equation of state
Enthalpy	H	h	J	J/kg	$U + P\mathcal{V}$	Based on first law and calculated from calorific equation of state
Constant-volume specific heat	—	c_v	—	J/kg·K	$(\partial u/\partial T)_v$	Appears in calorific equation of state
Constant-pressure specific heat	—	c_p	—	J/kg·K	$(\partial h/\partial T)_p$	Appears in calorific equation of state
Specific-heat ratio	—	γ	—	*Dimensionless*	c_p/c_v	—
Entropy	S	s	J/K	J/kg·K	—	Based on second law of thermodynamics
Gibbs free energy (or Gibbs function)	G	g	J	J/kg	$H - TS$	Based on second law of thermodynamics
Helmholtz free energy (or Helmholtz function)	A	a	J	J/kg	$U - TS$	Based on second law of thermodynamics

* Molar intensive properties are obtained by substituting the number of moles, N, for the mass and changing the units accordingly. For example, $s = S/M$ [=] J/kg·K becomes $\bar{s} \equiv S/N$ [=] J/kmol·K.

2.3a Properties Related to the Equation of State

MASS

As one of our fundamental dimensions (see Chapter 1), **mass**, like time, cannot be defined in terms of other dimensions. Much of our intuition of what mass is follows from its role in Newton's second law of motion,

$$F = Ma. \tag{2.6}$$

In this relationship, the force F required to produce a certain acceleration of a particular body is proportional to its mass M. In 2019, the SI mass standard (a platinum–iridium cylinder) was replaced by a definition of mass based solely on three physical constants: Planck's constant, the speed of light, and the hyperfine transition frequency of cesium-133.

NUMBER OF MOLES

In some applications, such as reacting systems, the number of moles N comprising the system is more useful than the mass. In 2019, the **mole** was formally redefined as the amount of substance in a system that contains exactly $6.02214076 \times 10^{23}$ elementary entities. The elementary entities may be atoms, molecules, ions, electrons, etc. The abbreviation for the SI unit for a mole is mol, and kmol refers to 10^3 mol.

The **atomic** or **molecular weight** $\mathcal{M}$; is the ratio of the system mass to the number of moles in the system:

$$\mathcal{M} = M/N, \tag{2.7}$$

where $\mathcal{M}$ has units of g/mol or kg/kmol. (Note that the "atomic weight" is not a weight at all but is the relative atomic mass.) The **Avogadro constant** $\mathcal{N}_{AV}$ defines the number of particles (atoms, molecules, etc.) in a mole:

$$\mathcal{N}_{AV} \equiv \begin{cases} 6.02214076 \times 10^{23} \text{ particles/mol,} \\ 6.02214076 \times 10^{26} \text{ particles/kmol.} \end{cases} \tag{2.8}$$

For example, we can use the Avogadro constant and the known atomic weight of carbon-12 (12.0 g/gmol) to determine the mass of a single ^{12}C atom:

$$\begin{aligned} M_{12_C} &= \frac{M(1\,\text{mol}^{12}\text{C})}{\mathcal{N}_{AV} N_{12_C}} \\ &= \frac{0.012}{6.02214076 \times 10^{23}(1)} = 1.9926469 \times 10^{-26} \\ &[=] \frac{\text{kg}}{(\text{atom/mol})\text{mol}} = \frac{\text{kg}}{\text{atom}}. \end{aligned}$$

Although not an SI unit, one-twelfth of the mass of a single ^{12}C atom is sometimes used as a mass standard and is referred to as the **unified atomic mass unit**, defined as

$$1\, m_u \equiv (1/12) M_{12_C} = 1.66053907 \times 10^{-27} \text{kg}.$$

VOLUME

The familiar property of **volume** is formally defined as the amount of space occupied in three-dimensional space. The SI unit of volume is cubic meters (m^3).

DENSITY

Consider the small volume $\Delta\mathcal{V}$ $(= \Delta x \Delta y \Delta z)$ as shown in Fig. 2.3. We formally define the density to be the ratio of the mass of this element to the volume of the element, under the condition that the size of the element shrinks to the continuum limit, that is,

$$\rho \equiv \lim_{\Delta\mathcal{V}\to\mathcal{V}_{\text{continuum}}} \frac{\Delta M}{\Delta\mathcal{V}} \; [=] \, \text{kg/m}^3. \tag{2.9}$$

What is meant by the **continuum limit** is that the volume is very small but yet sufficiently large that the number of molecules within the volume is essentially constant and unaffected by any statistical fluctuations. For a volume smaller than the continuum limit, the number of molecules within the volume fluctuates with time as molecules randomly enter and exit the volume. As an example, consider a volume such that, on average, only two molecules are present. Because the volume is so small, the number of molecules within may fluctuate wildly, with three or more molecules present at some times, and one or none at other times. In this situation, the volume is below the continuum limit. In most practical systems, however, the continuum limit is quite small, and the density, and other thermodynamic properties, can be considered to be smooth functions in space and time. Table 2.2 provides a quantitative basis for this statement.

TABLE 2.2 Size of Gas Volume Containing *n* Molecules at 25 °C and 1 atm

Number of molecules (n)	Volume (mm^3)	Size of equivalent cube (mm)
10^5	4.06×10^{-12}	0.00016
10^6	4.06×10^{-11}	0.0003
10^{10}	4.06×10^{-7}	0.0074
10^{12}	4.06×10^{-5}	0.0344

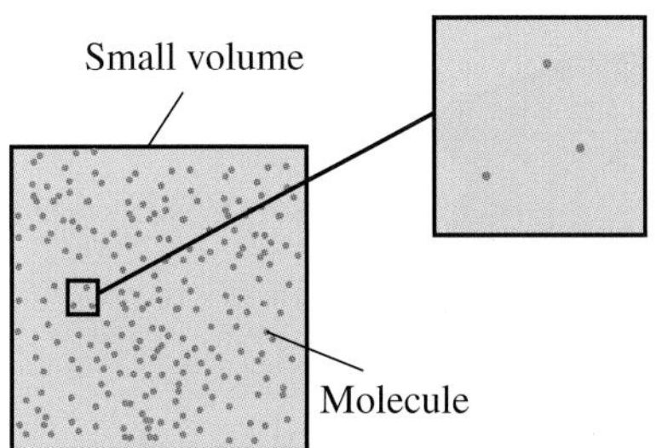

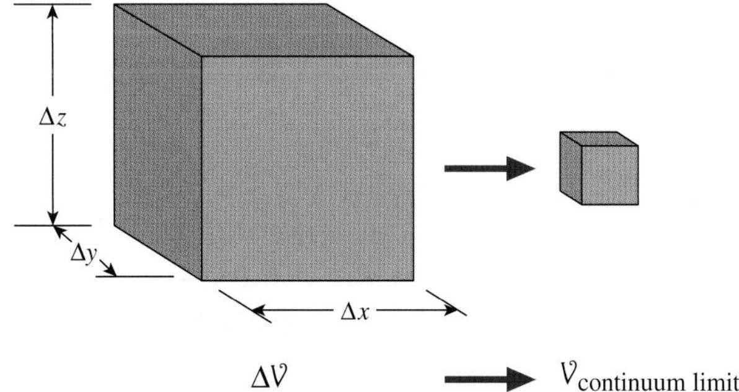

FIGURE 2.3 A finite volume element shrinks to the continuum limit to define macroscopic properties. Volumes smaller than the continuum limit experience statistical fluctuations in properties as molecules enter and exit the volume.

SPECIFIC VOLUME

The **specific volume** is the inverse of the density, that is,

$$v \equiv \frac{1}{\rho} \; [=] \, \text{m}^3/\text{kg}. \tag{2.10}$$

Physically this is interpreted as the amount of volume per unit mass associated with a volume at the continuum limit. The specific volume is most frequently used in thermodynamic applications, whereas density is more commonly used in fluid mechanics and heat transfer. You should be comfortable with both properties and immediately recognize their inverse relationship.

PRESSURE

For a fluid (liquid or gaseous) system, the pressure is defined as the magnitude of the normal force exerted by the fluid on a solid surface or a neighboring fluid element, per unit area, as the area shrinks to the continuum limit, that is,

$$P \equiv \lim_{\Delta A \to A_{\text{continuum}}} \frac{F_{\text{normal}}}{\Delta A}. \tag{2.11}$$

This definition assumes that the fluid is in a state of equilibrium at rest. It is important to note that the pressure is a scalar quantity, having no direction associated with it. Figure 2.4 illustrates this concept, showing that the pressure at a point is independent of orientation. Throughout this book, the idea of *a point* is interpreted in light of the continuum limit, that is, a point is of some small dimension rather than of zero dimension, as required by the mathematical definition of a point. That the pressure force is always directed normal to a surface (real or imaginary) is a consequence of a fluid being unable to sustain any tangential force without movement. If a tangential force is present, the fluid layers simply slip over one another.

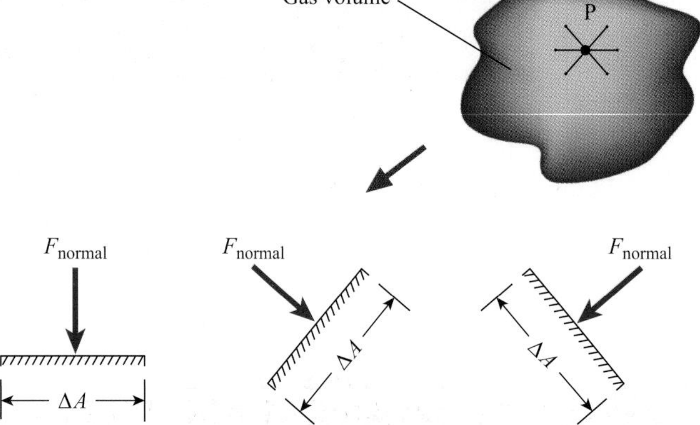

FIGURE 2.4 The pressure at a point P is a scalar quantity independent of orientation. Regardless of the orientation of ΔA, application of the defining relationship (Eq. 2.11) results in the same value of the pressure.

From a microscopic (molecular) point of view, the pressure exerted by a gas on the walls of its container is a measure of the rate at which the momentum of the molecules colliding with the wall changes.

The SI unit for pressure is a *pascal,* defined by

$$P[=]\,\text{Pa} \equiv 1\ \text{N/m}^2. \tag{2.12a}$$

Since one pascal is typically a small number in engineering applications, multiples of 10^3 and 10^6 are employed, which result in the usage of the kilopascal, kPa (10^3 Pa), and the megapascal, MPa (10^6 Pa). Also commonly used is the **bar,** which is defined as

$$1\ \text{bar} \equiv 10^5\ \text{Pa}. \tag{2.12b}$$

Pressure is also frequently expressed in terms of a standard atmosphere:

$$1\ \text{standard atmosphere (atm)} \equiv 101{,}325\ \text{Pa}. \tag{2.12c}$$

FIGURE 2.5 A Bourdon gage contains a coiled, flattened tube that uncoils slightly when it is connected to a fluid having a pressure greater than the local atmospheric pressure. The uncoiling tube is connected through gears and levers to cause a corresponding rotation of the needle on the dial. In addition to SI units, many other units for pressure are commonly employed. Most of these units originate from the application of a particular measurement method. For example, the use of manometers results in pressures expressed in inches of water or millimeters of mercury. American (or British) customary units (pounds-force per square inch or psi) are frequently appended with a "g" or an "a" to indicate gage or absolute pressures, respectively (i.e., psig and psia). Conversion factors for common pressure units are provided at the front of this book (Image Source / Getty Images).

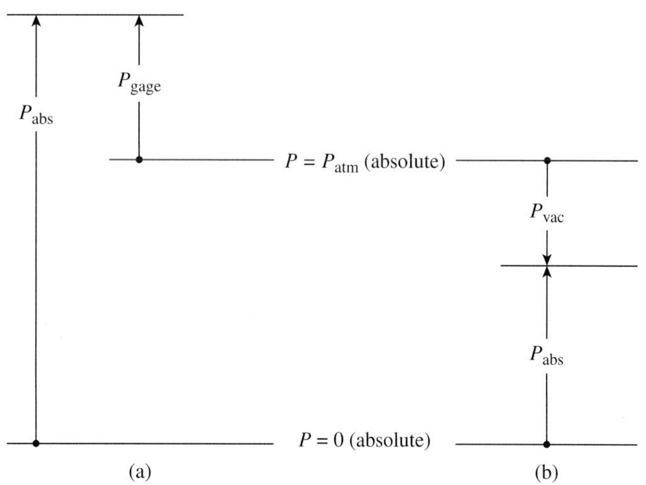

FIGURE 2.6 The absolute pressure is zero in a perfectly evacuated space; gage pressure is measured relative to the local atmospheric pressure.

Because some practical devices measure pressures relative to the local atmospheric pressure (see Fig. 2.5), we distinguish between **gage pressure** and **absolute pressure.** Gage pressure is defined as

$$P_{gage} \equiv P_{abs} - P_{atm,abs}, \tag{2.13}$$

where the absolute pressure P_{abs} is that as defined in Eq. 2.11. In a perfectly evacuated space, the absolute pressure is zero. Figure 2.6 illustrates the relationship between gage and absolute pressures. The term vacuum or vacuum pressure is also employed in engineering applications (and leads to confusion if one is not careful) and is defined as

$$P_{vacuum} = P_{atm,abs} - P_{abs}. \tag{2.14}$$

This relationship is also illustrated in Fig. 2.6.

Example 2.1 Gage and Absolute Pressure

A pressure gage is used to measure the inflation pressure of a tire. The gage reads 35 psig in San Francisco when the barometric pressure is 28.5 inches of mercury. What is the absolute pressure in the tire in kPa and psia?

(Credit: Paul Bradbury / Caiaimage / Getty Images.)

Solution

Known $P_{tire,\,g}$, $P_{atm,\,abs}$

Find $P_{tire,\,abs}$

Sketch

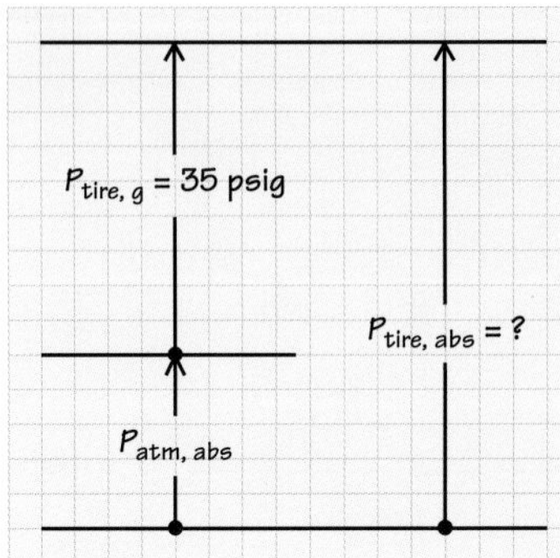

Analysis From the sketch and from Eq. 2.13, we know that

$$P_{\text{tire, abs}} = P_{\text{tire, g}} + P_{\text{atm, abs}}.$$

We need to deal with the mixed units given in order to apply this relationship. Applying the conversion factor from the front of this book to express the atmospheric pressure in units of psia yields

$$P_{\text{atm, abs}} = (28.5\,\text{in Hg})\left[\frac{14.70\,\text{psia}}{29.92\,\text{in Hg}}\right] = 14.0\,\text{psia}.$$

Thus,

$$P_{\text{tire, abs}} = 35 + 14.0 = 49\ \text{psia}.$$

Converting this result to units of kPa yields

$$P_{\text{tire, abs}} = 49\,\text{lb}_\text{f}/\text{in}^2\left[\frac{1\,\text{Pa}}{1.4504\times10^{-4}\,\text{lb}_\text{f}/\text{in}^2}\right]\left[\frac{1\,\text{kPa}}{1000\ \text{Pa}}\right] = 337.8\,\text{kPa}.$$

Comments Note the use of three different units to express pressure: Pa (or kPa), psi (or $\text{lb}_\text{f}/\text{in}^2$), and inches of Hg. Other commonly used units are millimetres of Hg and inches of H_2O. You should be comfortable working with any of these. Note also the usage psia and psig to denote absolute and gage pressures, respectively, when working with $\text{lb}_\text{f}/\text{in}^2$ units.

Self-Test 2.1

☑ A car tire suddenly goes flat and a pressure gage indicates zero psig. Is the absolute pressure in the tire also zero psia?

(Answer: No. A zero gage reading indicates that the absolute pressure in the tire is atmospheric pressure.)

Example 2.2 SI Engine Application

A vintage automobile has an intake manifold vacuum gage in the dashboard instrument cluster. Cruising at 20.1 m/s (45 mph), the gage reads 14 inches of Hg vacuum. If the local atmospheric pressure is 99.5 kPa, what is the absolute intake manifold pressure in kPa?

Solution

Given $P_{\text{man, vac}}$, $P_{\text{atm, abs}}$

Find $P_{\text{man, abs}}$

Sketch

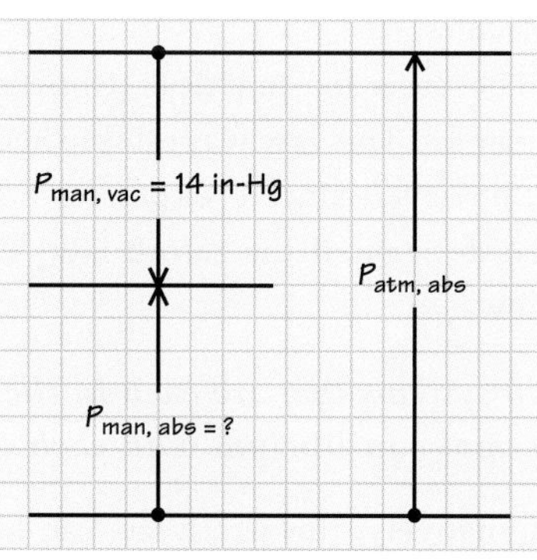

Analysis From the sketch and a rearrangement of Eq. 2.14, we find the intake manifold absolute pressure to be

$$P_{\text{man,abs}} = P_{\text{atm,g}} - P_{\text{man,vac}}.$$

Substituting numerical values and converting units yields

$$\begin{aligned} P_{\text{man,abs}} &= 99.5\,\text{kPa} - (14\,\text{in Hg})\left[\frac{1\ \text{atm}}{29.92\,\text{in Hg}}\right]\left[\frac{101.325\,\text{kPa}}{1\ \text{atm}}\right] \\ &= 99.5\,\text{kPa} - 47.4\ \text{kPa} = 52.1\,\text{kPa}. \end{aligned}$$

Comments The pressure drops across the throttle plate of an SI engine (see Chapter 8). This lower pressure, in turn, results in a decreased air density and a smaller quantity of air entering the cylinder than would occur without the throttle. The throttle thus controls the power delivered by the engine.

Self-Test 2.2

 A technician performs a compression test on a vehicle engine and finds that the maximum pressure in one cylinder is 105 psig while the minimum pressure in another is 85 psig. What is the absolute pressure difference between the two cylinders?

(Answer: 20 psia)

TEMPERATURE

Like mass, length, and time, temperature is a fundamental dimension and, as such, eludes a simple and concise definition. Nevertheless, we all have some experiential notion of temperature when we say that some object is hotter than another, that is, that some object has a greater temperature than another. (The perception of temperature, however, is a complex phenomenon involving the physical properties of the object and perceiver, the rate of heat transfer between the object and perceiver, and the actual temperature.) From a macroscopic point of view, we define temperature as the property that is shared by two systems, initially at different states, after they have been placed in thermal contact and allowed to come to thermal equilibrium. Although this definition may not be very satisfying, it is the best we can do from a macroscopic viewpoint. For the special case of an ideal gas, the microscopic (molecular) point of view may be somewhat more satisfying. Here the temperature is directly proportional to the square of the mean molecular speed. A higher temperature means faster-moving molecules.

The basis for practical temperature measurement is the zeroth law of thermodynamics. This law was established after the first and second laws of thermodynamics; however, since it expresses a concept logically preceding the other two, it has been designated the zeroth law. The **zeroth law of thermodynamics** is stated as follows:

Two systems that are each in thermal equilibrium with a third system are in thermal equilibrium with each other.

Alternatively, the zeroth law can be expressed explicitly in terms of temperature:

When two systems have equality of temperature with a third system, they in turn have equality of temperature with each other.

This law forms the basis for thermometry. A thermometer measures the same property, temperature, independently of the nature of the system subject to the measurement. A temperature of 20 °C measured for a block of steel means the same thing as 20 °C measured for a container of water. Putting the 20 °C steel block in the 20 °C water results in no temperature change to either.

As a result of the zeroth law, a *practical temperature scale* can be based on a *thermometric* substance. Such a substance expands as its temperature increases; mercury is a thermometric substance. The height of the mercury column in a glass tube can be calibrated against standard *fixed points* of reference. For example, the original Celsius scale (i.e., prior to 1954) defines 0 °C to be the temperature at **ice point** and 100 °C to be the temperature at **steam point. Ice point** is the temperature at which an ice and water mixture is in equilibrium with water vapor-saturated air at one atmosphere, and the **steam point** is the temperature at which steam and water are in equilibrium at one atmosphere. The modern Celsius scale assigns a temperature of 0.01 °C to the **triple point** of water and the size of a single degree equal to that from the absolute, or Kelvin, temperature scale, as will be discussed in Chapter 6. The **triple point** is the temperature at which ice, liquid water, and steam all coexist in equilibrium. With the adoption of the International Temperature Scale of 1990 (ITS-90), the ice point is still 0 °C, but the steam point is now 99.974 °C. For practical purposes, the original and modern Celsius scales are identical.

Four temperature scales are in common use today: the Celsius scale and its absolute counterpart, the Kelvin scale, and the Fahrenheit scale and its absolute counterpart, the Rankine scale. Both absolute scales start at absolute zero, the lowest temperature possible. The conversions among these scales are shown in Table 2.3.

TABLE 2.3 Temperature Scales

Temperature scale	Unit*	Relation to other scales
Celsius	degree Celsius (°C)	$T(^\circ\mathrm{C}) = T(\mathrm{K}) - 273.15$
		$T(^\circ\mathrm{C}) = \frac{5}{9}[T(\mathrm{F}) - 32]$
Kelvin	kelvin (K)	$T(\mathrm{K}) = T(^\circ\mathrm{C}) + 273.15$
		$T(\mathrm{K}) = \frac{5}{9}T(\mathrm{R})$
Fahrenheit	degree Fahrenheit (F)	$T(\mathrm{F}) = T(\mathrm{R}) - 459.67$
		$T(\mathrm{F}) = \frac{9}{5}T(^\circ\mathrm{C}) + 32$
Rankine	degree Rankine (R)	$T(\mathrm{R}) = T(\mathrm{F}) + 459.67$
		$T(\mathrm{R}) = \frac{9}{5}T(\mathrm{K})$

* Note that capital letters are used to refer to the units for each scale. The degree symbol (°), however, is used only with the Celsius unit to avoid confusion with the coulomb. Note also that the SI Kelvin scale unit is the kelvin; thus, we say that a temperature, for example, is 100 kelvins (100 K), not 100 degrees Kelvin.

Example 2.3 Temperature Conversion

On a hot day in Boston, a high of 97 degrees Fahrenheit is reported on the nightly news. What is the temperature in units of °C, K, and R?

(Credit: DenisTangney Jr / iStock / Getty Images Plus.)

Solution

Known $T(\mathrm{F})$

Find $T(^\circ\mathrm{C})$, $T(\mathrm{K})$, $T(\mathrm{R})$

Analysis We apply the temperature-scale conversions provided in Table 2.3 as follows:

$$
\begin{aligned}
T\,(^\circ\mathrm{C}) &= \tfrac{5}{9}[T(\mathrm{F}) - 32] \\
&= \tfrac{5}{9}(97 - 32) = 36.1\,^\circ\mathrm{C}, \\
T\,(\mathrm{R}) &= T(\mathrm{F}) + 459.67 \\
&= 97 + 459.67 = 556.7\,\mathrm{R}, \\
T\,(\mathrm{K}) &= \tfrac{5}{9}T(\mathrm{R}) \\
&= \tfrac{5}{9}(556.7) = 309.3\,\mathrm{K}.
\end{aligned}
$$

Comments Except for the Rankine scale, it is likely that you are familiar with the conversions in Table 2.3. Note that the size of the temperature unit is identical for the Fahrenheit and Rankine scales. Similarly, the Celsius and Kelvin units are of identical size, and each is 9/5 (or 1.8) times the size of the Fahrenheit or Rankine unit.

Self-Test 2.3

On the same hot day in Boston, the air conditioning keeps your room at 68 degrees Fahrenheit. Find the temperature difference between the inside and the outside air in (a) R, (b) °C, and (c) K.

(Answer: (a) 29 R, (b) 16.1 °C, (c) 16.1 K)

2.3b Properties Related to the First Law and Calorific Equation of State

INTERNAL ENERGY

In this section, we introduce the thermodynamic property **internal energy**. Further discussion of internal energy is presented in Chapter 4, which focuses on the many ways that energy is stored and transferred.

Internal energy has its origins with the microscopic nature of matter; specifically, we define internal energy as the energy associated with the motions of the microscopic particles (atoms, molecules, electrons, etc.) comprising a macroscopic system. For simple monatomic gases (e.g., helium and argon), the internal energy is associated only with the translational kinetic energy of the atoms (Fig. 2.7a). If we assume that a gas can be modeled as a collection of point-mass hard spheres that collide elastically, the translational kinetic energy associated with n particles is

$$U_{\mathrm{trans}} = n\tfrac{1}{2}M_{\mathrm{molec}}\overline{V^2}, \tag{2.15}$$

where $\overline{V^2}$ is the mean-square molecular speed. Using kinetic theory (see, for example, Ref. [2]), the translational kinetic energy can be related to temperature as

$$U_{\mathrm{trans}} = n\tfrac{3}{2}k_{\mathrm{B}}T, \tag{2.16}$$

where k_{B} is the Boltzmann constant,

$$k_{\mathrm{B}} = 1.380649 \times 10^{-23}\,\mathrm{J/K}$$

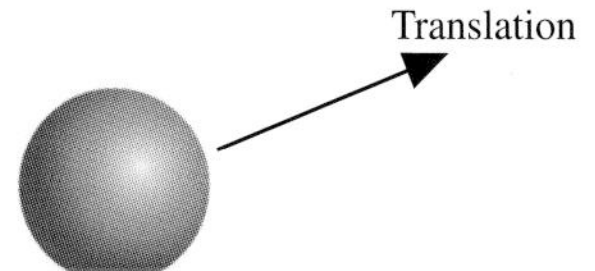

(a) Monatomic species

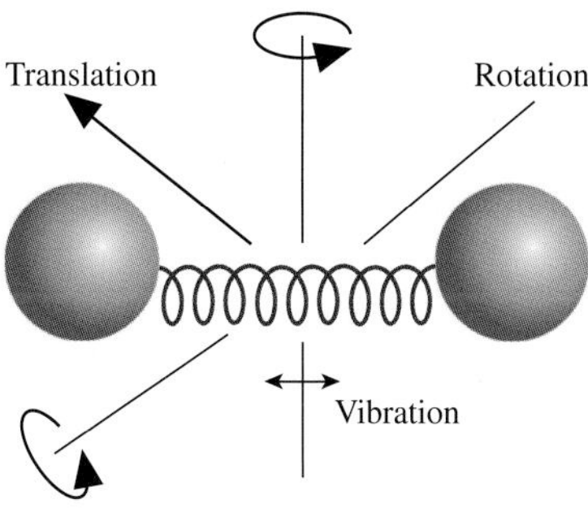

(b) Diatomic species

FIGURE 2.7 (a) The internal energy of a monatomic species consists only of translational (kinetic) energy. **(b)** In a diatomic species, the internal energy results from translation together with energy from vibration (potential and kinetic) and rotation (kinetic). Image from *An Introduction to Combustion: Concepts and Applications*, 3rd edn, Stephen Turns. © McGraw-Hill Education.

and T is the absolute temperature in kelvins. (By comparing Eqs. 2.15 and 2.16, we see the previously mentioned microscopic interpretation of temperature, i.e., $T \equiv M_{\text{molec}}\overline{V^2}/[3k_B]$.)

For molecules more complex than single atoms, internal energy is stored in vibrating molecular bonds and the rotation of the molecule about two or more axes, in addition to the translational kinetic energy. Figure 2.7b illustrates this model of a diatomic species. In general, the internal energy is expressed

$$U = U_{\text{trans}} + U_{\text{vib}} + U_{\text{rot}}, \tag{2.17}$$

where U_{vib} is the vibrational kinetic and potential energy, and U_{rot} is the rotational kinetic energy. The amount of energy that is stored in each mode varies with temperature and is described by quantum mechanics. One of the fundamental postulates of quantum theory is that energy is quantized; that is, energy storage is modeled by discrete bits (quanta) rather than continuous functions. The translational kinetic energy states are very close together, so that, for practical purposes, quantum states need not be considered and the continuum result, Eq. 2.15, is a useful model. For vibrational and rotational states, however, quantum behavior is important. We will see the effects of this later in our discussion of specific heats.

Another form of internal potential energy is that associated with chemical bonds and their rearrangements during chemical reactions (Fig. 2.8). Similarly, internal energy is associated with nuclear bonds. We address the topic of chemical energy storage in Chapters 12 and 13; nuclear energy storage, however, lies beyond our scope.

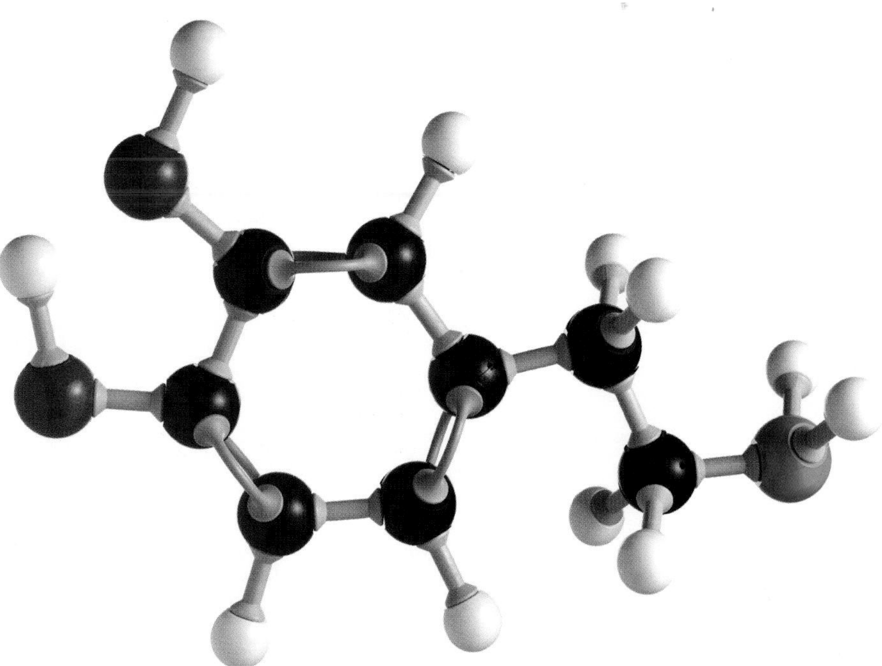

FIGURE 2.8 For reacting systems, chemical bonds make an important contribution to the system internal energy (Science Photo Library / Getty Images).

The SI unit for internal energy is the joule (J); for mass-specific internal energy, it is joules per kilogram (J/kg) and for molar-specific internal energy, it is joules per kilomole (J/kmol).

ENTHALPY

Enthalpy is a useful property defined by the following combination of more common properties:

$$H \equiv U + P\mathcal{V}. \tag{2.18}$$

On a mass-specific basis, the enthalpy of a system involves the specific volume or the density, that is,

$$h \equiv u + Pv \tag{2.19a}$$

or

$$h \equiv u + P/\rho. \tag{2.19b}$$

Enthalpy first appears in conservation of energy for systems in Example 5.4.

Conservation of energy for open systems (Eq. 5.15) uses enthalpy to replace the combination of flow work (see Chapter 4) and internal energy.

Enthalpy has the same units as internal energy (i.e., J or J/kg). Molar-specific enthalpies are obtained by the application of Eq. 2.4.

The usefulness of the enthalpy concept will become clear during our discussion of the first law of thermodynamics (the principle of energy conservation) in Chapter 5. There we will see that the combination of properties $u + Pv$ arises naturally in analyzing systems at constant pressure and in analyzing open systems. In the former, the Pv term results from expansion and/or compression work; for the latter, the Pv term is associated with the work needed to push the fluid into or out of the open system, that is, flow work. Later in this chapter, we discuss internal energy and enthalpy further.

SPECIFIC HEATS AND SPECIFIC-HEAT RATIO

Here we deal with two intensive properties,

$$c_v \equiv \text{constant-volume specific heat}$$

and

$$c_p \equiv \text{constant-pressure specific heat.}$$

These properties mathematically relate to the specific internal energy u and enthalpy h, as follows:

$$c_v \equiv \left(\frac{\partial u}{\partial T}\right)_v \tag{2.20a}$$

and

$$c_p \equiv \left(\frac{\partial h}{\partial T}\right)_p. \tag{2.20b}$$

Similar defining relationships relate the molar-specific heats and molar-specific internal energy and enthalpy. Physically, the constant-volume specific heat is the slope of the internal energy versus temperature curve for a substance undergoing a process conducted at constant volume. Similarly, the constant-pressure specific heat is the slope of the enthalpy versus temperature curve for a substance undergoing a process conducted at constant pressure. These ideas are illustrated in Fig. 2.9. It is

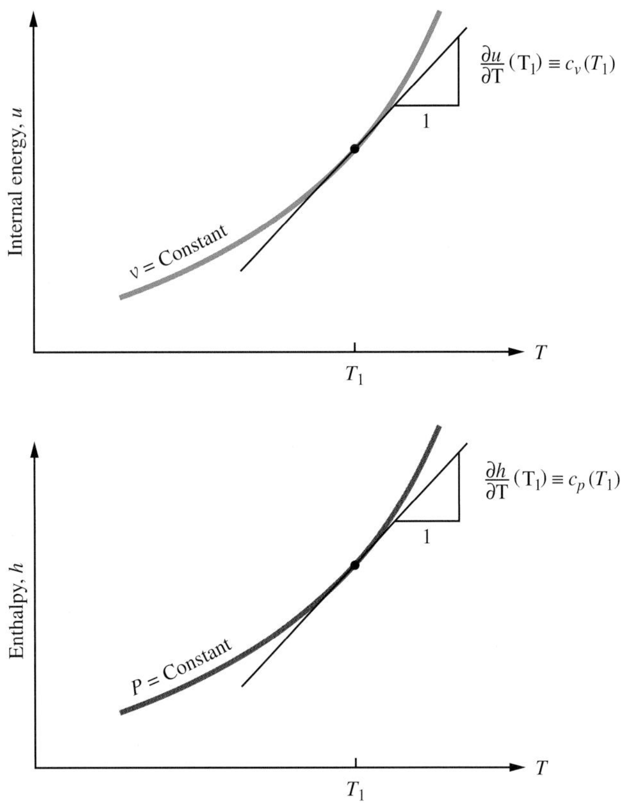

FIGURE 2.9 The constant-volume specific heat c_v is defined as the slope of u versus T for a constant-volume process (top). Similarly, c_p is the slope of h versus T for a constant-temperature process (bottom). Generally, both c_v and c_p are functions of temperature, as suggested by these graphs.

important to note that, although the definitions of these properties involve constant-volume and constant-pressure processes, c_v and c_p can be used in the description of any process regardless of whether the volume (or pressure) is held constant.

For solids and liquids, specific heats generally increase with temperature, in a way that is essentially uninfluenced by pressure. A notable exception to this is mercury, which exhibits a decreasing constant-pressure specific heat with temperature. Values of specific heats for selected liquids are presented in Appendix G.

For both real (nonideal) and ideal gases, the specific heats c_v and c_p are functions of temperature. The specific heats of nonideal gases also possess a pressure dependence. For gases, the temperature dependences of c_v and c_p are a consequence of the fact that the internal energy of a molecule consists of three components – translational, vibrational, and rotational – and the fact that the vibrational and rotational energy storage modes become increasingly active as temperature increases, as described by quantum theory. As discussed previously, Fig. 2.7 schematically illustrates these three energy storage modes by contrasting a monatomic species, whose internal energy consists solely of translational kinetic energy, and a diatomic molecule, which stores energy in a vibrating chemical bond, represented as a spring between the two nuclei, and by rotation about two orthogonal axes, as well as possessing kinetic energy from translation. With these simple models (Fig. 2.7), we expect the specific heats of diatomic molecules to be greater than those of monatomic species, which is indeed true. In general, the more complex the molecule, the greater is its molar specific heat. This can be seen clearly in Fig. 2.10, where molar constant-pressure specific heats for a number of gases are shown as functions of temperature. As a group, the triatomics have the greatest specific heats, followed by the diatomics, and lastly, the monatomics. Note that the triatomic molecules are also more temperature dependent than the diatomics, a consequence of the greater number of

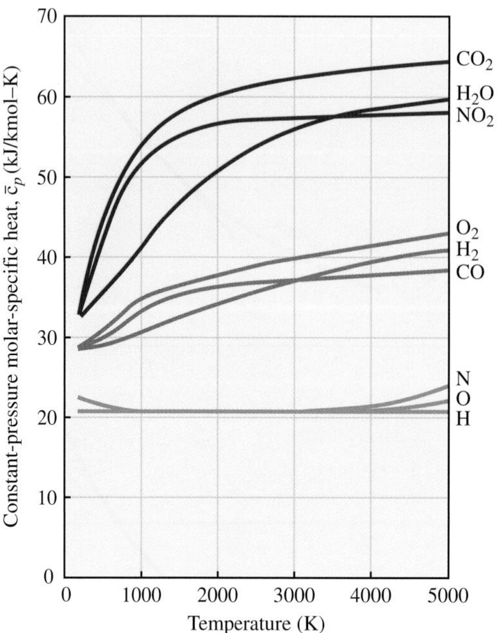

FIGURE 2.10 Molar constant-pressure specific heats as functions of temperature for monatomic (H, N, and O), diatomic (CO_2, H_2, and O_2), and triatomic (CO_2, H_2O, and NO_2) species. Values are from Appendix D. Image from *An Introduction to Combustion: Concepts and Applications*, 3rd edn, Stephen Turns. © McGraw-Hill Education.

vibrational and rotational modes that are available to become activated as the temperature is increased. In comparison, the monatomic species have nearly constant specific heats over a wide range of temperatures; in fact, the specific heat of monatomic hydrogen is constant ($c_p = 20.786$ kJ/kmol·K) from 200 K to 5000 K.

Constant-pressure molar-specific heats are tabulated as a function of temperature for various ideal-gas species in Tables D.l to D.12 in Appendix D. Also provided in Appendix D are curve-fit coefficients taken from the Chemkin thermodynamic database [3], which were used to generate the tables. These coefficients can be easily used with spreadsheet software to obtain $\bar{c}_p$ values at any temperature within the given temperature ranges.

Values of c_v and c_p for a number of substances are also available from the National Institute of Standards and Technology (NIST) WebBook [6] and various software packages (e.g., the *miniREFPROP*, *REFPROP*, and Matlab toolboxes). We discuss the use of these important resources later in this chapter.

The ratio of specific heats, γ, is another commonly used property and is defined by

$$\gamma \equiv \frac{c_p}{c_v} = \frac{\bar{c}_p}{\bar{c}_v}. \tag{2.21}$$

The specific-heat ratio is frequently denoted by k or r, as well as gamma (γ), our choice here.

Example 2.4 Molar-Specific Heat and Mass-Specific Heat

Compare the values of the constant-pressure specific heats for hydrogen (H_2) and carbon monoxide (CO) at 3000 K using the ideal-gas molar-specific values from the tables in Appendix D. How does this comparison change if mass-specific values are used?

Solution

Known H_2 and CO at T

Find $\bar{c}_{p,H_2}$, $\bar{c}_{p,CO}$, c_{p,H_2}, $c_{p,CO}$

Modeling, Premises and Assumptions Ideal-gas behavior

Analysis To answer the first question requires only a simple table look-up. The molar constant-pressure specific heats found in Table D.3 for H_2 and in Table D.l for CO are as follows:

$$\bar{c}_{p,H_2}(T = 3000\,K) = 37.112\ kJ/kmol{\cdot}K.$$
$$\bar{c}_{p,CO}(T = 3000\,K) = 37.213\ kJ/kmol{\cdot}K.$$

The difference between these values is 0.101 kJ/kmol·K, or approximately 0.3%.

Using the molecular weights of H_2 and CO found in Tables D.3 and D.1, we can calculate the mass-based constant-pressure specific heats using Eq. 2.5 as follows:

$$\begin{aligned} c_{p,H_2} &= \bar{c}_{p,H_2}/\mathcal{M}_{H_2} \\ &= \frac{37.213\,kJ/kmol{\cdot}K}{2.016\,kg/kmol} \\ &= 18.409\ kJ/kg{\cdot}K \end{aligned}$$

and

$$\begin{aligned} c_{p,CO} &= \bar{c}_{p,CO}/\mathcal{M}_{CO} \\ &= \frac{37.112\,kJ/kmol{\cdot}K}{28.010\,kg/kmol} \\ &= 1.329\ kJ/kg{\cdot}K. \end{aligned}$$

Comments We first note that, on a molar basis, the specific heats of H_2 and CO are nearly identical. This result is consistent with Fig. 2.10, where we see that the molar-specific heats are similar for the three diatomic species. On a mass basis, however, the constant-pressure specific heat of H_2 is almost 14 times greater than that of CO, which results from the molecular weight of CO being approximately 14 times that of H_2.

Self-Test 2.4

Calculate the specific heat ratios for (a) H_2, (b) CO, and (c) air using the data from Table E.l in Appendix E.

(Answer: (a) 1.402, (b) 1.398, (c) 1.400)

2.3c Properties Related to the Second Law

Several properties of importance to our study have their origins in the second law of thermodynamics: **entropy** (S), **Gibbs free energy** or **Gibbs function** (G), **Helmholtz free energy** (A), and **availability** or **exergy** ($\mathcal{A}$). Entropy is particularly useful in determining the spontaneous direction of a process and for establishing maximum possible efficiencies, for example. The Gibbs free energy is important in determining chemical equilibrium and phase equilibrium at constant temperature and pressure.

The Helmholtz free energy plays a similar role at fixed temperature and volume. Availability (exergy) is a measure of the quality of energy, expressing how much energy in a system, or flow stream, is available to produce useful work. We discuss these second law properties in Chapters 7 and 13 and, for the time being, just note their existence here.

2.4 Concept of State Relationships

2.4a State Principle

An important concept in thermodynamics is the **state principle (or state postulate):**

In dealing with a simple compressible substance, the* thermodynamic state *is completely defined by specifying two independent intensive properties.

The state principle allows us to define **state relationships** among the various thermodynamic properties. Before developing such state relationships, we examine what is meant by independent properties.

The concept of independent properties is particularly important in dealing with substances when more than one phase is present. For example, temperature and pressure are not independent properties when water (liquid) and steam (vapor) coexist. As you are well aware, water at one atmosphere boils at a specific temperature (i.e., 100 °C). Increasing the pressure results in an increase in the boiling point, the principle upon which the pressure cooker is based. One cannot change the pressure and keep the temperature constant: A fixed relationship exists between temperature and pressure; hence, they are not independent. We will examine this concept of independence in detail later when we study the properties of substances that exist in multiple phases.

2.4b *P–v–T* Equations of State

What is generally known as an **equation of state** is the mathematical relationship among the following three intensive thermodynamic properties: pressure P; specific volume v, and temperature T. The state principle allows us to determine any one of the three properties from knowledge of the other two. In its most general and abstract form, we can write the P–v–T equation of state as

$$f_1(P, v, T) = 0. \tag{2.22}$$

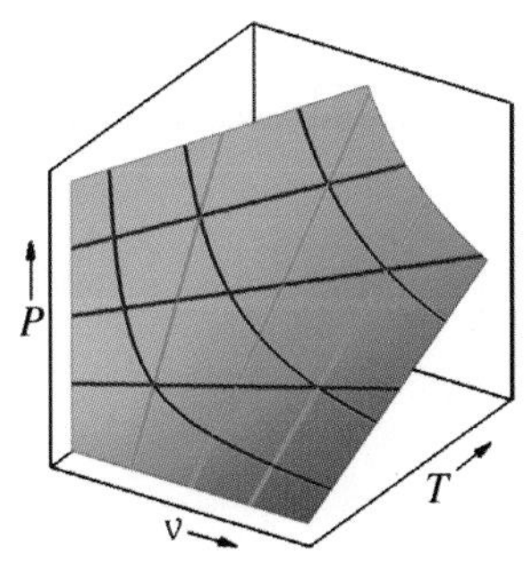

FIGURE 2.11 Ideal gas P–v–T surface (with permission from ARL).

Equation 2.22 describes a three-dimensional surface in P–v–T space as illustrated in Fig. 2.11. In the following sections, we explore explicit functions relating P, v, and T for various substances, starting with an ideal gas, a concept with which you should already have some familiarity.

2.4c Calorific Equations of State

A second type of state relationship connects energy-related thermodynamic properties to pressure, temperature, and specific volume. The state principle also applies here;

thus, for a simple compressible substance, a knowledge of any two intensive properties is sufficient to determine any of the others. The most common **calorific equations of state** relate specific internal energy u to v and T, and, similarly, enthalpy h to P and T, that is,

$$f_2(u, T, v) = 0, \tag{2.23a}$$

or

$$f_3(h, T, P) = 0. \tag{2.23b}$$

These ideas are developed in more detail for various substances in the sections that follow.

2.5 Ideal Gases as Pure Substances

In this section, we define all the useful state relationships for a class of pure substances known as ideal gases. We begin with the definition of an ideal gas.

2.5a Ideal-Gas Definition

The following definition of an **ideal gas** is tautological in that it uses a state relationship to define what is meant by an ideal gas:

An ideal gas is a gas that obeys the relationship $Pv = RT$.

In this definition, P and T are the absolute pressure and absolute temperature, respectively, and R_i is the **particular gas constant,** a physical constant. Figure 2.12 illustrates how the ideal gas law can be applied to thermometry. The particular gas constant depends on the molecular weight of the gas as follows:

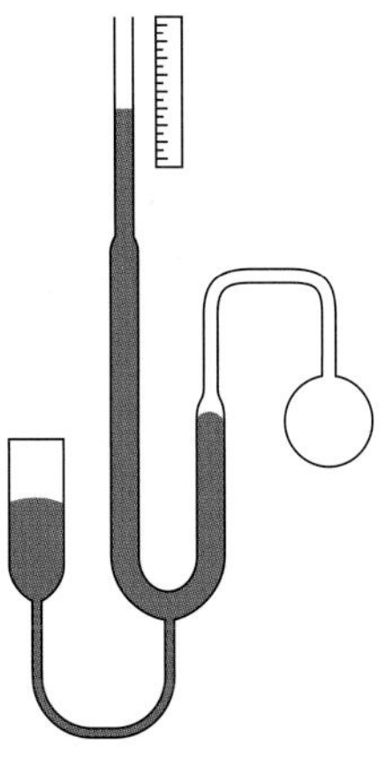

FIGURE 2.12 An ideal-gas thermometer consists of a sensing bulb filled with an ideal gas (right), a movable closed reservoir (left), and a liquid column (center). The height of the liquid column is directly proportional to the temperature of the gas in the bulb when the reservoir position is adjusted to maintain a fixed volume for the ideal gas.

$$R_i \equiv R_u/\mathcal{M}_i \, [=] \, \mathrm{J/kg{\cdot}K}, \tag{2.24}$$

where the subscript i denotes the species of interest, and R_u is the **universal gas constant,** defined by

$$R_u \equiv 8314.472 \, [=] \, \mathrm{J/kmol{\cdot}K}. \tag{2.25}$$

For example, if we are dealing with helium, the particular gas constant would be denoted R_{He}. Frequently, the symbol R without a subscript is used to represent the particular gas constant when there is no confusion as to what gas is being considered. The universal gas constant, however, will always be denoted using a subscript, i.e., R_u.

This definition of an ideal gas can be made more satisfying by examining what is implied from a molecular, or microscopic, point of view. Kinetic theory predicts that $Pv = RT$, first, when the molecules comprising the system are infinitesimally small, hard, round spheres occupying negligible volume and, second, when no forces exist among these molecules except during collisions. Qualitatively, these conditions imply a gas at *low* density. What we mean by low density will be discussed in later sections.

2.5b Ideal-Gas Equation of State

Formally, the P–v–T equation of state for an ideal gas is expressed as

$$Pv = RT. \tag{2.26a}$$

Alternative forms of the ideal-gas equation of state arise in various ways. First, by recognizing that the specific volume is the reciprocal of the density ($v = 1/\rho$), we get

$$P = \rho RT. \tag{2.26b}$$

Second, expanding the definition of the specific volume ($v = \mathcal{V}/M$) yields

$$P\mathcal{V} = MRT. \tag{2.26c}$$

Third, expressing the mass in terms of the number of moles and molecular weight of the particular gas of interest ($M = N\mathcal{M}_i$) yields

$$P\mathcal{V} = NR_u T. \tag{2.26d}$$

Finally, by employing the molar specific volume $\bar{v} = v\mathcal{M}_i$, we obtain

$$P\bar{v} = R_u T. \tag{2.26e}$$

We summarize these various forms of the ideal-gas equation of state in Table 2.4 and encourage you to become familiar with these relationships by performing the various conversions on your own (see Problem 2.52).

TABLE 2.4 Various Forms of the Ideal-Gas Equation of State

$Pv = RT$	Eq. 2.26a
$P = \rho RT$	Eq. 2.26b
$P\mathcal{V} = MRT$	Eq. 2.26c
$P\mathcal{V} = NR_u T$	Eq. 2.26d
$P\bar{v} = R_u T$	Eq. 2.26e

Example 2.5 Mass of Ideal Gas

A compressed-gas cylinder contains N_2 at room temperature (25 °C). A gage on the pressure regulator attached to the cylinder reads 120 psig. A mercury barometer in the room in which the cylinder is located reads 750 mm Hg. What is the density of the N_2 in the tank in units of kg/m^3? Also, determine the mass of the N_2 contained in the steel tank if its volume is 1.54-ft^3.

Solution

Known T_{N_2}, $P_{N_2,g}$, P_{atm}, $\mathcal{V}_{N_2}$

Find ρ_{N_2}, M_{N_2}

Modeling, Premises and Assumptions Ideal-gas behavior

Analysis To find the density of the N_2, we apply the ideal-gas equation of state (Eq. 2.26b, Table 2.4). Before doing so we must determine the particular gas constant for N_2 and perform several unit conversions for the given information.

From Eq. 2.24 and Eq. 2.25, we find the particular gas constant,

$$R_{N_2} = \frac{R_u}{\mathcal{M}_{N_2}} = \frac{8314.47\,\text{J/kmol·K}}{28.013\,\text{kg/kmol}}$$
$$= 296.8\,\text{J/kg·K},$$

where the molecular weight for N_2 is found in Table D.7.

The absolute pressure in the tank is (Eq. 2.13)

$$P_{N_2} = P_{N_2,g} + P_{\text{atm, abs}},$$

where

$$P_{N_2,g} = 120\frac{\text{lb}_f}{\text{in}^2}\left[\frac{39.370\,\text{in}}{1\text{ m}}\right]^2\left[\frac{1\text{ N}}{0.224809\text{ lb}_f}\right]$$
$$= 827{,}367\,\text{Pa (gage)}$$

and

$$P_{\text{atm, abs}} = (750\text{ mm Hg})\left[\frac{1\text{ atm}}{760\text{ mm Hg}}\right]\left[\frac{101{,}325\,\text{Pa}}{1\text{ atm}}\right]$$
$$= 99{,}992\text{ Pa}.$$

Thus, the absolute pressure of the N_2 is

$$P_{N_2} = 827{,}367\,\text{Pa (gage)} + 99{,}992\text{ Pa} = 927{,}359\text{ Pa},$$

which rounds off to

$$P_{N_2} = 927{,}000\,\text{Pa}.$$

The absolute temperature of the N_2 is

$$T_{N_2} = 25^\circ\text{C} + 273.15 = 298.15\text{ K}.$$

To obtain the density, we now apply the ideal-gas equation of state (Eq. 2.26b)

$$\rho_{N_2} = \frac{P_{N_2}}{R_{N_2}T_{N_2}}$$
$$= \frac{927{,}000}{296.8\,(298.15)}$$
$$= 10.5$$
$$[=]\frac{\text{Pa}\left[\frac{1\text{ N/m}^2}{\text{Pa}}\right]}{\frac{\text{J}}{\text{kg·K}}\left[\frac{1\,\text{N·m}}{\text{J}}\right]\text{K}} = \text{kg/m}^3.$$

Note that we have set aside the units and unit conversions to ensure their proper treatment. Unit conversion factors are always enclosed in square brackets. We obtain the mass from the definition of density (Eq. 2.9):

$$\rho_{N_2} \equiv \frac{M_{N_2}}{V_{N_2}}$$

or

$$M_{N_2} = \rho_{N_2} V_{N_2}.$$

The tank volume is

$$V_{N_2} = 1.54\,\text{ft}^3 \left[\frac{1\,\text{m}}{3.2808\,\text{ft}}\right]^3 = 0.0436\,\text{m}^3.$$

Thus,

$$M_{N_2} = 10.5\frac{\text{kg}}{\text{m}^3}(0.0436\,\text{m}^3) = 0.458\,\text{kg}.$$

Comments Note that, although the application of the ideal-gas law to find the density is straightforward, the unit conversions and calculations of absolute pressures and temperatures make the calculation nontrivial.

Self-Test 2.5

The valve of the tank in Example 2.5 is slowly opened and 0.1 kg of N_2 escapes. Calculate the density of the remaining N_2 and find the final gage pressure in the tank, assuming the temperature remains at 25 °C.

(Answer: 8.2 kg/m³, 626.6 kPa)

Example 2.6 Density of Air

It is a cold, sunny day in Merrill, WI. The temperature is −10 F, the barometric pressure is 100 kPa, and the humidity is nil. Estimate the outside air density. Also, estimate the molar density, N/V, of the air.

Solution

Known T_{air}, P_{air}

Find ρ_{air}

Modeling, Premises and Assumptions

i. Air can be treated as a pure substance.
ii. Air can be treated as an ideal gas.
iii. Air is dry.

(Credit: AccretionPoint / iStock / Getty Images Plus.)

Analysis With these assumptions, we can use the data in Appendix C together with the ideal-gas equation of state (Eq. 2.26b) to find the air density. First, we convert the temperature to SI absolute units:

$$T_{air}(\text{K}) = \frac{5}{9}(-10 + 459.67) = 249.8\,\text{K}.$$

The density is thus

$$\rho_{air} = \frac{P_{air}}{R_{air}T_{air}} = \frac{100,000\,\text{Pa}}{287.0\,\text{J/kg·K}\,(249.8\,\text{K})} = 1.395\,\text{kg/m}^3,$$

where R_{air}, the particular gas constant for air, was obtained from Table C.1 in Appendix C. The treatment of units in this calculation is the same as detailed in the previous example.

The molar density is the number of moles per unit volume. We calculate this quantity by dividing the mass density (ρ_{air}) by the apparent molecular weight of the air, that is,

$$\frac{N_{air}}{\mathcal{V}_{air}} = \frac{\rho_{air}}{\mathcal{M}_{air}} = \frac{1.395\,\mathrm{kg/m^3}}{28.97\,\mathrm{kg/kmol}} = 0.048\,\mathrm{kmol/m^3}.$$

Comments The primary purpose of this example is to introduce the approximation of treating air, a mixture of gases (see Table C.1 for the composition of dry air), as a simple substance that behaves as an ideal gas. Note the introduction of the apparent molecular weight, $\mathcal{M}_{air} = 28.97$ kg/kmol, and the particular gas constant, $R_{air} = R_u/\mathcal{M}_{air} = 287.0$ J/kg·K. Ideal-gas thermodynamic properties for dry air are also tabulated in Appendix C. In our study of air conditioning (Chapter 11), we investigate the influence of moisture in air.

Self-Test 2.6

Calculate the mass of air in an uninsulated, unheated 10 ft × 15 ft × 12 ft garage on this same cold day.

(Answer: 71.1 kg)

2.5c Processes in *P*–v–*T* Space

Plotting thermodynamic processes using *P*–v or other thermodynamic property coordinates (Fig. 2.13) is very useful in analyzing many thermal systems. In this section, we introduce this topic by illustrating common processes in *P*–v and *T*–v coordinates, restricting our attention to ideal gases. Later in this chapter, we add the complexity of a phase change.

We begin by examining *P*–v coordinates. Rearranging the ideal-gas equation of state (Eq. 2.26a) to a form in which *P* is a function of v yields the hyperbolic relationship

$$P = (RT)\frac{1}{v}. \tag{2.27}$$

By considering the temperature to be a fixed parameter, Eq. 2.27 can be used to create a family of hyperbolas for various values of *T*, as shown in Fig. 2.14 for nitrogen. Increasing the temperature moves the isotherms further from the origin.

Graphs such as Fig. 2.14 are quite useful; from them one can immediately visualize how properties must vary for a particular thermodynamic process. For example, consider the constant-pressure expansion process shown in Fig. 2.14, where the initial and final states are designated as points 1 and 2, respectively. Knowing the arrangement of the constant-temperature lines allows us to see that the temperature must increase in the process 1–2. For the values given, the temperature increases from 300 to 600 K. The important point here, however, is not this quantitative result, but the

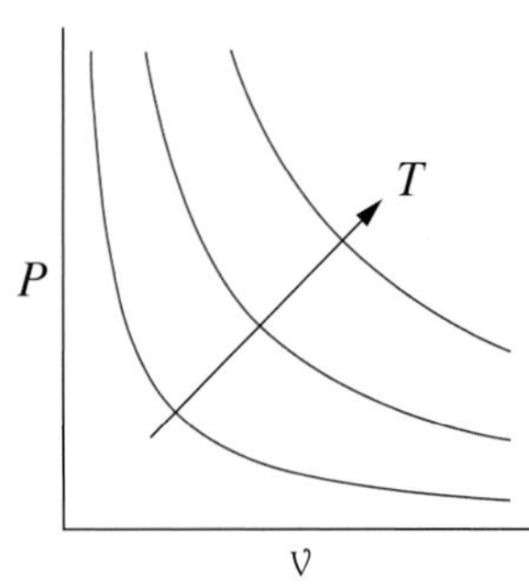

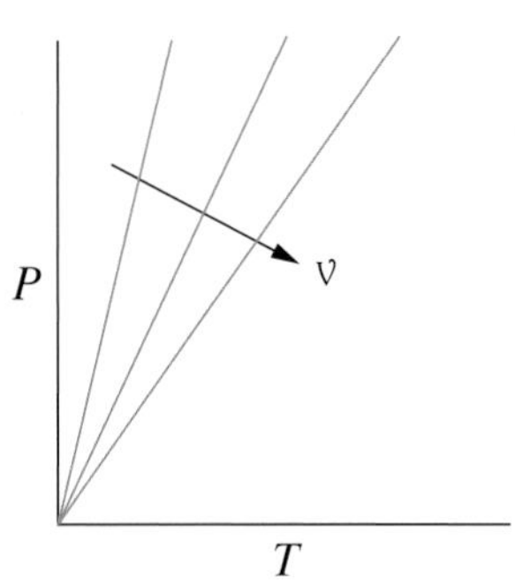

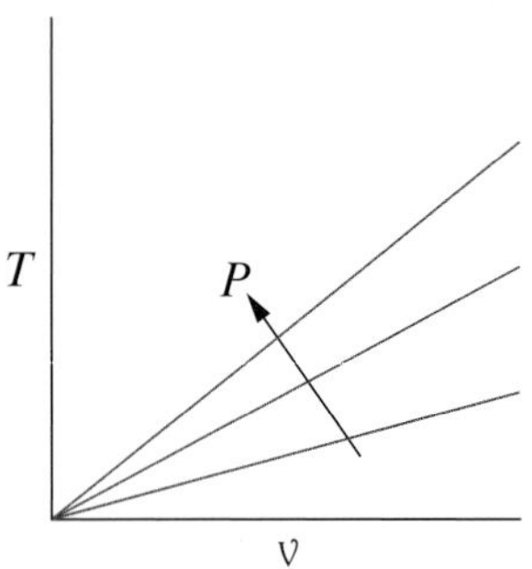

FIGURE 2.13 Cutting planes through the P–v–T surface of Fig. 2.11 yield two-dimensional plots of any two thermodynamic properties (P vs. v, P vs. T, and T vs. v) with the third property (T, v, and P) as a parameter (with permission from ARL).

qualitative information available from plotting processes on P–v coordinates. Also shown in Fig. 2.14 is a constant-volume process (assuming that we are dealing with a system of fixed mass). In going from state 2 to state 3 at constant volume, we immediately see that both the temperature and pressure must fall.

In choosing a pair of thermodynamic coordinates to draw a graph, one usually selects those that allow given constant-property processes to be shown as straight lines. For example, P–v coordinates are the natural choice for systems involving either constant-pressure or constant-(specific)volume processes. If, however, one is interested in a constant-temperature process, then T–v coordinates may be more useful. In this case, we can rearrange the ideal-gas equation of state to yield

$$T = \left(\frac{P}{R}\right)v. \tag{2.28}$$

Treating pressure as a fixed parameter, this relationship yields straight lines with slopes P/R. Higher pressures result in steeper slopes.

Figure 2.15 illustrates the ideal-gas T–v relationship. Consider a constant-temperature compression process going from state 1 to state 2. For this process, we immediately see from the graph that the pressure must increase. Also shown on Fig. 2.15 is a constant- (specific) volume process going from state 2 to state 3 for conditions of decreasing temperature. Again, we immediately see that the pressure falls during this process.

Plotting processes using thermodynamic coordinates develops understanding and aids in problem solving. Experts frequently employ plots to solve thermodynamics problems, whereas students – to their detriment – tend to avoid creating useful plots. Whenever possible, we will use such diagrams in examples throughout the book. In addition, many homework problems are designed to foster development of your ability to draw and use such plots.

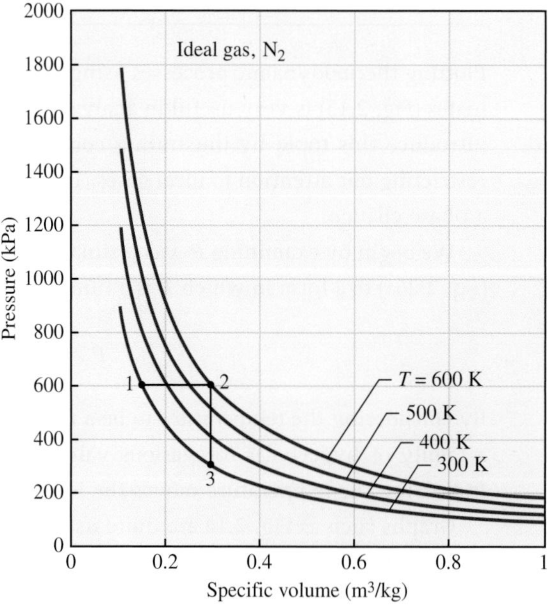

FIGURE 2.14 Constant-pressure (1–2) and constant-volume (2–3) processes are shown on P-v coordinates for an ideal gas (N_2). Lines of constant temperature are hyperbolic, following the ideal-gas equation of state, $P = (RT)/v$.

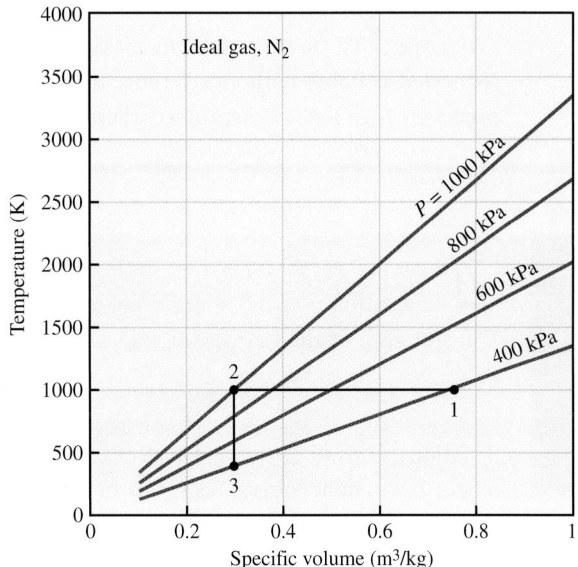

FIGURE 2.15 Constant-temperature (1–2) and constant-volume (2–3) processes are shown using T–v coordinates for an ideal gas (N_2). Lines of constant pressure are straight lines, following the ideal-gas equation of state, $T = (P/R)v$.

Example 2.7 Ideal Gas Cycle

An ideal gas system undergoes a thermodynamic cycle composed of the following processes:

1–2: constant-pressure expansion,
2–3: constant-temperature expansion,
3–4: constant-volume return to the state-1 temperature, and
4–1: constant-temperature compression.

Sketch these processes (a) on P–v coordinates and (b) on T–v coordinates.

Solution

The sequences of processes are shown on the sketches.

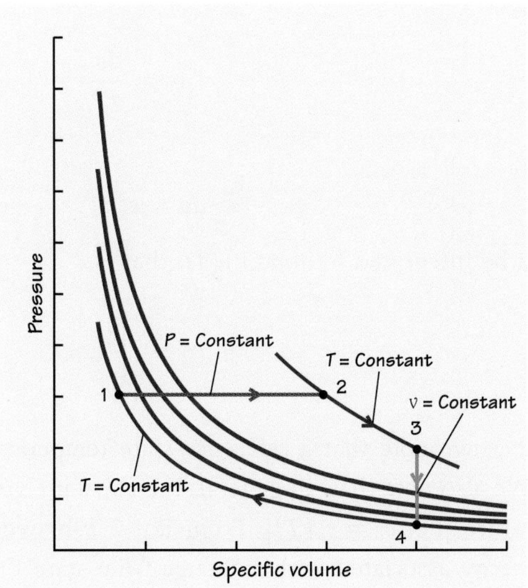

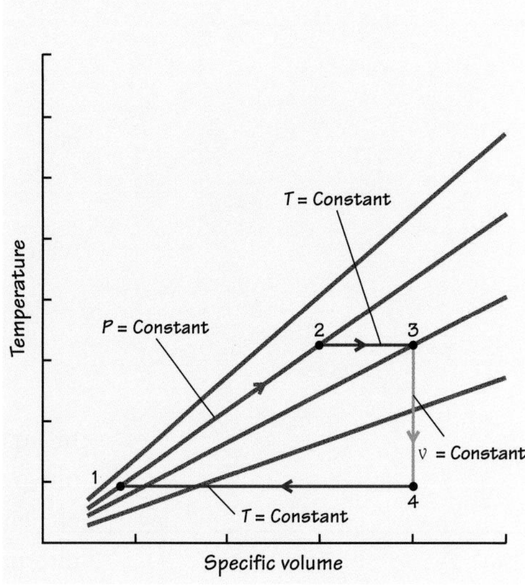

Comments To develop skill in making such plots, the reader should redraw the requested sketches without reference to the solutions given. Note that the sequence of processes 1–2–3–4–1 constitutes a thermodynamic cycle.

Self-Test 2.7

Looking at the sketches in Example 2.7, state whether the pressure P, temperature T, and specific volume v increase, decrease, or remain the same for each of the four processes.

(Answer: 1–2: P constant, T increases, v increases; 2–3: P decreases, T constant, v increases; 3–4: P decreases, T decreases, v constant; 4–1: P increases, T constant, v decreases)

2.5d Ideal-Gas Calorific Equations of State

From kinetic theory we predict that the internal energy of an ideal gas will be a function of temperature only. That the internal energy is independent of pressure follows from the neglect of any intermolecular forces in the model of an ideal gas. In real gases, molecules do exhibit repulsive and attractive forces, which results in a pressure dependence of the internal energy. As in our discussion of the P–v –T equation of state, the ideal gas approximation, however, is quite accurate at sufficiently low densities, and the result that $u = u(T \text{ only})$ is quite useful.

Ideal-gas specific internal energies can be obtained from experimentally or theoretically determined values of the constant-volume specific heat. Starting with the general definition (Eq. 2.20a)

$$c_v = \left(\frac{\partial u}{\partial T}\right)_v,$$

we recognize that the partial derivative becomes an ordinary derivative when $u = u\,(T$ only); thus

$$c_v = \frac{du}{dT} \quad \textbf{(2.29a)}$$

or

$$du = c_v dT, \quad \textbf{(2.29b)}$$

which can be integrated to obtain $u(T)$, that is,

$$u(T) = \int_{T_{\text{ref}}}^{T} c_v dT. \quad \textbf{(2.29c)}$$

In Eq. 2.29c, we note that a reference-state temperature is required to evaluate the integral. We also note that, in general, the constant-volume specific heat is a function of temperature [i.e., $c_v = c_v(T)$]. From Eq. 2.29b, we can easily find the change in internal energy associated with a change from state 1 to state 2:

$$u_2 - u_1 = u(T_2) - u(T_1) = \int_{T_1}^{T_2} c_v dT. \tag{2.29d}$$

If the temperature difference between the two states is not too large, the constant-volume specific heat can be treated as a constant, $c_{v,\mathrm{avg}}$; thus,

$$u_2 - u_1 = c_{v,\mathrm{avg}}(T_2 - T_1). \tag{2.29e}$$

We will return to Eq. 2.29 after discussing the calorific equation of state involving enthalpy.

To obtain the h–T–P calorific equation of state for an ideal gas, we first show that the enthalpy of an ideal gas, like the internal energy, is a function of only the temperature [i.e., $h = h(T \text{ only})$]. We start with the definition of enthalpy (Eq. 2.19),

$$h = u + Pv,$$

and replace the Pv term using the ideal-gas equation of state (Eq. 2.26a). This yields

$$h = u + RT. \tag{2.30}$$

Since $u = u\,(T \text{ only})$, we see from Eq. 2.30 that h, too, is a function only of temperature for an ideal gas. With $h = h(T \text{ only})$, as before the partial derivative becomes an ordinary derivative; thus, we have

$$c_p \equiv \left(\frac{\partial h}{\partial T}\right)_p = \frac{dh}{dT}. \tag{2.31a}$$

From this, we can write

$$dh = c_p dT, \tag{2.31b}$$

which can be integrated to yield

$$h(T) = \int_{T_{\mathrm{ref}}}^{T} c_p dT. \tag{2.31c}$$

The enthalpy difference for a change in state mirrors that for the internal energy, that is,

$$h_2 - h_1 = h(T_2) - h(T_1) = \int_{T_1}^{T_2} c_p dT, \tag{2.31d}$$

and, if the constant-pressure specific heat does not vary much between states 1 and 2,

$$h_2 - h_1 = c_{p,\mathrm{avg}}(T_2 - T_1). \tag{2.31e}$$

All these relationships (Eqs. 2.29–2.31) can be expressed in a molar basis simply by substituting molar-specific properties for mass-specific properties. Note that then R becomes R_u.

Before proceeding, we will obtain some useful auxiliary ideal-gas relationships by differentiating Eq. 2.30 with respect to temperature, obtaining

$$\frac{dh}{dT} = \frac{du}{dT} + R.$$

Recognizing the definitions of the ideal-gas specific heats (Eqns. 2.29 and 2.31a), we then have

$$c_p = c_v + R, \tag{2.32a}$$

or

$$c_p - c_v = R. \tag{2.32b}$$

Because property data sources sometimes only provide values or curve fits for c_p, one can use Eq. 2.32 to obtain values for c_v.

Figure 2.16 provides graphical interpretations of Eqs. 2.29c and 2.31c, our ideal-gas calorific equations of state. Here we see that the area under the c_v vs. T curve represents the internal energy change for a temperature change from T_{ref} to T_1. A similar interpretation applies to the $c_p(T)$ curve, where the area now represents the enthalpy change.

In the same spirit in which we graphically illustrated the ideal-gas equation of state using P–v and T–v plots (Figs. 2.14 and 2.15, respectively), we now illustrate the ideal-gas calorific equations of state using u–T and h–T plots. Figure 2.17 illustrates these u–T and h–T relationships for N_2. Because the specific internal energy and specific enthalpy of ideal gases are functions of neither pressure nor specific volume, each relationship is expressed by a single curve. From this, we conclude that u and T are

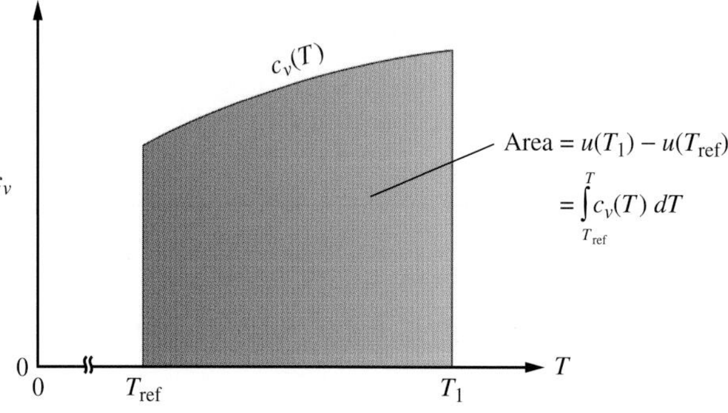

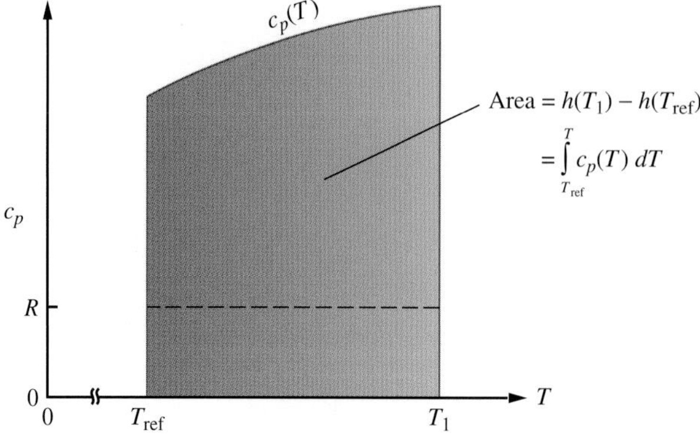

FIGURE 2.16 The area under the c_v vs. T curve is the internal energy (top), and the area under the c_p vs. T curve is the enthalpy. Note that the difference in the areas, $[h(T_1)-h(T_{ref})] - [u(T_1-u(T_{ref})]$, is $R(T_1-T_{ref})$, the area below the dashed line.

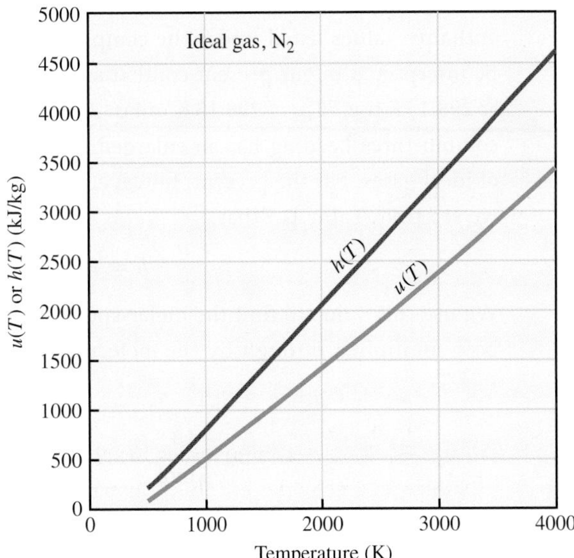

FIGURE 2.17 For ideal gases, the specific internal energy and specific enthalpy are functions of temperature only. The concave upward curvature in these nearly straight line plots results because the specific heats (c_v and c_p) for N_2 increase with temperature.

not independent properties for ideal gases; likewise, h and T are not independent properties. Thus, some other property is needed to define the state of an ideal gas. This result (Fig. 2.17) contrasts with the P–v (Fig. 2.14) and T–v (Fig. 2.15) plots generated from the equation of state, where families of curves are required to express these state relationships. Being able to sketch processes in u–T and h–T coordinates, as well as in P–v and T–v coordinates, greatly aids problem solving.

Numerical values for molar specific enthalpies $\bar{h}$ (and molar constant-pressure specific heats $\bar{c}_p$) are available from tables contained in Appendix D for a number of gaseous species where ideal-gas behavior is assumed. Note the reference temperature is 298.15 K. Curve fits for $\bar{c}_p$ are also provided in Appendix D for these same gases. Although air is a mixture of gases, it can be treated practically as a pure substance (see Example 2.6). Properties of air are provided in Appendix C. Note that the reference temperature used in these air tables is 78.903 K, not 298.15 K. The following examples illustrate the use of some of the information available in the appendices.

Example 2.8 Specific Internal Energy of an Ideal Gas

Determine the specific internal energy u for N_2 at 2500 K.

Solution

Known N_2, T

Find u

Modeling, Premises and Assumptions Ideal-gas behavior

Analysis We use Table D.7 to determine u. Before that can be done a few preliminaries are involved. First, we note that molar-specific enthalpies, not mass-specific internal energies, are provided in the table; however, Eq. 2.30 relates u and h for ideal gases and the conversion to a mass basis is straightforward. A second issue is how to interpret the third column in Table D.7, the column containing enthalpy data. The enthalpy values listed under the complex column-three heading, $\bar{h}^\circ(T) - \bar{h}_f^\circ(T_{ref})$, can be interpreted in our present context as simply $\bar{h}\,(T)$, where $\bar{h}$ is assigned a zero value at 298.15 K (i.e., $T_{ref} = 298.15$ K in Eq. 2.31c). As we will see in Chapters 10 and 12, the column-three heading has an enlarged meaning when dealing with reacting mixtures of ideal gases. For the present, however, this meaning need not concern us. At 2500 K, we see from Table D.7 that

$$\bar{h} = 74{,}305 \text{ kJ/kmol}.$$

We use this value to find the molar-specific internal energy from Eq. 2.30, which has been multiplied through by the molecular weight to yield

$$\begin{aligned}\bar{u} &= \bar{h} - R_u T \\ &= 74{,}305 \frac{\text{kJ}}{\text{kmol}} - 8.31447 \frac{\text{kJ}}{\text{kmol·K}} 2500\,\text{K} \\ &= 74{,}305\,\text{kJ/mol} - 20{,}786\,\text{kJ/mol} = 53{,}519\,\text{kJ/kmol}.\end{aligned}$$

Converting the molar-specific internal energy to its mass-specific form (Eq. 2.5) yields

$$\begin{aligned}u &= \bar{u}/\mathcal{M}_{N_2} \\ &= \frac{53{,}519 \text{ kJ/kmol}}{28.013 \text{ kg/kmol}} \\ &= 1910.5 \text{ kJ/kg}.\end{aligned}$$

Comments Several very simple, yet very important, concepts are illustrated by this example: (1) the conversions between mass-specific and molar-specific properties, (2) the use of the enthalpy data in Tables D.1–D.12 for nonreacting ideal gases, (3) the calculation of ideal-gas internal energies from enthalpies, and (4) practical recognition of the use of reference states for enthalpies (and internal energies).

Self-Test 2.8

Determine the specific enthalpy h and internal energy u for air at 1500 K and 1 atm.

(*Answer: 1762.24 kJ/kg, 1331.46 kJ/kg*)

Example 2.9 Enthalpy and Internal Energy of an Ideal Gas

Determine the changes in the mass-specific enthalpy and the mass-specific internal energy for air, for a process that starts at 300 K and ends at 1000 K. Also show that $\Delta h - \Delta u = R\Delta T$, and compare a numerical evaluation of this with tabulated data.

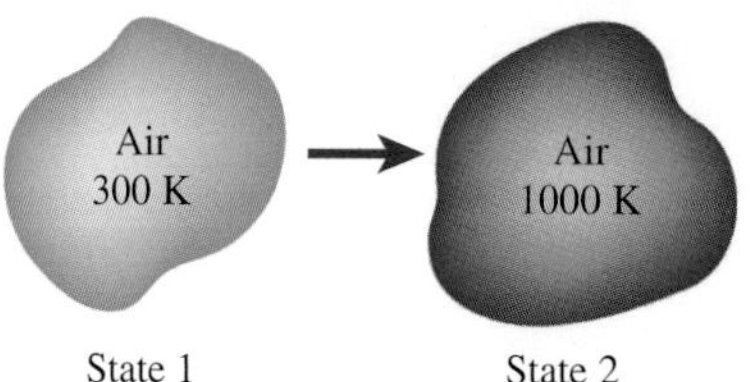

Solution

Known air, T_1, T_2

Find $\Delta h\ [= h(T_2) - h(T_1)]$, $\Delta u\ [= u(T_2) - u(T_1)]$, $\Delta h - \Delta u$

Sketch

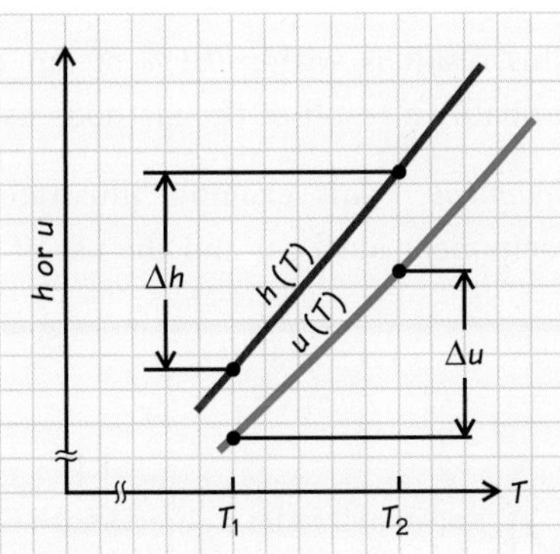

Modeling, Premises and Assumptions Air behaves as a single-component ideal gas.

Analysis From Table C.2, we obtain the following values for h and u:

T(K)	h (kJ/kg)	u (kJ/kg)
300	426.04	339.93
1000	1172.43	885.22

Using these data, we calculate

$$\begin{aligned}\Delta h &= h(T_2) - h(T_1)\\ &= h(1000) - h(300)\\ &= 1{,}172.43\,\text{kJ/kg} - 426.04\,\text{kJ/kg}\\ &= 746.39\,\text{kJ/kg}\end{aligned}$$

and

$$\begin{aligned}\Delta u &= 885.22\,\text{kJkg} - 339.93\,\text{kJ/kg}\\ &= 545.29\,\text{kJ/kg}.\end{aligned}$$

Applying Eq. 2.30, we can relate Δh and Δu, that is,

$$h_2 = u_2 + RT_2$$

and

$$h_1 = u_1 + RT_1.$$

Subtracting these yields

$$h_2 - h_1 = u_2 - u_1 + R(T_2 - T_1).$$

or

$$\Delta h - \Delta u = R\Delta T.$$

We evaluate this equation using the particular gas constant for air ($R = R_u/\mathcal{M}_{air}$), giving us

$$\begin{aligned}\Delta h - \Delta u &= \frac{8.31447\,\text{kJ/mol·K}}{28.97\,\text{kg/kmol}}(1000 - 300)\,\text{K}\\ &= 200.90\,\text{kJ/kg}.\end{aligned}$$

This compares with the value obtained from the Table C.1 data as follows:

$$(\Delta h - \Delta u)_{\text{tables}} = 746.39\,\text{kJ/kg} - 545.29\,\text{kJ/kg} \\ = 201.10\,\text{kJ/kg}.$$

This result is within 0.1% of our calculation. Failure to achieve identical values is probably a result of the methods used to generate the tables (i.e., curve fitting).

Comment This example illustrates once again the treatment of air as a single-component ideal gas and the use of Appendix C for air properties.

Self-Test 2.9

Recalculate the values for the change in mass-specific enthalpy and internal energy for the process in Example 2.9 using Eqs. 2.29e and 2.31e and an appropriate average value of c_v and c_p. Compare your answer with that of Example 2.9.

(Answer: 751.8 kJ/kg, 550.2 kJ/kg)

2.6 Nonideal-Gas Properties

2.6a State (*P*–*v*–*T*) Relationships

In this section, we will look at three ways to determine P–v–T state relations for gases that do not necessarily obey the ideal-gas equation of state (Eq. 2.26): (1) the use of tabular data (or various computer applications), (2) the use of the van der Waals equation of state, and (3) the application of the concept of generalized compressibility. Before discussing these methods, however, we define the thermodynamic **critical point,** a concept important to all three methods (see also Fig. 2.18):

The critical point is the point in P–v–T space defined by the highest possible temperature and the highest possible pressure for which distinct liquid and gas phases can be observed.

At the critical point, we designate the thermodynamic variables with the subscript c, so that

$$\text{critical temperature} \equiv T_c,$$
$$\text{critical pressure} \equiv P_c,$$

and

$$\text{critical specific volume} \equiv v_c.$$

Critical properties for a number of substances are given in Table E.1 in Appendix E. We elaborate on the significance of the critical point later in this chapter in our discussion of substances existing in multiple phases. One immediate use of critical point properties, however, is to establish a rule of thumb for determining when a gas

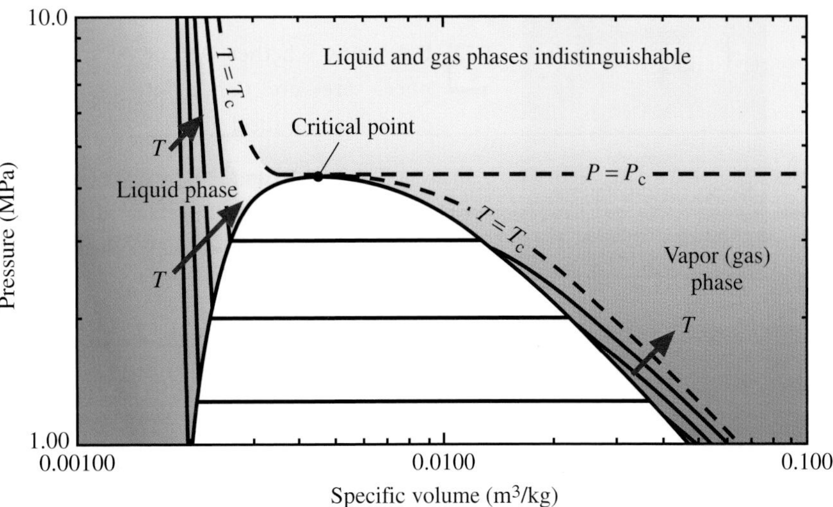

FIGURE 2.18 The bold dot indicates the critical point on this P–v–T plot for propane. When both temperature and pressure are greater than their respective critical values T_c and P_c, the liquid and gas phases are indistinguishable.

can be considered ideal: a real gas approaches ideal-gas behavior when $P \ll P_c$. In a later section, we will see other conditions where ideal-gas behavior is approached.

Example 2.10 Validity of the Ideal-Gas Approximation

Determine whether N_2 is likely to approximate ideal-gas behavior at 298 K and 1 atm.

Solution

The critical properties for N_2 from Table E.1 are

$$T_c = 126.2\,\text{K},$$
$$P_c = 3.39\,\text{MPa}.$$

Comparing the given properties with the critical properties, we have

$$\frac{T}{T_c} = \frac{298\,\text{K}}{126.2\,\text{K}} = 2.36$$

and

$$\frac{P}{P_c} = \frac{101{,}325\,\text{Pa}}{3.39 \times 10^6\,\text{Pa}} = 0.030.$$

Clearly, $P \ll P_c$; thus, the ideal-gas equation of state is likely to be a good approximation to the true state relation for N_2 at these conditions. Moreover, T is greater than T_c, which is also in the direction of ideal-gas behavior.

Comment Since P_c is quite high (33 atm) and T_c is quite low (126.2 K), ideal-gas behavior is likely to be a good approximation for N_2 over a wide range of conditions covering many applications. However, caution should be exercised. The methods described in the following allow us to estimate quantitatively the deviation of real-gas behavior from ideal-gas behavior.

Self-Test 2.10

 Determine whether O_2 can be considered an ideal gas at 298 K and atmospheric pressure. Can air at room temperature and atmospheric pressure be considered an ideal gas?

(Answer: $T/T_c = 1.93$ and $P/P_c = 0.020$, so ideal-gas behavior is a good approximation for O_2. Since the main constituents of air (N_2 and O_2) both behave as ideal gases, ideal-gas behavior for air is a good approximation.)

For conditions near the critical point or for applications requiring high accuracy, alternatives to the ideal-gas equation of state are needed. The following subsections present three approaches.

TABULATED PROPERTIES

Accurate P–v –T data in tabular or curve-fit form are available for a number of gases of engineering importance. Because of the importance of steam in electric power generation, a large database is available for this fluid, and tables are published in a number of sources (e.g., Refs. [4] and [5]). Thermodynamic data are also available from MATLAB toolboxes provided by various third-party sources. Particularly useful and convenient sources of thermodynamic properties are available from NIST. A downsized version of the NIST database, *miniREFPROP*, is available as freeware and can be downloaded from NIST at http://trc.nist.gov/refprop/MINIREF/MINIREF.HTM. (See also the publisher's website associated with this book.) The NIST Chemistry WebBook [6], an internet resource, is also quite useful. Table 2.5 lists the fluids for which properties are available from these sources. We will use the NIST resources throughout this book, and the reader is encouraged to become familiar with these valuable sources of thermodynamic data. A tutorial on the use of *miniREFPROP* is provided later in this chapter. Selected tabular data for steam are also provided in Appendix B.

TABLE 2.5 Fluid Properties: Fluids Included in *miniREFPROP* Software and NIST WebBook [6]

Fluid	*miniREFPROP*	NIST WebBook
Air	X	
Water*	X	X
Nitrogen	X	X
Carbon dioxide	X	X
Dodecane	X	X
Methane	X	X
Propane	X	X
Ethane, 1,1,1,2-tetrafluoro- (R-134a)	X	X
68 additional fluids, including most modern refrigerants		X

* See also Appendix B.

Example 2.11 Deviation from Ideal-Gas Conditions

Determine the deviation from ideal-gas behavior associated with the following gases and conditions:

N_2 at 200 K from 1 atm to P_c,
CO_2 at 300 K from 1 to 40 atm, and
H_2O at 600 K from 1 to 40 atm.

Solution

To quantify the deviation from ideal-gas behavior, we define the factor $Z = Pv/RT$. For an ideal gas, Z is unity for any temperature or pressure. We employ the NIST Web-Book [6] to obtain values of specific volume for the conditions specified. These data are then used to calculate Z values. For example, the specific volume of CO_2 at 300 K and 20 atm is 0.025001 m^3/kg; thus,

$$Z = \frac{Pv}{RT} = \frac{20\,\text{atm}(101{,}325\,[\text{Pa/atm}])\,0.025001\,\text{m}^3/\text{kg}}{\left(\dfrac{8314.47\,\text{J/kmol}\cdot\text{K}}{44.011\,\text{kg/kmol}}\right)300\,\text{K}} \times \frac{\left[\dfrac{1\,\text{N/m}^2}{\text{Pa}}\right]}{\left[\dfrac{1\,\text{N}\cdot\text{m}}{\text{J}}\right]}$$

$$= 0.8939\,(\text{dimensionless})$$

A few selected results are presented in the following table, and all the results are plotted in Fig. 2.19.

	v (m^3/kg)			$Z = Pv/RT$		
P (atm)	N_2	CO_2	H_2O	N_2	CO_2	H_2O
1	0.8786	0.5566	2.7273	0.9998	0.9951	0.9980
10	0.8773	0.5309	0.2676	0.9983	0.9491	0.9792
20	0.4381	0.0250	0.1308	0.9972	0.8939	0.9572
30	0.0292	0.0155	0.0851	0.9965	0.8330	0.9341
40	—	0.0107	0.0621	—	0.7642	0.9097

Comments From the table and from Fig. 2.19, we see that N_2 behaves essentially as an ideal gas over the entire range of pressures from 1 atm to the critical pressure (P_c = 33.46 atm): its Z values deviate less than 0.4% from the ideal-gas value of unity. Both CO_2 and water vapor, in contrast, exhibit significant departures from ideal-gas behavior, which become larger as the pressure increases. These large departures indicate the need for caution in applying the ideal-gas equation of state, $Pv = RT$.

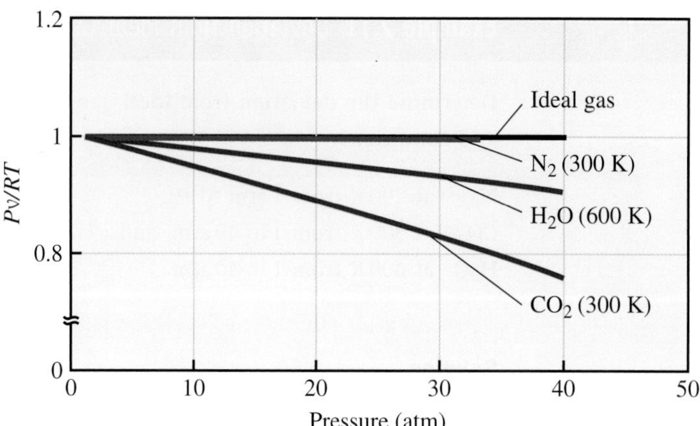

FIGURE 2.19 Deviations from ideal-gas behavior can be seen by the extent to which $Z\,(=Pv/RT)$ deviates from unity. At 1 atm, CO_2, N_2, and H_2O all have Z values near unity. At higher pressures, CO_2 and H_2O deviate significantly from ideal-gas behavior, whereas N_2 still behaves essentially as an ideal gas.

Self-Test 2.11

Using the ideal-gas equation of state, calculate the specific volume for H_2O at 40 atm and 600 K. Compare your answer with the value from the table in Example 2.11 and determine the calculation error.

(Answer: 0.0683 m³kg, 10%)

Tutorial 1 – How to Interpolate

Relationships among properties are frequently presented in tables. For example, internal energy, enthalpy, and specific volume for steam are often tabulated as functions of temperature at a fixed pressure. Table B.3 illustrates such tables and employs temperature increments of 20 K. If the temperature of interest is one of those tabulated, a simple look-up is all that is needed to retrieve the desired properties. In many cases, however, the temperature of interest will fall somewhere between the tabulated temperatures. To estimate property values for such a situation, you can apply **linear interpolation.** We illustrate this procedure with the following concrete example.

Given: Steam at 835 K and 10 MPa.
Find: Enthalpy using Table B.3P.

T (K)	h (kJ/kg)
820	3494.1
835	?
840	3543.9

From the 10-MPa pressure table we see that the given temperature lies between the tabulated values.

Linear interpolation assumes a straight-line relationship between h and T for the interval $820 \leq T \leq 840$. For the given data, we see that the desired temperature, 835 K, lies three-fourths of the way between 820 and 840 K, that is,

$$\frac{835-820}{840-820}=\frac{15}{20}=0.75.$$

With the assumed linear relationship, the unknown h value must also lie three-fourths of the way between the two tabulated enthalpy values, that is,

$$\frac{h(835\,\text{K})-h(820\,\text{K})}{h(840\,\text{K})-h(820\,\text{K})}=\frac{h(835\,\text{K})-3494.1}{3543.9-3494.1}=0.75.$$

We now solve for h (835 K), obtaining

$$\begin{aligned} h(835\,\text{K}) &= 0.75\,(3543.9-3494.1)\,\text{kJ/kg}+3494.1\ \text{kJ/kg} \\ &= 37.4\,\text{kJ/kg}+3494.1\ \text{kJ/kg}=3531.5\ \text{kJ/kg}. \end{aligned}$$

To generalize, we express this procedure as follows, denoting $h(T_1)$ as h_1, $h(T_2)$ as h_2, and $h(T_3)$ as h_3:

$$\frac{h(T_3)-h(T_1)}{h(T_2)-h(T_1)}=\frac{h_3-h_1}{h_2-h_1}=\frac{T_3-T_1}{T_2-T_1}$$

or

$$h_3=\left(\frac{T_3-T_1}{T_2-T_1}\right)(h_2-h_1)+h_1.$$

For any property pair $Y(X)$, this can be expressed as

$$\frac{Y(X_3)-Y(X_1)}{Y(X_2)-Y(X_1)}=\frac{Y_3-Y_1}{Y_2-Y_1}=\frac{X_3-X_1}{X_2-X_1}$$

or

$$Y_3=\left(\frac{X_3-X_1}{X_2-X_1}\right)(Y_2-Y_1)+Y_1.$$

Rather than remembering or referring to any equations, knowing the physical interpretation of linear interpolation allows you to create the needed relationships. The following sketch provides a graphical aid for this procedure.

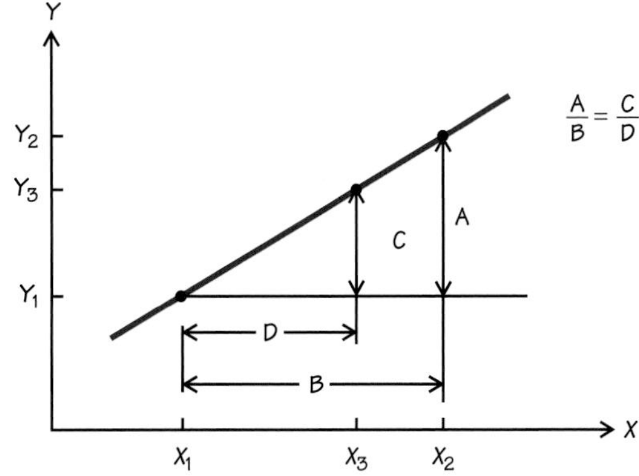

Although the use of computer-based property data may minimize the need to interpolate, some data may only be available in tabular form. Moreover, the ability to interpolate is a generally useful skill and should be mastered.

OTHER EQUATIONS OF STATE

The kinetic theory model of a gas that leads to the ideal-gas equation of state is based on two assumptions, which break down when the density of the gas is sufficiently high. The first of these is that the molecules themselves occupy a negligible volume compared to the volume of gas under consideration. Clearly, as a gas is compressed, the average spacing between molecules becomes less and the fraction of the macroscopic gas volume occupied by the microscopic molecules increases. The **van der Waals equation of state** accounts for the finite volume of the molecules by subtracting a molecular volume from the macroscopic volume; thus, the molar-specific volume $\bar{v}$ in the state equation is replaced with $\bar{v} - b$, where b is a constant for a particular gas.

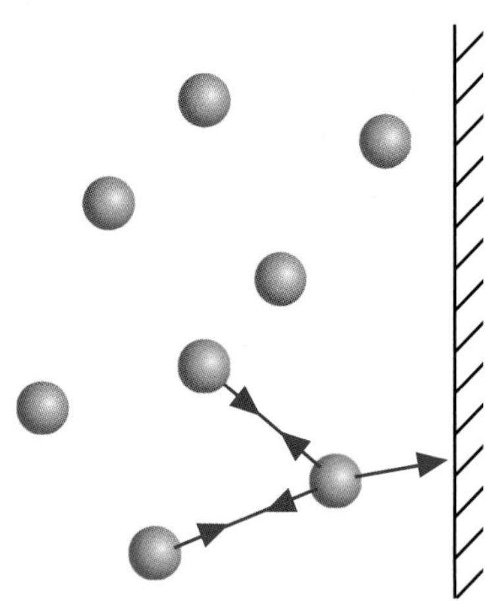

FIGURE 2.20 Intermolecular attractive forces result in a gas pressure less than would result from an ideal gas in which there are no long-range intermolecular forces.

The second assumption that is violated at high densities is that the long-range forces between molecules are negligible. Figure 2.20 illustrates how the existence of such forces can affect the equation of state. Since the pressure of a gas is a manifestation of the momentum transferred to the wall by molecular collisions, long-range attractive forces will "pull back" molecules as they approach the wall, thus decreasing the momentum exchange. This effect, in turn, results in a decrease in pressure. In the van der Waals equation of state, this effect of intermolecular attractive forces is accounted for by replacing the pressure P with $P + a/\bar{v}^2$, where a is a constant, again dependent upon the particular molecular species involved. The reciprocal of $\bar{v}^2$ appears because the molecular collision frequency with the wall is proportional to the molar gas density $(1/\bar{v})$ and the intermolecular attractive force is also proportional to $1/\bar{v}$.

With these two corrections to the ideal-gas model, we write the van der Waals equation of state as follows:

$$\left(P + \frac{a}{\bar{v}^2}\right)(\bar{v} - b) = R_u T. \tag{2.33}$$

Values for the constants a and b for a number of gases are provided in Appendix E.

Example 2.12 van der Waals Equation of State

Use the van der Waals equation of state to evaluate $Z\ (= P\bar{v}/R_u T)$ for CO_2 at 300 K and 30 atm. Compare this result with the value calculated using data from the NIST WebBook in Example 2.11.

Solution

Known CO_2, T, P

Find Z

Sketch See Fig. 2.19.

Modeling, Premises and Assumptions van der Waals gas

Analysis We use the van der Waals equation of state (Eq. 2.33) to find the molar-specific volume $\bar{v}$. This value of $\bar{v}$ is then used to calculate Z. We rearrange Eq. 2.33,

$$\left(P + \frac{a}{\bar{v}^2}\right)(\bar{v} - b) = R_u T,$$

to the following cubic form:

$$\bar{v}^3 - \left(\frac{R_u T}{P} + b\right)\bar{v}^2 + \frac{a}{P}\bar{v} - \frac{ab}{P} = 0.$$

Before solving for $\bar{v}$, we calculate the coefficients using values for a and b from Table E.2:

$$a = 3.643 \times 10^5\,\mathrm{Pa}\cdot(\mathrm{m}^3/\mathrm{kmol})^2,$$
$$b = 0.0427\,\mathrm{m}^3/\mathrm{kmol}.$$

The coefficients are thus

$$\left(\frac{R_u T}{P} + b\right) = \frac{8314.47\,\mathrm{J/kmol\cdot K}\,(300\,\mathrm{K})}{30\,\mathrm{atm}\left[101{,}325\left(\frac{\mathrm{N/m^2}}{\mathrm{atm}}\right)\right]}\left[\frac{\mathrm{N\cdot m}}{1\,\mathrm{J}}\right] + 0.0427\,\mathrm{m}^3/\mathrm{kmol}$$
$$= 0.863274\,\mathrm{m}^3/\mathrm{kmol},$$

$$\frac{a}{P} = \frac{3.643 \times 10^5\,\mathrm{Pa}\cdot(\mathrm{m}^3/\mathrm{kmol})^2}{30\,\mathrm{atm}\,(101{,}325\ \mathrm{Pa/atm})}$$
$$= 0.119845\,(\mathrm{m}^3/\mathrm{kmol})^2,$$

and

$$\frac{ab}{P} = \frac{3.643 \times 10^5\,\mathrm{Pa}\cdot(\mathrm{m}^3/\mathrm{kmol})^2(0.0427)(\mathrm{m}^3/\mathrm{kmol})}{30 \times 101{,}325\,\mathrm{Pa}}$$
$$= 0.005117\,(\mathrm{m}^3/\mathrm{kmol})^3.$$

Substituting these coefficients into the cubic van der Waals equation of state yields

$$\bar{v}^3 - 0.863274\bar{v}^2 + 0.119845\bar{v} - 0.005117 = 0.$$

Many methods exist to find the useful root of this polynomial, including the use of a hand-held calculator. Using spreadsheet software to implement the iterative Newton–Raphson method, with an initial guess of $\bar{v} = R_u T/P$ ($= 0.8206\,\mathrm{m}^3/\mathrm{kmol}$), results in the following converged value for $\bar{v}$ after four iterations:

$$\bar{v}\,(30\,\mathrm{atm}, 300\,\mathrm{K}) = 0.7032\,\mathrm{m}^3/\mathrm{kmol}.$$

With this value of $\bar{v}$, we evaluate Z:

$$Z = \frac{P\bar{v}}{R_u T} = \frac{30(101{,}325)0.7032}{8314.47(300)}$$
$$= 0.8570.$$

The reader should verify that this is dimensionless.

This value of Z is only 2.9% higher than the value 0.8330 calculated from the NIST WebBook (or software) in Example 2.15.

Comments For this particular example, note that the van der Waals model did an excellent job of predicting the specific volume for conditions far from the ideal-gas regime. Although the procedure used to calculate $\bar{v}$ was straightforward, we still needed to employ the power of a calculator/computer to solve the cubic van der Waals equation of state.

Self-Test 2.12

 Repeat Example 2.12 for H_2O at 40 atm and 600 K.

(Answer: $Z = 0.930$)

Other equations of state have been developed that provide more accuracy than the van der Waals equation. Discussion of these is beyond the scope of this book. For more information, we refer the interested reader to Ref. [1].

GENERALIZED COMPRESSIBILITY

As we will see later, the $T = T_c$ isotherm in P–v space exhibits an inflection point at the critical point. One can relate the constants a and b in the van der Waals equation of state to the properties at the critical state by recognizing that both the slope and the curvature of the $T = T_c$ isotherm are zero; thus,

$$a = \frac{27}{64}\frac{R^2 T_c^2}{P_c}, \tag{2.34a}$$

$$b = \frac{RT_c}{8P_c}, \tag{2. 34b}$$

$$Z = \frac{P_c v_c}{RT_c}\frac{3}{8}. \tag{2. 34c}$$

That a and b depend only on the critical pressure and critical temperature suggests that a generalized state relationship exists when actual pressures and temperatures are normalized by their respective critical values. More explicitly, defining the reduced pressure and temperature as

$$P_R \equiv \frac{P}{P_c} \tag{2.35a}$$

and

$$T_R \equiv \frac{T}{T_c}, \tag{2. 35b}$$

respectively, we expect

$$Z = Z(P_R, T_R) \tag{2. 35c}$$

to be a single "universal" relationship. This idea is known as the **principle of corresponding states,** and Z is called the **compressibility factor.** Figure 2.21 shows data and the best fit for this relationship, where Z is presented as a function of P_R using T_R as a parameter. For the gases chosen, individual data points are quite close to the curve fits, illustrating the "universal" nature of Eq. 2.35c. Plots such as Fig. 2.21 are known as **generalized compressibility charts.** Figure 2.22 is a working generalized compressibility chart for reduced pressures up to 10 and reduced temperatures up to 15.

In reality, Eq. 2.35c is not truly universal but rather a useful approximation. The best accuracy is obtained when gases are grouped according to shared characteristics. For example, a single plot constructed for polar compounds, such as water and alcohols (see Fig. 2.23), provides a tighter "universal" relationship than is obtained

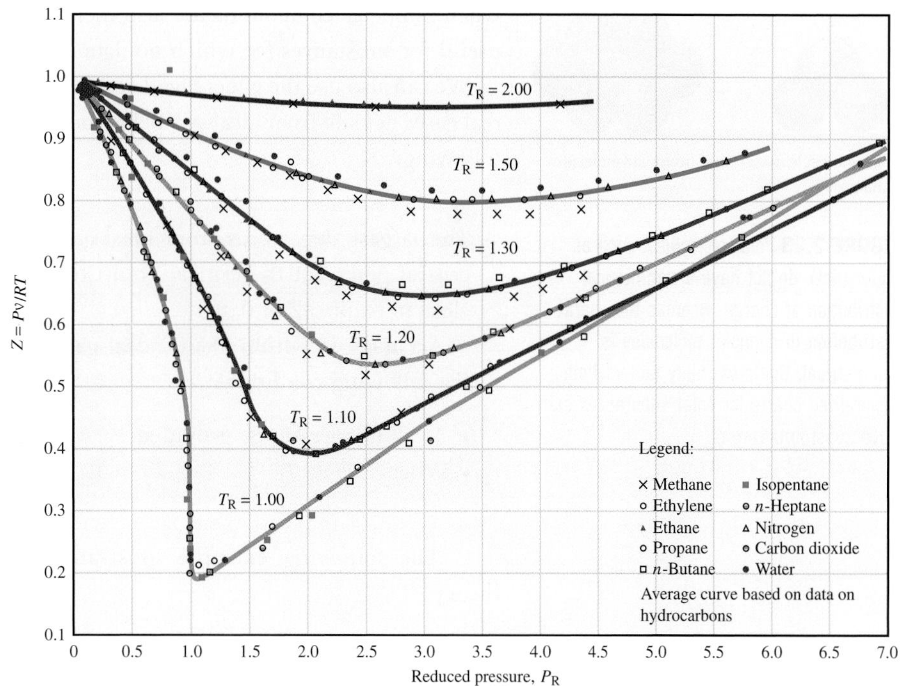

FIGURE 2.21 The compressibility factor Z can be correlated using reduced properties (i.e., P/P_c and T/T_c). Note how the data for various substances collapse when plotted in this manner. Adapted from Ref. [7] with permission.

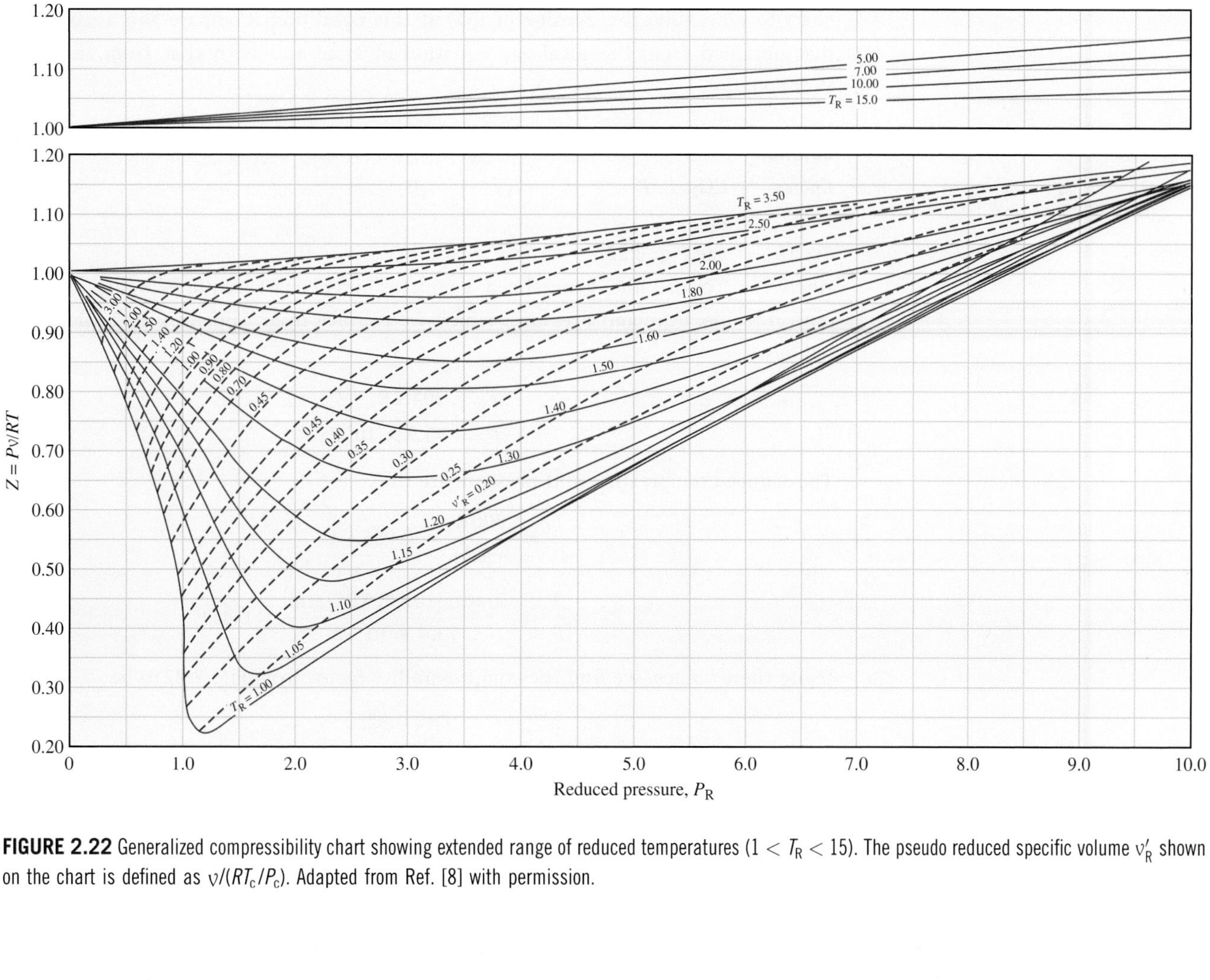

FIGURE 2.22 Generalized compressibility chart showing extended range of reduced temperatures ($1 < T_R < 15$). The pseudo reduced specific volume v'_R shown on the chart is defined as $v/(RT_c/P_c)$. Adapted from Ref. [8] with permission.

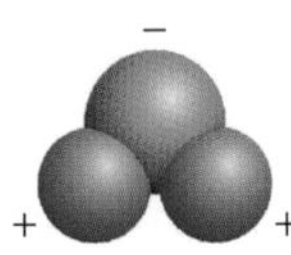

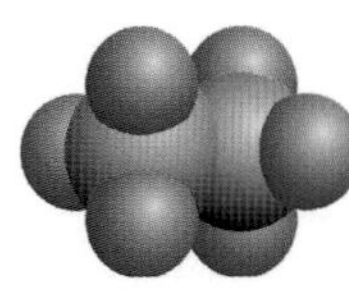

Polar molecule (water) and nonpolar molecule (ethane).

FIGURE 2.23 Polar molecules, such as water (left), do not have a symmetrical distribution of charge, whereas the charge distribution in nonpolar molecules is symmetrical. Compressibility factors from generalized charts for polar substances can have substantial errors.

when nonpolar compounds are also included. Generalized compressibility charts are useful for substances for which no data are available other than critical properties.

We can also use the generalized compressibility chart to ascertain conditions where real gases deviate from ideal-gas behavior. Using Fig. 2.22 as our guide, we observe the following:

> The largest departures from ideal-gas behavior occur at conditions near the critical point. At the critical point, the ideal-gas equation of state is not at all close to reality ($Z \approx 0.3$).
>
> Accuracy to within 5% of ideal-gas behavior ($0.95 < Z < 1.05$) is found for the following conditions:
>
> - At all temperatures, provided $P_R < 0.1$, **or**
> - When $1.95 < T_R < 2.4$ or $T_R > 15$, both for $P_R < 7.5$.

The following example illustrates the use of generalized compressibility theory.

Example 2.13 Generalized Compressibility Chart

Once again, consider CO_2 at 30 atm and 300 K. Use the generalized compressibility chart to determine the density of CO_2 at this condition. Compare this result with that obtained from the ideal-gas equation of state and with that from the NIST WebBook [6].

Solution

Known CO_2, P, T

Find ρ_{CO_2}

Sketch See Fig. 2.22.

Analysis To use the generalized compressibility chart requires values for the critical temperature and pressure, which we retrieve from Appendix E, Table E.l, as follows:

$$T_c = 304.2\,\text{K},$$
$$P_c = 73.9 \times 10^5\,\text{Pa}.$$

The reduced temperature and pressure are thus

$$T_R = \frac{T}{T_c} = \frac{300\,\text{K}}{304.2\,\text{K}} = 0.986,$$

$$P_R = \frac{P}{P_c} = \frac{30\,\text{atm}\,[101{,}325\,\text{Pa/atm}]}{73.9 \times 10^5\,\text{Pa}} = 0.411.$$

Using these values, we find the compressibility factor from Fig. 2.22 to be

$$Z \approx 0.82.$$

Applying the definition of Z, we obtain the density:

$$Z \equiv \frac{Pv}{RT} = \frac{P}{\rho RT}$$

or

$$\rho = \frac{P}{ZRT} = \frac{P\mathcal{M}_{CO_2}}{ZR_u T}$$
$$= \frac{1}{0.82}\frac{30(101{,}325)44.01}{8314.47(300)}\,\text{kg/m}^3$$
$$= \frac{53.63}{0.82}\,\text{kg/m}^3 = 65\ \text{kg/m}^3.$$

The reader should verify the units in this calculation. Note that the ideal-gas density is the numerator in the last line of this calculation. Using the results of Example 2.11, we make the requested comparisons.

Method	Z	ρ (kg/m^3)
Fig. 2.22	0.82	65
NIST WebBook	0.833	64.51
Ideal gas	1.000	53.63

Comments Even though reading Z from the chart is an approximate procedure, excellent agreement is found between the density calculated from the chart and that from the NIST WebBook.

Self-Test 2.13

 Repeat Example 2.13 for H_2O at 40 atm and 600 K.

(Answer: $Z \cong 0.95$, $\rho = 15.41$ kg/m^3)

FIGURE 2.24 The invisible steam (vapor) escaping this relief valve condenses into liquid drops made visible by their ability to reflect light (Danita Delimont / Gallo Images / Getty Images).

2.6b Calorific Relationships

To obtain values of enthalpy h, internal energy u, and specific heats c_v and c_p as functions of temperature and pressure for nonideal gases, we recommend the use of the NIST WebBook or other software. (Most thermodynamic property applications employ the NIST database as their source of property data.) Prior to the widespread use of computer-based methods to obtain thermodynamic properties, using generalized corrections to ideal-gas properties was one method to deal with nonideal gas calorific properties.

2.7 Pure Substances Involving Liquid and Vapor Phases

In this section, we investigate the thermodynamic properties of simple, compressible substances that frequently exist in both liquid and vapor states. Because of its engineering importance, we focus on water (Fig. 2.24). When discussing water, we use the common designations ice, water, and steam to refer to the solid, liquid, and vapor states, respectively, and use the chemical designation H_2O when we wish to be general and not designate any particular phase. Some textbooks distinguish a gas from a vapor using the critical pressure as a criterion. In this usage, gases exist above P_c, whereas vapors exist below P_c. However, we use the terms *vapor* and *gas* interchangeably in this book.

2.7a State (P–v–T) Relationships

PHASE BOUNDARIES

We begin our discussion of multiphase properties by conducting a thought experiment. Consider a kilogram of H_2O enclosed in a piston–cylinder arrangement as shown in the sketch in Fig. 2.25. The weight on the piston fixes the pressure in the cylinder as we add energy to the system. At the initial state, the H_2O is liquid, and remains liquid, as we add energy by heating from A to B. You might envision that a Bunsen burner is used as the energy source, for example. The added energy results in an increase in the temperature of the water, while the water expands slightly, causing a small increase in the specific volume. The temperature and the specific volume increase until state B is reached. At this point, further addition of energy results in a portion of the water becoming vapor or steam, at a fixed temperature. At 1 atm, this phase change occurs at 373.12 K (99.97 °C). (With the adoption of the International Temperature Scale of 1990 (ITS-90), the normal boiling point of water is not exactly 100 °C, but rather 99.974 °C.) At state C, we see that most of the H_2O is still in the liquid phase. With continued heating, more and more water is converted to steam. At state D, the transformation is nearly complete; and at state E, all of the liquid has

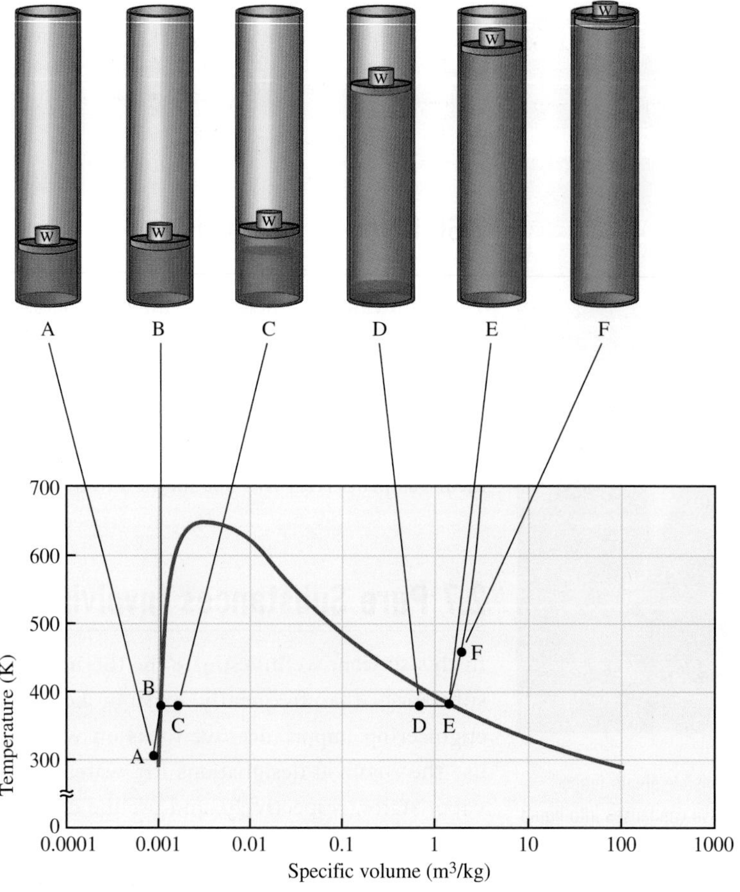

FIGURE 2.25 Heating water at constant pressure follows the path A–B–C–D–E–F on a temperature specific-volume diagram. Note the huge increase in volume in going from the liquid state (B) to the vapor state E. The volumes shown in the piston-cylinder sketches are not to scale.

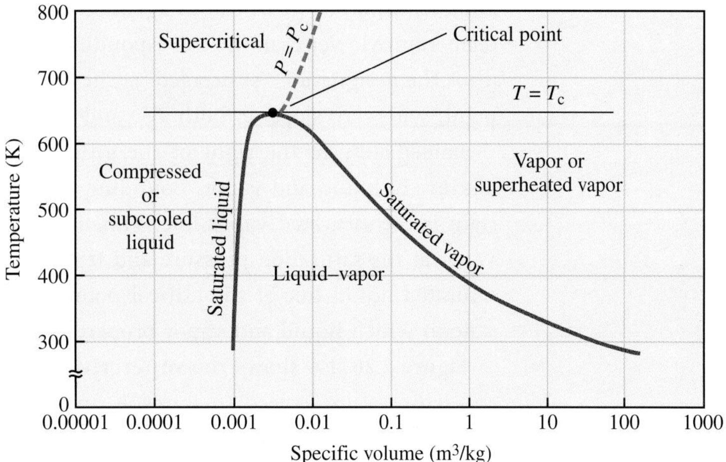

FIGURE 2.26 On T–v coordinates, the compressed liquid region lies to the left of the saturated liquid line, the liquid–vapor region between the saturated liquid and saturated vapor lines, and the vapor or superheat region lies to the right of the saturated vapor line.

turned to vapor. Adding energy beyond this point results in an increase in both the temperature and specific volume of the steam, creating what is known as superheated steam. The point at F designates a superheated state. Assuming that the heating was conducted quasi-statically, the path A–B–C–D–E–F is a collection of equilibrium states following an isobar (a line of constant pressure) in T–v space. We could repeat the experiment with lesser or greater weight on the piston (i.e., at lower or higher pressures), and map out a phase diagram showing a line where vaporization just begins (states like B) and a line where vaporization is complete (states like E). The bold line in Fig. 2.25 summarizes these thought experiments. Although we considered H_2O, similar behavior is observed for many other fluids, for example, the various fluids used as refrigerants.

Figure 2.26 presents a temperature–specific-volume (T–v) diagram showing the designations of the various regions and lines associated with the liquid and vapor states of a fluid. Although the numerical values shown apply to H_2O, the general ideas presented in this figure and discussed in the following apply to many liquid–vapor systems.

Consider the region in T–v space that lies below the critical temperature but above temperatures at which a solid phase forms. Here we have three distinct regions, corresponding to

- Compressed or subcooled liquid,
- Liquid–vapor or saturation, and
- Vapor or superheated vapor.

We also have two distinct lines, which join at the critical point:

- The saturated liquid line and
- The saturated vapor line.

The saturated liquid line is the locus of states at which the addition of energy at constant pressure results in the formation of vapor (state B in Fig. 2.25). The temperature and pressure associated with states on the saturated liquid line are called the **saturation temperature** and **saturation pressure**, respectively. To the left of the saturated liquid line is the compressed liquid region, which is also known as the subcooled liquid region. These designations arise from the fact that at any given point in this region, the pressure is higher than the corresponding saturation pressure at the same temperature (thus the designation *compressed*) and, similarly, any point is at a temperature lower than the corresponding saturation temperature at the same pressure (thus the designation "subcooled"). The saturated liquid line terminates at the critical point, with zero slope on both T–v and P–v coordinates.

Immediately to the right of the saturated liquid line are states that consist of a mixture of liquid and vapor. Bounding this liquid–vapor or saturation region on the right is the saturated vapor line. This line is the locus of states that consist of 100% vapor at the saturation pressure and temperature. The saturated vapor line joins the saturated liquid line at the critical point. This junction, the critical point, defines a state in which liquid and vapor properties are indistinguishable.

Figure 2.26 also shows the **supercritical region.** Within this region, the pressure and temperature exceed their respective critical values, and there is no distinction between a liquid and a gas.

To the right of the saturated vapor line lies the vapor or superheated vapor region. The use of the word *superheated* refers to the idea that within this region the temperature at a particular pressure is greater than the corresponding saturation temperature.

The liquid–vapor or saturation region is frequently referred to as the **vapor dome** because of its shape on a T–log v plot. When dealing with water, the term *steam dome* is sometimes applied. Because two phases coexist in the saturation region, defining a thermodynamic state here is a bit more complex than in the single-phase regions. We will examine this issue in the next section.

The three-dimensional P–v–T surface (Fig. 2.27) provides the most general view of a simple substance that exhibits multiple phases. From Fig. 2.27, we see that the T–v, P–v, and P–T plots are simply projections of the three-dimensional surface onto these respective planes.

A NEW PROPERTY – QUALITY

In our discussion of gases, whether ideal or real, we saw that simultaneously specifying the temperature and pressure defines the thermodynamic state. A knowledge of P and T is sufficient to define all other properties. In the liquid–vapor region, however, this situation no longer holds, although we still require two independent properties to define the thermodynamic state. In Fig. 2.25, we see that the temperature and pressure are both fixed for all states lying between point B and point E; thus, in the two-phase region, the temperature and pressure are not independent properties. If one is known, the other is therefore given. To define the state within the liquid–vapor mixture region thus requires another property in addition to P or T. For example, we see from Fig. 2.25 that specifying the specific volume together with the temperature can define all states along B–E.

To assist in defining states in the saturation region, a property called **quality** is used. Formally, the quality (x) is defined as the mass fraction of the mixture existing in the vapor state (see Fig. 2.28); that is,

$$x \equiv \frac{M_{\text{vapor}}}{M_{\text{mix}}}. \tag{2.36a}$$

Conventional notation uses the subscript g to refer to the vapor (gas) phase and the subscript f to refer to the liquid phase. The subscripts f and g are actually from the

FIGURE 2.27 A P–v–T surface for a substance that contracts upon freezing. After Ref. [9].

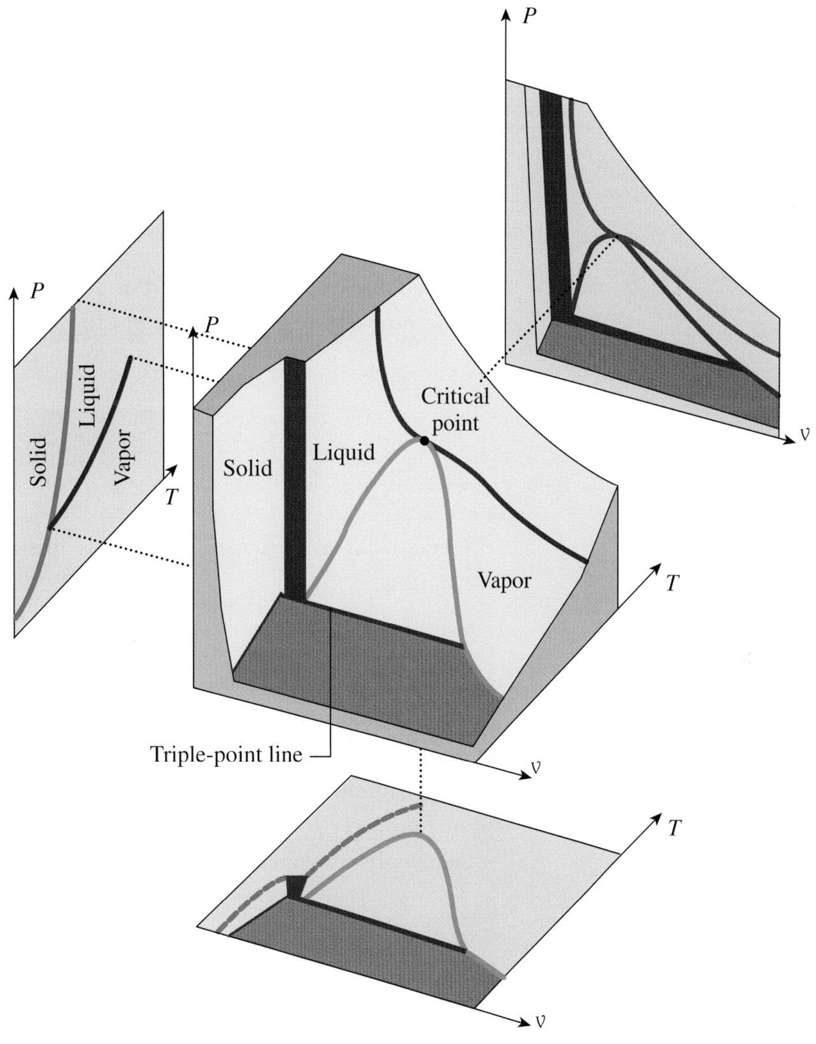

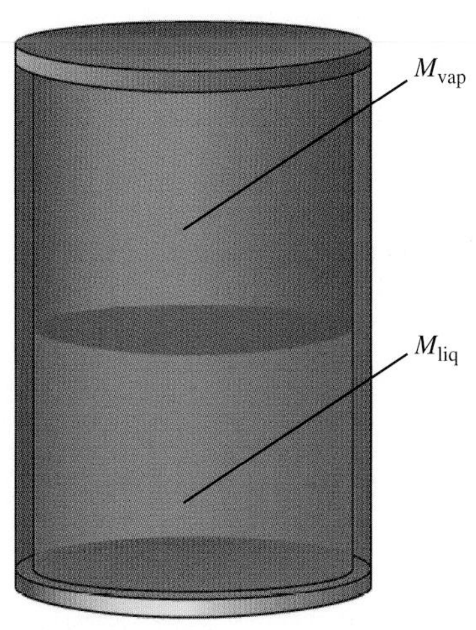

FIGURE 2.28 The quality property, defined in the text as x, is the mass fraction of a liquid–vapor mixture that is vapor.

German words *Flussigkeit* (liquid) and *Gaszustand* (gaseous state), respectively. Using this notation, the quality is defined as

$$x \equiv \frac{M_g}{M_g + M_f}. \quad \textbf{(2.36b)}$$

The quality for states lying on the saturated liquid line is zero, whereas for states lying on the saturated vapor line, the quality is unity, or 100% when expressed as a percentage. Since the actual mass in each phase is not usually known, the quality is usually derived from, or related to, other intensive properties. For example, we can relate the quality to the specific volume as follows:

$$\mathrm{v} = (1 - x)\mathrm{v}_f + x\mathrm{v}_g, \quad \textbf{(2.37a)}$$

or

$$x = \frac{\mathrm{v} - \mathrm{v}_f}{\mathrm{v}_g - \mathrm{v}_f}. \quad \textbf{(2.37b)}$$

That this is true is easily seen by expressing the intensive properties v and x in terms of their defining extensive properties, $\mathcal{V}$ and M, that is,

$$
\begin{aligned}
v &= \frac{M_f}{M_f + M_g}\left(\frac{\mathcal{V}_f}{M_f}\right) + \frac{M_g}{M_f + M_g}\left(\frac{\mathcal{V}_g}{M_g}\right) \\
&= \frac{\mathcal{V}_f}{M_f + M_g} + \frac{\mathcal{V}_g}{M_f + M_g} = \frac{\mathcal{V}_f + \mathcal{V}_g}{M_f + M_g} \\
&\equiv \frac{\mathcal{V}_{\text{mixture}}}{M_{\text{mixture}}}.
\end{aligned}
$$

The relationship expressed by Eq. 2.37a can be generalized, in that any mass-specific property (e.g., u, h, and s) can be used in place of the specific volume, so that

$$\beta = (1 - x)\beta_f + x\beta_g, \tag{2.37c}$$

or

$$x = \frac{\beta - \beta_f}{\beta_g - \beta_f}, \tag{2.37d}$$

where β represents any mass-specific property. A physical interpretation of Eq. 2.37d is easy to visualize by referring to Fig. 2.25. Here we interpret the quality at state C as the length of the line B–C divided by the length of the line B–E, where a linear rather than logarithmic scale is used for v. This same interpretation applies to any mass-specific property β plotted in a similar manner.

Commonly employed rearrangements of Eqs. 2.37a and 2.37b are

$$v = v_f + x(v_g - v_f) \tag{2.38a}$$

and

$$\beta = \beta_f + x(\beta_g - \beta_f). \tag{2. 38b}$$

Defining v_{fg} and β_{fg},

$$v_{fg} \equiv v_g - v_f \tag{2. 38c}$$

and

$$\beta_{fg} \equiv \beta_g - \beta_f, \tag{2. 38d}$$

we can rewrite Eqs. 2.38a and 2.38b more compactly as

$$v = v_f + xv_{fg} \tag{2.38e}$$

and

$$\beta = \beta_f + x\beta_{fg}. \tag{2.38f}$$

Equations 2.38e and 2.38f are useful when one is looking up tables that provide values for v_{fg} and similar properties β_{fg}.

Example 2.14 Water Mixture

A rigid tank contains 3 kg of H_2O (liquid and vapor) at a quality of 0.6. The specific volume of the liquid is 0.001041 m^3/kg and the specific volume of the vapor is 1.859 m^3/kg. Determine the mass of the vapor, the mass of the liquid, and the specific volume of the mixture.

Solution

Known M_{mix}, x, v_f, v_g

Find M_g, M_f, v_{mix}

Sketch See Fig. 2.28.

Modeling, Premises and Assumptions Equilibrium prevails.

Analysis We apply the definition of quality (Eq. 2.36b) to find the mass of H_2O in each phase:

$$\begin{aligned} M_g &= x(M_g + M_f) = xM_{mix} \\ &= 0.6(3\,\text{kg}) = 1.8\,\text{kg} \end{aligned}$$

and

$$\begin{aligned} M_f &= M_{mix} - M_g \\ &= 3\,\text{kg} - 1.8\,\text{kg} = 1.2\,\text{kg}. \end{aligned}$$

Using the given quality value of 0.6 and the given values for the specific volumes of the saturated liquid and saturated vapor, we can calculate the mixture specific volume (Eq. 2.36a):

$$\begin{aligned} v_{mix} &= (1-x)v_f + xv_g \\ &= (1-0.6)\,0.001041\,\text{m}^3/\text{kg} + 0.6\,(1.859\,\text{m}^3/\text{kg}) \\ &= 1.1158\,\text{m}^3/\text{kg}. \end{aligned}$$

Comments This example is a straightforward application of definitions. Knowledge of definitions is important in the solution of complex problems.

Self-Test 2.14

☑ Repeat Example 2.14 with $x = 0.8$ and find the volume of the tank.

(Answer: $M_g = 2.4\,kg$, $M_f = 0.6\,kg$, $v_{mix} = 1.4874\,m^3/kg$, $V = 4.46\,m^3$)

PROPERTY TABLES AND DATABASES

Tabulated properties are available for H_2O and many other fluids of engineering interest (see Table 2.5). Typically, data are presented in five forms:

- Saturation properties for convenient temperature increments,
- Saturation properties for convenient pressure increments,
- Superheated vapor properties at various fixed pressures with convenient temperature increments,

- Compressed (subcooled) liquid properties at various fixed pressures with convenient temperature increments, and
- Saturated solid–vapor properties for convenient temperature increments.

Tables in Appendix B provide data in these forms for H_2O.

Tables are instrumental to developing a conceptual understanding of the P–v–T behavior of substances (see Tutorial 3 below). After establishing a firm grasp of principles using tables, you should then develop skill in using computer-based methods to obtain properties. Computerized and online databases are making the use of printed tables obsolete. Using such databases also eliminates the need to interpolate, which can be tedious. Throughout this book, we illustrate the use of both tabulated data (Appendix B) and the NIST resources [6]. Until the use of computers is commonplace in the classroom and for examinations, printed tables will still be useful for thermodynamics instruction. Therefore, you should develop proficiency with both sources of properties. Note, also, that mastering the use of the NIST resources will be useful in fluid mechanics, heat transfer, and other disciplines.

Figures 2.29–3.32 illustrate the use of the NIST online database. Figure 2.29 shows one of the input menus for creating your own tables and graphs. After selecting the fluid of interest and choosing the units desired (not shown), you have the choice of accessing saturation data in temperature or pressure increments (Figs. 2.30 and 2.31, respectively) or accessing data as an isotherm or as an isobar (Fig. 2.32). Figure 2.30 presents saturation data for temperatures from 300 to 310 K in increments of 10 K; Fig. 2.31 shows similar data but for a range of saturation pressures from 0.1 to 0.2 MPa in increments of 0.1 MPa. Figure 2.32 presents data for the 10-MPa isobar for a 800–840 K temperature range in 20-K increments. Note that the data in Fig. 2.32 all lie within the superheated vapor region, as indicated in the final column of the table. Tables generated from the database can also be downloaded. Figures 2.33 and 2.34, for example, were generated using NIST data with spreadsheet software.

The *miniREFPROP* software provides expanded capability to that available online. We cannot overstress the importance of learning to become proficient with this invaluable tool. Tutorials 2 and 3 describe the use of this software.

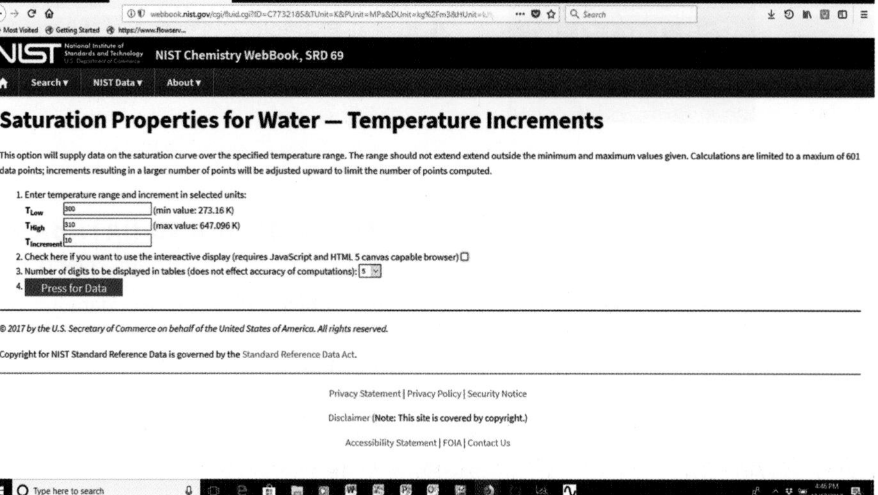

FIGURE 2.29 Graphical user interface for the NIST online thermophysical property database (NIST WebBook [6]). See Table 2.5 for the fluids for which properties are available.

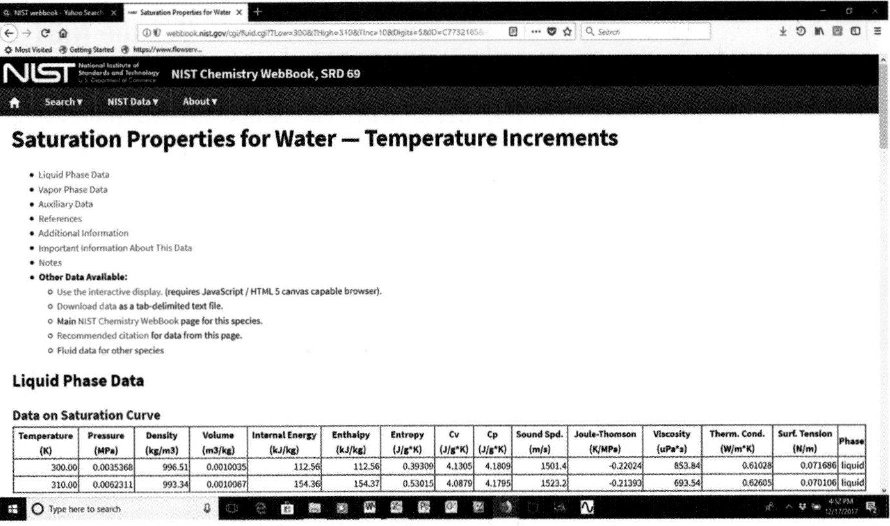

NIST Chemistry WebBook, SRD 69

Search ▾ NIST Data ▾ About ▾

Saturation Properties for Water — Temperature Increments

- Liquid Phase Data
- Vapor Phase Data
- Auxiliary Data
- References
- Additional Information
- Important Information About This Data
- Notes
- **Other Data Available:**
  - Use the interactive display. (requires JavaScript / HTML 5 canvas capable browser).
  - Download data as a tab-delimited text file.
  - Main NIST Chemistry WebBook page for this species.
  - Recommended citation for data from this page.
  - Fluid data for other species

Liquid Phase Data

Data on Saturation Curve

Temperature (K)	Pressure (MPa)	Density (kg/m3)	Volume (m3/kg)	Internal Energy (kJ/kg)	Enthalpy (kJ/kg)	Entropy (J/g*K)	Cv (J/g*K)	Cp (J/g*K)	Sound Spd. (m/s)	Joule-Thomson (K/MPa)	Viscosity (uPa*s)	Therm. Cond. (W/m*K)	Surf. Tension (N/m)	Phase
300.00	0.0035368	996.51	0.0010035	112.56	112.56	0.39309	4.1305	4.1809	1501.4	-0.22024	853.84	0.61028	0.071686	liquid
310.00	0.0062311	993.34	0.0010067	154.36	154.37	0.53015	4.0879	4.1795	1523.2	-0.21393	693.54	0.62605	0.070106	liquid

FIGURE 2.30 Saturation properties for water from NIST WebBook [6]: 300–310 K in 10-K increments. Vapor phase properties (not shown) follow those of the liquid phase.

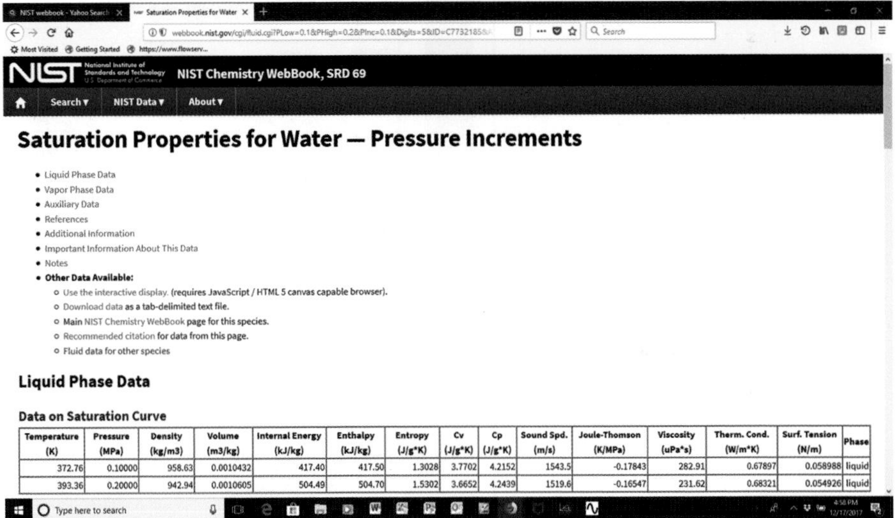

NIST Chemistry WebBook, SRD 69

Search ▾ NIST Data ▾ About ▾

Saturation Properties for Water — Pressure Increments

- Liquid Phase Data
- Vapor Phase Data
- Auxiliary Data
- References
- Additional Information
- Important Information About This Data
- Notes
- **Other Data Available:**
  - Use the interactive display. (requires JavaScript / HTML 5 canvas capable browser).
  - Download data as a tab-delimited text file.
  - Main NIST Chemistry WebBook page for this species.
  - Recommended citation for data from this page.
  - Fluid data for other species

Liquid Phase Data

Data on Saturation Curve

Temperature (K)	Pressure (MPa)	Density (kg/m3)	Volume (m3/kg)	Internal Energy (kJ/kg)	Enthalpy (kJ/kg)	Entropy (J/g*K)	Cv (J/g*K)	Cp (J/g*K)	Sound Spd. (m/s)	Joule-Thomson (K/MPa)	Viscosity (uPa*s)	Therm. Cond. (W/m*K)	Surf. Tension (N/m)	Phase
372.76	0.10000	958.63	0.0010432	417.40	417.50	1.3028	3.7702	4.2152	1543.5	-0.17843	282.91	0.67897	0.058988	liquid
393.36	0.20000	942.94	0.0010605	504.49	504.70	1.5302	3.6652	4.2439	1519.6	-0.16547	231.62	0.68321	0.054926	liquid

FIGURE 2.31 Saturation properties from NIST WebBook [6]: 0.1–0.2 MPa in 0.1-MPa increments. Vapor phase properties (not shown) follow those of the liquid phase.

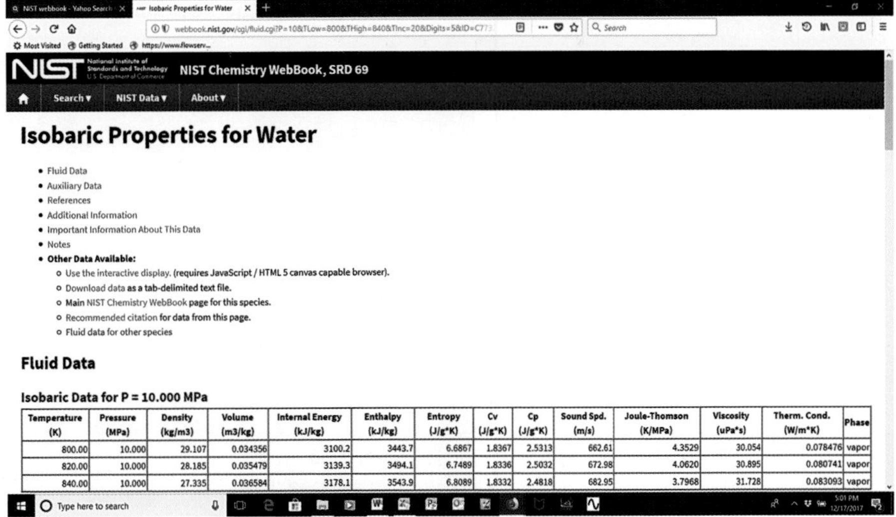

NIST Chemistry WebBook, SRD 69

Search ▾ NIST Data ▾ About ▾

Isobaric Properties for Water

- Fluid Data
- Auxiliary Data
- References
- Additional Information
- Important Information About This Data
- Notes
- **Other Data Available:**
  - Use the interactive display. (requires JavaScript / HTML 5 canvas capable browser).
  - Download data as a tab-delimited text file.
  - Main NIST Chemistry WebBook page for this species.
  - Recommended citation for data from this page.
  - Fluid data for other species

Fluid Data

Isobaric Data for P = 10.000 MPa

Temperature (K)	Pressure (MPa)	Density (kg/m3)	Volume (m3/kg)	Internal Energy (kJ/kg)	Enthalpy (kJ/kg)	Entropy (J/g*K)	Cv (J/g*K)	Cp (J/g*K)	Sound Spd. (m/s)	Joule-Thomson (K/MPa)	Viscosity (uPa*s)	Therm. Cond. (W/m*K)	Phase
800.00	10.000	29.107	0.034356	3100.2	3443.7	6.6867	1.8367	2.5313	662.61	4.3529	30.054	0.078476	vapor
820.00	10.000	28.185	0.035479	3139.3	3494.1	6.7489	1.8336	2.5032	672.98	4.0620	30.895	0.080741	vapor
840.00	10.000	27.335	0.036584	3178.1	3543.9	6.8089	1.8332	2.4818	682.95	3.7968	31.728	0.083093	vapor

FIGURE 2.32 Properties of water for the 10-MPa isobar from the NIST WebBook [6]: 800–840 K in 20-K increments.

Tutorial 2 — How to Use the NIST Software *miniREFPROP*

The National Institute of Science and Technology (NIST) provides *miniREFPROP* as freeware that can be downloaded from NIST at http://trc.nist.gov/refprop/MINIREF/MINIREF.HTM. This software – an invaluable resource for the thermodynamic and transport properties of several pure substances and air – is very easy to use: You can learn to use it in only a few minutes and can be an expert in less than an hour. The purpose of this tutorial is to acquaint you with the basic capabilities of this powerful package and to encourage you to become an expert user.

What fluids are contained in miniREFPROP?

From the "Substance" menu a selection of seven pure fluids is available. A listing is shown in the figure. Air is also available as a pseudo-pure fluid from the substance menu.

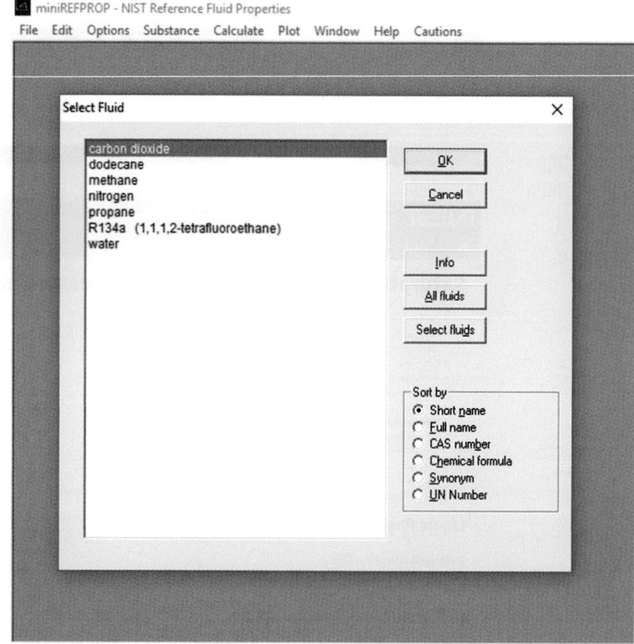

What properties are available?

All the properties used in this book and many others are provided. "Properties" can be selected from the "Options" menu as illustrated here.

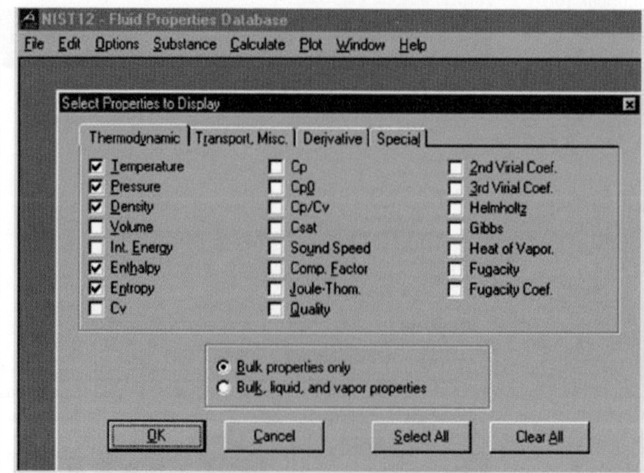

Which units can be used?

From the "Options" menu, the user can select SI, US customary, and other units using pull-down selections. Mixed units can be used if desired and either mass- or molar-specific quantities can be selected.

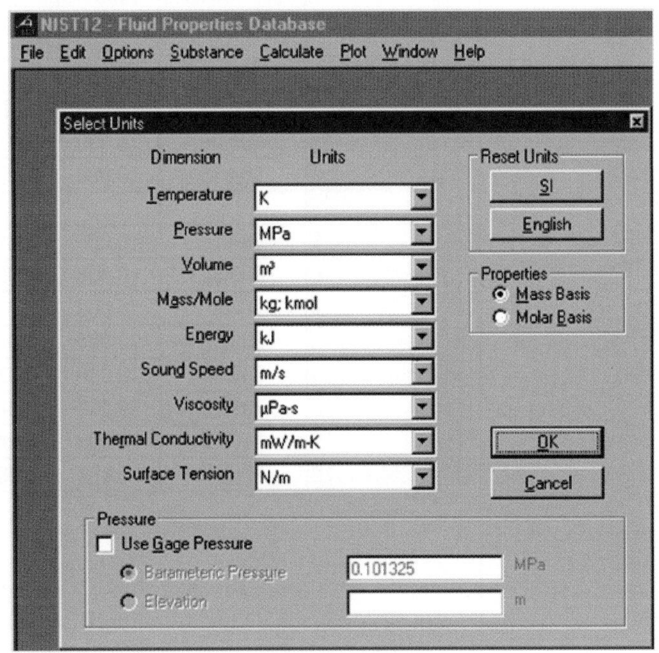

Can tables be generated?

Under the "Calculate" menu, the user has the option to generate two types of tables: saturation tables and isoproperty tables. The creation of an isoproperty for table is illustrated here (for P fixed).

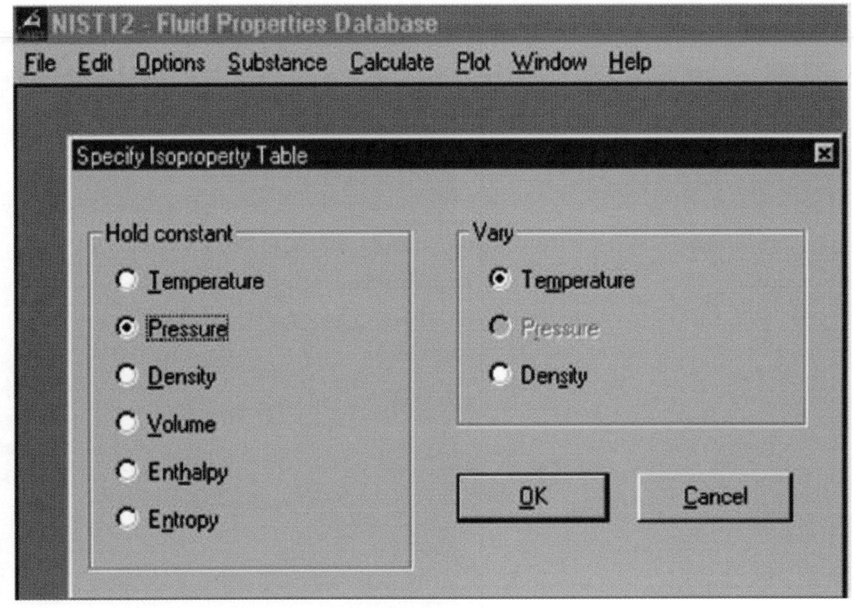

NIST12 - Fluid Properties Database

File Edit Options Substance Calculate Plot Window Help

1: water: p = 10.0 MPa

	Temperature (K)	Pressure (MPa)	Volume (m^3/kg)	Int. Energy (kJ/kg)	Enthalpy (kJ/kg)	Entropy (kJ/kg-K)
1	300.00	10.000	0.00099905	111.74	121.73	0.39029
2	400.00	10.000	0.0010611	529.06	539.67	1.5921
3	500.00	10.000	0.0011933	965.25	977.18	2.5669
4	584.15	10.000	0.0014526	1393.5	1408.1	3.3606
5	584.15	10.000	0.018030	2545.2	2725.5	5.6160
6	600.00	10.000	0.020091	2619.1	2820.0	5.7756
7	700.00	10.000	0.028285	2894.5	3177.4	6.3305
8	800.00	10.000	0.034356	3100.2	3443.7	6.6867
9	900.00	10.000	0.039804	3293.5	3691.6	6.9787
10	1000.0	10.000	0.044963	3485.8	3935.5	7.2357
11	1100.0	10.000	0.049959	3681.0	4180.6	7.4693
12	1200.0	10.000	0.054854	3880.6	4429.2	7.6855

What if I want properties at a single state point?

From the "Calculate" menu the user can select "Specified State Points" or "Saturation Points."

What plots can be generated with the software?

All the standard thermodynamic coordinates (*P*–*v*, *T*–*s*, *h*–*s*, etc.) and others are available from the "Plot" menu. Plots with iso-lines, as shown in the figure, are easy to create. The option "Other Diagrams" allows the user to select any pair of thermodynamic coordinates. The *T*–*v* diagram here illustrates this option.

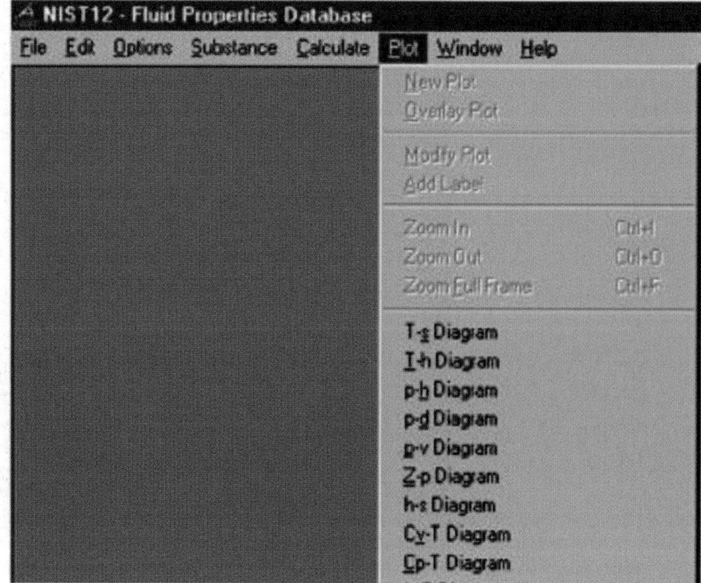

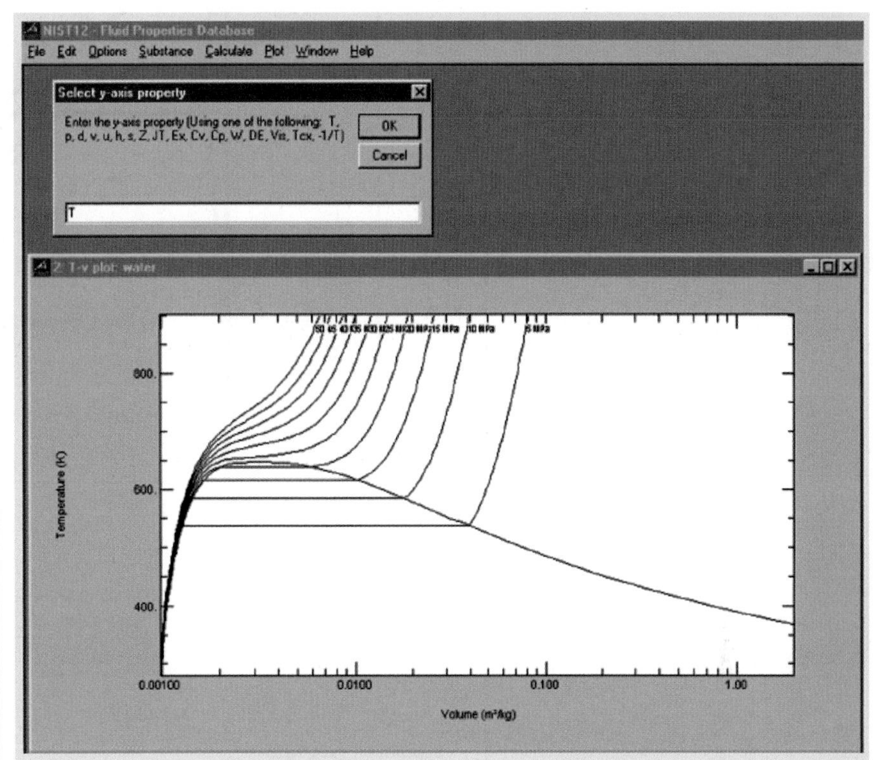

Is the NIST software compatible with other common software?

Tables generated using *miniREFPROP* can be easily pasted into spreadsheets. Similarly, plots can be pasted into word-processing documents.

The reader is encouraged to use this software to solve the homework problems provided in this book.

Tutorial 3 — How to Define a Thermodynamic State

Whether employing tabular data, the NIST WebBook, or other computer applications, one needs to become proficient at identifying and defining thermodynamic states. What we mean by that is the following: given a pair of properties (e.g., P_1, T_1), can you determine the region, or region boundary, in or on which the state lies? Specifically, is the state of the substance a compressed liquid, saturated liquid, liquid–vapor mixture, saturated vapor, or superheated vapor? Or does the state lie at the critical point or within the supercritical region? This tutorial offers hints to help you answer such questions efficiently and with confidence.

Given an arbitrary temperature T_1 and pressure P_1 we desire to identify the state and to determine the specific volume v_1. To aid our analysis, we employ the following P–v diagram for H_2O.

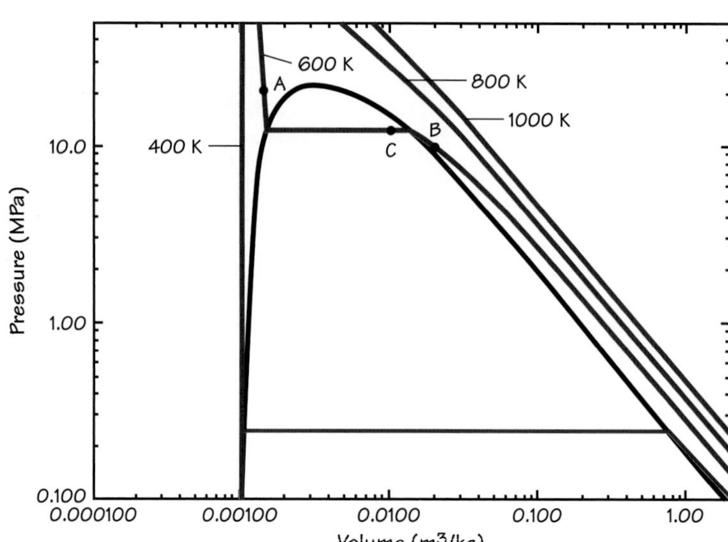

We begin with a few concrete examples using H_2O, with all states having the same temperature, 600 K. The 600-K isotherm is one of the several shown on our P–v plot.

Given: $T = 600$ K, $P = 20$ MPa

Find: Region, v

The key to defining the region is determining the relationship of the given properties to those of (1) the critical point and (2) the saturation states. For H_2O, the critical point is 22.064 MPa and 647.1 K. Since the given pressure is less than the critical pressure, we can rule out the supercritical region; we can easily verify this by inspecting the P–v diagram. Next we determine the saturation states corresponding to the given P and T. Using Tables B.1 and B.2, respectively (or the NIST WebBook), we find that

$$P_{sat}(600\,\mathrm{K}) = 12.34\,\mathrm{MPa} \quad (\mathrm{Table\,B.1}),$$

$$T_{sat}(20\,\mathrm{MPa}) = 638.9\,\mathrm{K} \quad (\mathrm{Table\,B.2}).$$

From these values, we see that the given temperature, 600 K, is less than the saturation value, 638.9 K [i.e., T (20 MPa) $< T_{sat}$ (20 MPa)]. The state thus lies in the subcooled region. Alternatively, we can see that the given pressure, 20 MPa, is greater than the saturation value, 12.34 MPa [i.e., P (600 K) $< P_{sat}$ (600 K)]; the state thus lies in the compressed-liquid region. The compressed-liquid and subcooled-liquid regions are synonymous. Point A on the P–v plot denotes this state. Now, knowing the region, we turn to the compressed-liquid table, Table B.4D, to find the specific volume. No interpolation is needed and we can read the value directly:

$$v(20\,\mathrm{MPa}, 600\,\mathrm{K}) = 0.001481\,\mathrm{m^3/kg}.$$

A simpler, but less instructive, procedure is to use the NIST software. By selecting "Specified State Points" from the "Calculate" menu, we obtain the density [= 675.11 kg/m^3 = $1/v$ = 1/(0.001481 m^3/kg)].

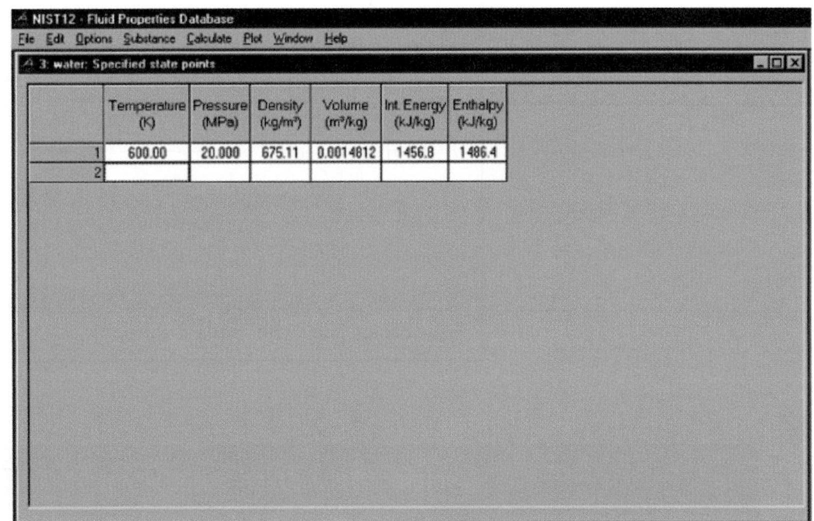

Alternatively, the NIST WebBook can be employed to find the specific volume. To do this, we select "Isothermal Properties" for water and continue as prompted.

Note that finding the region in which the state lies must be a separate step because the "Specified State Point" calculation does not identify the state location. Being able to locate a state relative to the liquid, liquid–vapor-mixture, and vapor boundaries is an essential skill.

Given $T = 600$ K, $P = 10\,\text{MPa}$
Find Region, v

Again, we use Tables B.1 and B.2 to find the corresponding saturation values:

$$P_{\text{sat}}(600\,\text{K}) = 12.34\,\text{MPa} \quad (\text{Table B.1}),$$
$$T_{\text{sat}}(10\,\text{MPa}) = 584.15\,\text{K} \quad (\text{Table B.2}).$$

Because T (10 MPa) $> T_{\text{sat}}$ (10 MPa) (i.e., 600 K $>$ 584.15 K), the state lies in the superheated vapor region. We could come to the same conclusion by recognizing that P (600 K) $< P_{\text{sat}}$ (600 K). Point B on the P–v plot denotes this state. To find the specific volume, we employ the superheated vapor table for 10 MPa, Table B.3P, which directly yields

$$v(10\text{ MPa}, 600\text{ K}) = 0.020091\text{ m}^3/\text{kg}.$$

The given values were deliberately selected to avoid any interpolation. Finding properties in the superheated region may be complicated by the need to interpolate both temperature and pressure. In such cases, we recommend using the NIST WebBook or other software to avoid this tedious process.

Given $T = 600\,\text{K}$, $v = 0.01\,\text{m}^3/\text{kg}$
Find Region, P

Since the specific volume is given, we determine the relationship of this value to those of the saturated liquid and saturated vapor at the same temperature in order to define the region. From Table B.1, we obtain

$$v_{\text{sat liq}}(600\,\text{K}) = 0.0015399\,\text{m}^3/\text{kg},$$
$$v_{\text{sat vapor}}(600\,\text{K}) = 0.014984\,\text{m}^3/\text{kg}.$$

Because the given specific volume falls between these two values, the state must lie in the liquid–vapor mixture region. The pressure must then be the saturation value, $P_{sat}(600\,\text{K}) = 12.34\,\text{MPa}$. If the given specific volume were less than $v_{sat\ liq}$, we would have to conclude that the state lies in the compressed-liquid region. Similarly, if the given specific volume were greater than $v_{sat\ vapor}$, the state would then lie in the superheated-vapor region. Point C denotes this state on the P–v diagram.

Although we determined the regions in which various states lie using T, P, and v, other properties can be employed as well. The key, regardless of the particular properties given, is to find their relationships to the corresponding saturation state boundaries (i.e., the "steam dome").

Example 2.15 Turbine Outflow State

A turbine expands steam to a pressure of 0.01 MPa and a quality of 0.92. Determine the temperature and the specific volume at this state.

Solution

Known P, x

Find T, v

Sketch

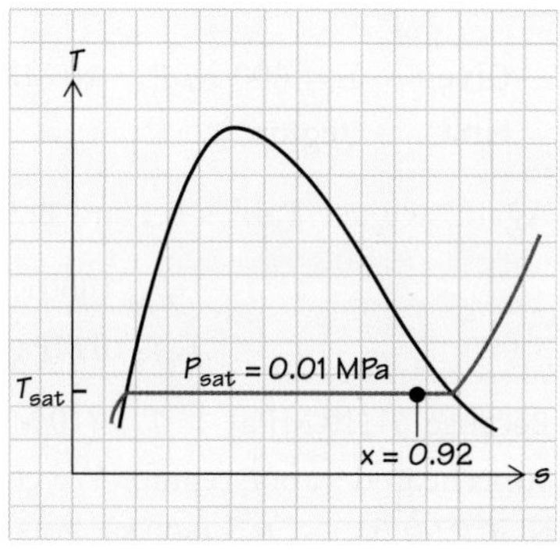

Analysis The given state lies in the liquid–vapor region on the 0.01-MPa isobar, as shown in the sketch. Knowing this, we use Table B.2, Saturation Properties of Water and Steam – Pressure Increments, to obtain the following data:

$$T = T_{sat}(P = P_{sat} = 0.01\,\text{MPa}) = 318.96\text{ K},$$
$$v_f = 0.0010103\,\text{m}^3/\text{kg},$$
$$v_g = 14.670\,\text{m}^3/\text{kg}.$$

Using the given quality and these values for the specific volumes of the saturated liquid and saturated vapor, we calculate the specific volume from Eq. 2.37a as follows:

$$\begin{aligned} v &= (1 - x)v_f + xv_g \\ &= (1 - 0.92)\,0.0010103\,\text{m}^3/\text{kg} + 0.92\,(14.670\,\text{m}^3/\text{kg}) \\ &= 13.496\,\text{m}^3/\text{kg}. \end{aligned}$$

Comments The keys to this problem are, first, recognizing the region in which the given state lies, and, second, choosing the correct table to evaluate properties. The reader should verify the values obtained from Table B.2 for this example and also verify that the values agree with the NIST WebBook.

Self-Test 2.15

Calculate the values of the specific internal energy u and the specific enthalpy h for the conditions given in Example 2.15.

(Answer: $u = 2257.57\ \mathrm{kJ/kg}$, $h = 2392.53\ \mathrm{kJ/kg}$)

T–v DIAGRAMS

Figure 2.33 presents a T–v diagram for H_2O showing several lines of constant pressure. In the liquid region, lines of constant pressure hug the saturated-liquid line quite closely. We illustrate this fact more clearly in the expanded view of the liquid region provided in Fig. 2.34, where the 10-MPa constant-pressure line is shown together with the saturated-liquid line. All the constant-pressure lines for pressures less than 10 MPa, but greater than P_{sat} at any given temperature, are crowded between the two lines shown in Fig. 2.34; thus, we see that the specific volume is much more sensitive to temperature than to pressure. As a consequence, values for the specific volume for compressed liquids can be approximated to those at the saturated-liquid state at the same temperature; that is,

$$v_{liq}(T) \cong v_f(T), \tag{2.39}$$

where v_f is the specific volume of the saturated liquid at T. This rule of thumb can be used when comprehensive compressed-liquid data are not available. We discuss compressed-liquid properties in more detail in a later section.

Returning to Fig. 2.33 we now examine the behavior of the constant-pressure lines on the other side of the steam dome. We first note the significant dependence of the specific volume on pressure along the saturated-vapor line, where an order-of-magnitude increase in pressure results in approximately an order-of-magnitude decrease in specific volume. If the saturated vapor could be treated as an ideal gas, the relationship would follow $v^{-1} \propto P$. In the vapor region, specific volume increases with temperature along a line of constant pressure, and pressure increases with temperature along a line of constant volume. Although somewhat obscured by the use of a logarithmic scale for specific volume, the vapor region of Fig. 2.33 looks much like the ideal-gas T–v plot shown earlier (cf. Fig. 2.15). Of course, the real-gas behavior of steam affects the details of such a comparison.

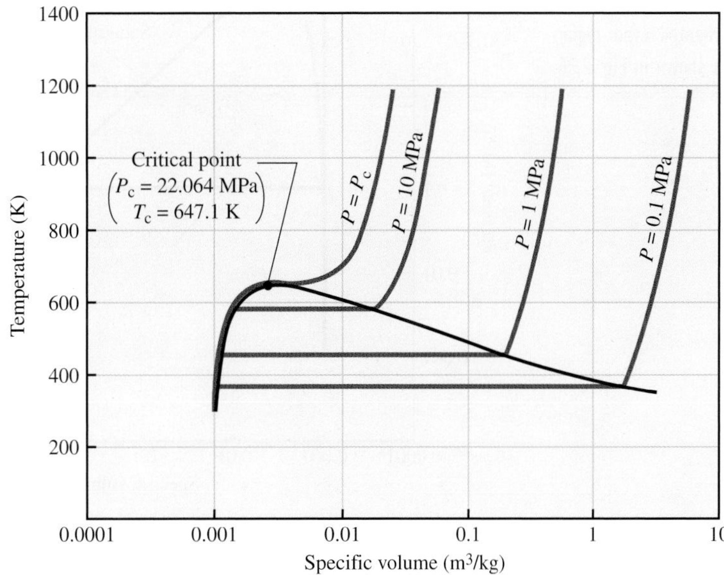

FIGURE 2.33 Temperature–specific-volume diagram for H_2O showing isobars for 0.1, 1, and 10 MPa. Also shown is the critical pressure isobar.

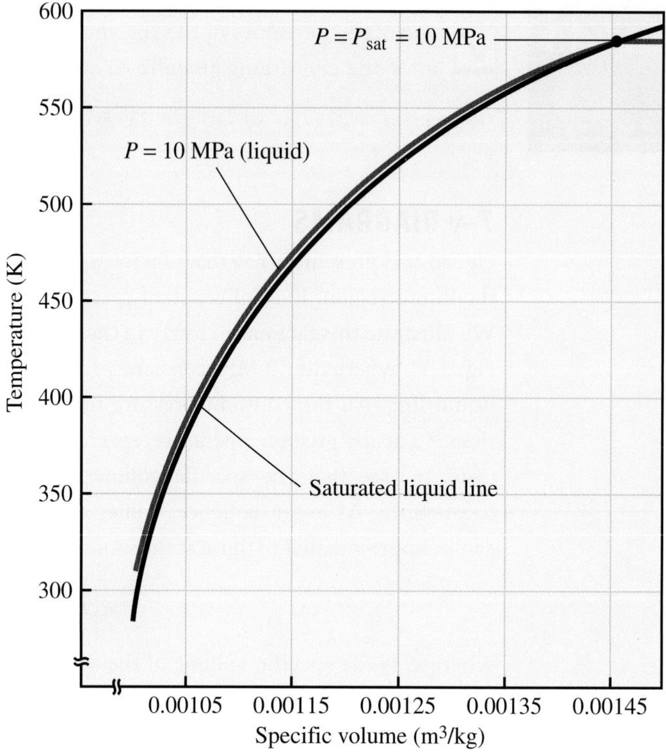

FIGURE 2.34 Isobars in T–v space in the compressed-liquid region for H_2O lie close to the saturated-liquid line. The 10-MPa isobar crosses the saturated-liquid line when $P_{sat} = 10$ MPa and then proceeds horizontally across the steam dome (a small part of the line is shown).

P–v DIAGRAMS

Figure 2.35 presents a P–v diagram for H_2O showing a single isotherm ($T = 373.12$ K or 99.97 °C). The corresponding saturation pressure is 101,325 Pa (1 atm). The saturation condition thus corresponds to the normal boiling point of water. Note that both axes in Fig. 2.35 are logarithmic. Isotherms for temperatures greater than 373.12 K lie above

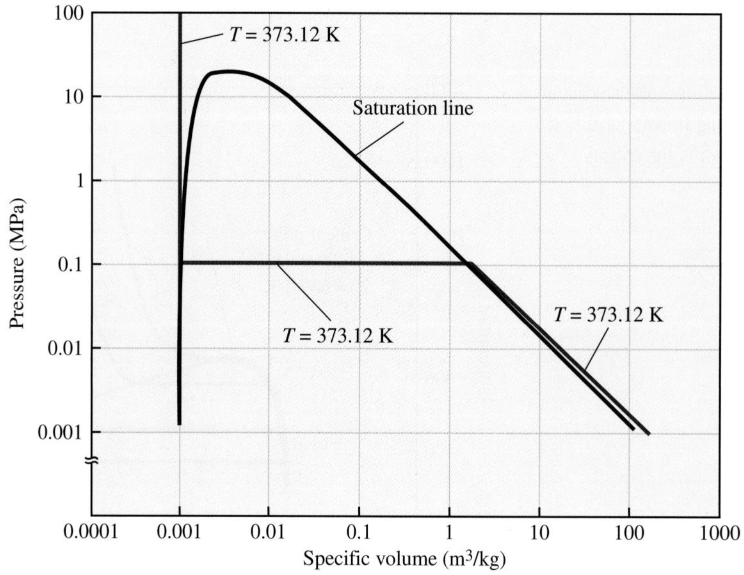

FIGURE 2.35 P–v diagram for H_2O showing the $T = 373.12$ K (99.97 °C) isotherm. An expanded view of several isotherms in the superheated-vapor region ($P = 0.01$–0.1 MPa) is shown in Fig. 2.36.

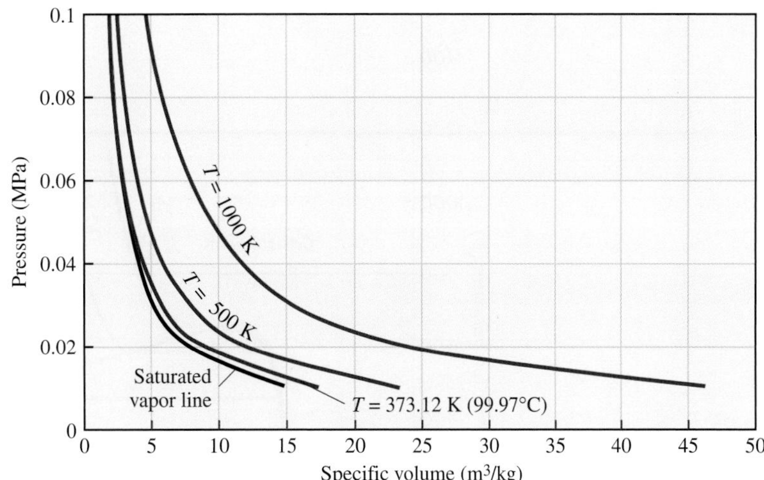

FIGURE 2.36 Isotherms in P–v space for the superheated region of H_2O exhibit a hyperbolic-like behavior (cf. the ideal-gas behavior in Fig. 2.14).

and to the right of the isotherm shown. In the superheated-vapor region, the isotherms lie relatively close to the saturated-vapor line in the log–log coordinates of Fig. 2.35. Figure 2.36 presents an expanded view of the region $0.01 < P < 0.1$ MPa, showing several isotherms in linear P–v space.

Example 2.16 Isochoric Vapor Process

A 0.5-kg mass of steam at 800 K and 1 MPa is contained in a rigid vessel. The steam is cooled to the saturated-vapor state. Plot the process in both T–v and P–v coordinates. Determine the volume of the vessel and the final temperature and pressure of the steam.

Solution

Known Steam, M, P_1, T_1, saturated vapor at state 2

Find P_2, T_2, V

Modeling, Premises and Assumptions

i. Rigid tank (given)

ii. Simple compressible substance

Analysis Before we can create T–v and P–v sketches, we need to determine the region in which the state-1 point lies. Since the state-1 temperature (800 K) is greater than the critical temperature ($T_c = 647.27$ K) and the state-1 pressure (1 MPa) is less than the critical pressure ($P_c = 22.064$ MPa), state 1 must lie in the superheated-vapor region. With this information, we can now construct the following T–v and P–v plots.

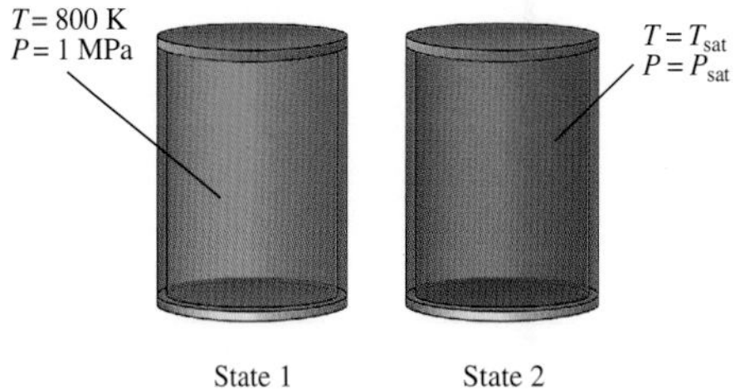

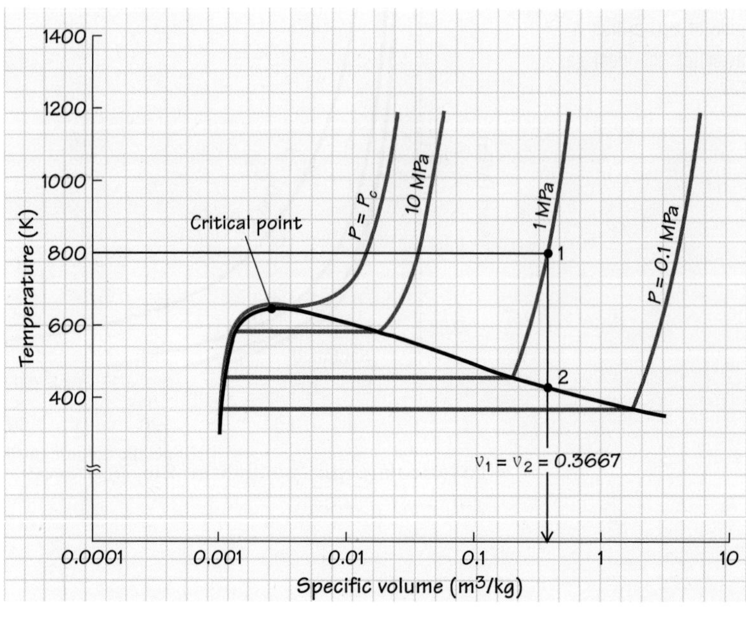

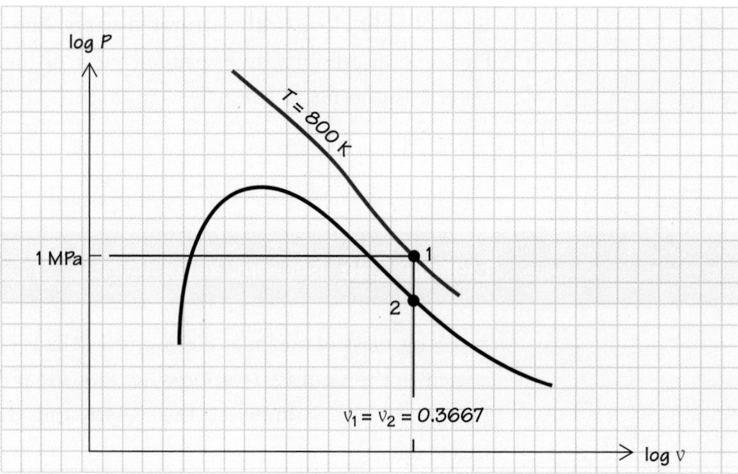

We note that the cooling of the steam from state 1 to state 2 is a constant-volume process because the tank is rigid. Because the mass is also constant, the specific volume at state 1 equals the specific volume at state 2, as shown on the sketches. To find P_2, T_2, and $\mathcal{V}$ requires a numerical value for v_2 $(= v_1)$. Using data from Table B.3, we find the specific volume v_1:

T (K)	v (m^3/kg)
780	0.35734
$800 = T_1$	$0.36677 = v_1$
820	0.37618

The NIST WebBook yields a value identical to the tabulated value. Since state 2 lies on the saturated-vapor line, we know that $v_g(T_2, P_2) = v_2 = v_1$. From Table B.2, we see that the state-2 specific volume is between those of saturated vapor at 0.5 and 0.6 MPa; thus, we interpolate to find P_2 and T_2 as follows.

v_{sat} (m^3/kg)	P_{sat} (MPa)	T_{sat} (K)
0.37481	0.5	424.98
0.36677 = v_2	0.51 = P_2	425.9 = T_2
0.31558	0.6	431.98

Using the NIST software provides a more accurate result:

$$P_2 = 0.511\,\text{MPa},$$
$$T_2 = 425.8\,\text{K}.$$

To calculate the volume of the rigid vessel containing the steam, we apply the definition of specific volume,

$$v \equiv \mathcal{V}/M$$

or

$$\mathcal{V} = Mv.$$

Thus,

$$\mathcal{V} = (0.5\,\text{kg})\,0.36677\,\text{m}^3/\text{kg} = 0.1834\,\text{m}^3.$$

Comments This example illustrates some of the thought processes involved in determining the region in which a state point lies. We also see the importance of recognizing that here the process involved is a constant-volume process.

Self-Test 2.16

The system of Example 2.16 is further cooled to a final temperature of 350 K. Determine the final pressure and the quality at this state.

(Answer: $P = P_{sat}$ (350 K) = 41.68 kPa, x = 0.095)

2.7b Calorific and Second-Law Properties

The sources available for calorific properties – specific enthalpies and specific internal energies – and the second-law property, entropy – are the same as for state properties: Appendix B, the NIST WebBook [6], and NIST software. In the NIST sources, both enthalpies and internal energies are provided; however, many sources provide only enthalpies, leaving it to the user to obtain internal energies from the definition

$$u(T, P) = h(T, P) - Pv.$$

Also frequently tabulated is the **enthalpy of vaporization,** h_{fg}, which is defined as

$$h_{fg} \equiv h_{vap} - h_{liq} = h_g - h_f. \tag{2.40}$$

Physically, this represents the amount of energy required to vaporize a unit mass of liquid at constant pressure. This quantity is also sometimes referred to as the latent heat of vaporization, a term coined during the reign of the caloric theory. See Chapter 4 for a brief history of the caloric theory and how it was superseded by modern constructs of energy.

T–s DIAGRAMS

Of equal importance to *T–v* and *P–v* diagrams is the temperature–entropy, or *T–s*, diagram. We introduce the temperature–entropy diagram here; Chapter 7 provides a more expansive discussion.

Figure 2.37 presents a temperature–entropy diagram for water and shows an isobar traversing the compressed liquid region, across the steam dome, and up into the superheat region. Without an expanded scale, the 1-MPa isobar in the compressed-liquid region is indistinguishable from the saturated-liquid line; however, it does lie above and to the left of the saturated-liquid line. Isobars for pressures greater than 1 MPa lie above the isobar shown, and those for lower pressures lie below. Although not shown, isochors (constant-volume lines) in the superheat region have steeper slopes than the isobars. Note that Fig. 2.37 employs linear scales for both temperature and entropy, rather than the semilog and log–log scales previously used in the analogous *T–v* (Fig. 2.33) and *P–v* (Fig. 2.35) diagrams, respectively.

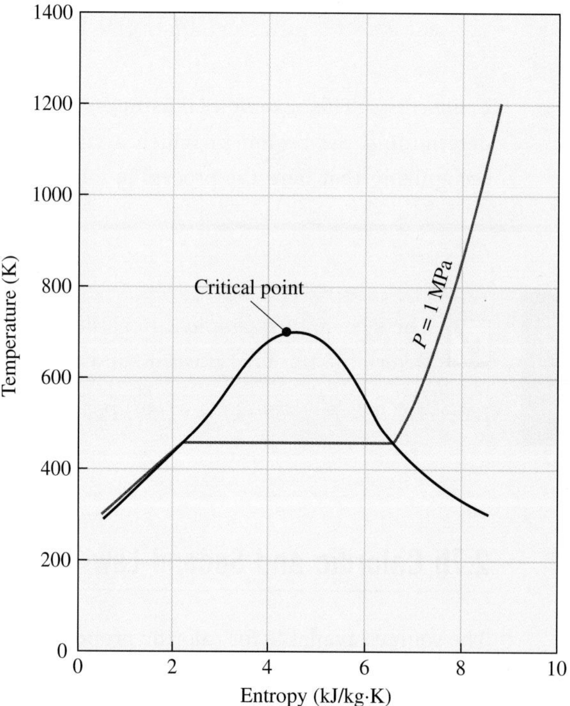

FIGURE 2.37 Temperature–entropy (*T–s*) diagram for water showing the liquid–vapor saturation lines and the 1-MPa isobar.

Example 2.17 Isothermal Vapor Process

Consider an isothermal expansion of steam from an initial state of saturated vapor at 3 MPa to a pressure of 1 MPa. Plot the process in *T–s* and *P–v* coordinates and determine the initial and final specific volumes.

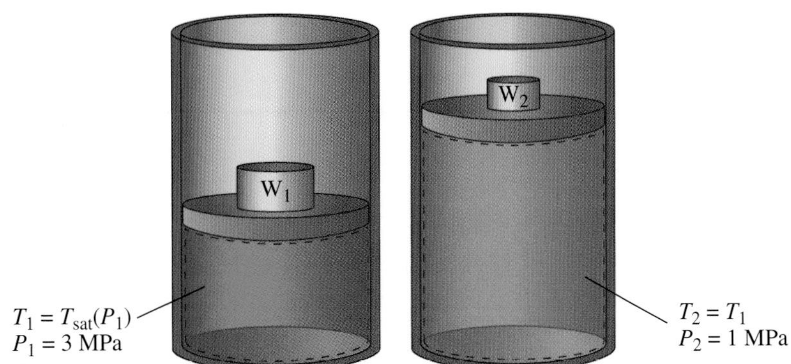

Solution

Known Saturated vapor at P_1, T_1 (= T_2), P_2

Find $v_1\,(= v_g)$, v_2

Modeling, Premises and Assumptions Simple compressible substance at equilibrium

Analysis We begin by drawing the P_1 (= 3 MPa) and P_2 (= 1 MPa) isobars on a T–s diagram as shown. State 1 is identified on this diagram as a saturated vapor. For the isothermal process, a horizontal line is extended from the state-1 point. The location where this isotherm, $T = T_{sat}$ (3 MPa), crosses the 1-MPa isobar identifies the state-2 point. State 2 is in the superheated vapor region. In P– v coordinates, we draw the same $T = T_{sat}$ (3 MPa) isotherm. Where it crosses the 1-MPa isobar identifies the state-2 point in P–v space.

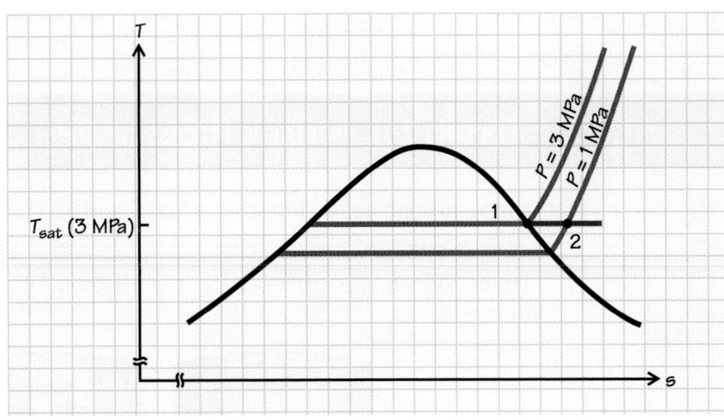

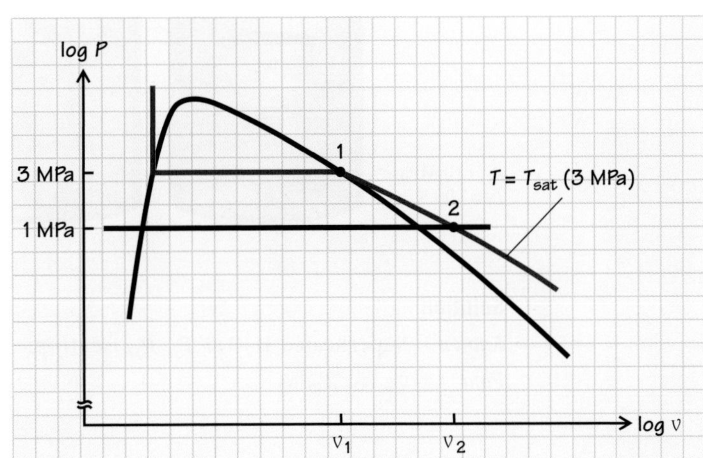

From Table B.2, we obtain the following properties:

$$P_1 = 3\,\text{MPa},$$
$$T_1 = 507.00\,\text{K},$$
$$v_1 = 0.066664\,\text{m}^3/\text{kg}.$$

To obtain the state-2 properties, we can interpolate to find v_2 in the superheated-vapor table for $P = 1$ MPa (see Table B.3). Alternatively, the NIST WebBook can be used directly to determine v_2 by generating isothermal data ($T = 507.00$ K) with pressure increments containing $P = 1$ MPa. The result is

$$v_2 = 0.22434\,\text{m}^3/\text{kg}.$$

Comments We note the utility of defining constant-property lines on T–s and P–v diagrams. The reader should verify the state-2 property determinations using Table B.3 and the NIST WebBook and/or software.

Self-Test 2.17

The system of Example 2.17 is allowed to expand isothermally until the final specific volume is 0.32 m³/kg. Find the final pressure.

(Answer: $P \cong 0.70$ MPa)

Example 2.18 Isentropic Process for Water

Superheated steam, initially at 700 K and 10 MPa, is expanded to 1 MPa isentropically, i.e., at constant specific entropy s. Determine the values of v, h, T, and x at the final state. Sketch the process on a T–s diagram.

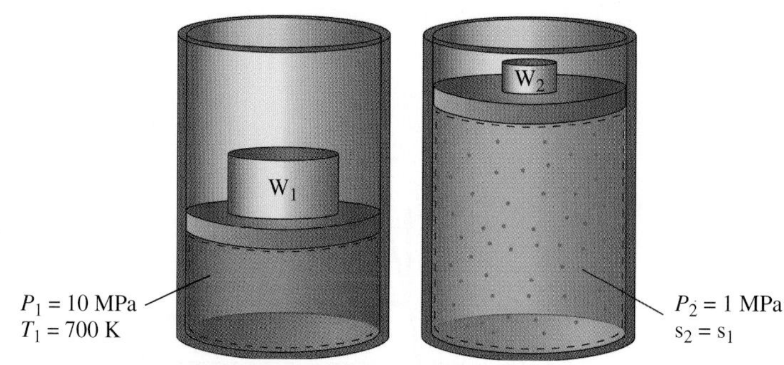

Solution

Known Superheated H_2O at T_1, P_1, isentropic expansion to P_2

Find v_2, h_2, T_2, x_2

Sketch

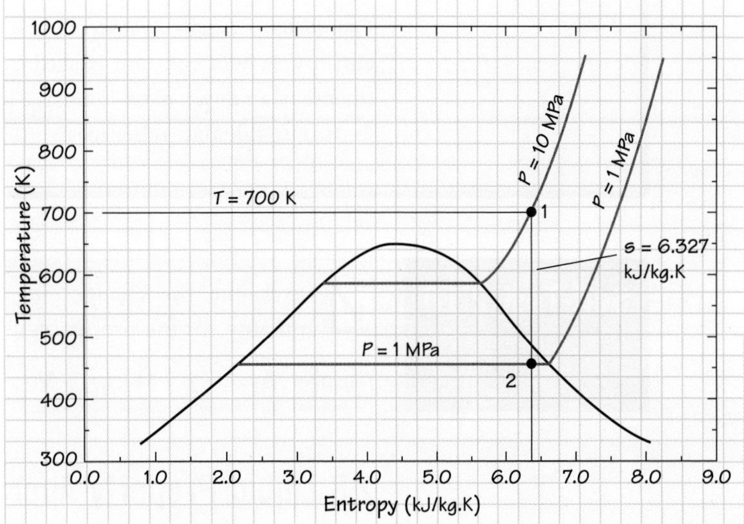

Modeling, Premises and Assumptions Simple compressible substance at equilibrium

Analysis Our strategy here is to locate state 1 in T–s space, that is, find s_1 given T_1 and P_1. State 2 is then determined by the given pressure ($P_2 = 1$ MPa) and the fact that the process is isentropic (i.e., that $s_2 = s_1$.) The relationship between state 1 and state 2 is shown in the sketch. Since we know two independent properties at state 2 (P_2, s_2), all other properties can be obtained.

To find s_1, we employ Table B.3 for the superheated steam at 10 MPa, which gives

$$s_1(T_1 = 700\,\text{K}, P_1 = 10\,\text{MPa}) = 6.3305\,\text{kJ/kg·K}.$$

Since $s_2 = s_1$ (= 6.3305 kJ/kg·K), we see from Table B.2 that state 2 must be in the liquid–vapor mixture region, since

$$s_f(P_{\text{sat}} = 1\,\text{MPa}) < s_2 < s_g(P_{\text{sat}} = 1\,\text{MPa}),$$

that is,

$$2.1381 < 6.3305 < 6.585.$$

With this knowledge, we can calculate the quality from Eq. 2.37d as follows:

$$x = \frac{s_2 - s_f}{s_g - s_f}.$$

Numerically evaluating this expression yields

$$x = \frac{6.3305 - 2.1381}{6.585 - 2.1381} = 0.9428.$$

This value of quality is now used to determine v_2 and h_2 using Eq. 2.37c. The following table summarizes these calculations. Also shown are the saturated liquid and saturated vapor properties at 1 MPa.

x	s (kJ/kg·K)	v (m^3/kg)	h (kJ/kg)
0	2.1381	0.0011272	762.52
0.9421	6.3305 = s_2	0.18330 = v_2	2661.8 = h_2
1	6.585	0.19436	2777.1

Comments Again, we see how a process in which one property, in this case, the entropy, is held constant is used to define the final state. This example also illustrates the use of a known mass-intensive property (s) to determine the quality, which we use, in turn, to calculate other mass-intensive properties (specific volume v and enthalpy h).

Self-Test 2.18

Determine the changes in the specific internal energy and the specific enthalpy for the process described in Example 2.18.

(Answer: $\Delta u = -415.61$ kJ/kg, $\Delta h = -515.13$ kJ/kg)

2.8 Liquid Property Approximations

Although accurate thermodynamic property data are available for many substances in the compressed-liquid region (see Table 2.5), we now discuss some approximations. These approximations for liquid properties are useful to simplify some analyses, and they can be used when detailed compressed-liquid data are not available.

For most liquids, the specific volume v and specific internal energy u are nearly independent of pressure; hence, they depend only on temperature, so that,

$$v(T,P) \cong v(T),$$

$$u(T,P) \cong u(T).$$

Since saturation properties are available for many liquids, the specific volume v and specific internal energy u can be approximated using the corresponding saturation values, where the saturation state is evaluated at the temperature of interest. That is,

$$v(T,P) \cong v_f(T_{sat} = T) \quad \textbf{(2.41a)}$$

and

$$u(T,P) \cong u_f(T_{sat} = T). \quad \textbf{(2.41b)}$$

The enthalpy can also be approximated using these relationships, together with the definition $h = u + Pv$; thus,

$$h(T,P) \cong h_f(T_{sat} = T) + (P - P_{sat})v_f(T_{sat} = T). \quad \textbf{(2.41c)}$$

Example 2.19 Validity of Mildly Compressed Liquid Approximation

Determine the error associated with the evaluation of the specific volume and specific enthalpy from Eqs. 2.41a–2.41c for water at 10 MPa and 300 K.

Solution

Known H_2O (liquid), P, T

Find v and h (approximations and "exact" values)

Sketch

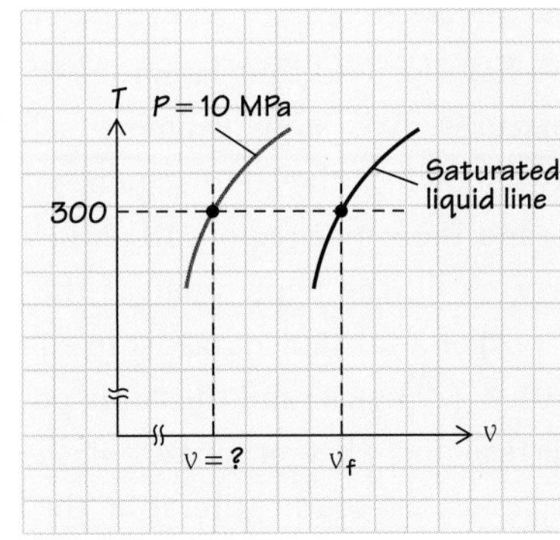

Modeling, Premises and Assumptions Water is approximately incompressible

Analysis Using the NIST WebBook or software, we obtain the following values for the compressed liquid at 10 MPa and 300 K:

$$v = 0.00099905 \text{ m}^3/\text{kg},$$
$$h = 121.73 \text{ kJ/kg}.$$

To obtain the approximate specific volume, we employ Eq. 2.41a together with Table B.1, which gives

$$v(T, P) \cong v_f(T_{sat} = T) = v_f(300 \text{ K}) = 0.0010035 \text{ m}^3/\text{kg}.$$

We also employ values from Table B.2 to evaluate Eq. 2.41c for the approximate enthalpy:

$$\begin{aligned} h(T, P) &\cong h_f(\text{T}) + [P - P_{sat}(T)]v_f(T) \\ &= 112.56 \times 10^3 \text{ J/kg} + (10 \times 10^6 \text{Pa} - 3.5369 \times 10^3 \text{ Pa})\, 0.0010035 \text{ m}^3/\text{kg} \\ &= 122.59 \times 10^3 \text{J/kg or } 122.59 \text{ kJ/kg}. \end{aligned}$$

Verification of the units is left to the reader. The errors associated with the approximations are given in the following table:

Property	NIST Value	Approximation	Error
v (m^3/kg)	0.00099905	0.0010035	+ 0.44%
h (kJ/kg)	121.73	122.59	+0.71%

Comments We see that the approximate values are quite close to the exact values, with errors of less than 1%. We also note that the fact that the water is at a high pressure (10 MPa) makes a significant contribution to the enthalpy.

Self-Test 2.19

A beaker contains H_2O at atmospheric pressure and 10 °C. Find the specific volume, specific internal energy, and specific enthalpy of the H_2O.

(Answer: $v = 0.001$ m^3/kg, $u = 41.391$ kJ/kg, $h = 41.491$ kJ/kg)

2.9 Solids

We have focused thus far on the thermodynamic properties of gases, liquids, and their mixtures. We now examine solids. Figure 2.38 presents a phase diagram for H_2O showing: the solid region, which is designated I; the liquid region, designated II; and the vapor region, III. A key feature of this diagram is the triple point. As you may recall, the triple point is the state at which all three phases (solid, liquid, and vapor) of a substance coexist in equilibrium. For water, the triple point temperature is 0.01 °C (273.16 K) and the triple point pressure is 0.00604 atm (0.6117 kPa). The nearly vertical line that originates at the triple point, the **solidification** or **fusion line**, separates the solid region from the liquid region. Another line, the **sublimation line**, separates the solid from the vapor phase and ends at the triple point. A third line, which starts at the triple point and continues up to the critical point, comprises the saturation states that were defined in our discussion of liquid–vapor mixtures. The normal (i.e., 1-atm) freezing point is also indicated on the phase diagram and corresponds to a temperature of 0 °C (273.15 K).

Important properties associated with solid–liquid and solid–vapor phase changes are the enthalpy of fusion, h_{fusion}, and the enthalpy of sublimation, h_{sublim}, respectively:

$$h_{\text{fusion}} = h_{\text{liq}} - h_{\text{solid}}, \tag{2.42a}$$

$$h_{\text{sublim}} = h_{\text{vap}} - h_{\text{solid}}. \tag{2.42b}$$

Values of these quantities for H_2O at the triple point are 333.4 kJ/kg for h_{fusion} and 2834.3 kJ/kg for h_{sublim}. Table B.5 provides vapor-phase properties along the saturated solid–vapor line (200–273.16 K) for H_2O. Sublimation is an important mechanism for drying clothing outdoors in icy weather and in many industrial processes. Iodine is a substance that can be observed to sublimate at room temperature (Fig. 2.39), not because of a high vapor pressure at room temperature (it is only about 40 Pa) but because of its optical properties. Carbon dioxide (dry ice) can only exist as a solid or a vapor at 1 atm; hence, a solid chunk of dry ice will sublime without melting when exposed to a room-temperature, 1-atm ambient environment.

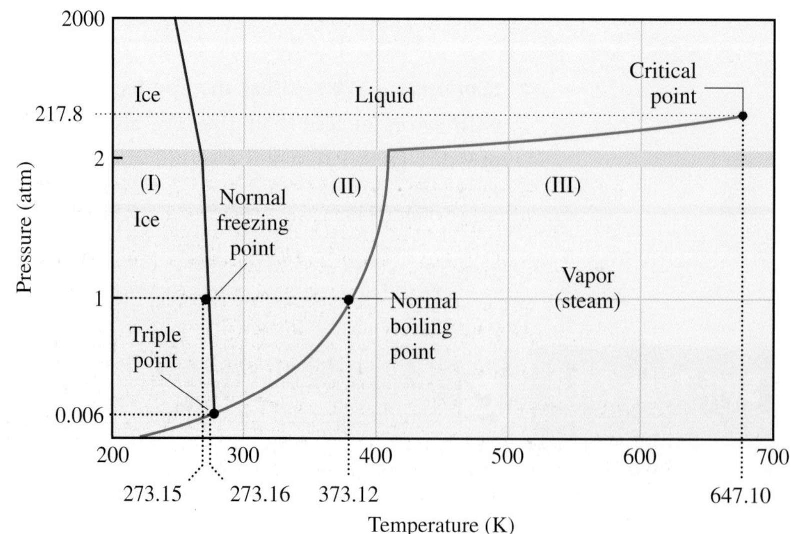

FIGURE 2.38 Phase diagram for water showing solid (I), liquid (II), and vapor (III) regions. Indicated on the diagram are the critical point ($P_c = 217.8$ atm, $T_c = 647.10$ K), the normal boiling point (steam) point ($P_{\text{boil}} = 1$ atm, $T_{\text{boil}} = 373.12$ K), the normal freezing (ice) point ($P_{\text{freeze}} =$ 1 atm, $T_{\text{freeze}} = 273.15$ K), and the triple point ($P_{\text{triple}} = 0.006$ atm, $T_{\text{triple}} = 273.16$ K). Note the scale change above 2 atm. Adapted from Ref. [9] with permission of Oxford University Press.

FIGURE 2.39 Iodine sublimates from a bluish-black, metallic looking solid to a purple vapor. The vapor pressure of solid iodine at 90 °C is 3.57 kPa and at room temperature is about 40 Pa (age fotostock / Alamy Stock Photo).

In most thermal–science applications, the thermodynamic properties of interest for solids are the density (reciprocal specific volume) and the specific heats. The dependence of these properties on pressure is very slight over a wide range; in fact, the effect of pressure is so small that solid properties are usually assumed to be functions of temperature alone, that is,

$$\rho = \rho(T\,\text{only}) \tag{2.43a}$$

and

$$c_p = c_p(T\,\text{only}). \tag{2.43b}$$

Densities and constant-pressure specific heats are tabulated in Ref. [10] for a number of solids of engineering interest.

To simplify the thermal analysis of solid systems, we frequently assume that the solid is an **incompressible** substance. With this assumption, the density (or specific volume) is a constant, independent of both pressure and temperature; that is,

$$\rho = 1/v = \text{constant}. \tag{2.44a}$$

Furthermore,

$$c_p = c_v \equiv c = \text{constant}. \tag{2.44b}$$

When invoking the incompressible-substance approximation, properties are usually evaluated at an appropriate average temperature.

Example 2.20 Heating of a Solid

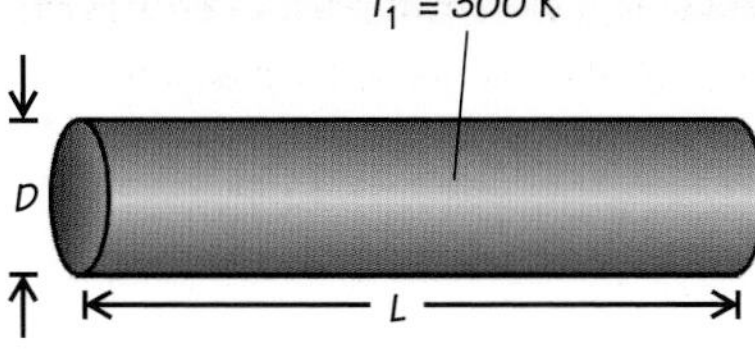

A pure copper rod of diameter $D = 25$ mm and length $L = 150$ mm is initially at a uniform temperature of 300 K. The rod is heated at 1 atm to a uniform temperature of 450 K. Estimate the change in internal energy U ([=] J) of the rod in going from the initial to the final state.

Solution

Known Pure Cu, L, D, T_1, T_2

Find $U_2 - U_1 \equiv \Delta U$

Sketch

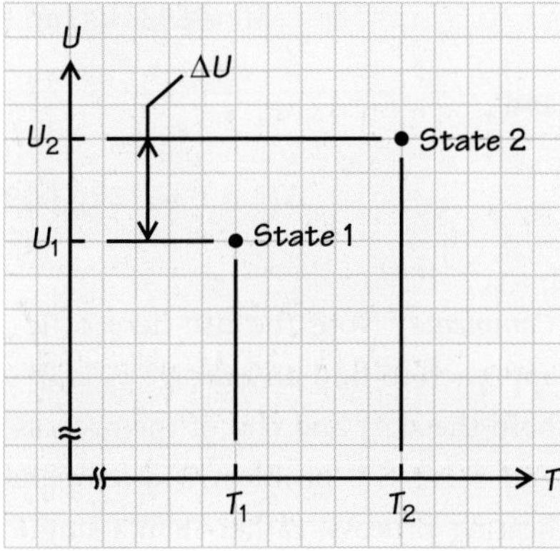

Modeling, Premises and Assumptions Incompressible solid

Analysis We develop the required calorific equation of state for copper by starting with the definition of the constant-volume specific heat (Eq. 2.20a),

$$c_v \equiv \left(\frac{\partial u}{\partial T}\right)_v .$$

With the assumption of incompressibility, this becomes an ordinary derivative, which can be separated and integrated as follows:

$$\Delta u = \int_{u_1}^{u_2} du = \int_{T_1}^{T_2} c_v dT = c(T_2 - T_1).$$

For a system of mass M, the internal energy change is thus

$$\Delta U = M\Delta u = Mc(T_2 - T_1).$$

We find the mass of the rod as follows:

$$M = \rho \mathcal{V} \rho = \frac{\pi D^2}{4} L.$$

We can use Ref. [10] to obtain values for c and ρ. Because values of c are given there for several temperatures, we interpolate for an average temperature T [$= (T_1 + T_2)/2$] of 375 K:

T(K)	c_p (J/kg · K)
200	356
375	392
400	397

Because only a single value of the density is given in Ref. [10], we use that value:

$$\rho(300\,\text{K}) = 8933\,\text{kg/m}^3 \cong \rho(375\,\text{K}).$$

Evaluating the mass and internal energy change yields

$$M = 8933\,\text{kg/m}^3 \frac{\pi(0.025\ \text{m})^2}{4} 0.150\,\text{m} = 0.6577\,\text{kg}$$

and

$$\Delta U = 0.6577\,\text{kg}\left(392\,\frac{\text{J}}{\text{kg}\cdot\text{K}}\right)(450 - 300)\,\text{K} = 38{,}700\,\text{J}.$$

Comments Note the use here of a c_p value based on an average temperature. This approach to find an average c_p is frequently used. That the temperature is uniform at both the start and end of the process also greatly simplifies this problem. (How would you solve the problem if a temperature distribution were given for the end of the heating process, rather than a single uniform temperature?)

Self-Test 2.20

 Redo Example 2.20 for an iron rod of the same dimensions. Compare your answer with that of Example 2.20 and discuss your results.

(Answer: $\Delta U = 41.4\,kJ$. Even though the density of iron is lower than that of copper, more energy is required for the same process because iron has a higher c_p value.)

SUMMARY

This chapter introduced the reader to the myriad thermodynamic and thermophysical properties that are used throughout this book. The chapter also showed how these properties relate to one another through equations of state and calorific equations of state. The concept of an ideal gas was presented, and methods were given to obtain properties for substances that do not behave as ideal gases. The properties of H_2O in both its liquid and vapor states were emphasized. You should be familiar with the use of both tables and computer-based resources to obtain property data for a wide variety of substances. You should also be proficient at sketching simple processes on thermodynamic coordinates. A more detailed summary of this chapter can be obtained by reviewing the learning objectives presented at the outset. It is likely that you will revisit this chapter many times in the course of your study of later chapters.

KEY EQUATIONS

Review the most important equations presented in this chapter (i.e., those boxed with a yellow background). What physical principles do they express? What restrictions apply?

CHAPTER 2 KEY CONCEPTS AND DEFINITIONS CHECKLIST

Answer the Questions and solve the Problems following the arrows to demonstrate mastery of the listed concepts and definitions.

2.2 Key Definitions

- ☐ Pure substance ➔ Question 2.1
- ☐ Simple compressible substance ➔ Question 2.1
- ☐ Extensive and intensive properties ➔ Problem 2.1
- ☐ Mass- and molar-specific properties ➔ Question 2.3

2.3 Frequently Used Thermodynamic Properties

- ☐ Common thermodynamic properties (list)
- ☐ Continuum limit ➔ Question 2.2, Problem 2.45
- ☐ Absolute, gage, and vacuum pressures ➔ Problems 2.5, 2.8
- ☐ Zeroth law of thermodynamics ➔ Question 2.7
- ☐ Internal energy ➔ Problem 2.48
- ☐ Enthalpy ➔ Problem 2.39
- ☐ Constant-volume and constant-pressure specific heats ➔ Question 2.6
- ☐ Specific heat ratio ➔ Question 2.8

2.4 Concept of State Relationships

- ☐ State principle ➔ Problems 2.47, 2.108
- ☐ P–v–T relationships ➔ Questions 2.4, 2.5
- ☐ Calorific relationships ➔ Questions 2.4, 2.5

2.5 Ideal Gases as Pure Substances

- ☐ Ideal gas definition ➔ Problems 2.49, 2.52
- ☐ Ideal-gas equation of state (Table 2.4) ➔ Problems 2.49, 2.52
- ☐ Particular gas constant ➔ Problem 2.52
- ☐ P–v and T–v diagrams ➔ Questions 2.9, 2.10, Problem 2.83
- ☐ u, c_v, T relationships (Eqs. 2.29a–2.29e) ➔ Problem 2.74
- ☐ h, c_p, T relationships (Eqs. 2.31a–2.31e) ➔ Problems 2.69, 2.70
- ☐ u–T and h–T diagrams ➔ Problem 2.74

2.6 Nonideal-Gas Properties

- ☐ Critical point ➔ Question 2.13
- ☐ Use of tables and NIST resources ➔ Problems 2.94, 2.95, 2.96
- ☐ Van der Waals equation of state ➔ Problem 2.99
- ☐ Generalized compressibility ➔ Problem 2.94

2.7 Pure Substances Involving Liquid and Vapor Phases

- ☐ Regions and phase boundaries ➔ Problems 2.107, 2.108
- ☐ T–v and P–v diagrams ➔ Problems 2.107, 2.108
- ☐ Quality and liquid–vapor-mixture properties ➔ Question 2.14, Problems 2.138, 2.139
- ☐ Use of tables and NIST resources ➔ Problems 2.114, 2.115, 2.119

2.8 Liquid Property Approximations

- ☐ Specific volume, internal energy, and enthalpy approximations ➔ Problem 2.160

2.9 Solids

- ☐ Fusion and sublimation properties ➔ Question 2.16

REFERENCES

1. Reid, R. C., Prausnitz, J. M., and Poling, B. E., *The Properties of Gases and Liquids*, 4th edn, McGraw-Hill, New York, 1987.
2. Halliday, D., and Resnick, R., *Physics*, combined 3rd edn, Wiley, New York, 1978.
3. Kee, R. J., Rupley, F. M., and Miller, J. A., "The Chemkin Thermodynamic Data Base," Sandia National Laboratories Report SAND 87–8215 B, March 1991.
4. Keenan, J. H., Keyes, F. G., Hill, P. G., and Moore, J. G., *Steam Tables: Thermodynamic Properties of Water Including Vapor, Liquid & Solid Phases*, Krieger, Melbourne, FL, 1992.
5. Irvine, T. E., Jr., and Hartnett, J. P. (Eds.), *Steam and Air Tables in SI Units*, Hemisphere, Washington, DC, 1976.
6. Linstrom, P., and Mallard, W. G. (Eds.), *NIST Chemistry WebBook*, Thermophysical Properties of Fluid Systems, National Institute of Standards and Technology, Gaithersburg, MD, 2000, http://webbook.nist.gov/chemistry/fluid.
7. Su, G.-J., "Modified Law of Corresponding States," *Industrial Engineering Chemistry*, 38:803 (1946).
8. Nelson, L. C., and Obert, F. E., "Generalized Compressibility Charts," *Chemical Engineering*, 61:203 (1954).
9. Atkins, P. W., *Physical Chemistry*, 6th edn, Oxford University Press, Oxford, 1998.

10. Incropera, F. P., and DeWitt, D. P., *Fundamentals of Heat and Mass Transfer*, 4th edn, Wiley, New York, 1996.
11. Myers, G. E., *Engineering Thermodynamics*, Prentice Hall, Englewood Cliffs, NJ, 1989.

Some end-of-chapter problems were adapted with permission from the following:

Look, D. C., Jr., and Sauer, H. J., Jr., *Engineering Thermodynamics*, PWS, Boston, 1986.
Myers, G. E., *Engineering Thermodynamics*, Prentice Hall, Englewood Cliffs, NJ, 1989.

QUESTIONS

2.1 Distinguish between a pure substance and a simple compressible substance.

2.2 Explain the continuum limit to a classmate.

2.3 How do the properties u and $\bar{u}$ differ? The properties v and $\bar{v}$?

2.4 Explain the distinction between an equation of state and a calorific equation of state.

2.5 What are the three types of thermodynamic state relationships? What properties are typically used in each?

2.6 Distinguish between constant-volume and constant-pressure specific heats.

2.7 What is the practical significance of the zeroth law of thermodynamics?

2.8 Define the specific-heat ratio. Which Greek symbol is used to denote this ratio?

2.9 Sketch two isotherms on a P–v plot for an ideal gas. Label each, taking $T_2 > T_1$.

2.10 Sketch two isobars on a T– v plot for an ideal gas. Label each, taking $P_2 > P_1$.

2.11 List fluids that you know are used as working fluids in thermal-fluid devices.

2.12 Compare your list of fluids from Question 2.11 with the fluids for which property data are available from the NIST online database.

2.13 What is the physical significance of the critical point? Locate the critical point on a P–v diagram.

2.14 Write out a physical interpretation of the thermodynamic property "quality".

2.15 Explain the meaning of quality to a classmate.

2.16 Distinguish between the enthalpy of fusion and the enthalpy of sublimation.

Chapter 2 Problem Subject Areas

2.1–2.48	State and calorific properties: definitions, units, and conversions
2.49–2.81	Ideal gases: Equation of state and calorific equations of state
2.82–2.91	Ideal gases: Process relationships
2.92–2.106	Real gases: Tabulated properties, generalized compressibility, and van der Waals equation of state
2.107–2.160	Pure substances with liquid and vapor phases
2.161–2.162	Solids
2.163–2.167	EES/NIST problems
2.168–2.177	FE problems

PROBLEMS

2.1–2.48 State and calorific properties: Definitions, units, and conversions

2.1 Create a table with the following symbols as the first column: T, ρ, $\mathcal{V}$, P, and v. In the second column, write out in words the precise meaning of these symbols. Be very specific about using appropriate adjectives as needed. In the third column, indicate whether the symbol represents an extensive or an intensive property. In the fourth column, provide the usual SI units associated with the quantities.

2.2 Consider 2.1 kg of water vapor. The specific volume of water vapor at 150 kPa and 120 °C is 1.188 m^3/kg.

A. Determine the molar-specific volume and the density of the water vapor.

B. Determine the volume of the water vapor.

C. Determine the number of kmols of the water vapor.

D. Determine the number of water vapor molecules.

2.3 Consider 0.65 kg of water vapor. The specific volume of the water vapor at 215 kPa and 140 °C is 0.8687 m^3/kg.

A. Determine the molar-specific volume and the density of the water vapor.

B. Determine the volume of the water vapor.

C. Determine the number of kmols of water vapor.

D. Determine the number of water vapor molecules.

2.4 Consider 0.83 kg of water vapor. The specific volume of the water vapor at 850 kPa and 500 °C is 0.4171 m^3/kg.

A. Determine the molar-specific volume and the density of the water vapor.

B. Determine the volume of the water vapour.

C. Determine the number of kmols of water vapor.

D. Determine the number of water vapor molecules.

2.5 A. An air compressor fills a tank to a gage pressure of 100 psi. The barometric pressure is 751 millimeters of mercury. What is the absolute pressure in the tank in kPa?

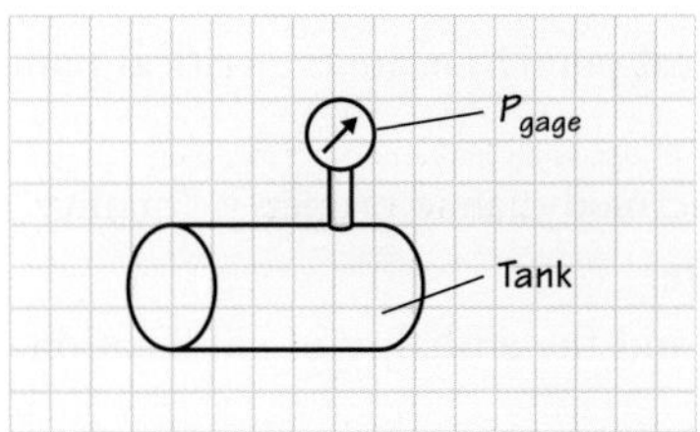

B. The air in the tank is bled out through a valve, the valve is closed, and the tank and its contents sit out overnight. The temperature drops, while the barometric pressure increases to 775 millimeters of mercury. A vacuum gage on the tank in the morning reads 1.2 inches of mercury. What is the absolute pressure in the tank in kPa?

2.6 A. An air compressor fills a tank to a gage pressure of 125 psi. The barometric pressure is 760 millimeters of mercury. What is the absolute pressure in the tank in kPa?

B. The air in the tank is bled out through a valve, the valve is closed, and the tank and its contents sit out overnight. The temperature drops, while the barometric pressure decreases to 755 millimeters of mercury. A vacuum gage on the tank in the morning reads 0.5 inches of mercury. What is the absolute pressure in the tank in kPa?

2.7 A. An air compressor fills a tank to a gage pressure of 132 psi. The barometric pressure is 755 millimeters of mercury. What is the absolute pressure in the tank in kPa?

B. The air in the tank is bled out through a valve, the valve is closed, and the tank and its contents sit out overnight. The temperature drops, while the barometric pressure increases to 762 millimeters of mercury. A vacuum gage on the tank in the morning reads 0.7 inches of mercury. What is the absolute pressure in the tank in kPa?

2.8 An instrument used to measure the concentration of the pollutant nitric oxide uses a vacuum pump to create a vacuum of 28.3 inches of Mercury in a reaction chamber. What is the absolute pressure in the chamber if the barometric pressure is 1 standard atmosphere? Express your result in psia, millimeters of mercury, and Pa.

2.9 An instrument used to measure the concentration of the pollutant nitric oxide uses a vacuum pump to create a vacuum of 28.0 inches of mercury in a reaction chamber. What is the absolute pressure in the chamber if the barometric pressure is 1.05 atm? Express your result in psia, millimeters of mercury, and Pa.

2.10 An instrument used to measure the concentration of the pollutant nitric oxide uses a vacuum pump to create a vacuum of 27.5 inches of mercury in a reaction chamber. What is the absolute pressure in the chamber if the barometric pressure is 0.98 atm? Express your result in psia, millimeters of mercury, and Pa.

2.11 Find the specific volume (in both ft^3/lb_m and m^3/kg) of 45 lb_m of a substance of density $10\,kg/m^3$, where the acceleration due to gravity is $30\,ft/s^2$.

2.12 Determine the specific volume of a spherical volume of air weighing 100 N. The diameter of the volume is 2.5 m. Assume standard earth gravity and a barometric pressure of 1 atm.

2.13 Determine the specific volume of a cubic volume of argon gas weighing 1.17 N. The length of a side of the cube is 0.5 m. The local gravitational acceleration is 9.85 m^2/s, and the barometric pressure is 0.98 atm.

2.14 Water is contained in a cylindrical tank having a diameter of 0.3 m. The water depth is 0.35 m, and the water weighs 920 N. Determine the specific volume of the water. Assume standard earth gravity and a barometric pressure of 1.05 atm.

2.15 An unknown liquid is contained in a cylindrical tank. The tank has diameter 0.35 m, and the liquid depth is 0.6 m. The liquid weighs 1920 N. Determine the specific volume of the liquid. Assume standard earth gravity and a barometric pressure of 98 kPa.

2.16 Determine the weight of oil contained in a vertical cylindrical tank. The tank has a diameter of 0.6 m, and the liquid depth is 0.5 m. The density of the oil is 870 kg/m^3. The local gravitational acceleration is 9.79 m^2/s, and the barometric pressure is 100 kPa.

2.17 A pressure gage connected to a water line reads 227.5 kPa. The local barometric pressure reading is 26.27 inches of mercury. Calculate the absolute pressure in units of psia, psfa (pounds-force per square foot absolute), and atm.

2.18 A pressure gage connected to a compressed air tank reads 325 kPa. The local barometric pressure reading is 28.3 inches of mercury. Calculate the absolute pressure in units of psia, psfa (pounds-force per square foot absolute), and atm.

2.19 A pressure gage connected to a natural gas pipeline reads 150 kPa. The local barometric pressure reading is 29.5 inches of mercury. Calculate the absolute pressure in units of psia, psfa (pounds-force per square foot absolute), and atm.

2.20 A pressure gage connected to a water line reads 60 psig. The local barometric pressure reading is 27.8 inches of mercury. Calculate the absolute pressure in units of psia, psfa (pounds-force per square foot absolute), kPa, and atm.

2.21 The pressure in a partially evacuated enclosure is 26.8 inches of mercury when the local barometer reads 29.5 inches of mercury. Determine the absolute pressure in units of inches of mercury, psia, and atm.

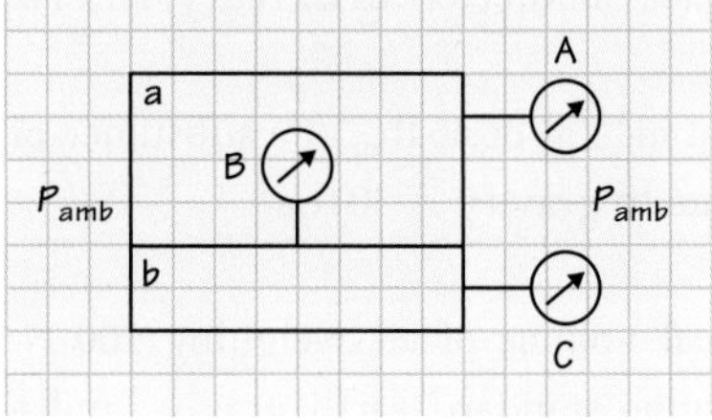

2.22 The accompanying sketch shows a compartment divided into two sections *a* and *b*. The ambient barometric pressure reading is 30.0 inches of mercury (absolute). Gage *C* reads 620,528 Pa and gage *B* reads 275,790 Pa. Determine the reading of gage *A* and convert this reading to an absolute value. **HINT**: For gage B, the ambient (surroundings) pressure (analogous to the atmospheric pressure for gages A and C) is the absolute pressure in compartment *a*.

2.23 The sketch for Problem 2.22 shows a compartment divided into two sections *a* and *b*. The ambient barometric pressure reading is 30.0 inches of mercury (absolute). Gage *C* reads 615 kPa and gage *B* reads 312 kPa. Determine the reading of gage *A* and convert this reading to an absolute value. **HINT**: For gage *B*, the ambient (surroundings) pressure (analogous to the atmospheric pressure for gages *A* and *C*) is the absolute pressure in compartment *a*.

2.24 The sketch for Problem 2.22 shows a compartment divided into two sections *a* and *b*. The ambient barometric pressure reading is 29.0 inches of mercury (absolute). Gage *A* reads 435 kPa and gage *B* reads 150 kPa. Determine the reading of gage *C* and convert this reading to an absolute value. **HINT**: For gage *B*, the ambient (surroundings) pressure (analogous to the atmospheric pressure for gages A and C) is the absolute pressure in compartment *a*.

2.25 The sketch for Problem 2.22 shows a compartment divided into two sections *a* and *b*. The ambient barometric pressure reading is 28.0 inches of mercury (absolute). Gage *A* reads 5 kPa and gage *B* reads 150 kPa. Determine the reading of gage *C* and convert this reading to an absolute value. **HINT**: For gage B, the ambient (surroundings) pressure (analogous to the atmospheric pressure for gages A and C) is the absolute pressure in compartment *a*.

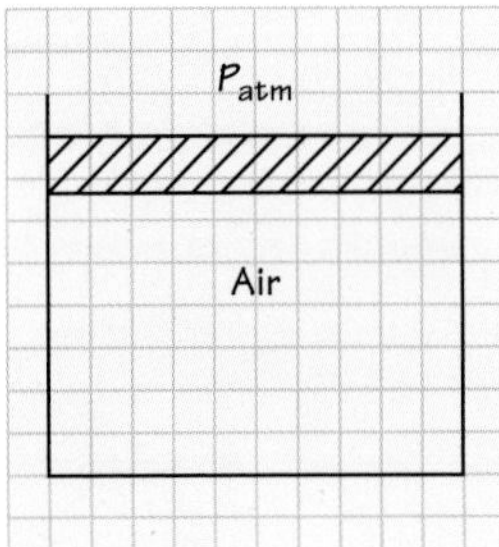

2.26 A vertical cylinder containing air is fitted with a freely moving (frictionless) piston of 68 lb_m and cross-sectional area 35 in^2. The ambient pressure outside the cylinder is 14.6 psia, and the local acceleration due to gravity is 31.1 ft/s^2. Determine both the absolute and gage pressures (psi) of the air inside the cylinder. **HINT**: This is a statics problem with a thermodynamics twist. Be sure to start by drawing a free-body diagram.

2.27 A cylinder containing a gas is fitted with a piston (see the above sketch). The cross-sectional area of the piston is 0.029 m^2. Assume that there is no frictional force at the piston–cylinder wall interface. The atmospheric pressure is 0.1035 MPa and the acceleration due to gravity is 30.1 ft/s^2. To produce an absolute pressure in the gas of 0.1517 MPa, what mass (kg) of the piston is required? **HINT**: This is a statics problem with a thermodynamics twist. Be sure to start by drawing a free-body diagram.

2.28 A cylinder containing a gas is fitted with a piston having a diameter of 0.12 m. (See the sketch in Problem 2.26.) Assume that there is no frictional force at the piston–cylinder wall interface. The atmospheric pressure is 100 kPa and the local acceleration due to gravity is 9.81 m/s^2. To produce an absolute pressure in the gas of 200 kPa, what mass (kg) of the piston is required? **HINT**: This is a statics problem with a thermodynamics twist. Be sure to start by drawing a free-body diagram.

2.29 A cylinder containing a gas is fitted with a piston as illustrated above. Assume that there is no frictional force at the piston–cylinder-wall interface. The mass of the piston is 200 kg. The atmospheric pressure is 99 kPa, and the local acceleration due to gravity is 9.81 m/s^2. The absolute pressure of the gas is 130 kPa. Determine the diameter of the piston. **HINT**: This is a statics problem with a thermodynamics twist. Be sure to start by drawing a free-body diagram.

2.30 Perform the temperature conversions as requested.

A. For safety, cans of whipped cream (with propellant) should not be stored above 120 F. What is this temperature expressed on the Rankine, Celsius, and Kelvin temperature scales?

B. Albuterol inhalers, used for the control of asthma, are to be stored and used between 15 °C and 30 °C. What is the acceptable temperature range on the Fahrenheit scale?

2.31 How fast, on average, do nitrogen molecules travel at room temperature (25 °C)? How does this speed compare to the average speed of a modern jet aircraft that travels 2500 miles in 5 hours? **HINT**: See Eqns. 2.15 and 2.16.

2.32 A thermometer reads 72 F. Specify the temperature in °C, K, and R.

(Credit: Stockbyte / Getty Images.)

2.33 Convert the following Celsius temperatures to Fahrenheit: (a) –30 °C, (b) –10 °C, (c) 0 °C, (d) 200 °C, and (e) 1050 °C.

2.34 At what temperature are temperatures expressed in Fahrenheit and Celsius numerically equal? Solve using algebra.

2.35 Create a table with the following symbols as the first column: R_u, $\mathcal{M}$, P, u, and $\bar{h}$. In the second column, write out in words the precise meaning of these symbols. Be very specific, using appropriate adjectives as needed. In the third column, indicate whether the symbol represents an extensive or an intensive property. In the fourth column, provide the usual SI units associated with the quantities.

2.36 Create a table with the following symbols as the first column: R, γ, U, h, and c_v. In the second column, write out in words the precise meaning of these symbols. Be very specific, using appropriate adjectives as needed. In the third column, indicate whether the symbol represents an extensive or an intensive property. In the fourth column, provide the usual SI units associated with the quantities.

2.37 Create a table with the following symbols as the first column: e, $\bar{c}_p$, N, x, and Z. In the second column, write out in words the precise meaning of these symbols. Be very specific, using appropriate adjectives as needed. In the third column, indicate whether the symbol represents an extensive or an intensive property. In the fourth column, provide the usual SI units associated with the quantities.

2.38 Create a table with the following symbols as the first column: E, M, $\bar{v}$, c_p, and s. In the second column, write out in words the precise meaning of these symbols. Be very specific, using appropriate adjectives as needed. In the third column, indicate whether the symbol represents an extensive or an intensive property. In the fourth column, provide the usual SI units associated with the quantities.

2.39 At 0.3 MPa, the mass-specific internal energy and enthalpy of a substance are 3313.6 J/kg and 3719.2 J/kg, respectively. Determine the density of the substance at these conditions.

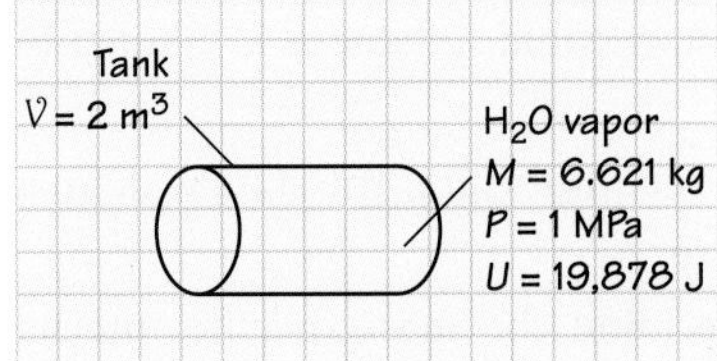

2.40 A tank having a volume of $2\ m^3$ contains 6.621 kg of water vapor at 1 MPa. The internal energy of the water vapor is 19,878 J. Determine the density, the mass-specific internal energy, and the mass-specific enthalpy of the water vapor, using the given information.

2.41 The constant-volume molar-specific heat of nitrogen (N_2) at 1000 K is 24.386 kJ/kmol·K and the specific-heat ratio γ is 1.3411. Determine the constant-pressure molar-specific heat and the constant-pressure mass-specific heat of the N_2.

2.42 ***Conceptual Problem.*** A. What is the microscopic interpretation of internal energy for a monatomic gas?

B. What is the microscopic interpretation of internal energy for a gas comprising diatomic molecules?

2.43 For temperatures between 300 and 1000 K and at 1 atm, the molar specific enthalpy of O_2 is expressed by the following polynomial:

$$\begin{aligned}\bar{h}_{O_2} = R_u(&3.697578\,T + 3.0675985 \times 10^{-4}T^2\\ &-4.19614 \times 10^{-8}T^3 + 4.4382025 \times 10^{-12}T^4\\ &-2.27287 \times 10^{-16}T^5 - 1233.9301),\end{aligned}$$

where $\bar{h}$ is expressed in kJ/kmol and T in kelvins.

Determine the constant-pressure molar-specific heat $\bar{c}_p$ at 500 K and at 1000 K. Compare the magnitudes of the values at the two temperatures and discuss your results. Also determine the constant-pressure mass-specific heat c_p.

2.44 Plot a graph of the molar specific enthalpy for O_2 given in Problem 2.43 for the temperature range 300–1000 K (i.e., plot $\bar{h}_{O_2}$ versus T). (Spreadsheet software is recommended.) Use your plot to graphically determine $\bar{c}_p$ at 500 K and at 1000 K. Use a pencil and ruler to perform this operation. How do these estimated values for $\bar{c}_p$ compare with your computations in Problem 2.43?

2.45 ***Conceptual Problem.*** Demonstrate your knowledge of the following thermodynamics concepts by writing brief paragraphs as directed:

A. Discuss the relationships among properties, states, and processes.

B. Describe the continuum limit. Indicate why it is important to your study of thermodynamics.

C. Explain the difference between extensive and intensive properties. List five (or more) properties of each type.

2.46 ***Conceptual Problem.*** Write out the state principle for a simple compressible substance and illustrate this principle with a specific example.

2.47 ***Conceptual Problem.*** Which of the following statements are true? Justify your choices by writing a sentence or two.

A. For a simple compressible substance, the only forces that can act are pressure and gravitational body forces.

B. The thermodynamic state of a simple compressible substance is defined by two extensive properties.

C. For a mixture of liquid water and steam in equilibrium, the state is defined by knowing the temperature and the pressure.

D. For an ideal gas, it is possible to determine the mass-specific internal energy if the pressure and the specific volume are known.

2.48 At 600 K and 0.10 MPa, the mass-specific internal energy of some water vapor is 2852.4 kJ/kg and the specific volume is 2.7635 m^3/kg. Determine the density and mass-specific enthalpy of the water vapor. Also, determine the molar-specific internal energy and enthalpy.

2.49–2.81 Ideal gases: Equation of state and calorific equations of state

2.49 What is the mass of a cubic meter of air at 25 °C and 1 atm? How many times greater is the mass of a cubic meter of liquid water? Assume the density of water is approximately 1000 kg/m^3.

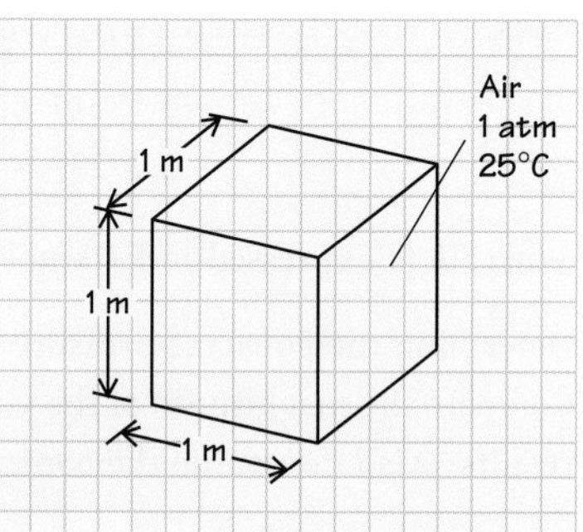

2.50 A. Determine the density of air at Coors Stadium in Denver, Colorado, on a warm summer evening when the temperature is 78 F. The barometric pressure is 85.1 kPa.

B. Assume to a first approximation that the drag force exerted on a well-hit baseball arcing to the outfield, or beyond, is proportional to the air density. Discuss the implications of this for balls hit at the stadium in Denver versus balls hit in Yankee Stadium in New York (which is at sea level).

(Credit: Blaine Harrington III / Alamy Stock Photo,)

2.51 A compressor pumps air into a tank until a pressure gage reads 120 psi. The temperature of the air is 85 F, and the tank is a 0.3-m-diameter cylinder 0.6 m long. Determine the mass of the air contained in the tank in grams. What common items (pencils, apples, etc.) have about the same mass?

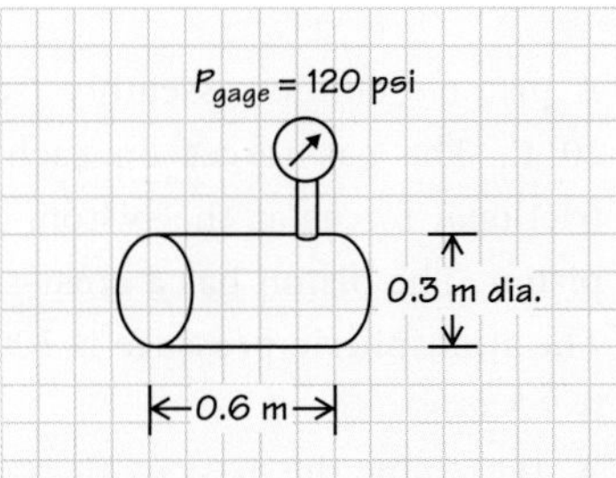

2.52 Starting with Eq. 2.26c, derive all the other forms of the ideal-gas law, (i.e., Eqs. 2.26a, 2.26b, 2.26d, and 2.26e). Discuss when you might prefer to use each of these rather than the others.

2.53 Determine the number of kmols of carbon monoxide contained in a 0.027-m^3 compressed-gas cylinder at 200 psia and 72 F. Assume ideal-gas behavior. Also, determine the weight of the gas, assuming standard gravitational acceleration. What common items (bananas, apples, etc.) have about the same weight?

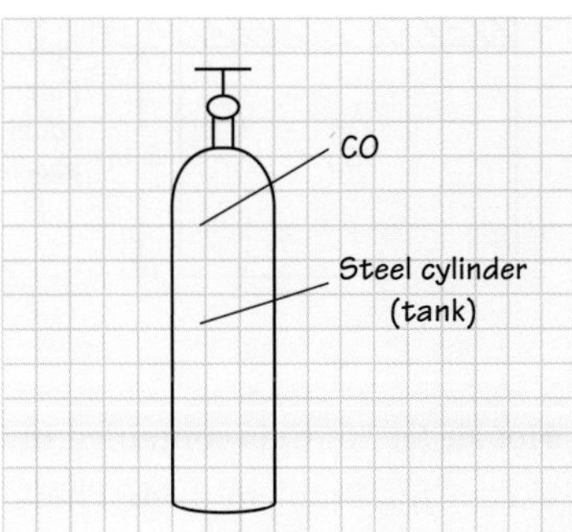

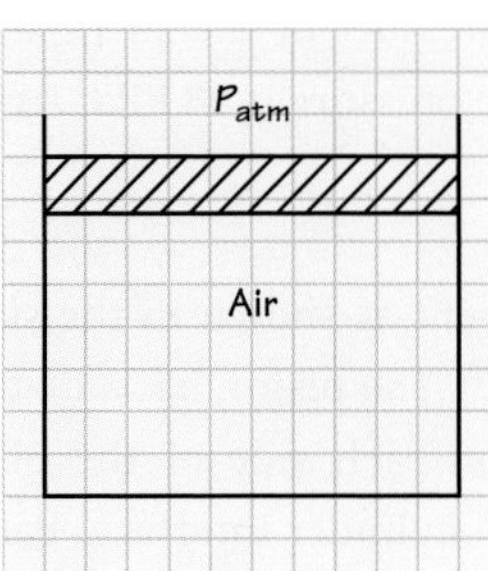

2.54 A piston–cylinder assembly (see sketch) contains 0.005 m^3 air at 29.4 °C. The gas forces on each side of the piston, assuming there are no frictional forces at the piston-cylinder interface, balance the weight of the piston. The piston has a cross-sectional area of 20.2 in^2 and a mass of 160.6 kg. The local gravitational acceleration is 9.144 m/s^2. Determine the mass of air trapped within the cylinder. The atmospheric pressure is 101.35 kPa.

2.55 A piston–cylinder assembly (see the sketch) contains 0.003 m^3 air at 25 °C. The gas forces on each side of the piston, assuming there are no frictional forces at the piston-cylinder interface, balance the weight of the piston. The piston has a cross-sectional area of 0.02 m^2 and a mass of 150 kg. The atmospheric pressure is 101.35 kPa, and the local gravitational acceleration is 9.8 m/s^2.

A. Determine the mass of air trapped within the cylinder.

B. Determine the distance the piston moves in millimeters when a 450-N weight is placed on top of the piston. Assume the new equilibrium temperature is still 25 °C.

C. In what way is the gas like a mechanical spring?

2.56 A piston–cylinder assembly (see the above sketch) contains 0.006 m^3 air at 20 °C. The gas forces on each side of the piston, assuming there are no frictional forces at the piston–cylinder interface, balance the weight of the piston. The piston has a cross-sectional area of 100 cm^2 and a mass of 300 kg. The atmospheric pressure is 100 kPa, and the local gravitational acceleration is 9.8 m/s^2.

A. Determine the mass of air trapped within the cylinder.

B. Determine the distance the piston moves after a 600-N weight is placed on top of the piston. Assume the new equilibrium temperature is still 20 °C.

C. In what way is the gas like a mechanical spring?

2.57 Consider the piston–cylinder arrangement shown in the sketch below. The gas forces on each side of the piston, assuming there are no frictional forces at the piston–cylinder interface, balance the weight of the piston. Determine the absolute pressure of the air (in psia) and the mass of air in the cylinder (in lb_m). The atmospheric pressure is 14.6 psia.

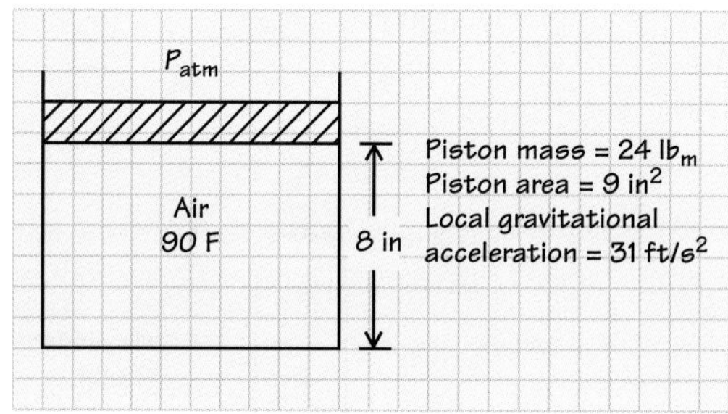

2.58 Hot air (at 50 °C) is contained in a piston–cylinder arrangement as shown in the sketch for Problem 2.54. The volume of the air is 250 cm^3. The piston has a diameter of 4 cm and a mass of 49 kg. The gas forces on each side of the piston, assuming there are no frictional forces at the piston–cylinder interface, balance the weight of the piston. Determine the absolute pressure of the air (kPa) and the mass of air in the cylinder (kg). The atmospheric pressure is 98 kPa. Assume standard gravity.

2.59 Hot air is contained in a piston–cylinder arrangement as shown in the sketch for Problem 2.54. The density of the air is 3.6 kg/m^3. The piston has a diameter of 5 cm and a mass of 110 kg. The gas forces on each side of the piston, assuming there are no frictional forces at the piston–cylinder interface, balance the weight of the piston. Determine the temperature (in °C) of the air in the cylinder. The atmospheric pressure is 100 kPa. Assume standard gravity.

2.60 Hot air is contained in a piston–cylinder arrangement as shown in the sketch for Problem 2.54. The specific volume of the air is 0.25 m^3/kg. The piston has a diameter of 6 cm and a mass of 95 kg. The gas forces on each side of the piston, assuming there are no frictional forces at the piston–cylinder interface, balance the weight of the piston. Determine the temperature (in °C) of the air in the cylinder. The atmospheric pressure is 102 kPa. Assume standard gravity.

2.61 Determine the mass of air in a room that is 15 m by 15 m by 2.5 m if the temperature and pressure are 25 °C and 1 atm, respectively. How does this mass compare with your body mass?

(Credit: Klaus Vedfelt / DigitalVision / Getty Images.)

2.62 Pure carbon dioxide flows through a pipe in a vacuum process at 49 °C and 0.8 kPa. Determine the specific volume (m^3/kg) of the CO_2. How does this value compare to that for air at the same temperature and pressure?

2.63 A piston–cylinder arrangement in an air separation plant contains nitrogen at 21 °C and 1.379 MPa. The piston moves and compresses the nitrogen from 98 to 82 cm^3. The final (equilibrium) temperature is 27 °C. Determine the final pressure (in kPa).

2.64 Consider the following gases for conditions under which ideal-gas behavior occurs: argon, helium, hydrogen, nitrogen, and oxygen.

A. Rank the gases from that with the greatest density to that with the least density for fixed temperature and pressure.

B. Rank the pressures associated with the five gases when the temperatures and volumes have the same fixed values for each gas.

C. Rank the masses associated with the five gases when the temperatures, pressures, and volumes have the same fixed values for each gas.

2.65 The temperature of an ideal gas remains constant while the pressure changes from 101 kPa to 827 kPa. If the initial volume is 0.08 m^3, what is the final volume?

2.66 Nitrogen (3.2 kg) at 348 °C is contained in a vessel having a volume of 0.015 m^3. Use the ideal-gas equation of state to determine the pressure of the N_2. If the N_2 is replaced by the same mass of hydrogen at the same temperature, does the pressure increase, decrease, or remain the same. Explain.

2.67 Consider three 0.03-m^3 tanks filled, respectively, with N_2, Ar, and He. Each tank is filled to a pressure of 400 kPa at room temperature, 298 K. Determine the mass of gas contained in each tank. Also determine the number of moles of gas (kmol) in each tank.

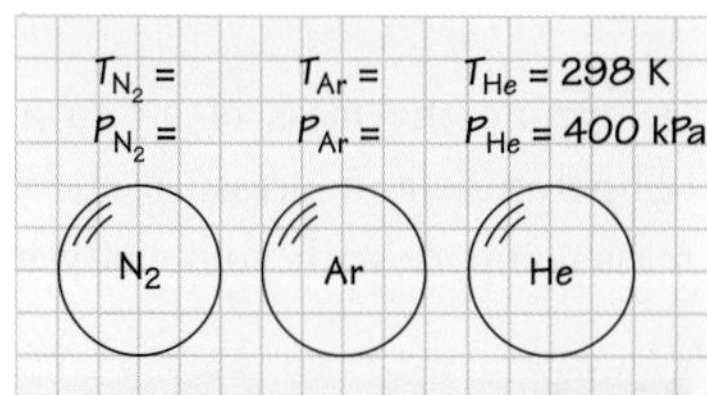

2.68 The constant-pressure specific heat of a gas is 0.24 Btu/lb_m·R at room temperature. Determine the specific heat in units of kJ/kg·K.

2.69 Compute the mass-specific enthalpy change associated with N_2 that is undergoing a change in state from 400 K to 800 K. Assume the constant-pressure specific heat is constant for your calculation. Use the arithmetic average of the values at 400 K and 800 K from Table D.7.

2.70 Compare the specific enthalpy change calculated in Problem 2.69 with the change determined directly from the ideal-gas tables (Table D.7). Also compare these values with that obtained from the NIST software or online database at 1 atm.

2.71 Use Table C.2 to calculate the mass-specific enthalpy change for air undergoing a change of state from 300 K to 1000 K. How does this value compare with that estimated using the constant-pressure specific heat at the average temperature, $T_{avg} = (300 + 1000)/2$?

2.72 Determine the mass-specific internal energy for O_2 at 900 K for a reference-state temperature of 298.15 K. Also determine the constant-volume mass-specific heat.

2.73 Hydrogen is compressed in a cylinder from 101 kPa and 15 °C to 5.5 MPa and 121 °C. Determine the changes in specific volume (Δv), mass-specific internal energy (Δu), and mass-specific enthalpy (Δh).

2.74 As air flows across the cooling coil of an air conditioner at a rate of 3856 kg/hr, the temperature of the air drops from 26 °C to 12 °C. Determine the average internal energy rate of change in kJ/hr. Sketch the process in mass-specific enthalpy–temperature (h–T) coordinates.

2.75 Air cools an electronics compartment by entering at 300 K and leaving at 340 K. The pressure is essentially constant at 101.3 kPa. Determine (a) the change in the mass-specific internal energy of the air as it flows through the compartment and (b) its change in specific volume.

(Credit: Adrian Sherratt / Alamy Stock Photo.)

2.76 In a closed (fixed-mass) system, an ideal gas undergoes a process from an initial state with pressure 517 kPa and volume of $0.1416\,\text{m}^3$ to a final state with pressure 172.3 kPa and volume of $0.2741\,\text{m}^3$. For this process, the enthalpy change ΔH is –65.4 kJ and the average constant-volume specific heat $c_{v,\text{avg}}$ is 3.157 kJ/kg·K. Determine (a) the internal energy change ΔU, (b) the average constant-pressure specific heat $c_{p,\text{avg}}$, and (c) the specific gas constant R.

2.77 In a fixed-mass system, 1.8 kg of air is heated at constant pressure (200 kPa) from 278 K to 333 K. Determine the change in internal energy (kJ) for this process assuming an average constant-volume molar specific heat $\bar{c}_{v,\text{avg}}$ of 20.76 kJ/kmol·K. The specific heat ratio γ is 1.4.

2.78 Calculate the mass-specific volume and enthalpy changes for air undergoing a change of state from 400 K and 2 atm to 800 K and 7.2 atm.

2.79 Air is compressed in the compressor of a turbojet engine. The air enters the compressor at 270 K and 58 kPa and exits the compressor at 465 K and 350 kPa. Determine the mass-specific volume, enthalpy, and internal energy changes associated with the compression process.

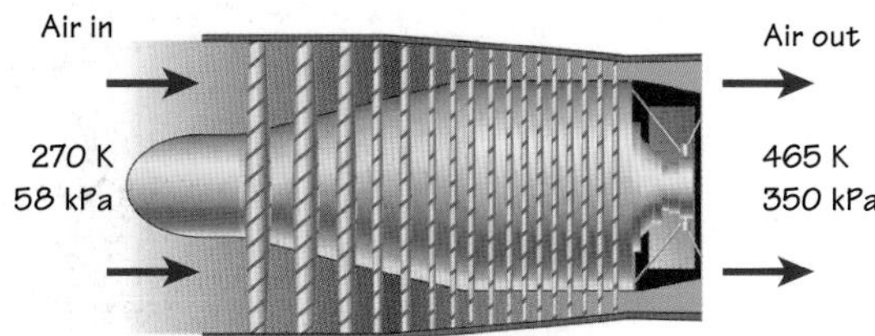

2.80 Air is compressed in a piston–cylinder system having an initial volume of 80 in^3. The initial pressure and temperature are 20 psia and 140 F. The final volume is one-eighth of the initial volume at a pressure of 175 psia. Determine the following:

A. The final temperature (F)
B. The mass of air (lb_m)
C. The change in internal energy (Btu)
D. The change in enthalpy (Btu)

2.81 Determine the enthalpy change for air undergoing a process that causes a change of state from 300 K and 1 atm to 1000 K and 5 atms. Do this in two ways: (1) Assume ideal gas behavior and use Table C.2, and (2) use the NIST software, which incorporates real gas behavior. How do the two results compare? Why do the results differ? One objective of this problem is to acquaint you with ways to find the calorific properties of air. **HINT**: To use the NIST software, select "Pseudo-Pure Fluid" from the "Substance Menu."

2.82–2.91 Ideal-gases: Process relationships

2.82 ***Conceptual Problem.*** Using P–v coordinates, sketch a process in which the product of the pressure and specific volume is constant from state 1 to state 2. Assume an ideal gas. Also assume $P_1 > P_2$. Show this same process on P–T and T– v diagrams.

2.83 ***Conceptual Problem.*** Consider the five processes a–b, b–c, c–d, d–a, and a–c shown below, as sketched in P–v coordinates. Show the same processes in P–T and T–v coordinates assuming ideal-gas behavior.

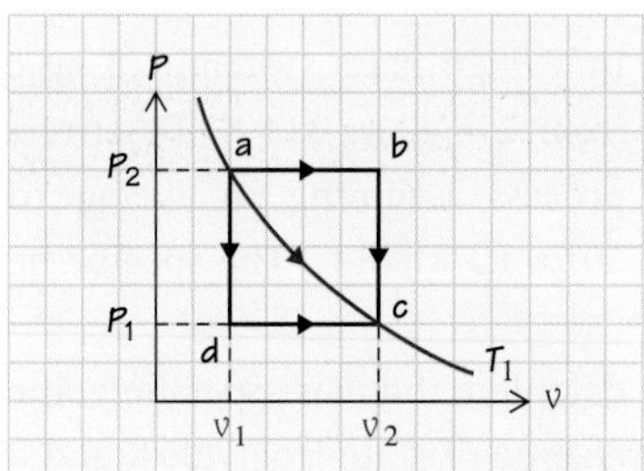

2.84 Nitrogen in a cylinder slowly expands from an initial volume of 0.025 m^3 to 0.05 m^3 at a constant pressure of 400 kPa. Determine the final temperature if the initial temperature is 500 K. Plot the process in P–v and T–v coordinates using spreadsheet software.

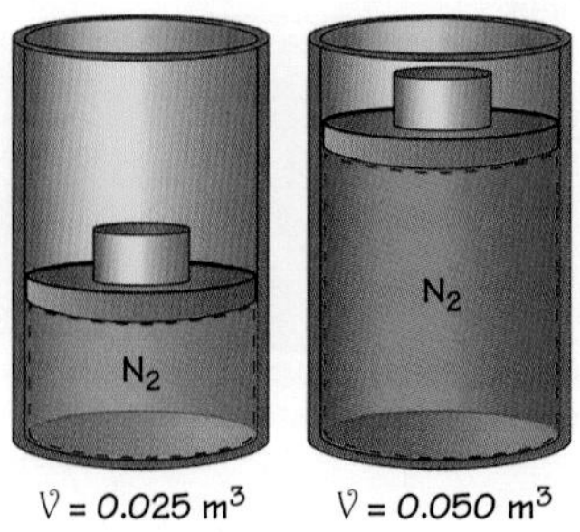

2.85 Nitrogen in a cylinder slowly expands from an initial volume of $0.025\,\mathrm{m}^3$ to $0.05\,\mathrm{m}^3$ at a constant temperature of 500 K. Determine the final pressure if the initial pressure is 600 kPa. Plot the process in P–v and T–v coordinates using spreadsheet software.

2.86 Nitrogen in a cylinder is slowly compressed from an initial volume of $0.06\,\mathrm{m}^3$ to $0.03\,\mathrm{m}^3$ at a constant temperature of 400 K. Determine the final pressure if the initial pressure is 500 kPa. Plot the process in P–v and T–v coordinates using spreadsheet software.

2.87 ***Conceptual Problem.*** Consider air (an ideal gas) within a piston–cylinder assembly as a thermodynamic system. We wish to compare three different quasi-equilibrium expansion processes for this system by plotting the pressure as a function of volume. Each process starts from the same initial condition: $P =$ 100 kPa and $\mathcal{V} = 1\,\mathrm{m}^3$. The final volume for each process is $\mathcal{V} = 3\,\mathrm{m}^3$. The plots are to made by connecting three points ($\mathcal{V} = 1$, 2, and $3\,\mathrm{m}^3$) with hand-drawn lines. The first process (which is trivial) is conducted at constant pressure. The second process is conducted at constant temperature. The third process is conducted such that $P\mathcal{V}^{1.4}$ is maintained constant. Note that here only the volume is raised to the 1.4 power, not the pressure–volume product.

A. Complete the table below by calculating the six values for the pressure.

B. Plot the results on the figure or on graph paper. Connect the data points with a smooth line.

C. As you will see in Chapter 4, the work done by a gas undergoing a quasi-equilibrium process can be expressed as the area under a pressure vs. volume plot, i.e., $\int_{\mathcal{V}_1}^{\mathcal{V}_2} P d\mathcal{V}$. Using your plots, rank the quantity of work done by the three processes from the least to the most.

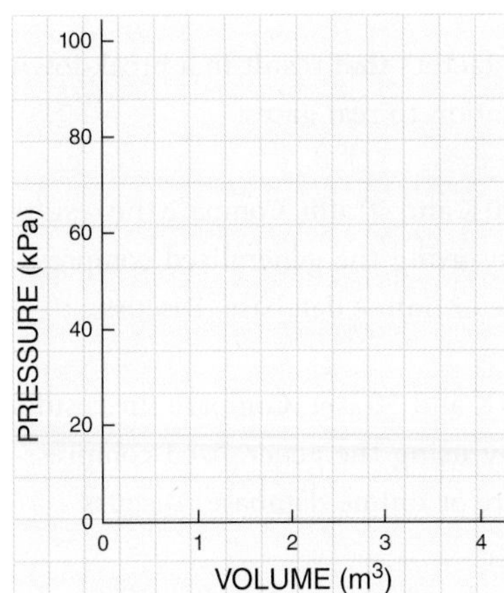

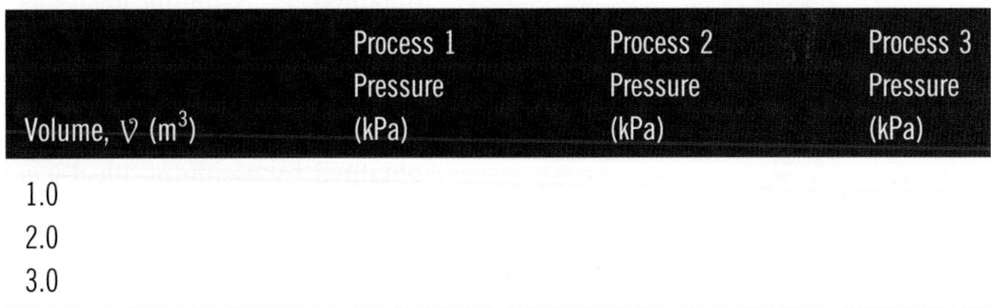

Volume, $\mathcal{V}$ (m^3)	Process 1 Pressure (kPa)	Process 2 Pressure (kPa)	Process 3 Pressure (kPa)
1.0			
2.0			
3.0			

2.88 ***Conceptual Problem.*** Consider an ideal gas. Indicate whether the following thermodynamic properties depend on pressure when the temperature is fixed: density, specific volume, molar-specific internal energy, mass-specific enthalpy, constant-volume specific heat, and constant-pressure specific heat. Explain your reasoning.

2.89 Create the requested plots for air, assuming ideal-gas behavior.

A. Sketch isotherms on a pressure–specific-volume (P–v) plot for temperatures of 360 and 520 K with specific volumes ranging from 0.2 to $1.0\,\mathrm{m}^3/\mathrm{kg}$. Label the isotherms.

B. Sketch two isobars on a temperature–specific-volume (T–v) plot for pressures of 520 and 680 kPa with specific volumes ranging from 0.2 to $1.0\,\mathrm{m}^3/\mathrm{kg}$. Label the isobars.

2.90 Create the requested plots for air, assuming ideal-gas behavior.

A. Sketch isotherms on a pressure–specific-volume (P–v) plot for temperatures of 750 and 1000 K with specific volumes ranging from 0.5 to 2.5 m^3/kg. Label the isotherms.

B. Sketch two isobars on a temperature–specific-volume (T–v) plot for pressures of 180 and 230 kPa with specific volumes ranging from 0.5 to 2.5 m^3/kg. Label the isobars.

2.91 Create the requested plots for air, assuming ideal-gas behavior.

A. Sketch isotherms on a pressure–specific-volume (P–v) plot for temperatures of 610 and 830 K with specific volumes ranging from 0.2 to 1.0 m^3/kg. Label the isotherms.

B. Sketch two isobars on a temperature–specific-volume (T–v) plot for pressures of 825 and 1100 kPa with specific volumes ranging from 0.2 to 1.0 m^3/kg. Label the isobars.

2.92–2.106 Real gases: Tabulated properties, generalized compressibility, and van der Waals equation of state

2.92 ***Conceptual Problem.*** Explain the principle of corresponding states. How does this principle relate to the use of the generalized compressibility chart?

2.93 ***Conceptual Problem.*** List and explain the factors that result in a breakdown of the application of the ideal-gas approximation to real gases.

2.94 Determine the density of methane (CH_4) at 300 K and 40 atm. Compare the results obtained by assuming ideal-gas behavior, by using the generalized compressibility chart, and by using the NIST software or online database. Discuss.

2.95 Determine the density of oxygen (O_2) at 250 K and 50 atm. Compare the results obtained by assuming ideal-gas behavior, by using the generalized compressibility chart, and by using the NIST software or online database. Discuss.

2.96 Determine the density of propane (C_3H_8) at a fixed temperature (300 K) for the following pressures: 200 kPa, 400 kPa, and 600 kPa. Also determine the compressibility factors Z for these conditions. Use the NIST software or online database. Discuss.

2.97 Compare the values of the specific volume of superheated steam at 2 MPa and 500 K found in Appendix B with those calculated assuming ideal-gas behavior. Discuss.

2.98 Consider steam. Plot the 500-K isotherm in P–v space using the NIST online database as your data source. Also plot on the same graph the 500-K isotherm assuming ideal-gas behavior. Start (actually, end) the isotherm at the saturated vapor line. Use a sufficiently large range of pressures that the real isotherm approaches the fictitious ideal-gas isotherm at large specific volumes. Use spreadsheet software for your calculations and plot.

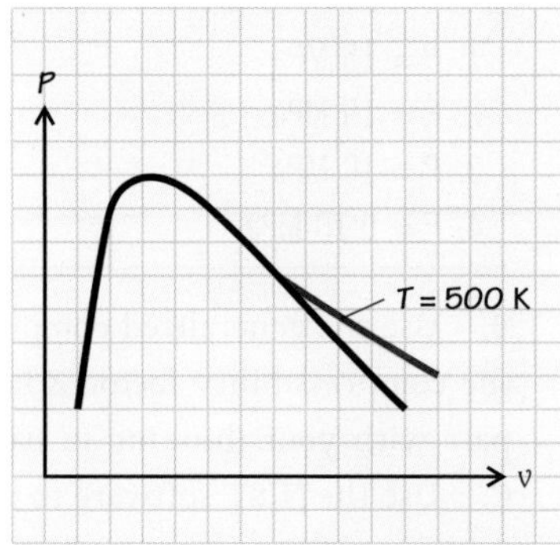

2.99 Use the van der Waals equation of state to determine the density of propane (C_3H_8) at 400 K and 12.75 MPa. How does this value compare with that obtained from the NIST software or online database?

2.100 Use the van der Waals equation of state to determine the density of propane (C_3H_8) at 300 K and 800 kPa. How does this value compare with that obtained from the NIST software or online database?

2.101 Use the van der Waals equation of state to determine the density of methane (CH_4) at 300 K and 5 MPa. How does this value compare with that obtained from the NIST software or online database?

2.102 Carbon dioxide (CO_2) is heated in a constant-pressure process from 15 °C and 101.3 kPa to 86 °C. Determine, per unit mass, the changes in (a) enthalpy, (b) internal energy, and (c) volume, all in SI units. Use the NIST software or online database to find these quantities.

2.103 A large tank contains nitrogen at −65 °C and 91 MPa. Can you assume that this nitrogen is an ideal gas? What is the specific-volume error in this assumption?

2.104 The principal components of air are N_2 (about 78%), O_2 (about 21%), and Ar (about 0.9%). The NIST miniREFPROP software allows you to select air as a pseudo-pure fluid under the substance menu. Review *Tutorial 2 – How to Use the NIST Software* to see the location of the substance menu bar. In addition, look specifically at the section *What plots can be generated with this software?* Use the NIST software to plot the mass-specific internal energy u (kJ/kg) versus temperature T for nitrogen N_2. Select the "Other diagrams" under the "Plot" menu to create your u–T plot. Use the default settings (500 kPa to 7000 kPa, step 500 kPa) to create a series of isobars. Set the u axis to range from $-200 < u < 1000$ kJ/kg and $100 < T < 1000$ K. Repeat for O_2 using the same isobars and scale settings. What does this plot tell you about the ideal-gas calorific behavior of air? Discuss and comment.

2.105 Explore the properties of steam (H_2O) at the four conditions below:

A. Calculate the reduced pressure and the reduced temperature. Based on these values, decide whether ideal-gas behavior is reasonable, i.e., whether $Pv = RT$ within +/−5%.

1. $P = 15$ MPa	$T = 1200$ K
2. $P = 15$ MPa	$T = 800$ K
3. $P = 10$ MPa	$T = 600$ K
4. $P = 0.10$ MPa	$T = 600$ K

B. Use the generalized compressibility chart (Fig. 2.22) to estimate a value of the compressibility factor Z for these conditions. How do these values compare with your thinking in part A?

C. Use the NIST software to determine the compressibility factor Z for steam at the conditions above. HINT: Under the "Options" menu, select "Properties." Check the box "Comp. Factor" to get values for the compressibility factor Z. How "ideal" is steam at these various conditions? How do these Z-values compare with your estimates from part B?

D. Calculate the specific volume for the steam at the conditions above using $v = ZRT/P$ and compare these calculations with the values given by the NIST software. **Note**: The purpose of this problem is to illustrate that there are conditions for which $H_2O(g)$ is far from ideal-gas behavior.

2.106 Is steam at 10 MPa and 500 °C close to behaving as an ideal gas? Discuss.

2.107–2.160 Pure substances with liquid and vapor phases

2.107 ***Conceptual Problem.*** Draw from memory (or familiarity) a temperature–specific-volume (T–v) diagram for H_2O. Show an isobar that begins in the compressed-liquid region and ends in the superheated-vapor region. Label all lines and regions and indicate the critical point.

2.108 ***Conceptual Problem.*** Draw from memory (or familiarity) a pressure–specific-volume (P–v) diagram for H_2O. Show an isotherm that begins in the compressed-liquid region and ends in the superheated-vapor region. Label all lines and regions and indicate the critical point.

2.109 Review *Tutorial 1 – How to Interpolate*. Apply interpolation to the property data in Table B.3 to determine the following properties of superheated steam:

A. The specific volume at 523 K and 1.0 MPa

B. The mass-specific enthalpy at 815 K and 0.3 MPa

C. The mass-specific internal energy at 620 K and 4.5 MPa

2.110 Review *Tutorial 1 – How to Interpolate*. Apply interpolation to the property data in Table B.3 to determine the following properties of superheated steam:

A. The specific volume at 727 K and 1.5 MPa

B. The mass-specific enthalpy at 823 K and 10.0 MPa

C. The mass-specific internal energy at 815 K and 0.15 MPa

2.111 Review *Tutorial 1 – How to Interpolate*. Apply interpolation to the property data in Table B.3 to determine the following properties of superheated steam:

A. The specific volume at 637 K and 700 kPa

B. The mass-specific enthalpy at 752 K and 13 MPa

C. The mass-specific internal energy at 780 K and 0.13 MPa

2.112 Review *Tutorial 1 – How to Interpolate*. Apply interpolation to the property data in Table B.1 to determine the following properties of saturated liquid H_2O:

A. The specific volume at 300.7 K

B. The saturation pressure at 282.6 K

C. The mass-specific enthalpy at 514 K

2.113 Review *Tutorial 1 –How to Interpolate*. Apply interpolation to the property data in Table B.2 to determine the following properties of saturated vapor H_2O:

A. The specific volume at 50.8 kPa

B. The saturation temperature at 8.4 MPa

C. The mass-specific internal energy at 3.33 MPa

2.114 Review *Tutorial 3 – How to Define a Thermodynamic State*. Given the following property data for H_2O, designate the region, line, or point in T–v or P–v space (i.e., compressed liquid, liquid–vapor mixture, superheated vapor, etc.) and find the value of the requested property. If the state is a saturated mixture, also find the quality. Use the tables in Appendix B and indicate which table was used. Provide units with your answers.

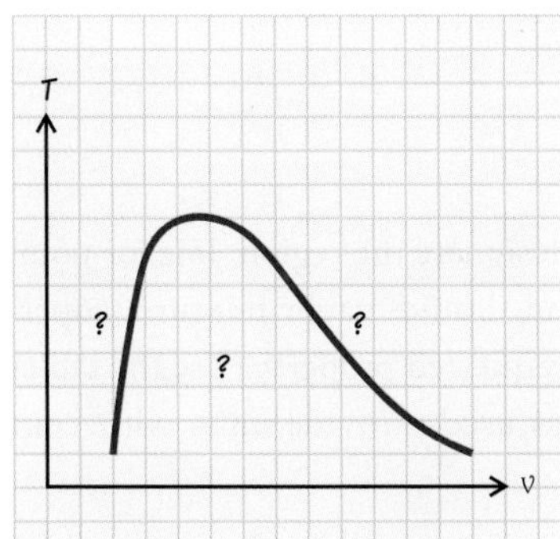

A. $T = 320\,\text{K}$, $v = 13.954\,\text{m}^3/\text{kg}$
Table? Region, etc.? $P = ?$

B. $T = 320\,\text{K}$, $v = 12.5\,\text{m}^3/\text{kg}$
Table? Region, etc.? $P = ?$

C. $T = 320\,\text{K}$, $P = 5.0\,\text{MPa}$
Table? Region, etc.? $u = ?$

D. $T = 320\,\text{K}$, $P = 6\,\text{kPa}$
Table? Region, etc.? $h = ?$

E. $T = 647.096\,\text{K}$, $P = 22.064\,\text{MPa}$
Table? Region, etc.? $v = ?$

F. $T = 800\,\text{K}$, $P = 28\,\text{MPa}$
Table? Region, etc.? $\rho = ?$

G. $T = 800\,\text{K}$, $P = 5\,\text{MPa}$
Table? Region, etc.? $\rho = ?$

2.115 Review *Tutorial 3 – How to Define a Thermodynamic State*. Given the following property data for H_2O, designate the region, line, or point in T–v or P–v space (i.e., compressed liquid, liquid–vapor mixture, superheated vapor, etc.) and find the value of the requested property. If the state is a saturated mixture, also find the quality. Use the tables in Appendix B and indicate which table was used. Provide units with your answers.

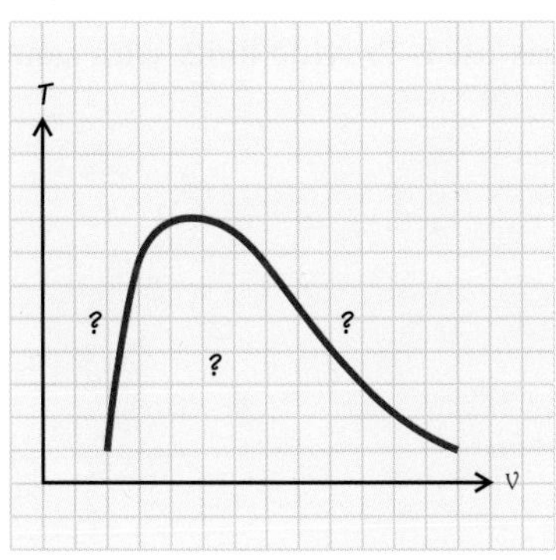

A. $P = 10\,\text{MPa}$, $v = 0.01803\,\text{m}^3/\text{kg}$
Table? Region, etc.? $T = ?$

B. $P = 10\,\text{MPa}$, $v = 0.0014526\,\text{m}^3/\text{kg}$
Table? Region, etc.? $u = ?$

C. $P = 10\,\text{MPa}$, $v = 0.01\,\text{m}^3/\text{kg}$
Table? Region, etc.? $T = ?$

D. $P = 10\,\text{MPa}$, $T = 400\,\text{K}$
Table? Region, etc.? $h = ?$

E. $P = 22.064$ MPa, $T = 647.096$ K
Table? Region, etc.? $v = ?$

F. $P = 10$ MPa, $T = 800$ K
Table? Region, etc.? $\rho = ?$

2.116 Review *Tutorial 3 – How to Define a Thermodynamic State*. Given the following property data for H_2O, designate the region, line, or point in T–v or P–v space (i.e., compressed liquid, liquid–vapor mixture, superheated vapor, etc.) and find the value of the requested property. If the state is a saturated mixture, also find the quality. Use the tables in Appendix B and indicate which table was used. Provide units with your answers.

A. $T = 600$ K, $v = 1.8405\ m^3/kg$
Table? Region, etc.? $P = ?$

B. $T = 600$ K, $v = 0.010\ m^3/kg$
Table? Region, etc.? $P = ?$

C. $T = 600$ K, $P = 6.0$ MPa
Table? Region, etc.? $v = ?$

D. $T = 600$ K, $P = 6$ kPa
Table? Region, etc.? $h = ?$

E. $P = 6$ kPa, $h = 2784.6$ kJ/kg
Table? Region, etc.? $v = ?$

2.117 Given the following property data for H_2O, designate the region, line, or point in T–v or P–v space (i.e., compressed liquid, liquid–vapor mixture, superheated vapor, etc.) and find the value of the requested property. Use the tables in Appendix B and indicate which table was used. Provide units with your answers.

A. $T = 600$ K, $h = 2677.8$ kJ/kg
Table? Region, etc.? $v = ?$

B. $T = 600$ K, $h = 2000$ kJ/kg
Table? Region, etc.? $x = ?$

C. $T = 600$ K, $v = 0.0015399\ m^3/kg$
Table? Region, etc.? $h = ?$

D. $T = 600$ K, $h = 2400$ kJ/kg
Table? Region, etc.? $v = ?$

2.118 Given the following property data for H_2O, designate the region, line, or point in T–v or P–v space (i.e., compressed liquid, liquid–vapor mixture, superheated vapor, etc.) and find the value of the requested property. Use the tables in Appendix B and indicate which table was used. Provide units with your answers.

A. $T = 500$ K, $P = 1.0$ MPa
Table? Region, etc.? $h = ?$

B. $T = 500$ K, $h = 1000$ kJ/kg
Table? Region, etc.? $P = ?$

C. $T = 500$ K, $h = 1000$ kJ/kg
Table? Region, etc.? $x = ?$

D. $T = 500$ K, $h = 1000$ kJ/kg
Table? Region, etc.? $u = ?$

2.119 Review *Tutorial 3 – How to Use the NIST Software*. Given the following property data for H_2O, designate the region, line, or point in T–v or P–v space (i.e., compressed liquid, liquid–vapor mixture, superheated vapor, etc.) and find the value of the requested property. Use the NIST software *miniREFPROP* that you can download from the NIST website http://trc.nist.gov/refprop/MINIREF/MINIREF.HTM. Provide units with your answers.

A. $T = 515\,\text{K}$, $P = 1.0$ MPa
Region, etc.? $h = ?$

B. $T = 515\,\text{K}$, $x = 0.45$
Region, etc.? $u = ?$

C. $T = 515\,\text{K}$, $v = 0.02\ \text{m}^3/\text{kg}$
Region, etc.? $h = ?$

D. $T = 156\,^\circ\text{C}$, $P = 4.6$ atm
Region, etc.? $u = ?$

E. $T = 474$ F, $P = 230$ psia
Region, etc.? $u = ?$ (Btu/lb_m and kJ/kg)

2.120 Review *Tutorial 2 – How to Use the NIST Software*. Given the following property data for H_2O, designate the region, line, or point in T–v or P–v space (i.e., compressed liquid, liquid–vapor mixture, superheated vapor, etc.) and find the value(s) of the requested property or properties. Use the NIST *miniREFPROP* software. Provide units with your answers.

A. $T = 452\,\text{K}$, $P = 5.0$ MPa
Region, etc.? $h = ?$

B. $T = 452\,\text{K}$, $P = 0.10$ MPa
Region, etc.? $u = ?$

C. $T = 452\,\text{K}$, $x = 0$
Region, etc.? $h = ?$

D. $T = 452$ K, $u = 1200$ kJ/kg
Region, etc.? $u = ?$

E. $P = 200$ kPa, $h = 2500$ kJ/kg
Region, etc.? $T = ?$, $x = ?$

2.121 Review *Tutorial 3 – How to Use the NIST Software*.

A. At room temperature (25 °C), what pressure (in both kPa and psi) is required to liquefy propane (C_3H_8)?

B. Determine values of the specific volume of the saturated vapor and saturated liquid. Also determine their ratios.

C. Determine the enthalpy of vaporization h_{fg} for these same conditions.

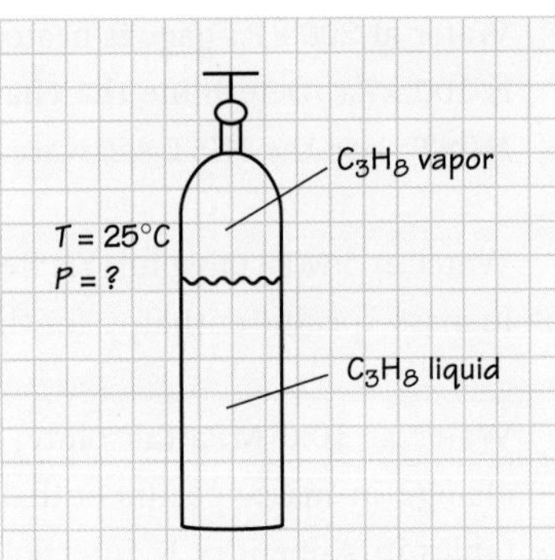

2.122 Plot the 1-atm isobar for H_2O in T–v coordinates. Start in the compressed-liquid region, continue across the liquid–vapor dome, and end well into the superheated-vapor region. **HINT**: Use a log scale for specific volume.

2.123 Given the following property data for H_2O, designate the region (or line or point) in T–v or P–v space (i.e., compressed liquid, liquid–vapor mixture, superheated vapor, saturated liquid, etc.) and find the value of the requested property. Use the tables in Appendix B and then verify your results using the NIST software or the NIST online database. Indicate the table used, e.g., Table B.3K, etc.

A. $T = 440$ K, $P = 0.733679$ MPa, $v = 0.26090$ m^3/kg: Region, etc. = ? h = ? Table = ?

B. $T = 440$ K, $P = 5.00$ MPa: Region, etc. = ? u = ? Table = ?

C. $T = 440$ K, $P = 0.100$ MPa: Region, etc. = ? v = ? Table = ?

D. $T = 440$ K, $P = 0.733679$ MPa, $v = 0.200$ m^3/kg: Region, etc. = ? h = ? Table = ?

E. $T = 647.096$ K, $P = 22.064$ MPa: Region, etc. = ? v = ? Table = ?

F. $T = 700$ K, $P = 32$ MPa: Region, etc. = ? ρ = ? Table = ?

G. $T = 605$ K, $P = 6.0$ MPa: Region, etc. = ? u = ? Table = ?

2.124 Determine the temperature and/or quality of H_2O for the following states. Use the Appendix B tables, not the NIST software. Some single interpolation may be required (but no double interpolation). Indicate the tables used.

A. $T = 500$ K; $u = 1500$ kJ/kg

B. $P = 1.2$ MPa; $u = 1000$ kJ/kg

C. $P = 3$ MPa; $v = 0.10$ m^3/kg

D. $P = 3$ MPa; $v = 0.03$ m^3/kg

E. $T = 400$ K; $v = 0.06$ m^3/kg

F. $P = 3$ MPa; $\rho = 11$ kg/m^3

2.125 A piston–cylinder arrangement contains 0.05 m^3 of liquid water and 2.0 m^3 of water vapor in equilibrium at 1.5 MPa. Heat crosses the boundary from the surroundings to the H_2O at constant pressure until the temperature reaches 700 K. Use tables, not the NIST software, to solve this problem.

A. Sketch the process in both T–v and P–v coordinates.

B. Determine the initial temperature of the H_2O.

C. Determine the total mass of the H_2O.

D. Determine the final specific volume v and total volume V.

2.126 Water at 200 kPa-gage is heated from 17 °C to 46°C. The atmospheric pressure is 100 kPa. Determine the change in mass-specific enthalpy for this process. **HINT**: Use the NIST software or WebBook.

2.127 Water at 5 MPa (absolute) is heated from 320 K to 340 K. Determine the change in mass-specific enthalpy for this process. Use the tables found in Appendix B.

2.128 Water at 10.0 MPa (absolute) is heated from 350 K to 380 K. Determine the change in mass-specific enthalpy for this process in three ways. (1) Use the tables in Appendix B, interpolating as required. (2) Use the NIST software or online database. (3) Use the liquid property approximation, Eqn. 2.41c. Compare your results and discuss.

2.129 Water at 5.0 MPa (absolute) is heated from 300 K to 350 K. Determine the change in mass-specific enthalpy for this process in three ways. (1) Use the tables in Appendix B, interpolating as required. (2) Use the NIST software or online database. (3) Use the liquid property approximation, Eqn. 2.41c. Compare your results and discuss.

2.130 Water at 500 kPa (absolute) and 310 K exists in the compressed liquid state. Estimate the specific volume, the mass-specific internal energy, and the mass-specific enthalpy for this state using Eqns. 2.41a–2.41c. Compare your approximate results with the values obtained from the NIST software or WebBook and discuss.

2.131 Review *Tutorial 2 – How to Use the NIST Software*, paying particular attention to how to use the various units. Water flows through a hot water heater at 2.0 gal/min, entering at 50 F and 40 psig. The water leaves the heater at 160 F and 39 psig. The barometric pressure is one standard atmosphere. Determine the change in enthalpy in Btu/lb_m.

2.132 Water at 6.90 MPa and 95 °C enters the steam-generating unit of a power plant and leaves the unit as steam at 6.90 MPa and 850 °C. Determine the following properties in SI units using the NIST software or WebBook:

Inlet			Outlet		
v =	?	m^3/kg	v =	?	m^3/kg
h =	?	kJ/kg	h =	?	kJ/kg
u =	?	kJ/kg	u =	?	kJ/kg
s =	?	kJ/kg·K	s =	?	kJ/kg·K
Region:	?		*Region:*	?	

2.133 In a proposed automotive steam engine, the steam after expansion would reach a state at which the pressure is 200 kPa (gage) and the specific volume is 0.2997 m^3/kg. Atmospheric pressure is 100 kPa. In what region does this state lie? Determine the temperature and mass-specific internal energy at this state.

2.134 A water heater operating under steady-flow conditions delivers 10 liters/min at 75 °C and 370 kPa. The input conditions are 10 °C and 379 kPa. What are the corresponding changes in internal energy and enthalpy per kilogram of water supplied?

2.135 Water at 3.4 MPa is pumped through pipes embedded in the concrete of a large dam. The water, in picking up the heat of hydration of the concrete curing, increases in temperature from 10 °C to 40 °C. Determine (a) the change in enthalpy (kJ/kg) and (b) the change in entropy of the water (kJ/kg·K).

2.136 Steam enters the condenser of a modern power plant with temperature 32 °C and quality 0.98. The condensate (water) leaves at 7 kPa and 27 °C. Determine the change in specific volume between the inlet and outlet of the condenser in m^3/kg. Use the NIST software or online database.

2.137 What is the temperature or quality of H_2O in the following states?

A:	20 °C, 2000 kJ/kg (u)
B:	2 MPa, 0.1 m^3/kg
C:	140 °C, 0.5089 m^3/kg
D:	4 MPa, 25 kg/m^3
E:	2 MPa, 0.111m^3/kg

2.138 Wet steam exits a turbine at 50 kPa with quality 0.83. Determine the following properties of the wet steam: T, v, h, u, and s. Give the units.

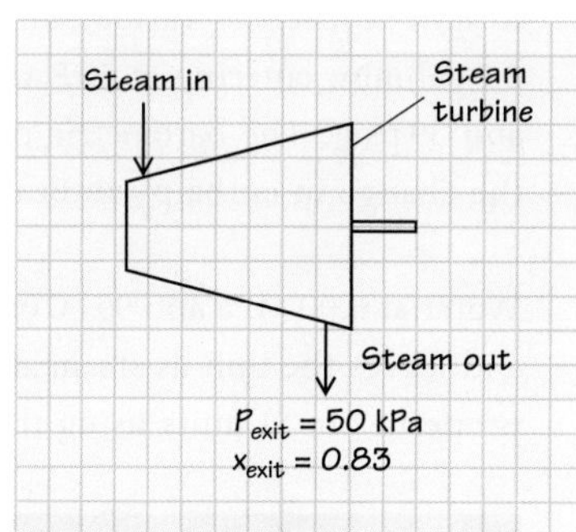

2.139 Wet steam at 375 K has a specific enthalpy of 2600 kJ/kg. Determine the quality of the mixture and the specific internal energy u.

2.140 Consider H_2O at 1 MPa. Plot the following properties as a function of the quality x: T (K), h (kJ/kg), s (kJ/kg·K), and v (m^3/kg). Discuss.

2.141 Steam is condensing in the shell of a heat exchanger at 305 K under steady conditions. The volume of the shell is 2.75 m^3. Determine the mass of the liquid in the shell if the specific enthalpy of the mixture is 2500 kJ/kg.

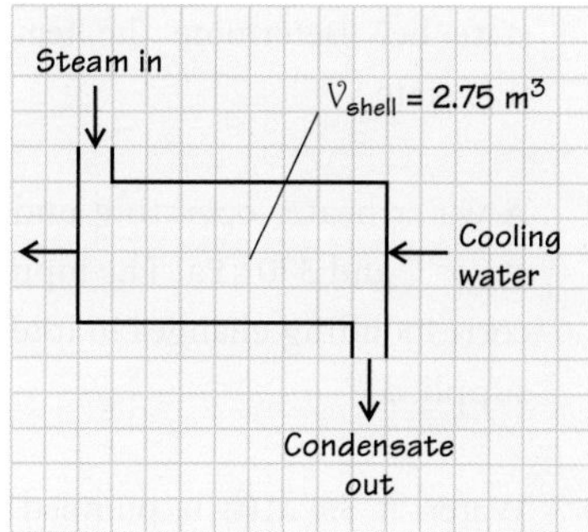

2.142 A 0.6-m^3 tank contains 0.2 kg of H_2O at 350 K. Determine the pressure in the tank and the enthalpy of the H_2O.

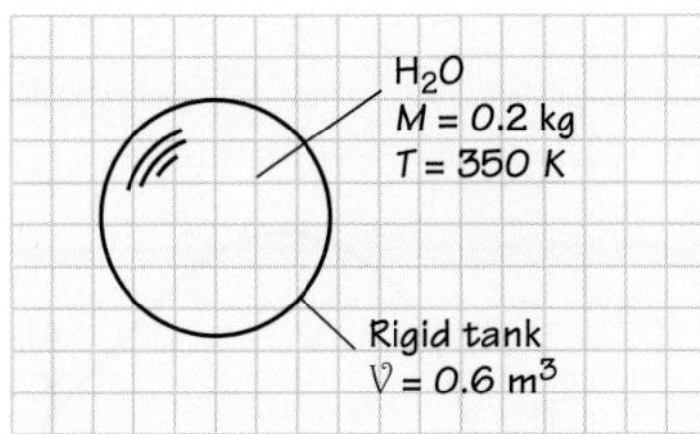

2.143 A 7.57-m^3 rigid tank contains 0.546 kg of H_2O at 37.8 °C. The H_2O is then heated to 204.4 °C. Determine (a) the initial and final pressures of the H_2O in the tank (MPa) and (b) the change in internal energy (kJ). **HINT**: Use the NIST software or online database.

2.144 A 2.7-kg mass of H_2O is in a 0.566-m^3 container (liquid and vapor in equilibrium) at 700 kPa. Calculate (a) the volume and mass of liquid and (b) the volume and mass of vapor in the container.

2.145 Consider 0.25 kg of steam contained in a rigid container at 600 K and 4 MPa. The steam is cooled to 300 K. Determine the entropy change of the H_2O associated with this cooling process. Note: Find $S_2 - S_1$, not $s_2 - s_1$. **HINT**: Use the NIST software or online database.

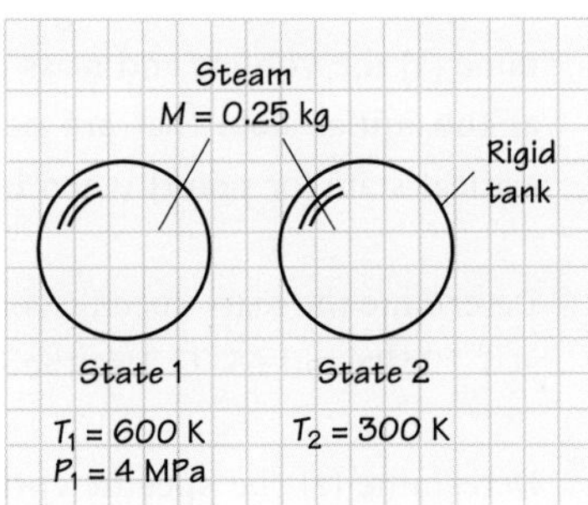

2.146 Steam expands isentropically (i.e., at constant entropy s) from 2 MPa and 500 K to a final state in which the steam has become saturated vapor. What is the temperature of the steam at the final state?

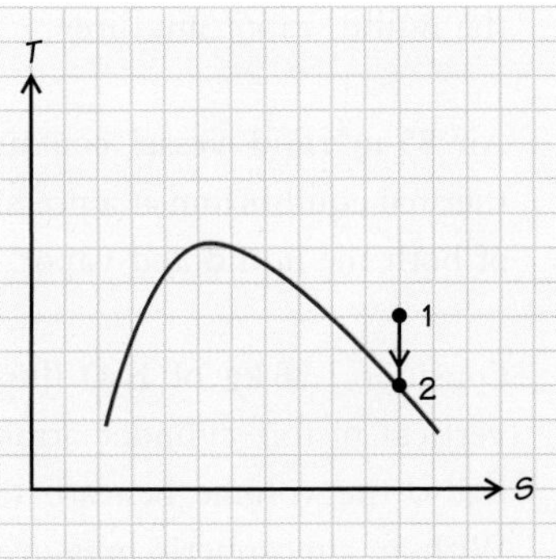

2.147 Steam expands isentropically (i.e., at constant entropy s) from 2 MPa and 500 K to a final state in which the quality is 0.90. Determine the final-state temperature, pressure, and specific volume.

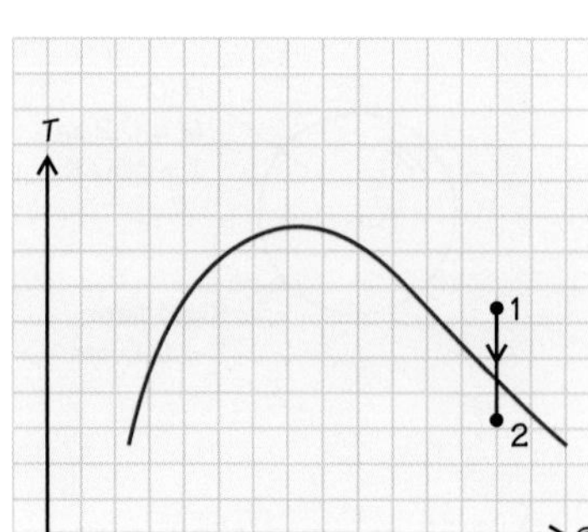

2.148 Steam expands isentropically (i.e., at constant entropy s) from 6 MPa and 1000 K to 1 MPa. Determine the temperature at the final state. Also determine the specific enthalpy change for the process, (i.e., $h_2 - h_1$).

2.149 A piston–cylinder assembly contains steam initially at 0.965 MPa and 315.6 °C. The steam then expands in an isentropic process (i.e., at constant entropy s) to a final pressure of 0.138 MPa. Determine the change in mass-specific internal energy. Use the NIST software or online database for property evaluations.

2.150 A rigid vessel contains 1 kg of saturated R-134a refrigerant at 15.6 °C. Determine (a) the volume and mass of liquid and (b) the volume and mass of vapor at the initial state that are necessary to make the R-134a pass through the critical state (or point) when heated.

2.151 Determine the mass-specific enthalpy (Btu/lb$_m$) of superheated ammonia vapor at 1.3 MPa and 65 °C. Use the NIST WebBook.

2.152 Determine (a) the specific enthalpy of evaporation of steam (kJ/kg) and (b) the ratio of the vapor to liquid specific volumes at standard atmospheric pressure.

2.153 A liquid–vapor mixture of H_2O at 2 MPa is heated in a constant-volume process. The final state is the critical point. Determine the initial quality of the liquid–vapor mixture.

2.154 An 85-m^3 rigid vessel contains 10 kg of water (both liquid and vapor, in thermal equilibrium at a pressure of 0.01 MPa). Calculate the volume and mass of both the liquid and vapor.

2.155 Consider 1.36 kg of H_2O (liquid and vapor in equilibrium) contained in a vertical piston–cylinder arrangement at 50°C, as shown in the sketch. Initially, the volume beneath the 113.4-kg piston (area 11.15 cm^2) is 0.03 m^3. With an atmospheric pressure of 101.325 kPa (g = 9.14 m/s^2), the piston rests on the stops (indicated in black). Energy is transferred to this arrangement until there is only saturated vapor inside.

A. Show this process on a T–v diagram.

B. What is the temperature of the H_2O when the piston first rises from the stops?

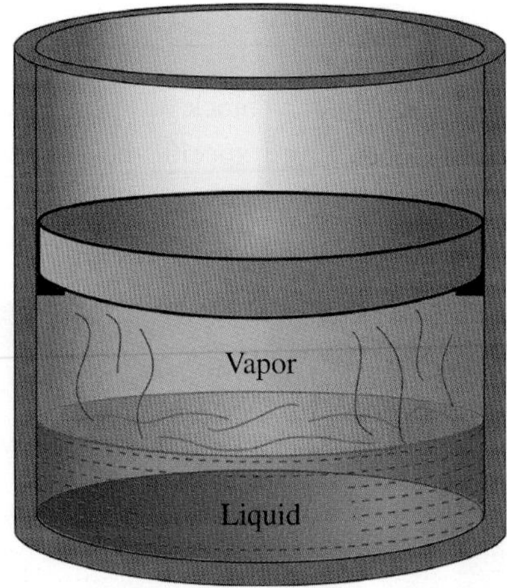

2.156 Consider 0.15 kg of H_2O (liquid and vapor in equilibrium) contained in a vertical piston–cylinder arrangement at 350 K, as shown in the sketch for problem 2.155. Initially, the volume beneath the 150-kg piston (area 10 cm^2) is 0.001306 m^3. With an atmospheric pressure of 100 kPa, the piston rests on the stops. The local gravitational acceleration is 9.807 m/s^2. Energy is transferred to this arrangement until there is only saturated vapor inside.

A. Show this process on a T–$\mathcal{V}$ diagram.

B. What is the temperature of the H_2O when the piston first rises from the stops?

2.157 In a steam turbine, H_2O expands isentropically (i.e., at constant entropy) through the turbine from inlet conditions of 700 kPa and 820 K to an exit pressure of 8 kPa. Determine the specific volume and specific enthalpy at both the inlet and outlet conditions. Use the Appendix B tables.

2.158 Fifty kilograms of H_2O liquid and vapor in equilibrium at 570 K occupy a volume of 1.0 m^3. What is the liquid fraction, that is, the moisture content, $1 - x$?

2.159 Consider water at 400 K and a pressure of 5 atm. Estimate the enthalpy of the water using the saturated liquid tables from Appendix B. Compare this result with the value obtained from the NIST software or online database. Repeat for the density. **HINT**: See Eqns. 2.41a–2.41c.

2.160 Water exits a pump at 15 MPa and 440 K. Use Eqs. 2.41a–2.41c to estimate the specific volume, internal energy, and enthalpy of the water. Compare with the exact values obtained from the NIST software (or Appendix B tables) and calculate the percentage error associated with each.

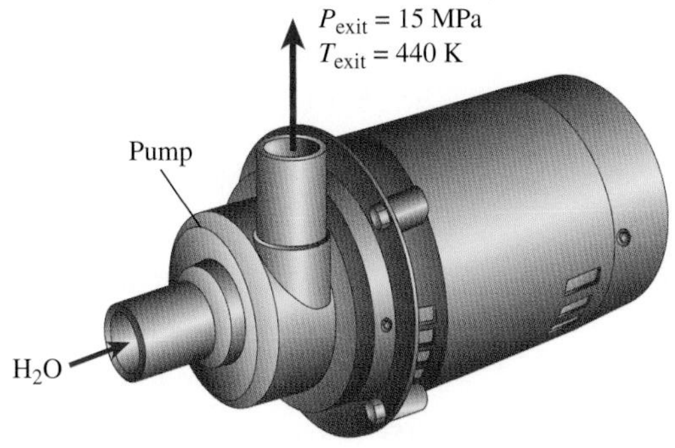

2.161–2.162 Solids

2.161 Consider a block of pure aluminum measuring $25 \times 300 \times 200\,\text{mm}^3$. The density and specific heat of aluminium are approximately 2702 kg/m^3 and 903 J/kg·K, respectively. Estimate the change in internal energy ΔU associated with a temperature change from 600 K to 400 K.

(Credit: Evgeny Dergachev / Alamy Stock Photo.)

2.162 Consider identically sized blocks ($30 \times 250 \times 200\,\text{mm}^3$) of pure aluminium, pure copper, and pure iron. Equal amounts of energy are added to each block such that the internal energy increases by 600 kJ in the process. Estimate the final temperature of each block if the initial temperature is 300 K. **HINT**: You will need to obtain property data from appropriate sources.

2.163–2.167 EES/NIST problems

2.163 (Section 2.7) Use EES or NIST to generate a T–v diagram for water that includes:

A. Constant pressure curves at 100 kPa (approximately atmospheric), 1 MPa, and 10 MPa that extend from compressed liquid to superheated vapor;

B. the vapor dome, showing the saturated liquid and vapor conditions;

C. a linear scale for the temperature and specific volume axes.

D. Now generate the same T–v diagram but using a linear scale for the temperature axes and a logarithmic scale for the specific volume axes;

E. label the axes and isobars.

In your discussion,

F. Compare your figures with Figure 2.33 in the text.

G. From your plot, at what temperature does water at 100 kPa boil? (That is, become a saturated mixture.)

H. Use your plot to identify how the saturation temperature increases with pressure.

2.164 (Section 2.7) Use EES or NIST to generate a P–v diagram for water that includes:

A. Constant temperature curves at 300 K, 373 K, and 450 K that extend from compressed liquid to superheated vapor;

B. the vapor dome showing the saturated liquid and vapor conditions;

C. a linear scale for the pressure and specific volume axes.

D. Now generate the same P–v diagram but using a logarithmic scale for the pressure and specific volume axes;

E. label the axes and isotherms.

In your discussion,

F. Compare your figures with Figure 2.35 in the text.

G. From your plot, at what pressure does water at 373 K boil? (That is, become a saturated mixture.)

H. Use your plot to identify how the saturation pressure increases with temperature.

2.165 (Section 2.7) Use EES or NIST to generate a T–v diagram for compressed liquid water that includes:

A. Constant pressure curves at 5 MPa, 10 MPa, and 20 MPa from 400 K to saturation conditions;

B. the vapor dome showing the saturated liquid conditions from 400 K to 650 K;

C. a linear scale for the temperature and specific volume axes.

D. Mark the following three states: (1) 10 MPa, 450 K; (2) 450K saturated liquid; (3) 10 MPa, saturated liquid.

E. Label the axes and isobars.

In your discussion,

F. Compare your figures with Figure 2.34 in the text.

G. Use your plot to explain why the specific volume of a mildly compressed liquid is approximated as the specific volume of a saturated liquid at the same temperature as stated in Eq. 2.39: $v_{\text{liq}}(T) \cong v_{\text{f}}(T)$ but not the same pressure.

2.166 (Section 2.7) Use EES or NIST to plot the following states as a cycle on a P–v diagram and a T–s diagram. Include the vapor dome on each plot. (This cycle is the simple Rankine cycle shown in Figure 1.1 and studied in Chapter 9.)

1. Saturated liquid water at 3 MPa
2. Saturated vapor water at 3 MPa
3. Saturated mixture at 30 kPa and the same entropy as state 2 ($s_3 = s_2$)
4. Saturated liquid water at 30 kPa
5. Compressed liquid water at 3 MPa and the same entropy as state 4 ($s_5 = s_4$)

2.167 (Section 2.7) Use EES or NIST to plot the following states as a cycle on a P–v diagram and a T–s diagram. Include the vapor dome on each plot. (This cycle is the Rankine cycle with superheat shown in Figure 1.1 and to be studied in Chapter 9.)

1. Saturated liquid water at 3 MPa
2. Saturated vapor water at 3 MPa
3. Superheated water at 3 MPa and 640 K
4. Saturated mixture at 30 kPa and the same entropy as state 3 ($s_4 = s_3$)
5. Saturated liquid water at 30 kPa
6. Compressed liquid water at 3 MPa and the same entropy as state 5 ($s_6 = s_5$)

2.168–2.177 FE problems

2.168 What is the volume of 3 kg saturated water mixture at a temperature of 420 K and quality of 80%? a. 0.34 m^3, b. 1.02 m^3, c. 0.425 m^3, d. 0.0012 m^3.

2.169 What is the mass of air in a 4 m × 5 m × 3 m room that is at 23 °C and 101 kPa? a. 918 kg, b. 0.92 kg, c. 71.3 kg, d. 23.8 kg.

2.170 A fixed volume of air is initially at 300 kPa and 70°C. If the air is heated to 250°C, what is the pressure? a. 457 kPa, b. 197 kPa, c. 1071 kPa, d. 84 kPa.

2.171 Compressed air in a rigid tank is at a pressure of 3 MPa and 200 °C. If half the air is released and the air temperature is 50 °C, what is the pressure? a. 4.4 MPa, b. 2.04 MPa, c. 1.5 MPa, d. 1.02 MPa.

2.172 Five kg of air at 600 kPa and 300 K expands to three times the volume in an isothermal process. What is the final pressure of the air? a. 200 kPa, b. 1.8 MPa, c. 600 kPa, d. 100 kPa.

2.173 What is the mass of water in a 3 m^3 tank at 500 kPa and 460 K? a. 1.24 kg, b. 7.29 kg, c. 3.2 kg, d. 0.81 kg.

2.174 What is the pressure of 5 kg water at 400 K when the volume is 20 m^3? a. 100 kPa, b. 500 kPa, c. 25 kPa, d. 3.55 kPa.

2.175 A 40%-quality water mixture (boiling water) is in a pressure cooker at 100 kPa (atmospheric). The pressure cooker is closed, allowing the pressure to increase during a constant volume process. What is the quality of the mixture when the pressure is 200 kPa? a. 0.21, b. 0.42, c. 0.77, d. 0.81.

2.176 What is the volume of 4 kg water at 500 kPa and 560 K? a. 2.04 m^3, b. 8.74 m^3, c. 4.01 m^3, d. 5.24 m^3.

2.177 Five m^3 of air at 200 kPa and 30 °C is heated to 200 °C in an isobaric (constant-pressure) process. What is the final volume of the air? a. 3.2 m^3, b. 7.8 m^3, c. 33.3 m^3, d. 12.3 m^3.

CHAPTER 3

Conservation of Mass

LEARNING OBJECTIVES

After studying Chapter 3, you should:

- Be able to define and calculate volume and mass flow rates for simple situations.
- Be able to express the conservation of mass principle for closed systems and to apply this principle to analyze practical situations.
- Be able to express the conservation of mass principle for open systems (control volumes) for both steady and unsteady flows and to apply this principle to analyze practical situations.
- Have an improved ability to determine states and properties of ideal gases and two-phase substances.

CHAPTER 3 OVERVIEW

WITH THE DISCLAIMER that we will consider neither situations involving nuclear transformations nor situations where relativistic effects are important, this chapter presents general and specific statements of mass conservation. We consider, first, closed thermodynamic systems. Before extending the mass conservation principle to open systems (control volumes), the concept of a flow rate and its relationship to the average velocity of a flowing fluid is introduced. The concept of steady-state, steady flow for open systems is presented. The principle of mass conservation is then applied to both steady and unsteady flows for single and multiple streams into and out of the system.

Figure 3.1 introduces the first conservation principle essential to our study of thermodynamics: the conservation of mass. This figure illustrates the key role that Chapter 3 plays in our study. Although the chapter is relatively short, the ideas that it presents form one of the pillars of engineering analysis.

<table>
<tr><td colspan="3">DESIGN AND ANALYSIS OF PRACTICAL DEVICES AND SYSTEMS</td></tr>
<tr><td>CONSERVATION OF MASS</td><td>CONSERVATION OF ENERGY (First Law of Thermodynamics)</td><td>SECOND LAW OF THERMODYNAMICS (Entropy)</td></tr>
<tr><td colspan="3">PROPERTIES OF MATTER</td></tr>
<tr><td colspan="3">FRAMEWORKS FOR ANALYSIS, KEY CONCEPTS, AND DEFINITIONS</td></tr>
</table>

FIGURE 3.1 Hierarchical arrangements of the topics in our study of engineering thermodynamics. Chapter 3 deals with the fundamental principle of mass conservation.

Historical Context

(Credit: pictore / iStock / Getty Images Plus.)

Early Greek philosophers postulated basic concepts of mass conservation with the notion that "nothing comes from nothing" and similar philosophical ideas about the nature of matter. Most pertinent to this chapter, however, is **Leonardo da Vinci's** (1452–1519) understanding of mass conservation of a flowing fluid. Being an astute observer of nature, **Leonardo's** observations of flowing fluids led him to be the first to express a clear and concise statement of mass conservation (continuity) for incompressible flows. The following statement, and others, reveals his quantitative understanding of this principle [1]:

A river of uniform depth will have a more rapid flow at the narrower section than at the wider, to the extent that the greater surpasses the lesser.

For reacting systems, a correct statement of mass conservation had to wait until 1798, when **Antoine Laurent Lavoisier** (1743–1794) presented his results from careful experiments on closed thermodynamic systems [2], which we discuss in Chapter 12. **Albert Einstein's** (1879–1955) special theory of relativity added a new dimension to the concept of mass conservation, with the idea that energy crossing a system boundary also carries mass with it. From the perspective of our study, relativistic effects on classical mass conservation are so small that they can be neglected.

3.1 Generic Balance Principle

Here we introduce a generic balance principle that can be transformed into any of the various conservation principles used in engineering. This balance principle can also be used to create useful expressions for quantities that are not conserved, specifically entropy. For our study, we employ this principle to develop expressions for the conservation of mass (this chapter), expressions for the conservation of energy (Chapter 5), and entropy balances (Chapter 7). We start by expressing the generic balance principle, first, for a finite time interval and, second, for an instant:

For a finite time interval, $\Delta t = t_2 - t_1$,

$$X_{in} - X_{out} + X_{generated} = \Delta X_{stored} \equiv X(t_2) - X(t_1). \tag{3.1}$$

X_{in}	X_{out}	$X_{generated}$	ΔX_{stored}
Quantity of X crossing boundary and passing into system	Quantity of X crossing boundary and passing out of system	Quantity of X generated within system	Change of quantity of X stored within system during time interval

For our study, the generic quantity X can be mass, energy, or entropy.

Instantaneously, we have

$$\dot{X}_{in} - \dot{X}_{out} + \dot{X}_{generated} = \dot{X}_{stored} \tag{3.2}$$

$\dot{X}_{in}$	$\dot{X}_{out}$	$\dot{X}_{generated}$	$\dot{X}_{stored}$
Time rate of X crossing boundary and passing into system	Time rate of X crossing boundary and passing out of system	Time rate of generation of X within system	Time rate of storage of X within system

3.2 Mass Conservation for a Closed System

We now explicitly transform the generic conservation principles expressed in Eqs. 3.1 and 3.2 to statements of mass conservation for closed systems. Equation 3.1 thus becomes

Equations with yellow backgrounds express key concepts and are the most important relationships in the chapter.

$$M_{\text{in}} - M_{\text{out}} + M_{\text{generated}} = \Delta M_{\text{stored}} \equiv M_{\text{sys}}(t_2) - M_{\text{sys}}(t_1). \tag{3.3a}$$

Quantity of mass crossing boundary and passing into system	Quantity of mass crossing boundary and passing out of system	Quantity of mass generated within system	Change of quantity of mass stored in system during time interval $\Delta t = t_2 - t_1$

By definition, no mass crosses the closed system boundaries, when relativistic effects are ignored; thus both M_{in} and M_{out} must be zero. Furthermore, if no nuclear transformations occur, no mass is generated (or destroyed) within the system; thus, all terms on the left-hand side of Eq. 3.3a are zero. We formally conclude the obvious:

$$0 = M(t_2) - M(t_1),$$

or

Central to our study of mass conservation are the thermodynamic properties M, $\mathcal{V}$, and ρ or M/v. See Eq. 2.9 and related material in Chapter 2.

$$M(t_1) = M(t_2) = M = \text{constant}. \tag{3.3b}$$

Furthermore, since the system mass is a constant, its time derivative must be zero, that is,

$$\frac{dM}{dt} = 0. \tag{3.3c}$$

Note that, had we started with Eq. 3.2, the rate form of the generic conservation principles, we would have arrived at Eq. 3.3c directly.

For a system in which the density is uniform (i.e., it has the same value at every location), we can express Eqs. 3.3b and 3.3c as

$$\rho\mathcal{V} = \text{constant}\ (= M), \tag{3.4a}$$

or

$$\frac{d(\rho\mathcal{V})}{dt} = 0, \tag{3.4b}$$

where ρ is the mass density and $\mathcal{V}$ is the volume.

Some systems have boundaries that move with time. A familiar example is the gas trapped in the combustion chamber of an internal combustion engine, suggested by the sketch in Fig. 3.2. To ensure that we have a closed system, both intake and exhaust valves must be tightly closed, and no gas can leak past the piston rings. If we assume that the gas in the system has a uniform but time-varying density, Eq. 3.4b can be expanded using the product rule for differentiation to give

$$\rho\frac{d\mathcal{V}}{dt} + \mathcal{V}\frac{d\rho}{dt} = 0. \tag{3.5}$$

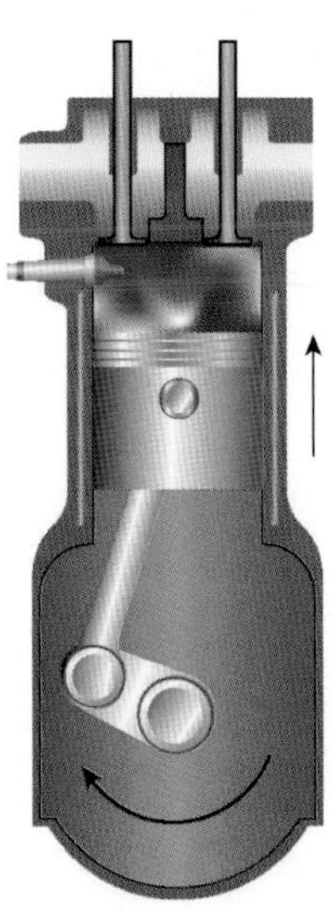

FIGURE 3.2 The gas trapped in the combustion chamber is the closed system. The motion of the piston creates a moving boundary and a time-varying volume. Note that the system is closed only when the valves are closed.

To apply Eq. 3.5, say, to a simulation of an engine, $\mathcal{V}$ and $d\mathcal{V}/dt$ can be determined purely from geometric and kinematic relationships (see Appendix 3A). Obtaining the density and density time derivative, however, requires the application of additional thermal science concepts: an equation of state (see Chapter 2), conservation of energy (Chapter 5), and expressions for heat-transfer rates (Chapter 4).

The assumption of a uniform density is an approximation because the temperature of the gas within a real combustion chamber is not uniform. The uniform-property approximation is probably most reasonable during the compression and power strokes, but it is far from reality during the combustion event, when a flame propagates through a relatively cool unburned mixture and produces hot products. Dealing with nonuniform properties is beyond the scope of our study.

Example 3.1 Isochoric Ideal-Gas Process

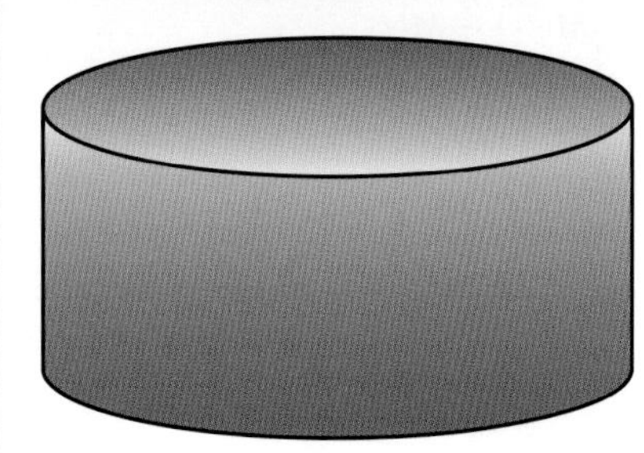

Gaseous nitrogen is trapped in a cylinder having a diameter D of 75 mm and height L of 40 mm. The initial uniform temperature and pressure are 1250 K and 500 kPa, respectively. Heat transfer from the nitrogen results in a new equilibrium temperature of 500 K. Determine the mass of the nitrogen and the final pressure.

Solution

Known P_1, T_1, T_2, D, L

Find M_{N_2}, P_2

Sketch

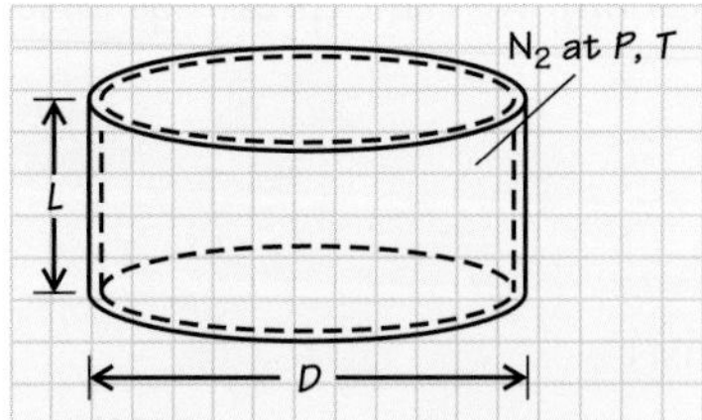

Modeling, Premises and Assumptions

i. N_2 behaves as an ideal gas.
ii. Gravity does not affect the pressure.
iii. The cylinder is rigid and leak free.

Analysis The solution to this is quite straightforward. We first calculate the mass of the nitrogen and then calculate the final pressure.

To determine the mass, we use the given geometry to calculate the volume of the N_2, then apply the ideal-gas equation of state (Eq. 2.26c) as follows:

$$\mathcal{V} = (\pi D^2/4)L = \pi(0.075\,\text{m})^2(0.040\,\text{m})/4 = 1.767 \times 10^{-4}\,\text{m}^3$$

and

$$P\mathcal{V} = MR_{N_2}T.$$

Solving for M and recognizing that the gas constant $R_{N_2} \equiv R_u/\mathcal{M}_{N_2}$ yields

$$M = \frac{P\mathcal{V}\mathcal{M}_{N_2}}{R_u T}.$$

Evaluating numerically, we get

$$M = \frac{500 \times 10^3\,\text{Pa}\,(1.767 \times 10^{-4}\,\text{m}^3)\,28.013\,\text{kg/kmol}}{8314.47\,\text{J/kmol}\cdot\text{K}\,(1250\,\text{K})}\left[\frac{\text{N/m}^2}{\text{Pa}}\right]\left[\frac{\text{J}}{\text{N}\cdot\text{m}}\right]$$

$$= 2.38 \times 10^{-4}\,\text{kg}.$$

To obtain the final pressure, we explicitly start with the concept that the mass is conserved during the cooling process that takes the system from its initial to its final state:

$$M_1 = M_2\,(= M).$$

Applying the ideal-gas equation of state yields

$$M = \frac{P_1\mathcal{V}_1}{RT_1} = \frac{P_2\mathcal{V}_2}{RT_2}.$$

Because the cylinder is rigid, $\mathcal{V}_1 = \mathcal{V}_2$ and the volumes cancel. Solving the above for P_2 and evaluating the result numerically yields

$$P_2 = P_1\frac{T_2}{T_1} = 500\,\text{kPa}\,\frac{500\,\text{K}}{1250\,\text{K}} = 200\,\text{kPa}.$$

Comments Note the importance of the ideal-gas law and the explicit use of mass conservation in solving this nearly trivial problem. That the final pressure is less than the initial pressure meets our expectations for a constant-volume cooling process.

Self-Test 3.1

☑ A sliding piston in the cylinder in Example 3.1 compresses the nitrogen into a final volume with a height of 15 mm. Determine the final density of the N_2.

(Answer: $\rho = 3.59\,\text{kg/m}^3$)

Example 3.2 SI Engine Compression

(Credit: Courtesy of Compact Radial Engines, Inc.)

Consider a spark-ignition engine. The temperature and pressure are assumed to be uniform within the cylinder with values of 345 K and 184 kPa, respectively, and the properties of the fuel–air mixture can be treated as those of air. Geometric parameters are defined and kinematic relationships for the instantaneous volume $\mathcal{V}(\theta)$ and its time derivative $d\mathcal{V}(\theta)/dt$ are given in Appendix 3A (see the end of this chapter). For the geometrical parameters defined there,

$$B = 70\,\text{mm}, \qquad CR = 8,$$
$$S = 70\,\text{mm}, \qquad \ell/a = 3.5,$$

determine the instantaneous value of $d\rho/dt$ during the compression stroke at a crank angle of $\theta = 270^\circ$ ($3\pi/2$ rad) for a rotational speed of 2000 rpm.

Solution

Known $P, T, \theta, N, B, S, CR, \ell/a$

Find $d\rho/dt$

Sketch

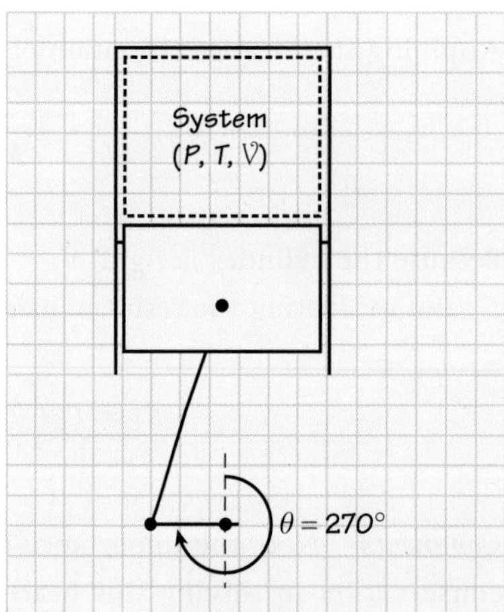

Modeling, Premises and Assumptions

i. Quasi-static equilibrium
ii. Uniform properties
iii. Closed system with no leakage past rings or valves
iv. Air properties approximate those of the mixture (ideal gas)

Analysis Assumptions i, ii, and iii allow us to express conservation of mass for the system shown in the sketch using Eq. 3.5. Solving Eq. 3.5 for $d\rho/dt$ yields

$$\frac{d\rho}{dt} = \frac{\rho}{\mathcal{V}}\frac{d\mathcal{V}}{dt}.$$

We evaluate the density from the ideal-gas equation of state (Eq. 2.26b); $\mathcal{V}$ and $d\mathcal{V}/dt$ are evaluated from the relationships in Appendix 3A:

$$\rho = \frac{P}{R_{\text{air}}T} = \frac{P\mathcal{M}_{\text{air}}}{R_u T}$$
$$= \frac{184 \times 10^3\,\text{Pa}\,(28.97\,\text{kg/kmol})}{8314.47\,\text{J/kmol·K}(345\,\text{K})}\left[\frac{\text{N/m}^2}{\text{Pa}}\right]\left[\frac{\text{J}}{\text{N·m}}\right] = 1.858\,\text{kg/m}^3.$$

To evaluate $\mathcal{V}(\theta)$ from Eq. 3A.4 requires a value for $\mathcal{V}_{\text{TC}}$. Using the definition of compression ratio and displacement presented in Appendix 3A at the end of the chapter, we solve Eq. 3A.3 for $\mathcal{V}_{\text{TC}}$ as follows:

$$\mathcal{V}_{\text{TC}} = \frac{\mathcal{V}_{\text{disp}}}{CR - 1},$$

where

$$\mathcal{V}_{\text{disp}} = S\pi B^2/4.$$

Thus,

$$\mathcal{V}_{\text{TC}} = \frac{S\pi B^2/4}{CR - 1}$$
$$= \frac{(0.07\text{ m})\pi(0.07\text{ m})^2/4}{8 - 1} = 3.848 \times 10^{-5}\text{m}^3.$$

We can now evaluate $\mathcal{V}(\theta)$ and $d\mathcal{V}(\theta)/dt$ using expressions from Appendix 3A:

$$\mathcal{V}(\theta) = \mathcal{V}_{\text{TC}}\left\{1 + \tfrac{1}{2}(CR - 1)\left[\tfrac{\ell}{a} + 1 - \cos\theta - \left(\tfrac{\ell^2}{a^2} - \sin^2\theta\right)^{1/2}\right]\right\},$$

where

$$\mathcal{V}(270^\circ) = 3.848 \times 10^{-5}\text{m}^3\left\{1 + \tfrac{1}{2}(8-1)\left[3.5 + 1 - \cos 270^\circ - (3.5^2 - \sin^2(270^\circ))^{1/2}\right]\right\}$$
$$= 3.848 \times 10^{-5}\text{m}^3\left\{1 + \tfrac{7}{2}[4.5 - 3.3541]\right\} = 1.928 \times 10^{-4}\text{m}^3,$$

and

$$\frac{d\mathcal{V}(\theta)}{dt} = SN\frac{\pi B^2}{4}\pi\sin\theta\left[1 + \frac{\cos\theta}{(\ell^2/a^2 - \sin^2\theta)^{1/2}}\right].$$

Recognizing that cos (270°) = 0, we see that the term in brackets becomes unity; thus,

$$\frac{d\mathcal{V}}{dt}(270^\circ) = 0.07\,\text{m}\left(\frac{2000\,\text{rev}}{\text{min}}\right)\left[\frac{1\text{ min}}{60\,\text{s}}\right]\frac{\pi(0.07)^2\text{m}^2}{4}\pi\sin(270^\circ)(1) = -0.0282\,\text{m}^3/\text{s}.$$

We now evaluate $d\rho/dt$ from Eq. 3.5:

$$\frac{d\rho}{dt} = -\frac{\rho}{\mathcal{V}(270^\circ)}\frac{d\mathcal{V}}{dt}(270^\circ)$$
$$= -\frac{1.858\,\text{kg/m}^3(-0.0282)\text{m}^3/\text{s}}{1.928 \times 10^{-4}\,\text{m}^3}$$
$$= +271.8\,\text{kg/m}^3\text{·s},$$

where the positive sign indicates that the mixture is being compressed, as expected.

Comments This example illustrates how the purely mechanical relationships that describe the geometry and kinematics of a reciprocating engine relate to the thermodynamics within the engine cylinder.

3.3 Flow Rates and Average Velocity

Before we can apply conservation of mass to open systems, we need to define and explore these two concepts.

3.3a Flow Rates

UNIFORM VELOCITY

Consider a fluid with a uniform and steady velocity V crossing an imaginary (i.e., nonmaterial), planar surface of area A, as shown in Fig. 3.3. The fluid velocity is perpendicular to the surface and aligned with the x-direction. In a time interval Δt, a fluid particle will travel a distance $\Delta x = V\Delta t$ after crossing the surface. The volume of fluid crossing the surface in this same time interval is simply the product of the flow area A and the distance traveled by the first fluid particle to cross the surface at the start of the time interval, that is,

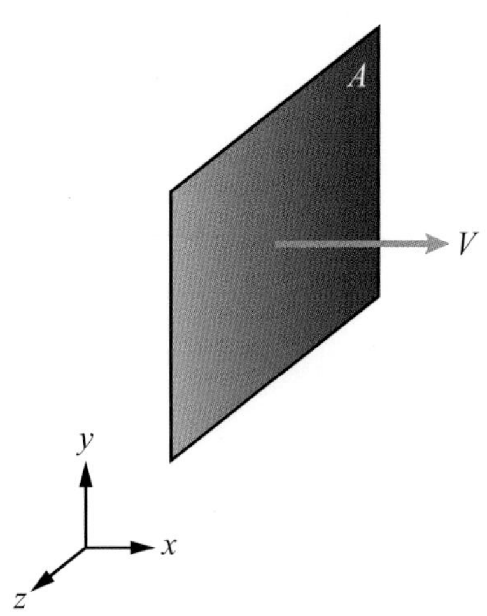

FIGURE 3.3 Fluid with uniform velocity V crosses surface A with the velocity everywhere perpendicular to the plane of A.

$$\Delta\mathcal{V} = A\Delta x.$$

Substituting for Δx and rearranging yields

$$\frac{\Delta\mathcal{V}}{\Delta t} = VA.$$

The quantity $\Delta\mathcal{V}/\Delta t$ we define as the **volumetric flow rate** $\dot{\mathcal{V}}$. Shrinking the time interval to a very small value, but not going below the limit required to maintain a continuum, allows us to interpret VA as the instantaneous volumetric flow rate, removing the restriction that the velocity be steady (i.e., that it does not vary with time). Thus,

$$\dot{\mathcal{V}} = VA, \tag{3.6}$$

where we emphasize that V is uniform over A and is everywhere perpendicular to the plane of A. This particular condition of uniform velocity is sometimes called **plug** or **slug flow**.

If the fluid density also is uniform over A, the amount of mass crossing the plane in Δt is just

$$\Delta M = \rho\Delta\mathcal{V} = \rho AV\Delta t.$$

We thus define the **mass flow rate** $\dot{m}$ to be $\Delta M/\Delta t$ or, instantaneously,

$$\dot{m} = \rho VA. \tag{3.7}$$

The SI units for mass flow rate are kg/s.

DISTRIBUTED VELOCITY

In flows within pipes, tubes, and channels, the velocity distribution is not uniform. In these flows, the fluid sticks to the walls of the flow passage (see Fig. 3.4). This results in a velocity distribution with a zero value at the walls and a maximum value some distance from the wall. That the fluid velocity is zero at a surface is termed the **no-slip condition**. For pipe or tube flows, velocity profiles are commonly parabolic (**laminar**

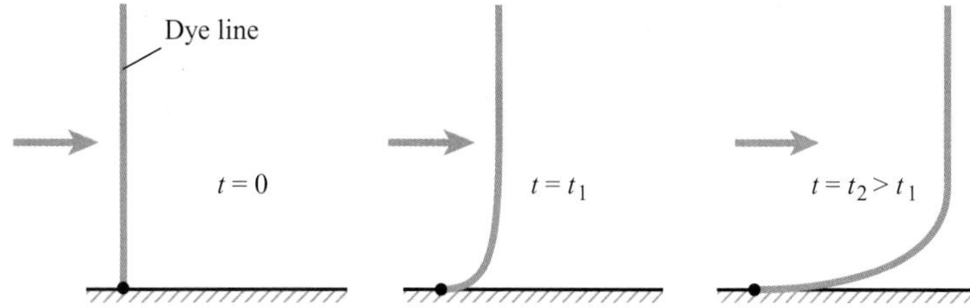

FIGURE 3.4 The flow from left to right transports a dye line downstream. The distance traveled by each segment of the line is proportional to the local velocity. The dye line sticks to the wall as the velocity there is zero.

flow) or obey a power law (**turbulent flow**). Details and further development of these concepts can be found in textbooks dealing with fluid flow [3–5]. Evaluating flow rates when the velocity is not uniform over the flow area is beyond the scope of this book. Our approach is to assume that the uniform velocity we use is the appropriate average velocity for any real flow.

3.3b Average Velocity

In many situations, an area-averaged velocity is employed to characterize the flow. An average velocity is particularly useful in expressing conservation of mass in flows where the density is uniform.

As you may recall from your study of mathematics, the definition of the average of a function $f(s)$ over the interval from s_1 to s_2 is

$$f_{avg} \equiv \frac{1}{s_2 - s_1} \int_{s_1}^{s_2} f(s)ds, \tag{3.8}$$

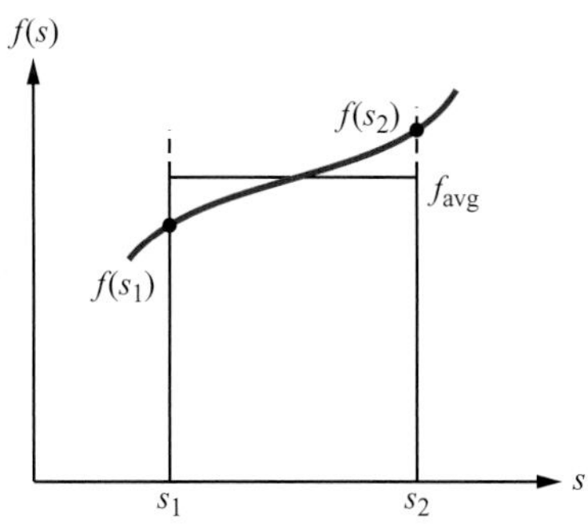

FIGURE 3.5 The average value of the function f over the interval s_2 to s_1 multiplied by the length of the interval $(s_2 - s_1)$ yields the area of a rectangle that has the same area as the area under the curve $f(s)$.

as illustrated in Fig. 3.5. For our purposes, f is identified as the velocity distribution and s is the cross-sectional flow area perpendicular to V, $A_{x\text{-sec}}$; that is,

$$V_{avg} \equiv \frac{1}{A_{x\text{-sec}}} \int_A VdA. \tag{3.9}$$

By comparing Eq. 3.9 with Eq. 3.6, we see that the volumetric flow rate $\dot{V}$ is just the average velocity times the flow area:

$$\dot{V} = \int_A VdA = V_{avg}A_{x\text{-sec}} \tag{3.10}$$

If the density is uniform over the flow area, the mass flow rate also simplifies to

$$\dot{m} = \rho V_{avg}A_{x\text{-sec}}. \tag{3.11}$$

Table 3.1 shows some typical average velocities associated with pipe flows in a steam power plant. Note that the typical velocities for steam flows are much greater than those for flows of liquid water. This difference results from the fact that much more power is required to pump liquids, compared to gases, for a given velocity.

TABLE 3.1 Typical Average Velocities for Selected Pipe Flows*

Fluid	Application	Velocity (m/s)
Steam	Superheated process steam	45–100
	Auxiliary heat steam	30–75
	Saturated and low-pressure steam	30–50
Water	Centrifugal pump suction lines	0.9–1.5
	Feedwater	2.4–4.6
	General service	1.2–3.1
	Potable water	Up to 2.1

* Adapted from Department of the Army, TM 5–810-15, Central Steam Boiler Plants, August 1995.

Example 3.3 Solar Collector Heating Water

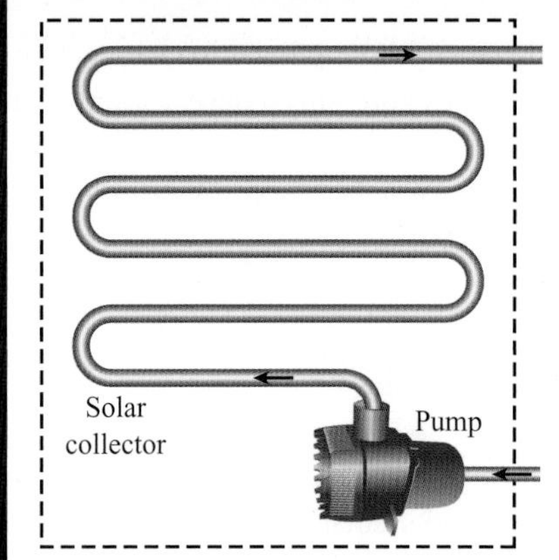

A solar collector for heating water consists of a 292.6-m length of black EPDM tubing, as shown in the sketch. The outside diameter of the tubing is 8.9 mm and the inside diameter is 5.7 mm. Water is pumped through the tubing at a steady volumetric flow rate of $3.9 \times 10^{-5}\,\mathrm{m^3/s}$. The water temperature is approximately 365 K and the pressure in the tubing is nominally 2 atm. Determine the average velocity V_{avg} and the mass flow rate of the water flowing through the collector.

Solution

Known $\dot{\mathcal{V}}$, D, T

Find V_{avg}

Sketch

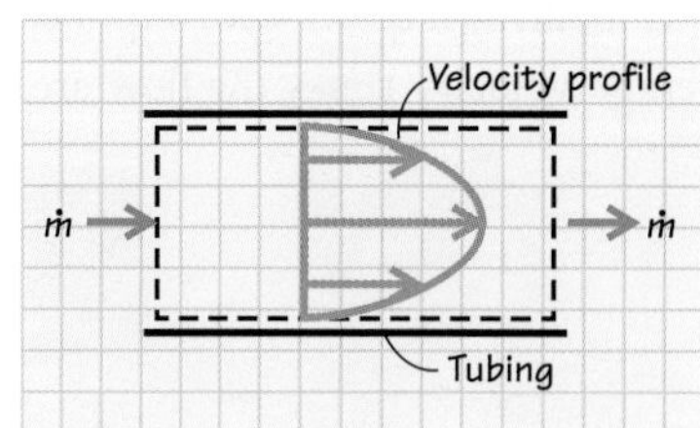

Modeling, Premises and Assumptions Uniform water properties are evaluated at 365 K and 2 atm.

Analysis The average velocity is calculated from the straightforward application of Eq. 3.10, where the appropriate diameter is the inside diameter. Solving for V_{avg} yields

$$V_{avg} = \frac{\dot{\mathcal{V}}}{A_{x\text{-sec}}},$$

or

$$V_{\text{avg}} = \frac{\dot{V}}{\pi D^2/4}.$$

Substituting numerical values gives

$$V_{\text{avg}} = \frac{4(3.9 \times 10^{-5}\,\text{m}^3/\text{s})}{\pi(0.0057\,\text{m})^2} = 1.528\,\text{m/s}.$$

To calculate the mass flow rate, we need to determine the density of the water and apply Eq. 3.11. From the NIST WebBook, we find the density to be 964.1 kg/m^3. Thus,

$$\begin{aligned}\dot{m} &= \rho V_{\text{avg}} A_{\text{x-sec}} \\ &= \rho \dot{V} \\ &= 964.1\,\text{kg/m}^3(3.9 \times 10^{-5}\text{m}^3/\text{s}) = 0.0376\,\text{kg/s}.\end{aligned}$$

Comments Equations 3.10 and 3.11 are key relationships, which we will use many times throughout this book.

Self-Test 3.2

Calculate the average velocity and the mass flow rate for the solar collector in Example 3.3 if the tubing diameter is doubled with all other factors unchanged.

(Answer: $V_{avg} = 0.382\,m/s$, $\dot{m} = 0.0376\,kg/s$)

3.4 Mass Conservation for an Open System (Control Volume)

We now consider conservation of mass applied to an **open system** (a **control volume**). In this section, we develop various mathematical statements of this conservation principle for the most simple and restricted cases as well as for more complex and general situations.

3.4a General View of Mass Conservation for Open Systems

With the knowledge of how to express and calculate mass flow rates, we are now able to write explicit mass conservation statements for open systems (control volumes). In the following sections, we develop the simplest statements and then add complexity, carefully stating the restrictions that apply. We start by transforming the instantaneous form of the generic balance principle presented at the beginning of the chapter into a general statement of mass conservation.

We rewrite the instantaneous generic balance principle expressed in Eq. 3.2, choosing the quantity of interest X to be mass M, as follows:

$$\dot{m}_{\text{in}} - \dot{m}_{\text{out}} + \dot{m}_{\text{generated}} = \frac{dM_{\text{cv}}}{dt} \quad (3.12)$$

$\dot{m}_{\text{in}}$	$\dot{m}_{\text{out}}$	$\dot{m}_{\text{generated}}$	$\frac{dM_{\text{cv}}}{dt}$
Time rate of mass crossing boundary and passing into the open system, i.e., the mass flow rate in	Time rate of mass crossing boundary and passing out of the open system, i.e., the mass flow rate out	Time rate of mass generated within the open system	Time rate of storage of mass within the open system

Because we limit all our analyses to situations in which relativistic effects are negligible and no nuclear reactions occur, we can eliminate the generation term; thus, Eq. 3.12 simplifies to

$$\dot{m}_{\text{in}} - \dot{m}_{\text{out}} = \frac{dM_{\text{cv}}}{dt}. \quad (3.13)$$

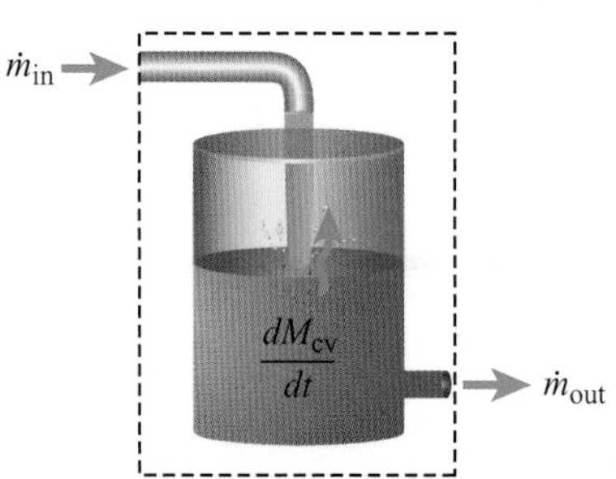

FIGURE 3.6 Mass accumulates within an open system when the inlet flow rate exceeds the exit flow rate. The system boundary is denoted by the dashed line.

Figure 3.6 illustrates this relationship. In words, Eq. 3.13 states that the net rate at which mass flows across the boundary into an open system ($\dot{m}_{\text{in}} - \dot{m}_{\text{out}}$) must equal the rate at which mass accumulates, or is stored, within the boundaries of the open system (dM_{cv}/dt). Using different terminology to refer to the same physics, we see that this storage term is equivalently the time rate of change of mass within the open system. Note that the storage rate is thus expressed mathematically as a time derivative and refers to what is happening *within the system*, whereas the mass flow rates are not expressed as derivatives and refer to what is happening *at the boundaries* of the system. We reinforce these distinctions by showing an arrow interior to the open system (storage) and arrows that stop or start at the system boundaries (flow rates) as illustrated in Fig. 3.6.

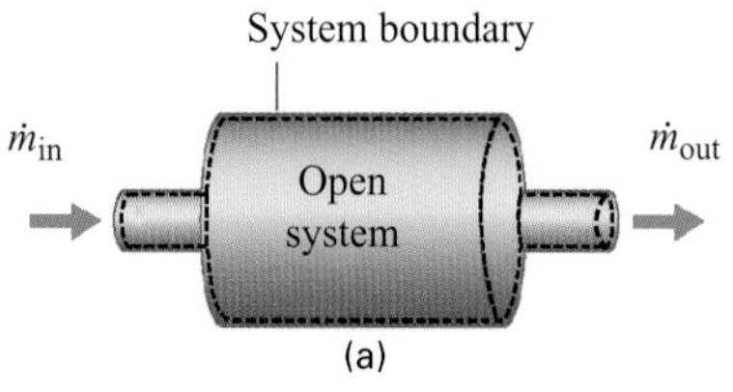

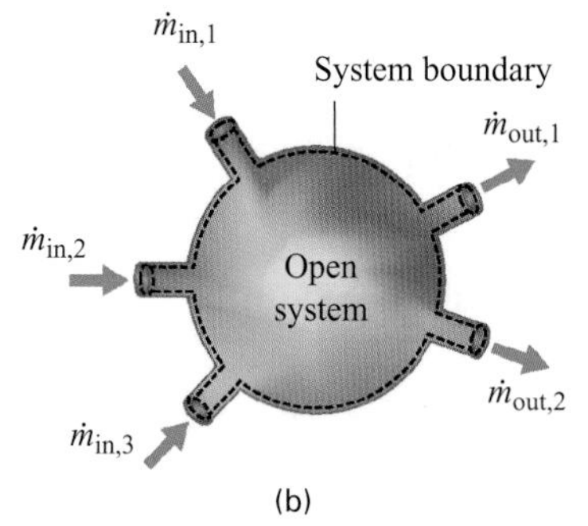

FIGURE 3.7 The boundary of the open system is indicated by the dashed line for open systems with **(a)** a single inlet and single outlet and **(b)** multiple inlets and outlets.

3.4b Steady-State, Steady Flow

We start with the simplest case of a **steady-state, steady flow** for a large-scale (i.e., integral) control volume possessing a single inlet and a single outlet stream, as illustrated in Fig. 3.7a. The restriction to a **steady state** requires that all thermodynamic properties at all locations within the control volume and on the boundary do not vary with time. The assumption of a **steady flow** requires that the velocity at each location where the fluid enters or exits the control volume does not change with time. With these restrictions, conservation of mass is expressed by

$$\dot{m}_{\text{in}} = \dot{m}_{\text{out}}, \quad (3.14a)$$

where the mass flow rates relate to the inlet and outlet velocities, as previously discussed (e.g., Eq. 3.11).

If more than one stream enters or exits the control volume (Fig. 3.7b), the total mass flow entering must equal the total mass flow exiting, that is,

$$\sum_{j=1}^{N\text{inlets}} \dot{m}_{\text{in},j} = \sum_{k=1}^{M\text{outlets}} \dot{m}_{\text{out},k}. \quad (3.14b)$$

Practical applications of Eqs. 3.14a, b abound. Chapter 8 (Thermal-Fluid Analysis of Steady-Flow Devices) focuses on these applications, which include pumps, turbines, and heat exchangers, among others. The following examples illustrate applications of steady-flow mass conservation and reinforce ideas from Chapter 2.

Example 3.4 Solar Collector Heating Air

Air at nearly atmospheric pressure (100 kPa) enters a solar collector at 60 °C and exits at 89 °C, as shown in the sketch. The width of the collector (perpendicular to the page in the right-hand figure) is 1 m. The air mass flow rate is 0.056 kg/s. Determine the average velocity of the air at the entrance and at the exit of the collector.

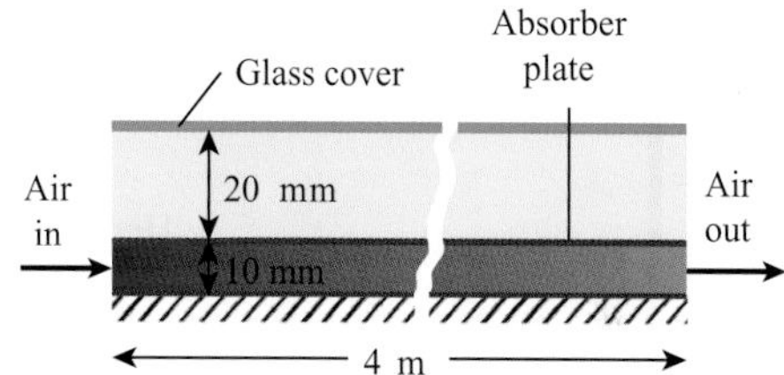

(Credit: Zoonar GmbH / Alamy Stock Photo.)

Solution

Known $\dot{m}$ T_{in}, T_{out}, P, *geometry*

Find $V_{avg,in}$, $V_{avg,out}$

Sketch

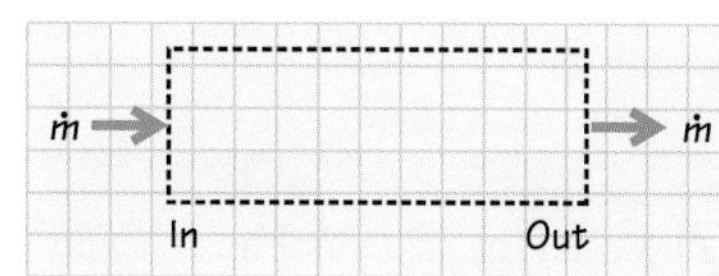

Modeling, Premises and Assumptions

i. Steady flow
ii. Negligible pressure drop from inlet to outlet
iii. Ideal-gas behavior
iv. Uniform density at inlet and outlet

Analysis With the assumption of a steady flow, the conservation of mass for the control volume shown is given by Eq. 3.14a:

$$\dot{m}_{in} = \dot{m}_{out} = \dot{m}.$$

Since the air density is assumed to be uniform across the inlet and outlet, we apply Eq. 3.11 as follows:

$$\dot{m} = \rho_{in} V_{avg,\,in} A_{x\text{-}sec} = \rho_{out} V_{avg,\,out} A_{x\text{-}sec},$$

where the cross-sectional area is the product of the air-channel height, d, and the width of the collector, W, that is,

$$A_{x\text{-}sec} = dW = (0.010\ \text{m})(1.0\ \text{m}) = 0.010\ \text{m}^2.$$

The density is calculated from the ideal-gas equation of state (Eq. 2.26b), taking care to use the absolute temperature:

$$\rho_{\text{in}} = \frac{P_{\text{in}}}{R_{\text{air}} T_{\text{in}}} = \frac{100 \times 10^3\,\text{Pa}}{287.0\,\text{J/kg·K}(273 + 60)\,\text{K}} \left[\frac{\text{N/m}^2}{\text{Pa}}\right]\left[\frac{\text{J}}{\text{N·m}}\right] = 1.046\,\text{kg/m}^3.$$

Thus,

$$V_{\text{avg, in}} = \frac{\dot{m}}{\rho_{\text{in}} A_{\text{x-sec}}} = \frac{0.056\,\text{kg/s}}{1.046\,\text{kg/m}^3(0.01)\text{m}^2} = 5.35\,\text{m/s}.$$

Because both $\dot{m}$ and $A_{x\text{-sec}}$ are constants, the average velocity changes only as a result of the reduced density at the outlet; thus,

$$V_{\text{avg, out}} = V_{\text{avg, in}} \frac{T_{\text{out}}}{T_{\text{in}}}$$

$$= 5.35\text{m/s} \frac{(273 + 89)\,\text{K}}{(273 + 60)\,\text{K}}\text{m/s} = 5.82\ \text{m/s}.$$

Here we make use of our assumption that $P_{\text{in}} \cong P_{\text{out}}$.

Comments A more rigorous calculation might include a determination of the pressure drop from the inlet to the outlet. Good design practice, however, seeks to keep this pressure drop small to minimize the power required to push the air through the collector, so our assumption is probably reasonable. Note that we employed the approximation that air can be treated as an ideal gas. You should use the NIST software and compare the densities to see how good this approximation is for the conditions here.

Self-Test 3.3

Consider the system of Example 3.4, which had a constant mass flow rate of 0.056 kg/s. Is the volumetric flow rate for this system constant also?

(Answer: If the mass flow rate of an incompressible flow is constant, then the volumetric flow rate will also be constant because the density is fixed. For a gas, however, even though the mass flow rate is constant, the volumetric flow rate will not necessarily be constant as the density may vary along the flow path.)

Example 3.5 Water Duct Flow

Consider a flow of room-temperature water through a duct of circular cross section and constant taper. The diameters of the duct at the entrance and at the exit are 50 mm and 80 mm, respectively. The length of the duct is 0.75 m. The average velocity at the entrance is 3 m/s. Determine the mass flow rate through the duct and the average velocity at the exit.

Solution

Known $V_{avg,in}$, D_{in}, D_{out}, T

Find $\dot{m}$, $V_{avg,out}$

Sketch

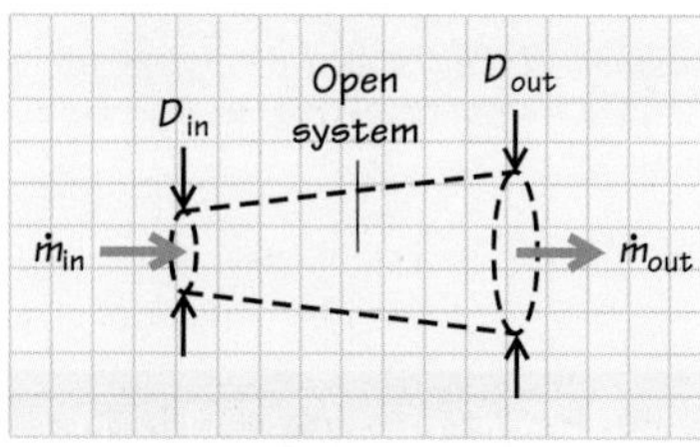

Modeling, Premises and Assumptions

i. One-dimensional flow
ii. Steady-state, steady flow
iii. Incompressible flow (i.e., constant density)
iv. $\rho \approx \rho\ (T_{sat} = T)$

Analysis Because the density is constant, we can apply Eq. 3.11 to find the mass flow rate at the inlet:

$$\dot{m} = \rho_{in} V_{avg,in} \frac{\pi D_{in}^2}{4}.$$

From Appendix B, we find the room-temperature (298 K) value for the water density to be 997 kg/m^3; thus,

$$\dot{m} = 997\,\text{kg/m}^3 (3.0\,\text{m/s}) \pi (0.050)^2\,\text{m}^2/4$$
$$\dot{m} = 5.87\,\text{kg/s}.$$

Steady-state conservation of mass for our integral control volume is expressed by Eq. 3.14a as

$$\dot{m}_{in} = \dot{m}_{out}$$

or

$$\rho V_{avg,in} \frac{\pi D_{in}^2}{4} = \rho V_{avg,out} \frac{\pi D_{out}^2}{4}.$$

Solving for $V_{avg,out}$ yields

$$V_{avg,out} = V_{avg,in} \frac{D_{in}^2}{D_{out}^2}$$
$$= 3.0\,\text{m/s} \frac{(0.050)^2\,\text{m}^2}{(0.080)^2\,\text{m}^2}$$
$$= 1.17\,\text{m/s}.$$

Nozzles and diffusers are important components of jet engines. See Example 9.10 and related material in Chapter 9.

Comment The tapered duct described in this example is called a **diffuser**. These passive devices are deliberately used to slow the flow and thus increase the pressure. We will discuss the operation of diffusers and their opposite-taper counterparts, **nozzles**, in Chapters 5 and 8.

Self-Test 3.4

Consider a household plumbing tee that has one 3/4-in inlet and two 1/2-in outlets. If 0.01 m^3/s of water flows steadily into the tee at 10 °C, what are the outlet mass flow rates and velocities?

(Answer: 5.0 kg/s and 39.5 m/s for each outlet).

Example 3.6 Steam Turbine Flow

A chemical processing plant burns by-product gases to generate electricity for the plant using a simple Rankine cycle, as shown in the diagram. The working fluid (H_2O) flow rate is 11.6 kg/s. Steam enters the turbine at 10 MPa and 800 K through a 0.15-m diameter pipe. The steam exits at 6 kPa with a quality of 0.95. The cross-sectional area of the duct connecting the turbine to the condenser is 2 m^2. Determine the velocities in the inlet pipe and the turbine-to-condenser duct.

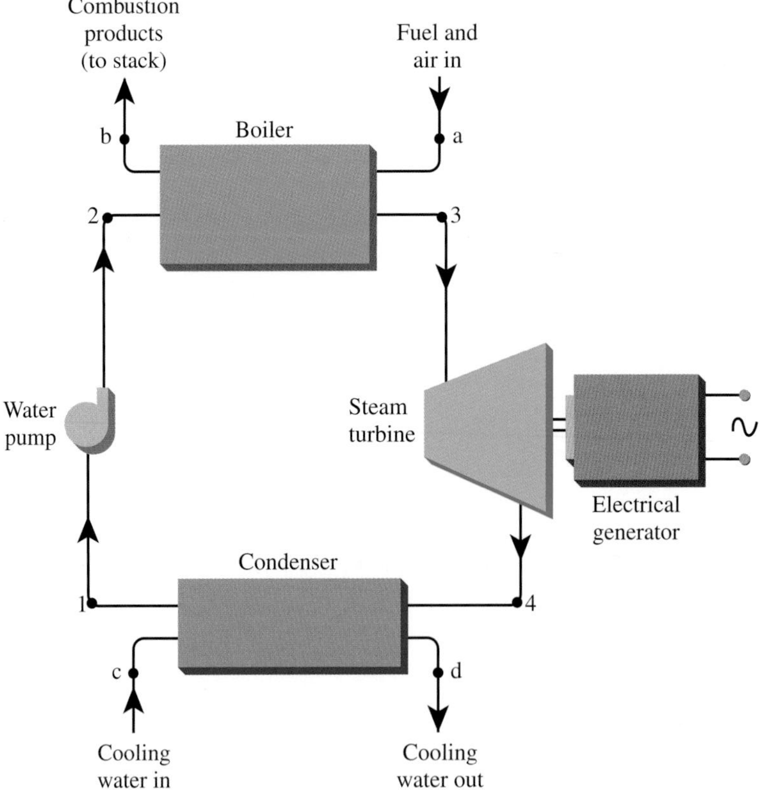

Solution

Known H_2O, T_3, P_3, $\dot{m}$, D_3, P_4, x_4, A_4

Find V_3, V_4

Sketch

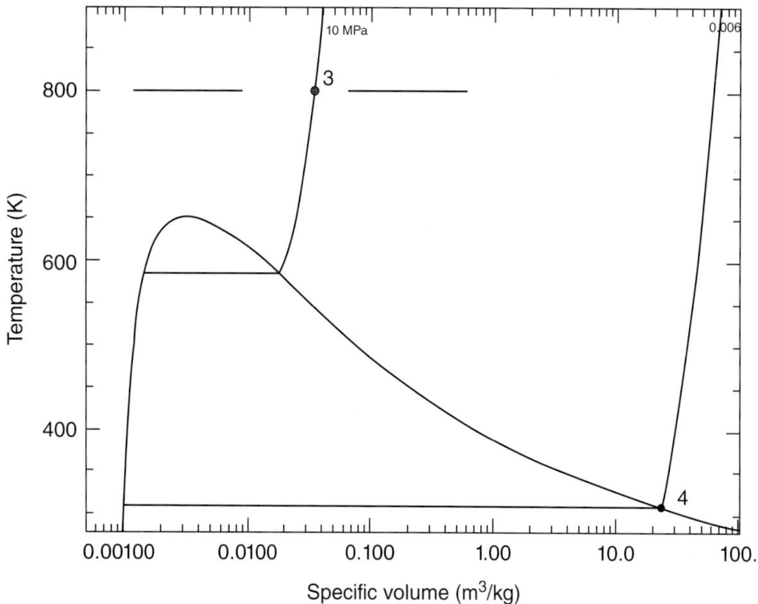

Modeling, Premises and Assumptions

i. No steam leaks from the turbine between the inlet and the outlet.
ii. The thermodynamic properties of the steam in the interconnecting duct are the same as at the turbine exit.
iii. Steady-state and steady flow conditions prevail.
iv. Properties are uniform across the inlet and outlet.

Analysis Because the inlet flow rate is known and we assume there are no leaks, we can write mass conservation (Eq. 3.14a) as

$$\dot{m}_{\text{in}} = \dot{m}_{\text{out}}$$

or

$$\dot{m}_3 = \dot{m}_4.$$

To find the velocities, we apply the definition of the mass flow rate (Eq. 3.11):

$$\dot{m} = \rho V_{\text{avg}} A_{\text{x-sec}}.$$

Thus,

$$V_3 = \frac{\dot{m}}{\rho_3 A_3} = \frac{\dot{m}}{\rho_3(\pi D_3^2/4)},$$

$$V_4 = \frac{\dot{m}}{\rho_4 A_4}.$$

We now only have to find the inlet and exit densities to find the average inlet and outlet velocities. Because the temperature of the steam entering the turbine (800 K) is above the corresponding (10 MPa) saturation temperature (584.15 K), the inlet steam is superheated. This state is denoted with a labeled dot on the pressure–specific-volume plot in the sketch. We can use Table B.3P or the NIST WebBook or software to find the state-3 density:

$$\rho_3 = 29.107 \text{ kg/m}^3.$$

We will use the Appendix B tables to find the exit density to (i) reinforce understanding and use of the tables, and (ii) to illustrate that the quality (x) relates mass-based properties, not volume-based properties. Jumping immediately to use the NIST software, we might miss this second important point. Because the quality ($x_4 = 0.95$) is given, we know that the exiting steam is a wet mixture. For a saturation pressure (P_4) of 6 kPa, we can find the values for the saturated-liquid and saturated-vapor specific volumes from Table B.2:

$$v_f = 0.0010065 \text{ m}^3/\text{kg},$$
$$v_g = 23.733 \text{ m}^3/\text{kg}.$$

We apply Eq. 2.37c to find the specific volume of the wet mixture, v_4:

$$v_4 = (1 - x_4)v_f + x v_g$$

Numerically evaluating yields

$$v_4 = (1 - 0.95)\ 0.0010065 \text{ m}^3/\text{kg} + 0.95(23.733 \text{ m}^3/\text{kg}) = 0.034356 \text{ m}^3/\text{kg}.$$

The exit density is thus

$$\rho_4 = 1/(0.034356 \text{ m}^3/\text{kg}) = 0.044352 \text{ kg/m}^3.$$

We can now find the exit velocities as indicated at the beginning of our analysis:

$$V_3 = \frac{\dot{m}}{\rho_3 A_3} = \frac{\dot{m}}{\rho_3(\pi D_3^2/4)} = \frac{11.6 \text{ kg/s}}{29.107 \text{ kg/m}^3\left[\pi(0.15 \text{ m})^2/4\right]} = 22.6 \text{ m/s}$$

and

$$V_4 = \frac{\dot{m}}{\rho_4 A_4} = \frac{11.6 \text{ kg/s}}{0.044352 \text{ kg/m}^3\,(2.0 \text{ m}^2)} = 131 \text{ m/s}.$$

Comments As alluded to in our solution, the expression $\beta_4 = (1-x)\beta_f + x\beta_g$ (Eq. 2.37a) applies only to mass-specific properties, i.e., β must be expressed per kilogram. Using the density in this equation does not provide the correct density for the wet mixture. The reader should verify this for this example. We also note that our assumption of no leakage is only an approximation, as in a real steam turbine some steam will leak past the turbine blade tips, and smaller amounts at the shaft.

Example 3.7 Steam Boiler Flow

A simple Rankine cycle is used to generate electricity as shown in the drawing in Example 3.6. Saturated steam from the steam generator at 8 MPa enters the superheater section of the boiler, where the steam is heated at constant pressure to 700 K. The superheater consists of a bundle of tubes with equal flow rates in each tube. The total mass flow rate through the superheater section is 10 kg/s. The average velocity through the individual superheater tubes is restricted to be everywhere less than 25 m/s. The inside diameter of the tubes is 38 mm. Determine the minimum number of tubes required to meet the maximum-velocity restriction.

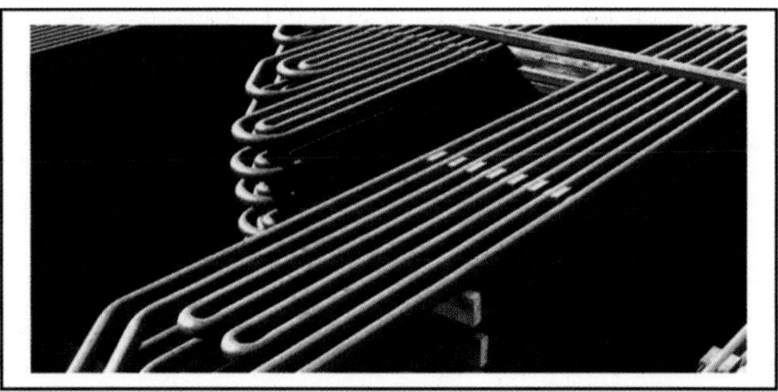

(Image courtesy of National Boiler Service, Inc.)

Solution

Known H_2O, $P_{in} = P_{sat}$ (= 8 MPa), T_{out}, D_{tube}, $\dot{m}_{tot}\, V_{max}$

Find Number of tubes

Sketch

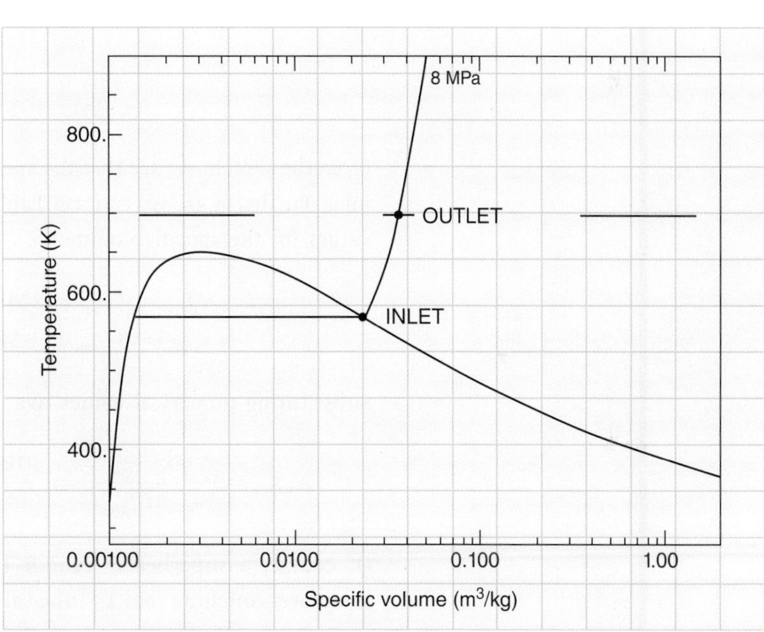

Modeling, Premises and Assumptions

i. Maximum velocity is average maximum velocity.
ii. No pressure drop from inlet to exit.
iii. Steady-state steady flow.
iv. Flow is divided equally among all the tubes.
v. Uniform properties over flow cross-section.

Analysis Overall mass conservation (Eq. 3.14a) for the superheater tells us that the total inlet and outlet mass flow rates are equal, i.e.,

$$\dot{m}_{in,tot} = \dot{m}_{out,tot} = \dot{m}_{tot}.$$

The total flow is the sum of the flows through each individual tube,

$$\dot{m}_{tot} = N_{tubes}\dot{m}_{tube},$$

where N_{tubes} is the number of tubes in the superheater. We apply the definition of the mass flow rate (Eq. 3.11) to relate the maximum velocity to the flow rate in an individual tube:

$$\dot{m}_{tube} = \rho V_{max} A_{tube} = \frac{V_{max} \pi D_{tube}^2/4}{v}.$$

Substituting this expression into the previous one and solving for the number of tubes yields

$$N_{tubes} = \frac{\dot{m}_{tot}}{\dot{m}_{tube}} = \frac{\dot{m}_{tot} v}{V_{max} \pi D_{tube}^2/4}.$$

To evaluate the above requires us to select the appropriate specific volume v associated with the flow, i.e., at the inlet or the exit. From the definition of the mass flow rate, we see that the maximum velocity must occur at the exit where the specific volume is the greatest:

$$V_{max} = \frac{\dot{m}_{tube} v}{\pi D_{tube}^2/4}.$$

From the sketch, we see that the specific volume at the outlet is much larger than at the inlet. Furthermore, we can use Tables B.2 and B.30, respectively, to obtain numerical values for the specific volumes:

$$v_{in} = v_g(P_{sat} = 8\,\text{MPa}) = 0.023526\,\text{m}^3/\text{kg},$$
$$v_{out} = v(P_{out} = 8\,\text{MPa}, T_{out} = 700\,\text{K}) = 0.036455\,\text{m}^3/\text{kg}.$$

Substituting numerical values, we obtain the number of tubes:

$$N_{tubes} = \frac{\dot{m}_{tot} v_{out}}{V_{max} \pi D_{tube}^2/4} = \frac{10\,\text{kg/s}\,(0.036455\,\text{m}^3/\text{kg})}{(10\,\text{m/s})\,\pi[(0.038)^2/4]\,\text{m}^2} = 12.85\,(\text{dimensionless}).$$

Of course, a superheater cannot be built with a fractional number of tubes; we, therefore, conclude that 13 tubes are required to meet the maximum velocity requirement. We verify this by calculating the maximum velocity:

$$V_{max,13} = \frac{\left(\frac{\dot{m}_{tot}\,\text{kg/s}}{13}\right) v_{out}\,\text{m}^3/\text{kg}}{\frac{\pi D_{tube}^2\,\text{m}^2}{4}} = 24.7\,\text{m/s}.$$

Comments We note the importance of recognizing that the maximum velocity occurs at the outlet, where the specific volume is a maximum. Had we chosen the inlet specific volume to determine the number of tubes (which would be 8.29, or 9 whole tubes), the outlet velocity would have been 35.7 m/s, which greatly exceeds the design requirement of 25 m/s. Note that being able to sketch the states on a P–v diagram allows us to make this choice before we evaluate any properties.

Example 3.8 Feedwater Pump Flow

A simple Rankine cycle is used to generate electricity at a manufacturing plant (the diagram at the start of Example 3.6 is reproduced below). Saturated steam enters the turbine at 5 MPa (state point 3 in the diagram) with a flow rate of 6.8 kg/s. The steam is condensed at 0.01 MPa and exits the condenser as a liquid (state point 1). This condensate flows to the pump. The H_2O exits the pump at 5 MPa and 335 K through an exit pipe (state point 2). Determine the minimum pump exit-pipe diameter such that the maximum average velocity in the pipe does not exceed 10 m/s.

Solution

Known H_2O, P, T, $\dot{m}$, V_{max}

Find D_{pipe}

Sketch

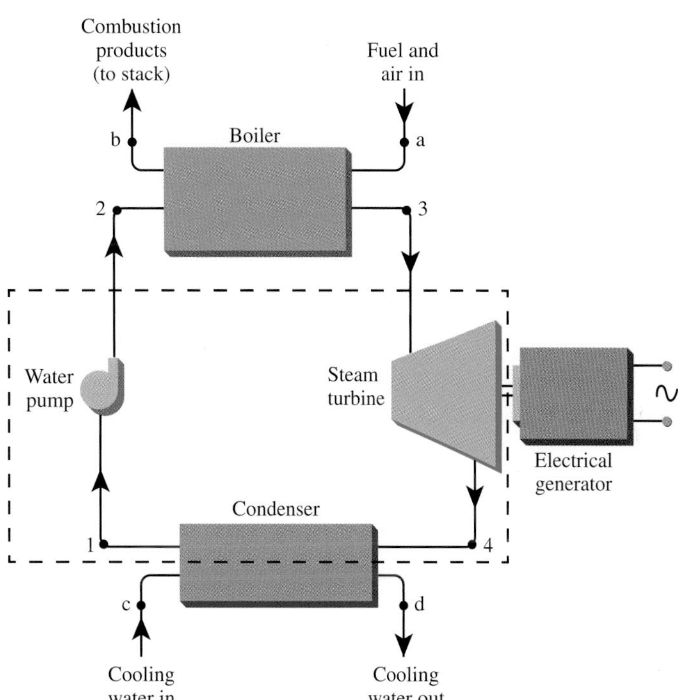

Modeling, Premises and Assumptions

i. Maximum velocity is average maximum velocity.
ii. Steady-state steady flow.
iii. Uniform properties over flow cross-section.

Analysis We choose an open system (control volume) that includes the turbine, the condenser, and the pump. The system boundary, the dashed line overlaying the diagram, is shown in the sketch. Mass enters the system at point 3 and exits at point 2. Because the flow is steady, the incoming and outgoing mass flow rates must be equal (Eq. 3.14a), i.e.,

$$\dot{m}_3 = \dot{m}_1 \left(= \dot{m}_{pipe}\right) = 6.8 \text{ kg/s}.$$

To find the minimum tube diameter, we apply the definition of mass flow rate in the pipe (Eq. 3.11):

$$\dot{m}_{\text{pipe}} = \rho V_{\max} A_{\text{pipe}} = \frac{V_{\max} \pi D_{\text{pipe}}^2 / 4}{v}.$$

Solving for the pipe diameter yields

$$D_{\text{pipe}} = \left(\frac{4 \dot{m}_{\text{pipe}} v}{\pi V_{\max}} \right)^{1/2}.$$

The problem is now reduced to determining the specific volume of the water. We will use the Appendix B tables for this task, and then confirm our result using the NIST software. We begin by determining the saturation pressure corresponding to the given temperature (T_1 = 335 K). From Table B.1, we obtain P_{sat} (T = 335 K) = 0.021719 MPa. The given pressure at point 1 (P_1 = 5 MPa) is much greater than the saturation pressure, so we conclude that state 1 is in the compressed liquid region. The following temperature–specific-volume plot illustrates this reasoning: we see that state 2 lies to the left of the saturated liquid line in the compressed liquid region. Also shown for reference on the sketch is the 0.02 MPa isobar, the pressure that prevails from state 4 to state 1.

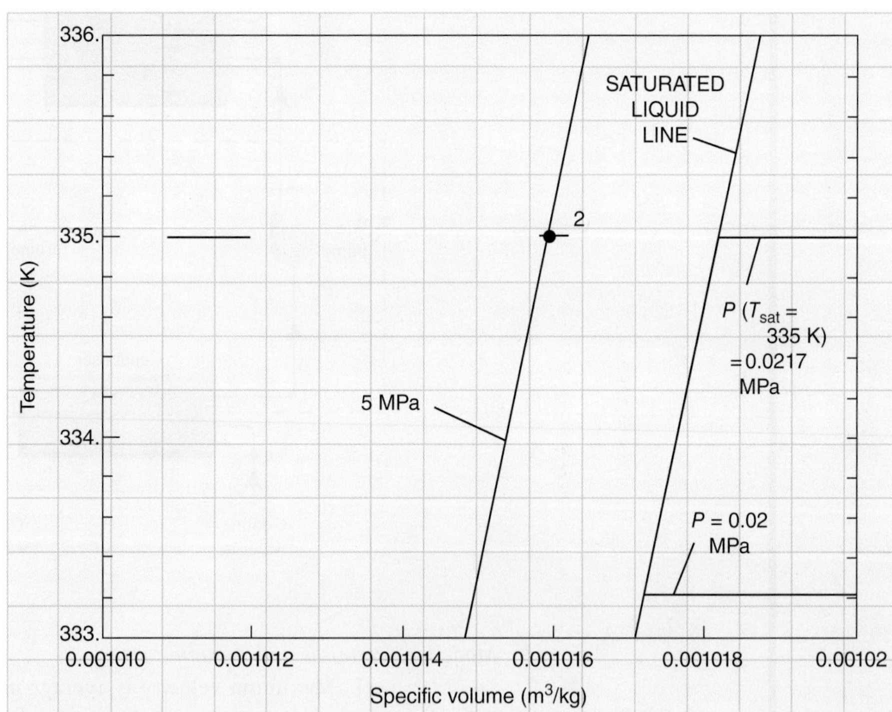

Having identified state 2 as lying in the compressed liquid region, we can find the specific volume from the compressed-liquid tables (Table B.4). Because the value at 335 K is not tabulated, we interpolate using the values at 320 K and 340 K as follows:

$$\begin{aligned} v_2\,(P_2 &= 5\,\text{MPa},\ T_2 = 335\,\text{K}) \\ &= 0.001009\,\text{m}^3/\text{kg} + \left(\frac{335\ \text{K} - 320\,\text{K}}{340\,\text{K} - 320\,\text{K}} \right) (0.001019 - 0.001009)\,\text{m}^3/\text{kg} \\ &= 0.0010165\,\text{m}^3/\text{kg}. \end{aligned}$$

Using the NIST software to evaluate the specific volume directly yields a value of 0.0010159 m^3/kg. This difference is negligible as the error is only 0.06%. The minimum required pipe diameter is thus

$$D_{\text{pipe}} = \left[\frac{4\,(6.8\,\text{kg/s})\,0.0010165\,\text{m}^3/\text{kg}}{\pi\,10\,\text{m/s}}\right]^{1/2} = 0.0297\,\text{m}.$$

In practice, a standard pipe size close to this value, without exceeding it, would be chosen.

Comments We note how our choice of the system boundary enabled us to determine easily the flow rate in the pipe. Note also how we employed our Chapter 2 knowledge to find the specific volume and to create a useful temperature–volume plot.

3.4c Unsteady Flows

In the preceding analyses time played no role. Our assumption of a steady state and steady flow eliminated time as a variable. Although many, many situations can be treated as steady to a good approximation, time plays a key role in a variety of important problems. For example, almost all engineering devices undergo a **transient**, that is, a time-dependent, start-up and/or shutdown. An interesting example illustrated in Fig. 3.8 is the ignition and start-up of the engines and solid rocket boosters for the Space Shuttle. The timing of events and the transient build-up of thrust are critical to a successful lift-off.

FIGURE 3.8 The ignition and start-up of the engines and solid rocket boosters for the Space Shuttle involved many unsteady flows (courtesy NASA).

Some of the most challenging engineering problems arise in the control of transients. By their very nature, the thermal-fluid processes associated with reciprocating internal combustion engines never achieve a steady state. Other, less complicated, unsteady flows involve the emptying and filling of tanks and pressure vessels. We now present open-system mass conservation statements that include time as an explicit independent variable.

For simplicity, we start with an open system that has a single inlet flow and a single exit flow, as illustrated in Fig. 3.9. What makes this different from the situation described in Fig. 3.7a is that now the inlet and exit flow rates are not equal so that the mass within the open system will be increasing with time, if $\dot{m}_{\text{in}}$ exceeds $\dot{m}_{\text{out}}$, or decreasing with time, if $\dot{m}_{\text{out}}$ exceeds $\dot{m}_{\text{in}}$. Our general statement of mass conservation (Eq. 3.13) applies without further simplifications:

$$\underbrace{\dot{m}_{\text{in}}}_{\text{Mass flow into the control volume}} - \underbrace{\dot{m}_{\text{out}}}_{\text{Mass flow out of the control volume}} = \underbrace{\frac{dM_{\text{cv}}}{dt}}_{\text{Rate of change of mass within the control volume}}. \quad \text{(3.15a)}$$

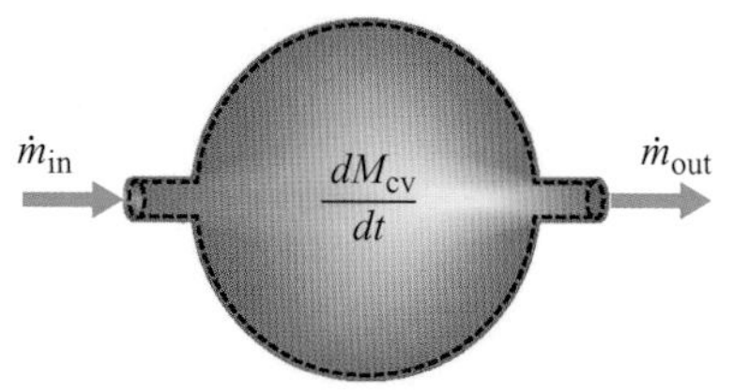

FIGURE 3.9 In an unsteady flow, the mass within the open system can vary with time. Note that the system boundary is not necessarily fixed and can move with time.

Because Eq. 3.11 expresses an instantaneous quantity, it can be used to evaluate the flow rates in Eq. 3.15a.

Equation 3.15a is easily extended to open systems with multiple inlets and outlets by writing

$$\sum_{j=1}^{N_{inlets}} \dot{m}_{in,j} - \sum_{k=1}^{M_{outlets}} \dot{m}_{out,k} = \frac{dM_{cv}}{dt}. \tag{3.15b}$$

Since dM_{cv}/dt is a new term in our analysis, it is useful to explore its physical meaning in greater detail. For simplicity, we assume that the density is uniform throughout the open system, that is, ρ takes on the same value irrespective of location. However, we allow the density ρ to be a function of time. With this modeling premise, the mass within the control volume at any instant is just $M_{cv} = \rho\mathcal{V}$. Applying the product rule to the differentiation of $\rho\mathcal{V}$ gives

$$\frac{dM_{cv}}{dt} = \rho\frac{d\mathcal{V}}{dt} + \mathcal{V}\frac{d\rho}{dt}. \tag{3.16}$$

From Eq. 3.16, we see that the mass within the open system can increase either by increasing the volume (i.e., $d\mathcal{V}/dt$ is positive) or increasing the density (i.e., $d\rho/dt$ is positive). Practical examples of either possibility are easy to visualize. Consider the filling of a bathtub (Figure 3.10a) where the open-system boundary includes only the water in the tub. In this case, the density of the water is constant but the open system boundary continually expands as long as the faucet remains open. Just the opposite occurs in the filling of an air-compressor storage tank (Figure 3.10c). In this case, the open system volume remains fixed, whereas the density of the air in the tank continually increases as air is pumped in. A common example in which both the volume and the density vary with time is the inflation of a rubber balloon (Figure 3.10b).

(a)

(b)

(c)

FIGURE 3.10 Examples of unsteady flows: **(a)** the filling of a bathtub, **(b)** the inflation of a toy balloon, and **(c)** the filling of a compressed air tank.

We illustrate these concepts now with an example.

Example 3.9 Compressed Air Tank

A steel, compressed-gas cylinder is filled with air at a fixed rate of 5 g/s for 30 s. The volume of the cylinder is 0.1 m^3. The initial pressure in the cylinder is 100 kPa, and the initial temperature is 24 °C. After the filling process is complete, the cylinder and its contents sit until they reach thermal equilibrium with the surroundings at the initial temperature. What is the final pressure in the cylinder?

Solution

Known: $\mathcal{V}$, P_1, T_1 $(= T_2)$, $\dot{m}_{\text{in}}$, Δt

Find: P_2

Sketch

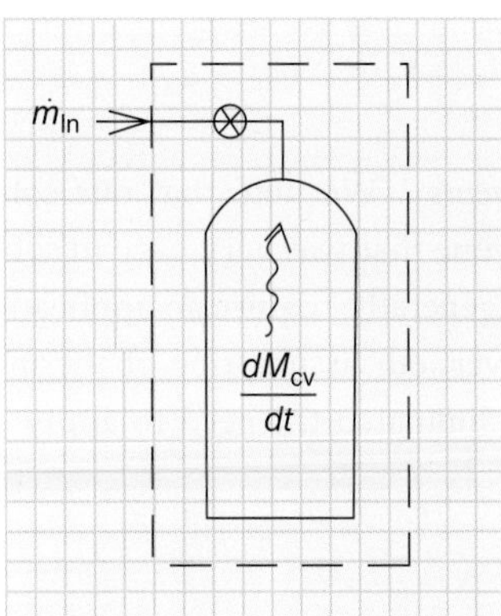

Modeling, Premises and Assumptions

i. Air behaves as an ideal gas.
ii. Rigid cylinder ($\mathcal{V}_2 = \mathcal{V}_1$) with no leaks.
iii. Heat transfer results in final temperature equaling initial temperature (given).

Analysis Our strategy here is to (i) use the ideal-gas equation of state to find the initial mass, (ii) apply conservation of mass to find the final mass, and then (iii) use the ideal-gas equation of state again to find the final pressure.

To find the initial mass, we apply Eq. 2.26c with the value of the gas constant for air from Appendix C:

$$P_1\mathcal{V}_1 = M_1RT_1$$

or

$$M_1 = \frac{P_1 \Psi_1}{RT_1} = \frac{100 \cdot 10^3 \text{ Pa} (0.1 \text{ m}^3)}{287 \frac{\text{J}}{\text{kg}\cdot\text{K}} (24 + 273)\text{K}} \left[\frac{\text{N/m}^2}{\text{Pa}}\right]\left[\frac{\text{J}}{\text{N}\cdot\text{m}}\right] = 0.117 \text{ kg}.$$

Mass conservation for this unsteady flow is expressed by Eq. 3.15a. Setting $\dot{m}_{\text{out}}$ to zero results in

$$\frac{dM_{\text{cv}}}{dt} = \dot{m}_{\text{in}}.$$

This simple ordinary differential equation can be separated and integrated as follows:

$$\int_{M_1}^{M_2} \frac{dM_{\text{cv}}}{dt} dt = \int_{t_1=0}^{t_2=\Delta t} \dot{m}_{in} dt$$

Evaluating with the limits as shown and rearranging, the final mass can be found as

$$\begin{aligned} M_2 &= M_1 + \dot{m}_{\text{in}} \Delta t \\ &= 0.117 \text{ kg} + 0.005 \text{ kg/s} (30 \text{ s}) = 0.267 \text{ kg}. \end{aligned}$$

Our last step is to find the final pressure. With the final temperature equal to the initial temperature, the ideal-gas equation of state yields

$$P_2 = \frac{M_2 R T_2 (= T_1)}{\Psi_2 (= \Psi_1)} = \frac{0.267 \text{ kg} (287 \text{ J/kg}\cdot\text{K})(24 + 273)\text{K}}{0.1 \text{ m}^3} \left[\frac{\text{N}\cdot\text{m}}{\text{J}}\right] = 228{,}000 \frac{\text{N}}{\text{m}^2} \text{ or Pa.}$$

Comments We note that our solution involved an ordinary differential equation, albeit one that was very easy to solve. Also, tank filling/emptying problems involving gases generally require an application of the conservation of energy for their solution; however, our assumption of thermal equilibrium with the surroundings for the final state eliminated the need to apply energy conservation.

Example 3.10 Water Tank Fill

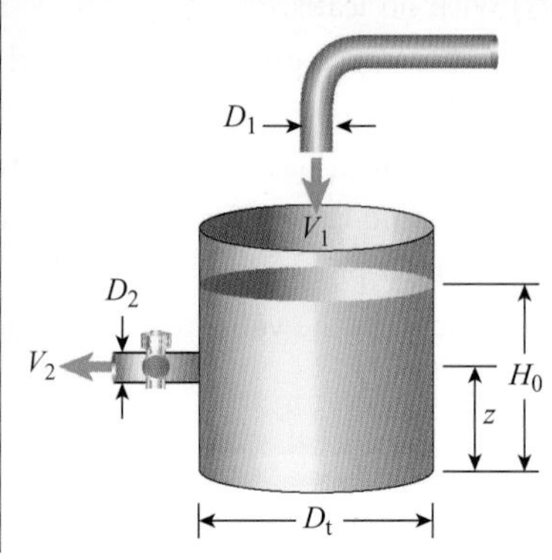

Consider the tank and water supply system as shown in the drawing. The diameter of the supply pipe D_1 is 20 mm, and the average incoming velocity V_1 is 0.595 m/s. A shut-off valve is located at $z = 0.1$ m, and the exit pipe diameter D_2 is 10 mm. The tank diameter D_t is 0.3 m. The water density is 997 kg/m^3.

A. Determine the time taken to fill the tank to a depth of 1 m ($\equiv H_0$) assuming the tank is initially empty and the shut-off valve is closed. Neglect the volume associated with the short pipe connecting the tank to the shut-off valve.

B. At the instant the water level reaches $H_0 = 1$ m, the shut-off valve is opened. The instantaneous average velocity of the outflow depends on the water depth above z, that is, $H(t) - z$, and is given by

$$V_2 = 0.85[g(H(t) - z)]^{1/2},$$

where g is the gravitational acceleration. Determine whether the tank continues to fill or begins to empty immediately after the valve is opened.

C. Determine the steady-state value of the water depth.

Solution

Known D_1, D_2, D_t, z, H_0, V_1, expression for V_2

Find t_0, H_{ss}, t_{ss}

Sketch See the sketches that follow for each part of the problem.

Modeling, Premises and Assumptions

i. Incompressible flow

ii. Outlet velocity instantaneously adjusts to changes in H

Analysis (Part A) For this part, we select an expanding control volume that contains all the water in the tank, as shown in the sketch.

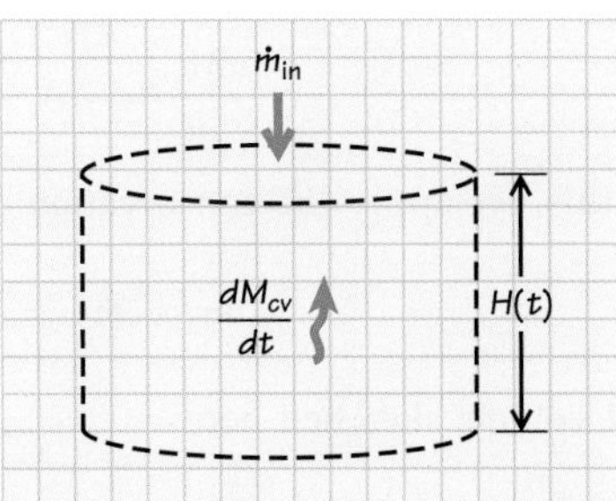

Conservation of mass for the unsteady filling process is expressed by Eq. 3.15a, where $\dot{m}_{out} = 0$:

$$\dot{m}_{in} = \dot{m}_1 = \frac{dM_{cv}}{dt}.$$

The inlet mass flow rate is expressed as (Eq. 3.11)

$$\dot{m}_1 = \rho V_1 A_1,$$

and the instantaneous mass within the control volume is simply the product of the density and the instantaneous volume of water within the tank, that is,

$$M_{cv} = \rho \mathcal{V}(t) = \rho A_t H(t),$$

where $A_t \left(= \pi D_t^2/4\right)$ is the cross-sectional area of the tank. Substituting these expressions for $\dot{m}_1$ and M_{cv} into Eq. 3.15a yields

$$\rho A_t \frac{dH(t)}{dt} = \rho V_1 A_1.$$

This first-order ordinary differential equation is easily integrated from the initial condition $H(t = 0) = 0$ to $H(t_0) = H_0$ as follows:

$$\int_0^{H_0} dH(t) = \int_0^{t_0} V_1 \frac{A_1}{A_t}\,dt$$

$$H_0 = V_1 \frac{A_1}{A_t} t_0.$$

Solving for the unknown t_0 yields

$$t_0 = \frac{H_0 A_t}{V_1 A_1}.$$

Since the ratio A_t/A_1 is the ratio D_t^2/D_1^2, we evaluate this as

$$\begin{aligned} t_0 &= \frac{H_0 D_t^2}{V_1 D_1^2} \\ &= \frac{1.0\,\text{m}(0.3\,\text{m})^2}{0.595\,\text{m/s}\,(0.020\,\text{m})^2} = 378\,\text{s} \;\text{ or }\; 6.30\text{ min}. \end{aligned}$$

Analysis (Part B) To determine whether the water level continues to increase or begins to fall when the valve is opened, we need to know whether the mass in the tank is increasing or decreasing since

$$\frac{dM_{cv}}{dt} = \rho A_t \frac{dH(t_0)}{dt}.$$

To determine this, we apply the conservation of mass principle,

$$\dot{m}_1 - \dot{m}_2 = \frac{dM_{cv}}{dt},$$

for the control volume sketched here:

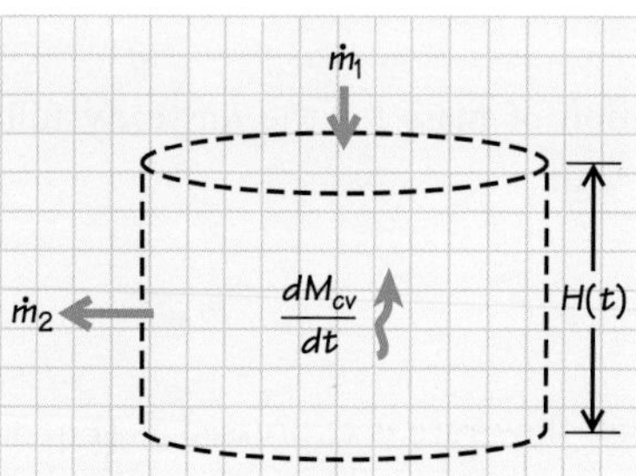

The instantaneous outlet mass flow rate $\dot{m}_2$ is expressed as

$$\begin{aligned} \dot{m}_2 &= \rho V_2 A_2 \\ &= \rho\left(0.85[g(H(t_0) - z)]^{1/2}\right)\frac{\pi D_2^2}{4}. \end{aligned}$$

We evaluate $\dot{m}_2(t_0)$ and $\dot{m}_1(t_0)$ as follows:

$$\dot{m}_2 = 997\text{ kg/m}^3\left[0.85\left(9.81\,\text{m/s}^2(1.0 - 0.1)\,\text{m}\right)^{1/2}\right]\frac{\pi(0.01\,\text{m})^2}{4} = 0.198\,\text{kg/s}$$

and

$$\begin{aligned} \dot{m}_1 &= \rho V_1 \pi D_1^2/4 \\ &= 997\text{ kg/m}^3(0.595\text{ m/s})\pi(0.020\text{ m})^2/4 \\ &= 0.186\,\text{kg/s}. \end{aligned}$$

Thus,

$$\frac{dM_{cv}}{dt} = 0.186 \text{ kg/s} - 0.198 \text{ kg/s}$$
$$= -0.012 \text{ kg/s},$$

where the negative sign indicates that the water level must start to fall when the valve is opened.

Analysis (Part C) For a steady state, dM_{cv}/dt is zero. The appropriate sketch and mathematical expression for mass conservation are as follows:

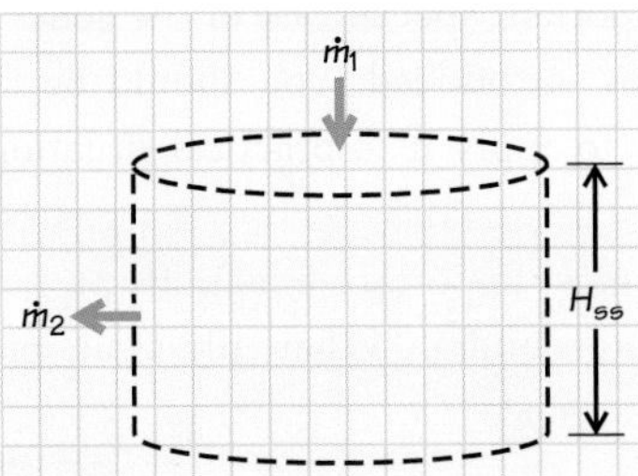

and

$$\dot{m}_1 = \dot{m}_2.$$

The inlet mass flow rate is as previously calculated (0.186 kg/s), whereas $\dot{m}_2$ is expressed in terms of the steady-state water level H_{ss}. Thus,

$$\dot{m}_1 = \dot{m}_2 = \rho\left(0.85[g(H_{ss} - z)]^{1/2}\right)\frac{\pi D_2^2}{4}.$$

Solving for H_{ss} yields

$$H_{ss} = \frac{1}{g}\left(\frac{4\dot{m}_1}{0.85\,\rho\pi D_2^2}\right)^2 + z$$
$$= \frac{1}{9.81}\left(\frac{4(0.186 \text{ kg/s})}{0.85 \text{ m/s}^2(997 \text{ kg/m}^3)\pi(0.010 \text{ m})^2}\right)^2 + 0.1 \text{ m}$$
$$= 0.90 \text{ m}.$$

Comments This example illustrates the use of both steady and unsteady expressions of mass conservation for an open system (control volume). Note that, in all parts of the problem, we chose a boundary that contained only the water in the tank; thus, our system expanded or contracted with time. An alternative, but more complex, choice would have been to choose a larger fixed volume containing some air above the water. Our original choice is clearly superior because of its simplicity. We also note that formulation of the unsteady problem generated an ordinary differential equation, a common result for this class of problems.

Self-Test 3.5

Water at a constant flow rate of 3 kg/s is entering a partially full bathtub measuring 6 ft by 2 ft. The drain plug is removed and at one particular instant the water level in the tub is decreasing at 0.5 in/min. Determine the exit mass flow rate at this instant.

(Answer: 3.236 kg/s).

SUMMARY

This chapter focused on the concept of mass conservation – one of the foundational principles of engineering analysis. For closed systems, the statement of this principle is straightforward. For open systems, the mathematical statement of this principle is more interesting, and requires introducing volume and mass flow rates. To evaluate flow rates simply, we developed the concept of an average velocity. Although we presented several mass-conservation expressions, all are based on the general concept that the rate at which mass accumulates within an open system (control volume) equals the difference between all incoming mass flow rates and all outgoing mass flow rates. As an important special case of the general principle, the concept of steady-state steady flow was introduced. You should be familiar with this idea and understand when and how to apply it to practical situations.

KEY EQUATIONS

Review the most important equations presented in this chapter (i.e., those with yellow backgrounds). What physical principles do they express? What restrictions apply?

CHAPTER 3 KEY CONCEPTS AND DEFINITIONS CHECKLIST

Answer the Questions and solve the Problems following the arrows to demonstrate mastery of the listed concepts and definitions.

3.2 Mass Conservation for a Closed System ➔ Question 3.1, Problems 3.2, 3.11

3.3 Flow Rates and Average Velocity ➔ Problems 3.18, 3.19

3.4 Mass Conservation for an Open System

3.4b Steady-State Steady Flow ➔ Questions 3.4A, B, Problems 3.30, 3.32, 3.65

3.4c Unsteady Flows ➔ Questions 3.4C, D, Problems 3.67, 3.68

REFERENCES

1. Rouse, H., and Ince, S., *History of Hydraulics*, Iowa Institute of Hydraulic Research, State University of Iowa, 1957.
2. Lavoisier, A. L., *Elements of Chemistry, in a New Systematic Order, Containing All of the Modern Discoveries*, translated by R. Kerr, Dover, New York, 1965.
3. Turns, S. R., *Thermal-Fluid Sciences: An Integrated Approach*, Cambridge University Press, New York, 2006.
4. White, F. M., *Fluid Mechanics*, 5th ed., McGraw Hill, New York, 2003.
5. Fox, R. W., McDonald, A. T., and Pritchand, P. J., *Introduction to Fluid Mechanics*, 6th ed., Wiley, New York, 2004.
6. Heywood, J. B., *Internal Combustion Engine Fundamentals*, McGraw-Hill, New York, New York, 2004.

Some end-of-chapter problems were adapted with permission from the following:

Look, D. C, Jr., and Sauer, H. J., Jr., *Engineering Thermodynamics*, PWS, Boston, 1986.

Myers, G. E., *Engineering Thermodynamics*, Prentice Hall, Englewood Cliffs, NJ, 1989.

Pnueli, D., and Gutfinger, C., *Fluid Mechanics*, Cambridge University Press, Cambridge, England, 1992.

QUESTIONS

3.1 Express the conservation of mass for a closed system for the following conditions:

A. For a change in state with a uniform system density

B. At an instant with a uniform system density

3.2 Although they share the same units (kg/s), the mass flow rate, $\dot{m}$, and the unsteady term, dM_{cv}/dt, in a general statement of mass conservation are quite different. Explain (discuss) the physical difference between these two quantities.

3.3 Consider the quotation from Leonardo da Vinci given in the historical context section. Rewrite this passage so that it both sounds modern and is precise in an engineering sense.

3.4 Write down from memory (or familiarity) the following expressions of mass conservation applied to an open system (control volume):

A. Steady flow with one inlet and one outlet

B. Steady flow with two inlets and three outlets

C. Unsteady flow with a single inlet and no outlet

D. Unsteady flow with a single inlet and a single outlet

3.5 For flow through a tube, what is the physical interpretation of the average velocity?

3.6 How do typical velocities for steam flowing through pipes compare to those for water? Why are they different?

Chapter 3 Problem Subject Areas

3.1–3.15	Mass conservation: closed systems
3.16–3.27	Volume and mass flow rates
3.28–3.46	Mass conservation for open systems with steady flow – single stream
3.47–3.65	Mass conservation for open systems with steady flow – multiple streams
3.66–3.75	Mass conservation for open systems with unsteady flow
3.76–3.77	Computer problems
3.78–3.86	FE problems

PROBLEMS

3.1–3.15 Mass conservation: Closed systems

3.1 A 18-cm-diameter spherical balloon contains air at 1.2 atm and 25 °C. Determine the mass of the air in the balloon.

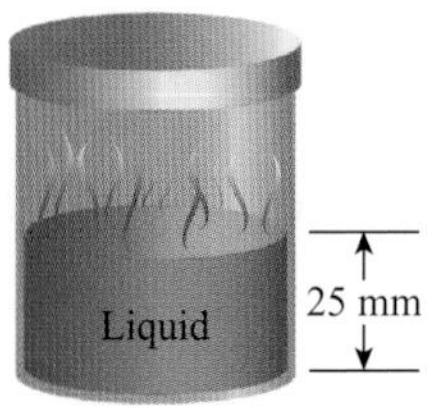

3.2 A vertical, closed, cylindrical tank 0.4 m high with 0.3 m diameter contains a saturated mixture of water and steam at 400 K. The depth of the liquid is 25 mm. Determine the mass and quality of the mixture contained in the tank.

3.3 During the expansion process at a crank angle $\theta = 400^\circ$, the temperature and pressure in the cylinder of a spark-ignition engine are 2400 K and 1.8 MPa, respectively. (Note: Top center is 360°.) The geometric properties of the engine are the following:
Bore, $B = 90$ mm,
Stroke, $S = 80$ mm,
Compression ratio, $CR = 8.5$, and
Connecting rod length to crank radius ratio, $\ell/a = 3.5$.

A. Determine the mass of the combustion products in the cylinder assuming the products' molecular weight is 28.5 kg/kmol.
B. Determine the time rate of change of the products' density, $d\rho/dt$, when the engine is operating at 1600 rpm.
HINT: See Appendix 3A for $\mathcal{V}(\theta)$ and $d\mathcal{V}(\theta)/dt$ relationships.

3.4 Determine the mass (lb_m) of air in a classroom that is 30 ft wide, 50 ft long, and 15 ft high. The pressure is 14.7 psia and the temperature is 70 F. Convert units to SI before performing your calculation.

(Credit: adventtr / iStock / Getty Images Plus.)

3.5 A gas is confined in a 0.057-m^3 rigid tank at 57 °C and 690 kPa. The mass of the gas is 0.23 kg.
A. Determine the apparent molecular weight (mass) of this gas in kg/kmol.
B. After a heating process, the gas temperature is 445 K. Determine the pressure (kPa).

3.6 Initially, a rigid 0.00246-m^3 tank is filled with saturated liquid water at 100 kPa. A cover is placed on the tank, and the water then cools to 294 K. The diameter of the opening underneath the cover is 63.5 mm. The atmospheric pressure outside the tank is 100 kPa.

A. Determine the mass (kg) of water initially in the tank.

B. Determine the final volume (m^3) of liquid in the tank.

C. Determine the force (N) that must be applied to lift the cover from the tank at the final state. Neglect the mass of the cover.

3.7 With the valves open, the radiator of a steam heating system has a volume of 0.06 m^3 and contains dry saturated vapor ($x = 1.0$) at 0.14 MPa. After the valves are closed on the radiator, the pressure drops to 0.07 MPa as a result of heat transfer to the room.

A. Determine the total mass (kg) contained in the radiator.

B. Determine the final mass (kg) and volume (m^3) of vapor.

C. Determine the final mass (kg) and volume (m^3) of liquid.

(Credit: northlightimages / iStock / Getty Images Plus.)

3.8 A rigid 0.0283-m^3 tank initially containing water at 380 K and 0.070 MPa cools until it reaches 294 K.

A. Determine the final pressure (MPa) in the tank.

B. Determine the final liquid mass (kg) and volume (m^3).

C. Determine the final vapor mass (kg) and volume (m^3).

3.9 A rigid tank initially contains a mixture of liquid water and water vapor at a pressure of 100 kPa. Determine the proportions by volume of liquid and vapor necessary to make the mixture pass through the critical point when heated.

3.10 An uninsulated rigid 0.8-m^3 tank initially contains steam at 860 K and 0.9 MPa. The tank then cools until equilibrium is established at a final temperature of 40 °C.

A. Determine the pressure (MPa) at the final state.

B. Determine the mass (kg) of liquid water at the final state.

3.11 Initially, a rigid 6-m^3 tank contains helium (assumed to be an ideal gas) at 300 K and 0.10 MPa. The tank is then heated until the temperature reaches 400 K. A pressure-relief valve at the top of the tank opens when the pressure reaches

0.15 MPa, and helium flows out, preventing the pressure from ever exceeding 0.15 MPa. Determine the final mass (kg) of helium in the tank.

3.12 Air is contained in a vertical cylinder fitted with a frictionless piston and a set of stops, as shown in the sketch. The cross-sectional area of the piston is 0.05 m^2. The air is initially at 400 °C and 0.2 MPa. The air is then cooled by heat flow to the surroundings.

A. Determine the temperature (°C) of the air inside when the piston reaches the stops.

B. Determine the final pressure (MPa) if the cooling is continued until the temperature reaches 20 °C.

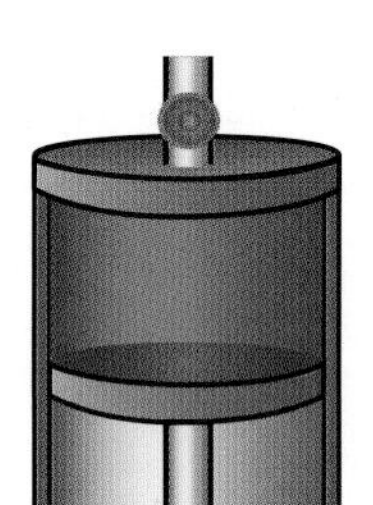

3.13 A piston–cylinder device contains air at 20 °C and 0.14 MPa, as shown in the sketch. Initially the volume is 0.3 m^3. The piston is then slowly pushed upward until the volume reaches 0.06 m^3. Heat transfer with the surroundings maintains the temperature at 20 °C during this process. The pressure-relief valve at the top of the cylinder opens when the pressure reaches 0.6 MPa to allow mass to escape, thus preventing the pressure from ever exceeding 0.6 MPa.

A. Determine the final pressure (MPa), temperature (°C), and specific volume (m^3/kg) of the air in the cylinder.

B. Determine the amount of mass (kg) that flows through the valve.

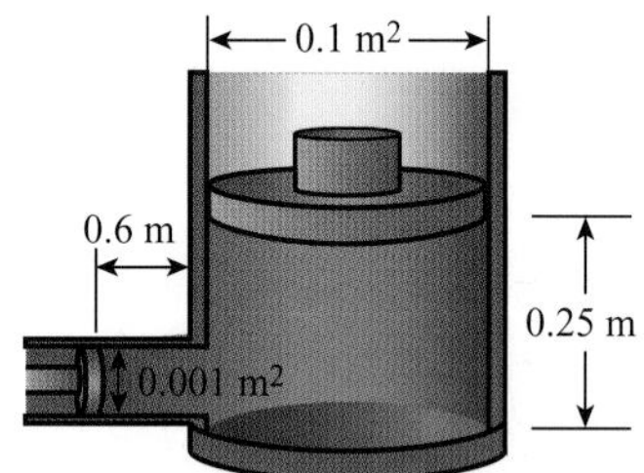

3.14 Superheated steam initially at 420 K and 0.3 MPa (state 1) is held in the device shown in the sketch. At state 1, the plunger is 0.6 m away from the cylinder, and the piston is 0.25 m above the bottom of the cylinder. The plunger moves toward the cylinder until it just reaches the cylinder (i.e., the plunger moves 0.6 m). At this condition (state 2), the steam temperature is 480 K. Determine the distance (m) and direction in which the piston moves.

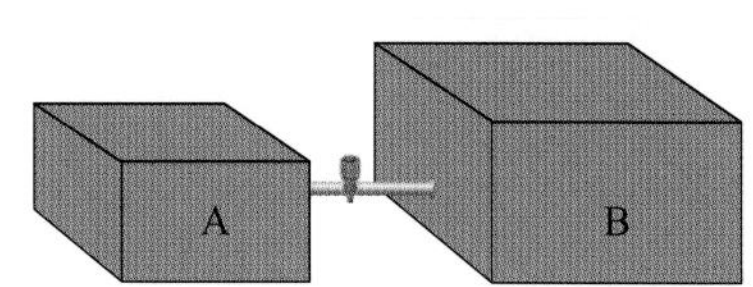

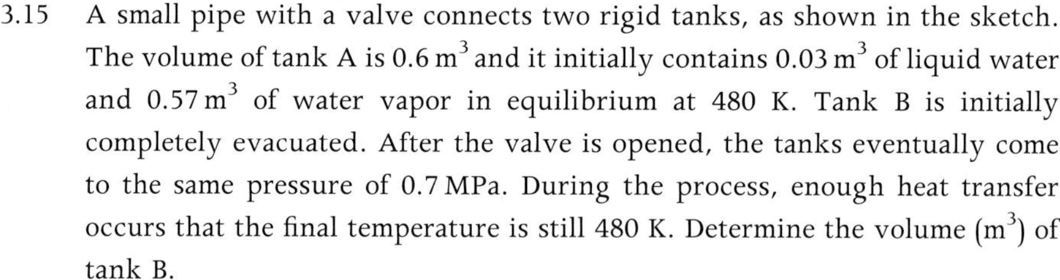

3.15 A small pipe with a valve connects two rigid tanks, as shown in the sketch. The volume of tank A is 0.6 m^3 and it initially contains 0.03 m^3 of liquid water and 0.57 m^3 of water vapor in equilibrium at 480 K. Tank B is initially completely evacuated. After the valve is opened, the tanks eventually come to the same pressure of 0.7 MPa. During the process, enough heat transfer occurs that the final temperature is still 480 K. Determine the volume (m^3) of tank B.

3.16–3.27 Volume and mass flow rates

3.16 Water with a density of 990 kg/m^3 is discharged by a pump at a rate of 3×10^5 cm^3/min from a pipe. Determine the mass flow rate in units of kg/min and lb_m/hr.

3.17 Water exits a nozzle into the atmosphere at 0.1 MPa. The flow has an essentially uniform velocity of 0.7 m/s. The nozzle outlet diameter is 15 mm, and the temperature of the water is 305 K. Determine the volume flow rate and the mass flow rate of the water. Discuss your choice of the value for the density (or specific volume). Give units.

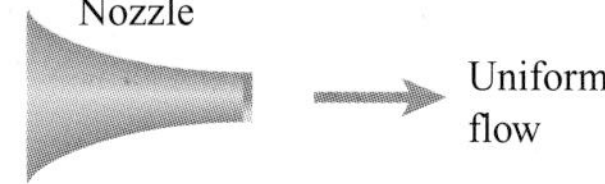

3.18 A feedwater pump for a 30-MW steam power plant pumps water, which enters as a saturated liquid at 0.78 MPa with a flow rate of 40.0 kg/s. Determine the minimum inlet pipe diameter required if the maximum allowed average velocity in the inlet pipe is 4.6 m/s. Use the NIST WebBook or software to evaluate properties.

(Credit: Flowserve Corporation.)

3.19 Natural gas enters a 150-km-long pipeline (0.30-m inner diameter) at 6.5 MPa (gage) and exits at 3.0 MPa (gage). The natural gas flow rate is 24.06 kg/s. The temperature of the natural gas is 17 °C (it is an isothermal process), and the atmospheric pressure is 100 kPa. Assuming the natural gas to be methane (CH_4), determine the maximum average velocity in the pipeline. Use the NIST WebBook or software to evaluate properties. How would your result differ if you were to assume ideal-gas behavior? Discuss.

(Credit: Prykhodov / iStock / Getty Images Plus.)

3.20 Liquid water at 380 K and 5 MPa flows at a rate of 1 gal/min in a nominal 1-in, schedule-40 pipe (of 1.049-in inside diameter). Determine the average velocity (m/s) of the water.

3.21 Steam at 580 K and 0.70 MPa flows through a 3-in-diameter pipe at 15 m/s. Determine the mass flow rate in kg/hr.

3.22 Air at 540 °C and 140 KPa flows through a 2-ft-inside-diameter pipe at a mass flow rate of 1.05 kg/s. Determine the average velocity (m/s) of the air. Assume ideal-gas behavior.

3.23 Air at 100 °C and 0.2 MPa flows at an average velocity of 30 m/s through a pipe whose cross-sectional flow area is 0.2 m^2. Determine the mass flow rate of the air in kg/s.

3.24 Compressed air is introduced into the ballast tank of a submarine and drives the water out of the tank at the rate of 2 m^3/s when the submarine is at a depth of 10 m. The pressure at this depth is 198.1 kPa. What is the volume flow rate of the ejected water when the same mass flow of compressed air is introduced at a depth of 100 m, where the pressure is 1081 k Pa? Justify your use of the assumption of ideal-gas behavior. **HINT**: See Table E.1.

3.25 Moisture in a steam pipeline can result in corrosion, scaling, and other problems; hence, it is desirable to remove suspended water droplets from the flow. One way to accomplish this is to use a cyclonic separator. This separator removes the moisture from the flow using the centrifugal action of a swirling flow to cause the droplets to impact the walls of the separator. The liquid then drains from the bottom of the separator, and dry steam exits at the top. Wet steam at 410 K with a quality of 0.98 enters a cyclonic separator with a flow rate of 0.50 kg/s. The inlet diameter of the separator is 100 mm. Determine the velocity of the steam entering the separator.

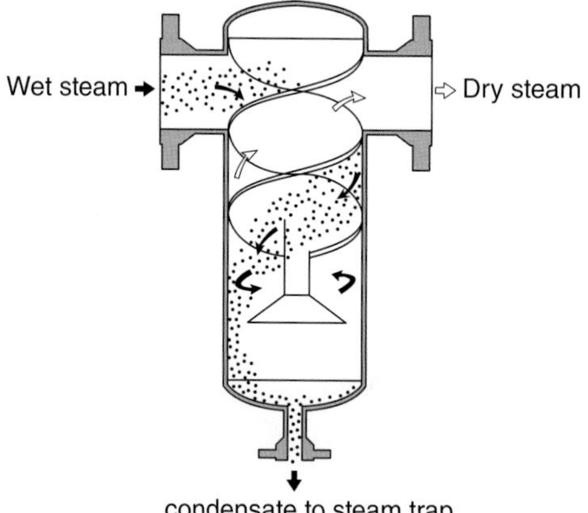

(Drawing is copyright, remains the intellectual property of Spirax Sarco, and has been used with their full permission.)

3.26 Wet steam at 0.40 MPa enters a cyclonic separator (see Problem 3.25) with a quality of 0.96 at a velocity of 30 m/s. The diameter of the separator inlet is 150 mm. Determine the volumetric and mass flow rates of the wet steam.

3.27 Wet steam at 380 K enters a cyclonic separator (see Problem 3.25) at 0.059 kg/s. The inlet diameter is 62.5 mm. What is the maximum allowable quality such that the inlet velocity does not exceed 25 m/s?

3.28–3.46 Mass conservation for open systems with steady flow – single stream

3.28 A steam boiler is fed with water at the rate of 1 kg/s. Steam exits the boiler at 100 kPa and 380 K. The average velocity of the steam in the pipe is 10 m/s. What is the pipe diameter?

3.29 Water enters a garden-hose nozzle with an average velocity of 1.8 m/s. The diameter at the nozzle entrance is 15 mm. The open nozzle is such that the

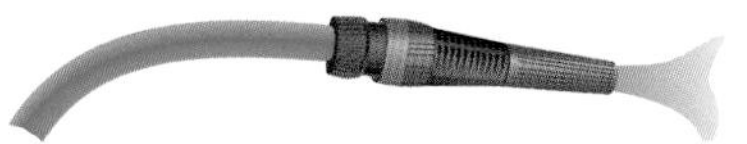

effective flow area at the exit is an annulus with outer and inner diameters of 6 mm and 5 mm, respectively. What is the average water velocity at the exit annulus? Comment on any assumptions you made regarding the water density.

3.30 Air enters a constant-diameter tube with an average velocity of 2 m/s at a temperature of 400 K and a pressure of 0.12 MPa. The air exits the tube at a temperature of 350 K and a pressure of 0.10 MPa. Determine the average velocity of the air at the exit.

3.31 Steam enters a 60-mm-inner-diameter tube with an average velocity of 4 m/s. The specific volume of the entering steam is 0.127 m^3/kg. The steam exits the tube with a density of 9.94 kg/m^3. Determine the average velocity of the steam at the tube exit, assuming steady flow.

3.32 A saturated liquid–vapor H_2O mixture enters a 25-mm-inner-diameter tube at a temperature of 490 K with a quality of 0.95. The mixture is heated as it travels through the tube and exits at 620 K and 2 MPa. The average velocity of the fluid exiting the tube is 17 m/s. Show the entering and exit states on a T–v diagram. Before you perform any numerical calculations, consider whether you expect the velocity at the entrance to be greater than, less than, or the same as the exit velocity. Determine the average velocity of the liquid–vapor mixture at the tube entrance.

3.33 Steam enters a 100-mm-inner-diameter pipe at 760 K and 4.0 MPa. The steam cools as it travels through the pipe and exits at 4 MPa with a quality of 0.93. The average velocity of the fluid exiting the pipe is 25 m/s. Show the entering and exit states on a T–v diagram. Before you perform any numerical calculations, consider whether you expect the velocity at the entrance to be greater than, less than, or the same as the exit velocity. Determine the average velocity of the steam at the pipe entrance.

3.34 A saturated liquid–vapor H_2O mixture enters a 30-mm-inner-diameter tube at a temperature of 425 K with a quality of 0.97. The mixture is heated as it travels through the tube and exits at 520 K and 0.5 MPa. The average velocity of the fluid exiting the tube is 20 m/s. Show the entering and exit states on a T–v diagram. Before you perform any numerical calculations, consider whether you expect the velocity at the entrance to be greater than, less than, or the same as the exit velocity. Determine the average velocity of the liquid–vapor mixture at the tube entrance.

3.35 A saturated liquid–vapor H_2O mixture enters a 25-mm-inner-diameter tube at a temperature of 490 K with a quality of 0.95. The mixture is heated at constant pressure as it travels through the tube and exits as a saturated vapor. The average velocity of the mixture entering the tube is 10 m/s. Show the entering and exit states on a T–v diagram. Determine the average velocity of the steam at the tube exit.

3.36 A steady-flow boiler is essentially a long, heated, variable-area duct with liquid water coming in at one end and steam coming out at the other end. Water enters at 420 K and 20 MPa and leaves at 880 K and 20 MPa. Determine the ratio of the outlet area to the inlet area if the outlet velocity is equal to the inlet velocity.

3.37 The average air velocity in the intake duct of an air conditioner is 3 m/s. The intake air temperature is 35 °C. The air comes out of the air conditioner at 20 °C and flows through a duct, also at 3 m/s. The ambient pressure is 100 kPa, which can be assumed to be the condition at both the inlet and exit. What is the ratio of the cross-sectional areas of the ducts assuming no condensation of moisture? Discuss how moisture condensation affects the velocity in the outlet duct.

3.38 Air with a density of 1.20 kg/m^3 enters a steady-flow control volume through a 0.3-m-diameter duct with an average velocity of 3 m/s. The air leaves with a specific volume of 0.311 m^3/kg through a 0.1-m-diameter duct. Determine (a) the mass flow rate (kg/hr) and (b) the average outlet velocity (m/s).

3.39 A gas enters a steady-flow system through a 5-cm-diameter tapering duct with an average velocity of 3.5 m/s and a density of 1.20 kg/m^3. It leaves the duct with a specific volume of 0.31 m^3/kg through a 1.6-cm-diameter constriction. Determine (a) the mass flow rate (kg/hr) and (b) the average outlet velocity (m/s).

3.40 A hot-water heater has 2.0 gal/min entering at 50 F and 40 psig. The water leaves the heater at 160 F and 39 psig. Determine the flow rate of the exiting water in gal/min. The local barometric pressure is 14.7 psia. Use the NIST software to obtain the water properties in the given units.

3.41 Air is heated as it flows through a 75-mm-diameter pipe at a rate of 0.045 kg/s. At station 1, the air has an average velocity of 5.5 m/s and a temperature of 311 K. Downstream, at station 2, the air has a temperature of 389 K and a pressure of 130 kPa. Determine (a) the average velocity (m/s) at section 2 and (b) the pressure (kPa) at station 1.

3.42 Air is heated as it flows through a constant-diameter tube in steady flow. The air enters the tube at 345 kPa and 300 K and has an average velocity of 3.0 m/s at the entrance. The air leaves at 310 kPa and 397 K.

A. Determine the average velocity of the air (m/s) at the exit.

B. If 10 kg/min of air is to be heated, what diameter (mm) tube is required?

3.43 A variable-diameter duct has a flow cross-sectional area of 0.006 m^2 at state 1 and 0.002 m^2 at state 2. Steam at 680 K and 6.0 MPa enters at state 1 at a velocity of 30 m/s, flows steadily through the duct, and leaves at 580 K and 3.0 MPa at state 2. (a) Determine the velocity (m/s) at state 2, and (b) use the NIST software to plot the 6.0 and 3.0 MPa isobars on a T–v plot and indicate the inlet and outlet states using bold dots. Add the dots by hand.

3.44 A reversed Brayton cycle can be used to provide refrigeration. The components of an air standard reversed Brayton cycle are shown in the diagram below. This cycle is used to cool the passenger cabin of a commercial aircraft using air as the working fluid. The cycle comprises the following steady-flow processes:

1–2 (compressor): Ideal compression from $T_1 = 289$ K and $P_1 = 68.9$ kPa to $P_2 = 550$ kPa. The process follows the path $T^{1.4}P^{-0.4}$ = constant. Substitution of the ideal-gas law shows that this process also follows the path $Pv^{1.4}$ = constant.

2–3 (heat exchanger): Constant-pressure heat rejection with $T_3 = 311$ K.

3–4 (turbine): Ideal expansion from T_3 and P_3 to P_4 $(=P_1) = 68.9$ kPa. Like the compression process, this process also follows the paths $T^{1.4}P^{-0.4}$ = constant and $Pv^{1.4}$ = constant.

4–1 (heat exchanger): Constant-pressure heat transfer from the passenger compartment to the working fluid raising the air temperature to the initial temperature ($T_1 = 289$ K).

The working fluid (air) flows at a steady rate of 0.213 kg/s. Tubes having a 75-mm diameter connect each of the components. (a) Create a table showing the pressure, temperature, and specific volume at each state point; (b) justify your use of the ideal-gas equation of state (see Appendix E for the equivalent critical properties of air); (c) sketch the cycle in P–v coordinates; and (d) determine the average velocity of the air at each state in the cycle.

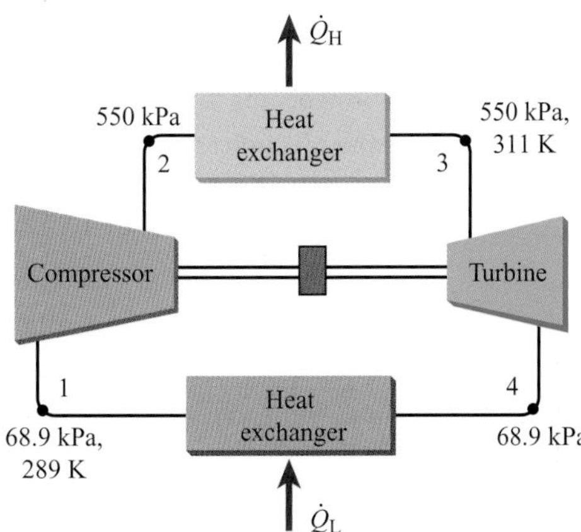

3.45 A reversed Brayton cycle can be used to provide refrigeration. The components of an air-standard reversed Brayton cycle are shown in Problem 3.44. This cycle is used to cool the passenger cabin of a commercial aircraft using air as the working fluid. The cycle comprises the following steady-flow processes.

1–2 (compressor): Ideal compression from $T_1 = 288$ K and $P_1 = 70.0$ kPa to $P_2 = 490$ kPa. The process follows the path $T^{1.4}P^{-0.4}$ = constant. Substitution of the ideal-gas law shows that this process also follows the path $Pv^{1.4}$ = constant.

2–3 (heat exchanger): Constant-pressure heat rejection with $T_3 = 315$ K.

3–4 (turbine): Ideal expansion from T_3 and P_3 to P_4 $(=P_1) = 70.0$ kPa. Like the compression process, this process also follows the paths $T^{1.4}P^{-0.4}$ = constant and $Pv^{1.4}$ = constant.

4–1 (heat exchanger): Constant-pressure heat transfer from the passenger compartment to the working fluid raising the air temperature to the initial temperature ($T_1 = 288$ K).

The working fluid (air) flows at a steady rate of 0.230 kg/s. The diameters of the tubes connecting each of the components were selected so that the velocity was always 20 m/s in each tube. (a) Create a table showing the pressure, temperature, and specific volume at each state point; (b) justify your use of the ideal-gas equation of state (see Appendix E for the equivalent critical properties of air); (c) sketch the cycle in P–v coordinates; and (d) determine the diameter of the four interconnecting tubes.

3.46 A reversed Brayton cycle can be used to provide refrigeration. The components of an air-standard reversed Brayton cycle are shown in Problem 3.44. This cycle is used to cool the passenger cabin of a commercial aircraft using air as the working fluid. The cycle comprises the following steady-flow processes.

1–2 (compressor): Ideal compression from $T_1 = 290$ K and $P_1 = 70.0$ kPa to $P_2 = 630$ kPa. The process follow the path $T^{1.4}P^{-0.4} =$ constant. Substitution of the ideal-gas law shows that this process also follows the path $Pv^{1.4} =$ constant.

2–3 (heat exchanger): Constant-pressure heat rejection with $T_3 = 320$ K.

3–4 (turbine): Ideal expansion from T_3 and P_3 to $P_4\ (=P_1) = 70.0$ kPa. Like the compression process, this process also follows the paths $T^{1.4}P^{-0.4} =$ constant and $Pv^{1.4} =$ constant.

4–1 (heat exchanger): Constant-pressure heat transfer from the passenger compartment to the working fluid, raising the air temperature to the initial temperature ($T_1 = 290$ K).

The working fluid (air) flows at a steady rate of 0.250 kg/s. Tubes having 120-mm diameters connect the components on the low-pressure side of the cycle, and tubes having 80-mm diameters connect the components on the high-pressure side of the cycle. (a) Create a table showing the pressure, temperature, and specific volume at each state point; (b) justify your use of the ideal-gas equation of state (see Appendix E for the equivalent critical properties of air); (c) sketch the cycle on P–v coordinates; and (d) determine the average velocity of the air in each of the interconnecting tubes.

3.47–3.65 Mass conservation for open systems with steady flow – multiple streams

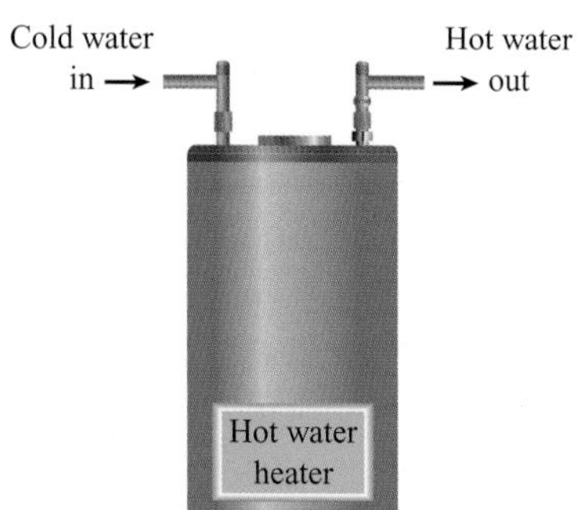

3.47 Hot water at 120 F (48.9 °C) from a water heater is used to supply simultaneously a shower with 2.5 gal/ min, an automatic dishwasher with 6.5 gal/ min, and a washing machine with 9.4 gal/min. Cold water at 60 F (15.5 °C) is supplied to the water heater through a copper tube of 25-mm inner diameter. Determine the average velocity of the incoming water. Comment on any assumptions you made regarding the water density.

3.48 While preparing a bath you have both the hot- and cold-water faucets turned on. For the conditions indicated, what is the necessary volume flow rate at the exit for the total mass in the tub to remain constant?

Cold-water inlet	Hot-water inlet	Exit
$T_1 = 60$ F	$T_2 = 140$ F	$\rho_3 = 61\ \mathrm{lb_m/ft^3}$
$\rho_1 = 62.4\ \mathrm{lb_m/ft^3}$	$v_2 = 0.01666\ \mathrm{ft^3/lb_m}$	
$\dot{m}_1 = 2\ \mathrm{lb_m/s}$	$\dot{m}_2 = 1.5\ \mathrm{lb_m/s}$	

3.49 While preparing a bath you have both the hot- and cold-water faucets on. The ambient pressure is 100 kPa. This pressure can also be assumed to be the pressure associated with each stream. The cold- and hot-water streams enter, respectively, at 289 K with a flow rate of 1.00 kg/s and at 335 K with a flow rate of 0.75 kg/s. The temperature of the water exiting the drain is 307 K. What is the necessary volume flow rate at the drain for the total mass in the tub to remain constant? **HINT**: Use the approximations for compressed liquids given in Chapter 2 to determine any needed properties (Eq. 2.41a).

3.50 While preparing a bath you have both the hot- and cold-water faucets on. A steady state has been reached with the water level in the tub constant. The ambient pressure is 100 kPa. This pressure can also be assumed to be the pressure associated with each stream. The cold-water stream enters at 286 K with a volumetric flow rate of 8.5051×10^{-3} m^3/s. Water exits the drain at 308 K with a volumetric flow rate of 1.8108×10^{-4} m^3/s. The hot water enters at 330 K. Determine the mass and volumetric flow rates of the incoming hot-water stream. **HINT**: Use the approximations for compressed liquids given in Chapter 2 to determine any needed properties (Eq. 2.41a).

3.51 For the mixing chamber shown in the sketch, determine the unknown mass flow rate assuming there is no loss or gain of fluid in the chamber.

3.52 For the mixing chamber shown in the sketch, determine the unknown mass flow rate assuming there is no loss or gain of fluid in the chamber.

3.53 Consider the mixing chamber shown in the sketch. Two streams of fluid enter the chamber and two streams exit. Determine whether the system is in steady state, and, if not, determine the magnitude (kg/s) and direction (in or out) of a fifth stream that would be required to achieve steady state.

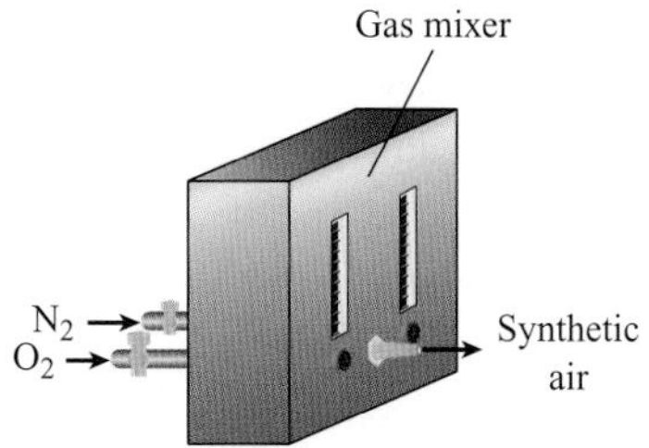

3.54 Synthetic air is made by blending pure O_2 and pure N_2. All streams are at 100 kPa and 298 K. The synthetic air has a composition of 79% N_2 and 21% O_2 by volume with an apparent molecular weight of 28.85 kg/kmol. (i) Determine the mass flow rates of O_2 and N_2 required to produce a 0.2 kg/s flow of synthetic air. (ii) How would your results change if the pressure or temperature changed?

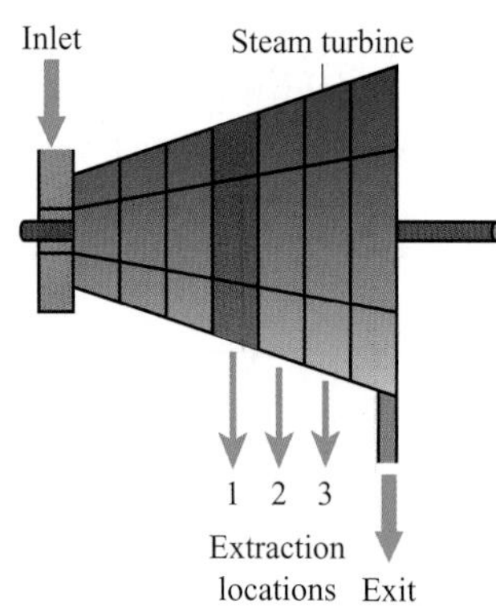

3.55 Consider a steam turbine in which steam enters at 10.0 MPa and 780 K with a flow rate of 39.0 kg/s. As shown in the sketch, a portion of the steam is extracted from the turbine after partial expansion at three different locations. The extracted steam is then led to various heat exchangers. The mass flow rates and the temperatures and pressures at each extraction point are given in the following table:

Extraction location	$\dot{m}$ (kg/s)	P (MPa)	T (K) or x
1	4.3	3.0	640
2	4.3	0.30	440
3	2.9	0.10	$x = 0.99$

The remaining wet steam exits the turbine at 12.0 kPa with a quality of 0.88. (i) Determine the minimum pipe diameters required to restrict the maximum inlet velocity to 40 m/s, the maximum extraction velocities each to 65 m/s, and the turbine outlet velocity to 120 m/s. Use the tables in Appendix B to obtain any necessary steam properties. (ii) Sketch a temperature–specific-volume plot showing the various state points and their associated isobars.

3.56 Consider a steam turbine in which steam enters at 10.45 MPa and 780 K with a flow rate of 38.739 kg/s. As shown in the sketch for Problem 3.55, a portion of the steam is extracted from the turbine after partial expansion at three different locations. The extracted steam is then led to various heat exchangers. The mass flow rates and the temperatures and pressures at each extraction point are given in the following table

Extraction location	$\dot{m}$ (kg/s)	P (MPa)	T (K) or x
1	4.343	3.054	620
2	4.345	0.332	482
3	2.871	0.136	$x = 0.949$

The remaining wet steam exits the turbine at 11.5 kPa with a quality of 0.88. (i) Determine the minimum pipe diameters required to restrict the maximum inlet velocity to 50 m/s, the maximum extraction velocities each to 75 m/s, and the turbine outlet velocity to 130 m/s. Use the NIST software to obtain any necessary steam properties. (ii) Use the NIST software to create a temperature–specific-volume plot showing the various state points and their associated isobars. You will have to locate the state points by hand, but the software can create the isobars.

3.57 Consider a steam turbine in which steam enters at 1500 psia and 900 F with a flow rate of 85.4 lb_m/s. As shown in the sketch for Problem 3.55, a portion of the steam is extracted from the turbine after partial expansion at three different locations. The extracted steam is then led to various heat exchangers. The mass flow rates and the temperatures and pressures at each extraction point are given in the following table.

Extraction location	$\dot{m}$ (lb_m/s)	P (psia)	T (F) or x
1	9.5	450	650
2	9.6	60	340
3	6.3	20	$x = 0.9405$

The remaining wet steam exits the turbine at 2.0 psia with a quality of 0.8564. (i) Determine the minimum pipe diameters in inches required to restrict the maximum inlet velocity to 160 ft/s, the maximum extraction velocities each to 240 ft/s, and the turbine outlet velocity to 350 ft/s. (ii) Sketch a temperature–specific-volume plot showing the various state points and their associated isobars.

3.58 Moisture in a steam pipeline can result in corrosion, scaling, and other problems; hence, it is desirable to remove suspended water droplets from the flow. One way to accomplish this is to use a cyclonic separator. This separator removes the moisture from the flow using the centrifugal action of a swirling flow to cause the droplets to impact the walls of the separator. The liquid then drains from the bottom of the separator, and dry steam exits at the top. Wet steam at 410 K with a quality of 0.98 enters a cyclonic separator with a flow rate of 0.50 kg/s. Saturated (dry) steam (410 K) exits the separator. Determine the mass flow rates of the dry steam and the condensate.

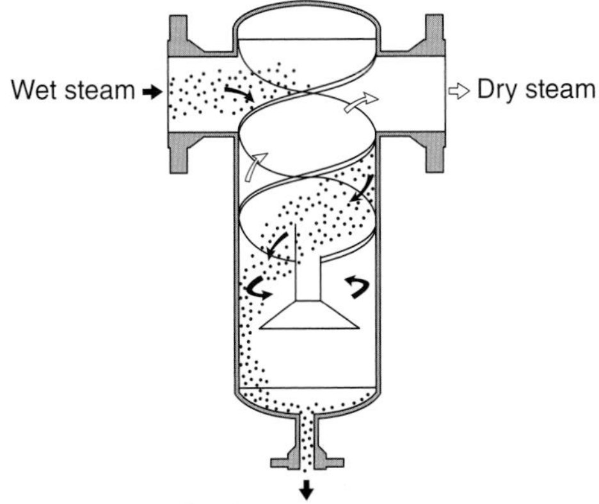

(Drawing is copyright, remains the intellectual property of Spirax Sarco, and has been used with their full permission.)

3.59 Geothermal power plants utilize steam produced at locations where the hot core of the earth is quite close to the surface. Yellowstone National Park is such a location, with many hot springs and geysers. Geothermal steam is typically wet and available only at relatively low temperatures. At a particular geothermal power plant, wet steam is available at 450 K with a quality of 0.90. Before the steam enters the turbine a cyclonic separator removes the moisture from the steam. (See the sketch in Problem 3.58.) The dry-steam flow rate to the turbine is 300 kg/s. Determine the mass flow rates of the wet steam entering the separator and of the exiting condensate.

(Credit: Bartek Wrzesniowski / Alamy Stock Photo.)

3.60 Process steam is extracted from a turbine in a paper mill power plant. Moisture is removed from the extracted steam using a cyclonic separator before it enters a distribution line. (See the sketch from Problem 3.58.) The steam enters the separator at 0.40 MPa with a quality of 0.97 and a flow rate of 12 kg/s. Determine the individual mass flow rates of the dry steam and the condensate exiting the separator.

3.61 Process steam is extracted from a turbine in a paper mill power plant. Moisture is removed from the extracted steam using a cyclonic separator before it enters a distribution line. (See the sketch from Problem 3.58.) The steam enters the separator at 60 psia with a quality of 0.98 and a flow rate of 25 $\mathrm{lb_m/s}$. Determine the individual mass flow rates of the dry steam and the condensate exiting the separator.

3.62 Consider a rectangular sheet-metal duct carrying air at 60 F at 100 kPa. The volume flow rate at the entrance to the duct is 1200 $\mathrm{ft^3/min}$. A portion of the air exits through a grille, as shown in the sketch. On the basis of noise considerations, the average velocity in the duct is to be limited to 600 ft/min. The height of the duct is 12 in.

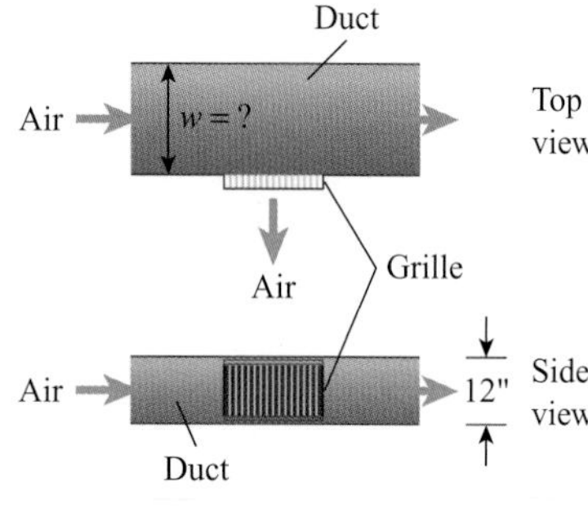

A. Determine the minimum duct width required, subject to the constraint that the duct is only available with dimensions in even increments of 2 inches (e.g., 14, 16, 18 inches, etc.).

B. Determine the volumetric flow rate through the grille if the velocity of the air entering the room is 35 ft/min, the maximum value for comfort. The dimensions of the grille are 10 inches by 15 inches.

C. Determine the mass flow rate through the duct downstream of the grille. Assume that all pressure drops in the duct system are sufficiently small that a uniform pressure is a reasonable approximation.

3.63 A system of pipes is shown in the sketch below. All the pipes have the same diameter of 0.1 m. The average flow velocity V_1 in pipe 1 is 10 m/s to the right.

In pipe 2, the velocity V_2 is 6 m/s to the right. (i) Determine V_3. (ii) If the flow is reversed in pipe 2, i.e., V_2 is to the left, what is the velocity V_3 in pipe 3?

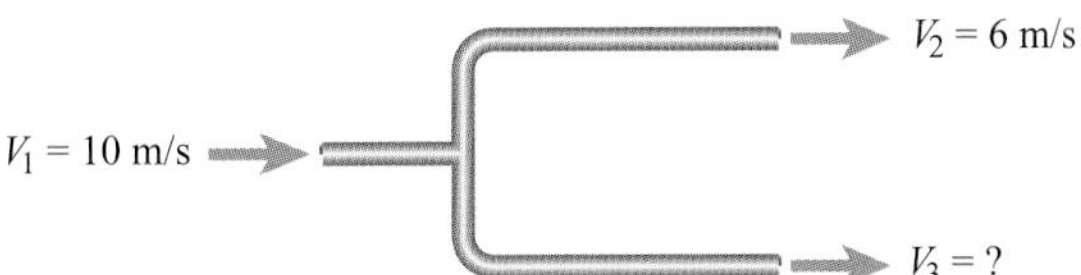

3.64 At steady state, fuel oil is supplied at 0.01 gal/min to an oil furnace for home heating. (Note: 1 gal $= 3.785 \times 10^{-3}\,\text{m}^3$.) The fuel oil density is $870\,\text{kg/m}^3$. Combustion air at 65 F and 1 atm enters the furnace with a volumetric flow rate of $15.2\,\text{ft}^3/\text{min}$. (i) Determine the mass flow rate of combustion product gases flowing up the chimney flue. (ii) Also determine the average velocity of the flue gas if the gas temperature is 225 F and the flue diameter is 9 inches. Assume that the flue gases behave as an ideal gas with an apparent molecular weight of 29 kg/kmol. Also assume that the pressure in the flue is essentially 1 atm (absolute).

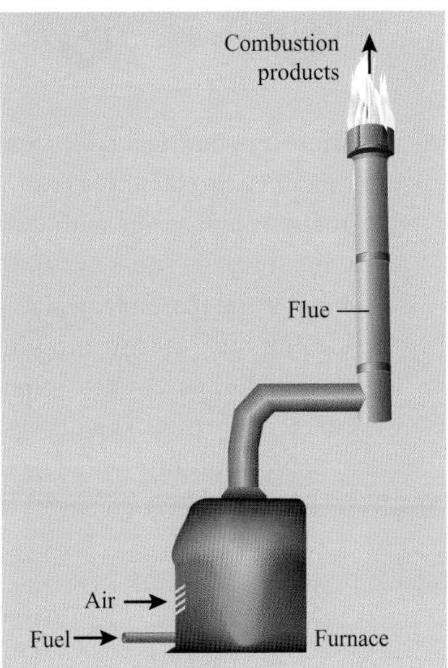

3.65 At steady state, fuel oil is supplied at $0.50\,\text{cm}^3/\text{s}$ to an oil furnace for home heating. (See the illustration in Problem 3.64.) The fuel oil density is $877\,\text{kg/m}^3$. Combustion air at 19 °C and 100 kPa enters the furnace with a mass flow rate 16 times that of the fuel mass flow rate. The pressure in the chimney flue is approximately 98 kPa, and the flue diameter is 0.20 m. (i) Determine the mass flow rate of combustion product gases flowing up the chimney flue. (ii) Determine the volumetric flow rate of the air and the ratio of that to the fuel volumetric flow rate. (iii) Determine the average velocity of the combustion products in the flue if the gas temperature is 108 °C. Assume that the flue gases behave as an ideal gas with an apparent molecular weight of 29 kg/kmol.

3.66–3.75 Mass conservation for open systems (control volumes) with unsteady flow

3.66 The water faucet on a stoppered kitchen sink was accidentally left slightly open. The faucet volumetric flow rate is 0.12 gal/min. The ambient pressure is 100 kPa. Formally apply the principle of the conservation of mass to answer the following questions. (i) If the sink is a rectangular basin 45 cm by 56 cm and 22 cm deep, how long will it take the sink to overflow? (ii) How long will it take to flood the 350-ft^2 apartment to a depth of 1 inch if there is no leakage through the floor or baseboards?

3.67 To fill an initially empty 275-gal tank with fuel oil requires 12 min. If the flow from the filling nozzle is steady, estimate the velocity of the fuel oil exiting the 20-mm-diameter filling nozzle. Formally apply the principle of the conservation of mass to solve this problem.

3.68 Consider a cylindrical tank with a diameter D of 1.0 m and a height H of 1.5 m as shown in the sketch below. Water ($\rho = 1000\,\text{kg/m}^3$) enters the tank at the bottom through a tube with diameter d of 5 mm at a constant average velocity of 4 m/s. Initially (time $t = 0$), the tank is half full as shown; water continues to enter until the tank begins to overflow at time t_{full}. Selecting the water inside the tank at any instant as the control volume of interest, write or develop a formal explicit expression of mass conservation for this control volume. Your result should be in the form of an ordinary differential equation. Solve your differential equation for t_{full} and determine a numerical value for it using the data given.

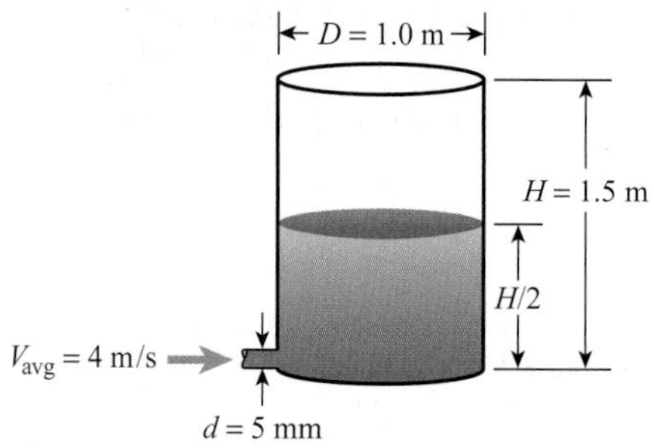

3.69 Consider the stepped cylindrical tank shown in the sketch below. The tank is initially filled with water ($\rho = 997\,\text{kg/m}^3$) to a height of 0.7 m. A plug in the bottom of the tank is quickly removed and water begins to flow from the 6-mm-diameter hole. The velocity of the exiting water relates to the instantaneous height of the water in the tank, $h(t)$, as $V = 0.98\,[2gh(t)]^{1/2}$, where g is the gravitational acceleration (9.807 m/s^2). Estimate the time required to drain the tank.

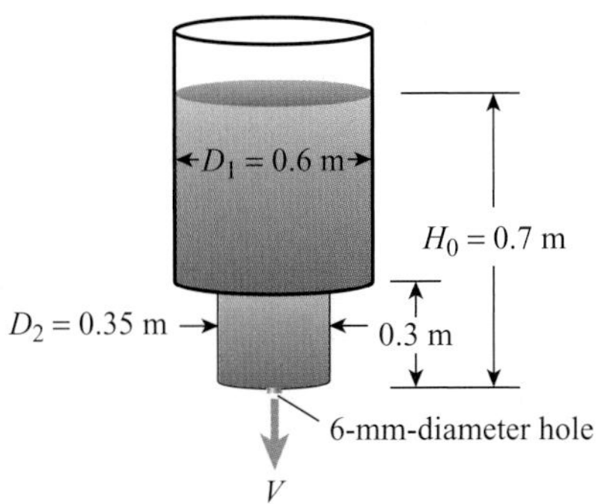

3.70 An automobile is cruising along a level highway steadily at 62 miles/hr. The on-board computer indicates a fuel economy of 26 miles/gal. The engine is operating at an air–fuel ratio of 14.7 (mass basis). The density of the fuel is 860 kg/m^3. (i) Assuming the exhaust gases have a molecular weight of 28.5 kg/kmol and a temperature of 580 K, determine their velocity in the 50-mm-diameter exhaust pipe. The exhaust pressure is nominally 100 kPa. (ii) Considering the automobile and its contents to be the open system (control volume) of interest, determine the time rate of change of the system mass, dM_{cv}/dt. (iii) Estimate the change in weight of the automobile and its contents in the time it takes to travel 100 miles.

3.71 Oil enters an open system (control volume) at a rate of 100 kg/s through pipe 1. Oil leaves the open system (control volume, through pipe 2 at 50 kg/s and through pipe 3 at 70 kg/s. Determine the rate of change of mass (kg/s) within the system.

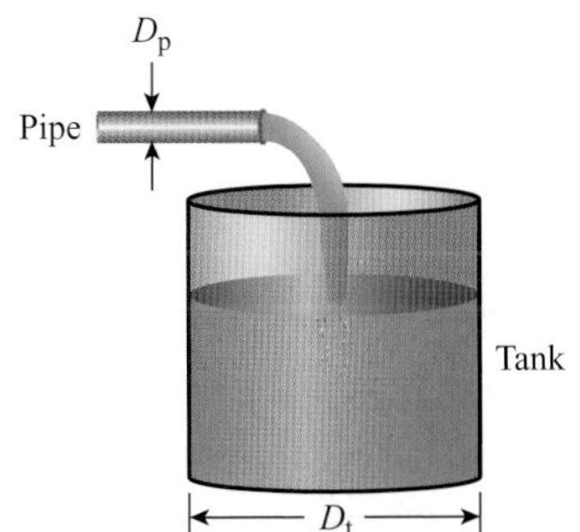

3.72 A 1-m-diameter tank (empty at time zero) is being filled by liquid water flowing through a 0.025-m-diameter pipe at an average velocity of 30 m/s with a temperature of 100 °C. Evaporation is negligible.

A. Derive a differential equation relating the water level z in the tank to the time t.

B. Integrate this differential equation to determine the water level (m) in the tank at the end of 2 minutes.

C. Discuss any assumptions you made regarding the water density.

3.73 A rigid tank (volume $\mathcal{V}$) containing an ideal gas is initially at T_1 and P_1. At time zero, an exit pipe (area A) is opened and gas flows out of the tank at velocity $V = K(P - P_{atm})^{1/2}$, where P is the absolute pressure in the tank, P_{atm} is the pressure of the atmosphere outside the tank, and K is a constant. The temperature of the gas in the tank is maintained at T_1 during the process. The pressure and the temperature of the gas exiting the tank are P_{atm} and T_1, respectively. Assume that, inside the tank, the pressure and specific volume do not vary with position.

A. Derive a differential equation for the tank pressure P as a function of time t.

B. Determine the time required for the tank pressure to reach P_{atm}.

3.74 Consider the draining of a liquid from a tank. The level of the liquid above the bottom of the tank at any instant of time is z. The cross-sectional area of the tank is A_t. The flow area at the tank exit is A_e. The exit velocity V is given by $V = Cz^{1/2}$, where C is a constant.

A. Derive a differential equation for $z(t)$.

B. Integrate the differential equation to determine the time taken to drain the tank from level z_1 to a lower level z_2.

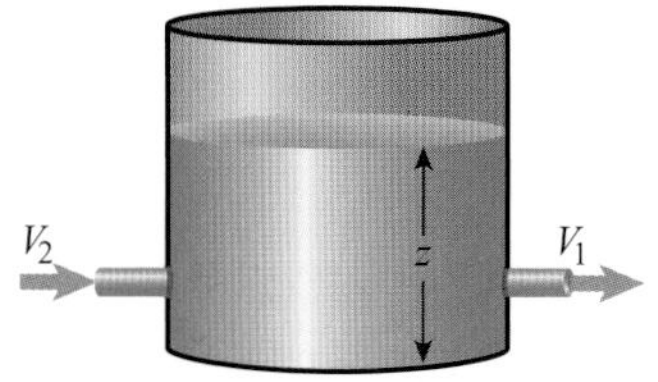

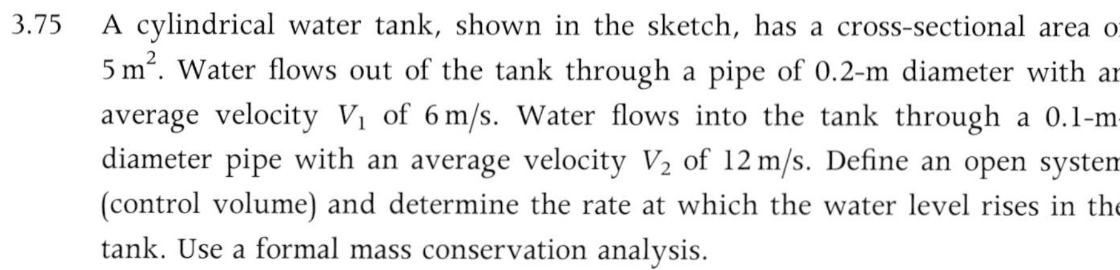

3.75 A cylindrical water tank, shown in the sketch, has a cross-sectional area of $5\,\text{m}^2$. Water flows out of the tank through a pipe of 0.2-m diameter with an average velocity V_1 of 6 m/s. Water flows into the tank through a 0.1-m-diameter pipe with an average velocity V_2 of 12 m/s. Define an open system (control volume) and determine the rate at which the water level rises in the tank. Use a formal mass conservation analysis.

3.76–3.77 Computer problems

3.76 Search online for the following information and convert it to each of the units requested. Set up a spreadsheet to show your results and to convert the speeds. You will find conversion factors in the front matter of the textbook. Include the web address where you found each speed value and highlight the value in your reference.

A. Typical walking speed in mph, ft/s, m/s
B. Typical bicycle speed in mph, ft/s, m/s
C. Automobile highway speed in mph, ft/s, m/s
D. Cessna 172, private plane, cruising speed in mph, ft/s, m/s
E. Boeing 737, commercial plane, cruising speed in mph, ft/s, m/s
F. Space Shuttle orbiting speed in mph, ft/s, m/s
G. Speed of sound at sea level in mph, ft/s, m/s

In future problems, you can compare calculated velocities to the values in this table as part of your discussion.

3.77 Using a four-cup (one-quart) measuring cup, find the time that it takes to fill four cups of water from your kitchen (or bathroom) sink and the shower. (You can also use a one-gallon jug or other container of known volume.) You must time the filling of four cups or more. A calibrated milk jug with the top cut off works well. Do not use a one- or two- cup measurement. Use a spreadsheet for the calculations below and include a printed copy of the spreadsheet with your homework. Write the equations that you used for each calculation on the spreadsheet.

A. Make three measurements from the sink and average. Make three measurements from the shower and average. List all data in your homework solution. You could make these measurements with one other person. Indicate the location where you made the measurements (name of dormitory or apartment). If you make measurements with another person, include their name.

B. Find the volume flow rate from the sink and the volume flow rate from the shower in units of gal/min, ft^3/s, and m^3/s.

C. Calculate the mass flow rate in kg/s and lbm/s.

D. If the inflow pipe is 1/2″ id (inner diameter) find the average velocity in each pipe in units of mph, ft/s, and m/s.

E. Compare these water velocities to your values in Problem 3.76 (if assigned). Is the water velocity in the pipe similar to walking speeds, bicycle speeds, or highway speeds?

3.78–3.86 FE problems

3.78 Water with a density of $990\,\text{kg/m}^3$ is discharged from a pipe with a volume flow rate of $0.4\,\text{m}^3/\text{min}$. Determine the mass flow rate in units of kg/s. a. 396 kg/s, b. 6.6 kg/s, c. 0.4 kg/s, d. 3.96 kg/s.

3.79 Water is flowing through a 4-cm diameter pipe at a volume flow rate of $0.3\,\text{m}^3/\text{min}$. What is the flow velocity in the pipe? a. 12.5 m/s, b. 24.0 m/s, c. 1.0 m/s, d. 4.0 m/s.

3.80 Water with a density of $990\,\text{kg/m}^3$ is flowing through a 3-cm diameter pipe at a mass flow rate of 5 kg/s. What is the flow velocity in the pipe? a. 7.1 m/s, b. 1.8 m/s, c. 22.4 m/s, d. 0.21 m/s.

3.81 Air at 300 kPa and 350 K flows through a 6-cm diameter pipe at a velocity of 15 m/s. What is the mass flow rate? a. 12.7 kg/s, b. 0.51 kg/s, c. 0.13 kg/s, d. 0.04 kg/s.

3.82 Air at 200 kPa and 50 °C flows through a 9-cm diameter pipe at a mass flow rate of 0.15 kg/s. What is the flow velocity? a. 12.7 m/s, b. 10.9 m/s, c. 15.7 m/s, d. 7.8 m/s.

3.83 Water at 300 kPa and 440 K flows through a 12-cm diameter pipe at a velocity of 16 m/s. What is the mass flow rate? a. 0.273 kg/s, b. 0.12 kg/s, c. 1.09 kg/s, d. 0.09 kg/s.

3.84 Air at 150 kPa and 350 K enters a compressor with a velocity of 30 m/s. If the air exits the compressor at 2 MPa and 600 K, what is the outflow velocity when the inflow and outflow duct diameters are the same? a. 233 m/s, b. 120 m/s, c. 9.2 m/s, d. 3.9 m/s.

3.85 Water enters a nozzle with a velocity of 0.4 m/s. If the nozzle inflow diameter is 2 cm and the outflow diameter is 0.7 cm, what is the outflow velocity? Assume that the water is incompressible. a. 3.27 m/s, b. 1.14 m/s, c. 0.14 m/s, d. 2.34 m/s.

3.86 Water at 500 kPa and 600 K flows through a 15-cm-diameter pipe at a mass flow rate of velocity of 0.2 m^3/s. What is the flow velocity? a. 1.5 m/s, b. 6.2 m/s, c. 19.5 m/s, d. 12.4 m/s.

Appendix 3A Spark-Ignition Engine Geometry

Since we will be revisiting the spark-ignition (SI) engine, it is useful to define a few engine-related geometrical terms at this point, rather than repeating them in a variety of locations. These are:

B: The **bore** is the diameter of the cylinder.

S: The **stroke** is the distance traveled by the piston in moving from top center to bottom center, or vice versa.

$\mathcal{V}_c$ (or $\mathcal{V}_{TC}$): The **clearance volume** is the combustion chamber volume when the piston is at top center.

$\mathcal{V}_d$ (or $\mathcal{V}_{disp}$): The **displacement** is the difference in the volume at bottom center and at top center, that is,

$$\mathcal{V}_{disp} = \mathcal{V}_{BC} - \mathcal{V}_{TC}. \tag{3A.1}$$

The displacement is also equal to the product of the stroke and the cross-sectional area of the cylinder:

$$\mathcal{V}_{disp} = S\pi B^2/4. \tag{3A.2}$$

CR: The **compression ratio** is the ratio of the volume at bottom center to the volume at top center, that is,

$$CR = \frac{\mathcal{V}_{BC}}{\mathcal{V}_{TC}} = \frac{\mathcal{V}_{disp}}{\mathcal{V}_{TC}} + 1. \tag{3A.3}$$

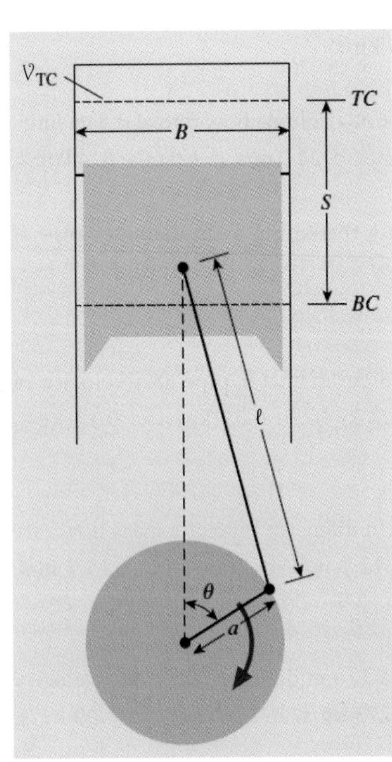

FIGURE 3A.1 Definition of geometrical parameters for reciprocating engines. After Ref. [6].

The following geometrical parameters associated with a reciprocating engine are illustrated in Fig. 3A.1:

B = bore
S = stroke
ℓ = connecting-rod length
a = crank radius
θ = crank angle
$\mathcal{V}_{TC}$ = volume at top center
$\mathcal{V}_{BC}$ = volume at bottom center
CR = compression ratio $(=\mathcal{V}_{BC}/\mathcal{V}_{TC})$

From geometric and kinematic analyses [6], the instantaneous volume $\mathcal{V}(\theta)$ and its time derivative $d\mathcal{V}(\theta)/dt$ are given by

$$\mathcal{V}(\theta) = \mathcal{V}_{\mathrm{TC}}\left\{1 + \frac{1}{2}(CR - 1)\left[\frac{\ell}{a} + 1 - \cos\theta - \left(\frac{\ell^2}{a^2} - \sin^2\theta\right)^{1/2}\right]\right\} \tag{3A.4}$$

and

$$\frac{d\mathcal{V}(\theta)}{dt} = SN\left(\frac{\pi B^2}{4}\right)\pi\sin\theta\left[1 + \frac{\cos\theta}{(\ell^2/a^2 - \sin^2\theta)^{1/2}}\right]. \tag{3A.5}$$

where N is the crank rotational speed in rev/s.

CHAPTER 4
Energy and Energy Transfer

LEARNING OBJECTIVES

After studying Chapter 4, you should:

- Understand the differences between the bulk (macroscopic) and internal (microscopic) energies possessed by closed and open systems.
- Understand the formal definitions of both heat and work and be able to state these precisely.
- Be able to identify various forms of work in practical situations.
- Be able to write the formal definitions of compression and expansion work, i.e., moving-boundary work, for a simple compressible substance.
- Be able to calculate compression and expansion work for cases where it is possible to relate the pressure to the volume of a system as the system executes a thermodynamic process.
- Be able to define a polytropic process and to use this definition to calculate compression and expansion work.
- Be able to identify the three modes of heat transfer in practical situations.

CHAPTER 4 OVERVIEW

IN THIS CHAPTER, we review the concept of energy and the various ways in which closed and open systems can possess energy at both microscopic (molecular) and macroscopic levels. We also carefully define heat and work, which are boundary interactions and, therefore, not properties of a system or control volume. The chapter concludes with a brief examination of the rate laws that govern heat transfer. Figure 4.1 illustrates how this chapter relates to other thermodynamics topics. We begin with a brief historical overview of our subject matter.

DESIGN AND ANALYSIS OF SYSTEMS		
ANALYSIS OF PRACTICAL DEVICES		
CONSERVATION OF MASS	CONSERVATION OF ENERGY (First Law of Thermodynamics)	SECOND LAW OF THERMODYNAMICS (Entropy)
PROPERTIES OF MATTER		
FRAMEWORKS FOR ANALYSIS, KEY CONCEPTS, AND DEFINITIONS		

FIGURE 4.1 Chapter 4 deals with energy and the transfer of energy across system boundaries. Here we define and explore in detail the concepts of heat and work, concepts that are essential to applying the principle of conservation of energy. This figure illustrates how Chapter 4 relates to other thermodynamic topics.

Historical Context

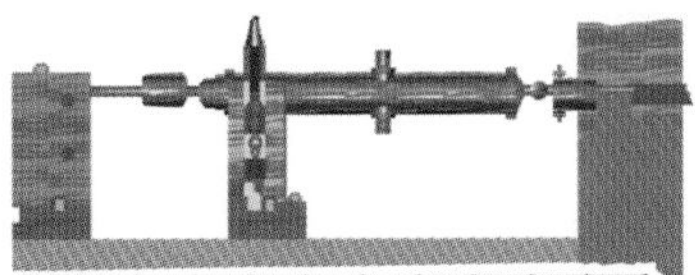

Thompson's canon boring experiments connected the concepts of heat and work.

Benjamin Thompson (1753–1814) (Credit: The Granger Collection / Alamy Stock Photo.)

With the word *energy* being commonplace, you may find it difficult to believe that the scientific underpinning of this concept as we know it today is just 200 years old. In 1801, Thomas Young (1773–1829) (of Young's modulus fame) presented the idea that *the energy of a system is the capacity to do work* [1]. In the early 1800s, various forms of energy had yet to be defined in useful ways. For example, the concept of heat was a muddle of the concepts that we now distinguish as temperature, internal energy, and heat. The discovery that energy is conserved – a premier conservation principle of classical physics – had to wait until Robert Mayer's (1814–1878) statement of the theory of conservation of energy in 1842. One of the keys to the recognition of this law was that work could be converted to heat and vice versa. Benjamin Thompson (aka Count Rumford) (1753–1814), a traitor to the colonists in the Revolutionary War between Great Britain and her American colonies, discovered in 1798, by a series of carefully planned and conducted experiments, that friction produces an inexhaustible supply of heat. Exactly what heat is, however, was not clear in 1798.

The invention and development of the steam engine spurred theoretical development of the thermal sciences at the end of the eighteenth and the beginning of the nineteenth centuries. Ironically, the steam engine's very low thermal efficiency (a few percent) may have prevented **Sadi Carnot** (1796–1832), the discoverer of the **second law of thermodynamics,** from also discovering the **first law of thermodynamics** (i.e., the conservation of energy principle). With early steam engines converting such a small percentage of the heat supplied to work, it appeared to Carnot that there was no conversion of heat to work, but merely a heat addition at high temperature and an equal heat rejection at a low temperature. For the interested reader, a timeline of important lives and events in the history of the thermal sciences is presented in Appendix A.

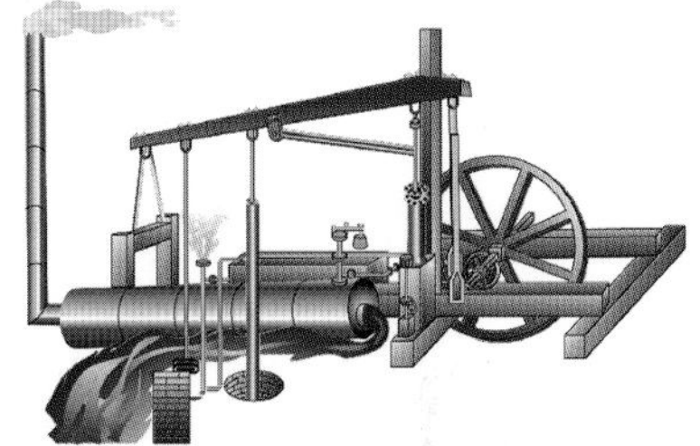

A 1893 illustration of Oliver Evans' Columbian steam engine of 1812. (Library of Congress. Published in the USA before 1923 and in the public domain in the USA.)

4.1 Closed and Open System Energy

A closed system or an open system (control volume) can possess energy, first, as a consequence of its bulk motion or position within a force field, such as gravity, and second, as a consequence of the motion of its constituent molecules and their interactions (i.e., the internal energy). We can therefore express the total closed-system energy as the sum of these two energy components:

Equations with yellow backgrounds express key concepts and are the most important relationships in the chapter.

$$E_{\text{sys}} = E_{\text{bulk}} + U_{\text{internal}}. \tag{4.1a}$$

The same relationship holds for an open system (control volume):

$$E_{\text{cv}} = E_{\text{bulk,cv}} + U_{\text{internal,cv}}. \tag{4.1b}$$

We now examine each of the terms on the right-hand sides of Eqs. 4.1a, b.

4.1a Energy Associated with a System as a Whole

A system (closed or open) can possess energy by virtue of its bulk motion. The most common forms of this bulk energy in engineering applications are kinetic energy *KE* and potential energy *PE*. We can then write

$$E_{\text{bulk}} = (KE)_{\text{bulk}} + (PE)_{\text{bulk}}. \tag{4.2}$$

Assuming a rigid system with no relative motion of the subsystem elements, we can easily relate the kinetic and potential energies to the linear and angular velocities of the system and its position as follows:

$$E_{\text{bulk}} = \frac{1}{2}MV^2 + \frac{1}{2}I\omega^2 + Mg(z - z_{\text{ref}}), \tag{4.3}$$

where we assume that the only contribution to the potential energy is the action of gravity. The first two terms on the right-hand side of Eq. 4.3 are the translational and rotational contributions to the system kinetic energy, where V is the magnitude of the bulk translational velocity of the system center of mass, I is the system rotational moment of inertia, ω is the system angular velocity, and z_{ref} is some reference elevation. Implicitly, the reference kinetic energies are zero (i.e., $V_{\text{ref}} = 0$ and $\omega_{\text{ref}} = 0$). Equation 4.3 should be familiar to you from your previous study of physics and rigid-body mechanics. If other external fields exist in addition to gravity (e.g., magnetic or electrostatic fields), additional contributions to the potential energy can result, depending on the nature of the matter within the system. Discussion of these effects is beyond the scope of this book. Additional information can be found in more advanced texts such as Ref. [2].

The specification of a single translational velocity and a single angular velocity in Eq. 4.3 is possible only for a rigid system. In a control volume, the bulk velocity is likely to vary with position; thus, the kinetic energies are obtained by integrating over the control volume:

$$(KE)_{\text{bulk}} = \frac{1}{2}\int_{M_{\text{CV}}} V^2 dM, \tag{4.4a}$$

or

$$(KE)_{\text{bulk}} = \frac{1}{2}\int_{\text{CV}} \rho V^2 d\mathcal{V}, \tag{4.4b}$$

where the differential mass $dM = \rho d\mathcal{V}$. The application of Eq. 4.4 to a rigid translating and rotating system captures the two kinetic energy terms of Eq. 4.3; thus, Eq. 4.4 applies equally well to open and closed systems.

Figure 4.2 shows several examples that illustrate the energies associated with bulk motions. The spinning baseball in Fig. 4.2a is a closed system of fixed mass. At any instant in time, the baseball possesses a particular translational velocity V, a particular angular velocity ω, and a particular elevation relative to a datum, $z - z_{\text{ref}}$. Knowing the mass of the ball and its moment of inertia, we could easily calculate the instantaneous values of the terms contributing to E_{bulk}.

A similar, but more complex, example is the system defined by the bicycle (Fig. 4.2b), where we have deliberately excluded the rider from our system. In this case, the translational kinetic energy and the potential energy are easy to determine. Since the entire system does not rotate with a single velocity about a common axis, the

(a)

(b)

(c)

FIGURE 4.2 Examples of closed systems and an open system (control volume) having macroscopic energies: **(a)** a baseball thrown with spin imparted (Mike Watson Images / moodboard / Getty Images), **(b)** a bicycle rolling down a hill (Ascent/PKS Media Inc. / DigitalVision / Getty Images), and **(c)** the Space Shuttle with solid-fuel booster engines (Stocktrek / DigitalVision / Getty Images).

rotational kinetic energy would have to be determined by subdividing the system to treat each rotating part separately. For example, each wheel makes a separate contribution to the system rotational kinetic energy.

A third example (Fig 4.2c) is an open system (control volume) containing a solid-fuel rocket. We assume that the rocket contains no internal macroscopic moving parts, such as pumps, as would be found in a liquid-fuel rocket. With this assumption, evaluating the control-volume kinetic energy is simplified. Nevertheless, evaluating the integral in Eq. 4.4 remains a daunting task. To do this requires knowledge of the local velocity and density at every point within the complex flow created by the burning of the solid propellant. Evaluating the potential energy is much easier. If the control-volume center of mass does not change as the propellant is consumed, or if we ignore any small change in its location, the potential energy is simply $M_{cv}g\,(z - z_{ref})$.

The three examples of Fig. 4.2 were deliberately chosen to illustrate situations in which macroscopic system energies can be important. In many thermal-science applications, however, these macroscopic system energies can be neglected because they, or their changes, are small compared to the heat and work exchanges at the boundaries and to the changes in internal (molecular) energy during a process. Alternatively, we may select system boundaries to exclude some bulk energies. Nevertheless, being able to identify all the various energies and energy exchanges is crucial in analyzing thermal systems. In our future theoretical developments and examples, we will be careful to point out when we are neglecting the macroscopic energies. Such practice should help you develop critical thinking in your approach to thermal-science problems.

Rereading the section on internal energy in Chapter 2 is useful at this point.

4.1b Energy Associated with Matter at a Microscopic Level

As defined in Chapter 2, internal energy is the energy associated with the motion of the microscopic particles (atoms, molecules, electrons, etc.) comprising a closed system or control volume. Seeing how macroscopic properties relate to the microscopic structure of matter can be intellectually satisfying, although an understanding of microscopic behavior is not necessary to solve most engineering problems in the thermal sciences.

(a)

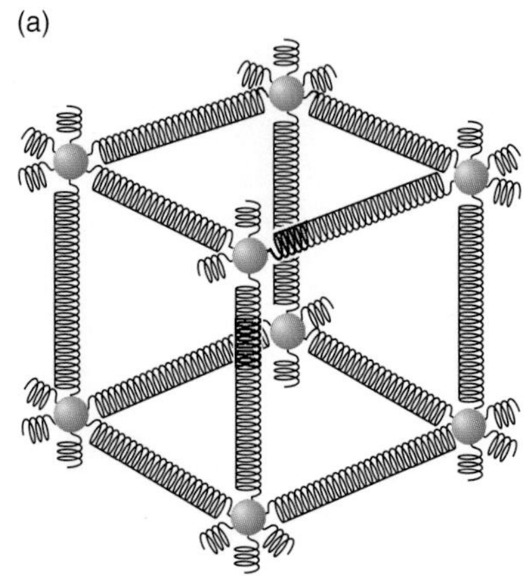

(b)

FIGURE 4.3 (a) Energy storage in the vibrating lattice of a solid. **(b)** False-color scanning tunneling microscope (STM) image of the surface of pyrolytic graphite reveals the regular pattern of individual carbon atoms (Colin Cuthbert/Science Photo Library).

(Recall that the subjects of classical thermodynamics and heat transfer predate modern ideas concerning the microscopic nature of matter.) Interestingly, many of the challenges associated with the microminiaturization of engineering devices require a detailed understanding of microscopic behavior. See, for example, Refs. [3, 4].

In Chapter 2, we saw that gas molecules possess energy in ways analogous to our macroscopic systems: translational kinetic energy, rotational kinetic energy, and vibrational kinetic and potential energies. As the temperature is increased, not only is more energy associated with some storage modes (e.g., translational kinetic energy) but new states become accessible to the molecule (e.g., vibrational kinetic and potential energies). Figure 2.7 illustrated these "degrees of freedom" and energy storage modes; their impact on specific heats was shown in Fig. 2.9.

For solids (Fig. 4.3), the internal energy is associated with lattice vibrations that give rise to vibrational kinetic and potential energies. Figure 4.3a illustrates in cartoon fashion the basic structure of a solid, comprising masses (molecular centers) connected by springs (intermolecular forces). The internal energy associated with liquids has its origin in the relatively close-range interactions among the molecules making up the liquid. A simplified view is to consider the structure of a liquid as lying between the extremes of a disorganized gas and a well-ordered crystalline solid.

4.2 Energy Transfer across Boundaries

4.2a Heat

DEFINITION

The recognition that heat is not a property of a system – not something that the system possesses – but, rather, an exchange of energy from one system to another, or to the surroundings, was a breakthrough in thermodynamics. Our formal definition of heat [3] is the following:

Heat *is energy transferred, without transfer of mass, across the boundary of a system (or across a control surface) because of a temperature difference between the system and the surroundings or a temperature gradient at the boundary.*

Because of the importance of this definition, let us elaborate some of the important implications. First, heat transfer occurs at the boundary of a system; that is, it is a boundary phenomenon. As a consequence, a system cannot *contain* heat. The addition of heat to a system with all else held constant, however, increases the energy of the system, and, conversely, heat removal from a system decreases the energy of the system. Another important element in this definition of heat is that the "driving force" for this energy exchange is a temperature difference or temperature gradient. Other energy transfers across a system boundary may occur, but only heat is controlled solely by a temperature difference. We can gain some physical insight into this boundary energy exchange by examining the molecular processes involved. The energy exchange is carried out by the collision of molecules, in which the higher kinetic energy molecules, in general, impart some of their energy to the lower kinetic energy molecules. Because the higher kinetic energy molecules are at a higher temperature than the lower kinetic energy molecules, the direction of the energy exchange is from high temperature to low temperature. A more rigorous treatment of this process would involve the subject of irreversible thermodynamics [6–8] and is beyond the scope of this book.

SEMANTICS

In spite of our attempts here to be very precise about the definition of heat, some semantic problems arise out of traditions and nomenclature developed prior to the advent of modern thermodynamic principles. The science of heat transfer was developed in the early 1800s [9], prior to our understanding that heat is not possessed by a body and to the development of energy conservation. Even today, many scientific disciplines still use the word heat to mean internal energy, e.g., the heat content of the ocean. When Fourier published his theory of heat transfer (1811–1822), the common wisdom was that *caloric* (or heat) was a material substance; despite the fact that Benjamin Thompson in 1798 had shown that friction produces an inexhaustible supply of heat. Regardless of the fundamental nature of heat, Fourier's mathematical analyses describing temperature distributions in solids are accurate descriptions and stand yet today as the foundation of heat-transfer theory.

Rates of heat transfer and rates per unit area are also of importance and need to be distinguished. We adopt the following symbols to denote these heat interactions:

Q = heat (or heat transfer) [=] J,

$\dot{Q}$ = rate of heat transfer [=] J/s or W,

q = heat transfer per unit mass [=] J/kg,

$\dot{Q}''$ = heat flux or heat transfer rate per unit area [=] W/m^2.

We use the word **adiabatic** to describe a process in which there is no heat transfer. Later in this chapter we will elaborate on the principles of heat transfer.

4.2b Work

DEFINITION

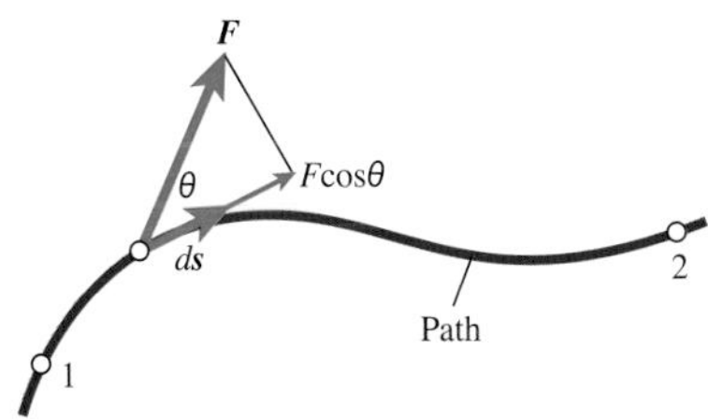

FIGURE 4.4 An increment of work δW is express as the dot product of a force acting through the distance $\boldsymbol{ds}$.

Another fundamental transfer of energy across a system boundary is work. All forms of work, regardless of their origin, are fundamentally expressions of a force acting through a distance (Fig. 4.4),

$$\delta W = \boldsymbol{F} \cdot \boldsymbol{ds}; \tag{4.5}$$

that is, an increment of work is the scalar product of a vector force and the displacement vector $\boldsymbol{ds} = \hat{\boldsymbol{i}}\,dx + \hat{\boldsymbol{j}}\,dy + \hat{\boldsymbol{k}}\,dz$, where $\hat{\boldsymbol{i}}, \hat{\boldsymbol{j}}$, and $\hat{\boldsymbol{k}}$ are the unit vectors in a Cartesian coordinate system. We adopt the notation δW to indicate that the incremental work done depends upon the path taken. For a particular process, Eq. 4.5 can be integrated following the process path from position 1 to position 2 to obtain the total work done:

$${}_1W_2 \equiv \int_C \boldsymbol{F} \cdot \boldsymbol{ds}, \tag{4.6}$$

where the $\int_C$ symbol indicates a path integral along the curve C. Equations 4.5 and 4.6 should be familiar to you from your previous study of physics.

We adopt the particular notation ${}_1W_2$ to emphasize that this quantity is the work done in going from point 1 to point 2, which cannot be represented as a difference in the values of a properly of the system, since work, like heat, is not a property of a system. Mathematically, work is a **path function** (corresponding to an **inexact** differential δW). In contrast, thermodynamic properties are **point functions** (corresponding to **exact** differentials such as dE (see below)). To illustrate the differences between point and path functions, consider a hypothetical example involving work (a path function) and elevation (a point function): You can bicycle to a nearby town taking a short path that is relatively flat, or you can bicycle to the same location by taking a circuitous and hilly path. Clearly, you will perform more work if you take the circuitous, hilly path. However, the change in elevation between your initial and final positions is the same regardless of which path you took. Elevation (the height above a datum plane) is a point function; its change associated with going from your present location to the nearby town is independent of the path traveled.

All thermodynamic properties (pressure, temperature, specific volume, internal energy, entropy, etc.) are point functions. Thus, property changes associated with a system undergoing a process are expressed as differences. For example, the change in system energy for a process that takes the system from state 1 to state 2 is expressed as $\int_{\text{State 1}}^{\text{State 2}} dE = \int_{E_1}^{E_2} dE = E_2 - E_1 \equiv \Delta E$. To write a similar expression for work would be nonsense, as there is no such thing as W_1 or W_2 the work associated with state 1 or state 2.

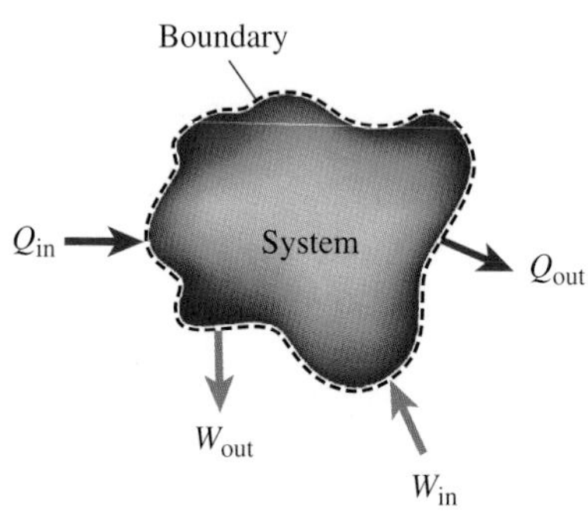

FIGURE 4.5 Heat and work are boundary phenomena. Arrows representing work or heat start or stop at the system boundary.

The force appearing in Eqs. 4.5 and 4.6 may be a purely mechanical one, or it may have other origins, such as the force acting on a charge moving through an electric or magnetic field or the force on a particle with a magnetic moment in a magnetic field. In mechanical engineering applications, work is most commonly associated with mechanical and electrical forces. In this book, we deal only with work arising from these two forces, although we define a mechanical force fairly broadly.

In our study of thermal sciences, it is important to emphasize that work occurs *only* at the boundary of a system or a control volume. Like heat, work is not possessed by a thermodynamic system or a control volume but is just the name of a particular form of energy transfer from a system to the surroundings, or vice versa. For this reason we draw arrows representing work or heat that start or stop at the system boundary without crossing it (Fig. 4.5). In this context, we offer the following formal definition of work:

Work ***is the transfer of energy across a system or control-volume boundary, exclusive of energy carried across the boundary by a flow, and is not the result of a temperature gradient at the boundary or a difference in temperature between the system and the surroundings.***

Before presenting examples of work, it is useful to convert Eq. 4.5 to a form expressing the rate at which work is done. The time rate of doing work is called power, defined as

$$\mathcal{P} \equiv \dot{W} = \lim_{\Delta t \to 0} \frac{\delta W}{\Delta t} = \lim_{\Delta t \to 0} \frac{\boldsymbol{F} \cdot d\boldsymbol{s}}{\Delta t} = \boldsymbol{F} \cdot \frac{d\boldsymbol{s}}{dt}, \quad \textbf{(4.7a)}$$

or

$$\mathcal{P} = \boldsymbol{F} \cdot \boldsymbol{V}, \quad \textbf{(4.7b)}$$

where we recognize that $d\boldsymbol{s}/dt$ *is the velocity vector* $\boldsymbol{V}$. From this definition, we see that power enters or exits a system or control volume wherever a component of a force is aligned with the velocity at the boundary.

TYPES

Some common types of work are listed in Table 4.1. In many situations, the power, or rate of working, is the important quantity; therefore, expressions to evaluate the power are also shown.

TABLE 4.1 Common Types of Work

Type	Expression for W	Expression for $\dot{W}$ or $\mathcal{P}$*
Expansion or compression (moving boundary) work	$\delta W = Pd\mathcal{V}$ ${}_1W_2 = \int_1^2 Pd\mathcal{V}$	$\dot{W} = P\dfrac{d\mathcal{V}}{dt}$
Viscous work	${}_1W_2 = \int_{t_1}^{t_2} \tau_{\text{visc}} AVdt$	$\dot{W} = \tau_{\text{visc}} AV$
Shaft work	${}_1W_2 = \int_{\Omega_1}^{\Omega_2} \mathcal{T} d\Omega$	$\dot{W} = \mathcal{T}\omega$
Electrical work	${}_1W_2 = \int_{t_1}^{t_2} i\Delta\mathfrak{V} dt$	$\dot{W} = i\Delta\mathfrak{V}$
Flow work	${}_1W_2 = \int_{t_1}^{t_2} \dot{m}Pv dt$	$\dot{W} = \dot{m}Pv$

* Rate of work, $\dot{W}$, and power, $\mathcal{P}$, are used synonymously throughout this book.

Expansion (or Compression) Work

In systems or control volumes where a boundary moves, work is performed *by* the system if it expands, whereas work is done *on* the system if the system is compressed. Concomitantly, work is done *on* the surroundings by an expanding system, and work is done *by* the surroundings when the system contracts. As an example of this type of work, consider the expansion of a gas contained in a piston–cylinder assembly, as shown in Fig. 4.6. We use this specific example to illustrate application of the fundamental definition of work and some of the subtleties that need to be considered. Examining the entire boundary of the system, we see that work can only be done at the portion of the boundary that is in contact with the piston, as this is the only part of the boundary where there is motion. For a force to act over a distance (Eq. 4.5), the boundary must move. In our examination of the boundary, we also note that the only force present is that due to the pressure of the gas in the system, where we have ignored any possible viscous forces created by friction between the cylinder wall and the moving gas. For our simple geometry, the magnitude of the force exerted by the gas on the piston is given by

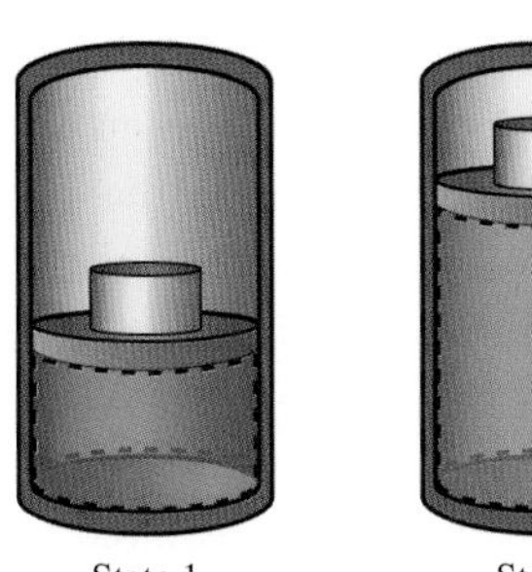

FIGURE 4.6 Work is done as a gas expands and pushes back the surroundings in a piston–cylinder assembly.

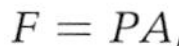

$$F = PA,$$

where A is the cross-sectional area of the piston and P is the pressure. The pressure force acts vertically upward in the same direction as the piston motion; thus, the dot product $\boldsymbol{F} \cdot \boldsymbol{ds}$ in our definition reduces to the product of the magnitude of the force and the vertical displacement (i.e., Fdx). The incremental work done is then

$$\delta W = PAdx.$$

For our simple cylindrical geometry, we immediately recognize that Adx is the volume displaced; that is,

$$d\mathcal{V} = Adx.$$

The incremental moving boundary work then is

$$\delta W = Pd\mathcal{V}, \tag{4.8a}$$

and so the total moving-boundary work done in going from state 1 to state 2 is

$$_1W_2 = \int_1^2 P d\mathcal{V} \tag{4.8b}$$

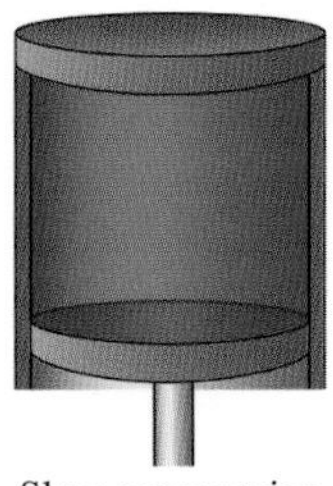

Slow compression

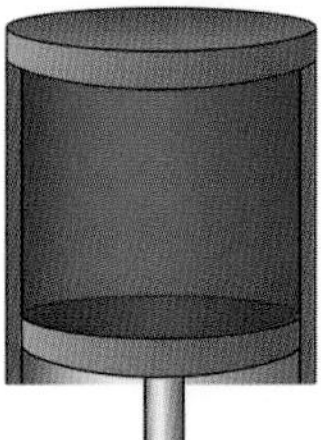

Rapid compression

FIGURE 4.7 Slow compression (top) results in the pressure being uniform throughout the system as represented by the uniform blue shading. Rapid compression (bottom) results in a pressure gradient with a higher pressure near the piston face.

Because of the relationship between pressure and volume appearing in Eqs. 4.8a and 4.8b, moving boundary work is frequently called P–$d\mathcal{V}$ work. We can also express the instantaneous power produced as

$$\mathcal{P} = \dot{W} = P\frac{d\mathcal{V}}{dt}. \tag{4.8c}$$

These expressions for work and power (Eqs. 4.8a–4.8c) represent the simple reversible work mode associated with a simple compressible substance.

At this juncture it is important to ask, what assumptions are built into Eqs. 4.8 that might restrict their use? First, as suggested in our development, we assume that the only force acting at the system boundary is that resulting from pressure. Second, we require that the pressure be a meaningful thermodynamic property of the system as a whole. For this to be true, the motion of the piston must be sufficiently slow so that there is enough time for a sufficient number of molecular collisions to cause the pressure to be uniform within the gas volume (Fig. 4.7). A characteristic time to achieve mechanical (pressure) equilibrium is of the order of the height of the volume divided by the speed of sound in the gas. As an example, consider room-temperature air and a volume height of 150 mm (~6 in). For this situation, approximately 0.4 ms are required for the change at the moving boundary to be communicated to the gas molecules at the bottom of the cylinder. Thus, our theoretical restriction is that the process must be quasi-static, where the practical meaning of quasi-static is determined by the time scale $t_c \equiv L_c/a$, where L_c is the characteristic length and a is the speed of sound.

Equilibrium and quasi-equilibrium processes are discussed in Chapter 1. You may find a review of that material useful here.

Our piston–cylinder example is also useful to illustrate that motion is required to produce work, that is, to transfer energy from the gas molecules to the surroundings in the form of work. Consider throwing a ball against a stationary wall in which the ball rebounds in a perfectly elastic manner, as suggested in Fig. 4.8. If, say, you threw the ball at 100 m/s, the ball would rebound back at 100 m/s. The kinetic energy of the ball, $MV^2/2$, is thus the same before and after the collision. The ball experiences no loss of energy. We now allow the wall to move. What then happens to the magnitude of the rebound velocity? In this case, the ball will return with a velocity less than its incoming velocity. For example, when the wall moves at 25 m/s, the rebound velocity

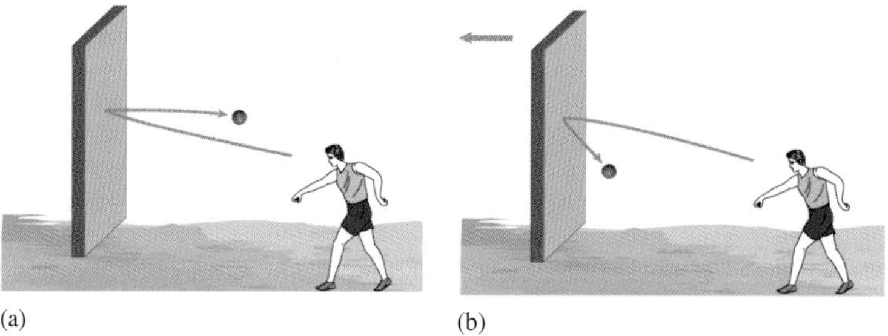

FIGURE 4.8 (a) A ball thrown at a stationary wall rebounds with the same speed that it strikes the wall, assuming an elastic collision. **(b)** If the wall is moving away from the incoming ball, the rebound velocity is less than the approach velocity.

is 50 m/s, a value substantially less than the incoming velocity of 100 m/s. The kinetic energy of the ball undergoes a considerable reduction after the collision with the moving wall. Analogous to our ball-throwing example, gas molecules lose energy in their collisions with a receding boundary. If the boundary is advancing, the molecules, of course, gain energy. This example highlights the fundamental idea that work is an energy transfer across a boundary.

Although the expression defining moving-boundary work is simple – work equals the integral of pressure with respect to volume – thermodynamics students frequently have difficulty applying this expression for processes other than the straightforward constant-pressure process. The following examples illustrate how the expression (Eq. 4.8) can be applied to a variety of processes, starting with the trivial constant-pressure case. In all cases, the key task is to relate the pressure as a function of volume.

Example 4.1 Piston–Cylinder Processes with Air

Consider a piston–cylinder arrangement containing 5.057×10^{-4} kg of dry air. For the following two quasi-static processes, determine the quantity of work performed by or on the air in the cylinder:

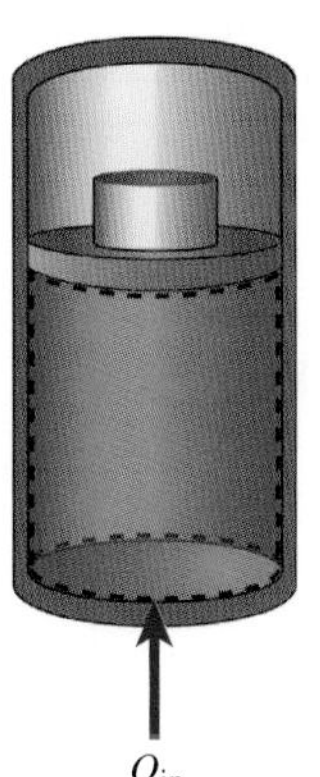

A. Constant-pressure heat-addition process
B. Isothermal expansion process

Also completely define the final state (i.e., P_2, $\mathcal{V}_2$, and T_2) and sketch the process in P–$\mathcal{V}$ coordinates. For both processes, the following conditions apply:

At initial state	At final state
$\mathcal{V}_1 = 2.54 \times 10^{-4}$ m^3 $T_1 = 350$ K	$\mathcal{V}_2 = 5.72 \times 10^{-4}$ m^3

Solution (Part A)

Known Constant-pressure process, M, $\mathcal{V}_1$, T_1, $\mathcal{V}_2$

Find ${}_1W_2$, P_2, T_2

Sketch

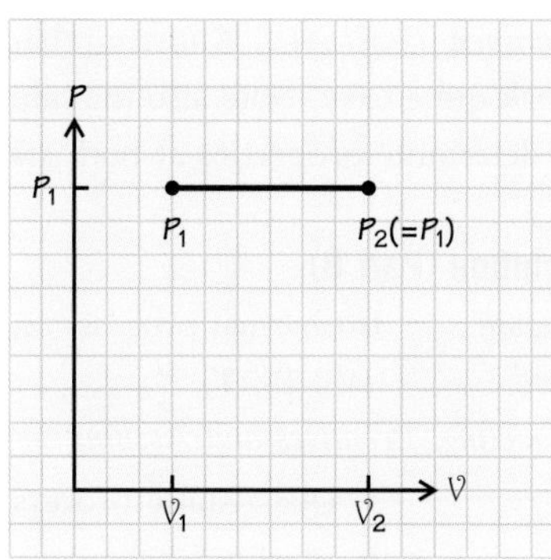

Modeling, Premises and Assumptions

i. Quasi-static process (given)
ii. Ideal-gas behavior

Analysis Because the process is quasi-static, we can apply Eq. 4.8b, which is easily integrated since the pressure is constant, so

$$ {}_1W_2 = \int_1^2 P d\mathcal{V} = P \int_1^2 d\mathcal{V} = P(\mathcal{V}_2 - \mathcal{V}_1). $$

To evaluate this equation, we only need to find the pressure, as both $\mathcal{V}_1$ and $\mathcal{V}_2$ are given. To find P, we apply the ideal-gas equation of state (Eq. 2.26c) at state 1:

$$ P_1\mathcal{V}_1 = MRT_1 $$

or

$$ P_1 = \frac{MRT_1}{\mathcal{V}_1} $$

where $R(\equiv R_u/\mathcal{M} = 287.0\ \text{J/kg·K})$ is the gas constant for air (Appendix C). Substituting numerical values, we get

$$ P_1 = \frac{5.057 \times 10^{-4}\text{kg}\,(287.0\ \text{J/kg·K})\,350\,\text{K}}{2.54 \times 10^{-4}\,\text{m}^3}\left[\frac{\text{N·m}}{\text{J}}\right]\left[\frac{\text{Pa}}{\text{N/m}^2}\right] = 200 \times 10^3\,\text{Pa}. $$

The work is then

$$ {}_1W_2 = 200 \times 10^3\,\text{Pa}(5.72 \times 10^{-4} - 2.54 \times 10^{-4})(\text{m}^3)\left[\frac{\text{N/m}^2}{\text{Pa}}\right]\left[\frac{\text{J}}{\text{N·m}}\right] = 63.6\ \text{J}. $$

We now define the final state. If we know two independent intensive properties, the state principle tells us that all other properties can be found. Since P is constant, we know $P_2 = P_1 = 200\,\text{kPa}$, and $\mathcal{V}_2$ and M are given. With this information, we again apply the ideal-gas equation of state (Eq. 2.26c), this time to find T_2:

$$ \begin{aligned} T_2 &= \frac{P_2\mathcal{V}_2}{MR} \\ &= \frac{200 \times 10^3\,\text{Pa}(5.72 \times 10^{-4}\,\text{m}^3)}{5.057 \times 10^{-4}\,\text{kg}(287.0\text{J/kg·K})}\left[\frac{\text{N/m}^2}{\text{Pa}}\right]\left[\frac{\text{J}}{\text{N·m}}\right] = 788.2\,\text{K}. \end{aligned} $$

Comment (Part A) Knowing that the pressure is constant made the calculation of the work quite easy. Note also the importance of the use of the ideal-gas equation of state to determine properties at both state 1 and state 2.

Solution (Part B)

Known Isothermal process, M, $\mathcal{V}_1$, T_1, $\mathcal{V}_2$

Find ${}_1W_2$, P_2, T_2

Modeling, Premises and Assumptions

i. Quasi-static process
ii. Ideal-gas behavior

Analysis We delay drawing a P–$\mathcal{V}$ sketch until the appropriate mathematic relationship between P and $\mathcal{V}$ is determined. We appeal again to the ideal-gas law to do this:

$$P = MRT\left(\frac{1}{\mathcal{V}}\right).$$

Here we recognize that (1) MRT is a constant, since an isothermal process is one carried out at constant temperature, and (2) the P–$\mathcal{V}$ relation is hyperbolic ($P \sim \mathcal{V}^{-1}$). The work can now be found from Eq. 4.8b as

$$\begin{aligned} {}_1W_2 &= \int_1^2 P d\mathcal{V} = MRT_1 \int_1^2 \frac{d\mathcal{V}}{\mathcal{V}} \\ &= MRT_1[\ln \mathcal{V}]_1^2 = MRT_1(\ln \mathcal{V}_2 - \ln \mathcal{V}_1) = MRT_1 \ln \frac{\mathcal{V}_2}{\mathcal{V}_1}. \end{aligned}$$

Substituting numerical values, we obtain

$${}_1W_2 = 5.057 \times 10^{-4}\,\text{kg}(287.0\,\text{J/kg·K})(350\,\text{K}) \ln\left[\frac{5.72 \times 10^{-4}\,\text{m}^3}{2.54 \times 10^{-4}\,\text{m}^3}\right] = 41.2\,\text{J}.$$

To completely define state 2, we now need P_2. Following the same procedure as in Part A, we apply the ideal-gas equation of state (Eq. 2.26c):

$$P_2 = \frac{MRT_2}{\mathcal{V}_2}.$$

Since $T_2 = T_1 = 350\,\text{K}$,

$$P_2 = \frac{5.057 \times 10^{-4}\,\text{kg}(287.0\,\text{J/kg·K})350\,\text{K}}{5.72 \times 10^{-4}\,\text{m}^3}\left[\frac{\text{N·m}}{\text{J}}\right]\left[\frac{\text{Pa}}{\text{N/m}^2}\right] = 88.8 \times 10^3\,\text{Pa}.$$

We can now plot this process in P–$\mathcal{V}$ coordinates:

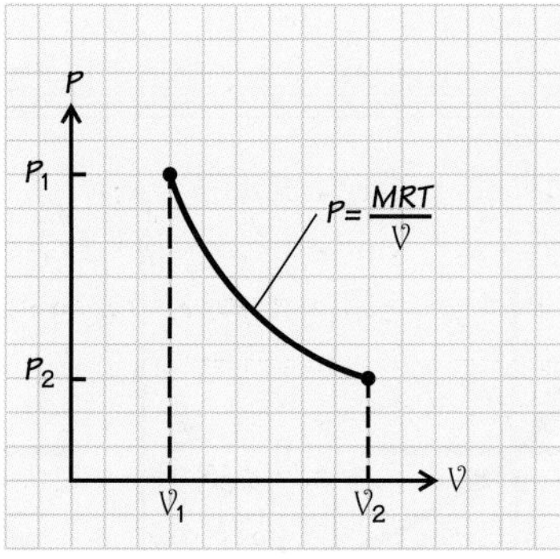

Comment (Part B) Note that the area under this curve is the work. Comparing this graph with that of Part A, we immediately see that less work is performed in the isothermal process, which is consistent with our calculations. Being able to sketch a process in P–$\mathcal{V}$ coordinates is particularly useful in dealing with thermodynamic systems. Such sketches immediately show the work (area under the curve), provided the process is carried out quasi-statically, a requirement for ${}_1W_2 \equiv \int P d\mathcal{V}$ to hold. Note also that in the solutions to both Parts A and B, we employed only fundamental definitions and the state principle.

Self-Test 4.1

Determine whether the work in Example 4.1 is performed on or by the system and on or by the surroundings.

(Answer: Since the numerical values for the work done by the system are positive, the work is being performed by the system. A commensurate quantity of work is performed on the surroundings.)

The preceding example illustrates the calculation of expansion/compression (moving boundary) work for two very specific cases: ideal gases undergoing (i) constant-pressure and (ii) constant-temperature expansions. Here we introduce the concept of the polytropic process and apply it to the calculation of compression/expansion work. The polytropic process relates pressure and volume in a generalized way, independent of the working fluid. For example, the fluid does not have to be an ideal gas. Moreover, liquids can follow a polytropic process. We define a polytropic process as any process that obeys the property relation

$$Pv^n = \text{constant}, \tag{4.9a}$$

or

$$P\mathcal{V}^n = \text{constant}. \tag{4.9b}$$

Here the **polytropic exponent** n is a known constant. For certain values of n, we recover relationships previously developed for ideal gases. These are shown in Table 4.2. For example, when n is unity, a constant-temperature (isothermal) process is described, that is,

$$Pv^1 = \text{constant } (= RT).$$

Figure 4.9 graphically illustrates these special-case, ideal-gas, polytropic processes in pressure–volume, P–v, and temperature–entropy, T–s, coordinates.

TABLE 4.2 Special Cases of Polytropic Processes – Ideal Gases

Process	Constant property	Polytropic exponent (n)
Isobaric	P	0
Isothermal	T	1
Isentropic	s or S	γ
Isochoric	v or $\mathcal{V}$	$\pm\infty$

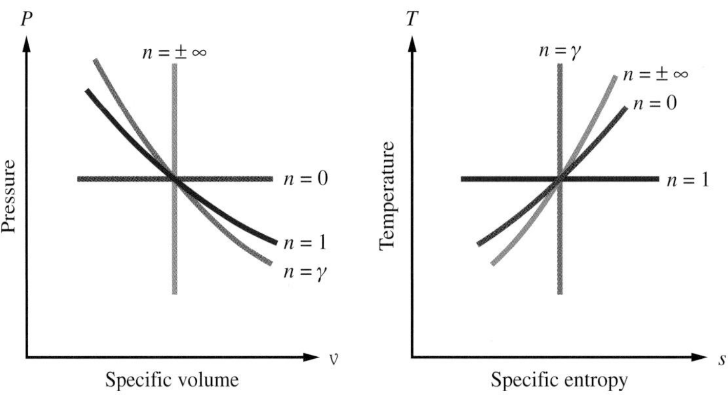

FIGURE 4.9 P–v and T–s diagrams for ideal gases illustrating polytropic paths for special cases of constant pressure ($n = 0$), constant temperature ($n = 1$), constant entropy ($n = \gamma$), and constant volume ($n = +/- \infty$).

A polytropic process is often used to simplify and model complex processes. A common use is modeling compression and expansion processes when heat-transfer effects are present, as illustrated in the following example.

Example 4.2 SI Engine Work

The compression and expansion processes associated with spark-ignition engines can be modeled crudely as quasi-static (reversible) adiabatic (no heat transfer) processes. We will see in Chapter 7 that such processes are isentropic (constant-entropy) processes. For compression and expansion processes described in this way, neither the pressure nor the temperature will remain constant during the process (cf. Example 4.1). Furthermore, we assume that the working fluid is dry air, rather than a mixture of fuel and air (compression) or combustion products (expansion). This set of assumptions forms the basis for the air standard Otto cycle, which we explore in detail in Chapter 9. With these assumptions, the compression and expansion processes obey

$$P\mathcal{V}^{\gamma} = \text{constant},$$

where γ $(= c_p/c_v)$ is the ratio of specific heats and has a value of 1.4 for air over a wide range of temperatures. Using this model, determine the compression work ${}_1W_2$, the expansion work ${}_3W_4$, and the net work associated with the cycle defined by the following processes:

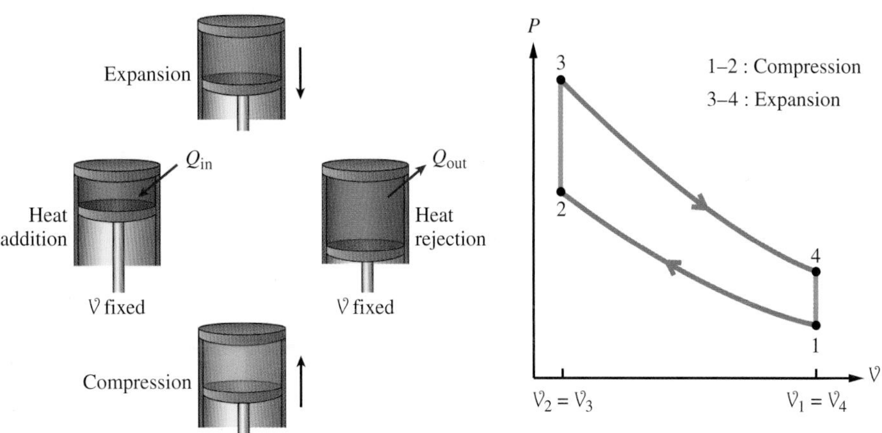

State	1	2	3	4
P (kPa)	100	—	—	—
T (K)	300	—	2800	—
$\mathcal{V}$ (m^3)	6.543×10^{-4}	0.8179×10^{-4}	0.8179×10^{-4}	6.543×10^{-4}

Solution

Known Adiabatic reversible (isentropic) processes ($P\mathcal{V}^{\gamma}$ = constant); selected properties at states 1, 2, 3, and 4; working fluid is air.

Find ${}_1W_2$, ${}_3W_4$, net work

Sketch

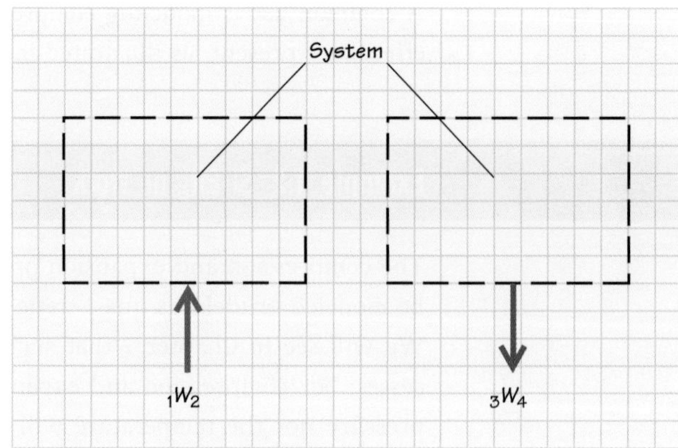

Modeling, Premises and Assumptions

i. Closed system
ii. Reversible (quasi-static) processes
iii. Adiabatic processes
iv. Ideal-gas behavior

Analysis First we recognize that we are dealing with a closed thermodynamic system, the air trapped in the cylinder, and not an open system. This fact, combined with the assumption of quasi-static compression and/or expansion, allows us to use Eq. 4.8b to evaluate the work done, ${}_1W_2$ and ${}_3W_4$:

$${}_1W_2 = \int_1^2 Pd\mathcal{V}$$

and

$${}_3W_4 = \int_3^4 Pd\mathcal{V}.$$

The functional relationship between P and $\mathcal{V}$ is given by $P\mathcal{V}^\gamma =$ constant. Knowing both P and $\mathcal{V}$ at state 1, we express the process from state 1 to state 2 as

$$P\mathcal{V}^\gamma = P_1\mathcal{V}_1^\gamma,$$

or

$$P = \frac{P_1\mathcal{V}_1^\gamma}{\mathcal{V}^\gamma}.$$

As noted in the previous example, obtaining the pressure P as a function of the volume $\mathcal{V}$ is key to solving for the work. Substituting this into Eq. 4.8b and integrating yields

$$\begin{aligned} {}_1W_2 &= \int_1^2 Pd\mathcal{V} = P_1\mathcal{V}_1^\gamma \int_1^2 \frac{d\mathcal{V}}{\mathcal{V}^\gamma} \\ &= P_1\mathcal{V}_1^\gamma \left[\frac{\mathcal{V}^{1-\gamma}}{1-\gamma}\right]_{\mathcal{V}_1}^{\mathcal{V}_2} \\ &= \frac{P_1\mathcal{V}_1^\gamma}{1-\gamma}\left(\mathcal{V}_2^{1-\gamma} - \mathcal{V}_1^{1-\gamma}\right). \end{aligned}$$

Because all the quantities on the right-hand side are known, we can numerically evaluate ${}_1W_2$ as follows:

$$\begin{aligned}{}_1W_2 &= \frac{100\times10^3\text{Pa}(6.543\times10^{-4}\text{m}^3)^{1.4}}{-0.4}\\ &\quad\times\left[(0.8179\times10^{-4}\text{m}^3)^{-0.4}-(6.543\times10^{-4}\text{m}^3)^{-0.4}\right]\left[\frac{\text{N/m}^2}{\text{Pa}}\right]\left[\frac{\text{J}}{\text{N}\cdot\text{m}}\right]\\ &= -212.2\text{ J}.\end{aligned}$$

Note that the sign of ${}_1W_2$ is negative since work is done *on* the air. Note also the treatment of the fractional powers in dealing with the units: $(\text{m}^3)^{1.4}\times(\text{m}^3)^{-0.4}=\text{m}^3$.

Our analysis of the expansion process is similar; however, it is complicated by our not knowing the pressure at state 3. We need this to evaluate the constant associated with $P\mathcal{V}^\gamma = \text{constant} = P_3\mathcal{V}_3^\gamma$. To find P_3 we recognize that, by definition of a closed system, $M_1 = M_2 = M_3 = M_4 = M$. We thus apply the ideal-gas equation of state (Eq. 2.26) twice: once to find M, using state 1 properties, and a second time to obtain P_3. We express these operations mathematically as

$$M = \frac{P_1\mathcal{V}_1}{RT_1}$$

and

$$P_3 = M\frac{RT_3}{\mathcal{V}_3} = \left(\frac{P_1\mathcal{V}_1}{RT_1}\right)\frac{RT_3}{\mathcal{V}_3},$$

which simplifies to

$$P_3 = P_1\left(\frac{\mathcal{V}_1}{\mathcal{V}_3}\right)\left(\frac{T_3}{T_1}\right).$$

Substituting numerical values, we obtain

$$P_3 = 100\times10^3\text{Pa}\left(\frac{6.543\times10^{-4}\text{m}^3}{0.8179\times10^{-4}\text{m}^3}\right)\left(\frac{2800\text{ K}}{300\text{ K}}\right) = 7.466\times10^6\text{ Pa}.$$

Following the same procedures as for the compression process, we integrate Eq. 4.8b to obtain

$${}_3W_4 = \frac{P_3\mathcal{V}_3^\gamma}{1-\gamma}\left[\mathcal{V}_4^{1-\gamma}-\mathcal{V}_3^{1-\gamma}\right],$$

which is numerically evaluated as

$$\begin{aligned}{}_3W_4 &= \frac{7.466\times10^6\text{ Pa}(0.8179\times10^{-4}\text{ m}^3)^{1.4}}{-0.4}\\ &\quad\times\left[(6.543\times10^{-4}\text{ m}^3)^{-0.4}-(0.8179\times10^{-4})^{-0.4}\text{ m}^3\right]\left[\frac{\text{N/m}^2}{\text{Pa}}\right]\left[\frac{\text{J}}{\text{N}\cdot\text{m}}\right]\\ &= +862.1\text{ J}.\end{aligned}$$

The plus sign here emphasizes that work is done *by* the air during the expansion process.

Because there is no volume change for the heat-interaction processes 2–3 and 4–1, no work is done in either process; thus, the net work for the cycle 1–2–3–4–1 is

$$\begin{aligned}W_{\text{net}} &= {}_1W_2 + {}_3W_4 = -212.2\text{ J} + 862.1\text{ J}\\ &= +649.9\text{ J}.\end{aligned}$$

Comments We first note that the net work done is positive, which meets our expectations that engines produce work. It is also important to point out how our crude model differs from the actual processes in a real spark-ignition engine: First, heat transfer is present in the real engine; in particular, there is a substantial heat loss from the hot gases to the cylinder walls during the expansion process. Also affecting the actual net work is the timing of the combustion process, which begins before the piston reaches top center and ends somewhat after the piston begins its descent during the expansion. Both of these factors affect the P–$\mathcal{V}$ relationship; nevertheless, if we were to measure P versus $\mathcal{V}$ and apply Eq. 4.8b, this would yield a close approximation to the work performed. In fact, experimental P–$\mathcal{V}$ data are used in just this way in engine research (see the image). Heat losses and a finite combustion time result in the actual work being less than that predicted by our crude model.

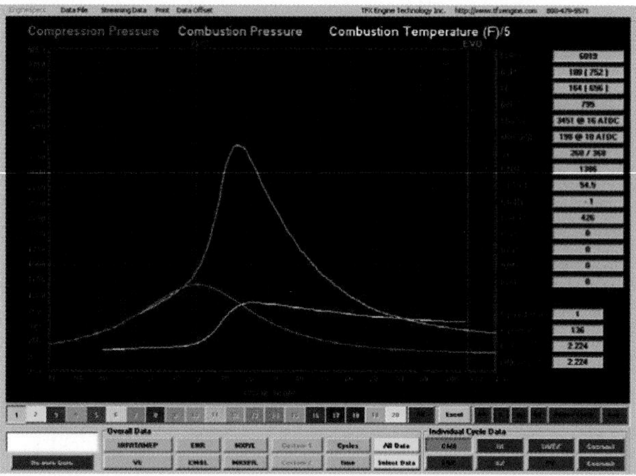

Pressure transducers mounted within the combustion chamber of an engine provide a record of pressure versus crank angle, i.e., time. The crank angle record can be related to the instantaneous combustion chamber volume. Courtesy of TFX Engine Technology Inc. (www.tfxengine.com).

Self-Test 4.2

A piston–cylinder device contains 5 kg of saturated liquid water at a pressure of 100 kPa. Heat is added until a saturated vapor state exists. Determine the work performed and whether it is done by or on the system.

(Answer: 846.4 kJ, done by the system)

We summarize and generalize the results of Examples 4.1 and 4.2 in Table 4.3. Note the restriction of the constant-temperature process to ideal gases. All processes are assumed to be quasi-static (quasi-equilibrium) processes.

Shaft Work[1]

In many practical devices, power is transmitted across a control surface via a rotating shaft. Figure 4.10 shows gas-turbine and diesel engines, fans, and a propeller-driven aircraft, all of which rely on shaft power for their operation. A control surface that

[1] The rate of work $\dot{W}$, or power $\mathcal{P}$, is frequently implied by the use of the word *work*, as here. The context usually makes clear whether the reference is to W or $\dot{W}$ (i.e., $\mathcal{P}$).

TABLE 4.3 Closed-System Moving-Boundary (P–$d\mathcal{V}$) Work: Some Special Cases

Process	$({}_1W_2)_{out}$	
Constant pressure	$P(\mathcal{V}_2 - \mathcal{V}_1)$	T4.3a
Constant temperature (Ideal gases)	$P_1\mathcal{V}_1 \ln \frac{\mathcal{V}_2}{\mathcal{V}_1}$	T4.3b
	or	
	$MRT \ln \frac{\mathcal{V}_2}{\mathcal{V}_1}$	T4.3c
Polytropic	$\frac{P_2\mathcal{V}_2 - P_1\mathcal{V}_1}{1-n}$	T4.3d

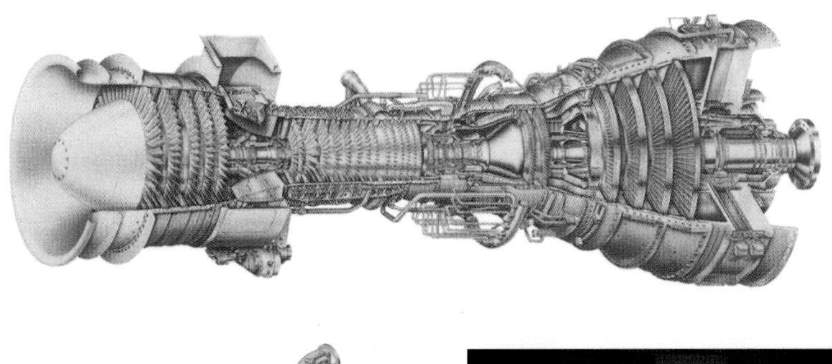

FIGURE 4.10 Examples of devices in which shaft work or power is important. Stationary gas-turbine engine (top), diesel engine (left), wind tunnel fans and propeller-driven aircraft (right).

Drawings and photographs courtesy of General Electric Co., Scania, and NASA, respectively.

cuts through a shaft exposes a force acting over a distance. From a formal analysis of the forces within the shaft and the application of Eq. 4.7, the instantaneous work rate or shaft power is expressed as

$$\dot{W}_{\text{shaft}} = \mathcal{P}_{\text{shaft}} = \mathcal{T}\omega \tag{4.10}$$

where $\mathcal{T}$ is the torque and ω is the angular velocity of the shaft. In many of the applications in this book, $\dot{W}_{\text{shaft}}$ (or $\mathcal{P}_{\text{shaft}}$) will be a given quantity or a quantity derived from, usually, a conservation of energy expression; thus, we seldom refer to either the torque or the angular velocity.

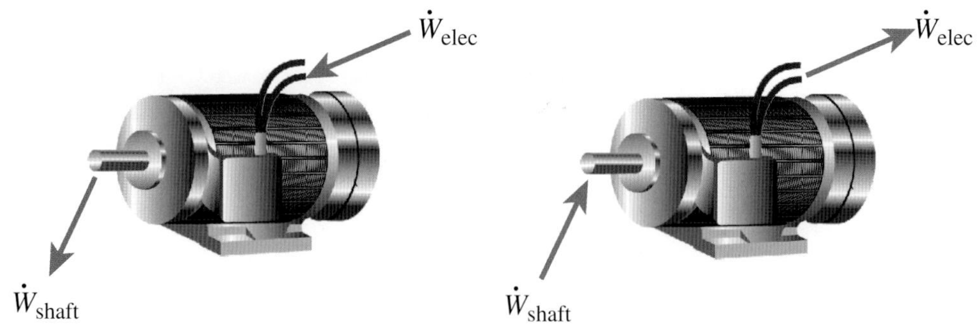

FIGURE 4.11 Electric motors and electrical generators interconvert electrical power and shaft power.

Electrical Work

The flow of an electrical current across the boundary of either a closed or an open system results in a flow of energy. This transfer of energy corresponds to work (over a time interval) or power (at an instant). See Fig. 4.11. That this is true can be seen from a careful application of our definition of work (Eq. 4.5) to the electrical forces and the motion of electrons through a conductor. For our purposes, it is sufficient to know how the electrical work and power relate to voltage and current:[2]

$$W_{elec} = \int_{t_1}^{t_2} i\Delta\mathcal{V}\,dt \tag{4.11a}$$

and

$$\dot{W}_{elec} = \mathcal{P}_{elec} = i\Delta\mathcal{V}. \tag{4.11b}$$

Note that electrical power flows into the system when $\Delta\mathcal{V}(=\mathcal{V}_{in} - \mathcal{V}_{out})$ is positive and, conversely, flows out when $\Delta\mathcal{V}(=\mathcal{V}_{in} - \mathcal{V}_{out})$ is negative.

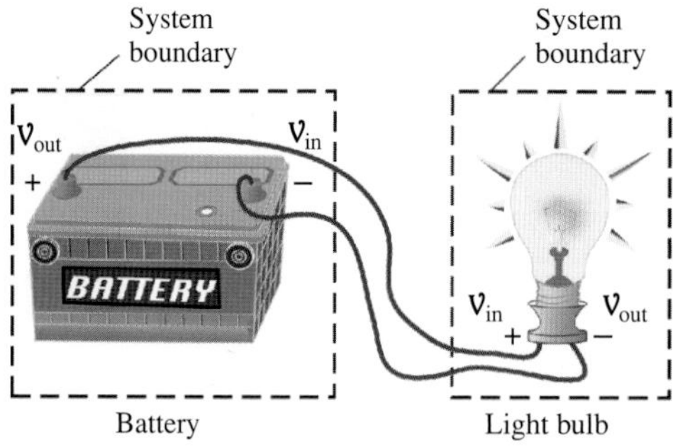

FIGURE 4.12 Power exits the system defined as the battery, whereas power enters the system defined as the light bulb.

Figure 4.12 shows examples of electrical work. Choosing a boundary to select just the battery as a thermodynamic system, we see that electrical power is delivered across the boundary from the system to the surroundings, since $\mathcal{V}_{out} > \mathcal{V}_{in}$. In contrast, selecting the light bulb to be a system, we see that electrical power is now delivered in the opposite direction (i.e., from the surroundings to the system). Here $\mathcal{V}_{in}$ is greater

[2] Implicit in our discussion of electrical work, current, and voltage is that we are dealing with DC circuits or with AC circuits containing only resistance elements. A treatment of AC circuits, nonresistive loads, and power factors is beyond the scope of this book.

than $\mathcal{V}_{out}$. If we were to choose a system boundary that enclosed *both* the battery and the light bulb, no work interaction would exist.

As another example, consider the steam power plant, one of our integrating applications. Our choice of the open-system boundary shown in Fig. 4.13 results in electrical work (power) crossing to the surroundings, while high-pressure, high-temperature steam enters the system and low-pressure, low-temperature steam exits the system. Many systems of practical importance involve the conversion of electrical work to other forms of work (see Fig. 4.14).

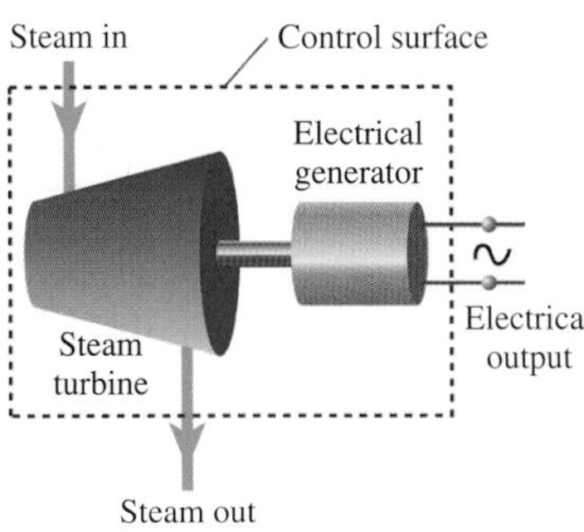

FIGURE 4.13 The open system shown includes a steam turbine and an electrical generator (left). Steam enters and exits the system, and electrical work (power) exits the system. The photographs show an electrical generator for a 322.5 MW_e power plant (middle) and electrical generator windings (right) (AGEfotostock).

FIGURE 4.14 Multiple energy transformations are involved in pumping water. Electrical power drives a motor; the shaft power from the motor drives the pump; and the pump transfers energy to the water, resulting in flow work ($P_{out} > P_{in}$). (AGEfotostock).

Flow Work

The work associated with moving a fluid into and out of an open system (control volume) is called **flow work**. Pressure forces acting over flow inlets and exits produce this work. To evaluate the flow work, we appeal to our fundamental definition, Eq. 4.7b, which requires the identification of the appropriate forces.

When ascertaining what forces act *on* a system, our point of view is always from a position *outside* the system. Forces due to pressure always act perpendicular to the boundary (control surface) and are directed to the interior of the system (control volume). As an example, consider the flow in a pipe illustrated in Fig. 4.15. We focus our attention on the inlet and exit areas, designated as stations 1 and 2. Assuming a uniform pressure distribution at stations 1 and 2, we can calculate the pressure forces as simply the products of the respective pressures and areas as indicated.

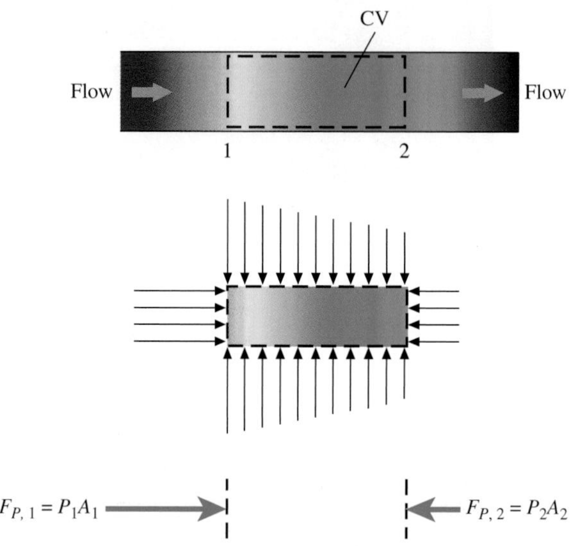

FIGURE 4.15 Pressure forces associated with an open system (control volume, CV) inside a pipe with a flowing fluid.

Having identified the forces at the inlet and exit, we now can evaluate the flow work done *on* the fluid at the inlet surface designated 1. Since the velocity over the inlet area A_1 has the same direction as $F_{P,1}$ the dot product in Eq. 4.7b is simply

$$\dot{W}_{\text{flow},1} = P_1 A_1 V_1,$$

where, for the time being, we have assumed that V_1 is uniform over A_1. The simple mathematical manipulation of multiplying and dividing by the density ρ yields

$$\dot{W}_{\text{flow},1} = \rho_1 A_1 V_1 \frac{P_1}{\rho_1}.$$

In this equation, we recognize that $\dot{m}_1 = \rho_1 A_1 V_1$ and that $1/\rho_1$ is just v_1, the specific volume. Thus,

$$\dot{W}_{\text{flow},1} = \dot{m}_1 P_1 v_1, \tag{4.12a}$$

where we emphasize that this quantity is the rate of work done *on* the system by the surroundings. It is a simple matter to show that Eq. 4.12 applies even if the velocity distribution is not uniform; of course, both P_1 and v_1 must be uniform over A_1.

A similar derivation can be applied to obtain the flow work at the open system exit (i.e., at station 2). In this case, however, the pressure force is directed opposite to the velocity; thus, the dot product $\mathbf{F}_{P,2}\cdot\mathbf{V}_2 = F_{P,2}\,V_2\cos(180°) = -F_{P,2}\,V_2$. The rate of flow work performed on the fluid is then

$$\dot{W}_{\text{flow},2} = -\dot{m}_2 P_2 v_2. \quad \textbf{(4.12b)}$$

We end this development by noting that flow work only occurs when a fluid crosses a boundary. If there is no flow, there cannot be any component of velocity aligned with the pressure force.

To review enthalpy, see Eq. 2.18 and the associated discussion in Chapter 2.

Foreshadowing the development of the conservation of energy principle in Chapter 5, we point out that the pressure–specific-volume product Pv in the flow work is frequently grouped with the internal energy of the entering or exiting fluid (i.e., $u + Pv$). You may recall that this particular grouping of thermodynamic properties is termed the enthalpy.

4.3 Sign Conventions and Units

In the next chapter, we will examine the principle of energy conservation and the many ways that this principle can be expressed. Since writing a conservation of energy expression is analogous to maintaining an accountant's ledger, we need to know whether various energy terms are credits or debits to our energy account. In this brief section, we present a consistent set of sign conventions for heat and work interactions.

Heat transfer and its time rate are regarded as *positive* when the direction of the energy exchange is *from the surroundings to the system*. Conversely, heat transfer from a system to the surroundings is a negative quantity. Work and power are defined to be *positive* when they are delivered *from the system to the surroundings* and negative when the converse is true. Thus, the work associated with an expanding gas is positive, whereas the work associated with compression is negative, as we saw in Example 4.2 for the spark-ignition engine. In a steam turbine, the steam expands and produces positive power, as suggested by Fig. 4.13. In contrast, a water pump requires a power input to operate.

We note that sign conventions can cause confusion – because they are just that, conventions, rather than physically intuitive constructs. Moreover, different books may adopt different conventions. Confusion can be avoided by using known directions for both heat and work interactions (*into* or *out of* the system) in writing energy balances (applying the first law of thermodynamics); thus, energy quantities (both heat and work) *in* minus those *out* will always equal the energy change within the system. We discuss this in greater detail in Chapter 5.

The SI unit associated with energy, heat, and work is the joule, which is abbreviated as J. The joule is derived from the definition of work and relates to the fundamental units as follows:

$$\text{joule} = \text{newton}\cdot\text{meter} = \frac{\text{kilogram}\cdot\text{meter}}{\text{second}^2}\cdot\text{meter},$$

or

$$\text{J} = \text{kg}\cdot\text{m}^2/\text{s}^2.$$

Power, the time rate of doing work, is expressed in watts (W), as is the heat transfer rate, $\dot{Q}$. The watt is expressed in terms of the fundamental units as

$$\text{watt} = \frac{\text{joule}}{\text{second}},$$

or

$$\mathrm{W} = \frac{\mathrm{kg{\cdot}m^2}}{\mathrm{s^3}}.$$

In the United States, other units also are used for heat work and power. The British thermal unit (Btu) is frequently used for energy, heat, and work; and horsepower is used for mechanical power. Conversions to SI units are as follows:

$$1 \text{ British thermal unit (Btu)} = 1055.056 \text{ joules (J)},$$
$$1 \text{ horsepower(hp)} = 745.7 \text{ watts (W)}.$$

Other units are used as well. An extensive list of unit conversions is provided in the front matter of this book. Unfortunately, a wide variety of units is commonly used in commerce and industry. You should be comfortable and proficient in converting units. Some end-of-chapter problems are designed for you to practice these conversions.

Example 4.3 Energy Transfer in a Steam Power Plant

(Credit: aaron007 / iStock / Getty Images Plus.)

(Credit: Flowserve Corporation.)

Consider the simple steam power plant cycle shown in Fig. 4.16. Using the indicated control volumes, identify all the energy transfers (i.e., heat and work interactions) for each of the components.

Solution

We start at the water (feedwater) pump and proceed around the loop.

Water Pump We have redrawn the control volume for the pump in Fig. 4.17a. Here we identify a work input from an electric motor (i.e., shaft work in) and flow work in and out. We assume that the casing of the pump will be hotter than the surroundings, so a heat loss is also indicated. We will see in Chapter 5 that this heat loss is negligible compared to the other energy transfers.

Boiler In our simplified boiler, water enters the water chamber (drum) and steam exits the steam drum. Water is converted to steam in a series of tubes connecting the

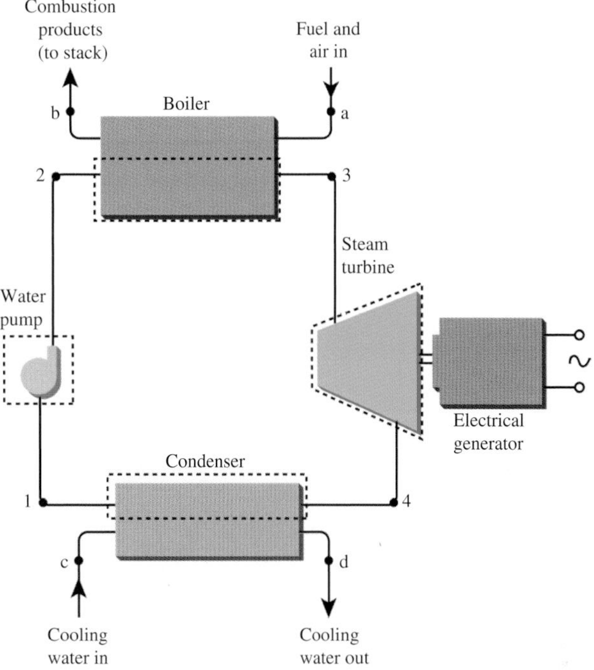

FIGURE 4.16 Rankine cycle schematic for Example 4.3 with individual open systems defined for each component.

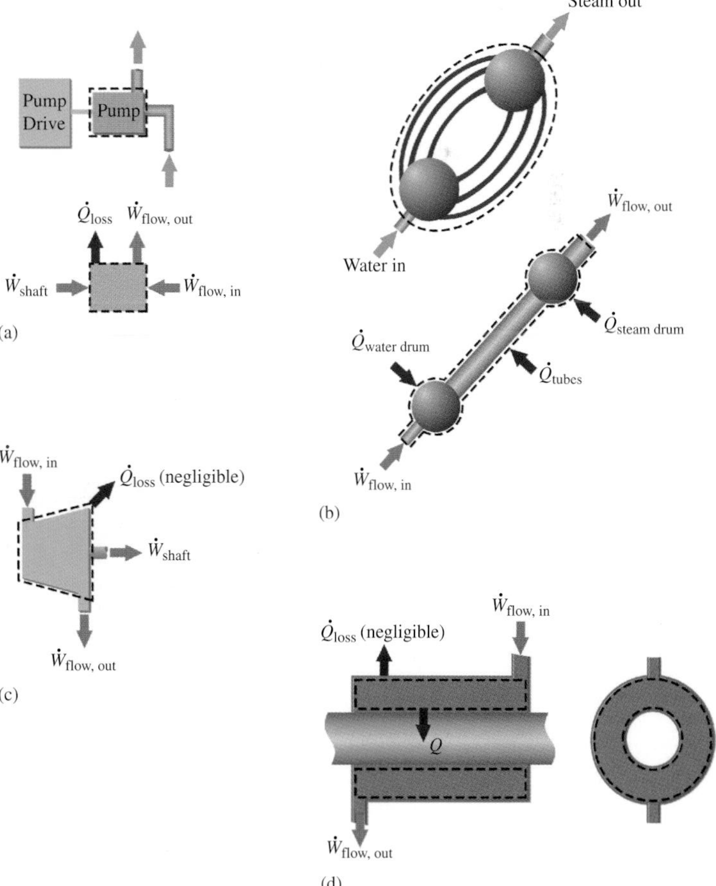

FIGURE 4.17 Open systems showing heat and work interactions for (a) feedwater pump, (b) boiler, (c) steam turbine, and (d) condenser.

water drum and the steam drum. The outside surfaces of these tubes are exposed to hot products of combustion. Figure 1.6 in Chapter 1 shows a cutaway view of such a boiler for a marine application, and Fig. 4.18 shows a photograph of a biomass boiler. Our system boundaries (Fig. 4.17b, upper) include the drums and tubes, along with the water and the steam that they contain. Our system is simplified and represented as a single tube in Fig. 4.17b (lower). The hot combustion products are external to our system. Because the tube and drum geometries are significantly different, we have arbitrarily divided the heat transfer from the combustion products to our open system into three components. Again, flow work exists where the system boundary cuts through the flowing fluid. Note that we could have just as easily chosen a system that contains only the water and steam. Sometimes an analysis is simplified by either including or excluding the hardware surrounding the working fluid.

FIGURE 4.18 Steam is generated in these boilers by the combustion of residues from lumber waste and pulp paper (biomass). Photograph courtesy of NREL.

Steam Turbine We now consider the steam turbine. Our control surface here (Fig. 4.17c) cuts through the inlet steam line, exposing the flow work at the inlet, and similarly at the outlet. The control surface also cuts through the shaft connecting the turbine to the electrical generator and, thus, shaft work occurs at this location. Since the outer casing of the turbine is likely to be hotter than the surroundings, there will be some heat transfer from the control volume. Although this heat loss is shown in Fig. 4.17c for completeness, the loss is quite small in comparison to all the other energy transfers and is usually neglected in thermodynamic analyses.

FIGURE 4.19 Condensers and cooling towers at The Geysers geothermal power plant in California. Photograph courtesy of NREL.

Condenser In the condenser (see Fig. 4.19), all the entering high-quality steam condenses to liquid water. Figure 4.17d schematically shows the condenser in which the steam is the shell-side fluid occupying the annular control volume, while cold water flows through the tube. The single tube in this schematic represents all the tubes in a real condenser (see Fig. 1.5). The primary heat transfer is from the condensing steam to the cold water. Again, there is a small, and usually negligible, heat loss to the surroundings. Again, flow work is present as the fluid must be pushed into and out of the control volume.

Comment Note that we have identified the small heat losses that occur in all these real devices. Although these are usually neglected in applying the conservation of energy principle to these devices, it is important that you become skillful in identifying *all* heat and work interactions. You can always discard a term as you proceed, but if you missed an important term at the beginning of an analysis, there is no later recourse.

It is also important to point out that in conservation of energy analyses the flow work terms identified in all these devices are conventionally grouped with the rate of internal energy flowing into or out of the control volume as the rate of enthalpy flow (i.e., $\dot{m}u + \dot{W}_{\text{flow}} = \dot{m}u + \dot{m}Pv = \dot{m}h$). We will deal with this at some length in the next chapter.

Self-Test 4.3

Write an expression for the net work and net heat transfer for the steam power plant of Fig. 4.16.

(Answer: $\dot{W}_{net} = {}_1\dot{W}_2 + {}_3\dot{W}_4 = \dot{W}_{turb} - \dot{W}_{pump}$, $\dot{Q}_{net} = {}_2\dot{Q}_3 + {}_4\dot{Q}_1 = \dot{Q}_{boil} - \dot{Q}_{cond}$. *Note that the directions of work and heat transfer are explicitly defined in Fig. 4.16.)*

Example 4.4 Net Flow Work in a Steam Power Plant

For the Rankine cycle illustrated in Fig. 4.16, calculate and compare the net flow work associated with the feedwater pump and the steam turbine for the following conditions:

State 1	State 2	State 3	State 4
Saturated liquid $P_1 = 5$ kPa	Compressed liquid $P_2 = 1$ MPa	Saturated vapor $P_3 = 1$ MPa	Wet mixture $P_4 = 5$ kPa, $x_4 = 0.9$

These conditions are typical for an oil-fired industrial power plant producing approximately 1000 kW electrical power [10].

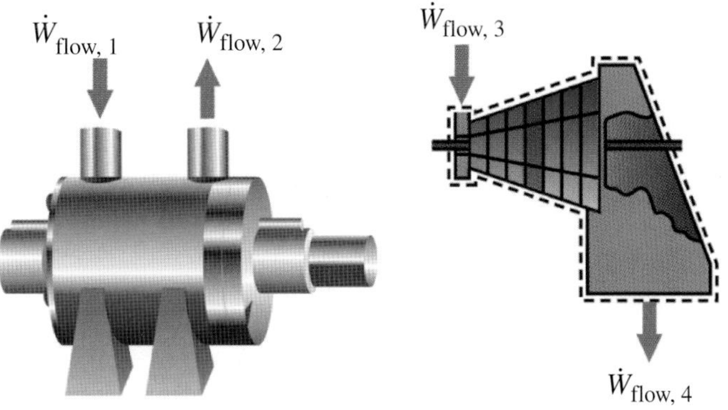

Working fluid (water/steam) flow rate: 2.36 kg/s.

Solution

Known $\dot{m}$, physical states, P_1, P_2, P_3, P_4, x_4

Find $\dot{m}_{\text{flow, net}}$ for pump and turbine

Sketch

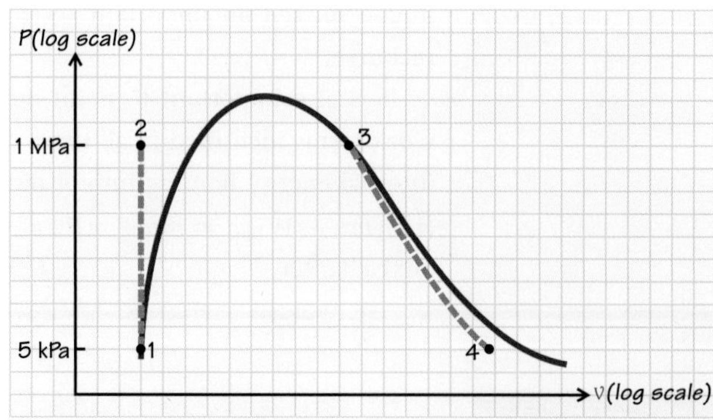

Modeling, Premises and Assumptions

i. The specific volume of the compressed liquid at state 2 is approximately equal to that of the saturated liquid at state 1.
ii. Properties are uniform over the inlet and outlet stations.

Analysis We calculate the flow work from a straightforward application of Eqs. 4.12a and 4.12b, recognizing that the net work is the sum of the flow work in and flow work out and with careful attention being paid to signs. Since all the required pressures are given, we need only determine the specific volumes. Starting with the feedwater pump, we find the inlet specific volume from the saturated steam tables (NIST sources or Appendix B):

$$v_1 = v_f(P_{sat} = 5\,\text{kPa}) = 0.0010053\ \text{m}^3/\text{kg}.$$

The specific volume at state 2 can be approximated as being the same as that at state 1. If we knew another property at state 2, we could use tabulated (or computer-based) data in the compressed liquid region to obtain a precise value. Applying the definition of flow work (Eq. 4.12), we calculate

$$\begin{aligned}\dot{W}_{\text{flow},1} &= \dot{m}P_1 v_1\\ &= 2.36(\text{kg/s})(5\times10^3\,\text{N/m}^2)0.0010053\,\text{m}^3/\text{kg}\left[\frac{\text{J}}{\text{N}\cdot\text{m}}\right]\left[\frac{\text{W}}{\text{J/s}}\right]\\ &= 11.9\ \text{W},\end{aligned}$$

$$\begin{aligned}\dot{W}_{\text{flow},2} &= \dot{m}P_2 v_2\\ &= -2.36(1\times10^6)0.0010053\,\text{W}\\ &= -2372.5\,\text{W},\end{aligned}$$

and

$$\begin{aligned}\dot{W}_{\text{flow, pump net}} &= \dot{W}_{\text{flow},1} + \dot{W}_{\text{flow},2}\\ &= 11.9\,\text{W} - 2372.5\,\text{W}\\ &= -2360.6\,\text{W}.\end{aligned}$$

For the steam turbine, the inlet specific volume is found from the NIST WebBook or Appendix B to be

$$v_3 = v_g(P_{sat} = 1\,\text{MPa}) = 0.19436\,\text{m}^3/\text{kg}.$$

Since the outlet condition lies in the wet region, we use the quality (x_4) to determine v_4, that is,

$$v_4 = (1 - x_4)v_f + x_4 v_g,$$

where v_f and v_g are the specific volumes for the saturated liquid and vapor at P_4 (= 5 kPa), respectively. Using values for v_f and v_g from the NIST sources or Appendix B, we calculate

$$\begin{aligned} v_4 &= 0.1(0.0010053\,\text{m}^3/\text{kg}) + 0.9(28.185\,\text{m}^3/\text{kg}) \\ &= 25.37\,\text{m}^3/\text{kg}. \end{aligned}$$

The flow work can now be calculated (treating the units as above):

$$\begin{aligned} \dot{W}_{\text{flow},3} &= \dot{m}P_3 v_3 \\ &= 2.36(1 \times 10^6)0.19436\,\text{W} \\ &= 458{,}690\,\text{W} \end{aligned}$$

and

$$\begin{aligned} \dot{W}_{\text{flow},4} &= -\dot{m}P_4 v_4 \\ &= 2.36(5 \times 10^3)23.37\,\text{W} \\ &= 299{,}370\,\text{W}. \end{aligned}$$

Thus, the net turbine flow work is

$$\begin{aligned} \dot{W}_{\text{flow, turbine net}} &= \dot{W}_{\text{flow},3} + \dot{W}_{\text{flow},4} \\ &= 458{,}690\,\text{W} - 299{,}370\,\text{W} \\ &= 159{,}320\,\text{W}. \end{aligned}$$

Comment In comparing the net flow work for the pump and turbine, we note, first, that the magnitude of the turbine flow work is about 70 times that of the pump, and, second, that the signs differ between the pump and turbine flow work. The turbine flow work is large because the specific volume of the steam is much larger than that of the liquid water. In the next chapter, we will return to the sign issue in our first-law (conservation of energy) analysis of pumps and turbines.

Self-Test 4.4

Calculate the change in enthalpy of the working fluid for the turbine of Example 4.4 by using (a) the definition of enthalpy (i.e., $h = u + Pv$) and (b) tabulated values for h.

(Answer: (a) –454.32 kJ/kg, (b) –459.32 KJ/kg. Note: The relatively large discrepancy (1%) results from interpolation for properties at 5 kPa. Using NIST data at 5 kPa without interpolation yields no discrepancy.)

Example 4.5 Solar Heating of Domestic Hot Water

(Credit: electra kay-smith / Alamy Stock Photo.)

Solar-heated domestic hot water is provided using the system shown schematically in Fig. 4.20 [11]. The solar collector consists of a 292.6-m length of pliable black plastic tubing (EPDM) through which water is continuously pumped. An electric motor drives the pump. The solar-heated water is returned to the solar storage tank. The high-pressure (~400 kPa) domestic water is physically separated from the solar-heated water, allowing the solar circuit to operate close to atmospheric pressure. For the control volume indicated by the dashed line in Fig. 4.20, identify all the heat and work interactions.

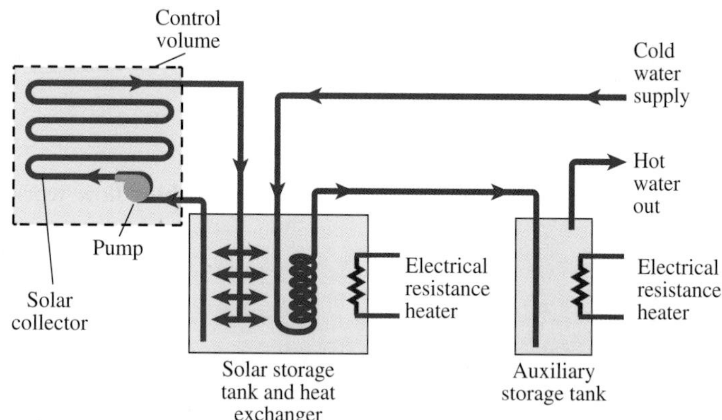

FIGURE 4.20 System for solar heating of domestic hot water [11] considered in Example 4.5.

Solution

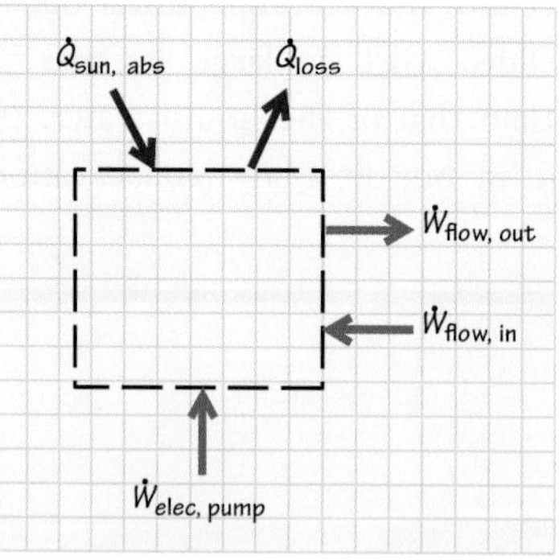

We begin by redrawing the control volume. We identify a work (power) input associated with the pump, and flow work in and out where the control surface cuts through the tubes. The heat transfer is a bit more complicated, and we will define two components rather than just considering a single net rate. A portion of the radiant energy from the sun is absorbed by the control volume, which we designate as $\dot{Q}_{\text{sun,abs}}$. Because the collector will be at a higher temperature than the surroundings (air, ground, etc.), there will be heat transferred to the surroundings, $\dot{Q}_{\text{loss}}$. These two heat-transfer components are shown as arrows pointing in the known directions.

Comments The details of the heat-transfer processes depend critically on the specific geometry, the characteristics of the solar radiation, and the properties of the surroundings (e.g., the ambient temperature and wind speed). Such details are the subject of courses and books on heat transfer.

Self-Test 4.5

Consider the solar storage tank and heat exchanger in Fig. 4.20. Identify the energy transfers associated with the cold-water coil in the tank.

(Answer: Flow work in, flow work out, and heat transfer in)

Section 4.4 is not essential to our study of thermodynamics; however, you may find it interesting and useful to your understanding of *heat*.

4.4 Rate Laws for Heat Transfer

Recall our definition of heat (or heat transfer) as the transfer of energy across a system or control-volume boundary resulting from a difference in temperature or a temperature gradient. In this section, we present the basic relationships that allow us to calculate the heat transfer, knowing either temperature differences or temperature gradients.

There are two physical mechanisms for heat transfer: (1) the transfer of energy resulting from molecular collisions, lattice vibrations, and unbound-electron flow, that is, **conduction**, and (2) the net exchange of electromagnetic radiation, that is, **radiation** (see Fig. 4.21). Since conduction depends on interactions among neighboring particles (i.e., a local exchange of energy), the process is said to be **diffusional** in nature and is driven by **temperature gradients.** In contrast, no medium is required for the transfer of energy by electromagnetic radiation and **temperature differences**. Actually, the driving potential for radiation is a difference of the fourth power of the absolute temperature [i.e., $\Delta(T^4)$]. In the case of flowing fluids, **convection** is treated as a third mode of heat transfer; however, conduction is still the only fundamental mechanism for energy exchange at the boundary in a convection problem. We will discuss this later in more detail.

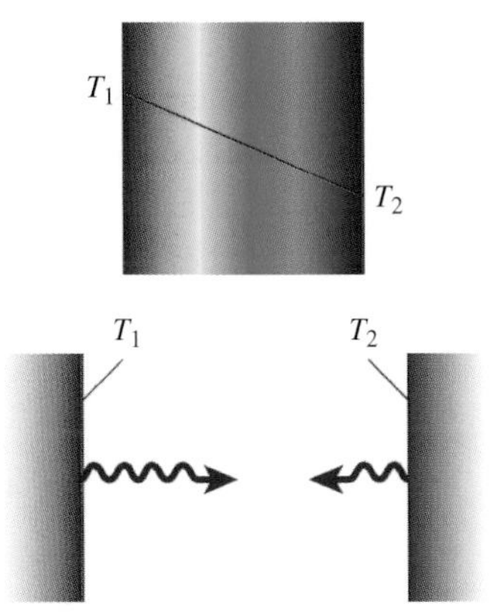

FIGURE 4.21 A temperature gradient within a medium drives energy transfer by conduction (top), whereas no medium is required for energy transfer by radiation (bottom).

Examples of heat transfer abound in both our natural and synthetic environments. A typical home in the United States is chock full of devices that involve heat transfer: light bulbs, furnaces, space heaters, toasters, hair dryers, computers, stoves, ovens, air conditioners, heat pumps, etc. Keeping your body at an appropriate temperature is a close-to-home and very practical example of heat transfer.

4.4a Conduction

The fundamental rate law governing conduction heat transfer in solids and stagnant fluids is Fourier's law, which is expressed as

$$\dot{Q}_{\text{cond}} = -kA\frac{dT}{dx} \tag{4.13a}$$

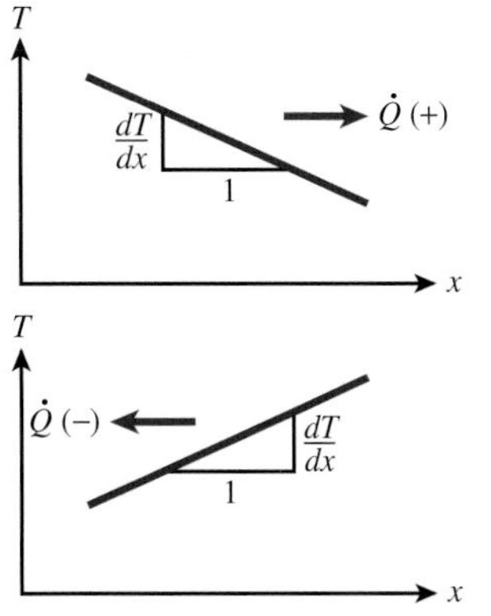

FIGURE 4.22 The temperature gradient *dT/dx* determines the direction of heat transfer. A negative *dT/dx* (top) results in a positive heat transfer (i.e., from left to right), whereas a positive *dT/dx* (bottom) results in a negative heat transfer (i.e., from right to left).

for a one-dimensional Cartesian system. In Eq. 4.13a, k is the thermal conductivity of the conducting medium, A is the area perpendicular to the direction of heat transfer (i.e., perpendicular to the x-direction for this one-dimensional case), and dT/dx is the temperature gradient. The negative sign appearing in Eq. 4.13a is the result of the fact that heat transfer has an associated direction. Heat transfer is directed in the negative x-direction when the temperature gradient is positive, as illustrated in Fig. 4.22. That heat transfer has a natural direction from a high-temperature region to a low-temperature region is a consequence of the second law of thermodynamics, the primary subject of Chapter 6. You can find thermal conductivity values for selected fluids from the NIST resources, and many sources exist for thermal conductivity data for many other substances, e.g., [12].

Applying Eq. 4.13a to the special case of steady-state conduction through a wall with fixed temperatures on the wall faces, we obtain

$$\dot{Q}_{\text{cond}} = kA(T_{\text{hot wall}} - T_{\text{cold wall}}). \tag{4.13b}$$

Fourier's law can be applied to obtain conduction heat-transfer rates and temperature distributions for any configuration, provided the medium is stagnant.

4.5b Convection

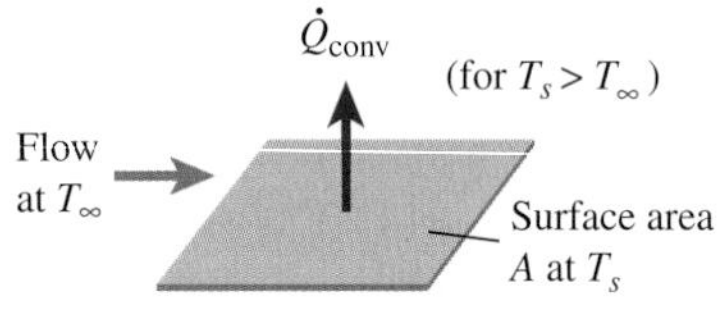

FIGURE 4.23 Convection heat transfer occurs when a fluid flows over a surface. When the surface temperature T_s is greater than the fluid temperature T_∞, the direction of the heat transfer is from the surface to the fluid.

Convection heat transfer is defined as the heat-transfer process that occurs at the interface between a solid surface and a flowing fluid and is not associated with radiation (Fig. 4.23). As discussed in Chapter 3, the fluid velocity at a fluid–solid interface is zero. As a result of this no-slip condition, Fourier's law applies within the fluid immediately adjacent to the wall. We can explicitly write Eq. 4.13a for the fluid at the wall as

$$\dot{Q}_{\text{conv}} = -k_{\text{f}} A \left. \frac{dT}{dy} \right|_{\text{wall}}, \tag{4.14}$$

where k_{f} is the fluid thermal conductivity and $dT/dy|_{\text{wall}}$ is the temperature gradient in the fluid. What complicates the evaluation of this simple relation is the need to know the temperature gradient in the fluid at the wall. The motion of the fluid away from the wall determines the temperature distribution throughout the flow, including the near-wall region, and, hence, the temperature gradient at the wall. In some relatively simple situations, Eq. 4.14 can be evaluated from theory [13, 14].

Because of the complexity associated with convection heat transfer, an empirical approach is frequently adopted. Such empiricism has its origins with Sir Isaac Newton but has been greatly refined in modern times. A simple empirical rate law is given by

$$\dot{Q}_{\text{conv}} = \bar{h}_{\text{conv}} A(T_{\text{s}} - T_\infty), \tag{4.15}$$

where $\bar{h}_{\text{conv}}$ is the heat-transfer coefficient averaged over the surface area A, T_{s} is the surface temperature, and T_∞ is the fluid temperature far from the wall in the bulk of the fluid. In this expression, a temperature difference is the driving potential for the heat transfer. Some typical values for heat-transfer coefficients are presented in Table 4.4 for a wide range of situations.

The heat-transfer coefficient is not a thermo-physical property of the fluid but, rather, a proportionality coefficient relating the heat flow $\dot{Q}_{\text{conv}}$ to a driving potential difference $T_{\text{s}} - T_\infty$; thus, Eq. 4.15 can be considered to be the definition of $\bar{h}_{\text{conv}}$ rather than a statement of some physical law. Heat-transfer coefficients depend on the

TABLE 4.4 Typical Values of Convective Heat-Transfer Coefficients [12]

Process	Heat-transfer coefficient, $\bar{h}_{conv}$ (W/m^2·K or °C)
Free convection	
Gases	2–25
Liquids	50–1000
Forced convection	
Gases	25–250
Liquids	50–20,000
Convection with phase change	
Boiling or condensation	2500–100,000

specific geometry, the flow velocity, and several thermo-physical properties of the fluid (e.g., the thermal conductivity, viscosity, and density). The wide range of values for $\bar{h}_{conv}$ shown in Table 4.4 results from the variation of these factors. Note that the temperature units associated with the heat-transfer coefficient are interchangeably K or °C, since the units refer to a temperature difference rather than a temperature, as can be seen from Eqs. 4.15.

Table 4.4 also categorizes convection heat transfer as: free convection, forced convection, or convection with phase change. **Free** (or **natural**) **convection** refers to flows that are driven naturally by buoyancy, whereas **forced convection** employs some external agent to drive the flow. The convection heat transfer from an incandescent light bulb and rising air currents (thermals) in the atmosphere are examples of free convection; the cooling of circuit boards in a computer with a fan and both air-side and coolant-side heat transfer in an automobile radiator are examples of forced convection.

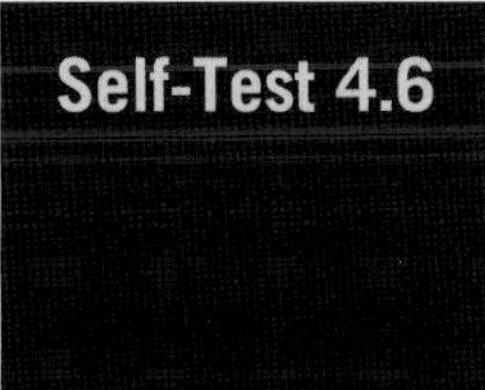

Considering the above discussion of free versus forced convection, explain why a person would seek relief on a hot day by standing in front of a fan.

(Answer: Since a forced-convection heat-transfer coefficient can be many times larger than that associated with free convection, standing in front of the fan results in a much greater heat-transfer rate from the body to the air.)

4.4c Radiation

Of the three modes of heat transfer, radiation is the most difficult to describe in general terms. This difficulty arises, in part, because of the wavelength dependence of both radiant energy and the radiant properties of substances. Also contributing is the action-at-a-distance nature of radiation. Unlike conduction, which depends only on the local material conductivity and local temperature gradient (see Eq. 4.13), radiant energy can be exchanged without any intervening medium and at distances that may be long relative to the dimensions of the device under study. A fortunate example of this is the radiant energy that travels 93 million miles from the sun to the earth through outer space, which is essentially a vacuum.

We can define a simplified rate law for radiation heat transfer for a solid surface that exchanges radiant energy with its surroundings. The net radiation heat transfer from the surface to the surroundings can be expressed as

$$\dot{Q}_{rad} = \varepsilon_s A_s \sigma \left(T_s^4 - T_{surr}^4\right), \tag{4.16}$$

where A_s is the area of the surface, and T_s and T_{surr} are the absolute temperatures of the surface and surroundings, respectively. The emissivity of the surface ε_s is a property of the surface and represents the ratio of the actual radiant energy emitted by the surface to that of a blackbody at the same temperature. Thus, values of emissivities range between zero and unity.

Use of Eq. 4.16 to estimate radiation heat-transfer rates relies on the following conditions being met:

1. The surface and the surroundings are, respectively, isothermal; thus, one single temperature characterizes the surface and another single temperature characterizes the surroundings.
2. All the radiation that leaves A_s is incident upon the surroundings; that is, there are no intervening surfaces that intercept a portion of the radiation from A_s, and there is no radiation leaving any one region of A_s that is incident upon any other region of A_s. Furthermore, any medium between the two surfaces (e.g., air) neither absorbs nor emits any radiation.
3. The surroundings behave as a blackbody *or* the area of the surroundings is much greater than A_s.
4. Surface A_s is a diffuse-gray emitter and reflector, which means that the radiation properties of the surface do not depend on direction or wavelength. This restriction results in the emissivity having a single fixed value.

Any examples or problems involving radiation that appear in this book are such that these four conditions are met or approximated. To deal with more complex situations, we refer the interested reader to Ref. [12]. The following two examples illustrate the application of the simplified rate expressions for radiation (Eq. 4.16) and convection (4.15).

Example 4.6 Radiation Heat Transfer from a Thermocouple

A spherical (0.75-mm-diameter) thermocouple with an emissivity of 0.41 is used to measure the temperature of hot combustion products flowing through a duct, as shown in the sketch. The walls of the duct are cooled and maintained at 52 °C. The thermocouple temperature is 1162 °C. Ignoring the presence of the thermocouple lead wires, determine the rate of radiation heat transfer from the thermocouple to the duct walls.

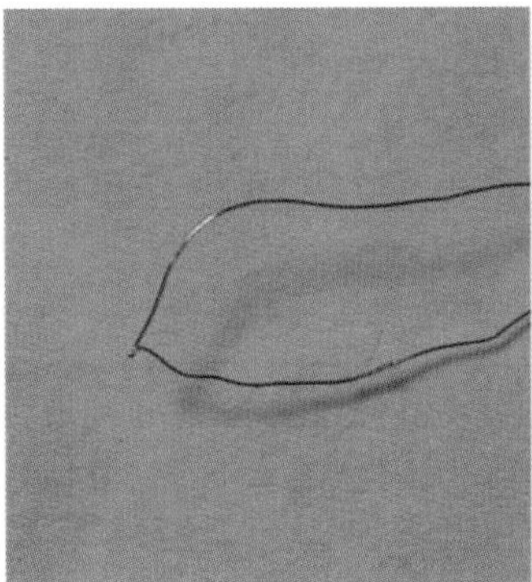

Fine-wire Pt–Rh thermocouple (small sphere at left).

Solution

Known D_{TC}, ε_{TC}, T_{wall}, T_{TC}

Find $\dot{Q}_{rad,TC-wall}$

Sketch

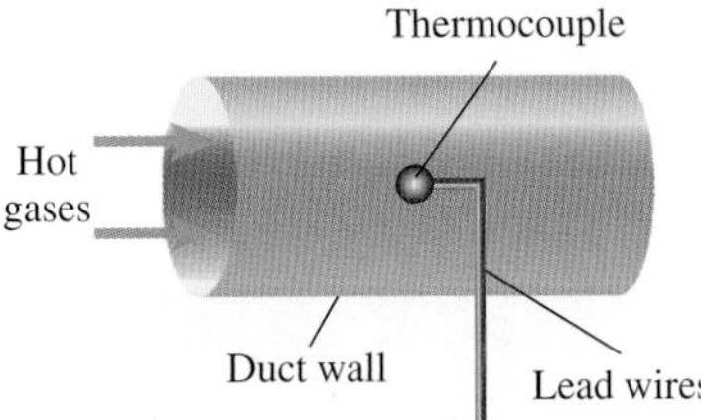

Modeling, Premises and Assumptions

i. The effects of the lead wires are negligible.
ii. All the restrictions associated with Eq. 4.16 are met.
iii. The combustion product gases neither absorb nor emit any radiation.

Analysis We apply Eq. 4.16, recognizing that T_s is the thermocouple temperature, T_{TC}, and that T_{surr}, the temperature of the surroundings, is identified with the duct wall temperature, that is, $T_{surr} = T_{wall}$. Since absolute temperatures are required for Eq. 4.16, we need to convert the given temperatures:

$$T_{wall} = 52 + 273 = 325\,\text{K},$$
$$T_{TC} = 1162 + 273 = 1435\,\text{K}.$$

The radiant heat-transfer rate is calculated from

$$\dot{Q} = \varepsilon_{TC} A_{TC} \sigma \left(T_{TC}^4 - T_{wall}^4\right),$$

where A_{TC} is the surface area of the spherical thermocouple ($= \pi D^2$). Substituting numerical values, we obtain

$$\begin{aligned}\dot{Q} &= 0.41\,\pi(0.00075\,\text{m})^2 5.67 \times 10^{-8}\,\text{W}/(\text{m}^2 \cdot \text{K}^4)\left[(1435\,\text{K})^4 - (325\,\text{K})^4\right] \\ &= 0.174\,\text{W}.\end{aligned}$$

Comments We note that differences in temperatures to the fourth power are not the same when temperatures are expressed in °C (or F) rather than K (or R), so care must be used in numerically evaluating the radiation rate expression (Eq. 4.16). (However, in the rate expression for convection (Eq. 4.15), temperature differences are the same regardless of the temperature scales, i.e., °C vs. K.)

In our analysis, we assumed that the hot gases themselves do not participate in the exchange of radiation and that the only exchange occurs between the two solid surfaces, the thermocouple and the duct wall. This is an excellent assumption for diatomic molecules such as N_2 and O_2, the primary components of air. Asymmetrical molecules, such as CO, CO_2, and H_2O, which are found in combustion products, can also participate in the radiation process; thus our analysis is a first approximation to the actual radiation exchange associated with the thermocouple. Treatments involving the participating media (i.e., gas molecules and particulate matter), can be found in advanced textbooks [15, 16].

Self-Test 4.7

Consider a cast-iron stove with emissivity $\varepsilon = 0.44$ and surface area $A_s = 16\ \text{ft}^2$ located in a large room. If the stove surface temperature is 500 F and the room wall temperature is 60 F, determine the radiation heat-transfer rate to the room walls.

(Answer: $\dot{Q}_{rad} = 9380\ \text{Btu/h}$)

Example 4.7 Thermocouple Temperature Measurement

Consider the same situation as that described in Example 4.6. For a steady state to be achieved, the heat lost by the thermocouple must be balanced by an energy input. This input comes from convection from the hot gases to the thermocouple, so

$$\dot{Q}_{\text{conv}} = \dot{Q}_{\text{rad}}.$$

Assuming an average heat-transfer coefficient of $595\ \text{W/m}^2\cdot\text{K}$, estimate the temperature of the hot flowing gases.

Solution

Known $\dot{Q}_{\text{rad}}$ (Ex. 4.6), $\bar{h}_{\text{conv}}$, D_{TC}

Find T_{gas}

Sketch

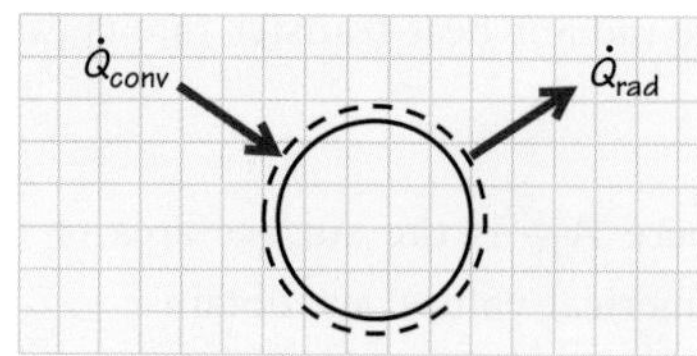

Modeling, Premises and Assumptions

i. The situation is steady state.
ii. The effects of the lead wires are negligible.
iii. The combustion products neither absorb nor emit any radiation.

Analysis Given that

$$\dot{Q}_{\text{conv}} = \dot{Q}_{\text{rad}},$$

we substitute the convection rate law (Eq. 4.15) and write

$$\bar{h}_{conv}A_{TC}\left(T_{gas} - T_{TC}\right) = \dot{Q}_{rad}.$$

Here we recognize that the gas must be hotter than the thermocouple for the heat transfer to be in the direction shown; thus, the appropriate temperature difference is $T_{gas} - T_{TC}$. Solving for T_{gas} yields

$$T_{gas} = \frac{\dot{Q}_{rad}}{\bar{h}_{conv}A_{TC}} + T_{TC}.$$

Also recognizing that the surface area of the thermocouple is πD_{TC}^2, we numerically evaluate this as

$$\begin{aligned} T_{gas} &= \frac{0.174\,\text{W}}{595\ \text{W}/(\text{m}^2\cdot\text{K})\,\pi(0.00075\ \text{m})^2} + 1435\ \text{K} \\ &= 1600\ \text{K}. \end{aligned}$$

Comments This example foreshadows the power of the principle of conservation of energy, from which the statement that $\dot{Q}_{conv}$ equals $\dot{Q}_{rad}$ was derived. We also see the inherent difficulty in making temperature measurements in high-temperature environments, as the presence of radiation from the thermocouple to the relatively cold wall results in a substantial error in the hot-gas temperature measurement. In this specific example, the error is 165 K.

Self-Test 4.8

The air temperature of a house is kept at 22 °C all the year round. Determine the rate of radiation heat transfer from a person ($A_s = 1.2\,\text{m}^2$) to the walls of a room (a) during summer when the walls are at 28 °C and (b) during winter when the walls are at 18 °C. Discuss.

(Answer: (a) 66.4 W, (b) 133.6 W. With more heat loss by radiation during the winter than in the summer, a person will feel colder, even though the room air temperature is the same.)

SUMMARY

This chapter began by considering the various ways in which energy is stored in systems. We saw the importance of distinguishing between the bulk system energy, which is associated with the system as a whole, and the internal energy, which is associated with the constituent molecules. We also carefully defined heat and work and their time rates. We saw that both of these quantities are energy transfers that occur only at the boundaries of closed and open systems. Neither is possessed by, or a property of, a system. Various types of work were discussed along with the three principal modes of heat transfer: conduction, convection, and radiation. To consolidate your knowledge further, reviewing the learning objectives presented at the beginning of this chapter and the checklist below is recommended.

KEY EQUATIONS

Review the most important equations presented in this chapter (i.e., those boxed with a yellow background). What physical principles do they express? What restrictions apply?

CHAPTER 4 KEY CONCEPTS AND DEFINITIONS CHECKLIST

Answer the Questions and solve the Problems following the arrows to demonstrate mastery of the listed concepts and definitions.

4.1 Closed and Open System Energy

- ☐ Bulk versus internal energies ➔ Question 4.2, Problems 4.9, 4.12
- ☐ Linear and rotational kinetic energies ➔ Problems 4.6, 4.11
- ☐ Potential energy ➔ Problem 4.10
- ☐ Molecular description of internal energy ➔ Question 4.3

4.2 Energy Transfer Across Boundaries

4.2a Heat

- ☐ Definition of heat (or heat transfer) ➔ Question 4.4
- ☐ Driving force for heat transfer ➔ Question 4.5
- ☐ Distinction between heat and heat-transfer rate ➔ Questions 4.6, 4.7
- ☐ Identification of heat interactions in real devices and systems ➔ Problem 4.75 (heat interactions only)

4.2b Work

- ☐ Definition of work ➔ Question 4.8
- ☐ Definition of power ➔ Question 4.12
- ☐ Five common types of work (list) ➔ Question 4.13
- ☐ P–$d\mathcal{V}$ (compression and expansion) work ➔ Problems 4.14, 4.15, 4.46, 4.56
- ☐ Shaft work ➔ Problem 4.63
- ☐ Electrical work ➔ Problem 4.66
- ☐ Flow work ➔ Problem 4.72
- ☐ Identification of work interactions in real devices and systems ➔ Problem 4.75 (work interactions only)

4.3 Sign Conventions and Units

- ☐ Units for energy and energy rates ➔ Problems 4.1, 4.3

4.4 Rate Laws for Heat Transfer

4.4a Conduction ➔ Problem 4.82

4.4b Convection ➔ Problem 4.83

4.4c Radiation ➔ Question 4.92

REFERENCES

1. Young, Thomas, "Energy", in *The World of Physics*, Vol. 1 (J. H. Weaver, Ed.), Simon & Schuster, New York, 1987.
2. Hatsopoulis, G. N., and Keenan, J. H., *Principles of General Thermodynamics*, Wiley, New York, 1965.
3. Obert, E. F., *Thermodynamics*, McGraw-Hill, New York, 1948.
4. Carey, V. P., *Statistical Thermodynamics and Microscale Thermophysics*, Cambridge University Press, New York, 1999.
5. Tien, C.-L. (Ed.), *Microscale Thermophysical Engineering*, Taylor & Francis, Philadelphia, published quarterly.
6. Eu, B. C., *Generalized Thermodynamics: The Thermodynamics of Irreversible Processes and Generalized Hydrodynamics*, Kluwer, Boston, 2002.
7. Sieniutycz, S., and Salamon, P. (Eds.), *Flow, Diffusion, and Rate Processes*, Taylor & Francis, New York, 1992.

8. Stowe, K., *Introduction to Statistical Mechanics and Thermodynamics*, Wiley, New York, 1984.
9. Fourier, J., *The Analytical Theory of Heat*, translation by A. Freeman, Cambridge University Press, Cambridge, England, 1878. (Original, *Théorie Analytique de la Chaleur*, published in 1822.)
10. Goodall, R. P., *The Efficient Use of Steam*, IPC Science and Technology Press, Surrey, England, 1980.
11. Smith, T. R., Menon, A. B., Burns, P. J., and Hittle, D. C, "Measured and Simulated Performances of Two Solar Domestic Hot Water Heating Systems," Colorado State University Engineering Report.
12. Incropera, F. P., and DeWitt, D. P., *Fundamentals of Heat and Mass Transfer*, 5th edn, Wiley, New York, 2002.
13. Kays, W. M., *Convection Heat and Mass Transfer*, McGraw-Hill, New York, 1966.
14. Burmeister, L. C., *Convective Heat Transfer*, 2nd edn, Wiley, New York, 1993.
15. Modest, M. E., *Radiative Heat Transfer*, 2nd edn, Academic Press, San Diego, 2003.
16. Siegel, R., and Howell, J. R., *Thermal Radiation Heat Transfer*, 3rd edn, McGraw-Hill, New York, 1992.

Some end-of-chapter problems were adapted with permission from the following:

Chapman, A. J., *Fundamentals of Heat Transfer*, Macmillan, New York, 1987.
Look, D. C., Jr., and Sauer, H. J., Jr., *Engineering Thermodynamics*, PWS, Boston, 1986.
Myers, G. E., *Engineering Thermodynamics*, Prentice Hall, Englewood Cliffs, NJ, 1989.

QUESTIONS

4.1 Name three pioneers in the thermal-fluid sciences and indicate their major contributions. Provide an approximate date for these contributions.

4.2 Distinguish between the bulk and internal energies associated with a thermodynamic system. Give a concrete example of a system that possesses both.

4.3 From a molecular point of view, how is energy stored in a volume of gas composed of monatomic species? How is energy stored in diatomic gas molecules? In triatomic gas molecules?

4.4 From memory, write out a formal definition of heat or heat transfer. Compare your definition with that in the text.

4.5 What thermodynamic property "drives" a heat-transfer process?

4.6 How does heat (or heat transfer) Q differ from the heat-transfer rate $\dot{Q}$?

4.7 List the units for Q and $\dot{Q}$.

4.8 From memory, write out a formal definition of work. Compare your definition with that in the text.

4.9 What distinguishes work from internal energy? What distinguishes heat from internal energy? Discuss.

4.10 What distinguishes work from heat? Discuss.

4.11 What distinguishes a path function from a point function? Give an example and discuss.

4.12 How does work W differ from power $\dot{W}$ (or $\mathcal{P}$)? What units are associated with each?

4.13 List, define, and discuss five types of work (or power). For example, one of these should be shaft work (power).

4.14 List the three modes of heat transfer. Compare and contrast these modes.

Chapter 4 Problem Subject Areas

4.1–4.4	Definitions, symbols, and units
4.5–4.13	Macroscopic and microscopic energies
4.14–4.61	Moving boundary (expansion/compression) work
4.14–4.24	Conceptual problems for ideal gases – no numerical results required
4.25–4.54	Numerical problems – ideal gases
4.55–4.61	Numerical problems – two-phase substances
4.62–4.65	Shaft work and power
4.66–4.70	Electrical work and power
4.72–4.74	Flow work (power)
4.75–4.79	Identifying heat and work interactions
4.80–4.93	Heat-transfer modes
4.94–4.99	EES problems
4.100–4.108	FE problems

PROBLEMS

4.1–4.4 Definitions, symbols, and units

4.1 Provide the names and units associated with the following symbols: E, u, W, $\dot{W}$, Q, and q.

4.2 Indicate whether each of the following statements is true or false. Justify your choice.

A. The electrical power associated with a battery always moves from the battery (system) to the surroundings.

B. The rate of enthalpy flow across an open-system boundary $\dot{m}\ h$is the sum of the rate of internal energy flow across the boundary and the rate of flow work at the boundary.

C. Flow work (rate) is another name for shaft work (rate).

D. The pressure force at inlets and outlets is responsible for flow work (rate).

4.3 Write out in words the precise meaning of the following symbols. Be very specific, using appropriate adjectives as needed. Also provide the usual SI units associated with the quantities. For the terms that are products of symbols, give a single word or short phrase describing the grouping (see example).

	Symbol	Meaning (words only)	Usual SI units
Example:	$\rho\mathcal{V}$	mass	kg
a.	$\dot{Q}$		
b.	V (not v or $\mathcal{V}$)		
c.	$\dot{m}Pv$		
d.	$u + Pv$		
e.	$\dot{m}V^2/2$		
f.	$\dot{W}_{shaft}$		

4.4 Show in detail that the units associated with the term $V^2/2$ are J/kg. Be sure to show how basic unit definitions are required to obtain the final result; for example, $1\ N \equiv 1 kg{\cdot}m/s^2$.

4.5–4.13 Macroscopic and microscopic energies

4.5 Determine the instantaneous kinetic and potential energies (kJ) of a 90-kg object moving at a velocity of 10 m/s, 10 m above the surface of a planet where $g = 7\,\text{m/s}^2$.

4.6 Determine the instantaneous kinetic and potential energies (kJ) of a 860-kg automobile crossing the Rocky Mountains on Trail Ridge Road traveling at 35 mph at an elevation of 9700 feet above sea level. Use sea level as the reference state for the potential energy.

4.7 The Chelyabinsk meteor exploded over Russia in 2013 with a blast whose brightness exceeded that of the sun. The meteor entered the atmosphere with a speed of approximately 60,000 km/h and exploded at an altitude of approximately 29.7 miles. The estimated mass of the meteor was 13,000 metric tons (1 metric ton = 1000 kg).

A. Determine the meteor's kinetic energy at the entry velocity and its potential energy at the point of explosion.

B. How many 55-kg bicycle riders traveling at 20 mph would be required to result in the same kinetic energy as the Chelyabinsk meteor?

(Credit: Carlos Fernandez / Moment / Getty Images.)

4.8 Initially, 1 lb_m of water is at rest at 14.7 psia and 70 F. The water then undergoes a process where the final state is 30 psia and 700 F with a velocity of 100 ft/s and an elevation of 100 ft above the starting location. Determine the increases in internal energy, potential energy, and kinetic energy of the water in Btu/lb_m. Compare the increases in potential energy and kinetic energy individually to the change in the internal energy. HINT: Use the NIST software or online WebBook to evaluate thermodynamic properties.

4.9 At a particular instant, a hot-air balloon is ascending at 5 m/s. In the preceding time period, the air in the balloon was heated from 100 °C to 110 °C. The initial volume of the balloon is 2800 m^3 and the pressure of the contained gas is 103 kPa. Treating the gas as air, and assuming a constant mass of gas in the balloon, calculate and compare the kinetic energy and the internal energy change of the air in the balloon.

4.10 A 100-kg man wants to lose weight by increasing his energy consumption through exercise. He proposes to do this by climbing a 6-m staircase several times a day. His current food consumption provides him with an energy input of 4000 Calories/day. A Calorie (Cal) is a unit for measuring the energy released by food when oxidized in the body; 1 Cal = 1 kcal. Determine the vertical distance (m) the man must climb to convert his food intake into potential energy. What is your advice regarding exercise versus a reduced food intake to effect a weight loss?

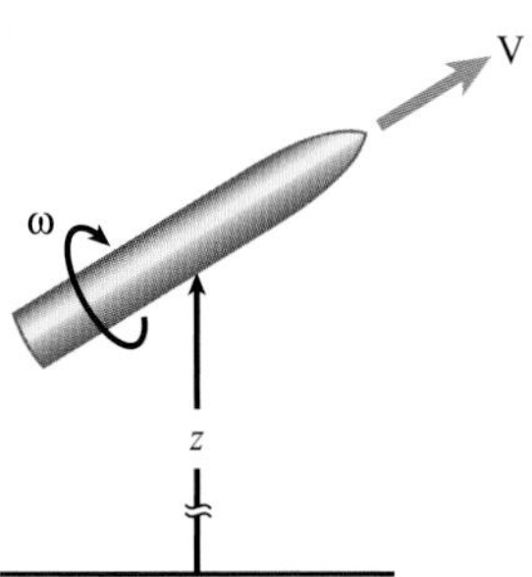

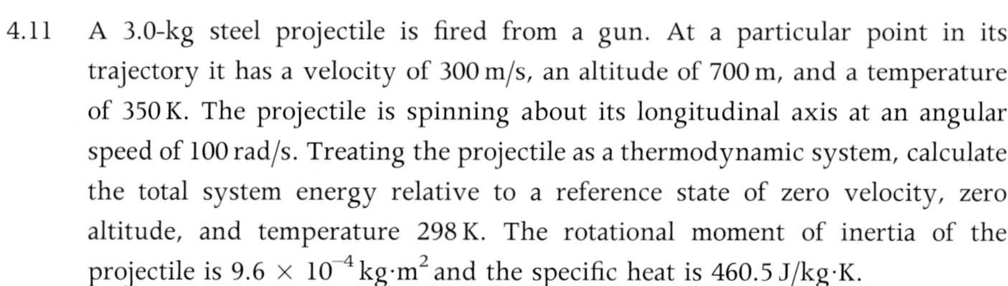

4.11 A 3.0-kg steel projectile is fired from a gun. At a particular point in its trajectory it has a velocity of 300 m/s, an altitude of 700 m, and a temperature of 350 K. The projectile is spinning about its longitudinal axis at an angular speed of 100 rad/s. Treating the projectile as a thermodynamic system, calculate the total system energy relative to a reference state of zero velocity, zero altitude, and temperature 298 K. The rotational moment of inertia of the projectile is 9.6×10^{-4} kg·m^2 and the specific heat is 460.5 J/kg·K.

4.12 Consider a 25-mm-diameter steel ball that has a density of 7870 kg/m^3 and a constant-volume specific heat of the form

$$c_v(\text{J/kg}\cdot\text{K}) = 447 + 0.4233[T(\text{K}) - 300].$$

What velocity would the ball have to achieve so that its bulk kinetic energy equals its internal energy at 500 K? Use an internal energy reference state of 300 K.

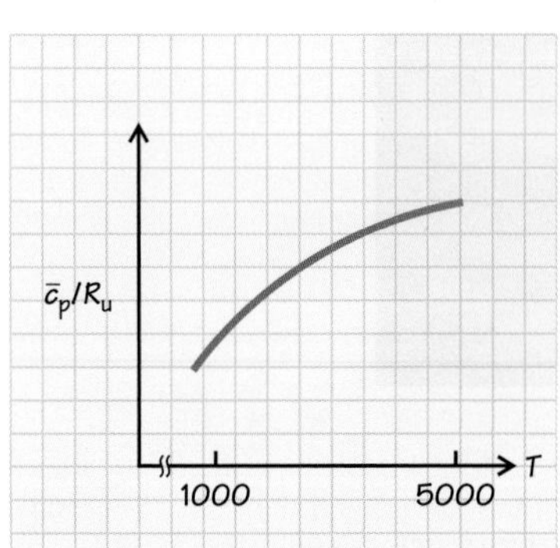

4.13 For the temperature range 1000–5000 K, the ratio of the molar constant-pressure specific heat to the universal gas constant is given as follows for CO_2:

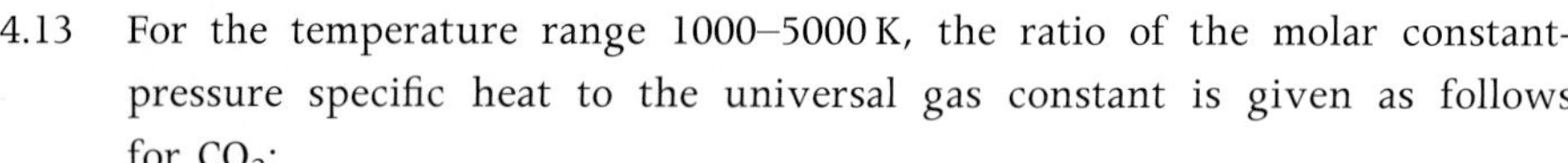

$$\begin{aligned}\bar{c}_p/R_u = {} & 4.45362 + (3.140168 \times 10^{-3})T - (1.278410 \times 10^{-6})T^2 \\ & + (2.393996 \times 10^{-10})T^3 - (1.6690333 \times 10^{-14})T^4,\end{aligned}$$

where T [=] K. For 1 kg of CO_2, plot $U(T) - U(1000\text{ K})$ versus T up to 5000 K. Discuss.

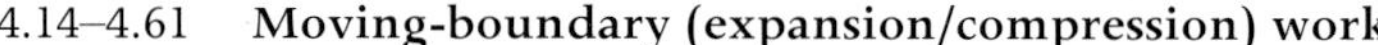

4.14–4.61 Moving-boundary (expansion/compression) work

4.14–4.24 Conceptual problems for ideal gases – no numerical results required

4.14 ***Conceptual problem***. Assuming simple straight-line motion, show how the definition of work as the integral of a force with respect to distance, $\int F dx$, is identical to the integral of pressure with respect to volume $\int P d\mathcal{V}$ for a simple compressible substance.

4.15 ***Conceptual problem***. Consider an ideal-gas fixed-mass system undergoing a quasi-equilibrium (reversible) process. Which of the following statements are true? Explain your answer in each case.

A. On a pressure–volume plot, a constant-temperature process is represented by a hyperbola.

B. On a pressure–volume plot, a constant-temperature process is represented by a straight line with a positive slope.

C. On a pressure–volume plot, the area under the line representing the process is the work done on or by the system.

D. On a temperature–volume plot, a constant-pressure process is represented by a hyperbola.

E. On a temperature–volume plot, a constant-pressure process is represented by a straight line with a positive slope.

F. The moving-boundary work associated with a constant-volume process is zero.

4.16 ***Conceptual Problem***. Consider an ideal gas contained in a piston–cylinder arrangement as a closed thermodynamic system. The gas undergoes a constant-pressure expansion process from state 1 to state 2.

A. Plot the process in pressure–volume coordinates. Label the initial and final states.

B. Starting with the definition of moving-boundary work $\int Pd\mathcal{V}$ derive an expression to evaluate the work in terms of the pressures and volumes at the initial and final states.

C. Illustrate the work on your plot from part A.

D. Is the work for this process into or out of the system? Why?

4.17 ***Conceptual Problem.*** Consider an ideal gas contained in a piston–cylinder arrangement as a closed thermodynamic system. The gas undergoes a constant-pressure compression process from state 1 to state 2.

A. Plot the process in pressure–volume coordinates. Label the initial and final states.

B. Starting with the definition of moving-boundary work $\int Pd\mathcal{V}$ derive an expression to evaluate the work in terms of the pressures and volumes at the initial and final states.

C. Illustrate the work on your plot from part A.

D. Is the work for this process into or out of the system? Why?

4.18 ***Conceptual Problem***. Consider an ideal gas contained in a sealed rigid tank as a closed thermodynamic system. The gas undergoes a constant-volume process from state 1 to state 2.

A. Plot the process in pressure–volume coordinates. Label the initial and final states.

B. Use the definition of moving-boundary work $\int Pd\mathcal{V}$ to evaluate the work.

C. Is it possible to illustrate the work on your plot from part A? Discuss.

4.19 ***Conceptual Problem***. Consider an ideal gas contained in a piston–cylinder arrangement as a closed thermodynamic system. The gas undergoes a constant-temperature expansion process from state 1 to state 2.

A. Plot the process in pressure–volume coordinates. Label the initial and final states.

B. Starting with the definition of moving-boundary work $\int Pd\mathcal{V}$ derive an expression to evaluate the work in terms of the pressures and volumes at the initial and final states.

C. Illustrate the work on your plot from part A.

D. Is the work for this process into or out of the system? Why?

4.20 ***Conceptual Problem***. Consider an ideal gas contained in a piston–cylinder arrangement as a closed thermodynamic system. The gas undergoes a constant-temperature compression process from state 1 to state 2.

A. Plot the process in pressure–volume coordinates. Label the initial and final states.

B. Starting with the definition of moving-boundary work $\int P d\mathcal{V}$ derive an expression to evaluate the work in terms of the pressures and volumes at the initial and final states.

C. Illustrate the work on your plot from part A.

D. Is the work for this process into or out of the system? Why?

4.21 ***Conceptual Problem***. Consider an ideal gas contained in a piston–cylinder arrangement as a closed thermodynamic system. The gas expands following a polytropic process $P\mathcal{V}^n = P_1\mathcal{V}_1^n$ from state 1 to state 2. The polytropic exponent n is a constant.

A. Starting with the definition of moving-boundary work $\int P d\mathcal{V}$ derive an expression to evaluate the work in terms of the pressures and volumes at the initial and final states.

B. Is the work for this process into or out of the system? Why?

4.22 ***Conceptual Problem***. Consider an ideal gas contained in a piston–cylinder arrangement as a closed thermodynamic system. The gas expands from an initial volume at state 1 to a larger volume $\mathcal{V}_2$. The expansion process can be carried out as either a constant-pressure process or an isothermal process. It is desired to maximize the work done in the process. Determine which process will maximize the work and justify your choice. The initial states and the final volumes are the same for both processes.

4.23 ***Conceptual Problem***. Consider an ideal gas contained in a piston–cylinder arrangement as a closed thermodynamic system. The gas is compressed from an initial volume at state 1 to a smaller volume $\mathcal{V}_2$. The compression process can be carried out as either a constant-pressure process or an isothermal process. It is desired to minimize the work required to achieve the desired compression. Determine which process will minimize the work input and justify your choice. The initial states and the final volumes are the same for both processes.

4.24 ***Conceptual Problem***. A polytropic process is one that obeys the relationship $P\mathcal{V}^n = \text{constant}$, where n may take on various values.

A. Define the type of process associated with each of the following values of n, assuming the working fluid is an ideal gas (i.e., determine which state variable is constant):

i. $n = 0$

ii. $n = 1$

iii. $n = \gamma\ (= c_p/c_v)$

B. Starting from the same initial state, which process results in the most amount of work performed by the gas when the gas is expanded quasi-statically to a final volume twice as large as the initial volume? Which process results in the least amount of work?

C. Sketch the three processes on P–$\mathcal{V}$ coordinates.

4.25–4.54 Numerical problems – ideal gases

4.25 Consider air (ideal gas) within a piston–cylinder assembly as a thermodynamic system. We wish to compare three different quasi-equilibrium expansion processes for this system by plotting the pressure as a function of volume. Each process starts from the same initial condition: $P = 100$ kPa and $\mathcal{V} = 1$ m^3. The final volume for each process is $\mathcal{V} = 3$ m^3. The plots are to be made by connecting three points ($\mathcal{V} = 1$, 2, and 3 m^3) with hand-drawn lines. The first process (trivial) is conducted at constant pressure. The second process is conducted at constant temperature. The third process is conducted such that $P\mathcal{V}^{1.4}$ is maintained constant. (Note that only the volume here is raised to the 1.4 power, not the pressure–volume product.)

A. Complete the table below by calculating the six values for the pressure.

B. Plot the results on the graph axes given here or on graph paper. Connect the data points with a smooth line.

C. Using your plots, rank the quantity of work done by the three processes from the least to the most.

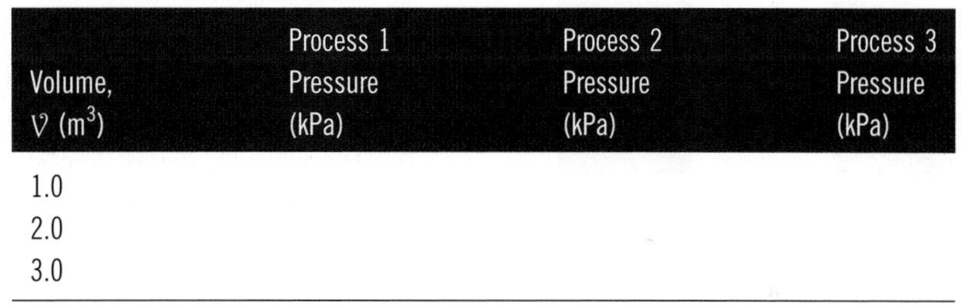

Volume, $\mathcal{V}$ (m^3)	Process 1 Pressure (kPa)	Process 2 Pressure (kPa)	Process 3 Pressure (kPa)
1.0			
2.0			
3.0			

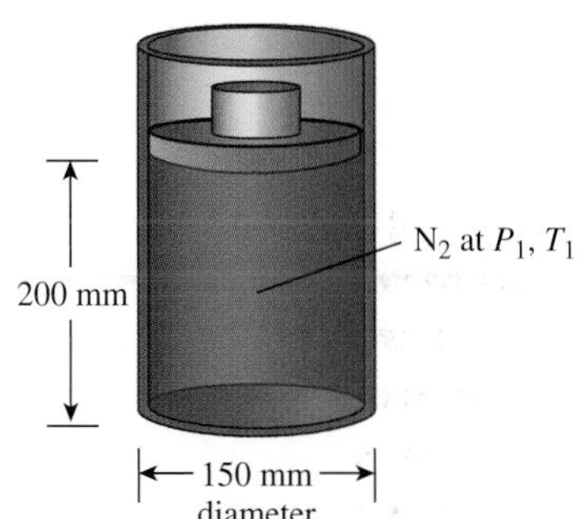

4.26 Consider nitrogen gas contained in a piston-cylinder arrangement as shown in the sketch. The piston is free to move without friction. The cylinder diameter is 150 mm. At the initial state the temperature of the N_2 is 500 K, the pressure is 100 kPa, and the height of the gas column within the cylinder is 200 mm. Heat is removed from the N_2 by a quasi-static, constant-pressure process until the temperature of the N_2 is 300 K. Calculate the work done by the N_2 gas.

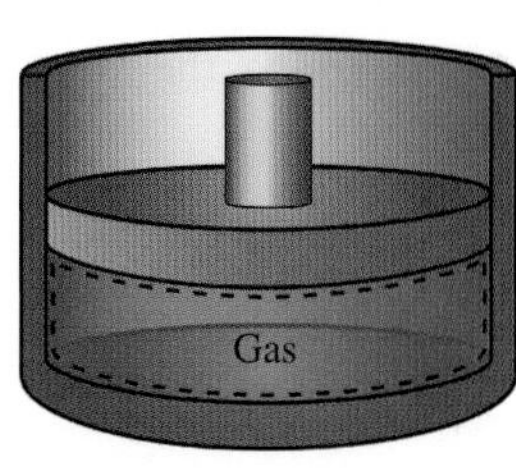

4.27 A gas is trapped in a cylinder–piston arrangement as shown in the sketch. The initial pressure and volume are 13,789.5 Pa and 0.02832 m^3, respectively. Determine the work (kJ) assuming that the volume is increased to 0.08496 m^3 in a constant-pressure process.

4.28 For the same conditions as in Problem 4.27, determine the work done if the process is polytropic with $n = 1$ (i.e., $P\mathcal{V}$ = constant).

4.29 For the same conditions as in Problem 4.27, determine the work done if the process is polytropic with $n = 1.4$ (i.e., $P\mathcal{V}^{1.4}$ = constant).

4.30 Air is compressed in a piston–cylinder assembly by a constant-pressure process with a work input of 225 J. The initial volume is 0.001 m^3, and the final volume is one-fourth of the initial volume. Determine the gas pressure during the process.

4.31 A 0.0065-kg mass of air expands in a piston–cylinder assembly following a constant-pressure path with a work output of 1.3 kJ. The initial specific volume of the air is 0.25 m^3/kg. The final volume is three times that of the initial volume.

A. Sketch the process in P–$\mathcal{V}$ and T–$\mathcal{V}$ coordinates.
B. Determine the gas pressure during the process.
C. Determine the initial and final temperatures of the air.

4.32 Initially, 0.05 kg of air is contained in a piston–cylinder device at 200 °C and 1.6 MPa. The air then expands at constant temperature to a pressure of 0.4 MPa. Assume the process occurs slowly enough that the acceleration of the piston can be neglected. The ambient pressure is 101.35 kPa.

A. Determine the work (kJ) performed by the air in the cylinder on the piston.
B. Determine the work (kJ) performed by the piston on the ambient environment. Neglect the cross-sectional area of the connecting rod.
C. Determine the work transfer (kJ) from the piston to the connecting rod. Neglect friction between the piston and cylinder. Assume that no heat transfer to or from the piston occurs and that the energy of the piston does not change.
D. Discuss why the assumption of negligible piston acceleration was made.
E. Which of these work interactions might be useful for driving a car? Why?

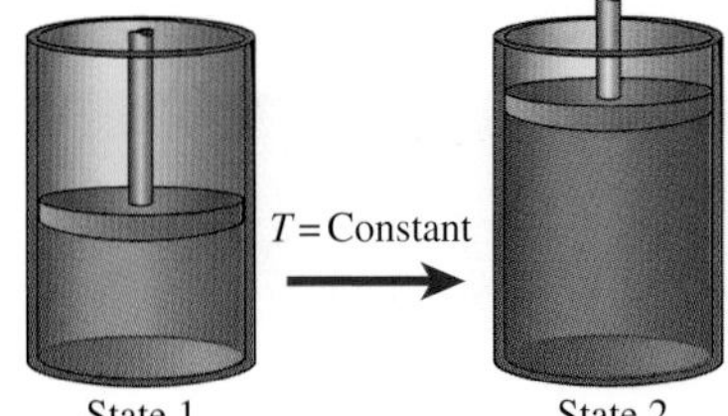

4.33 One cubic meter of an ideal gas expands in an isothermal process from 760 to 350 kPa. Determine these work done by this gas in kJ and Btu.

4.34 Air is compressed reversibly in a piston–cylinder assembly. The 0.12 lb_m of air in the cylinder is initially at 15 psia and 80 F, and the compression process takes place isothermally to 120 psia.

A. Determine the work required to compress the air (Btu).
B. Determine the final pressure of the air (psia).

4.35 Air is compressed reversibly and isothermally in a piston–cylinder assembly. The 0.05 kg of air in the cylinder is initially at 110 kPa and 300 K and is compressed to 830 kPa. Assuming ideal-gas behavior, determine the work required to compress the air.

4.36 A 1-kg mass of air expands in a piston–cylinder assembly following a constant-temperature path with a work output of 120 kJ. The final volume is 2.5 times that of the initial volume.

A. Sketch the process in P–$\mathcal{V}$ and T–$\mathcal{V}$ coordinates.
B. Determine the gas temperature during the process.
C. What additional information do you need to determine the initial and final pressures of the air?

4.37 Starting with the definition of reversible moving-boundary work, develop an expression for the work done by an ideal gas in a reversible adiabatic expansion from an initial temperature T_1 to a final temperature T_2. Note: $n = \gamma$.

4.38 Consider the two processes shown in the sketch: *ac* and *abc*. Determine the work done by an ideal gas in executing these reversible processes if $P_2 = 2P_1$ and $v_2 = 2v_1$. Assume you are dealing with a closed system and express your answer in terms of the gas constant R and the temperature T_1.

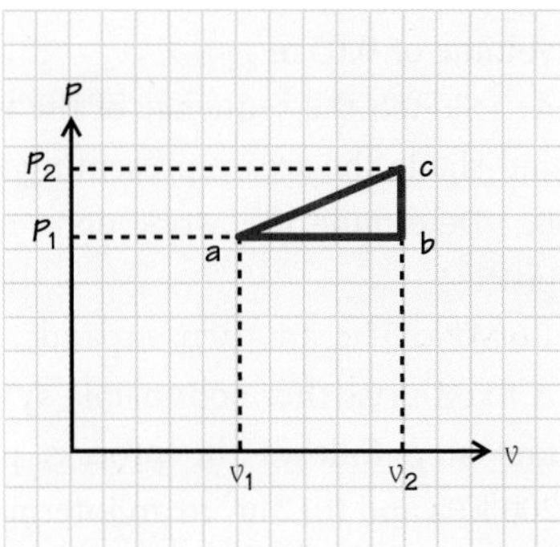

4.39 Initially, 0.5 kmol of an ideal gas at 20 °C and 2 atm is contained in a piston–cylinder device. The piston is held in place by a pin. The pin is then removed and the gas expands rapidly. After a new equilibrium is attained, the gas is at 20 °C and 1 atm. Determine the work transfer by the piston to the atmosphere. Explicitly list and discuss the assumptions that are required to perform this calculation.

4.40 A spherical balloon is inflated from a diameter of 10 in to a diameter of 20 in as air is forced into the balloon with a tire pump. The initial pressure inside the balloon is 20 psia. During the process the pressure is proportional to the balloon diameter.

A. Determine the amount of work transferred (Btu) from the air inside the balloon to the balloon.

B. Determine the amount of work transferred (Btu) from the balloon to the surrounding atmosphere.

4.41 A spherical balloon is inflated from a diameter of 0.15 m to a diameter of 0.30 m as air is forced into the balloon with a tire pump. The initial pressure inside the balloon is 1.5 atm. During the process the pressure is proportional to the balloon diameter.

A. Determine the amount of work transferred from the air inside the balloon to the balloon.

B. Determine the amount of work transferred from the balloon to the surrounding atmosphere.

4.42 A fixed-mass thermodynamic system is compressed in a two-step process. The first step follows the process path PV^2 = constant from an initial state at 15.0 psia and 3.00 ft^3 to an intermediate state at 45.0 psia. From this intermediate state, the compression follows a constant-pressure path to a final state having a volume of 1.00 ft^3.

A. Sketch the sequence of the two processes on pressure–volume coordinates and shade the area representing the work.

B. Determine the work done on the system (Btu) for each step of this two-step process.

4.43 A fixed-mass thermodynamic system is compressed in a two-step process. The first step follows the process path $P\mathcal{V}^2$ = constant from an initial state at 125 kPa and 0.15 m^3 to an intermediate state at 300 kPa. From this intermediate state, the compression follows a constant-pressure path to a final state having a volume of 0.05 m^3.

A. Sketch the sequence of the two processes in pressure–volume coordinates and shade the area representing the work.

B. Determine the work done on the system for each step of this two-step process.

4.44 A fixed-mass thermodynamic system is compressed in a two-step process. The first step follows the process path $P\mathcal{V}^3$ = constant from an initial state at 200 kPa and 0.12 m^3 to an intermediate state at 380 kPa. From this intermediate state, the compression follows a constant-pressure path to a final state having a volume of 0.03 m^3.

A. Sketch the sequence of the two processes in pressure–volume coordinates and shade the area representing the work.

B. Determine the work done on the system for each step of this two-step process.

4.45 A fixed-mass thermodynamic system (an ideal gas) expands in a piston–cylinder assembly in a two-step process. The first step follows an isothermal path from an initial state at 375 kPa and 0.020 m^3 to an intermediate state at 200 kPa. From this intermediate state, the expansion follows the process path $P\mathcal{V}^{0.5}$ = constant to a final state having a volume of 0.045 m^3.

A. Sketch the sequence of the two processes in pressure–volume coordinates and shade the area representing the work.

B. Determine the work done by the system for each step of this two-step process.

C. Is it possible to determine the temperature of the gas during the isothermal expansion? Explain.

4.46 A mass of air initially at 300 K and 100 kPa is contained in a piston–cylinder arrangement. The initial volume is 0.2 m^3. The air undergoes the following sequence of two processes:

Process 1–2: Constant-pressure expansion with a doubling of the volume.

Process 2–3: Isothermal expansion from state 2 to a volume of 0.6 m^3.

A. Sketch the sequence of processes in pressure–volume coordinates and shade the area representing the work.

B. Determine the work done by the system for each step of this two-step process.

4.47 A mass of air is trapped in a piston–cylinder arrangement initially at 500 K and 100 kPa (absolute). The volume is 0.15 m^3. Consider the air to be the thermodynamic system of interest. The system undergoes the following sequence of three processes:

Process 1–2: Constant-temperature compression to a volume of 0.075 m^3.
Process 2–3: Constant-volume heat addition to a temperature of 1000 K.
Process 3–4: Constant-pressure expansion to a volume of 0.15 m^3.

A. Sketch the sequence of processes in pressure–volume coordinates and shade the area representing the work.
B. Determine the work done on or by the system for each step of this three-step process.
C. Determine the net work transfer for the system and indicate whether this is work done on or by the system.

4.48 Air is the working fluid in a thermodynamic cycle for the production of power. The air undergoes the following processes in a reciprocating piston–cylinder assembly:

State 1–State 2: Isothermal compression from an initial volume (0.0003 m^3) to a volume one-third of the initial volume. The initial pressure and temperature are 100 kPa and 400 K, respectively.
State 2–State 3: Constant-pressure expansion until the volume reaches the initial (state-1) volume.
State 3–State 4: Isothermal expansion until the pressure reaches the initial (state-1) pressure.
State 4–State 1: Constant-pressure compression until the volume reaches the initial (state-1) volume.

A. Sketch the cycle on pressure–volume coordinates and shade the area representing the work.
B. Determine the net work of the cycle.

4.49 Air is the working fluid in a thermodynamic cycle for the production of power. The air undergoes the processes listed below in a reciprocating piston–cylinder assembly.

State 1–State 2: Isothermal compression from an initial volume (0.0008 m^3) to a volume one-eighth of the initial volume. The initial pressure and temperature are 100 kPa and 300 K, respectively.
State 2–State 3: Constant-volume heat addition until the temperature reaches 1200 K ($= T_3$).
State 3–State 4: Constant-pressure expansion until the volume reaches three-eighths of the initial (state-1) volume.
State 4–State 5: Isothermal expansion until the volume reaches the initial (state-1) volume.
State 5–State 1: Constant-volume heat rejection to return to the initial state (state-1).

A. Sketch the cycle in pressure–volume coordinates, label the state points (1, 2, etc.), and shade the area representing the work.
B. Determine the net work of the cycle.

4.50 Consider a power cycle in which air is the working fluid. The air is contained in a piston–cylinder assembly and undergoes the following processes:

State 1–State 2: Isentropic compression from the maximum volume $\forall_{max}$ to the minimum volume $\forall_{min}$, where $\forall_{max} = 10\forall_{min}$. The initial state pressure, temperature, and volume are 100 kPa, 300 K, and 0.001 m^3, respectively. Note: This process can be described as $P\forall^{\gamma}$ = constant, where γ is the specific-heat ratio.

State 2–State 3: Constant-volume heat addition until the temperature T_3 reaches 1000 K.

State 3–State 4: Constant-pressure heat addition until the volume equals $3\forall_{min}$.

State 4–State 5: Isentropic expansion from $\forall_4$ (= $3\forall_{min}$) to the maximum volume $\forall_{max}$. See the note for process 1–2.

State 5–State 1: Constant-volume heat rejection to return to the initial state.

A. Carefully sketch the cycle in pressure–volume coordinates. Label the state points 1, 2, 3, etc.

B. Assume the specific heats can be treated as constants with the following values: $c_p = 1.008$ kJ/kg·K, and $c_v = 0.721$ kJ/kg·K. Determine the work transfer associated with each process and the net work for the cycle.

4.51 An air-standard Otto cycle is often used as a very simplified model of a spark-ignition engine. The following sequence of processes applied to a fixed mass of air in a piston–cylinder assembly constitutes the Otto cycle:

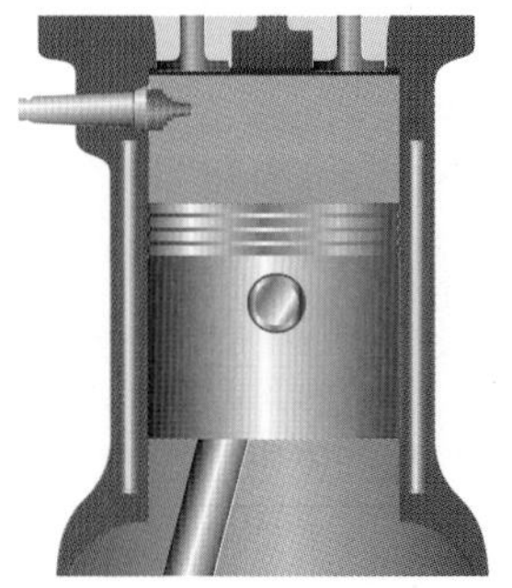

(Photograph courtesy of Scania.)

Process 1–2: Isentropic compression from the maximum volume $\forall_{max}$ to the minimum volume $\forall_{min}$. Note: This process can be described as $P\forall^{\gamma}$ = constant, where γ is the specific-heat ratio.

Process 2–3: Constant-volume heat addition to a peak cycle temperature T_3.

Process 3–4: Isentropic expansion from $\forall_{min}$ to $\forall_{max}$. See the note for process 1–2.

Process 4–1: Constant-volume heat rejection to return to the initial state.

A. Sketch this cycle in P–$\forall$ coordinates and label each state point (1, 2, 3, and 4).

B. The following properties are associated with a particular execution of the Otto cycle. The initial state (state 1) temperature, pressure, and volume are 320 K, 120 kPa, and 0.001 m^3, respectively. The compression ratio, $\forall_{max}/\forall_{min}$, is 9.0. The maximum temperature is 3000 K (state 3). Assume the specific heats can be treated as constants with the following values: $c_p = 1.008$ kJ/kg·K, and $c_v = 0.721$ kJ/kg·K. Determine the work transfer associated with each process and the net work for the cycle.

4.52 Consider a fixed mass of air trapped in a piston–cylinder assembly. The air-standard diesel cycle is defined by the following sequences of processes applied to this system:

Process 1–2: Isentropic compression from the maximum volume $\forall_{max}$ to the minimum volume $\forall_{min}$. Note: This process can be described as $P\forall^{\gamma}$ = constant, where γ is the specific-heat ratio.

Process 2–3: Constant-pressure heat addition to a peak cycle temperature T_3. Note that the volume at state 3 lies between the minimum and maximum volumes, i.e., $\mathcal{V}_{min} < \mathcal{V}_3 < \mathcal{V}_{max}$.

Process 3–4: Isentropic expansion to $\mathcal{V}_{max}$. See the note for process 1–2.

Process 4–1: Constant-volume heat rejection to return to the initial state.

A. Sketch this cycle on P–$\mathcal{V}$ coordinates and label each state point (1, 2, 3, and 4).

B. The following properties are associated with a particular execution of the diesel cycle: The initial state (state 1) temperature, pressure, and volume are 310 K, 105 kPa, and 0.0012 m^3, respectively. The compression ratio, $\mathcal{V}_{max}/\mathcal{V}_{min}$, is 15.0. The maximum temperature is 3200 K (state 3). Assume the specific heats can be treated as constants with the following values: $c_p = 1.008$ kJ/kg·K, and $c_v = 0.721$ kJ/kg·K. Determine the work transfer associated with each process and the net work for the cycle.

4.53 Consider the Otto cycle described in Problem 4.51. The initial state (state 1) temperature, pressure, and volume are 90 F, 14.7 psia, and 40 in^3, respectively. The compression ratio, $\mathcal{V}_{max}/\mathcal{V}_{min}$, is 8.5. The maximum temperature is 4500 F (state 3). Assume the specific heat ratio is constant with a value of 1.4. Determine the work transfer associated with each process and the net work for the cycle.

4.54 Consider the Diesel cycle described in Problem 4.52. The initial state (state 1) temperature, pressure, and volume are 95 F, 20.0 psia, and 60 in^3, respectively. The compression ratio, $\mathcal{V}_{max}/\mathcal{V}_{min}$, is 14.5. The maximum temperature is 4200 F (state 3). Assume the specific heat ratio is constant with a value of 1.4. Determine the work transfer associated with each process and the net work for the cycle.

4.55–4.61 Numerical problems – two-phase substances

4.55 Saturated steam at 2 MPa is contained in a piston–cylinder arrangement as shown in the sketch below. The initial volume is 0.25 m^3. Heat is removed from the steam during a quasi-static constant-pressure process such that 197.6 kJ of work is done by the surroundings on the steam. Determine the quality of the steam at the final state.

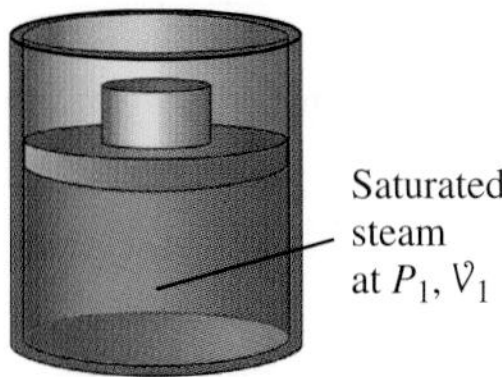

4.56 One kilogram of wet steam at 450 K with a quality of 0.75 is contained in a piston–cylinder assembly. Heat is added to the steam during a quasi-static isothermal process until the steam becomes saturated vapor. Sketch the process in P–$\mathcal{V}$ coordinates and determine the work done by the steam on the surroundings.

4.57 A 0.5-kg mass of saturated steam at 1 MPa is contained in a piston–cylinder assembly. Heat is removed from the steam during a quasi-static constant-pressure process such that 65 kJ of work is done by the surroundings on the steam. Sketch the process in P–$\mathcal{V}$ coordinates and determine the quality of the steam at the final state.

4.58 Determine the work done in ft·lb$_f$ by a 2-lb$_m$ steam system as it expands slowly in a cylinder–piston arrangement from the initial conditions of 324 psia and 12.44 ft^3 to the final volume of 25.256 ft^3 in accordance with the following relations: (a) $P = 20\mathcal{V} + 75.12$, where $\mathcal{V}$ and P are expressed in units of ft^3 and psia, respectively; and (b) $P\mathcal{V} = \text{constant}$.

4.59 Consider 0.15 kg of H_2O (liquid and vapor in equilibrium) contained in a vertical piston–cylinder arrangement at 350 K, as shown in the sketch. Initially, the piston rests on the stops, and the volume beneath the 142.8-kg piston (area of 10 cm^2) is 0.001306 m^3. The atmospheric pressure is 100 kPa, and the local gravitational acceleration is 9.80 m/s^2. Energy is transferred to this arrangement until there is only saturated vapor inside. The piston moves without friction.

A. Show this process on T–$\mathcal{V}$ and P–$\mathcal{V}$ diagrams.
B. Determine the moving boundary work done by the H_2O system.

4.60 A vertical cylinder–piston arrangement contains 0.3 lb$_m$ of H_2O (liquid and vapor in equilibrium) at 120 F, as shown in the sketch for the previous problem. Initially, the piston rests on the stops, and the volume beneath the 250-lb$_m$ piston is 1.054 ft^3. The piston area is 120 in^2. The atmospheric pressure is 14.7 lb$_f$/in^2, and the gravitational acceleration is 30.0 ft/s^2. Heat is transferred to the arrangement until only saturated vapor exists inside. **HINT**: Use the NIST software or online WebBook to obtain thermodynamic properties.

A. Show this process on T–$\mathcal{V}$ and P–$\mathcal{V}$ diagrams.
B. Determine the moving boundary work done (Btu) by the H_2O system.

4.61 Saturated steam is condensed to saturated liquid in a piston–cylinder assembly at constant pressure. The 721-kg piston has a diameter of 15 cm and moves freely without friction during the process. The work transfer to the steam during this process is 56 kJ. The atmospheric pressure is 100 kPa, and the local gravitational acceleration is 9.80 m/s^2. Determine the mass of the H_2O.

4.62–4.65 Shaft work and power

4.62 An automobile drive shaft rotates at 3000 rev/min and delivers 75 kW of power from the engine to the wheels. Determine the torque (N–m and ft·lb$_f$) in the drive shaft.

4.63 The output shaft of a gas-turbine engine in a locomotive rotates steadily at 3000 rev/min. The torque in the shaft is 3500 N·m. Determine (i) the instantaneous power delivered (kW) and (ii) the amount of work transfer for one minute of operation.

4.64 The torque in the shaft of a water pump is 15 N·m. How much work is performed after 1000 revolutions of the shaft?

4.65 The shaft of an electric motor in a drone helicopter delivers 52.5 W of power to a propeller. The 3.17-mm shaft spins at 10,500 rev/min. Determine the torque in the motor shaft.

4.66–4.70 Electrical work and power

4.66 A toaster draws 9.6 amperes of current from a 120-volt wall socket when operating. Bread is toasted to a nice golden brown after 2 1/2 minutes. Determine (i) the instantaneous electrical power supplied to the toaster and (ii) the energy used to toast the bread. Discuss the fate of the electrical energy supplied to the toaster during the process.

4.67 Determine all the combinations of the following 120-volt kitchen appliances that can be operated simultaneously without tripping a 15-ampere circuit breaker: blender (300 W); clock radio (70 W); coffee maker (1200 W); food processor (200 W); microwave oven (1450 W); toaster oven (1150 W). Would you recommend having more than one 15-amp circuit to accommodate these appliances? Would you recommend a dedicated circuit for any of these appliances?

4.68 A somewhat hard-to-start car engine requires the starter motor to crank for approximately 5 seconds. The current drain from the 12-volt battery during cranking is 116 amperes. If an equivalent amount of the energy is used to operate a 43-W CFL lightbulb, for how many minutes would the lightbulb be lit?

4.69 You are charged with the design of the electrical circuits for a small workshop. A design criterion is that a circular saw (1200 W) and a drill press (1100 W) are able to operate simultaneously. Choices of circuit breakers are 15 amperes and 20 amperes. How many circuits do you propose and at what current rating(s)? Discuss (defend) your proposed design.

4.70 A 40-ohm resistor is rated at 5 watts. Would this resistor be safe to use (i) in a computer with a 5-volt drop, (ii) in an automobile with a 12-volt drop, or (iii) in a home appliance with a 120-volt drop?

4.71 You like to play very loud music on your car stereo such that your stereo system draws about 4 amperes more or less continuously. At what average rate would you have to eat donuts to consume food energy at the same rate as your stereos' electricity consumption? A donut contains 260 food calories (1 food calorie = 4184 J). Assume your vehicle operates with a 12-volt system. Express your result in donuts per day.

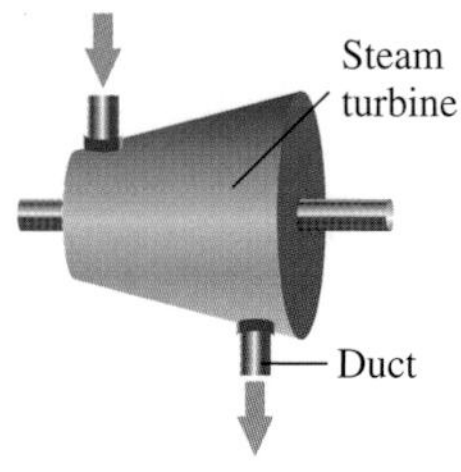

4.72–4.74 **Flow work rate**

4.72 Steam exits a turbine and enters a 1.435-m-diameter duct. The steam is at 4.5 kPa and has a quality of 0.90. The mean velocity of the steam through the duct is 40 m/s. Determine the flow work rate required to push the steam into the duct.

4.73 Natural gas enters a 12-in-diameter pipeline at 300 psig and 68 F. The average velocity of the entering gas is 30 m/s. Determine the flow work rate required to push the gas into the duct. Express your result in ft·lb_f/s, Btu/hr, horsepower, and kW. Assume the natural gas has the properties of methane. The atmospheric pressure is 14.7 psia. **HINT**: Use the NIST software for any properties you need.

4.74 Natural gas (methane) is compressed from 100 kPa at 290 K to 500 kPa at 415 K. The gas flows at 0.2 kg/s. Determine the flow work rates into and out of the compressor. Use the NIST software for methane properties.

4.75–4.79 **Identifying heat and work interactions**

4.75 Consider the electrical circuit shown in the sketch, in which a light bulb and a motor are wired in series with a battery. The motor drives a fan. Analyze the following thermodynamic systems for all heat and work interactions. Sketch the system boundaries and use labeled arrows for the various values of $\dot{Q}$ and $\dot{W}$. Be sophisticated enough in your analysis to include small interactions. Use subscripts to denote the type of work and make sure that your arrows point in the correct directions.

A. The light bulb alone

B. The light bulb and motor, with the system boundary cutting through the motor shaft

C. The light bulb and motor, but the system boundary is now contiguous with the fan blade rather than cutting through the shaft

D. The battery alone

E. The battery, light bulb, motor, and fan

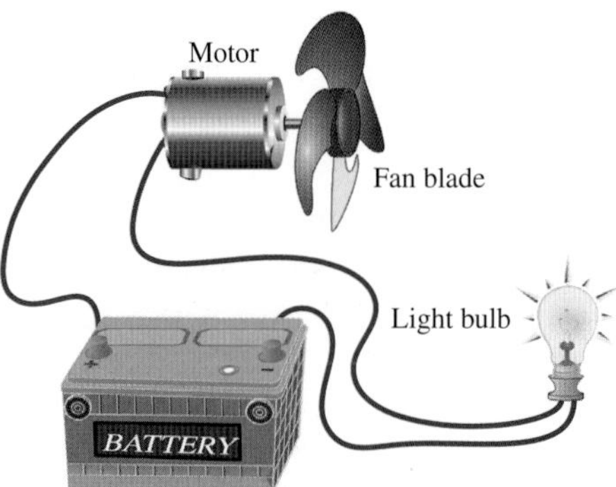

4.76 Consider the Rankine cycle power plant, as shown in the sketch below. Using a control volume that includes *only* the working fluid (steam/water), draw and label a sketch showing all the heat and work interactions associated with the cycle control volume.

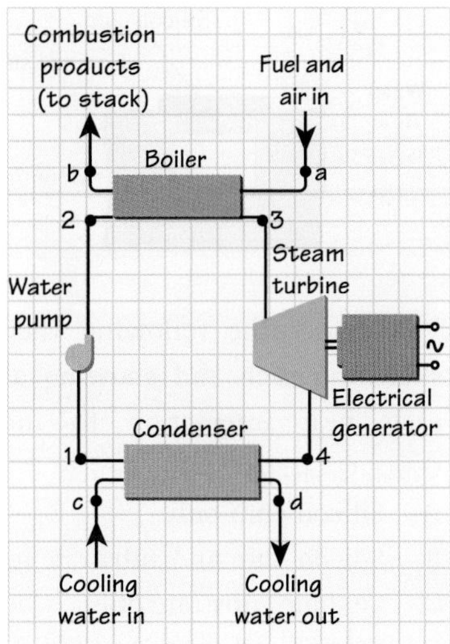

4.77 Consider the solar water heating scheme illustrated in Fig. 4.20. To analyze this scheme, choose a control surface that includes all the components shown and cuts through the cold water supply and hot-water-out lines. Assume hot water is being drawn for a shower. Sketch the control volume and show using labeled arrows all the heat and work interactions (i.e., all values of $\dot{Q}$ and $\dot{W}$). Use appropriate subscripts to denote the type of work or heat exchanges.

4.78 To conduct a laboratory test, a multicylinder spark-ignition engine is connected to a dynamometer (a device that measures the shaft power output), as shown in the sketch.

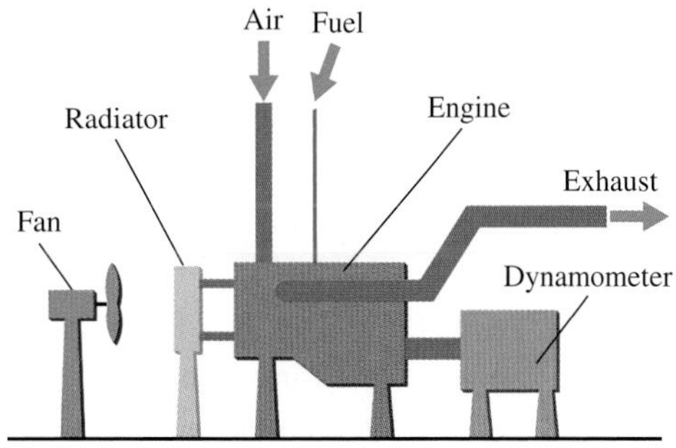

For the following control volumes, draw a sketch indicating all heat and work interactions (i.e., all values of $\dot{Q}$ and $\dot{W}$):

A. The control volume contains only the engine, and the control surface cuts through the hoses to and from the radiator, through the air and fuel supply lines, through the exhaust pipe, and through the shaft connecting the engine and dynamometer.

B. Same as in part A, but now the radiator is included within the control volume.

4.79 In a high-performance computer, silicon chips are actively cooled by water flowing through the ceramic substrate, as shown in the sketch. The chip assembly is inside a case through which air freely circulates.

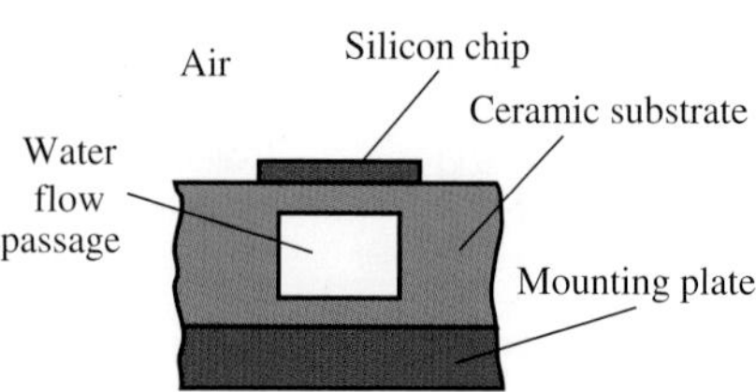

Consider the following thermodynamic systems (or control volumes), sketching each and showing all the heat interactions with arrows. Label each heat interaction to indicate the mode of heat transfer (i.e., $\dot{Q}_{cond}$, $\dot{Q}_{conv}$, or $\dot{Q}_{rad}$):

A. Silicon chip only
B. Silicon chip and substrate (excluding the water)
C. Ceramic substrate only
D. Ceramic substrate and the flowing water

4.80–4.93 Heat-transfer modes

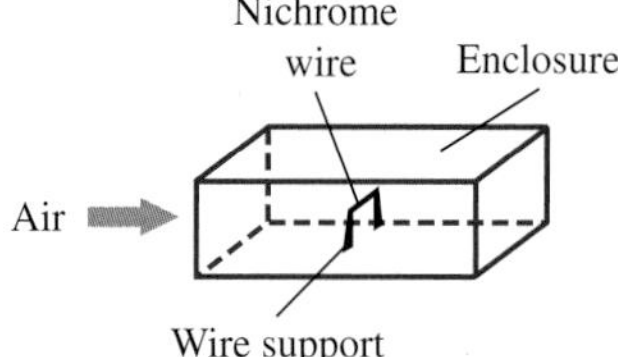

4.80 A 35-mm length of 1-mm-diameter nichrome (80% Ni, 20% Cr) wire is heated to 915 K by the passage of an electrical current through the wire (Joule heating). The wire is in an enclosure, the walls of which are maintained at 300 K. Warm air at 325 K flows slowly through the enclosure. The emissivity of the wire is 0.5 and the convective heat-transfer coefficient has a value of 54.9 W/m^2·K. Determine the heat-transfer rate from the nichrome wire.

4.81 Consider the same physical situation as in Problem 4.80, except that the electrical current is reduced by an amount such that the nichrome wire temperature is now 400 K. This temperature reduction also results in a reduction of the heat-transfer coefficient associated with forced convection, which now has a value of 36.9 W/m^2·K. Compare the value of the fraction of the total heat transfer due to radiation for this situation to the value from Problem 4.80. Discuss.

4.82 Air at 300 K ($\equiv T_\infty$) flows over a flat plate maintained at 400 K ($\equiv T_s$). At a particular location on the plate, the temperature distribution in the air measured from the plate surface is given by

$$T(y) = (T_s - T_\infty)\exp[-22{,}000y(\text{m})] + T_\infty,$$

where y is the perpendicular distance from the plate, as shown in the sketch.

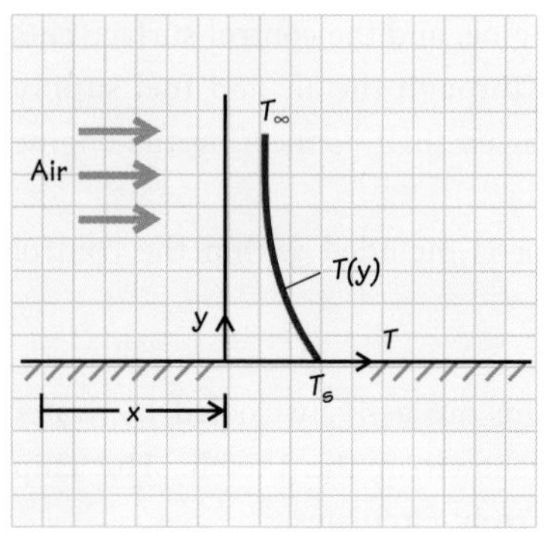

A. Determine the conduction heat flux (the heat-transfer rate per unit area) through the air at the plate surface where the no-slip condition applies (i.e., at $y = 0^+$).

B. The physical mechanism of convection heat transfer starts with conduction in the fluid at the wall, as described in part A. The details of the flow, however, determine the temperature distribution through the fluid. Use your result from Part A to determine a value for the local heat-transfer coefficient $h_{conv,x}$ at the location x where $T(y)$ is given.

4.83 Determine the instantaneous rate of heat transfer from a 1.5-cm-diameter ball bearing with a surface temperature of 150 °C submerged in an oil bath at 75 °C if the surface convective heat transfer coefficient is 850 W/m^2·°C.

4.84 Air at 480 F flows over a 20-in by 12-in flat surface. The surface is maintained at 75 F and the convective heat-transfer coefficient at the surface is 45 Btu/hr·ft^2·F. Find the rate of heat transfer to the surface.

4.85 A convective heat flux of 20 W/m^2 (with free convection) is observed between the walls of a room and the ambient air. What is the heat-transfer coefficient when the air temperature is 32 °C and the wall temperature is 35 °C?

4.86 A 30-cm-diameter, 5-m-long steam pipe passes through a room where the air temperature is 20 °C. If the exposed surface of the pipe is a uniform 40 °C, find the rate of heat loss from the pipe to the air if the surface heat-transfer coefficient is 8.5 W/m^2·°C.

4.87 A 10-ft-diameter spherical tank is used to store petroleum products. The products in the tank maintain the exposed surface of the tank at 75 F. Air at 60 F blows over the surface, resulting in a heat-transfer coefficient of 10 Btu/hr·ft^2·F. Determine the rate of heat transfer from the tank.

4.88 The heat-transfer coefficient for water flowing normal to a cylinder is measured by passing water over a 3-cm-diameter, 0.5-m-long electric resistance heater. When water at a temperature of 25 °C flows across the cylinder with a velocity of 1 m/s, 30 W of electrical power are required to maintain the heater surface temperature at 60°C. Estimate the heat-transfer coefficient at the heater surface. What is the significance of the water velocity in determining the answer?

4.89 A cylindrical electric resistance heater has a diameter of 1 cm and a length of 0.25 m. When water at 30 °C flows across the heater, a heat-transfer coefficient of 15 W/m^2·°C exists at the surface. If the electrical input to the heater is 5 W, what is the surface temperature of the heater?

4.90 A typical value of the heat-transfer coefficient for water boiling on a flat surface is 900 Btu/hr·ft^2·F. Estimate the heat flux on such a surface when the surface is maintained at 222 F and the water is at 212 F.

4.91 Find the rate of radiant energy emitted by an ideal blackbody having a surface area of 10 m^2 when the surface temperature is maintained at (a) 50 °C, (b) 100 °C, and (c) 500 °C.

4.92 A surface is maintained at 100 °C and is enclosed by very large surrounding surfaces at 80°C. What is the net radiant flux (W/m^2) from this surface if its emissivity is (a) 1.0 or (b) 0.8?

4.93 Repeat Problem 4.92 for a surface temperature of 150 °C, with all other data remaining unchanged.

4.94–4.99 EES problems

4.94 Air initially at 500 kPa and 2 m^3 is expanded to a volume of 7 m^3 in an isothermal process. Use EES or a spreadsheet to plot the process on a $P–V$ diagram. At least ten points should be used from the initial to final state, in order to produce a smooth curve.

4.95 Air initially at 500 kPa and 2 m^3 is expanded to a volume of 7 m^3 in an isentropic process. (For an ideal gas, an isentropic process is a polytropic process with $n = \gamma$.) Use EES or a spreadsheet to plot the process on a $P–V$ diagram. At least ten points should be used from the initial to final state, in order to produce a smooth curve.

4.96 Plot problems 4.94 and 4.95 on the same $P–V$ diagram. Also plot the isobaric process from the initial state to the final volume. Which process produces the greatest $P–V$ work? Explain how you used the $P–V$ diagram to answer this question.

4.97 7 kg water initially at 500 kPa and 2 m^3 is expanded to a volume of 5 m^3 in an isothermal process. Use EES or NIST and a spreadsheet to plot the process on a $P–V$ diagram. At least ten points should be used from the initial to final state, in order to produce a smooth curve.

4.98 7 kg water initially at 500 kPa and 2 m^3 is expanded to a volume of 5 m^3 in an isentropic process. Use EES or NIST and a spreadsheet to plot the process on a $P–V$ diagram. At least ten points should be used from the initial to final state, in order to produce a smooth curve.

4.99 Plot Problems 4.97 and 4.98 on the same $P–V$ diagram. Also plot the isobaric process from the initial state to the final volume. Which process produces the greatest P–V work? Explain how you used the $P–V$ diagram to answer this question.

4.100–4.108 FE problems

4.100 Air ($\gamma = 1.4$) expands at a constant pressure of 300 kPa from a volume of 0.1 m^3 to 0.4 m^3. What is the work done by the air? a. 9 kJ, b. 30 kJ, c. 90 kJ, d. 120 kJ.

4.101 Steam expands at a constant pressure of 300 kPa from a volume of 0.1 m^3 to 0.4 m^3. What is the work done by the steam? a. 9 kJ, b. 30 kJ, c. 90 kJ, d. 120 kJ.

4.102 What is the shaft work when a torque of 130 N·m is required to rotate a pump shaft at 300 rpm? a. 245 MW, b. 16.3 MW, c. 156 MW, d. 2.6 MW.

4.103 Air ($\gamma = 1.4$) initially at 500 kPa and 400 K expands at constant temperature from a volume of 0.3 m^3 to 0.6 m^3. What is the work done by the air? a. 124 kJ, b. 104 kJ, c. 90 kJ, d. 180 kJ.

4.104 What is the rate of electrical power required if a current of 5 A passes through an electric heater having a resistance of 30 Ω? a. 200 W, b. 1.6 kW, c. 500 W, d. 5.0 kW.

4.105 Air ($\gamma = 1.4$) initially at 300 kPa and 400 K expands in an isentropic process from a volume of 0.3 m^3 to 0.6 m^3. What is the work done by the air? a. 54.5 kJ, b. 63.4 kJ, c. 90 kJ, d. 120 kJ.

4.106 Air ($\gamma = 1.4$) initially at 600 kPa and 400 K expands in a polytropic process where $P\mathcal{V}^{1.1}$ = constant What is the work done by the air as it expands from a volume of 0.8 m^3 to 1.2 m^3? a. 120 kJ, b. 180 kJ, c. 95 kJ, d. 191 kJ.

4.107 A hot steel rod at 600 K is cooled by quenching it in cold water at 300 K. If the rod is 0.5 m long and 3 cm in diameter, what is the initial heat transfer rate when the average convection coefficient is 80 $W/m^2 \cdot K$? a. 24 kW, b. 360 kW, c. 1.13 kW, d. 300 W.

4.108 3 kg steam expands from saturated liquid to saturated vapor at a constant temperature of 400 K. What is the work done by the steam? a. 538 kJ, b. 179 kJ, c. 875 kJ, d. 60 kJ.

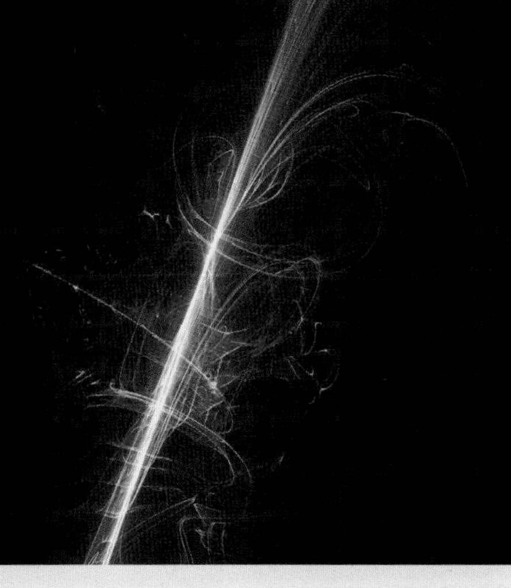

CHAPTER 5
First Law of Thermodynamics

LEARNING OBJECTIVES

After studying Chapter 5, you should:

- Be able to write from memory (or intimate familiarity) the conservation of energy (the first law of thermodynamics) applied to a closed system for an incremental change in state, for a change from state 1 to state 2, and at an instant.
- Be able to write down from memory (or intimate familiarity) both steady and unsteady forms of energy conservation (first law) for an open system with a single inlet and a single outlet.
- Be able to explain the physical significance of each term in the various first-law expressions.
- Understand when and how to apply the first-law statements to practical situations.
- Be able to explain the physical origins of enthalpy, in the first-law analysis of both closed and open systems.
- Be able to simplify and apply the integral form of the first law for steady-flow devices. (This objective links to Chapter 8.)
- Be able to solve first-law problems.

CHAPTER 5 OVERVIEW

IN THIS CHAPTER, we apply the fundamental principle of energy conservation to both closed and open thermodynamic systems. In our analyses of closed (fixed-mass) systems, we will express energy conservation for incremental and finite changes in state and at an instant. In dealing with open systems, we again follow a hierarchical development by starting with simple, steady-state, steady-flow cases and then adding detail and complexity to arrive at more general statements of energy conservation. We apply the energy conservation principle to analyses of steady-flow devices, linking our theoretical developments to practical applications. Figure 5.1 illustrates the key role that this chapter plays in our study.

FIGURE 5.1 Chapter 5 is key to our study of thermodynamics. Previous chapters laid the groundwork to enable us to explore and apply the principle of energy conservation to systems of engineering interest. This figure illustrates the key role that Chapter 5 plays in our study. Chapter 5 is also the jumping off point that allows us to begin to study practical devices and systems.

Historical Context

The development of an energy conservation principle depended critically on the evolving idea that work was convertible to heat, and vice versa. **Benjamin Thompson** (1753–1814) (Count Rumford) in 1798 [1], **Julius Robert Mayer** (1814–1878) in 1842 [2], and **James Prescott Joule** (1818–1889) in 1849 [3] published results of their research on the mechanical equivalent of heat and presented numerical values. The issue here was to define the specific number of foot-pounds of work that is equivalent to the heat required to raise the temperature of 1 lb_m of water by 1 F. (The accepted value today is 778.16 ft·lb_f and defines the British thermal unit or Btu.) Mayer was the first person to state a formal conservation of energy principle. In 1842, he wrote [2]: "Forces (energies) are therefore indestructible, convertible, and (in contradistinction to matter) imponderable objects. . . a force (energy) once in existence, cannot be annihilated." (The parenthetical material has been added; in the mid 1800s, the word *force* commonly meant *energy* [4, 5].) **Hermann Helmholtz** (1821–1894), who produced an array of scientific achievements, extended the conservation of energy principle in his 1847 paper, "On the Conservation of Force," to include all known forms of energy: mechanical, thermal, chemical, electrical, and magnetic. (See Ref. [4].) **Rudolf Clausius** (1822–1888) wrote in 1850 one of the most succinct and modern-sounding statements of the energy conservation principle [6]: "The energy of the universe is constant." (Clausius also named the property **entropy** and presented clear statements of the second law of thermodynamics. We will consider these concepts in Chapters 6 and 7.)

(Credit: ZU_09 / DigitalVision Vectors / Getty Images.)

(Credit: Oprea Nicolae)

(Credit: Alamy Stock Photo.)

5.1 Energy Conservation for a Closed System

To clarify what we mean by the **first law of thermodynamics**, we offer this quotation from Max Planck's thermodynamics textbook [7]: *The first law of thermodynamics is nothing more than the principle of the conservation of energy applied to phenomena involving the production or absorption of heat.* Throughout this book, we will use the terms conservation of energy and the first law of thermodynamics synonymously, in accord with Planck's definition.

We begin our study of the conservation of energy principle by considering a closed system (one with fixed mass). We start by explicitly transforming the generic balance principles introduced in Chapter 3 to statements of energy conservation. We repeat the generic balance principle (Eq. 3.1), first, for a finite time interval, and second, for an instant:

> At this point, you may find it useful to review the detailed discussion of closed and open systems in Chapter 1.

For a finite time interval, $\Delta t = t_2 - t_1$,

$$X_{\text{in}} - X_{\text{out}} + X_{\text{generated}} = \Delta X_{\text{stored}} \equiv X(t_2) - X(t_1).$$

X_{in}	X_{out}	$X_{\text{generated}}$	$\Delta X_{\text{stored}} \equiv X(t_2) - X(t_1)$
Quantity of X crossing boundary and passing into system	Quantity of X crossing boundary and passing out of system	Quantity of X generated within system	Change of quantity of X stored within system during time interval

For our study, the generic quantity X can be mass, energy, or entropy.

Instantaneously, we have

$$\dot{X}_{\text{in}} - \dot{X}_{\text{out}} + \dot{X}_{\text{generated}} = \dot{X}_{\text{stored}}$$

$\dot{X}_{\text{in}}$	$\dot{X}_{\text{out}}$	$\dot{X}_{\text{generated}}$	$\dot{X}_{\text{stored}}$
Time rate of X crossing boundary and passing into system	Time rate of X crossing boundary and passing out of system	Time rate of generation of X within system	Time rate of storage of X within system

By defining X to be the energy E, Eq. 3.1, which applies to the time interval $t_2 - t_1$, becomes

> Equations with yellow backgrounds express key concepts and are the most important relationships in the chapter.

$$E_{\text{in}} - E_{\text{in}} + E_{\text{generated}} = \Delta E_{\text{stored}} \equiv E_{\text{sys}}(t_2) - E_{\text{sys}}(t_1). \tag{5.1}$$

E_{in}	E_{in}	$E_{\text{generated}}$	$\Delta E_{\text{stored}} \equiv E_{\text{sys}}(t_2) - E_{\text{sys}}(t_1)$
Quantity of energy crossing boundary and passing into system	Quantity of energy crossing boundary and passing out of system	Quantity of energy generated within system	Change in quantity of energy stored within system during time interval $\Delta t = t_2 - t_1$

Similarly, by defining $\dot{X}$ to be the time rate of energy $\dot{E}$, Eq. 3.2, which applies to an instant, becomes

$$\dot{E}_{\text{in}} - \dot{E}_{\text{out}} + \dot{E}_{\text{generated}} = \dot{E}_{\text{stored}}. \tag{5.2}$$

$\dot{E}_{\text{in}}$	$\dot{E}_{\text{out}}$	$\dot{E}_{\text{generated}}$	$\dot{E}_{\text{stored}}$
Time rate of energy crossing boundary and passing into system	Time rate of energy crossing boundary and passing out of system	Time rate of energy generated within system	Time rate of energy stored within system

FIGURE 5.2 Energy enters (E_{in}) and exits (E_{out}) **across** the system boundaries; energy is generated ($E_{\text{generated}}$) or stored (ΔE_{stored}) **within** the system.

Figure 5.2 shows a system with superimposed arrows representing the various terms in these equations.

Our task now is to associate each term in these conservation of energy equations with particular forms of energy for various physical situations. We will consider rather general statements of energy conservation, which include all forms of energy and their interconversions with the exception of nuclear energy and its transformations. Because we include *all* but nuclear forms of energy in our energy accounting, no generation terms appear, i.e., $\dot{E}_{\text{generated}} = 0$. The proper treatment of nuclear transformations and the relationship between mass and energy are beyond the scope of this book.

Consider the general thermodynamic system illustrated in Fig. 5.3. The system here is a macroscopic (i.e., an integral) system. Several forms of energy may be associated with the system. For example, the system will always possess internal energy – which can be further subdivided into thermal and chemical contributions – and it may also possess bulk kinetic and potential energies. Thus, we can write

$$E_{\text{sys}} = U_{\text{sys}} + (KE)_{\text{sys}} + (PE)_{\text{sys}}. \tag{5.3}$$

Heat and work only have meaning as energy crossing a boundary.
Heat and work do not exist within a system or control volume.

We cannot emphasize too strongly that the energies expressed in Eq. 5.3 are contained *within* the system boundary. The system possesses these energies. The system may lose or gain energy, however, by energy transfers *across* the system boundary. As we saw in Chapter 4, these transfers occur only because of heat and/or work interactions with the surroundings. Thus, the energy of the system can be increased either by heat transfer to the system from the surroundings or by the surroundings performing work on the system. These transfers of energy across the system boundary correspond to E_{in} in Fig. 5.2. Conversely, the system may lose energy by heat transfer from the system to the surroundings or by the system performing work on the surroundings (i.e., E_{out} in Fig. 5.2). With only three quantities involved, writing a mathematical expression of conservation of energy is easy to do and intuitively satisfying. The system energy is analogous to your bank account balance, where heat and work interactions are the credits and debits. Note that $E_{\text{generated}}$ is zero since we include in our energy balance all forms of energy, excluding nuclear reactions. We now apply these ideas to several situations.

5.1a For an Incremental Change

Consider the transfer of incremental quantities of energy across the system boundary, either into the system, δE_{in}, or out of the system, δE_{out} (the reason why we use δ rather than d for these quantities will become clear later), during the time interval dt. These transfers result in an incremental change in the amount of energy stored within the system, δE_{sys}, as indicated in Fig. 5.3a. Conservation of energy is then expressed as

$$\underbrace{\delta E_{\text{in}}}_{\substack{\text{Incremental energy}\\ \text{crossing the system}\\ \text{boundary from the}\\ \text{surroundings}\\ \text{to the system}}} - \underbrace{\delta E_{\text{out}}}_{\substack{\text{Incremental energy}\\ \text{crossing the system}\\ \text{boundary from the}\\ \text{system to the}\\ \text{surroundings}}} = \underbrace{\delta E_{\text{sys}},}_{\substack{\text{Incremental change}\\ \text{in energy stored}\\ \text{within the}\\ \text{system boundary}}} \tag{5.4a}$$

where

$$\delta E_{\text{in}} \equiv \begin{cases} \delta W_{\text{in}}, \text{ work performed on the} \\ \text{system by the surroundings, plus} \\ \delta Q_{\text{in}}, \text{ heat transfer from the} \\ \text{surroundings to the system,} \end{cases} \tag{5.4b}$$

and

$$\delta E_{\text{out}} \equiv \begin{cases} \delta W_{\text{out}}, \text{ work performed on the} \\ \text{surroundings by the system, plus} \\ \delta Q_{\text{out}}, \text{ heat transfer from the} \\ \text{system to the surroundings.} \end{cases} \tag{5.4c}$$

Substituting the definitions from Eqs. 5.4b and 5.4c back into Eq 5.4a yields

$$(\delta W_{\text{in}} + \delta Q_{\text{in}}) - (\delta W_{\text{out}} + \delta Q_{\text{out}}) = \delta E_{\text{sys}}, \tag{5.4d}$$

Note that the arrows representing Q_{in} and W_{in} end precisely at the boundary. Similarly, Q_{out} and W_{out} arrows begin precisely at the boundary. Q and W have no meaning inside the boundary.

Equation 5.4d preserves the *in minus out equals stored* formulation, while introducing heat and work. Traditionally, the heat in and out and work in and out are combined; thus,

$$(\delta Q_{\text{in}} - \delta Q_{\text{out}}) - (\delta W_{\text{out}} - \delta W_{\text{in}}) = \delta E_{\text{sys}}, \tag{5.5a}$$

or

$$\delta Q_{\text{in, net}} - \delta W_{\text{out, net}} = \delta E_{\text{sys}}. \tag{5.5b}$$

This grouping of terms establishes a sign convention for heat and work, where Q is conventionally treated as a credit (i.e., $Q(\equiv Q_{\text{in, net}})$ is from the surroundings to the system), whereas work W is conventionally treated as a debit (i.e., $W(\equiv W_{\text{out, net}})$ is performed by the system on the surroundings). In some situations, we will adopt this convention; however, to avoid any ambiguity or confusion, we employ the subscript "net" to emphasize that both positive and negative heat and work interactions are implied within a single term as in Eq. 5.5b. Regardless of any sign convention, you should always be prepared to think through any new or ambiguous situation to ensure that heat and work interactions are properly credited or debited in your energy account as expressed by Eq. 5.4d.

We now consider a system that undergoes a finite change in state.

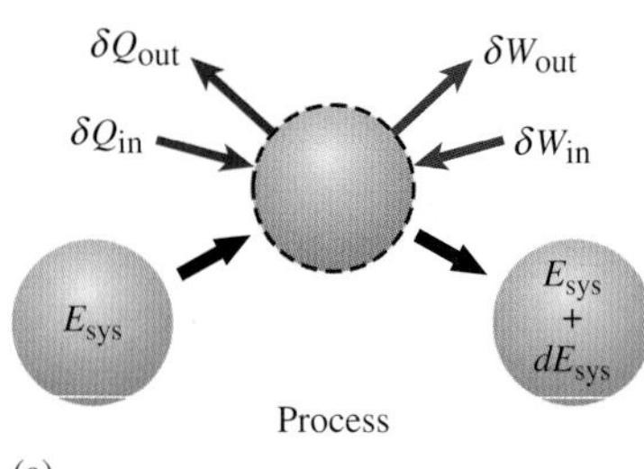

(a)

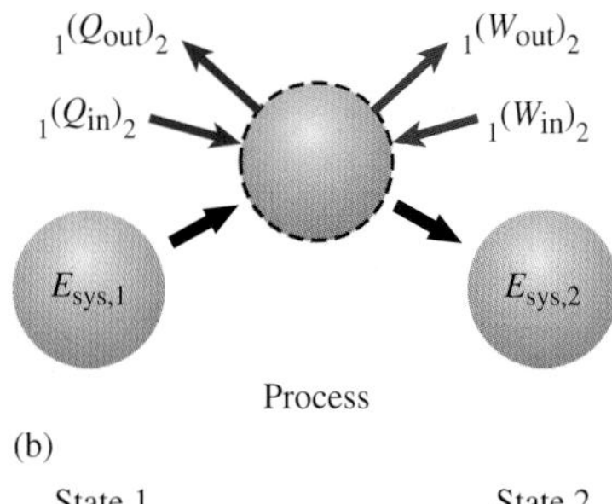

(b)

State 1 State 2

FIGURE 5.3 Heat and work interactions at the system boundaries result in a change in the amount of energy possessed by the system for **(a)** incremental interactions and **(b)** finite interactions.

5.1b For a Change in State

Conservation of energy is expressed for a process in which the system state changes from state 1 to state 2 by integrating Eqs. 5.5a and 5.5b, with the result that

$$\left[{}_1(Q_{\text{in}})_2 + {}_1(W_{\text{in}})_2\right] - \left[{}_1(Q_{\text{out}})_2 + {}_1(W_{\text{out}})_2\right] = \Delta E_{\text{sys}}, \tag{5.6a}$$

or

$${}_1(Q_{\text{in, net}})_2 - {}_1(W_{\text{out, net}})_2 = \Delta E_{\text{sys}}, \tag{5.6b}$$

where

$$\Delta E_{\text{sys}} \equiv \Delta E_{\text{sys, 2}} - \Delta E_{\text{sys, 1}}. \tag{5.7}$$

Recall from Chapter 4 that the notation ${}_1Q_2$ and ${}_1W_2$ does not represent a change from state 1 to 2 but rather refers to the integration of δQ and δW over particular paths. However, ΔE_{sys} does represent a change in the system energy as defined in Eq. 5.7. Mathematically, Q and W are **path functions**, and δQ and δW are referred to as **inexact differentials**; whereas E is a **state function**, and dE is referred to as an **exact differential.**

Equation 5.6 is an extremely important relationship. You should become very familiar with both its physical interpretation and its application. The following tutorial and examples should assist you in this endeavor.

Tutorial – First Law for Ideal-Gas Processes

A series of six short video presentations with three downloadable exercises have been developed to consolidate and integrate your thermodynamics knowledge [8]. These videos and exercises focus on ideal gases, making property evaluations simple and thus allowing you to focus on thermodynamic principles rather than procedure. The videos and exercises can be accessed at http://www.cambridge.org/thermo. Using the worksheets (exercises) with the videos is an essential aspect of the learning experience. See also end-of-chapter Problem 5.4.

MATRIX I – IDEAL-GAS PROPERTIES & PROCESSES

	Column 1	Column 2	Column 3	Column 4	Column 5	Column 6
	PROCESS	Relation of *P* to *V*	Relation of *T* to *V*	*P-V* Plot	*T-V* Plot	Moving Boundary Work
Row P	**Constant Pressure Expansion**					
Row V	**Constant Volume with Increase in Pressure**					
Row T	**Constant Temperature Expansion**					

Example 5.1 Drop Velocity

A 450-kg instrument package is dropped in NASA's 132-m drop tower facility. To minimize the drag force acting on the falling package, the air has been evacuated from the tower ($P = 10^{-2}$ torr). The package is dropped from rest. Determine the velocity of the package at the bottom of the tower just before it impacts the decelerator mechanism. Also, determine the kinetic energy of the package at this same instant.

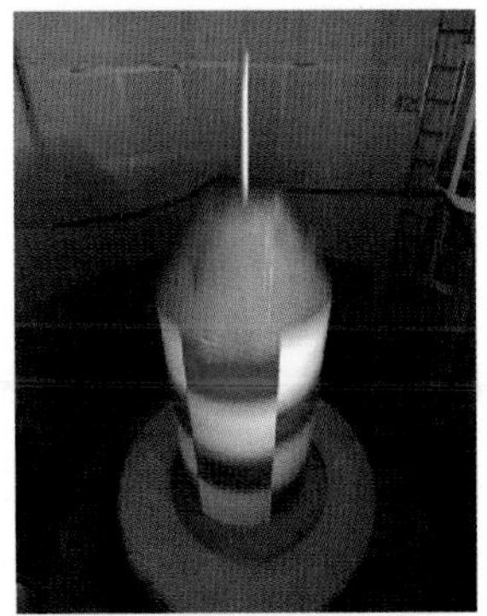

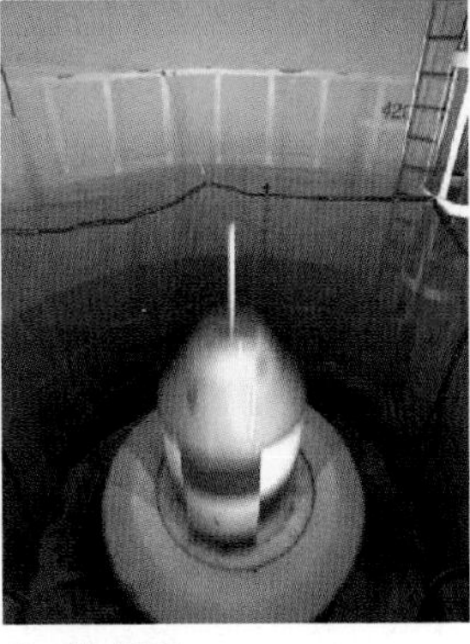

Credit: NASA.

Solution

Known M, $z_2 - z_1$, V_1

Find V_2, KE

Sketch

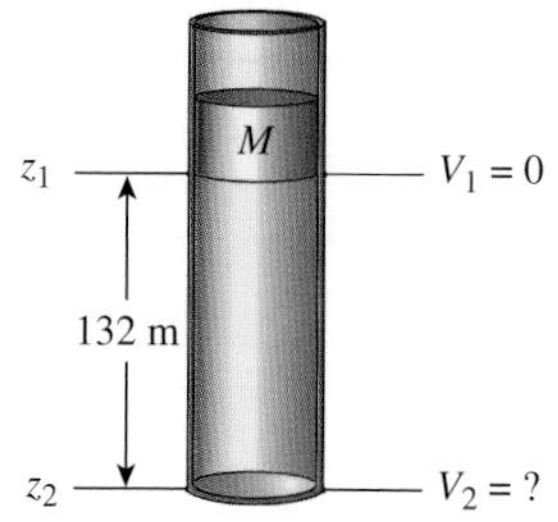

z_1 — $V_1 = 0$; M; 132 m; z_2 — $V_2 = ?$

Modeling, Premises and Assumptions

i. No frictional or drag forces act on the package.

ii. The process is adiabatic.

Analysis We begin by identifying the instrument package as a thermodynamic system. This system undergoes a process from the initial state ($V_1 = 0$, z_1) to the final state (V_2, z_2). The appropriate conservation of energy expression is given by Eq. 5.6:

$$_1(Q_{\text{in, net}})_2 - {_1(W_{\text{out, net}})_2} = E_{\text{sys, 2}} - E_{\text{sys, 1}}.$$

The net heat transfer is zero, as is the net work. That the work is zero follows from the fact that no boundary forces act on the package during its fall:

$$W = \int_1^2 F \cdot ds = 0.$$

Thus,

$$0 - 0 = E_{\text{sys, 2}} - E_{\text{sys, 1}},$$

or

$$E_{\text{sys, 2}} - E_{\text{sys, 1}}.$$

The system energy is given by Eq. 4.3. Assuming no rotational kinetic energy, we write

$$U_2 + \tfrac{1}{2}MV_2^2 + Mg(z_2 - z_{\text{ref}}) = U_1 + \tfrac{1}{2}MV_1^2 + Mg(z_1 - z_{\text{ref}}).$$

With our assumptions, there is no mechanism to change the system temperature; hence, $U_2 = U_1$. Recognizing that $V_1 = 0$, we can simplify the previous equation to

$$\tfrac{1}{2}MV_2^2 = Mg(z_1 - z_2).$$

Solving for V_2 yields

$$V_2 = [2g(z_1 - z_2)]^{1/2},$$

which is numerically evaluated as

$$V_2 = \left[2\left(9.8067\frac{\text{m}}{\text{s}^2}\right)(132\,\text{m})\right]^{1/2}$$
$$= 50.88\,\text{m/s}.$$

The kinetic energy is also evaluated as

$$\begin{aligned}\tfrac{1}{2}MV_2^2 &= Mg(z_1 - z_2)\\ &= 450\,\text{kg}\left(9.8067\frac{\text{m}}{\text{s}^2}\right)(132\,\text{m})\left[\frac{\text{N}}{\text{kg}\cdot\text{m/s}^2}\right]\left[\frac{\text{J}}{\text{N}\cdot\text{m}}\right]\\ &= 5.825\times 10^5\,\text{J}.\end{aligned}$$

Comment Note how the thermal energy terms drop out of this problem, leaving only the conversion between system potential and kinetic energy. We note also the substantial impact velocity of the package at the bottom of the tower (50.88 m/s = 113.8 mph). A catch basin of polystyrene beads is used to decelerate the package in NASA's 5.2-s drop tower.

Self-Test 5.1

A 0.2-kg projectile is launched vertically upward from the surface of the moon ($g = 1.62\,\text{m/s}^2$) with an initial velocity of 30 m/s. Determine (a) the initial kinetic energy of the projectile and (b) the maximum altitude it attains.

(Answer: (a) 90 J, (b) 277.8 m)

Example 5.2 Piston–Cylinder Isothermal Process with Air

Consider a piston–cylinder assembly containing 0.14 kg of air initially at a pressure of 350 kPa. Heat is added during a quasi-equilibrium isothermal process in which the volume increases from an initial value of 0.1 m^3 to a final value of 0.3 m^3. Sketch the process in T–$\mathcal{V}$ and P–$\mathcal{V}$ coordinates and calculate the amount of heat added during the process.

Solution

Known *Quasi-equilibrium isothermal process, M, P_1, $\mathcal{V}_1$, $\mathcal{V}_2$*

Find ${}_1Q_2$

Sketch

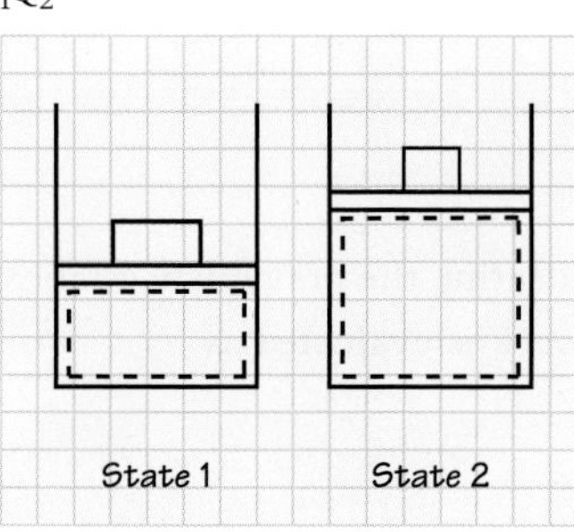

Modeling, Premises and Assumptions

i. Ideal-gas behavior
ii. No system kinetic energy
iii. Negligible change in system potential energy

Analysis We first identify the air in the cylinder as the thermodynamic system of interest, as indicated by the dashed lines in the sketch. Since the process is stated to be isothermal, we know that the temperature remains constant during the process and the T–$\mathcal{V}$ plot is a simple horizontal line as shown in the left-hand sketch here. With our assumption of ideal-gas behavior (Eq. 2.26c), we also know the relationship between P and $\mathcal{V}$ for a fixed T, that is,

$$P = (MRT)\frac{1}{\mathcal{V}}.$$

Since MRT is a constant, the P–$\mathcal{V}$ relationship is a simple hyperbola, as shown in the sketch on the right.

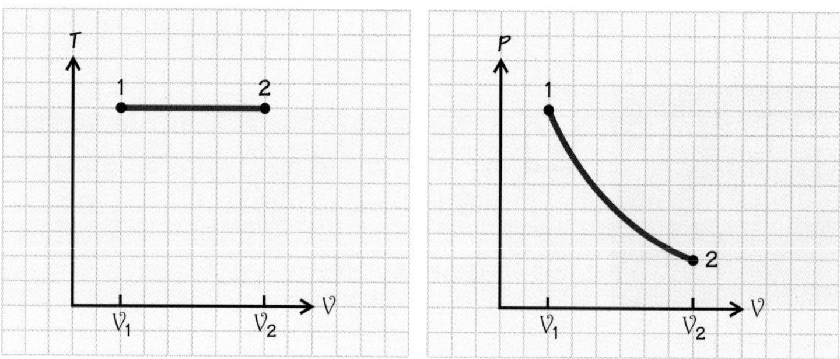

To determine the heat added, we write down the conservation of energy for the process (Eq. 5.6), that is,

$${}_1(Q_{in})_2 - {}_1(W_{out})_2 = U_2 - U_1.$$

For an ideal gas, the internal energy is a function of temperature only; thus, the internal energy change for the process must be zero since the temperature is constant. More formally, we evaluate Eq. 2.29d as

$$U_2 - U_1 = M\Delta u = M \int_{T_1}^{T_2=T_1} c_v dT = 0.$$

With the assumption of a quasi-equilibrium process, we can evaluate the work done by the system from Eq. 4.8b:

$${}_1W_2 = \int_1^2 P d\mathcal{V}.$$

Substituting the previously established relationship between P and $\mathcal{V}$ for this process, the work is evaluated as

$${}_1W_2 = \int_1^2 (MRT)\frac{1}{\mathcal{V}} d\mathcal{V}.$$

$$= MRT\int_1^2 \frac{d\mathcal{V}}{\mathcal{V}} = MRT \ln \frac{\mathcal{V}_2}{\mathcal{V}_1}.$$

We also know that $P_1\mathcal{V}_1 = MRT$; thus,

$$
\begin{aligned}
{}_1W_2 &= P_1\mathcal{V}_1 \ln \frac{\mathcal{V}_2}{\mathcal{V}_1} \\
&= 350 \times 10^3(0.1) \ln \left(\frac{0.3}{0.1}\right) \\
&= 38{,}450 \\
&[=] \frac{\text{N}}{\text{m}^2}\text{m}^3 = \text{N·m} = \text{J}.
\end{aligned}
$$

We now obtain the heat added from energy conservation:

$$ {}_1(Q_{\text{in}})_2 - {}_1(W_{\text{out}})_2 = 0 $$

or

$$
\begin{aligned}
&{}_1(Q_{\text{in}})_2 - {}_1(W_{\text{out}})_2 \\
&= 38{,}450\,\text{J}.
\end{aligned}
$$

Comment This example illustrates the combined use of a state equation, a calorific equation of state, and the conservation of energy. We also note, first, that the requirement that the process is conducted in a quasi-equilibrium manner is critical to our evaluation of the work and, second, that this work can be visualized as the area under the curve on our P–$\mathcal{V}$ sketch.

Self-Test 5.2

☑ Repeat Example 5.2 for a constant-pressure process, finding T_1, T_2, ${}_1W_2$, and ${}_1Q_2$.

(Answer: 871.1 K, 2613.2 K, 70 kJ, 300.2 kJ using c_v of 0.944 kJ/kg·K)

Example 5.3 Isochoric Process with Water

Steam is contained in a rigid tank at 500 K and 1 MPa. A cooling coil removes energy from the steam as heat until a final temperature of 400 K is reached. The tank volume is 0.15 m^3. Determine the amount of heat removed.

Solution

Known T_1, P_1, $\mathcal{V}_1 (= \mathcal{V}_2)$, T_2

Find ${}_1Q_2$

Sketch

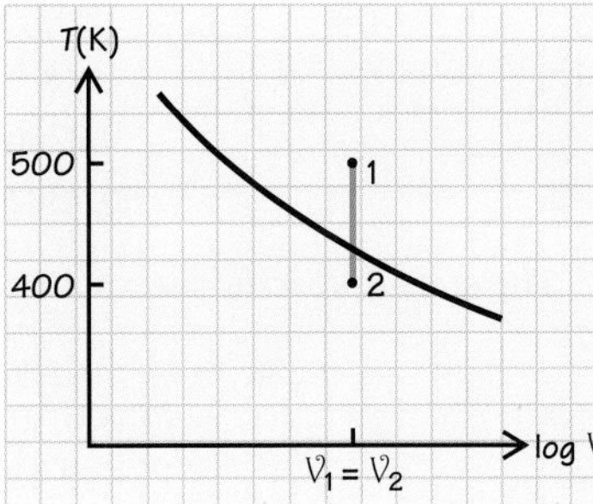

Modeling, Premises and Assumptions

i. Equilibrium prevails at initial and final states.
ii. There are no changes in system kinetic or potential energies.

Analysis We select the steam as the system of interest. The heat removed can then be determined from application of the conservation of energy for a constant-volume process, together with knowledge of the thermodynamic properties at the initial and final states. Since the tank is rigid no expansion or compression work is performed; hence, Eq. 5.6a can be simplified as follows:

$$_1(Q_{\text{in}})_2 + {}_1(W_{\text{in}})_2 = {}_1(Q_{\text{out}})_2 - {}_1(W_{\text{out}})_2 = E_{\text{sys},2} - E_{\text{sys},1}$$
$$\Rightarrow 0 + 0 - {}_1(Q_{\text{out}})_2 - 0 = U_2 - U_1$$

or

$$-_1(Q_{\text{out}})_2 = M(u_2 - u_1).$$

Since two independent thermodynamic properties are known at the initial state, all other properties can be determined. For the given conditions, the steam is superheated; hence, its properties can be obtained from the NIST software or WebBook or Table B.3 as follows:

$$v_1 = 0.22064 \text{ m}^3/\text{kg},$$
$$u_1 = 2670.6 \text{ kJ/kg}.$$

The system mass is obtained as

$$M = \frac{\mathcal{V}_1}{v_1} = \frac{0.15\,\text{m}^3}{0.22064\,\text{m}^3/\text{kg}} = 0.6798\,\text{kg}.$$

Because the process occurs at constant volume with a fixed mass, v_2 equals v_1. This knowledge and the given state-2 temperature allow us to determine the state-2 specific internal energy. From the NIST resources or Table B.1 we see that, at 400 K,

$$v_f(400\,\text{K}) < v_2 < v_g(400\,\text{K}),$$

so

$$0.0010667 < 0.22064 < 0.73024.$$

State 2 thus lies in the liquid–vapor mixture region, as shown in the sketch at the start of the example.

The useful saturation properties from the NIST resources or Table B.1 are as follows:

$$v_f = 0.0010667\,\text{m}^3/\text{kg}, \quad v_g = 0.73024\,\text{m}^3/\text{kg},$$
$$u_f = 532.69\,\text{kJ/kg}, \quad u_g = 2536.2\,\text{kJ/kg}.$$

To find u_2, we must first find the quality at state 2, x_2. Applying Eq. 2.37b yields

$$x_2 = \frac{v_2(= v_1) - v_{f,2}}{v_{g,2} - v_{f,2}}$$
$$= \frac{0.22064\,\text{m}^3/\text{kg} - 0.0010667\,\text{m}^3/\text{kg}}{0.73024\,\text{m}^3/\text{kg} - 0.0010667\,\text{m}^3/\text{kg}} = 0.3011.$$

The state-2 specific internal energy is obtained as (Eq. 2.37c)

$$u_2 = (1 - x_2)u_{f,2} + x_2 u_{g,2}$$
$$= (1 - 0.3011)\,532.69\,\text{kJ/kg} + (0.3011)\,2536.2\,\text{kJ/kg}$$
$$= 1135.9\,\text{kJ/kg}.$$

Thus, we find the heat removed to be

$$
\begin{aligned}
{}_1(Q_{out})_2 &= -M(u_2 - u_1) \\
&= -0.6798\,\text{kg}\,(1135.9 - 2670.6)\,\text{kJ/kg} \\
&= 1043.3\,\text{kJ}.
\end{aligned}
$$

Comments Recognizing that a constant-volume process occurs is important, as the work is then simply evaluated as zero. This example emphasizes the use of the quality property in dealing with liquid–vapor mixtures.

Self-Test 5.3

A piston–cylinder device originally contains 5 kg of saturated liquid water at 100 kPa. Determine the heat addition required to bring the fluid to a saturated vapor state.

(Answer: 11,287 kJ)

Example 5.4 Piston–Cylinder Isobaric Process with Air

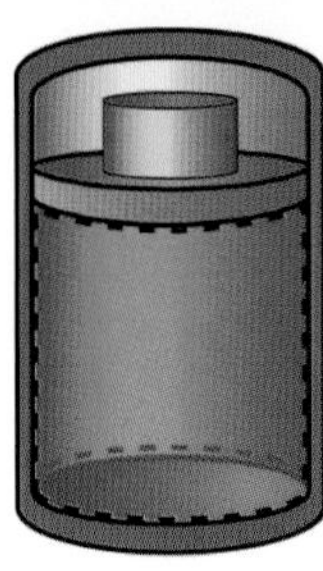

Air is contained in a piston–cylinder arrangement, as shown in the sketch. A weight placed on the piston keeps the pressure constant (at 120 kPa) inside the cylinder as 11,820 J of energy is added to the air by heat transfer. The initial temperature is 300 K, and the initial volume is 0.12 m^3. Determine the temperature at the end of the heat-addition process.

Solution

Known T_1, V_1, ${}_1(Q_{in})_2$, $P_1 = P_2 =$ constant

Find T_2

Sketch

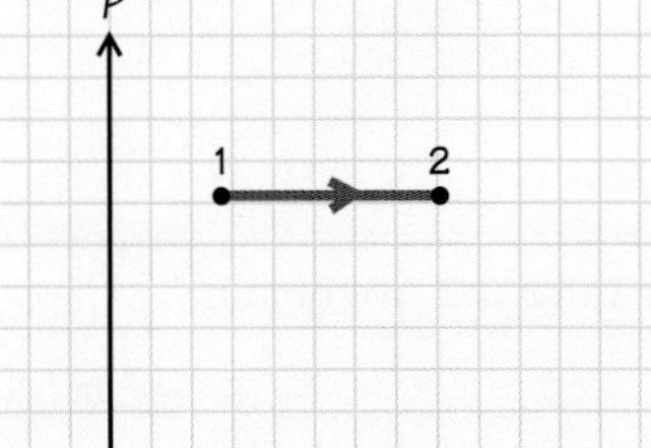

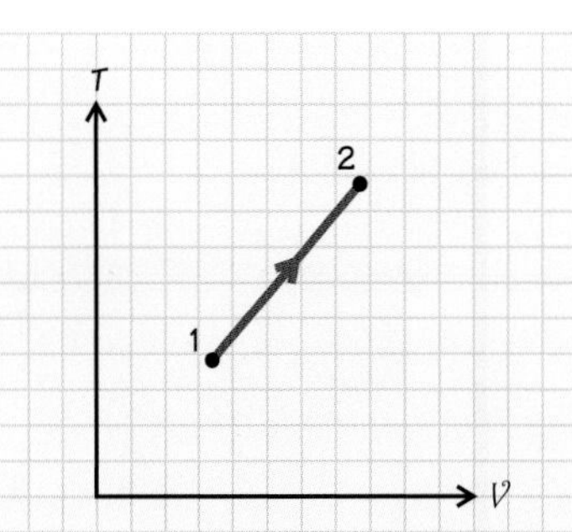

Modeling, Premises and Assumptions

i. Quasi-equilibrium process
ii. Ideal-gas behavior
iii. Negligible change in system potential energy

Analysis We define the air contained in the cylinder as the thermodynamic system of interest and apply the conservation of energy (Eq. 5.6b) to the process, as follows:

$$_1(Q_{\text{in}})_2 - {}_1(W_{\text{out}})_2 = E_{\text{sys},2} - E_{\text{sys},1}.$$

Since the pressure is constant, the work out is simply evaluated as

$$_1W_2 = \int_1^2 Pd\mathcal{V} = P(\mathcal{V}_2 - \mathcal{V}_1)$$

or

$$_1W_2 = MP(v_2 - v_1).$$

We also recognize that the only system energy is the internal energy u. Upon rearrangement, our conservation of energy expression thus becomes

$$\begin{aligned} _1(Q_{\text{in}})_2 &= M(u_2 - u_1) + MP(v_2 - v_1) \\ &= M(u_2 + Pv_2 - u_1 - Pv_1) \\ &= M(h_2 - h_1) \end{aligned}$$

or

$$\frac{_1(Q_{\text{in}})_2}{M} = h_2 - h_1.$$

The mass is evaluated from the ideal-gas equation of state (Eq. 2.26c) as

$$\begin{aligned} M &= \frac{P\mathcal{V}_1}{RT_1} \\ &= \frac{120 \times 10^3 \dfrac{\text{N}}{\text{m}^2}(0.12\,\text{m}^3)}{287\dfrac{\text{J}}{\text{kg·K}}(300\,\text{K})}\left[\frac{\text{J}}{\text{N·m}}\right] = 0.1672\,\text{kg}, \end{aligned}$$

where $R = R_{\text{air}} = 287\,\text{J/kg·K}$ (Table C.l). To find T_2, we recognize that h_1 and h_2 depend only on temperature and that Table C.2 provides the necessary values of h_1 and h_2, that is, it is a tabular form of the calorific equation of state for air. Using the given value of the heat added, we evaluate the enthalpy difference as

$$\begin{aligned} h_2 - h_1 &= \frac{11,820\,\text{J}}{0.1672\,\text{kg}} = 70,694\,\text{J/kg} \\ &= 70.69\,\text{kJ/kg}. \end{aligned}$$

From Table C.2, we obtain

$$h_1 = h_1(300\,\text{K}) = 426.04\,\text{kJ/kg}.$$

Thus,

$$\begin{aligned} h_2 &= h_1 + (h_2 - h_1) \\ &= 426.04\,\text{kJ/kg} + 70.69\,\text{kJ/kg} \\ &= 496.73\,\text{kJ/kg}. \end{aligned}$$

We see that this value is just slightly greater than the tabulated value for 370 K. Interpolating data from Table C.2 yields

$$T_2 = 370.1\,\text{K}.$$

See Eq. 2.31e in Chapter 2.

Comments We note that enthalpy appears as a useful thermodynamic property for this constant-pressure process. Note that we could also have solved this problem using an average value of c_p rather than tabular values for the enthalpy. The reader should verify this.

Self-Test 5.4

 A well-insulated (i.e., adiabatic) rigid tank contains 2 kg of air at 300 K and 150 kPa. An electric resistance heater in the tank is turned on and receives 200 kJ of electrical work. Find the final temperature and pressure of the air.

(Answer: 439.3 K, 219.7 kPa)

5.1c At an Instant

We now consider the conservation of energy at an instant. The general rate form of energy conservation, Eq. 5.2, can be used directly for this situation by identifying

$$\dot{E}_{\text{in}} \equiv \dot{W}_{\text{in}} + \dot{Q}_{\text{in}}, \tag{5.8a}$$

$$\dot{E}_{\text{out}} \equiv \dot{W}_{\text{out}} + \dot{Q}_{\text{out}}, \tag{5.8b}$$

$$\dot{E}_{\text{generated}} \equiv 0, \tag{5.8c}$$

and

$$\dot{E}_{\text{stored}} \equiv \frac{dE_{\text{sys}}}{dt}, \tag{5.8d}$$

where

$$\dot{Q} \equiv \lim_{\Delta t \to 0} \frac{\delta Q}{\Delta t}, \tag{5.9a}$$

$$\dot{W} \equiv \lim_{\Delta t \to 0} \frac{\delta W}{\Delta t}, \tag{5.9b}$$

$$\frac{dE_{\text{sys}}}{dt} = \lim_{\Delta t \to 0} \frac{E_{\text{sys}}(t+\Delta t) - E_{\text{sys}}(t)}{\Delta t}. \tag{5.9c}$$

Equation 5.2 is thus reassembled as

Equation 5.10 is a key conservation of energy expression for closed systems.

$$\underbrace{(\dot{W}_{\text{in}} + \dot{Q}_{\text{in}})}_{\substack{\text{Time rate of}\\ \text{energy crossing}\\ \text{boundary into}\\ \text{system}}} - \underbrace{(\dot{W}_{\text{out}} + \dot{Q}_{\text{out}})}_{\substack{\text{Time rate of}\\ \text{energy crossing}\\ \text{boundary out}\\ \text{of system}}} = \underbrace{\frac{dE_{\text{sys}}}{dt}}_{\substack{\text{Time rate of}\\ \text{change of energy}\\ \text{stored within}\\ \text{system}}}, \tag{5.10a}$$

$\dot{Q}_{in,\,net}$ $\dot{W}_{out,\,net}$

$\frac{dE_{sys}}{dt}$

$$\frac{dE_{sys}}{dt} = \dot{Q}_{in,\,net} - \dot{W}_{out,\,net}$$

FIGURE 5.4 The rate at which the energy of a system increases equals the difference between the net rate at which energy enters the system by heat transfer across the boundary and the net rate at which energy exits by the work done by the system.

or alternatively,

$$\underbrace{\dot{Q}_{in.net}}_{\substack{\text{Net heat transfer}\\ \text{rate to the system}}} - \underbrace{\dot{W}_{out.net}}_{\substack{\text{Net rate of work done}\\ \text{by the system, that is,}\\ \text{net power produced}}} = \underbrace{\frac{dE_{sys}}{dt}}_{\substack{\text{Time rate of change}\\ \text{of energy stored}\\ \text{within system}}}. \qquad \textbf{(5.10b)}$$

This relationship is shown schematically in Fig. 5.4. Equation 5.10, like Eq. 5.6, is a key relationship, and again its meaning and application should become firmly fixed in your storehouse of knowledge. The following examples help to illuminate the usefulness of this expression.

Example 5.5 System Energy Balance

Consider the electrical circuit shown in the sketch in which a light bulb and electric motor are wired in series with a battery. The motor drives a fan. For the following thermodynamic systems, write an appropriate conservation of energy expression:

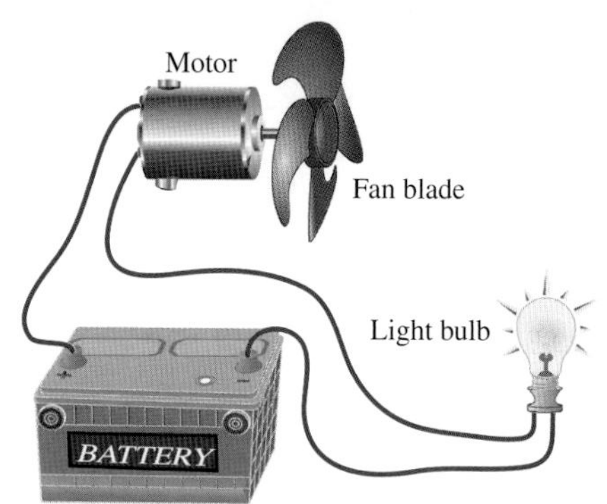

A. For the light bulb alone

B. For the battery alone

Solution

The nature of this problem suggests that we depart from the standard solution format. We begin by drawing a system boundary around the light bulb and identifying all the energy flows:

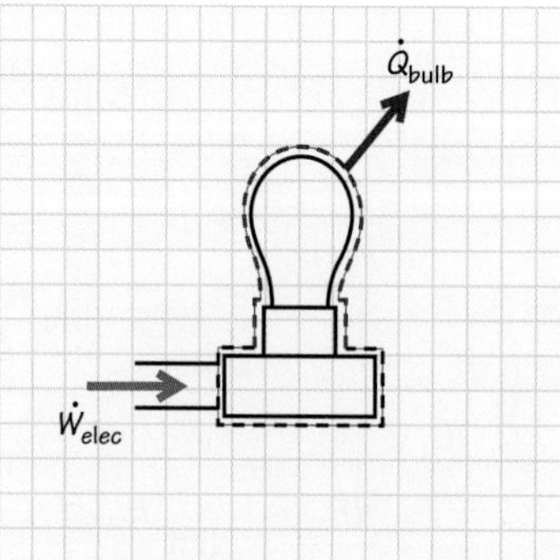

Here we identify a rate of electrical work into the system, $\dot{W}_{elec}$, and a rate of heat transfer out of the system, $\dot{Q}_{bulb}$. Assuming a steady state, there are no other energy terms to consider. We now apply Eq. 5.10a,

$$(\dot{W}_{in} + \dot{Q}_{in}) - (\dot{W}_{out} + \dot{Q}_{out}) = \frac{dE_{sys}}{dt},$$

which simplifies to

$$(\dot{W}_{elec} + 0) - (0 + \dot{Q}_{bulb}) = 0,$$

or

$$\dot{W}_{\text{elec}} = \dot{Q}_{\text{bulb}}.$$

From this, we formally conclude that all the electrical power input is converted to a heat-transfer rate from the light bulb to the surroundings. Note that visible radiation is included in this heat-transfer rate.

We now consider the battery alone as our system of interest. The most obvious flow of energy is the net electrical power out from the battery terminals, $\dot{W}_{\text{elec}}$. Furthermore, if the battery has some small internal resistance, the battery will be warmer than the surroundings, resulting in a small heat transfer to the surroundings, $\dot{Q}_{\text{batt}}$. Since both $\dot{W}_{\text{elec}}$ and $\dot{Q}_{\text{batt}}$ are outflows of energy, we identify the rate of change of the internal energy of the battery, dU_{batt}/dt, as their source. The physical mechanism for this change in internal energy entails the rearrangement of chemical bonds within the battery. We indicate these energy flows on a sketch:

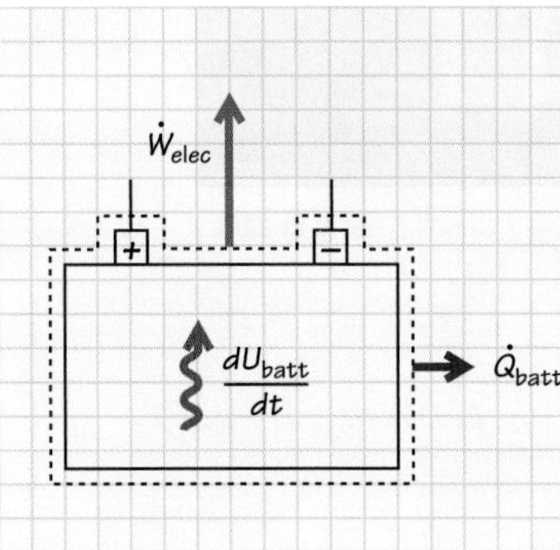

Conservation of energy is again expressed by Eq. 5.10a,

$$(\dot{W}_{\text{in}} + \dot{Q}_{\text{in}}) - (\dot{W}_{\text{out}} + \dot{Q}_{\text{out}}) = \frac{dE_{\text{sys}}}{dt},$$

which simplifies to yield

$$(0 + 0) - (\dot{W}_{\text{elec}} + \dot{Q}_{\text{batt}}) = \frac{dU_{\text{batt}}}{dt},$$

Here we assume that the energy of the battery (E_{sys}) comprises only internal energy ($E_{\text{syst}} \equiv U_{\text{batt}}$); that is, there are no kinetic and potential energies. We note that the formal application of Eq. 5.10a indicates that dU_{batt}/dt is negative; that is, the internal energy decreases with time, which is consistent with our experience that batteries "run down" with use.

Self-Test 5.5

Consider a 100-W light bulb in a well-insulated 34-m^3 room with the air initially at 100 kPa and 25 °C. Determine (a) the rate of change of the air temperature and (b) the temperature of the air after 10 h.

(Answer: (a) 0.210 K/min, (b) 151 °C)

Example 5.6 Light Bulb Filament Temperature

Assume that the filament in a 100-W light bulb can be modeled as a cylinder 1 mm in diameter and 25 mm long, as shown in the sketch. Estimate the steady-state operating temperature of the filament assuming that it loses energy by both radiation and convection heat transfer. The emissivity of the wire is 0.25, and the surrounding glass bulb has a temperature of 400 K. The average convective heat-transfer coefficient at the filament surface is approximately 15 W/m^2·K, and the ambient gas temperature inside the bulb is approximately 425 K.

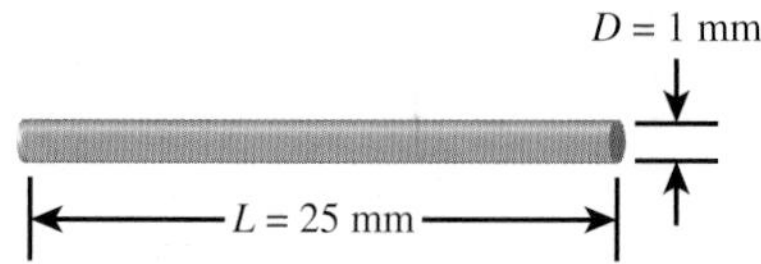

(Credit: BlackJack3D / iStock / Getty Images.)

Solution

Known $\dot{W}_{\text{elec}}, L, D, \varepsilon, h_{\text{conv}}, T_{\text{surr}}, T_{\infty}$

Find T_{f}

Sketch

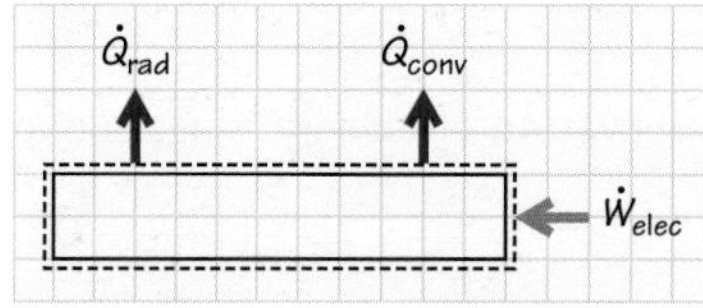

Modeling, Premises and Assumptions

i. Steady state
ii. Coiled filament can be modeled as a simple cylinder
iii. Uniform surface temperature
iv. Negligible heat transfer from ends
v. Glass bulb absorbs and emits as a blackbody (i.e., the fraction of radiation transmitted is small)

Analysis We begin by identifying the filament as our thermodynamic system of interest and indicating all the energy (rate) terms, as shown in the sketch. We show the electric power input as a single arrow indicating the net effect of the current flowing in and out of the system (see Eq. 4.11b). Since we are dealing with energy rates (J/s = W), the conservation of energy is expressed by Eq. 5.10:

$$(\dot{W}_{\text{in}} + \dot{Q}_{\text{in}}) - (\dot{W}_{\text{out}} + \dot{Q}_{\text{out}}) = \frac{dE_{\text{sys}}}{dt}.$$

Since we assume steady-state operation, the energy storage term, dE_{sys}/dt, is zero. Using the previous sketch as a guide, this relationship becomes

$$(\dot{W}_{\text{elec}} + 0) - (0 + \dot{Q}_{\text{conv}} + \dot{Q}_{\text{rad}}) = 0$$

or

$$\dot{Q}_{conv} + \dot{Q}_{rad} = \dot{W}_{elec}.$$

Since $\dot{W}_{elec}$ is given as 100 W, the right-hand side is fixed. The heat-transfer rates depend on the filament temperature T_f and can be expressed using the "rate laws" presented in Chapter 4 (Eqs. 4.15 and 4.16). These are expressed for the present problem as

$$\dot{Q}_{conv} = h_{conv}\pi DL(T_f - T_\infty)$$

and

$$\dot{Q}_{rad} = \varepsilon_f \pi DL\sigma\left(T_f^4 - T_{surr}^4\right),$$

where we identify T_∞ as the gas temperature within the bulb and T_{surr} as the temperature of the bulb itself. Substituting these expressions and dividing through by the surface area, πDL, yields

$$h_{conv}(T_f - T_\infty) + \varepsilon_f\sigma\left(T_f^4 - T_{surr}^4\right) = \frac{\dot{W}_{elec}}{\pi DL}.$$

Conceptually, we are done since the only unknown quantity in this relationship is the filament temperature T_f. We substitute numerical values as follows:

$$15(T_f - 425) + 0.25(5.67 \times 10^{-8})\left(T_f^4 - 400^4\right) = 100/[\pi(0.001)0.025],$$

where each term has units of W/m^2. (The reader should verify this.) Performing the indicated arithmetic and rearranging yields the following polynomial:

$$(1.4175 \times 10^{-8})T_f^4 + 15T_f - 1.279978 \times 10^6 = 0.$$

The final task is to find the particular root of this equation that satisfies our problem. (There are multiple roots.) Any number of root-solving methods can be applied. The final result obtained by application of the Newton–Raphson iterative method is

$$T_f = 3054.6\,\text{K}.$$

Comment This result is reasonable, although it is somewhat higher than the 2800 K value frequently used to characterize standard tungsten incandescent lights.

Example 5.7 Cooling Rate of an Orange

Consider a 6-cm-diameter orange growing on a tree. A cold front causes the ambient temperature to drop rapidly from 50 F to 30 F. Estimate the initial rate of temperature change (dT/dt) of the orange if the convective heat-transfer coefficient is approximately 1.5 W/m^2·K. Assume the orange is initially at 50 F. The density and specific heat of the orange are approximately 850 kg/m^3 and 3770 J/kg · K, respectively [9].

(Credit: Vinson Motas / EyeEm / Getty Images.)

Solution

Known $D, \rho, c, T_i, T_\infty, h_{\text{conv}}$

Find dT/dt

Sketch

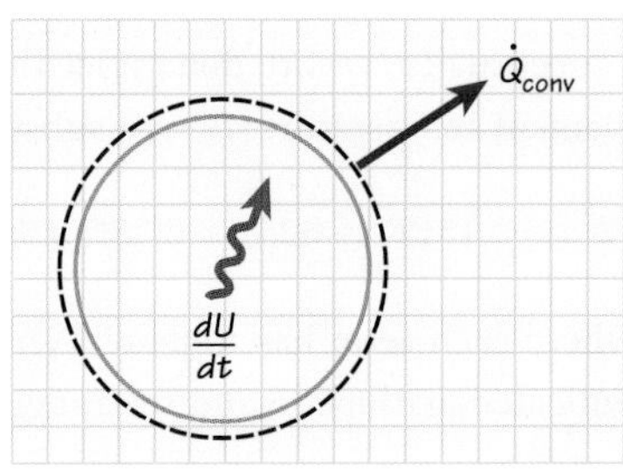

Modeling, Premises and Assumptions

i. The orange cools with uniform temperature.
ii. The orange is an incompressible solid ($c_p = c_v = c$; see Eq. 2.44b).
iii. No changes in kinetic or potential energies.

Analysis We denote the orange as our thermodynamic system and apply conservation of energy by simplifying Eq. 5.10:

$$-\dot{Q}_{\text{conv}} + 0 = \frac{dU_{\text{sys}}}{dt},$$

noting that $\dot{Q}_{\text{conv}}$ has a negative sign since it is a transfer *out* of the system. To introduce temperature as an explicit variable in our conservation of energy expression, we recognize that

$$\frac{dU}{dt} = M\frac{du}{dt},$$

and apply the chain rule and the definition of the specific heat (Eq. 2.20a) to yield

$$\frac{du}{dt} \equiv \left(\frac{\partial u}{\partial T}\right)\frac{dT}{dt} = c_v\frac{dT}{dt}.$$

With our assumption of incompressibility, $c_p = c_v = c$ (Eq. 2.44b), energy conservation is expressed as

$$-\dot{Q}_{\text{conv}} = M_{sys}c = \frac{dT_{\text{orange}}}{dt}.$$

We use Eq. 4.15 to express the convection heat-transfer rate as

$$Q_{\text{conv}} = h_{\text{conv}} A\left(T_{\text{orange}} - T_{\infty}\right),$$

where A is the surface area ($= \pi D^2$). The mass of the orange is the product of the density and the volume,

$$M_{\text{sys}} = \rho \forall = \rho \pi D^3/6.$$

Substituting these expressions into our energy balance and solving for dT_{orange}/dt yields

$$\frac{dT_{\text{orange}}}{dt} = -\frac{h_{\text{conv}}\left(T_{\text{orange}} - T_{\infty}\right)}{(D/6)(\rho c)_{\text{orange}}}.$$

Converting the given temperatures to SI units, we numerically evaluate this as

$$\frac{dT_{\text{orange}}}{dt} = -\frac{1.5\,\text{W/m}^2\cdot\text{K}(283.15 - 272.0)\text{K}}{(0.06\,\text{m}/6)850\text{kg/m}^3(3770\,\text{J}/(\text{kg}\cdot\text{K}))}\left[\frac{\text{J/s}}{\text{W}}\right] = 5.2 \times 10^{-4}\text{K/s},$$

or

$$= 1.9\,\text{K/hr}.$$

Comment To determine the temperature–time relationship describing the cooling of the orange requires the solution of the original energy conservation equation, an ordinary differential equation, and not simply its algebraic rearrangement as was done here to obtain the initial cooling rate.

Self-Test 5.6

Show that $d(T - T_{\infty})/dt = dT/dt$ and then perform the integration to develop an expression for the temperature of the orange in Example 5.7 as a function of time.

(Answer: $(T(t)_{orange} - T_{\infty})/(T_{orange,init} - T_{\infty}) = \exp\left[-h_{conv}/((D/6)(\rho c)_{orange} t)\right]$*)*

Example 5.8 Heating Rate of Light Bulb

Consider the tungsten light-bulb filament as modeled in Example 5.6. Estimate the initial rate of temperature rise of the filament at the instant after the current starts to flow. The current is 0.9 A and the voltage drop is 110 V. Assume the filament is initially in equilibrium with its surroundings at 25 °C. The density and heat capacity of tungsten are 19,300 kg/m^3 and 132 J/kg·K, respectively.

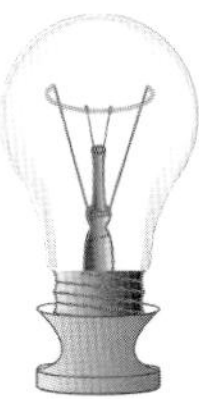

$t = 0$

$t = t$

$t = \infty$

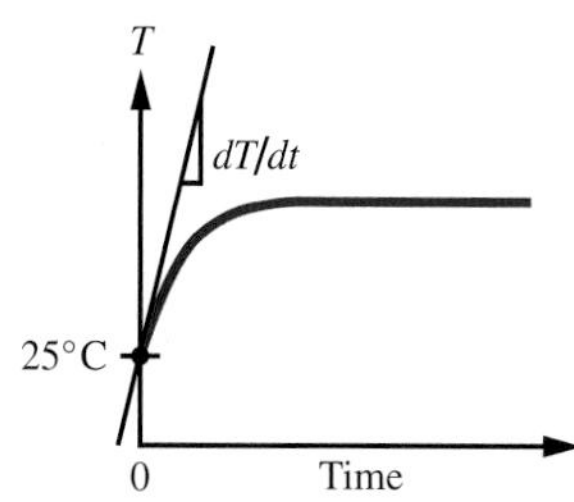

Solution

Known $D, L, i, \mathcal{V}, T_{\text{init}}, \rho, c$

Find dT_{f}/dt

Sketch

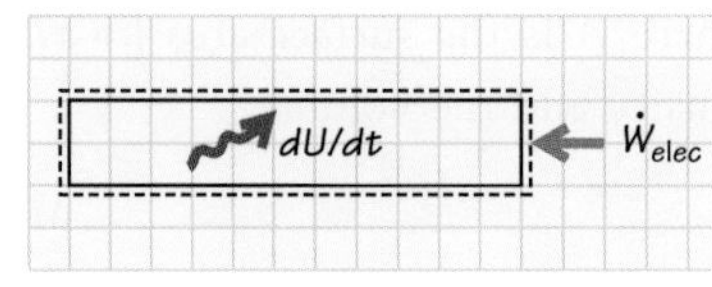

Modeling, Premises and Assumptions

i. Incompressible solid ($c_p = c_v = c$)
ii. Filament heats with uniform temperature rise
iii. Kinetic and potential energies negligible ($E_{\text{sys}} = U_{\text{sys}}$)

Analysis The filament is chosen as the system of interest. Since the filament is initially at the same temperature as the surroundings, there is no heat transfer. (After the filament heats, however, the heat transfer will be significant.) As shown in the sketch, all the electrical power is used to increase the internal energy of the filament. We simplify Eq. 5.10 as follows:

$$(\dot{W}_{\text{in}} + \dot{Q}_{\text{in}}) - (\dot{W}_{\text{out}} + \dot{Q}_{\text{out}}) = \frac{dE_{\text{sys}}}{dt} = \frac{dU_{\text{sys}}}{dt}$$
$$(\dot{W}_{\text{elec}} + 0) - (0 + 0) = \frac{dU_{sys}}{dt}.$$

From Example 5.7, we find

$$\frac{dU_{\text{sys}}}{dt} = M\frac{du}{dt} = Mc\frac{dT}{dt}.$$

Thus,

$$\dot{W}_{\text{elec}} = Mc\frac{dT_{\text{f}}}{dt},$$

which we can solve for the time rate of temperature change:

$$\frac{dT_{\text{f}}}{dt} = \frac{\dot{W}_{\text{elec}}}{Mc},$$

where

$$M = \rho\mathcal{V} = \rho\pi D^2 L/4$$
$$= 19{,}300\pi(0.001)^2 0.025/4 = 3.789 \times 10^{-4}\text{kg}.$$

The electrical power is the product of the current flow and voltage drop (Eq. 4.11b), so

$$\dot{W}_{\text{elec}} = i\Delta\mathcal{V}$$
$$= 0.9\,\text{A}(110\,\text{V})\left[\frac{1\,\text{W}}{\text{V}\cdot\text{A}}\right] = 99\,\text{W}.$$

Numerically evaluating dT_{f}/dt yields

$$\frac{dT_{\text{f}}}{dt} = \frac{99\,\text{W}}{3.789 \times 10^{-4}\text{kg}(132\text{J/kg}\cdot\text{K})}\left[\frac{1\,\text{J/s}}{\text{W}}\right] = 1979\,\text{K/s}.$$

Comment Consistently with our experience, the filament heats very quickly. The complete heating problem requires the solution of Eq. 5.10 for $T_{\text{f}}(t)$ with the inclusion of both convection and radiation heat-transfer rates.

5.2 Energy Conservation for Open Systems

Our treatment here parallels the development of the mass conservation principle for control volumes from Chapter 3. We start by developing steady-state expressions of energy conservation for large-scale (integral) open systems and follow by developing the general unsteady case.

5.2a Open Systems with Steady Flow

Recall from Chapter 3 the idea that for steady-state and steady flow, properties do not change with time within the open system (control volume) or within the inlet and outlet streams. Furthermore, for the single-inlet, single-exit, open system shown in Fig. 5.5, the mass flow rate in must equal the mass flow rate out (Eq. 3.14a); that is,

$$\dot{m}_{\text{in}} = \dot{m}_{\text{out}} = \dot{m}.$$

For simplicity, we assume that the fluid properties are uniform where the flow crosses the control surface, or that appropriately corrected average values characterize the inlet and outlet streams. In most engineering applications of steady-state and steady flow, the energy rates are of importance (e.g., $\dot{W}$and $\dot{Q}$ expressed in joules per second or watts).

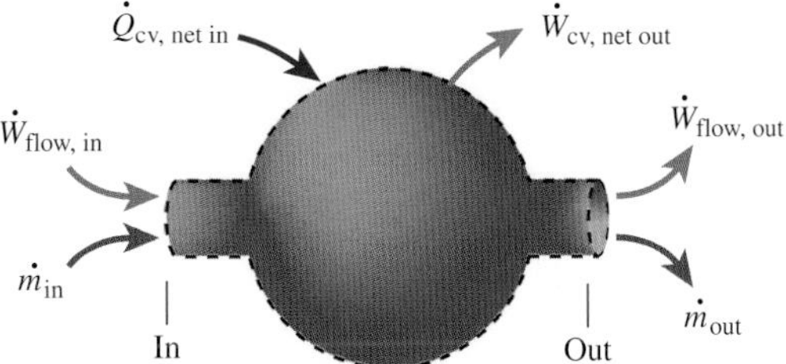

FIGURE 5.5 Open system with one inlet and one outlet for steady-state, steady-flow analysis of energy conservation.

We begin our analysis by identifying all the energy interactions around the open system boundary: The inlet stream carries energy into the system, and the exit stream carries energy out of the system. The rates of these energy transfers are expressed as the product of the mass flow rate and the mass-specific energy of each stream, i.e., $\dot{m}e_{\text{in}}$ and $\dot{m}e_{\text{out}}$. Flow work is performed by the surroundings to push the fluid into the system at a rate $\dot{W}_{\text{flow,in}}$, whereas the system performs work to push the exiting fluid from the system at a rate $\dot{W}_{\text{flow,out}}$. In addition, the open system itself may perform work that crosses the boundary at locations other than the fluid inlet and exit. The rate of this work we denote $\dot{W}_{\text{cv,net,out}}$. The most common example of this is shaft work rate or power. Another example would be electrical power, if electricity flowed across the system boundary. Finally, a net rate of heat transfer into the control volume ($\dot{Q}_{\text{cv,net,in}}$) will exist if there are temperature gradients and/or differences at the boundary. The subscript cv stands for *control volume*, which is synonymous with the term *open system*.

Having identified all the energy interactions at the boundary – heat, work, and mass flow (there can be no others) – we now apply the following broad conservation of energy principle:

In a steady state, the net rate at which energy crosses the control surface must be zero; that is, the rate at which energy enters the control volume must equal the rate at which energy exits.

This is the most basic steady-state, steady-flow statement of energy conservation.

This principle is mathematically expressed as

$$\dot{E}_{B,\text{in}} = \dot{E}_{B,\text{out}}, \tag{5.11}$$

where $\dot{E}_B$ is the energy transfer rate across the entire boundary of the open system. Using this as the starting point, and making some rearrangements and substitutions, we arrive at a standard form of energy conservation for steady-state, steady flow. Substituting the particular energy rates yields

$$\dot{Q}_{\text{cv,in}} + \dot{W}_{\text{cv,in}} + \dot{W}_{\text{flow,in}} + \dot{m}e_{\text{in}} = \dot{Q}_{\text{cv,out}} + \dot{W}_{\text{cv,out}} + \dot{W}_{\text{flow,out}} + \dot{m}e_{\text{out}}. \tag{5.12}$$

Rearranging, we write

$$(\dot{Q}_{\text{cv,in}} - \dot{Q}_{\text{cv,out}}) + (\dot{W}_{\text{cv,in}} - \dot{W}_{\text{cv,out}}) = \dot{m}(e_{\text{out}} - e_{\text{in}}) + (\dot{W}_{\text{flow,out}} - \dot{W}_{\text{flow,in}}). \tag{5.13}$$

The specific energy (energy per unit mass) of the flowing streams can be expanded to show explicitly the mass-specific internal, mass-specific kinetic, and mass-specific potential energies:

$$e = u + (KE) + (PE). \tag{5.14}$$

Furthermore, the flow work, as derived in Chapter 4 (Eq. 4.12), is expressed as

$$\dot{W}_{\text{flow}} = \dot{m}P\text{v}.$$

Substituting Eqs. 5.14 and 4.12 into Eq. 5.13 yields

$$(\dot{Q}_{\text{cv,in}} - \dot{Q}_{\text{cv,out}}) + (\dot{W}_{\text{cv,in}} - \dot{W}_{\text{cv,out}}) = \dot{m}\left[(u + P\text{v})_{\text{out}} - (u + P\text{v})_{\text{in}} + (ke)_{\text{out}} - (ke)_{\text{in}} + (pe)_{\text{out}} - (pe)_{\text{in}}\right].$$

In this expression, we recognize the specific enthalpy h (= $u + P\text{v}$). With this, and the substitution of the definitions of the mass-specific kinetic and potential energies [i.e., $ke \equiv V^2/2$ and $pe \equiv g(z - z_{\text{ref}})$], we can write our final steady-state, steady-flow conservation of energy expression:

$$(\dot{Q}_{\text{cv,in}} - \dot{Q}_{\text{cv,out}}) + (\dot{W}_{\text{cv,in}} - \dot{W}_{\text{cv,out}}) = \dot{m}\left[(h_{\text{out}} - h_{\text{in}}) + \tfrac{1}{2}\left(V_{\text{out}}^2 - V_{\text{in}}^2\right) + g(z_{\text{out}} - z_{\text{in}})\right]. \tag{5.15a}$$

Equation 5.15a is one of those key relationships whose meaning and applications should be deeply integrated in your knowledge base. A slightly more compact form that may be easier to recall is

Equations 5.15a and 5.15b are key relationships.

$$\dot{Q}_{\text{cv, net in}} - \dot{W}_{\text{cv, net out}} = \dot{m}[\Delta h + \Delta(ke) + \Delta(pe)], \tag{5.15b}$$

where $\Delta \equiv (\ )_{\text{out}} - (\ )_{\text{in}}$ and

$$\dot{Q}_{\text{cv, net in}} \equiv \dot{Q}_{\text{cv, in}} - \dot{Q}_{\text{cv, out}} \tag{5.15c}$$

and

$$\dot{W}_{\text{cv, net out}} \equiv \dot{W}_{\text{cv, out}} - \dot{W}_{\text{cv, in}}. \tag{5.15d}$$

The reader should work to become familiar with Eqs. 5.15a–5.15d as these are powerful tools and can be used to solve many practical problems.

If multiple streams enter and/or exit an open system, Eq. 5.15 is easily extended to treat such a case by writing

In spite of its apparent complexity, Eq. 5.16 simply states that the rate of all energy entering the system $\dot{E}_{\text{cv, in}}$ must equal the rate of all energy exiting the system $\dot{E}_{\text{cv, out}}$, i.e., $\dot{E}_{\text{cv, in}} = \dot{E}_{\text{cv, out}}$.

$$\begin{aligned}\dot{Q}_{\text{cv, net in}} - \dot{W}_{\text{cv, net out}} = &\sum_{k=1}^{M\text{ outlets}} \dot{m}_{\text{out},k}\left[h_k + \tfrac{1}{2}V^2_{\text{avg},k} + g(z_k - z_{\text{ref}})\right] \\ &- \sum_{j=1}^{N\text{ inlets}} \dot{m}_{\text{in},j}\left[h_j + \tfrac{1}{2}V^2_{\text{avg},j} + g(z_j - z_{\text{ref}})\right].\end{aligned} \tag{5.16}$$

Several practical examples of the application of steady-state, steady-flow energy conservation are presented in Section 5.3, following a discussion of energy conservation for unsteady flow in Section 5.2b.

5.2b Open Systems with Unsteady Flow

Although many engineering applications involve steady-flow processes, unsteady (i.e., time-dependent) flows can be important in some situations. Transient start-ups and shutdowns fall in this category, as do the emptying or filling of tanks and pressure vessels. At this point, it is a relatively simple matter to extend our previous analyses to handle unsteady flows. Figure 5.6 shows the addition of the term needed to deal with the unsteady case: dE_{cv}/dt, indicated with the squiggly arrow inside the system (cf. Fig. 5.5). With the addition of this single term to the previously identified energy interactions at the control surface, we can state the conservation of energy as follows:

The net rate at which energy from the surroundings crosses the system boundary into the system must equal the time rate of increase of energy within the system.

Equations 5.17 express a key concept in a very compact way.

Symbolically, we express this most simply as

$$\dot{E}_{\text{cv, in}} - \dot{E}_{\text{cv, out}} = \frac{dE_{\text{cv}}}{dt}, \tag{5.17a}$$

or

$$\dot{E}_{\text{cv, net in}} = \frac{dE_{\text{cv}}}{dt}. \tag{5.17b}$$

From this point, our analysis follows directly from the previous development for the steady-flow case (i.e., Eqs. 5.11–5.16). Therefore, we need only state the final results. First, for a control volume with a single inlet and outlet,

$$\dot{Q}_{\text{cv, net in}} - \dot{W}_{\text{cv, net out}} + \dot{m}_{\text{in}}\left[h_{\text{in}} + \tfrac{1}{2}V_{\text{in}}^2 + g(z_{\text{in}} - z_{\text{ref}})\right] - \dot{m}_{\text{out}}\left[h_{\text{out}} + \tfrac{1}{2}V_{\text{out}}^2 + g(z_{\text{out}} - z_{\text{ref}})\right] = \frac{dE_{\text{cv}}}{dt}, \tag{5.18}$$

and, second, for control volumes with multiple inlets and/or outlets,

$$\dot{Q}_{\text{cv, net in}} - \dot{W}_{\text{cv, net out}} = \sum_{j=1}^{N\text{ inlets}} \dot{m}_{\text{in},j}\left[h_j + \tfrac{1}{2}V_j^2 + g(z_j - z_{\text{ref}})\right] - \sum_{k=1}^{M\text{ outlets}} \dot{m}_{\text{out},k}\left[h_k + \tfrac{1}{2}V_k^2 + g(z_k - z_{\text{ref}})\right] = \frac{dE_{\text{cv}}}{dt}. \tag{5.19}$$

It is important to emphasize that Eqs. 5.18 and 5.19 are *instantaneous* expressions of energy conservation. The examples in the next section illustrate the application of these expressions.

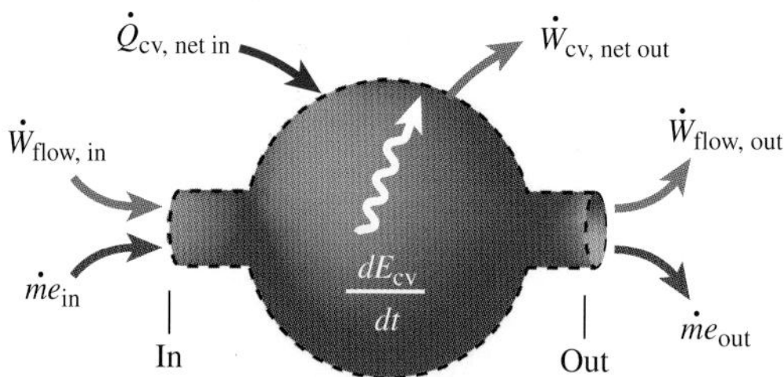

FIGURE 5.6 Open system (control volume) with one inlet and one outlet for unsteady analysis of energy conservation.

5.3 Applications of the First Law to Open Systems

This section presents several examples of the application of energy conservation (the first law) for open systems. Table 5.1 lists important devices that are amenable to a simple steady-flow analysis, and Chapter 8 explores these devices further. Combinations of these devices create complex systems such as power plants, propulsion systems, and systems for heating and cooling, the subject of Chapter 9.

TABLE 5.1 Important Devices Amenable to Steady-Flow Analysis*

Nozzle	Fan
Diffuser	Turbine
Throttle	Heat exchanger
Pump	Furnace or boiler
Compressor	Combustor

* Simplified analyses of each of these devices are presented in Chapter 8.

A key step in analyzing practical devices is deciding which terms in our general first-law expression (Eq. 5.16 or 5.19) should be discarded, either because they are small compared to all the other terms, or because they are nonexistent. As a beginning learner, you may have to rely on commonly invoked assumptions. These may be stated explicitly, or implicitly, in a problem statement (in textbook problems). These assumptions are usually good starting points in real-world situations. In Chapter 8, we list these common assumptions for many devices. As your problem-solving skills develop, you should become proficient in making and justifying your assumptions used to simplify the first law. The examples below should help you develop these skills.

Example 5.9 Jet Engine Diffuser

In a jet engine, the incoming air is compressed before it enters the combustor. Located between the compressor outlet and the combustor inlet is a diffuser. The purpose of the diffuser is to slow the high-velocity air stream exiting the compressor to velocities sufficiently low to allow combustion to take place within the combustor. To accomplish this velocity change, the flow area of the passive diffuser increases in the flow direction as shown in the lower drawing. For a particular engine, the velocity and temperature at the diffuser entrance are 30 m/s and 450 K, respectively. The velocity at the diffuser exit is 4 m/s. Assuming adiabatic operation, estimate the temperature at the diffuser exit.

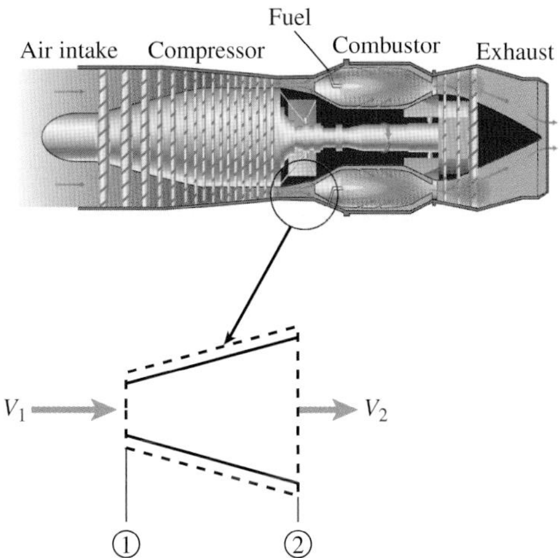

Solution

Known V_1, V_2, T_1

Find T_2

Sketch See system definition in sketch above.

Modeling, Premises and Assumptions

i. Steady state
ii. $\dot{Q}_{CV} = 0$ (adiabatic)
iii. $\dot{W}_{CV} = 0$
iv. $\Delta pe = 0$
v. Uniform velocity profiles

Analysis Because the diffuser is a passive device (with no moving parts and no moving boundaries), no work is done. We also neglect the change in potential energy because elevation changes are likely to be small, say, 5–10 cm. With these assumptions, and designating the inlet as state 1 and the outlet as station 2, we simplify the basic steady-flow, steady-state conservation of energy expression (Eq. 5.15) as follows:

$$\dot{Q}_{cv} - \dot{W}_{cv} = \dot{m}\left(h_2 - h_1 + \frac{V_2^2 - V_1^2}{2} + g(z_2 - z_1)\right),$$

or

$$0 - 0 = \dot{m}\left(h_2 - h_1 + \frac{V_2^2 - V_1^2}{2} + 0\right).$$

Dividing by $\dot{m}$ and rearranging, we obtain

$$h_2 - h_1 = \frac{V_1^2 - V_2^2}{2}.$$

Treating air as an ideal gas, we know that finding $h_2 - h_1$ is tantamount to finding the temperature difference since $h = h(T \text{ only})$. Substituting numerical values yields

$$h_2 - h_1 = \frac{(30\,\text{m/s})^2 - (4\,\text{m/s})^2}{2}\left[\frac{1\,\text{J}}{\text{N·m}}\right]\left[\frac{1\,\text{N}}{\text{kg·m/s}^2}\right] = 442\,\text{J/kg}.$$

To determine the temperature change, we could employ the calorific equation of state represented by the air property table (Table C.2). Because the enthalpy difference is small ($\Delta h = 0.442\,\text{kJ/kg}$), a simpler approach is to use the constant-pressure specific heat at 450 K from Table C.3 and the approximate relationship (Eq. 2.29e) $\Delta h = c_{p,avg}\Delta T$; thus,

$$T_2 = T_1 + \frac{h_2 - h_1}{c_p}$$
$$= 450\,\text{K} + \frac{0.442\,\text{kJ/kg}}{1.021\,\text{kJ/kg·K}} = 450.4\,\text{K}.$$

Comments We note first the procedure used to apply the open-system conservation of energy equation: listing reasonable assumptions and then using these to simplify Eq. 5.15. We also note that, even though the diffuser reduces the velocity by a factor of 7.5, the temperature rise is quite small. This calls into question our original assumption of adiabaticity, since heat-transfer effects might easily produce a temperature change of the same order as was found in our adiabatic analysis. The diffuser in most jet engines is an annulus surrounding the rotating components in the engine core.

Self-Test 5.7

Water flowing at 38.7 kg/s enters a pump as a saturated liquid at 320 K and exits at 325 K and 10 MPa. Neglecting any heat losses and potential and kinetic energy changes, determine the work rate associated with the pump.

(Answer: $\dot{W} = -1140\,\text{kW}$, where the negative sign indicates work is done on the system)

Example 5.10 Steam Turbine Power Output

A steam turbine is used to generate electricity at a pulp and paper manufacturing plant. Superheated steam (10 MPa and 780 K) enters the turbine with a flow rate of 38.7 kg/s. The steam exits the turbine at 320 K with a quality of 0.88. The turbine operation is nearly adiabatic, and changes in both the kinetic and potential energies of the entering and exiting steam are negligible. Estimate the shaft power delivered by the turbine to the generator.

(Credit: James Hardy / PhotoAlto Agency RF Collections / Getty Images.)

Solution

Known $\dot{m}, T_1, P_1, T_2, x_2$

Find $\dot{W}_{turb}$

Sketch

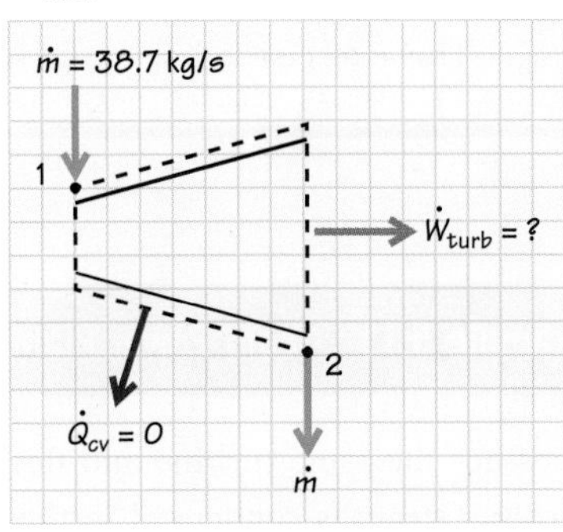

Modeling, Premises and Assumptions

i. Steady-state steady flow
ii. $\dot{Q}_{cv} = 0$ (adiabatic)
iii. $\Delta ke = \Delta pe = 0$
iv. Uniform velocity profiles at inlet and outlets

Analysis We begin with the steady-state, steady-flow conservation of energy expression for an open system having a single inlet and a single outlet (Eq. 5.15):

$$\dot{Q}_{cv,\,net\,in} - \dot{W}_{cv,\,net\,out} = \dot{m}\left(h_2 - h_1 + \frac{V_2^2 - V_1^2}{2} + g(z_2 - z_1)\right),$$

which simplifies to the following on applying the assumptions:

$$0 - \dot{W}_{turb} = \dot{m}(h_2 - h_1 + 0 + 0)$$

or

$$\dot{W}_{turb} = \dot{m}(h_1 - h_2).$$

The problem is now reduced to finding the values of h_1 and h_2. To find the inlet enthalpy, h_1, we use the NIST resources or Table B.3. For steam at 10 MPa and $T = 780$ K, we find

$$h_1 = 3392.4\,\text{kJ/kg}.$$

Given the outlet state ($T_2 = 320$ K, $x_2 = 0.88$), we employ data from the NIST WebBook (or Table B.l) with Eq. 2.37c, which expresses mass-specific properties in the liquid–vapor region, to find h_2:

$$\begin{aligned} h_2 &= (1 - x_2)h_{f,2} + x_2 h_{g,2} \\ &= (1 - 0.88)\,196.17\,\text{kJ/kg} + 0.88\,(2585.7\,\text{kJ/kg}) \\ &= 2299.0\,\text{kJ/kg}. \end{aligned}$$

The turbine power is thus

$$\dot{W}_{\text{turb}} = 38.7\frac{\text{kg}}{\text{s}}(3392.4 - 2299.0)\frac{\text{kJ}}{\text{kg}}\left[\frac{\text{kW}}{\text{kJ/s}}\right]$$
$$= 42{,}315\,\text{kW}.$$

Comments We can now justify our original assumptions that the kinetic and potential energies of the steam (or their changes) are negligible. The National Electrical Manufacturers Association (NEMA) has set standards for maximum inlet and outlet velocities for steam turbine applications: 175 ft/s (53.3 m/s) and 250 ft/s (76.2 m/s), respectively. The kinetic energy rate associated with the 53.3-m/s velocity is $\dot{m}V^2/2 = 38.7(53.3)^2/2 = 55.0 \times 10^3$ W or 55.0 kW. This energy rate is 0.13% of the power delivered by the turbine. The kinetic energy difference, that is, $\left|\dot{m}\left(V_2^2 - V_1^2\right)/2\right|$, associated with these standard maximum velocities is equally small, amounting to about 0.14% of the power delivered by the turbine. Similar estimates can be made to show that the potential energy differences are also much less than $\dot{W}_{\text{turb}}$. This exercise is left to the reader.

Example 5.11 Steam Power Plant Condenser

Consider the same turbine and flow condition as in Example 5.10. After exiting this turbine, steam is condensed in the shell side of a condenser, as shown in the sketch. The state of the steam entering the condenser (station 1) is essentially the same as the turbine exit state defined in Example 5.10 ($x = 0.88$, $T = 320$ K). The condensate exits the condenser as saturated liquid at 320 K (station 2). Cooling water is pumped through the tubes of the condenser at a rate of 2119.4 kg/s and a pressure of 0.3 MPa. The water enters the condenser at 305 K (station 3). Determine the outlet temperature of the cooling water (station 4).

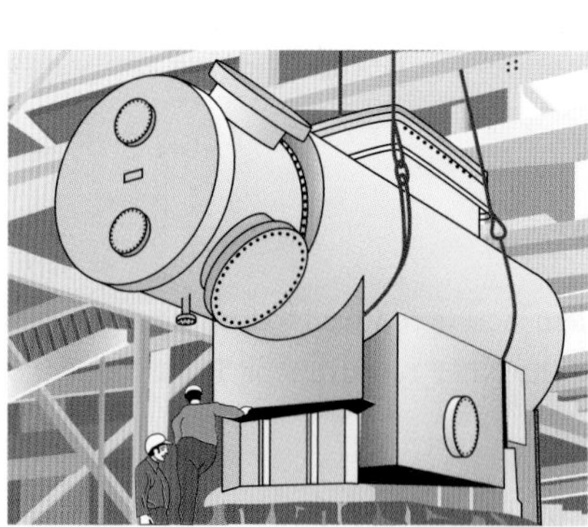

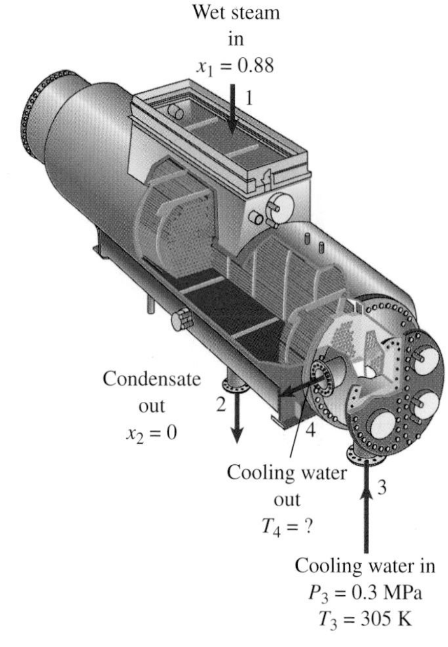

Steam condenser. Images used with permission from SPX Heat Transfer LLC for educational and informational purposes only.

Solution

Known $\dot{m}_1$, $\dot{m}_3$, T_1, T_2, x_1, x_2, P_3, T_3

Find T_4

Sketch

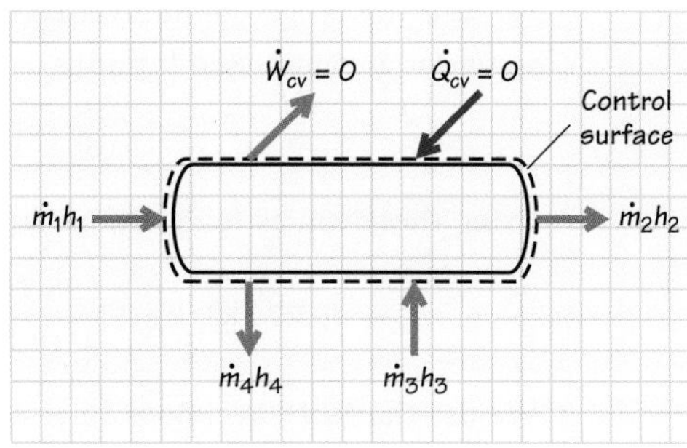

Modeling, Premises and Assumptions

i. Steady-state, steady flow
ii. $\dot{Q}_{cv} = 0$ (adiabatic operation)
iii. Negligible pressure drops ($P_2 = P_1$ and $P_4 = P_3$)
iv. Negligible kinetic and potential energies
v. Uniform velocity profiles at inlets and outlets

Analysis We apply conservation of mass and conservation of energy to the condenser for the control volume shown in the sketch. Since the condensing steam and the cooling water streams remain separated as they pass through the condenser, mass conservation requires

$$\dot{m}_2 = \dot{m}_1 \equiv \dot{m}_{\text{hot}}$$

and

$$\dot{m}_4 = \dot{m}_3 \equiv \dot{m}_{\text{cold}},$$

where we denote the two flow rates, respectively, as $\dot{m}_{\text{hot}}$ and $\dot{m}_{\text{cold}}$. For a control volume with two inlets and two outlets, energy conservation is expressed by Eq. 5.16, which we simplify as follows by applying the assumptions listed, so that

$$\dot{Q}_{\text{cv, net in}} - \dot{W}_{\text{cv, net out}} = \sum \dot{m}_{\text{out},k}\left(h_k + \tfrac{1}{2}V_k^2 + g(z_k - z_{\text{ref}})\right) - \sum \dot{m}_{\text{in},j}\left(h_j + \tfrac{1}{2}V_j^2 + g(z_j - z_{\text{ref}})\right)$$

becomes

$$0 + 0 = \dot{m}_2(h_2 + 0 + 0) + \dot{m}_4(h_4 + 0 + 0) - \dot{m}_1(h_1 + 0 + 0) - \dot{m}_3(h_3 + 0 + 0)$$

or

$$0 = \dot{m}_{\text{hot}}(h_2 - h_1) + \dot{m}_{\text{cold}}(h_4 - h_3).$$

We solve this for h_4 to get

$$h_4 = h_3 + \frac{\dot{m}_{\text{hot}}}{\dot{m}_{\text{cold}}}(h_1 - h_2).$$

From Example 5.10, we have

$$h_1 = 2299.0 \text{ kJ/kg}$$

and

$$h_2 = h_f(320\,\text{K}) = 196.17\ \text{kJ/kg}.$$

We use the NIST online WebBook to determine the enthalpy of the entering cold water (as a compressed liquid):

$$h_3(0.3\,\text{MPa}, 305\,\text{K}) = 133.74\ \text{kJ/kg}.$$

Using these values to determine h_4 yields

$$h_4 = 133.74\ \text{kJ/kg} + \frac{38.7\,\text{kg/s}}{2119.4\ \text{kg/s}}(2299 - 196.17)\,\text{kJ/kg} = 172.14\,\text{kJ/kg}.$$

We use the NIST WebBook once again now to find

$$T_4(0.3\,\text{MPa}, 172.14\ \text{kJ/kg}) = 314.2\ \text{K}.$$

The cooling water thus experiences a temperature increase of $\Delta T = 9.2$ $(= 314.2 - 305)$ K.

Comments Note the large flow rate used for the cooling water (2119.4 kg/s) to condense a much smaller amount of steam (38.7 kg/s). Maintaining a small ΔT for the cooling water mitigates thermal pollution for an open system that discharges its water into a river or other natural body of water.

Self-Test 5.8

Consider the steady flow of hot and cold water through a mixing faucet. Neglecting any heat losses and potential and kinetic energy changes, find the final temperature of the flow when 0.03 kg/s of water at 52 °C mixes with 0.045 kg/s of water at 10 °C.

(Answer: $T \cong 300\,K$ (27 °C))

Example 5.12 Air Tank Filling

A 0.002-m^3 steel tank is rapidly filled with air for 10 s. The air enters the tank at 300 K with a mass flow rate of 0.00017 kg/s. The tank initially contains air at 100 kPa and 300 K. Determine the temperature and pressure of the air in the tank immediately at the end of the 10-s filling process.

Solution

Known $T_{in} = T_1$, P_1, $\dot{m}_{in}$, Δt

Find T_2 and P_2

Sketch

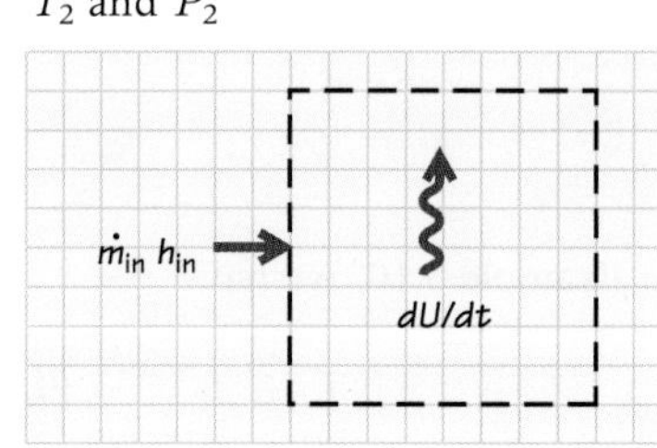

Modeling, Premises and Assumptions

i. The air behaves as an ideal gas.
ii. Air enters with constant properties and constant flow rate.
iii. The kinetic energy of the entering air is negligible (i.e., $V_{\text{in}}^2/2 \approx 0$).
iv. Properties are uniform within the tank owing to the jet stirring action.
v. The filling process is sufficiently rapid that there is no time for heat transfer to occur from the tank, i.e., the process is adiabatic ($\dot{Q}_{\text{cv}} \approx 0$).

Analysis We immediately recognize that we will want to write an appropriate energy conservation expression because our desired quantity is a temperature. If we can find the final internal energy of the air in the tank, we can easily determine the temperature. Our first step to solving this problem is to draw an energy diagram for our open system, the tank (see above). Here the energy entering with the flow $\dot{m}_{\text{in}}h_{\text{in}}$ equals the time rate of increase in the total system internal energy, dU_{cv}/dt. Thus,

$$\dot{m}_{\text{in}}h_{\text{in}} = \frac{dU_{\text{cv}}}{dt}.$$

Since both $\dot{m}_{\text{in}}$ and h_{in} are constants, we can easily integrate this expression as

$$\int_{t_1}^{t_2} \dot{m}_{\text{in}}h_{\text{in}}dt = \int_{U_1}^{U_2} dU_{\text{cv}},$$

to obtain, upon evaluating limits and rearranging,

$$U_{\text{cv},2} - U_{\text{cv},1} = \dot{m}_{\text{in}}h_{\text{in}}(t_2 - t_1).$$

Since we assume the properties within the control volume are uniform, $U_{\text{cv}} = Mu$; thus,

$$M_2u_2 - M_1u_1 = \dot{m}_{\text{in}}h_{\text{in}}\Delta t,$$

where $\Delta t \equiv t_2 - t_1$, our 10-s filling time.

We now apply mass conservation (Eq. 3.15a) to find M_2: $dM_{\text{cv}}/dt = \dot{m}_{\text{in}} - h_{\text{out}}$, and

$$\int_{M_1}^{M_2} dM_{\text{cv}} = \int_{t_1}^{t_2} \dot{m}_{\text{in}}dt.$$

Performing the integration and solving for M_2 yields

$$M_2 = M_1 + \dot{m}_{\text{in}}\Delta t.$$

Substituting this expression for M_2 into our energy conservation expression and solving for u_2 results in

$$u_2 = \frac{\dot{m}_{\text{in}}h_{\text{in}}\Delta t + M_1u_1}{M_2}.$$

We apply the ideal-gas equation of state to determine the initial (state-1) mass in the tank:

$$M_1 = \frac{P_1\mathcal{V}_1}{R_{\text{air}}T_1} = \frac{(100\ \text{kN/m}^2)(0.002\ \text{m}^3)}{(0.287\ \text{kJ/kg·K})(300\ \text{K})}\left[\frac{\text{kJ}}{\text{kN·m}}\right] = 0.002323\ \text{kg}.$$

With this value we can find the final (state-2) mass:

$$M_2 = M_1 + \dot{m}_{in}\Delta t = 0.002323\,\text{kg} + (0.00011\,\text{kg/s})(10\text{ s}) = 0.004023\text{ kg}.$$

From Table C.2 (Appendix C), we can obtain values for h_{in} and u_1:

$$h_{in}(300\text{ K}) = 426.04\text{ kJ/kg, and}$$

$$u_1(300\text{ K}) = 339.93\text{ kJ/kg}.$$

Substituting all our now known values into our conservation of energy expression, we obtain the final mass-specific internal energy of the air in the tank:

$$\begin{aligned} u_2 &= \frac{\dot{m}_{in}h_{in}\Delta t + M_1u_1}{M_2} \\ &= \frac{(0.00017\,\text{kg/s})(426.04\,\text{kJ/kg})10\text{ s} + 0.002323\,\text{kg}(339.93\,\text{kJ/kg})}{0.004023\,\text{kg}} \\ &= 376.3\,\text{kJ/kg}. \end{aligned}$$

We now interpolate values in Table C.2 to find the final temperature:

$$\begin{aligned} u_2(T_2) &= 376.3\text{ kJ/kg} = u_2(350.5\text{ K}),\text{ i.e.,} \\ T_2 &= 350.5\text{ K}. \end{aligned}$$

The final pressure is obtained from the ideal-gas law as follows:

$$\begin{aligned} P_2 &= \frac{M_2R_{air}T_2}{\mathcal{V}_2(=\mathcal{V}_1)} \\ &= \frac{0.004023\text{ kg}(0.287\,\text{kJ/kg·K})\ 350.5\,\text{K}}{0.002\text{ m}^3}\left[\frac{\text{kN·m}}{\text{kJ}}\right]\left[\frac{\text{kPa}}{\text{kN/m}^2}\right] \\ &= 202.4\text{ kPa}. \end{aligned}$$

Comments We note the many layers of knowledge employed to solve this problem: conservation of energy, conservation of mass, ideal-gas equation of state, ideal-gas calorific equation of state (via Table C.2), and the ability to perform simple integrations. Also, assuming the process was adiabatic was crucial to obtaining a solution.

Example 5.13 Heating Milk

Specialty coffee beverages are made by adding steamed milk to espresso coffee. In the preparation of the steamed milk (see sketch), saturated water vapor at 100 kPa jets into a pitcher containing 0.17 kg of skim milk, initially at 278 K (~40 F). If the steam flow rate is 4 g/min, estimate how long it takes to heat the milk to 322 K (~120 F). Assume that the process effectively occurs at constant pressure (100 kPa) and that the steam jet provides a good stirring action.

(Credit: J-Roman / iStock / Getty Images.)

(Credit: Image Source / Getty Images.)

Solution

Known M_1, T_1, T_2, $P_1 = P_2 = P_{in}$, $\dot{m}_{in}$

Find $\Delta t\ (= t_2 - t_1)$

Steam
Steel tube
Milk

Sketch

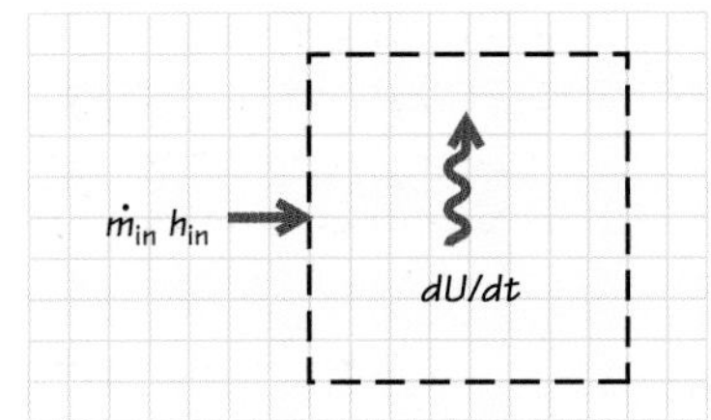

Modeling, Premises and Assumptions

i. Steam enters with constant properties and constant flow rate.
ii. The process occurs at constant pressure (i.e., pressure variations within the milk can be neglected).
iii. The kinetic energy of the entering steam is negligible (i.e., $V_{\text{in}}^2/2 \approx 0$).
iv. Properties are uniform within the control volume from the jet stirring action.
v. $\dot{W}_{\text{cv}} \approx 0$ [i.e., the boundary work rate $(Pd\mathcal{V}/dt)$ is negligibly small].
vi. $\dot{Q}_{\text{cv}} \approx 0$ (i.e., the process is adiabatic).
vii. Kinetic and potential energies of the control volume are negligible.
viii. The properties of the milk are approximated by those of water.
ix. There is no evaporation into the surroundings.

Analysis To solve this problem requires the application of the unsteady forms of both the conservation of energy and the conservation of mass, similarly to the previous example. Here we proceed more formally, beginning with the conservation of energy, Eq. 5.18:

$$\dot{Q}_{\text{cv, in}} - \dot{W}_{\text{cv, out}} + \dot{m}_{\text{in}}\left[h_{\text{in}} + V_{\text{in}}^2/2 + g(z_{\text{in}} - z_{\text{ref}})\right]$$
$$- \dot{m}_{\text{out}}\left[h_{\text{out}} + V_{\text{out}}^2/2 + g(z_{\text{out}} - z_{\text{ref}})\right] = \frac{dE_{\text{cv}}}{dt}.$$

Applying our assumptions and noting that $\dot{m}_{\text{out}} = 0$, this simplifies to

$$0 - 0 + \dot{m}_{\text{in}}h_{\text{in}} - 0 = \frac{dU_{\text{cv}}}{dt},$$

or

$$\frac{dU_{\text{cv}}}{dt} = \dot{m}_{\text{in}}h_{\text{in}}.$$

Since both $\dot{m}_{\text{in}}$ and h_{in} are constants, we can easily integrate this expression and, as a result, introduce the unknown time difference as follows:

$$\int_{U_1}^{U_2} dU_{\text{cv}} = \int_{t_1}^{t_2} \dot{m}_{\text{in}}h_{\text{in}}dt,$$

and

$$U_{\text{cv},2} - U_{\text{cv},1} = \dot{m}_{\text{in}}h_{\text{in}}(t_2 - t_1).$$

Since we are assuming the properties within the control volume are uniform, $U_{\text{cv}} = Mu$; thus,

$$M_2u_2 - M_1u_1 = \dot{m}_{\text{in}}h_{\text{in}}\Delta t,$$

where $\Delta t \equiv t_2 - t_1$.

We now apply mass conservation (Eq. 3.15a) to find M_2:

$$dM_{cv}/dt = \dot{m}_{in} - \dot{m}_{out},$$

and

$$\int_{M_1}^{M_2} dM_{cv} = \int_{t_1}^{t_2} \dot{m}_{in} dt.$$

Performing the integration and solving for M_2 yields

$$M_2 = M_1 + \dot{m}_{in}\Delta t.$$

Substituting this expression for M_2 into our energy conservation expression and solving for Δt results in

$$\Delta t = \frac{M_1(u_2 - u_1)}{\dot{m}_{in}(h_{in} - u_2)}.$$

Values for u_1, u_2, h_{in} are easily obtained from the NIST database:

$$u_1(100\,\text{kPa}, 278\,\text{K}) = 20.388\ \text{kJ/kg},$$
$$u_2(100\,\text{kPa}, 322\,\text{K}) = 204.51\ \text{kJ/kg},$$
$$h_{in} = h_g(100\,\text{kPa}) = 2674.9\ \text{kJ/kg}.$$

Before substituting numerical values we need to express the given flow rate in SI units:

$$\dot{m}_{in} = 4\text{g/min}\left[\frac{1\,\text{kg}}{1000\,\text{g}}\right]\left[\frac{1\ \text{min}}{60\,\text{s}}\right] = 6.67 \times 10^{-5}\,\text{kg/s}.$$

Our final result is thus

$$\Delta t = \frac{0.17\,\text{kg}\ (204.51 - 20.388)\ \text{kJ/kg}}{(6.67 \times 10^{-5}\ \text{kg/s})(2674.9 - 204.51)\ \text{kJ/kg}} = 190\,\text{s}.$$

Comments The final result (slightly more than 3 min) seems quite reasonable. Note the many simplifying assumptions used to solve this problem. The application of reasonable assumptions to complex problems is an important part of the art of engineering.

SUMMARY

Although several different expressions for the conservation of energy were presented in this chapter, it should be clear that they all express a single principle. You should be quite familiar with the meanings of the simpler expressions for closed systems, in particular Eqs. 5.5, 5.6, and 5.10, and for open systems, Eqs. 5.11, 5.15, and 5.18. You should also be able to select from these several expressions the most appropriate one for any particular problem or analysis. As a higher-level alternative, you should be able to start with the most general expression of energy conservation (Eq. 5.19) and derive the specific expression needed for any particular case.

KEY EQUATIONS

Review the most important equations presented in this chapter (i.e., those boxed with a yellow background). What physical principles do they express? What restrictions apply?

CHAPTER 5 KEY CONCEPTS AND DEFINITIONS CHECKLIST

Answer the questions and solve the problems following the arrows to demonstrate mastery of the listed concepts and definitions.

5.1 Energy Conservation for Closed Systems

- ☐ Energy conservation for a time interval (Eq. 5.1) ➔ Question 5.2, Problem 5.2B
- ☐ Energy conservation at an instant rate form (Eq. 5.2) ➔ Problems 5.2C, 5.72, 5.73
- ☐ Creating closed-system boundaries with dashed lines and representing energy transfers or changes with arrows
- ☐ Various system energies ➔ Problem 5.1
- ☐ First law for incremental change in state ➔ Question 5.1
- ☐ First law for finite change in state ➔ Question 5.2, Problems 5.3, 5.5, 5.16
- ☐ Systems undergoing constant-P, constant-v, or other simple processes, for ideal-gases or two-phase substances ➔ Problems 5.13, 5.18, 5.22, 5.24
- ☐ Application of the first law to systems involving ideal gases ➔ Problems 5.25, 5.28, 5.31
- ☐ Application of the first law to systems involving two-phase substances ➔ Problems 5.46, 5.49, 5.50, 5.69

5.2 Energy Conservation for Open Systems

5.2a Open Systems with Steady Flow

- ☐ Creating open-system boundaries with dashed lines and representing energy transfer rates with arrows
- ☐ First law to analyze steady-flow devices ➔ Problems 5.81, 5.82, 5.89, 5.96, 5.103, 5.113
- ☐ Control volumes with multiple inlets and outlets ➔ Problems 5.128, 5.129

5.2b Open Systems with Unsteady Flow ➔ Problems 5.134, 5.136

REFERENCES

1. Thompson, B. (Count Rumford), "An Inquiry Concerning the Source of Heat Which Is Excited by Friction," in *The Complete Works of Count Rumford*, American Academy of Arts and Sciences, Boston, 1870, pp. 471–491. (Original publication 1798.)
2. Mayer, J. R., "The Forces of Inorganic Nature," in *The Correlation and Conservation of Forces*, E. L. Youmans (Ed.), Appleton, New York, pp. 251–258, 1865. (Original publication 1842.)
3. Joule, J. P., "On the Mechanical Equivalent of Heat," *Philosophical Transactions of the Royal Society*, **140**: 61–82 (1850).

4. Helmholtz, H., "Interaction of Natural Forces," in *The Correlation and Conservation of Forces*, E. L. Youmans (Ed.), Appleton, New York, 1865, pp. 211–247. (Original publication 1854.)
5. Mayer, J. R., "The Mechanical Equivalent of Heat," in *The Correlation and Conservation of Forces*, E. L. Youmans (Ed.), Appleton, New York, 1865, pp. 316–355. (Original publication 1851.)
6. Clausius, R., "The Second Law of Thermodynamics," in *A Source Book of Physics*, W. F. Magie (Ed.), Harvard University Press, Cambridge, MA, 1963, pp. 228–236. (Original publication 1850.)
7. Planck, M., *Treatise on Thermodynamics*, Longmans, Green, and Co., London, 1903, p. 38.
8. Van Meter, P. N., Firetto, C. M., Turns, S. R., *et al.*, "Improving Students' Conceptual Reasoning by Prompting Cognitive Operations," *Journal of Engineering Education*, 105:245–277 (2016).
9. Johnson, A. T, *Biological Process Engineering: An Analogical Approach to Fluid Flow, Heat Transfer, and Mass Transfer Applied to Biological Systems*, Wiley, New York, 1998.

Some end-of-chapter problems were adapted with permission from the following:

Chapman, A. J., *Fundamentals of Heat Transfer*, Macmillan, New York, 1987.
Look, D. C, Jr., and Sauer, H. J., Jr., *Engineering Thermodynamics*, PWS, Boston, 1986.
Myers, G. E., *Engineering Thermodynamics*, Prentice Hall, Englewood Cliffs, NJ, 1989.

QUESTIONS

5.1 Write out the first law of thermodynamics for an incremental change in state. Include heat, work, and energy terms in your expression.

5.2 Write out the first law of thermodynamics for a finite change in state. Include heat, work, and energy terms in your expression.

5.3 Explain the relationship between electrical work, or power, and Joule heating.

Chapter 5 Problem Subject Areas

5.1–5.12	Closed system processes and the first law of thermodynamics (little or no numerical calculation required)
5.13–5.17	General closed-system first-law problems involving a finite change in state
5.18–5.45	Energy conservation applied to closed systems with finite state changes (ideal gases)
5.46–5.69	Energy conservation applied to closed systems with finite state changes (H_2O and other two-phase substances)
5.70–5.71	Energy conservation applied to closed systems for a finite change in state (solids)
5.72–5.79	Energy conservation applied to closed systems at an instant (rate processes)
5.80–5.88	Steady-flow open systems (not device-specific)
5.89–5.95	Steady-flow open systems (*ke*/*pe* featured)
5.96–5.102	Steady-flow open systems (heating/cooling)
5.103–5.110	Steady-flow open systems (turbine)
5.111–5.117	Steady-flow open systems (pump/compressor)

5.118–5.124	Steady-flow open systems (nozzle)
5.125–5.126	Steady-flow open systems (throttle)
5.127–5.133	Steady-flow open systems (multiple streams)
5.134–5.140	Unsteady open systems
5.141–5.143	System analysis
5.144–5.147	EES problems
5.148–5.155	FE problems

PROBLEMS

5.1–5.12 Closed system processes and the first law of thermodynamics (little or no numerical calculation required)

5.1 ***Conceptual problem.*** Distinguish between bulk system energy and internal energy. Illustrate your discussion with three examples.

5.2 ***Conceptual problem.*** Without reference to the text, write symbolic expressions for the conservation of energy (the first law of thermodynamics) for a system, for the following conditions:

A. A change from state 1 to state 2

B. An incremental change (use δ and d as appropriate)

C. At an instant

Below each term, write its meaning in words.

5.3 ***Conceptual problem.*** Consider a closed thermodynamic system undergoing a finite change in state.

A. During a certain process, 6 kJ of energy enters the system as a heat interaction. No energy leaves or is generated. The energy change for the system is (positive, negative, zero). Give the numerical value. The energy of the system at the end of the process (can or cannot) be determined from the given information. Give the numerical value, if possible.

B. During a certain process, 9 kJ of energy exits the system as a work interaction. No energy enters or is generated. The energy change for the system is (positive, negative, zero). Give the numerical value. The energy of the system at the end of the process (can or cannot) be determined from the given information. Give the numerical value, if possible.

C. At the start of a certain process, the system energy is 14 kJ. During the process, 2 kJ of energy enters the system as work and 5 kJ of energy exits as a heat interaction. No energy is generated. The energy change for the system is (positive, negative, zero). Give the numerical value. The energy of the system at the end of the process (can or cannot) be determined from the given information. Give the numerical value, if possible.

D. At the start of a certain process, the system energy is 150 kJ. During the process, 15 kJ of energy enters the system as work and 15 kJ of energy exits as a heat interaction. No energy is generated. The energy change for the system is (positive, negative, zero). Give the numerical value. The energy of the system at the end of the process (can or cannot) be determined from the given information. Give the numerical value, if possible.

5.4 ***Conceptual problem.*** View the videos and complete the three exercises discussed in the tutorial *First Law for Ideal-Gas Processes*. These can be found at http://www.cambridge.org/thermo

5.5 ***Conceptual problem.*** A piston–cylinder assembly contains a fixed mass of air. The air is the system of interest and undergoes a finite process from state 1 to state 2. The piston moves such that the final (state-2) volume of the air is several times greater than the initial volume (state 1). The process occurs at constant pressure.

A. Does the temperature increase or decrease? Explain your answer.

B. Is there any work associated with the process? If so, is the work done on or by the system? Explain your answer.

C. Does the internal energy of the system increase, decrease, or remain the same? Explain your answer.

D. Is there any heat interaction associated with the process? If so, is the heat interaction into or out of the system? Explain your answer.

5.6 ***Conceptual problem.*** A piston–cylinder assembly contains a fixed mass of air. The air is the system of interest and undergoes a finite process from state 1 to state 2. The piston moves such that the final (state-2) volume of the air is several times smaller than the initial volume (state 1). The process occurs at constant pressure.

A. Does the temperature increase or decrease? Explain your answer.

B. Is there any work associated with the process? If so, is the work done on or by the system? Explain your answer.

C. Does the internal energy of the system increase, decrease, or remain the same? Explain your answer.

D. Is there any heat interaction associated with the process? If so, is the heat interaction into or out of the system? Explain your answer.

5.7 ***Conceptual problem.*** A piston–cylinder assembly contains a fixed mass of air. The air is the system of interest and undergoes a finite process from state 1 to state 2. The piston moves such that the final (state-2) volume of the air is several times larger than the initial volume (state 1). The process occurs at constant temperature.

A. Does the pressure increase or decrease? Explain your answer.

B. Is there any work associated with the process? If so, is the work done on or by the system? Explain your answer.

C. Does the internal energy of the system increase, decrease, or remain the same? Explain your answer.

D. Is there any heat interaction associated with the process? If so, is the heat interaction into or out of the system? Explain your answer.

5.8 ***Conceptual Problem.*** A piston–cylinder assembly contains a fixed mass of air. The air is the system of interest and undergoes a finite process from state 1 to state 2. The piston moves such that the final (state-2) volume of the air is several times smaller than the initial volume (state 1). The process occurs at constant temperature.

A. Does the pressure increase or decrease? Explain your answer.

B. Is there any work associated with the process? If so, is the work done on or by the system? Explain your answer.

C. Does the internal energy of the system increase, decrease, or remain the same? Explain your answer.

D. Is there any heat interaction associated with the process? If so, is the heat interaction into or out of the system? Explain your answer.

5.9 ***Conceptual problem.*** A rigid tank contains a fixed mass of air. The air is the system of interest and undergoes a finite process from state 1 to state 2. The process results in the final (state-2) pressure of the air being much greater than the initial pressure (state 1).

A. Does the temperature increase or decrease? Explain your answer.

B. Is there any work associated with the process? If so, is the work done on or by the system? Explain your answer.

C. Does the internal energy of the system increase, decrease, or remain the same? Explain your answer.

D. Is there any heat interaction associated with the process? If so, is the heat interaction into or out of the system? Explain your answer.

5.10 ***Conceptual problem.*** A rigid tank contains a fixed mass of air. The air is the system of interest and undergoes a finite process from state 1 to state 2. The process results in the final (state-2) pressure of the air being much less than the initial pressure (state 1).

A. Does the temperature increase or decrease? Explain your answer.

B. Is there any work associated with the process? If so, is the work done on or by the system? Explain your answer.

C. Does the internal energy of the system increase, decrease, or remain the same? Explain your answer.

D. Is there any heat interaction associated with the process? If so, is the heat interaction into or out of the system? Explain your answer.

5.11 ***Conceptual problem.*** A rigid tank contains a fixed mass of air. The air is the system of interest and undergoes a finite process from state 1 to state 2. The process results in the final (state-2) temperature of the air being much higher than the initial temperature (state 1).

A. Does the pressure increase or decrease? Explain your answer.

B. Is there any work associated with the process? If so, is the work done on or by the system? Explain your answer.

C. Does the internal energy of the system increase, decrease, or remain the same? Explain your answer.

D. Is there any heat interaction associated with the process? If so, is the heat interaction into or out of the system? Explain your answer.

5.12 ***Conceptual problem.*** A rigid tank contains a fixed mass of air. The air is the system of interest and undergoes a finite process from state 1 to state 2. The process results in the final (state-2) temperature of the air being much lower than the initial temperature (state 1).

A. Does the pressure increase or decrease? Explain your answer.

B. Is there any work associated with the process? If so, is the work done on or by the system? Explain your answer.

C. Does the internal energy of the system increase, decrease, or remain the same? Explain your answer.

D. Is there any heat interaction associated with the process? If so, is the heat interaction into or out of the system? Explain your answer.

5.13–5.17 General closed-system first-law problems involving a finite change in state

5.13 A closed system rejects 25 kJ of energy in a heat interaction while experiencing a volume change of 0.1 m^3 (0.15 to 0.05 m^3). Assuming a reversible constant-pressure process at 350 kPa, determine the change in internal energy.

5.14 A tank contains a fluid that is stirred by a paddle wheel. The work input to the paddle wheel is 4309 kJ. The heat transferred from the tank is 1371 kJ. Considering the tank and the fluid as a closed system, determine the change in the internal energy (kJ) of the system.

5.15 Front-wheel-drive cars do not distribute the work involved in stopping the car equally among all wheels. The front-wheel brakes dissipate about 60% of the energy transfer involved in braking, while the rear wheels take care of the remaining 40%. If you are in such a car, traveling at 55 miles/hr, and if you slow the car down to 20 miles/hr by applying the brakes, determine the amount of kinetic energy (kJ) dissipated in the front brakes. Neglect rolling resistance and aerodynamic drag. Assume that the car and driver have a total mass of 1087 kg.

5.16 An insulated container, filled with 10 kg of liquid water at 20°C, is fitted with a stirrer. The stirrer is made to turn by lowering a 25-kg object outside the container a distance of 10 m using a frictionless pulley system. The local acceleration of gravity is 9.7 m/s^2. Assume that all work done by the object is transferred to the water and that the water is incompressible.

A. Determine the work transfer (kJ) to the water.

B. Determine the increase in internal energy (kJ) of the water.

C. Determine the final temperature (°C) of the water. HINT: Assume that the temperature change is small enough that a constant value of the specific heat is a good approximation.

D. Determine the heat transfer (kJ) from the water required to return the water to its initial temperature.

5.17 A frictionless 3000-N piston maintains a gas at constant pressure in a cylinder. The cross-sectional area of the piston is 52 cm^2 and the atmospheric pressure acting on this area is 0.1 MPa. A process occurs in which a paddle wheel transfers 6800 N·m of work to the gas, 10 kJ of heat is transferred from the gas, and the internal energy of the gas decreases by 1 kJ. Determine the distance (cm) moved by the piston.

5.18–5.45 Energy conservation applied to closed systems for a finite change in state (ideal gases)

5.18 Consider a piston–cylinder assembly containing 0.1 kg of dry air initially at 300 K and 200 kPa (state 1). Energy is added to the air (by heat transfer) at

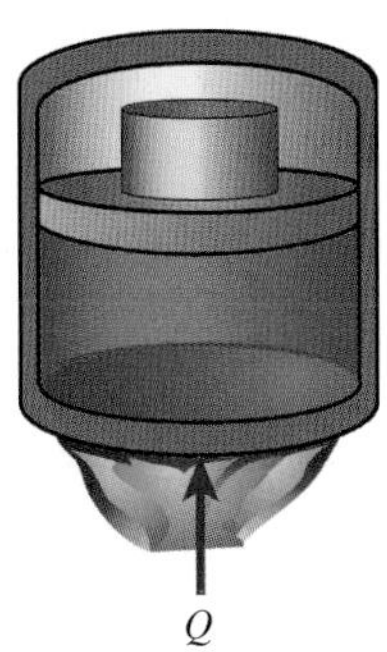

constant pressure until the final temperature is 450 K at state 2. Plot the process in P–v and T–v coordinates and determine the following quantities: ${}_1W_2$, ΔU, ΔH, and ${}_1Q_2$. Give units.

5.19 Consider a piston–cylinder assembly containing 0.12 kg of nitrogen at 300 K and 1 atm (state 1). The nitrogen is heated at constant pressure to 700 K (state 2). Plot the process in P–v and T–v coordinates. Assuming a constant value of the constant-pressure specific heat of 1.067 kJ/kg·K for the N_2, determine the following quantities: ${}_1W_2$, ΔU, ΔH, and ${}_1Q_2$. Give units.

5.20 Repeat Problem 5.19 but use tabulated values of $\bar{h}$ (Table B.7), rather than the given constant specific heat, to find the desired quantities. Give units.

5.21 ***Conceptual Problem.*** Show in detail how the first law reduces to ${}_1Q_2 = \Delta H$ for a system comprising a compressible substance undergoing a constant-pressure process. What assumptions are required?

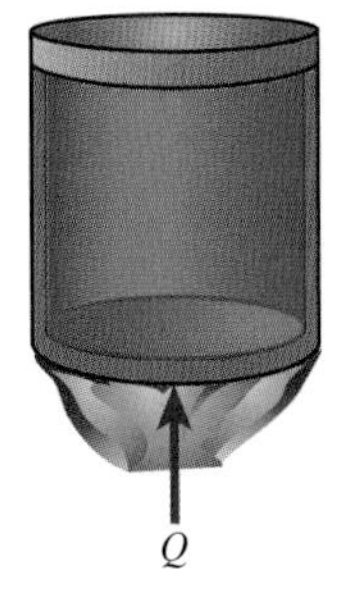

5.22 Consider a sealed rigid tank containing 0.15 kg of dry air at 300 K and 100 kPa (state 1). The air is heated to 600 K (state 2). Plot the process in P–v and T–v coordinates and determine the following quantities: ${}_1W_2$, ΔU, ΔH, and ${}_1Q_2$. Give units.

5.23 Consider a sealed rigid tank containing 0.12 kg of N_2 at 300 K and 1 atm (state 1). The N_2 is heated to a temperature of 700 K (state 2). Plot the process in P–v and T–v coordinates and determine the following quantities: ${}_1W_2$, ΔU, ΔH, and ${}_1Q_2$. Compare these results with the results obtained in Problems 5.19 or 5.20. Discuss.

5.24 Consider a piston–cylinder assembly containing 0.12 kg of nitrogen initially at 700 K and 300 kPa (state 1). The piston moves such that the final volume of the nitrogen (state 2) is three times larger than the initial volume (state 1). The process occurs at constant temperature. Plot the process on P–v and T–v coordinates. Assuming a constant value of the constant-pressure specific heat of 1.067 kJ/kg·K for the N_2, determine the following quantities: ${}_1W_2$, ΔU, ΔH, and ${}_1Q_2$. Give units.

5.25 Consider a piston–cylinder assembly containing 0.10 kg of air initially at 300 K and 300 kPa (state 1). The piston moves such that 35.9 kJ of work is done on the air by the surroundings. The process occurs adiabatically, i.e., there is no heat interaction. Assuming an average value of the constant-pressure specific heat of 1.005 kJ/kg·K for the air, determine the final temperature of the air. Is it possible to find the final pressure of the air? Explain your answer.

5.26 Consider a piston–cylinder assembly containing 0.25 kg of air initially at 750 K and 200 kPa (state 1). The piston moves such that 20 kJ of work is done on the air by the surroundings. The process occurs at constant pressure. Determine the magnitude of the heat interaction associated with this process. Indicate the direction of the heat interaction (into or out of the system).

5.27 Consider a piston–cylinder assembly containing 0.14 kg of air initially at 500 K and 250 kPa (state 1). The piston moves such that 5 kJ of work is done on the air by the surroundings. The process occurs at constant pressure. Determine the magnitude of the heat interaction associated with this process. Indicate the direction of the heat interaction (into or out of the system).

5.28 Consider a piston–cylinder assembly containing 0.08 kg of air initially at 950 K and 400 kPa (state 1). The piston moves such that 30 kJ of work is done by the air on the surroundings. The process occurs at constant pressure. Determine the magnitude of the heat interaction associated with this process. Indicate the direction of the heat interaction (into or out of the system).

5.29 Consider a piston–cylinder assembly containing 0.09 kg of air initially at 600 K and 350 kPa (state 1). The piston moves such that 18 kJ of work is done by the air on the surroundings. The process occurs at constant pressure. Determine the magnitude of the heat interaction associated with this process. Indicate the direction of the heat interaction (into or out of the system).

5.30 Consider a sealed rigid tank containing 0.1 kg of argon at 300 K and 100 kPa (state 1). A heat interaction occurs, with 15 kJ transferred from the surroundings to the argon. The molecular weight of argon is 39.948 kg/kmol. Assuming an average molar constant-pressure specific heat of 20.83 kJ/kmol·K for the process, determine the final pressure of the argon.

5.31 Consider a sealed rigid tank containing 0.23 kg of air at 800 K and 350 kPa (state 1). A heat interaction occurs, with 25 kJ transferred from the air to the surroundings. Determine the final pressure of the air.

5.32 Consider a piston–cylinder assembly containing 0.06 kg of air initially at 350 K and 200 kPa (state 1). The piston moves such that the final pressure of the air is 400 kPa. The process occurs at constant temperature. Determine the magnitude of the heat interaction associated with this process. Indicate the direction of the heat interaction (into or out of the system).

5.33 Consider a piston–cylinder assembly containing 0.10 kg of N_2 initially at 425 K and 150 kPa (state 1). The piston moves such that the final pressure of the air is 350 kPa. The process occurs at constant temperature. Determine the magnitude of the heat interaction associated with this process. Indicate the direction of the heat interaction (into or out of the system).

5.34 Consider a piston–cylinder assembly containing 0.85 kg of air initially at 400 K and 620 kPa (state 1). The piston moves such that the final pressure of the air is 300 kPa. The process occurs at constant temperature. Determine the magnitude of the heat interaction associated with this process. Indicate the direction of the heat interaction (into or out of the system).

5.35 Consider a piston–cylinder assembly containing 0.15 kg of N_2 initially at 515 K and 800 kPa (state 1). The piston moves such that the final pressure of the air is 175 kPa. The process occurs at constant temperature. Determine the magnitude of the heat interaction associated with this process. Indicate the direction of the heat interaction (into or out of the system).

5.36 A hydrogen storage tank containing 30 kg of H_2 is exposed to a fire such that 128.5 MJ is transferred to the H_2 in a heat interaction. The initial temperature and pressure in the tank are 300 K and 500 kPa, respectively. The maximum safe pressure for the tank is 1.5 MPa.

A. Is there a danger that the tank will rupture? Treat the hydrogen as an ideal gas for your analysis. Be sure to list your assumptions. **HINT**: Use the NIST software or Table E.1 in Appendix E to obtain approximate values for specific heats.

B. Is the assumption of ideal gas behavior reasonable? Justify your answer.

5.37 Air is contained in a piston–cylinder assembly at 2 MPa and 400 K (state 1). The 0.15-m-diameter piston is locked in place by stops at the 0.2-m position, and a 2-kg steel block sits on top of the piston, as shown in the sketch. The mass of the piston is 0.1 kg. The stops are removed and the air rapidly expands until the piston comes to rest at the upper position (x = 0.4 m), while the block continues upward as shown in the sketch on the right. The temperature is 310 K (state 2). Assume that the process is so rapid that it can be considered adiabatic and that the constant-pressure specific heat is constant (c_p= 1.013 kJ/kg·K). The atmospheric pressure is 100 kPa.

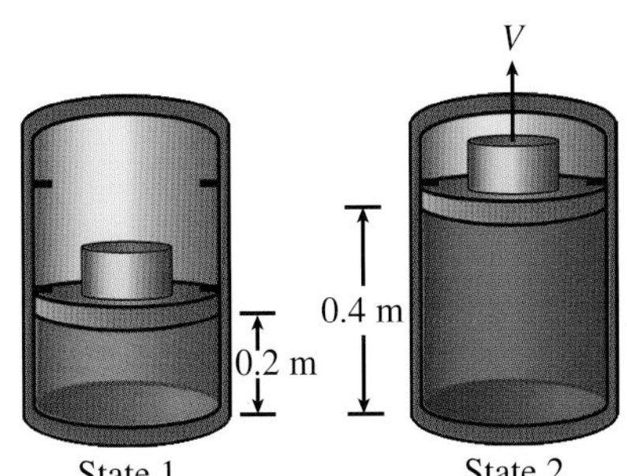

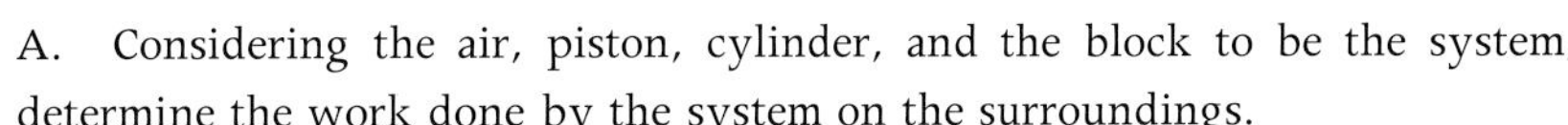

A. Considering the air, piston, cylinder, and the block to be the system, determine the work done by the system on the surroundings.

B. Determine the velocity of the steel block immediately after the piston comes to rest.

C. Neglecting drag, determine the maximum height achieved by the block measured from the bottom of the cylinder.

5.38 During the compression stroke in an automobile engine, air initially at 95 kPa and 305 K is compressed reversibly according to $P\mathcal{V}^{1.48}$ = constant. The engine compression ratio (= $\mathcal{V}_1/\mathcal{V}_2$) is 7.5. Determine (a) the work and (b) the heat transfer, each in kJ/kg.

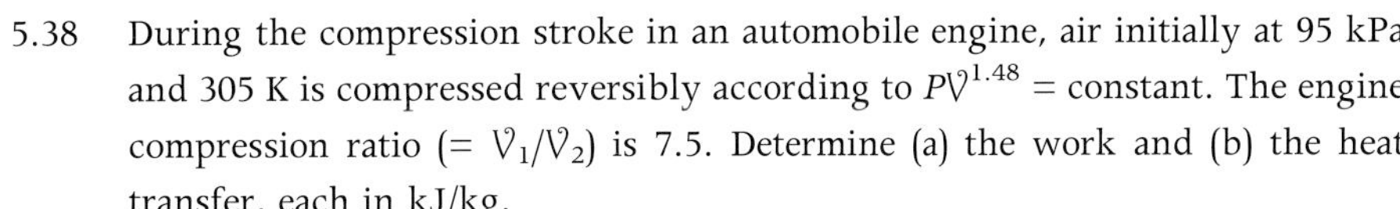

5.39 Consider 5 kg of air initially at 101.3 kPa and 38°C. Heat is transferred to the air until the temperature reaches 260°C. Determine the change of internal energy, the change in enthalpy, the heat transfer, and the work done for (a) a constant-volume process and (b) a constant-pressure process.

5.40 While trapped in a cylinder, 2.27 kg of air is compressed isothermally (a water jacket being used around the cylinder to maintain constant temperature) from initial conditions of 101 kPa and 16°C to a final pressure of 793 kPa. For this process, determine (a) the work required and (b) the heat removed, both in units of kJ.

5.41 Consider a thermodynamic system executing the two reversible processes shown in the sketch below (*ab* and *adb*), where $P_2 = 2P_1$ and $v_2 = 2v_1$. The working fluid is air with $P_1 = 100$ kPa and $v_1 = 0.2\ \text{m}^3/\text{kg}$. Determine the heat transfer per unit mass.

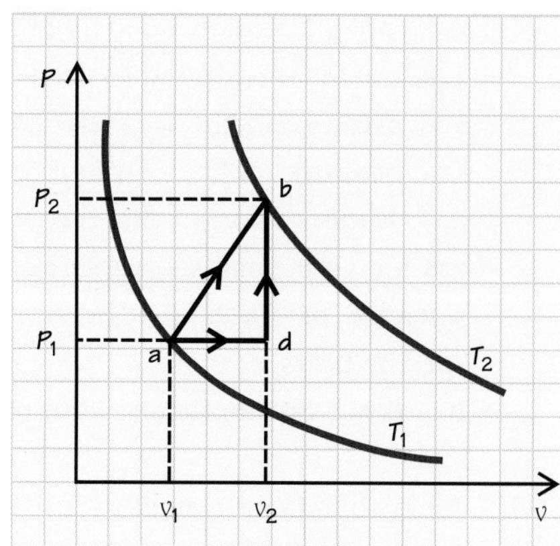

5.42 Consider a thermodynamic system consisting 3 kg of air. The air is compressed frictionlessly and adiabatically from 100 kPa and 300 K to 400 kPa. The process follows the path PV^{γ} = constant ($= P_1V_1^{\gamma} = P_2V_2^{\gamma}$). Compute (a) the initial volume, (b) the final volume, (c) the final temperature, and (d) the work.

5.43 A balloon at sea level contains 2 kg of helium at 30°C and 1 atm. The balloon then rises to 1500 m above sea level. At this height, the helium temperature is 6°C. Determine the change in internal energy (kJ) of the helium.

5.44 Initially, 1 kg of air at 101 kPa and 700 K is contained in a rigid tank. The tank is not insulated and cools down, as a result of heat transfer to the atmosphere (T_{atm} = 25°C, P_{atm} = 101 kPa), until it reaches thermal equilibrium. Determine the final pressure (kPa) and the heat transfer (kJ) for the process.

5.45 Air, initially at 205°C and 3.45 MPa, expands isothermally in a piston–cylinder device until its volume is 100 times larger than its initial volume. Determine the following:

A. The increase in internal energy (kJ/kg) of the air
B. The work transfer (kJ/kg) and direction into or out of the air
C. The heat transfer (kJ/kg) and direction into or out of the air

5.46–5.69 Energy conservation applied to closed systems (H_2O and other two-phase substances) for a finite change in state

5.46 Consider a piston–cylinder assembly containing 0.1 kg of saturated steam at 300 kPa (state 1). Energy is added to the steam by a heat interaction at constant pressure until a temperature of 600 K is reached (state 2). Plot the process in P–v and T–v coordinates and determine the following quantities: ${}_1W_2$, ΔU, ΔH, and ${}_1Q_2$. Include the steam dome on your plots. Give units.

5.47 Superheated steam is contained in a sealed rigid tank at 600 K and 0.3 MPa (state 1). Energy is removed from the steam by heat transfer until a temperature

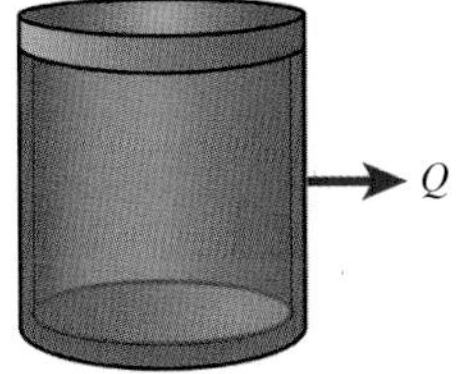

of 350 K (state 2) is reached. Plot the process in P–v and T–v coordinates. Include the steam dome on your plots. Determine the following quantities for this process: ${}_1W_2/M$, Δu, Δh, and ${}_1Q_2/M$.

5.48 Consider a piston–cylinder assembly containing 0.09 kg of steam having a quality of 0.9 at 500 kPa (state 1). Energy is added to the steam by a heat interaction at constant pressure until a temperature of 700 K is reached (state 2). Plot the process in P–v and T–v coordinates and determine the following quantities: ${}_1W_2$, ΔU, ΔH, and ${}_1Q_2$. Include the steam dome on your plots. Give units.

5.49 Saturated steam is contained in a sealed rigid tank at 600 K (state 1). Energy is removed from the steam by heat transfer until a temperature of 400 K (state 2) is reached. Plot the process in P–v and T–v coordinates. Include the steam dome on your plots. Determine the following quantities for this process: ${}_1W_2/M$, Δu, Δh, and ${}_1Q_2/M$.

5.50 Consider a piston–cylinder assembly containing 0.125 kg of wet steam with a quality of 0.45 at 300 kPa (state 1). Energy is added to the steam at constant temperature by a heat interaction until the volume of the steam is doubled (state 2). Determine the heat transfer. Give units.

5.51 Consider a piston–cylinder assembly containing 0.25 kg of wet steam with a quality of 0.85 at 400 kPa (state 1). Energy is removed from the steam at constant temperature by a heat interaction until the final volume (state 2) of the steam is one-third that of the original volume. Determine the heat transfer. Give units.

5.52 Saturated steam is contained in a sealed rigid tank at 500 K (state 1). Energy is removed from the steam by heat transfer until a temperature of 350 K (state 2) is reached. Determine the heat transfer.

5.53 Consider a piston–cylinder assembly containing 0.12 kg of superheated steam at 700 kPa and 800 K (state 1). Energy is removed from the steam by heat transfer until a temperature of 400 K is reached (state 2). The process is conducted at constant pressure. Determine the work and heat transfer.

5.54 Steam (13.2 kg) at 1.5 MPa with an 80% quality comprises a thermodynamic system. The steam is heated in a frictionless process until the temperature is 540 K.

A. Calculate the heat transfer (kJ) if the process occurs at constant pressure.
B. What is the heat transfer if the process is carried out at constant volume?

5.55 Consider 1.1 kg of steam initially at 0.15 MPa and 400 K in a closed, rigid container. This steam is heated to 440 K. Determine the amount of heat added to the system in kJ. HINT: Use NIST software to avoid double interpolation.

5.56 Initially, 1 kg of water (liquid and/or vapor) at 0.1 MPa is contained in a rigid 1.5-m^3 tank. The H_2O is then heated and equilibrates at 300°C. Determine the final pressure (MPa) and the heat transfer (kJ). Plot the process in P–v and T–v coordinates. **HINT**: Use the NIST software to obtain properties.

5.57 Old homes are sometimes heated with so-called steam radiators. In operation, steam flows through the metal radiator causing the metal surface to become hot, which, in turn, results in heat transfer to the room air. A particular steam radiator has a volume of 0.1416 m^3 and initially contains dry, saturated vapor at 200 kPa. The inlet and outlet valves on the radiator are then closed, and, as a result of heat transfer to the room, the radiation temperature drops to 311 K. Determine the heat transfer (kJ) from the radiator to the room.

5.58 An 2.27-m^3 steam boiler initially contains 1.7025 m^3 of liquid water and 0.5675 m^3 of water vapor in equilibrium at 100 kPa. The boiler is fired up and the liquid and vapor in the boiler are heated. Somehow the valves on the inlet and discharge of the boiler are both left closed. The relief valve lifts when the pressure reaches 4.0 MPa. Determine the heat transfer (kJ) to the boiler before the relief valve lifts.

5.59 The temperature of 150 liters of liquid water, initially at 10°C, is increased to 60°C by a 2500-W electric heater.

A. Determine the total energy (MJ) required.

B. Determine the length of time (hr) required.

C. If electricity can be purchased for 6 cents per kW·hr, determine the cost (cents).

5.60 A piston–cylinder assembly initially contains 0.45 kg of water at 8.0 MPa and a quality of 0.25. The water is then heated to a temperature of 760 K. The piston moves freely and maintains the cylinder contents at constant pressure. Determine the work transfer (kJ) and the heat transfer (kJ).

5.61 One kilogram of water (liquid and/or vapor) at 140 kPa is contained in a piston–cylinder device. The initial volume is 0.7 m^3. The water is then heated until its temperature reaches 700 K. The piston is free to move up or down unless it reaches a set of stops, which prevent piston motion. When the piston is against the stops, the cylinder volume is 1.4 m^3. **HINTS**: Use the NIST software for water properties, and be sure to draw a T–v diagram.

A. Determine the initial internal energy (kJ) of the water.

B. Determine the final internal energy (kJ) of the water.

C. Determine the work transfer (kJ) from the water.

D. Determine the heat transfer (kJ) to the water.

5.62 Initially, 1 kg of water at 420 K with a quality of 0.60 is contained in a piston–cylinder device. The piston is then slowly moved until the water is entirely changed into liquid. Heat transfer maintains the temperature at 420 K. Determine the following:

A. The change in volume of the cylinder

B. The work transfer (J)

C. The heat transfer (J)

5.63 Initially, saturated liquid water at 475 K is contained in a piston-cylinder device. The water then expands isothermally until its volume is 100 times larger than its initial volume. **HINT**: Be sure to draw a T–v diagram. Determine the following:

A. The increase in energy (kJ/kg) of the water
B. The work transfer (kJ/kg) and its direction into or out of the water
C. The heat transfer (kJ/kg) and its direction into or out of the water

5.64 A piston–cylinder device has an initial volume of $0.003\ m^3$ and contains dry, saturated water vapor at 200°C. A connecting rod is attached to the piston and transfers work to (or from) the device. The piston then moves out until the volume reaches $0.015\ m^3$. The final pressure of the steam is 0.25 MPa, and the atmospheric pressure acting on the outside of the piston is 101.3 kPa. The process occurs rapidly enough that it may be assumed to be adiabatic. Draw a T–v diagram to guide your solution. Use the NIST software to obtain properties and determine the following:

A. The work transfer (kJ) from the steam
B. The work transfer (kJ) to the atmosphere
C. The work transfer (kJ) from the piston to the connecting rod

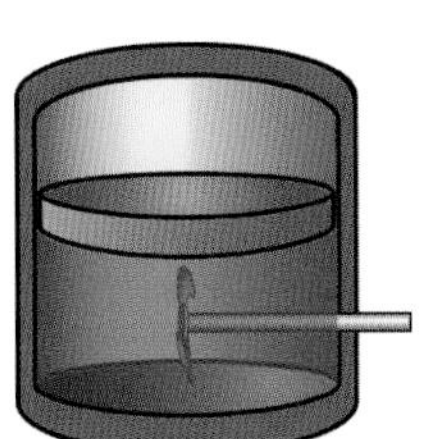

5.65 Initially, 0.45 kg of steam at 0.7 MPa with a quality of 0.90 is held in an adiabatic cylinder fitted with a freely floating piston and a paddle wheel. The paddle wheel is then turned on for 1 min. During this time interval, 200 kJ of work is transferred to the steam from the paddle wheel. Use the NIST software to obtain properties. Determine the following:

A. The final pressure (MPa)
B. The final temperature (K)
C. The final volume (m^3)
D. The final internal energy (kJ) of the steam

5.66 Initially, 0.5 kg of dry, saturated steam at 115°C is contained inside a spherical elastic balloon whose internal pressure is proportional to its diameter. Heat is then transferred to the steam until the steam pressure reaches 0.2 MPa. Determine the final temperature (°C) and the heat transfer in kJ. Use the NIST software to obtain properties.

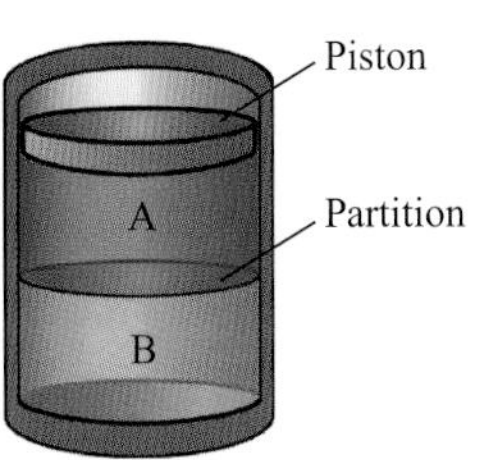

5.67 A rigid cylinder fitted with a freely floating piston is divided into two parts (A and B) by a rigid metal partition. Initially, part A contains 0.91 kg of water (liquid and/or vapor) at 1.7 MPa and part B contains 0.45 kg of dry, saturated water vapor at 600 kPa. The piston and the sides of A and B are perfectly insulated. The partition is a good heat conductor and allows enough heat transfer to keep the temperature of part A always equal to the temperature of part B. The bottom of part B is then heated until the pressure of part A equals the pressure of part B. Determine the heat transfer (kJ) into the bottom of part B. **HINT**: Use NIST or other software to avoid double interpolation for final state properties.

CNG/LPG Problems

5.68 Compressed natural gas (CNG) is contained in a rigid tank near room temperature (300 K) and with a pressure of 6.75 MPa. The CNG is cooled until it just starts to condense. Assuming the CNG can be treated as methane (CH_4), answer the following:

A. In what thermodynamic region does the initial state lie?

B. What are the final state temperature and pressure?

C. What is the heat transfer (per kg of methane) required for this process?

5.69 A small propane tank ($\mathcal{V} = 0.0185\ m^3$), used to fuel a barbeque grill, is approximately half full and contains 10 pounds (4.55 kg) of propane. A fire breaks out enveloping the tank in flames. There is no pressure relief system to prevent the pressure from building up as the tank heats. If the pressure in the tank reaches 1000 psia (6.89 MPa), there is a good chance that the tank will rupture. How much energy must be transferred to the propane for this pressure to be reached? Is this likely to occur if the average flame temperature is 1200 K?

5.70–5.71 Energy conservation applied to closed systems for a finite change in state (solids)

5.70 A 1-kg piece of iron, initially at 700°C, is quenched by dropping it into an insulated tank containing 2 kg of liquid water. The initial temperature of the water is 20°C. Determine the final temperature of the iron, assuming no water is vaporized.

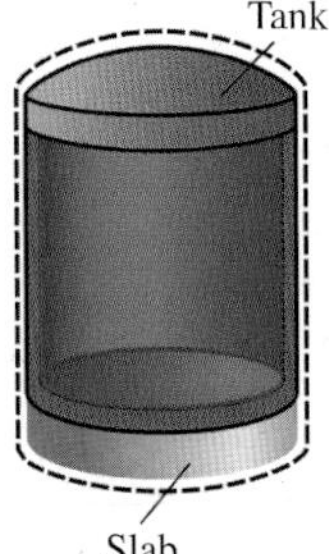

5.71 A closed, rigid steel tank has an inner volume of $0.02832\ m^3$ and has a mass of 22.7 kg when empty. Initially, the tank contains 0.45 kg of water (liquid plus vapor), and the tank and the water are both at 294 K. The tank containing the water is then placed on a 45 kg slab of steel, which is at another temperature. Assume the tank–water–slab system comes to equilibrium with no heat transfer to the surroundings. The final temperature of the tank–water–slab system is 555 K. The specific heat of the steel in both the tank and the slab is 0.419 kJ/kg·K. Determine the initial slab temperature (K). **HINT**: To avoid double interpolation, use the NIST software.

5.72–5.79 Energy conservation applied to closed systems at an instant (rate processes)

5.72 ***Conceptual problem.*** Consider the following four (A–D) closed thermodynamic systems at an instant. The systems consist of a single-phase substance. Answer the questions posed for the following scenarios:

A. At a particular instant, energy enters a system as a heat interaction at a rate of 14 kJ/s. No energy leaves. Is the rate of change of the system energy positive, negative, or zero? Give the numerical value. Is the system temperature increasing, decreasing, or constant? Explain your answers.

B. At a particular instant, energy exits a system as a work interaction at a rate of 7 kJ/s. No energy enters. Is the rate of change of the system energy positive, negative, or zero? Give the numerical value. Is the system temperature increasing, decreasing, or constant? Explain your answers.

C. At an instant, energy enters a system as a heat interaction at a rate of 20 kJ/s and energy exits the system as a work interaction at a rate of 15 kJ/s. Is the rate of change of the system energy positive, negative, or zero? Give the numerical value. Is the system temperature increasing, decreasing, or constant? Explain your answers.

D. At an instant, energy enters a system as a heat interaction at a rate of 16 kJ/s, and energy exits the system as a work interaction at a rate of 30 kJ/s. Is the rate of change of the system energy positive, negative, or zero? Give the numerical value. Is the system temperature increasing, decreasing, or constant? Explain your answers.

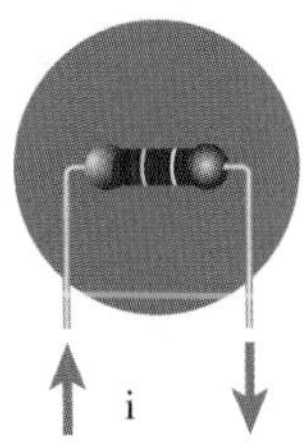

5.73 An electric current of 0.25 A flows through a 5-ohm electrical resistor placed in an evacuated chamber. The resistor is a 30-mm-long cylinder with a diameter of 5 mm. Estimate the steady-state surface temperature of the resistor if it has an emissivity of unity (i.e., is black) and the temperature of the chamber walls is 300 K. Neglect any conduction heat transfer through the lead wires.

5.74 Consider the same situation as in Problem 5.73 except that now air at 300 K fills the chamber. Does the steady-state temperature of the resistor increase or decrease? Explain. Estimate the temperature of the resistor if the average convective heat-transfer coefficient is 2.5 $W/m^2{\cdot}K$.

5.75 Pulverized coal particles at 300 K are injected into hot furnace gases at 1500 K. Estimate the initial heating rate of a 70 μm-diameter particle. Express your result in kelvins per second. The convective heat-transfer coefficient is approximately 1500 $W/m^2{\cdot}K$. The specific heat and density of the coal are 1.3 kJ/kg · K and 1650 kg/m^3, respectively.

5.76 Perform a first-law analysis of the earth to determine the temperature of the earth with and without an atmosphere.

Use the following simplifying assumptions to determine these temperatures. Note that the actual physics is quite complex and that these assumptions greatly oversimplify performing an energy balance of the earth.

i. Steady-state conditions prevail, i.e., the earth is in equilibrium with its surroundings.

ii. The incoming solar radiation (heat interaction) is distributed evenly over the earth's surface. (Can you offer a justification for this assumption?) The apparent incoming energy rate per unit surface area of the earth is $\dot{Q}''_{app} = 341.75\,W/m^2$.

iii. We assume that the earth behaves as a blackbody. This means that 100% of the incoming solar energy is absorbed and that the Earth emits the maximum possible radiation. For this assumption, the outgoing radiation from the earth per unit surface area is σT^4 [= W/m^2], where σ is the Stefan–Boltzmann constant (= 5.67×10^{-8} $W/m^2{\cdot}K^4$). See Chapter 4 in for more details about this.

Consider two cases: (1) there is no atmosphere, and (2), there is an atmosphere that prevents 5% of the outgoing radiation from reaching the top of the atmosphere and being emitted into space but does not affect the incoming radiation. Ignore the radiation from the atmosphere itself.

5.77 The heat-transfer rate to the surroundings from a person at rest is about 100 W. Consider an auditorium containing 2000 people in which the ventilation system fails. Answer the following assuming that there is no heat transfer through the auditorium walls.

(Credit: Caiaimage/Martin Barraud / OJO+ / Getty Images.)

A. Determine the increase in internal energy of the air in the auditorium during the first 20 min after the ventilation system fails. Express your result in MJ.
B. Considering the auditorium and all the people as a system, determine the increase in internal energy (MJ) of the system. How do you explain the fact that the temperature of the air increases?

5.78 A car battery is charged by applying a current of 40 A at 12 V for 30 min. During the charging process, there is a heat transfer of 200 Btu from the battery to the surroundings. Determine the increase in internal energy (Btu) of the battery.

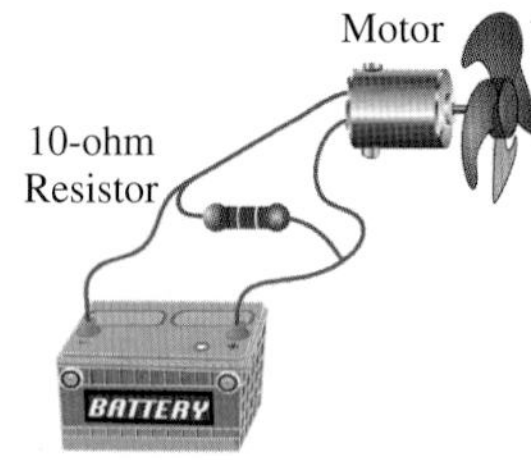

5.79 An electric circuit consists of a 12-V storage battery, a 10-ohm resistor, and an electric motor connected in parallel. If the motor produces 100 mW of power, estimate the rate of energy depletion from the battery.

5.80–5.88 Steady-flow open systems (not device-specific)

5.80 ***Conceptual problem.*** Which of the following are true statements? Justify your responses, true or false, with a sentence or two.
A. Enthalpy appears in the conservation of energy equation for open systems because open systems involve constant-pressure processes.

B. Enthalpy appears in the conservation-of-energy equation for open systems because the internal energies of the inlet and outlet fluid streams are negligible.

C. Flow work (rate) is another name for shaft work (rate).

D. The pressure force at the inlets and outlets is responsible for the flow work (rate).

5.81 Write out in words the precise meaning of the following symbols. Be very specific, using appropriate adjectives as needed. Also provide the usual SI units associated with the quantities. If the quantity is dimensionless, write "dimensionless." For the terms that are products of symbols, give a single word or short phrase describing the grouping (see example).

	Symbol	Meaning (words only)	Usual SI units
Example:	$\rho \mathcal{V}$	mass	kg
A.	$\dot{Q}$		
B.	V (not v or $\mathcal{V}$)		
C.	$\dot{m} P v$		
D.	$u + Pv$		
E.	$\dot{m} V^2/2$		
F.	$\dot{W}_{shaft}$		

5.82 ***Conceptual problem.*** Without reference to the text, write a symbolic expression for the conservation of energy (the first law of thermodynamics) for a open system having a single inlet and a single outlet. Assume steady state. Below each term, write its meaning in words.

5.83 ***Conceptual problem.*** Repeat Problem 5.82, eliminating the assumption of a steady state.

5.84 ***Conceptual problem.*** Explain the origin of enthalpy in the expression of the conservation of energy for an open system.

5.85 Show in excruciating detail that the units associated with the term $V^2/2$ are J/kg.

5.86 Mass flows through a control volume (open system) at 1 kg/s. The enthalpy, velocity, and elevation at entrance are 200 kJ/kg, 30 m/s, and 100 m, respectively. At the exit, these quantities are 198 kJ/kg, 0.3 m/s, and −3 m. Heat is transferred to the system at a rate of 5 kJ/s. How much work is done by this system (a) per kilogram of fluid and (b) per minute? (c) What is the rate of working (power) in kilowatts?

5.87 The inlet and outlet streams of an open system are described by the following information.

	Inlet	Outlet
Velocity	36.58 m/s	12.19 m/s
Elevation	30.48 m	54.86 m
Enthalpy	2791.2 kJ/kg	2795.9 kJ/kg
Mass rate	0.756 kg/s	0.756 kg/s

If the net work rate out is 4.101 kW, what is the heat-transfer rate?

5.88 Saturated steam at 0.3 MPa flows through a 5.08-cm-inside-diameter pipe at a rate of 700 kg/hr. Determine the specific kinetic energy of the steam in kJ/kg.

5.89–5.95 Steady-flow open systems (*ke*/*pe* featured)

5.89 Steam at 580 K and 1.5 MPa enters a well-insulated steady-state device through a standard 3-inch pipeline (inside diameter = 3.068 in) at 10 ft/s. The exhaust from the device flows through a standard 10-in pipeline (inside diameter = 10.02 in) at 360 K and 35 kPa. Determine the power output of the device.

5.90 As a fluid flows steadily past a turbine blade with friction present, the fluid velocity drops from 400 m/s to 100 m/s while the fluid enthalpy increases by 25 kJ/kg. Assuming the process is adiabatic, determine the specific work transfer (kJ/kg) from the fluid to the blade.

5.91 An ideal gas passes steadily through a device that increases the gas velocity from 5 m/s to 300 m/s without transfer of heat or work.

A. Determine the increase in specific enthalpy (kJ/kg) of the gas.

B. If the specific heat of the gas is a linear function of temperature given by c_p [kJ/kg·K] = 1.00 + 0.01T [K], determine the increase in temperature of the gas if it enters the device at 20°C.

5.92 Quite some time ago, Frank Lloyd Wright designed a one-mile-high building. Suppose that, in such a building, steam for the heating system enters a pipe at ground level as dry, saturated vapor at 200 kPa. On the top floor of the building, the pressure in the pipe is 70 kPa. The heat transfer from the steam as it flows up the pipe is 116 kJ/kg of steam. Taking the one-mile-high pipe as a control volume (open system), determine the quality of the steam at the top of the pipe.

5.93 Consider a waterfall having a drop of 84.7 m.

A. Determine the specific potential energy (J/kg) of the water at the top of the waterfall with respect to the base of the waterfall.

B. Assuming no energy is exchanged with the surroundings, determine the velocity (m/s) of the water just before it reaches the bottom.

C. What happens to the kinetic energy of the water after it reaches the bottom?

5.94 Water (assumed to be incompressible) is pumped at a constant rate of 22.7 kg/min through a pipeline that has an internal diameter of 50.8 mm. The pipe discharges through a nozzle that has a diameter of 25.4 mm at the exit and is at an elevation of 30 m above the inlet to the pump. At the inlet to the pump the water is at 294 K and 137.9 kPa. At the exit of the nozzle, the water is at 294 K and 101.3 kPa. Neglecting friction and heat losses, determine the power that must be supplied to the pump.

5.95 The manager of an amusement park at the bottom of Niagara Falls wants to install a water turbine to produce 100 kW. Water (assumed to be incompressible) would enter the pipeline leading to the turbine at 20°C and 0.10135 MPa at the top of the falls, 51 m above the turbine exit, with a velocity of 3 m/s. The water should leave the turbine at 20°C and 0.10135 MPa. Assume the pipeline and the turbine are both adiabatic.

A. Determine the mass flow rate of the water in kg/min.

B. Determine the diameter of the pipeline.

5.96–5.102 Steady-flow open systems (heating/cooling)

5.96 Water at 5 MPa and 400 K enters the steam-generating unit of a power plant and leaves the unit as steam at 5 MPa and 1000 K. The water mass flow rate is 14,000 kg/hr. Determine the capacity of the steam-generating unit in units of kJ/hr and MW.

5.97 Electrical heating elements heat a 5-gal/min flow of water at 30 psig from 62 F to 164 F. The barometric pressure is one standard atmosphere. Determine (a) the wattage required and (b) the current in amperes if a single-phase 220-V circuit is used. Use the NIST software to obtain properties.

5.98 Air, entering at 16°C, is used to cool an electronic compartment. The maximum allowable air temperature is 38°C. If the equipment in the compartment dissipates 3600 W of energy to the air, determine the necessary air flow rates at the inlet in (a) kg/hr and (b) m^3/min.

5.99 Cool air enters a classroom at 12.8°C. The cool air is heated by the students in the room, by the lights in the room, and by heat transfer through the classroom walls. The total heat-transfer rate to the air is 22,156 kJ/hr. The air leaves the room at 25.5°C. Determine the following:

A. The air mass flow rate (kg/hr)

B. The air volumetric flow rate (m^3/hr) at inlet conditions

C. The duct diameter (m) for an entering air velocity of 3.1 m/s

5.100 Dry, saturated steam from a turbine enters a condenser at 4 kPa and exits as saturated liquid, also at 4 kPa. In the condenser, energy is removed from the steam by heat transfer to a stream of lake water. The lake water enters at 5°C and is then returned to the lake at 10°C (the maximum allowed by local regulations). Determine the mass of lake water required per mass of steam condensed.

5.101 Saturated liquid water at 0.3 MPa enters a steady-flow boiler. Inside the boiler, the water is heated at constant pressure to 580 K. The potential and kinetic energies are negligible.

A. Determine the heat input in kJ/kg.

B. If the exit area is 0.2 m^2 and the average exit velocity is 30 m/s, determine the mass flow rate in kg/s.

5.102 Air undergoes a steady-flow, constant-pressure heating process at 150 kPa from 20°C to 150°C. Determine the increase in the specific internal energy of the air and the mass-specific heat addition, both in kJ/kg.

5.103–5.110 Steady-flow open systems (turbine)

(Credit: Michael Dwyer / Alamy Stock Photo.)

5.103 Superheated steam (8 MPa, 900 K) enters a turbine with a flow rate of 0.16 kg/s. The steam exits at 16 kPa. Determine the power produced by the turbine if the expansion process is isentropic, i.e., the mass-specific entropy of the exiting steam equals the mass-specific entropy of the entering steam. Neglect all heat losses and changes in kinetic and potential energies.

5.104 In a coal-fired power plant, steam enters a turbine at 800 K and 4 MPa at a rate of 189 kg/s. The steam expands and exits the turbine at 100 kPa with a quality of 0.97. Determine the power produced by the turbine in kilowatts.

5.105 Superheated steam at 780 K and 3 MPa enters a steady-state turbine with a mass flow rate of 500 kg/s. The steam leaves the turbine at 560 K and 0.7 MPa. The turbine is not well insulated, resulting in a heat-transfer rate of 660 kW to the atmosphere.

A. Determine the power delivered by the turbine in MW.

B. Determine the ratio of the heat loss rate to the shaft power expressed as a percentage.

C. Solve the problem again, including the kinetic energies associated with the inlet and exit streams, where $V_1 = 85$ m/s and $V_2 = 25$ m/s. Determine the percentages of the shaft power represented by the kinetic energy rates at the inlet and outlet, respectively.

5.106 Air enters an adiabatic steady-flow gas turbine at 650 K and 550 kPa and leaves at 430 K and 100 kPa. The mass flow rate of the air is 45.5 kg/s. Treat the air as an ideal gas and use average specific heats.
A. Neglecting the potential energy and kinetic energy at the inlet and outlet, determine the power (kW) produced by the turbine.
B. If the air enters the turbine through a pipe with a flow area of 0.07 m^2 and exits through a 0.37-m^2 pipe, determine the increase in kinetic energy rate (kW) of the air.

5.107 Air at 400°C and 0.4 MPa is steadily supplied to an adiabatic gas turbine. The air leaves the turbine at 200°C and 0.10135 MPa. Neglecting changes in potential energy and kinetic energy, determine the following:
A. The ratio of the exit flow area to the inlet flow area required for an exit velocity equal to the inlet velocity
B. The specific work (kJ/kg), assuming constant specific heats

5.108 Steam enters an adiabatic turbine at 800 K and 6.0 MPa and leaves at 700 K and 3.0 MPa. The mass flow rate is 450 kg/hr. The potential and kinetic energies can be neglected. Determine the following:
A. The power delivered (kW)
B. The ratio of the outlet flow area to the inlet flow area needed to keep the exit velocity equal to the inlet velocity

5.109 During the operation of a steam power plant, the steam flow rate is 227,000 kg/hr with turbine inlet conditions of 4 MPa and 800 K and turbine exhaust (condenser inlet) conditions of 8 kPa and 90% quality. The H_2O exits the condenser as a saturated liquid. Determine (a) the turbine output in kW and (b) the condenser heat-rejection rate in kW.

5.110 During the operation of a steam power plant, steam enters the turbine with a flow rate of 230,000 kg/hr at 3.5 MPa and 550°C. The turbine exhaust (condenser inlet) condition is 0.01 MPa with 85% quality. The steam (water) exits the condenser as a saturated liquid at 0.01 MPa. Determine (a) the turbine output in kW and (b) the condenser heat-rejection rate in kJ/hr. Use the NIST software to obtain fluid properties.

5.111–5.117 Steady-flow open systems (pump/compressor)

5.111 What is the minimum-power motor (hp) that would be necessary to operate a pump that handles 85 gal/min of city water while increasing the water pressure from 15 to 90 psia? The water is at room temperature. Be sure to list your assumptions. **HINT**: Assume that the water is incompressible and that the temperature rise across the pump is negligible.

(Credit: Baloncici / iStock / Getty.)

5.112 A pump is used to remove water after a flood. Estimate the mass-specific work (kJ/kg) required to operate this pump, assuming the process is adiabatic and steady-flow. The water enters the pump at 0.0689 MPa and 292 K and leaves at 3.516 MPa. The temperature increase across the pump is 0.05 kelvins. Assume no changes in the kinetic or potential energies. Use NIST or other software to obtain water properties.

5.113 A pump is used to supply water at a rate of 3 kg/s to a spray-cooling system used in a heat-treating operation. Estimate the power required to operate this pump, assuming the process is adiabatic and steady-flow. The water enters the pump at 200 kPa and 295 K and leaves at 800 kPa. The temperature increase across the pump is 0.01 kelvins.

A. Estimate the minimum power required to operate the pump. Assume changes in the kinetic and potential energies are small compared to the power required. Use NIST or other software to obtain water properties.

B. Assuming the inlet velocity is negligible, what would the outlet velocity have to be if the exiting kinetic energy rate were 1 percent of the pump power?

(Credit: loraks / iStock / Getty Images.)

5.114 Estimate the power required to compress a steady flow of air ($\dot{m} = 0.02$ kg/s) from 100 kPa and 300 K to 598 kPa at 500 K. Neglect changes in the kinetic and potential energies of the air stream.

5.115 Air at a rate of 18 kg/s is drawn into the compressor of a jet engine at 55 kPa and −23°C and is compressed reversibly and adiabatically to 276 kPa. For an adiabatic reversible compression, the inlet state and exit state entropies are the same, i.e., $s_1 = s_2$. Determine the required power input to the compressor expressed in horsepower. Use the NIST software to obtain air properties.

5.116 Air enters a steady-flow adiabatic compressor at 15°C and 0.1 MPa at 2 m^3/s. The air leaves at 150°C and 0.385 MPa. Neglecting changes in potential energy and kinetic energy, determine the power (kW) required to drive the compressor. Use the NIST software to obtain air properties.

5.117 Steam flowing at 450 kg/hr is compressed from 480 K and 0.5 MPa to 760 K and 1.5 MPa in an adiabatic steady-flow process. Determine the input power required in kilowatts. Neglect changes in potential energy and kinetic energy.

5.118–5.124 Steady-flow open systems (nozzle)

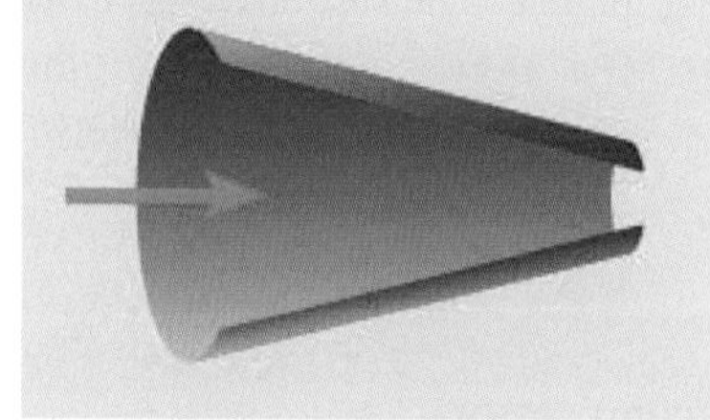

5.118 Air enters a nozzle at 300 K with a velocity of 10 m/s. Determine the temperature at the nozzle exit, where the velocity is 250 m/s. Work this problem first assuming a constant specific heat c_p of 1.005 kJ/kg·K. Repeat using the air tables in Appendix C.

5.119 Low-velocity steam (with negligible kinetic energy) enters an adiabatic nozzle at 580 K and 3 MPa. The steam leaves the nozzle at 2 MPa with a velocity of 400 m/s. The mass flow rate is 0.4 kg/s. Determine the quality and temperature (K) of the steam leaving the nozzle and the exit area of the nozzle in mm^2.

5.120 After flowing from the combustion chamber and expanding through the turbine of a jet engine, combustion products at low velocity and 1150 K enter the jet nozzle. Determine the maximum velocity (m/s) that can be obtained at the nozzle exit if the products discharge at 640 K. Approximate the thermodynamic properties of the combustion products as those of air.

5.121 Steam at 0.7 MPa and 480 K enters a rigid, insulated nozzle with a velocity of 70 m/s. The steam leaves as a saturated vapor at a pressure of 0.15 MPa. Determine the exit velocity.

5.122 Steam at 1.5 MPa and 600 K enters a rigid, insulated nozzle with a velocity of 90 m/s. The steam leaves at a pressure of 0.3 MPa and a temperature of 440 K. Determine the exit velocity.

5.123 The mass flow rate of air into a nozzle is 100 kg/s. The discharge pressure and temperature are 0.1 MPa and 270 °C, respectively, and the inlet conditions are 1.4 MPa and 800 °C. The inlet velocity is negligible, and the nozzle is well insulated. Determine the outlet diameter of the nozzle in meters.

5.124 A gas expands through an adiabatic nozzle. During the expansion there is a decrease in specific enthalpy of 116 kJ/kg from entrance to exit.

A. If the initial velocity of the gas entering the nozzle is nearly zero, determine the exit velocity in m/s.

B. If the initial velocity of the gas entering the nozzle is 30 m/s, determine the exit velocity in m/s.

5.125–5.126 Steady-flow open systems (throttle)

5.125 When the pressure in a steam line reaches 0.7 MPa, the pressure-relief safety valve opens and releases steam to the atmosphere (100 kPa) in a constant-enthalpy process across the valve. The temperature of the escaping steam (after the valve) is measured as 880 K. Determine the temperature (K) and the specific volume (m^3/kg) of the steam in the line. Also identify the state region (e.g., superheated vapor, etc.) of the steam in the line.

5.126 H_2O is throttled through an expansion valve from a saturated liquid state at 375 K to a wet mixture at a temperature of 325 K. What is the quality of the steam after passing through the expansion valve? Assume the process occurs adiabatically with no work interaction. Also neglect changes in kinetic and potential energy changes across the valve.

5.127–5.133 Steady-flow open systems (multiple streams)

5.127 Steam enters a turbine at a rate of 90,720 kg/hr with an enthalpy of 3038.6 kJ/kg. It leaves the turbine at two exits. The mass flow rate at the first exit is 22,680 kg/hr with an enthalpy of 2748.6 kJ/kg. At the second exit, the enthalpy is 2357.9 kJ/kg. The only other energy transfer is work. There is no storage of mass or energy within the turbine. Determine the power (kW) delivered by the turbine.

5.128 Consider a steam turbine in which steam enters at 10.45 MPa and 780 K with a flow rate of 38.739 kg/s. A portion of the steam is extracted from the turbine after partial expansion at three different locations, as shown in the sketch. The extracted steam is then led to various heat exchangers. The mass flow rates and the temperatures and pressures at each extraction point are given in the following table.

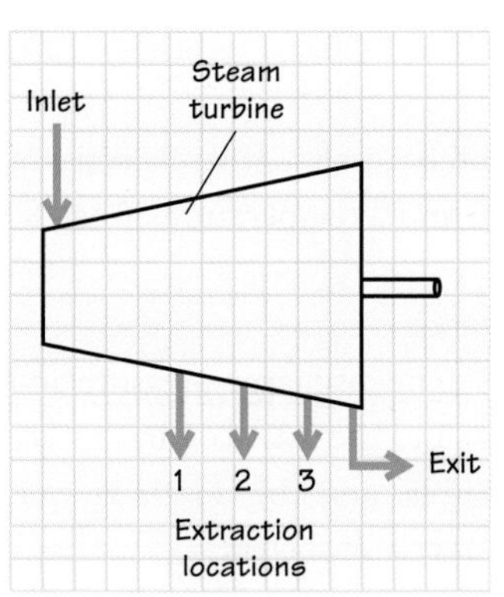

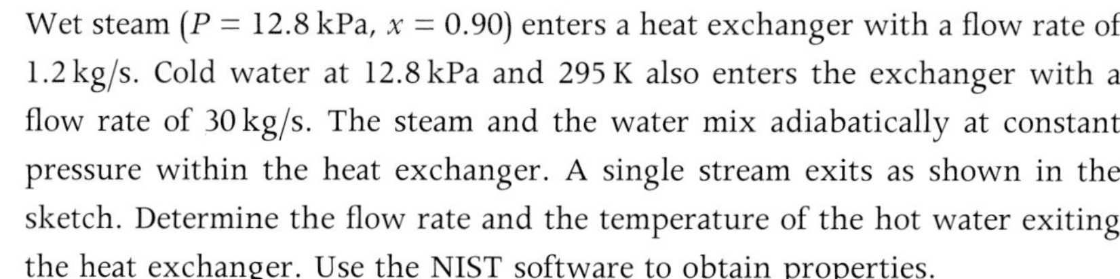

Extraction Location	$\dot{m}$ (kg/s)	P (MPa)	T (K)
1	4.343	3.054	620
2	4.345	0.332	482
3	2.871	0.136	$x = 0.949$

The remaining wet steam exits the turbine at 11.5 kPa with a quality of 0.88. Estimate the power produced by the turbine assuming adiabatic operation and negligible kinetic and potential energies for all streams. Use the NIST software to obtain steam properties.

5.129 Wet steam ($P = 12.8$ kPa, $x = 0.90$) enters a heat exchanger with a flow rate of 1.2 kg/s. Cold water at 12.8 kPa and 295 K also enters the exchanger with a flow rate of 30 kg/s. The steam and the water mix adiabatically at constant pressure within the heat exchanger. A single stream exits as shown in the sketch. Determine the flow rate and the temperature of the hot water exiting the heat exchanger. Use the NIST software to obtain properties.

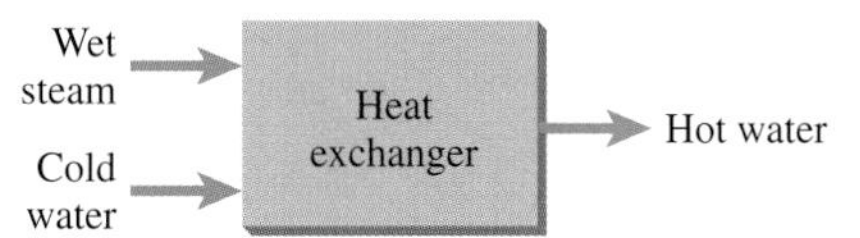

5.130 Water at 366.5 K and 206.8 kPa flows into a rigid, insulated tank through pipe 1 at a rate of 45.4 kg/s. Steam at 477.6 K and 206.8 kPa flows into the same tank through pipe 2 at a rate of 90.7 kg/s. The two flows mix together within the tank and leave through pipe 3 at 206.8 kPa. This is a steady-flow process with negligible potential and kinetic energies. Determine the temperature (and quality, if applicable) of the flow leaving through pipe 3. Use the NIST software for properties.

5.131 Superheated steam at 400 °C and 1.6 MPa flows steadily into a control volume (open system) at a rate of 0.2 kg/s. A second stream (dry, saturated water vapor at 1.6 MPa) enters the control volume steadily at 0.1 kg/s. The control volume is adiabatic and the pressure at the only exit stream is 1.6 MPa. The exit flow is steady. Determine the mass flow rate (kg/s) and temperature (°C) of the exit flow. Use the NIST software for properties.

5.132 Water at 283K and 3.45 MPa enters a steady-flow control volume (open system) at a rate of 0.91 kg/s. Wet steam at 101.325 kPa with a quality of 0.25 also enters the control volume but at a rate of 0.45 kg/s. Steam at 645 K and 4.14 MPa leaves the control volume. The work rate into the control volume is 2640 kW. Determine the rate of heat transfer (kW). Is the heat flow into or out of the control volume? Use the NIST software for properties.

5.133 In certain situations when only superheated steam is available, a need for saturated steam may arise. This need can be met in an adiabatic desuperheater, in which liquid water is sprayed into the superheated steam in such amounts that dry, saturated steam leaves the desuperheater. The following data apply to such a steadily operating desuperheater. Superheated steam at 580 K and 3 MPa enters the desuperheater at 0.25 kg/s. Liquid water enters the desuperheater at 320 K and 5 MPa. Dry, saturated vapor leaves at 3 MPa. Determine the mass flow rate (kg/s) of liquid water.

5.134–5.140 Unsteady open systems

5.134 A well-insulated, 80-gal (0.3028 m^3), electric hot-water heater initially contains water at 322 K (120 F). Hot water is withdrawn from the tank at 3 gal/min, while cold water at 283 K (50 F) enters to keep the tank full. The cold water mixes completely with the water in the tank at each instant. Because of a power outage, there is no energy input to heat the incoming cold water. Estimate the time required for the temperature of the outgoing water to reach 299.8 K (80 F). Assume constant water properties with $\rho = 994\ \text{kg/m}^3$ and $c_p = c_v = 4149\ \text{J/kg}\cdot\text{K}$.

5.135 A tank having a volume of 5.66 m^3 (200 ft^3) contains saturated water vapor at 138 kPa (20 psia). A line is attached to the tank in which vapor flows at 689 kPa (100 psia) and 478 K (400 F). Steam from this line enters the tank until the pressure is 689 kPa (100 psia). Calculate the mass of steam that enters the tank assuming that the process is adiabatic and that the heat capacity of the tank is negligible. Use the NIST software for properties.

5.136 A 216-in^3 tank initially contains saturated water vapor at 0.143 MPa. A line is attached to the tank in which vapor flows at 0.7 MPa and 200°C. Steam from this line enters the tank until the pressure in the tank is 0.7 MPa. Calculate the mass of steam that enters the tank assuming that the process is adiabatic and that the heat capacity of the tank is negligible. Use the NIST software for properties.

5.137 A rigid tank contains water (liquid and vapor) at 589 K (600 F). Liquid is withdrawn from the bottom at a slow rate. The cross-sectional area of the tank is 323 cm^2 (50 in^2) and the liquid level drops 15.2 cm (6 in). During this time heat transfer maintains the temperature at 589 K (600 F). Neglecting changes in potential energy, determine the heat transfer (kJ). Use the NIST software for properties.

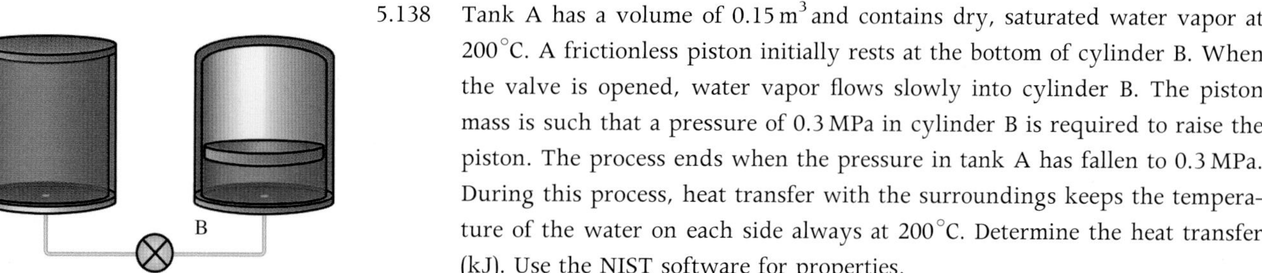

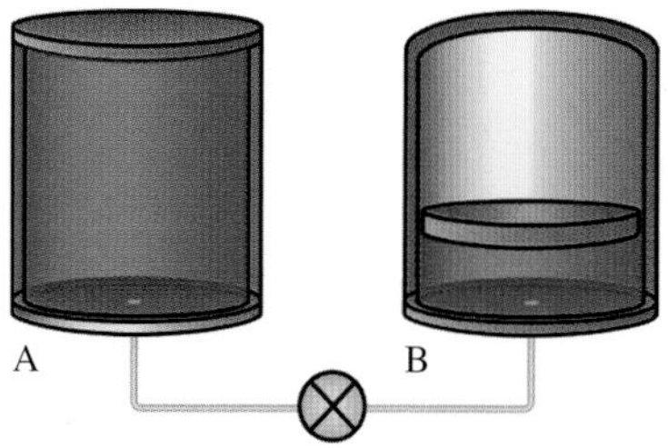

5.138 Tank A has a volume of 0.15 m^3 and contains dry, saturated water vapor at 200°C. A frictionless piston initially rests at the bottom of cylinder B. When the valve is opened, water vapor flows slowly into cylinder B. The piston mass is such that a pressure of 0.3 MPa in cylinder B is required to raise the piston. The process ends when the pressure in tank A has fallen to 0.3 MPa. During this process, heat transfer with the surroundings keeps the temperature of the water on each side always at 200°C. Determine the heat transfer (kJ). Use the NIST software for properties.

5.139 Consider the process of filling a scuba diving tank with air. The tank is initially at T_1 and P_1. A large reservoir can supply air at $T_r > T_1$ and $P_r > P_1$. The scuba tank is connected to the reservoir and quickly filled to P_r. Assume the filling process is adiabatic. The tank temperature returns to T_1 as a result of heat transfer with the surroundings. Assuming air is an ideal gas with constant specific heats (c_v and c_p), derive an expression for the final pressure in the tank in terms of known quantities.

5.140 Initially, 0.1 kg of water at 40°C and 0.2 MPa is contained in an insulated piston–cylinder device. During a time interval of 35 s, 0.1 kg of steam is added to the contents of the cylinder through the valve from a source at 250°C and 0.5 MPa. Assume the pressure within the cylinder is constant during the process. Determine the final volume (m^3) of the cylinder contents.

5.141–5.143 System analysis

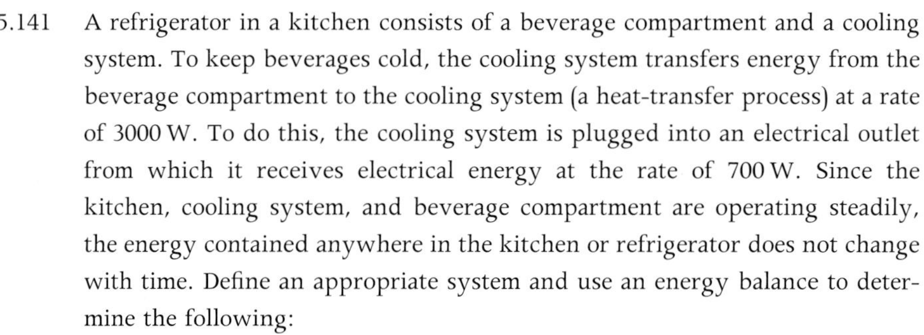

5.141 A refrigerator in a kitchen consists of a beverage compartment and a cooling system. To keep beverages cold, the cooling system transfers energy from the beverage compartment to the cooling system (a heat-transfer process) at a rate of 3000 W. To do this, the cooling system is plugged into an electrical outlet from which it receives electrical energy at the rate of 700 W. Since the kitchen, cooling system, and beverage compartment are operating steadily, the energy contained anywhere in the kitchen or refrigerator does not change with time. Define an appropriate system and use an energy balance to determine the following:

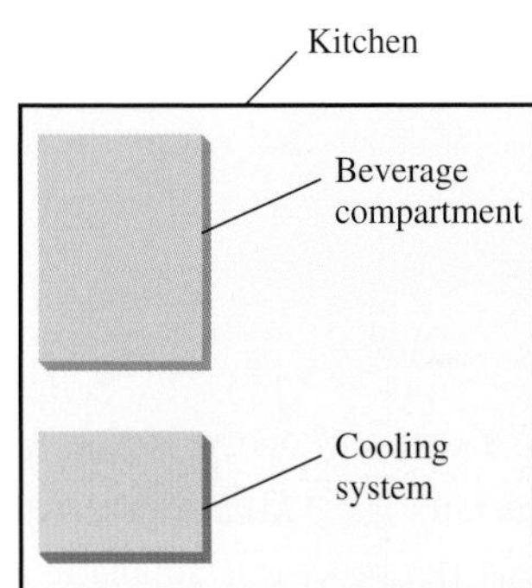

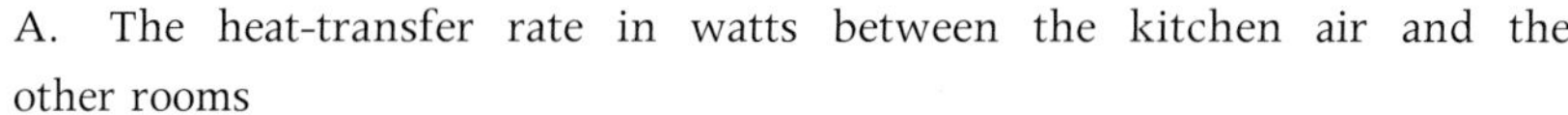

A. The heat-transfer rate in watts between the kitchen air and the other rooms

B. The heat-transfer rate in watts between the cooling system and the kitchen air

C. The heat-transfer rate in watts between the beverage compartment and the kitchen air

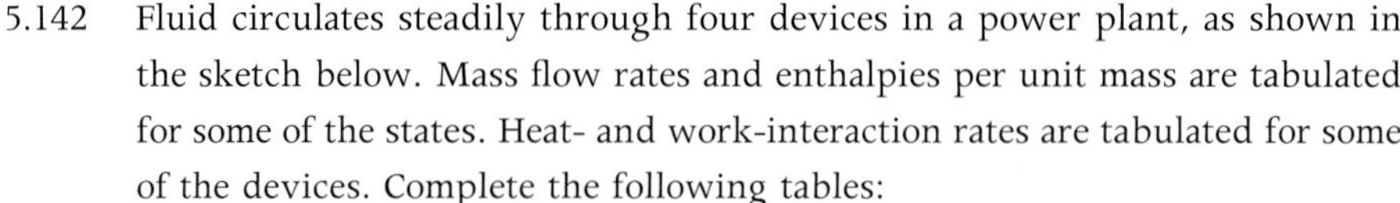

5.142 Fluid circulates steadily through four devices in a power plant, as shown in the sketch below. Mass flow rates and enthalpies per unit mass are tabulated for some of the states. Heat- and work-interaction rates are tabulated for some of the devices. Complete the following tables:

State	$\dot{m}$ (kg/s)	h (J/kg)
1		15
2		13
3	25	9
4		
5	5	

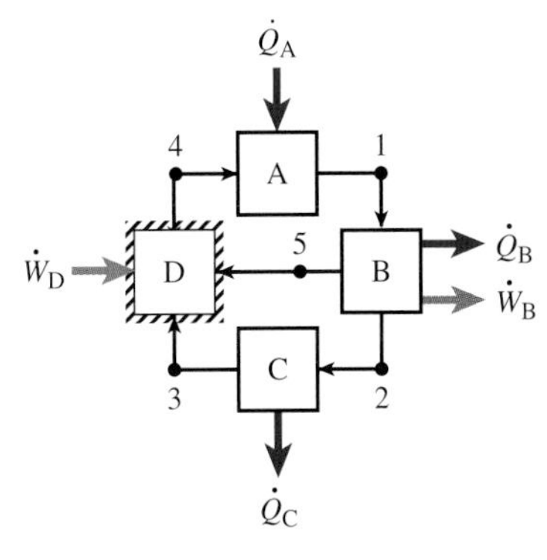

Device	$\dot{Q}$(W)	$\dot{W}$(W)
A	150	0
B	30	
C		0
D	0	5

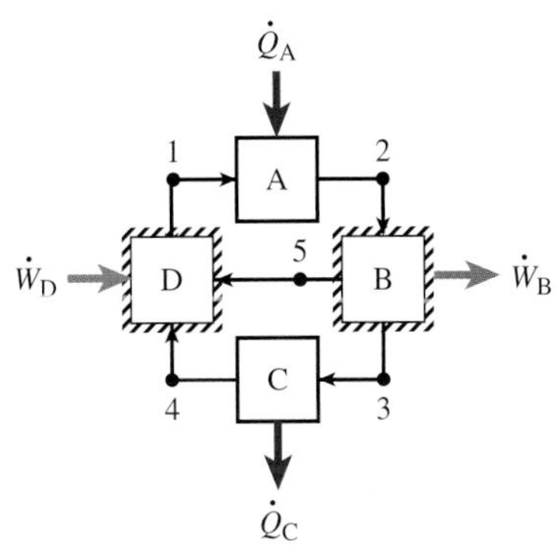

5.143 Fluid circulates steadily through four devices in a power plant, as shown in the sketch alongside. All transfers of heat and work are indicated. The mass flow rates at states 1 and 5 are 45.5 kg/s and 9.07 kg/s, respectively. The fluid enthalpies per unit mass at states 1 through 5 are 9.30, 23.25, 11.6, 2.33, and 16.28 kJ/kg, respectively. Determine the power delivered by device B in kW.

See Chapter 8 for additional problems dealing with steady-flow devices.

5.144–5.147 EES problems

5.144 Steam is throttled from saturated liquid at 500 K to a temperature of 300 K. Use EES or NIST tables to find the quality of the steam after passing through the expansion valve. Assume $\Delta PE = \Delta KE = Q = W = 0$.

5.145 Use your EES program from Problem 5.144 to solve for the steam outflow quality for eight outflow temperatures from 500 K to 300 K. Use these steam conditions to plot a smooth curve showing the process on T–v and T–s diagrams. Include the vapor dome in your plots.

5.146 A refrigeration cycle includes an expansion valve (throttle) of saturated liquid R-134a at 20°C to a temperature of −20°C. (See Figures 9.57 and 9.58 for the complete refrigeration cycle.) Use EES or NIST tables to find the quality of the refrigerant after passing through the expansion valve. Assume $\Delta PE = \Delta KE = Q = W = 0$.

5.147 Use your EES program from Problem 5.146 to solve for the R-134a outflow quality for eight outflow temperatures from 20°C to −20°C. Use these R-134a conditions to plot a smooth curve showing the process on T–v and T–s diagrams. Include the vapor dome in your plots.

5.148–5.155 FE problems

5.148 2 kg water in a cylinder is heated from saturated vapor to a temperature of 520 K during a constant-pressure process at 150 kPa. What is the heat transfer during the process? a. 288 kJ, b. 547 kJ, c. 144 kJ, d. 273 kJ.

5.149 An adiabatic turbine with a flow rate of 4 kg/s has an inflow air temperature of 700 K and outflow air temperature of 500 K. What is the power output from the turbine? a. 574 kW, b. 144 kW, c. 201 kW, d. 804 kW.

5.150 A 0.2 m^3 volume of saturated water vapor at 500 kPa expands to twice its volume in an isobaric process. What is the heat transfer to the system during this process? a. 241 kJ, b. 441 kJ, c. 2200 kJ, d. 341 kJ.

5.151 An adiabatic compressor with a flow rate of 6 kg/s has an inflow air temperature of 350 K and outflow air temperature of 670 K. What is the power required to drive the compressor? 1.93 MW, b. 1.38 MW, c. 322 kW, d. 230 kW.

5.152 The exit of a jet engine contains a nozzle to accelerate the flow from 67 m/s to 90 m/s. If the air entering the nozzle is at 600 K, what is the temperature of the air exiting the nozzle? a. 602 K, b. 598 K, c. 580 K, 540 K.

5.153 A steam turbine with a flow rate of 0.4 kg/s has inflow conditions of 660 K and 1.0 MPa. If the outflow from the turbine is saturated vapor at 70 kPa, determine the power output assuming negligible heat transfer. a. 231 kW, b. 577 kW, c. 237 kW, 124 kW.

5.154 A 0.5 m^3 volume of air in a piston at 250 kPa is compressed to one-third of its volume in an isothermal process. What is the heat transfer from the system during this process? a. 83 kJ, b. 53 kJ, c. 137 kJ, d. 274 kJ.

5.155 4 kg of saturated water vapor at 350 K in a rigid tank is heated to 500 K. What is the heat transfer during the process? a. 267 kJ, b. 222 kJ, c. 785 kJ, d. 889 kJ.

CHAPTER 6
Second Law of Thermodynamics and Some of Its Consequences

LEARNING OBJECTIVES

After studying Chapter 6, you should be able to:

- State in words the Kelvin–Planck statement of the second law of thermodynamics and at least one other meaningful statement of the second law.
- State two or more ways in which the second law is useful in engineering applications.
- Explain the difference between a reversible and an irreversible process and illustrate it with a concrete example.
- List four or more irreversible processes.
- Draw a sketch of a heat engine that includes labeling the direction of the energy transfer terms to the high- and low-temperature reservoirs.
- Draw a sketch of a heat pump that includes labeling the direction of the energy transfer terms to the high- and low-temperature reservoirs.
- Calculate the thermal efficiencies of reversible heat engines
- Calculate the coefficient of performance for a heat pump.
- Calculate the coefficient of performance for an air conditioner.
- Relate temperatures and heat transfers for a reversible heat engine.
- Relate temperatures and heat transfers for a reversible air conditioner.
- Enjoy the beauty of the second law of thermodynamics.

CHAPTER 6 OVERVIEW

THE FIRST LAW of thermodynamics can be mathematically expressed in a variety of ways. All these expressions, however, are easily viewed as rearrangements of the statement that energy can neither be created nor destroyed but only converted from one form to another. In contrast, there is no single universally agreed upon statement of the second law of thermodynamics. Kline [1] indicates that many seemingly different statements have been accepted as the second law, all of which, however, can be shown to be equivalent after a careful and sometimes subtle application of logic. This multiplicity of apparently disparate statements can lead to confusion in understanding the second law. In this chapter, we examine several statements of the second law and discuss the consequences of each.

To understand these various second-law statements and their consequences, we define and develop many concepts in this chapter. These include various devices that execute thermodynamic cycles (heat engines, heat pumps, and refrigerators), the distinctions between reversible and irreversible processes, the thermodynamic temperature scale, and second-law-based maximum theoretical efficiencies for cycles and individual components.

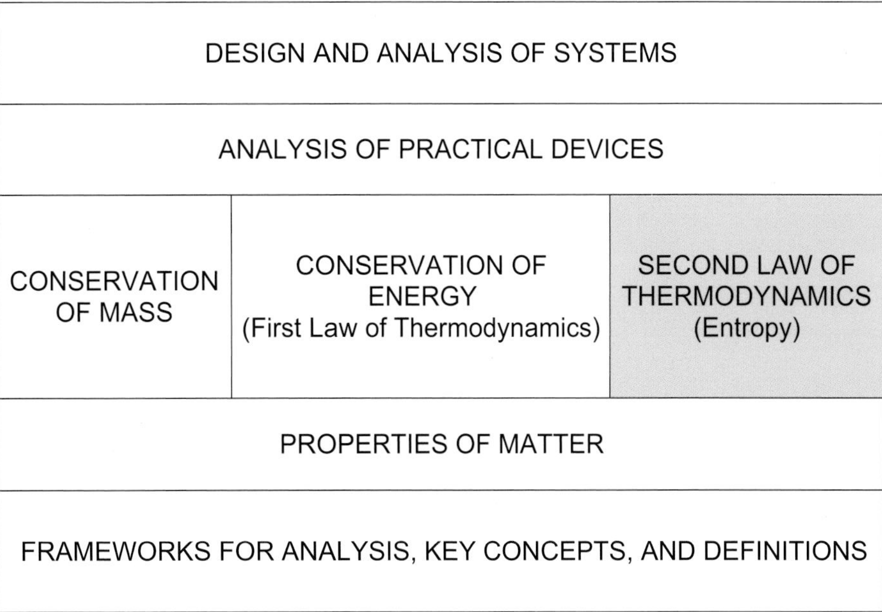

FIGURE 6.1 Hierarchical arrangements of the topics in our study of engineering thermodynamics. Chapter 6 deals with the second law of thermodynamics.

There are two common approaches to introducing second-law analysis. These two different approaches are covered in two different chapters of this text. Chapter 6 applies the second law to cycle analysis and the Kelvin–Planck statement. Chapter 7 introduces the definition of entropy, Gibbs free energy, and availability. These two approaches are presented in two separate chapters so that an instructor can choose the order of presentation. Chapter 7 was written to be independent of Chapter 6 so that it can be used to begin second-law analysis using the definition of entropy. If Chapter 7 is covered first, it is suggested that the reader return to Chapter 6 after Section 7.4d, "A Closed System Undergoing a Cycle".

Figure 6.1 illustrates the key role that Chapter 6 plays in our study. Although this chapter is relatively short, the ideas that it presents form some of the pillars of engineering analysis.

Historical Context

Sadi Carnot (1796–1832). (Credit: gameover / Alamy Stock Photo.)

Rudolf Clausius (1822–1888). (Credit: A. R. Thayer, from a portrait by Bailly, courtesy AIP Emilio Segre Visual Archive, Physics Today Collection.)

Sadi Carnot (1796–1832), a French military engineer during the rise and fall of Napoleon, worked to understand the relationship between the heat supplied to an engine and the work it produced. Steam engines during Carnot's time converted only about six percent of the energy in the fuel to useful work [2]. In 1824, Carnot published his now celebrated paper in which he set forth the idea that a reversible cycle is the most efficient of any cycle and that the efficiency of the conversion of heat to work depends only on the temperatures at which an engine receives and rejects heat. Although Carnot worked under the misconception of the caloric theory, his ideas concerning the second law have stood the test of time as fundamental principles. Today we still use the *Carnot efficiency* as a limiting case for practical devices.

Rudolf Clausius (1822–1888) recognized the importance of Carnot's work, and building upon this, he presented a clear statement of the second law, a statement that we will explore in some detail. Although others made significant contributions to the development of the second law, Clausius is frequently regarded as its discoverer because of his naming of the property *entropy* [2]. Because the word *energy* is central to the first law of thermodynamics, he chose a like-sounding

William Thomson (1824–1907), Baron Kelvin of Largs, Scottish mathematician and physicist. (Credit: Photos.com / Getty Images.)

Ludwig Boltzmann (1844–1906). (Credit: A. R. Thayer, from a portrait by Bailly, courtesy AIP Emilio Segre Visual Archive, Segre Collection.)

word, entropy, to designate the property that is central to the second law. Other key figures in the development of the second law are the following.

William Thomson (Lord Kelvin) (1824–1907) used second-law concepts to define an absolute thermodynamic temperature scale, and **Josiah Willard Gibbs** (1839–1903) a Yale professor, rigorously generalized and extended thermodynamic concepts to reacting systems. **Ludwig Boltzmann** (1844–1906), one of the originators of statistical mechanics, derived a physical meaning for the entropy of gases from the behavior of assemblies of molecules. His famous equation $S = k \ln W$, which states that the entropy of a system is a measure of the probability of its state, is engraved on his tombstone. **Max Planck** (1858–1947), in addition to winning the Nobel Prize for his quantum theory of energy, advanced the understanding of the second law and entropy by approaching the subject from a macroscopic point of view, that is, from a point of view that does not take into account the detailed behavior of the atoms and molecules comprising a system. **Joseph Keenan** (1900–1977) is often credited as a modern codifier of second-law concepts for engineers [3] and with introducing irreversibility as a thermodynamic property [1].

6.1 Usefulness of the Second Law

To provide a firm focus for the second law, we set out the following two ways in which the second law is particularly useful to engineers:

- The second law establishes the theoretical limits of performance of cycles, engines, and other energy conversion devices – over and above those imposed by energy conservation – and provides a means to quantitatively compare real devices with these theoretical ideals. This approach is explored in Chapter 6.
- The second law determines the direction for any spontaneous change and, furthermore, can be used to determine the equilibrium state of any system. This approach is applied in Chapter 7.

The second-law analysis topics can be started from Chapter 6 or Chapter 7.

The second law is useful in many other ways as well; for example, it provides a means to define a thermodynamic temperature scale that is independent of the properties of any substance. Although we discuss this and other uses, a major objective of these chapters is for the reader to develop an appreciation of the two particular uses just stated.

6.2 One Fundamental Statement of the Second Law

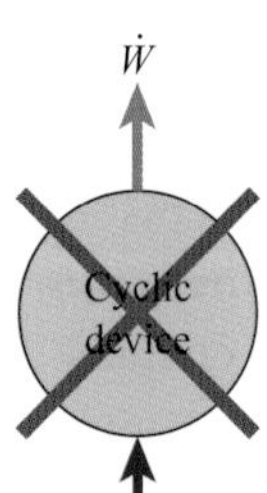

FIGURE 6.2 Statement IB of the second law: It is impossible to construct a cyclically operating device for which the sole effect is the exchange of heat with a single reservoir and the creation of an equivalent amount of work.

As indicated in the chapter overview, the second law of thermodynamics can be stated in many ways. We choose the following statement to be a useful starting point for engineering purposes:

Statement IA: Although all work can be converted completely to heat, heat cannot be completely and *continuously* converted into work.

This statement paraphrases the often quoted and more precise **Kelvin–Planck statement** of the second law [4]:

Statement IB: It is impossible to construct a cyclically operating device for which the sole effect is the exchange of heat with a single reservoir and the creation of an equivalent amount of work (Fig. 6.2).

The immediate and obvious consequence of statement IA is that, in effect, not all forms of energy are equally valuable. Here we see that work is inherently a more valuable form of energy than is heat because we can always convert all work to heat, but not vice versa. The first law of thermodynamics, the conservation of energy principle, puts no limits on interconversion, only that energy cannot appear or disappear, that is, our energy accounting ledger must always balance.

Chapter 4 presents precise definitions of heat and work.

We could gain some insight into this qualitative difference between heat and work by returning to our discussion of these two energy transfer modes in Chapter 4. There we saw that, in a gas, heat transfer by conduction is a result of the random collisions of molecules and requires no macroscopic organization of these molecules to cause the energy exchange that we call heat transfer; however, for work to be extracted from a volume of gas requires a macroscopic organization superimposed on the random molecular motion. This macroscopic organization is the macroscopic flow velocity resulting from a moving boundary. As illustrated in Fig. 4.4, the system boundary must be moving for energy to be extracted from the molecules that collide with it. No such organized motion is required for heat transfer to take place. That energy has quality (value) as well as quantity has important implications for engineering design.[1]

See Chapter 1 to review the definitions of thermodynamic processes and cycles.

A second consequence of statements IA and IB is that they place no restriction on the exchange of heat for work in a *process;* that is, all heat can be converted to work in a process. They do, however, place a severe restriction on devices that execute a thermodynamic cycle, that is, a series of processes that returns the working fluid to its initial state. This restriction to cycles is explicit in the Kelvin–Planck statement (statement IB) and implicit in statement IA by the use of the word *continuously*.

[1] Quality as used here should not be confused with the quality (x) used to describe a liquid–vapor mixture.

Example 6.1 Isothermal Ideal Gas in a Piston–Cylinder

Consider a piston–cylinder arrangement filled with an ideal gas. Show that, for a quasi-static, isothermal, heat-addition process, all the heat added is converted to work.

Ideal gas

Solution

Sketch

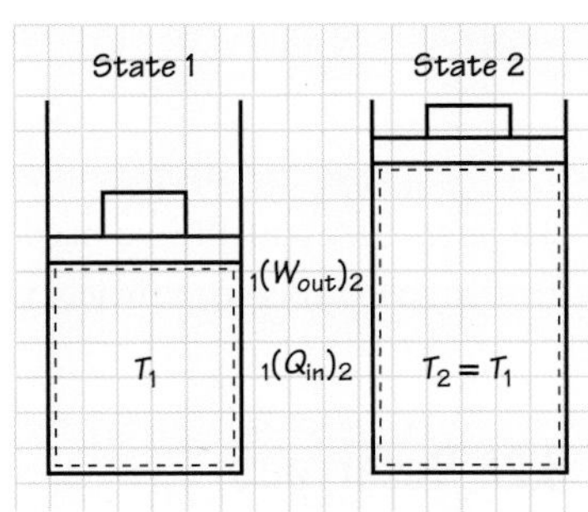

Modeling, Premises and Assumption Changes in kinetic and potential energies are negligible.

Analysis We define the gas in the cylinder to be the thermodynamic system of interest. Conservation of energy relates the heat added and the work performed by the gas (Eq. 5.6) as follows:

$$_1(Q_{\text{in, net}})_2 - {}_1(W_{\text{out, net}})_2 = \Delta E_{\text{syst}}.$$

With negligible changes in kinetic and potential energies, the change in the system energy for the process is expressed as

$$\Delta E_{\text{syst}} = U_2 - U_1 = m(u_2 - u_1).$$

For an ideal gas, the internal energy for a process is expressed by Eq. 2.31d:

$$u_2 - u_1 = \int_{T_1}^{T_2} c_v dT.$$

As the given process is isothermal (i.e., $T_1 = T_2$), we conclude that

$$u_2 - u_1 = 0.$$

Energy conservation is thus

$$_1(Q_{\text{in}})_2 - {}_1(W_{\text{out}})_2 = 0,$$

or

$$_1(W_{\text{out}})_2 = {}_1(Q_{\text{in}})_2.$$

We thus conclude that, for this particular process, all the heat added is converted to work.

Comment The purpose of this example is to illustrate that for a thermodynamic process it is indeed possible to convert all heat added to work. Later in this chapter, we see that, for a continuously operating device, the second law restricts how much of the heat added can be converted to work. The important distinction here is between a thermodynamic process and a thermodynamic cycle. We discuss this later.

Self-Test 6.1

Consider a piston–cylinder device and a rigid tank, both containing saturated liquid water at room temperature. Heat is added to both devices until a saturated vapor state exists. Is heat converted entirely to work in both cases?

(Answer: For the piston-cylinder device, boundary work is performed, but the internal energy of the system has also increased because $\Delta U = m(u_g - u_f) \neq 0$; for the rigid tank, no work is performed and all the heat transfer results in an increase in the internal energy of the water.)

To explore the deeper meaning and significance of the second law given by statements IA and IB, we need to define formally what is meant by heat reservoirs, heat engines, thermal efficiency, and reversibility.

6.2a Reservoirs

A **heat reservoir** is a source of heat energy that is sufficiently large that the extraction of any desired amount of energy as heat does not change the temperature of the reservoir. For most practical purposes, the earth's atmosphere, oceans, lakes, and rivers can be considered heat reservoirs or thermal reservoirs (Fig. 6.3). Large amounts of heat can be added or removed from these without any significant change in their temperatures. For example, an air conditioning unit exchanges heat with the atmosphere without affecting the temperature of the atmosphere.[2] Frequently a boiling or condensation phase change has the practical effect of acting as a constant-temperature reservoir.

6.2b Heat Engines

The Kelvin–Planck statement explicitly mentions a *cyclically operating device,* which can be construed as a **heat engine.** Many consequences of the various statements of the second law involve heat engines. Figure 6.4 illustrates this concept. Here we see that energy is transported as heat from a high-temperature reservoir to the engine. The engine converts a portion of this heat energy to work and rejects the remainder to the low-temperature reservoir. Because the engine operates in a thermodynamic cycle – with the working fluid always returning to its initial state – the device is capable of continual operation. The first law of thermodynamics, the energy conservation principle, relates the heat supplied and rejected during the cycle to the net work produced. For any incremental portion of the cycle, Eq. 5.5b applies:

$$\delta Q - \delta W = dU. \tag{6.1}$$

We integrate over the cycle to yield

$$\oint \delta Q - \oint \delta W = \Delta U = 0, \tag{6.2}$$

[2] At intermediate scales, however, thermal pollution and heat-island effects can come into play such that local regions of the atmosphere are affected. For example, the huge energy released at night from roads, parking lots, and buildings affects nighttime temperatures in Phoenix, Arizona.

FIGURE 6.3 Oceans, large lakes, and the earth's atmosphere serve as thermal reservoirs (PetroGraphy / iStock / Getty Images Plus; Thorsten Fritz / EyeEm / Getty Images; Dobresum / iStock / Getty Images Plus).

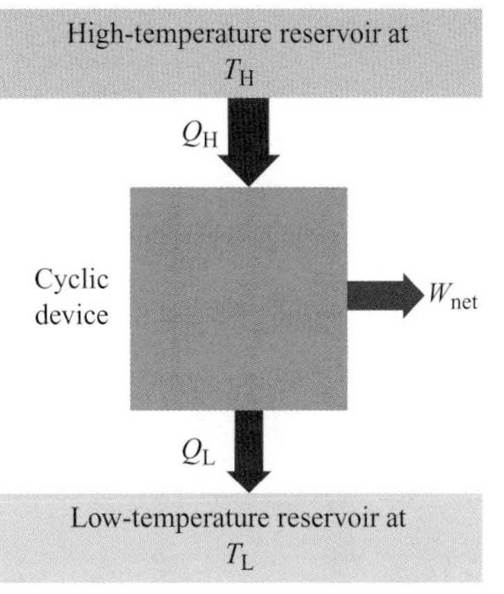

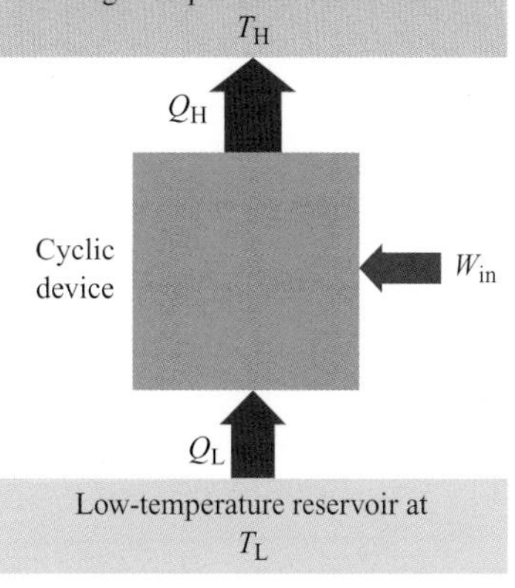

FIGURE 6.4 A heat engine **(a)** receives energy from a high-temperature reservoir, some of which is converted to work, while the remainder is rejected to the low-temperature reservoir. Reversing the cycle converts the heat engine into a heat pump or refrigerator **(b)**. Work is used to transfer energy from the low-temperature to the high-temperature reservoir.

where the internal energy change is zero since the initial and final states are identical. The path integrals of the heat and work are just $Q_H - Q_L$ and W_{net}, respectively; thus,

$$(Q_H - Q_L) - W_{net} = 0 \tag{6.3a}$$

or

$$W_{net} = Q_H - Q_L. \tag{6.3b}$$

A practical example of a heat engine is the closed-loop portion of a Rankine-cycle steam power plant (Fig. 6.5). Figure 6.6 shows the heat addition to the steam in the boiler. In a Rankine cycle, not all the heat is added at a single temperature, however, as the water from the feedwater pump is first heated to the saturation temperature before boiling and the saturated steam may then be superheated. In the Rankine cycle, the heat removal in the condenser typically occurs at a fixed temperature at a low-pressure, saturation condition. The net work crossing the boundary is that from the

FIGURE 6.5 The generation of electrical power from steam is an important application of the thermal-fluid sciences (see Chapter 1). The high-temperature heat source in the boiler can be from coal or gas combustion, or a nuclear reactor (zhuyongming / Moment / Getty).

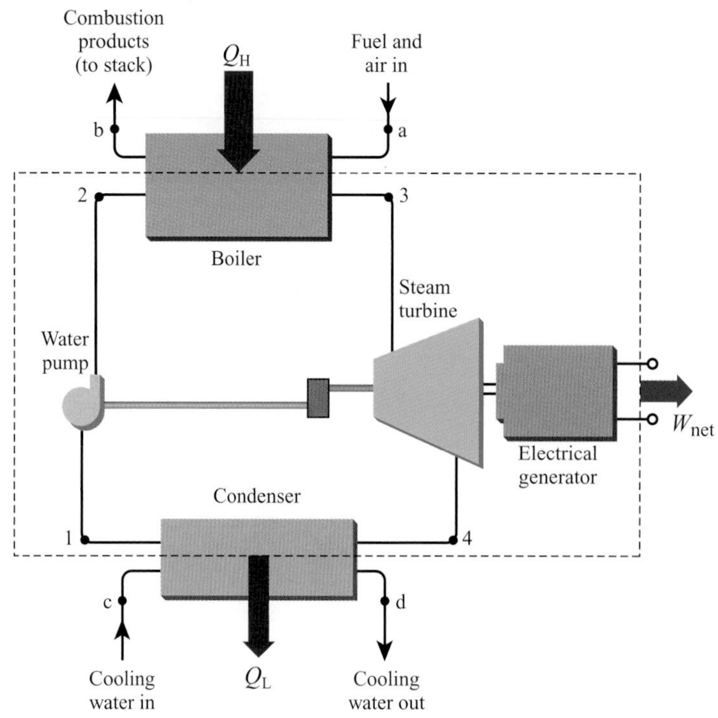

FIGURE 6.6 Steam power plant (Rankine cycle) schematically represented as a heat engine. The dashed line contains the components corresponding to the cyclic device of Fig. 6.4a.

turbine-driven electrical generator. No other work crosses the boundary. Note that the pump is driven by the steam turbine *within* the closed system (Fig. 6.6) so that we do not consider this work interaction in W_{net}.

A heat engine can be reversed by providing an input of work from the surroundings, as shown in Fig. 6.4b. In this reversed engine, energy from the low-temperature reservoir is delivered to the high-temperature reservoir. Reversed-cycle engines are called heat pumps, refrigerators, or air conditioners depending on their application. A **heat pump** removes energy from the atmosphere, the ground, or a body of water, and delivers energy to provide residential heating, for example (see Fig. 6.7a). The desired effect for a heat pump is to supply energy to the high-temperature reservoir (e.g., a building interior). The desired effect for a **refrigerator or air conditioner** is the removal of energy from the low-temperature reservoir. In a household refrigerator (see Fig. 6.7b), energy is removed from the interior of the refrigerator where food items are stored, and energy is rejected to the surroundings, typically through a heat exchanger located on the back of the appliance. As a result of this heat rejection, operating a refrigerator helps to heat your home in the winter and provides an extra cooling load for the summer air conditioning. An air conditioner removes heat from the cool inside of a building and rejects heat to the outside.

FIGURE 6.7 The refrigeration cycle can be used for different objectives. **(a)** Two 36-ton geothermal heat pumps used at the College of Southern Idaho (courtesy US Department of Energy). **(b)** Home refrigerator (JazzIRT / E+ / Getty Images).

Applying the conservation of energy principle (Eq. 6.2) to a reversed cycle, we relate the desired energy to the work input and the secondary heat transfer process as follows:

For a heat pump,

$$Q_H = W_{in} + Q_L. \tag{6.4}$$

For a refrigerator,

$$Q_L = Q_H - W_{in}. \tag{6.5}$$

See Fig. 9.40 and Examples 9.12–9.14 in Chapter 9.

In Eqs. 6.4 and 6.5, Q_H, Q_L, and W_{in} are all numerically positive, in agreement with the direction of the arrows shown in Fig. 6.2b.

Our current purpose is to use these devices to develop an understanding of the second law. A detailed and practical analysis of heat pumps and refrigerators is presented in Chapter 9.

6.2c Thermal Efficiency and Coefficients of Performance

In general, efficiency is the ratio of the desired output to the cost of operating the system. For heat engines, the desired output is the net work output which can be used to drive an electrical generator. The cost of operating the heat engine is the heat transfer in the high-temperature boiler, which is produced from some fuel source, such as natural gas, coal, or nuclear fuel. We define a **thermal (or first-law) efficiency** for heat engines as the ratio of the net useful work produced to the heat energy supplied:

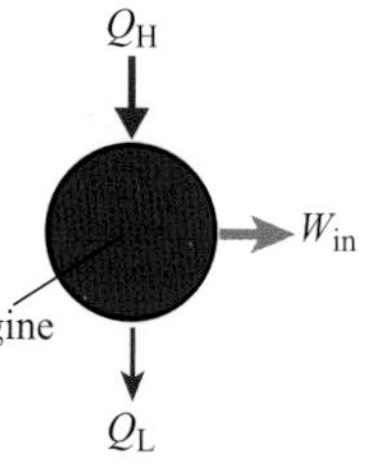

Q_H

W_{in}

heat engine

Q_L

FIGURE 6.8 A heat engine uses heat transfer from a high-temperature source to produce work.

$$\eta_{th} \equiv \frac{\text{Useful work produced}}{\text{Energy supplied}}. \tag{6.6}$$

Using Eq. 6.3, this is expressed as

$$\eta_{th} = \frac{W_{net}}{Q_H} = \frac{Q_H - Q_L}{Q_H}, \tag{6.7a}$$

or

$$\eta_{th} = 1 - \frac{Q_L}{Q_H}. \tag{6.7b}$$

Equations 6.6 and 6.7 are quite general and are not restricted to devices that receive or reject heat at fixed temperatures. All that is required is for Q_H, Q_L, and W_{net} to be determined by a proper integration over the cycle, that is,

$$Q_H \equiv \oint_{\text{Cycle}} \delta Q_H,$$

where the temperature at which heat is added can vary throughout the cycle. Thus, we could apply Eq. 6.7 to calculate the thermal efficiency of a real Rankine-cycle steam power plant were we able to measure any two of the quantities Q_H, Q_L, or W_{net}.

The **coefficient of performance,** or ***COP,*** defines a measure of performance for a reversed cycle. The following definition applies to both the heat pump and refrigerator:[3]

[3] Note that the right-hand side of Eq. 6.8 also defines the thermal efficiency for a heat engine (cf. Eq. 6.6).

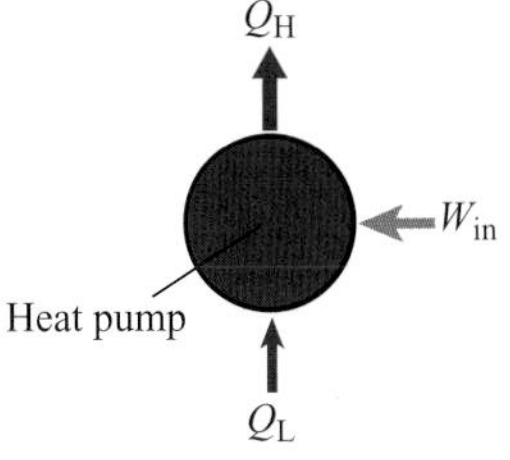

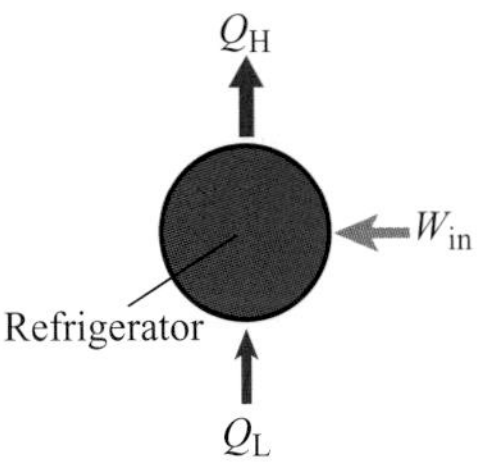

FIGURE 6. 9 The heat pump and refrigerator use the same thermodynamic cycle but have different objectives.

$$COP \equiv \beta \equiv \frac{\text{Desired energy}}{\text{Energy that costs}}. \tag{6.8}$$

For the heat pump, the desired energy is that delivered to the high-temperature reservoir, and the "energy that costs" is the work input; thus, the coefficient of performance for a heat pump is given by

$$\beta_{\text{heat pump}} = \frac{Q_H}{W_{in}}. \tag{6.9}$$

For a refrigerator or air conditioner, the desired energy is the energy removed from the cold space, and the energy that costs, once again, is the work input; thus,

$$\beta_{\text{refrig}} = \beta_{AC} = \frac{Q_L}{W_{in}}. \tag{6.10}$$

Unlike the thermal efficiency of a heat engine (Eq. 6.7), the coefficient of performance exceeds unity. Typical values for practical devices range, say, from 2 to 5.

Example 6.2 Stirling Engine

In a Stirling engine, heat is supplied to the working fluid, typically helium, from an external combustor. (See the schematic diagram.) A particular Stirling engine continuously produces 5 kW of shaft power with a thermal efficiency of 0.24. Determine the rate at which heat is added and the rate at which heat is rejected by this engine.

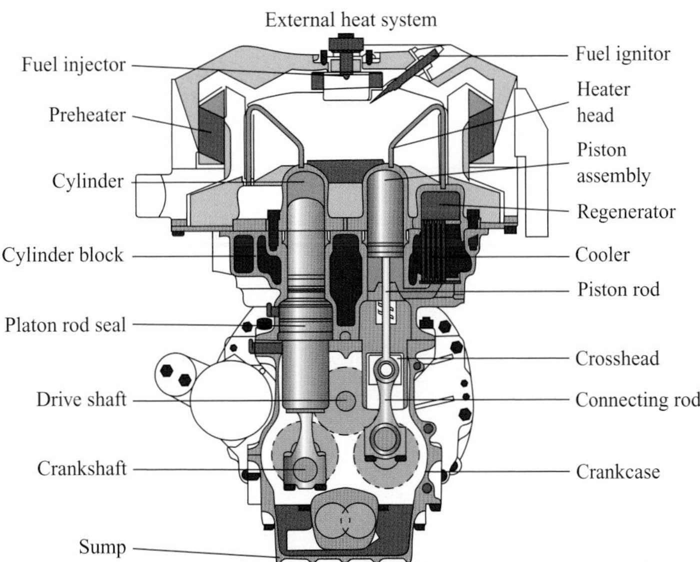

A Stirling engine consists of displacer pistons, power pistons, and a complex drive mechanism. The displacer piston(s) shuttle the working gas (shown in orange) back and forth between the heated space above the piston and the cooled space below the piston through a regenerator. (Schematic diagram courtesy of Kockums AB, Sweden.)

Solution

Known $\dot{W}_{net}, \eta_{th}$

Find $\dot{Q}_H, \dot{Q}_L$

Sketch

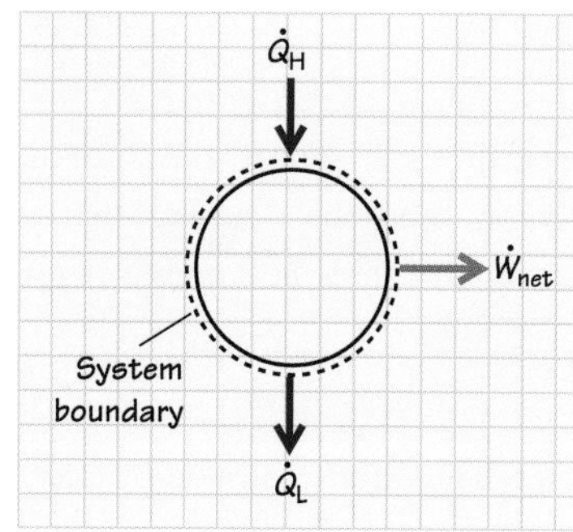

Analysis Actual heat engines execute a thermodynamic cycle many times over each minute to produce an essentially continuous power output; therefore, relationships involving the heat and work interactions, W_{net}, Q_H, and Q_L, apply equally well when these quantities are expressed on a rate basis (i.e., $(\dot{W}_{net}, \dot{Q}_H, \text{and } \dot{Q}_L)$ To find the heat addition rate, we apply the definition of thermal efficiency for a heat engine (Eq. 6.7a),

$$\eta_{th} = \dot{W}_{net}/\dot{Q}_H,$$

or

$$\dot{Q}_H = \dot{W}_{net}/\eta_{th}$$
$$= \frac{5\,\text{kW}}{0.24} = 20.8\,\text{kW}.$$

We obtain the heat-rejection rate from the conservation of energy (Eq. 6.3b):

$$\dot{Q}_L = \dot{Q}_H - \dot{W}_{net}$$
$$= 20.8\text{ kW} - 5\text{ kW} = 15.8\text{ kW}.$$

Comments Note the relatively low value of the thermal efficiency for this real heat engine; only 24% of the supplied energy is converted to useful work.

Self-Test 6.2

A heat engine receives 85 kW of heat from a high-temperature source and rejects 50 kW to a low-temperature sink. Determine the work produced and the thermal efficiency.

(Answer: $\dot{W}_{net} = 35\,kW \eta_{th} = 41.2\%$)

Example 6.3 Heat Pump

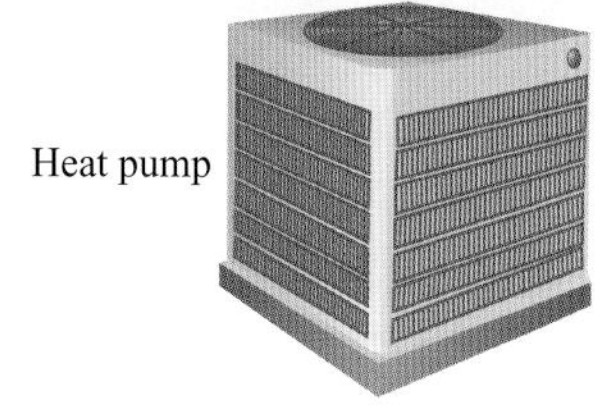

Heat pump

A heat pump operates with a coefficient of performance of 3 and requires a power input of 3.5 kW.

A. Determine the rate of heat delivered to the high-temperature reservoir.

B. The heat pump is reconfigured to operate as an air conditioner with the same power input and the same operating conditions as in part A. Determine the rate of heat removal from the low-temperature reservoir and the coefficient of performance of this air conditioner.

Solution

Known $\beta_{\text{heat pump}}, \dot{W}_{\text{in}}$

Find $\dot{Q}_{\text{H}}, \dot{Q}_{\text{L}}, \beta_{\text{refrig}}$

Sketch

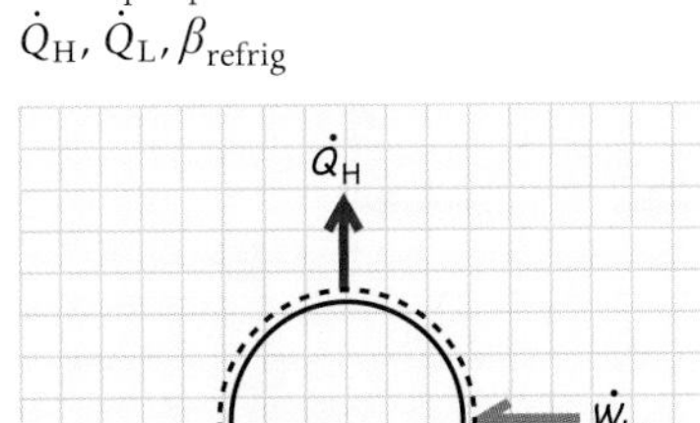

Analysis We use the definition of the coefficient of performance for a heat pump (Eq. 6.9) and a cycle energy balance (Eq. 6.3b) to find $\dot{Q}_{\text{H}}$ (Part A):

$$\beta_{\text{heat pump}} = \dot{Q}_{\text{H}}/\dot{W}_{\text{in}},$$

so

$$\begin{aligned}\dot{Q}_{\text{H}} &= \beta_{\text{heat pump}}\dot{W}_{\text{in}}\\ &= 3(3.5\,\text{kW}) = 10.5\,\text{kW}\end{aligned}$$

To find the coefficient of performance for the device reconfigured as a refrigerator, we first solve for $\dot{Q}_{\text{L}}$ from an energy balance,

$$\begin{aligned}\dot{Q}_{\text{L}} &= \dot{Q}_{\text{H}} - \dot{W}_{\text{in}}\\ &= 10.5\,\text{kW} = -3.5\,\text{kW} = 7\,\text{kW},\end{aligned}$$

and apply Eq. 6.10 for a refrigerator,

$$\begin{aligned}\beta_{\text{AC}} &= \dot{Q}_{\text{L}}/\dot{W}_{\text{in}}\\ &= \frac{7\,\text{kW}}{3.5\,\text{kW}} = 2.0.\end{aligned}$$

Comment Note that the same device operating under the same conditions has a different coefficient of performance when used as a heat pump rather than an air conditioner. The general definition of the coefficient of performance, Eq. 6.8, has the desired heat transfer in the numerator. For a heat pump, the desired heat transfer is to the high-temperature reservoir, for example, the inside of a house on a cold day. For an air conditioner, the desired heat transfer is from the inside of a cool house on a warm day.

Manipulating the definitions of the coefficient of performance for the heat pump and the refrigerator shows that $\beta_{\text{heat pump}}$ always exceeds $\beta_{AC}\left(\text{or } \beta_{\text{refrig}}\right)$ by one unit, provided the energy quantities are all the same, that is,

$$\beta_{\text{heat pump}} = \frac{W_{\text{H}}}{W_{\text{in}}} = \frac{Q_{\text{L}} + W_{\text{in}}}{W_{\text{in}}} = \frac{Q_{\text{L}}}{W_{\text{in}}} + 1,$$

so that

$$\beta_{\text{heat pump}} = \beta_{\text{AC}} + 1.$$

We can understand the purpose of the heat pump or air conditioner by considering the house as our system. In the cold winter, there are heat losses, $\dot{Q}_{\text{loss}}$ from the walls,

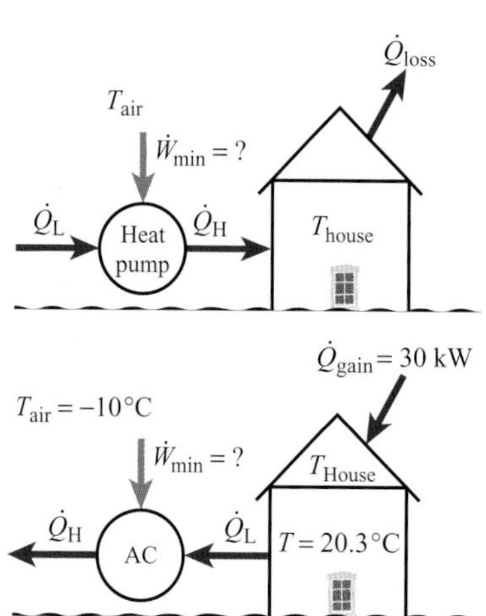

windows, and roof of the house to the surrounding cold air. To maintain the inside of the house at a steady warm temperature, the heat pump delivers heat to the house, $\dot{Q}_H = \dot{Q}_{loss}$. In the warm summer, there is heat transfer at the perimeter of the house from the warm surrounding air to the cool inside of the house. To maintain the inside of the house at a steady cool temperature, the air conditioner removes heat from the house at the same rate, $\dot{Q}_L = \dot{Q}_{gain}$.

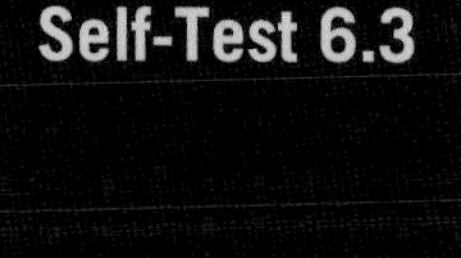

Heat pump A has a coefficient of performance of 2.5 and heat pump B has a coefficient of performance of 3.2. If both heat pumps provide 85 kW of heat, determine $\dot{Q}_L$ and the work required for each. Which one is more efficient?

(Answer: $\dot{W}_A = 34$ kW. $\dot{Q}_{LA} = 51$ kW, $\dot{W}_B = 26.6$ kW, $\dot{Q}_{L,B} = 58.4$ kW. Heat pump B is more efficient because it extracts more heat from the low-temperature reservoir and provides the same heating for less work input.)

6.2d Reversibility

Many of the consequences of the second law require an understanding of a **reversible process,** and its counterpart, an **irreversible process.** We define a reversible process as follows:

> **A reversible process *is one such that the system and all parts of the surroundings can be restored to their initial states.***

FIGURE 6.10 The thermal heating produced by friction in the hand-driven drill is sufficient to ignite easily inflammable tinder.

From this definition, we see that the effects of a reversible process can be undone, so that there is no evidence of the process ever having occurred. Reversing a reversible process leaves no trace in *either* the system *or* the surroundings.

A true reversible process is a piece of fiction (an idealization) since all real processes have some aspect that prevents them from being reversible. We often find that a movie played in reverse is funny because the actions cannot occur. For example, a child does not suddenly move up a slide, water does not suddenly jump from the ground to a glass, or bowling pins do not suddenly right themselves. The reason that these actions only occur in one direction is due to irreversibilities in the processes. Table 6.1 lists common irreversible processes, several of which are illustrated in Figs. 6.10–6.14.

Friction is a source of irreversibility that is present in essentially all mechanical processes (i.e., processes involving forces and motion). Friction is most obviously present in dry sliding, as you can easily demonstrate by firmly pressing your hands together and then sliding one over the other. The organized macroscopic energy resulting from the relative motion of your hands ($\dot{W} = F_{fric}V$) is converted

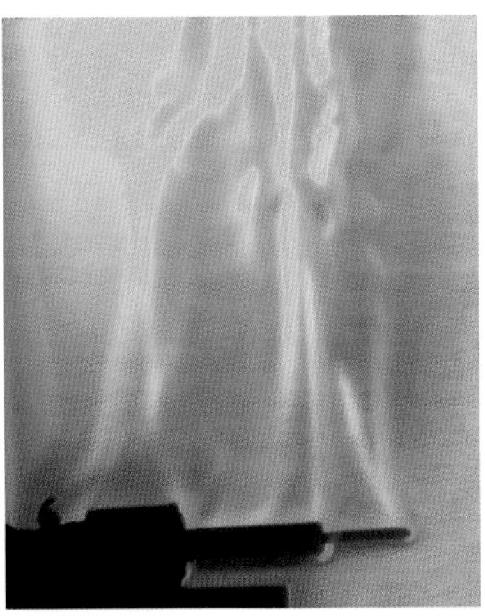

FIGURE 6.11 Electrical current flowing through a soldering iron tip results in a heating of the tip (i.e., Joule heating). This illustrates the irreversible conversion of work (the flow of electricity; see Chapter 4) to heat. The schlieren optical technique makes visible the free convection currents set up by the hot soldering iron. (Photograph courtesy of Gary Settles.)

FIGURE 6.12 The mixing of a dye jet with a reservoir fluid is an irreversible process, as are all mixing processes that involve two or more different substances.

In Chapter 7, we define the "head loss" that describes the effects of flow friction.

TABLE 6.1 Common Irreversible Processes

Any process involving friction (Fig. 6.10)
Heat transfer across a finite temperature difference
Unrestrained expansion of a gas to a low pressure
Joule i^2R heating (Fig. 6.11)
Plastic deformation of a solid
Mixing of two different substances (Fig. 6.12)
Turbulent flow
Shock waves (Fig. 6.13)
Spontaneous chemical reaction (Fig. 6.14)
Magnetic hysteresis
Freezing of a subcooled liquid
Condensation of a supersaturated vapor

to the randomized thermal motion of molecules, which you sense as a heating up of your hands. There is no way for you to move your hands back to their original position that can recover the thermal energy generated by the friction. A "fire-producing" drill (Fig. 6.10) utilizes the irreversible effects of friction for a useful purpose.

Friction also occurs within fluids. At a system boundary, the effect of viscosity is manifest as work (shear work) if the boundary is moving. The viscous forces acting internally to an open system cause a loss of mechanical energy, usually as reduced pressure from inflow to outflow, that results in an increase in the internal energy of the fluid. If we are interested in the mechanical energy of the flow, the effects of fluid viscosity are described as "viscous losses" or "head losses" in the system. If we are describing the thermal energy changes in the system, the effects of fluid viscosity are described as "viscous heating." We will look at the effects of viscosity in more detail in Chapter 7. In an ideal frictionless flow (a reversible process), the viscous losses would be zero, and no work would be required to keep the fluid flowing at steady-state conditions. We will explore the connection between the second law and viscous loss in greater detail later in Chapter 7.

By examining the other irreversible processes shown in Table 6.1, it is easy to see that reversing any of the processes would leave a significant history in the surroundings, the absence of such a history being the requirement for a process to be reversible. For example, consider a metal rod plastically deformed in a tensile test machine. A plastically deformed rod increases in temperature and will not spontaneously return to its original shape, since it has been stretched beyond the elastic limits. It is also unlikely that any mechanical process could restore the rod to its original state short of melting and refabrication. If one were to reconstruct the rod, a permanent change most certainly would occur in the surroundings.

The electrical heating of an object (Fig. 6.11) uses electrical work to increase the temperature (and internal energy) of the object. This process is irreversible; it can only occur in one direction. It is not possible to generate electrical work from a single heat reservoir at high temperature as indicated by Statement IA of the second law.

The mixing of two different liquids (Fig. 6.12) is also an irreversible process that can only occur spontaneously in one direction. The two different fluids can be easily

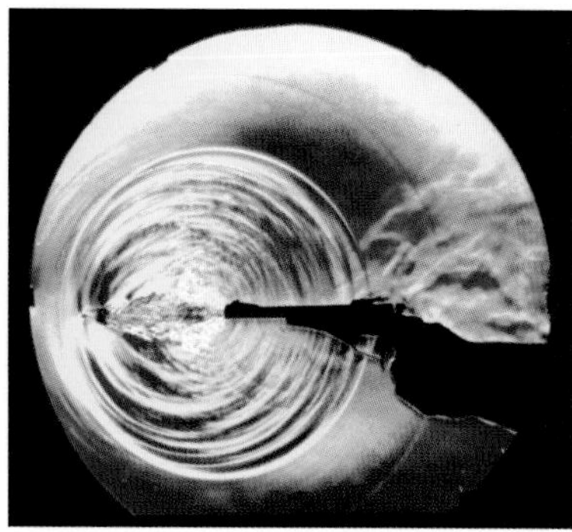

FIGURE 6.13 The schlieren optical technique makes visible the shock waves created by a bullet. A shock wave is a very thin region in a gas over which the gas properties change dramatically. Hearing a sonic boom is the result of the remnants of a shock wave passing by your ear. (Photograph courtesy of Gary Settles.)

FIGURE 6.14 Combustion is a highly irreversible process. The products of combustion cannot be converted back to their original reactant state without a large expenditure of work (Richard Drury / DigitalVision / Getty).

mixed but they cannot be separated again without work through a distillation process or chemical reaction. The combustion process shown in Fig. 6.14 also has irreversibilities and the process cannot occur in the reverse direction to "unburn" the wood.

One purpose of the preceding discussion is to show that a reversible process is an *ideal* process and that all *real* processes involve some degree of irreversibility. We will see in the next section how the reversible process is a standard to which real processes can be compared and how such comparisons affect engineering choices. In Chapter 7 we will look again at reversible and irreversible processes using the thermodynamic property called entropy.

6.3 Consequences of the Kelvin–Planck Statement

At this point, you may be asking how all this relates to the second law of thermodynamics. Table 6.2 presents some answers to this question, listing some practical consequences of the second law.

The first entry in Table 6.2, that work is inherently more valuable than heat, has already been discussed; the remaining entries, however, need some elaboration. The second entry in Table 6.2, that the thermal efficiency of a heat engine can never be 100%, follows from the Kelvin–Planck statement (IB) of the second law stating that it is impossible to convert heat from a *single* reservoir continuously to work. Implicit is the requirement for a *second* reservoir and the rejection of some heat. The idea here is that it is impossible to close the cycle (i.e., bring the working fluid back to its original state) without rejecting some heat in the course of performing some work. There is no possible cycle path that does not include a heat rejection. Theoretical maximum

TABLE 6.2 Consequences of the Kelvin–Planck Statement of the Second Law

No.	Consequence
1	Energy transfer by work is more valuable than energy transfer by heat.
2	All heat engines must reject a portion of the heat energy supplied; thus, their thermal efficiency can never be 100%.
3	The energy contained in the earth's heat reservoirs (atmosphere, oceans, rivers, ground, etc.) cannot be utilized to produce a continuous supply of work unless a second reservoir exists at a lower temperature.
4	For any engine working between two reservoirs having given high and low temperatures, a reversible engine will have the greatest thermal efficiency.
5	All reversible heat engines have the same thermal efficiency when operating between the same two reservoirs.
6	A thermodynamic temperature scale can be defined that is independent of the thermometric properties of any substance; the attainment of negative absolute temperatures is impossible; and it is impossible for a finite system to attain a zero value on the absolute scale.

thermal efficiencies will therefore fall far short of 100%. The accuracy of this statement will also be verified in Chapter 7 using an entropy analysis for the system. (See Section 7.4d.)

The third entry in Table 6.2 is a way of restating the Kelvin–Planck statement that makes the importance of the second law obvious. If one could construct a device that converted low-grade heat from the earth's atmosphere, oceans, etc. to work, we would enjoy an essentially inexhaustible supply of useful energy with no cost other than that of the conversion device. Such a device is called a **perpetual-motion machine**[4] of the *second kind* and violates the second law of thermodynamics. (A perpetual-motion machine of the *first kind* violates the first law of thermodynamics, that is, the conservation of energy principle.) Several examples of perpetual-motion machines of the second kind are shown in Fig. 6.15. In these drawings, the marbles move down and back up through the cycle. The only way this can occur is if the kinetic energy of the marble rolling or falling downward can be completely converted back to potential energy at the top elevation. An actual cycle, however, will have frictional losses as the marble rolls and the paddles rotate. These frictional losses are similar to the viscous losses that occur in a fluid cycle where the kinetic energy cannot be completely converted back to potential energy.

The fourth and fifth consequences shown in Table 6.2 provide standards for thermal efficiency to which all real heat engines can be compared. Recall that the Rankine-cycle steam power plant is just such a real heat engine. As we will show shortly, the thermal efficiency of a reversible engine is solely determined by the temperatures of the hot and cold reservoirs. Carnot was the first to discover this, and his name is associated with the efficiency of a reversible engine.

Before exploring this efficiency in greater detail, we show how the fourth and fifth consequences (Table 6.2) do indeed follow from the Kelvin–Planck statement of the

[4] For an interesting history of the search for a perpetual-motion machine, the reader is referred to Ref. [5]. Figure 6.15 whimsically presents some ideas of perpetual motion.

FIGURE 6.15 Perpetual motion machines violate the second law of thermodynamics (Dorling Kindersley / Getty Images).

second law. Figure 6.16 illustrates the logic applied. In Fig. 6.16a, an irreversible engine and a reversible engine operate between the two reservoirs at T_H and T_L. As a counter-example of the Kelvin–Planck statement, we postulate that the irreversible engine has a greater thermal efficiency than the reversible engine; that is, for the same heat addition, this engine produces more work than the reversible engine. These conditions are expressed mathematically as

$$Q_{H,\,irrev} = Q_{H,\,rev},$$
$$W_{irrev} > W_{rev},$$

and it follows from the application of the first law to the two cycles that

$$Q_{L,\,irrev} < Q_{L,\,rev}.$$

We now choose to operate the reversible engine in reverse (Fig. 6.16b), that is, like a heat pump, in that work is supplied (W_{rev}), heat is taken from the low-temperature reservoir ($Q_{L,rev}$), and heat is added to the high-temperature reservoir ($Q_{H,rev}$). That the cycle is reversed in no way violates either the first or second laws of thermodynamics since this is a *reversible* cycle, meaning that it can be operated in the reverse direction with the same magnitudes of the heat transfer terms $Q_{H,rev}$ and $Q_{L,rev}$.

Since the heat added to the high-temperature reservoir by the reversible engine (heat pump) has the same magnitude as the heat taken from the high-temperature reservoir by the irreversible engine, we can replace this heat exchange by a direct path not involving a reservoir, as shown in Fig. 6.16c. We also use some of the work produced by the irreversible engine to drive the heat pump, as indicated by the arrow labeled W_{rev} in Fig. 6.16c. Since the magnitude of the work produced by the irreversible engine exceeds that required to drive the reversible heat pump, a net amount of work $W_{excess} = W_{irrev} - W_{rev}$ is produced by the combined cycles. To complete our analysis, we make use of the fact that $|Q_{L,\,irrev}| < |Q_{L,\,rev}|$ to replace these two arrows with a single arrow directed *from* the low-temperature reservoir, as shown in Fig. 6.16d. What have we now created? We clearly have a device that directly violates the Kelvin–Planck statement, a device that produces work continuously from a single heat reservoir. Since such a device is impossible to construct, our original postulate that the thermal efficiency of the irreversible engine exceeds that of the reversible

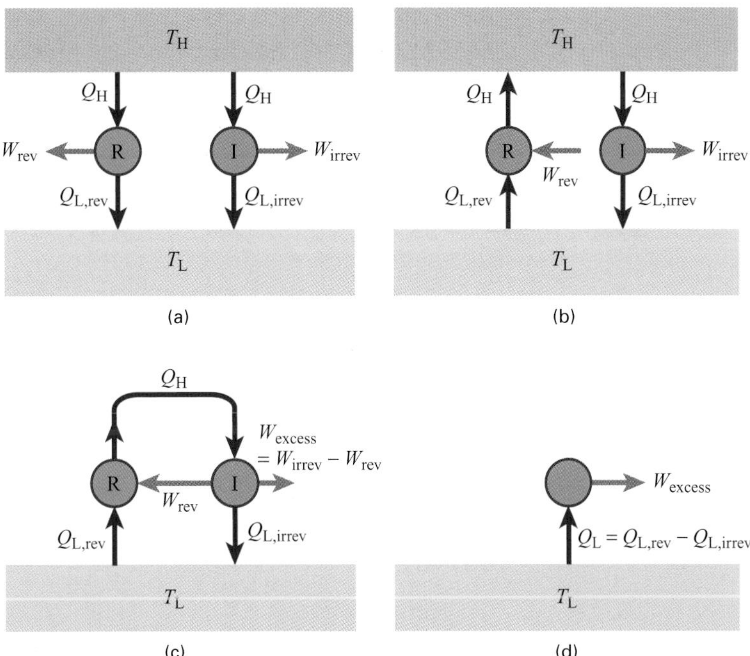

FIGURE 6.16 Logical sequence illustrating that no engine can have a thermal efficiency greater than that of a reversible engine without violating the Kelvin–Planck statement of the second law.

engine must be false. We therefore conclude that the fourth consequence listed in Table 6.2 must be true. Similar arguments can be mustered to show that the fifth consequence is true as well.

Having just shown that a heat engine operating in a reversible cycle is the most efficient of any engine, we seek to quantify this maximum efficiency. Such quantification is invaluable to engineering analyses as it provides hard and fast standards to which real devices can be compared. Imagine the wasted effort of an engineer seeking to improve the thermal efficiency of a power plant to 60% when the second law indicates that 55% is the theoretical maximum achievable. To achieve our goal of quantifying the thermal efficiency of a reversible cycle, we must first define an absolute thermodynamic temperature scale (see item 6 in Table 6.2).

6.3a Kelvin's Absolute Temperature Scale

We have just shown the validity of the statement that, for any engine working between two reservoirs having respectively a given high temperature and a given low temperatures, a reversible engine will have the greatest thermal efficiency. William Thomson, Lord Kelvin, used this statement to define a temperature scale that is independent of any substance or particular measuring instrument, that is, an **absolute temperature scale**[5] [6]. We now explore how this was done by mathematically expressing this statement as

$$\eta_{\text{rev}} = f(T_{\text{H}}, T_{\text{L}}), \quad \textbf{(6.11)}$$

where f indicates an arbitrary function of the two variables T_{H} and T_{L}. Equation 6.11 implies that the *only* factors affecting the thermal efficiency of a reversible cycle are

[5] An absolute temperature scale is also referred to as a **thermodynamic temperature scale**.

the temperatures of the two reservoirs. We combine this second-law conclusion, Eq. 6.11, with the first-law definition of thermal efficiency, Eq. 6.7b, as follows:

$$\eta_{\text{rev}} = 1 - \frac{Q_{\text{L}}}{Q_{\text{H}}} = f(T_{\text{H}}, T_{\text{L}}). \tag{6.12}$$

Although there are several choices that can be made for the function *f*, Kelvin's choice [9] was to set

$$f(T_{\text{H}}, T_{\text{L}}) \equiv 1 - \frac{T_{\text{L}}}{T_{\text{H}}}, \tag{6.13a}$$

or

$$\frac{Q_{\text{L}}}{Q_{\text{H}}} \equiv \frac{T_{\text{L}}}{T_{\text{H}}}. \tag{6.13b}$$

Equation 6.13b is then used to create an absolute thermodynamic temperature scale by arbitrarily assigning a numerical value to one of the reservoir temperatures, so that

$$T = T_{\text{fixed}} \left(\frac{Q_T}{Q_{T_{\text{fixed}}}} \right)_{\text{rev}}. \tag{6.14a}$$

This definition of temperature states that the thermodynamic temperature is directly proportional to the ratio of the heat received by a reversible heat engine at the temperature of interest to the heat rejected at a known fixed temperature. The fixed point for the kelvin scale adopted in 1954 by the Conférence Générale des Poids et Mesures (CGPM) is the triple point of water and is assigned the thermodynamic temperature of 273.16 K, that is,

$$T_{\text{fixed}} \equiv 273.16\,\text{K}.$$

Equation 6.14a then becomes

$$T(\text{K}) = 273.16 \left(\frac{Q_{T(\text{K})}}{Q_{273.16}} \right)_{\text{rev}}. \tag{6.14b}$$

Figure 6.17 graphically illustrates this absolute thermodynamic temperature scale.

Since one cannot in reality measure the ratio of heat received and rejected in a reversible heat engine, Eq. 6.14b and Fig. 6.17 are theoretical constructs. Practical temperature scales are defined to allow accurate approximations to this thermodynamic ideal using laboratory instruments.

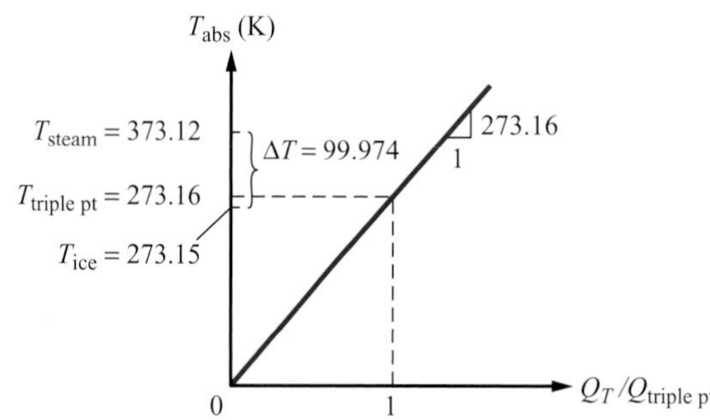

FIGURE 6.17 Kelvin's absolute thermodynamic temperature scale is independent of the properties of any substance and depends only on the ratio of the heat transferred from a high-temperature reservoir to the low-temperature reservoir in a reversible heat engine. The slope of the Kelvin scale is set by choosing the triple point of water as a reference condition and assigning to it a temperature of 273.16 K.

6.3b The Carnot Efficiency

With the establishment of a thermodynamic temperature scale and its relation to a practical means of measurement, we have a way to actually quantify the ideal (maximum) thermal efficiency of a heat engine operating between two heat reservoirs:

$$\eta_{rev} = 1 - \frac{T_L}{T_H}, \tag{6.15a}$$

where T_L and T_H are absolute temperatures. The maximum thermal efficiency defined by Eq. 6.15a is also named the **Carnot efficiency** in honor of Carnot's contributions. From Eq. 6.15a, we see that increasing the temperature of the heat source increases the ideal efficiency, as does decreasing the temperature of the heat sink. To approach an efficiency of unity (100%) requires either that the temperature of the high-temperature reservoir approach infinity ($T_H \to \infty$) or that the low-temperature reservoir approach absolute zero ($T_L \to 0$).

For any real device, maximum temperatures are usually dictated by the consideration of material properties. For example, the strength of metal parts decreases as the temperature increases, resulting in metallurgical limits. For highly stressed boiler tubes and steam turbine blades [10], this metallurgical limit is approximately 900 K, whereas for modern gas-turbine engines, the maximum turbine blade temperature for nickel-based alloys is approximately 1100 K [11]. The high-pressure turbine blades downstream of the combustor encounter high temperatures that can cause the blades to fail, as shown in Fig. 6.18. Cooling the blades allows working fluid temperatures to be significantly higher than the metallurgical limits. Advanced gas-turbine systems permit the use of turbine inlet temperatures above 1775 K. The use of ceramic materials provides the hope of extending metallurgical limits by several hundred kelvins [12]. If we consider the sun as an extreme high-temperature source, energy is available at approximately 5800 K. The ambient temperatures of the atmosphere, or a body of water into which heat is rejected fix the practical minimum temperatures for heat rejection. In North America, typical ambient temperatures range, say, from about 279 K (40 F) to 300 K (80 F). To create a reservoir at a temperature less than that of ambient requires the use of refrigeration, which, in turn, requires a work input. This work input more than cancels any improvement in efficiency gained by lowering the sink temperature of the original cyclic device.

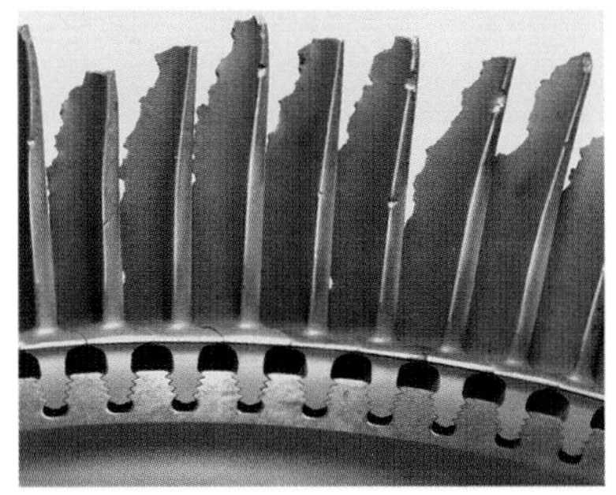

FIGURE 6.18 Turbine rotor blade failure from excessive temperatures. Photograph courtesy of Rolls-Royce (The Jet Engine).

See Chapter 9 for analyses of power plants, jet engines, and other systems.

Table 6.3 presents the Carnot efficiencies associated with some of the practical temperatures just discussed. Here we see that for all the cycles constrained by considerations of materials ($T < 1775$ K), thermal efficiencies are well below unity, with approximately 20%–30% of the supplied heat rejected to the cold reservoir. It should be kept in mind that real cyclic devices (steam power plants, for example) do not operate between two fixed-temperature reservoirs; nevertheless, the theoretical maximum thermal efficiencies shown are useful approximations to real devices receiving energy on the average at such temperatures. Older steam power plants operate with actual thermal efficiencies on the order of 40%, whereas modern dual-cycle power plants that employ both steam turbines and gas turbines can achieve actual efficiencies of approximately 60% [13].

TABLE 6.3 Carnot Efficiencies for Various High-Temperature Reservoir Temperatures*

	T_H(K)	$\eta_{rev} = 1 - \frac{T_L}{T_H}$
Temperature of sun	~5800	0.949
Advanced gas-turbine inlet gas temperatures	~1775	0.835
Metallurgical limit (Hastelloy X, etc.)	~1100	0.734
Metallurgical limit (steels)	~900	0.674

* The low-temperature reservoir is fixed at 293 K (68 F).

Equation 6.13b can also be use to define ideal (Carnot) coefficients of performance for heat pumps and refrigerators or air conditioners, as follows:

$$\beta_{\text{rev, heat pump}} = \frac{T_H}{T_H - T_L}, \tag{6.15b}$$

$$\beta_{\text{rev, refrig}} = \beta_{\text{rev, AC}} = \frac{T_L}{T_H - T_L}. \tag{6.15c}$$

Example 6.4 Heat Engine

A reversible heat engine rejects heat at 25 °C and has a thermal efficiency of 50%. At what temperature is heat added to the engine cycle?

Solution

Known T_L, η_{rev}

Find T_H

Analysis We apply the definition of thermal efficiency (Eq. 6.15a) for a reversible heat engine to find the temperature of the high-temperature reservoir as follows:

$$\eta_{rev} = 1 - \frac{T_L}{T_H},$$

which is rearranged to yield

$$T_H = \frac{T_L}{1 - \eta_{rev}}$$

$$= \frac{(25 + 273.15)\,\text{K}}{1 - 0.5} = 569\,\text{K or } 323\,^\circ\text{C}.$$

Comment Note that absolute temperatures must be used when defining the thermal efficiency.

Self-Test 6.4

A refrigerator operates between two reservoirs, one at 275 K and the other at 310 K. Determine the maximum coefficient of performance for the refrigerator. If $\dot{Q}_L = 10\,kW$ for this maximum COP, determine the input work required.

(Answer: $\beta_{refrig} = 7.86$, $\dot{W}_{in} = 1.27\,kW$)

6.4 Alternative Statements of the Second Law

As discussed at the outset of this chapter, the second law can be expressed in a number of ways, any of which can be used (ultimately) to generate the others. Table 6.4 presents four classes of second-law statements. We have already considered the class I statements in some detail.

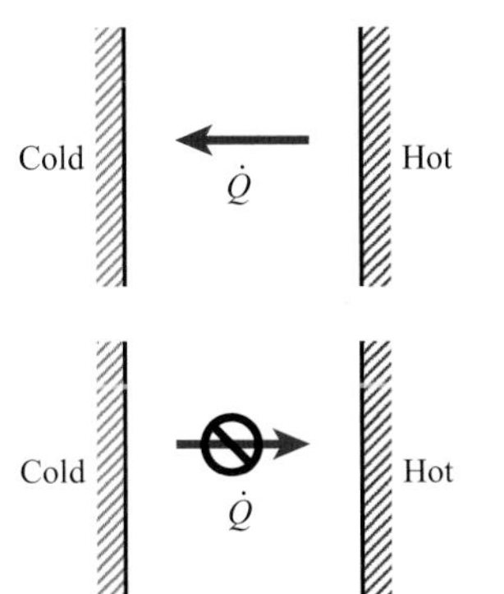

FIGURE 6.19 Heat transfer occurs spontaneously from a hot reservoir to a cold reservoir but not conversely.

TABLE 6.4 Various Statements of the Second Law of Thermodynamics

Designation	Statement
IA: Paraphrase of Kelvin–Planck statement	Although all work can be converted completely to heat, heat cannot be completely and continuously converted into work.
IB: Kelvin–Planck statements [4]	It is impossible to construct a cyclically operating device for which the sole effect is the exchange of heat with a single reservoir and the creation of an equivalent amount of work.
II: Clausius (original) statements [7]	It is impossible to operate a cyclic device such that the sole effect external to the device is the transfer of heat from one energy reservoir to another at a higher temperature. Heat flows spontaneously from high to low temperature, but not conversely.
III: Clausius entropy statement [6]	The entropy of a system and the environment with which it is in contact increases, or, in the limit of a reversible process, remains constant [1]. The entropy of the universe tends toward a maximum [8].
IV A, B, C: Equilibrium statements	A. Equilibrium occurs when the entropy is a maximum for a simple closed system at constant internal energy and volume. B. Equilibrium occurs when the Gibbs free energy is a minimum for a simple closed system at constant pressure and temperature. C. Equilibrium occurs when the Helmholtz free energy is a minimum for a simple closed system at constant temperature and volume.

The original statements of Clausius (II) are frequently invoked in discussions of the second law. The following provides the spontaneous direction for heat flow (see also Fig. 6.19):

Heat flows spontaneously from high to low temperature, but not conversely.

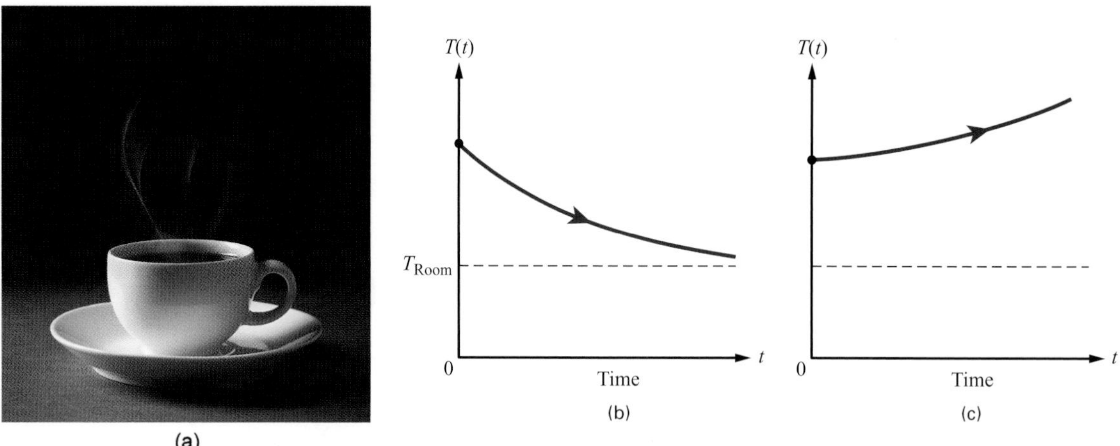

FIGURE 6.20 The second law of thermodynamics provides unambiguous criteria for the direction of spontaneous change. **(a)** For the coffee cup shown, **(b)** the coffee cools spontaneously. **(c)** The spontaneous heating of coffee in cooler surroundings is impossible (hdere / E+ / Getty).

The first law provides no guideline for the direction of spontaneous processes, only stipulating that energy must be conserved. One of the particularly useful aspects of the second law is that it *does* provide clear guidelines for what processes will occur naturally. Clausius' statement clearly asserts that one will never observe a cup of coffee spontaneously getting hotter at the expense of energy drawn from the colder atmosphere (Fig. 6.20). Here the use of the word **spontaneous** refers to the idea that no work is applied. Clearly, a heat pump could be used to heat the coffee at the expense of the atmosphere. Clausius' original statements can be deduced from the Kelvin–Planck statements (I). These deductions are left as exercises for the reader.

Both the class III and IV statements focus on the ways in which the second law is useful in determining the spontaneous direction for real processes and the conditions required for equilibrium. This usefulness is one of the two aspects emphasized at the beginning of this chapter. As we see from Table 6.4, an understanding of the class III and IV statements requires, in turn, an understanding of entropy and the Gibbs and Helmholtz free energies. Engineering analyses use these thermodynamic properties in quantitative ways. For example, we use the Gibbs free energy, or Gibbs function, to calculate the detailed composition of the products of combustion at high temperatures where chemical dissociation occurs. We will investigate this particular application in detail in Chapter 13.

Before we can continue our discussion of the class III and IV second-law statements, we need to define and explore entropy and related thermodynamic properties.

SUMMARY

In this chapter, we introduced the second law of thermodynamics and indicated its usefulness in establishing theoretical performance limits for real devices: the Carnot efficiency for cyclically operating devices, such as the closed-loop steam power plant, and the isentropic efficiency for noncyclic devices, such as turbines, compressors, pumps, and nozzles. Achieving an understanding of these performance limits requires a fairly large number of concepts and definitions. Among these, you should be familiar with: thermal reservoirs; heat engines and their reversed-cycle counterparts, heat pumps and refrigerators; the distinctions between reversible and irreversible processes; and Kelvin's absolute temperature scale. At the chapter outset, we also indicated the usefulness of the second law in establishing the direction of spontaneous change and defining thermodynamic equilibrium.

An organizing theme of these chapters is the idea that the second law can be expressed in a number of ways – all of which are ultimately equivalent. We started with the Kelvin–Planck statement, with which you should be quite familiar, and ended with the various equilibrium statements that involve entropy and its companion properties. From this presentation, the reader should now be able to recognize the various forms of the second law and be comfortable with the idea that no single statement is universally recognized as *the* second law of thermodynamics. As a further summary of this chapter, reviewing the learning objectives presented at the outset of this chapter is recommended.

KEY EQUATIONS

Review the most important equations presented in this chapter (i.e., those boxed with a yellow background). What physical principles do they express? What restrictions apply?

CHAPTER 6 KEY CONCEPTS AND DEFINITIONS CHECKLIST

Numbers following arrows refer to Questions and Problems at the end of the chapter.

6.1 Usefulness of the Second Law

- ☐ Performance limits
- ☐ Direction for spontaneous change

6.2 One Fundamental Statement of the Second Law

- ☐ Kelvin–Planck statement ➔ Question 6.2
- ☐ Heat reservoir ➔ Question 6.5
- ☐ Heat engine ➔ Questions 6.6, 6.7
- ☐ Heat pump ➔ Questions 6.8, 6.9
- ☐ Refrigerator ➔ Question 6.10
- ☐ Thermal (first-law) efficiency ➔ Questions 6.12, 6.14
- ☐ Coefficients of performance ➔ Questions 6.12, 6.13
- ☐ Reversible and irreversible processes ➔ Questions 6.15, 6.16, 6.18, 6.19

6.3 Consequences of the Kelvin–Planck Statement

- ☐ Six consequences (Table 6.2) ➔ Question 6.21
- ☐ Absolute temperature scale ➔ Question 6.24
- ☐ Carnot efficiency ➔ Problems 6.54, 6.57
- ☐ Stirling cycle ➔ Problem 6.1

6.4 Alternative Statements of the Second Law

- ☐ Original Clausius statements ➔ Question 6.23
- ☐ Clausius entropy statement ➔ Question 6.23

REFERENCES

1. Kline, S. J., *The Low-Down on Entropy and Interpretive Thermodynamics*, DCW Industries, La Cañada, CA, 1999.
2. Weaver, J. H. (Ed.), *World of Physics*, Simon & Schuster, New York, 1987, p. 734.
3. Keenan, J. H., *Thermodynamics*, Wiley, New York, 1957.
4. Planck, M., *Treatise on Thermodynamics*, translated by Alexander Ogg, 3rd edn, Dover, New York, 1945.
5. Ord-Hume, A. W. J. G., *Perpetual Motion: The History of an Obsession*, St. Martin's Press, New York, 1977.
6. Thomson, W. (Lord Kelvin), "On an Absolute Thermometric Scale Founded on Carnot's Theory of the Motive Power of Heat, and Calculated from Regnault's Observations," *Cambridge Philosophical Society Proceedings*, June 5, 1848.
7. Clausius, R. J. E., "Ueber die bewegende Kraft der Wärme," *Annalen der Physik und Chemie*, **79**:368 (1850); translated excerpts in *A Source Book of Physics*, W. F. Magie, McGraw-Hill, New York, 1935, pp. 228–233.
8. Clausius, R. J. E., "Ueber verschiedene für die Anwendung bequeme Formen der Hauptgleichungen der mechanischen Wärmetheorie," *Annalen der Physik und Chemie*, **125**:353 (1865); translated excerpts in *A Source Book of Physics*, W. F. Magie, McGraw-Hill, New York, 1935, pp. 234–236.
9. Thomson, W. (Lord Kelvin), "On the Dynamical Theory of Heat, with Numerical Results Deduced from Mr. Joule's Equivalent of a Thermal Unit, and M. Regnault's Observations on Steam," *Transactions of the Royal Society of Edinburgh*, March, 1851.
10. Goodall, P. M., *The Efficient Use of Steam*, IPC Science and Technology Press, Surrey, England, 1980.
11. Lefebvre, A. H., *Gas Turbine Combustion*, Taylor & Francis, Bristol, PA, 1983.
12. Ohhashi, I., and Arakawa, S., "Development of 300 kW Class Ceramic Gas Turbine (CGT 303)," *Journal of Engineering for Turbines and Power–Transactions of the ASME*, **117**:777–782 (1995).
13. Chase, D. L., and Kehoe, P. T., "GE Combined-Cycle Product Line and Performance," GE Power Systems, GER-3574G, October, 2000.

Some end-of-chapter problems were adapted with permission from the following:

Look, D. C., Jr., and Sauer, H. J., Jr., *Engineering Thermodynamics*, PWS, Boston, 1986.
Myers, G. E., *Engineering Thermodynamics*, Prentice Hall, Englewood Cliffs, NJ, 1989.

CONCEPTUAL QUESTIONS

6.1 Review the most important equations presented in this chapter, i.e., those with a yellow background. What physical principles do they express? What restrictions apply?

6.2 State in words the Kelvin–Planck statement of the second law of thermodynamics.

6.3 Write a paragraph explaining the second law of thermodynamics for your friends not majoring in engineering- or science-related fields.

6.4 List two or more ways in which the second law is useful to engineers.

6.5 Define a heat reservoir.

6.6 Explain the concept of a heat engine. Of what significance is this concept to the study of thermodynamics?

6.7 What is the primary objective of a heat engine?

6.8 Explain the thermodynamic concept of a heat pump. Also explain the term "refrigerator".

6.9 What is the primary objective of a heat pump?

6.10 What is the primary objective of an air conditioner or refrigerator?

6.11 What is the difference between a heat pump and air conditioner?

6.12 Write in words the definitions of thermal efficiency and coefficient of performance for cyclic devices.

6.13 Explain the difference between a reversible and an irreversible process. List three or more examples of irreversible processes.

6.14 Consider a reversible heat engine operating between two heat reservoirs. How does the thermal efficiency of the engine relate to the temperatures of the reservoirs?

6.15 Will a reversible or irreversible heat engine between a given T_H and T_L have a greater power output for a given heat transfer from the high-temperature reservoir?

6.16 Will a reversible or irreversible heat engine between a given T_H and T_L have a greater thermal efficiency?

6.17 Consider a reversible heat pump operating between two heat reservoirs. How does the heat pump coefficient of performance relate to the temperatures of the reservoirs? How does your answer change if the device is operated as a refrigerator rather than a heat pump?

6.18 Will a reversible or irreversible heat pump between a given T_H and T_L require a greater power input to deliver a desired heat transfer to the high-temperature reservoir?

6.19 Will a reversible or irreversible heat pump between a given T_H and T_L have a greater coefficient of performance?

6.20 Write an equation that relates the coefficient of performance of a heat pump to the coefficient of performance of the same device operating as an air conditioner.

6.21 List as many consequences of the Kelvin–Planck statement of the second law as you can remember. Compare your list with Table 6.2.

6.22 Explain the difference between a perpetual-motion machine of the first kind and a perpetual-motion machine of the second kind.

6.23 State in words two statements of the second law of thermodynamics other than the Kelvin–Planck statement.

6.24 How is the absolute temperature scale defined using the Kelvin–Plank statement?

Chapter 6 Problem Subject Areas

6.1–6.26	Thermal efficiency and coefficient of performance
6.27–6.67	Carnot cycle, ideal efficiencies, and *COP*s
6.68–6.80	FE exam problems

PROBLEMS

6.1–6.26 Thermal efficiency and coefficients of performance

6.1 During each cycle, the working fluid of a Stirling engine receives 2.67 kJ of energy in the heat interaction with its combustor. The engine operates with a thermal efficiency of 0.20 and delivers 4 kW of shaft power. Determine (a) the number of cycles executed per minute and (b) the energy rejected as heat to the low-temperature reservoir during each cycle.

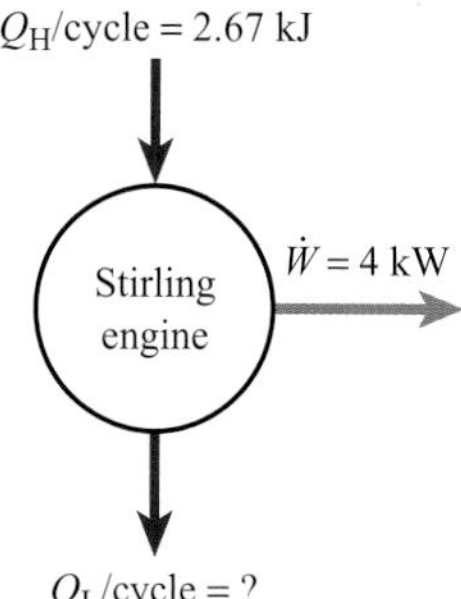

6.2 Energy is transferred to the working fluid (H_2O) in the boiler of a Rankine-cycle power plant at a rate of 131 MW. Energy is rejected in the condenser at a rate of 97 MW. Determine the net power produced and the thermal efficiency of this power plant.

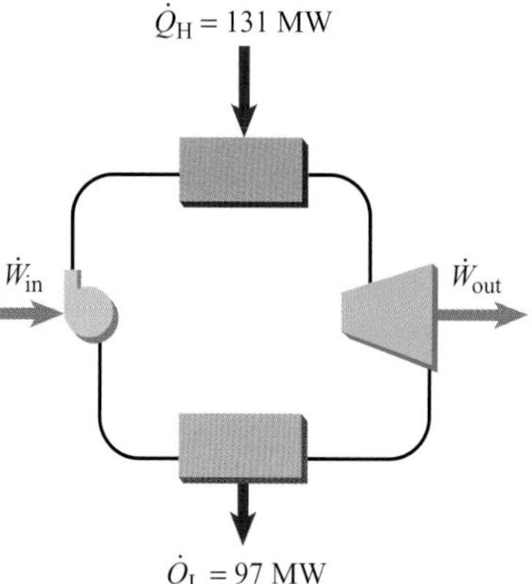

6.3 A steam power plant with a net power output of 87 MW has a thermal efficiency of 40%. What is the heat transfer to the water in the boiler? What is the heat transfer from the water in the condenser?

6.4 A steam power plant is 35% efficient and has a heat rejection of 95 MW. What is the heat transfer to the water in the boiler? What is the net power output from the power plant?

6.5 A steam power plant provides 120 MW of heat to the cycle in the boiler. If the power plant has a thermal efficiency of 38%, what is the net power delivered? What is the heat transfer in the condenser?

6.6 What is the efficiency of a steam power plant that delivers 45 MW of power and has a heat transfer of 80 MW in the condenser? What heat transfer is required in the boiler to operate this power plant?

6.7 A gas turbine operates as a heat engine with the fuel combustion producing an effective heat transfer of 180 MW from a high-temperature reservoir. If the gas turbine produces a net power output of 80 MW, what is the thermal efficiency of the gas turbine?

6.8 A gas turbine with a thermal efficiency of 35% produces a net power output of 50 MW. What is the heat transfer in the high-temperature reservoir from the fuel combustion?

6.9 Using atmospheric air as the heat source, a heat pump delivers 43,000 Btu/hr to heat a home. The coefficient of performance is 3.2. Determine (a) the electrical power required to operate the heat pump and (b) the rate at which energy is removed from the outside air.

6.10 A ground-source heat pump has a coefficient of performance of 4.2. What power is required to operate the heat pump if the desired heating of a house is 13 kW?

6.11 A heat pump with a coefficient of performance of 3.9 requires 4 kW of power to maintain the inside temperature of a house. What is the heat loss from the exterior of the house that needs to be added by the heat pump?

6.12 An air conditioner with a coefficient of performance of 3.9 requires 4 kW of power to maintain the inside temperature of a house. What is the heat gain from the exterior of the house that needs to be removed by the air conditioner? What is the heat rate from the air conditioner to the outside air?

6.13 If the air conditioner in Problem 6.12 is operated in the winter as a heat pump, what is the coefficient of performance? Why does the same device operating as a heat pump or air conditioner have different coefficients of performance?

6.14 A residential heat pump uses the ground as an energy source. At a particular operating condition, the heat pump delivers energy at a rate of 13.37 kW to the heated interior while simultaneously removing energy from the ground at a rate of 10.05 kW. Determine (a) the electrical power required to operate the heat pump and (b) the coefficient of performance.

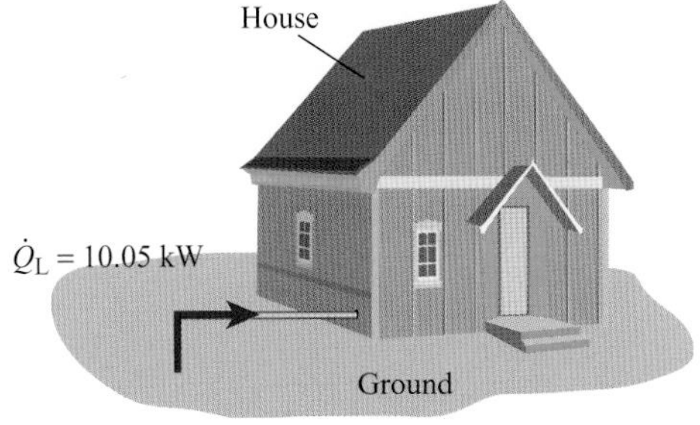

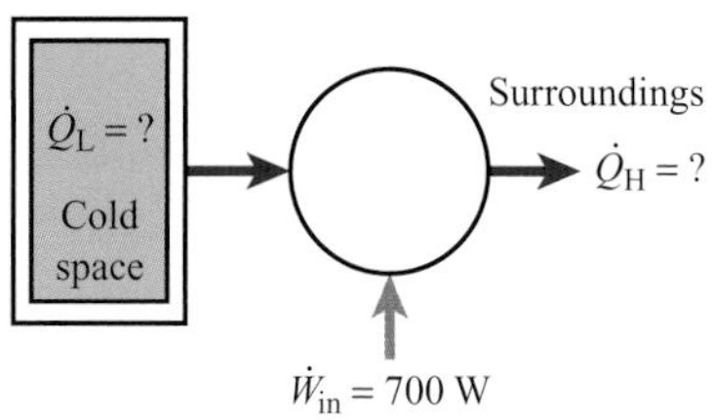

6.15 A refrigerator operates with a coefficient of performance of 4.2 and requires an electrical input of 700 W. Determine (a) the rate of heat transfer from the cold space of the refrigerator and (b) the rate at which the refrigerator rejects heat energy to the surroundings.

(Credit: Katrina Wittkamp / DigitalVision / Getty Images.)

6.16 What power is required to remove 1500 W of heat from a freezer if the coefficient of performance is 3.7?

6.17 A 10,000 BTU/hr room air conditioner requires 1080 W to operate. What is the coefficient of performance? (Note the mixed units in this problem!)

6.18 An ideal gas in a piston–cylinder device is compressed isothermally at T_L from state 1 to state 2. The gas is then heated at constant volume to state 3. Next the gas expands isothermally at T_u to state 4 and is then cooled at constant volume, returning to state 1. Derive an expression for the thermal efficiency of this cycle in terms of T_H, T_L, $\mathcal{V}_1/\mathcal{V}_2$, the gas constant R, and the constant-volume specific heat c_v.

6.19 A cycle using helium in a piston–cylinder device starts at $T_1 = 300\,\text{K}$ and $P_1 = 0.2\,\text{MPa}$. The helium is then compressed isothermally until $\mathcal{V}_2 = \mathcal{V}_1/10$. Heat addition at constant volume then raises the temperature to $T_3 = 1500\,\text{K}$. An isothermal expansion followed by a constant-volume cooling process returns the helium to the initial state 1. Assume that all processes are carried out slowly, so that equilibrium is always maintained.

A. Sketch (to scale) this cycle on a P–$\mathcal{V}$ diagram.

B. Choosing the helium to be your system, determine the work and heat transfer (kJ/kg) for each process in the cycle.

C. Determine the thermal (first-law) efficiency of this cycle.

6.20 (EES) Use EES or other software to plot the cycle described in Problem 6.19 on a P–$\mathcal{V}$ diagram. If you are using EXCEL to plot the processes, use at least five points across the compression and expansion processes to define a smooth curve.

6.21 (EES) Repeat Problem 6.19 using EES or other software to determine the work and heat transfer (kJ/kg) for each process in the cycle and the thermal efficiency of the cycle.

6.22 (EES) Using your computer solution from Problem 6.21, vary the temperature T_3 from 500 to 1500 K in increments of 100 K and calculate the thermal efficiency of the cycle at each value of T_3. Show your solutions in a table and as a plot. Plot the curve without symbols, label the axes, and include a plot title. Explain why the thermal efficiency changes as observed.

6.23 A heat engine of 30% thermal efficiency drives a refrigerator having a coefficient of performance of 4. Determine the ratio of the heat input to the engine to the heat removed from the cold body by the refrigerator.

6.24 During the month of January, a house is found to have an average heat loss of 14 kW. If the cost of electricity is 5.4 cents per kW·hr,

A. Find the cost of heating the house for a month using electrical resistance heating.

B. Find the cost of heating the house for a month using a heat pump with coefficient of performance of 3.7.

C. If the house currently has electric resistance heating, would you recommend that a heat pump be installed? In your analysis, assume that the total annual heating is three times the January heating and that the heat pump will cost $5000 to install. How many years of cost savings are required to pay for the cost of the heat pump installation?

6.25 Two heat engines (A and B) operate in series. Engine A receives heat Q_A from a source at T_A, produces work W_A, and transfers heat Q_B at temperature T_B to engine B. Engine B produces work W_B and rejects heat Q_0 to the environment at T_0. Determine the thermal efficiency η of the combined engine (A + B) in terms of the thermal efficiencies η_A and η_B of the individual engines.

6.26 A computer laboratory houses 16 computers that each generate 400 W of heat and eight 40-W fluorescent bulbs lighting the room. When fully occupied, there are 16 students each generating 100 W of heat. Find the power required to operate an air conditioner with $COP = 1.8$ to maintain a constant temperature in the computer lab.

6.27–6.67 Carnot efficiency, ideal efficiencies, and *COPs*

6.27 Energy from a high-temperature reservoir at 750 K flows at a rate of 2.1 kW to a heat engine. If the engine rejects heat to a low-temperature reservoir at 300 K, what is the maximum possible power the engine can deliver? What is the rate of heat rejection?

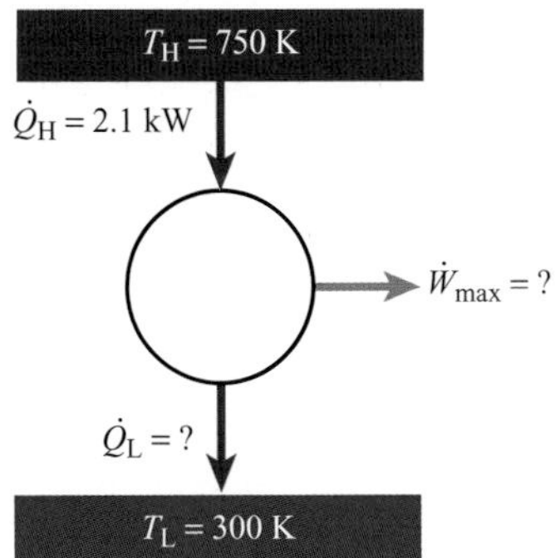

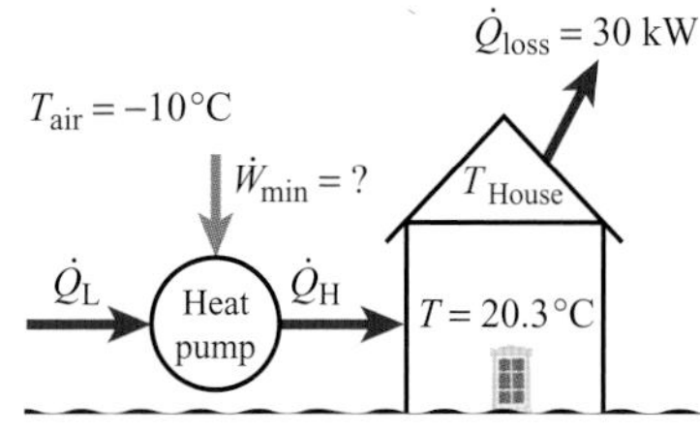

6.28 A heat pump is used to heat a building. The outside air temperature is −10 °C, and the desired inside air temperature is 20.3 °C. The heat loss through the walls, ceiling, etc., is 30 kW. Determine the ideal coefficient of performance and the minimum power required to operate the heat pump in order to maintain the heated space at a constant temperature.

6.29 Derive and expression for the maximum *COP* for a heat pump operating between T_H and T_L.

6.30 A reversible heat engine using air as the working fluid receives equal amounts of heat from two different sources. One source is at an absolute temperature of T_1 and the other source is at an absolute temperature of T_2. The heat engine rejects heat to a sink at an absolute temperature of T_3. Determine an expression for the thermal efficiency of this heat engine in terms of T_1, T_2, T_3.

6.31 A reversible heat engine operating between a 1500-F source and the atmosphere at 0 F produces work to drive a reversible heat pump. The heat pump is connected to the 0-F atmosphere and supplies the heating required to maintain a home at 70 F. The rate of heat transfer from the source is 100 kBtu/hr.

A. Determine the power delivered by the heat engine (kBtu/hr).

B. Determine the heat-transfer rate (kBtu/hr) to the home.

C. Is there any potential advantage of this system compared to simply using the 1500-F source to supply the home heating needs directly? Explain.

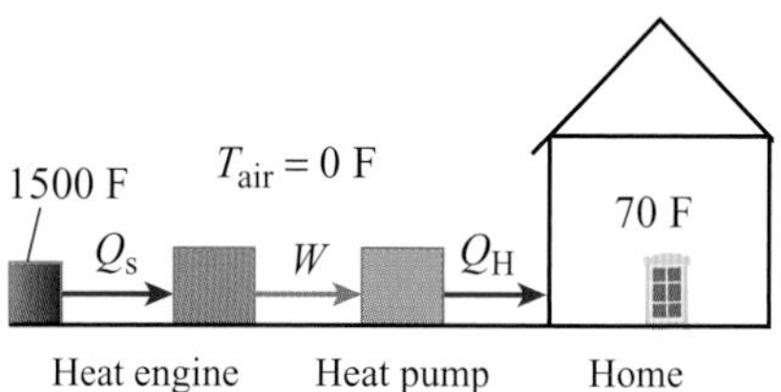

6.32 The maximum allowable temperature of a working fluid is usually determined by metallurgical considerations. In a certain power plant this temperature is 700 °C. A nearby river has a water temperature of 10 °C. Determine the maximum possible thermal efficiency for this power plant.

6.33 It is proposed to heat a home using a heat pump. The rate of heat transfer from the house to the environment is 15 kW. The home is to be maintained at 20 °C. The outside air is at −20 °C. Determine the minimum power (kW) required to drive the heat pump.

6.34 A heat source is maintained at 700 F and the environment is at 70 F. The working fluid of a reversible heat engine receives 1000 Btu of energy in a heat-transfer process at 700 F. After the fluid performs work, the fluid rejects heat at 70 F. Determine the heat rejected (Btu) at 70 F and the thermal efficiency of the device.

6.35 To maintain the interior of a structure at 20 °C when the outside temperature is 45 °C, one must provide 15 kW of cooling. Determine the minimum possible power (kW) required by an air conditioner to handle this load.

6.36 It is desired to produce refrigeration at −25 F. A heat source is available at 500 F and the ambient temperature is 80 F. A heat engine can produce work operating between the 500-F source and the ambient surroundings. This work in turn can be used to drive the refrigerator. Assuming all processes are reversible, determine the ratio of the heat transfer from the source to the heat transfer from the refrigerated space.

6.37 A heat pump is to be used to heat a house in winter and then reversed to cool the house in summer. The interior temperature is to be maintained at 70 F. Heat transfer through the walls and roof is estimated to be 1400 Btu/hr per degree temperature difference between the inside and outside.

A. If the outside temperature in winter is 35 F, determine the minimum power (Btu/hr) required to drive the heat pump.

B. If the power input is the same as in part A, determine the maximum outside temperature (F) for which the inside can be maintained at 70 F.

C. If the outside temperature in winter is 35 F, and the house is to be heated with electric resistance heating, determine the power (kW) required. Discuss the difference between your solutions in parts A and C.

6.38 Two reversible engines A and B operate in series. Engine A receives heat at 600 °C and rejects heat to a reservoir at temperature T. Engine B receives the heat rejected by the first engine and, in turn, rejects heat to a thermal reservoir at 30 °C. Determine the temperature T (°C) for the following situations:

A. The work outputs of the two engines are equal.

B. The thermal efficiencies of the two engines are equal.

6.39 An inventor claims to have developed a refrigeration unit that maintains a cold space at −10 °C while operating in a room where the temperature is 25 °C, and which has a coefficient of performance of 7.5. Determine whether constructing such a unit is possible.

(Credit: JazzIRT / E+ / Getty.)

6.40 An inventor claims to have developed a refrigeration unit that maintains a refrigerated space at 20 F while operating in a room where the temperature is 80 F. Determine the maximum possible *COP*.

6.41 An inventor claims to have developed a device that converts heat transferred to it at 810 °C into work, with a thermal efficiency of 75%. Heat is rejected to the 20 °C surroundings. Is this possible?

6.42 Determine the applicable efficiency or coefficient of performance for each of the following:

A. An ideal heat pump using refrigerant-12 and operating between pressures of 35.7 and 172.4 psia

B. A refrigerator providing 4500 Btu/hr of cooling while drawing 585 W

C. A heat engine to recover the thermal energy in the ocean by operating between the warm surface water (82 F) and the colder water (45 F) at a depth of 1200 feet

6.43 A heat pump is used to heat a house. When the outside air temperature is 10 F and the inside is maintained at 70 F, the heat loss from the house is 60,000 Btu/hr. Determine the minimum electric power (kW) required to operate the heat pump.

6.44 A heat pump is used to heat a house. When the outside air temperature is −10 °C and the inside is maintained at 21 °C, the heat loss from the house is 200 kW. Determine the minimum electric power (kW) required to operate the heat pump.

6.45 Assuming that the temperature of the surroundings remains at 60 F, determine the minimum increase in operating temperature (ΔT_H) needed to increase the thermal efficiency of a Carnot heat engine from 30% to 40%.

6.46 We wish to use a heat pump from the outside air to the interior of a house to maintain a house temperature of 67 F. What are the maximum coefficients of performance when the outside air is at 40 F and when the outside air is at 10 F?

6.47 (EES) Repeat Problem 6.46 using EES or other software to determine the maximum coefficient of performance when the outside air is at 40 F.

6.48 (EES) Using your computer solution from Problem 6.47, vary the outside temperature from −10 F to 50 F in increments of 5 F. Show your solutions in a table and as a plot. Plot the curve without symbols, label the axes, and include a plot title. Explain why the coefficient of performance changes as observed.

6.49 Solar energy is used to warm a large collector plate. This energy in turn is transferred as heat to a fluid in a heat engine. The engine rejects energy as heat to the atmosphere. Experiments indicate that about 200 Btu/hr·ft^2 of energy can be collected when the plate is operating at 190 F. Estimate the minimum collector area (ft^2) required for a plant to produce 1 kW of useful shaft power when the atmospheric temperature is 70 F.

6.50 A Carnot engine operates between a heat source at 1200 F and a heat sink at 70 F. The engine delivers 200 hp. Compute the heat supplied (Btu/hr), the heat rejected (Btu/hr), and the thermal efficiency of the heat engine.

6.51 The efficiency of a Carnot engine discharging heat to a cooling pond at 80 F is 30%. If the cooling pond receives 800 Btu/min, what is the power output of the engine? What is the source temperature?

6.52 A Carnot refrigerator is used for making ice. Water freezing at 32 F is the cold reservoir. Heat is rejected to a river at 72 F. Determine the work required to freeze 2000 lb_m of ice? (The latent heat of fusion of ice is 144 Btu/lb_m.)

6.53 In Problem 6.52, determine the required power input in kW and hp if this operation is to be carried out in one hour.

6.54 A Carnot engine operating between 750 and 300 K produces 100 kJ of work. Determine (a) the thermal efficiency and (b) the heat supplied (kJ).

6.55 (EES) Repeat Problem 6.54 using EES or other software to determine the thermal efficiency and heat supplied.

6.56 (EES) Using your computer solution from Problem 6.55, vary the low-temperature reservoir from 250 K to 400 K in increments of 10 K. Generate a table showing listing the thermal efficiency and heat supplied at each value of the low temperature. Also plot the thermal efficiency as a function of the low temperature. Plot the curve without symbols, label the axes, and include a plot title. Explain why the thermal efficiency changes as observed.

6.57 A reversed Carnot cycle operating between −20 °C and 30 °C receives 126.375 kJ of heat. If this cycle is operating as a refrigerator, determine (a) the coefficient of performance and (b) the heat rejected (kJ).

6.58 Rework Problem 6.55 but assume the device is a heat pump.

6.59 Calculate the thermal efficiency of a Carnot-cycle heat engine operating between 1051 F and 246 F. What would the coefficient of performance of this device be if it were reversed to run as a heat pump? As a refrigerator?

6.60 A Carnot engine operates between a source at 800 F and a sink at 100 F. If 200 Btu is rejected each minute to the sink, determine the power output.

6.61 A Carnot engine receives 15 Btu/s from a source at 900 F and delivers 6000 $ft \cdot lb_f/s$ of power. Determine the engine efficiency and the temperature (F) of the low-temperature reservoir.

6.62 A Carnot heat engine receives heat from a high-temperature reservoir at 527 °C. The heat rejected from this engine is supplied to a second Carnot engine. This second engine rejects heat to a low-temperature reservoir at 17 °C. The first engine rejects 400 kJ to the second engine. If both engines have the same efficiency, determine the following:

A. The temperature of the high-temperature reservoir for the second engine (i.e., the temperature of the low-temperature reservoir of the first engine)

B. The energy received by the first engine from the 527 °C source

C. The work done by each engine

D. The efficiency of each engine

6.63 Rework parts A and D of Problem 6.62. When the two engines now deliver the same work instead of having the same efficiency.

6.64 A refrigerator operates on a Carnot cycle between thermal reservoirs at −6 °C and 22 °C. Calculate the coefficient of performance. Also determine the refrigeration effect and the heat rejected to the high-temperature reservoir, both per kJ of work supplied.

6.65 The low-temperature reservoir of a Carnot heat engine is at 10 °C. To increase the efficiency of this heat engine from 40% to 55%, by how many degrees must the temperature of the high-temperature reservoir be increased?

6.66 The efficiency of a Carnot heat engine is $\eta_{Carnot} = (T_H - T_L)/T_H$. To increase the efficiency is it better to increase T_H or decrease T_L? **HINT**: Determine $d\eta$.

6.67 Consider four Carnot engines in series thermally: Q_1 is absorbed by the first engine; Q_2 is rejected by the first engine into an intermediate reservoir at T_2; Q_2 is absorbed by the second Carnot engine, and so on. Each engine produces the same work. Show that for these conditions the temperature differences across the engines are equal.

6.68–6.80 FE exam problems

6.68 A steam power plant with a power output of 105 MW has a thermal efficiency of 35%. What is the heat transfer to the water in the boiler? a. 300 MW, b. 450 MW, c. 195 MW, d. 104 MW.

6.69 A ground-source heat pump has a coefficient of performance of 3.8. What power is required to operate the heat pump if the desired heating of a house is 15 kW? a. 57 kW, b. 4.0 kW, c. 11 kW, d. 15 kW.

6.70 An air conditioner with a coefficient of performance of 3.2 requires 5 kW of power to maintain the inside temperature of a house. What is the heat gain from the exterior of the house that is to be removed by the air conditioner? a. 21 kW, b. 5 kW, c. 16 kW, d. 26 kW.

6.71 What is the maximum thermal efficiency for a heat engine operating between heat reservoirs at 650 K and 300 K? a. 54%, b. 26%, c. 46%, d. 73%.

6.72 What is the maximum coefficient of performance for an air conditioner operating between 23 °C and 50 °C? a. 3.2, b. 5.7, c. 12, d. 11.

6.73 What is the maximum coefficient of performance for a heat pump operating between 20 °C and 5 °C? a. 18.5, b. 19.5, c. 1.05, d. 2.5.

6.74 To maintain the interior of a structure at 20 °C when the outside temperature is 35 °C, one must provide 12 kW of cooling. Determine the minimum possible power (kW) required by an air conditioner to handle this load. a. 12.6 kW, b. 1.2 kW, c. 0.6 kW, d. 11.4 kW.

6.75 A Carnot engine operates between a heat source at 1200 F and a heat sink at 70 F. The engine delivers 200 hp. Determine the heat supplied (Btu/hr). a. 51,000 Btu/hr, b. 24,000 Btu/hr, c. 52,000 Btu/hr, d. 75,000 Btu/hr.

6.76 A Carnot engine operates between a heat source at 1200 F and a heat sink at 70 F. The engine delivers 200 hp. Determine the thermal efficiency of the heat engine. a. 68%, b. 56%, c. 32%, d. 47%.

6.77 What is the maximum coefficient of performance for a freezer with an interior temperature of –8 °C that is in a room at 21 °C? a. 0.9, b. 9.14, c. 10.14, d. 1.1.

6.78 A freezer with an interior temperature of –8 °C is in a room at 21 °C. What power is required to remove 3 kW of heat from the freezer? a. 3.3 kW, b. 0.5 kW, c. 0.1 kW, d. 0.33 kW.

6.79 Determine the coefficient of performance of a reversible heat pump that heats the inside of a house to 70 F when the outside temperature is 15 F. a. 8.6, b. 9.6, c. 1.1, d. 3.5.

6.80 What power is required to provide 51,000 Btu/hr to the inside of a house at 70 F when the outside temperature is 15 F? a. 17.7 hp, b. 2.1 hp, c. 22 hp, d. 18.5 hp.

CHAPTER 7
Entropy and Availability

LEARNING OBJECTIVES

After studying Chapter 7, you should be able to:

- State two or more ways in which the second law is useful in engineering applications.
- Explain the difference between a reversible and an irreversible process and illustrate it with a concrete example.
- List four or more irreversible processes.
- Draw the processes comprising the Carnot cycle in pressure–volume and temperature–entropy coordinates.
- State in words and write out symbolically the macroscopic (Clausius) definition of entropy.
- Be proficient in illustrating ideal and real processes in P–v, T–s, and h–s coordinates for the following common devices: turbines, compressors, pumps, heat exchangers, throttles, and nozzles.
- Explain the increase of entropy principle and show how it is an indicator of spontaneous change.
- Calculate equilibrium constants from tabulations of Gibbs function of formation data.

CHAPTER 7 OVERVIEW

THE FIRST LAW OF THERMODYNAMICS can be mathematically expressed in a variety of ways. All these expressions, however, are easily viewed as rearrangements of the statement that energy can neither be created nor destroyed but only converted from one form to another. In contrast, there is no single universally agreed statement of the second law of thermodynamics. Kline [1] indicates that many seemingly different statements have been accepted as the second law, all of which, however, can be shown to be equivalent after careful and sometimes subtle application of logic. This multiplicity of apparently disparate statements can lead to confusion in understanding the second law.

Chapter 6 considered the thermal efficiency of cycles and used a second-law analysis to determine the best possible thermal efficiency. This best possible thermal efficiency (or coefficient of performance) is achieved in a reversible Carnot cycle. All other cycles have a thermal efficiency (or coefficient of performance) that is less than this optimum value for a cycle operating between given high- and low-temperature reservoirs.

In this chapter, we will define the entropy of a system and examine both reversible and irreversible processes for closed and open systems. In doing this, several statements of the second law will be verified. The second law will allow us to determine the

<table>
<tr><td colspan="3">DESIGN AND ANALYSIS OF SYSTEMS</td></tr>
<tr><td colspan="3">ANALYSIS OF PRACTICAL DEVICES</td></tr>
<tr><td>CONSERVATION OF MASS</td><td>CONSERVATION OF ENERGY (First Law of Thermodynamics)</td><td>SECOND LAW OF THERMODYNAMICS (Entropy)</td></tr>
<tr><td colspan="3">PROPERTIES OF MATTER</td></tr>
<tr><td colspan="3">FRAMEWORKS FOR ANALYSIS, KEY CONCEPTS, AND DEFINITIONS</td></tr>
</table>

FIGURE 7.1 Hierarchical arrangements of the topics in our study of engineering thermodynamics. Chapters 6 and 7 explore the second law of thermodynamics.

direction of any spontaneous change and can be used to determine the equilibrium state of any system.

The introduction of second-law analysis has been included in two separate chapters (Chapters 6 and 7) to give the course instructor flexibility in how this material is presented. The chapters are written so that the entropy analysis in Chapter 7 can be presented first before the cycle analysis in Chapter 6, if desired.

Figure 7.1 illustrates the key role of Chapters 6 and 7 in any thermodynamic analysis. Later chapters will apply first- and second-law analysis to devices and systems.

Historical Context

(Credit: Bettmann / Contributor / Getty Images.)

Josiah Willard Gibbs (1839–1903) was an American mathematical physicist who developed the important thermodynamics equations later called the Gibbs relationships (Eqs. 7.6 and 7.7). Gibbs was born in New Haven, Connecticut. He graduated from Yale College in 1858 and continued there for a Ph.D. in 1863. In 1871 he was appointed as a professor of mathematical physics in Yale College and held that position for the rest of his life. Gibbs developed the concepts of the free energy of substances and the tendency of a substance to lower its energy and increase its entropy. In 1873, Dr. Josiah Gibbs published his derivation of two equations that relate entropy to other state properties. Gibbs studied phase changes and chemical reactions using the concept of the chemical potential, which he developed [2]. Gibbs was honored in 2005 on a United States stamp.

7.1 Entropy

Chapter 6 introduced the second law of thermodynamics. In this chapter, we will consider the second law in a different way by defining the property called entropy. Entropy can be defined most generally from a macroscopic viewpoint and, in more restrictive ways, from a microscopic (molecular) viewpoint. We will focus on the macroscopic definition in this chapter.

7.1a Definition[1]

The entropy change of a system is defined by considering the heat transfer to or from the system during an internally reversible process. The entropy change, dS, for a macroscopic thermodynamic system is defined as:

$$dS \equiv \left(\frac{\delta Q}{T}\right)_{\substack{\text{int}\\ \text{rev}}}. \tag{7.1a}$$

To properly implement this definition, we enforce the convention that heat *into* the system is positive and heat *out* is negative. Integrating Eq. 7.1a for any process involving a change from state 1 to state 2 yields

[1] This definition from classical thermodynamics is usually attributed to Clausius [3].

$$\Delta S(\equiv S_2 - S_1) \equiv \int_1^2 \left(\frac{\delta Q}{T}\right)_{\substack{\text{int}\\ \text{rev}}}. \quad \textbf{(7.1b)}$$

A reversible process, as defined in Section 6.3d, is a process after which the system and all parts of the surroundings can be restored to their initial state. Irreversibilities in the system or irreversibilities between the system and surroundings will prevent a process from occurring in the opposite direction, however. Since entropy is a property of the system and not the surroundings, the definition of entropy specifies only that irreversiblities do not occur within the system.

Appendix 7A provides a microscopic interpretation of entropy for a gas.

Equation 7.1b is curious. To employ it to calculate a change in entropy, ΔS, a *reversible* path must first be defined between states 1 and 2, and then the heat transferred divided by the temperature at which the heat exchange takes place must be integrated over this path. This definition of entropy also is quite abstract and begs for a physical interpretation. Unfortunately, classical thermodynamics offers none. Definitions, however, are free to stand on their own as part of a logical framework. Fortunately, microscopic (statistical) thermodynamics does lend some physical insight into the meaning of entropy for some systems. (See Appendix 7A at the end of the chapter.)

We can use the definition of entropy in Eq 7.1a to determine the heat transfer to or from a closed system during a reversible process:

$$_1Q_2 = \int_1^2 T\,dS|_{\substack{\text{int}\\ \text{rev}}}. \quad \textbf{(7.1c)}$$

If a reversible process is shown on a T–S diagram, the area under the curve will be the heat transfer that occurs during the process, as shown in Figure 7.2. This is similar to the visual interpretation of P–V work using the P–V diagram. (See Example 4.1.)

7.1b Isothermal Heat Transfer

One process that we will consider in entropy analysis is heat transfer that occurs at constant temperature. There are several common conditions where this can occur. Isothermal heat transfer occurs when a saturated mixture boils or condenses at constant pressure. A constant-pressure boiling process is shown in Fig. 2.19 from states B–C–D–E. A constant-pressure condensation process would go from states E–D–C–B. Isothermal heat transfer can also occur when the system (or region of interest) is very large. A large region that can accept or reject heat transfer at constant temperature is called a **heat reservoir**. For an isothermal heat transfer process, Eq. 7.1a can be used across an internally reversible process to give the total change in entropy as

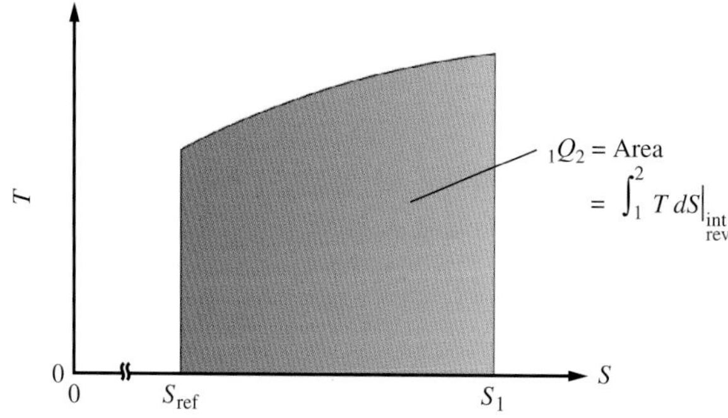

FIGURE 7.2 The area under a T–s curve is the net heat transfer during a reversible process.

$$\Delta S = S_2 - S_1 = \frac{{}_1Q_2}{T}\bigg|_{\text{int rev}}. \tag{7.2a}$$

If the change in entropy is known in an internally reversible, isothermal process the heat transfer can be found as:

$${}_1Q_2 = T(S_2 - S_1)\big|_{\text{int rev}}. \tag{7.2b}$$

Example 7.1 Closed Isothermal Reversible System

Consider 500 kJ of energy added to a closed system as heat transfer in a reversible isothermal process at 320 K. What is the change in entropy of the system?

Solution

Known ${}_1Q_2$ and T in an isothermal, reversible process

Find $S_{\text{final}} - S_{\text{init}}$

Sketch

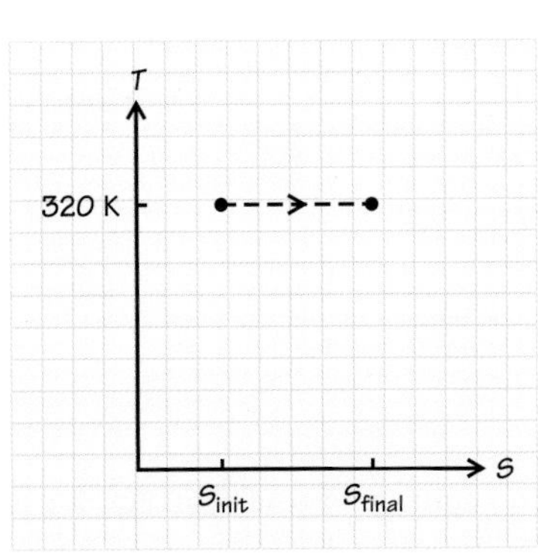

Modeling, Premises and Assumptions

i. Isothermal process
ii. Internally reversible process

Analysis For an isothermal reversible process, the change in entropy of the system is found using Eq. 7.2:

$$\Delta S = S_2 - S_1 = \frac{{}_1Q_2}{T}\bigg|_{\substack{\text{int}\\\text{rev}}}$$

$$\Delta S = S_2 - S_1 = \frac{500\,\text{kJ}}{320\,\text{K}} = 1.563\ \text{kJ/K}.$$

Comments The entropy of the closed system increases when heat is transferred to the system. Note that the mass or type of fluid of the system was not specified.

Self-Test 7.1

Consider 500 kJ of energy is removed from a closed system as heat transfer in a reversible isothermal process at 420 K. What is the change in entropy of the system?

(Answer: –1.19 kJ/K)

(This section can be omitted without loss of continuity.)

7.1c Derivation of Entropy as a Property

That entropy, as defined in Eq. 7.1a, is a thermodynamic property is not obvious. To prove that entropy is a thermodynamic property requires that we demonstrate that the total entropy change in a cycle is zero. Substituting for the definition of entropy gives:

$$\oint_{\text{cycle}} dS = \oint_{\text{cycle}} \left(\frac{\delta Q}{T}\right)_{\substack{\text{int}\\ \text{rev}}} = 0. \tag{7.3}$$

Recall that for *any* quantity to be a thermodynamic property requires that, upon executing a thermodynamic cycle, the value of the quantity returns to its initial value, that is,

$$\oint_{\text{cycle}} d(\text{property}) \equiv 0. \tag{7.4}$$

It is in this sense that energy is a thermodynamic property, since from the first law,

$$\oint_{\text{cycle}} (\delta Q - \delta W) = 0,$$

or

$$\oint_{\text{cycle}} dE = 0.$$

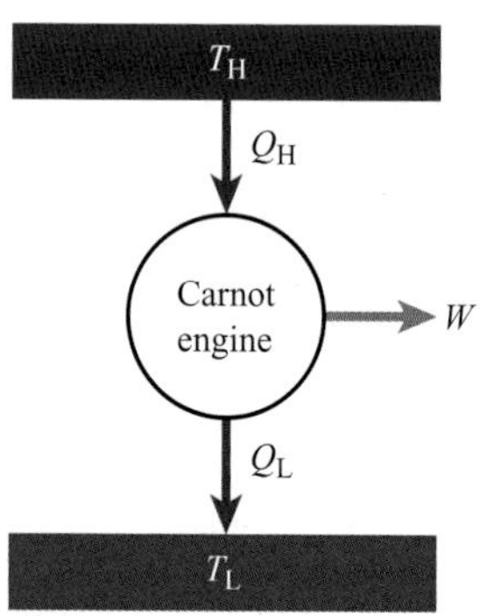

FIGURE 7.3 Ideal Carnot engine operating between high- and low-temperature reservoirs.

To begin our proof that Eq. 7.3 is true, we consider a Carnot heat engine operating between a high-temperature reservoir at T_H and a low-temperature reservoir at T_L, as shown in Fig. 7.3. We choose a Carnot engine because the heat transfers at T_H and T_L are reversible processes, as are all the processes in the engine's execution of a cycle. For the Carnot cycle, the integral of $(\delta Q/T)_{rev}$ has two components, one for the heat exchange at T_H and one for the heat exchange at T_L:

$$\oint_{\substack{\text{Carnot}\\ \text{cycle}}} \left(\frac{\delta Q}{T}\right)_{\text{rev}} = \frac{Q_H}{T_H} - \frac{Q_L}{T_L}, \tag{7.5a}$$

where the minus sign indicates that Q_L is rejected during the cycle. We also know from the definition of the thermodynamic temperature, Eq. 6.13b, that

$$\frac{Q_H}{T_H} = \frac{Q_L}{T_L}. \tag{7.5b}$$

Substituting this result into Eq. 7.5a yields

$$\oint_{\substack{\text{Carnot}\\ \text{cycle}}} \left(\frac{\delta Q}{T}\right)_{\text{rev}} = \frac{Q_L}{T_L} - \frac{Q_L}{T_L} = 0, \tag{7.5c}$$

which provides the desired proof. To generalize to *any* reversible cycle, we note that such cycles can always be modeled as an appropriate assembly of Carnot engines.

7.1d Specific Entropy, a State Property

The thermodynamic property called **entropy (S)** originates from the second law of thermodynamics.[2] This property is particularly useful in determining the spontaneous direction of a process and for establishing maximum possible efficiencies, for example. The entropy per mass is the specific entropy, $s = S/M$.

The entropy property can be interpreted from both macroscopic and microscopic (molecular) points of view. One way of describing the entropy of matter is that:

Entropy is a measure of the unavailability of thermal energy to do work in a closed system.

This definition implies that two identical quantities of energy are not able to produce the same amount of useful work. Entropy is valuable in quantifying the usefulness of energy.

The following informal definition presents a microscopic (molecular) interpretation of entropy:

Entropy is a measure of the microscopic randomness associated with a closed system

To help understand this statement, consider the physical differences between water existing as a solid (ice) and as a vapor (steam), as shown in Fig. 7.4. In a piece of ice, the individual H_2O molecules are locked in relatively rigid positions, with the individual hydrogen and oxygen atoms vibrating within well-defined domains. In contrast, in steam the individual molecules are free to move within any containing vessel. Thus, we say that the state of the steam is more disordered than that of the ice and that the steam has a greater entropy per unit mass. It is this idea, in fact, that leads to the **third law of thermodynamics**, which states that all perfect crystals have zero entropy at a temperature of absolute zero. For the case of a perfectly ordered crystal at absolute zero, there is no molecular motion, and there are no imperfections in the lattice; thus, there is no uncertainty about the microscopic state (because there is no disorder or randomness) and the entropy is zero. A more detailed discussion of the microscopic interpretation of entropy is presented in the appendix to this chapter.

The SI units for entropy S, mass-specific entropy s, and molar-specific entropy $\bar{s}$, are J/K, J/kg·K, and J/kmol·K, respectively. Tabulated values of entropies for H_2O, air, and ideal gases are found in Appendices B, C and D, respectively. Entropies for selected substances are also available from the NIST software and WebBook.

7.1e Gibbs Relationships

Sections 7.1a–d considered entropy from several different perspectives: the change in entropy in a reversible process was related to the heat transfer to the system and entropy was also shown to be a state property of a fluid. Since entropy is a state property of the fluid, we can relate it to other state properties: temperature, pressure, internal energy, and enthalpy. These equations are called the Gibbs relationships.

[2] Rudolf Clausius (1822–1888) chose *entropy*, a Greek word meaning transformation, because of its root meaning and because it sounded similar to *energy*, a closely related concept [3].

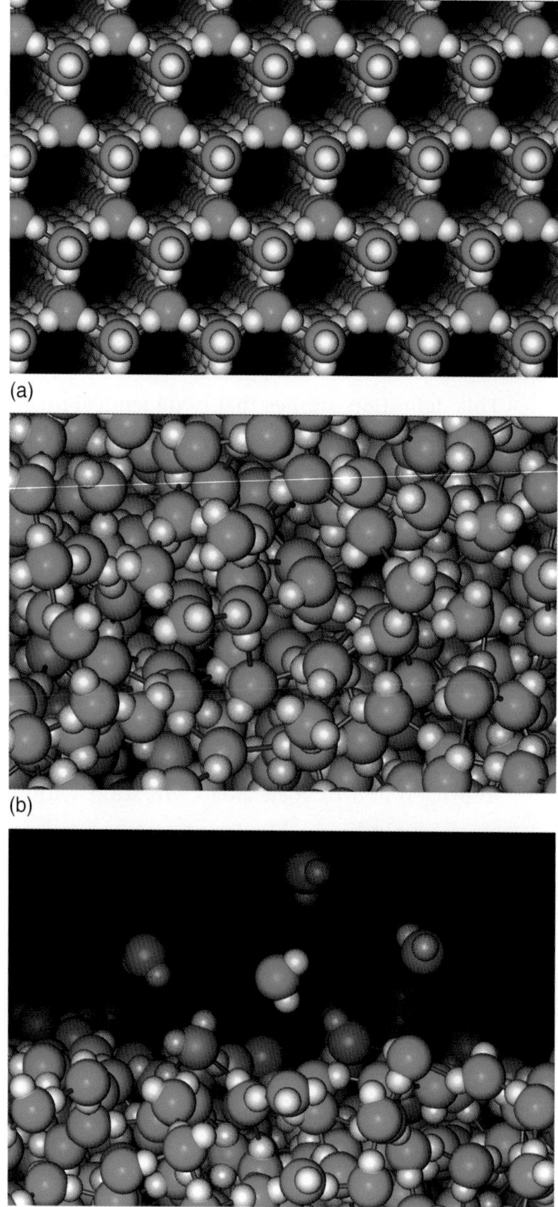

FIGURE 7.4 Molecular interactions for water in three phases. **(a)** Hexagonal crystal structure of ice. The open structure causes ice to be less dense than liquid water. **(b)** Structure of liquid water. **(c)** Evaporating water molecules. (Credits: Clive Freeman/ Biosym Technologies/Science Photo Library.)

To obtain the desired temperature–entropy relationships for an ideal gas, we apply concepts associated with the first and second laws of thermodynamics. These concepts are developed in Chapter 5 and in Section 7.1a, respectively.

We begin by considering a simple compressible closed system undergoing an internally reversible process that results in an incremental change in state. For such a process, the first law of thermodynamics is expressed (Eq. 5.5) as

$$\delta Q_{\text{rev}} - \delta W_{\text{rev}} = dU.$$

The incremental heat interaction δQ_{rev} is related directly to the entropy change through the formal definition of entropy from Section 7.1a (i.e., Eq. 7.1a),

$$dS \equiv \left(\frac{\delta Q}{T}\right)_{\text{rev}},$$

or

$$\delta Q_{\text{rev}} = TdS.$$

For a simple compressible substance, the only reversible work mode is compression and/or expansion, that is,

$$\delta W_{\text{rev}} = Pd\mathcal{V}.$$

Substituting these expressions for δQ_{rev} and δW_{rev} into the first-law statement yields

$$TdS - Pd\mathcal{V} = dU.$$

We rearrange this result slightly and write

$$TdS = dU + Pd\mathcal{V}, \tag{7.6a}$$

which can also be expressed on a per-unit-mass basis as

$$Tds = du + Pdv. \tag{7.6b}$$

Equation 7.6 is the first of the so-called Gibbs or T–ds equations. We can obtain a second Gibbs equation by employing the definition of enthalpy (Eq. 2.17), that is,

$$U = H - P\mathcal{V},$$

which can be differentiated to yield

$$dU = dH - Pd\mathcal{V} - \mathcal{V}dP.$$

Substituting this expression for dU into Eq. 2.35a and simplifying yields

$$TdS = dH - \mathcal{V}dP, \tag{7.7a}$$

or on a per-unit-mass basis

$$Tds = dh - vdP. \tag{7.7b}$$

Note that Eqs. 7.6 and 7.7 apply to any simple compressible substance, not just an ideal gas; furthermore, these relationships apply to any incremental process, not just an internally reversible one. The Gibbs relationships can be used to find the incremental change in entropy during an incremental process. The equations can be integrated to determine the change in entropy during a finite process from state 1 to state 2. Since the Gibbs relationships include only state properties, they can be used for reversible and irreversible processes.

7.2 Ideal-Gas Properties and Processes

This section will introduce several ways to find entropy changes for an ideal gas and then look at isentropic processes of an ideal gas. This section is a continuation of the treatment of state relationships for ideal gases found in Section 2.5.

7.2a Ideal-Gas Temperature–Entropy (Gibbs) Relationships

Gibbs relationships can be used to find the change in entropy of an ideal gas, using the ideal-gas equation of state and the ideal-gas calorific equations of state found in

Section 2.5. Starting with Eq. 7.6b, we substitute $v = RT/P$ (Eq. 2.26a) and $dh = c_p dT$ to yield

$$Tds = c_p dT - \frac{RT}{P} dP,$$

which, upon dividing through by T, becomes

$$ds = c_p \frac{dT}{T} - R \frac{dP}{P}. \tag{7.8}$$

Similarly, Eq. 7.7b is transformed using the substitutions $P = RT/v$ and $du = c_v dT$ to yield

$$ds = c_v \frac{dT}{T} + R \frac{dv}{v}. \tag{7.9}$$

To evaluate the entropy change in going from state 1 to state 2, we integrate Eqs. 7.8 and 7.9, that is,

$$s_2 - s_1 = \int_1^2 ds = \int_1^2 c_p \frac{dT}{T} - \int_1^2 R \frac{dP}{P}$$

and

$$s_2 - s_1 = \int_1^2 ds = \int_1^2 c_v \frac{dT}{T} + \int_1^2 R \frac{dv}{v}.$$

The second term on the right-hand side of each of these expressions can be easily evaluated since R is a constant, and so

$$s_2 - s_1 = \int_1^2 c_p \frac{dT}{T} - R \ln \frac{P_2}{P_1} \tag{7.10a}$$

and

$$s_2 - s_1 = \int_1^2 c_v \frac{dT}{T} + R \ln \frac{v_2}{v_1}. \tag{7.10b}$$

To evaluate the integrals in Eqs. 7.10a and 7.10b requires knowledge of $c_p(T)$ and $c_v(T)$. Curve-fit expressions for $c_p(T)$ are readily available for many species (e.g., Table D.14); $c_v(T)$ can be determined using the c_p curve fits and the ideal-gas relationship, $c_v(T) = c_p(T) - R$ (Eq. 2.32).

Since $\int_1^2 c_p dT/T$ is a function of temperature only for an ideal gas, it can be tabulated. This expression can be found as $s^\circ(T)$ for air in Table C.2 and for other ideal gases in Tables D. The change in entropy from state 1 to state 2 is then

$$s_2(T_2, P_2) - s_1(T_1, P_1) = s^\circ(T_2) - s^\circ(T_1) - R \ln \frac{P_2}{P_1}. \tag{7.11}$$

In many engineering applications, an average value of c_p (or c_v) over the temperature range of interest can be used to evaluate $s_2 - s_1$ with reasonable accuracy:

$$s_2 - s_1 = c_{p,\text{avg}} \ln \frac{T_2}{T_1} - R \ln \frac{P_2}{P_1} \tag{7.12a}$$

and

$$s_2 - s_1 = c_{v,\text{avg}} \ln \frac{T_2}{T_1} + R \ln \frac{v_2}{v_1}. \tag{7.12b}$$

Note that we must always find the difference in specific entropy using one of these methods. We must never use different methods or tables to find s_1 and s_2 since the reference value may be different in these different methods or tables.

Molar-specific forms of Eqs. 7.7–7.12 are formed by substituting molar-specific properties ($\bar{v}$, $\bar{u}$, $\bar{h}$, $\bar{s}$, $\bar{c}_p$, and $\bar{c}_v$) for their mass-specific counterparts (v, u, h, s, c_p, and c_v) and substituting R_u for R.

Example 7.2 Closed Isochoric System

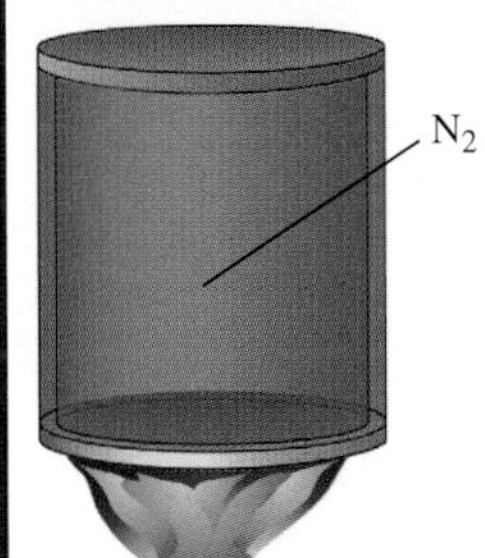

A rigid tank contains 1 kg of N_2 initially at 300 K and 1 atm. Energy is added to the gas until a final temperature of 600 K is reached. Calculate the entropy change of the N_2 associated with this heating process.

Solution

Known M_{N_2}, P_1, T_1, T_2

Find $s_2 - s_1$

Sketch

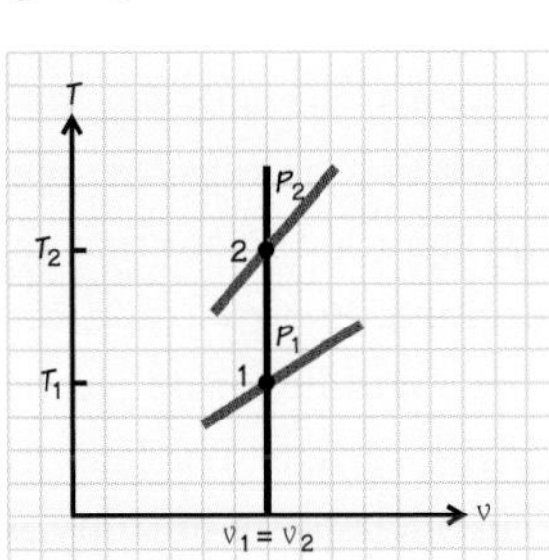

Modeling, Premises and Assumptions

i. The tank is leak free and perfectly rigid; therefore, $v_2 = v_1$.
ii. The N_2 behaves like an ideal gas.

Analysis Recognizing that this is a constant-volume process (see sketch and assumptions) makes the calculation of the specific entropy change straightforward. To accomplish this, we simplify Eq. 7.12b, which applies to an ideal gas. Starting with

$$s_2 - s_1 = c_{v,\text{avg}} \ln \frac{T_2}{T_1} + R \ln \frac{v_2}{v_1},$$

we recognize that the second term on the right-hand side is zero since $v_2 = v_1$ and $\ln(v_2/v_1) = \ln(1) = 0$; thus,

$$s_2 - s_1 = c_{v,\text{avg}} \ln \frac{T_2}{T_1}.$$

To evaluate $c_{v,\,\text{avg}}$, we use data from Table D.7 together with Eq. 2.32 as follows:

$$\bar{c}_{p,\text{avg}} = \frac{\bar{c}_p(300) + \bar{c}_p(600)}{2} = \frac{29.075 + 30.086}{2}\ \text{kJ/kmol}\cdot\text{K} = 29.581\ \text{kJ/mol}\cdot\text{K}.$$

Converting to a mass basis,

$$\begin{aligned} c_{p,\text{avg}} &= \bar{c}_{p,\text{avg}}/\mathcal{M}_{N_2} \\ &= \frac{29.581\,\text{kJ/kmol}\cdot\text{K}}{28.013\,\text{kg/kmol}} = 1.0560\ \text{kJ/kg}\cdot\text{K}, \end{aligned}$$

and calculating the specific gas constant for N_2,

$$\begin{aligned} R &= R_u/\mathcal{M}_{N_2} \\ &= \frac{8.31447\,\text{kJ/kmol}\cdot\text{K}}{28.013\,\text{kg/kmol}} = 0.2968\ \text{kJ/kg}\cdot\text{K}, \end{aligned}$$

we obtain $c_{v,\,\text{avg}}$ from Eq. 2.32, that is,

$$\begin{aligned} c_{v,\text{avg}} &= c_{p,\text{avg}} - R \\ &= 1.0560\ \text{kJ/kg}\cdot\text{K} - 0.2968\ \text{kJ/kg}\cdot\text{K} \\ &= 0.7592\ \text{kJ/kg}\cdot\text{K}. \end{aligned}$$

We now calculate the entropy change from the simplified version of Eq. 7.12b that we just derived:

$$\begin{aligned} s_2 - s_1 &= (0.7592\ \text{kJ/kg}\cdot\text{K}) \ln\left(\frac{600\,\text{K}}{300\,\text{K}}\right) \\ &= 0.5262\ \text{kJ/kg}\cdot\text{K}. \end{aligned}$$

Comments First, we note the importance of recognizing that the volume is constant for the given process. Discovering that a property is constant is frequently the key to solving problems. Second, we note that, because both the initial and final temperatures were given, finding an average value for the specific heat is accomplished without any iteration. Note that the method used to add energy to the closed system was not specified. Energy could be added by heat transfer to the system, by mixing of the gas using a paddle, or by electrical heating. All these methods cause an increase in entropy for the system. The heat transfer process can occur as an internally reversible process if the increase in entropy is due to heat transfer to the closed system. Mixing of the gas or electrical heating would be an internal irreversibility.

Self-Test 7.2

Repeat Example 7.2 using Equation 7.12a. Do you expect that your answer will be the same as in Example 7.2? Why or why not?

(Answer: 0.5262 kJ/kg·K. Yes, because entropy is a property of the system and is defined by the states.)

7.2b Ideal-Gas Isentropic-Process Relationships

For an **isentropic process,** a process in which the initial and final entropies are identical, Eqs. 7.12a and 7.12b can be used to develop some useful engineering relationships. Setting $s_2 - s_1$ equal to zero in Eq. 7.12a gives

$$0 = c_{p,\mathrm{avg}} \ln \frac{T_2}{T_1} - R \ln \frac{P_2}{P_1}.$$

Dividing by R and rearranging yields

$$\ln \frac{P_2}{P_1} = \frac{c_{p,\mathrm{avg}}}{R} \ln \frac{T_2}{T_1},$$

and removing the logarithm by exponentiation, we obtain

$$\frac{P_2}{P_1} = \left(\frac{T_2}{T_1}\right)^{c_{p,\mathrm{avg}}/R},$$

The ratio c_p/R can be rewritten by introducing the specific heat ratio $\gamma = c_p/c_v$ and the fact that $c_p - c_v = R$ (Eq. 2.32b), and performing a little algebra,

$$\frac{c_{p,\mathrm{avg}}}{R} = \frac{c_{p,\mathrm{avg}}}{c_{p,\mathrm{avg}} - c_{v,\mathrm{avg}}} = \frac{\dfrac{c_{p,\mathrm{avg}}}{c_{v,\mathrm{avg}}}}{\dfrac{c_{p,\mathrm{avg}} - c_{v,\mathrm{avg}}}{c_{v,\mathrm{avg}}}} = \frac{\gamma}{\gamma - 1};$$

thus,

$$\frac{P_2}{P_1} = \left(\frac{T_2}{T_1}\right)^{\gamma/\gamma-1}, \tag{7.13a}$$

or, more generally,

$$T_1^{\gamma} P_1^{1-\gamma} = T_2^{\gamma} P_2^{1-\gamma} = \text{constant.} \tag{7.13b}$$

Equation 7.12b can be similarly manipulated to yield

$$\frac{T_2}{T_1} = \left(\frac{v_1}{v_2}\right)^{\gamma-1}, \tag{7.14a}$$

or

$$T_1 v_1^{\gamma-1} = T_2 v_2^{\gamma-1} = \text{constant.} \tag{7.14b}$$

Equating the expression for the temperature ratio in Eq. 7.13a and Eq. 7.14a allows us to combine the two expressions into one relationship:

$$\frac{P_2}{P_1} = \left(\frac{v_1}{v_2}\right)^{\gamma} = \left(\frac{T_2}{T_1}\right)^{\gamma/(\gamma-1)}. \tag{7.15}$$

If we look only at the pressure and specific volume expressions in Eq. 7.15, we have a third and final ideal-gas isentropic-process relationship:

$$\frac{P_2}{P_1} = \left(\frac{v_1}{v_2}\right)^{\gamma}, \tag{7.16a}$$

TABLE 7.1 Ideal-Gas Isentropic-Process Relationships

General form	State 1 to State 2	Equation reference
$Pv^{\gamma} = \text{constant}$	$\frac{P_2}{P_1} = \left(\frac{v_1}{v_2}\right)^{\gamma}$	Eq. 7.16a
$Tv^{\gamma-1} = \text{constant}$	$\frac{T_2}{T_1} = \left(\frac{v_1}{v_2}\right)^{\gamma-1}$	Eq. 7.14a
$T^{\gamma}P^{1-\gamma} = \text{constant}$	$\frac{P_2}{P_1} = \left(\frac{T_2}{T_1}\right)^{\gamma/(\gamma-1)}$	Eq. 7.13a
P–v work for a closed system	${}_1W_2 = \frac{P_2\mathcal{V}_2 - P_1\mathcal{V}_1}{1-\gamma}$ or ${}_1W_2 = \frac{MR(T_2 - T_1)}{1-\gamma}$	Eq. 7.17

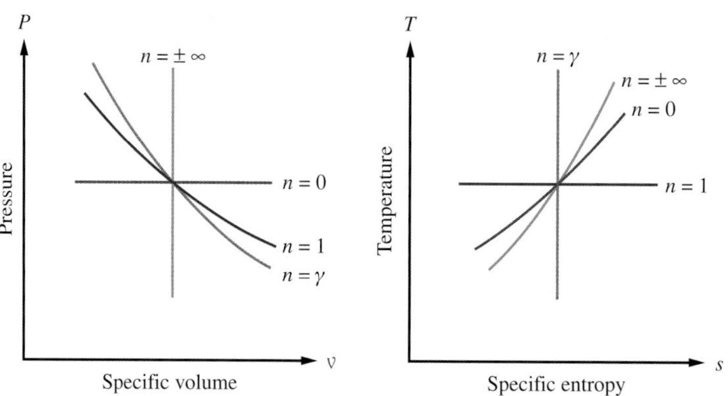

FIGURE 7.5 P–v and T–s diagrams illustrating polytropic process paths for special cases: constant pressure ($n = 0$), constant temperature ($n = 1$), constant entropy ($n = \gamma$), and constant volume ($n = \pm\infty$).

or

$$P_1 v_1^{\gamma} = P_2 v_2^{\gamma} = \text{constant}. \tag{7.16b}$$

This last relationship (Eq. 7.16b) is easy to remember; it is a polytropic process with $n = \gamma$. All the other isentropic relationships are easily derived from this by applying the ideal-gas equation of state. Figure 7.5 compares some ideal-gas polytropic processes. Several of these processes were considered in Section 4.3b. We have now added isentropic processes (for $n = \gamma$) to Table 4.3.

Using Eq. T4.3d for a polytropic process with exponent $n = \gamma$, the P–v work for an ideal gas undergoing a reversible expansion/compression in a closed system is

$$_1W_2 = \frac{P_2\mathcal{V}_2 - P_1\mathcal{V}_1}{1-\gamma} = \frac{MR(T_2 - T_1)}{1-\gamma}. \tag{7.17}$$

The ideal-gas isentropic-process relationships are summarized in Table 7.1 for convenient future reference. One use of these relationships is to model ideal compression and expansion processes in internal combustion engines, air compressors, and gas-turbine engines, for example. Example 7.3 illustrates this use in this chapter; other examples are found throughout the book. (See also Fig. 7.6.)

FIGURE 7.6 Pipeline compressor station for the distribution of natural gas (CreativeNature_nl / iStock / Getty Images Plus).

Example 7.3 SI Engine Compression

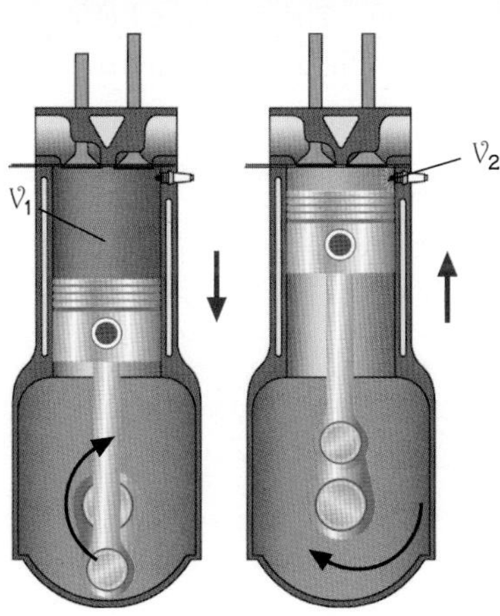

Consider a spark-ignition engine in which the effective compression ratio (CR) is 8:1. Compare the pressure at the end of the compression process for an isentropic compression ($n = \gamma = 1.4$) with that for a polytropic compression with $n = 1.3$. The initial pressure in the cylinder is 100 kPa, a wide-open-throttle condition.

Solution

Known P_1, CR, γ, n

Find $P_2(n = \gamma), P_3(n = 1.3)$

Sketch

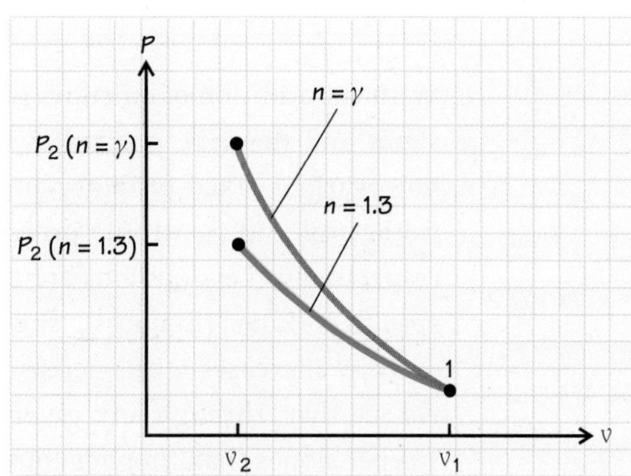

Modeling, Premises and Assumption The working fluid can be treated as air with $\gamma = 1.4$.

Analysis We apply the polytropic process relationship, Eq. 4.9a, for the two processes, both starting at the same initial state. From Appendix 3A, we know that the compression ratio is defined as $\mathcal{V}_1/\mathcal{V}_2$. Since the mass is fixed, the compression ratio also equals the ratio of specific volumes, v_1/v_2. For the isentropic compression,

$$\begin{aligned} P_2(n=\gamma) &= P_1(v_1/v_2)^{\gamma} \\ &= 100\,\text{kPa}\,(8)^{1.4} \\ &= 1840\ \text{kPa}. \end{aligned}$$

For the polytropic compression,

$$\begin{aligned} P_3(n=1.3) &= P_1(v_1/v_2)^{1.3} \\ &= 100\,\text{kPa}\,(8)^{1,3} \\ &= 1490\ \text{kPa}. \end{aligned}$$

Comments We see that the isentropic compression pressure is approximately 20% higher than the polytropic compression pressure. The ideal compression process is adiabatic and reversible, and, hence, isentropic. In an actual engine, energy is lost from the compressed gases by heat transfer through the cylinder walls, and frictional effects are also present. The combined effects result in a compression pressure lower than the ideal value. The use of a polytropic exponent is a simple way to account for these effects.

Self-Test 7.3

Given an initial temperature of 400 K, calculate the final temperature for each process in Example 7.3.

(Answer: 920 K, 745 K)

7.2c Processes in *T–s* and *P–v* Space

With the addition of entropy, we now have four properties (P, v , T, and s) to define states and describe processes. We will now investigate T–s and P–v space to see how various fixed-property processes appear on these coordinates. In T–s space, isothermal and isentropic processes are by definition horizontal and vertical lines, respectively. Less obvious are the lines of constant pressure (isobars) and constant-specific-volume (isochors), shown in Fig. 7.7. For the isobars and isochors, the entropy increases with increasing temperature and both are curved upward. Here we see that, through any given state point, lines of constant specific volume have steeper slopes than those of constant pressure. It is important to note that the isobars diverge with increasing temperature and entropy. That is, the temperature difference between the P_1 and P_2 isobars will increase as the entropy increases. This will be an important characteristic when we look at gas turbines in Chapter 8. The divergence of the isobars allows the gas turbine engine to generate a net power

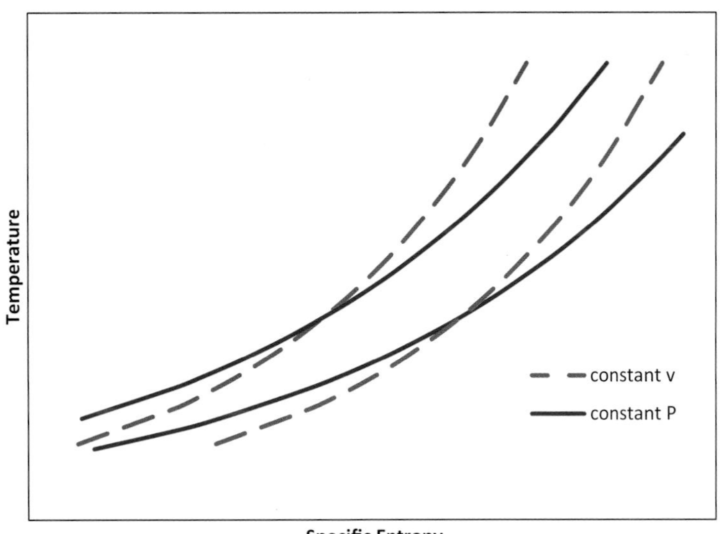

FIGURE 7.7 On a *T*–s diagram, lines of constant pressure (isobars) and lines of constant specific volume (isochors) both exhibit positive slopes. At a particular state, a constant-volume line has a greater slope than a constant-pressure line.

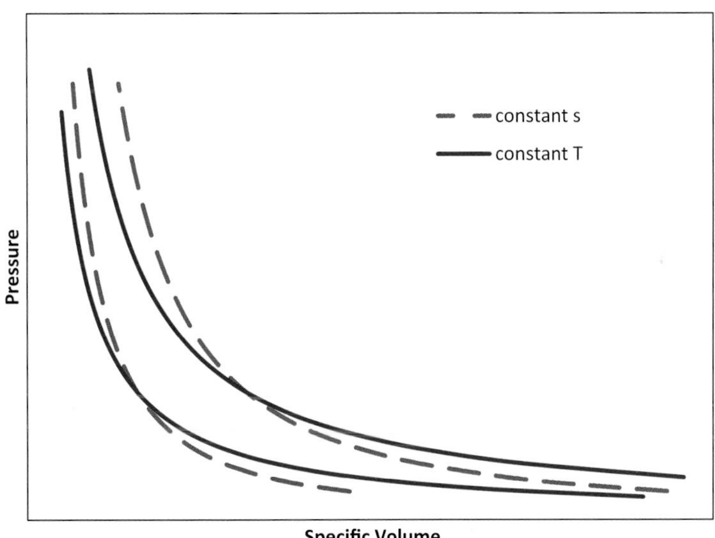

FIGURE 7.8 On a *P*–v diagram, lines of constant temperature (isotherms) and lines of constant entropy (isentropes) have negative slopes. At a particular state, an isentrope is steeper (has a greater negative slope) than an isotherm.

output. The isochors have a similar characteristic of diverging with increasing temperature and entropy.

In *P*–v space, constant-pressure and constant-volume processes are, again by definition, horizontal and vertical lines, whereas constant-entropy and constant-temperature processes follow curved paths, as shown in Fig. 7.8. Here we see that lines of constant entropy are steeper (i.e., have a greater negative slope) than those of constant temperature.

You should become familiar with these characteristics of *T*–*s* and *P*–v diagrams. Sketching the isotherms, isobars, isochors, or isentropes is often the first step in analysing an ideal gas process.

Example 7.4 Carnot Cycle

Sketch the following set of processes on T–s and P–v diagrams:

Ideal Carnot Cycle

State Change Process:
1–2: isothermal expansion,
2–3: isentropic expansion,
3–4: isothermal compression, and
4–1: isentropic compression.

Solution

We begin with the T–s diagram since the temperature is fixed for two of the four processes, whereas the entropy is fixed for the remaining two. We recognize that in an expansion process v_2 is greater than v_1 and, therefore, s_2 is also greater than s_1. Starting with Fig. 7.7 and knowing that state 2 must lie to the right of state 1, we draw process 1–2 as a horizontal line from left to right. Because the volume continues to increase in process 2–3, we know that the temperature must be decreasing, as indicated from Eq. 7.14a for an isentropic process. We therefore draw our isentrope as a vertical line downward from state 2 to state 3. In the compression process from state 3 to state 4, the entropy decreases, as the volume decreases in an isothermal process as indicated by Eq. 7.12b. We therefore draw the state-3 to state-4 isotherm from right to left, stopping when s_4 is equal to s_1. An upward vertical line completes the cycle and returns the fluid to state 1. The cycle completes a rectangle in T–s coordinates.

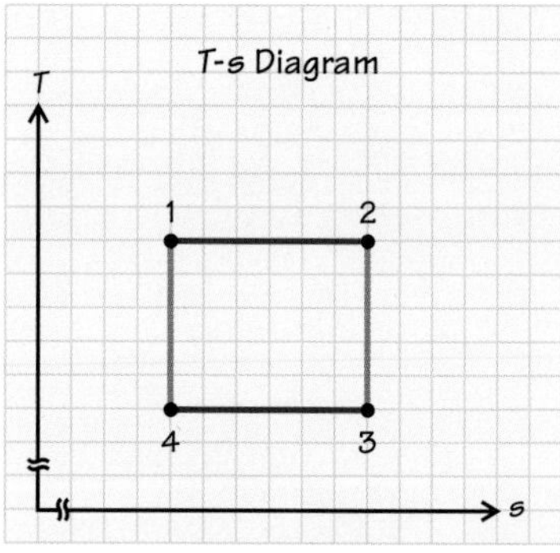

To draw the P–v plot, we refer to Fig. 7.8, which shows lines of constant temperature and lines of constant entropy. Because the volume increases in the process 1–2, we draw an isotherm directed downward and to the right. The sketch shows us that the pressure decreases as the volume increases in an isothermal process for an ideal gas. This can also be observed from the ideal gas law, which results in $P_1v_1 = P_2v_2$ when $T_1 = T_2$. The expansion continues from state 2 to state 3, but the process line now follows the steeper isentrope to the lower isotherm. The sketch shows that pressure also decreases as the volume increases in an isentropic process. This is consistent with Eq. 7.16a. The cycle is completed by following this lower isotherm upward and to the left until it intercepts the isentrope that passes through the initial state, 1.

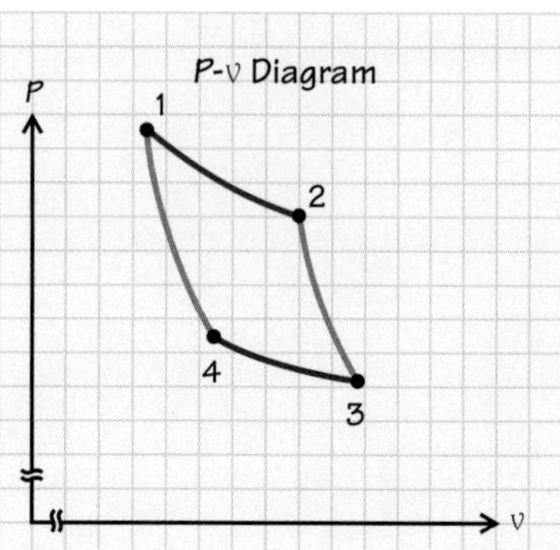

Comments This sequence of processes is the famous Carnot cycle for an ideal gas, which is discussed at length in Chapter 6.

Self-Test 7.4

Referring to Example 7.4, for which process(es) are the equations in Table 7.1 valid? Why or why not?

(Answer: 2–3 and 4–1 only. The equations in Table 7.1 are for isentropic processes only.)

Example 7.5 SI Otto Cycle

(Credit: ineb1599 / iStock / Getty Images Plus.)

The following processes constitute the air-standard Otto cycle. Plot these processes in P– v and T–s coordinates:

1–2: constant-volume energy addition,
2–3: isentropic expansion,
3–4: constant-volume energy removal, and
4–1: isentropic compression.

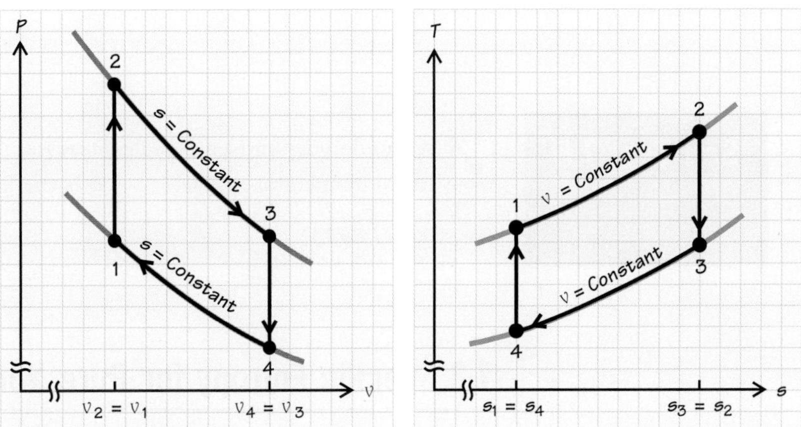

Solution

Since there are two constant-volume processes and two constant-entropy processes, we begin by drawing two isentropes in P–v coordinates and two isochors in T–s

coordinates. These are shown in the sketches. Because we expect the pressure to increase with energy addition at constant volume, we draw a vertical line upward from state 1 to state 2 on the P–v diagram. Because process 2–3 is an expansion, with state 3 lying below state 2, we thus know that process 1–2 follows the upper constant-volume line on the T–s diagram. Having established the relative locations of states 1, 2, and 3 on each plot, completing the cycle is straightforward.

Also see Example 4.2 in Chapter 4.

During the intake stroke of a real spark-ignition engine, fresh fuel and air enter the cylinder and mix with the residual gases.

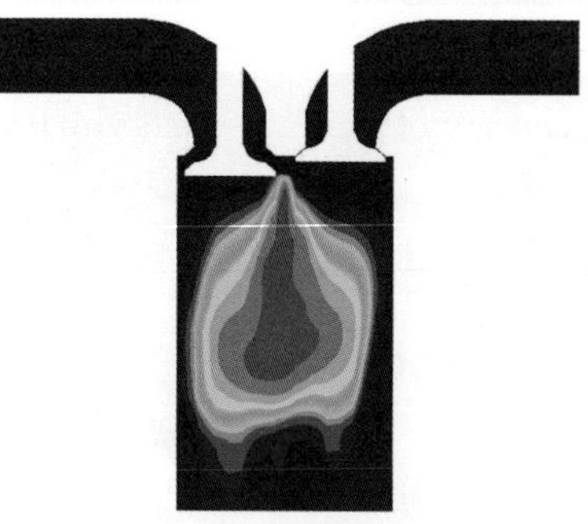

Image courtesy of Eugene Kung and Daniel Haworth.

Comments The cycle illustrated in this example using air as the working fluid is often used as a starting point for understanding the thermodynamics of spark-ignition (Otto cycle) engines. In the real engine, a combustion process, rather than energy addition from the surroundings, causes the pressure to rise from states 1 to 2. Furthermore, the real "cycle" is not closed because the gases exit the cylinder and are replaced by a fresh charge of air–fuel mixture in each mechanical cycle. Nevertheless, this air-standard cycle does capture the effect of the compression ratio on thermal efficiency and can be used to model other effects as well. Some of these are shown in examples throughout this book.

Self-Test 7.5

☑ Air undergoes an isentropic compression process from 100 to 300 kPa. If the initial temperature is 30 °C, what is the final temperature?

(Answer: 414.9 K)

7.3 Specific Entropy for Pure Nonideal Substances

When the fluid is not an ideal gas, property tables can be used to find the specific entropy.

7.3a Vapor State

We can use the textbook tables or the NIST resources to find the specific entropy as a function of T and P for fluids in the vapor state. For water, Tables B.3 list the specific entropy. The superheated vapour tables can be interpolated between different pressure or temperatures to approximate the property at conditions that are not listed.

7.3b Saturated Mixture

The specific entropy of a saturated mixture is found using the quality of the saturated mixture and the specific entropies of the saturated liquid and saturated vapour at the specified pressure and temperature. Equations 2.37c and 2.38b can be used with $\beta = s$ to find the specific entropy of the mixture:

$$s = (1 - x)s_{\mathrm{f}} + xs_{\mathrm{g}} = s_{\mathrm{f}} + x\left(s_{\mathrm{g}} - s_{\mathrm{f}}\right). \tag{7.18a}$$

If the entropy of the mixture is known, the quality can be determined in the same way as in Eq. 2.37b:

$$x = \frac{s - s_{\mathrm{f}}}{s_{\mathrm{g}} - s_{\mathrm{f}}}. \tag{7.18b}$$

7.3c Compressed Liquid

The specific entropy for a compressed liquid can be found using the compressed liquid tables (see Table B.4 for H_2O) is or the NIST resources. It is noted that the lowest pressure tabulated in Table B.4, Compressed Liquid Water, is 5.0 MPa. For pressures below 5.0 MPa, water is considered a mildly compressed liquid and the approximations in Eq. 2.41 can be used. That is, the specific volume and specific internal energy are approximated as being those of saturated liquid at the same temperature. The same is done for approximating the specific entropy of a mildly compressed liquid:

$$s(T, P) \cong s_{\mathrm{f}}(T_{\mathrm{sat}} = T). \tag{7.19}$$

It can be shown that this approximation results from the Gibbs relationship Eq. 7.6b. If $v(T, P) \cong v_{\mathrm{f}}(T_{\mathrm{sat}} = T)$ and $u(T, P) \cong u_{\mathrm{f}}(T_{\mathrm{sat}} = T)$, then the entire right-hand side of the Gibbs relationship is only a function of temperature. This indicates that entropy changes only slightly with pressure in the compressed liquid region. As an example, we can look at the conditions in Example 2.19 and find that the error between s_f (T_{sat} = 300 K) = 0.39309 kJ/kg·K and the value from the NIST WebBook s(300 K, 10 MPa) = 0.39029 kJ/kg·K is only 0.72%.

7.3d Incompressible Solids and Liquids

To find the entropy change for an incompressible solid or liquid, we will start with the Gibbs relationship

$$Tds = du + Pdv.$$

For an incompressible solid or liquid, $dv = 0$ and the specific internal energy is described using the specific heat:

$$du = c_v dT.$$

Substituting into the Gibbs relationship yields

$$Tds = c_v dT.$$

We can solve for ds and integrate from state 1 to state 2 to give

$$\Delta s = s_2 - s_1 = \int_1^2 c_v \frac{dT}{T}$$

Approximating as a constant the specific heat from T_1 to T_2 and integrating gives

$$\Delta s = s_2 - s_1 = c_v \ln\left(\frac{T_2}{T_1}\right). \quad (7.20)$$

7.3e *T–s* Diagrams

Of equal importance to the T–v and P–v diagrams is the temperature–entropy, or T–s, diagram. For a reversible process, the integral of Tds provides the energy added as heat to a closed system; thus, the area under a reversible process line in T–s coordinates represents the energy added or removed by heat interactions.

Figure 7.9 presents a T–s diagram for water and shows an isobar traversing the compressed liquid region, across the steam dome, and up into the superheat region. Without an expanded scale, the 1-MPa isobar in the compressed liquid region is indistinguishable from the saturated liquid line; however, it does lie above and to the left of the saturated liquid line. Isobars for pressures greater than 1 MPa lie above the isobar shown, and those for lower pressures lie below. Although not shown, isochors (constant-volume lines) in the superheated region have steeper slopes than the isobars (as in Fig. 7.7). Note that Fig. 7.9 employs linear scales for both temperature and entropy, rather than the semilog and log–log scales previously used in the analogous T–v (Fig. 2.33) and P–v (Fig. 2.35) diagrams, respectively.

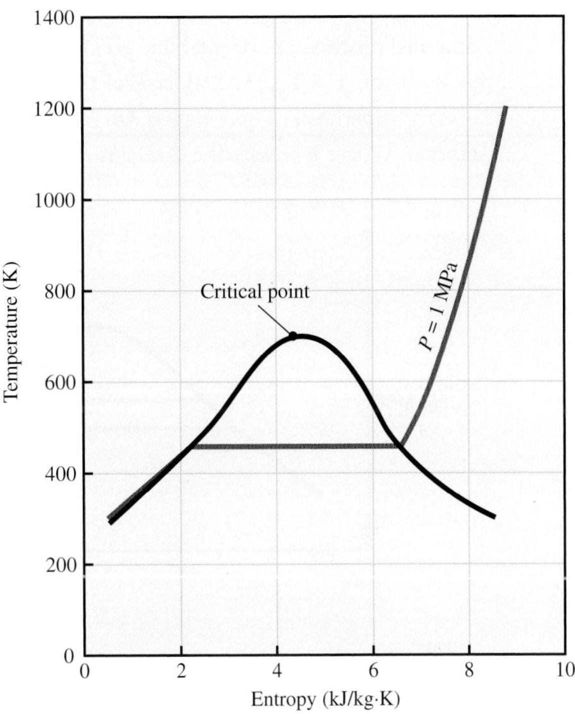

FIGURE 7.9 Temperature–entropy (*T*–*s*) diagram for water showing liquid–vapor saturation lines and the 1-MPa isobar.

Example 7.6 Isothermal Expansion

Consider an internally reversible, isothermal expansion of 2 kg of steam from an initial state of saturated vapor at 3 MPa to a pressure of 1 MPa. Sketch the process in *T*–*s* and *P*–*v* coordinates and determine the initial and final specific volumes, specific internal energies, and specific entropies. Determine the heat transfer and *P*–*v* work for this internally reversible process.

Solution

Known Saturated vapor at P_1, T_1 $(= T_2)$, P_2

Find v_1 $(= v_g)$, v_2, u_1 $(= u_g)$, u_2, s_1 $(= s_g)$, s_2, ${}_1Q_2$, and ${}_1W_2$.

Sketch

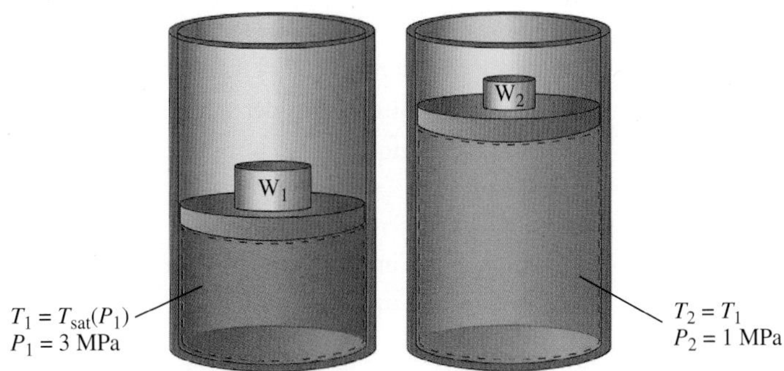

Modeling, Premises and Assumptions

i. Simple compressible substance in a quasi-equilibrium process.
ii. The closed system is not moving so the bulk kinetic energy is zero and the change in potential energy is zero.

Analysis We begin by drawing the P_1 (= 3 MPa) and P_2 (= 1 MPa) isobars on a *T*–*s* diagram as shown. State 1 is identified on this diagram as a saturated vapor. For the isothermal process, a horizontal line is extended from the state-1 point. The location where this isotherm, $T = T_{sat}$ (3 MPa), crosses the 1-MPa isobar identifies the state-2 point. State 2 is in the superheated vapor region. On P–v coordinates, we draw the same $T = T_{sat}$ (3 MPa) isotherm. Where it crosses the 1-MPa isobar identifies the state-2 point in P–v space.

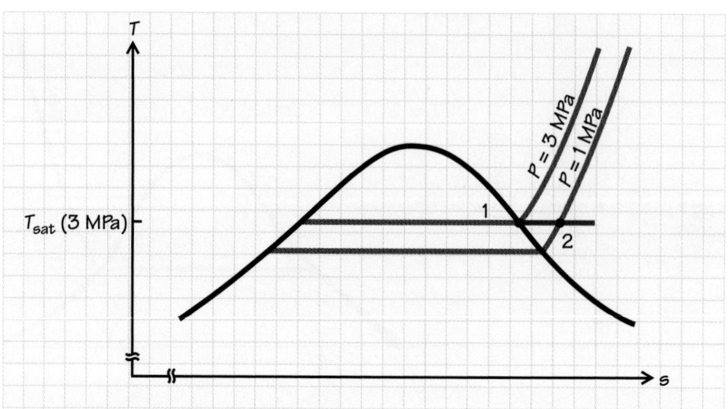

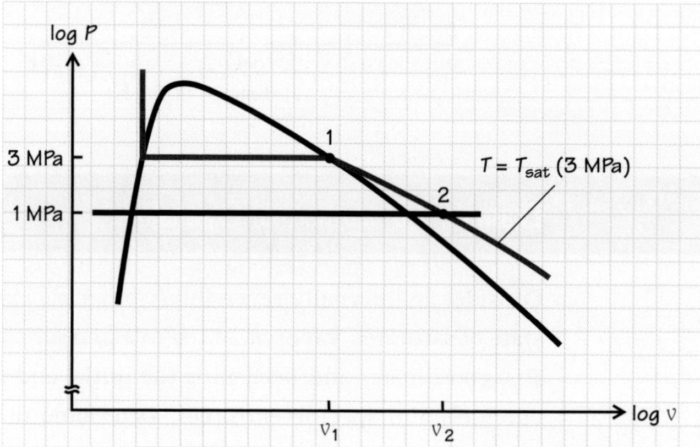

From Table B.2, we obtain the following properties:

$$P_1 = 3\,\text{MPa},\ T_1 = 507.00\,\text{K},$$
$$v_1 = 0.066664\,\text{m}^3/\text{kg},\quad u_1 = 2603.2\,\text{kJ/kg},\quad s_1 = 6.1856\,\text{kJ/kg·K}.$$

To obtain the state-2 properties, we can interpolate to find v_2 in the superheated-vapor table for $P = 1$ MPa (see Table B.3). Alternatively, the NIST resources can be used directly to determine v_2 by generating isothermal data ($T = 507.00$ K) with pressure increments containing P = 1 MPa. The result is

$$v_2 = 0.22434\,\text{m}^3/\text{kg},\quad u_2 = 2682.7\,\text{kJ/kg},\quad s_2 = 6.8565\,\text{kJ/kg·K}.$$

The heat transfer to the closed system in this internally reversible isothermal process can be found using Eq. 7.2b:

$$_1Q_2 = T(S_2 - S_1)\Big|_{\substack{\text{int}\\\text{rev}}} = MT(s_2 - s_1)\Big|_{\substack{\text{int}\\\text{rev}}}$$
$$_1Q_2 = (2\,\text{kg})(507\,\text{K})(6.8565 - 6.1856)\,\text{kJ/kg·K} = 680.3\,\text{kJ}.$$

The energy equation can now be used to find the P–v work during this process:

$$_1Q_2 - {}_1W_2 = M(u_2 - u_1)$$
$$_1W_2 = {}_1Q_2 + M(u_1 - u_2)$$
$$= 680.3\,\text{kJ} + (2\,\text{kg})(2603.2 - 2682.7)\,\text{kJ/kg} = 521.3\,\text{kJ}.$$

Comments In an internally reversible process, heat transfer to the closed system causes an increase in the system entropy. Since the specific volume increases from state 1 to state 2, we know that the P–v work will also be positive, as seen in our answer. We note the utility of sketching constant-property lines on T–s and P–v diagrams to understand the process being studied.

Self-Test 7.6

The closed system of Example 7.6 is now allowed to expand isothermally until the final specific volume is 0.32 m^3/kg. Find the final pressure.

(Answer: $P \cong 0.70$ MPa)

7.3f *h–s* Diagrams

Figure 7.10 illustrates an enthalpy–entropy diagram for water. Unlike all the other previously shown property diagrams, the steam dome is skewed because both enthalpy and entropy increase during the liquid–vapor phase change. Note that the critical point does not lie at the topmost point on the saturation line in h–s space. Also, the constant pressure curves are smooth and do not suddenly change slope at the vapor dome. In the analysis of many processes and devices, enthalpy and entropy are key properties and, hence, h–s diagrams can be helpful. For example, consider the turbine analysis in Chapter 5. The energy analysis for an open system (control volume)

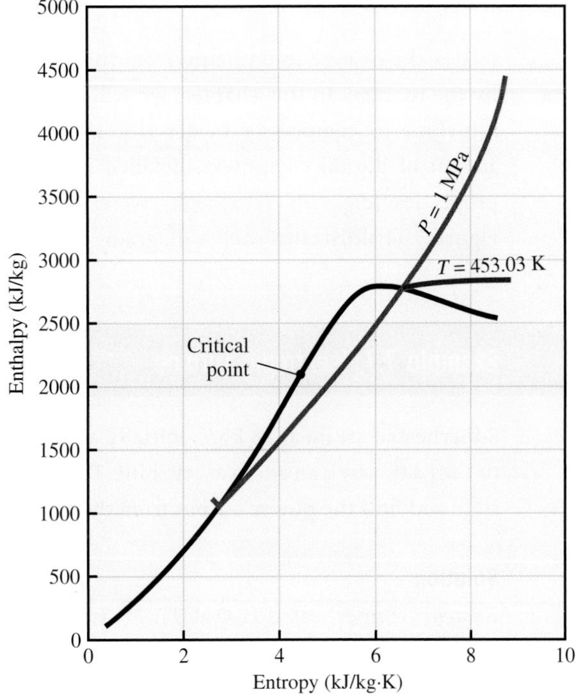

FIGURE 7.10 Enthalpy–entropy (h–s) diagram for water showing the liquid and vapor saturation lines and the 1-MPa isobar. Also shown is the 453.03-K isotherm, where 453.03 K is the saturation temperature at 1 MPa.

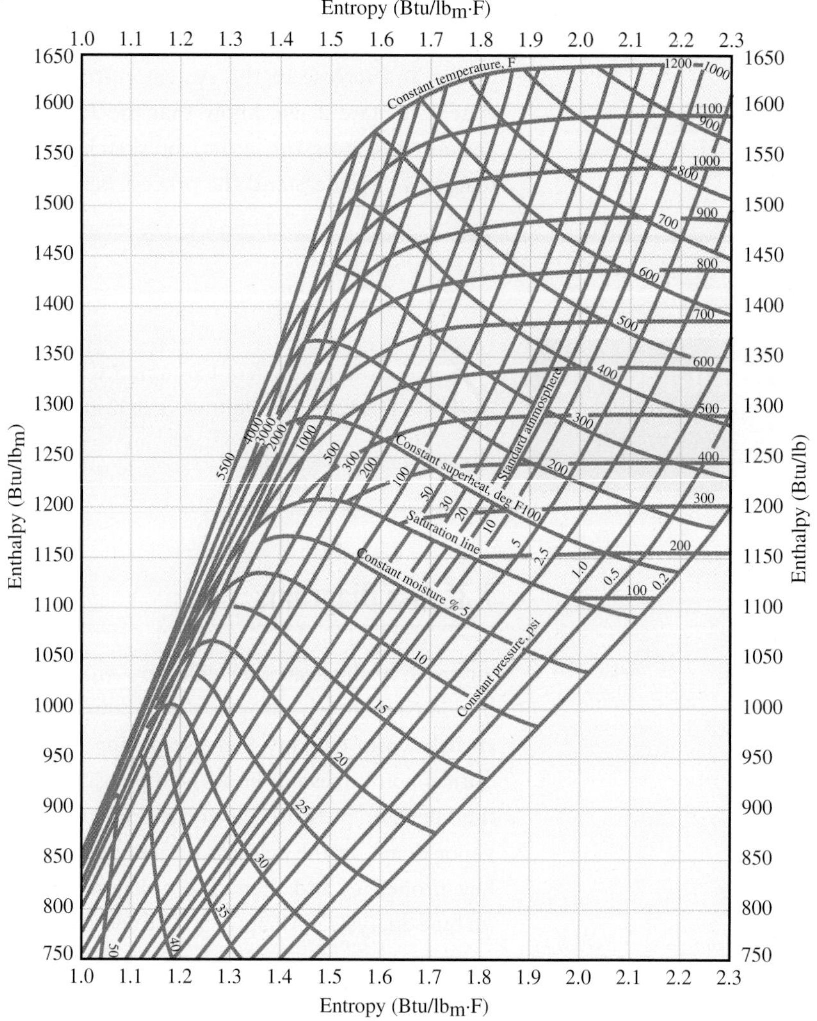

FIGURE 7.11 Mollier *h–s* diagram for water. Courtesy of The Babcock & Wilcox Company.

relates the change in enthalpy from the inflow to the outflow to the power delivered by the turbine. In this chapter, we will use entropy changes from inflow to outflow of a turbine or compressor to describe the performance of these devices. Prior to the advent of digital computers, detailed *h–s* or **Mollier diagrams** showing numerous properties (e.g., P, T, x, etc.) were routinely used in thermodynamic analyses. Figure 7.11 illustrates such a diagram.

Example 7.7 Isentropic Turbine

Superheated steam at 18 kg/s, initially at 700 K and 10 MPa, is isentropically expanded to 1 MPa through an adiabatic turbine. Determine the values of s, h, T, and x at the outlet state and find the power output from the turbine. Sketch the process on a T–s diagram.

Solution

Known Superheated H_2O at T_1, P_1, isentropic expansion to P_2

Find v_2, h_2, T_2, x_2, $\dot{W}_{turbine}$

Sketch

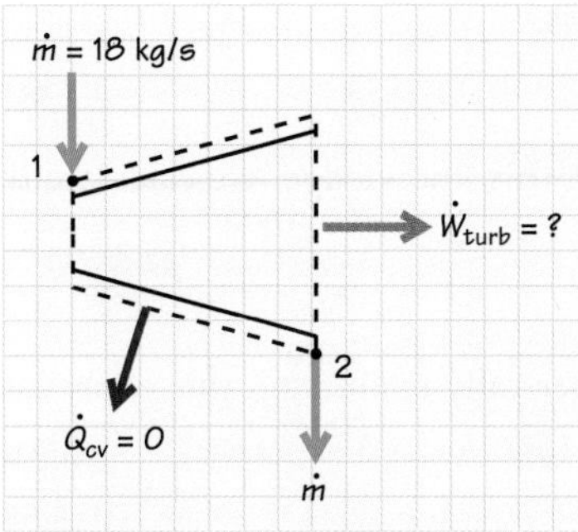

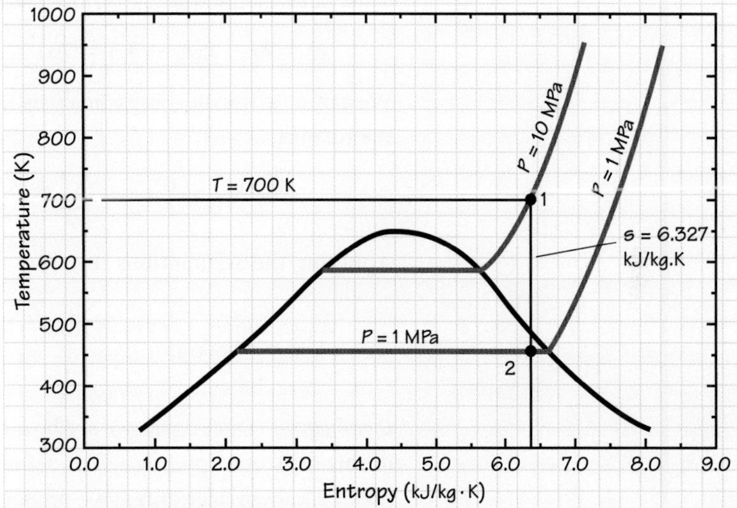

Modeling, Premises and Assumptions

i. Simple compressible substance in quasi-equilibrium through turbine
ii. Adiabatic turbine, $\dot{Q}_{CV} = 0$
iii. Isentropic process
iv. Neglect changes in kinetic and potential energy

Analysis Our strategy here is to locate state 1 in T–s space, that is, to find s_1 given T_1 and P_1. State 2 is then determined by the given pressure ($P_2 = 1$ MPa) and the fact that the process is isentropic (i.e., that $s_2 = s_1$.) The relationship between state 1 and state 2 is shown in the sketch. Since we know two independent properties at state 2 (P_2, s_2), all other properties can be obtained.

To find h_1 and s_1, we employ Table B.3 for the superheated steam at 10 MPa, which gives

$$h_1(T_1 = 700\,\text{K}, P_1 = 10\,\text{MPa}) = 3177.4\,\text{kJ/kg},$$
$$s_1(T_1 = 700\,\text{K}, P_1 = 10\ \text{MPa}) = 6.3305\,\text{kJ/kg}\cdot\text{K}.$$

Since $s_2 = s_1$ (= 6.3305 kJ/kg·K), we see from Table B.2 that state 2 must be in the liquid–vapor mixture region since the entropy is between that of saturated liquid and saturated vapour state at 1 MPa:

$$s_f(P_{sat} = 1\,\text{MPa}) < s_2 < s_g(P_{sat} = 1\,\text{MPa}),$$

that is,

$$2.1381 < 6.3305 < 6.585.$$

With this knowledge, we calculate the quality from Eq. 7.18b as follows:

$$x = \frac{s_2 - s_f}{s_g - s_f}.$$

Numerically evaluating this expression yields

$$x = \frac{6.3305 - 2.1381}{6.585 - 2.1381} = 0.9428.$$

This value of the quality is now used to determine h_2 using Eq. 2.37c.

$$h_2 = h_f + x(h_g - h_f) = 762.52 + 0.9428(2777.1 - 762.52) = 2661.9 \text{ kJ/kg}.$$

We can now use an energy analysis for an open system to find the power output from the turbine:

$$\dot{Q}_{CV} - \dot{W}_{CV} = \sum_{out} \dot{m}\left[h + \tfrac{1}{2}V^2 + gz\right] - \sum_{in} \dot{m}\left[h + \tfrac{1}{2}V^2 + gz\right].$$

Applying the assumptions listed for the adiabatic turbine,

$$\dot{W}_{CV} = \dot{m}(h_1 - h_2) = (18 \text{ kg/s})(3177.4 - 2661.9) \text{ kJ/kg}\left[\frac{\text{kW}}{\text{kJ/s}}\right] = 9.279 \text{ MW}.$$

Comments Once again we see how a process in which one property is held constant, in this case, entropy, is used to define the final state. This example also illustrates the use of a known mass-intensive property (s) to determine the quality, which, in turn, is used to calculate another mass-intensive property (h). We will see in a later section that the isentropic turbine gives the maximum power output for a turbine between a given initial state and final pressure.

Self-Test 7.7

Determine the changes in the specific internal energy and the specific enthalpy for the process described in Example 7.7.

(Answer: $\Delta u = -415.61$ kJ/kg, $\Delta h = -515.13$ kJ/kg)

7.4 Entropy Balances for a Closed System

7.4a Systems Undergoing a Change in State

We now develop an entropy balance for a system, adopting the generic conservation principle introduced in Chapter 3 (Eq. 3.1) that

$$\Delta X_{\text{stored}} = X_{\text{in}} - X_{\text{out}} + X_{\text{generated}},$$

where X is the conserved quantity of interest. For a process taking a closed system from state 1 to state 2, we write

$$\underset{\substack{\text{Change of}\\ \text{system}\\ \text{entropy}}}{S_2 - S_1} = \underset{\substack{\text{Net transfer of}\\ \text{entropy into the}\\ \text{system by heat}\\ \text{interactions}}}{\int_1^2 \frac{\delta Q}{T}} + \underset{\substack{\text{Entropy generated}\\ \text{by irreversibilities}\\ \text{within the system}}}{\int_1^2 \delta \mathbb{S}_{gen}}. \quad (7.21)$$

The first term on the right-hand side of this entropy balance corresponds to $X_{in} - X_{out}$ and can be positive or negative depending on the direction of the heat interaction, δQ. The second term is always positive because irreversibilities always generate an increase in entropy; for a reversible process, this term is zero. In the same sense that heat is path dependent and not a property of the system, the entropy generation is also a path function and not a system property; that is,

$$\int_1^2 \delta Q \equiv {}_1Q_2,$$

the magnitude of which is path dependent, and

$$\int_1^2 \delta \mathbb{S}_{gen} \equiv {}_1\mathbb{S}_{gen2},$$

the magnitude of which is path dependent. We use the symbol $\mathbb{S}_{gen}$ to emphasize that $\mathbb{S}_{gen}$ is not a property and is not identical to the entropy S, which is a property.

7.4b Entropy Change for an Isolated System

An **isolated system** is defined as a closed system that has no interactions with the surroundings. An isolated system will not have any energy transfer to or from the system to the surroundings. This means that there will be no heat transfer or work done by the closed system. Since a change in volume for a system would create $P - \mathsf{V}$ work, an isolated system cannot change in volume. For an isolated system that is stationary, the energy equation

$${}_1Q_2 - {}_1W_2 = M(u_2 - u_1) + \tfrac{1}{2}M\left(V_1^2 - V_1^2\right) + Mg(z_2 - z_1)$$

simplifies to

$$\Delta U = U_2 - U_1 = M(u_2 - u_1) = 0.$$

There will be no change in the internal energy for an isolated system. The entropy balance

$$\underset{\substack{\text{Change of}\\ \text{system}\\ \text{entropy}}}{S_2 - S_1} = \underset{\substack{\text{Net transfer of}\\ \text{entropy into the}\\ \text{system by heat}\\ \text{interactions}}}{\int_1^2 \frac{\delta Q}{T}} + \underset{\substack{\text{Entropy generated}\\ \text{by irreversibilities}\\ \text{within the system}}}{\int_1^2 \delta \mathbb{S}_{gen}}$$

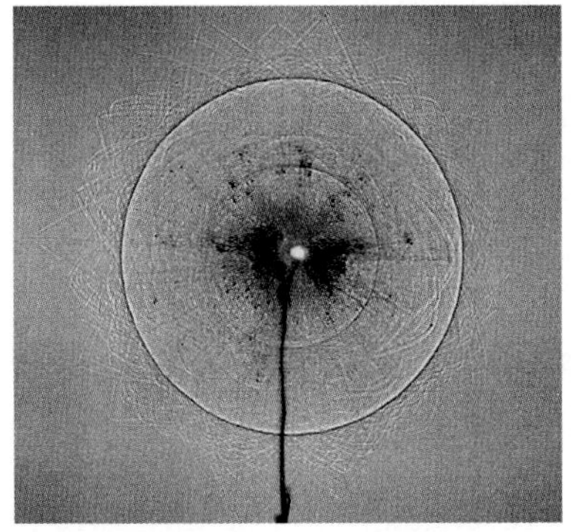

FIGURE 7.12 Explosions of all sorts are highly irreversible processes. This instantaneous image shows an explosion of one gram of triacetone triperoxide. Image courtesy of Gary Settles.

is also simplified since there is no heat transfer to or from the isolated system. This gives

$$\underbrace{S_2 - S_1}_{\substack{\text{Change of}\\ \text{system}\\ \text{entropy}}} = \underbrace{\int_1^2 \delta \mathbb{S}_{gen}}_{\substack{\text{Entropy generated}\\ \text{by irreversibilities}\\ \text{within the system}}}.$$

Since irreversibilities will always cause an increase in entropy of the isolated system, we can say that

$$\Delta S_{\substack{\text{isolated}\\ \text{system}}} = (S_2 - S_1)_{\substack{\text{isolated}\\ \text{system}}} \geq 0. \tag{7.22a}$$

OR

$$S_2 \geq S_1 \tag{7.22b}$$

for an isolated system.

The equality will apply when there are no internal irreversibilities. If there are internal irreversbilities, the inequality will apply. Table 6.1 listed many common irreversibilities. We will look at three of them now; see also Fig. 7.12. Ireversibilities can be caused by:

a. The free expansion of a gas into an evacuated space

b. The mixing of two dissimilar gases

c. Heat transfer from a hot body to a cold body

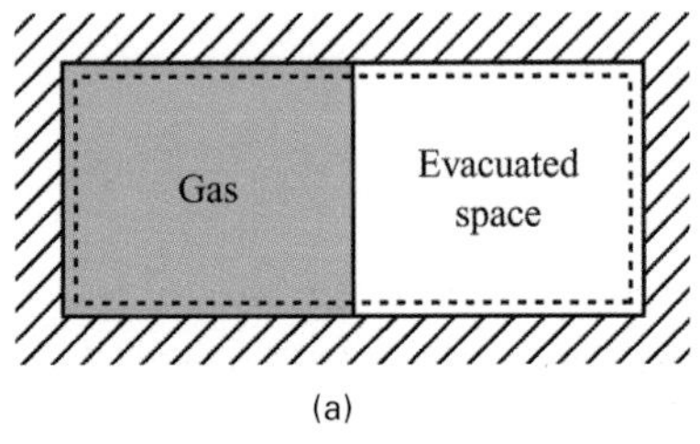

(a)

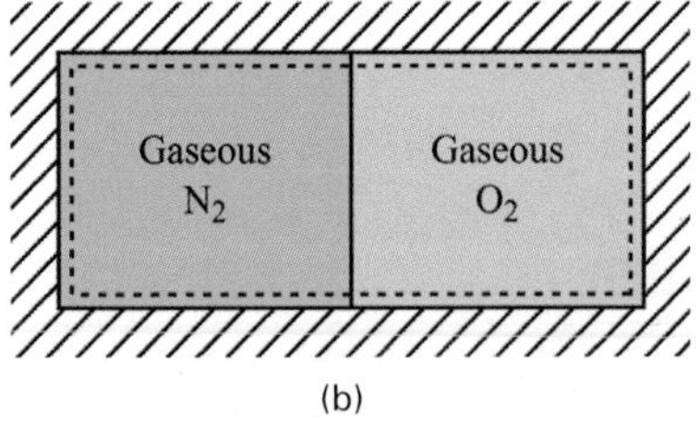

(b)

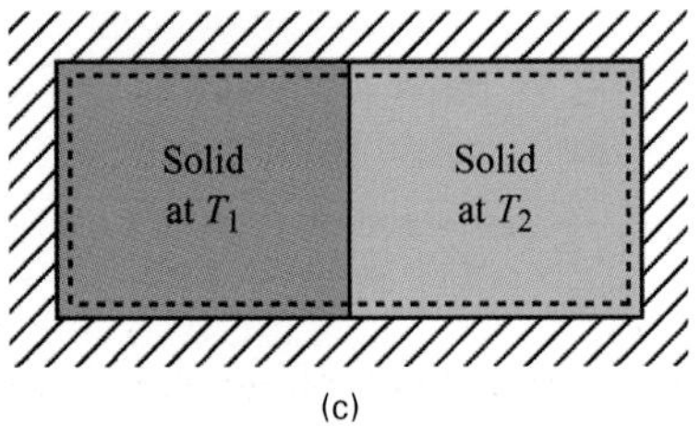

(c)

FIGURE 7.13 Three examples of spontaneous processes: **(a)** the free expansion of a gas into an evacuated space, **(b)** the mixing of two dissimilar gases, and **(c)** the transfer of heat from a hot body to a cold body.

To calculate ΔS for ideal gases, use Eqs. 7.10, 7.11 and 7.12. For many other substances, use the NIST resources.

We will look at each of these cases in an example of a spontaneous process. **Spontaneity** or a **spontaneous process** is defined as follows:

A process is spontaneous if, in the absence of any interactions with the surroundings (i.e., heat or work), a system undergoes a change in state.

What is meant by a spontaneous process can be made clear by considering the simple examples shown in Fig. 7.13: the free expansion of a gas into an evacuated space, the mixing of two dissimilar gases, and the transfer of heat from a hot body to a cold body. Experience tells us that, all three of these processes have a preferred, or spontaneous, direction. We expect that, after the partition separating the N_2 and O_2 gases is removed, the gases will mix and, given enough time, the composition in the combined space will become uniform. However, it is inconceivable that, starting with an N_2–O_2 mixture, the N_2 molecules will gather in the left half of the enclosure while the O_2 molecules congregate in the right. The same is true for the other two processes. The evacuated space will spontaneously fill after the removal of the partition, but not the reverse. A hot body and cold body in contact will spontaneously exchange energy until they achieve a common uniform temperature, but two bodies initially in contact at the same temperature will not spontaneously change temperature. In all three examples, the first law (conservation of energy) does not preclude the reverse processes from occurring; the second law, however, does. Moreover, a quantitative evaluation of spontaneity is possible by considering the second-law statements involving entropy. This is the beauty of the entropy property.

Example 7.8 Free Expansion of a Gas

A perfectly insulated tank has two sections. The first section has a volume of 0.2 m^3 and contains 0.5 kg of air at 350 K. The second section of the tank has a volume of 0.7 m^3 and is totally evacuated (no air). A wall is removed to allow the air to expand to fill both sections of the tank. Determine the change in entropy for the process.

Known air, $T_1, \mathcal{V}_1$

Find $S_{\text{final}} - S_{\text{init}}$

Sketch

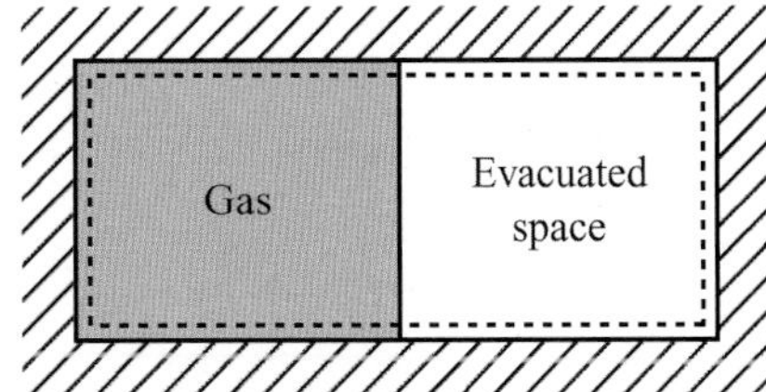

Modeling, Premises and Assumptions

i. Ideal-gas behavior, constant specific heat
ii. Adiabatic process
iii. Rigid tank, no change in volume of the total tank
iv. Stationary closed system, no kinetic energy or change in potential energy

Analysis Assuming a constant specific heat, we can find the change in entropy using the ideal gas relationship, Eq. 7.10b:

$$\Delta S = M(s_2 - s_1) = M\left[c_v \ln\left(\frac{T_2}{T_1}\right) + R\ln\left(\frac{\mathcal{V}_2}{\mathcal{V}_1}\right)\right].$$

We will use the energy equation to determine the temperature at state 2.

$${}_1Q_2 - {}_1W_2 = M(u_2 - u_1) + \tfrac{1}{2}M\left(v_1^2 - v_1^2\right) + Mg(z_2 - z_1).$$

Simplifying, for the assumptions above:

$$0 = M(u_2 - u_1).$$

For an ideal gas, the internal energy is only a function of temperature. If the specific internal energy does not change during the unrestrained expansion then the temperature does not change, $T_2 = T_1 = 350$ K. We can now find the change in entropy

$$\Delta S = M(s_2 - s_1) = M\left[c_v \ln\left(\frac{T_2}{T_1}\right) + R\ln\left(\frac{\mathcal{V}_2}{\mathcal{V}_1}\right)\right]$$

$$\Delta S = M\left[0 + R\ln\left(\frac{\mathcal{V}_2}{\mathcal{V}_1}\right)\right] = (0.5\,\text{kg})\left[(0.287\,\text{kJ/kg}\cdot\text{K})\ln\left(\frac{0.2+0.7}{0.2}\right)\right] = 0.2158\,\text{kJ/k}.$$

Comment We find that the entropy of the isolated system increases in an unrestrained expansion. An unrestrained expansion is an internal irreversibility. This process will not spontaneously occur in the reverse direction. That is, the air molecules will not move to the smaller volume and leave part of the tank evacuated without work being done on the system.

Example 7.9 Mixing of Two Dissimilar Gases

Partition

Consider 1 kmol of O_2 and 2 kmol of N_2, both at 298 K and 1 atm, separated by a partition. The partition is removed and the O_2 and N_2 mix. Determine the entropy change associated with this mixing process.

Solution

Known $N_{O_2}, N_{N_2}, T_1, P_1$

Find $S_{\text{final}} - S_{\text{init}}$

Sketch

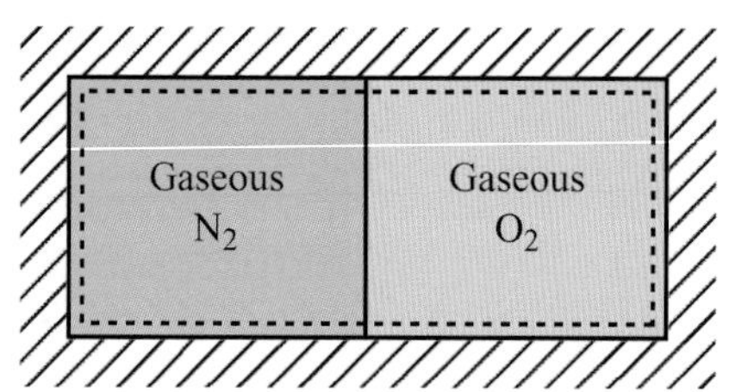

Modeling, Premises and Assumptions

i. Ideal-gas behavior
ii. Adiabatic process
iii. Rigid tank, no change in volume of the total tank
iv. Stationary closed system, no kinetic energy or change in potential energy

Analysis For the mixing of ideal gases with no heat or work interactions, conservation of energy tells us that the initial and final temperatures are equal. To calculate the entropy change, we therefore need to be concerned only with the changing partial pressures. This example is solved in detail in Chapter 10, "Ideal-Gas Mixtures". The entropy analysis gives

$$\Delta S = 15.88 \text{ kJ/K}.$$

Comment We find that the entropy increases in the mixing of two dissimilar gases in an isolated system. The mixing of two dissimilar gases is an internal irreversibility. This process will not spontaneously occur in the reverse direction. That is, the oxygen and nitrogen molecules will not move to separate regions in the tank without work being done on the system.

What if we were to perform the same mixing process, but using a single constituent, say, O_2? With no change in pressure, ΔS is then zero. We can view this process as reversible since we could put the partition back in place after the mixing takes place and retrieve the initial state, treating the O_2 molecules as indistinguishable.

Example 7.10 Heat Transfer from a Hot Body to a Cold Body

A perfectly insulated container has two solid blocks. The first block is 2 kg copper at a temperature of 400 K. The second block is 3 kg aluminum at a temperature of 250 K.

Determine the equilibrium temperature of the two blocks and the change in entropy during the process.

Known copper, M_A, M_B, T_{A1}, T_{B1}

Find $S_{\text{final}} - S_{\text{init}}$

Sketch

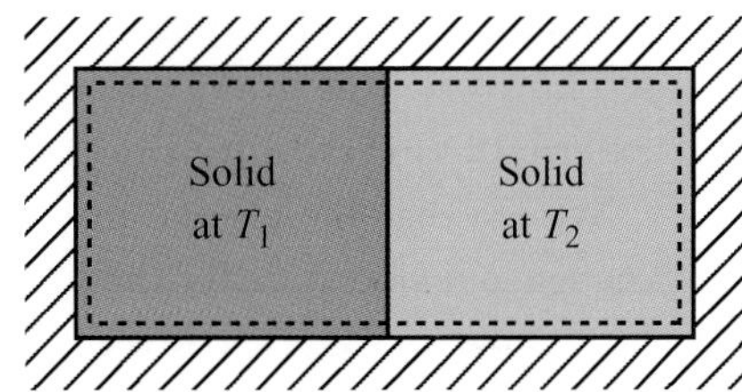

Assumptions

i. Incompressible solid, constant specific heat
ii. Adiabatic process
iii. Stationary closed system, no kinetic energy or change in potential energy

Properties The specific heat for copper and aluminum is given as:
$c(\text{copper}) = 385$ J/kg·K, $c(\text{aluminum}) = 903$ J/kg·K

Analysis We will use the energy equation to determine the final temperature of the copper blocks. At equilibrium, the two blocks will be at the same final temperature:

$$ {}_1Q_2 - {}_1W_2 = \sum_{blocks} \left[M(u_2 - u_1) + \tfrac{1}{2}M\left(V_2^2 - V_1^2\right) + Mg(z_2 - z_1) \right]. $$

Applying the assumptions in this problem simplifies the energy equation to

$$ 0 = M_A(u_{A2} - u_{A1}) + M_B(u_{B2} - u_{B1}). $$

Assuming a constant specific heat for the copper gives

$$ \begin{aligned} 0 &= M_A c_A (T_{A2} - T_{A1}) + M_B c_B (T_{B2} - T_{B1}), \\ (M_A c_A + M_B c_B) T_2 &= M_A c_A T_{A1} + M_B c_B T_{B1} \\ T_2 &= \frac{M_A c_A T_{A1} + M_B c_B T_{B1}}{(M_A c_A + M_B c_B)} \\ T_2 &= \frac{(2\,\text{kg})(385\,\text{J/kg}\cdot\text{K})(400\,\text{K}) + (3\,\text{kg})(903\,\text{J/kg}\cdot\text{K})(250\,\text{K})}{(2\,\text{kg})(385\,\text{J/kg}\cdot\text{K}) + (3\,\text{kg})(903\,\text{J/kg}\cdot\text{K})} \\ &= 283.2\ \text{K}. \end{aligned} $$

Assuming a constant specific heat, we can find the change in entropy using Eq. 7.20 for a solid:

$$ \Delta S = \Delta S_A + \Delta S_B = M_A(s_{A2} - s_{A1}) + M_B(s_{B2} - s_{B1}) = M_A c_A \ln\left(\frac{T_{A2}}{T_{A1}}\right) + M_B c_B \ln\left(\frac{T_{B2}}{T_{B1}}\right), $$

$$ \Delta S = (2\,\text{kg})(385\,\text{J/kg·K}) \ln\left(\frac{283.2\,\text{K}}{400\,\text{K}}\right) + (3\,\text{kg})(903\,\text{J/kg·K}) \ln\left(\frac{283.2\,\text{K}}{250\,\text{K}}\right) $$

$$ \Delta S = (-265.9 + 337.8)\,\text{J/K} = 71.9\,\text{J/K} $$

Comments We find that the entropy of the system (copper+aluminium blocks) increases because of the heat transfer from a hot body to a cold body. This example confirms one statement of the second law (Clausius II from Table 6.4):

Heat flows spontaneously from high to low temperature but not conversely.

This process will not spontaneously occur in the reverse direction. That is, the hotter block will not increase in temperature while the colder block decreases in temperature unless work is done by the surroundings. In a refrigerator, an input of power at the compressor is required to remove heat from the cold temperature reservoir.

In this example, the hot and cold blocks were both included in the closed system being analyzed. The heat transfer from a hot region to a cold region in a system is an **internal irreversibility**. If only one of the blocks was defined as the system and the second block was in the surroundings, we would say that this irreversibility is an **external irreversibility** since it occurs between the system and the surroundings.

We see that the change in temperature of the aluminium block is less than the change in temperature for the copper block because the aluminium block has a greater **heat capacity** ($M_A c_A < M_B c_B$).

7.4c Entropy Change for a Closed System with Energy Exchange with the Surroundings

In many cases, a closed system is not isolated from the surroundings but has some energy exchange with the surroundings as work or heat. In these cases, a second-law entropy analysis needs to consider both the closed system and the surroundings. The Clausius entropy statement (Table 6.4) describes this as:

The entropy of a system and the environment with which it is in contact increases, or, in the limit of a reversible process, remains constant [1].

OR

The entropy of the universe tends toward a maximum [5].

This is the celebrated **increase in entropy principle.** This can be written as:

$$dS_{\text{universe}} \equiv dS_{\text{sys}} + dS_{\text{surr}} \geq 0. \tag{7.23}$$

Here we have divided the universe into two regions: the region that we want to study (the system) and the rest of the universe, which may affect the process but is not our focus (the surroundings). We cannot emphasize too strongly the idea that these entropy-based second-law statements (Eqs. 7.22 and 7.23) rest firmly on the Kelvin–Planck statement of the second law, as a review of their derivation will show.

Example 7.11 Entropy Changes Due to Heat Transfer

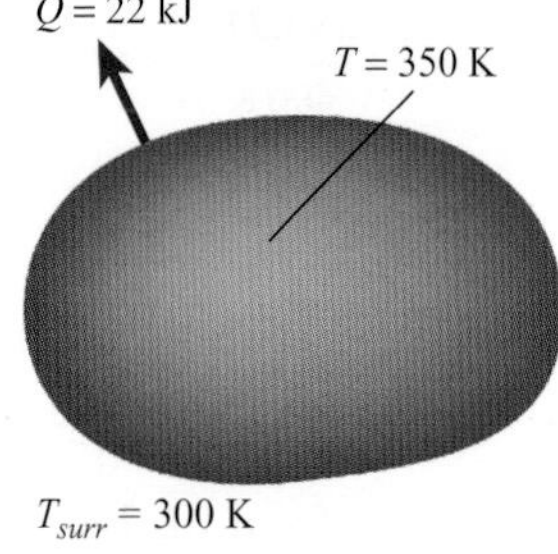

During an internally reversible process, 22 kJ of energy is removed as heat, from a thermodynamic system at a constant temperature of 350 K to the surroundings at 300 K.

a. Determine the entropy change of the closed system, surroundings, and universe.
b. Determine the entropy change of the system for the same heat removal when there are also internal irreversibilities.

Solution

Known T_{sys}, T_{surr}, Q

Find ΔS

Sketch

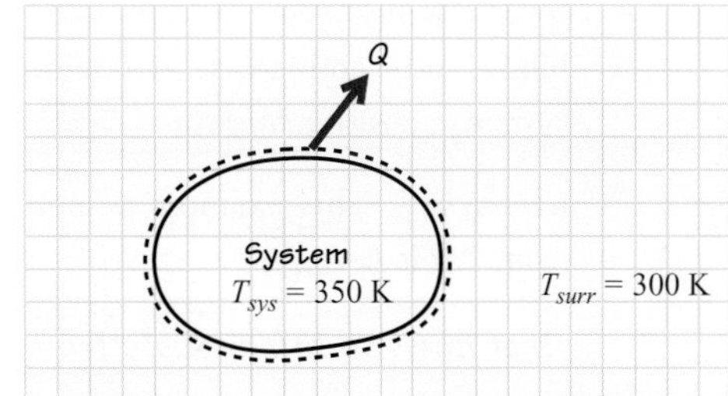

Analysis

a. Because the temperature of the closed system is constant during the heat transfer process, we can use Eq. 7.12a to find the change in the entropy of the system during an internally reversible process:

$$\Delta S_{\text{sys}} = (S_2 - S_1)_{\text{sys}} = \left.\frac{{}_1Q_2}{T_{\text{sys}}}\right|_{\substack{\text{int}\\\text{rev}}} = \frac{-22\text{ kJ}}{350\text{ K}} = -62.9\text{ J/K}.$$

Since energy is *removed* from the closed system as heat, Q is negative and the entropy of the system decreases during the internally reversible process.

We will now find the entropy change of the surroundings. Since the surroundings are large, they act as a heat reservoir and the temperature of the surroundings does not change during the heat transfer process. For this calculation, the surroundings are now our region of interest and the heat transfer is positive, into the region of interest.

$$\Delta S_{\text{surr}} = (S_2 - S_1)_{\text{surr}} = \left.\frac{{}_1Q_2}{T_{\text{surr}}}\right|_{\substack{\text{int}\\\text{rev}}} = \frac{22\text{ kJ}}{300\text{ K}} = 73.3\text{ J/K}.$$

The entropy of the surroundings increases owing to heat transfer to the surroundings.

We have divided the universe into two regions, the system and the surroundings. The entropy change of the universe is found by adding the entropy changes of these two regions:

$$dS_{\text{universe}} \equiv dS_{\text{sys}} + dS_{\text{surr}} = -6.29\text{ J/K} + 73.3\text{ J/K} = 10.4\text{ J/K} \geq 0.$$

We find that the entropy of the universe increases, as described by the Clausius entropy statement.

Inequalities containing negative quantities can be confusing. Exercise care in their interpretation.

b. To find the entropy change of the system when there are internal irreversibilities, we can use Eq. 7.21:

$$\underset{\substack{\text{Change of}\\\text{system}\\\text{entropy}}}{S_2 - S_1} = \underset{\substack{\text{Net transfer of}\\\text{entropy into the}\\\text{system by heat}\\\text{interactions}}}{\int_1^2 \frac{\delta Q}{T}} + \underset{\substack{\text{Entropy generated}\\\text{by irreversibilities}\\\text{within the system}}}{\int_1^2 \delta \mathcal{S}_{\text{gen}}}.$$

The term $\int_1^2 \delta Q/T$ is the change in entropy of the system in an internally reversible process. This was found in part a to be –62.9 J/K. Any irreversibilities in the system, $\int_1^2 \delta\mathcal{S}_{\text{gen}}$, will always give rise to a positive term. For a reversible process, $\Delta S_{\text{sys,rev}} = -62.9$ J/K; thus, the entropy change for the irreversible process must be greater than this. Because reversible heat removal causes the entropy of the system to decrease, whereas irreversibilities cause an increase, we cannot determine whether the actual entropy change for the system is positive, negative, or zero. The best we can say is that $\Delta S_{sys} > -62.9$ J/K.

Comment This example demonstrates one statement of the second law (Clausius II from Table 6.4):

Heat flows spontaneously from high to low temperature but not conversely.

This example is similar to Example 7.10 in that there is heat transfer across a finite temperature difference. In Example 7.10, the heat transfer occurred between two regions within the defined system and there was no energy transfer to the surroundings. In this example we have heat transfer between the system (the region of interest) and the surroundings. In both cases, the total entropy in the universe increases owing to heat transfer across a finite temperature difference. Heat transfer between two regions in the system is an internal irreversibility. Heat transfer between the system and surroundings is an external irreversibility.

Note the importance of the sign convention that heat added to a system is positive and that heat removed is negative.

Most engineering applications involve closed systems that are not isolated from their surroundings. Table 7.2 summarizes the ways in which the entropy of such closed systems can change. Here we see that there are only two possible ways for the entropy to remain unchanged after a process: (1) if the process is both adiabatic and reversible, and (2) if the process entails heat removal in an amount that balances exactly the entropy increase generated by irreversibilities. These are the first and last entries in Table 7.2. Note that the increase-of-entropy principle does not require the entropy change for any process to be positive. Heat removal, in the absence of any irreversibilities, always results in an entropy decrease.

We can summarize these results by saying that the change in entropy for a closed system undergoing an irreversible process will be greater than the change in entropy due to heat transfer to or from the system.

The equality applies when the process is internally reversible. Equation 7.24 is an alternative way to state the second law.

$$dS_{\text{sys}} \geq \left(\frac{\delta Q}{T}\right)_{\text{irrev}}. \tag{7.24}$$

TABLE 7.2 Entropy Changes for a Closed System*

Process	ΔS_{rev}	$\int_1^2 \delta \mathbb{S}_{\text{gen}}$	Entropy Change of System, $\Delta S_{\text{sys}} = \Delta S_{\text{rev}} + \int_1^2 \delta \mathbb{S}_{\text{gen}}$
Adiabatic and reversible	0	0	$\Delta S = 0$
Adiabatic and irreversible	0	positive	$\Delta S > 0$
Reversible heat addition	positive	0	$\Delta S > 0$
Irreversible heat addition	positive	positive	$\Delta S > 0$
Reversible heat removal	negative	0	$\Delta S < 0$
Irreversible heat removal	negative	positive	$\Delta S \gtreqless 0$

* The results shown here follow from the application of Eqs. 7.1b and 7.21.

7.4d A Closed System Undergoing a Cycle

In some cases, we will want to analyze an arbitrary closed thermodynamic system undergoing a thermodynamic cycle that includes both heat and work interactions. One example is the cycle of a fluid in a piston cylinder in Example 4.2, shown below in Fig 7.14a. The complete cycle includes four separate processes: expansion, heat rejection, compression, and heat addition. The cycle is called a **heat engine** since energy transfer into the closed system as heat produces a net work output. In Example 4.2, the P–$\mathcal{V}$ work was found for the expansion and compression processes. The heat transfer at the heat rejection and heat addition processes can be found using an energy balance. The change in entropy for the closed system can also be calculated for each process in this cycle using the methods described in the previous sections of this chapter. In this section, we will analyze the entire cycle as shown in Fig. 7.14b. For this analysis, the heat transfer and work terms Q_H, Q_L, Q_{net} are all positive values in our analysis, which the magnitude of these terms only; the direction of heat transfer and work will be included explicitly in the equations. In general, the heat transfer terms Q_H, Q_L may occur at uniform temperature or at a varying temperature.

A second type of cycle that we will consider is a continuously operating cycle that has several steady-state devices. An example is the steam power cycle shown in Fig 4.16 and repeated here as Fig 7.15a. This cycle is also called a **heat engine** since energy transfer as heat produces a net work output. In Examples 4.4, 5.10, and 5.11, individual devices using the steam power cycle were studied. In this section, we will analyze the entire cycle, as shown in Fig. 7.15b. For this analysis, the heat transfer and work terms $\dot{Q}_H$, $\dot{Q}_L$, and $\dot{W}_{net}$ are all positive values; the direction of heat transfer and work will be included explicitly in the equations. The heat transfer terms $\dot{Q}_H$ and $\dot{Q}_L$ may occur at uniform temperature or at a varying temperature. Since the cycle components operate continuously, the energy terms are written on a rate basis as $\dot{Q}_H$, $\dot{Q}_L$, and $\dot{W}_{net}$ or as energy per mass, q_H, q_L, and w_{net}.

The four-process piston–cylinder cycle and the steam power cycle can be analyzed in a similar way. The only difference is that the total energy terms are used in the piston–cylinder cycle and the rate of energy transfer is used for the steam power cycle. Both of these devices are called heat engines.

In the following discussion, the piston-cylinder cycle is considered and all terms are written as the total heat transfer or the total entropy change during the process.

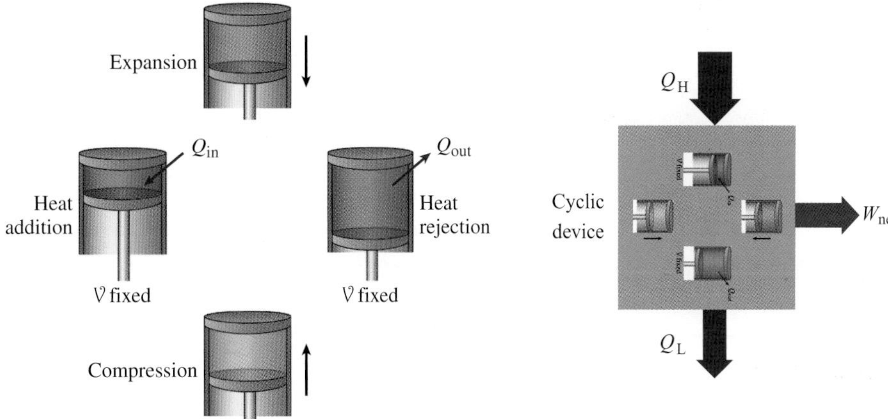

FIGURE 7.14 A thermodynamic cycle with four piston–cylinder processes: **(a)** four sequential processes; **(b)** energy transfer for entire cycle.

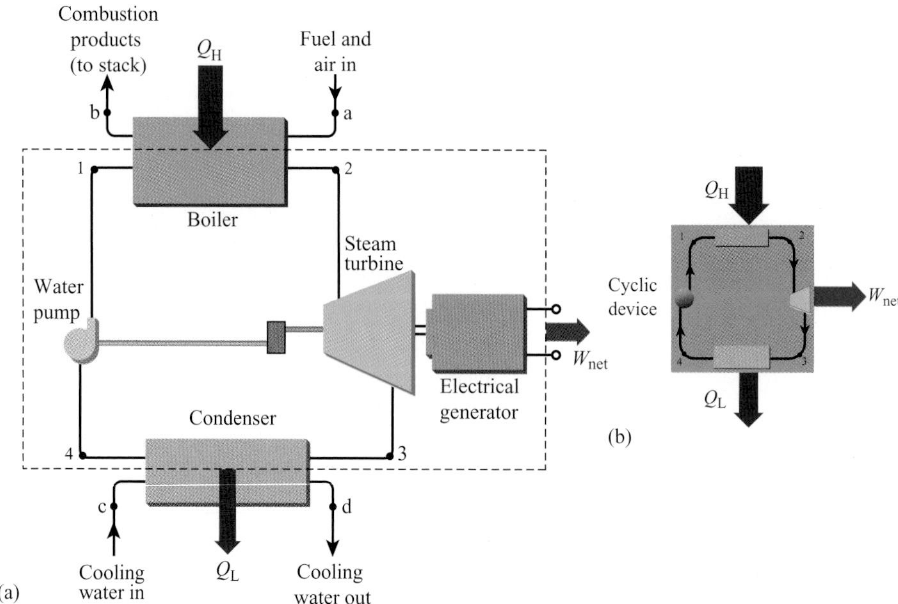

FIGURE 7.15 A thermodynamic steam power cycle: **(a)** four steady-state devices; **(b)** energy transfer for entire cycle.

For the entire cycle, there is no change in entropy since the closed system returns to its original state, as shown in Section 7.1c and stated as Eq. 7.3:

$$\oint_{\text{cycle}} dS_{\text{sys}} = 0.$$

We can substitute this into Eq. 7.23, $dS_{\text{universe}} \equiv \oint_{\text{cycle}} dS_{\text{sys}} + dS_{\text{surr}} \geq 0$, to give

$$dS_{\text{surr}} \geq 0 \tag{7.25}$$

for a cycle (or $d\dot{S}_{\text{surr}} \geq 0$ for a steady-state cycle). For a reversible cycle, $dS_{\text{surr}} = 0$. For a cycle with irreversibilities in the closed system (internal irreversibilities) or between the closed system and surroundings, owing to heat transfer across a finite temperature difference (external irreversibilities), we find that $dS_{\text{surr}} \geq 0$.

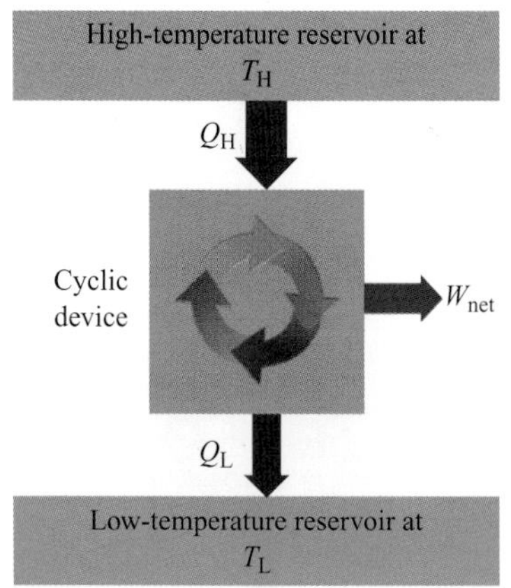

FIGURE 7.16 Heat engine with isothermal heat transfer.

CYCLE WITH ISOTHERMAL HEAT TRANSFER

We can apply the second-law analysis for a cycle, Eq. 7.25, to a heat engine where the heat transfer occurs at constant temperature (isothermally) as shown in Fig. 7.16. The change in entropy of the surroundings is due to the heat transfer from the high-temperature reservoir to the closed system and from the closed system to the low-temperature reservoir. The heat reservoirs are assumed to be very large, so that the heat transfer to a reservoir occurs at constant temperature with no internal irreversibilities. Since the heat transfer occurs at constant temperature, we can use Eq 7.2a to calculate the change in entropy for the surroundings:

$$\Delta S_{\text{surr}} = \left.\frac{Q_L}{T_L}\right|_{\text{int rev}} - \left.\frac{Q_H}{T_H}\right|_{\text{int rev}} \geq 0.$$

Note that the surroundings are now the region of interest and the negative Q_H term describes heat transfer out of the region of interest, which will reduce the entropy of that region. The Q_L term is positive since heat is added to the surroundings at the low-temperature reservoir.

In order for heat transfer between the closed system and the surroundings to occur at a reasonable rate for an actual heat engine, in fact there must be a finite temperature difference between the closed system and the surroundings. This is an external irreversibility, which causes an increase in the entropy of the surroundings during the cycle and gives the inequality above.

Solving for Q_L gives

$$Q_L \geq \frac{T_L}{T_H} Q_H.$$

An energy balance for the closed system gives: $Q_L = Q_H - W_{net}$. Substituting this into the equation above and solving for W_{net} yields

$$W_{net} \leq \left(1 - \frac{T_L}{T_H}\right) Q_H. \tag{7.26}$$

In general, the thermal efficiency of a cycle is defined as the desired output divided by the cost. For a heat engine, the desired output is W_{net} and the cost is Q_H, the heating of the closed system by some external source. The thermodynamic efficiency of the heat engine with isothermal heat transfer is

$$\eta_{HE} = \frac{W_{net}}{Q_H} \leq \left(1 - \frac{T_L}{T_H}\right). \tag{7.27}$$

If there is only a very small (differential) temperature difference between the closed system and surroundings, there will be no external irreversibilities and no net change in the entropy of the surroundings. A differential temperature difference between the closed system and surroundings would produce a very low rate of heat transfer and would not be practical. But we will look at this case as exhibiting ideal reversible conditions; it is called the Carnot cycle. If there are no irreversibilities in the universe, there is no change in entropy of the surroundings: $\Delta S_{surr} = 0$. The derivation above then leads to the thermal efficiency for an ideal reversible Carnot heat engine:

$$\eta_{rev,HE} = \frac{W_{net}}{Q_H} = \left(1 - \frac{T_L}{T_H}\right).$$

This is Eq. 6.15a for an ideal Carnot heat engine, now derived using the definition of the entropy change in an isothermal process. The Carnot cycle gives the maximum work output and the maximum thermal efficiency for a cycle operating between given high- and low-temperature reservoirs. Any actual cycle operating between these two temperatures will produce less work, as indicated by Eq. 7.26.

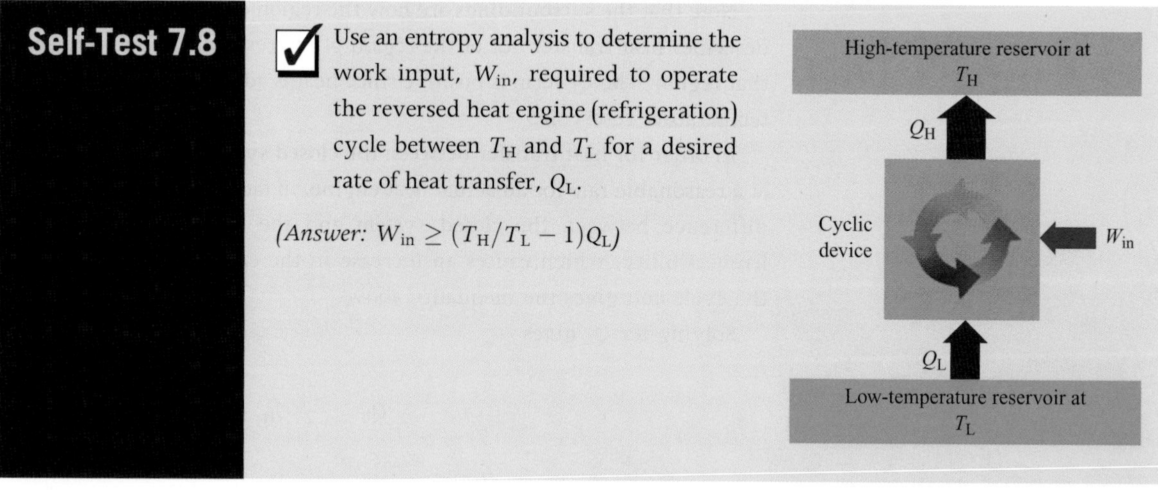

CLAUSIUS STATEMENT

We now look at a closed system undergoing a cycle in order to write another form of the second law, called the Clausius statement. We will consider an actual heat engine cycle where there are internal and external irreversibilities, as shown in Fig. 7.17. For our analysis, we will look at a steam power cycle where the heat addition at the boiler and the heat removal at the condenser occur at constant temperature. For heat transfer in the boiler to occur at a practical rate, the temperature of the steam in the boiler, T_{boiler}, will be less than the temperature of the reservoir, T_H. Likewise, for a finite rate heat transfer in the condenser, the temperature of the steam in the condenser, T_{cond}, will be greater than the temperature of the reservoir, T_L.

We will first start with Eq. 7.3, which states that there is no net change in entropy for a closed system undergoing a complete cycle:

$$\oint_{cycle} dS_{sys} = 0.$$

The entropy change in the system cycle is due to the heat transfer from the reservoirs and irreversibilities in the closed system:

$$\oint_{cycle} d\dot{S}_{sys} = \oint_{cycle} \left(\frac{\delta \dot{Q}}{T}\right)_{irrev} + \dot{\mathcal{S}}_{gen} = 0. \tag{7.28}$$

For a reversible process, there is no entropy generated within the cycle, $\dot{\mathcal{S}}_{gen} = 0$. For an irreversible process, the entropy generation is always positive. From this, we know that

$$\oint_{cycle} \left(\frac{\delta \dot{Q}}{T}\right)_{irrev} \leq 0, \tag{7.29a}$$

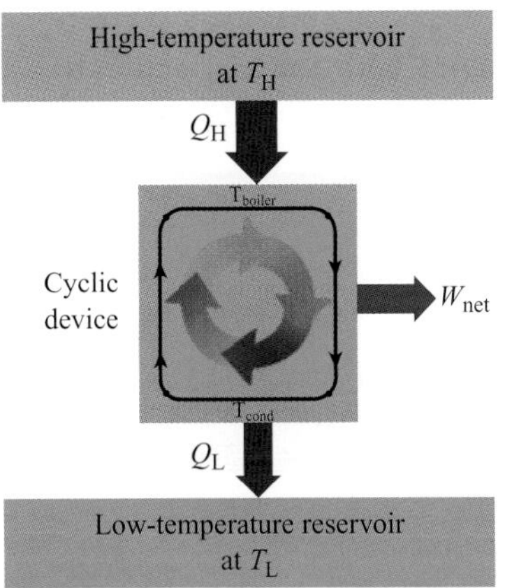

FIGURE 7.17 Heat engine with isothermal heat transfer.

where the equality applies for a reversible cycle and the inequality applies for an irreversible cycle. This equation is called the Clausius inequality for a continuously operating cycle. For a system enclosed by a piston, such as that in Fig. 7.14, The Clausius inequality is written with the heat transfer in each process of the cycle instead of the rate of heat transfer:

$$\oint_{\text{cycle}} \left(\frac{\delta Q}{T}\right)_{\text{irrev}} \leq 0. \tag{7.29a}$$

The Clausius inequality can be written as the following statement:

Whenever a closed system executes a complete cyclic process, the integral of $\delta Q/T$ around the cycle is less than zero, or in the limit is equal to zero.

Example 7.12 Entropy Change of a Heat Engine

The heat engine shown in Fig. 7.17 operates between two heat reservoirs at 550 K and 300 K. The temperature of the saturated water in the boiler is 520 K and the temperature of the saturated water in the condenser is 320 K. The rates of heat transfer in the boiler and condenser are 5.0 MW and 3.3 MW respectively. Determine the entropy generation due to internal and external irreversibilities in the cycle. What is the thermal efficiency of this heat engine and what is the ideal efficiency for a Carnot heat engine operating between the high- and low-temperature reservoirs?

Solution

Known T_H, T_L, Q_H and Q_L

Find $\dot{\mathcal{S}}_{gen}$ in the closed system and between the system and surroundings

To find the entropy generated within the closed system, we use Eq. 7.28 and solve for $\dot{\mathcal{S}}_{gen}$:

$$\left(\dot{\mathcal{S}}_{gen}\right)_{int} = -\oint_{\text{cycle}} \left(\frac{\delta \dot{Q}}{T}\right)_{\text{irrev}} = -\left(\frac{Q_H}{T_{\text{boiler}}} - \frac{Q_L}{T_{\text{cond}}}\right)$$

$$\left(\dot{\mathcal{S}}_{gen}\right)_{int} = -\left(\frac{5.0\,\text{MW}}{520\,\text{K}} - \frac{3.3\,\text{MW}}{320\,\text{K}}\right) = 0.697\,\text{MW/K}$$

When the closed system is the region of interest, Q_H is positive and Q_L is negative.

We can find the total entropy generation by calculating the change in entropy for the surroundings:

$$\left(\dot{\mathcal{S}}_{gen}\right)_{total} = \int \left(\frac{\delta \dot{Q}}{T}\right)_{\text{surr}} = \frac{Q_L}{T_L} - \frac{Q_H}{T_H},$$

$$\left(\dot{\mathcal{S}}_{gen}\right)_{total} = \frac{3.3\,\text{MW}}{300\,\text{K}} - \frac{5.0\,\text{MW}}{550\,\text{K}} = 1.909\,\text{kW/K}.$$

When the surroundings is the region of interest, Q_H is negative and Q_L is positive.

The total entropy generation in this cycle is due to both internal and external irreversibilities. The external irreversibilities can now be found:

$$\left(\dot{S}_{gen}\right)_{total} = \left(\dot{S}_{gen}\right)_{int} + \left(\dot{S}_{gen}\right)_{ext},$$
$$\left(\dot{S}_{gen}\right)_{ext} = \left(\dot{S}_{gen}\right)_{total} - \left(\dot{S}_{gen}\right)_{int},$$
$$\left(\dot{S}_{gen}\right)_{ext} = 1.909\,\text{kW/K} - 0.697\,\text{kW/K} = 1.212\,\text{kW/K}.$$

The thermal efficiency of the heat engine is

$$\eta_{actual,\,HE} = \frac{W_{net}}{Q_H} = \frac{Q_H - Q_L}{Q_H} = \frac{5.0\,\text{MW} - 3.3\,\text{MW}}{5.0\,\text{MW}} = 0.340.$$

The thermal efficiency of the ideal Carnot heat engine is

$$\eta_{rev,\,HE} = \frac{W_{net}}{Q_H} = 1 - \frac{T_L}{T_H} = 1 - \frac{300\,\text{K}}{550\,\text{K}} = 0.455.$$

Comment The internal irreversibilities are due to frictional losses and mixing within the closed system. The external irreversibilities are due to the finite temperature difference through which heat transfer occurs between the system and surroundings. As expected, the actual heat engine has a thermal efficiency that is less than an ideal Carnot cycle operating between the high- and low-temperature reservoirs.

AN ADDITIONAL PROOF OF THE KELVIN–PLANCK STATEMENT

We can use a cycle analysis to verify two statements of the second law listed in Table 6.4 called the Kelvin–Planck statements:

Statement IA: Although all work can be converted completely to heat, heat cannot be completely and *continuously* converted into work.

Statement IB: It is impossible to construct a cyclically operating device for which the sole effect is exchange of heat with a single reservoir and the creation of an equivalent amount of work.

To challenge this statement, we will consider a hypothetical cycle as a counter-example that would violate the Kelvin–Planck statement. If the counter-example is true, then the Kelvin–Planck statement is disproved. This hypothetical cycle has heat transfer to the closed system from a high-temperature reservoir and converts that energy completely to work.

For a cycle, there is no change in entropy of the closed system. The only change in entropy of the universe will be within the surroundings (see Eq. 7.25)

$$\Delta S_{surr} \geq 0.$$

The change in entropy of the surroundings for isothermal heat transfer to the hypothetical closed system gives

$$\Delta S_{surr} = -\frac{Q_H}{T_H} < 0.$$

Since the heat transfer is out of the surroundings (the region of interest in this calculation), the entropy of the surroundings will decrease. But we know that the entropy of the universe cannot decrease during any process. Therefore, the process described cannot occur, demonstrating the validity of the Kelvin–Planck statement that heat cannot be completely converted into work (see Fig. 7.18).

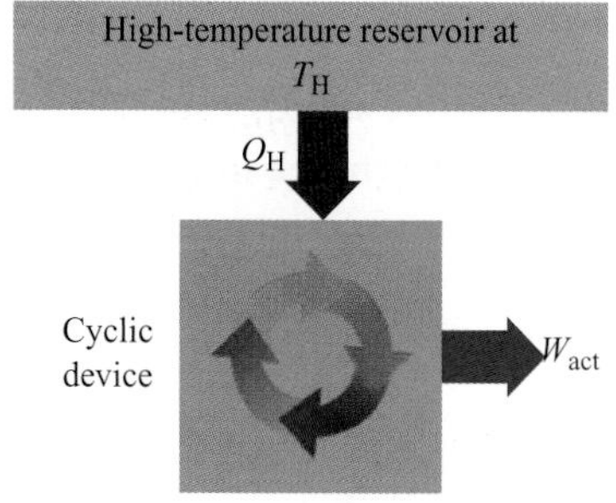

FIGURE 7.18 An impossible cycle used to prove the Kelvin–Planck statement that heat cannot be converted completely to work.

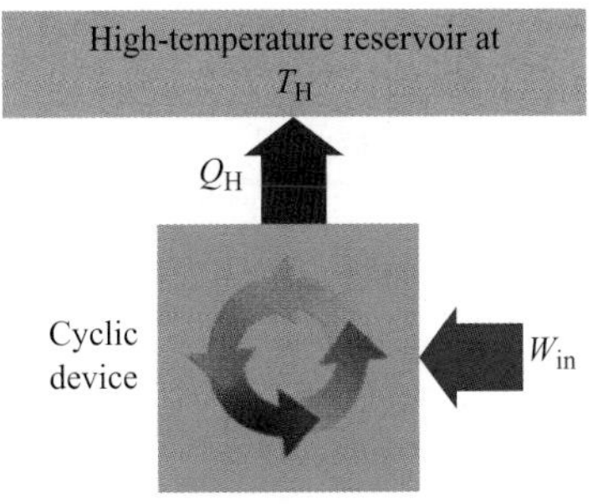

FIGURE 7.19 A cycle demonstrating that all work can be converted to heat, as described in the Kelvin–Planck statement 1A.

As a second example, we will reverse the direction of the heat transfer and work, as shown in Fig. 7.19. In this example, the work into the closed system is converted to heat, such as in an electrical heater or a mixing process. Again there is no change in entropy for the closed system since it operates in a complete cycle. The change in entropy of the surroundings is

$$\Delta S_{surr} = \frac{Q_H}{T_H} > 0.$$

This cycle is possible since the entropy of the world will increase during the process. This example demonstrates the first Kelvin–Planck statement that "All work can be converted completely to heat."

As described in Section 7.1d, heat transfer to a system causes an increase in the randomness (entropy) of the system. Energy transfer in the form of work does not increase the randomness (entropy) of the system since work is an organized form of energy transfer. For example, consider shaft work, where all the molecules of the shaft are rotating together, or P–$\mathcal{V}$ work, where the entire mass of the piston is moving at the same speed.

WORK AND HEAT FOR REVERSIBLE CYCLE

For a complete cycle, as shown in Figs. 7.14 and 7.15, the state of the fluid through the cycle can be tracked on P–$\mathcal{V}$ and T–s diagrams. Since the state of the fluid is the same at the beginning and the end of the cycle, all the state properties will also be the same at the beginning and the end of the cycle. Plotting a complete cycle on the P–$\mathcal{V}$ and T–s diagrams will give a closed curve, as we found in Examples 7.4 and 7.5. Figure 7.20 repeats the P–$\mathcal{V}$ and T–s diagrams that we found for the Carnot cycle in Example 7.4. We can use these figures to determine the net heat and work for the cycle.

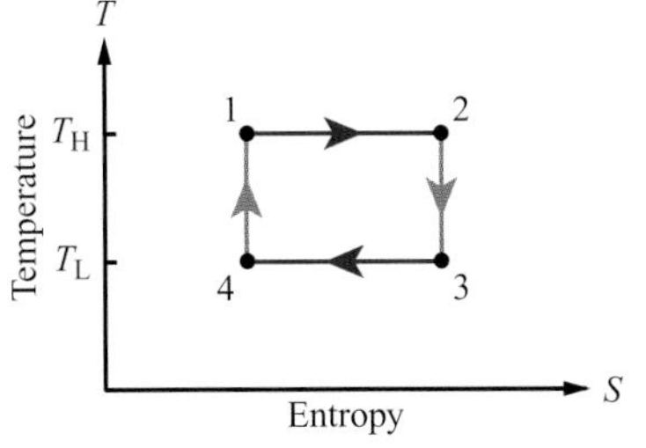

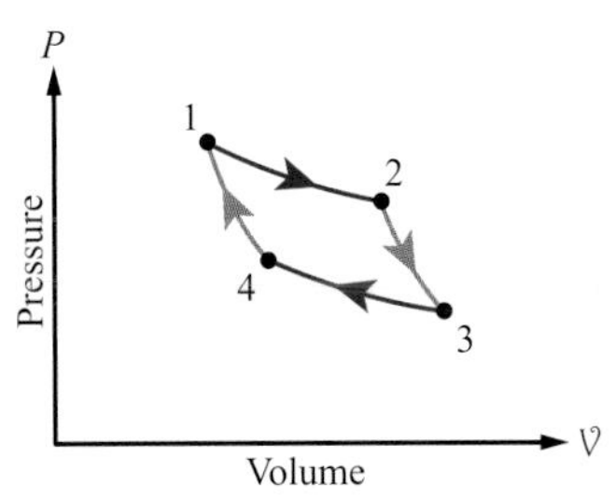

FIGURE 7.20 The four reversible processes of the Carnot cycle are presented using (top) temperature–entropy (T–S) and (bottom) pressure–volume (P–$\mathcal{V}$) thermodynamic coordinates.

In Chapter 4, we found that the work done per mass by the fluid in a cylinder during a reversible expansion or compression is (see Eq. 4.8b)

$$_{a}w_{b} = \int_{a}^{b} P d\mathcal{V}.$$

For the process to be reversible, we assumed that it occurs slowly (quasi-steady) so that the fluid properties remain uniform through the cylinder. We also assumed that there were no irreversibilities such as friction in the cylinder. The P–$\mathcal{V}$ work done during a process is the area under the P–$\mathcal{V}$ curve. Since P is always positive, we note that the P–$\mathcal{V}$ work will be positive for an expansion process ($\mathcal{V}_2 > \mathcal{V}_1$) and negative for a compressible process ($\mathcal{V}_2 < \mathcal{V}_1$) .

For a cycle, such as the Carnot cycle in Figure 7.20, we can determine the net P–$\mathcal{V}$ work per mass done by the fluid during the entire cycle by adding the P–$\mathcal{V}$ work per mass in each step of the cycle and substituting the expression for the P–$\mathcal{V}$ work for a reversible process, Eq. 4.8b:

$$\begin{aligned} w_{net} &= {}_1w_2 + {}_2w_3 + {}_3w_4 + {}_4w_1 \\ &= \int_1^2 Pd\mathcal{V} + \int_2^3 Pd\mathcal{V} + \int_3^4 Pd\mathcal{V} + \int_4^1 Pd\mathcal{V}, \\ w_{net} &= \oint Pd\mathcal{V}. \end{aligned} \tag{7.30}$$

To review plotting processes in thermodynamic coordinates, see Examples 2.11 and 2.12 in Chapter 2; also see Fig. 2.13.

This equation tells us mathematically that the area enclosed in the P–v_2 diagram represents the net work produced by the cycle.

We can now repeat this process using the T–s diagram and Eq. 7.1c:

$$_1q_2 = \int_1^2 T\,ds\bigg|_{\text{int rev}}.$$

For a reversible process, the heat transfer per mass can be found knowing the change in entropy of the fluid with temperature. On the T–s diagram, the area under the T–s curve is the heat transferred during the process. Since T is always positive, we note that an increase in entropy $(s_2 > s_1)$ is related to the heat into the system and a decrease in entropy $(s_2 < s_1)$ is related to the heat out of the system.

For a cycle, such as the Carnot cycle in Fig. 7.20, we can determine the net heat transfer to the fluid during the entire cycle by adding the heat in each step of the cycle and substituting Eq. 7.1c:

$$\begin{aligned} q_{\text{net}} &= {_1q_2} + {_2q_3} + {_3q_4} + {_4q_1} \\ &= \int_1^2 T\,ds + \int_2^3 T\,ds + \int_3^4 T\,ds + \int_4^1 T\,ds, \\ q_{\text{net}} &= \oint T\,ds. \end{aligned} \tag{7.31}$$

The area enclosed in the T–s diagram, therefore, represents the net heat per mass into the cycle.

We can now use a first-law analysis for the cycle:

$$\oint \delta q = \oint \delta w + \Delta u.$$

Recall from Section 4.2b that we use "δ" in δq and δw to denote path dependent changes. Changes in state properties use "Δ", as in Δu.

Since the fluid returns to its original state at the end of the cycle, there is no net change in internal energy from the beginning to the end of the cycle, $\Delta u = 0$. This gives

$$\oint \delta q = \oint \delta w,$$

or

$$q_{\text{net}} = w_{\text{net}} = \oint T\,ds \tag{7.32}$$

for the cycle. We can therefore use the area beneath the P–v curve or beneath the T–s curve to find the net work or heat during a cycle. We prefer to use a simple curve, such as a rectangle, when possible. For example, in the Carnot cycle shown in Figure 7.12, the cycle forms a rectangle on a T–s diagram. It is therefore simplest to find the net work and heat in the cycle in terms of T and s:

$$q_{\text{net}} = w_{\text{net}} = (T_H - T_L)(s_2 - s_1).$$

We have written these equations on a per mass basis. The net work and heat transfer in a piston cycle, such as in Fig. 7.9 can be found as $W_{\text{net}} = Mw_{\text{net}}$ and $Q_{\text{net}} = Mq_{\text{net}}$, respectively. For a continuous cycle with flow passing through multiple devices, such as in Fig. 7.10, the net power output and the net rate of heat transfer can be found as $\dot{W}_{\text{net}} = \dot{m}w_{\text{net}}$ and $\dot{Q}_{\text{net}} = \dot{m}q_{\text{net}}$.

7.4e Some Reversible Cycles

In this section, we discuss two ideal cycles that are of particular practical and historical interest: the Carnot cycle and the Stirling cycle. In both of these cycles, a cyclic device exchanges heat with two constant-temperature reservoirs through a series of reversible processes, thus meeting the requirements of our previous analysis. Exploring these ideal cycles provides insight into the conversion of heat to work, in general, and also provides a framework for understanding practical energy-conversion devices.

CARNOT CYCLE

In his 1824 publication, Carnot defined an ideal cycle that now bears his name. This Carnot cycle is illustrated in Fig. 7.21 for a single-phase working fluid.

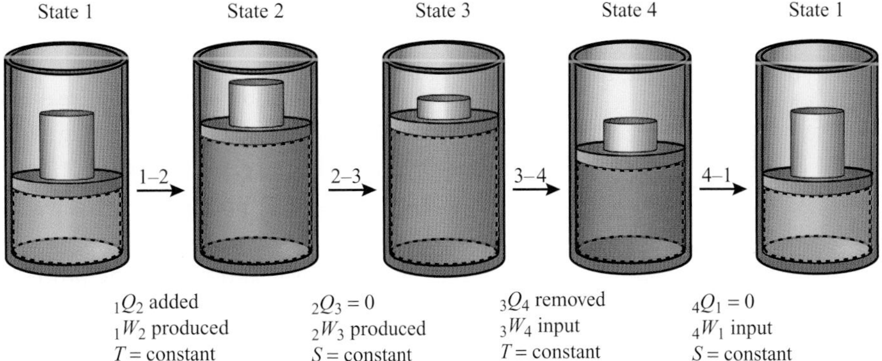

FIGURE 7.21 This sequence of states illustrates the Carnot cycle applied to a fixed mass of gas. The size of the weights placed on top of the piston indicates the relative pressure levels of the gas within the cylinder.

Here the closed thermodynamic system under consideration is the gas within the piston–cylinder assembly. The reversible processes that comprise the Carnot cycle are the following:

Ideal Carnot Cycle

State change	Process
1–2:	Reversible heat addition at constant temperature
2–3:	Reversible adiabatic (isentropic) expansion
3–4:	Reversible heat rejection at constant temperature
4–1:	Reversible adiabatic (isentropic) compression

These four processes are conveniently illustrated using temperature–entropy (T–S) and pressure–volume (P–$V\!\!\!/$) coordinates, as previously shown in Fig. 7.20 and Example 7.4 .[3]

The two reversible and adiabatic processes in the cycle are isentropic (they occur at constant entropy) and appear as vertical lines on the T–S diagram. The other two processes are isothermal (they occur at constant temperature) and appear as horizontal

[3] The emblem of the Mechanical Engineering Honor Society, Pi Tau Sigma, uses the outline of the Carnot cycle in P–$V\!\!\!/$ coordinates in honor of Sadi Carnot's contribution to the field of mechanical engineering.

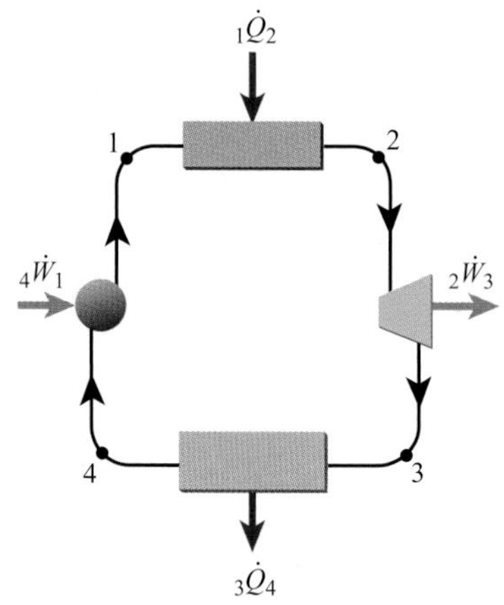

FIGURE 7.22 Steady-flow Carnot cycle: 1–2, heat addition at constant temperature; 2–3, reversible adiabatic expansion; 3–4, heat rejection at constant temperature; and 4–1, reversible adiabatic compression.

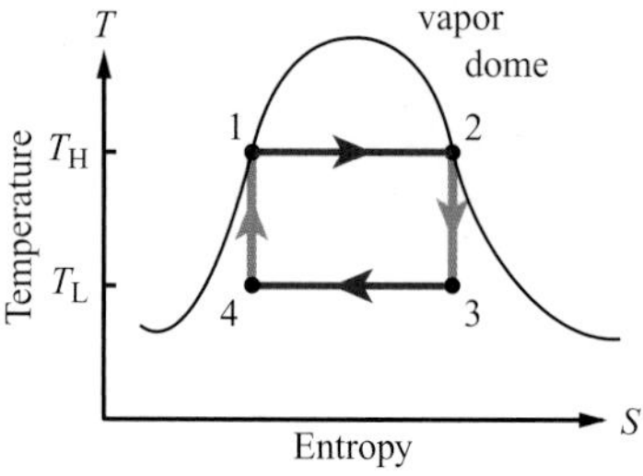

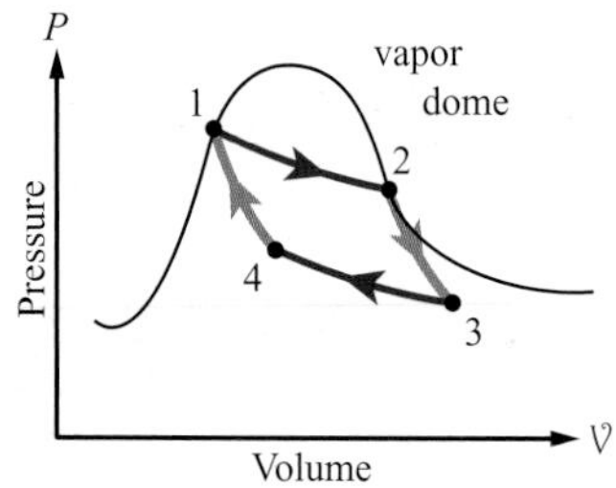

FIGURE 7.23 The four reversible processes of the Carnot cycle are presented using (top) temperature–entropy (T–S) and (bottom) pressure–volume (P–$\mathcal{V}$) thermodynamic coordinates.

lines on the T–S diagram. Thus, the Carnot cycle is described by a rectangle in T–S coordinates. In P–$\mathcal{V}$ coordinates, the isothermal and isentropic processes follow curved paths for an ideal gas, with isotherm slopes that are not as steep as those of the isentropes.[4]

The Carnot cycle can also be applied to a sequence of steady-flow processes, as illustrated in Fig. 7.22. This steady-flow cycle is a key building block in our analysis of the steam power plant (Rankine cycle). The Carnot cycle used in a steam power plant would occur under the vapor dome, as shown in Fig. 7.23. For the saturated steam in the power cycle, the isothermal heating (1–2) results in boiling of the saturated liquid in a constant-pressure process. The isothermal cooling (3–4) that condenses saturated steam to saturated liquid also occurs at constant pressure. The isothermal heating and cooling of saturated water appear as horizontal lines in the P–$\mathcal{V}$ diagram.

STIRLING CYCLE

Another reversible cycle is that derived by Robert Stirling (1790–1878). The Stirling cycle is approximated by a number of real engines. Stirling engines fill a niche for quiet, small engines capable of operating with a wide variety of fuels [6]. (See Fig. 7.24.) Typically, light gases (H_2 and He) are used as the working fluid in Stirling engines. Example 6.2 presented performance data for a commercially available engine.

Note that there is heat transfer in all processes in the Stirling cycle. Heat transfer to the closed system occurs during processes 4–1 and 1–2. To find Q_H we must add these two heat transfer terms: $Q_H = {}_4Q_1 + {}_1Q_2$. The heat transfer from process 1–2 occurs at constant temperature and can be found using Eq. 7.2b: ${}_1Q_2 = T(S_2 - S_1)|_{\text{int rev}}$. The heat transfer from process 4–1 can be found using the energy equation: ${}_4Q_1 = {}_4W_1 + (U_2 - U_4)$. Since the closed system does not change volume during process 4–1, there is no P–v work and ${}_4W_1 = 0$. We now have the total heat addition to the closed system as:

$$Q_H = {}_4Q_1 + {}_1Q_2,$$
$$Q_H = (U_1 - U_4) + T(S_2 - S_1)|_{\text{int rev}}$$

For an ideal gas, the change in entropy can be found using Eq. 7.12a:

$$s_2 - s_1 = c_{p,\text{avg}} \ln \frac{T_2}{T_1} - R \ln \frac{P_2}{P_1}.$$

Substituting this into the expression for Q_H gives

$$Q_H = {}_4Q_1 + {}_1Q_2,$$
$$Q_H = M(u_1 - u_4) + MT\left(c_{p,\text{avg}} \ln \frac{T_2}{T_1} - R \ln \frac{P_2}{P_1}\right).$$

The net work done by the closed system can be found as the difference between the P–v expansion and compression work, both at constant temperature: $W_{\text{net}} = {}_1W_2 + {}_3W_4$

[4] As discussed in Chapter 2, isotherms refer to lines of constant temperature and isentropes to lines of constant entropy.

FIGURE 7.24 Mirrors focus sunlight onto a thermal receiver of a Stirling engine. The engine drives a 25 kW electrical generator. Photograph courtesy of Bill Timmerman (NREL).

Note that ${}_3W_4$ will be negative since the system volume decreases during the compression process. For an isothermal expansion or compression of an ideal gas, we use Eq. T4.3b:

$$ {}_1W_2 = P_1\mathcal{V}_1 \ln \frac{\mathcal{V}_2}{\mathcal{V}_1} = mRT_1 \ln \frac{\mathcal{V}_2}{\mathcal{V}_1} \quad \textbf{(T4.3b)} $$

The net work done by the Stirling cycle is

$$ W_{\text{net}} = {}_1W_2 + {}_3W_4, $$
$$ W_{\text{net}} = P_1\mathcal{V}_1 \ln \frac{\mathcal{V}_2}{\mathcal{V}_1} + P_3\mathcal{V}_3 \ln \frac{\mathcal{V}_4}{\mathcal{V}_3}. $$

The second P–$\mathcal{V}$ term in W_{net} will be negative since $\mathcal{V}_4 < \mathcal{V}_3$. The thermal efficiency for the ideal Stirling cycle can be found by substituting for the heat transfer at high temperature and the net work term:

$$ \eta_{\text{HE}} = \frac{W_{\text{net}}}{Q_{\text{H}}} = \frac{MP_1\mathcal{V}_1 \ln \frac{\mathcal{V}_2}{\mathcal{V}_1} + MP_3\mathcal{V}_3 \ln \frac{\mathcal{V}_4}{\mathcal{V}_3}}{M(u_1 - u_4) + MT\left(c_{\text{p,avg}} \ln \frac{T_2}{T_1} - R \ln \frac{P_2}{P_1}\right)} $$
$$ = \frac{P_1\mathcal{V}_1 \ln \frac{\mathcal{V}_2}{\mathcal{V}_1} + P_3\mathcal{V}_3 \ln \frac{\mathcal{V}_4}{\mathcal{V}_3}}{(u_1 - u_4) + T\left(c_{\text{p,avg}} \ln \frac{T_2}{T_1} - R \ln \frac{P_2}{P_1}\right)}. $$

Note that the thermal efficiency of the ideal Stirling engine is different than that for the ideal Carnot engine. The ideal Stirling cycle consists of the following processes:

IDEAL STIRLING CYCLE

State change	Process
1–2:	Reversible heat addition at constant temperature
2–3:	Reversible heat rejection at constant volume
3–4:	Reversible heat rejection at constant temperature
4–1:	Reversible heat addition at constant volume

If you have not covered Chapter 6, this is a good place to go back to Chapter 6 material.

These processes are shown in P–$\mathcal{V}$ and T–S coordinates in Fig. 7.25.

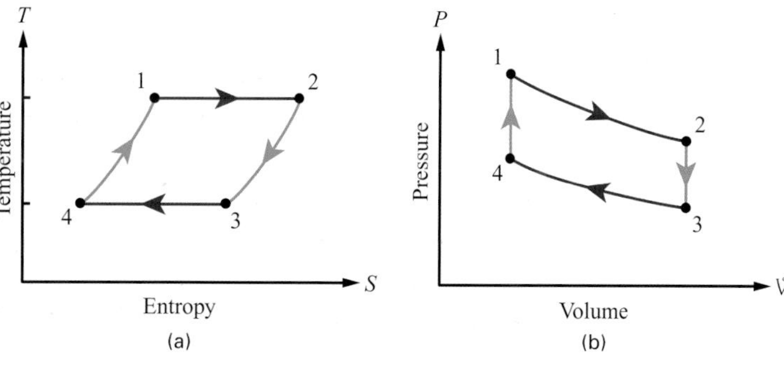

FIGURE 7.25 The ideal Stirling cycle consists of four reversible processes, shown here using **(a)** T–S coordinates and **(b)** P–$\mathcal{V}$ coordinates.

7.5 Entropy Balances for an Open System (Control Volume)

For a control volume, we may have a process that occurs over a finite time interval or a process that is operating continuously at steady state. For a process that occurs over a finite time interval, we will apply Eq. 3.1 to the entropy change:

$$(S_2 - S_1)_{CV} = \int_1^2 \frac{\delta Q}{T} + \sum M_{\text{in}}\, s_{\text{in}} - \sum M_{\text{out}}\, s_{\text{out}} + \mathbb{S}_{\text{gen}}. \tag{7.33a}$$

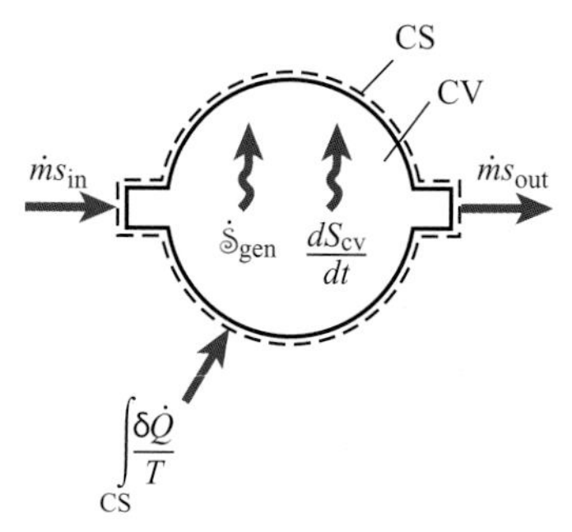

FIGURE 7.26 Entropy terms for a control volume.

The terms of this equation are shown in Fig. 7.26. The entropy in the control volume will increase owing to heat transfer to the control volume, inflows of mass carrying entropy, and irreversibilities in the control volume. Outflows of mass from the control volume will also carry entropy out of the control volume. For many control volume problems, we will analyze a device that is running in steady state, for example, a turbine or compressor. For steady-state devices, we adopt the rate form of the generic conservation principle Eq. 3.2.

$$dX_{\text{cv}}/dt = \dot{X}_{\text{in}} - \dot{X}_{\text{out}} + \dot{X}_{\text{gen}}.$$

For our entropy balance, this becomes

$$\frac{dS_{\text{cv}}}{dt} = \int_{CS} \frac{\delta \dot{Q}}{T} + \sum \dot{m}_{\text{in}}\, s_{\text{in}} - \sum \dot{m}_{\text{out}}\, s_{\text{out}} + \dot{\mathbb{S}}_{\text{gen}}. \tag{7.33b}$$

For a steady-state device with only one inflow and one outflow, $\dot{m} = \dot{m}_{\text{in}} = \dot{m}_{\text{out}}$ and Eq. 7.33b becomes

$$\frac{dS_{\text{cv}}}{dt} = \int_{CS} \frac{\delta \dot{Q}}{T} + \dot{m}(s_{\text{in}} - s_{\text{out}}) + \dot{\mathbb{S}}_{\text{gen}}. \tag{7.33c}$$

The left-hand term is the rate at which the entropy within the control volume increases. For a steady state, this term is zero. Here we see that entropy enters (or leaves) the control volume by a heat interaction, the first term on the right-hand side in Eq. 7.33b, or is carried in ($\dot{m}s_{\text{in}}$) or out ($\dot{m}s_{\text{out}}$) by the flow. The term $\dot{\mathbb{S}}_{\text{gen}}$ represents the rate of entropy production within the control volume by irreversibilities. A common source of $\dot{\mathbb{S}}_{\text{gen}}$ in real systems is fluid friction, either at the walls of a device or within the moving fluid itself. The thermodynamic properties appearing in this entropy balance are the specific entropies s_{in} and s_{out}, the total entropy within the control volume, S_{cv}, and the temperature T; again $\delta \dot{Q}$ and $\dot{\mathbb{S}}_{\text{gen}}$ are process-dependent quantities and are not thermodynamic properties.

7.5a Isentropic Efficiency of a Turbine

In addition to entropy's usefulness in predicting the direction of spontaneous change and establishing equilibrium conditions (statements IVA–C of Table 6.4), we also use entropy to quantify the performance of practical devices. Our discussions of entropy thus reinforce the two ways in which the second law is particularly useful to engineers, as outlined at the outset of this chapter. In an earlier section, we saw that the second law establishes theoretical limits to thermodynamic *cycles* (Carnot efficiency); in an analogous manner, the second law also establishes theoretical limits to *processes*. For example, consider the expansion of steam through a turbine. The maximum work will be obtained from this process when it occurs adiabatically – with no loss of potential to do work through heat transfer – and reversibly – with no loss of potential to do work through friction or other irreversibilities. For all the steady-flow devices we wish to consider (i.e., turbines, compressors, pumps, and nozzles; see Chapter 8), the reversible adiabatic, or isentropic, process is the standard to which real processes can be compared. The isentropic efficiency for a turbine is thus defined as

$$\eta_{s,\text{turbine}} \equiv \frac{\dot{W}_{\text{actual}}}{\dot{W}_{\text{ideal}}}. \tag{7.34}$$

In Eq. 7.34, both the actual power $\dot{W}_{\text{actual}}$ and the isentropic power $\dot{W}_{\text{ideal}}$ are evaluated using the same inlet states and for the same pressures at the exit states.

Example 7.13 Power Output from an Actual Steam Turbine

Example 5.10 in Chapter 5 provides useful background for this example.

Consider a steam turbine in which the steam enters as superheated vapor at 800 K and 6 MPa and exits at 0.1 MPa. The flow rate of the steam is 15 kg/s, and the isentropic efficiency of the turbine is 90%. Determine the outlet state of the steam and the power produced by the turbine.

Solution

Known $T_1, P_1, P_2, \eta_{s,t}$

Find State-2 properties, $\dot{W}_{\text{act}}$

Sketch

A 660-MW steam turbine. Photograph courtesy of General Electric Company.

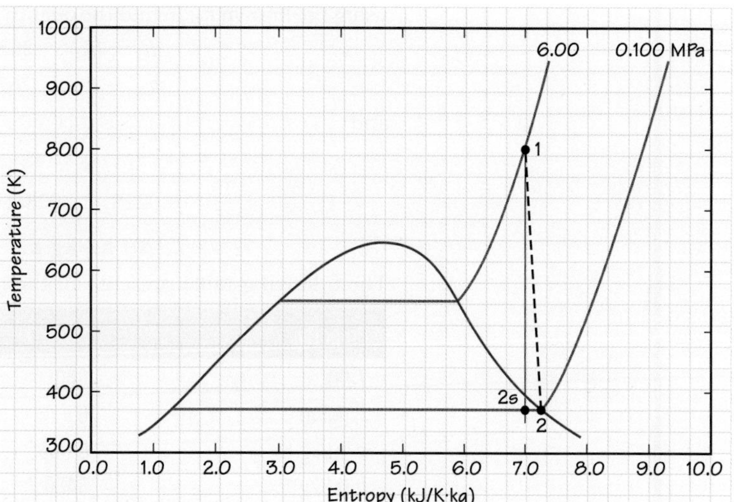

Modeling, Premises and Assumptions

i. Steady state

ii. Negligible ΔKE and ΔPE

Analysis Using the NIST WebBook, we find that $s_1 = 6.9635$ kJ/kg·K for the given inlet temperature and pressure. For an isentropic expansion from state 1 to state 2, we see that s_{2s} falls between s_{2f} and s_{2g} (the subscripts "f" and "g" refer to "liquid" and "gas") at 0.1 MPa (i.e., $1.3028 < 6.9635 < 7.3588$ kJ/kg·K). The actual state-2 entropy will lie to the right of s_{2s}, somewhere on the 0.1-MPa isobar. To determine the actual state 2, we apply the definition of isentropic efficiency (Eq. 7.34):

$$\eta_{s,t} = \frac{\dot{W}_{\text{actual}}}{\dot{W}_{\text{ideal}}}.$$

Here both the actual power $\dot{W}_{\text{actual}}$ and the isentropic power $\dot{W}_{\text{ideal}}$ are evaluated using the same inlet states and with the same exit-state pressure. With our assumptions, the turbine power is expressed as $\dot{W}_t = \dot{m}(h_{\text{in}} - h_{\text{out}})$; thus,

$$\eta_{s,t} = \frac{\dot{m}(h_1 - h_2)}{\dot{m}(h_1 - h_{2s})} = \frac{h_1 - h_2}{h_1 - h_{2s}}.$$

The enthalpy at state 1 is found from the NIST WebBook and is listed in the table that follows this discussion. To find h_{2s}, we use Eq. 7.18b to find the quality x_{2s} using known entropies:

$$x_{2s} = \frac{s_{2s} - s_{2f}}{s_{2g} - s_{2f}} = \frac{6.9635\,\text{kJ/kg·K} - 1.3028\,\text{kJ/kg·K}}{7.3588\,\text{kJ/kg·K} - 1.3028\,\text{kJ/kg·K}} = 0.9347$$

and

$$\begin{aligned} h_{2s} &= (1 - x_{2s})h_{2f} + x_{2s}h_{2g} \\ &= 0.0653(417.5\,\text{kJ/kg}) + 0.9347(2674.9\,\text{kJ/kg}) \\ &= 2527.5\,\text{kJ/kg}, \end{aligned}$$

where s_{f2}, s_{2g}, h_{2f}, and h_{2g} are found from the NIST WebBook for $P_2 = P_{\text{sat}} = 0.1$ MPa. Using the given value of $\eta_{s,t}(= 0.90)$, we calculate h_2 as follows:

$$0.90 = \frac{3486.7\,\text{kJ/kg} - h_2}{3486.7\,\text{kJ/kg} - 2527.5\,\text{kJ/kg}},$$

or

$$h_2 = 2623.4\,\text{kJ/kg}.$$

The enthalpy at state 2 is less than the enthalpy of the saturated vapor at the outflow pressure. We therefore know that the outflow from the actual turbine is a saturated mixture. To completely define the properties at state 2, we again apply Eq. 2.37d, now using the known h_2, to find the quality x_2. This quality is used in turn to find s_2 (Eq. 2.37c):

$$x_2 = \frac{h_2 - h_{2f}}{h_{2g} - h_{2f}} = \frac{2623.4\,\text{kJ/kg} - 417.5\,\text{kJ/kg}}{2674.9\,\text{kJ/kg} - 417.5\,\text{kJ/kg}} = 0.9772$$

and

$$\begin{aligned} s_2 &= (1 - x_2)s_{2f} + x_2 s_{2f} \\ &= (0.0228)(1.3028\,\text{kJ/kg·K}) + 0.9772(7.3588\,\text{kJ/kg·K}) \\ &= 7.2207\,\text{kJ/kg·K}. \end{aligned}$$

The following table presents all the properties:

Property	State 1	State 2s	State 2
T (K)	800	372.76	372.76
P (MPa)	6	0.10	0.10
s (kJ/kg·K)	6.9635	6.9635	7.2207
h (kJ/kg)	3486.7	2527.5	2623.4
x	—	0.9347	0.9772

The actual turbine power is

$$\begin{aligned}\dot{W}_{\text{actual}} &= \dot{m}(h_1 - h_2)\\ &= (15\ \text{kg/s})(3486.7\ \text{kJ/kg} - 2623.4\ \text{kJ/kg})\left[\frac{\text{kW}}{\text{kJ/s}}\right]\\ &= 12{,}949\ \text{kW}\end{aligned}$$

Comment We note that the final state still lies within the liquid-vapor dome on our T–s diagram, to the right of the 2s state, as expected for an irreversible, nearly adiabatic expansion. Note also the importance of the thermodynamic property "quality" in the solution of this problem.

Self-Test 7.9

Steam enters a turbine at 10 MPa and 800 K and exits at a quality of 0.91 at 100 kPa. Determine the isentropic efficiency of the turbine.

(Answer: $\eta_{isen,t} = 95.4\%$)

7.5b Isentropic Efficiency of a Pump or Compressor

For devices that consume power (e.g., pumps and compressors) one desires to minimize the power input; thus, the isentropic efficiency is now defined as

$$\eta_{\text{s, p or c}} \equiv \frac{\dot{W}_{\text{ideal}}}{\dot{W}_{\text{actual}}}. \tag{7.35}$$

Again, the same inlet conditions and identical outlet pressures apply to the evaluation of both the isentropic and the actual power.

Example 7.14 Power Input for an Actual Feedwater Pump

A feedwater pump operates with a flow rate of 464 kg/s. The inlet pressure is 689 kPa and the outlet pressure is 26 MPa. The water enters the pump at 422 K. Assuming an isentropic efficiency of 85%, estimate the power required to drive the pump.

Solution

Known $\dot{m}, P_1, P_2, T_1, \eta_{s,p}$

Find $\dot{W}_{\text{in, act}}$

Sketch

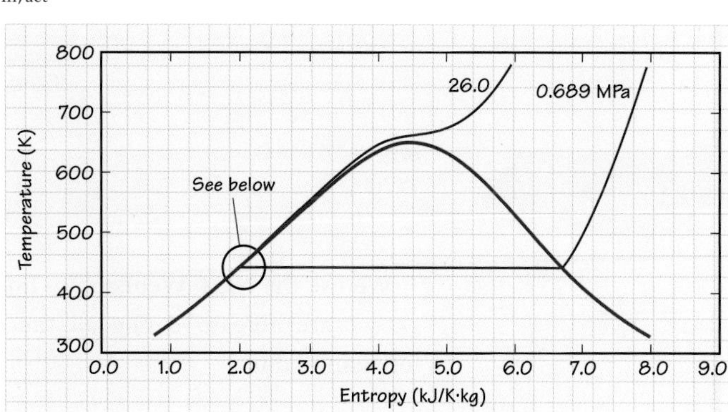

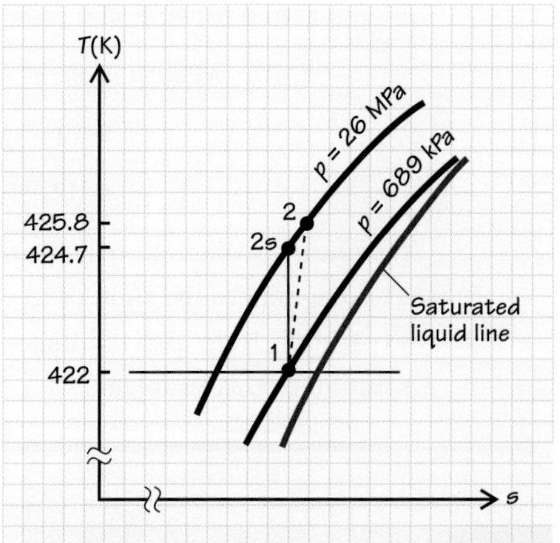

Feedwater pump.
Photograph courtesy of Flowserve Corporation.

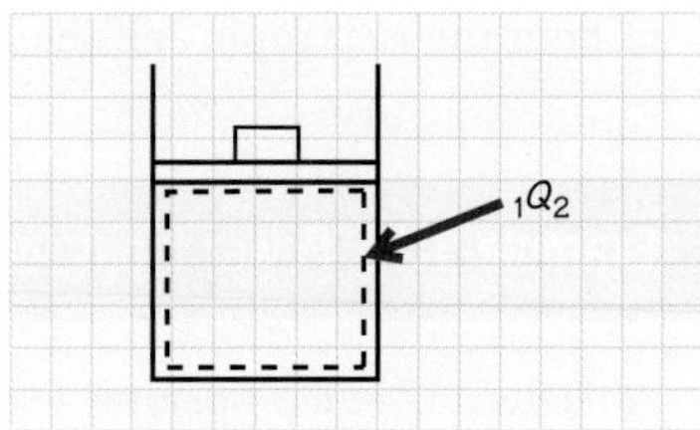

Modeling, Premises and Assumptions

i. Adiabatic process
ii. Negligible changes in kinetic and potential energies

Analysis To find the pump power, we apply the definition of isentropic efficiency (Eq. 7.26) and the simplified first-law analysis of the pump (Eq. 5.15b), that is,

$$\eta_{s,\,pump} = \frac{\dot{W}_{ideal}}{\dot{W}_{actual}},$$

where

Also see Table 8.1

$$\dot{W}_{ideal} = \dot{m}(h_{2s} - h_1).$$

We use the NIST WebBook to find the properties at state 1. Recognizing that $s_{2s} = s_1$, we are able to define all the properties at state 2. These are shown in the following table.

Property	State 1	State 2s
P(MPa)	0.689	26
T(K)	422	424.67
s(kJ/kg·K)	1.8298	1.8298
h(kJ/kg)	627.36	654.74

With the inlet and exit properties now known, we evaluate $\dot{W}_{\text{in, actual}}$ as follows:

$$\dot{W}_{\text{in, actual}} = \frac{\dot{m}(h_{2s} - h_1)}{\eta_{\text{s, pump}}}$$

$$= \frac{(464\,\text{kg/s})(654.74\,\text{kJ/kg} - 627.36\,\text{kJ/kg})}{0.85}\left[\frac{\text{kW}}{\text{kJ/s}}\right] = 14.95\,\text{MW}.$$

Comments The irreversibilities in the pump, such as friction, convert some of the input power to thermal energy. To find the additional temperature rise associated with the irreversibilities, we first calculate the actual state-2 enthalpy:

$$h_2 = \frac{\dot{W}_{\text{actual}}}{\dot{m}} + h_1$$
$$= 14{,}950\,\text{kW}/(464\,\text{kg/s}) + 627.36\,\text{kJ/kg} = 659.58\,\text{kJ/kg}.$$

With h_2 and P_2 (= 26 MPa) defining the state, we find that $T_2 = 425.8$ K using the NIST WebBook. This value is about 1.1 K greater than T_{2s} and is shown in the sketch.

Self-Test 7.10

Air enters a compressor at 100 kPa and 300 K and exits at 600 kPa and 550 K. Assuming constant specific heats at 300 K and neglecting potential and kinetic energy changes, determine the isentropic efficiency of the compressor.

(Answer: $\eta_{isen,c}$ = 80%)

Isentropic efficiencies are also illustrated in Example 8.4.

7.5c Isentropic Efficiency of a Nozzle

For a nozzle (n), the outlet kinetic energy is the quantity to be maximized; thus, the isentropic efficiency for this device is

$$\eta_{\text{s, n}} \equiv \frac{KE_{\text{actual}}}{KE_{\text{ideal}}}. \tag{7.36}$$

Table 7.3 shows typical ranges of isentropic efficiency for the steady-flow devices discussed here.

TABLE 7.3 Typical Isentropic Efficiencies for Turbines, Compressors, Pumps, and Nozzles

Device	Isentropic Efficiency (%)
Turbine	70–90
Compressor	75–85
Pump	75–85
Nozzle	>95

Isentropic efficiencies are also illustrated in Examples 8.7–8.11.

7.6 Availability (or Exergy)

Earlier in this chapter, we stated that energy is characterized both by *quantity* and *quality*. For example, consider the same quantity of energy stored in two systems: a mass of gas with a temperature slightly higher than the temperature of the surroundings and a mass of gas at a temperature much greater than the surroundings. To convert the same quantity of energy to useful work from either the warm or the hot gas, we employ a reversible heat engine operating between the gas system at $T_0 + \Delta T$ and the surroundings at T_0. The Carnot efficiency (Eqn. 6.15a) establishes the maximum possible conversion of energy from the gas to useful work, i.e.,

$$W = \eta_{\text{rev}} Q_{\text{H}}.$$

or

$$W = \left(1 - \frac{T_0}{T_0 + \Delta T}\right) Q_{\text{H}}.$$

For a small value of ΔT (warm gas), the Carnot efficiency η_{rev} is quite small; hence, very little energy from the system (transferred as heat) can be converted to work. For the hot gas, however, a larger proportion of the energy transferred from the system is converted to useful work.

Thus, 10 MJ of energy stored in a gas at 1000 K is much more valuable than 10 MJ stored in a gas at 310 K. The thermodynamic property **availability** is one way to quantify the *quality* of energy. This property is also known as **exergy**. The property availability quantifies how much energy in a system, or in a flow stream, is potentially *available* to produce useful work. In the sections below, we formally define availability and develop availability balances for both closed (fixed-mass) systems and open systems (control volumes).

7.6a Definitions

Before defining availability, we first revisit the basic concepts of *system* and *surroundings*. Figure 7.27 illustrates a closed thermodynamic system (with fixed mass) separated from its surroundings by a dashed line contiguous with the physical boundary of the system. The surroundings are further subdivided into two entities: the *immediate surroundings* and the *environment*. In the interaction of the system with the surroundings, the thermodynamic properties of the immediate surroundings may

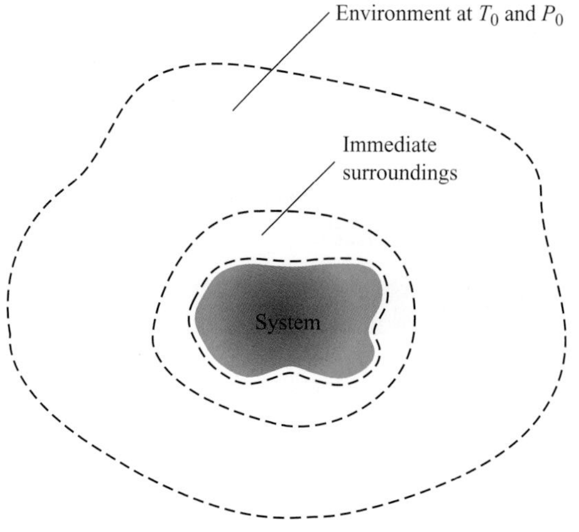

FIGURE 7.27 The surroundings associated with any thermodynamics system can be subdivided into two parts: the immediate surroundings and the environment.

be different from that of the environment. For example, the air close to a steam turbine may be much hotter than the air further away from the turbine. For our analysis of availability, we assume that the environment temperature, T_0, and pressure, P_0, are unaffected by any energy transfers between the system and its surroundings. These properties of the environment are usually referred to as the **dead state**. The physical interpretation of the dead state is that, when a system is in equilibrium with its surroundings, the system no longer has any potential to produce useful work. The system is *dead* or is said to be at the *dead state*. Although the actual dead state pressure and temperature may be different for different situations, they usually are assigned values close to one atmosphere and room temperature (say, 25°C), respectively.

We now formally define the availability $\mathcal{A}$, a thermodynamic property of a system, as follows:

The Availability $\mathcal{A}$ is the maximum theoretical work obtainable as a system interacts with its environment until they are in equilibrium.

We now relate this new property to analyses of closed systems and open systems (control volumes) in the following sections.

7.6b Closed System Availability

To relate the availability $\mathcal{A}$ to other thermodynamic properties, we answer the question: how much useful work can a closed system produce in going from an arbitrary initial state 1 to the dead state? Consider a system with energy E_1 at the initial state, i.e.,

$$E_1 = M\left(u_1 + V_1^2/2 + gz_1\right).$$

As there are no first- or second-law limits on completely converting kinetic and potential energies to useful energy, we have

$$(KE)_1 - (KE)_{\text{dead state}} = MV_1^2/2 - 0 = MV_1^2/2 = W_{\text{useful},\,KE} \tag{7.37a}$$

and

$$(PE)_1 - (PE)_{\text{dead state}} = Mgz_1 - 0 = Mgz_1 = W_{\text{useful},\,PE}. \tag{7.37b}$$

To assess the amount of useful work associated with the system internal energy, we first apply conservation of energy to the system:

$$dU = -\delta Q_{\text{out}} - \delta W_{\text{out}}. \tag{7.38}$$

Here we explicitly assume that the energy transfers are *from* the system *to* the surroundings and designate them with the subscript "out". The incremental work of the system, δW_{out}, can be split into two terms: the useful work associated with the moving boundary and the non-useful work used to push back the surroundings, i.e.,

$$\delta W_{\text{out}} = Pd\mathcal{V} = \delta W_{\text{useful, MB}} + P_0 d\mathcal{V}.$$

Rearranging, we obtain the incremental useful work as

$$\delta W_{\text{useful, MB}} = (P - P_0)d\mathcal{V}. \tag{7.39}$$

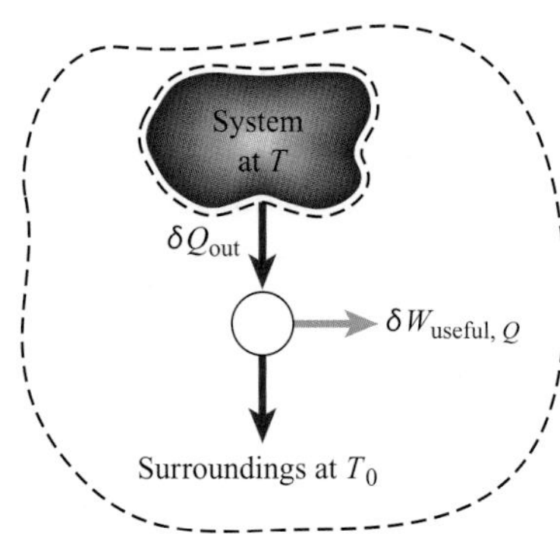

FIGURE 7.28 A hypothetical ideal heat engine delivers the maximum useful work associated with a heat-transfer process between a closed system and its surroundings at T_0.

Assessing the maximum possible useful work associated with heat transfer from the closed system to the surroundings is interesting and requires the application of second-law concepts. To obtain the maximum possible useful work associated with heat transfer, we replace the heat-transfer process with the operation of an ideal heat engine, as shown in Fig. 7.28. Thus,

$$\delta W_{\text{useful},\,Q} = \eta_{\text{Carnot}}\delta Q_{\text{out}} = \left(1 - \frac{T_0}{T}\right)\delta Q_{\text{out}}, \tag{7.40}$$

where the ideal Carnot thermal efficiency has been expressed using the temperature of the system, T, and that of the surroundings, T_0 (Eq. 6.15a). Using the definition of entropy (Eq. 7.1a),

$$dS \equiv \left(\frac{\delta Q}{T}\right)_{\text{rev}} = -\frac{\delta Q_{\text{out}}}{T},$$

we transform Eq. 7.40 to

$$\delta W_{\text{useful},\,Q} = \delta Q_{\text{out}} + T_0 dS. \tag{7.41}$$

We return now to Eq. 7.38 to finally assess the maximum useful work associated with a change in internal energy from an arbitrary state 1 to the dead state. Substituting Eqs. 7.39 and 7.41 into Eq. 7.38 and rearranging yields

$$\begin{aligned} \delta W_{\text{useful},\,U} &\equiv \delta W_{\text{useful, MB}} + \delta W_{\text{useful},\,Q} \\ &= -dU - P_0 d\mathcal{V} + T_0 dS. \end{aligned} \tag{7.42a}$$

Integrating this expression from state 1 to the dead state yields the following:

$$\begin{aligned} W_{\text{useful},\,U} &= \int_1^0 (\delta W_{\text{useful, MB}} + \delta W_{\text{useful},\,Q}) \\ &= \int_1^0 (-dU - P_0 d\mathcal{V} + T_0 dS) \\ &= U_1 - U_0 + P_0(\mathcal{V}_1 - \mathcal{V}_0) - T_0(S_1 - S_0). \end{aligned} \tag{7.42b}$$

We now formally define the availability $\mathcal{A}$ by combining the maximum useful work associated with the three forms of system energy (internal, kinetic, and potential), i.e., Eqs. 7.42b, 7.37a, and 7.37b:

$$\mathcal{A}_1 \equiv U_1 - U_0 + P_0(\mathcal{V}_1 - \mathcal{V}_0) - T_0(S_1 - S_0) + M\text{V}_1^2/2 + Mgz_1 \tag{7.43a}$$

or

$$\mathcal{A}_1 = E_1 - U_0 + P_0(\mathcal{V}_1 - \mathcal{V}_0) - T_0(S_1 - S_0). \tag{7.43b}$$

The availability can also be expressed as an intensive thermodynamic property by dividing by the system mass:

$$a_1 = \frac{\mathcal{A}_1}{M} = e_1 - u_0 + P_0(v_1 - v_0) - T_0(s_1 - s_0). \tag{7.43c}$$

Note that once values are assigned to the dead state properties, the availability is a only function of the state of the system; hence, the availability a is a thermodynamic property of the system, just like any of the more common intensive properties, P, T, v, u, etc.

Equations 7.43b and 7.43c can be used to evaluate the change in availability for a system that undergoes a change from one state to another, e.g.,

$$\Delta\mathcal{A} \equiv \mathcal{A}_2 - \mathcal{A}_1 = E_2 - E_1 + P_0(\mathcal{V}_2 - \mathcal{V}_1) - T_0(S_2 - S_1) \tag{7.44a}$$

or

$$\Delta a \equiv a_2 - a_1 = e_2 - e_1 + P_0(v_2 - v_1) - T_0(s_2 - s_1). \tag{7.44b}$$

The following example illustrates the use of Eqs. 7.44.

Example 7.15 Availability Change in an Ideal Gas Piston–Cylinder Expansion

Here we revisit Example 5.4 from Chapter 5. Air is contained in a piston–cylinder arrangement initially at 120 kPa and 300 K with a volume of 0.12 m^3. Energy as heat (11,820 J) is transferred to the air in a quasi-equilibrium constant-pressure process, to yield a final temperature of 370.2 K. The piston moves without friction. Assuming constant specific heats (c_p = 1.009 kJ/kg·K, c_v = 0.720 kJ/kg·K), determine the availability change for the process. The reference environment is at 298 K and 1 atm.

Solution

Known: T_1, P_1, $\mathcal{V}_1$, T_2, P_2 $(=P_1)$, ${}_1Q_2$

Find: $\Delta\mathcal{A}$

Sketch

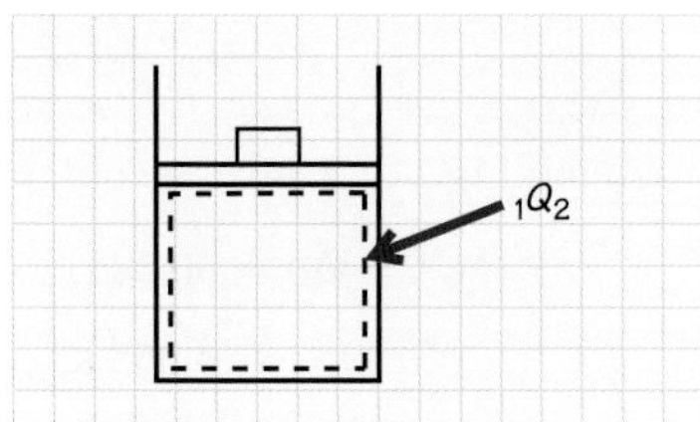

Modeling, Premises and Assumptions

i. Ideal-gas behavior
ii. Quasi-equilibrium (given)
iii. Frictionless piston (given)
iv. System kinetic energy zero
v. Negligible change in system potential energy
vi. Constant c_p, c_v (given)

Analysis: The mass-specific availability is given by Eq. 7.44b. With our assumptions, $e_2 - e_1$ is simply $u_2 - u_1$; thus, Eq. 7.44b becomes

$$\Delta a = u_2 - u_1 + P_0(v_2 - v_1) - T_0(s_2 - s).$$

With the assumptions of an ideal gas with constant specific heats,

$$u_2 - u_1 = c_v(T_2 - T_1) = \left(0.720\frac{\text{kJ}}{\text{kg·K}}\right)(370.2\,\text{K} - 300\,\text{K}) = 50.544\ \text{kJ/kg}.$$

To evaluate the second term requires values for the system mass and the volume at the final state. We apply the ideal-gas equation of state to obtain

$$M = \frac{P_1 \mathcal{V}_1}{RT_1} = \frac{120\,\mathrm{kN/m^2}(0.12\,\mathrm{m^3})}{0.287\,\mathrm{kJ/kg{\cdot}K}(300\,\mathrm{K})}\left[\frac{\mathrm{kJ}}{\mathrm{kN{\cdot}m}}\right] = 0.1672\ \mathrm{kg}$$

and

$$\mathcal{V}_2 = \mathcal{V}_1 \frac{P_1}{P_2}\frac{T_2}{T_1}$$

$$= (0.12\,\mathrm{m^3})(1)\frac{370.2\,\mathrm{K}}{300\,\mathrm{K}} = 0.14808\,\mathrm{m^3}.$$

Thus,

$$P_0(v_2 - v_1) = \frac{P_0}{M}(\mathcal{V}_2 - \mathcal{V}_1)$$

$$= \frac{101.25\,\mathrm{kN/m^2}}{0.1672\,\mathrm{kg}}(0.14808\,\mathrm{m^3} - 0.12\,\mathrm{m^3})\left[\frac{\mathrm{kJ}}{\mathrm{kN{\cdot}m}}\right] = 17.017\ \mathrm{kJ/kg}.$$

To evaluate the final term, we use Eq. 7.12a to find the entropy change for the process:

$$s_2 - s_1 = c_p \ln\frac{T_2}{T_1} - R\ \ln\frac{P_2}{P_1}$$

$$= 1.009\frac{\mathrm{kJ}}{\mathrm{kg{\cdot}K}}\ \ln\left(\frac{370.2\ \mathrm{K}}{300\ \mathrm{K}}\right) - 0.287\frac{\mathrm{kJ}}{\mathrm{kg{\cdot}K}}\ \ln\left(\frac{120\ \mathrm{kPa}}{120\ \mathrm{kPa}}\right)$$

$$= 0.2121\,\frac{\mathrm{kJ}}{\mathrm{kg{\cdot}K}} - 0\,\frac{\mathrm{kJ}}{\mathrm{kg{\cdot}K}} = 0.2121\,\frac{\mathrm{kJ}}{\mathrm{kg{\cdot}K}}.$$

Thus,

$$T_0(s_2 - s_1) = 298\,\mathrm{K}\left(0.2121\frac{\mathrm{kJ}}{\mathrm{kg{\cdot}K}}\right) = 63.206\frac{\mathrm{kJ}}{\mathrm{kg}}.$$

Reassembling Eq. 7.44b, we obtain the availability change:

$$\Delta a = 50.544\,\mathrm{kJ/kg} + 17.017\,\mathrm{kJ/kg} - 63.206\ \mathrm{kJ/kg}$$

$$= 4.355\ \mathrm{kJ/kg}$$

or

$$\Delta\mathcal{A} = M\Delta a = 0.1672\,\mathrm{kg}(4.355\,\mathrm{kJ/kg}) = 0.728\ \mathrm{kJ}.$$

Comment This example is a straightforward application of the defining relationships for availability and a good review of the ideal-gas property relationships from Section 7.2. The work produced by the expansion of the air is 3.370 kJ [$= P(\mathcal{V}_2 - \mathcal{V}_1)$]; hence, the availability change represents 21.6% of the work delivered.

7.6c Closed System Availability Balance

Consider an arbitrary fixed-mass closed thermodynamic system, as shown in Fig. 7.27. By combining an energy balance (first law) with an entropy balance (second law) for this system, we can obtain an availability balance. The final result is consistent with the generic balance principle presented in Chapter 3 (Eqn. 3.1) where we now replace the generic variable X with $\mathcal{A}$ and note that availability is always destroyed and never generated:

$$\mathcal{A}_{\mathrm{in}} - \mathcal{A}_{\mathrm{out}} - \mathcal{A}_{\mathrm{destroyed}} = \Delta\mathcal{A} \equiv \mathcal{A}_2 - \mathcal{A}_1. \tag{7.45}$$

From Eq. 5.5b, we express energy conservation for a change in state as

$$\int_1^2 \delta Q - \delta W = E_2 - E_1. \tag{7.46}$$

The corresponding entropy balance is obtained from Eq. 7.21 as

$$\int_1^2 \frac{\delta Q}{T} + \mathcal{S}_{\text{gen}} = S_2 - S_1, \tag{7.47}$$

where $\mathcal{S}_{\text{gen}}$ is the entropy generated by irreversibilities within the closed system. In both Eqs. 7.46 and 7.47, we adopt the usual convention that Q is positive for transfer from the surroundings to the system and W is positive for work transferred from the system to the surroundings. We also note that the temperature in Eq. 7.47 is that of the boundary at which the heat transfer takes place. To arrive at our final result, several operations are required. We first multiply Eq. 7.47 by T_0 and subtract the result from Eq. 7.46, to yield

$$\int_1^2 \left(1 - \frac{T_0}{T}\right) \delta Q - W - T_0 \mathcal{S}_{\text{gen}} = E_2 - E_1 - T_0(S_2 - S_1).$$

We now add $P_0 (\mathcal{V}_2 - \mathcal{V}_1)$ to both sides of this result and recognize that the right-hand side is simply $\mathcal{A}_2 - \mathcal{A}_1$ (cf. Eq. 7.44a), i.e.,

$$\underbrace{\int_1^2 \left(1 - \frac{T_0}{T}\right) \delta Q}_{\text{Net availability transfer in by heat}} - \underbrace{[W - P_0(\mathcal{V}_2 - \mathcal{V}_1)]}_{\text{Net availability transfer out by work}} - \underbrace{T_0 \mathcal{S}_{gen}}_{\text{Availability destroyed}} = \underbrace{\mathcal{A}_2 - \mathcal{A}_1 = \Delta \mathcal{A}}_{\text{Availability change for process}} \tag{7.48}$$

Note that whether the net availability transfer by heat is positive or negative depends not only on the direction of the heat transfer but also on the relative magnitude of the boundary temperature T and the dead state temperature T_0. For $T > T_0$, the term $1 - T_0/T$ is positive, while for $T < T_0$ this term is negative.

Example 7.16 A Second Method to Find the Change in Availability

Consider the situation described in Example 7.15. Evaluate each term in the availability balance given by Eq. 7.48 to obtain the availability change $\Delta \mathcal{A}$ for the process. Compare the result with that obtained in Example 7.15. Assume the heat transfer occurs with a boundary temperature of 500 K.

Solution

Known T_b and, from Example. 7.13, P_1, T_1, $\mathcal{V}_1$, P_2, T_2, $\mathcal{V}_2$, ${}_1Q_2$

Find $\mathcal{A}_{Q,\text{in}}$, $\mathcal{A}_{W,\text{out}}$, $\mathcal{A}_{\text{destroyed}}$, $\Delta \mathcal{A}$

Sketch

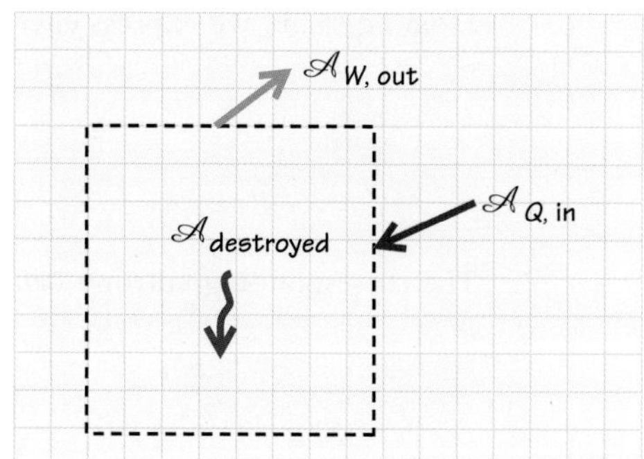

Modeling, Premises and Assumptions All as in Example 7.15

Analysis For a process, the fixed-mass system availability balance is given by (Eq. 7.45)

$$\mathcal{A}_{Q,\,\text{in}} - \mathcal{A}_{W,\,\text{out}} - \mathcal{A}_{\text{destroyed}} = \Delta\mathcal{A}.$$

From Eq. 7.48, we see that the availability transfer by heat is

$$\mathcal{A}_{Q,\,\text{in}} = \int_1^2 \left(1 - \frac{T_0}{T}\right)\delta Q.$$

For a constant boundary temperature ($T = T_\text{b}$), this becomes

$$\mathcal{A}_{Q,\,\text{in}} = \left(1 - \frac{T_0}{T_\text{b}}\right){}_1Q_2$$

$$= \left(1 - \frac{298\,\text{K}}{500\,\text{K}}\right)11.820\ \text{kJ} = 4.775\ \text{kJ}.$$

From Eq. 7.48, we see that the availability transfer by work is

$$\mathcal{A}_{w,\,\text{out}} = {}_1W_2 - P_0(\mathcal{V}_2 - \mathcal{V}_1),$$

which for a constant-pressure process simplifies to

$$\begin{aligned}\mathcal{A}_{w,\,\text{out}} &= (P_1 - P_0)(\mathcal{V}_2 - \mathcal{V}_1)\\ &= (120\,\text{kN/m}^2 - 101.325\ \text{kN/m}^2)(0.14808\,\text{m}^3 - 0.12\,\text{m}^3)\left[\frac{1\,\text{kJ}}{\text{kN}\cdot\text{m}}\right]\\ &= 0.5244\ \text{kJ}.\end{aligned}$$

The availability destruction is given by

$$\mathcal{A}_{\text{destroyed}} = T_0\mathcal{S}_{\text{gen}}.$$

To find the entropy generated by irreversibilities, we employ an entropy balance (Eq. 7.33) as follows:

$$S_2 - S_1 = \frac{{}_1Q_2}{T_\text{b}} + \mathcal{S}_{\text{gen}}$$

or

$$\begin{aligned}\mathcal{S}_{\text{gen}} &= M(s_2 - s_1) - \frac{{}_1Q_2}{T_\text{b}}\\ &= 0.1672\,\text{kg}\left(0.2121\,\frac{\text{kJ}}{\text{kg}\cdot\text{K}}\right) - \frac{11.82\,\text{kg}}{500\,\text{K}} = 0.0011823\ \text{kJ/K}.\end{aligned}$$

where values for the mass and entropy change $s_2 - s_1$ are taken from Example 7.15. The availability destruction is thus

$$\mathcal{A}_{\text{destroyed}} = 298\,\text{K}(0.001823\,\text{kJ/K}) = 3.523\,\text{kJ}.$$

We now reassemble the availability balance to calculate $\Delta\mathcal{A}$:

$$\mathcal{A}_{Q,\text{in}} - \mathcal{A}_{W,\text{out}} - \mathcal{A}_{\text{destroyed}} = \Delta\mathcal{A}$$

$$4.775\,\text{kJ} - 0.5244\,\text{kJ} - 3.523\,\text{kJ} = 0.728\,\text{kJ}.$$

This is the same result as that obtained in Example 7.15 for $\Delta\mathcal{A}$.

Comment We note the large value of the availability destroyed in this process. The availability destroyed is 73.8% of the availability transferred in by heat.

7.6d Open System (Control Volume) Availability

In addition to the availability transfers associated with heat and work, availability transfers are also associated with, first, the flow work[5] and, second, the energy of the entering and exit streams (Fig. 7.29).

In Chapter 4, we derived the flow work as (Eqs. 4.12)

$$\dot{W}_{\text{flow}} \equiv \dot{m}Pv.$$

The rate of availability associated with this flow work is the flow work minus $\dot{m}P_0v$, i.e.,

$$\dot{\mathcal{A}}_{\text{flow work}} \equiv \dot{m}(Pv - P_0v). \quad \textbf{(7.49)}$$

The rate of availability associated with the energy of the entering or exiting stream is simply the product of the mass flow rate and the mass-specific availability:

$$\dot{\mathcal{A}}_{\text{flow}} = \dot{m}a, \quad \textbf{(7.50)}$$

where a is defined by Eq. 7.43c. We now combine Eqs. 7.49 and 7.50 to obtain the total rate of availability transfer associated with the flow crossing the open system boundary,

$$\dot{\mathcal{A}}_{\text{flow work}} + \dot{\mathcal{A}}_{\text{flow}} = \dot{m}(Pv - P_0v) + \dot{m}[(e - u_0) + P_0(v\text{-}v_0) - T_0(s - s_0)]. \quad \textbf{(7.51)}$$

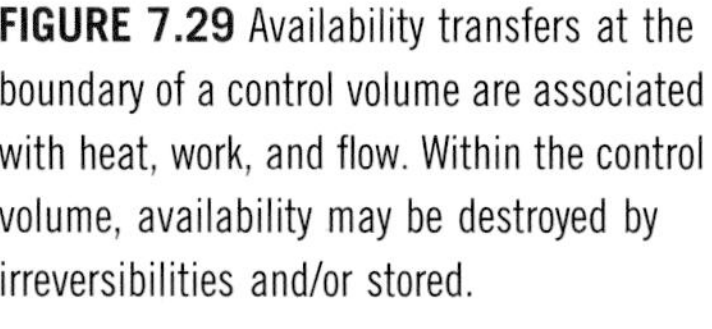

FIGURE 7.29 Availability transfers at the boundary of a control volume are associated with heat, work, and flow. Within the control volume, availability may be destroyed by irreversibilities and/or stored.

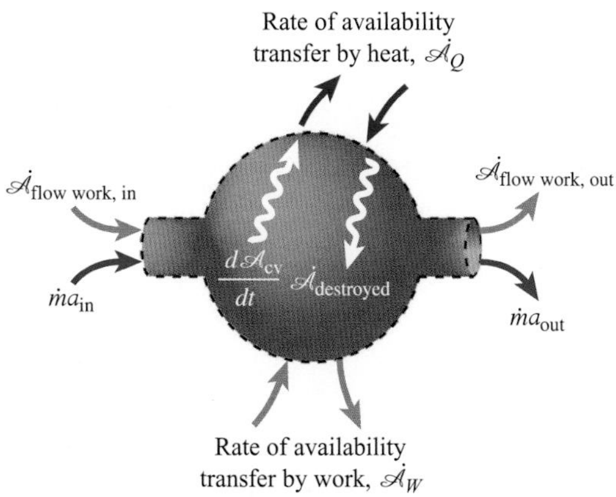

[5] Here we use the word *work* synonymously with *time rate of work*.

Recognizing that

$$e = u + \frac{V^2}{2} + gz$$

and

$$h = u + Pv,$$

Eqn. 7.51 becomes

$$\dot{\mathscr{A}}_{\text{flow work}} + \dot{\mathscr{A}}_{\text{flow}} = \dot{m}\left[(h - h_0) + \frac{V^2}{2} + gz - T_0(s - s_0)\right]. \tag{7.52}$$

With this expression, we are now able to formulate an availability balance for an open system (control volume).

7.6e Open System (Control Volume) Availability Balance

Consider an open system (control volume) with a single inlet and a single outlet stream, as shown in Fig. 7.29. We furthermore assume that the kinetic and potential energies of the open system itself are negligible. The rate form of the generic balance expression applied to the availability is simply

$$\sum \dot{\mathscr{A}}_{\text{in}} - \sum \dot{\mathscr{A}}_{\text{out}} - \dot{\mathscr{A}}_{\text{cv, destroyed}} = \frac{d\mathscr{A}_{\text{cv}}}{dt}. \tag{7.53}$$

The $\dot{\mathscr{A}}_{\text{in}}$ and $\dot{\mathscr{A}}_{\text{out}}$ terms comprise the rates of availability transfer by heat, work, and flow (including flow work). The net availability transfer rate into the open system (control volume) by heat can be expressed as

$$\dot{\mathscr{A}}_{\text{in}, Q} = \int_{\text{cv}} \left(1 - \frac{T_0}{T_j}\right) \delta\dot{Q} \tag{7.54a}$$

or, assuming heat transfer across the various fixed-temperature regions of the open system boundary, as

$$\dot{\mathscr{A}}_{\text{in}, Q} = \sum_j \left(1 - \frac{T_0}{T_j}\right) \dot{Q}_j. \tag{7.54b}$$

The net availability transfer rate out of the control volume by work can be expressed as

$$\dot{\mathscr{A}}_{\text{out}, W} = \dot{W}_{\text{cv}} - P_0 \frac{d\mathcal{V}_{\text{cv}}}{dt}. \tag{7.54c}$$

Lastly, the availability destruction rate within the open system can be expressed as

$$\dot{\mathscr{A}}_{\text{cv, destroyed}} = T_0 \dot{\mathbb{S}}_{\text{gen}}, \tag{7.54d}$$

where $\dot{\mathbb{S}}_{\text{gen}}$ is the entropy generation rate by irreversibilities within the open system (see Eqs. 7.33b). We now substitute Eqs. 7.54a–7.54d, together with expressions for the inlet and outlet stream availability flows from Eq. 7.52, into our availability balance (Eq. 7.53) to yield

$$\sum_j \left(1 - \frac{T_0}{T_j}\right)\dot{Q}_j - \left(\dot{W}_{cv} - P_0 \frac{d\mathcal{V}_{cv}}{dt}\right)$$
$$+ \dot{m}\left[(h_{in} - h_{out}) + \frac{V_{in}^2 - V_{out}^2}{2} + g(z_{in} - z_{out}) - T_o(s_{in} - s_{out})\right]$$
$$-T_0\dot{\mathcal{S}}_{gen} = \frac{d\mathcal{A}_{cv}}{dt}. \tag{7.55}$$

Before simplifying this expression, we restate the assumptions embodied in this result:

- Single inlet and outlet streams
- Uniform properties at the inlet and outlet
- Heat transfer subdivided over portions of the open system boundary at their respective temperatures T_j.

By adding the assumption of steady-state and steady flow, the two time derivatives appearing in Eqn. 7.55 become zero. With this additional restriction, our availability balance becomes

$$\sum_j \left(1 - \frac{T_0}{T_j}\right)\dot{Q}_j - \dot{W}_{cv}$$
$$+\dot{m}\left[(h_{in} - h_{out}) + \frac{V_{in}^2 - V_{out}^2}{2} + g(z_{in} - z_{out}) - T_0(s_{in} - s_{out})\right] - T_0\dot{\mathcal{S}}_{gen} = 0. \tag{7.56}$$

The following example illustrates the application of availability analysis to a simple steady-flow device.

Example 7.17 Availability Change in a Steam Turbine

Steam enters a well-insulated turbine at 800 °C and 10 MPa at a flowrate of 2.5 kg/s. The steam exits at 50 kPa. The isentropic efficiency of the turbine is 0.9332. Assuming a reference environment at 25 °C and 1 atm, determine (a) the rate at which availability enters the turbine with the flow, including that associated with flow work, and (b) the availability destruction rate, from an availability balance.

Solution

Known T_1, P_1, P_2, $\dot{m}$, $\eta_{isen,t}$

Find $\dot{\mathcal{A}}_{in}\ (= \dot{\mathcal{A}}_{flow\ work} + \dot{\mathcal{A}}_{flow})$, $\dot{\mathcal{A}}_{destoyed}$

Sketch

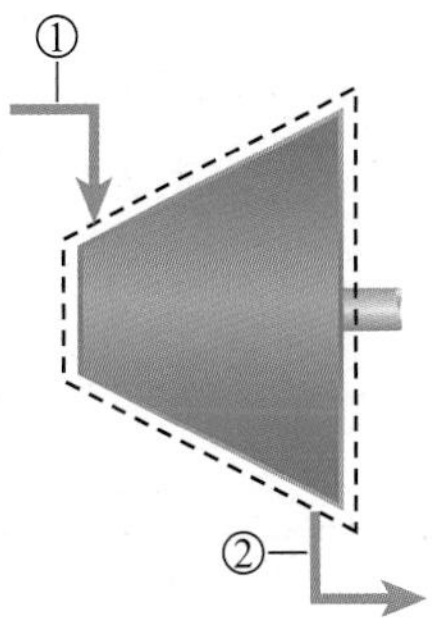

①

②

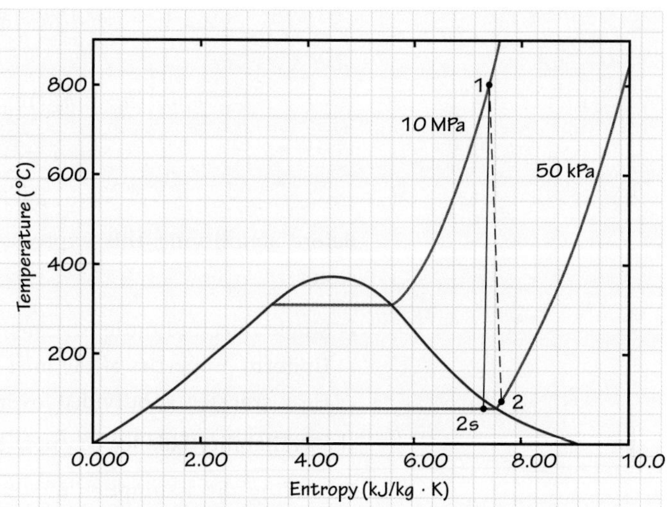

Modeling, Premises and Assumptions:

i. Steady-state and steady flow
ii. Adiabatic $(\dot{Q}_{cv} = 0)$
iii. Negligible kinetic and potential energies

Analysis: The rate of availability entering the turbine at the inlet (station 1) is given by Eq. 7.52 as

$$\dot{\mathscr{A}}_{in} = \left(\dot{\mathscr{A}}_{flow\ work} + \dot{\mathscr{A}}_{flow}\right)_1 = \dot{m}[(h_1 - h_0) - T_0(s_1 - s_0)]$$

The properties h_1, s_1, h_0, and s_0 are obtained from the NIST database for 800 °C and 10 MPa and for 25 °C and 1 atm, respectively. Using these values yields

$$\dot{\mathscr{A}}_{in} = \left(2.5\frac{\mathrm{kg}}{\mathrm{s}}\right)\left[(4114.8 - 104.92)\frac{\mathrm{kJ}}{\mathrm{kg}} - (298\,\mathrm{K})(7.4077 - 0.3672)\frac{\mathrm{kJ}}{\mathrm{kg\cdot K}}\right]\left[\frac{\mathrm{kW}}{\mathrm{kJ/s}}\right]$$
$$= 4780\,\mathrm{kW}.$$

To obtain $\dot{\mathscr{A}}_{destoyed}$, we simplify and rearrange the availability balance expressed by Eq. 7.56 as follows:

$$\dot{\mathscr{A}}_{destoyed}\left(= T_0\,\dot{\mathbb{S}}_{gen}\right) = -\dot{W}_{cv} + \dot{m}[(h_1 - h_2) - T_0(s_1 - s_2)],$$

where the heat-transfer and kinetic and potential energy terms are assumed to be zero. To evaluate this expression, we need to completely define the outlet state. Following Example 7.13, we use the isentropic efficiency to determine the outlet enthalpy, h_2, as follows:

$$\eta_{isen,t} = \frac{\dot{m}(h_1 - h_2)}{\dot{m}(h_1 - h_{2s})} = \frac{h_1 - h_2}{h_1 - h_{2s}}$$

or

$$h_2 = \left(1 - \eta_{isen,t}\right)h_1 + \eta_{isen,t}h_{2s}.$$

To find the unknown h_{2s}, we first calculate the quality x_{2s} by recognizing that $s_{2s} = s_1$ (see sketch), i.e.,

$$x_{2s} = \frac{s_{2s} - s_{2f}}{s_{2g} - s_{2f}} = \frac{(7.4077 - 1.091)\,\mathrm{kJ/kg\cdot K}}{(7.5939 - 1.091)\,\mathrm{kJ/kg\cdot K}} = 0.9714$$

where s_{2f} and s_{2g} are the saturation values at 50 kPa from the NIST database. We can find h_{2s} as

$$\begin{aligned} h_{2s} &= (1 - x_{2s})h_{2f} + x_{2s}h_{2g} \\ &= 0.0286(340.49\,\text{kJ/kg}) + 0.9714(2645.9\,\text{kJ/kg}) \\ &= 2579.9\ \text{kJ/kg}. \end{aligned}$$

We now use this value to find h_2 from the previously developed expression involving the turbine efficiency, i.e.,

$$\begin{aligned} h_2 &= (1 - 0.9332)(4114.8\,\text{kJ/kg}) + 0.9332\,(2579.9\,\text{kJ/kg}) \\ &= 2682.4\,\text{kJ/kg}. \end{aligned}$$

Knowing h_2 and P_2, we find $s_2 = 7.6952$ kJ/kg·K using the NIST software. All terms in the availability balance can now be calculated:

$$\begin{aligned} \dot{W}_{cv} &= \dot{m}(h_1 - h_2) \\ &= 2.5\,\text{kg/s}(4114.8 - 2682.4)\,\text{kJ/kg}\left[\frac{\text{kW}}{\text{kJ/s}}\right] = 3581\,\text{kW} \end{aligned}$$

and

$$\begin{aligned} \dot{\mathcal{A}}_{in} - \dot{\mathcal{A}}_{out} &= \dot{m}[(h_1 - h_2) - T_0(s_1 - s_2)] \\ &= (2.5\ \text{kg/s})[(4114.8 - 2682.4)\text{kJ/kg} - (298\ \text{K})(7.4077 - 7.6952)\text{kJ/kg·K}]\left[\frac{\text{kW}}{\text{kJ/s}}\right] \\ &= 3795\ \text{kW}. \end{aligned}$$

Combining the above terms, we obtain our final result:

$$\begin{aligned} \dot{\mathcal{A}}_{destoyed} &= -\dot{W}_{cv} + \dot{\mathcal{A}}_{in} - \dot{\mathcal{A}}_{out} \\ &= -3581\,\text{kW} + 3795\,\text{kW} = 214\ \text{kW}. \end{aligned}$$

Comment The availability destruction rate can alternatively be determined by evaluating

$$\dot{\mathcal{A}}_{destoyed} = T_0\dot{\mathcal{S}}_{gen}$$

where $\dot{\mathcal{S}}_{gen}$ is obtained from an entropy balance (Eq. 7.33c) as $\dot{\mathcal{S}}_{gen} = \dot{m}(s_2 - s_1)$. Performing this calculation yields

$$\begin{aligned} \dot{\mathcal{A}}_{destoyed} &= T_0\dot{m}(s_2 - s_1) \\ &= (298\,\text{K})(2.5\,\text{kg/s})(7.6952 - 7.4077)\,\text{kJ/kg}\cdot\text{K}\left[\frac{\text{kW}}{\text{kJ/s}}\right]\cdot \\ &= 214\ \text{kW}. \end{aligned}$$

This result is identical to that obtained from the availability balance, as expected. The simplification of Eq. 7.33c to obtain $\dot{\mathcal{S}}_{gen}$ is left as an exercise for the reader.

SUMMARY

In this chapter we applied second-law analysis in terms of the entropy and the Gibbs relations for the system. Second-law analysis for a closed system allowed us to determine the direction of an irreversible process. The second-law analysis of an open system was used to determine the outflow conditions from an ideal, isentropic device. Using the isentropic efficiency, we can find the outflow conditions from an actual, nonisentropic device. Availability was used to find the maximum useful work that can be achieved in a process.

As a further summary of this chapter, reviewing the learning objectives presented at the outset of this chapter is recommended.

KEY EQUATIONS

Review the most important equations presented in this chapter (i.e., those boxed with a yellow background). What physical principles do they express? What restrictions apply?

CHAPTER 7 KEY CONCEPTS AND DEFINITIONS CHECKLIST

Numbers following arrows refer to Questions at the end of the chapter.

7.1 Entropy

- ☐ Formal definition of entropy (Eq. 7.1) ➔ Question 7.1, 7.6
- ☐ Isothermal heat transfer (Eq. 7.2) ➔ Question 7.4
- ☐ Entropy as a property 7.19, 7.61
- ☐ Gibbs relationships

7.2 Ideal Gas Properties and Processes

- ☐ Ideal-gas temperature–entropy (Gibbs) relationships
- ☐ Ideal-gas isentropic process relationships
- ☐ Processes in T–s and P–v space ➔ Question 7.17

7.3 Specific Entropy for Pure Nonideal Substances

7.4 Entropy Balances for a Closed System

- ☐ Systems undergoing a change in state ➔ Question 7.7
- ☐ Entropy change for an isolated system ➔ Questions 7.8, 7.12, 7.13
- ☐ Entropy change for a closed system with energy exchange with the surroundings ➔ Questions 7.9, 7.10, 7.11, 7.18
- ☐ A closed system undergoing a cycle ➔ Questions 7.5, 7.6
- ☐ Some reversible cycles, Carnot cycle, and Stirling cycle ➔ Questions 7.2, 7.3

7.5 Entropy Balances for an Open System (Control Volume)

- ☐ Isentropic efficiency of a turbine ➔ Question 7.15
- ☐ Isentropic efficiency of a pump or compressor ➔ Question 7.16
- ☐ Isentropic efficiency of a nozzle ➔ Question 7.14

7.6 Availability (or Exergy) Questions 7.152, 7.153

REFERENCES

1. Kline, S. J., *The Low-Down on Entropy and Interpretive Thermodynamics*, DCW Industries, La Cañada, CA, 1999.
2. "J. Willard Gibbs," APS Historic Site, Yale University, New Haven, CT. https://www.aps.org/programs/outreach/history/historicsites/gibbs.cfm.
3. Clausius, R. J. E., "Uber verschiedene für die Anwendung bequeme Formen der Hauptgleichungen der mechanischen Wärmetheorie," *Annalen der Physik und Chemie*, **125**:353 (1865); translated excerpts in *A Source Book of Physics*, W. F. Magie, McGraw-Hill, New York, 1935, pp. 234–236.
4. Lefebvre, A. H., *Gas Turbine Combustion*, Taylor & Francis, Bristol, PA, 1983.
5. Sears, F. W., *Thermodynamics, Kinetic Theory, and Statistical Thermodynamics*, Addision-Wesley, Reading, MA, 1975.

6. Podesser, E., "Electricity Production in Rural Villages with a Biomass Stirling Engine," *Renewable Energy*, **16**:1049–1052 (1999).
7. Wark, K., *Thermodynamics*, McGraw-Hill, New York, 1966.

Some end-of-chapter problems were adapted with permission from the following:

Look, D. C., Jr., and Sauer, H. J., Jr., *Engineering Thermodynamics*, PWS, Boston, 1986.
Myers, G. E., *Engineering Thermodynamics*, Prentice Hall, Englewood Cliffs, NJ, 1989.

QUESTIONS

7.1 Review the most important equations presented in this chapter i.e., those with a yellow background. What physical principles do they express? What restrictions apply?

7.2 List the processes comprising the Carnot cycle. Sketch the cycle on a T–s diagram.

7.3 List the processes comprising the Stirling cycle. Sketch the cycle on a T–s diagram.

7.4 Relate heat transfer to the change in entropy of a system.

7.5 State in words the Kelvin–Planck statement of the second law of thermodynamics.

7.6 Write symbolically and explain in words the macroscopic, or Clausius, definition of entropy.

7.7 Explain the increase of entropy principle. Be precise.

7.8 Explain how the increase of entropy principle is an indicator of spontaneous change. Illustrate with an example.

7.9 List the kinds of processes that produce an increase in entropy of a system.

7.10 Explain why heat transfer across a finite temperature difference is irreversible.

7.11 Why must heat transfer in an actual device be irreversible?

7.12 Why is a sudden expansion of a gas into an evacuated chamber an irreversible process?

7.13 Why is the mixing of two gases an irreversible process?

7.14 List the ways in which a process can be isentropic, that is, list the combined types of processes or conditions that are required for the overall process to be isentropic.

7.15 Explain how the isentropic efficiency of a turbine relates to the production of entropy by irreversibilities. Assume that the turbine operates adiabatically.

7.16 Explain how the isentropic efficiency of a pump relates to the production of entropy by irreversibilities. Assume that the fluid is incompressible and that the pump operates adiabatically.

7.17 On a T–s diagram, sketch lines of constant specific volume (isochors) and constant temperature (isobars) for an ideal gas. Which curves have a greater slope, $\partial T/\partial s$?

7.18 Consider the heating of an ideal gas from T_1 to T_2 ($T_2 > T_1$). For the particular application, the closed system can be heated in two ways: in a constant pressure process or a constant volume process. For each question, circle the appropriate answer.

   A. For which process will the P–$\mathcal{V}$ work done by the system be greater?
   The system will do more P–$\mathcal{V}$ work during a constant pressure process.

The system will do more P–$\mathcal{V}$ work during a constant volume process.
The system will do the same amount of P–$\mathcal{V}$ work during either process.

B. For which process will the heat added to the system be greater?
A constant pressure process will require a greater heat addition.
A constant volume process will require a greater heat addition.
Both processes require the same heat addition.

C. For which process will the entropy change of the gas be greater?
The entropy of the gas will increase more during a constant pressure process.
The entropy of the gas will increase more during a constant volume process.
The change in entropy of the gas will be the same for either process.

Chapter 7 Problem Subject Areas

7.1–7.14	Isothermal heat transfer and definition of entropy
7.15–7.52	Entropy for an ideal gas
7.53–7.74	Entropy for a pure nonideal substance
7.75–7.87	Entropy change for an isolated system
7.88–7.108	Entropy balance for a closed system with energy exchange
7.109–7.129	A closed system undergoing a cycle
7.130–7.151	Entropy balance for an open system
7.152–7.180	Availability (exergy)
7.181–7.201	FE exam problems

PROBLEMS

7.1–7.14 Isothermal heat transfer and definition of entropy

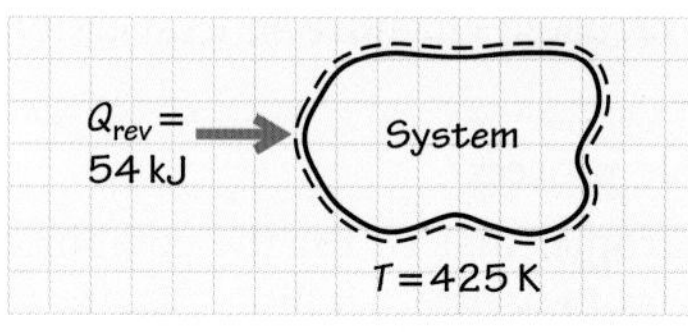

7.1 In a reversible heat interaction, 54 kJ of energy is transferred from the surroundings to a thermodynamic system. The process occurs isothermally at 425 K. Determine the entropy change of the system. How would the entropy of the system change if the process were irreversible rather than reversible?

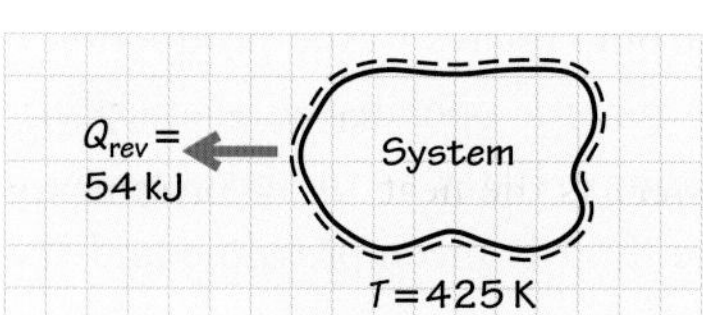

7.2 Repeat Problem 7.1 but for the case where 54 kJ of energy is removed from the system in a heat interaction. Also discuss how your result would change if the process were irreversible rather than reversible.

7.3 In a reversible heat interaction, 300 Btu of energy is transferred from the surroundings to a thermodynamic system. The process occurs isothermally at 200 F. Determine the entropy change of the system. How would the entropy of the system change if the process were internally irreversible rather than reversible?

7.4 An ideal gas contained in a piston–cylinder device expands isothermally at 20°C from state 1 to state 2.

A. If the expansion process is reversible and 100 kJ of heat is transferred from the surroundings to the gas, determine the increase in entropy (kJ/K) of the gas.

B. If the expansion process between these same two states had been internally irreversible, would the entropy increase be greater than, the same as, or less than the entropy increase found in part A? Explain.

C. For an irreversible process between the same two states, will the heat transfer to the gas be greater than, the same as, or less than the 100 kJ for the reversible process? Explain.

7.5 Consider the same isothermal process, but now include irreversible mixing in the system that increases the entropy by 80 J/K. Determine the entropy change of the system.

7.6 During a reversible process, 15 kJ of energy is added to a system at a constant temperature of 380 K. Determine the entropy change of the system. Consider the same isothermal process but now include irreversible mixing in the system that increases the entropy by 80 J/K. Determine the entropy change of the system.

7.7 An ideal gas contained in a piston–cylinder device expands isothermally at 80°C from state 1 to state 2.

A. If the expansion process is reversible and 100 kJ of heat is transferred from the surroundings to the gas, determine the increase in entropy (kJ/K) of the gas.

B. What is the P–$\mathcal{V}$ work done in this process?

C. If there is 0.5 kg of air in the piston-cylinder device initially at 100 kPa, find the initial and final volumes of the air.

7.8 An ideal gas contained in a piston–cylinder device expands isothermally at 120 F from state 1 to state 2.

A. If the expansion process is reversible and 20 Btu of heat is transferred from the surroundings to the gas, determine the increase in entropy (Btu/R) of the gas.

B. What is the P–$\mathcal{V}$ work done in this process?

C. If there is 1.3 lb_m of air in the piston-cylinder device initially at 15 psia, find the initial and final volumes of the air.

7.9 If a rigid system is at 350 K and the entropy of the system increases by 120 J/K in a reversible process, what is the heat transfer? Is the heat added or removed from the system?

7.10 If a rigid system is at 200 F and the entropy of the system decreases by 0.6 Btu/R in a reversible process, what is the heat transfer? Is the heat added or removed from the system?

7.11 Mixing of the fluid in a rigid system generates 0.05 kJ/K of entropy. What heat transfer is required to maintain the fluid temperature at 400 K during the mixing process?

7.12 Mixing of the fluid in a rigid system generates 0.03 Btu/R of entropy. What heat transfer is required to maintain the fluid temperature at 800 R?

7.13 Determine the coefficient of performance of the reversible cycle shown in the sketch.

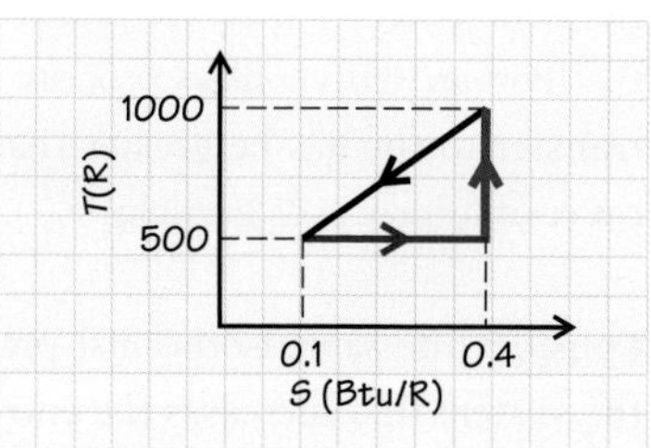

7.14 Determine the efficiency of the reversible cycle shown in the sketch.

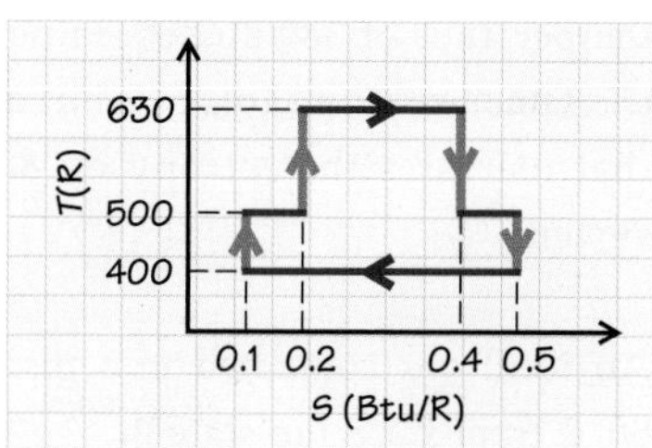

7.15–7.52 Entropy for an ideal gas

7.15 Nitrogen undergoes an isentropic process from an initial state at 425 K and 150 kPa to a final state at 600 K. Determine the density of the N_2 at the final state.

7.16 Starting with the relationship $P\text{–}v^{\gamma}$ = constant, derive Eq. 7.14a.

7.17 The following processes constitute the air-standard Diesel cycle:

1–2: isentropic compression,
2–3: constant-volume energy addition (T and P increase),
3–4: constant-pressure energy addition (v increases),
4–5: isentropic expansion, and
5–1: constant-volume energy rejection (T and P decrease).

Plot these processes in P–v and T–s coordinates. How does this cycle differ from the Otto cycle presented in Example 7.5?

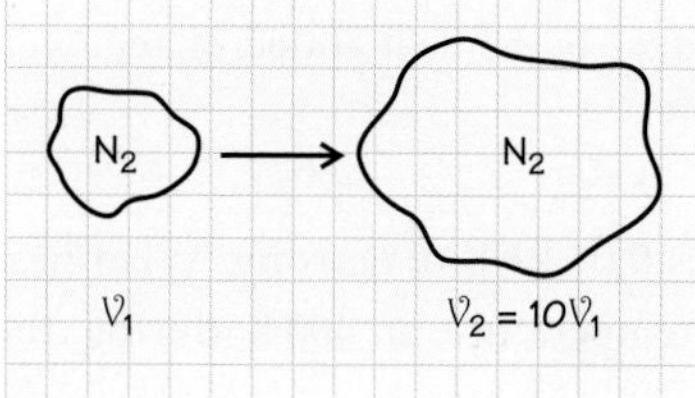

7.18 Consider the expansion of N_2 to a volume ten times larger than its initial volume. The initial temperature and pressure are 1800 K and 2 MPa, respectively. Determine the final-state temperature and pressure for (a) an isentropic expansion ($\gamma = 1.4$) and (b) a polytropic expansion with $n = 1.25$.

7.19 Consider a T–s diagram. Show that the constant specific volume line (v) must have a steeper slope than a line of constant pressure (P).

7.20 Air is drawn into the compressor of a jet engine at 55 kPa and –23°C. The air is compressed isentropically to 275 kPa. Determine (a) the temperature after compression (°C), (b) the specific volume before compression (m^3/kg), and (c) the change in specific enthalpy for the process (kJ/kg).

7.21 An automobile engine has a compression ratio (v_1/v_2) of 8.0. If the compression is isentropic and the initial temperature and pressure are 30°C and 101 kPa, respectively, determine (a) the temperature and the pressure after compression and (b) the change in enthalpy for the process.

7.22 Air is compressed in a piston–cylinder system having an initial volume of 80 in^3. The initial pressure and temperature are 20 psia and 140 F. The final volume is one-eighth of the initial volume at a pressure of 175 psia. Determine the following:

A. The final temperature (F)
B. The mass of air (lb_m)
C. The change in internal energy (Btu)
D. The change in enthalpy (Btu)
E. The change in entropy (Btu/lb_m·R)

7.23 An ideal gas expands in a polytropic process ($n = 1.4$) from 850 to 500 kPa. Determine the final volume if the initial volume is 100 m^3.

7.24 An ideal gas (3 lb_m) in a closed system is compressed in such a way that $\Delta s = 0$ from 14.7 psia and 70 F to 60 psia. For this gas, $c_p = 0.238$ Btu/lb_m·F, $c_v = 0.169$ Btu/ lb_m·F, and $R = 53.7$ ft lb_f/lb_m·R. Compute (a) the final volume if the initial volume is 40.3 ft^3 and (b) the final temperature.

7.25 Air expands from 172 kPa and 60°C to 101 kPa and 5°C. Determine the change in the mass-specific entropy s of the air (kJ/kg·K) assuming constant average specific heats.

7.26 A piston–cylinder assembly contains oxygen initially at 0.965 MPa and 315.5 °C. The oxygen then expands in such a way that the entropy s remains constant to a final pressure of 0.1379 MPa. Determine the change in internal energy per kg of oxygen.

7.27 During the compression stroke in an internal combustion engine, air initially at 41°C and 101 kPa is compressed isentropically to 965 kPa. Determine (a) the final temperature (°C), (b) the change in enthalpy (kJ/kg), and (c) the final volume (m^3/kg).

7.28 Air is heated from 49°C to 650°C at a constant pressure of 620 kPa. Determine the enthalpy and entropy changes for this process. Ignore any variation in specific heat and use the value at 27°C. Also determine the percentage error associated with the use of this constant value of specific heat.

7.29 Air expands through an air turbine from inlet conditions of 690 kPa and 538°C to an exit pressure of 6.9 kPa in an isentropic process. Determine the inlet specific volume, the outlet specific volume, and the change in specific enthalpy.

7.30 The following sketch illustrates three processes: ab, bc, and ac. Assuming constant specific heats, sketch these three processes on a T–S diagram. Assume that the working substance is an ideal gas.

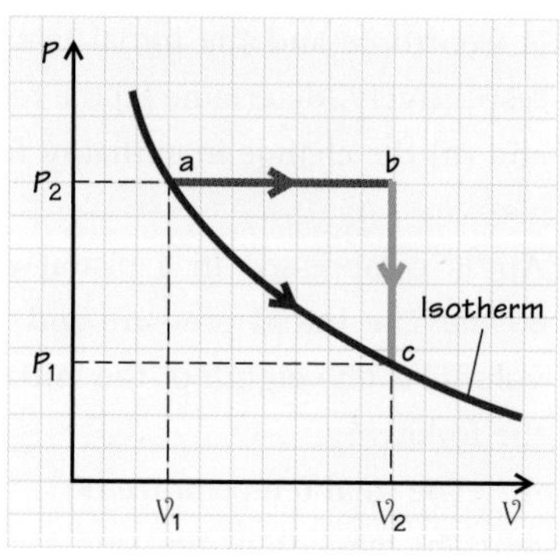

7.31 Is the adiabatic expansion of air from 175 kPa and 60°C to 101 kPa and 5°C possible? Assume the specific heats are constant.

7.32 Is the adiabatic compression of air from 150 kPa and 50°C to 400 kPa and 90°C possible? Assume the specific heats are constant.

7.33 For a simple compressible substance with $c_p = a(1 + bT)$, where a and b are constants, determine the entropy change for an isobaric process going from T_1 to T_2.

7.34 For the paths indicated in the sketch, show that the entropy change for an ideal gas by either path is the same.

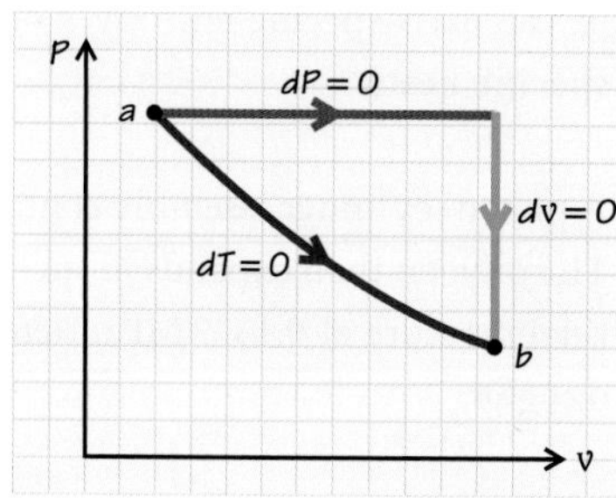

7.35 Consider a closed system (fixed mass) of air. For the following processes, indicate whether the entropy change of the system is zero, positive, negative, or indeterminate.

A. Reversible cooling at constant pressure
B. Irreversible cooling at constant pressure
C. Reversible heating at constant pressure
D. Irreversible heating at constant pressure

7.36 A 0.5-lb_m mass of air is compressed irreversibly from 15 psia and 40 F to 30 psia. During the process, 8.5 Btu of heat is removed from the air and 13 Btu of work is done on the air. Determine the entropy change of the air. **HINT**: Assume ideal-gas behavior with $c_p = 0.24\ \text{Btu/lb}_\text{m}{\cdot}\text{R}$ and $c_v = 0.17\ \text{Btu/lb}_\text{m}{\cdot}\text{R}$ and use the T–ds relationships.

7.37 In the cylinders of an internal combustion engine, air is compressed reversibly from 103.5 kPa and 23.9 °C to 793 kPa. Calculate the work per unit mass if the process is (a) adiabatic and (b) polytropic with $n = 1.25$.

7.38 Using computer software such as EES or a spreadsheet, plot the adiabatic and polytropic processes described in Problem 7.37 on a P–v diagram.

7.39 Air in a cylinder ($V_1 = 0.03\,\text{m}^3$, $P_1 = 100\,\text{kPa}$, $T_1 = 10\ °\text{C}$) is compressed reversibly at constant temperature to a pressure of 420 kPa. Determine the entropy change, the heat transferred, and the work done. Also sketch this process on T–S and P–V diagrams.

7.40 Air in a cylinder ($V_1 = 0.03\,\text{m}^3$, $P_1 = 400\,\text{kPa}$, $T_1 = 80\ °\text{C}$) expands reversibly at constant temperature to a pressure of 150 kPa. Determine the entropy change, the heat transferred, and the work done. Also sketch this process on T–S and P–V diagrams.

7.41 Consider a cylinder–piston arrangement trapping air at 630 kPa and 550 °C. Assume the air expands in a polytropic process ($PV^{1.3}$ = constant) to 100 kPa. What is the specific entropy change (kJ/kg·K)?

7.42 A 1.5-m^3 insulated rigid tank contains 2.7 kg of carbon dioxide at 100 kPa. As a paddle wheel turns in the tank the pressure rises to 120 kPa. Determine the entropy change of the carbon dioxide during this process. Assume constant specific heats of c_v=0.657 kJ/kg·K.

7.43 Air is compressed in a reversible, steady-state, steady-flow process from 15 psia and 80 F to 120 psia. The process is polytropic with $n = 1.22$. Calculate the work of compression, the change in entropy, and the heat transfer, all per unit mass (lb_m) of air compressed.

7.44 Air enters an expansion valve at 700 kPa and 200 °C. If the exit pressure is 150 kPa, determine the change in the specific entropy.

7.45 Air undergoes a steady-flow, reversible, adiabatic process through a turbine. The initial state is 200 psia and 1500 F, and the final pressure is 20 psia. Changes in kinetic and potential energy are negligible. Determine the following quantities:

A. The final temperature
B. The final specific volume
C. The enthalpy change per lb_m
D. The work per lb_m

7.46 Air undergoes a steady-flow, reversible, adiabatic process through a turbine. The initial state is 1400 kPa and 815 °C and the final pressure is 140 kPa. Changes in kinetic and potential energy are negligible. Determine the following quantities:

A. The final temperature
B. The final specific volume
C. The change in specific enthalpy
D. The specific work

7.47 Air undergoes a steady-flow, reversible, adiabatic process through a compressor. The initial state is 120 kPa and 315 °C and the final pressure is 1.2 MPa. Changes in kinetic and potential energy are negligible. Determine the following quantities:

A. The final temperature
B. The final specific volume
C. The change in specific enthalpy
D. The specific work

7.48 A 3.1-lb_m mass of air is trapped in a cylinder and compressed isothermally at 85 F from 15 psia to 100 psia. During the compression, 412 Btu of energy is removed in a heat-transfer process. Determine (a) the compression ratio ($= \mathcal{V}_1/\mathcal{V}_2$), (b) the work required (Btu), and (c) the entropy change (Btu/R).

7.49 Air at 15 psia and 20 F is compressed polytropically ($Pv^{1.35} = C$) in a cylinder to 50 psia. Determine the minimum work of compression and the corresponding heat transfer, both in Btu/lb_m.

7.50 A well-insulated 3-m^3 tank contains carbon monoxide at 500 K and 0.5 MPa. A valve is opened and gas is slowly bled out of the tank until the tank pressure is reduced to 0.2 MPa. Determine the temperature (K) of the gas remaining in the tank.

7.51 A well-insulated 1-m^3 tank contains carbon dioxide at 500 K and 0.4 MPa. A valve is opened and the gas is slowly bled out of the tank until the pressure in the tank is reduced to 0.1 MPa. Determine the temperature (K) of the gas remaining in the tank.

7.52 A 0.5-m^3 tank contains air at 300 K and 0.75 MPa. A valve is suddenly opened and air rushes out until the tank pressure drops to 0.15 MPa. The air in the tank at the end of the process may be assumed to have undergone a reversible adiabatic process. Determine the final temperature (K) and mass (kg) in the tank.

7.53–7.74 Entropy for a pure nonideal substance

7.53 A piston–cylinder device contains steam at 500 F and 100 psia. The piston is then slowly pushed in so that the steam undergoes a reversible, isothermal process until the pressure reaches 150 psia. Determine (a) the increase in entropy (Btu/lb_m·R), (b) the heat transfer (Btu/lb_m) from the steam, and (c) the work transfer (Btu/lb_m) from the piston to the steam. Sketch the process on a T–s diagram and include the vapor dome and isobars at 100 psia and 150 psia.

7.54 Is the adiabatic expansion of superheated steam from 850 °C and 2.0 MPa to 650 °C and 1.2 MPa possible? Explain.

7.55 Is the adiabatic compression of superheated steam from 440 K and 150 kPa to 680 K and 700 kPa possible? Explain.

7.56 Saturated water vapor in a piston–cylinder at 2.0 MPa is expanded to 300 kPa in an adiabatic reversible process. Determine the quality of the final state and the work done per mass of water. Sketch the process on a T–s diagram and include the vapor dome and isobars at 2.0 MPa and 300 kPa.

7.57 Superheated water (2.5 kg) in a piston–cylinder at 300 kPa and 500 K is compressed to saturated water vapor in a reversible isothermal process. Determine the work required to compress the water and the heat transfer during the process.

7.58 Saturated vapor water (1.2 kg) in a piston–cylinder at 420 K is expanded to 100 kPa in a reversible isothermal process. Determine the work done by the water and the heat transfer during the process.

7.59 Steam flows through an expansion valve. Conditions at the inlet are $P_1 =$ 700 kPa and $x_1 = 0.96$. The exit pressure P_2 is 350 kPa. Calculate the entropy change in kilojoules per kilogram of steam.

(Credit: AGEfotostock.)

7.60 Steam flows through an expansion valve. Conditions at the inlet are $P_1 =$ 100 psia and $x_1 = 0.96$. The exit pressure P_2 is 40 psia. Calculate the entropy change in Btu per lb_m of steam.

7.61 Sketch a T–s diagram for the changes of phase that occur between solid water and superheated steam. Assume the pressure is constant.

7.62 A rigid cylinder contains 0.5 m^3 of steam at 8 MPa and 350 °C. The steam is then cooled. The pressure at this state is 5 MPa. Calculate the heat rejected and sketch the process on a T–S diagram.

7.63 Steam in a piston–cylinder is expanded isentropically from 100 MPa and 375 °C to 1 MPa. Calculate the work done per unit mass.

7.64 Water in a piston–cylinder is expanded isentropically from saturated liquid at 1 MPa to 200 kPa. Calculate the work done per unit mass.

7.65 Three kilograms of saturated water in a piston–cylinder at 150 kPa with quality 0.6 are compressed isentropically to a pressure of 500 kPa. Determine the work done.

7.66 Consider a closed system (fixed mass) of superheated steam. For the following processes, indicate whether the entropy change of the system is zero, positive, negative, or indeterminate. Sketch the process on a *T–s* diagram.

A. Reversible cooling (reducing the steam temperature) at constant pressure
B. Irreversible cooling (reducing the steam temperature) at constant pressure
C. Reversible heating (increasing the steam temperature) at constant pressure
D. Irreversible heating (increasing the steam temperature) at constant pressure

7.67 A 0.5-lb_m mass of superheated steam is compressed irreversibly from 20 psia and 320 F to 40 psia. During the process, 7.5 Btu of heat is removed from the air and 39 Btu of work is done on the steam. Determine the entropy change of the water.

7.68 A 1.5-m^3 insulated rigid tank contains 0.82 kg of water at 100 kPa. As a paddle wheel turns in the tank the pressure rises to 150 kPa. Determine the entropy change of the water during this process and the shaft work required to turn the paddle wheel.

7.69 Superheated steam at a flow rate of 0.7 lb_m/s is expanded through a turbine in an adiabatic, reversible, steady-state steady-flow process from 200 psia and 550 F to 100 psia. Changes in kinetic and potential energy are negligible. Calculate the output power from the turbine and the change of entropy of the steam. Sketch the process on a *T–s* diagram showing the vapor dome and constant-pressure curves through the initial and final states.

7.70 Superheated steam undergoes a steady-flow, reversible, adiabatic process through a turbine. The initial state is 160 psia and 700 F, and the final pressure is 20 psia. Changes in kinetic and potential energy are negligible. Determine the following quantities:

A. The final temperature
B. The final specific volume
C. The work per lb_m

7.71 Superheated steam undergoes a steady-flow, reversible, adiabatic process through a compressor. The initial state is 150 kPa and 420 K and the final pressure is 1.0 MPa. Changes in kinetic and potential energy are negligible. Determine the following quantities:

A. The final temperature
B. The final specific volume
C. The specific work

7.72 Saturated water in a cylinder at 100 kPa with 40% quality is compressed in a reversible polytropic process ($Pv^{1.2} = C$) to 400 kPa. Determine the minimum work of compression and the corresponding heat transfer.

7.73 A well-insulated 0.012-m^3 tank contains 13.6 kg of R-22 refrigerant saturated mixture at 30 °C. A valve is opened to recharge an air-conditioning system.

A. Find the initial pressure, the quality of the refrigerant, and the initial volume of liquid in the tank.

B. Two kilograms of refrigerant are slowly (reversibly) bled out of the tank. (Note in the photo that the recharge tank is on a scale to measure the weight of the refrigerant.) Determine the tank temperature and pressure after the refrigerant bleed if the process is adiabatic. Also find the quality of the mixture in the tank.

C. After the bleed process, the temperature of the tank returns to 30 °C. Determine the final pressure. Also find the quality of the mixture and the final volume of liquid in the tank.

(Photo by Laura Pauley.)

7.74 A well-insulated 0.42-ft^3 tank contains 30 lb_m of R-22 refrigerant saturated mixture at 80 F. A valve is opened to recharge an air-conditioning system.

A. Find the initial pressure, the quality of the refrigerant, and the initial volume of liquid in the tank.

B. 5 lb_m of refrigerant is slowly (reversibly) bled out of the tank. (Note in the photo that the recharge tank is on a scale to measure the weight of the refrigerant.) Determine the tank temperature and pressure after the refrigerant bleed if the process is adiabatic. Also find the quality of the mixture in the tank.

C. After the bleed process, the temperature of the tank returns to 80 F. Determine the final pressure. Also find the quality of the mixture and the final volume of liquid in the tank.

7.75–7.87 Entropy change for an isolated system

7.75 An insulated rigid tank contains two chambers, each with a volume of 0.25 m^3. Air at 300 kPa and 25 °C is contained in one chamber of the insulated tank. The air is suddenly expanded to twice its volume when the wall dividing the two chambers is punctured.

A. Determine the final temperature and pressure of the air.

B. Determine the change in entropy of the air during the process.

C. Why does the entropy increase in this process?

7.76 An insulated rigid tank contains two chambers, each with a volume of 2.0 ft^3. Air at 45 psia and 80 F is contained in one chamber of the insulated tank. The air is suddenly expanded to twice its volume when the wall dividing the two chambers is punctured. Use $R = 0.3704$ psia·ft^3/(lb_m·R) for air.

A. Determine the final temperature and pressure of the air.

B. Determine the change in entropy of the air during the process.

C. Why does the entropy increase in this process?

7.77 An insulated rigid tank contains two chambers, each with a volume of 0.05 m^3. Steam at 1.0 MPa and 520 K is contained in one chamber of the insulated tank. The steam is suddenly expanded to twice its volume when the wall dividing the two chambers is punctured.

A. Determine the final temperature and pressure of the steam.
B. Determine the change in entropy of the steam during the process.
C. Why does the entropy increase in this process?

7.78 An insulated rigid tank contains two chambers, each with a volume of 0.25 m^3. Air at 300 kPa and 25 °C is contained in one chamber of the insulated tank. Air at 150 kPa and 25 °C is contained in the other chamber of the insulated tank. The air in the two chambers is suddenly mixed when the wall dividing the two chambers is punctured.

A. Determine the final temperature and pressure of the air.
B. Determine the change in entropy of the system during the process.
C. Why does the entropy increase in this process?

7.79 An insulated rigid tank contains two chambers, each with a volume of 0.25 m^3. Air at 400 kPa and 50 °C is contained in one chamber of the insulated tank. Air at 250 kPa and 200 °C is contained in the other chamber of the insulated tank. The air in the two chambers is suddenly mixed when the wall dividing the two chambers is punctured.

A. Determine the final temperature and pressure of the air.
B. Determine the change in entropy of the system during the process.
C. Why does the entropy increase in this process?

7.80 An insulated rigid tank contains two chambers, each with a volume of 3.0 ft^3. Air at 60 psia and 200 F is contained in one chamber of the insulated tank. Air at 35 psia and 250 F is contained in the other chamber of the insulated tank. The air in the two chambers is suddenly mixed when the wall dividing the two chambers is punctured. Use $R = 0.3704$ psia·ft^3/(lb_m·R) and $c_p = 1.2970$ psia·ft^3/(lbm·R) for air.

A. Determine the final temperature and pressure of the air.
B. Determine the change in entropy of the system during the process.
C. Why does the entropy increase in this process?

7.81 An insulated rigid tank contains two chambers, each with a volume of 0.1 m^3. Steam at 500 kPa and 560 K is contained in one chamber of the insulated tank. Steam at 300 kPa and 500 K is contained in the other chamber of the insulated tank. The steam in the two chambers is suddenly mixed when the wall dividing the two chambers is punctured.

A. Determine the final temperature and pressure of the steam.
B. Determine the change in entropy of the system during the process.
C. Why does the entropy increase in this process?

7.82 An insulated rigid tank contains two chambers. Steam at 300 kPa and 500 K is contained in one chamber with a volume of 0.9 m^3. Saturated liquid at 300 kPa is contained in the other chamber with a volume of 0.01 m^3. The steam in the two chambers is suddenly mixed when the wall dividing the two chambers is punctured.

A. Determine the final temperature and pressure of the steam.

B. Determine the change in entropy of the system during the process.

C. Why does the entropy increase in this process?

7.83 An insulated rigid tank with a volume of 9 m^3 contains air at 100 kPa and 25 °C and a 5-kg steel bar at 300 °C. (Note that the steel bar occupies part of the tank volume.) The steel bar cools until the insulated system reaches equilibrium. For the steel bar, use ρ = 7900 kg/m^3 and c = 477 J/kg·K.

A. Determine the final temperature and pressure of the air.

B. Determine the change in entropy of the system during the process.

C. Why does the entropy increase in this process?

7.84 An insulated rigid tank with a volume of 9 m^3 contains steam at 500 kPa and 580 K and a 5-kg steel bar at 300 K. (Note that the steel bar occupies part of the tank volume.) The steel bar heats until the insulated system reaches equilibrium. For the steel bar, use ρ = 7900 kg/m^3 and c = 477 J/kg·K.

A. Determine the final temperature and pressure of the steam. (**HINT**: Use the energy equation to derive an equation that includes the final temperature and specific internal energy of steam. Guess the final temperature and use the NIST resources to find the specific internal energy. Substitute this value into the energy equation to find an updated guess for the final temperature. Repeat this process until convergence.)

B. Determine the change in entropy of the system during the process.

C. Why does the entropy increase in this process?

7.85 An insulated rigid tank with a volume of 7 m^3 contains steam at 500 kPa and 480 K and a hot 5-kg steel bar. (Note that the steel bar occupies part of the tank volume.) The steel bar cools until the insulated system reaches an equilibrium temperature of 520 K. For the steel bar, use ρ = 7900 kg/m^3 and c = 477 J/kg·K.

A. Determine the initial temperature of the steel bar.

B. Determine the change in entropy of the system during the process.

C. Why does the entropy increase in this process?

7.86 A perfectly insulated chamber contains two solid blocks. The first block is 1 kg steel at a temperature of 400 K. The second block is 2 kg aluminium at a temperature of 250 K. Determine the equilibrium temperature of the two blocks and the change in entropy during the process. For the steel block, use ρ = 7900 kg/m^3 and c = 477 J/kg·K. For the aluminium block, use ρ = 2700 kg/m^3 and c = 903 J/kg·K.

7.87 An insulated rigid tank with a volume of 0.2 m^3 contains steam at 100 kPa and 400 K and a 4-cm^3 ice cube at 0°C. (Note that the ice cube occupies part of the tank volume.) Determine the equilibrium temperature of the water and the change in entropy during the process. For the ice, use v = 0.00111 m^3/kg, $h_{fusion} = h_{liq} - h_{solid} = 333.4$ kJ/kg and $\Delta s_{fusion} = h_{fusion}/T$. (See Section 2.9 for the enthalpy of fusion.)

7.88–7.108 Entropy balance for a closed system with energy exchange

7.88 The air in a piston–cylinder device is compressed from P_1 to P_2 at constant temperature. The reversible compression requires 12 kJ of P–V work. The air temperature is maintained constant at 25 °C during this process as a result of heat transfer to the surrounding medium at 10 °C. Determine:

A. The change in entropy of the ideal gas during this process.

B. The change in entropy of the surroundings during this process.

C. The change in entropy of the world (system + surroundings) during this process.

7.89 Air in a closed tank is mixed for 3 minutes with a paddle that requires 200 W to operate. The air temperature is maintained constant at 200 °C during this process as a result of heat transfer to the surrounding medium at 50 °C. Determine:

A. The change in entropy of the air during this process.

B. The change in entropy of the surroundings during this process.

C. The change in entropy of the world (system + surroundings) during this process.

7.90 Air in a piston–cylinder device expands from P_1 to P_2 at a constant temperature. The reversible expansion produces 30 kJ of P–V work. The air temperature is maintained constant at 40 °C during this process as a result of heat transfer from the surrounding medium at 90 °C. Determine:

A. The change in entropy of the ideal gas during this process.

B. The change in entropy of the surroundings during this process.

C. The change in entropy of the world (system + surroundings) during this process.

7.91 A closed office contains three computers each requiring 500 W to operate. The heat loss from the walls of the room to the surroundings gives a steady-state room temperature of 27 °C when the computers are operating. The surrounding air is at 10 °C. Determine:

A. The rate change in entropy of the ideal gas during this process.

B. The rate change in entropy of the surroundings during this process.

C. The rate change in entropy of the world (system + surroundings) during this process.

7.92 Isothermal heat transfer of 45 kJ of energy occurs from the surroundings at 120 °C to a system at 30 °C. For an isothermal process with no internal irreversibilities:

A. Determine the entropy change for the system. (Give to four digits.)

B. Determine the entropy change for the surroundings. (Give to four digits.)

C. Determine the entropy change for the world (system + surroundings).

7.93 Isothermal heat transfer of 2500 Btu of energy occurs from the surroundings at 200 F to the system at 100 F. For an isothermal process with no internal irreversibilities:

A. Determine the entropy change for the system. (Give to 4 digits.)
B. Determine the entropy change for the surroundings. (Give to 4 digits.)
C. Determine the entropy change for the world (system + surroundings).

7.94 The air in a piston–cylinder device initially has a volume of 0.2 m^3 at 100 kPa and 200 °C. The air is compressed reversibly from 100 kPa to 300 kPa at a constant temperature of 200 °C. The air temperature in the cylinder is maintained constant during this process as a result of heat transfer to the surrounding medium at 80 °C. Determine:
A. The P–$\mathcal{V}$ work during the process.
B. The change in entropy of the ideal gas during this process.
C. The change in entropy of the surroundings during this process.
D. The change in entropy of the world (system + surroundings) during this process.

7.95 The air in a piston–cylinder device initially has a volume of 0.2 m^3 at 100 kPa and 200 °C. The air is compressed from 100 kPa to 300 kPa at a constant temperature of 200 °C. During this process, a fan mixes the air for two minutes with a power of 100 W. The air temperature in the cylinder is maintained constant during this process as a result of heat transfer to the surrounding medium at 80 °C. Determine:
A. The P–$\mathcal{V}$ work during the process.
B. The change in entropy of the ideal gas during this process.
C. The change in entropy of the surroundings during this process.
D. The change in entropy of the world (system + surroundings) during this process.

7.96 Solve Problems 7.94 and 7.95 and compare your results. Why is the change in entropy of the world greater for Problem 7.95?

7.97 In Problem 7.94, there are two ways to calculate the change in entropy of the system during this reversible process. Find the entropy of the system using both of these methods.
A. Calculate the P–$\mathcal{V}$ work during the isothermal process and use an energy balance to find the heat transfer from the air in the cylinder. Then calculate the change in entropy for the reversible isothermal heat transfer process, using Eq. 7.2a.
B. Find the mass of air in the cylinder and use Eq. 7.12a to find the change in entropy of the air.
C. Do these two methods give the same answer?
D. Review Section 7.1e, "Gibbs Relationships", and explain why the two methods give the same answer for the change in entropy.

7.98 The air in a piston–cylinder device initially has a volume of 0.2 m^3 at 400 kPa and 120 °C. The air expands in a reversible process from 400 kPa to 150 kPa at a constant temperature of 120 °C. The air temperature in the cylinder is maintained constant during this process as a result of heat transfer to the surrounding medium at 250 °C. Determine:
A. The P–$\mathcal{V}$ work during the process.
B. The change in entropy of the ideal gas during this process.

C. The change in entropy of the surroundings during this process.

D. The change in entropy of the world (system + surroundings) during this process.

7.99 Steam in a piston–cylinder device initially has a volume of 0.2 m^3 at 300 kPa and 540 K. The steam is compressed reversibly from 300 kPa to 700 kPa at a constant temperature of 540 K. The steam temperature in the cylinder is maintained constant during this process as a result of heat transfer to the surrounding medium at 250 K. Determine:

A. The change in entropy of the steam during this process.

B. The heat transfer during this internally reversible process.

C. The P–$\mathcal{V}$ work during the process.

D. The change in entropy of the surroundings during the process.

E. The change in entropy of the world (system + surroundings) during this process.

7.100 Steam in a piston–cylinder device initially has a volume of 0.2 m^3 at 300 kPa and 540 K. The steam is compressed from 300 kPa to 700 kPa at a constant temperature of 540 K. During this process, a paddle mixes the steam for two minutes with a power of 100 W. The steam temperature in the cylinder is maintained constant during the process as a result of heat transfer to the surrounding medium at 250 K. Determine:

A. The change in entropy of the steam during this process.

B. The P–$\mathcal{V}$ work during a reversible compression process without the paddle.

C. The heat transfer during the actual process with mixing.

D. The change in entropy of the surroundings during this process.

E. The change in entropy of the world (system + surroundings) during this process.

7.101 Solve Problems 7.99 and 7.100 and compare your results. Why is the change in entropy of the world greater for Problem 7.100?

7.102 An insulated 2-m^3 rigid tank contains steam at 500 kPa and 580 K. A paddle mixes the steam for two minutes with a power of 50 W.

A. What is the final temperature and pressure of the steam?

B. What is the change in entropy of the system during this process?

C. What is the change in entropy of the world during this process?

7.103 A 2-m^3 rigid tank contains steam at 300 kPa and 520 K. A paddle mixes the steam for two minutes with a power of 50 W. The final temperature of the steam is 580 K.

A. What is the heat transfer during the process?

B. What is the change in entropy of the system during this process?

C. What is the change in entropy of the world (system + surroundings) during this process if the surroundings are at 620 K?

7.104 In the cylinders of a steam engine, superheated vapor is compressed reversibly from 300 kPa and 460 K to 500 kPa. Calculate the work per unit mass if the process is (a) adiabatic and (b) isothermal. For each case, sketch the process on P–$\mathcal{V}$ and T–s diagrams, showing the vapor dome. On the P–$\mathcal{V}$ diagram, include constant-temperature curves through the initial and final states. On the T–s diagram, include constant-pressure curves through the initial and final states. Which process requires more work to compress the steam?

7.105 Superheated steam in a cylinder ($\mathcal{V}_1 = 0.03\,\text{m}^3$, $P_1 = 500\,\text{kPa}$, $T_1 = 520$ K) is compressed reversibly at constant temperature to a pressure of 700 kPa. Determine the entropy change of the system, the heat transferred, and the work done. Also sketch this process on T–s and P–$\mathcal{V}$ diagrams.

7.106 Superheated steam in a cylinder ($\mathcal{V}_1 = 0.02\,\text{m}^3$, $P_1 = 300\,\text{kPa}$, $T_1 = 460$ K) expands in a reversible process at constant temperature to a pressure of 100 kPa. Determine the entropy change of the system, the heat transferred, and the work done. Also sketch this process on T–S and P–$\mathcal{V}$ diagrams.

7.107 Air in a cylinder ($\mathcal{V}_1 = 0.3\,\text{m}^3$, $P_1 = 400\,\text{kPa}$, $T_1 = 350$ K) is compressed reversibly at constant temperature to a pressure of 700 kPa. Determine the entropy change of the system, the heat transferred, and the work done. Also sketch this process on T–S and P– $\mathcal{V}$ diagrams.

7.108 A 3.0-lb_m mass of water is trapped in a cylinder and compressed isothermally at 360 F from 20 psia to 100 psia. Determine (a) the compression ratio ($= \mathcal{V}_1/\mathcal{V}_2$), (b) the entropy produced (Btu/R), (c) the heat transfer from the system if the compression occurs reversibly, (d) the work during the process (Btu) if 550 Btu of energy is removed in an irreversible heat-transfer process.

7.109–7.129 A closed system undergoing a cycle

7.109 Air in a piston–cylinder device undergoes the following processes to complete a reversible cycle:

1–2: a constant-pressure process at 50 psia to 700 F,

2–3: an adiabatic, constant-entropy process to 10 psia,

3–1: an isothermal process to 100 psia, and

4–1: a constant-volume process.

A. Carefully sketch (qualitatively, not to scale) these processes in P–$\mathcal{V}$ coordinates. Clearly indicate how you locate each of the four states.

B. For the gas as a system, complete the following table with *in*, *out*, or *none* to indicate the directions of heat and work transfer.

Process	Heat	Work
1–2		
2–3		
3–4		
4–1		

7.110 List the four processes involved in the ideal Carnot cycle and plot them in P–$\mathcal{V}$ and T–s coordinates for an ideal gas.

7.111 List the four processes involved in the ideal Stirling cycle and plot them in P–v and T–s coordinates for an ideal gas.

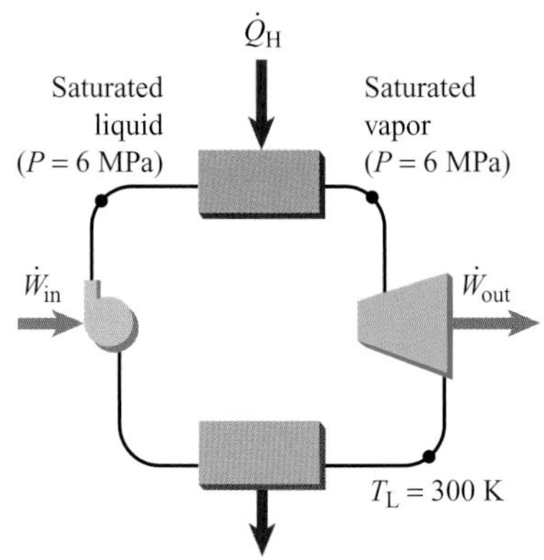

7.112 Consider a steady-flow ideal Carnot cycle, using steam as the working fluid, in which the high-temperature, constant-pressure heat-addition process starts with a saturated liquid and ends with a saturated vapor.

A. Plot this cycle in T–s coordinates showing the steam dome.

B. Calculate the thermal efficiency for this cycle if the pressure of the high-temperature steam is 6 MPa and the low-temperature heat-rejection process occurs at 300 K. Calculate the values of the quality at the beginning and end of the heat-rejection process.

C. Explain why it is difficult to operate a practical steam power plant on the ideal Carnot cycle of part A.

7.113 For the conditions given in part B of Problem 7.112, calculate the heat added, the heat rejected, and the net work performed (all per unit mass of steam).

7.114 Consider the situation described in Problem 7.112. How does the thermal efficiency change if the high-temperature heat-addition process is now conducted at 15 MPa instead of 6 MPa? Be quantitative. How does the net work produced per unit mass of steam ($\dot{W}_{net}/\dot{m}$) compare for the two cycles? (See Problem 7.113.)

7.115 Air in a piston–cylinder device undergoes an internally reversible cycle. That is, the air itself undergoes a reversible set of processes, but irreversible processes (e.g., heat transfer through a finite temperature difference) are possible outside the air in the cylinder. From state 1 at 100 °C and 0.75 MPa, the air expands isothermally to 0.15 MPa at state 2. The air then undergoes a constant-pressure process to state 3, from which the air returns to state 1 by an adiabatic process. Determine the thermal efficiency for the cycle and compare your answer to the thermal efficiency of a completely reversible cycle operating between the same maximum and minimum temperatures. Explain the difference.

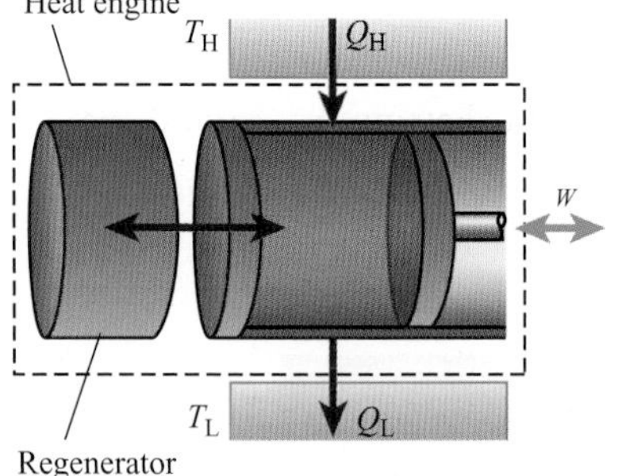

7.116 An ideal gas in a heat engine undergoes the following processes comprising a cycle in a piston–cylinder regenerator device:

1–2: isothermal heating at T_H from a heat source at T_H,
2–3: constant-volume cooling to T_L by the regenerator,
3–4: isothermal cooling at T_L by a heat sink at T_L, and
4–1: constant-volume heating to T_H by the regenerator.

A. Determine the thermal efficiency for the heat engine by considering the work and heat flow for each process.

B. Is the heat engine reversible? Why? Explain.

7.117 A Carnot heat engine receives 633 kJ from a reservoir at 650 °C while rejecting heat at 38 °C. Determine the work delivered (kJ) and the engine efficiency. Also determine the entropy change (kJ/K) associated with the high- and low-temperature reservoirs.

7.118 A Carnot heat engine between heat reservoirs at 550 °C and 50 °C delivers 420 kJ. Determine the heat transfer from the high-temperature reservoir (kJ) and the engine efficiency. Also determine the entropy change (kJ/K) associated with the high- and low-temperature reservoirs.

7.119 Complete the following equations by replacing the question mark with the proper equality sign or inequality sign.

A. For a closed system (any process):

i. $W ? \int P d\mathcal{V}$

ii. $\Delta S \; ? \; \int \delta Q/T$

iii. $T\,dS \; ? \; dU + P d\mathcal{V}$

iv. $T\,dS \; ? \; dU + \delta W$

B. For any cycle:

i. $W ? \oint P d\mathcal{V}$

ii. $\oint dS \; ? \; 0$

iii. $\oint dS \; ? \; \delta Q/T$

iv. $\oint \delta Q/T \; ? \; 0$

v. $\eta \; ? \; 1 - T_L/T_H$

vi. $\oint dH \; ? \; 0$

7.120 A Carnot-cycle heat engine operating between reservoirs at 950 K and 450 K receives 500 kW·hr of energy from a high-temperature source. Determine the following quantities:

A. The thermal efficiency

B. The work done

C. The entropy change of the high- and low-temperature reservoirs

D. The entropy change of the high- and low-temperature reservoirs, if the high-temperature reservoir temperature is changed to 1500 K, while energy still enters the heat engine at 950 K

E. The entropy change of the universe for both circumstances

7.121 Consider a steady-flow, ideal Carnot cycle in a closed piston–cylinder device, using steam as the working fluid, in which the high-temperature heat-addition process starts with a saturated liquid and ends with a saturated vapor. The working fluid is at 570 K during the heat-addition process and at 400 K during the heat removal process.

A. Sketch this cycle using *T–s* coordinates and showing the steam dome.

B. Calculate the quality at the beginning and end of the heat-rejection process.

C. Determine the heat addition and heat rejection per mass of steam (kJ/kg). Also calculate the net work per mass of steam (kJ/kg).

D. Calculate the thermal efficiency for this cycle using the values from part C.

7.122 Calculate the thermal efficiency of the cycle in Problem 7.121 using two different methods:

A. First calculate the heat addition and net work per mass of steam and find the thermal efficiency using $\eta_{\text{rev, HE}} = W_{\text{net}}/Q_{\text{H}}$.

B. Find the thermal efficiency using $\eta_{\text{rev, HE}} = (1 - T_L/T_H)$.

Do you find the same thermal efficiency?

7.123 In Problem 7.121, in order to have heat transfer at a finite rate, the high-temperature reservoir is at 600 K and the low-temperature reservoir is at 350 K. Find the change in entropy of the system and surroundings.

7.124 (EES) Solve Problem 7.121 using EES or other software to determine the net work and heat addition (kJ/kg) in the cycle and the thermal efficiency of the cycle.

7.125 (EES) Using your computer solution from Problem 7.124, vary the temperature of the working fluid in the heat-rejection process from 250 K to 400 K in 10-degree increments. Plot the net work and the thermal efficiency of the cycle as a function of the heat-rejection temperature. Why does the thermal efficiency increase as the heat-rejection temperature is decreased?

7.126 (EES) Using your computer solution from Problem 7.124, vary the temperature in the heat addition process from 500 K to 630 K in 10-degree increments. For each case, the high-temperature heat-addition process starts with a saturated liquid and ends with a saturated vapor. Plot the thermal efficiency of the cycle as a function of the heat addition temperature. On one plot, show the heat addition and net work as a function of the heat-addition temperature. Why does the thermal efficiency increase as the heat-addition temperature is increased?

7.127 Consider a steady-flow ideal Stirling cycle, using steam as the working fluid, in which the high-temperature heat-addition process starts with a saturated liquid and ends with a saturated vapor. The working fluid is at 570 K during the heat-addition process and at 400 K during the heat-removal process.

A. Sketch this cycle in T–s and P–$\mathcal{V}$ coordinates, showing the vapor dome.

B. Calculate the quality at the beginning and end of the heat-rejection process.

C. Determine the net work per mass of steam (kJ/kg). Also find the heat addition and heat rejection per mass of steam (kJ/kg).

D. Calculate the thermal efficiency for this cycle using the values from part C.

7.128 Using EES or other software, accurately plot the Carnot cycle from Problem 7.121 and the Stirling cycle from Problem 7.127 on T–s and $P-\mathcal{V}$ diagrams.

7.129 Calculate the thermal efficiencies for the Carnot cycle in Problem 7.121 and the Stirling cycle in Problem 7.127. Which cycle has a greater thermal efficiency? Explain the difference in the thermal efficiency of the two cycles using T–s and P–$\mathcal{V}$ diagrams.

7.130–7.151 Entropy balance for an open system

7.130 An isothermal steam turbine produces a power of 450 kW. Steam enters the turbine at 7 MPa and 320 °C and exits at 0.7 MPa. Assume that heat is added at a rate of 750 kW during this process. Determine the steam mass flow rate in kg/hr and the value of each of the following quantities:

$$\int \frac{\delta \dot{Q}}{T} [=] \text{kJ/(hr} \cdot \text{K)},$$

$$\sum_{\text{in}} (\dot{m}s) [=] \text{kJ/(hr} \cdot \text{K)},$$

$$\sum_{\text{out}} (\dot{m}s) [=] \text{kJ/(hr} \cdot \text{K)},$$

$$\dot{S}_{\text{gen}} [=] \text{kJ/(hr} \cdot \text{K)}.$$

7.131 An inventor claims to have developed an isothermal steady-flow turbine capable of producing 100 kW when operating with a steam flow rate of 10,600 lb_m/hr. The inlet conditions are 500 psia and 1000 F and the exit pressure is 14.7 psia. Heating takes place as the steam flows through the turbine to maintain an isothermal condition. Determine (a) the heat required and (b) the numerical value for each term in an entropy balance (Eq. 7.33b). Then evaluate the inventor's claim.

7.132 A contact feedwater heater operates on the principle of mixing steam and water. Steam enters the heater at 100 psia and 98% quality. Water enters the heater at 100 psia and 80 F. As a result, 25,000 lb_m/hr of water at 95 psia and 290 F leaves the heater. There is no heat transfer between the heater and the surroundings. Evaluate each term in the general entropy balance for the second law (Eq. 7.33).

7.133 Consider a steam turbine in which the steam enters at 23.26 MPa and 808 K with a flow rate of 17.78 kg/s. The steam exits at a pressure of 5.249 kPa with a quality of 0.9566. Determine the power produced by the turbine and the isentropic efficiency. Plot the process in T–s coordinates. Show the steam dome.

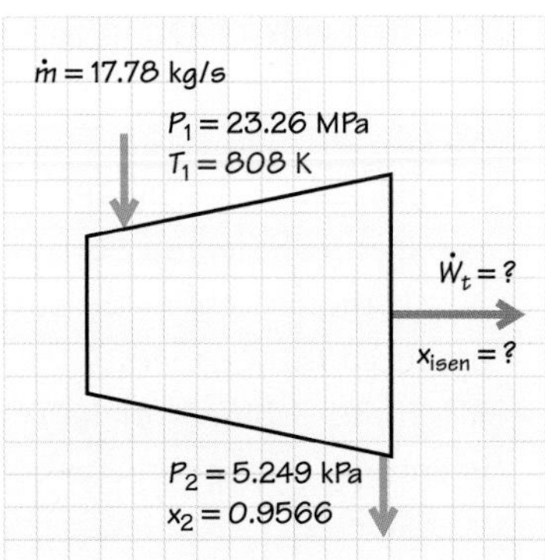

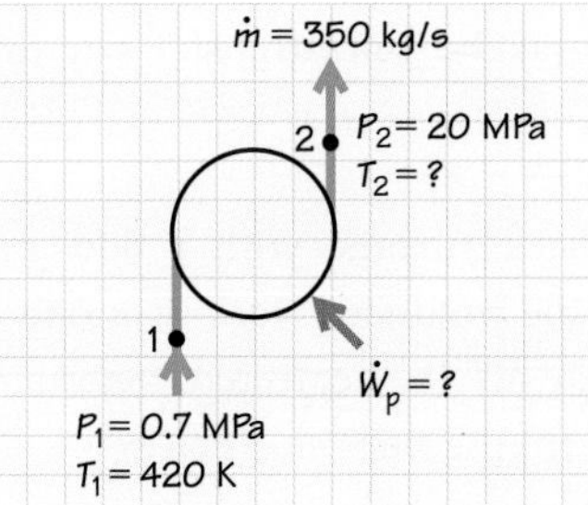

7.134 A feedwater pump operates with a flow rate of 350 kg/s. The water enters at 420 K. The inlet pressure is 0.7 MPa and the outlet pressure is 20 MPa. Determine the power needed to drive the pump if the pump isentropic efficiency is 87%. Also determine the outlet temperature of the water. Plot the process in T–s coordinates. Show the saturated-liquid line.

7.135 Consider the steam turbine and conditions described in Problem 7.133. Create a graph of isentropic efficiency as a function of the exit-steam quality. The exit quality should range from isentropic operation ($\eta_{\text{isen},\,t} = 100\%$) to the point where the steam exits as saturated vapor ($x = 100\%$).

7.136 Determine the rate of entropy production by irreversibilities associated with the pump and conditions described in Problem 7.134.

7.137 In a power plant, 1,500,000 lb_m/hr of steam enters a turbine at 1000 F and 500 psia. The steam expands adiabatically to 1 psia with 98% quality. Determine (a) the turbine power rating (MW) and (b) the turbine efficiency (%).

7.138 The power output of a steam turbine is 30 MW. Determine the rate of steam flow in the turbine for saturated steam entering at 0.1 MPa. The outlet pressure is 10 kPa and the expansion is reversible and adiabatic.

7.139 Gas enters a turbine at 550 °C and 500 kPa and leaves at 100 kPa. The entropy change is 0.174 kJ/kg·K (i.e., the turbine is only approximately adiabatic). What is the temperature of the gas leaving the turbine, assuming that the gas is ideal with $c_p = 1.11$ kJ/kg·K and $c_v = 0.835$ kJ/kg·K?

7.140 Steam enters a reversible adiabatic turbine at 250 °C and 1 MPa with a velocity of 60 m/s. The steam leaves at 0.2 MPa with a velocity of 180 m/s. Determine the specific work produced (kJ/kg).

7.141 Liquid water (assumed incompressible) at 70 F and 400 psia enters an adiabatic hydraulic turbine at the bottom of a mountain. The water leaves at 15 psia.

A. For a reversible turbine producing 1 MW of power, determine the water mass flow rate (Mlb_m/hr).

B. If the turbine is 80% efficient, with the same inlet state, exit pressure, and mass flow rate as in part A determine the exit temperature (F).

7.142 An adiabatic gas turbine receives a steady stream of air at 700 °C and 1 MPa and exhausts it at 0.15 MPa. The velocities at the inlet and outlet are negligible and specific heats can be assumed constant.

A. If the process in the turbine is isentropic, determine the specific work produced (kJ/kg).

B. If the turbine work is 70% of the isentropic value, determine the increase in entropy (kJ/kg·K) between inlet and outlet.

7.143 Air enters an adiabatic gas turbine at 1600 F and 40 psia and leaves at 15 psia. The turbine isentropic efficiency is 80%. The mass flow rate is 2500 lb_m/hr. Assuming constant specific heats, determine (a) the work-transfer rate (hp) and (b) the exit temperature (F).

7.144 Steam at 500 °C and 3 MPa enters an adiabatic turbine with a flow rate of 450 kg/s and leaves at 0.6 MPa. The turbine produces work at a rate of 177.7 MW. Potential energy and kinetic energy may be neglected. Determine (a) the exit temperature (°C) of the steam and (b) the turbine efficiency.

7.145 Steam at 400 °C and 1 MPa is steadily supplied to an adiabatic turbine at 500 kg/hr. The exhaust pressure of the turbine is 0.2 MPa. Determine the maximum rate of work produced (kW) by the turbine.

7.146 Air enters the compressor section of a turbojet engine at 15 °C and 0.1 MPa at a rate of 100 m^3/min. The air leaves at 0.5 MPa. Changes in potential energy and kinetic energy are negligible. If the process is reversible and adiabatic, determine the power (kW) required to drive the compressor.

7.147 The isentropic efficiency of an actual compressor with the same operating conditions as in Problem 7.146 is 78%. Determine (a) the compressor power input (kW) and (b) the temperature (K) of the air leaving the compressor.

7.148 A pump delivers 11,000 lb_m/hr of water from an elevation 30 ft below the pump to an elevation 50 ft above the pump through various-diameter pipes. At the lower elevation, the water is at 60 F and 10 psia and the flow area is 0.0233 ft^2. At the higher elevation, the water is at 80 psia and the flow area is 0.0060 ft^2. The pump and pipes are insulated. Determine the minimum pump power (Btu/hr) required.

7.149 Air at 80 F and 14.7 psia enters an adiabatic compressor at 125 lb_m/hr and leaves at 60 psia. The compressor isentropic efficiency is 80%. Determine the rate of power required to drive the compressor (hp).

7.150 Dry, saturated water vapor at 40 °C steadily enters a centrifugal compressor with a mass flow rate of 150 kg/hr. The vapor leaves at 200 °C and 0.04 MPa. During the process, heat is transferred from the vapor at the rate of 1 kW. Determine the following:

A. The power (kW) required to drive the compressor

B. The minimum power (kW) required for an adiabatic compression from the same initial state to the same final pressure

C. The compressor isentropic efficiency

7.151 Two steady flows of steam enter a rigid, adiabatic black box at 300 °C. One of these flows is at $P_1 = 2$ MPa and has a flow rate of 100 kg/hr. The other flow is at $P_2 = 0.5$ MPa and has a flow rate of 50 kg/hr. The box produces useful work and a single flow stream at 0.2 MPa leaves. Determine the maximum useful power (kW) that could be produced by this box.

7.152–7.180 Availability (Exergy)

7.152 Mass enters and leaves a device at a rate of 10 kg/s. It enters with an availability of 50 J/kg and leaves with an availability of 10 J/kg. A rotating shaft connected to the device transfers availability out of the device at the rate of 200 W. There is no transfer of availability due to heat transfer. Inside the device availability is being destroyed at the rate of 90 W. Determine the rate at which availability is being stored within the device.

7.153 The total energy supplied to a power plant during a time increment is 100 J. The availability supplied is 80 J. The energy rejected as heat transfer to the cooling water is 70 J. The amount of energy and availability contained in the power plant does not change during this time increment.

A. Determine the energy transferred as work out of the power plant during this time increment.

B. The work that this power plant does is completely useful. That is, the availability transferred out due to work transfer is equal to the work transfer. Determine the availability rejected by the heat transfer to the cooling water. Assume no availability is destroyed.

7.154 Mass enters a proposed energy system at the rate of 5 kg/s with an energy of 150 J/kg and an availability of 100 J/kg. Mass leaves the system at a rate of 8 kg/s with an energy of 50 J/kg and an availability of 10 J/kg. The spatial-average values of energy and availability of the mass contained in the system do not change with time and are given as 100 J/kg and 70 J/kg, respectively. The work-transfer rate out of the system is 500 W. The rate of availability transfer associated with the work transfer is also 500 W. The only other energy transfer to or from the system is heat transfer. The availability transfer associated with the heat transfer is zero.

A. Determine the magnitude and the direction (into or out of the system) of the heat-transfer rate.

B. Determine the availability-destruction rate within the system.

C. Is there any limit to the length of time for which this system could operate? Explain.

7.155 Fluid enters a power plant at the rate of 8 kg/s with an energy of 100 J/kg and an availability of 80 J/kg. The average energy of the fluid contained in the power plant is 75 J/kg and the average availability is 55 J/kg. These spatial-average values may be assumed to be constants. The mass flow rate leaving the plant is 6 kg/s. The fluid leaving has an energy of 50 J/kg. The heat transfer and the availability transfer due to heat transfer are both zero. The rate of availability destruction within this plant is zero.

A. Determine the work transfer rate.

B. Assuming that the availability that leaves the power plant due to work transfer is equal to the work transfer, determine the rate at which availability is carried out of the power plant by the fluid leaving it.

7.156 Initially a system contains 100 kg of fluid. Fluid then starts to enter the system at a rate of 2 kg/s with an energy of 10 J/kg and an availability of 8 J/kg. Fluid leaves at the rate of 2 kg/s with an energy of 6 J/kg and an availability of 3 J/kg. The energy transfer out of the system due to work is 15 W. The associated availability transfer out of the system due to work is also 15 W. The heat transfer and the availability transfer due to heat transfer are both zero. The energy per unit mass and the availability per unit mass of the fluid inside the system are always equal to each other during the process. For this system, complete the following table.

Quantity	Mass (kg/s)	Energy rate (W)	Availability rate (W)
Inflow			
Produced			
Outflow			
Stored			
Destroyed			

7.157 Fluid enters a steadily-operating device at the rate of 10 kg/s with an energy of 100 J/kg and an availability of 90 J/kg. Inside the device the flow is divided into two flow streams. One stream leaves the device with an energy of 50 J/kg and an availability of 60 J/kg. The other stream leaves with an energy of 175 J/kg and an availability of 120 J/kg. There are no transfers of work or heat. Determine the availability-destruction rate [W] within the device.

7.158 Fluid enters a system at a rate of 6 kg/s with an energy of 60 J/kg and an availability of 80 J/kg. A second stream of the same fluid enters the system at a rate of 4 kg/s with an energy of 55 J/kg and an availability of 35 J/kg. A heat transfer of 150 W leaves the system. The corresponding availability transfer is zero. The system does work, and the availability transfer due to the work is equal to the work. The system initially contains 20 kg of mass with an energy of 10 J/kg and an availability of 22 J/kg. After a 3-s time increment, the energy and availability within the system are 12 J/kg and 7 J/kg, respectively. No mass leaves the system during this time interval.

A. Determine the initial and final values of energy and availability.
B. Complete the table for the 3-s time increment:

Quantity Units	Mass	Energy	Availability
Inflow			
Produced			
Outflow			
Stored			
Destroyed			

7.159 Determine the availability of a 1-m^3 evacuated space by assuming that the vacuum is in a piston–cylinder device and then finding the maximum useful work that can be produced by an interaction between the vacuum and the atmosphere. The temperature and pressure of the reference atmosphere are 20 °C and 1 atm, respectively.

7.160 A piston–cylinder device contains 1 lb_m of air initially at 500 F and 50 psia. The air then interacts with the atmosphere by undergoing an adiabatic expansion process to 216.7 F and 14.7 psia, followed by a constant-pressure process to 70 F. The temperature and pressure of the reference atmosphere are 70 F and 14.7 psia, respectively. Determine the useful work produced.

7.161 Consider the process of inflating a balloon. Initially the air in the balloon is at 70 F and 14.7 psia and occupies 0.1 ft^3. After filling, the air in the balloon is at 70 F, 58.8 psia, and 1.0 ft^3. The temperature and pressure of the reference atmosphere are 70 F and 14.7 psia, respectively.

A. Determine the mass added to the balloon.

B. If the specific internal energy at the dead state is arbitrarily taken to be zero, determine the increase in energy inside the balloon.

C. Determine the increase in availability inside the balloon.

7.162 Determine the availability (kJ/kg) of flowing air at 400 °C and 3 MPa assuming constant specific heats. The temperature and pressure of the reference atmosphere are 20 °C and 1 atm, respectively.

7.163 Determine the availability (kJ/kg) of flowing steam at 400 °C and 3 MPa. The temperature and pressure of the reference atmosphere are 20 °C and 1 atm, respectively.

7.164 Air is compressed from 300 K, 0.1 MPa to 500 K, 0.4 MPa in a piston–cylinder device. Determine the increase in availability (kJ/kg). The temperature and pressure of the reference atmosphere are 20 °C and 1 atm, respectively.

7.165 Determine the maximum useful work that could be obtained from 0.6 m^3 of compressed air at 250 °C and 0.7 MPa. The temperature and pressure of the reference atmosphere are 20 °C and 1 atm, respectively.

7.166 Air at 800 F and 14.7 psia is contained in a rigid, uninsulated, 50-ft^3 tank. The tank is surrounded by an atmosphere at 70 F and 14.7 psia. Since the tank is not insulated, it cools until the contents reach 70 F because of heat transfer to the atmosphere. Determine (a) the heat transfer and (b) the availability destroyed.

7.167 Superheated steam at 800 F and 14.7 psia is contained in a rigid, uninsulated, 50-ft^3 tank. The tank is surrounded by an atmosphere at 70 F and 14.7 psia. Since the tank is not insulated, it cools until the contents reach 70 F because of heat transfer to the atmosphere. Determine (a) the heat transfer and (b) the availability destroyed.

7.168 A 0.6-m^3 steel tank initially contains steam at 200 °C and 0.2 MPa. The tank is surrounded by a reference atmosphere at 20 °C and 0.1014 MPa. The tank is uninsulated and cools down until the contents reach 20 °C. Determine the availability destruction.

7.169 An insulated and evacuated 0.01-m^3 vessel contains a capsule of water at 140 °C and 5 MPa. The volume of the capsule is 0.001 m^3. The capsule breaks and the contents then fill the entire volume. The temperature and pressure of the reference atmosphere are 20 °C and 1 atm, respectively. Determine the availability destroyed.

7.170 A rigid, uninsulated tank is divided into two parts by an interior partition. Initially, 0.003 m^3 of saturated liquid water at 250 °C is contained on one side of the partition and 0.3 m^3 of superheated water vapor at 250 °C and 0.7 MPa is contained on the other side. A small leak in the partition occurs, and the contents of the two sides slowly mix and reach a new equilibrium. The tank is surrounded by a heat source that maintains the tank at 250 °C. The temperature and pressure of the reference atmosphere are 20 °C and 1 atm, respectively. Determine (a) the final state of the contents in the tank and (b) the availability destruction inside the tank.

7.171 Two kilograms of steam are contained in a piston–cylinder device at 250 °C and 1.4 MPa. Determine the maximum possible useful work that could be done by the steam if it interacts with the atmosphere. The temperature and pressure of the reference atmosphere are 20 °C and 1 atm, respectively.

7.172 Steam at 500 F and 100 psia is contained in a piston–cylinder device. The piston is slowly pushed in so that the steam undergoes a reversible isothermal process until the pressure reaches 150 psia. The temperature and pressure of the reference atmosphere are 70 F and 14.7 psia, respectively. Determine the increase in availability of the steam (Btu/lb_m).

7.173 Steam initially at 300 F and 60 psia is compressed isothermally in a piston–cylinder device to 500 psia. The temperature and pressure of the reference atmosphere are 70 F and 14.7 psia, respectively. Determine the minimum useful work transfer (Btu/Ib_m) required.

7.174 Saturated water vapor at 200 psia enters an adiabatic valve. The exit pressure is 100 psia. The temperature and pressure of the reference atmosphere are 70 F and 14.7 psia, respectively. Determine the availability destruction (Btu/lb_m).

7.175 Nitrogen at 25 °C and 2 atm enters a partially open valve in an insulated pipe. The downstream pressure is 1 atm. Assume nitrogen behaves as an ideal gas with constant specific heats. The temperature and pressure of the reference atmosphere are 20 °C and 1 atm, respectively. Determine (a) the outlet temperature of the gas, (b) the increase in entropy (kJ/kg·K) of the nitrogen, and (c) the availability destruction (kJ/kg). (d) Is this process reversible?

7.176 Dry, saturated water vapor at 150 °C steadily enters an adiabatic valve at a rate of 1 kg/s. The exit pressure is 0.1 MPa. The temperature and pressure of the reference atmosphere are 20 °C and 1 atm, respectively. Determine the availability destruction in the valve.

7.177 Steam enters an adiabatic turbine at 300 °C and 1 MPa and leaves at 0.02 MPa. The work-transfer rate per unit mass flow rate produced by the turbine is 650 kJ/kg. The temperature and pressure of the reference atmosphere are 20 °C and 1 atm, respectively. Determine the rate of availability destruction rate per unit mass flow rate (kJ/kg).

7.178 Steam at 500 °C and 3 MPa enters an adiabatic turbine at 450 kg/s and leaves at 0.6 MPa. The turbine produces work at a rate of 177.7 MW. Potential energy and kinetic energy can be neglected. The temperature and pressure of the reference atmosphere are 20 °C and 1 atm, respectively. Determine (a) the exit temperature of the steam, (b) the turbine isentropic efficiency, and (c) the rate of availability destruction.

7.179 Water is heated in a steady-flow, constant-pressure process from 20 °C and 0.1 MPa to a temperature of 200 °C. The heat is supplied from a source at a constant temperature of 250 °C. The temperature and pressure of the reference atmosphere are 20 °C and 1 atm, respectively. Determine the availability destroyed (kJ/kg).

7.180 A 1-kg piece of iron, initially at a temperature of 500 °C, is quenched by dropping it into an insulated tank containing 1 kg of liquid water at 20 °C. The quenching process begins immediately after the iron enters the water and ends when the iron reaches its final temperature. Treat the water as incompressible. The temperature and pressure of the reference atmosphere are 20 °C and 1 atm, respectively. Determine (a) the final temperature of the iron and (b) the availability destroyed.

7.181–7.210 FE exam problems

7.181 During a reversible process, 35 kJ of energy is removed from a system at a constant temperature of 400 K. Find the entropy change of the system. a. 0.0875 kJ/K, b. –0.0875 kJ/K, c. 11.4 kJ/K, d. 1.29 kJ/K.

7.182 During a reversible process, 27 kJ of energy is added to a system at a constant temperature of 200 °C . Find the entropy change of the system. a. 0.074 kJ/K, b. –0.074 kJ/K, c. 0.175 kJ/K, d. –0.175 kJ/K.

7.183 During a reversible process, a system at a constant temperature of 250 °C has an entropy change of 0.12 kJ/K due to heat transfer. What is the heat transfer to the system? a. 30 kJ, b. –30 kJ, c. 62.76 kJ, d. –62.76 kJ.

7.184. Mixing of the fluid in a rigid vessel generates 0.07 kJ/K of entropy. What heat transfer is required to maintain the fluid temperature at 460 K during the mixing process? a. 6.57 MJ, b. 32.3 kJ, c. 6.57 kJ, d.–32.3 kJ.

7.185 Isothermal heat transfer of 65 kJ of energy occurs from a system at 240 °C to the surroundings at 50 °C. For an isothermal process with no internal irreversibilities, find the change in entropy of the world (system + surroundings) a. 0.075 kJ/K, b. –0.075 kJ/K, c. 0.554 kJ/K, d. –0.554 kJ/K.

7.186 What is the change in entropy of air during a reversible adiabatic expansion from 550 kPa and 60 °C to 180 kPa? Assume the specific heats are constant. a. 0.32 kJ/K, b. –0.32 kJ/K, c. 0.0 kJ/K, d. –0.554 kJ/K.

7.187 What is the change in entropy of air during a reversible isothermal expansion from 750 kPa and 220 °C to 150 kPa? Assume the specific heats are constant. a. 0.46 kJ/K, b. –0.46 kJ/K, c. 0.0 kJ/K, d. –0.554 kJ/K.

7.188 Water at 1.0 MPa and 540 K expands to 100 kPa in a reversible adiabatic process. What is the quality of the mixture at the end of the process? a. 0.85, b. 0.94, c. 1.0, d. 1.06.

7.189 What is the change in entropy of a 2-kg steel bar (c = 480 J/kg·K) that is heated from 300 K to 600 K? a. 332 J/K, b. −332 J/K, c. 665 J/K, d. –665 J/K.

7.190 An isolated system includes 0.5 kg of air at 300 K and 300 kPa in a closed chamber. A second chamber is initially evacuated as shown. When the partition between the two chambers is removed, the air expands to fill both volumes and the pressure drops to 100 kPa. Find the change in entropy of the air during this sudden expansion. Use c_p = 1.005 kJ/kg·K and R = 0.287 kJ/kg·K. a. 315 J/K, b. 552 J/K, c. −158 J/K, d. 158 J/K.

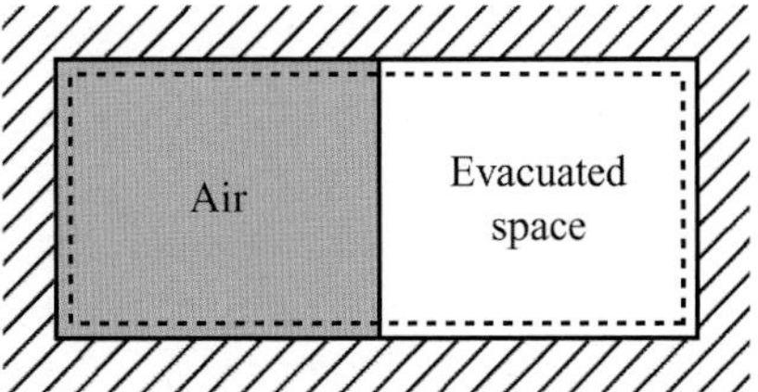

7.191 An isolated system includes 0.5 kg of steam at 500 K and 300 kPa in a closed chamber. A second chamber is initially evacuated as shown. When the partition between the two chambers is removed, the steam expands to fill both volumes and the pressure drops to 100 kPa. Find the change in entropy of the steam during this sudden expansion. a. 0.92 J/K, b. –0.92 J/K, c. 1.84 J/K, d. −1.84 J/K.

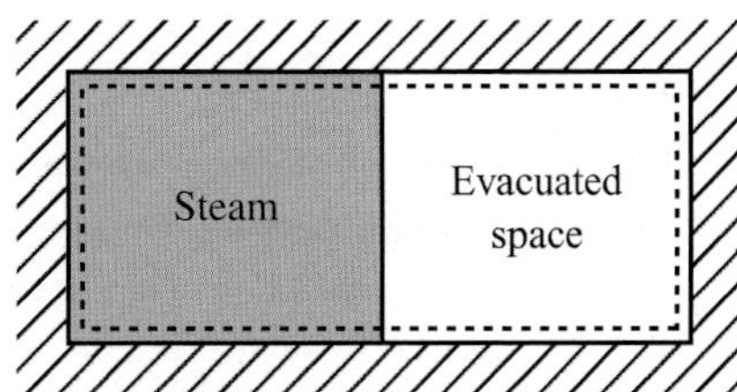

7.192 An isolated system includes 2.0 kg of saturated steam at 500 kPa in a closed chamber. A second chamber is initially evacuated as shown. When the partition between the two chambers is removed, the steam expands to fill both volumes and the pressure drops to 100 kPa. Find the change in entropy of the steam during this sudden expansion. a. 0.087 J/K, b. –0.087 J/K, c. 0.173 J/K, d. –0.173 J/K.

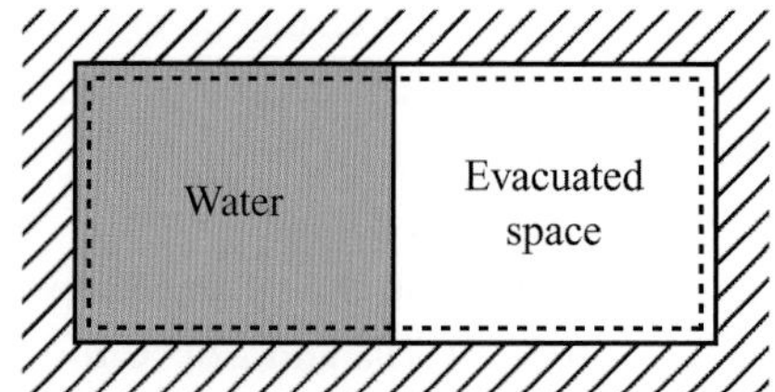

7.193 3 kg saturated steam in a rigid vessel is at 400 K. The steam is cooled to a temperature of 300 K. Determine the change in entropy of the water in the vessel. a. −6.51 kJ/K, b. −1.95 kJ/K, c. −15.4 kJ/K, d. −19.5 kJ/K.

7.194 3 kg saturated steam in a rigid vessel is at 400 K. The steam is cooled to a temperature of 300 K. Determine the heat transfer during the process. a. 7140 kJ to system, b. 7140 kJ from system, c. 2030 kJ to system, d. 2030 kJ from system.

7.195 0.5 kg of saturated water liquid at 400 kPa is expanded at constant pressure to saturated water vapor. What is the work done by the system? a. 92.3 kJ, b. 184.5 kJ, c. 0.231 kJ, d. 45.1 kJ.

7.196 0.5 kg of saturated water liquid at 400 kPa is expanded at constant pressure to saturated water vapor. What is the change in entropy of the system? a. 5.2 kJ/K, b. 2.6 kJ/K, c. 12.3 kJ/K, d. 6.8 kJ/K.

7.197 0.5 kg of saturated water liquid at 400 kPa is expanded at constant pressure to saturated water vapor. What is the change in entropy of the surroundings at 460 K? a. −2.32 kJ/K, b. −2.6 kJ/K, c. 2.32 kJ/K, d. 2.6 kJ/K.

7.198 Steam at a flow rate of 3 kg/s enters an isentropic turbine at 640K and 4 MPa. Determine the power output from the turbine if the outflow pressure is 300 kPa. a. 2.4 MW, b. 5.0 MW, c. 1.6 MW, d. 0.8 MW.

7.199 Steam at a flow rate of 0.7 kg/s at 660 K and 6 MPa enters a turbine having an isentropic efficiency of 85%. Determine the power output from the turbine if the outflow pressure is 100 kPa. a. 673 kW, b. 652 kW, c. 555 kW, d. 470 kW.

7.200 Air at a flow rate of 3 kg/s enters an isentropic compressor at 350 K and 200 kPa. Determine the power required to raise the outflow pressure to 2.0 MPa. a. 982 kW, b. 327 kW, c. 977 kW, d. 1.2 MW.

7.201 Air at a flow rate of 2 kg/s at 340 K and 150 kPa enters a compressor having an isentropic efficiency of 90%. Determine the power required to raise the outflow pressure to 2.0 MPa. a. 749 kW, b. 832 kW, c. 416 kW, d. 828 kW.

Appendix 7A Molecular Interpretation of Entropy

The purpose of this appendix is to introduce the reader to a molecular interpretation of entropy without getting bogged down in details. A rigorous development can be found in Ref. [6], for example, and a very readable elementary treatment is provided by Wark [7].

We begin by stating our final result, that the entropy S is given by the following expression:

$$S = k_B \ln W_{mp}, \tag{7A.1}$$

where k_B is the Boltzmann constant and W_{mp} is the probability of the most probable **thermodynamic macrostate,** a concept that requires some elaboration. Continuing to work backward, we define the thermodynamic probability as the number of microstates associated with a given macrostate divided by the total number of microstates. We are now at the heart of this issue: What do we mean by a microstate or by a macrostate? Answering these questions allows us to come full circle back to Eq. 7A. 1, our statistical definition of entropy.

To understand the concepts of microstates and macrostates, we consider an isolated group of N particles comprising our thermodynamic system. Furthermore, we assume that the individual particles in our system have various energy levels, ε_i, as dictated by quantum mechanics. Although any individual particle may have any particular allowed energy, the system energy must remain fixed and is constrained by

$$\sum N_i \varepsilon_i = U, \tag{7A.2}$$

where N_i is the number of particles that have the specific energy level ε_i. We continue now with a specific, but hypothetical, example in which our system contains only three particles, A, B, and C, and has a total system energy of six units, or quanta. Furthermore, we assume that the allowed energy levels are equally spaced intervals of one unit, beginning with unity. Thus, any individual particle can have energy of one, two, three, or four quanta. Clearly, energy levels above four are disallowed because, if one or more particles possessed this energy, the overall system constraint of six units would be violated. For example, if particle A possesses five units, the least possible total system energy is seven units, since the least energy particles B and C can possess is one unit $(5 + 1 + 1 = 7)$.

We now consider the number of ways in which the overall system can be configured, assuming particles A, B, and C are distinguishable. We can see that there is a total of ten ways that our three particles can be arranged while maintaining the total system energy at six units. Each one of these ten arrangements is identified as a microstate. We further note that some of these microstates are similar; if we remove the restriction that A, B, and C are distinguishable, there are six identical microstates in which one particle possesses three units of energy, another particle possesses two units of energy, and the third particle possesses one unit. The identification of a state purely by enumeration of the number of particles at each energy level without regard to the identification of the individual particles is defined as a **macrostate.** Employing this definition, we see that there are two other macrostates associated with our system: macrostate 2, in which one particle has four units of energy and two particles possess one unit, and macrostate 3, in which each particle has two units of energy. The probability of our system being found in a particular macrostate is related to the number of microstates comprising the macrostate. For our example, there are six internal arrangements that can be identified with macrostate 1, three for macrostate 2, and only one arrangement possible for macrostate 3.

Recall that one of our goals at the outset of this discussion was to understand the definition of thermodynamic probability as the number of microstates associated with a given macrostate. We come to closure on this goal by generalizing and defining the thermodynamic probability W for a system containing a total of N particles as follows:

$$W = \frac{N!}{N_1!N_2!\cdots N_M!} \tag{7A.3}$$

where N_i represents the number of particles having energies associated with the ith energy level. For our example, formal application of Eq. 7A.3 to the three macrostates yields (recognizing that $0! = 1$):

$$W_1 = \frac{3!}{1!1!1!0!} = 6,$$

$$W_2 = \frac{3!}{2!0!0!1!} = 3,$$

and

$$W_3 = \frac{3!}{0!3!0!0!} = 1.$$

If we normalize these results by the total number of microstates, a conventional concept of probability gives us:

$$P_1 = \frac{W_1}{W_1 + W_2 + W_3} = 0.6,$$

$$P_2 = \frac{W_2}{W_1 + W_2 + W_3} = 0.3,$$

and

$$P_3 = \frac{W_3}{W_1 + W_2 + W_3} = 0.1.$$

Thus we see that the probability of finding the system in macrostate 1 is 0.6, whereas the probabilities of finding the system in macrostates 2 and 3 are 0.3 and 0.1, respectively. As the number of particles increases, the probability of the most probable macrostate overwhelms that of all of the other possible macrostates.

We conclude our discussion of entropy with two observations. First, the practical evaluation of the defining relationship for entropy (Eq. 7A.1) requires a knowledge of how particles in a macroscopic system are distributed among all the allowed energy states (i.e., quantum levels) to determine the probability of the most probable thermodynamic macrostate. Various theories provide such distribution functions (e.g., Maxwell–Boltzmann, Bose–Einstein, and Fermi–Dirac statistics). Discussion of these is beyond the scope of this book and the interested reader is referred to Refs. [5,7], for example. Our second observation is that Boltzmann's definition of entropy (Eq. 7A.1) does indeed say something about "disorder" as being a favored condition.[6] By definition, the most probable macrostate is the one that has the greatest number of microstates. We can view this most probable macrostate as a state of maximum "disorder" – there is no other possible state that has as many different possible arrangements of particles.

---

[6] The idea of disorder is frequently invoked in nontechnical definitions of entropy. For example, one dictionary definition of entropy is a measure of the degree of disorder in a substance or a system.

CHAPTER 8
Thermal-Fluid Analysis of Steady-Flow Devices

LEARNING OBJECTIVES

After studying Chapter 8, you should be able to:

- Describe the use of steady-flow devices in practical applications and have a basic understanding of their operation and the important energy inputs and outputs associated with these devices.
- Apply the steady-flow forms of mass conservation, energy conservation, and as appropriate, to steady-flow devices.
- Simplify the general steady-flow expression for energy conservation to the standard forms used to describe each steady-flow device discussed in this chapter and understand the assumptions used in these simplifications.
- Plot in h–s, T–s, or other coordinates, the various thermodynamic processes associated with various ideal and actual steady-flow devices.
- Use the NIST resources to obtain thermodynamic properties needed to analyze steady-flow devices, in particular, those involving fluids existing in both liquid and gas phases.
- Describe the physical phenomena that result in loss of performance in various steady-flow devices and be able to calculate nonideal performances given isentropic efficiencies or other empirical information.

CHAPTER 8 OVERVIEW

IN THIS CHAPTER, we apply the basic conservation principles and other key concepts to analyze a number of important devices. Here we investigate typical components of more complex systems; these components include nozzles, diffusers, throttles, pumps, compressors, fans, turbines, and heat exchangers. In Chapter 9, we will combine these simple devices in more complex systems, which include steam power plants, jet engines, other power and propulsion cycles, heat pumps, refrigeration cycles, and air conditioning and humidification systems.

Chapter 8 will apply the mass and energy conservation principles and second-law considerations presented in earlier chapters to analyze steady-flow devices. Figure 8.1 illustrates the key role that Chapter 8 plays in our study. The device analysis is shown as bridging, or using, all the previous topics.

This chapter can be used simultaneously with earlier chapters; it does not stand alone and does not require complete mastery of the preceding seven chapters. Appropriate entry points have been indicated in Chapters 5 and 7. The analysis of each device treated in this chapter follows the sequence of presenting mass conservation, energy conservation, and entropy analysis, where appropriate. If the reader has yet to study isentropic efficiencies in Chapter 7, that portion of the analysis can be ignored

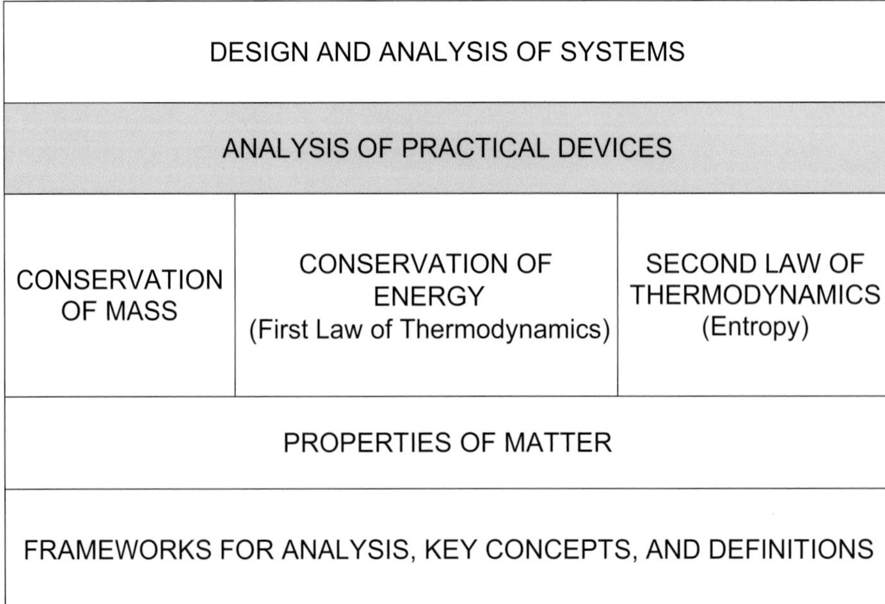

FIGURE 8.1 Hierarchical arrangements of the topics in our study of engineering thermodynamics. Chapter 8 applies the previous topics to steady-flow devices.

for the time being; however, mass and energy conservation principles are fundamental to every analysis presented here. One of the purposes of this chapter is to provide an opportunity to apply the many theoretical tools developed throughout this text.

8.1 Steady-Flow Devices

In the following sections, we investigate thermal-fluid devices that are traditionally called steady-flow devices. As indicated by this designation, the underlying assumption in analyzing all these devices is that steady state and steady flow prevail; hence, there are no time-dependent processes to consider. We will apply the mass and energy conservation equations and the entropy balance presented in earlier chapters to analyze these steady-flow devices.

As an overview, the particular steady-flow devices of interest are summarized in Table 8.1. These devices will be studied individually in this chapter and then combined together to form a thermodynamic cycle in Chapter 9. We have also considered such thermodynamic cycles in Chapter 6, where the cycles, efficiencies were defined for a steam power cycle and refrigeration/heat-pump cycle.

One can scan through the figures in Chapter 9 to see how individual steady-state devices will be combined to form a thermodynamic cycle. Figure 9.1 is a schematic of a steam power plant showing a cycle that includes pumps, turbines, and heat exchangers (boiler and condenser). The air power cycle used for the turbojet engine in Fig. 9.39 includes a diffuser, compressor, combustor, turbine, and nozzle. (The combustor will be

TABLE 8.1 Steady-Flow Devices

Device	Typical purpose(s)	Terms usually neglected in energy equation	Simplified energy conservation expression*
Nozzle	Creates high exit velocity	$\dot{Q}_{cv}, \dot{W}_{cv}, z_2 - z_1$	$h_2 - h_1 + \frac{V_2^2}{2} - \frac{V_1^2}{2} = 0$
Diffuser	Creates high outlet pressure, reduce velocity	$\dot{Q}_{cv}, \dot{W}_{cv}, z_2 - z_1$	$h_2 - h_1 + \frac{V_2^2}{2} - \frac{V_1^2}{2} = 0$
Throttle	Reduces pressure, control flow	$\dot{Q}_{cv}, \dot{W}_{cv}, +\frac{V_2^2}{2} - \frac{V_1^2}{2}, z_2 - z_1$	$h_2 - h_1 = 0$
Pump	Creates flow of a liquid, increases pressure	$\dot{Q}_{cv}, +\frac{V_2^2}{2} - \frac{V_1^2}{2}, z_2 - z_1$	$\dot{W}_{in,pump} = \dot{m}(h_2 - h_1)$
Compressor	Creates flow of a gas, increases pressure	$\dot{Q}_{cv}, \frac{V_2^2}{2} - \frac{V_1^2}{2}, z_2 - z_1$	$\dot{W}_{in,comp} = \dot{m}(h_2 - h_1)$
Fan†	Creates flow of a gas with minimal pressure change	$\dot{Q}_{cv}, z_2 - z_1$, other terms depending on open system (control volume) choice	$\dot{W}_{in,fan} = \dot{m}\left(h_2 - h_1 + \frac{V_2^2}{2} - \frac{V_1^2}{2}\right)$
Turbine	Produces power by expanding a fluid	$\dot{Q}_{cv}, \frac{V_2^2}{2} - \frac{V_1^2}{2}, z_2 - z_1$	$-\dot{W}_{out,turbine} = \dot{m}(h_2 - h_1)$
Heat exchanger or mixing chamber	Transfers energy from one fluid to another	$\dot{Q}_{cv}, \dot{W}_{cv}$, all kinetic and potential energy terms	$\sum_{\text{Inlets}} \dot{m}_i h_i = \sum_{\text{outlets}} \dot{m}_i h_i$

* The usual starting point for simplification is Eq. 8.4: $\dot{Q}_{cv,\text{net in}} = \dot{W}_{cv,\text{net out}} = \dot{m}\left(h_2 - h_1 + V_2^2/2 - V_1^2/2 + z_2 - z_1\right)$ where 1 and 2 designate inlet and outlet conditions, respectively. For devices with more than a single inlet and/or outlet stream, Eq. 5.16 is the starting point.

† There are several definitions of fans, depending upon the application. In industrial applications, the distinction between a fan and a compressor is that, for a fan, the density increase is less than 7% [1]; in propulsion applications, fans are specialized compressors with pressure ratios of less than 1.8 [2].

studied in Chapter 12.) The refrigeration cycle shown in Figure 9.58 includes a throttle, compressor, and two heat exchangers (condenser and evaporator).

FIGURE 8.2 In a subsonic nozzle, the flow area decreases in the flow direction resulting in a high-velocity exit stream.

8.2 Nozzles and Diffusers

A nozzle is a passive device having a flow area that varies in the flow direction such that the outlet velocity is higher than the inlet velocity. For subsonic flows, the area decreases in the flow direction (see Fig. 8.2). In practical applications, nozzles are purposefully used to produce a high-velocity fluid stream. Examples here include water

FIGURE 8.3 Nozzles are used in many applications. Here nozzles are used in: a pressurized water jet to clean an oil spill (image from Exxon Valdez Oil Spill Trustee Council shows the use of hot, pressurized water streams in cleaning up after the Exxon Valdez oil spill (NOAA) http://response.restoration.noaa.gov/oil-and-chemical-spills/significant-incidents/exxon-valdez-oil-spill/lessons-learned-exxon-valdez.html); a leaf blower (AdShooter / E+ / Getty Images); and a car wash (Brian Kelly / EyeEm / Getty Images).

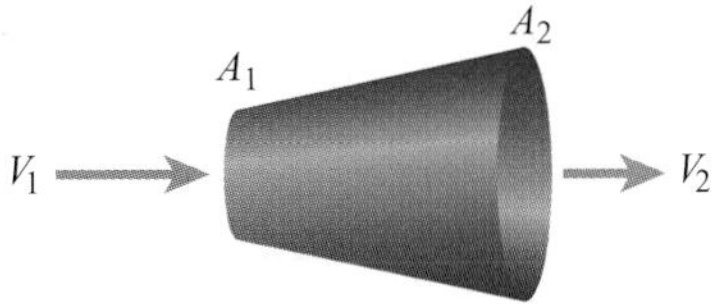

FIGURE 8.4 In a subsonic diffuser, the flow area increases in the flow direction, resulting in a higher pressure and lower velocity at the exit than at the inlet.

jets for fire fighting, boring holes in relatively soft rock with high-velocity water jets, cleaning operations using solvent or water jets, drying operations using air jets, and the use of liquid sprays in multitudinous applications. The thrust produced by rocket and jet engines depends critically on the high velocity generated by their exit nozzles.

A diffuser is a passive device that exchanges fluid kinetic energy for outlet flow work. Diffusers are used in applications where either high pressures or low velocities, or both, are desired. For subsonic flow through a diffuser, the flow area increases in the flow direction (Fig. 8.4). For such a device, the inlet velocity is greater than the outlet velocity, whereas the outlet pressure is greater than the inlet pressure. Applications of diffusers abound, as the following examples suggest. Diffusers are frequently used at the end of wind-tunnel test sections to slow the flow and, hence, decrease pumping requirements (Fig. 8.5a). Diffusers are also used in pumps and compressors to create a high outlet pressure. Figure 8.5b shows a curved diffuser section in a radial flow pump to further increase the pressure at the pump outflow. A diffuser section is used before the combustor in turbojet engines (Fig. 8.5c) to reduce the flow velocity during combustion. Diffusers have also been designed to improve the performance of wind turbines (Fig. 8.5d). The actual design and performance analysis of a diffuser is complicated by the need to consider some complex flow phenomena in this seemingly simple device. If a diffuser with a desired expansion ratio, A_2/A_1, is required to be made short, the walls of the diffuser will be expanding at a high diffuser angle. A high diffuser angle can result in a **stalled diffuser** where the outflow is unsteady and the outflow pressure is less than predicted and desired. Diffuser design is beyond the scope of this textbook but can be found in Ref. [4].

(a)

(b)

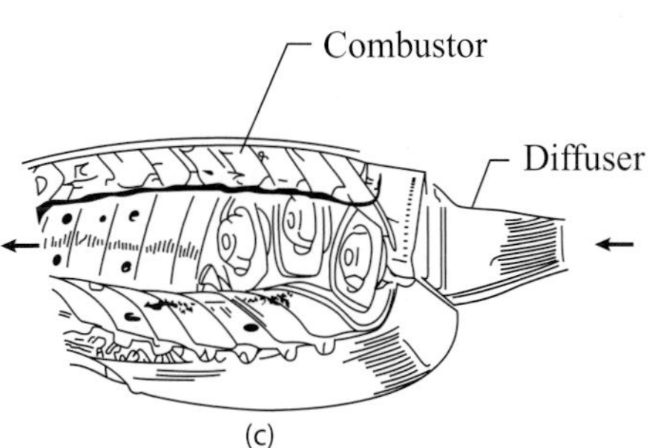

(c)

(d)

FIGURE 8.5 Various applications of diffusers include **(a)** decreasing the velocity after a wind-tunnel test section, **(b)** increasing the pressure at a pump outlet, **(c)** decreasing the velocity at the combustor inlet in a turbojet engine, and **(d)** improving the performance of a wind turbine (Vortec DAWT prototype). Images courtesy of dpa picture alliance / Alamy Stock Photo (a), S. R. Turns (b), Pratt & Whitney (c), and Arcadia Power (d).

8.2a General Analysis

We will look at a general analysis of nozzles and diffusers that applies to both incompressible and compressible flows. We will not consider the irreversibilities that can occur due to shock waves in supersonic flows.

We now apply mass and energy conservation to understand how the geometric variables (the inlet and outlet flow areas) affect the thermodynamic and flow properties.

MASS CONSERVATION

Consider the integral open system (control volume) shown in Fig. 8.6. Fluid enters the nozzle or diffuser at station 1 on the left and exits at station 2 on the right.

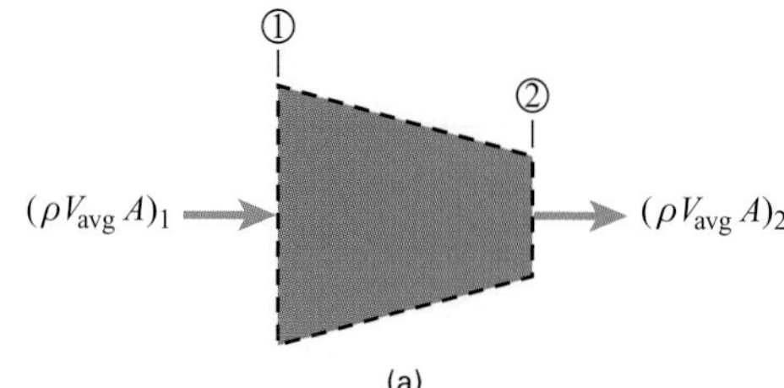

(a)

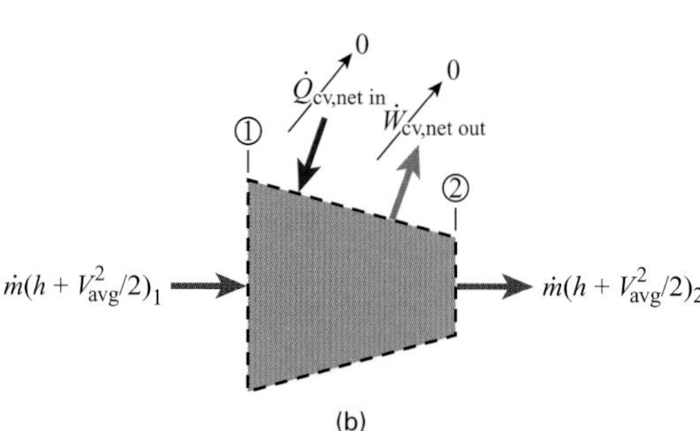

(b)

FIGURE 8.6 Open system for nozzle illustrating **(a)** mass flows and **(b)** energy flows. Steady flow and steady state are assumed.

We assume steady state and steady flow. For this open system, mass conservation expressed by Eq. 3.14a applies, so

$$\dot{m}_{in} = \dot{m}_{out}, \tag{8.1a}$$

or

$$\dot{m}_1 = \dot{m}_2. \tag{8.1b}$$

Since we are interested in the relationship between the flow velocity and flow area, we apply the definition of a flow rate (Eq. 3.11):

$$\rho_1 V_{avg,1} A_1 = \rho_2 V_{avg,2} A_2, \tag{8.2}$$

where V_{avg} is the velocity averaged over the flow cross-sectional area A (see Eq. 3.9). Throughout our analysis we will use the average flow velocity at a particular inflow or outflow area. For brevity, we will not use the subscript "avg" in our notation since the average velocity is always used. Solving for the exit velocity yields

$$V_2 = \frac{\rho_1 A_1}{\rho_2 A_2} V_1. \tag{8.3}$$

From Eq. 8.3, we see that for an incompressible flow (i.e., $\rho_1 = \rho_2 = \rho$) the exit velocity equals the inlet velocity multiplied by the area ratio A_1/A_2.

Example 8.1 Incompressible Water Nozzle

Water at 300 K enters a circular cross-section nozzle at an average velocity of 2 m/s. The inlet diameter is 25 mm and the exit diameter is 10 mm. Determine the exit velocity of the water.

Solution

Known T, V_1, D_1, D_2

Find V_2

Sketch

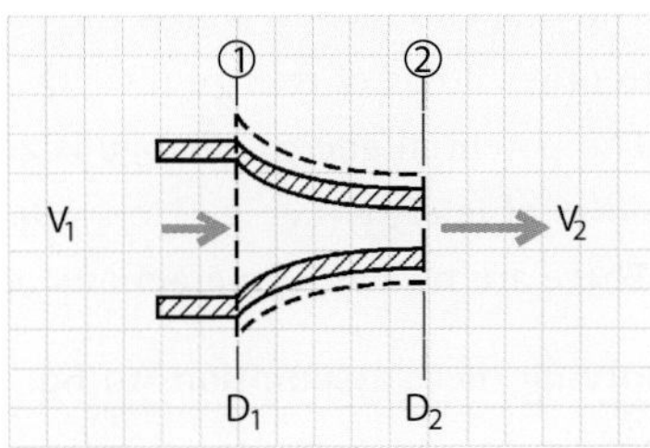

Modeling, Premises and Assumptions

i. One-dimensional flow
ii. Steady flow
iii. Incompressible fluid, $\rho_1 = \rho_2$
iv. No change in potential energy, $z_1 = z_2$

Analysis Assuming that the densities of the water at the inlet and exit are essentially equal, we apply mass conservation expressed by Eq. 8.3 directly:

$$V_2 = \frac{A_1}{A_2} V_1$$

where

$$A_1 = \pi D_1^2/4$$

and

$$A_2 = \pi D_2^2/4.$$

Thus,

$$\begin{aligned} V_2 &= \frac{D_1^2}{D_2^2} V_1 \\ &= \frac{(0.025\ \text{m})^2}{(0.010\ \text{m})^2}(2\ \text{m/s}) = 12.5\ \text{m/s}. \end{aligned}$$

Comments Since we assumed incompressible flow, no knowledge of the density at either the inlet or the exit was required to find the exit velocity.

Self-Test 8.1

Calculate the volumetric flow rate at the inlet and exit of the nozzle in Example 8.1.

(Answer: 9.817 × 10^{-4} m^3/s, 9.817 × 10^{-4} m^3/s)

ENERGY CONSERVATION

Consider again the integral open system (control volume) of Fig. 8.6. For this situation of steady state and steady flow with a single inlet and a single outlet, the conservation of energy as expressed by the first law of thermodynamics is given by Eq. 5.18, that is,

$$\dot{Q}_{\text{cv, net in}} - \dot{W}_{\text{cv, net out}} = \dot{m}\left[(h_2 - h_1) + \tfrac{1}{2}\left(V_2^2 - V_1^2\right) + g(z_2 - z_1)\right], \tag{8.4}$$

where z is the elevation. We can simplify Eq. 8.4 with the following assumptions:

- The heat interaction across the control surface is zero (adiabatic) or small compared to other flows of energy (i.e., $\dot{Q}_{\text{cv, net in}} = 0$).
- The potential energy change is zero (horizontal nozzle) or small compared to other flows of energy (i.e., $z_2 - z_1 = 0$).
- There are no work interactions other than flow work (i.e., $\dot{W}_{\text{cv, net out}} = 0$).

Applying these assumptions to Eq. 8.4 yields

$$0 - 0 = \dot{m}\left[(h_2 - h_1) + \tfrac{1}{2}\left(V_2^2 - V_1^2\right) + 0\right],$$

or

$$(h_2 - h_1) + \tfrac{1}{2}\left(V_2^2 - V_1^2\right) = 0. \tag{8.5}$$

Example 8.2 Pressure Drop in Water Nozzle

Estimate the pressure drop $P_1 - P_2$ for the nozzle and flow conditions given in Example 8.1. Assume that the flow is isothermal and neglect heat transfer.

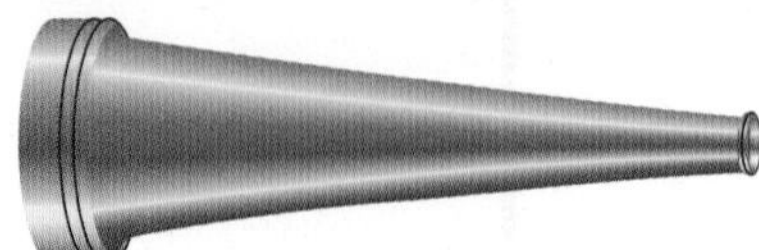

Solution

Known V_1, V_2, T

Find $P_1 - P_2$

Sketch

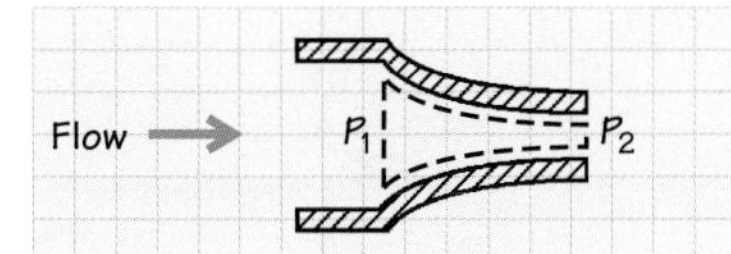

Modeling, Premises and Assumptions

i. One-dimensional flow
ii. Steady flow
iii. Incompressible fluid, $\rho_1 = \rho_2$
iv. Isothermal process, $T_1 = T_2$
v. No change in potential energy, $z_1 = z_2$

Analysis Although the pressure does not appear explicitly in Eq. 8.5, we recognize that it is buried in the enthalpy since $h \equiv u + P/\rho$. We again assume that the water is incompressible ($\rho_1 = \rho_2 = \rho$) and, furthermore, that $c_p = c_v = c$. Thus, we can rewrite Eq. 8.5 as

$$c(T_2 - T_1) + \frac{1}{\rho}(P_2 - P_1) + \frac{1}{2}\left(V_2^2 - V_1^2\right) = 0.$$

Since the flow is assumed to be isothermal, $T_2 - T_1 = 0$. Solving our previous equation for the pressure drop yields

$$P_1 - P_2 = \frac{\rho}{2}\left(V_2^2 - V_1^2\right) .$$

Approximating the density using the saturated-liquid value at 300 K (996.5 kg/m^3), we substitute numerical values as follows:

$$P_1 - P_2 = \frac{(996.5\ \mathrm{kg/m^3})}{2}\left[(12.5\ \mathrm{m/s})^2 - (2\ \mathrm{m/s})^2\right]\left[\frac{1\mathrm{N}}{\mathrm{kg{\cdot}m/s^2}}\right]\left[\frac{\mathrm{Pa}}{\mathrm{N/m^2}}\right]$$
$$= 75.86\ \mathrm{kPa}.$$

Comments This is a relatively large pressure drop compared to, say, the pressure losses resulting from friction in a 100-m run of pipe.

Example 8.3 Superheated Steam Nozzle

Superheated steam enters a nozzle at 0.13 MPa and 600 K and exits at 0.1 MPa. Heat interactions and frictional effects are both negligible. The inlet and outlet diameters of the nozzle are 20 and 10 mm, respectively. Determine the mass flow rate of the steam through the nozzle. Also find the inlet and outlet velocities.

Solution

Known T_1, P_1, P_2, D_1, D_2

Find $\dot{m}, V_1, V_2$

Sketch

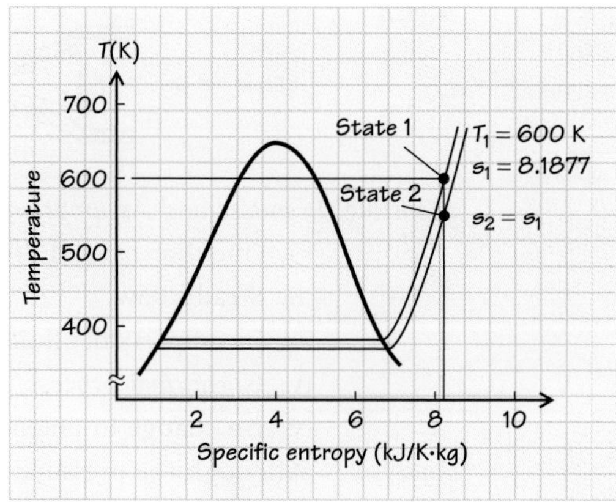

Modeling, Premises and Assumptions

i. The flow is adiabatic.
ii. The flow is frictionless.
iii. Local thermodynamic equilibrium prevails through the nozzle.
iv. $z_1 = z_2$ (i.e., no potential energy change).

Analysis Since neither the inlet nor outlet velocity is given, mass conservation alone will not allow us to find the mass flow rate. If we are able to determine the thermodynamic properties at the exit (state 2), then mass conservation (Eq. 8.2) and energy conservation (Eq. 8.5) can be combined to find the flow rate as follows: From Eq. 8.2,

$$V_2 = \frac{\dot{m}}{\rho_2 A_2} \quad \text{and} \quad V_1 = \frac{\dot{m}}{\rho_1 A_1}.$$

Rearranging the energy conservation equation (Eq. 8.5) and substituting these expressions yield

$$\begin{aligned} h_1 - h_2 &= \frac{1}{2}\left(V_2^2 - V_1^2\right) \\ &= \frac{\dot{m}^2}{2}\left[\frac{1}{(\rho_2 A_2)^2} - \frac{1}{(\rho_1 A_1)^2}\right]. \end{aligned}$$

Solving for $\dot{m}$, we obtain

$$\dot{m} = \left[\frac{2(h_1 - h_2)}{1/(\rho_2 A_2)^2 - 1/(\rho_1 A_1)^2}\right]^{1/2}.$$

To evaluate this expression, we require the properties h_1, h_2, ρ_1, and ρ_2. Since T_1 and P_1 are given, the thermodynamic state 1 is fully defined. We use the NIST WebBook to find h_1 and ρ_1 given T_1 and P_1:

$$\begin{aligned} h_1 &= 3128.1\,\text{kJ/kg}, \\ \rho_1 &= 0.47070\,\text{kg/m}^3 \end{aligned}$$

To define state 2, we assume that the flow process with negligible heat transfer and negligible friction approximates a reversible adiabatic process. With this assumption, the process is isentropic and $s_2 = s_1$. This process is shown as a vertical line on the T–s diagram in the sketch. Having a value for s_1 (= 8.1877 kJ/kg·K) defines state 2:

$$\begin{aligned} P_2 &= 0.1\,\text{MPa}, \\ s_2 &= 8.1877\,\text{kJ/kg·K}. \end{aligned}$$

We again use the NIST database to find h_2 and ρ_2 given P_2 and s_2:

$$h_2 = 3057.7 \text{ kJ/kg·K},$$
$$\rho_2 = 0.38461 \text{ kg/m}^3.$$

The inlet and exit areas are calculated from their respective diameters:

$$A_1 = \frac{\pi D_1^2}{4} = \frac{\pi(0.020\,\text{m})^2}{4} = 3.1416 \times 10^{-4}\text{m}^2,$$
$$A_2 = \frac{\pi D_2^2}{4} = \frac{\pi(0.010\,\text{m})^2}{4} = 7.854 \times 10^{-5}\text{m}^2.$$

With numerical values now available for all the quantities appearing in the expression derived for $\dot{m}$, we evaluate it as follows:

$$\dot{m} = \left[\frac{2(3128.1\times10^3 - 3057.7\times10^3)\text{J/kg}}{\left(\frac{1}{(0.38461\,\text{kg/m}^3)(7.854\times10^{-5}\,\text{m}^2)}\right)^2 - \left(\frac{1}{(0.4707\,\text{kg/m}^3)(3.1416\times10^{-4}\,\text{m}^2)}\right)^2} \times \left[\frac{\text{N·m}}{\text{J}}\right]\left[\frac{\text{kg·m/s}^2}{\text{N}}\right]\right]^{1/2} = 0.01158\,\text{kg/s}.$$

Now knowing the mass flow rate, we can calculate the velocities from mass conservation (Eq. 8.2):

$$V_1 = \frac{\dot{m}}{\rho_1 A_1} = \frac{(0.01158\text{ kg/s})}{(0.4707\text{ kg/m}^3)(3.1416 \times 10^{-4}\text{m}^2)} = 78.3\text{ m/s},$$
$$V_2 = \frac{\dot{m}}{\rho_2 A_2} = \frac{(0.01158\text{ kg/s})}{(0.38461\text{ kg/m}^3)(7.854 \times 10^{-5}\text{m}^2)} = 38.3\text{ m/s}.$$

Comments This problem brings together many thermal-fluid concepts: mass and energy conservation, the second law of thermodynamics, and thermodynamic state relations.

Self-Test 8.2

Calculate the volumetric flow rate at the inlet and exit of the nozzle in Example 8.3.

(Answer: $2.46 \times 10^{-2}\,m^3/s$, $3.01 \times 10^{-2}\,m^3/s$)

8.2b Incompressible Flow

The simplification of the general conservation relationships for incompressible nozzle and diffuser flows was illustrated in Examples 8.1 and 8.2. Application of mass conservation to an incompressible flow (see Example 8.1) is straightforward and needs no further discussion. We could say the same thing for energy conservation; however, we wish to explore some subtleties that we ignored in Example 8.2, to develop a more complete understanding of nozzle performance.

We begin with the general energy conservation expression (Eq. 8.5):

$$h_2 - h_1 + \tfrac{1}{2}\left(V_2^2 - V_1^2\right) = 0.$$

Applying the definition of enthalpy yields

$$(u_2 + P_2/\rho) - (u_1 + P_1/\rho) + \tfrac{1}{2}\left(V_2^2 - V_1^2\right) = 0,$$

or

$$u_2 - u_1 + \frac{P_2 - P_1}{\rho} + \tfrac{1}{2}\left(V_2^2 - V_1^2\right) = 0.$$

The internal energy difference is related to the fluid temperature by using the calorific equation of state for either an ideal gas (Eq. 2.29e) or an incompressible liquid as

$$u_2 - u_1 = c_v(T_2 - T_1),$$

where c_v is an appropriate average value of the constant-volume specific heat for the temperature range T_1 to T_2. With this substitution, energy conservation (the first law of thermodynamics) becomes

$$c_v(T_2 - T_1) + \frac{P_2 - P_1}{\rho} + \tfrac{1}{2}\left(V_2^2 - V_1^2\right) = 0. \tag{8.6}$$

Although we discarded the term $c_v(T_2 - T_1)$ in Example 8.2, this term appears as a result of the irreversible conversion of mechanical energy (flow work and kinetic energy) to thermal energy (internal energy). Thus, if the nozzle flow is not frictionless, as all real flows must be, then $c_v\,(T_2 - T_1)$ is not zero. It will be small, however, in many incompressible flow applications.

See Section 7.5c of Chapter 7.

8.2c Nozzle Efficiency

As discussed in Chapter 7, a frequently used measure of nozzle performance for high-speed flows is the isentropic efficiency (Eq. 7.36) given by

$$\eta_{\text{isen, n}} \equiv \frac{\dot{KE}_{\text{act}}}{\dot{KE}_{\text{isen}}}. \tag{8.7a}$$

In this definition, $\dot{KE}_{\text{act}}$ is the kinetic energy flow at the nozzle outlet for the actual process, and $\dot{KE}_{\text{isen}}$ is the theoretical kinetic energy flow at the nozzle outlet for a reversible adiabatic (isentropic) process starting at the same inlet state and ending at the same pressure as the actual process. Treating the flow as essentially one dimensional, we can express the isentropic efficiency as

$$\eta_{\text{isen, n}} = \frac{\left(\dot{m}V_2^2\right)_{\text{act}}}{\left(\dot{m}V_2^2\right)_{\text{isen}}} = \frac{\left(V_2^2\right)_{\text{act}}}{\left(V_2^2\right)_{\text{isen}}}. \tag{8.7b}$$

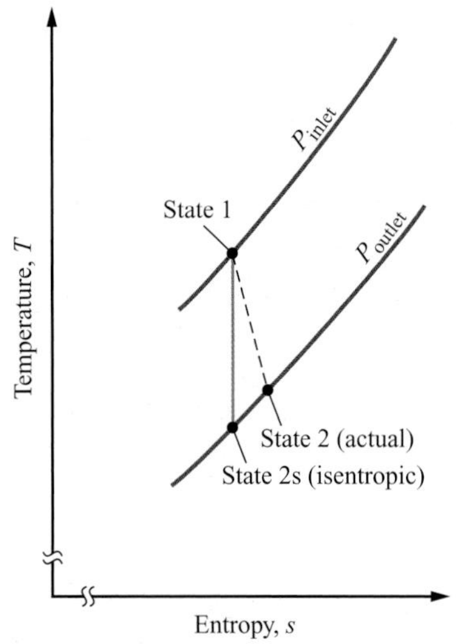

Temperature, T

State 1

P_{inlet}

P_{outlet}

State 2 (actual)

State 2s (isentropic)

Entropy, s

FIGURE 8.7 To define the nozzle efficiency, an ideal outlet state is defined by an isentropic expansion from the actual inlet state to the same outlet pressure as that for the actual process.

Figure 8.7 illustrates the actual and ideal processes in T–s coordinates. Because of irreversibilities, primarily due to fluid friction, the entropy of the outlet state is greater than the entropy at the inlet state.

Example 8.4 Turbojet Exit Nozzle

The exit nozzle of a turbojet engine expands the flow of exhaust products from 170 to 45 kPa. The products enter the nozzle at 800 K with a velocity of 225 m/s. The isentropic efficiency of the nozzle is 97%. Determine the jet exit velocity and the temperature of the products at the nozzle exit. Also determine the entropy change from inlet to outlet.

Solution

Given P_1, T_1, V_1, P_2, $\eta_{\text{isen, n}}$

Find V_2, T_2, $s_2 - s_1$

Sketch See Fig. 8.6.

(Credit: US Navy.)

Modeling, Premises and Assumptions

i. Steady, one-dimensional flow.
ii. The flow is adiabatic and isentropic (no shock).
iii. Local thermodynamic equilibrium prevails through the nozzle.
iv. The combustion products can be treated as an ideal gas with their properties approximated using those of air.
v. Constant values of c_p and γ evaluated at the average temperature, $(T_1 + T_2)/2$, can be used.

Analysis We begin by determining the temperature at the isentropic state 2s (Fig. 8.7) using the ideal-gas relationship for an isentropic process (Eq. 7.13a):

$$\frac{T_{2s}}{T_1} = \left(\frac{P_2}{T_1}\right)^{(\gamma-1/\gamma)}.$$

Solving for T_{2s} and evaluating using $\gamma = 1.37$ (calculated from $c_p\ (T_{\text{avg}}) \equiv c_{p,\text{avg}} =$ 1070 J/kg·K for $\text{T}_{\text{avg}} \approx 675$ K in Table C.3) yields

$$T_{2s} = T_1\left(\frac{P_2}{T_1}\right)^{(\gamma-1/\gamma)}$$

$$= 800\left(\frac{45}{170}\right)^{0.37/1.37} = 558.7 \text{ K}.$$

We now obtain the outlet velocity for the ideal (adiabatic and reversible) process by combining energy conservation (Eq. 8.5) with the ideal-gas calorific equation of state relating the enthalpy and temperature (Eq. 2.31e), as follows:

$$h_1 - h_{2s} = \tfrac{1}{2}\left(V_{2s}^2 - V_1^2\right)$$

and

$$c_{p,\mathrm{avg}}(T_1 - T_{2s}) = \tfrac{1}{2}\left(V_{2s}^2 - V_1^2\right) .$$

Solving for V_{2s} yields

$$\begin{aligned} V_{2s} &= \left[2c_{p,\mathrm{avg}}(T_1 - T_{2s}) + V_1^2\right]^{1/2} \\ &= \left[2\left(1070\frac{\mathrm{J}}{\mathrm{kg\cdot K}}\right)(800 - 558.7)\mathrm{K}\left[\frac{\mathrm{N\cdot m}}{1\mathrm{J}}\right]\left[\frac{\mathrm{kg\cdot m/s^2}}{1\mathrm{N}}\right] + (225\ \mathrm{m/s})^2\right]^{1/2} \\ &= 753\ \mathrm{m/s}. \end{aligned}$$

To find the actual exit velocity, we apply the definition of nozzle efficiency (Eq. 8.7b):

$$\begin{aligned} V_2 &= \left(\eta_{\mathrm{isen,n}} V_{2s}^2\right)^{1/2} \\ &= \left[0.97(753\ \mathrm{m/s})^2\right]^{1/2} = 741.6\ \mathrm{m/s}. \end{aligned}$$

To find the outlet temperature T_2, we apply energy conservation (Eq. 8.5) to the actual process, that is,

$$h_1 - h_2 = c_{p,\mathrm{avg}}(T_1 - T_2) = \tfrac{1}{2}\left(V_2^2 - V_1^2\right).$$

Solving for T_2 yields

$$\begin{aligned} T_2 &= T_1 - \frac{1}{2c_{p,\mathrm{avg}}}\left(V_2^2 - V_1^2\right) \\ &= 800\,\mathrm{K} - \frac{1}{2\left(1070\dfrac{\mathrm{J}}{\mathrm{kg\cdot K}}\right)\left[\dfrac{\mathrm{N\cdot m}}{1\,\mathrm{N}}\right]\left[\dfrac{\mathrm{kg\cdot m/s^2}}{1\,\mathrm{N}}\right]}\left(741.6^2 - 225^2\right)\mathrm{m^2/s^2} \\ &= 566.7\,\mathrm{K}. \end{aligned}$$

The verification of units is left as an exercise for the reader. To calculate the entropy change, we use the ideal-gas relationship from Chapter 7, Eq. 7.12a:

$$\begin{aligned} s_2 - s_1 &= c_{p,\mathrm{avg}} \ln\frac{T_2}{T_1} - R\ln\frac{P_2}{P_1} \\ &= \left(1070\,\frac{\mathrm{J}}{\mathrm{kg\cdot K}}\right)\ln\left(\frac{566.7\ \mathrm{K}}{800\ \mathrm{K}}\right) - \left(287\,\frac{\mathrm{J}}{\mathrm{kg\cdot K}}\right)\ln\left(\frac{45\ \mathrm{kPa}}{170\ \mathrm{kPa}}\right) \\ &= 12.54\ \mathrm{J/kg\cdot K}. \end{aligned}$$

Comments We see that the actual outlet temperature is 8 K (= 566.7 − 558.7 K) higher than the isentropic-process value. This increase in temperature results from the irreversible conversion of kinetic energy to thermal energy by frictional effects. The positive value for the entropy change $s_2 - s_1$ is consistent with this. Note also that our use of $T_{\mathrm{avg}} = 675$ K was reasonable since average temperatures based on calculations of T_{2s} and T_2 are quite close to this value (i.e., 679 K and 683 K, respectively).

Self-Test 8.3

Measurements show that the exit pressure and temperature of the nozzle in Example 8.4 are actually 50 kPa and 595 K, respectively. Recalculate the isentropic efficiency for this nozzle.

(Answer: 0.918)

8.3 Throttles

A throttling process is used to decrease the pressure and/or control the flow rate of a flowing fluid. Any narrow constriction in a flow can be effective as a throttle. In practice, valves, porous plugs, and fine-bore capillary tubes are all used (see Fig. 8.8). Throttling devices are essential components in refrigeration systems, which we will investigate later in Chapter 9. Throttles are also used to control the load in spark-ignition engines as shown in Fig. 8.9. When idling, the intake manifold pressure is reduced by the throttle from close-to-atmospheric pressure to about 0.4 atm, for example.

FIGURE 8.8 Globe valve (left) and photograph of a porous plug (right). Images courtesy Mike Kemp / Getty Images (left) and S. R. Turns (right).

FIGURE 8.9 Throttle body for spark-ignition engine (Lyroky / Alamy Stock Photo).

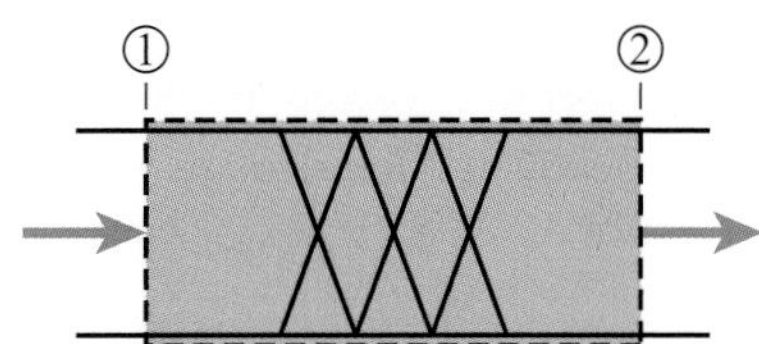

FIGURE 8.10 Open system for analysis of steady-flow throttling process. The throttling device proper (a porous plug, a valve, or other restriction) is schematically indicated by the Xs.

8.3a Analysis

To analyze the steady-flow throttling process, consider the open system (control volume) shown in Fig. 8.10. The upstream and downstream boundaries are chosen so that they are sufficiently far away from any localized high-velocity regions created by the throttling device. We now apply the fundamental conservation principles.

MASS CONSERVATION

Mass conservation (Eqs. 3.14 and 3.11) is straightforwardly expressed as

$$\dot{m}_1 = \dot{m}_2$$

or

$$\rho_1 V_1 A_1 = \rho_2 V_2 A_2.$$

ENERGY CONSERVATION

To apply energy conservation, we note, first, that there is no work interaction other than flow work since no power is delivered to or from the open system (control volume) by a shaft or any other means. Other simplifications result if we assume the following:

- Any heat interaction across the control surface is negligible (i.e., $\dot{Q}_{\text{cv, net in}} = 0$).
- The potential energy change from inlet to outlet is negligible (i.e., $z_2 - z_1 = 0$).
- The kinetic energy of the flow, or the change in the kinetic energy from inlet to outlet, is negligible [i.e., $\left(V_2^2 - V_1^2\right)/2 \approx 0$]. (See Table 8.2.)

Applying these facts and assumptions to the steady-flow energy equation (Eq. 5.15a),

$$\dot{Q}_{\text{cv, net in}} + \dot{W}_{\text{cv, net in}} = \dot{m}\left[(h_2 - h_1) + \tfrac{1}{2}\left(V_2^2 - V_1^2\right) + g(z_1 - z_2)\right],$$

TABLE 8.2 Conversion of Kinetic Energy to Thermal Energy for a Flow of Nitrogen

Velocity (m/s)	Specific kinetic energy (J/kg)	Equivalent temperature change* (K)
1	0.5	0.00048
10	50	0.048
20	200	0.19
50	1250	1.20
100	5000	4.8

$^*\Delta T_{\text{equiv}} \equiv \Delta h/c_p = V^2/2c_p$, where $c_p = c_{p,\text{N}_2}(300\,\text{K}) = 1041\,\text{J/Kg·K}$

yields

$$0 + 0 = \dot{m}[(h_2 - h_1) + 0 + 0],$$

or

$$h_2 - h_1 = 0. \tag{8.8}$$

For an ideal gas, the enthalpy is only a function of temperature. Since the enthalpy does not change across a throttle, the flow temperature also will not change across a throttle for an ideal gas:

$$\begin{aligned} h_2 - h_1 &= 0, \\ c_p(T_2 - T_1) &= 0, \\ T_1 &= T_2. \end{aligned} \tag{8.9}$$

8.3b Applications

The following examples illustrate some practical applications of throttling processes.

Example 8.5 Throttling Calorimeter

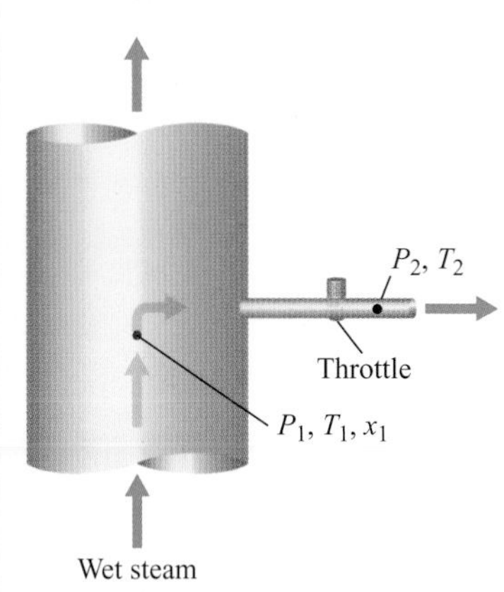

FIGURE 8.11 A throttling calorimeter is used to measure the quality of wet steam.

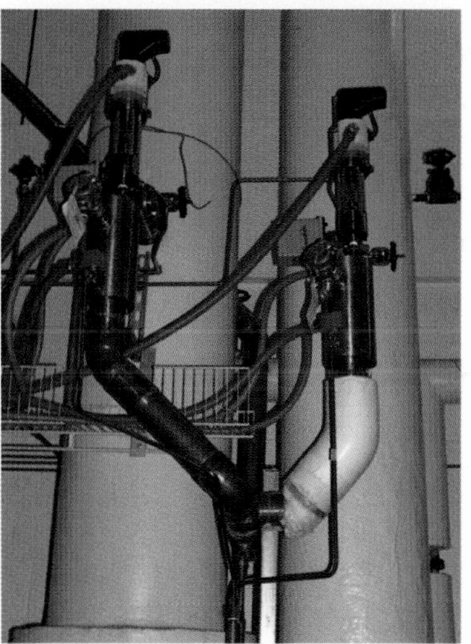

Throttling calorimeter measures steam quality. (Photograph courtesy of Croll Reynolds Co., Inc.)

A throttling calorimeter is a device used to measure the quality of wet steam. As shown in Fig. 8.11, a saturated liquid–vapor mixture at a known saturation temperature flows through a probe and is then throttled to a lower pressure such that the mixture becomes a superheated vapor. The downstream temperature and pressure are then measured. From these data (P_1, P_2, T_2) the quality x_1 can be determined.

Consider a throttling calorimeter used to measure the quality of steam entering a reheater at 3.0 MPa. The temperature and pressure measured in the calorimeter after the throttling process are 420 K and 0.1 MPa, respectively. Determine the quality of the mixture entering the reheater. Also determine the entropy change associated with the steam in going from state 1 to state 2.

Solution

Known P_1, P_2, T_2

Find x_1, $s_2 - s_1$

Sketch

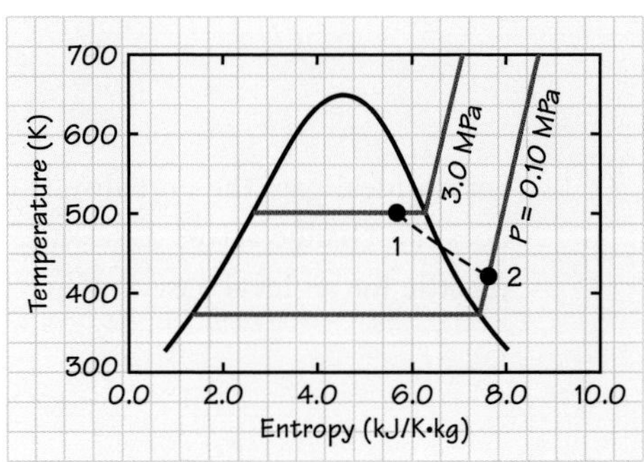

Modeling, Premises and Assumptions

i. Steady, one-dimensional flow through calorimeter
ii. Adiabatic calorimeter
iii. Negligible kinetic and potential energy changes

Analysis We choose an open system boundary that cuts across the entrance to the calorimeter tube, follows the exterior surface of the tube, and cuts across the tube where P_2 and T_2 are measured (see Fig. 8.11). Applying conservation of energy to this open system yields

$$h_1 = h_2.$$

Since we know two properties, we can determine h_2; that is, $h_2 = h_2(T_2, P_2)$. Using the superheated steam table B.3d, we find

$$h_2(420\text{ K}, 0.1\text{ MPa}) = 2770.3\text{ kJ/kg}$$

and

$$s_2(420\text{ K}, 0.1\text{ MPa}) = 7.5999\text{ kJ/kg}\cdot\text{K}.$$

The saturation properties at state 1 ($P_1 = P_{sat} = 3.0$ MPa) are determined from Table D.2:

$$h_{f,1} = 1008.3\text{ kJ/kg}, \quad s_{f,1} = 2.6455\text{ kJ/kg}\cdot\text{K},$$
$$h_{g,1} = 2803.2\text{ kJ/kg}. \quad s_{g,1} = 6.1856\text{ kJ/kg}\cdot\text{K}.$$

Using $h_1 = h_2$, we find the quality x_1 by applying its definition (Eq. 2.37d), that is,

$$x_1 = \frac{h_1 - h_{f,1}}{h_{g,1} - h_{f,1}}$$
$$= \frac{2770.3 - 1008.3}{2803.2 - 1008.3} = 0.9817.$$

We now use the quality to find s_1:

$$s_1 = (1 - x_1)s_{f,1} + x_1 s_{g,1}$$
$$= (1 - 0.9817)(2.6455) + 0.9817(6.1856)\text{ kJ/kg}\cdot\text{K}$$
$$s_1 = 6.1206\text{ kJ/kg}\cdot\text{K}.$$

Thus, the change in entropy is

$$\Delta s = s_2 - s_1 = 7.5999 - 6.1206 \text{ kJ/kg·K} = 1.4793\text{kJ/kg·K},$$

which, as expected, is a positive quantity for this highly irreversible process.

Comment Throttling calorimeters are used to measure the approximate quality of steam in geothermal wells for conditions where $x < 0.995$. Notice that the surroundings are not included in the second-law analysis. Since there is no heat transfer to or from the surroundings, there will be no change in entropy for the surroundings.

Self-Test 8.4

Refrigerant (R-134a) in an air-conditioning unit is throttled through several capillary tubes from a saturated-liquid condition at 0.8 MPa to a final pressure of 0.1 MPa. Determine the final quality x.

(Answer: 0.359)

Example 8.6 Pressurized Gas Flow

Nitrogen is flowing in a pipe at 170 atm and room temperature (298 K). A pressure regulator (throttle) is used to reduce the pressure to 2 atm in the pipe. Assuming the process is adiabatic, determine the temperature of the N_2 downstream of the regulator and the change in specific entropy. Solve this problem in two ways:

a. Use real-gas properties from the NIST resources.
b. Compare this result to that obtained assuming ideal-gas behavior.

Solution

Known N_2, P_1, P_2, T_1

Find T_2

Sketch

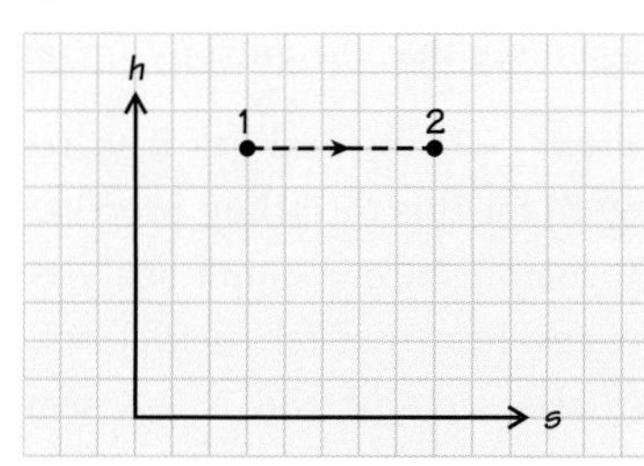

Modeling, Premises and Assumptions

i. Steady, one-dimensional flow
ii. Adiabatic process
iii. Negligible changes in kinetic and potential energies

Analysis We can use directly the open system sketched in Fig. 8.10, where station 1 is upstream of the regulator and station 2 downstream. With the assumption that heat interactions and kinetic energy changes are negligible, energy conservation is given by Eq. 8.8, so

$$h_2 = h_1.$$

Since both the temperature and pressure are known at state 1, we can find the enthalpy through the calorific equation of state for N_2. The NIST WebBook provides the needed information:

$$h_1(170\text{ atm}, 298\text{ K}) = 279.40\text{ kJ/kg}, \quad s_1(170\text{ atm}, 298\text{ K}) = 5.2104\text{ kJ/kg}.$$

From energy conservation, we know that

$$h_2\,(2\text{ atm}, T_2) = h_1 = 279.40\text{ kJ/kg}.$$

We employ the NIST database to find the state at 2 atm that has a specific enthalpy of 279.40 kJ/kg. Searching the database with a fixed pressure and incrementing the temperature, we quickly converge to the state-2 temperature:

$$T_2 = 269.74\text{ K}, \quad s_2(2\text{ atm}, 269.74\text{ K}) = 6.5244\text{ kJ/kg}.$$
$$\Delta s = s_2 - s_1 = (6.5244 - 5.2104)\text{ kJ/kg·K} = 1.3140\text{ kJ/kg·K}.$$

We again see that the entropy increases through the throttle.

For an ideal gas, the enthalpy is a function of temperature alone and is independent of pressure; thus,

$$h_2(T_2) = h(T_1)$$

and

$$T_2 = T_1(\text{ideal gas}) = 298\text{ K}.$$

Using the molecular weight from Table E.1, the gas constant can be found:

$$R = R_u/\mathcal{M} = (8314.472\text{ J/kmol·K})/(28.01\text{ kg/kmol}) = 296.83\text{ J/kg·K}$$

The change in entropy can be found using Eq. 7.12:

$$s_2 - s_1 = c_{p,\text{avg}} \ln\frac{T_2}{T_1} - R\ln\frac{P_2}{P_1},$$
$$s_2 - s_1 = (1040\text{J/kg·K})\ln\left(\frac{298\text{ K}}{298\text{ K}}\right) - (296.83\text{J/kg·K})\ln\left(\frac{2\text{ atm}}{170\text{ atm}}\right) = 1.3187\text{ kJ/kg·K}.$$

We again see that the entropy increases through the throttle.

Comments Using real-gas data results in a temperature drop of 28.3 K (= 298 − 269.69 K) for this throttling process. As a result, we expect that some heat interaction might complicate our simple analysis. A more detailed analysis would have to consider flow rates, heat-transfer coefficients, and the specific geometry.

We also see that the 28.3-K temperature drop is not captured in the ideal-gas analysis. From this, we generalize that care must be exercised in invoking the ideal-gas approximation in throttling processes. The **Joule–Thomson coefficient**, defined by

$$\mu_J \equiv \left(\frac{\partial T}{\partial P}\right)_h,$$

captures the effect of pressure on the temperature change for a throttling process. This thermodynamic property can be positive, negative, or zero, causing the temperature downstream of a throttling device to be less than, greater than, or the same as the upstream temperature, respectively. For an ideal gas, the Joule–Thomson coefficient is zero. Joule–Thomson coefficients for selected substances are available from the NIST resources.

8.4 Pumps, Compressors, and Fans

In this class of steady-flow devices, a power input is used to create a flow and/or increase the pressure of a fluid. You are likely to have some familiarity with these relatively common devices.

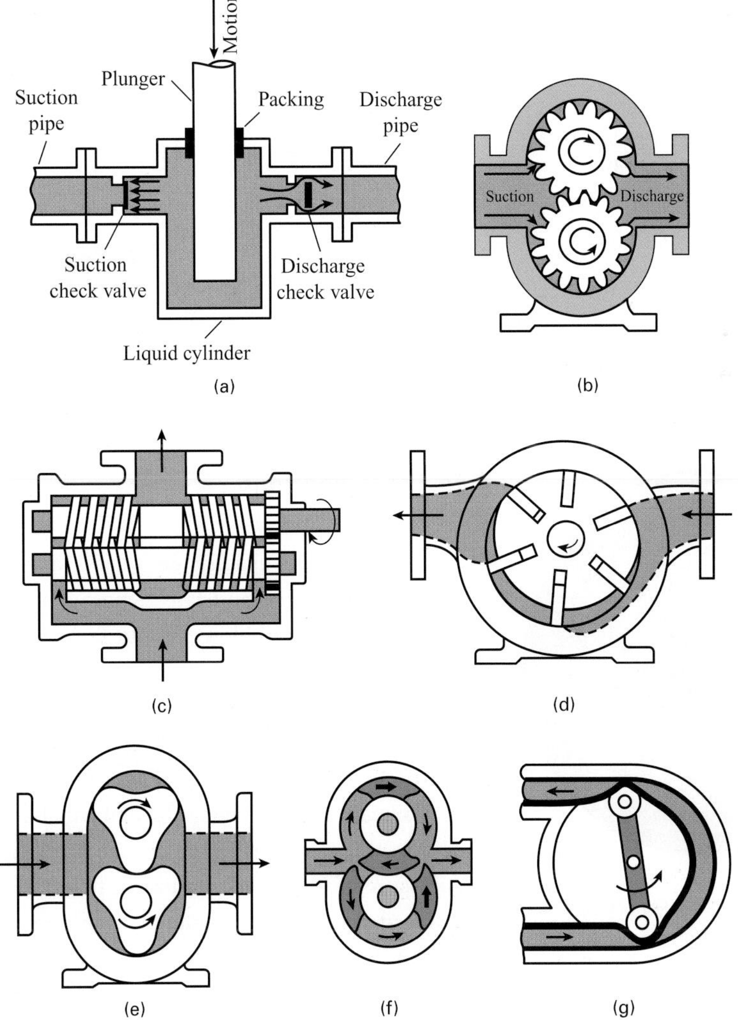

FIGURE 8.12 Various types of positive-displacement pumps: **(a)** reciprocating piston or plunger pump, **(b)** external gear pump, **(c)** double-screw pump, **(d)** sliding vane pump, **(e)** three-lobed pump, **(f)** double circumferential piston pump, and **(g)** flexible tube peristaltic pump. (Adapted from Ref. [3] with permission: Fluid Mechanics, 5th edn, Frank M. White. © McGraw-Hill Education.)

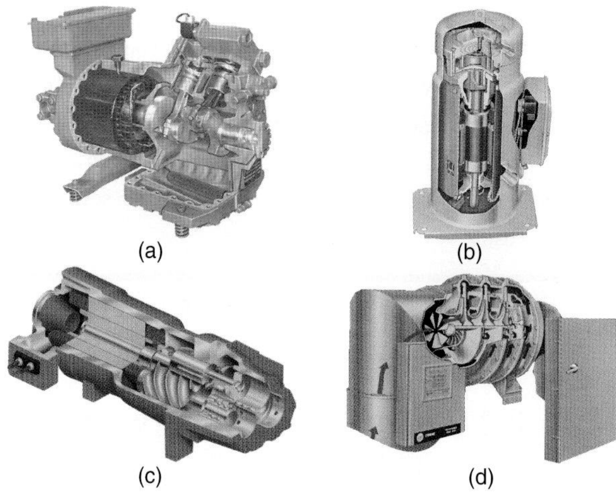

FIGURE 8.13 Refrigeration compressors. **(a)** Reciprocating compressor, **(b)** scroll compressor, **(c)** helical compressor, **(d)** centrifugal compressor. (Figures courtesy of Trane.)

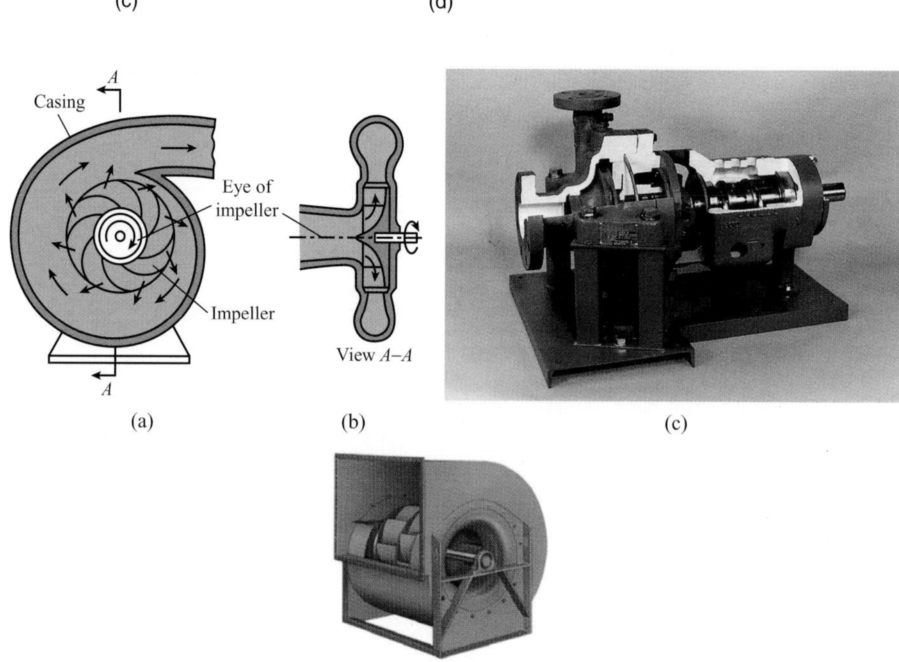

FIGURE 8.14 (a) and **(b)** Centrifugal pump schematic. Fluid enters the eye of the impeller, momentum is then imparted to the fluid by the impeller blades, and the fluid exits radially. The exit scroll functions as a diffuser. (Schematic adapted from Ref. [5] with permission.) **(c)** centrifugal pump cutaway view. (Photograph courtesy of Design Assistance Corporation (DAC).) **(d)** Centrifugal blower. (Photograph courtesy of New York Blower Company.)

FIGURE 8.15 Streamlines show fluid entering a centrifugal pump impeller in the axial direction and exiting in the radial direction. Shown in false color, the surface pressure distribution illustrates the increase in pressure from the inlet (blue) to the exit (red). (Courtesy of ARL (Penn State) Computational Mechanics.)

8.4a Classifications

The word "pump" usually designates a device that deals with liquids, whereas compressors, blowers, and fans denote devices that deal with gases. The distinction between compressors, blowers, and fans depends on the specific field of application (see the footnote to Table 8.1). In industrial applications, the distinction between a fan and a compressor is that, for a fan, the density increase is less than 7% [1]; in propulsion applications, fans are specialized compressors with pressure ratios of less than 1.8 [2]. The common use of the word "fan" denotes a device that creates a flow with minimal pressure change. A fan capable of delivering higher outlet pressures is sometimes referred to as a blower. The primary objective of a compressor is to increase the pressure of the flow from inflow to outflow.

There are many types of pumps, compressors, and fans; however, nearly all of these fall within two general categories: *positive-displacement* devices and *dynamic* devices. As the name suggests, a positive-displacement pump relies on a physical displacement (pushing) of the fluid by the device. A piston–cylinder arrangement in which a fluid enters through an open intake valve or port when the piston descends and is pushed out through an open exhaust valve or port when the piston ascends is a positive-displacement

FIGURE 8.16 Six fans drive the flow through the 80 × 120-ft^2 wind tunnel at NASA Ames Research Center. For a sense of scale, note the people in the photograph. (Photograph courtesy of NASA.)

device. In fact, a substantial part of the power produced by a throttled spark-ignition engine is used to pump the air and combustion products through the engine. Figure 8.12 illustrates several types of positive-displacement pumps. Figure 8.13 a–c shows several types of displacement compressors used in the refrigeration industry.

Dynamic devices rely on an exchange of momentum by spinning blades or vanes (rotodynamic devices) or on an exchange of momentum with a high-speed fluid stream (ejector or jet pump devices). A dynamic pump or compressor can be designed with an axial or radial outflow direction.

The centrifugal compressor in Fig. 8.13d is an example of a dynamic device used in the refrigeration industry. Figure 8.14 illustrates a centrifugal pump and a centrifugal air blower (a rotodynamic device) in which the fluid enters the pump axially through the eye of the impeller, follows the curved impeller passages, and exits radially. The exit scroll surrounding the impeller has an increasing area to convert the kinetic energy of the flow to an increase in pressure. Since pumps are typically used to increase the pressure of the flow, the flow velocity and therefore the kinetic energy are similar at the inflow and outflow of the pump. Figure 8.15 shows the pressure distribution on the surfaces of an impeller determined from a computer simulation of the pump operation. Centrifugal pumps are quite common and are used in many applications such as municipal water and furnace air blowers.

Figures 8.16 and 8.17 illustrate axial flow fans and compressors where the air flows in the same direction as the shaft. Fans can be small, as is the fan in a desktop computer or a room air fan. Fans are also used in large-scale applications such as the 80×120-foot wind tunnel at NASA Ames, shown in Fig. 8.16. (Note the people in the photograph for scale!) The propeller of an airplane and the first stage of a turbofan jet engine are also examples of axial flow fans. The primary objective of a fan is to increase the speed of the flow, causing a significant increase in the kinetic energy. As indicated in Table 8.1, the change in kinetic energy will be included in the energy equation when analysing fans. In contrast, the primary objective of an axial flow compressor, Fig. 8.17, is to increase the pressure of the flow from inflow to outflow; the change in kinetic energy through a compressor can often be neglected. Axial flow compressors are used in jet engines, natural gas power plants, and refrigeration systems. Axial flow compressors typically have multiple stages of rotating rotor blades with a row of stationary blades, called stators, between each

FIGURE 8.17 Multistage compressor blading for an aircraft engine. (Photograph courtesy of Pratt and Whitney.)

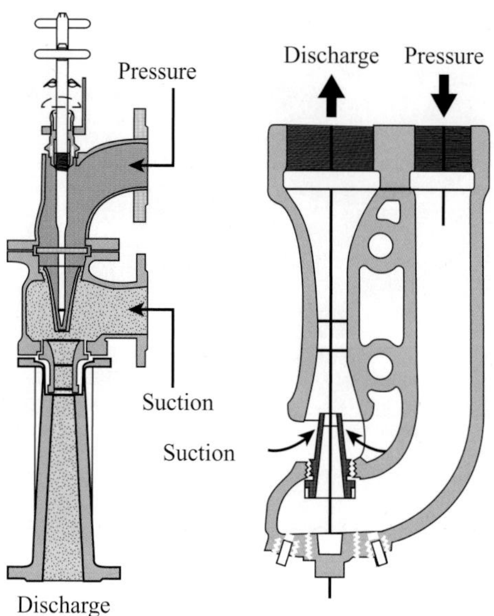

FIGURE 8.18 Cross-sectional views of eductors (i.e., liquid jet pumps). The eductor configuration shown on the right is used to pump sand and mud. (Adapted from Ref. [7] with permission: Pump Handbook, 4th edn, Karassik et al. © McGraw-Hill Education.)

rotor row. The stators reduce the swirl in the flow, thereby decreasing the kinetic energy and increasing the pressure.

The ejector (or eductor) pumps shown in Fig. 8.18 are submerged dynamic pumps that have no moving parts. The ejector pump has an inflow stream that travels through a nozzle section. The high-velocity flow at the exit of the nozzle is at low pressure and entrains the surrounding flow. This combined outflow is then expanded in a diffuser section to decrease the velocity and increase the pressure. Ejector pumps can be used to pump water, mud slurries, or sewage.

In the following, we will use the term "pumping device" to refer to pumps, compressors, and fans as a class of devices. When a distinction among these individual devices is required, the specific device name will be used.

8.4b Analysis

OPEN SYTEM (CONTROL VOLUME) CHOICE

See Eq. 4.7b in Chapter 4 for the general definition of open system power.

In dealing with pumping devices, several choices of open system (control volume) are possible. The particular choice depends on either the information provided as a given or on the information sought from the analysis. Three typical open-system choices are illustrated in Fig. 8.19. Let us examine these to see how they differ and the consequences of choosing one over another.

See Eq. 4.10 for the definition of shaft power.

Open system I contains only the fluid; no parts of the pumping device proper are included within the volume. The control surface follows the interior surfaces of the pumping device, as suggested in the sketch (Fig. 8.19a). The open system work $(\dot{W}_{cv})$ associated with this choice is the power delivered by the solid surface of the pumping device directly to the fluid and follows from integration of the product of the local force and velocity at the fluid–surface interface over the entire control surface. Since the force is distributed due to the pressure on the pump blades, and different parts of the blades are moving at different speeds, the work term is difficult to determine. Open system I might be used when one wants to isolate inefficiencies in the fluid itself from those produced in the pumping device proper. For example, losses from friction in the bearings of the pumping device would be *excluded* in an analysis involving open system I.

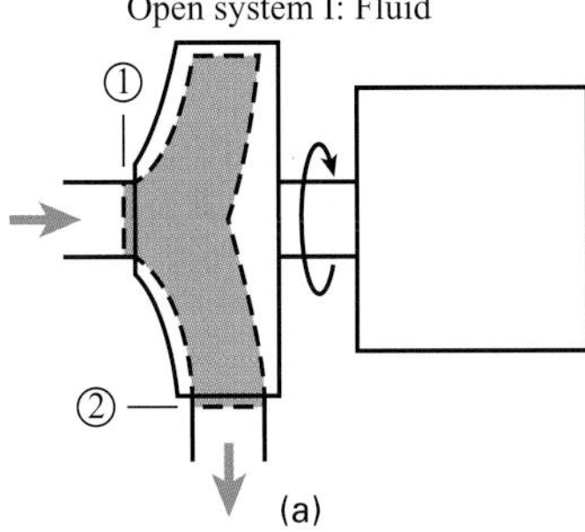

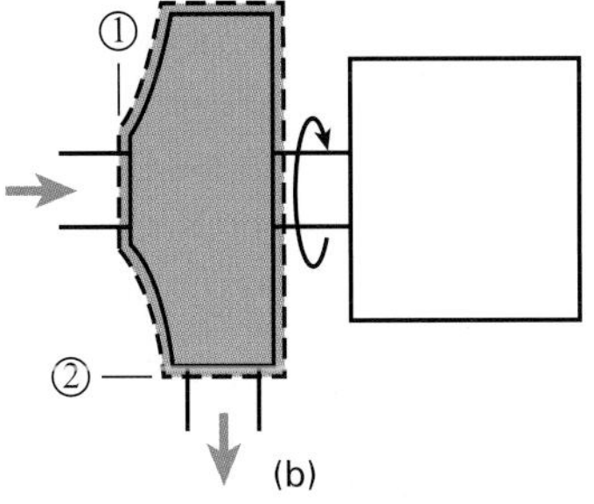

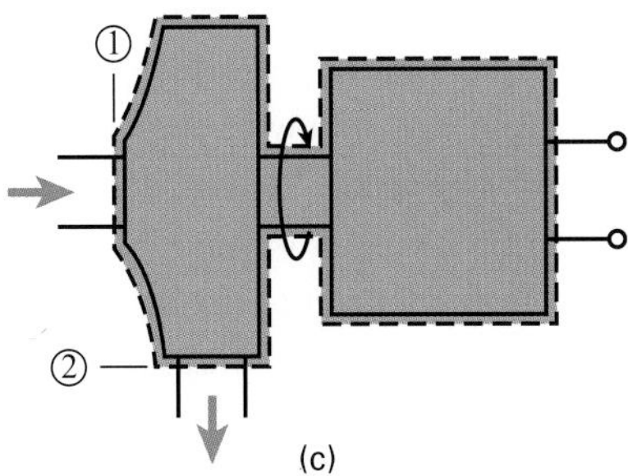

FIGURE 8.19 Various open systems (control volumes) that can be used to analyze the same pumping device.

If one desires to include bearing losses, open system II (Fig. 8.19b) may be an appropriate choice. Here the control surface cuts the inlet and exit flow pipes and surrounds the entire pumping device. The control surface also cuts through the shaft of the pumping device. For this choice of open system, the only open system work $(\dot{W}_{cv})$ is that associated with the spinning shaft. In choosing open system II, inefficiencies associated with friction in the fluid being pumped and the moving parts of the pumping device are all internal to the open system and cannot be determined separately.

A third open system choice (Fig. 8.19c) includes the unit that is driving the pumping device. This open system is frequently applied to electric-motor-driven pumping devices and is used to define an overall efficiency in converting the electrical power to useful fluid power. We will explore this concept later in our discussion of the various efficiencies used to characterize pumping devices.

APPLICATION OF CONSERVATION PRINCIPLES

We now apply the basic conservation principles to steady-flow, rotodynamic pumps and compressors.

Mass Conservation Mass conservation is represented in the same manner as in all of our previous analyses of steady-flow devices, that is, by Eq. 3.14a, which applies equally well to all the open systems of Fig. 8.19.

Energy Conservation We begin by considering the open system containing both the pump and its fluid (Fig. 8.19b) and invoking the following assumptions (see also Table 8.1):

- The heat interaction across the control surface, the pump external surface, is zero (adiabatic) or small compared to other flows of energy (i.e., $\dot{Q}_{cv,\,net\,in} = 0$).
- The kinetic energy of the flow, or the change in kinetic energy from inlet to outlet, is usually negligible compared to other flows of energy [i.e., $\left(V_2^2 - V_1^2\right)/2 \approx 0$]. Since pumps are typically used to increase the pressure and not the kinetic energy, they are usually designed to have a similar velocity at the inlet and exit. If we are analysing a fan, the change in kinetic energy will be included in our analysis.
- The potential energy change from inlet to outlet is negligible (i.e., $z_2 - z_1 = 0$).

Applying these assumptions to the steady-flow energy equation (Eq. 5.15a),

$$\dot{Q}_{cv,\,net\,in} - \dot{W}_{cv,\,net\,out} = \dot{m}\left[(h_2 - h_1) + \tfrac{1}{2}\left(V_2^2 - V_1^2\right) + g(z_2 - z_1)\right],$$

yields

$$0 - \dot{W}_{cv,\,net\,out} = \dot{m}\left[(h_2 - h_1) + 0 + 0\right].$$

Noting that power is always delivered to a pumping device (i.e., $\dot{W}_{cv,\,net\,out}$ is always negative), we define the pump shaft power as

$$\dot{W}_{pump} = \dot{W}_{shaft,\,in} \equiv -\dot{W}_{cv,\,net\,out}. \tag{8.10}$$

With this definition, our final simplified conservation of energy expression becomes

$$\dot{W}_{pump} = \dot{W}_{shaft,\,in} = \dot{m}(h_2 - h_1). \tag{8.11}$$

Note that Eq. 8.11 is quite general, subject to our assumptions, and applies to all rotodynamic pumping devices[1] for both incompressible and compressible flows (Fig. 8.20).

[1] Equation 8.11 also applies to positive-displacement devices if the pulsatile flow is treated as a time-averaged steady flow.

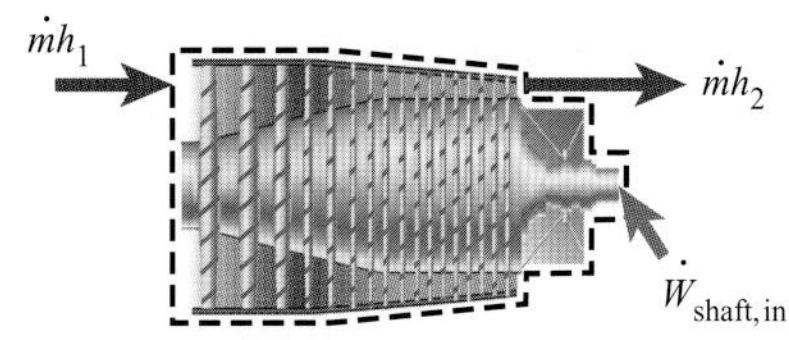

FIGURE 8.20 Open system analysis used to find the power required to operate the pumping device.

To better understand the sources of inefficiencies, we next consider a fluid-only open system (Fig. 8.19a). To apply energy conservation to this new open system and still be consistent with our previous analysis (Eq. 8.11) requires that we now include a heat-transfer term, $\dot{Q}_{\text{fluid, in}}$. This term results from the transfer of thermal energy from the frictionally heated parts of the pump, such as bearings, to the fluid. Energy conservation is now expressed as

$$\dot{Q}_{\text{fluid, in}} + \dot{W}_{\text{fluid, in}} = \dot{m}(h_2 - h_1), \tag{8.12}$$

where $\dot{W}_{\text{fluid, in}}$ is the power input to the fluid by the moving pump (or compressor) parts. Comparing Eqs. 8.11 and 8.12, we see that

$$\dot{W}_{\text{fluid, in}} = \dot{W}_{\text{shaft, in}} - \dot{Q}_{\text{fluid, in}}. \tag{8.13}$$

Thus the actual power delivered to the fluid is less than the power delivered by the shaft, as we expect. The difference between these power values is due to heat transfer to the open system. This heat-transfer process will cause an increase in the entropy of the universe and is an external irreversibility. We will return to this point later.

Momentum Conservation Generally, conservation of linear momentum is not particularly germane to an analysis of pumping devices. The conservation of angular momentum, however, is quite important to a detailed understanding of rotodynamic devices. Nevertheless, this topic reaches beyond the scope of this book and we refer the reader to other sources for further information [3, 5, 6, 8].

Reversible Steady-Flow Work For a pump or compressor, the reversible steady-flow work is the work required to raise the pressure from the inflow to the outflow conditions isentropically (i.e., by an adiabatic and reversible process). We will derive an expressions for the work performed in a reversible steady-flow process by applying first- and second-law principles to a parcel of fluid traveling from the inlet to the outlet of a steady-flow device in a reversible manner. We will start with the second Gibbs relationship, Eq. 7.7b:

$$Tds = dh - \text{v}dP.$$

For an isentropic process, there is no change in entropy. We then integrate the Gibbs relationship from the inflow to the outflow conditions:

$$h_{2s} - h_1 = \int_1^2 \text{v}dP. \tag{8.14}$$

Using the first-law analysis, we previously found that the change in specific enthalpy is equal to the work per mass required to drive the pump (Eq. 8.11) when the pump is assumed to be adiabatic and kinetic and when potential energy changes are neglected. Equation 8.14 uses the additional condition of a reversible process between states 1 and 2. We have included the subscript s on h to remind us that state 2 in this equation is reached by an isentropic process from state 1. Equating the two expressions for the change in specific enthalpy gives the reversible steady-flow work (power) delivered by the open system:

$$h_{2s} - h_1 = w_{\text{s, pump}} = \int_1^2 \text{v}dP. \tag{8.15a}$$

We also use the subscript s in the work terms to denote an isentropic process. The above expression can be written using the density of the fluid:

$$h_{2s} - h_1 = w_{\text{s, pump}} = \int_1^2 \frac{1}{\rho}dP. \tag{8.15b}$$

This expression allows us to find the work (or power) required to raise the pump pressure from state 1 to state 2 in an isentropic process where negligible kinetic and potential energy changes have been assumed. Evaluating the integral in Eq. 8.15 can be difficult because the variation of density with pressure depends on the details of the flow process; however, for simple ideal processes the integration can be performed. For example, for an ideal gas, we have

$$\mathcal{V} = \frac{RT}{P}.$$

and so, for an isothermal process,

$$h_{2s} - h = w_{s,\,\mathrm{comp}} = RT\int_1^2 \frac{1}{P}dP = RT\ln\frac{P_2}{P_1}. \tag{8.16}$$

Note that the subscript for the work term denotes the isentropic work per unit mass for a compressor since the flow is a gas. Equation 8.15 can also be evaluated for ideal gases undergoing idealized isentropic or polytropic processes. These work expressions are listed in Table 8.3. You are encouraged to derive them as practice problems. (See Problem 8.42.)

TABLE 8.3 Reversible Steady-Flow Work: Special Cases

Constant density	$\dfrac{P_2 - P_1}{\rho}$	Incompressible fluid	(8.20)
Constant temperature	$RT\ln\left(\dfrac{P_2}{P_1}\right)$	Ideal gas	(8.16)
Isentropic	$\dfrac{\gamma R(T_2 - T_1)}{\gamma - 1}$ or	Ideal gas with constant specific heats	(T8.3c)
	$\dfrac{\gamma RT_1}{\gamma - 1}\left[\left(\dfrac{P_2}{P_1}\right)^{(\gamma-1)/\gamma} - 1\right]$		(T8.3d)
Polytropic	$\dfrac{nR(T_2 - T_1)}{n - 1}$ or	Ideal gas	(T8.3e)
	$\dfrac{nRT_1}{n - 1}\left[\left(\dfrac{P_2}{P_1}\right)^{(n-1)/n} - 1\right]$		(T8.3f)

It may be helpful to contrast the work equation above (Eq. 8.16) with a similar equation that was derived in Chapter 4 and then examined again in Chapter 7. In Chapter 4, we derived an expression for the P-$\mathcal{V}$ work done by a system undergoing an isentropic (adiabatic and reversible) process using the definition of work, $\delta W = \boldsymbol{F}\cdot d\boldsymbol{s}$ (Eq. 4.5), and the concept of a quasi-equilibrium process (see Eq. 4.8b)

$$W_{\mathrm{sys,\,rev}} = \int_1^2 P d\mathcal{V}.$$

The above equation appears deceptively similar to Eq. 8.15a but must not be used interchangeably. Both equations were derived for an isentropic (adiabatic and reversible) process with kinetic and potential energy terms neglected. Equation 4.8b, however, applies to a *closed system* where there is no mass entering and leaving. State 1 is the initial state and state 2 is the final state of the system. In contrast, the pump work expression Eq. 8.15a, is used for an *open system* with a steady mass flow rate that enters at state 1 and leaves at state 2.

ISENTROPIC EFFICIENCY

The preceding discussion leads naturally to the consideration of pump efficiency. Various efficiencies can be defined. In keeping with previous analyses in this chapter, an isentropic efficiency can be defined for pumping devices (Eq. 7.35) by

$$\eta_{s,\text{pump}} \equiv \frac{\dot{W}_{s,\text{pump}}}{\dot{W}_{a,\text{pump}}} = \frac{w_{s,\text{pump}}}{w_{a,\text{pump}}}, \tag{8.17}$$

where the numerator is the ideal power input for the isentropic (adiabatic and reversible) process and the denominator is the actual shaft power input. For the ideal isentropic process, there are no flow losses or pump bearing losses so $\dot{W}_{\text{fluid,in}} = \dot{W}_{\text{shaft,in}}$. Note that the ratio of ideal to actual power input can be written with the total power flow rates, $\dot{W}_{s,\text{pump}}$ and $\dot{W}_{a,\text{pump}}$, or with the powers per mass flow rate, $w_{s,\text{pump}}$ and $w_{a,\text{pump}}$. Figure 8.21 illustrates the ideal (isentropic) process and the real process in *h–s* coordinates. Note that the real process, being irreversible, is shown as a dashed line in this sketch. Using our previous first-law statement (Eq. 8.11), we can relate the isentropic efficiency to the enthalpies associated with both the real and ideal processes as follows:

$$\eta_{s,\text{pump}} = \frac{h_{2s} - h_1}{h_2 - h_1}. \tag{8.18}$$

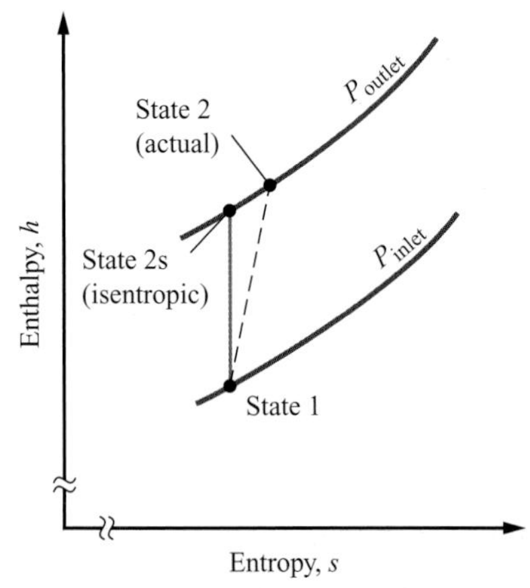

FIGURE 8.21 To define a pumping device efficiency, an ideal outlet state is defined by isentropic compression of a fluid from the actual inlet state to the same outlet pressure as that for the actual process.

Note that this relationship applies generally to all pumping devices (pumps, fans, blowers, and compressors) and is independent of whether the fluid is compressible or incompressible.

Pumping devices driven by electric motors frequently employ an overall efficiency [7], defined as

$$\eta_{OA,\text{pump}} = \frac{\dot{W}_{s,\text{pump}}}{\dot{W}_{\text{elec,in}}}, \eta_{s,\text{pump}}\eta_{\text{motor}}, \tag{8.19}$$

where $\dot{W}_{\text{elec,in}}$ is the electrical power delivered to the motor. The motor efficiency, $\eta_{\text{motor}} = \dot{W}_{a,\text{pump}}/\dot{W}_{\text{elec,in}}$, is defined as the ratio of the shaft power out (into the pump) of the motor to the electrical power delivered to the motor and includes irreversibilities within the motor such as Joule heating, eddy currents, fan losses (windage), bearing friction, etc. (See Fig. 8.19c.)

Example 8.7 Jet Engine Compressor

A dual-spool axial-flow compressor is used in a jet fighter engine. The maximum airflow through the engine is 84.4 kg/s for inlet conditions of 1 atm and 293 K. The compressor pressure ratio is 12.9:1. Determine (a) the ideal compressor power required if the isentropic efficiency is 94%, (b) the actual compressor power, and (c) the actual outlet temperature.

(Credit: NASA.)

Solution

Known $\dot{m}_{air}$, P_1, T_1, P_2/P_1, $\eta_{s,p}$

Find $\dot{W}_{ideal}$, $\dot{W}_{act}$, T_2

Sketch See Fig. 8.21.

Modeling, Premises and Assumptions

i. Steady one-dimensional flow
ii. Adiabatic process
iii. Negligible changes in kinetic and potential energies
iv. Ideal-gas behavior with constant c_p and γ

Analysis We start with the steady-state conservation of energy equation for an open system and apply assumptions ii and iii, which yields Eq. 8.11. For a compressor,

$$-\dot{W}_{cv,\,net\,in} = \dot{W}_{comp}.$$

$$\cancel{\dot{Q}_{cv,\,net\,in}} + \dot{W}_{comp} = \dot{m}\left[(h_2 - h_1) + \cancel{\frac{1}{2}(V_2^2 - V_1^2)} + \cancel{g(z_2 - z_1)}\right]$$

The ideal (minimum) power required to drive the compressor is for isentropic compression designated with subscript "s" in the equation; thus,

$$\dot{W}_{s,\,comp} = \dot{m}_{air}(h_{2s} - h_1).$$

Applying our fourth assumption allows us to express the enthalpy change as $c_{p,avg}(T_{2s} - T_1)$. With this substitution, the ideal power becomes

$$\dot{W}_{s,\,comp} = \dot{m}_{air}c_{p,avg}(T_{2s} - T_1).$$

For an isentropic process (cf. Eq. 7.13a),

$$\frac{T_{2s}}{T_1} = \left(\frac{P_2}{P_1}\right)^{(\gamma-1)/\gamma}.$$

Thus,

$$\dot{W}_{s,\,comp} = \dot{m}_{air}c_{p,avg}T_1\left[(P_2/P_1)^{(\gamma-1)/\gamma} - 1\right].$$

From Table C.3, we find c_p (293 K) = 1.007 kJ/kg; and with this value, we determine $\gamma = 1.4$ [$\equiv c_p/c_v = c_p/(c_p - R)$]. The ideal power is then evaluated as

$$\dot{W}_s = \left(84.4\frac{\text{kg}}{\text{s}}\right)\left(1.007\frac{\text{kJ}}{\text{kg}\cdot\text{K}}\right)(293\text{K})\left[(12.9)^{(1.4-1)/1.4} - 1\right]\left[\frac{\text{kW}}{\text{kJ/s}}\right]$$

$$= 26,800 \text{ kW}.$$

As an aside, direct calculation of T_{2s} from Eq. 2.41 yields 608.4 K.

From the definition of the isentropic efficiency (Eq. 8.18), we find the actual power required to drive the compressor to be

$$\dot{W}_a = \frac{\dot{W}_s}{\eta_{s,\,comp}} = \frac{26{,}800\,\text{kW}}{0.94} = 28{,}500\ \text{kW}.$$

To find the actual outlet temperature T_2, we now apply conservation of energy to the actual compressor, that is,

$$\dot{W}_a = \dot{m}_{air}(h_2 - h_1) = \dot{m}_{air} c_{p,\,avg}(T_2 - T_1).$$

Solving for T_2 yields

$$T_2 = T_1 + \frac{W_a}{\dot{m}_{air} c_{p,\,avg}} = 293 + \frac{28{,}500\text{kW}}{\left(84.4\frac{\text{kg}}{\text{s}}\right)\left(1.007\frac{\text{kJ}}{\text{kg}\cdot\text{K}}\right)}\left[\frac{\text{kJ/s}}{\text{kW}}\right] = 628.3\,\text{K}.$$

Comments We note that irreversibilities, primarily frictional effects, result in the actual outlet temperature being about 20 K ($\approx$ 628.3 K − 608.4 K) higher than that associated with an ideal isentropic compression.

Self-Test 8.5

The compressor of an air-conditioning unit compresses refrigerant (R-134a) from a saturated-vapor condition at 0.1 MPa to a final pressure and temperature of 0.8 MPa and 60 °C. Determine the isentropic efficiency of the compressor.

(Answer: 0.693)

Example 8.8 Feedwater Pump

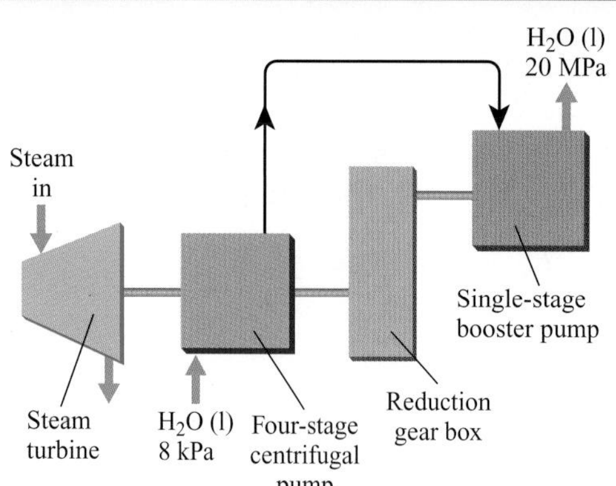

The main boiler feed pump for a 250-MW electrical generating unit consists of a four-stage centrifugal pump in series with a single-stage booster pump. A steam turbine with a 6-MW shaft output drives the pump system, as shown in the sketch. Saturated liquid water enters the pump at 8.0 kPa with a flow rate of 250 kg/s and exits at 20 MPa. Determine the isentropic efficiency of the pump.

Solution

Known $\dot{W}_{\text{shaft, in}}, P_1, P_2, \dot{m}$

Find $\eta_{s,p}$

Sketch

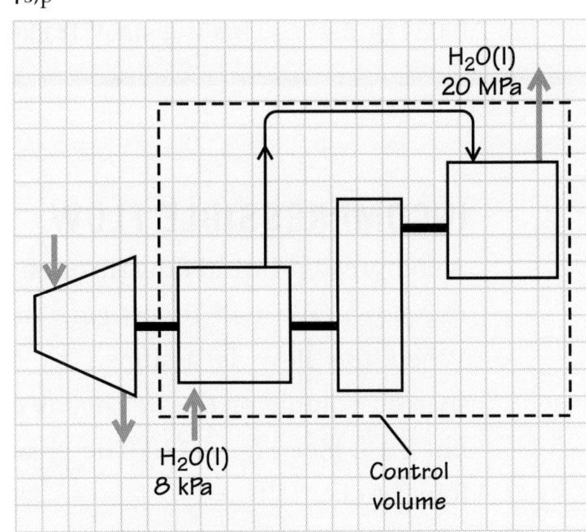

Material, Premises and Assumptions

i. Steady one-dimensional flow
ii. Adiabatic process
iii. Negligible kinetic and potential energy changes

Analysis We begin by selecting an open system that cuts through the turbine output/pump drive shaft as shown on the sketch. This open system includes both pumps and the interconnecting reduction gear box. We now need to determine the water properties at the inflow and outflow. The inflow conditions are saturated liquid water at 8 kPa. The properties can be found from Table B.2 or the NIST resources. At state 2, the pressure is at 20 MPa. For an ideal isentropic pump, the entropy will be the same at states 1 and 2s. With state 2s defined by the two independent properties of P_2 and s_{2s}, we can find the other properties at state 2s by interpolating Table B.4d or using the NIST resources. All needed properties are listed in the summary table below.

	P (kPa)	T (K)	ρ (kg/m^3)	h (kJ/kg)	s (kJ/kg·K)
State 1 sat liq	8.0 kPa	314.66	991.59	173.84	0.59249
State 2 isentropic	20 MPa	315.27	999.92	193.92	$s_1 = s_{2s}$

We start with the steady-state conservation of energy equation for an open system and apply assumptions ii and iii to give Eq. 8.11. For a pump, $-\dot{W}_{\text{cv, net out}} = \dot{W}_{\text{pump}}$.

$$\cancel{\dot{Q}_{\text{cv, net in}}} + \dot{W}_{\text{pump}} = \dot{m}\left[(h_2 - h_1) + \cancel{\frac{1}{2}(V_2^2 - V_1^2)} + \cancel{g(z_2 - z_1)}\right]$$

The power added to the fluid in the ideal isentropic pump, designated with subscript "s" in the equation, is found as

$$\dot{W}_{\text{s, pump}} = \dot{m}_{\text{air}}(h_{2s} - h_1) = \left(250\frac{\text{kg}}{\text{s}}\right)(193.92 - 173.84)\frac{\text{kJ}}{\text{kg}}\left[\frac{\text{kW}}{\text{kJ/s}}\right] = 5.02\ \text{MW}.$$

The isentropic efficiency of the pump can now be found using Eq. 8.17:

$$\eta_{s,p} \equiv \frac{\dot{W}_{\text{s, pump}}}{\dot{W}_{\text{a, pump}}} = \frac{5.02\ \text{MW}}{6.0\ \text{MW}} = 0.837 = 83.7\%.$$

Comments Since the data used in this example are from a real power plant, we have an opportunity to determine a realistic value for the ratio of the pumping power to the total power produced by the unit. This is simply $\dot{W}_{a,\,pump}/\dot{W}_{elec,\,out} = (6\text{ MW})/(250\text{ MW}) = 0.024$ or 2.4%. This is a small fraction, as expected.

INCOMPRESSIBLE FLOW

A compressed-liquid flow through a pump can often be approximated as an incompressible flow. (We will investigate the accuracy of this approximation in Example 8.9). When the flow is assumed to be incompressible ($v_1 = v_2$ or $\rho_1 = \rho_2$), then the specific volume or the density is constant and can be removed from the integrand in Eq. 8.15. (See Fig. 8.22.) The change in enthalpy across the isentropic pump is then found to be:

$$h_{2s} - h_1 = v\int_1^2 dP = v(P_2 - P_1) = \frac{P_2 - P_1}{\rho} \tag{8.20}$$

We will use this equation to analyze pumps found in a steam power cycle.

FIGURE 8.22 The pumping of water can often be treated as an incompressible process (AGEfotostock).

When changes in kinetic and potential energy across the pump are neglected, we can use Eq. 8.11 to find the power required to drive the ideal pump:

$$\dot{W}_{s,\,pump} = \dot{m}w_{s,\,pump} = \dot{m}(h_{2s} - h_1) = \frac{\dot{m}}{\rho}\int_1^2 dP = \frac{\dot{m}(P_2 - P_1)}{\rho}. \tag{8.21}$$

The power required to drive the adiabatic reversible (isentropic) pump, $\dot{W}_{s,\,pump}$, is the minimum power needed to raise the pressure of the flow from P_1 to P_2. An actual pump will have irreversibilities and will require a greater pumping power. Substituting Eq. 8.21 into the definition of isentropic efficiency (Eq. 8.17) yields the isentropic pump efficiency for incompressible flow:

$$\eta_{s,\,\text{pump}} = \frac{\dot{m}(P_2 - P_1)/\rho}{\dot{W}_{a,\,\text{pump}}} = \frac{(P_2 - P_1)/\rho}{w_{a,\,\text{pump}}}. \tag{8.22}$$

This is the working definition of efficiency used in pump testing [7] and is easily calculated by measuring the inlet and outlet pressures and the shaft torque and rotational speed (see Table 4.1). We also note that Eq. 8.22 is readily interpreted as the ratio of a useful output quantity, a pressure increase,[2] to an input quantity that must be paid for, the shaft power.

We can also substitute Eq. 8.22 into the definition of overall efficiency of the pumping system (pump and motor), Eq. 8.19, to find the overall efficiency of an incompressible pump system:

$$\eta_{\text{OA},\,\text{pump}} \equiv \frac{\dot{m}(P_2 - P_1)/\rho}{\dot{W}_{\text{elec, in}}}. \tag{8.23}$$

This is a direct expression of the general idea that an efficiency is the ratio of useful energy to energy that costs.

The following examples illustrate some of the concepts discussed here related to incompressible pumps.

Example 8.9 Water Pump

Repeat Example 8.8 using the incompressible flow approximation.

Modeling, Pemises and Assumptions

i. Steady one-dimensional flow
ii. Adiabatic process
iii. Negligible kinetic and potential energy changes
iv. Incompressible flow with $\rho = \rho(P_{\text{sat}} = P_1)$

Analysis We begin by selecting an open system that cuts through the turbine output/pump drive shaft as shown on the sketch for Example 8.8. This open system includes both pumps and the interconnecting reduction gear box.

For an incompressible fluid, the isentropic efficiency can be found using Eq. 8.22:

$$\eta_{s,\,\text{pump}} = \frac{\dot{m}(P_2 - P_1)/\rho}{\dot{W}_{\text{actual}}},$$

which we evaluate using a water density of 991.59 kg/m^3 [$= \rho(P_{\text{sat}} = 8.0\,\text{kPa})$] as follows:

[2] Rigorously, the useful output quantity is the net flow work (power), that is, $(\dot{m}P_2/\rho - \dot{m}P_1/\rho) = \dot{m}(P_2 - P_1)/\rho$.

$$\eta_{s,\text{pump}} = \frac{(250\ \text{kg/s})(20 \times 10^6 - 8000)(\text{N/m}^2)/(991.59\ \text{kg/m}^3)}{6 \times 10^6\ \text{W}} \left[\frac{\text{J}}{\text{N}\cdot\text{m}}\right]\left[\frac{\text{W}}{\text{J/s}}\right]$$
$$= \frac{5.035 \times 10^6\ \text{W}}{6 \times 10^6\ \text{W}}$$
$$= 0.840 = 84.0\%.$$

Comments Assuming an incompressible water flow gives a 0.41% error compared to the more accurate solution method used in Example 8.8. This is a very small error, similar in magnitude to the errors due to other assumptions in this problem. We also note that the calculation is much simpler when an incompressible flow is assumed. For these reasons, we will typically use an incompressible flow analysis for pumps with a compressed liquid. We can also determine the appropriateness of the incompressible approximation by comparing the densities at state 1 and state 2 tabulated in Example 8.8. The relative error between the densities at inflow and outflow is 0.84%, again demonstrating that the incompressible flow approximation can be used.

Example 8.10 Centrifugal Pump

A sales catalog presents the following data for a cast-iron centrifugal pump:

Flow rate (gal/min)	Pump head (ft)
50	15
30	75
10	100
0	54

A 1-hp (shaft) electrical motor drives the pump. Use these data to estimate the fluid power $(\dot{W}_{\text{fluid,in}})_{\text{isen}}$ and the pump efficiency η_{pump}, for pumping water. Plot η_{pump} and the pump head as a function of flow rate. The water density is 996 kg/m^3.

Solution

Known $\dot{W}_{\text{shaft,in}}$, $\dot{\forall}$, $\Delta P/\rho g(\equiv$ pump head$)$, ρ

Find $(\dot{W}_{\text{fluid,in}})_{\text{isen}}$, η_{pump}

Sketch

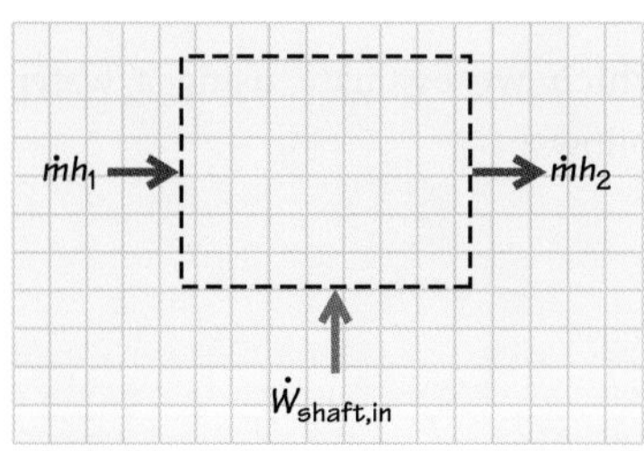

Modeling, Premises and Assumptions

i. Steady one-dimensional flow
ii. Incompressible liquid
iii. Adiabatic
iv. Negligible kinetic and potential energy changes

Analysis The standard pump performance data given as the head rise across the pump needs to be related to the pressure rise required in Eq. 8.22. The quantities will also be converted to metric units in our analysis. We will first look at the energy equation for a pump and substitute for the enthalpy, where $h = u + P/\rho$, to give

$$\dot{Q}_{\text{cv, net in}} + \dot{W}_{\text{comp}} = \dot{m}\left[(u_2 + P_2/\rho) - (u_1 + P_1/\rho) + \tfrac{1}{2}\left(V_2^2 - V_1^2\right) + g(z_2 - z_1)\right].$$

The given pump head is a measure of the pressure rise across the pump, expressed as an equivalent height of a column of the pumped fluid. The energy of the flow will increase as a result of an increase in pressure or potential head. Equating these forms of energy increase allows us to convert the head rise to an equivalent pressure rise across the pump:

$$\frac{(P_2 - P_1)_{\text{pump}}}{\rho} = g(\Delta z)_{\text{equiv, pump}},$$

where $(\Delta z)_{\text{equiv,pump}}$ is the pump head. Converting to SI units, we calculate ΔP $(= P_2 - P_1)$ for the first entry in the table at the start of this example as

$$(\Delta z)_{\text{equiv, pump}} = 15\,\text{ft}\,\frac{0.3048\,\text{m}}{\text{ft}} = 4.572\ \text{m}$$

and

$$\Delta P_{\text{pump}} = \rho g(\Delta z) = \left(996\,\frac{\text{kg}}{\text{m}^3}\right)\left(9.807\,\frac{\text{m}}{\text{s}^2}\right)(4.572\,\text{m})\left[\frac{\text{Pa}}{\text{N/m}^2}\right] = 44{,}700\,\text{Pa}.$$

However, we will not need to calculate actual pressure changes since the isentropic pump work for an incompressible fluid (Eq. 8.21) can be calculated as

$$\dot{W}_{\text{s, pump}} = \dot{m}\left(\frac{P_2 - P_1}{\rho}\right) = \dot{m}g(\Delta z)_{\text{equiv, pump.}}$$

The mass flow rate is the product of the density and the volume flow rate; that is, $\dot{m} = \rho\dot{\forall}$ (Eqs. 3.10 and 3.11). Thus,

$$\dot{W}_{\text{s, pump}} = \rho g\dot{\forall}(\Delta z)_{\text{equiv, pump.}}$$

The given flow data in gallons per minute are easily converted to SI units. For example,

$$50\,\frac{\text{gal}}{\text{min}}\left[\frac{3.785\times 10^{-3}\ \text{m}^3}{\text{gal}}\right]\left[\frac{1\,\text{min}}{60\ \text{s}}\right] = 3.15\times 10^{-3}\,\frac{\text{m}^3}{\text{s}}.$$

The fluid power for the first table entry is thus

$$\dot{W}_{s,\,pump} = \left(996\,\frac{kg}{m^3}\right)\left(9.807\frac{m}{s^2}\right)\left(3.15\times 10^{-3}\frac{m^3}{s}\right)(4.572\,m)\left[\frac{N}{kg\cdot m/s^2}\right]\left[\frac{J}{N\cdot m}\right]\left[\frac{W}{J/s}\right].$$
$$= 141\,W.$$

The pump efficiency, given by Eq. 8.22, is this fluid power divided by the shaft power. The shaft power is the actual power required to drive the pump with the given flow rate and pressure rise. Converting the shaft power to SI units yields

$$\dot{W}_{shaft,\,in} = 1\,hp\left[\frac{745.7\,W}{hp}\right] = 745.7\,W,$$

from which we calculate the efficiency as follows:

$$\eta_{pump} = \frac{\dot{W}_{s,\,pump}}{\dot{W}_{a,\,pump}} = \frac{141\,W}{745.7\,W}$$
$$= 0.19 \text{ or } 19\%.$$

Similar calculations can be performed for the other test conditions. The following table and the graph in Fig. 8.23 summarize the results.

$\dot{V}$			
(gal/min)	(m^3/s)	$\dot{W}_{s,\,pump}$(W)	η_{pump}(%)
50	3.15×10^{-3}	141	19
30	1.89×10^{-3}	422	57
10	0.63×10^{-3}	188	25
0	0	0	0

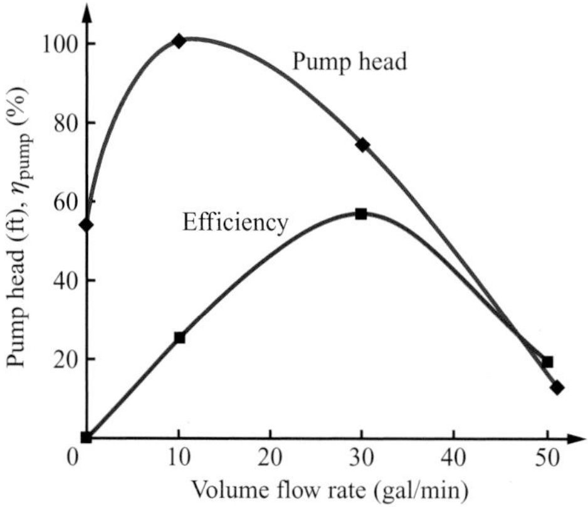

FIGURE 8.23 Centrifugal pump characteristics (Example 8.10)

Comments Our first observation from this example is the common use of non-SI units in the United States. Frequently needed conversion factors are provided at the front of this book. The second observation concerns the actual pump characteristics plotted in the figure above. In general, the efficiency of a centrifugal pump peaks within the overall flow range. In a well-designed pumping system, the pump should operate at or near the peak efficiency. We also note that the peak pressure (pump head) is delivered at a relatively low flow rate.

Although no information was provided in the example, one might like to know the overall efficiency of the pump–motor system. Data from Ref. [9] suggest an efficiency of 83% for a 1-hp motor. (Efficiency rises with motor size; e.g., it is 90% for a 10-hp motor and 96% for a 200-hp motor.) Using this value, we can estimate the peak overall efficiency (Eq. 8.23):

$$\begin{aligned}\eta_{\text{OA, pump}} &= \frac{\dot{W}_{\text{fluid}}}{\dot{W}_{\text{elec}}} = \frac{\dot{W}_{\text{fluid}}}{\dot{W}_{\text{shaft}}/\eta_{\text{motor}}} \\ &= \eta_{\text{pump}}\eta_{\text{motor}} = 0.57(0.83) = 0.47 \quad \text{or} \quad 47\%.\end{aligned}$$

Thus we conclude that, at best, less than half the electrical energy is used to perform the desired task. The remainder heats the fluid and surroundings in a non-useful way.

Self-Test 8.6

A 0.75-kW motor-driven pump is used to raise 5 kg/s of water through a total elevation of 10 m from a reservoir to a large open tank. Neglecting frictional losses in the pipe, determine the overall pump efficiency.

(Answer: 0.654)

8.5 Turbines

Turbines are steady-state flow devices that extract shaft power from the working fluid. This shaft power can be used to drive an electrical generator, a compressor, or other device.

8.5a Classifications and Applications

As shown in Table 8.4, we first classify turbines according to the working fluid (i.e., steam, gas, water, and wind). Figure 8.24 shows examples of gas turbine engines.

Within each working-fluid class, we can further classify turbines according to flow-passage types. For example, hydro (water) turbines are of two general types, which are then further subdivided as indicated in Table 8.5 and illustrated in Fig. 8.25. In the operation of **impulse turbines**, a free jet, or jets, strike the turbine wheel, also known as a runner (Fig. 8.25a and Fig. 8.26), whereas in reaction turbines the fluid is entirely enclosed (Figs. 8.25b, 8.25c, and 8.27). Reference [11] is a website devoted to small-scale hydropower and is an interesting starting point for those desiring more information in this area.

TABLE 8.4 Some Applications of Turbines

General classification	Application
Steam turbines	Utility and industrial power plants, electric generator drives, pump drives
Gas turbines	Turbochargers for internal combustion engines (trucks, buses, locomotives, and automobiles) Bus and truck engines Aircraft propulsion (turbojet, turbofan, turboshaft, and turboprop) Pipeline pump drives Ship propulsion Ship electric generator drives Rocket engine pump drives Stationary power generation
Water turbines	Hydroelectric power generation
Wind turbines	Small-scale power generation Large-scale power generation

TABLE 8.5 Hydro (Water) Turbine Types

	Application		
Type	High head	Medium head	Low head
Impulse turbines	Pelton Turgo*	Cross-flow† Multijet Pelton Turgo*	Cross-flow†
Reaction tubines	—	Francis	Propeller Kaplan

* The water jets of a Turgo turbine enter and exit obliquely from the sides, thus preventing interference of incoming and exiting jets.
† Water from a rectangular nozzle strikes a bladed cylindrical drum.

FIGURE 8.24 Gas turbine engines play an important role in transportation systems. Here are shown the 150-mile-per-hour JetTrain locomotive (left) and a hydrofoil ferry (right). Images from Wikicommons (https://commons.wikimedia.org/wiki/File:Bombardier_JetTrain.jpg) (left) and motive56 / iStock / Getty Images Plus (right).

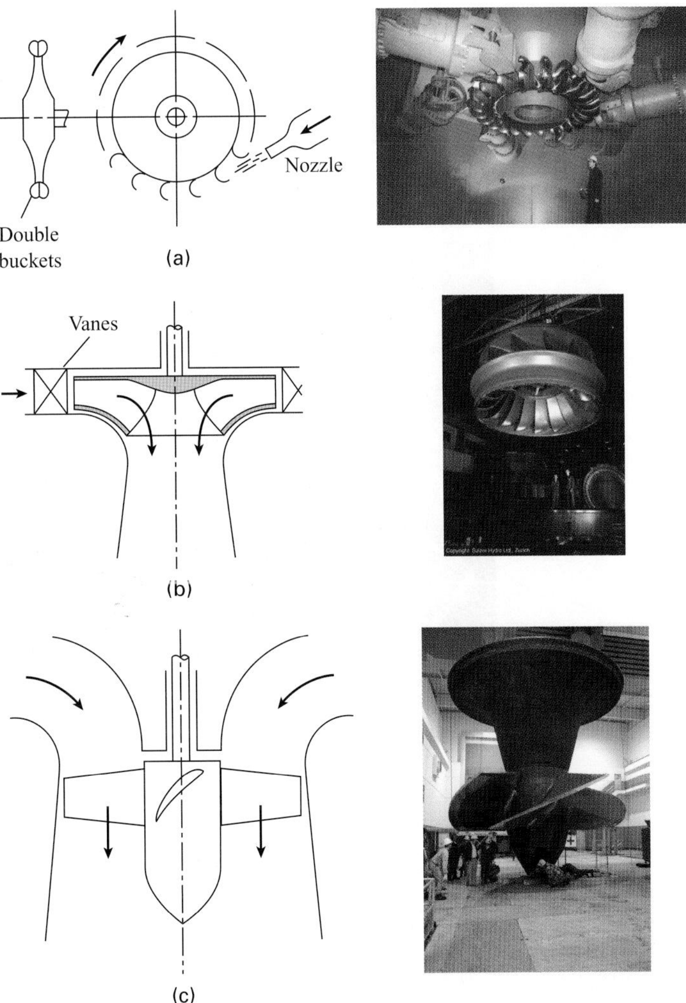

FIGURE 8.25 Hydro (water) turbine types: the impulse-type turbine **(a)** utilizes a jet of water in the open air striking the buckets or blades. The double-bucket configuration is known as a Pelton type; in reaction turbines, **(b)** and **(c)**, the fluid is enclosed. The fixed- or adjustable-vane radial inflow types **(b)** are called Francis turbines. Mixed radial- and axial-flow propeller types **(c)** may have fixed or variable blades. The variable-pitch type is also called a Kaplan turbine. Sketches reprinted from Ref. [10] with permission. Photographs courtesy of General Electric Company **(a)**, Sulzer Hydro Ltd. **(b)**, and US Army Corps of Engineers **(c)**.

FIGURE 8.26 The open housing of a Pelton-type turbine shows the turbine runner. This impulse turbine converts the kinetic energy from a water jet to shaft power. (See Fig. 8.25a.) Photograph courtesy of ABMS Consultants, Québec.

In steam and gas turbines, the working fluid is invariably contained within a casing. Small steam and gas turbines are frequently of the radial-inflow, or helical, configuration. Figure 8.28 schematically shows a radial-inflow gas turbine used in an automotive turbocharger. Larger steam and gas turbines are generally multistage, axial-flow machines. In this configuration, the general flow path follows an axial direction perpendicular to rows of spinning blades. Figure 1.4 in Chapter 1 shows a multistage steam turbine rotor for electrical power generation; Fig. 8.29 shows a large gas-turbine engine, also used in a power generation application. Multistage axial-flow gas turbines are invariably employed in aircraft engines. These are illustrated in the cutaway drawings of Fig. 8.30.

Windmills have been used for centuries to provide a local source of power. In the past few decades, however, wind turbines have been developed for both small- and large-scale power production. Figure 8.31 shows a wind farm near Great Basin, Nevada. Wind turbines can be classified into two types depending on their axis of revolution, either horizontal or vertical (Fig. 8.32). Horizontal-axis turbines dominate the current commercial market. The installed wind energy capacity in the United States in 2107 was 87,598 MW [28], Meeting approximately 6.390 of the country's total electricity demand [29]. The wind energy capacity has increased over five times in the last decade. In 2007, the installed wind capacity was 16,515 MW. The wind energy

FIGURE 8.27 The Grand Coulee Dam (left) is the largest concrete structure ever built in the United States. One of many radial inflow turbine wheels installed at the dam is shown on the right. The total installed generating capability is 6809 MW and the rated head is 330 ft. Photographs courtesy of US Bureau of Reclamation.

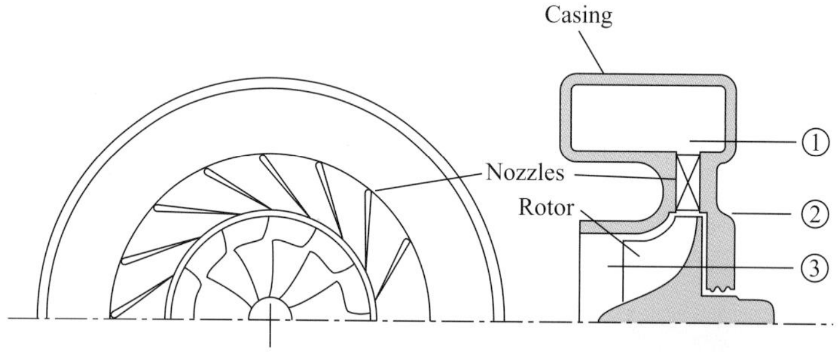

FIGURE 8.28 Schematic diagram of a radial-flow turbine used in an automotive turbocharger. Hot exhaust gases are accelerated through the nozzles (1–2) before entering the radial-inflow turbine rotor (2–3). Reprinted from Ref. [12] with permission: Figure 6.49 from Heywood, J. B., Internal Combustion Engine Fundamentals, McGraw-Hill, New York, 1988.

FIGURE 8.29 General Electric's 9H gas-turbine engine (480 MW) is used for stationary power generation. For a sense of scale, note the person standing on the frame. Photograph courtesy of General Electric Power Systems.

capacity is based on the "nameplate" or designed output if the wind turbine is operating at 100%. The capacity factor is the ratio of the actual average power output in a year to the nameplate power output. The capacity factor can range from 15% to 50% and depends on the wind turbine design and the wind conditions at the wind turbine farm. Although a renewable source of energy, wind power still poses environmental concerns such as the use of large tracts of land and the effect of wind turbines on bird populations.

8.5b Analysis

We will restrict our analysis to those turbines in which the fluid is entirely contained within flow passages defined by solid boundaries; impulse hydro turbines and wind turbines are thus excluded. Furthermore, we consider only conservation of mass and energy. Conservation of linear and angular momentum, although quite important to a thorough understanding of turbomachinery, is beyond the scope of this book.

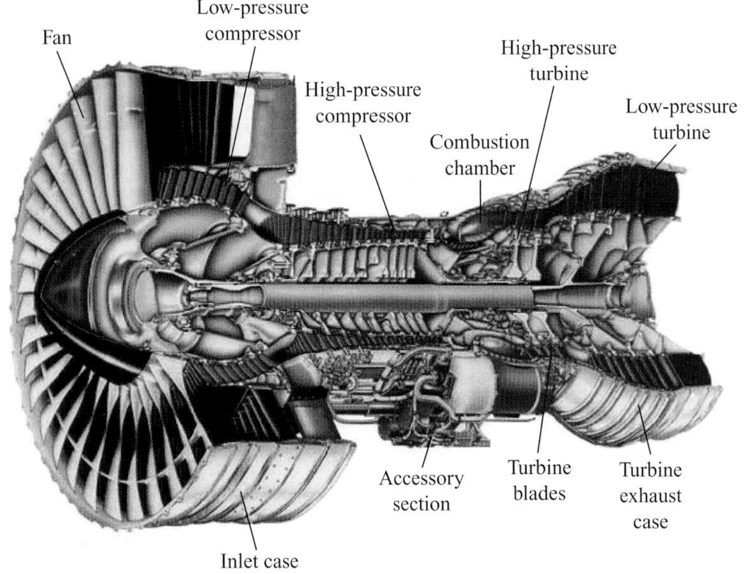

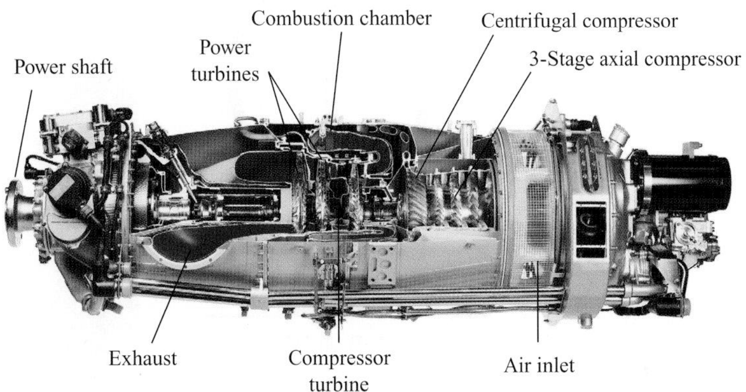

FIGURE 8.30 An aircraft turbofan engine (top) has a multistage, axial-flow, high-pressure turbine and a multistage, axial-flow, low-pressure turbine. The turboshaft engine (bottom) is configured similarly; however, the low-pressure turbine is not connected to the compressor but provides power only to the shaft. Turboshaft engines are used in helicopters, power generation, ship propulsion, and military tanks. Drawings courtesy of Pratt and Whitney.

FIGURE 8.31 Wind Farm at Great Basin, Nevada, contains 66 2.3-MW turbines with a total capacity of about 152 MW (Irina274 / iStock / Getty Images Plus).

We begin by considering the generic turbine open system (control volume) shown in Fig. 8.33. The control surface cuts through any inlet and outlet pipes, if they exist in a particular application, and is contiguous with the outer surface of the turbine housing. The open system also cuts through all shafts that transmit power from the

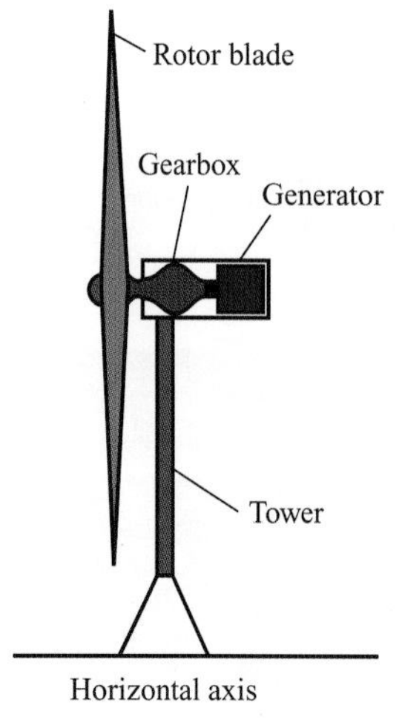

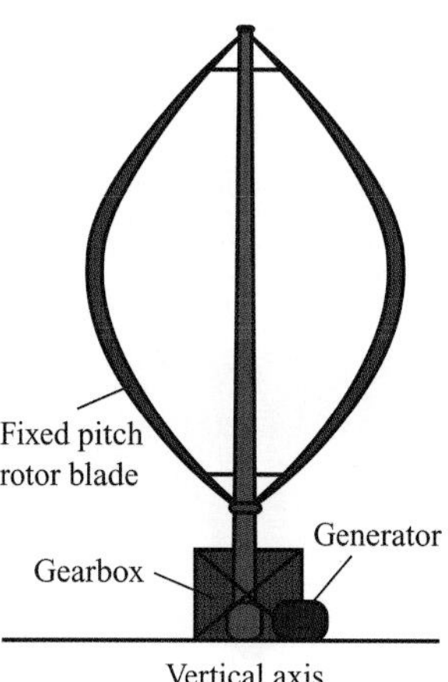

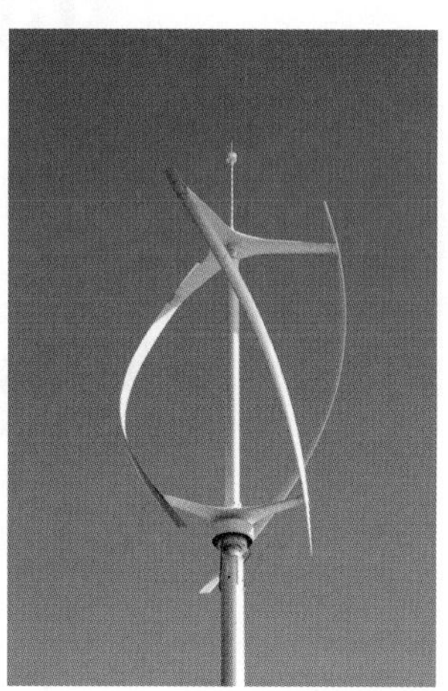

FIGURE 8.32 Wind turbines of the horizontal-axis type (top) and the vertical-axis type (bottom). Most commercial wind turbines are of the horizontal-axis type. (Photo credits: Hs Jo / EyeEm / Getty Images and ilbusca / E+ / Getty Images.)

turbine to a compressor or external load. In addition to the general requirement of steady-state and steady flow, we usually invoke the following simplifying assumptions for turbines:

- The heat interaction across the control surface is zero (in the adiabatic case) or small compared to other flows of energy (i.e., $\dot{Q}_{cv} = 0$).
- The potential energy change is zero or small compared to other flows of energy (i.e., $z_2 - z_1 = 0$).
- The kinetic energy of the inlet and outlet streams is small compared to other flows of energy, or the change in kinetic energy is small (i.e., $V_2^2/2 - V_1^2/2 = 0$).

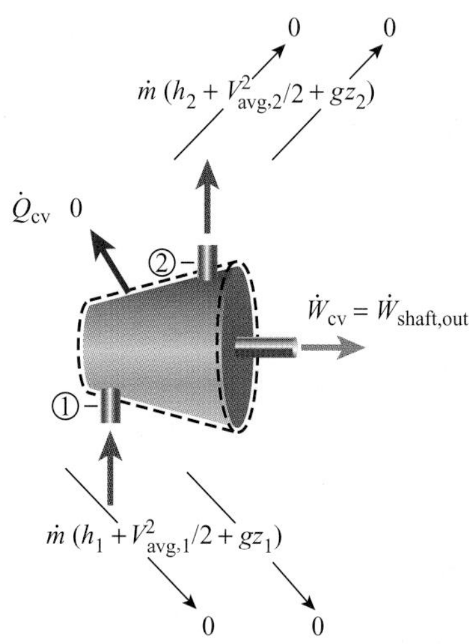

FIGURE 8.33 Open system for analysis of steam turbines, gas turbines, and reaction-type water turbines.

With steady flow through an open system (control volume) with a single inlet and a single outlet, mass conservation (Eqs. 3.11 and 3.14a) is expressed in the same way as for the previously discussed steady-flow devices:

$$\dot{m}_1 = \dot{m}_2, \tag{8.24a}$$

or

$$\rho_1 V_{avg,1} A_1 = \rho_2 V_{avg,2} A_2. \tag{8.24b}$$

With the assumptions listed, the conservation of energy (Eq. 5.15a) can be simplified to yield

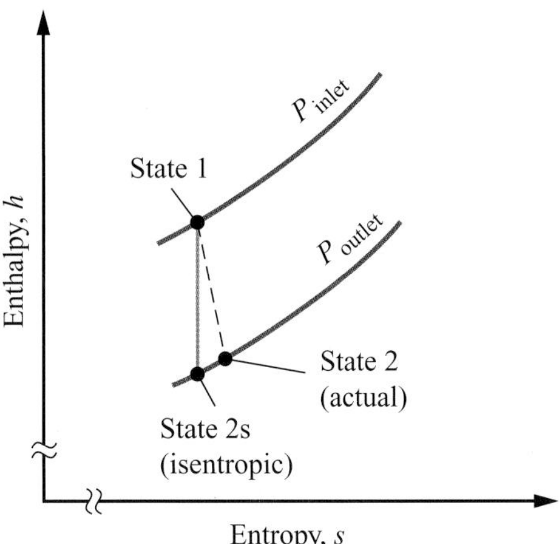

FIGURE 8.34 The maximum possible power is delivered by a turbine for the isentropic process (1–2s), whereas less power is delivered for the real process (1–2), in which irreversibilities (losses) are present. The use of a dashed line is a reminder that the real process does not follow a sequence of equilibrium states.

$$\cancel{\dot{Q}_{cv,\,\text{net in}}} - \dot{W}_{cv,\,\text{net out}} = \dot{m}\left[(h_2 - h_1) + \cancel{\frac{1}{2}(V_2^2 - V_1^2)} + \cancel{g(z_2 - z_1)}\right]$$

$$0 - \dot{W}_{cv,\,\text{net out}} = \dot{m}[(h_2 - h_1) + 0 + 0].$$

Identifying the turbine power out as $\dot{W}_{\text{turbine}} = \dot{W}_{cv,\,\text{net out}}$, this equation, upon rearrangement, becomes

$$\dot{W}_{\text{turbine}} = \dot{m}(h_1 - h_2). \tag{8.25}$$

The isentropic efficiency for a turbine is defined by Eq. 7.34. This definition, combined with our first-law expression for the actual turbine power, yields an expression involving only enthalpies;

$$\eta_{s,\,\text{turbine}} = \frac{\dot{W}_{\text{actual}}}{\dot{W}_{\text{ideal}}} = \frac{\dot{m}(h_1 - h_2)}{\dot{m}(h_1 - h_{2s})}, \tag{8.26a}$$

or

$$\eta_{s,\,\text{turbine}} = \frac{h_1 - h_2}{h_1 - h_{2s}}. \tag{8.26b}$$

Figure 8.34 illustrates on h–s coordinates the states and processes associated with Eqs. 8.26a, b (cf. Fig. 8.21 for pumping devices). Equations 8.26a, b apply equally well to incompressible and compressible flows.

Example 8.11 Steam Turbine

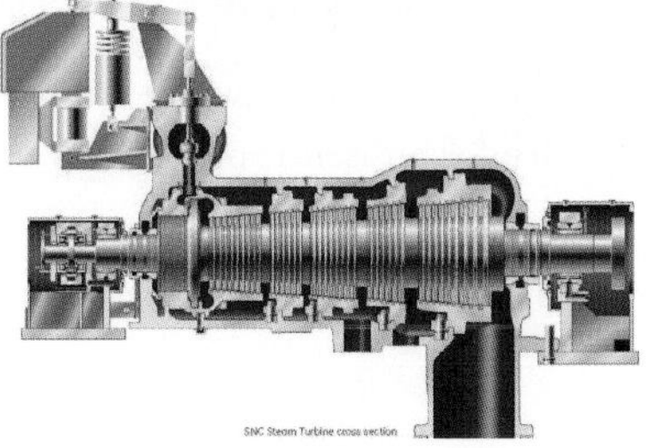

Drawing courtesy of General Electric Company.

A steam turbine is used to drive a feedwater pump of a large utility boiler. A 17.78 kg/s flow of supercritical steam at 808.3 K and 23.26 MPa is measured at the inflow to the turbine in this commercial power plant. The steam exits the turbine at 5.249 kPa with a quality of 0.9566. Determine the power produced by the turbine and the turbine isentropic efficiency.

Solution

Known $\dot{m}$, P_1, T_1, P_2, x_2

Find $\dot{W}_{act}$, $\eta_{isen,t}$

Sketch

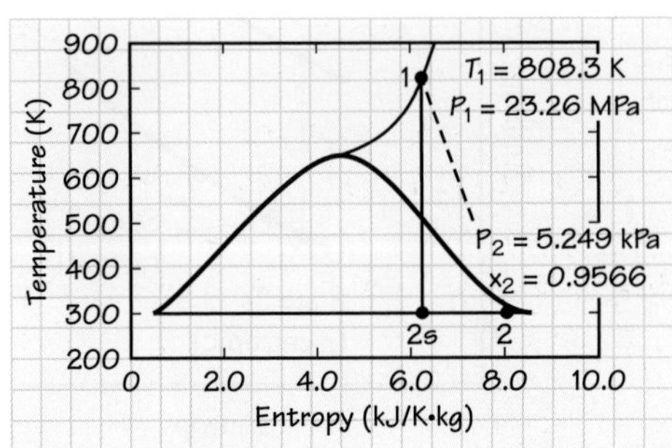

Modeling, Premises and Assumptions

i. Steady state
ii. Adiabatic process ($\dot{Q}_{cv} = 0$)
iii. Negligible kinetic and potential energy changes

Analysis We start with the steady-state energy equation, Eq. 5.15a, and apply the assumptions above:

$$\cancel{\dot{Q}_{cv}}_{,\text{net in}} - \dot{W}_{\text{CV, net out}} = \dot{m}\left[(h_2 - h_1) + \cancel{\frac{1}{2}(V_2^2 - V_1^2)} + \cancel{g(z_2 - z_1)}\right]$$

We determine the enthalpies of the entering and exiting steam from the NIST resources. Given T_1 and P_1, we get

$$h_1(23.26\text{ MPa},\ 808.3\text{ K}) = 3312.1\text{ kJ/kg},$$
$$s_1(23.26\text{ MPa},\ 808.3\text{ K}) = 6.1762\text{ kJ/kg·K}.$$

To find h_2, we apply the definition of quality (Eq. 2.37c) using the values for $h_{f,2}$ and h_{g2} at $P_2 = P_{sat} = 5.249$ kPa, that is,

$$\begin{aligned} h_2 &= (1 - x_2)h_{f2} + x_2 h_{g2} \\ &= [(1 - 0.9566)141.38 + 0.9556(2562.3)]\text{ kJ/kg} \\ &= 2457.2\text{ kJ/kg} \end{aligned}$$

The turbine power is thus

$$\begin{aligned} \dot{W}_{actual} &= \dot{m}(h_1 - h_2) \\ &= \left(17.78\frac{\text{kg}}{\text{s}}\right)(3312.1 - 2457.2)\frac{\text{kJ}}{\text{kg}}\left[\frac{\text{kW}}{\text{kJ/s}}\right] \\ &= 15{,}200\text{ kW}. \end{aligned}$$

To evaluate the isentropic efficiency, we apply its definition, Eq. 8.26b. Setting $s_{2s} = s_1$, we find the quality for an isentropic process using Eq. 2.37d:

$$\begin{aligned} x_{2s} &= \frac{s_{2s} - s_{f,2}}{s_{g,2} - s_{f,2}} \\ &= \frac{6.1762 - 0.48803}{8.3765 - 0.48803} \\ &= 0.72107\text{ (dimensionless)}. \end{aligned}$$

Using this value, we can find h_{2s}:

$$\begin{aligned} h_{2s} &= (1 - x_{2s})h_{f,2} + x_{2s}h_{g,2} \\ &= [(1 - 0.72107)141.38 + 0.72107(2562.3)]\ \text{kJ/kg} \\ &= 1887.0\ \text{kJ/kg} \end{aligned}$$

We now evaluate Eq. 8.26b:

$$\begin{aligned} \eta_{\text{isen, turbine}} &= \frac{h_1 - h_2}{h_1 - h_{2s}} \\ &= \frac{3312.1 - 2547.2}{3312.1 - 1887.0} \\ &= 0.60 \quad \text{or} \quad 60\% \end{aligned}$$

Comments We note the importance of the steam quality in determining both the actual and ideal (isentropic) enthalpies at the exit state. We also note the relatively low value for the isentropic efficiency of this steam turbine. Since the power required to drive the feedwater pump is a small fraction of the output power of the power plant, this low efficiency has a negligible effect on the overall efficiency of the power plant. The data from this example are from an actual power plant.

Self-Test 8.7

Helium, initially at 950 K and 800 kPa, is expanded through a turbine to a final state of 500 K and 100 kPa. Determine the isentropic efficiency of the turbine.

(Answer: 0.839)

Example 8.12 Gas-Turbine Engine

Consider a stationary gas-turbine engine used for electrical power generation. As shown in the sketch, the turbine drives both a compressor (part of the engine) and an electrical generator (external to the engine). The shaft power supplied to the generator from the turbine is 34,460 kW. Combustion products enter the turbine at 1530 K and exit at 720 K with a measured flow rate of 122.2 kg/s. Approximating the

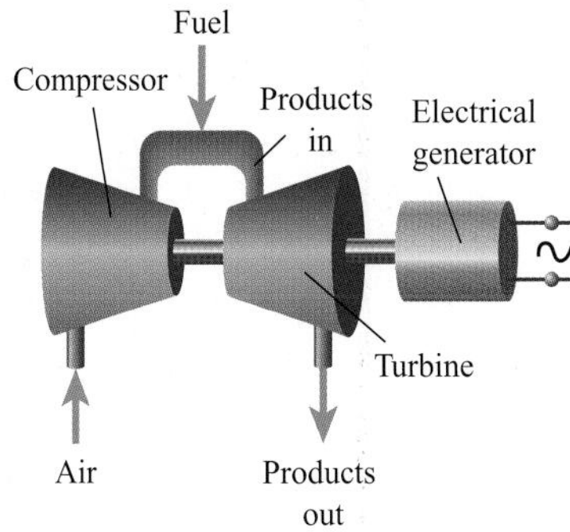

thermodynamic properties of the combustion products as those of air, estimate the fraction of the total turbine power used to generate electricity.

Solution

Known T_1, T_2, $\dot{m}$

Find $\dot{W}_{\text{t, gen}}/\dot{W}_{\text{t, tot}}$

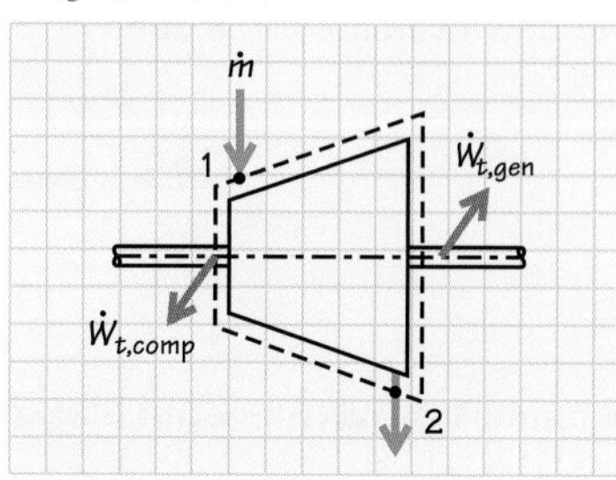

9HA.01 446 MW-class gas turbine engine. Image courtesy of General Electric Company.

Modeling, Premises and Assumptions

i. Steady state
ii. Air properties for combustion products
iii. Ideal-gas behavior
iv. Adiabatic process ($\dot{Q}_{\text{cv}} = 0$)
v. Negligible kinetic and potential energy changes

Analysis As shown in the sketch, we choose an open system that cuts through the shaft joining the turbine and the compressor and the shaft joining the turbine and the electrical generator. We start with the steady-state energy equation, Eq. 5.15a and apply the assumptions for this problem:

$$\cancel{\dot{Q}_{\text{cv, net in}}} - \dot{W}_{\text{turbine}} = \dot{m}\left[(h_2 - h_1) + \cancel{\frac{1}{2}(V_2^2 - V_1^2)} + \cancel{g(z_2 - z_1)}\right]$$

The power in this equation is the total shaft power from the turbine. Part of this power is used to drive the compressor and the remaining power drives the electrical generator:

$$\dot{W}_{\text{turbine}} = \dot{W}_{\text{t, comp}} + \dot{W}_{\text{t, gen}}.$$

Thus,

$$\dot{W}_{\text{t, comp}} + \dot{W}_{\text{t, gen}} = \dot{W}_{\text{turbine}} = \dot{m}(h_1 - h_2).$$

The enthalpies are evaluated from Table C.2:

$$h_1(1530\,\text{K}) = 1798.60\ \text{kJ/kg},$$
$$h_2(720\,\text{K}) = 861.11\ \text{kJ/kg}.$$

The total power produced is then

$$\dot{W}_{\text{shaft, out}} = \left(122.2\,\frac{\text{kg}}{\text{s}}\right)(1798.6 - 861.11)\frac{\text{kJ}}{\text{kg}}\left[\frac{\text{kW}}{\text{kJ/s}}\right]$$
$$= 114{,}600\,\text{kW},$$

and the fraction delivered to the generator is

$$\frac{\dot{W}_{\text{t, gen}}}{\dot{W}_{\text{shaft, out}}} = \frac{34,460\,\text{kW}}{114,600\,\text{kW}}$$

$$= 0.300 \quad \text{or} \quad 30\%.$$

Comment Note that 70% of the power produced by the turbine is used to drive the compressor. Gas-turbine engines typically use a large portion of the turbine output in this manner. Also note that we did not require values for the inlet and outlet pressures since we assumed ideal-gas behavior.

Example 8.13 Hydro Turbine

Hoover Dam, located on the Colorado River at the Arizona–Nevada border, creates Lake Mead in the lower Grand Canyon. Seventeen hydro turbines generate electricity at the dam. The total rated capacity of the power plant is 2074 MW. Water enters one of two intake towers (only one of which is shown in the sketch), from which it is distributed via a 9.1-m-diameter penstock (pipe) to several smaller 3.96-m-diameter penstocks, each supplying one of the main Francis-type hydro turbines.

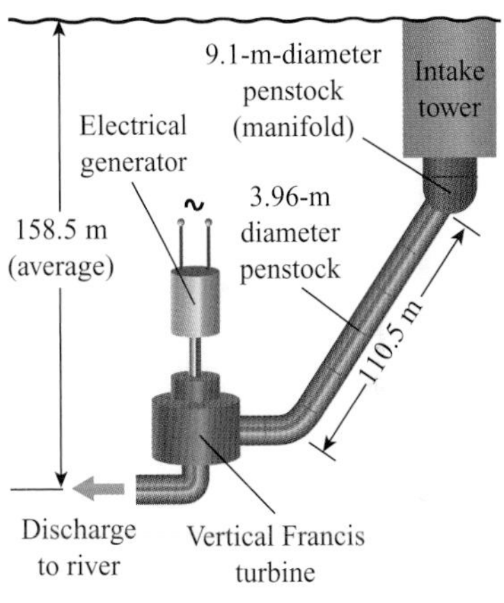

Water intakes at Hoover Dam. (Photo courtesy of the US Bureau of Reclamation.)

Consider a single main turbine with a volume flow rate of $79.4\,\text{m}^3/\text{s}$. Estimate the maximum possible power that this turbine can deliver using an average elevation change (Lake Mead surface to turbine discharge) of 158.5 m.

Engineers watch autonomous UAS testing in the Centerhill Dam penstock. Photo courtesy of the US Bureau of Reclamation; from https://www.usbr.gov/research/.

Solution

Known z_1, z_2, $\dot{\mathcal{V}}$, D_{penstock}, L_{penstock}

Find $\dot{W}_{\text{t, ideal}}$

Sketch

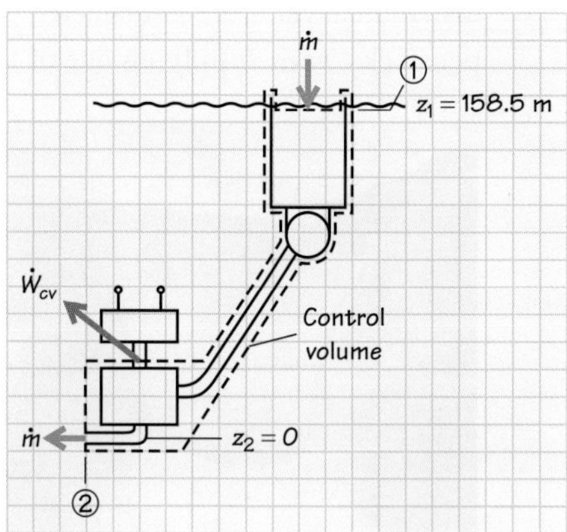

(Photo courtesy of the US Bureau of Reclamation.)

Modeling, Premises and Assumptions

i. Steady state
ii. Incompressible flow
iii. Adiabatic process ($\dot{Q}_{\text{cv}} = 0$)
iv. $P_1 = P_2 = P_{\text{atm}}$
v. $V_1 \approx 0$
vi. Uniform discharge velocity
vii. $D_2 = D_{\text{penstock}}$
viii. Isothermal fluid with $T = 25\,^{\circ}\text{C}$ (298 K)
ix. Ideal (frictionless) operation of all components (turbine, penstocks, etc.)
x. Water properties at 1 atm

Analysis We choose an open system that cuts through the turbine outlet shaft, so that the unknown power out is included in our analysis (see sketch). To find the maximum possible power (ideal), we can apply the conservation of energy for a steady-state device expressed by Eq. 5.15a. The only fluid flows crossing the control surface are the discharge at station 2 and the inlet at station 1. Since the area at the inlet is very large, we assume $V_1 = 0$. We also apply the adiabatic approximation.

$$\cancel{\dot{Q}_{ev}}_{\text{, net in}} - \dot{W}_{\text{turbine}} = \dot{m}\left[(h_2 - h_1) + \tfrac{1}{2}\left(V_2^2 - \cancel{V_1^2}\right) + g(z_2 - z_1)\right].$$

We employ the definition of enthalpy to yield:

$$h_2 - h_1 = u_2 - u_1 + \frac{P_2 - P_1}{\rho} = \frac{P_2 - P_1}{\rho}.$$

The pressure at the inflow and outflow locations are both atmospheric $(P_1 = P_2)$. We also apply the frictionless approximation $(u_2 - u_1 - Q_{ev}/\dot{m} = 0)$ of give

$$\dot{W}_{\text{turbine}} = \dot{m}\left[g(z_1 - z_2) - \tfrac{1}{2}V_2^2\right].$$

With the neglect of any losses, we recognize $\dot{W}_{\text{turbine}}$ to be the maximum possible, or ideal, power produced by the hydro turbine. To evaluate this expression, we apply mass conservation to obtain the discharge velocity V_2:

$$\dot{m} = \rho V_2 A_2 = \rho \dot{\forall},$$

or

$$V_2 = \frac{\dot{\forall}}{A_2} = \frac{\dot{\forall}}{\pi D_2^2/4}$$

$$= \frac{79.4\ \text{m}^3/\text{s}}{\pi(3.96\ \text{m})^2/4} = 6.45\ \text{m/s}$$

Thus, with the water density obtained from the NIST resources, we get

$$\dot{W}_{\text{turbine}} = \left(997.1\frac{\text{kg}}{\text{m}^3}\right)\left(79.4\frac{\text{m}^3}{\text{s}}\right)\left[\left(9.807\ \frac{\text{m}}{\text{s}^2}\right)(158.5 - 0)\text{m}\right.$$

$$\left. -0.5\left(6.45\frac{\text{m}}{\text{s}}\right)^2\right]\left[\frac{\text{N}}{\text{kg}\cdot\text{m/s}^2}\right]\left[\frac{\text{W}}{\text{N}\cdot\text{m/s}}\right] = 1.21 \times 10^8\ \text{W},$$

or

$$\dot{W}_{\text{turbine}} = 121\ \text{MW}.$$

Comments Note the importance of the choice of open system in this example, as this choice results in a relatively simple analysis.

8.6 Heat Exchangers

The purpose of a heat exchanger is to heat or cool one fluid stream at the expense of another. A common example is an automobile radiator. In this heat exchanger, energy is removed from the hot engine coolant (a mixture of ethylene glycol and water) by a

FIGURE 8.35 An automobile radiator is an example of a cross-flow heat exchanger. The engine coolant is pumped through the bank of horizontal tubes, perpendicular to the flow of air. Photograph courtesy of Delphi Automotive Systems.

flow of relatively cool ambient air (Fig. 8.35). Heat exchangers are typically passive devices with no moving parts.

8.6a Classifications and Applications

There are many types of heat exchangers. We can categorize them generally by the paths that the fluid streams take within the device, the following being the simplest:

- parallel flow,
- counterflow
- cross-flow.

These basic configurations are illustrated in Fig. 8.36. Many heat exchangers combine these path types. For example, the shell-and-tube heat exchanger shown in Fig. 8.37 exhibits elements of all three. The fluid in the first tube-pass generally runs counter to the shell fluid, whereas in the second tube-pass the fluid generally runs parallel to the shell fluid. The baffles, however, add a cross-flow component in both cases.

There are many, many applications of heat exchangers. Table 8.6 suggests a few of these and indicates the typical heat exchanger employed in the applications listed. Common heat exchangers are shown in Fig. 8.38.

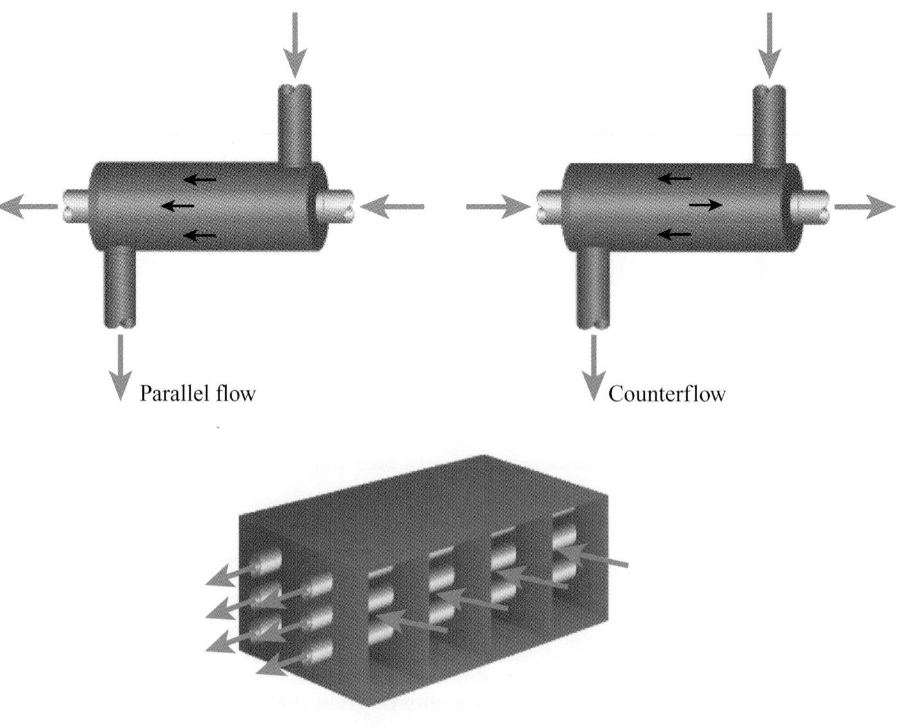

FIGURE 8.36 Three basic flow configurations for heat exchangers: parallel flow (top), counterflow (center), and cross-flow (bottom). Many heat exchangers operate with combinations of these three flow configurations.

TABLE 8.6 Some Applications of Heat Exchangers

Application	Typical heat exchanger configuration	Comment
Power plant steam condensers	Shell-and-tube*	See Figs. 8.37 and 8.38a
Nuclear power steam generators	Shell-and-tube*	See Figs. 8.37 and 8.38a
Oil coolers	Shell-and-tube*	See Figs. 8.37 and 8.38a
Process applications	Shell-and-tube*	See Figs. 8.37 and 8.38a
Chemical industry	Shell-and-tube*	See Figs. 8.37 and 8.38a
Refrigeration systems	Spiral tube	
Food processing	Gasketed plate	See Fig. 8.38e. Easily disassembled for cleaning
Sludges, viscous liquids, liquids with suspended solids, and slurries	Spiral plate	
Gas-to-gas exchanger	Plate-fin	See Fig. 8.38g. Applications include gas turbines, refrigeration, HVAC, waste heat recovery, and electronic cooling
Condensation and evaporation of refrigerants	Tube-fin	
Air preheaters in power plants and industrial processes	Rotary and fixed-matrix regenerators	Unsteady thermal energy storage devices
Cooling towers	Direct contact	See Chapter 9
Feedwater heaters	Shell-and-tube and direct contact	See Example 8.15

* More than 90% of the heat exchangers used in industry are of the shell-and-tube type [17].

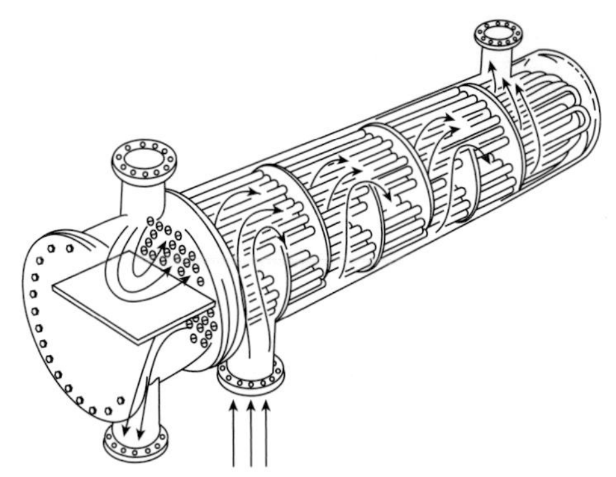

FIGURE 8.37 A U-tube shell-and-tube heat exchanger has two tube-passes. Baffles in the shell create a cross-flow component; thus the upper tube-pass is a cross-flow/parallel combination, whereas the lower tube-pass is a cross-flow/counterflow combination. Reprinted from Ref. [18] with permission.

8.6b Analysis

APPLICATION OF CONSERVATION PRINCIPLES

We consider three different open systems (control volumes) in our analysis of heat exchangers: (1) an overall open system that encloses the entire device and whose surface cuts through the inlets and outlets, (2) an open system that includes only the hot fluid, and (3) an open system containing only the cold fluid. These three open systems are sketched in Fig. 8.39.

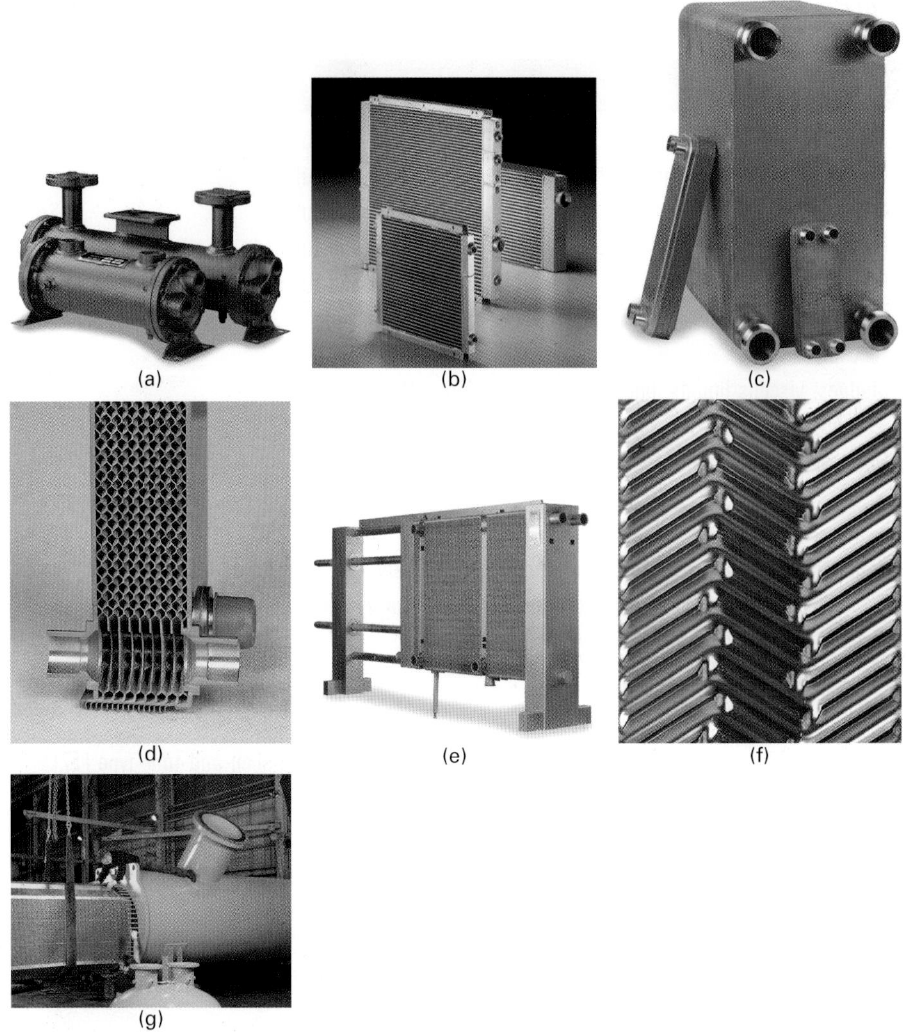

FIGURE 8.38 Various types of heat exchangers. **(a)** Shell-and-tube type. **(b)** Cross-flow type oil coolers, aftercoolers, and combination oil/aftercoolers of aluminum construction. **(c)** Brazed-plate type. **(d)** Cutaway view of flow channels in brazed-plate heat exchanger. **(e)** Plate-and-frame-type heat exchanger for food and beverage processing. **(f)** Flow channel pattern for heat-transfer plate. **(g)** Extended-surface, plate-fin heat exchanger for compressor intercooler and aftercooler applications. Photographs courtesy of API Heat Transfer, Inc.

Conservation of Mass Conservation of mass is simply expressed for each case as

$$\dot{m}_{\mathrm{H,in}} + \dot{m}_{\mathrm{C,in}} = \dot{m}_{\mathrm{H,out}} + \dot{m}_{\mathrm{C,out}}, \tag{8.27a}$$

$$\dot{m}_{\mathrm{H,in}} = \dot{m}_{\mathrm{H,out}} \equiv \dot{m}_{\mathrm{H}}, \tag{8.27b}$$

and

$$\dot{m}_{\mathrm{C,in}} = \dot{m}_{\mathrm{C,out}} \equiv \dot{m}_{\mathrm{C}}, \tag{8.27c}$$

where the subscripts H and C refer to the hot and cold streams, respectively.

Conservation of Energy For the overall open system (Fig. 8.39a), we simplify the following steady-flow statement of conservation of energy for open systems with multiple inlets and outlets (Eq. 5.16):

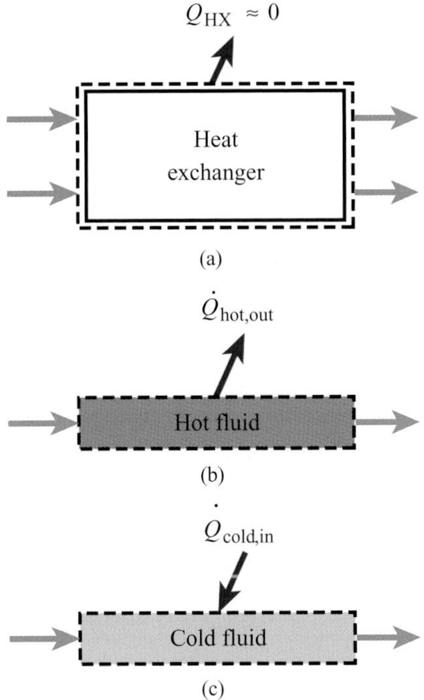

FIGURE 8.39 Open systems for heat exchanger analysis: **(a)** entire heat exchanger, **(b)** hot fluid stream only, and **(c)** cold fluid stream only.

$$\dot{Q}_{cv,\,net\,in} - \dot{W}_{cv,\,net\,in} = \sum_{k=1}^{M\,outlets} \dot{m}_{out,\,k}\left[h_k + \tfrac{1}{2}V_k^2 + g(z_k - z_{ref})\right] - \sum_{j=1}^{N\,inlets} \dot{m}_{in,\,j}\left[h_j + \tfrac{1}{2}V_j^2 + g(z_j - z_{ref})\right]. \quad \textbf{(8.28)}$$

We begin with the following facts and assumptions:

- No shaft power is associated with a heat exchanger; therefore, $\dot{W}_{cv}$ is zero.
- The heat interaction between the heat exchanger and its surroundings, $\dot{Q}_{cv}$, can be neglected. For heat exchangers employing fluids hotter than the ambient surroundings, there is likely to be some heat loss. This loss, however, is usually quite small compared to the enthalpy flows on the right-hand side of Eq. 8.28; thus, we will neglect it (i.e., $\dot{Q}_{cv} \approx 0$)
- Kinetic and potential energy changes between inlets and outlets for both the hot and cold streams are negligible.

With these provisions, Eq. 8.28 becomes

$$0 - 0 = \dot{m}_H h_{H,\,out} + \dot{m}_C h_{C,\,out} - \dot{m}_H h_{H,\,in} - \dot{m}_C h_{C,\,in},$$

which can be rearranged to yield

$$\underbrace{\dot{m}_H\left(h_{H,\,in} - h_{H,\,out}\right)}_{\text{Rate at which hot stream loses energy}} = \underbrace{\dot{m}_C\left(h_{C,\,out} - h_{C,\,in}\right)}_{\text{Rate at which cold stream gains energy}}. \quad \textbf{(8.29)}$$

In the analysis of heat exchangers in which a change of phase is not involved, Eq. 8.29 can be related to the inlet and outlet temperatures by assuming a constant specific heat in the calorific equation of state; that is, we assume that $\Delta h = c_{p,\text{avg}}\Delta T$ (see Eqs. 2.31e and 2.44b). With this assumption,

$$\dot{m}_\text{H}c_{p,\text{H}}(T_{\text{H,in}} - T_{\text{H,out}}) = \dot{m}_\text{C}c_{p,\text{C}}(T_{\text{C,out}} - T_{\text{C,in}}). \tag{8.30}$$

The physical interpretation of Eqs. 8.29 and 8.30 is that the rate at which the hot fluid loses energy equals the rate at which the cold fluid gains energy. This is physically satisfying, because it means that the open system itself neither loses nor gains energy at the expense of the surroundings (i.e., $\dot{Q}_\text{cv} = \dot{W}_\text{cv} = 0$). Frequently, the product of the flow rate and specific heat is called the **heat-capacity rate** C.

Examining either the hot or cold stream in isolation (Figs. 8.39b,c), we see a heat interaction across the control surface: For the hot stream, there is a loss of energy $\dot{Q}_{\text{H,out}}$, and for the cold stream, there is a gain of energy $\dot{Q}_{\text{C,in}}$. With this new wrinkle, but retaining all the other assumptions applied to the overall open system, we can express energy conservation for the hot fluid by simplifying Eq. 5.15a as follows:

$$-\dot{Q}_{\text{H,out}} + 0 = \dot{m}_\text{H}(h_{\text{H,out}} - h_{\text{H,in}} + 0 + 0),$$

or

$$\dot{Q}_{\text{H,out}} = \dot{m}_\text{H}(h_{\text{H,in}} - h_{\text{H,out}}). \tag{8.31a}$$

Similarly, energy conservation for the cold stream is expressed as

$$\dot{Q}_{\text{C,in}} = \dot{m}_\text{C}(h_{\text{C,out}} - h_{\text{C,in}}). \tag{8.31b}$$

Assuming constant specific heats, we can relate the heat-transfer rates in Eqs. 8.31a and 8.31b to the inlet and outlet temperatures as follows:

$$\dot{Q}_{\text{H,out}} = \dot{m}_\text{H}c_{p,\text{H}}(T_{\text{H,in}} - h_{\text{H,out}}). \tag{8.32a}$$

and

$$\dot{Q}_{\text{C,in}} = \dot{m}_\text{C}c_{p,\text{C}}(T_{\text{C,out}} - T_{\text{C,in}}). \tag{8.32b}$$

Comparing Eqs. 8.31a and 8.31b to our overall energy balance (Eq. 8.29), we can formally write

$$\dot{Q}_{\text{H,out}} = \dot{Q}_{\text{C,in}}. \tag{8.33}$$

The objective in the design of a heat exchanger is to define, in detail, a device that will transfer energy at the desired rate from one fluid to another. Details of the design include choosing the length of tubing, number of tubes, and the geometry of fins or extended surfaces to enhance (increase) the heat transfer from the hot stream to the cold stream. Heat exchanger design is beyond the scope of the present text. The interested reader can find details of heat exchanger design in Refs. [13–16].

Example 8.14 Oil/Water Heat Exchanger

A U-tube shell-and-tube heat exchanger (Fig. 8.37) is used to cool a 5.5-kg/s flow of hot oil from 380 to 320 K. The oil flows through the shell, while water entering at 280 K flows through a bundle of 16 tubes. The average velocity of the water in the 15-mm-inside-diameter tubes is 2 m/s. The average specific heats of the oil and water can be assumed to be 2.122 and 4.180 kJ/kg·K, respectively. Determine the outlet temperature of the water.

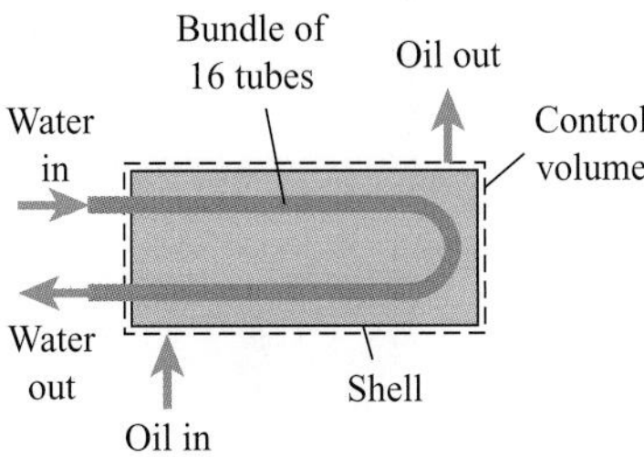

Solution

Known $\dot{m}_{\rm H}, c_{p,{\rm H}}, c_{p,{\rm C}}, T_{\rm H,in}, T_{\rm H,out}, T_{\rm C,in}, N, D_{\rm tube}, {\rm V}_{\rm avg}$

Find $T_{\rm C,out}$

Sketch

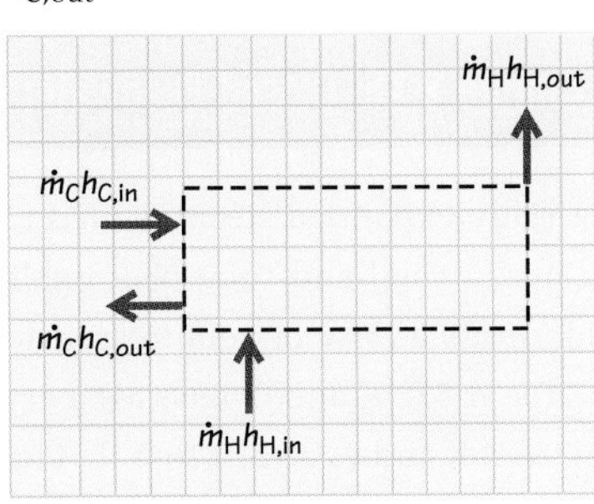

Modeling, Premises and Assumptions

i. Steady flow
ii. Constant (average) properties
iii. No residual heat loss ($\dot{Q}_{\rm cv} = 0$)
iv. Negligible kinetic and potential energy changes

Analysis For the open system shown in the sketch, the overall energy balance expressed by Eq. 8.30 applies. To use this equation to find $T_{\rm C,out}$ requires that we find the mass flow rate of the water. The mass flow rate through a single tube is given by

$$\dot{m}_{1\,{\rm tube}} = \rho_{\rm H_2O} V_{\rm avg} \pi D_{\rm tube}^2/4,$$

and for the bundle of N (= 16) tubes,

$$\begin{aligned} \dot{m}_{\rm C} &= N\dot{m}_{1\,{\rm tube}} = 16\rho_{\rm H_2O} V_{\rm avg}\pi D_{\rm tube}^2/4 \\ &= 16(999.1\,{\rm kg/m^3})(2.0\,{\rm m/s})\pi(0.015\,{\rm m})^2/4 \\ &= 5.65\,{\rm kg/s}, \end{aligned}$$

where the water density is obtained from the NIST WebBook for T = 280 K and P = 1 atm. (Although the pressure is unknown, using P = 1 atm should be reasonable since the liquid density does not vary much with pressure.) Isolating the unknown water outlet temperature in Eq. 8.30 yields

$$\begin{aligned} T_{\rm C,out} &= T_{\rm C,in} + \frac{\dot{m}_{\rm H} c_{p,{\rm H}}(T_{\rm H,in} - T_{\rm H,out})}{\dot{m}_{\rm C} c_{p,{\rm C}}} \\ &= 280\,{\rm K} + \frac{(5.5\,{\rm kg/s})(2.122\,{\rm kJ/kg\cdot K})(380 - 320){\rm K}}{(5.65\,{\rm kg/s})(4.180\,{\rm kJ/kg\cdot K})} = 310\,{\rm K}. \end{aligned}$$

Comment This example illustrates the application of energy conservation to find the one unknown temperature when three are given, a very common situation encountered in heat-exchanger design and analysis. We also see how the flow properties from a single tube are used to obtain the total flow rate associated with a tube bundle.

Self-Test 8.8

Oil flowing at 0.02 kg/s enters the inner channel of a parallel-flow heat exchanger at 400 K. Water at 290 K enters the outer channel of the heat exchanger. Neglecting any frictional effects, determine the minimum flow rate of the water such that the oil is cooled to a final temperature of 300 K.

(Answer: 0.101 kg/s)

Example 8.15 Closed Feedwater Heater

Partially expanded steam is bled from a steam turbine and fed into a closed feedwater heater, a shell-and-tube heat exchanger in which the condensing steam on the shell side heats a flow of compressed liquid water on the tube side. The steam enters the heater at 572 K and 0.4268 MPa at a flow rate of 9.38 kg/s. Cold water enters the tubes at 386.4 K and exits at 416.3 K. The water flow rate is 189 kg/s. Assuming the pressure in the shell is uniform, determine the temperature and state of the H_2O exiting the shell.

Solution

Known $\dot{m}_H$, $\dot{m}_C$, $T_{H,in}$, $P_{H,in}$, $(= P_{H,out})$, $T_{C,in}$, $T_{C,out}$

Find $T_{H,out}$, state (liquid, liquid–vapor mixture, or vapor)

Sketch

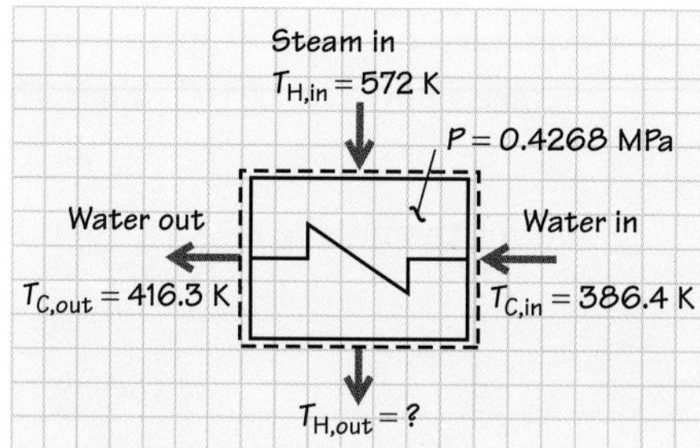

Modeling, Premises and Assumptions

i. Steady flow
ii. No residual heat loss ($\dot{Q}_{cv} = 0$)
iii. Negligible kinetic and potential energy changes
iv. Specific heat for cold-side water evaluated at mean temperature and arbitrary pressure

Analysis With our assumptions, application of conservation of energy to the open system shown in the sketch yields (Eq. 8.29)

$$\dot{m}_H(h_{H,in} - h_{H,out}) = \dot{m}_C(h_{C,out} - h_{C,in}).$$

For the cold-water stream, we express the enthalpy difference as $c_{p,C}\left(T_{C,out} - T_{C,in}\right)$, where an average value of the specific heat is employed; thus,

$$\dot{m}_H\left(h_{H,in} - h_{H,out}\right) = \dot{m}_C c_{p,C}\left(T_{C,out} - T_{C,in}\right).$$

Note that we cannot assume a constant specific heat for the hot stream since it enters as superheated steam and will change state during the process to a saturated mixture or compressed liquid. We use this energy expression to determine a value of $h_{H,out}$. Because we know the pressure at the outlet of the hot stream, we have two thermodynamic properties from which all other thermodynamic properties can be obtained; specifically,

$$T_{H,out} = T_{H,out}\left(h_{H,out}, P_{H,out}\right).$$

We proceed, first, by evaluating $c_{p,C}$. Since the pressure is unknown on the cold side, we will employ an arbitrary pressure of 5 atm to ensure that the cold stream enters and exits as a compressed liquid.[3] Our criterion for this choice of pressure is only that it exceed P_{sat} at the outlet temperature [i.e., $P_{sat}(T_{C,out}) = 3.9$ atm]. Using the NIST database, we find

$$c_p = (386.4\,\text{K}, 5\,\text{atm}) = 4.232\ \text{kJ/kg}\cdot\text{K},$$
$$c_p = (416.3\,\text{K}, 5\,\text{atm}) = 4.290\ \text{kJ/kg}\cdot\text{K},$$

the average of which is 4.261 kJ/kg·K. We also use the NIST database to evaluate $h_{H,in}$:

$$h_{H,in}(572\,\text{K}, 0.4268\,\text{Mpa}) = 3064.0\ \text{kJ/kg}.$$

The only unknown quantity left in our energy equation is $h_{H,out}$. Solving for $h_{H,out}$ yields

$$\begin{aligned} h_{H,out} &= h_{H,in} + \frac{\dot{m}_C c_{p,C}\left(T_{C,out} - T_{C,in}\right)}{\dot{m}_H} \\ &= 3064.0\ \text{kJ/kg} - \frac{(189\,\text{kg/s})(4.261\,\text{kJ/kg}\cdot\text{K})(416.3 - 386.4)\text{K}}{9.38\,\text{kg/s}} \\ &= 496.9\ \text{kJ/kg}. \end{aligned}$$

We can now find the outflow temperature for the hot stream using the "Specified State Points" option on the "calculate" menu of the NIST software. The pressure and specific enthalpy are entered and the other fluid properties are returned. The outflow temperature of the hot stream is: $T_{H,out}(0.4268\,\text{MPa}, 496.9\ \text{kJ/kg}) = 391.48$ K. The NIST table lists the state as "subcooled." In this textbook, we have referred to the "subcooled" state as "compressed liquid."

Comments The numerical values used in this example are taken from a real steam power plant. We will see in Chapter 9 that the purpose of a feedwater heater is to preheat the feedwater entering the boiler to increase the efficiency of the power cycle.

Self-Test 8.9

Heat from warm air drawn through the evaporator of an air-conditioning unit results in the evaporation of the refrigerant (R-134a) from $x = 0.359$ at 0.1 MPa to a saturated-vapor state. If the air is cooled from 27 °C to 17 °C at a flow rate 0.2 kg/s, determine the mass flow rate of the refrigerant.

(Answer: 0.0146 kg/s)

[3] Note that the choice of pressure affects the specific-heat value only slightly. It is recommended that the reader verify this.

SUMMARY

After studying this chapter, you should have a good understanding of the operation of a wide variety of practical steady-flow devices. Specifically, you should be able to formulate, simplify, and apply the basic conservation principles and thermodynamic property relationships to nozzles, diffusers, throttles, pumps, compressors, turbines, and heat exchangers. You should also have an appreciation of the losses and inefficiencies associated with these devices. As a further summary of this chapter, reviewing the learning objectives presented at the outset of the chapter is recommended.

KEY EQUATIONS

Review the most important equations presented in this chapter (i.e., those boxed with a yellow background). What physical principles do they express? What restrictions apply?

CHAPTER 8 KEY CONCEPTS AND DEFINITIONS CHECKLIST

Numbers following arrows refer to Questions at the end of the chapter.

8.1 Steady-Flow Devices

- ☐ Purpose of each device ➔ Question 8.2
- ☐ Typical simplifying assumptions ➔ Question 8.3

8.2 Nozzles and Diffusers

- ☐ Physical shapes ➔ Questions 8.4, 8.5
- ☐ Simplified mass and energy conservation ➔ Questions 8.1, 8.4
- ☐ Nozzle isentropic efficiency ➔ Questions 8.6, 8.7, 8.8, 8.11, 8.12

8.3 Throttles

- ☐ Physical configurations ➔ Question 8.9
- ☐ Throttling calorimeter ➔ Question 8.10
- ☐ Simplified mass and energy conservation ➔ Questions 8.11, 8.19

8.4 Pumps, Compressors, and Fans

- ☐ Positive-displacement versus dynamic devices ➔ Question 8.12
- ☐ Centrifugal and axial-flow devices ➔ Question 8.13
- ☐ Simplified mass and energy conservation ➔ Question 8.2
- ☐ Reversible, steady-flow work ➔ Questions 8.31, 8.42
- ☐ Isentropic efficiency ➔ Questions 8.14, 8.15, 8.16, 8.17, 8.38

8.5 Turbines

- ☐ Applications (Table 8.4) ➔ Question 8.18
- ☐ Types (Tables 8.4 and 8.5) ➔ Question 8.18
- ☐ Simplified mass and energy conservation ➔ Questions 8.2, 8.65
- ☐ Isentropic efficiency ➔ Questions 8.19, 8.20, 8.21, 8.67, 8.81

8.6 Heat Exchangers

- ☐ Parallel, counter-, and cross-flow ➔ Question 8.22
- ☐ Overall mass and energy conservation ➔ Questions 8.2, 8.92
- ☐ Heat-capacity rate ➔ Question 8.23

REFERENCES

1. Baumeister, T., and Marks, L. S. (Eds.), *Standard Handbook for Mechanical Engineers*, 7th edn, McGraw-Hill, New York, 1967, Chapter 14.

2. Cumpsty, N., *Jet Propulsion*, Cambridge University Press, New York, 1997.
3. White, F. M., *Fluid Mechanics*, 5th edn, McGraw-Hill, New York, 2003.
4. Runstadler, P. W., Jr., Dolan, F. X., and Dean, R. C., Jr., *Diffuser Data Book*, Technical Note TN-186, May 1975, Creare, Inc., Hanover, NH.
5. Roberson, J. A., and Crowe, C. T., *Engineering Fluid Mechanics*, 4th edn, Houghton Mifflin, Boston, 1990.
6. Shames, I. H., *Mechanics of Fluids*, 3rd edn, McGraw-Hill, New York, 1992.
7. Karassik, I. J., Krutzsch, W. C., Fraser, W. H., and Messina, J. P. (Eds.), *Pump Handbook*, 2nd edn, McGraw-Hill, New York, 1986.
8. Dixon, S. L., *Fluid Mechanics and Thermodynamics of Turbomachinery*, 4th edn, Butterworth-Heinemann, Woburn, MA, 1998.
9. Reliance Electric Co., "AC Motor Efficiency Guide," http://www.reliance.com/b7087_5/b7087_intro.htm, 1999.
10. Shepherd, D. G., *Elements of Fluid Mechanics*, Harcourt Brace & World, New York, 1965.
11. See http://www.microhydropower.net/index.php.
12. Heywood, J. B., *Internal Combustion Engine Fundamentals*, McGraw-Hill, New York, 1988.
13. Incropera, F. P., and DeWitt, D. P., *Fundamentals of Heat and Mass Transfer*, 5th edn, Wiley, New York, 2002.
14. Bowman, R. A., Mueller, A. C., and Nagle, W. M., "Mean Temperature Difference in Design," *Transactions ASME*, 62:283–294, 1940.
15. Van Dyke, M., *An Album of Fluid Motion*, Parabolic Press, Stanford, CA, 1982.
16. Kakaç, S., and Liu, H., *Heat Exchangers: Selection, Rating and Thermal Design*, 2nd edn, CRC Press, Boca Raton, FL, 2002.
17. Kuppan, T., *Heat Exchanger Design Handbook*, Marcel Dekker, New York, 2000.
18. Mills, A. F., *Heat and Mass Transfer*, Irwin, Chicago, 1995.

Some end-of-chapter problems were adapted with permission from the following:

Chapman, A. J., *Fundamentals of Heat Transfer*, Macmillan, New York, 1987.

Look, D. C., Jr., and Sauer, H. J., Jr., *Engineering Thermodynamics*, PWS, Boston, 1986.

Myers, G. E., *Engineering Thermodynamics*, Prentice Hall, Englewood Cliffs, NJ, 1989.

Pnueli, D., and Gutfinger, C., *Fluid Mechanics*, Cambridge University Press, Cambridge, England, 1992.

U.S. Department of Energy WINDExchange at http://apps2.eere.energy.gov/wind/windexchange/wind_installed_capacity.asp.

Energy Information Agency, U.S. Department of Energy, "Monthly Energy Review September 2103," http://www.eia.gov/totalenergy/data/monthly/index.cfm#electricity, Release Date: September 25, 2013.

QUESTIONS

8.1 Review the most important equations presented in this chapter (i.e., those with a yellow background). What physical principles do they express? What restrictions (approximations) apply?

8.2 List one or more purposes/applications for the following devices: nozzle, diffuser, throttle, pump, compressor, turbine, and heat exchanger.

8.3 What terms in the conservation of energy equation (Eqs. 5.15a or 5.16) are usually neglected in the thermal analysis of the following devices: nozzle, diffuser, throttle, pump, compressor, turbine, and heat exchanger?

8.4 Sketch an incompressible nozzle. How does the flow area change in the flow direction? How does the velocity change in the flow direction?

8.5 Repeat Question 8.4 for an incompressible diffuser.

8.6 Define the isentropic efficiency of a nozzle.

8.7 Sketch an isentropic and an actual nozzle process on a *T–s* diagram. Include the pressure contours that pass through the initial and final states.

8.8 Repeat Question 8.7 for a diffuser.

8.9 List several devices that can act as throttles.

8.10 Explain the operation and use of a throttling calorimeter.

8.11 Sketch a throttling process of an ideal gas on a *T–s* diagram. Include the pressure contours that pass through the initial and final states.

8.12 Distinguish between a positive-displacement pump and a dynamic (centrifugal) pump.

8.13 Create a sketch to illustrate the difference in the geometries of a centrifugal compressor and an axial-flow compressor.

8.14 Define the isentropic efficiency of a pump/compressor.

8.15 Sketch an isentropic and actual pumping process on a *T–s* diagram for a compressed liquid. Include the vapor dome and the pressure contours that pass through the initial and final states.

8.16 Sketch an isentropic and actual pumping process on a *T–s* diagram when the inlet flow is a saturated vapor. Include the vapor dome and the pressure contours that pass through the initial and final states.

8.17 Sketch an isentropic and actual pumping process on a *T–s* diagram for an ideal gas. Include the pressure contours that pass through the initial and final states.

8.18 List several applications of turbines. What type of turbine is typically used with the applications you list?

8.19 Define the isentropic efficiency of a turbine.

8.20 Sketch an isentropic and an actual turbine process on a *T–s* diagram when the inlet flow is a superheated vapor and the outflow is a saturated mixture. Include the vapor dome and the pressure contours that pass through the initial and final states.

8.21 Sketch an isentropic and an actual turbine process on a *T–s* diagram for an ideal gas. Include the pressure contours that pass through the initial and final states.

8.22 Distinguish among parallel-flow heat exchangers, counterflow heat exchangers, and cross-flow heat exchangers.

8.23 Define the heat-capacity rate.

Chapter 8 Problem Subject Areas

8.1–8.18	Nozzles and diffusers
8.19–8.30	Throttles
8.31–8.64	Pumps, compressors, and fans
8.65–8.88	Turbines
8.89–8.111	Heat exchangers
8.112–8.126	FE exam problems

PROBLEMS 8.1–8.18 Nozzles and diffusers

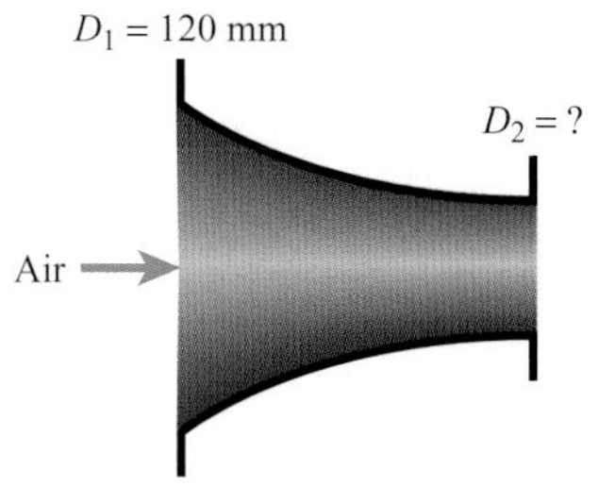

8.1 Air enters a nozzle at 1.30 atm and 25 °C with a velocity of 2.5 m/s. The nozzle entrance diameter is 120 mm. The air exits the nozzle at 1.24 atm with a velocity of 90 m/s. Determine the temperature of the exiting air and the nozzle exit diameter.

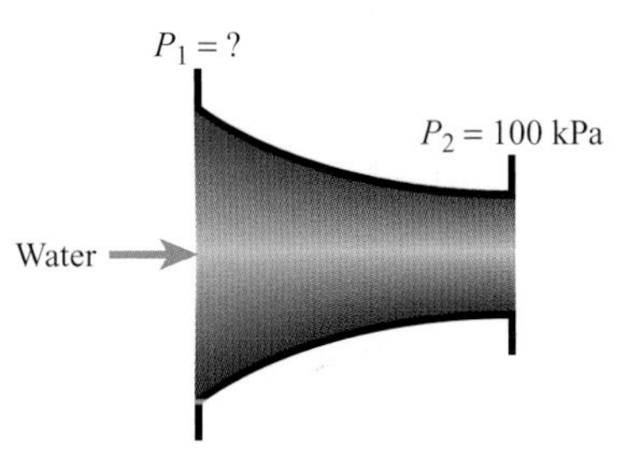

8.2 Water enters a nozzle with a velocity of 0.8 m/s and exits with a velocity of 5 m/s. The nozzle entrance diameter is 12 mm. Estimate the inlet pressure if the outlet pressure is 100 kPa and the water temperature is 300 K.

8.3 Consider the garden-hose nozzle described in Problem 3.29. For the flow conditions given, determine the pressure at the nozzle inlet. Assume that an ambient pressure of 100 kPa occurs at the nozzle exit.

(Credit: Bburdette / iStock / Getty Images Plus.)

8.4 Steam flows through a nozzle at 10^5 lb_m/min. The entering pressure and velocity are 250 psia and 400 ft/s, respectively; the exiting values are 1 psia and 4000 ft/s, respectively. Assuming the process is adiabatic, determine the specific enthalpy change of the steam (Btu/lb_m).

8.5 Steam at 100 lb_f/in^2 and 400 F enters a rigid, insulated nozzle with a velocity of 200 ft/s. The steam leaves at a pressure of 20 lb_f/in^2 and a velocity of 2000 ft/s. Assuming that the enthalpy at the entrance (h_i) is 1227.6 Btu/lb_m, determine the value of the enthalpy at the exit (h_e).

8.6 Air enters a diffuser at 100 kPa and 350 K with a velocity of 250 m/s. The air exits with a velocity of 30 m/s. Determine the temperature of the air at the outlet.

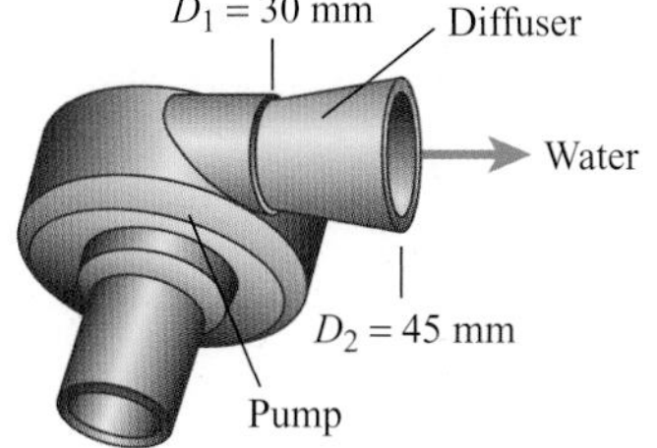

8.7 A diffuser is an integral part of a water pump. The diameter at the entrance to the diffuser section is 30 mm and the diameter at the exit is 45 mm. Water at 300 K flows through the pump at 2.2 kg/s. Determine the pressure rise in kPa and psi associated with the diffuser. Assume that the water is incompressible and use the specific volume and specific internal energy for saturated liquid water at 300 K.

8.8 Estimate the pressure drop $P_1 - P_2$ for the nozzle and flow conditions given in Example 8.1. Assume the flow is isothermal.

8.9 Steam flows through a nozzle from inlet conditions at 200 psia and 800 F to an exit pressure of 30 psia. The flow is reversible and adiabatic. For a flow rate of 10 lb_m/s, determine the exit area if the inlet velocity is negligible.

8.10 Steam at 400 psia and 600 F expands through a nozzle to 300 psia at a flow rate of 20,000 lb_m/hr. If the process occurs reversibly and adiabatically and the initial velocity is low, calculate (a) the velocity (ft/s) leaving the nozzle and (b) the exit area (in^2) of the nozzle.

8.11 Steam at 2 MPa and 290 °C expands to 1.400 MPa and 247 °C through a nozzle. If the entering velocity is 100 m/s, determine (a) the exit velocity and (b) the nozzle isentropic efficiency.

8.12 Consider a low-speed wind tunnel. At one point, the structure forms a nozzle with air inlet conditions of $V \approx 0$, $P = 14.7$ psia, and $T = 80$ F. If the nozzle isentropic efficiency is 90%, determine the air exit temperature when the exit pressure is 16 psia.

8.13 Steam enters a diffuser at 700 m/s, 200 kPa, and 200 °C. It leaves the diffuser at 70 m/s. Assuming reversible adiabatic operation, determine the final pressure and temperature.

8.14 Air enters a converging nozzle at 50 m/s, 3 atm, and 300 K. Assuming an adiabatic reversible process, find the outflow velocity and temperature when the nozzle exit pressure is 1 atm.

8.15 Following the combustion chamber of a jet engine, products of combustion and air enter the jet nozzle at low velocity at 100 psia and 1600 F. Determine the maximum velocity (ft/s) that can be obtained from the nozzle when the exit pressure is 11 psia. Assume the mixture properties are those of air. Also determine the exit diameter required to handle a flow of 23 lb_m/s.

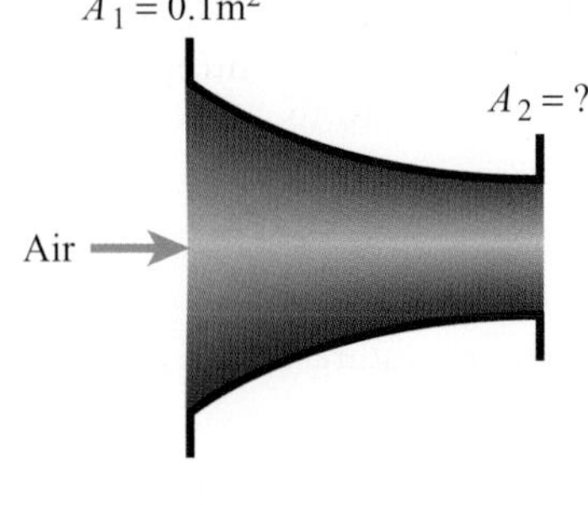

8.16 Air enters a converging nozzle at 10 m/s, 250 kPa, and 300 K. Assuming an adiabatic reversible process, find the outflow velocity and temperature when the nozzle exit pressure is 150 kPa. If the inflow nozzle area is 0.1 m^2, what is the area at the outflow of the nozzle?

8.17 Solve Problem 8.16 using EES or other software.

8.18 Using your computer solution from Problem 8.17, vary the outflow pressure from 150 kPa to 250 kPa (no nozzle). Plot T (ordinate) vs. $-A$ (abscissa). Also plot P (ordinate) vs. $-A$ (abscissa) and V (ordinate) vs. $-A$ (abscissa). For both of these plots, using the negative of the cross-sectional area on the horizontal axis means that the inflow conditions to the nozzle will be to the left and the outflow conditions will be to the right, similarly to the figure above for Problem 8.16.

8.19–8.30 Throttles

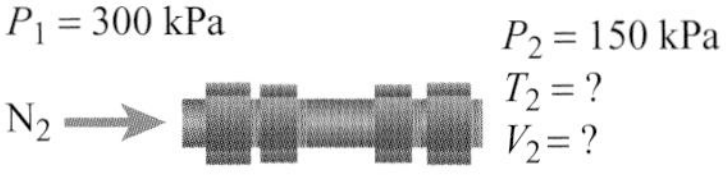

8.19 Nitrogen flows steadily through a 23-mm-diameter sintered metal filter. The pressure upstream of the filter is 300 kPa, the downstream pressure is 150 kPa. Determine the temperature and velocity of the N_2 downstream of the filter given an inlet temperature of 320 K and an inlet velocity of 2 m/s. The inlet and exit flow areas are identical.

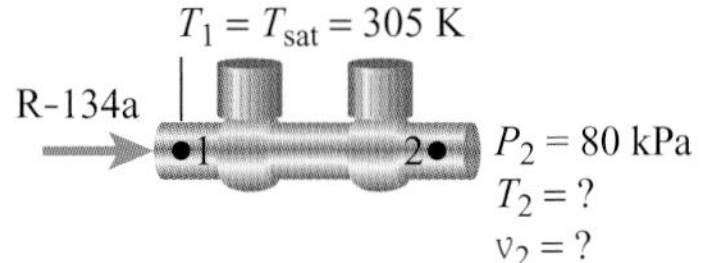

8.20 In a refrigerator, saturated liquid R-134a (a refrigerant) is throttled from an initial temperature of 305 K to a final pressure of 80 kPa. Determine the final temperature and specific volume of the R-134a.

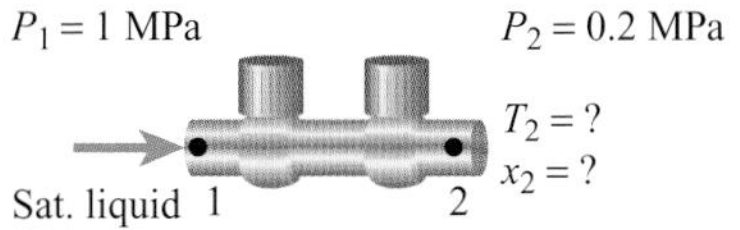

8.21 Saturated liquid water at 1 MPa enters a throttling device and exits at 0.2 MPa. Determine the temperature and quality of the exiting liquid–vapor mixture.

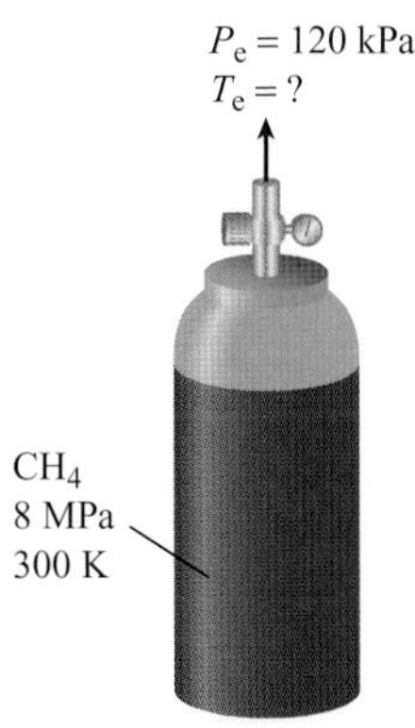

8.22 Methane stored in a tank at 8 MPa and 300 K is used as a fuel for a laboratory-scale gas-turbine combustor. The methane is throttled as it passes through a pressure regulator. The regulated pressure is 120 kPa. Determine the temperature of the methane at the exit of the pressure regulator. Assume the process is adiabatic and neglect any kinetic energy changes.

8.23 Water at 140 °C and 10 MPa is adiabatically throttled to a pressure of 0.2 MPa. Determine the quality after throttling. What is the change in specific entropy across the throttle? Sketch this process on a *T–s* diagram. Include the vapor dome and the pressure contours at the initial and final pressures.

8.24 Water is throttled (in a constant-enthalpy process) across a valve from 20.0 MPa and 260 °C to 0.143 MPa. Determine the temperature of the H_2O downstream of the valve. What is the physical state of the H_2O (i.e., superheated vapor, subcooled liquid, etc.)? Sketch the process on a *T–s* diagram. Include the vapor dome and the pressure contours at the initial and final pressures.

8.25 Air is throttled (in a constant-enthalpy process) across a valve from 20.0 MPa and 260 °C to 0.143 MPa. Determine the temperature and specific volume of the air downstream of the valve. Also determine the air specific entropy change across the valve (kJ/kg·K).

8.26 Superheated steam at 4 MPa and 660 K is adiabatically throttled to a pressure of 100 kPa. What is the temperature after throttling? What is the change in specific entropy across the throttle?

8.27 Saturated vapor at 2 MPa is adiabatically throttled to a pressure of 150 kPa. What is the temperature after throttling? What is the change in specific entropy across the throttle? Sketch the process on a *T–s* diagram. Include the vapor dome and the pressure contours at the initial and final pressures.

8.28 Air is throttled (in a constant-enthalpy process) across a valve from 5.0 MPa and 300 °C to 120 kPa. Determine the temperature and specific volume of the air downstream of the valve. Also determine the air specific entropy change across the valve (kJ/kg·K).

8.29 Solve Problem 8.28 using EES or other software.

8.30 Using your computer solution from Problem 8.29, vary the outflow pressure from 100 kPa to 5.0 MPa (no throttle). Plot a P–v diagram for this range of pressures. Compare your results to Figure 7.8.

8.31–8.64 Pumps, compressors, and fans

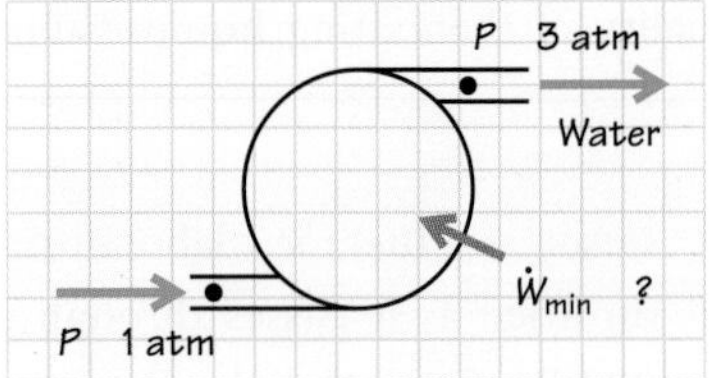

8.31 It is desired to pump water at 50 gal/min from 1 to 3 atm. The water temperature is 25 °C. Determine the minimum power input required, assuming ideal frictionless operation and incompressible flow.

8.32 Determine the minimum pump power required to pump water from one reservoir to another at an elevation of 110 m above the first. The desired flow rate is 0.2 kg/s. Assume the water is at room temperature (25 °C) and atmospheric pressure in each reservoir. Approximate the water as incompressible.

8.33 Consider the situation described in Problem 8.32. Calculate the actual power required by the pump if the pump efficiency is 85%.

8.34 A 1-hp electric motor drives a water pump. Water enters the pump at 90 kPa and 300 K. The volumetric flow rate is 2×10^{-3} m^3/s. Estimate the maximum possible outlet pressure of the pump when the water is considered incompressible.

Photograph courtesy of U.S. Department of Energy.

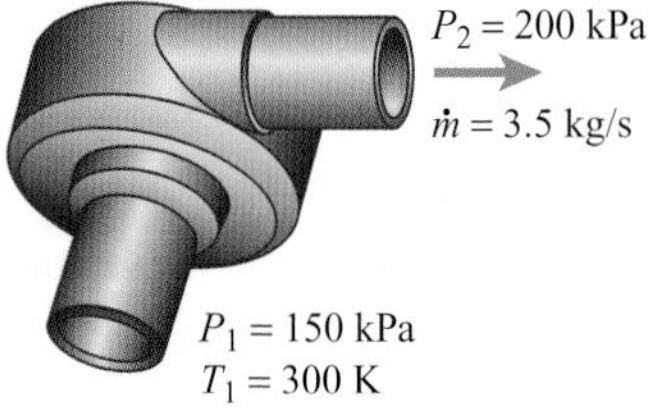

8.35 Water enters a pump at 300 K and 150 kPa and exits at 200 kPa with a flow rate of 3.5 kg/s. Estimate the shaft power required to pump the water at these conditions. Neglect all friction and heat-transfer effects. Assume the temperature of the water at the exit is approximately 300 K and regard the water as incompressible.

8.36 Consider the situation described in Problem 8.35. If the pump exit diameter is 50 mm, determine the ratio of the pump power to the kinetic energy rate of the exiting water.

8.37 Air initially at 100 kPa and 300 K is compressed to 350 kPa and 435 K in a multistage compressor. The air enters the 0.3-m-diameter inlet at 12 m/s. Determine (a) the work input per unit mass of air (i.e., $\dot{W}_{\text{shaft}}/\dot{m}$) and (b) the input power required to operate the compressor.

8.38 Consider the situation described in Problem 8.37. Determine the isentropic efficiency of the compressor for the given operating conditions.

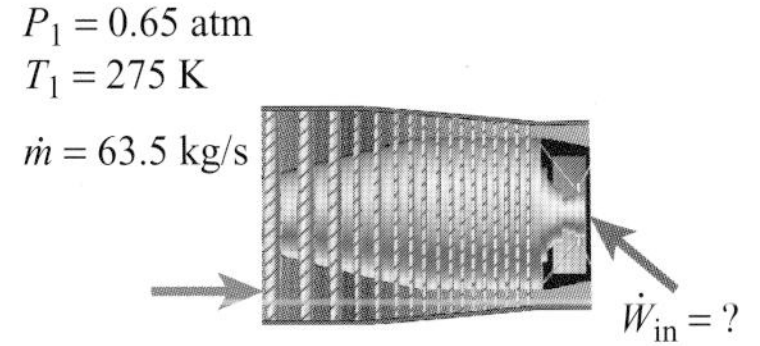

8.39 Air enters a multistage compressor of a jet engine at 0.65 atm and 275 K with a flow rate of 63.5 kg/s. The compressor has an overall pressure ratio of 24:1 and an isentropic efficiency of 94%. Determine the power required to drive the compressor and the temperature of the air at the compressor outlet.

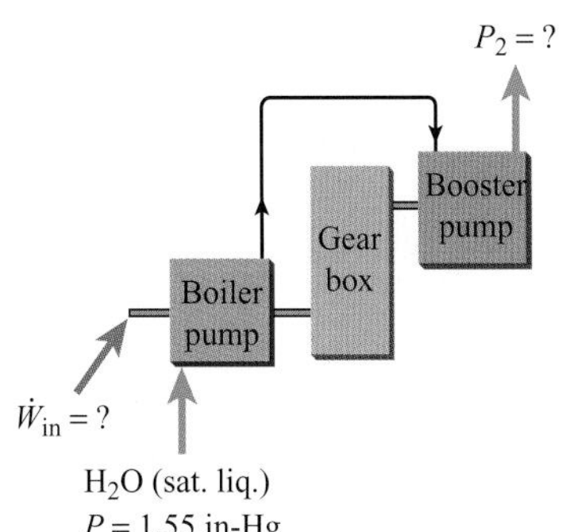

8.40 Consider a combination-boiler feed pump and booster pump similar to the arrangement shown in Example 8.8. The flow rate is 2,484,000 lb_m/hr, and the specific enthalpy increase from inlet to outlet is 16.79 Btu/lb_m. Saturated liquid water enters at a pressure of 1.55 in -Hg absolute. Estimate (a) the power to drive the pump in MW and (b) the outlet pressure in MPa.

8.41 Steam at 10 psia with a quality of 0.90 enters an adiabatic steady-flow compressor at a rate of 50 lb_m/s. The steam leaves at 400 F and 100 psia. Determine the horsepower required to drive the compressor.

8.42 Air is compressed in a steady-flow reversible process from 15 psia and 80 F to 120 psia. Determine the work and the heat transfer per pound of air compressed for each of the following types of process: (a) adiabatic, (b) isothermal, and (c) polytropic ($n = 1.25$).

8.43 The compressor at the inlet section of a jet engine has a diameter of 4 ft and receives air at 730 ft/s and 20 F. The air is compressed reversibly and adiabatically from 4 to 36 psia. The discharge velocity is negligible. Determine the horsepower required to operate the compressor.

8.44 Air is compressed through a pressure ratio of 4:1. Assume that the process is steady-flow and adiabatic and that the temperature increases by a factor of 1.65. Calculate the entropy change.

8.45 Determine the isentropic efficiency of a water pump when water enters as a saturated liquid at 96.5 kPa and exits at 5 MPa and 106 °C.

8.46 Water is pumped through pipes embedded in the concrete of a large dam. The water enters the pipes at 100 psia and exits at 20 psia. In picking up the heat of hydration of the curing concrete, the water increases in temperature from

50 to 100 F. During curing, the heat of hydration for a section of the dam is 140,000 Btu/hr. Determine (a) the required water flow rate for this section (lb_m/hr) and (b) the minimum size of motor needed to drive the pump (hp). Assume that the pump makes up the pressure lost through the pipes. Ignore changes in kinetic and potential energy.

8.47 A pump delivers 160 lb_m/hr of water at a pressure of 1000 psia when the inlet conditions are 15 psia and 100 F. Ignoring changes in kinetic and potential energies, find the minimum size of motor (hp) required to drive the pump.

8.48 The water table of a housing development is 400 ft below the surface. You have to install a well pump that will deliver 15 gal/min of water (8.33 lb_m/gal and 0.016 ft^3/lb_m) at a pressure of 30 psig at the surface. What horsepower motor should you use?

8.49 A booster pump is used to move water from the basement equipment room to the twelfth floor of an apartment building at the rate of 363 kg/min. The elevation change is 40 m between the basement and the twelfth floor. Determine the minimum size of pump (hp) required.

8.50 The discharge of a pump is 3 m above the inlet. Water enters at a pressure of 138 kPa and leaves at a pressure of 1.38 MPa. The specific volume of the water is 0.001 m^3/kg. If there is no heat transfer and no changes in kinetic or internal energies, what is the specific work (kJ/kg)?

8.51 A centrifugal pump receives liquid nitrogen at –340 F at the rate of 100 lb_m/s. The nitrogen enters the pump as a liquid at 15 psia. The discharge pressure is 400 psia. Estimate the minimum size of the motor (hp) needed to drive this pump.

8.52 Water enters a pump at 10 kPa and 35 °C and leaves at 5 MPa. For reversible adiabatic operation, calculate the work done per unit mass of working fluid and the exit temperature.

8.53 A compressor uses air as the working fluid. The air enters at 101 kPa and 16 °C and exits at 1.86 MPa and 775 °C. What is the compressor isentropic efficiency?

8.54 Air is compressed through a pressure ratio of 8:1 in a steady-flow process. For inlet conditions of 100 kPa and 25 °C, calculate the work, the heat transfer, and the entropy change per unit mass if the process is polytropic ($n = 1.25$). Sketch this process on T–s and P–v diagrams. Also sketch the process for an adiabatic compression using the same inlet conditions and pressure ratio.

8.55 Air is compressed from 101.3 kPa and 15 °C to 700 kPa. Determine the power required to process 0.3 m^3/min at the outlet if the operation is (a) polytropic ($n = 1.25$) and (b) isentropic.

8.56 A 200-ft^3/min flow of air at 14.7 psia and 60 F enters a fan with negligible inlet velocity. The fan discharge duct has a cross-sectional area of 3 ft^2. The process across the fan is isentropic (reversible and adiabatic). The fan discharge pressure is 14.8 psia. Determine (a) the velocity in the discharge duct (ft/min) and (b) the size of motor required to drive the fan (hp).

8.57 A fan is used to provide fresh air to the welding area in an industrial plant. The fan takes in outside air at 27 °C and 101 kPa at a rate of 34 m^3/min with negligible inlet velocity. In the 0.93-m^2 duct leaving the fan, the air pressure is 6.9 kPa gage. If the process is assumed to be reversible and adiabatic (isentropic), determine the size of motor (hp) needed to drive the fan.

8.58 Water flowing at 950,000 liters/min enters a pump in a power plant at 150 °C and 0.5 MPa. The pump increases the water pressure to 10 MPa (v = 0.063 m^3/kg). Determine (a) the mass flow rate (kg/hr), (b) the volume flow rate at discharge (liters/min), (c) the temperature change across the pump (°C), and (d) the enthalpy change across the pump (kJ/kg).

8.59 Air enters a multistage compressor of a jet engine at 50 kPa and 270 K with a flow rate of 40 kg/s. The compressor has an overall pressure ratio of 20:1. Determine the power required to drive the compressor and the temperature of the air at the compressor outlet. Use the following specific heat at the average temperature: $\bar{c}_p$ = 1.021 kJ/kg·K.

8.60 Solve Problem 8.59 using EES or other software.

8.61 Using your computer solution from Problem 8.60, vary the pressure ratio from 1 (no compressor) to 40. Plot the outflow temperature and the compressor power as a function of the pressure ratio. Put the pressure ratio on the abscissa (x-axis).

8.62 Determine the power required to isentropically pump 25 kg/s of water from saturated liquid at 50 kPa to 3.0 MPa. Assume incompressible flow.

8.63 Solve Problem 8.62 using EES or other software.

8.64 Using your computer solution from Problem 8.63, vary the outflow pressure from 300 kPa to 6 MPa. Plot the pump power as a function of the outflow pressure. Put pressure on the abscissa (x-axis) of the plot.

8.65–8.88 Turbines

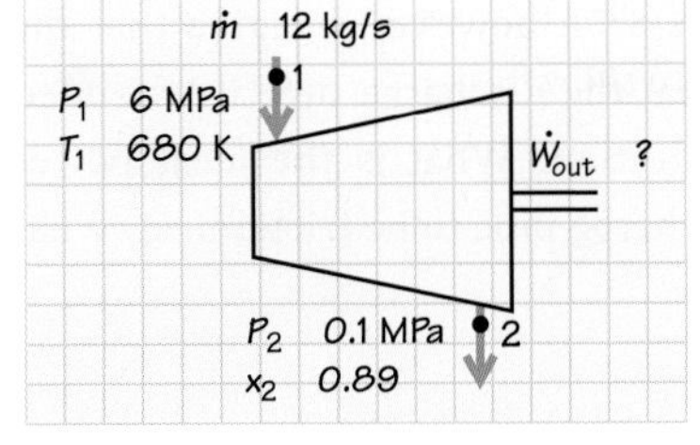

8.65 Steam enters a turbine superheated at 6 MPa and 680 K and exits the turbine at 0.1 MPa with a quality of 0.89. The steam flow rate is 12 kg/s. Determine the power delivered by the turbine.

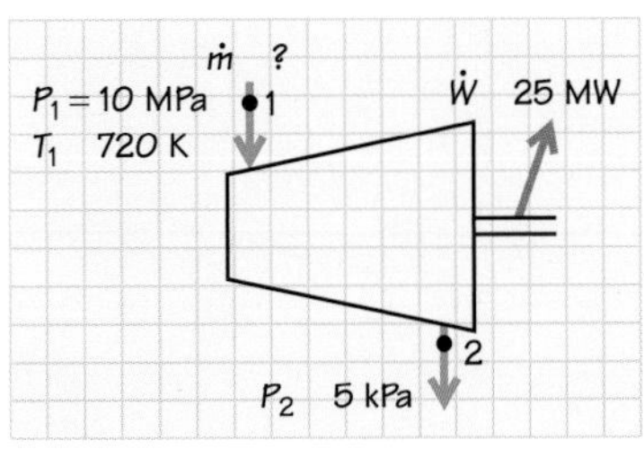

8.66 Determine the flow rate required to produce 25 MW of shaft power from a steam turbine in which the steam enters at 10 MPa and 720 K and exits at 5 kPa for the following cases: (a) an ideal turbine and (b) a turbine with an isentropic efficiency of 96%.

8.67 The steam flow rate in a power plant is 650,000 lb_m/hr. The steam enters the turbine at 500 psia and 1000 F. The turbine exhaust pressure is 1.0 psia and the quality of the exiting steam is 0.925. Determine (a) the turbine power output (kW) and (b) the isentropic efficiency of the turbine.

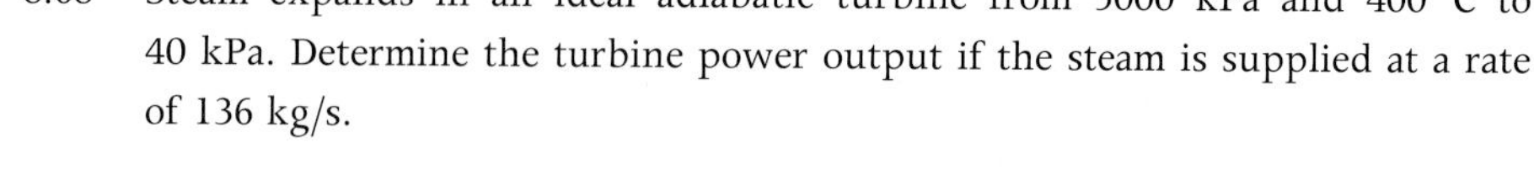
8.68 Steam expands in an ideal adiabatic turbine from 5000 kPa and 400 °C to 40 kPa. Determine the turbine power output if the steam is supplied at a rate of 136 kg/s.

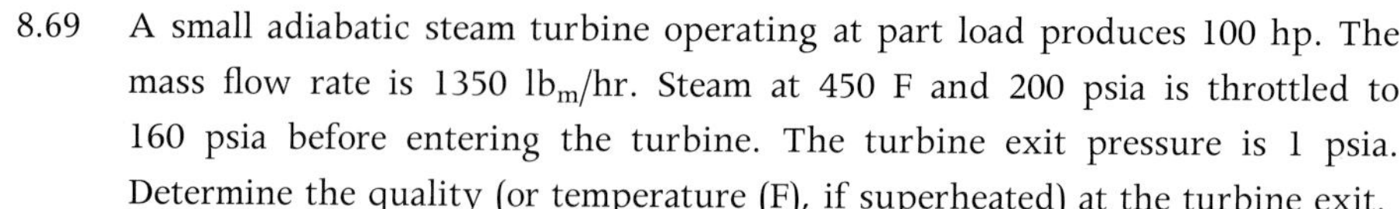
8.69 A small adiabatic steam turbine operating at part load produces 100 hp. The mass flow rate is 1350 lb_m/hr. Steam at 450 F and 200 psia is throttled to 160 psia before entering the turbine. The turbine exit pressure is 1 psia. Determine the quality (or temperature (F), if superheated) at the turbine exit.

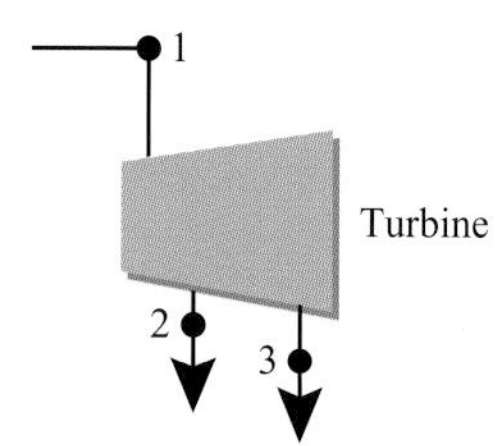

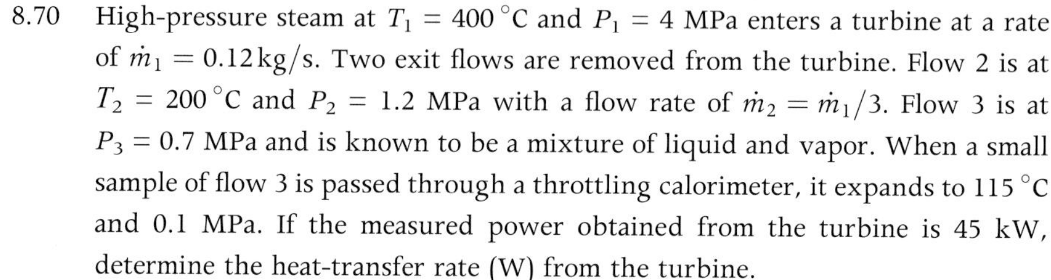
8.70 High-pressure steam at $T_1 = 400\,^\circ\text{C}$ and $P_1 = 4$ MPa enters a turbine at a rate of $\dot{m}_1 = 0.12\,\text{kg/s}$. Two exit flows are removed from the turbine. Flow 2 is at $T_2 = 200\,^\circ\text{C}$ and $P_2 = 1.2$ MPa with a flow rate of $\dot{m}_2 = \dot{m}_1/3$. Flow 3 is at $P_3 = 0.7$ MPa and is known to be a mixture of liquid and vapor. When a small sample of flow 3 is passed through a throttling calorimeter, it expands to 115 °C and 0.1 MPa. If the measured power obtained from the turbine is 45 kW, determine the heat-transfer rate (W) from the turbine.

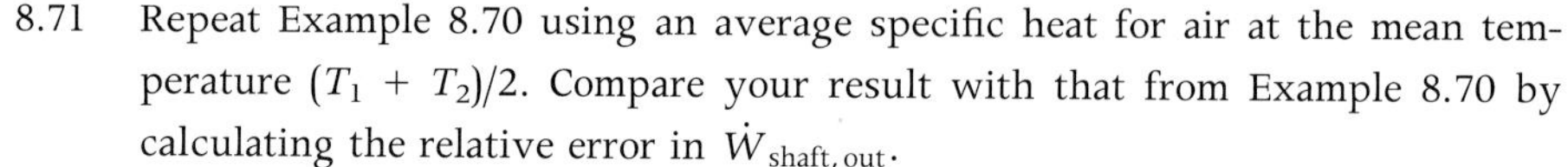
8.71 Repeat Example 8.70 using an average specific heat for air at the mean temperature $(T_1 + T_2)/2$. Compare your result with that from Example 8.70 by calculating the relative error in $\dot{W}_{\text{shaft, out}}$.

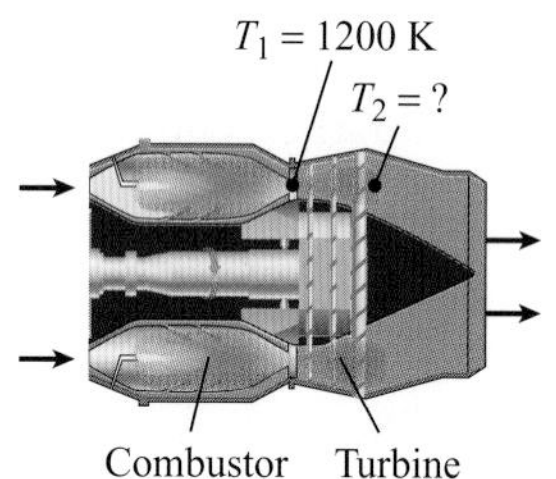

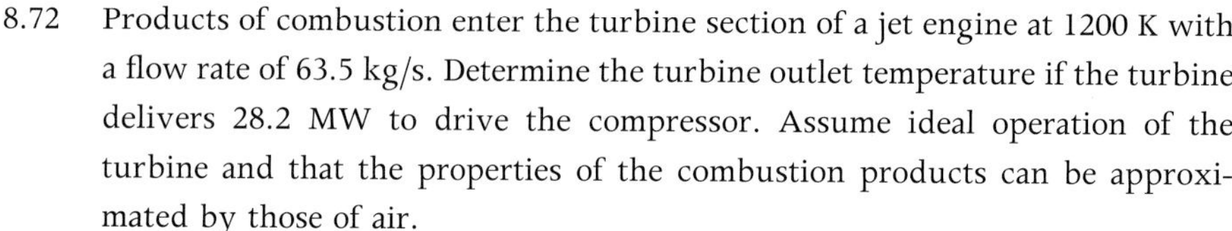
8.72 Products of combustion enter the turbine section of a jet engine at 1200 K with a flow rate of 63.5 kg/s. Determine the turbine outlet temperature if the turbine delivers 28.2 MW to drive the compressor. Assume ideal operation of the turbine and that the properties of the combustion products can be approximated by those of air.

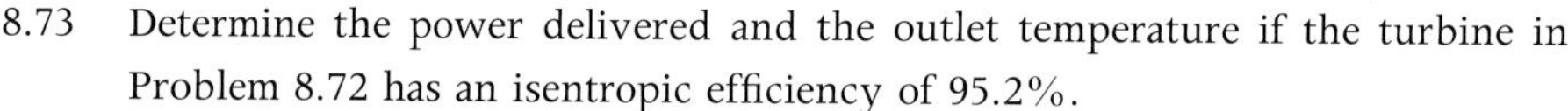
8.73 Determine the power delivered and the outlet temperature if the turbine in Problem 8.72 has an isentropic efficiency of 95.2%.

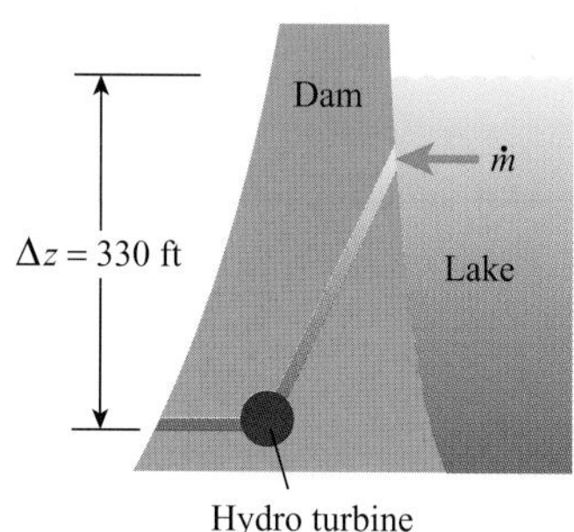

8.74 The elevation change available at the Grand Coulee Dam is 330 ft. What is the maximum possible power per unit mass flow rate that a hydro turbine can produce at this site? What mass flow rate of water is required to produce 100 MW of shaft power?

8.75 Consider the situation described in Example 8.74; however, imagine that the turbine is now located at one-half z_1 and that the discharge pipe is extended down to z_2 (= 0), the original discharge location. What is the ideal power produced by the turbine? Assuming the discharge pipe to be frictionless, what is the pressure at the turbine outlet? Discuss.

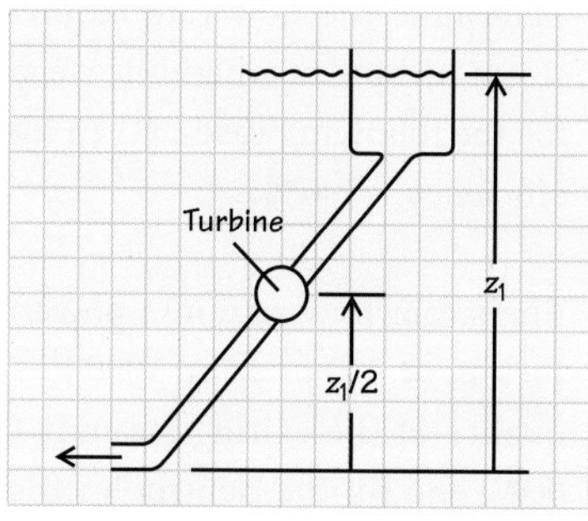

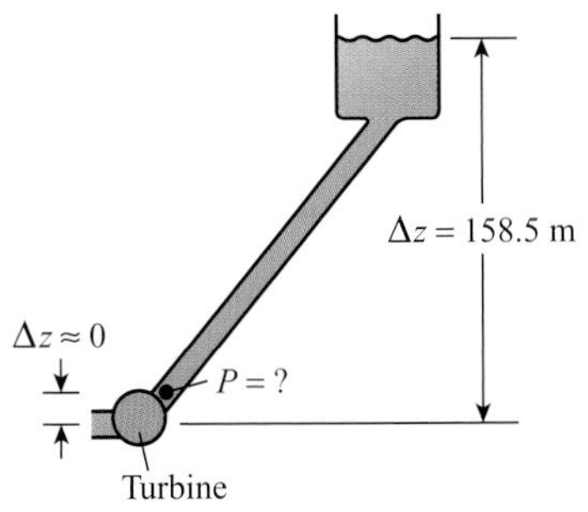

8.76 Using the results from Example 8.75, estimate the pressure at the hydro turbine inlet. Neglect any elevation change across the turbine.

8.77 Air at 50 psia and 90 F flows at the rate of 1.6 lb_m/s through an insulated ideal turbine. If the air delivers 11.5 hp to the turbine blades, at what temperature does the air leave the turbine?

8.78 A turbine receives steam at a pressure of 1000 psia and 1000 F and exhausts the steam at 3 psia. The velocity of the steam at the inlet is 50 ft/s. At the outlet, which is 10 ft higher, the velocity is 1000 ft/s. Assuming that the operation is reversible and adiabatic, determine the work produced per unit mass.

8.79 Steam flows through a turbine in a nuclear power plant at 1,230,000 lb_m/hr. The turbine inlet conditions are 500 psia and 1200 F, and the turbine outlet pressure is 5 psia. Determine the maximum turbine power output.

8.80 Air at 50 psia and 90 F flows through an expander (like a turbine) at the rate of 1.6 lb_m/s to an exit pressure of 14.7 psia. (a) What is the minimum temperature attainable at the expander exit if changes in kinetic energy are neglected? (b) If the inlet velocity is not to exceed 12 ft/s, what inlet diameter is required?

8.81 Steam enters a turbine as a saturated vapor at 2 MPa and exits at 101 kPa with a quality of 0.92. Determine the turbine isentropic efficiency.

8.82 For the turbine in Problem 8.81, how does the isentropic efficiency change if the exit pressure is decreased to 35 kPa with the same outflow quality of 0.92?

8.83 Steam flows at the rate of 12,000 lb_m/min through a turbine from 500 psia and 700 F to an exhaust pressure of 1 psia. Determine the ideal output of the turbine (hp). If the specific entropy increases between the inlet and the outlet of the turbine by 0.1 $Btu/lb_m{\cdot}R$, determine the isentropic turbine efficiency.

8.84 A high-speed turbine produces 1 hp while operating on compressed air. The inlet and outlet conditions are 70 psia and 85 F and 14.7 psia and −50 F, respectively. Assume changes in kinetic and potential energies are negligible. Determine the mass flow rate.

8.85 Heat is transferred from a steam turbine at a rate of 40,000 Btu/hr while the steam mass flow rate is 10,000 lb_m/hr. Using the following data for the steam entering and leaving the turbine, find the work rate. The gravitational acceleration is $g = 32.17\ ft/s^2$.

	Inlet Conditions	Outlet Conditions
Pressure	200 psia	15 psia
Temperature	700 F	—
Velocity	200 ft/s	600 ft/s
Elevation	16 ft	10 ft
Enthalpy, h	1361.2 Btu/lb_m	1150.8 Btu/lb_m

Also determine the fraction of the power contributed by each term of the first law compared to that associated with the change in enthalpy.

8.86 Steam at 600 °C and 12 MPa with a flow rate of 65 kg/s enters an isentropic turbine. What is the power output from the turbine if the outflow pressure is 70 kPa?

8.87 Solve Problem 8.86 using EES or other software.

8.88 Steam with a flow rate of 65 kg/s enters an isentropic turbine at 12 MPa and exits at 70 kPa. Using your computer solution from Problem 8.87, vary the inflow temperature from 500 °C to 650 °C. Plot the turbine power as a function of the inflow temperature. Put temperature on the abscissa (x-axis) of the plot.

8.89–8.111 Heat exchangers

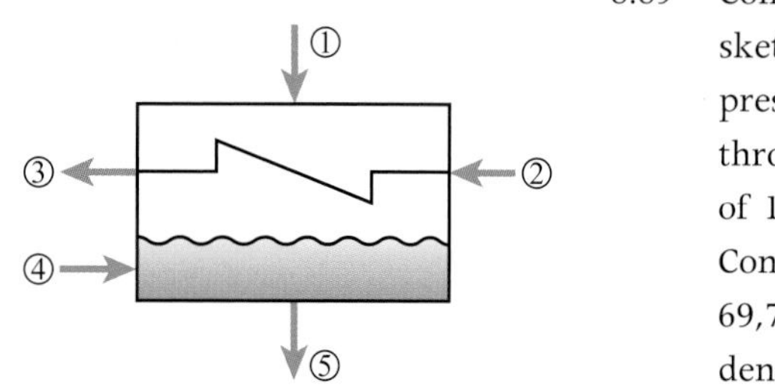

8.89 Consider a closed feedwater heater from a steam power plant, as shown in the sketch. Steam enters the heater at station 1 with a flow rate of 66,256 lb_m/hr at a pressure of 28.6 psia and an enthalpy of 1246.2 Btu/lb_m. The feedwater flows through the heater at 1,499,628 lb_m/hr, entering at station 2 with an enthalpy of 161.2 Btu/lb_m and exiting at station 3 with an enthalpy of 211.0 Btu/lb_m. Condensate from another feedwater heater enters at station 4 with a flow rate of 69,708 lb_m/hr and an enthalpy of 221.3 Btu/lb_m and is mixed with the condensing steam inside the heater.

Determine the temperature of the incoming steam (station 1) in degrees Fahrenheit and in kelvins and determine the mass flow rate and enthalpy of the liquid exiting at station 5 in both US customary and SI units. Also, estimate the temperature of this stream assuming that the mixing within the feedwater heater occurs at constant pressure (P_1).

Sat. liquid out
150 kPa
Outer tube (annulus)
Inner tube
Juice in
$T = 4$ °C
Juice out
$T = 93$ °C
Sat. steam in
150 kPa

8.90 In the production of orange juice, it is desired to heat the juice from 4 °C to 93 °C in a tubular heat exchanger. To accomplish this task, saturated steam enters the outer tube of the heat exchanger at 150 kPa and exits as a saturated liquid at the same pressure. For an orange juice flow rate of 2000 kg/hr ($\bar{c}_p$ = 3.9 kJ/kg·K), determine the required steam flow rate.

8.91 Steam enters the condenser of a modern power plant at a pressure of 1 psia and a quality of 0.98. The condensate leaves at 1 psia and 80 F. Determine (a) the heat rejected per pound and (b) the change in specific volume between inlet and outlet.

8.92 Water enters a heat exchanger at 180 F and 20 psia and leaves at 160 F and 19.8 psia. Air enters the heat exchanger at 70 F and 15 psia and leaves at 100 F and 14.7 psia. The mass flow rate of the water is 40 lb_m/min. Determine the heat-transfer rate (Btu/min) to the air and the air mass flow rate (lb_m/min).

8.93 Water is heated by air in a heat exchanger. The water enters at 420 K and 0.3 MPa and leaves at 560 K and 0.3 MPa. The mass flow rate of the water is 1.2 kg/s. The air enters at 620 K and 0.1 MPa at a rate of 2 kg/s. Determine the heat-transfer rate between the two fluids (kW) and the exit temperature (K) of the air.

8.94 In a shell-and-tube heat exchanger, cold water ($c_{p,\mathrm{C}} = 4.18$ kJ/ kg·°C) enters at 50 °C with a flow rate of 900 kg/hr. Hot water ($c_{p,\mathrm{H}} = 4.18$ kJ/ kg·°C) enters at 370 °C with a flow rate of 1260 kg/hr. The hot and cold streams are in parallel flow as shown in Figure 8.36a. If the heat exchanger area is infinitely large, the hot and cold streams will exit at the same temperature. Determine this outflow temperature and the heat transfer rate for this very long heat exchanger.

8.95 The lubricating oil cooler for a large gas turbine consists of a shell-and-tube heat exchanger. The oil ($c_p = 0.5$ Btu/lb$_\mathrm{m}$·F) enters at 320 F, flowing through the tubes at 7000 lb$_\mathrm{m}$/ hr. The water coolant ($c_p = 1.0$ Btu/lb$_\mathrm{m}$·F) enters at 80 F and flows through the shell at 9000 lb$_\mathrm{m}$/hr. The hot and cold streams are in parallel flow as shown in Figure 8.36a. If the heat exchanger area is infinitely large, the hot and cold streams will exit at the same temperature. Determine this outflow temperature and heat transfer rate for this very long heat exchanger.

8.96 A shell-and-tube heat exchanger has liquid water on the shell side and oil in the tubes. The hot-side water ($c_p \cong 4.18$ kJ/kg·°C) enters at 260 °C flowing at 9100 kg/hr. The cold-side oil ($c_p = 3.55$ kJ/kg·°C) enters at 30 °C flowing at 27,300 kg/hr. For a water outflow temperature of 150 °C, determine the outflow temperature of the oil and the heat transfer rate between the two streams.

8.97 A shell-and-tube heat exchanger uses 6800 kg/hr of water ($c_p = 4.18$ kJ/kg·°C) entering at 25 °C to cool 18,000 kg/h of oil ($c_p = 2.9$ kJ/kg·°C) entering at 100 °C. For a water outflow temperature of 58 °C, determine the outflow temperature of the oil and the heat-transfer rate between the two streams.

8.98 A shell-and-tube heat exchanger has liquid water on the shell side and oil in the tubes. The hot-side water ($c_p \cong 1.0$ Btu/lb$_\mathrm{m}$·F) flows at 18,000 lb$_\mathrm{m}$/hr and enters at 500 F. The tube-side oil ($c_p = 0.85$ Btu/lb$_\mathrm{m}$·F) flows at 50,000 lb$_\mathrm{m}$/hr and enters at 80 F. For a water outflow temperature of 210 F, determine the outflow temperature of the oil and the heat-transfer rate between the two streams.

8.99 A counterflow heat exchanger uses 6500 kg/hr of water ($c_p = 4.18$ kJ/kg·°C) entering at 25 °C to cool 13,000 kg/hr of oil ($c_p = 2.09$ kJ/kg·°C) entering at 100 °C. If the heat-transfer rate between the streams is 75 kW, determine the outflow temperature of the two streams.

8.100 A steam condenser is made with 115 parallel tubes each having an inner diameter of 0.584 inches. The cooling water ($c_p = 1.0$ Btu/lb$_\mathrm{m}$·F) enters the tubes at 70 F with an average velocity of 5 ft/s through each tube. Saturated-vapor steam at

5 psia condenses on the outside of the tubes. If the cooling water exits the condenser at 104 F, determine the condensation rate of the steam in lb_m/hr.

8.101 A steam condenser is made with 125 tube paths each having an inside diameter of 11.7 mm. Cooling water ($c_p = 4.18$ kJ/kg·°C) enters the tubes at 42 °C with an average velocity of 1.37 m/s through each tube. Saturated-vapor steam at 14 kPa condenses on the outside of the tubes. If the cooling water exits the condenser at 24 °C, determine the condensation rate of the steam in kg/hr.

8.102 A one-shell-pass, one-tube-pass, counterflow heat exchanger is used to cool oil ($c_p = 1.811$ kJ/kg·°C). The oil enters the shell at 120 °C flowing at 15,000 kg/hr. Cooling water (c_p= 4.18 kJ/kg ·°C) enters the tubes at 30 °C, flowing at 6500 kg/hr. If the cooling water exits the heat exchanger at 72 °C, find the outlet temperatures of the oil and the rate of heat transfer between the two flows.

8.103 A one-shell-pass, two-tube-pass heat exchanger is to be used to cool ethylene glycol with water. The ethylene glycol (c_p= 2.505 kJ/kg·°C) enters at 85 °C flowing at 1 kg/s; the water (c_p= 4.18 kJ/kg·°C) enters at 15 °C, flowing at 2 kg/s. If the cooling water exits the heat exchanger at 31 °C, find the outlet temperatures of the ethylene glycol and the rate of heat transfer between the two flows.

8.104 The lubricating oil in an engine is cooled by air in a cross-flow heat exchanger; both fluids are unmixed. Atmospheric air (c_p= 1.007 kJ/kg·°C) enters at 30 °C, flowing at 0.55 kg/s. The oil (c_p= 2.00 kJ/kg·°C) enters at 75 °C, flowing at 0.025 kg/s. Determine the exit temperature of the air if the oil exits at 46 °C.

8.105 Saturated steam condenses in a horizontal steam condenser at a pressure of 14 kPa. Each of the 180 parallel tubes in the condenser has an inside diameter of 12.6 mm and an average liquid water velocity of 1 m/s. If the cooling water (c_p= 4.18 kJ/kg·°C) enters at 20 °C and exits at 48 °C, find the steam condensation rate (kg/hr).

8.106 A one-shell-pass, two-tube-pass heat exchanger heats 27,000 kg/hr of water (c_p= 4.21 kJ/kg·°C) from 85 °C to 100 °C by the condensation of saturated steam at 345 kPa. Determine the condensation rate of the steam in kg/hr.

8.107 A counterflow heat exchanger heats 2160 kg/hr of water (c_p= 4.18 kJ/kg·°C) from 35 °C to 90 °C using oil ($c_p = 2.1$ kJ/kg·°C), which enters at 175 °C with a flow rate of 3240 kg/hr. Determine the outflow temperature of the oil.

8.108 A cross-flow heat exchanger is used to heat a 3-kg/s flow of water ($c_p = 4.18$ kJ/kg·°C) from 35 °C to 85 °C in a propane water heater. The hot combustion gases are approximated as air (1 atm), which enters the heat exchanger at 225 °C and exits at 100 °C. Determine the flow rate of the combustion gases that is required.

8.109 The combustion-air supplied to a boiler furnace is preheated in a cross-flow heat exchanger using hot exhaust gases from the furnace. The hot gases ($c_p = 1.075$ kJ/kg·°C) enter the exchanger at 825 °C, flowing at 15 kg/s, while the air ($c_p = 1.007$ kJ/kg·°C) enters at 25 °C and flows at 10 kg/s. Determine the outflow temperature of the combustion gases when the air is heated to 575 °C.

8.110 A one-shell-pass, one-tube-pass heat exchanger heats the water flow in the tubes using hot water on the shell side. Water (4.21 kJ/kg·°C) enters the shell side at 150 °C and leaves at 40 °C, flowing at 9000 kg/hr. The tube water (4.18 kJ/kg·°C) enters at 25 °C and leaves at 65 °C. Determine the flow rate of the water in the tubes.

8.111 Saturated steam at 350 K condenses within the shell of a shell-and-tube condenser. There are 150 brass tubes with inside diameters of 14.8 mm. The liquid cooling water (4.18 kJ/kg·°C) flows through the tubes at an average velocity of 1.52 m/s, entering at 25 °C and leaving at 50 °C. Determine the condensation rate of the steam in kg/hr.

8.112–8.126 FE exam questions

8.112 Incompressible water flow with velocity of 1.2 m/s enters a round nozzle with diameter of 4 cm. If the outflow diameter of the nozzle is 1 cm, what is the outflow velocity? a. 19.2 m/s, b. 4.8 m/s, c. 0.3 m/s, d. 38.4 m/s.

8.113 Superheated steam at 300 kPa and 440 K enters an adiabatic nozzle. If outflow pressure from the nozzle is 100 kPa, what is the outflow temperature? a. 440 K, b. 400 K, c. 372 K, d. 385 K.

8.114 Air at 400 kPa and 350 K enters an adiabatic nozzle. If outflow pressure from the nozzle is 100 kPa, what is the outflow temperature? a. 520 K, b. 350 K, c. 372 K, d. 236 K.

8.115 Air at 350 K has a velocity of 5 m/s when it enters a nozzle. What is the velocity at the outflow of the nozzle if the temperature is 300 K? a. 32 m/s, b. 11.2 m/s, c. 8.7 m/s, d. 121 m/s.

8.116 Air is adiabatically throttled from a pressure of 400 kPa to 150 kPa. If the inflow temperature is 350 K, what is the outflow temperature? a. 320 K, b. 380 K, c. 350 K, d. 300 K.

8.117 Steam at 1 MPa and 560 K is adiabatically throttled to 300 kPa. What is the outflow temperature? a. 560 K, b. 583 K, c. 535 K, d. 551 K.

8.118 Saturated liquid at 2 MPa is adiabatically throttled to 300 kPa. What is the outflow quality? a. 0.16, b. 0.92, c. 0.08, d. 0.23.

8.119 What power is required to isentropically pump 3 kg/s of saturated water liquid from 100 kPa to 3 MPa? a. 3.0 kW, b. 5.4 kW, c. 2.6 kW, d. 9.1 kW.

8.120 Saturated water at a flow rate of 15 kg/s is pumped from 50 kPa to 4 MPa. What power is required if the pump is 85% efficient? a. 52 kW, b. 72 kW, c. 3.5 kW, d. 61 kW.

8.121 Air at a flow rate of 20 kg/s at 300 K and 140 kPa enters an isentropic compressor. Determine the power required to raise the outflow pressure to 2.0 MPa. a. 6.9 MW, b. 690 kW, c. 3.5 MW, d. 350 kW.

8.122 Air at a flow rate of 30 kg/s at 300 K and 120 kPa enters a compressor having an isentropic efficiency of 90%. Determine the power required to raise the outflow pressure to 2.0 MPa. a. 12.9 MW, b. 13.2 MW, c. 10.7 MW, d. 441 kW.

8.123 Steam at a flow rate of 5 kg/s at 680 K and 6 MPa enters a turbine having an isentropic efficiency of 85%. Determine the power output from the turbine if the outflow pressure is 70 kPa. a. 736 kW, b. 867 kW, c. 3.7 MW, d. 4.3 MW.

8.124 Air at a flow rate of 15 kg/s at 700 K and 6 MPa enters a turbine having an isentropic efficiency of 85%. Determine the power output from the turbine if the outflow pressure is 300 kPa. a. 360 kW, b. 6.7 MW, c. 6.1 MW, d. 5.5 MW.

8.125 Saturated steam at 50 kPa is condensed to saturated liquid in a condenser. What is the rate of heat transfer needed to condense 10 kg/s of saturated steam? a. 23 MW, b. 2.3 MW, c. 26 MW, d. 520 kW.

8.126 Saturated steam at 100 kPa with a flow rate of 7 kg/s is condensed to saturated liquid in a condenser. What flow rate of river water at 290 K is required if the river water cannot be heated above 300 K? Use a specific heat of 4120 J/kg·K for the river water. a. 0.38 kg/s, b. 383 kg/s, c. 55 kg/s, d. 260 kg/s.

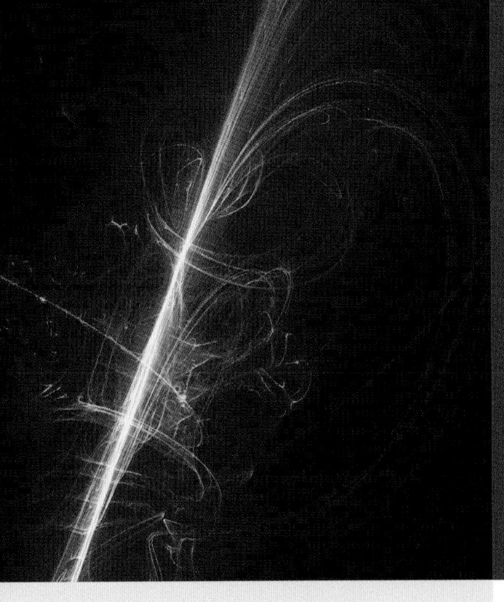

CHAPTER 9
Systems for Power Production, Propulsion, Heating, and Cooling

LEARNING OBJECTIVES

After studying Chapter 9, you should be able to:

- Understand how steady-flow devices are combined to form comprehensive systems for power production, propulsion, heating, and cooling.
- Sketch in *T–s*, *h–s*, or other useful coordinates, the thermodynamic cycles associated with the following systems: the basic steam power plant (Rankine cycle) and its improved-efficiency variants, turbojet engines, gas-turbine power systems (Brayton cycle), and simple heat pump and refrigeration systems.
- Perform a cycle analysis for all the various systems and calculate various performance measures such as thermal efficiency, specific thrust, and coefficient of performance.
- Understand the origins of the inefficiencies associated with system components and be able to include these in cycle analyses.
- Explain using words and sketches how superheat, reheat, and regeneration improve the efficiency of a steam power plant.
- Apply all knowledge gained in previous chapters to the analysis of complex thermal-fluid systems.

CHAPTER 9 OVERVIEW

IN THIS CHAPTER, we see how steady-flow devices combine to form complex systems for power production, propulsion, heating, and cooling. Not only are such systems important from an engineering perspective, but more generally, they are essential to everyday life in industrialized societies. Here we analyze these systems to understand their basic operation and to determine various performance measures. Thermodynamic cycle efficiency, first introduced in Chapter 6, provides a dominant theme for this chapter. Here we investigate in some detail the energy conversion efficiency of the various systems just listed. In some sense, this chapter is the culmination of all the preceding chapters. Chapter 9 not only provides an opportunity to integrate knowledge gained from previous chapters but also provides interesting applications that can be explored in parallel with earlier chapters. Figure 9.1 illustrates the key role that Chapter 9 plays in our study. The system analysis is shown as bridging, or using, all the previous topics.

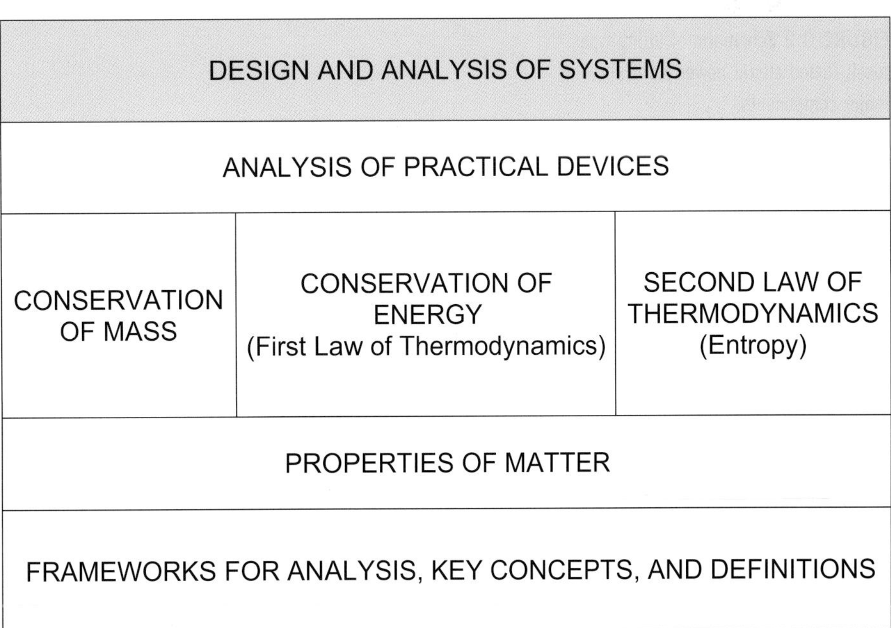

FIGURE 9.1 Hierarchical arrangements of the topics in our study of engineering thermodynamics. Chapter 9 applies the previous topics to thermodynamic cycles.

9.1 Steam Power Plants

The reader is encouraged to reread the appropriate material in Chapter 1 before proceeding.

Fossil-fueled steam power plants are enormously important in providing electricity in the United States and throughout the world. In Chapter 1, we discussed this importance (Table 1.1) and introduced the basic concept of a steam power plant. In the present chapter, we build upon this introduction. Chapter 9 combines the steady-flow device analyses from Chapter 8 to study complete power cycles. The steady-flow device analysis in Chapter 8 applied all previous tools developed in the text: thermodynamic properties, conservation of mass, conservation of energy, and second-law analysis. Figure 9.2 schematically illustrates the basic components and their arrangement in a fossil-fuel steam power plant. An actual power plant, such as seen in Fig. 1.7, is usually significantly more complex than suggested by the schematic in Fig. 9.2. Basic components include the following:

- a combustion chamber, or furnace, to create hot products of combustion; see Fig. 1.6. (Nuclear power plants use a nuclear reactor as the high-temperature source;)
- various heat exchangers (feedwater preheaters, economizers, boilers, and superheaters) to heat water and produce saturated and/or superheated steam as the working fluid for the turbines, see Fig. 8.37;
- pumps and pump drives (steam turbines or electric motors) to elevate the pressure of the working fluid, see Fig. 8.22;
- various steam turbines to extract the energy from the steam and transmit this power to electrical generators, see Fig. 9.3;
- electrical generators (dynamos) to convert shaft power to alternating-current electricity;
- heat-exchange equipment (condensers and cooling towers) to condense the expanded steam from the turbines, see Fig. 9.5;

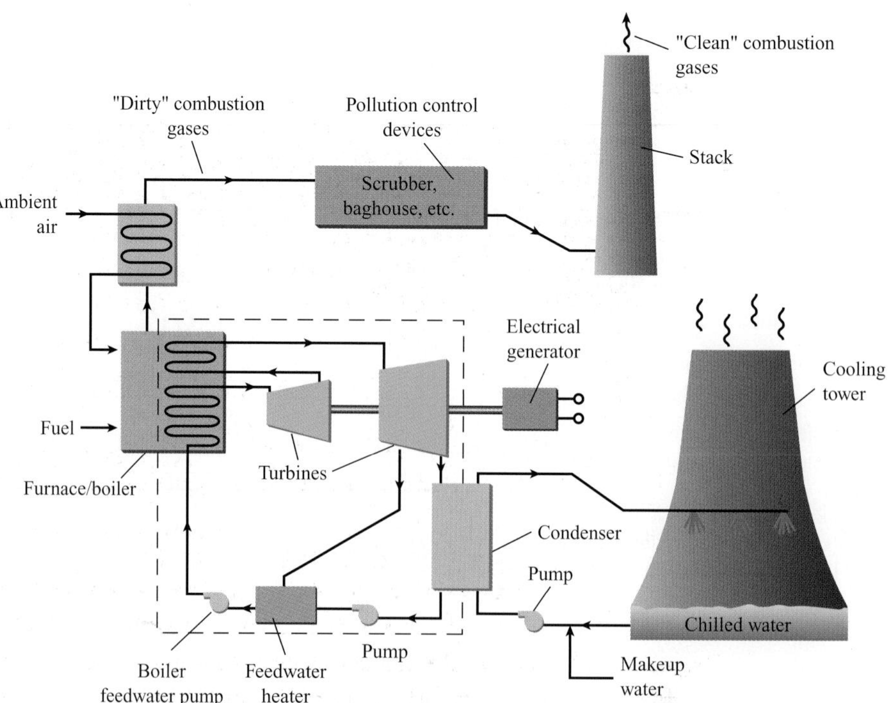

FIGURE 9.2 Schematic of utility-type fossil-fueled steam power plant showing major components.

FIGURE 9.3 Steam turbine assembly (Monty Rakusen / Cultura / Getty Images).

FIGURE 9.4 Power plant control room (Monty Rakusen / Cultura / Getty Images).

FIGURE 9.5 Coal-fired Navajo Generating Station near Page, Arizona. Photograph by J. C. Willett, courtesy of US Geological Survey.

- pollution-control equipment to remove particulate matter (baghouses), sulfur (scrubbers), and oxides of nitrogen (scrubbers and/or catalyst beds), see Fig. 9.6. (these are not present in a nuclear power plant) and
- control systems of various types to ensure the integrated operation of the various subsystems, see Fig. 9.4.

Before proceeding, we explicitly state our objectives for this section. These are:

- to illustrate the various physical arrangements used in typical fossil-fueled steam power plants (nuclear power plants use a similar power cycle);
- to determine the ideal thermodynamic efficiencies associated with various thermodynamic cycles of the working fluid;
- to explore how real processes deviate from ideal processes and quantitatively determine how these deviations affect thermal efficiency.

With these objectives in mind, our analysis first focuses on the working fluid (H_2O) and the components, or portions of components, in direct contact with the working fluid. A dashed line indicates this division in Fig. 9.2. A second focal point is the combustion-related equipment (or nuclear reactor). Using a thermodynamic analysis,

FIGURE 9.6 Electrostatic precipitators control the emission of fly ash from a coal-fired power plants. Photographs courtesy of Dr. H. Christopher Frey.

we explore the overall efficiency of converting fuel energy to electricity. Analysis of the electric generators proper is beyond the scope of this book.

9.1a Rankine Cycle Revisited

The basic Rankine cycle provides the foundation upon which an actual steam power plant cycle can be built. We introduced the Rankine cycle in Chapter 1, and we elaborate that discussion here using many of the concepts and tools developed in the intervening chapters. After exploring the basic Rankine cycle in detail, we add the complexity and features used in real power plants that either increase the overall efficiency or offer other practical advantages.

Figure 9.7 shows the basic Rankine cycle components and the corresponding thermodynamic state points on a T–s diagram for the ideal cycle. To achieve this *ideal cycle*, we assume that the pump and turbine operate adiabatically and reversibly (i.e., isentropic operation) and that there are no heat losses to the surroundings associated with heat-exchange equipment, or with any of the piping interconnecting the various components. Furthermore, all of the processes are reversible (e.g., the flow through all devices and pipes is assumed to be frictionless).

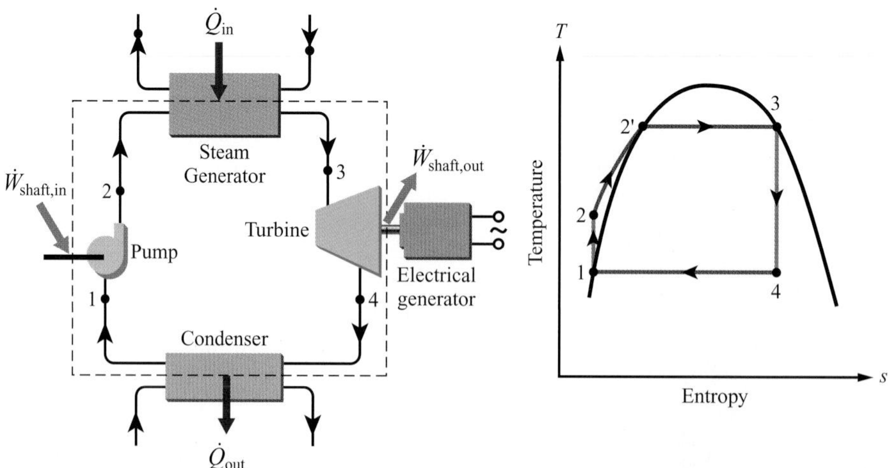

FIGURE 9.7 Components associated with a simple Rankine cycle (left) and the corresponding T–s diagram (right) for the cycle 1–2–3–4–1.

This ideal thermodynamic cycle consists of four processes:

- Process 1–2: A circulating pump boosts the pressure of the liquid water prior to its entering the boiler. To operate the pump, a power input is required, $\dot{W}_{shaft,in}$. The ideal processes is isentropic with $s_1 = s_2$. Note that saturated water enters the pump;

this increases the pressure to a compressed liquid state. This is an important difference between the Rankine cycle (Fig. 9.7) and the Carnot cycle for a saturated fluid. Since it is difficult or impossible to pump a two-phase mixture, the Rankine cycle compresses the one-phase liquid water.

- Process 2–3: Energy is added to the water in the boiler at constant pressure, resulting, first, in an increase in the water temperature and, second, in a phase change. Heat transferred from the hot products of combustion provide this energy, $\dot{Q}_{in}$. (In a nuclear power plant, heat is transferred from the hot reactor water to the working fluid in the steam power cycle.) The working fluid is all liquid at state 2 and all vapor (steam) at state 3. The phase change begins at 2′ on the T–s diagram. The ideal process is isobaric with $P_2 = P_3$.
- Process 3–4: Energy is removed from the high-temperature, high-pressure steam as it expands through a steam turbine. The ideal process is isentropic with $s_3 = s_4$. The shaft power of the turbine, $\dot{W}_{shaft,out}$, is delivered to an electrical generator for the production of electricity.
- Process 4–1: The low-pressure steam is returned to the liquid state as it flows through the condenser. This allows a one-phase liquid to be pumped in process 1–2. The ideal process is isobaric with $P_4 = P_1$. The energy from the condensing steam, $\dot{Q}_{out}$, is transferred to the cooling water.

As outlined here, the cycle consists of two constant-pressure (isobaric) processes and two constant-specific-entropy (isentropic) processes, as shown in the T–s diagram.

We now determine the thermodynamic efficiency of this cycle. In Chapter 6, we defined the thermal efficiency of a work-producing cycle as (Eq. 6.6)

$$\eta_{th} = \frac{\text{useful work produced}}{\text{energy supplied}}. \tag{9.1}$$

For the Rankine cycle, the rate at which useful work is produced is the difference between the output shaft power of the turbine and the input shaft power of the pump, that is,

$$\dot{W}_{net,\,out} = \dot{W}_{turbine,\,out} - \dot{W}_{pump,\,in}.$$

The rate at which energy is supplied to the cycle is the heat-transfer rate to the water in the steam generator, $\dot{Q}_{in}$. Thus, the Rankine cycle efficiency is given by

$$\eta_{th,\,Rankine} = \frac{\dot{W}_{turbine,\,out} - \dot{W}_{pump,\,in}}{\dot{Q}_{in}}. \tag{9.2}$$

From our previous component analyses, the quantities appearing in Eq. 9.2 can all be related to the thermodynamic states of the fluid at the inlets and outlets of the pump, turbine, and steam generator. Because the fluid kinetic and potential energy changes were neglected in our analyses, only the enthalpies are involved, so we have

$$\dot{W}_{turbine,\,out} = \dot{m}(h_3 - h_4), \tag{9.3a}$$

$$\dot{W}_{pump,\,in} = \dot{m}(h_2 - h_1), \tag{9.3b}$$

$$\dot{Q}_{in} = \dot{m}(h_3 - h_2). \tag{9.3c}$$

See Table 8.1 in Chapter 8.

Substituting these quantities into Eq. 9.2 yields

$$\eta_{\text{th, Rankine}} = \frac{(h_3 - h_4) - (h_2 - h_1)}{(h_3 - h_2)}, \tag{9.4a}$$

or, upon rearrangement,

$$\eta_{\text{th, Rankine}} = 1 - \frac{h_4 - h_1}{h_3 - h_1}. \tag{9.4b}$$

Since $\dot{Q}_{\text{out}} = \dot{m}(h_4 - h_1)$, Eq. 9.4b is also equivalent to

$$\eta_{\text{th, Rankine}} = 1 - \frac{\dot{Q}_{\text{out}}}{\dot{Q}_{\text{in}}}.$$

We note that Eqs. 9.3a–9.3c apply to real, as well as ideal, devices; thus, our thermal efficiency expressions (Eqs. 9.4a and 9.4b) apply equally well to real and ideal cycles.

EFFECTS ON CYCLE EFFICIENCY

In Section 7.1 we found that for an ideal reversible process the area under the T–s curve was the heat transfer (Eq. 7.1c).

$${}_1q_2 = \int_1^2 T ds \bigg|_{\substack{\text{int}\\ \text{rev}}}$$

In Section 7.4d, we found that for an ideal reversible cycle, the area inside the T–s curve was the net work output and the net heat transfer (Eq. 7.32)

$$q_{\text{net}} = w_{\text{net}} = \oint T ds.$$

We can use these expressions to better understand the effect of boiler and condenser temperatures on the thermal efficiency of the Rankine cycle. Recalling that the thermal efficiency of the cycle is the ratio of net work output to heat in, we can now find the thermal efficiency in terms of areas on the T–s diagram (see Fig. 9.8 and Eq. 6.7a)

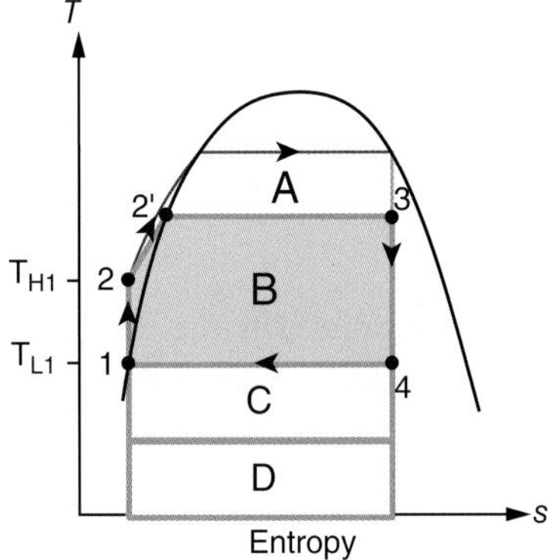

FIGURE 9.8 Graphical representation of W_{net} and Q_{H}.

$$\eta_{\text{th, 1}} = \frac{W_{\text{net}}}{Q_{\text{H}}} = \frac{B}{B + C + D}.$$

We will first look at the effects of decreasing the condenser temperature from T_{L1} to T_{L2}. The area in the cycle curve (W_{net}) increases while the area under the heat addition curve (2–2′-3) does not change (see Fig. 9.9).

The thermal efficiency is now:[1]

$$\eta_{\text{th, 2}} = \frac{W_{\text{net}}}{Q_{\text{H}}} = \frac{B + C}{B + C + D} > \eta_{\text{th, 1}}.$$

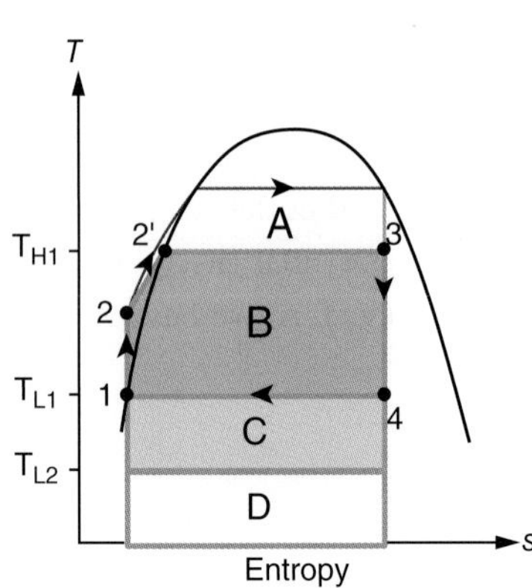

FIGURE 9.9 Graphical representation of W_{net} and Q_{H} when T_{L} is decreased.

We can see that decreasing the condenser temperature will increase the cycle efficiency. In order to decrease the temperature at which the steam condenses to liquid, we must decrease the pressure of the water in the condenser. For this reason, the pressure in the condenser of a steam power cycle is much lower than atmospheric pressure. Typical condenser pressures range from 30 kPa to 50 kPa. However, a low

[1] The small changes in the compressed liquid region are neglected in this discussion.

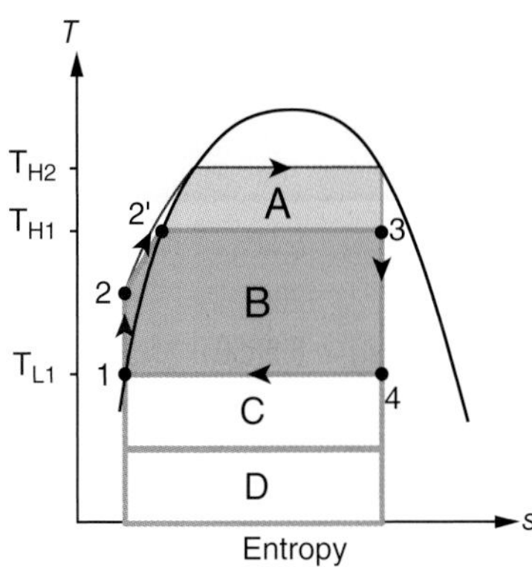

FIGURE 9.10 Graphical representation of W_{net} and Q_H when T_H is increased.

condenser pressure will cause air to leak into the condenser at any small opening. The entrained air must be removed to avoid corrosion of the pumps and heat exchangers. For this reason, a feedwater deaerator is typically included in the power plant.

For a second case, we will increase the temperature in the boiler from T_{H1} to T_{H2}. This increases the net work and the heat into the cycle (see Fig. 9.10).

The thermal efficiency of the cycle is now

$$\eta_{th,3} = \frac{W_{net}}{Q_H} = \frac{A+B}{A+B+C+D} > \eta_{th,1}.$$

Increasing the net work and heat in by the same amount will increase the overall thermal efficiency of the cycle. Therefore, it is desirable to increase the temperature at which heat is added at the boiler. The use of improved materials and surface cooling has allowed boiler temperatures to be increased.

Example 9.1 Ideal Rankine Cycle

Determine the maximum possible thermal efficiency for an industrial steam power plant operating with a high pressure of 1 MPa and a low pressure of 5 kPa. The power plant operates on a Rankine cycle with the steam entering the turbine as saturated vapor. Also determine the ratio of the pump power to the turbine power.

Oil refinery.
(Credit: Issarawat Tattong / Moment / Getty Images.)

Solution

Known Steam Rankine cycle with P_{low} ($= P_1 = P_4$) $= 5$ kPa and P_{high} ($= P_2 = P_3$) $= 1$ MPa.

Find η_{th} (ideal), $\dot{W}_{pump}/\dot{W}_{turbine}$

Sketch

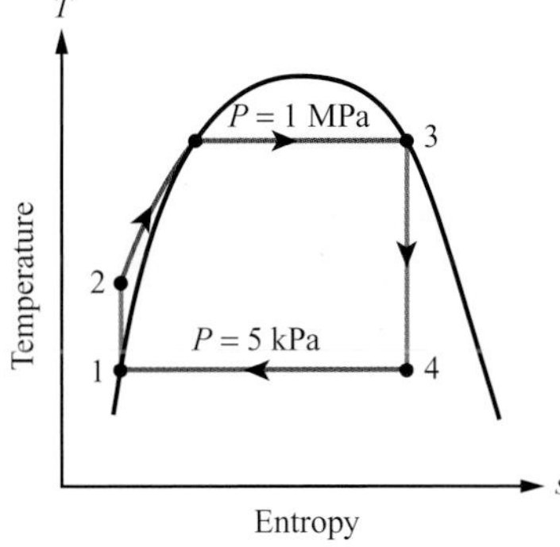

	1	2	3	4
Region	Saturated liquid	Compressed liquid	Saturated vapor	Liquid–vapor mixture
P	5 kPa	1 MPa	1 MPa	5 kPa
h (kJ/kg)	137.75	138.75	2777.1	2007.1
s (kJ/kg·K)	0.47620	0.47620 ($s_2 = s_1$)	6.5850 ($s_3 = s_4$)	6.5850

Modeling, Premises and Assumptions

i. Adiabatic, reversible (i.e., isentropic) operation of pump and turbine
ii. No frictional effects in steam generator and condenser
iii. No friction or heat losses associated with any interconnecting piping
iv. Changes in kinetic and potential energy can be neglected

Analysis The cycle thermal efficiency is calculated from a straightforward application of Eq. 9.1; this reduces the problem to determining the specific enthalpies at each state point in the cycle. To organize the required property data, we employ a table. The completed table is shown above; the steps required to fill the table are described in the following.

From the definition of the Rankine cycle, we can immediately identify the regions associated with states 1, 2, 3, and 4 (i.e., saturated liquid, compressed liquid, saturated vapor, and liquid–vapor mixture). Furthermore, states 1 and 3 are fully defined since they are saturated liquid ($P_{sat} = 5$ kPa) and saturated vapor ($P_{sat} = 1$ MPa) at the given pressures, respectively. From the NIST WebBook [1], we can retrieve the enthalpies h_1 and h_3. We will also retrieve the entropies s_1 and s_3, as they will be required to fully define states 2 and 4. We therefore have the following properties:

State 1 (Saturated Liquid)	State 3 (Saturated Vapor)
$P_1 = 5$ kPa (given)	$P_3 = 1$ MPa (given)
$T_1 = 306.02$ K	$T_3 = 453.03$ K
$h_1 = 137.75$ kJ/kg	$h_3 = 2777.1$ kJ/kg
$s_1 = 0.47620$ kJ/kg·K	$s_3 = 6.5850$ kJ/kg·K
$v_1 = 0.0010053$ m^3/kg	

State 2 is a compressed liquid at a pressure of 1 MPa. Since the pump process is isentropic, $s_2 = s_1$. Knowing two properties (P_2, s_2) allows us to find all other properties at state 2. Using the NIST software for a compressed liquid at $P_2 = 1$ MPa and $s_2 = 0.4762$ kJ/kg·K, we retrieve the following data using the "Specified State Point":

State 2 (Compressed Liquid)
$P_2 = 1$ MPa (known)
$T_2 = 306.05$ K
$h_2 = 138.75$ kJ/kg
$s_2 = 0.4762$ kJ/kg·K (known)

Since we only need to find the enthalpy at state 2, h_2, an alternative to the use of the NIST resources is to employ the incompressible approximation for the reversible pump work (Eq. 8.21):

$$\frac{\dot{W}_{\text{pump, rev}}}{\dot{m}} = h_2 - h_1 \cong v_1(P_2 - P_1),$$

or

$$\begin{aligned} h_2 &= h_1 + v_1(P_2 - P_1) \\ &= 137.75\frac{\text{kJ}}{\text{kg}} + \left(0.0010053\frac{\text{m}^3}{\text{kg}}\right)(1 \times 10^3 - 5)\frac{\text{kN}}{\text{m}^2}\left[\frac{1\text{k.J}}{\text{kN}\cdot\text{m}}\right] \\ &= 138.75\frac{\text{kJ}}{\text{kg}}. \end{aligned}$$

Both approaches yield the same result to five significant digits.

With the assumption that the turbine operates adiabatically and reversibly, we know that s_4 must equal s_3. Since P_4 is given, state 4 is defined; however, to find h_4 requires the calculation of the quality x_4. Using the known entropy s_4 $(= s_3)$, we calculate (Eq. 7.18b)

$$\begin{aligned} x_4 &= \frac{s_4 - s_f(P_{\text{sat}} = 5\text{ kPa})}{s_g(P_{\text{sat}} = 5\text{ kPa}) - s_f(P_{\text{sat}} = 5\text{ kPa})} \\ &= \frac{6.5850 - 0.47620}{8.3938 - 0.47620} = 0.7715, \end{aligned}$$

where s_f and s_g are obtained from the NIST resources. Knowing the quality, we perform the inverse of the previous operation (Eq. 2.37c) to find the unknown enthalpy h_4:

$$\begin{aligned} h_4 &= h_f + x_4(h_g - h_f) \\ &= 137.75 + 0.7715(2560.7 - 137.75)\text{kJ/kg} \\ &= 2007.1\text{ kJ/kg}, \end{aligned}$$

where h_f and h_g are determined in the same manner as s_f and s_g. Our property table above is now complete, and the ideal cycle efficiency can be calculated (Eq. 9.4):

$$\begin{aligned} \eta_{\text{th, Rankine}} &= 1 - \frac{h_4 - h_1}{h_3 - h_2} \\ &= 1 - \frac{2007.1 - 137.75}{2007.1 - 138.75} \\ &= 0.291, \end{aligned}$$

or

$$\eta_{\text{th, Rankine}} = 29.1\%.$$

We can also calculate the ratio of the pump to turbine power:

$$\begin{aligned} \frac{\dot{W}_{\text{pump}}}{\dot{W}_{\text{turbine}}} &= \frac{\dot{m}(h_2 - h_1)}{\dot{m}(h_3 - h_4)} = \frac{h_2 - h_1}{h_3 - h_4} \\ &= \frac{138.7.5 - 137.7.5}{2771.1 - 2007.1} = \frac{1}{764} = 0.0013, \end{aligned}$$

or

$$0.13\%.$$

Comments First, we note the low quality of the steam exiting the turbine (x_4 = 0.7715). Because liquid droplets erode turbine blades, many power plants operate with qualities in excess of 0.90. We explore this in a later example (Example 9.3).

We also note that when the cycle operates with the condensed steam at near-ambient temperatures, the steam side of the condenser operates at subatmospheric conditions. For this example, $T_4 = 306.02\,\text{K}$ when $P_4 = 5\,\text{kPa}$. This vacuum condition is one reason why the working fluid undergoes a closed cycle.

Third, we note that only 29.1% of the energy input to the working fluid is returned as useful work. If we were to consider all the irreversibilities and their attendant losses, the thermal efficiency would be even lower than this ideal value. The direct effect of turbine losses is considered in the next example. Various means to improve the ideal efficiency comprise the subjects of the next sections.

Self-Test 9.1

Repeat Example 9.1 for a high pressure of 1.5 MPa. Also determine the differences in the heat added (kJ/kg), the power (kJ/kg) required by the pump, and the power produced by the turbine.

(Answer: $\eta_{th,\,Rankine} = 31.1\%$. The higher pressure results in a small increase in pumping power (0.5 kJ/kg) and heat input (13.4 kJ/kg) and a substantial increase in the turbine power output (57.3 kJ/kg).)

Example 9.2 Rankine Cycle with Actual Turbine

Consider the same situation as that described in Example 9.1, except now the turbine has an isentropic efficiency of 0.90. Calculate the effect of the turbine efficiency on the quality of the steam exiting the turbine and on the cycle efficiency.

Solution

Known P_1, P_2, P_3, P_4, $\eta_{isen,t}$

Find x_4, η_{th}

Sketch

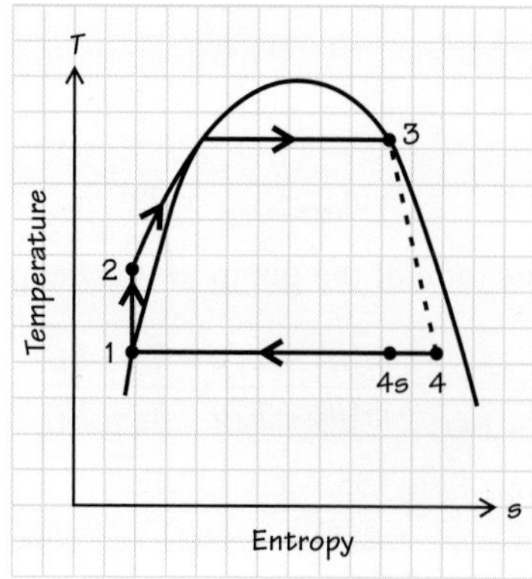

View of engine room in geothermal power station (Oscar Bjarnason / Cultura / Getty Images).

Modeling, Premises and Assumptions We invoke the same assumptions as applied in Example 9.1 *except* that turbine losses are considered, through $\eta_{isen,t}$.

Analysis States 1, 2, and 3 are unchanged by the addition of the turbine losses; only state 4 is affected. As indicated in the sketch, the entropy of the steam exiting the turbine (state 4) exceeds that of the corresponding isentropic-expansion state 4s. State 4s corresponds to state 4 in the previous example (i.e., $s_{4s} = 6.5850\,\text{kJ/kg·K}$). We can determine a value for h_4 by applying the definition of the isentropic efficiency for the turbine (Eq. 8.26b), that is,

$$\eta_{isen,t} = \frac{h_3 - h_4}{h_3 - h_{4s}}.$$

Solving for h_4 and evaluating, we get

$$\begin{aligned} h_4 &= h_3 - \eta_{isen,t}(h_3 - h_{4s}) \\ &= 2777.1 - 0.90(2777.1 - 2007.1)\text{kJ/kg} \\ &= 2084.1\,\text{kJ/kg}. \end{aligned}$$

where values for h_3 and h_{4s} are taken from Example 9.1 (see the table in Example 9.1). With this value of h_4, we can calculate the steam quality and the cycle efficiency as follows:

$$\begin{aligned} x_4 &= \frac{h_4 - h_f(P_{sat} = 5\,\text{kPa})}{h_g(P_{sat} = 5\,\text{kPa}) - h_f(P_{sat} = 5\,\text{kPa})} \\ &= \frac{2084.1 - 137.75}{2560.7 - 137.75} = 0.8033 \end{aligned}$$

and

$$\begin{aligned} \eta_{tn} &= 1 - \frac{h_4 - h_1}{h_3 - h_2} = 1 - \frac{2084.1 - 137.75}{2777.1 - 138.75} \\ &= 0.262 \text{ or } 26.2\%. \end{aligned}$$

Comments As expected, the less-than-ideal (less than 100%) turbine efficiency reduces the cycle efficiency. Since the turbine was the only nonideal component considered, $\eta_{th,Rankine} = \eta_{isen,t}\eta_{th,Rankine(ideal)}$ [i.e., $\eta_{th,Rankine} = (0.90)(0.2914) = 0.262$]. Thus, there is no need to directly calculate h_4 except to find the quality x_4. Note that the turbine inefficiency has the side benefit of improving the quality of the steam ($x_4 = 0.8033$ versus 0.7715), although the steam is still quite wet.

Self-Test 9.2

Repeat the analysis of Example 9.1 using a pump having an isentropic efficiency of 0.60. Comment on your results.

(Answer: $\eta_{th,Rankine} = 29.1\%$. Even though there is a significant increase in the required pump work, the overall contribution of this device is insignificant to the cycle efficiency.)

9.1b Rankine Cycle with Superheat and Reheat

In this section, we discuss two modifications to the basic Rankine cycle: the addition of superheat and the addition of reheat. All but the simplest systems employ superheat, whereas more sophisticated power plants employ both superheat and reheat.

SUPERHEAT

The addition of superheat is a relatively simple means to improve the basic Rankine cycle efficiency. Rather than allowing the steam to enter the turbine as saturated vapor, additional energy is supplied to the steam at constant pressure providing superheated vapor at the turbine inlet. A thought experiment illustrating the creation of superheated steam starting from a compressed liquid is shown in Fig. 9.11. The corresponding states are shown in the T–s diagram in Fig. 9.12b. The liquid water is first heated from the temperature at which it exits the pump (state 2 in Fig. 9.12b) to the saturated-liquid state (state 2′ in Fig. 9.12b). This occurs in the **economizer** section of the steam generator. Energy is added to the saturated liquid in the **boiler** section of the steam generator to produce saturated vapor (state 3 in Fig. 9.12b). The saturated vapor is then heated further in the **superheater** section of the boiler to state 3′ (Fig. 9.12b). Figure 9.12a shows these three sections of a steam generator as part of a power-plant schematic, and Fig. 9.13 shows a drawing of an actual small-scale marine steam generator. Note that the hot products of combustion strike the superheater first and then, sequentially, the boiler tubes and the economizer tubes. This flow path uses the hottest gases to superheat the steam because the temperature of the combustion products decreases as these gases pass through the steam generator.

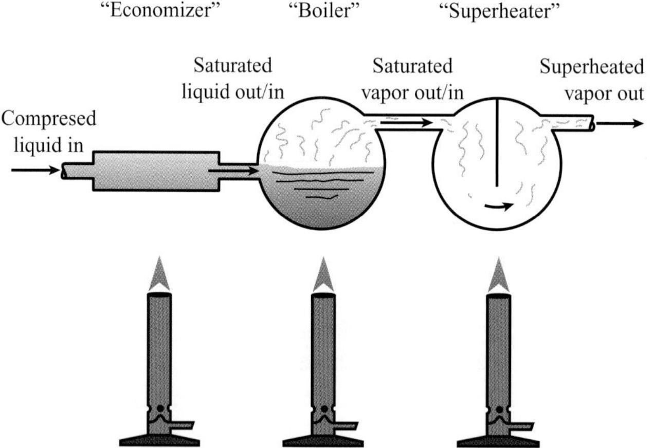

FIGURE 9.11 Thought experiment illustrating the processes involved in steam generation: At the left, cold water is heated to the saturated liquid state. The saturated liquid is then vaporized in the center vessel; and at the right, the vapor is superheated. The three vessels correspond to the economizer, the boiler, and the superheater sections used in real steam generators (see Figs. 9.12 and 9.13).

The use of superheat improves the cycle efficiency by adding energy to the working fluid at a higher mean temperature than would otherwise result at the same pressure. We saw in Chapter 6 (Eq. 6.15a) that the Carnot cycle efficiency increases with the temperature of the heat-addition reservoir. This idea carries through to the Rankine cycle as well. An additional benefit of superheat is improved steam quality at the turbine exit, thereby diminishing blade erosion from water droplets.

In the next example, we quantify these effects by adding superheat to the cycle analyzed in Example 9.1.

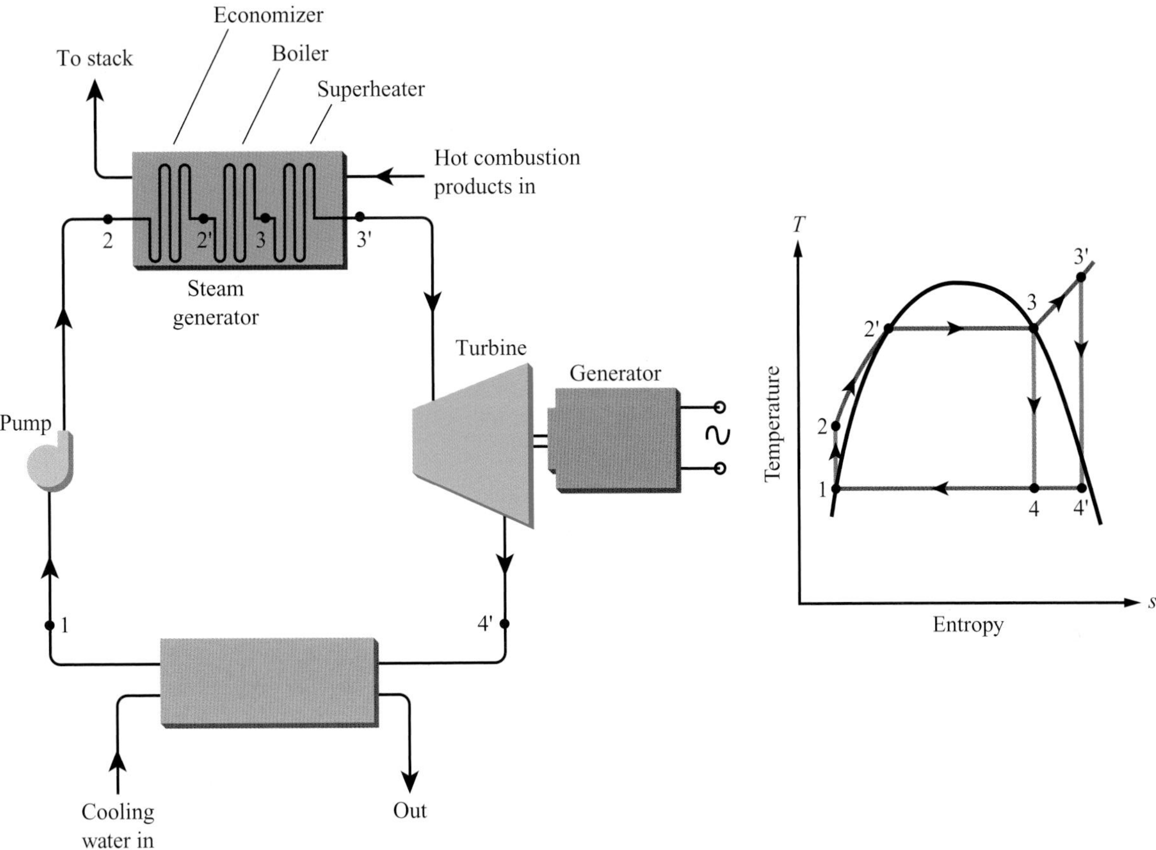

FIGURE 9.12 Components associated with a Rankine cycle with superheat (left) showing economizer, boiler, and superheater sections in a steam generator. The corresponding T–s diagram (right) shows the states for the cycle 1–2–3′–4′–1. Also shown on the T–s diagram is a Rankine cycle without superheat: 1–2–3–4–1.

FIGURE 9.13 Drawing of a steam generator for marine applications showing various subcomponents. Note that the hot combustion products produced by the burners pass through the superheater section first. Reprinted from Ref. [2] with permission (courtesy of the Babcock & Wilcox Company).

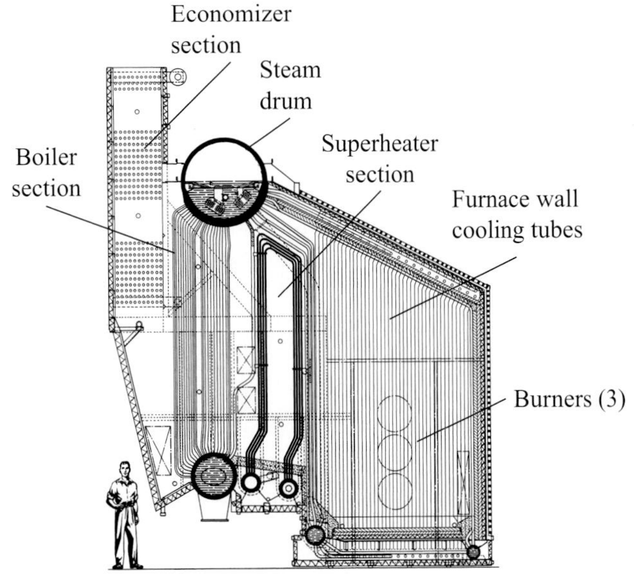

Example 9.3 Rankine Cycle with Superheat

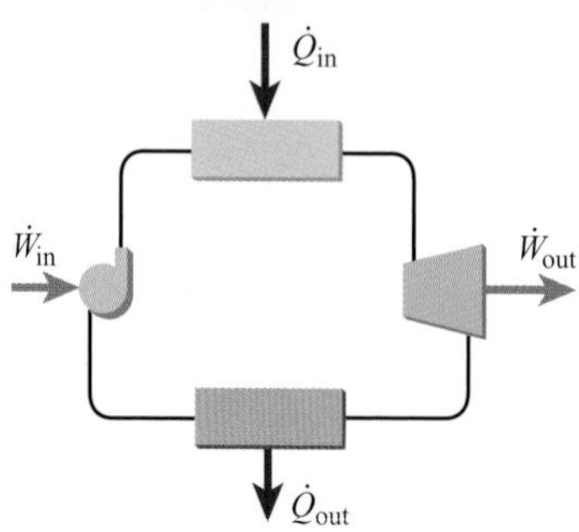

Calculate the ideal thermal efficiency for a Rankine cycle with superheat. The high and low pressures are the same as in Example 9.1, and the degree of superheat is such that the quality of the steam exiting the turbine is 0.90.

Solution

Known Superheat Rankine cycle with P_{low} $(= P_1 = P_4) = 5$ kPa, P_{high} $(= P_2 = P_3) =$ 1 MPa, and $x_{4'} = 0.90$

Find η_{tn} (ideal) for Rankine cycle with superheat.

Sketch See T–s diagram (Fig. 9.12b) for cycle 1–2–3′–4′–1. See also the figure here.

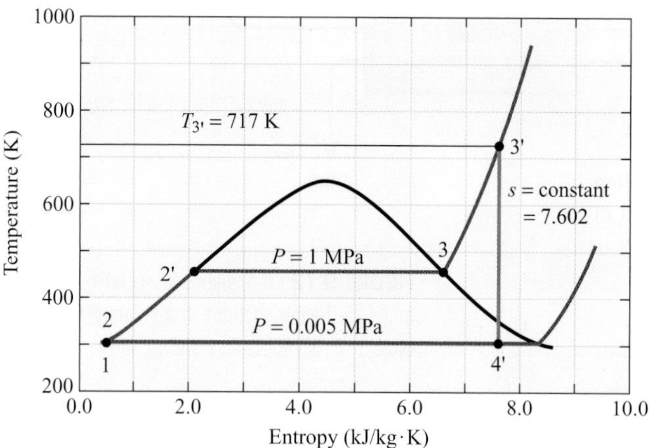

Modeling, Premises and Assumptions All processes are ideal (see the assumptions in Example 9.1).

Analysis To calculate the thermal efficiency from Eqs. 9.4 requires values for the enthalpy of the working fluid at states 1, 2, 3′, and 4′. States 1 and 2 are unchanged from Example 9.1, so we already have those values (see the table from Example 9.1):

$$h_1 = 137.75 \text{ kJ/kg},$$
$$h_2 = 138.75 \text{ kJ/kg}.$$

State 4′ is defined by the quality (0.90) and the saturation pressure (5 kPa); thus, the enthalpy and entropy at 4′ are calculated as

$$h_{4'} = h_f + x_{4'}\left(h_g - h_f\right)$$

and

$$s_{4'} = s_f + s_{4'}\left(s_g - s_f\right),$$

where the subscripts f and g refer to the saturated-liquid and saturated-vapor states at 5 kPa. Using the previously determined values for h_f, h_g, s_f, and s_g (see Example 9.1), $h_{4'}$ and $s_{4'}$ are determined as

$$h_{4'} = 137.75 + 0.90(2560.7 - 137.65)\text{kJ/kg}$$
$$= 2318.4 \text{ kJ/kg}$$

and

$$s_{4'} = 0.47620 + 0.90(8.3938 - 0.47620)\text{kJ/kg·K}$$
$$= 7.602\text{kJ/kg·K}.$$

Since we are assuming that the turbine operates ideally (adiabatically and reversibly), the process $3'$–$4'$ is isentropic, and so

$$s_{3'} = s_{4'} = 7.602 \text{ kJ/kg·K}$$

The state at $3'$ is thus defined: s_3' and P_3 are known. Using the NIST resources,[2] we then find

$$T_{3'} = 717.10 \text{ K},$$
$$h_{3'} = 3358.3 \text{ kJ/kg}.$$

The state points $3'$ and $4'$ are shown on the above T–s diagram, drawn to actual scale. Note that the maximum superheat temperature, $T_{3'} = 717.1$ K, is well above the critical temperature, $T_{cr} = 647.10$ K. With values for the enthalpies at each state (1, 2, $3'$, $4'$), we can apply Eq. 9.4 to calculate the ideal cycle efficiency:

$$\eta_{th} = 1 - \frac{h_{4'} - h_1}{h_{3'} - h_2}$$
$$= 1 - \frac{2318.4 - 137.75}{3358.3 - 138.75}$$
$$= 0.323,$$

or

$$\eta_{th} = 32.3.\%.$$

Comments The addition of superheat results in an increase in the ideal efficiency from 29.1% to 32.3%, a substantial improvement of 3.2 percentage points. Considering the relatively modest addition of hardware required to achieve this gain, we see why all but the smallest power plants employ superheat in practice.

Above, we illustrated this cycle on actual T–s coordinates to show the true positions of the various state points. Note, in particular, that the 1-MPa compressed-liquid isobar follows the saturation line so closely that it cannot be resolved on the scale shown. As a consequence, the separation between state 1 and state 2 cannot be resolved. In Example 9.1, we found the difference between T_2 and T_1 to be only 0.04 K (= 306.06 − 306.02 K).

This example combines cycle analysis with principles of internal flows from Chapter 10.

[2] Alternatively, the property tables in Appendix B can be used, and the desired properties can be obtained by interpolation.

Self-Test 9.3

Calculate the thermal efficiency and steam quality at the turbine exit for a Rankine cycle operating between the same pressures and with the same degree of superheating as in Example 9.3. The turbine has an isentropic efficiency of 0.92.

(Answer: $x = 0.934$, $\eta_{th,\,Rankine} = 29.7\%$)

REHEAT

Frequently, the simple Rankine cycle is modified with the addition of a **reheat** process. Reheat allows the use of relatively high boiler pressures while maintaining a sufficiently high value of the quality of the steam exiting the turbine. Figure 9.14a illustrates the basic component arrangement for a reheat cycle. Here we see that the simple cycle has been modified by replacing the single turbine with two: a high-pressure turbine and a low-pressure turbine. Furthermore, the steam exiting the high-pressure turbine returns to the boiler to be reheated before entering the low-pressure turbine. A T–s diagram illustrating an ideal reheat cycle is shown in Fig. 9.14b. Once again, by the word *ideal* we mean that no friction or other irreversibilities are associated with any of the components or interconnecting flow passages.

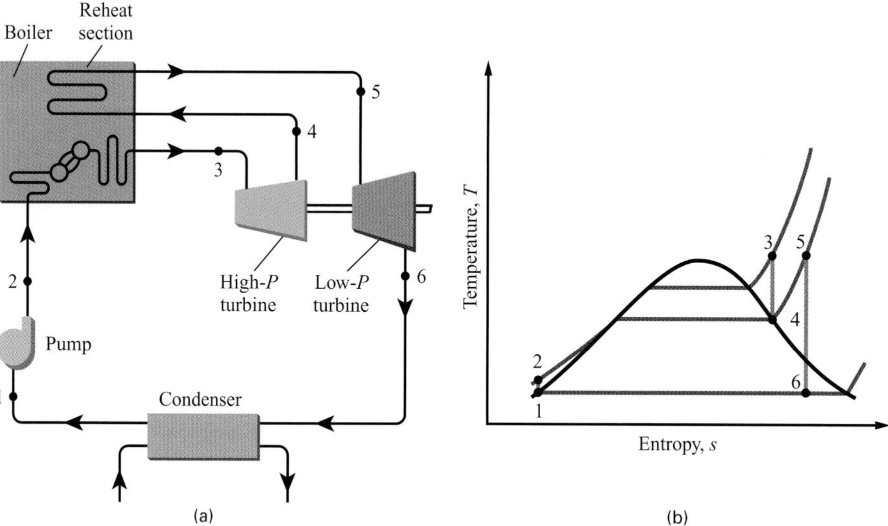

FIGURE 9.14 Rankine cycle with reheat: **(a)** component arrangement, **(b)** T–s diagram.

Analysis of the thermal efficiency of the reheat cycle follows that previously performed for the simple Rankine cycle with or without superheat. The primary differences are the inclusion of a second heat-addition process (state 4 to state 5) and power delivery by two turbines (state 3 to state 4 and state 5 to state 6). Figure 9.15 shows the high-pressure and low-pressure turbines in tandem driving one shaft. To evaluate the thermal efficiency defined by Eq. 9.1, we calculate the net useful work:

FIGURE 9.15 Tandem compound reheat steam turbine rotor displayed on half shell. Photograph courtesy of General Electric Power Company.

$$\dot{W}_{net,\ out} = \dot{W}_{hi\text{-}P\,turb,\,out} + \dot{W}_{low\text{-}P\,turb,\,out} + \dot{W}_{pump,\ in},$$

where

$$\dot{W}_{\text{hi-}P\,\text{turb, out}} = \dot{m}(h_3 - h_4), \tag{9.5a}$$

$$\dot{W}_{\text{low-}P\,\text{turb, out}} = \dot{m}(h_5 - h_6), \tag{9.5b}$$

and

$$\dot{W}_{\text{pump, in}} = \dot{m}(h_2 - h_1). \tag{9.5c}$$

The heat added is the sum of that for the steam generation process (2–3) and the reheat process (4–5):

$$\dot{Q}_{\text{boiler, in}} = \dot{m}(h_3 - h_2) \tag{9.5d}$$

and

$$\dot{Q}_{\text{reheated, in}} = \dot{m}(h_5 - h_4). \tag{9.5e}$$

The thermal efficiency is thus expressed as

$$\eta_{\text{th, reheat}} = \frac{\dot{W}_{\text{net, out}}}{\dot{Q}_{\text{in}}} = \frac{(h_3 - h_4) + (h_5 - h_6) - (h_2 - h_1)}{(h_3 - h_2) + (h_2 - h_4)}, \tag{9.6a}$$

which can be rearranged to yield

$$\eta_{\text{th, reheat}} = \frac{(h_5 + h_3 + h_1) - (h_6 + h_4 + h_2)}{(h_5 + h_3) - (h_2 + h_4)}. \tag{9.6b}$$

Note that in applying the conservation of energy (the first law) to each of the steady-flow components, no restrictions were imposed that the processes be ideal; thus, Eqs. 9.6 apply to both ideal and real cycles.

9.1c Rankine Cycle with Regeneration

Another Rankine-cycle modification that improves thermal efficiency is regeneration. **Regeneration** entails extracting a portion of the steam from the turbine at a pressure somewhere between that at the boiler outlet (state 3) and the condenser inlet (state 5). Figure 9.16a illustrates this **extraction** at state 4, between the high-pressure turbine stages and the low-pressure turbine stages. The extracted steam is at an intermediate pressure and temperature, between the values of steam exiting the boiler and entering the condenser. Since some steam is extracted after the high-pressure turbine stages, there will be a lower mass flow rate passing through the low-pressure turbine stages. The extracted steam then enters a **feedwater heater** where it heats the liquid water (the feedwater) pumped from the condenser before it enters the boiler.

The advantage of adding regeneration is that less heat needs to be added in the boiler since a portion of the working fluid has been preheated in the feedwater heater. The decreased power produced by the turbine resulting from extracting steam is more

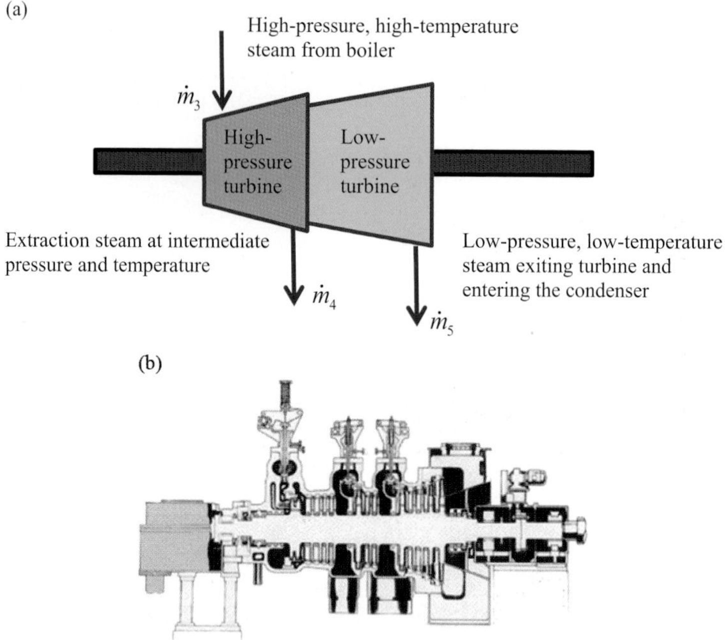

FIGURE 9.16 Steam turbine with extraction. **(a)** Schematic of the turbine with one extraction stream. **(b)** Steam turbine designed for steam to enter at 10 MPa, with two extractions at 1.19 and 0.59 MPa. The steam exits at 6 kPa. Drawing courtesy of Toshiba Corporation.

than offset by the diminished heat requirement in the boiler. Preheating the feedwater before it enters the boiler will raise the average temperature of heat transfer in the boiler and therefore increase the thermal efficiency of the cycle. In many power plants, several feedwater heaters are employed, with steam extracted at multiple locations from the turbine. Economic factors determine the optimum number of feedwater heaters for any particular power plant.

In an **open feedwater heater,** shown in Fig. 9.17, the extraction steam and feedwater are mixed together in the feedwater heater. In order to mix the two streams, the extraction steam and feedwater need to be at the same pressure. A second pump brings the total flow of working fluid up to the boiler pressure, and the cycle continues. Note that two pumps are required when an open feedwater heater is used. The open feedwater heater is often modified to serve also as a deaerator for the feedwater [9].

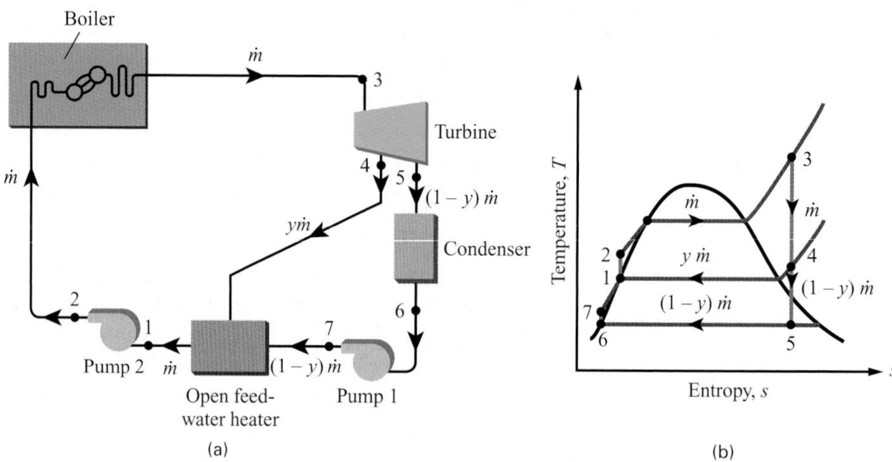

FIGURE 9.17 Regenerative Rankine cycle with a single open feedwater heater: **(a)** component arrangement, **(b)** T–s diagram for ideal components.

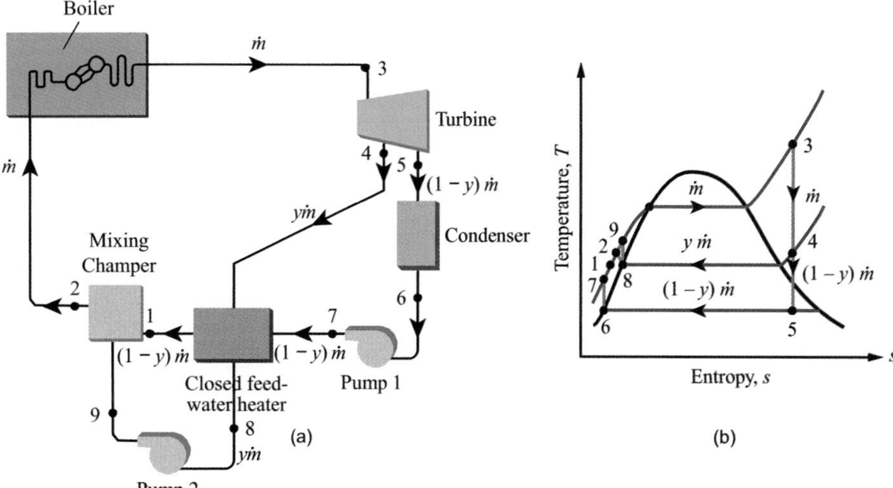

FIGURE 9.18 Regenerative Rankine cycle with a single closed feedwater heater and two pumps: **(a)** component arrangement, **(b)** *T–s* diagram for ideal components.

In a **closed feedwater heater**, shown in Fig. 9.18, the extraction steam and feedwater do not mix together in the feedwater heater. The closed feedwater heater is a heat exchanger with two separate inflow and outflow streams. Since the two streams do not mix, the pressure of the feedwater and extraction steam do not need to be the same, and only one feedwater pump is needed. After exiting the closed feedwater heater, the extraction steam is then added to the feedwater flow. The extraction steam (now saturated liquid) can be combined with the feedwater at two possible locations by:

1. pumping the extraction steam to a higher pressure and combining with the feedwater entering the boiler, as shown in Fig. 9.18, or
2. dropping the pressure of the extraction steam and adding it to the water exiting the condenser, as shown in Fig. 9.19. If there are several closed feedwater heaters, the outflow from one feedwater heater can be combined with the extraction steam in a lower-pressure feedwater heater. Note that only one pump is required when a closed feedwater heater is used and the extraction steam is combined with the outflow from the condenser.

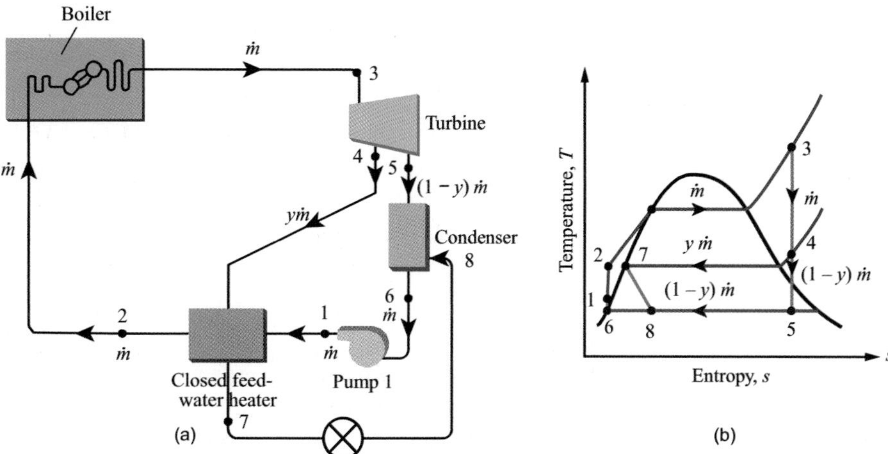

FIGURE 9.19 Regenerative Rankine cycle with a single closed feedwater heater and one pump: **(a)** component arrangement, **(b)** *T–s* diagram for ideal components.

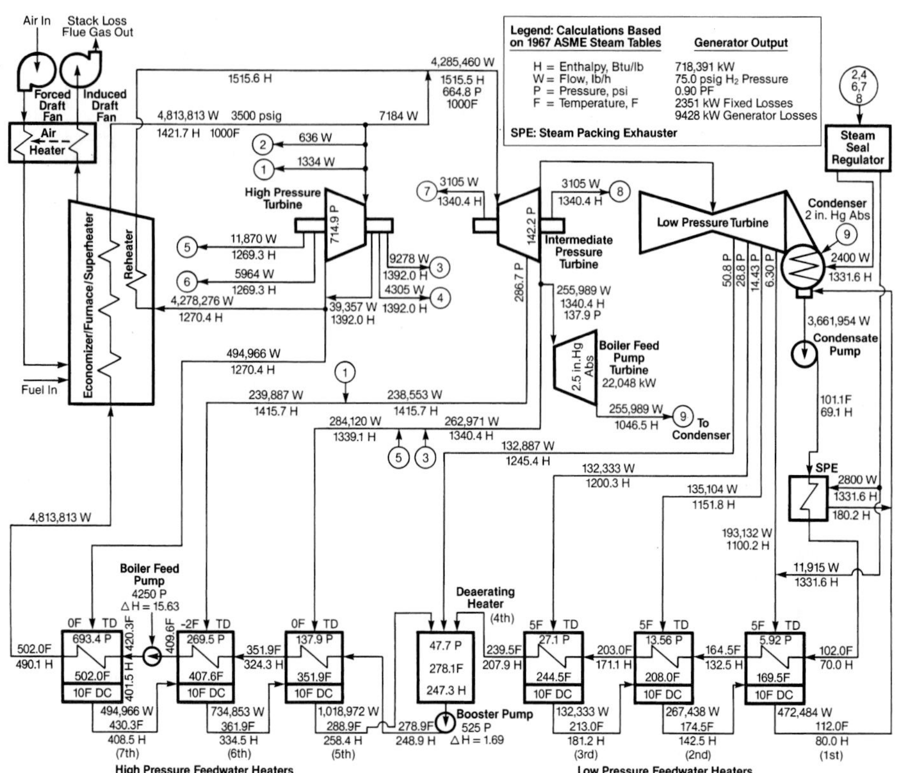

FIGURE 9.20 Steam power plant with five feedwater heaters. Reprinted from Ref. [2] with permission (courtesy of the Babcock & Wilcox Company).

Figure 9.20 gives an example of an actual industrial power plant, which produces 718 MW of power. Flow conditions through the cycle are labeled using pressure (P) in psu, flow rate (W) in lb_m/hr, and temperature (F) in F. The cycle extracts steam from the high-pressure, intermediate-pressure, and low-pressure turbines to be used in the feedwater heaters. We can follow the path of the extracted steam at each pressure. The steam extracted from the turbine at the lowest pressure of 6.30 psi passes through a closed feedwater heater (Heater 1 at right) and then is combined in the condenser with the turbine outflow. The extraction steam at 14.43 psi passes through a closed feedwater heater (Heater 2) and then is combined with the extraction steam in Heater 1. Feedwater heater 3 is also a closed feedwater heater and uses the extraction steam at 28.8 psi. Heater 4 is an open feedwater heater that combines the extraction steam from the low-pressure turbine at 50.8 psi with the feedwater at 47.7 psi and the outflow from the intermediate-pressure closed feedwater heater (Heater 5). Closed feedwater Heaters 5 and 6 use extraction steam from the intermediate turbine. Steam at the outflow of the high pressure turbine is extracted and used in the high-pressure closed feedwater heater (Heater 7). Most of the outflow from the high-pressure turbine passes through the reheater before entering the intermediate-pressure turbine. Note that there are two pumps in the cycle, one increasing the pressure at the outflow of the condenser and the other positioned after the open feedwater heater. Following the path of the water exiting the condenser shows that the feedwater temperature increases from 102.0 F, 164.5 F, 203.0 F, 239.5 F, 278.9 F, 351.9 F, 420.3 F, 502.0 F as it passes through the feedwater heaters before entering the steam generator (boiler) to exit as superheated steam and 1000 F and 3500 psi.

In the following, we utilize the principles of conservation of mass and energy to determine the thermal efficiency (Eq. 9.1) of a regenerative Rankine cycle employing an open feedwater heater.

OPEN FEEDWATER HEATER

We denote the fraction of the total mass flow rate extracted from the turbine as

$$y \equiv \frac{\dot{m}_{\text{extracted}}}{\dot{m}_{\text{total}}}. \tag{9.7}$$

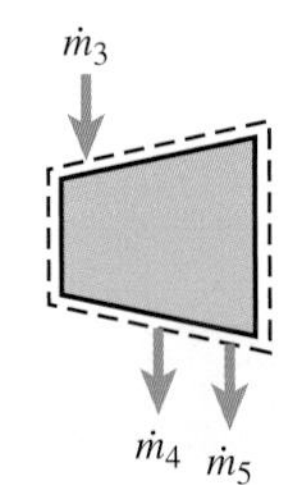

Mass conservation (Eq. 3.14b) to and from the turbine yields

$$\dot{m} = \dot{m}_4 + \dot{m}_5, \tag{9.8a}$$

or

$$\dot{m} = y\dot{m}_4 + (1 - y)\dot{m}, \tag{9.8b}$$

where $\dot{m} = \dot{m}_{\text{total}} (= \dot{m}_1 = \dot{m}_2 = \dot{m}_3)$. (See Figs. 9.21 and 9.17.) In the open feedwater heater, the two streams recombine; thus, mass conservation in this device is expressed as

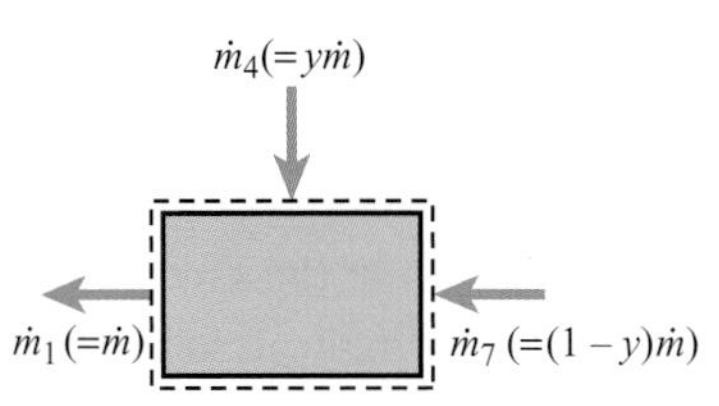

FIGURE 9.21 Open systems (control volumes) used for a turbine with extraction and for a feedwater heater.

$$\dot{m}_4 + \dot{m}_7 = \dot{m}_1, \tag{9.9a}$$

or

$$y\dot{m} + (1 - y)\dot{m} = \dot{m}. \tag{9.9b}$$

The extracted fraction y is an unknown quantity; the value of y is determined by requiring that the fluid exiting the feedwater heater be saturated liquid at the pressure of the extracted steam [i.e., $P_1 = P_{\text{sat}}(T_1) = P_4$]. Energy conservation is then employed to calculate y given the turbine inlet conditions (state 3), the extraction pressure $P_4 (= P_1)$, and the condenser pressure $P_5 (= P_6)$. See Fig. 9.17.

For the open feedwater heater, the steady-flow form of energy conservation is used for multiple flow streams (see Section 8.6b).

$$\dot{Q}_{\text{cv, net in}} - \dot{W}_{\text{cv, net out}} = \sum_{k=1}^{M \text{ outlets}} \dot{m}_{\text{out},k}\left[h_k + \tfrac{1}{2}V_k^2 + g(z_k - z_{\text{ref}})\right] - \sum_{j=1}^{N \text{ inlets}} \dot{m}_{\text{in},j}\left[h_j + \tfrac{1}{2}V_j^2 + g(z_j - z_{\text{ref}})\right]. \tag{8.40}$$

This equation is simplified by assuming that the kinetic and potential energy changes are small and can be neglected. The open feedwater heater is a mixing chamber that does no work. It is also assumed that the heat transfer from the open system (control volume) to the surroundings is small and can be neglected (see Table 8.1) to give:

$$\sum_{\text{inlets}} \dot{m}_i h_i = \sum_{\text{outlets}} \dot{m}_i h_i,$$

or

$$\dot{m}_4 h_4 + \dot{m}_7 h_7 = \dot{m}_1 h_1. \tag{9.10a}$$

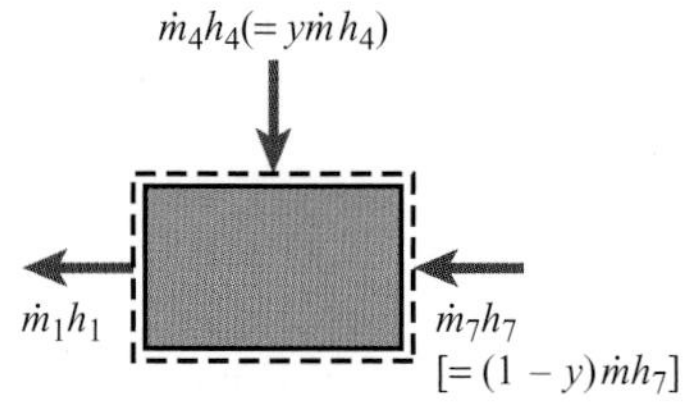

FIGURE 9.22 Energy analysis for an open feedwater heater.

Expressing the flow rates in terms of y yields

$$y\dot{m}h_4 + (1 - y)\dot{m}h_7 = \dot{m}h_7. \tag{9.10b}$$

From this expression of energy conservation, we can solve for y:

$$y = \frac{\eta - h_7}{h_4 - h_7}. \tag{9.11}$$

With the extracted mass fraction now determined, a thermodynamic analysis of the cycle can proceed in the same fashion as in our previous analyses. The following example illustrates this procedure.

Example 9.4 Open Feedwater Heater

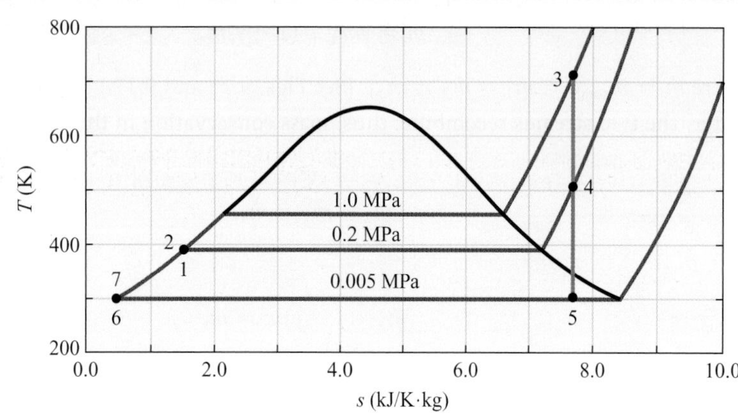

Calculate the ideal cycle efficiency for a regenerative Rankine cycle employing a single open feedwater heater, as shown in Fig. 9.22. A temperature–entropy diagram for the cycle is shown in the sketch. The low-pressure and high-pressure state points are the same as those in Example 9.3 (i.e., T_3 = 717.1 K and x_5 = 0.90). The extracted steam and the feedwater heater are at 0.2 MPa.

Solution

Known $P_1, P_2, P_3, P_4, P_6, P_7, T_3, x_5$

Find η_{th}

Modeling, Premises and Assumptions

i. The flow is steady.
ii. All processes and devices are ideal.

Analysis The cycle thermal efficiency is defined as the net useful power delivered divided by the rate at which heat is added to the working fluid. The heat added comes solely from the boiler, as the open feedwater heater is adiabatic with respect to the surroundings. Thus,

$$\eta_{th,regen} = \frac{\dot{W}_{net,out}}{\dot{Q}_{in}} = \frac{\dot{W}_{turbine} - \dot{W}_{pump1} - \dot{W}_{pump2}}{\dot{Q}_{boiler}}.$$

From an overall energy balance, we can also write

$$\begin{aligned}\dot{W}_{net,out} &= \dot{Q}_{in} - \dot{Q}_{out} \\ &= \dot{Q}_{boiler} - \dot{Q}_{condenser}\end{aligned}$$

Thus,

$$\begin{aligned}\eta_{th,regen} &= \frac{\dot{Q}_{boiler} - \dot{Q}_{condenser}}{\dot{Q}_{boiler}} \\ &= 1 - \frac{\dot{Q}_{condenser}}{\dot{Q}_{boiler}}.\end{aligned}$$

Steady-flow first-law analyses of the boiler and condenser yield

$$\dot{Q}_{\text{boiler}} = \dot{m}(h_3 - h_2)$$

and

$$\dot{Q}_{\text{condenser}} = (1 - y)\dot{m}(h_5 - h_6).$$

Substituting these back into the thermal efficiency expression, we have

$$\eta_{\text{th, regen}} = 1 - (1 - y)\frac{h_5 - h_6}{h_3 - h_2},$$

where y, the extracted fraction, is evaluated from Eq. 9.11, that is,

$$y = \frac{h_1 - h_7}{h_4 - h_7}.$$

From this, we see that enthalpy values are required at each of the seven state points. States 1 and 6 are saturated-liquid states, and the corresponding properties are easily obtained from the NIST resources; the properties for states 3 and 5 were previously evaluated in Example 9.3. These properties are summarized as follows:

Property	State 1	State 3	State 5	State 6
P (MPa)	0.20	1.00	0.005	0.005
T (K)	393.36	717	453.03	306.02
h (kJ/kg)	504.70	3358.1	2318.4	137.75
s (kJ/kg·K)	1.5302	7.602	7.602	0.4762
$v(\text{m}^3/\text{kg})$	0.0010605	—	—	0.0010053

The enthalpy at state 4 is determined from the NIST resources using the two known properties $P_4 = 0.2\,\text{MPa}$ and $s_4 = s_3 = 7.602\,\text{kJ/kg·K}$. This value is

$$h_4 = 2916.3\,\text{kJ/kg}.$$

The enthalpies at the pump outlets can be estimated by combining Eqs. 8.11 and 8.20:

$$\begin{aligned} h_7 &\cong h_6 + v_6(P_7 - P_6) \\ &= 137.75\frac{\text{kJ}}{\text{kg}} + \left(0.0010053\frac{\text{m}^3}{\text{kg}}\right)(200 - 5)\frac{\text{kN}}{\text{m}^2}\left[\frac{\text{kJ}}{\text{kN·m}}\right] = 137.95\frac{\text{kJ}}{\text{kg}}. \end{aligned}$$

and

$$\begin{aligned} h_2 &\cong h_1 + v_1(P_2 - P_1) \\ &= 504.70\frac{\text{kJ}}{\text{kg}} + \left(0.0010605\frac{\text{m}^3}{\text{kg}}\right)(1000 - 200)\frac{\text{kN}}{\text{m}^2}\left[\frac{\text{kJ}}{\text{kN·m}}\right] = 505.55\frac{\text{kJ}}{\text{kg}}. \end{aligned}$$

With values for all seven enthalpies, we can find the extracted fraction (Eq. 9.11) and the cycle thermal efficiency as follows:

$$y = \frac{h_1 - h_7}{h_4 - h_7} = \frac{504.70 - 137.95}{2916.3 - 137.95} = 0.1320$$

and

$$\begin{aligned} \eta_{\text{th, regen}} &= 1 - (1 - y)\frac{h_5 - h_6}{h_3 - h_2} \\ &= 1 - (1 - 0.1320)\frac{2318.4 - 137.75}{3358.1 - 505.55} \\ &= 0.3365 \quad \text{or} \quad 33.65\%. \end{aligned}$$

Comment Comparing the present result ($\eta_{th,regen} = 33.65\%$) with that for the same cycle without regeneration ($\eta_{th} = 32.3\%$), we see that a substantial improvement results from the addition of a single feedwater heater (i.e., the percentage improvement = [(33.65 − 32.3)/32.3] × 100% = 4.2%).

Self-Test 9.4

The desired net power output of the regenerative Rankine cycle of Example 9.4 is 20 MW. Determine the mass flow rate through the boiler, the power required by each pump, and the power generated by the turbine.

(*Answer*: $\dot{m} = 20.84\ kg/s$, $\dot{W}_{pump1} = 3.62\ kW$, $\dot{W}_{pump2} = 17.71\ kW$, $\dot{W}_{turbine} = 20.023\ MW$)

CLOSED FEEDWATER HEATER

The extraction from the turbine is the same for the open and closed feedwater heater, so the fraction of the total mass flow rate extracted from the turbine (Eq. 9.7) and mass conservation for the turbine (Eq. 9.8) can again be used.

The closed feedwater heater is a heat exchanger that does not mix the two streams, Fig. 9.20. We will look at the case where the extraction steam pressure is allowed to drop after the closed feedwater and the steam is combined with the water in the condenser (Fig. 9.19). Mass conservation can be written for the condenser where the two streams recombine as shown in Fig. 9.23:

$$\dot{m}_4 + \dot{m}_5 = \dot{m}_6. \tag{9.12a}$$

or

$$y\dot{m} + (1 - y)\dot{m} = \dot{m}. \tag{9.12b}$$

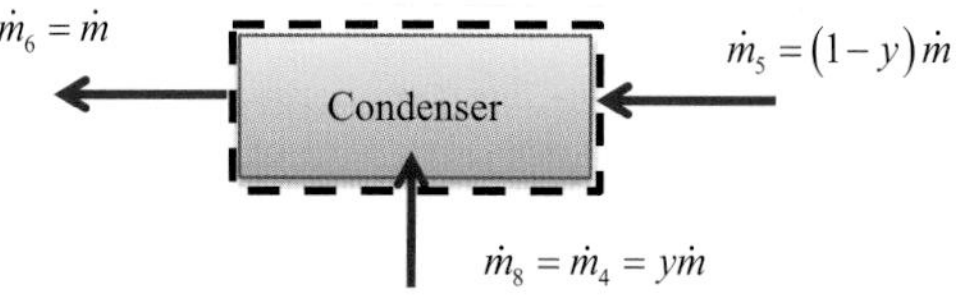

FIGURE 9.23 Mass analysis for the condenser in a regenerative cycle with a closed feedwater heater.

The extracted fraction y is an unknown quantity; the value of y is determined by requiring that the feedwater exiting the ideal closed feedwater heater be at the temperature of the extracted steam [i.e., $T_2 = T_{sat}(P_4) = T_7$]. Heating the feedwater to $T_2 = T_{sat}(P_4)$ would only occur for an ideal, very large, feedwater heat exchanger. An actual closed feedwater heater will not heat the feedwater to the saturation pressure of the extraction steam.

Energy conservation is now employed to calculate y given the turbine inlet conditions (state 3), the extraction pressure P_4, and the condenser pressure P_5+P_6.

For the closed feedwater heater, the steady-flow form of energy conservation is used for multiple flow streams and again simplifies to

$$\sum_{\text{inlets}} \dot{m}_i h_i = \sum_{\text{outlets}} \dot{m}_i h_i,$$

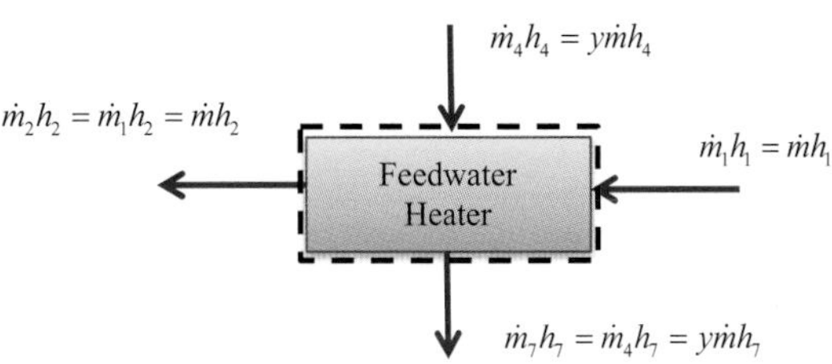

FIGURE 9.24 Energy analysis for the condenser in a regenerative cycle with a closed feedwater heater.

or

$$\dot{m}_4h_4 + \dot{m}_1h_1 = \dot{m}_7h_7 + \dot{m}_2h_2. \tag{9.13a}$$

Expressing the flow rates in terms of y yields

$$y\dot{m}h_4 + \dot{m}h_1 = y\dot{m}h_7 + \dot{m}h_2. \tag{9.13b}$$

From this expression of energy conservation, we can solve for y:

$$y = \frac{h_2 - h_1}{h_4 - h_7}. \tag{9.14}$$

With the extracted mass fraction now determined, a thermodynamic analysis of the cycle can proceed in the same fashion as our previous analyses. The following example illustrates this procedure.

Example 9.5 Closed Feedwater Heater

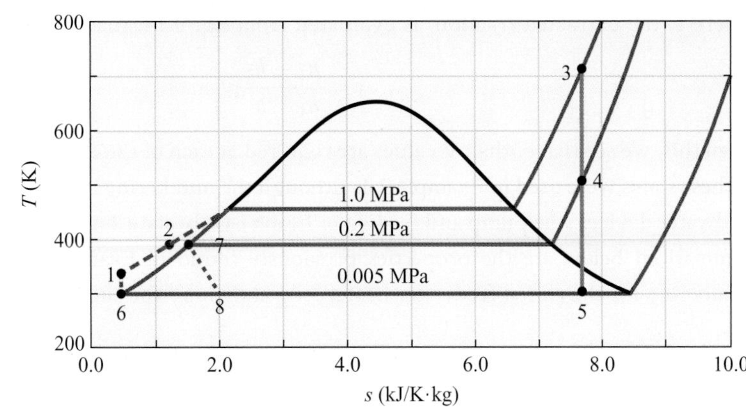

Calculate the ideal cycle efficiency for a regenerative Rankine cycle employing a single closed feedwater heater with the extraction steam being combined with the feedwater in the condenser, as shown in Fig. 9.19. A temperature–entropy diagram for the cycle is shown in the sketch. Note that state 6 is at 5 kPa and state 1 is at 1.0 MPa. Also, state 2 is at 1.0 MPa and state 7 is at 0.2 MPa. When plotted to scale on the T–s diagram, the states cannot be distinguished. The low-pressure and high-pressure state points are the same as those in Example 9.3 (i.e., $T_3 = 717.1$ K and $x_5 = 0.90$). The extracted steam and the feedwater heater are at 0.2 MPa.

Solution

Known P_1, P_2, P_3, P_4, P_6, P_7, T_3, x_5

Find η_{th}

Modeling, Premises and Assumptions

i. The flow is steady.

ii. All processes and devices are ideal.

Analysis The cycle thermal efficiency is defined as the net useful power delivered divided by the rate at which heat is added to the working fluid. The heat added comes solely from the boiler, as the closed feedwater heater is adiabatic with respect to the surroundings. Thus,

$$\eta_{\text{th, regen}} = \frac{\dot{W}_{\text{net, out}}}{\dot{Q}_{\text{in}}} = \frac{\dot{W}_{\text{turbine}} - \dot{W}_{\text{pump 1}} - \dot{W}_{\text{pump 2}}}{\dot{Q}_{\text{boiler}}}.$$

From an overall energy balance, we can also write

$$\dot{W}_{\text{net, out}} = \dot{Q}_{\text{in}} - \dot{Q}_{\text{out}}$$
$$= \dot{Q}_{\text{boiler}} - \dot{Q}_{\text{condenser}}.$$

Thus,

$$\eta_{\text{th, regen}} = \frac{\dot{Q}_{\text{boiler}} - \dot{Q}_{\text{condenser}}}{\dot{Q}_{\text{boiler}}}$$
$$= 1 - \frac{\dot{Q}_{\text{condenser}}}{\dot{Q}_{\text{boiler}}}.$$

Steady-flow first-law analyses of the boiler and condenser yield

$$\dot{Q}_{\text{boiler}} = \dot{m}(h_3 - h_2)$$

and

$$\dot{Q}_{\text{condenser}} = \dot{m}_5 h_5 + \dot{m}_8 h_8 - \dot{m}_1 h_1$$
$$= (1 - y)\dot{m} h_5 + y\dot{m} h_8 - \dot{m} h_1.$$

Substituting these back into the thermal efficiency expression, we have

$$\eta_{\text{th, regen}} = 1 - \frac{(1 - y)h_5 + yh_8 - h_1}{h_3 - h_2}$$

where y, the extracted fraction, is evaluated from Eq. 9.14, that is,

$$y = \frac{h_2 - h_1}{h_4 - h_7}.$$

From this, we see that enthalpy values are required at each of the eight state points. Most of these states were used in Example 9.4, although the numbering is different through the feedwater devices (the pump and feedwater heater). Only state 2 needs to be determined. Summarized below are the properties previously found in Example 9.4 but with the numbering for the closed feedwater heater loop shown by the subscripts in Fig. 9.19.

Property	State 3	State 4	State 5	State 6	State 7 (sat liq)
P (MPa)	1.00	0.20	0.005	0.005	0.20
T (K)	717		453.03	306.02	393.36
h (kJ/kg)	3358.1	2916.3	2318.4	137.75	504.70
s (kJ/kg·K)	7.602	7.602	7.602	0.4762	1.5302
v (m^3/kg)	—		—	0.0010053	0.0010605

The enthalpies at the pump outlet (state 1) can be estimated by combining Eqs. 8.11 and 8.20:

$$h \cong h_6 + v_6(P_1 - P_6)$$
$$= (137.75\ \text{kJ/kg}) + (0.0010053\ \text{m}^3/\text{kg})(1000 - 5)\text{kN/m}^2\left[\frac{\text{kJ}}{\text{kN·m}}\right] = 138.75\ \text{kJ/kg}.$$

The enthalpy for state 2 is that for compressed liquid at 1 MPa and the saturation temperature of the extraction steam, 393.36 K. From the NIST software we find:

$$h_2(1\ \text{MPa},\ 393.36\ \text{K}) = 505.27\ \text{kJ/kg}.$$

As the extraction steam passes through a trap (an expansion valve) between states 7 and 8, there is a drop in pressure but no change in enthalpy, $h_8 = h_7 = 504.70$ kJ/kg.

With values for all seven enthalpies, we find the extracted fraction (Eq. 9.14) and the cycle thermal efficiency as follows:

$$y = \frac{h_2 - h_1}{h_4 - h_7} = \frac{505.27 - 138.75}{2916.3 - 504.70} = 0.152$$

and

$$\begin{aligned}\eta_{\text{th, regen}} &= 1 - \frac{(1-y)h_5 + yh_7 - h_1}{h_3 - h_2}\\ &= 1 - \frac{(1-0.1520)\,2318.4 + 0.1520\,(504.7) - 138.75}{3358.1 - 505.27}\\ &= 0.3325 \quad \text{or} \quad 33.25\%.\end{aligned}$$

Comment The regeneration cycle with a closed feedwater heater requires a greater extraction steam flow rate and gives a lower thermal efficiency ($\eta_{\text{th, regen}} = 33.25\%$) compared to the open feedwater from Example 9.4 ($\eta_{\text{th, regen}} = 33.65\%$). Both regeneration cycles produce a significant improvement in the thermal efficiency compared to the same cycle without regeneration ($\eta_{\text{th}} = 32.3\%$).

Self-Test 9.5

The desired net power output of the regenerative Rankine cycle of Example 9.5 is 20 MW. Determine the mass flow rate through the boiler, the power required by the pump, and the power generated by the turbine.

(Answer: $\dot{m} = 21.08$ kg/s, $\dot{W}_{pump} = 21.08$ kW, $\dot{W}_{turbine} = 20.021$ MW)

9.1d Nuclear Power Plants

In 2012 there were 100 nuclear power plants operating in the United States providing 19% of the electricity. (See Table 1.1.) Nuclear power plants use the steam power cycle but the nuclear reactor serves as the high-temperature heat source. There are two different nuclear power plant designs used in the United States, the pressurized water reactor (PWR) and the boiling water reactor (BWR). There are currently 65 PWR units and 35 BWR units. In a boiling water reactor, the water is heated by the nuclear reactor to superheated conditions at approximately 7 MPa and then passes through the other components of the steam power cycle.

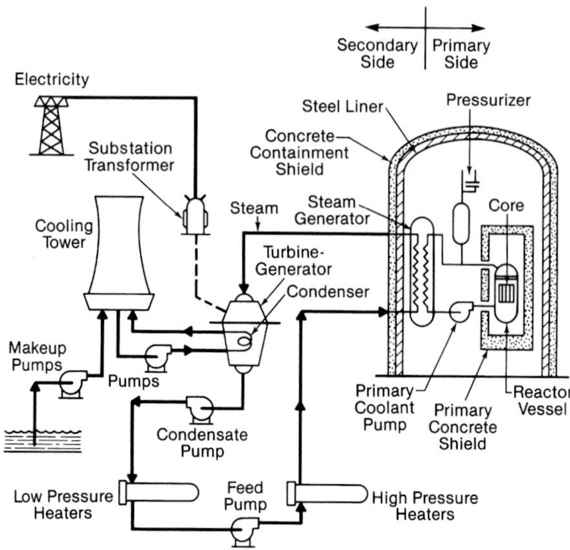

FIGURE 9.25 Schematic of a Pressured Water Reactor (PWR). Reprinted from Ref. [2] with permission (courtesy the Babcock & Wilcox Company).

In a pressurized water reactor (see Fig. 9.25), the primary reactor coolant is maintained at a high pressure but such that it does not boil. The primary coolant passes through the nuclear reactor and removes heat from the nuclear fuel rods at approximately 15 MPa. The primary coolant then passes to the steam generator, which is a large heat exchanger that uses the hot primary coolant to boil the water in the steam power cycle. The primary coolant is then pumped back to the reactor core. The primary reactor coolant follows a closed loop through the reactor, steam generator, and pumps, all housed in the containment building. The secondary flow in the steam generator is never in contact with the primary reactor coolant.

Typically there are two or four steam generators for each PWR. The secondary flow passes out of the containment building and is used in the steam power cycle. Figure 9.26 shows the flow paths for the primary and secondary flows in a Babcock & Wilcox once-through nuclear steam generator. The steam generator shown is a vertical straight tube and shell counterflow heat exchanger that is approximately 73 feet tall and 13 feet in diameter with over 15,000 tubes. The primary coolant from the nuclear reactor passes downward through the tubes and the secondary flow in the steam power cycle flows upward over the tubes. The steam generator produces superheated steam at 6.38 MPa and 313 °C at a flow rate of 680 kg/s [2].

9.2 Gas-Turbine Engines

Power cycles that produce electrical power can also be designed that use gases instead of water as the working fluid. In a gas-turbine engine, air enters and is compressed by a multistage compressor, fuel is added to this compressed air and burned, the hot products of combustion expand in a turbine, and the combustion products exit the engine at low velocity. Typically the gas power cycle is an open cycle with flows drawn from the environment and exiting into the environment. This eliminates the need for a low-temperature heat exchanger. A portion of the power produced by the turbine drives the compressor, while the remainder is delivered to a load through a shaft (Fig. 9.27). Many gas-turbine engines are more complex than is suggested by the basic arrangement in Fig. 9.27. For example, regeneration (using exhaust gases to heat

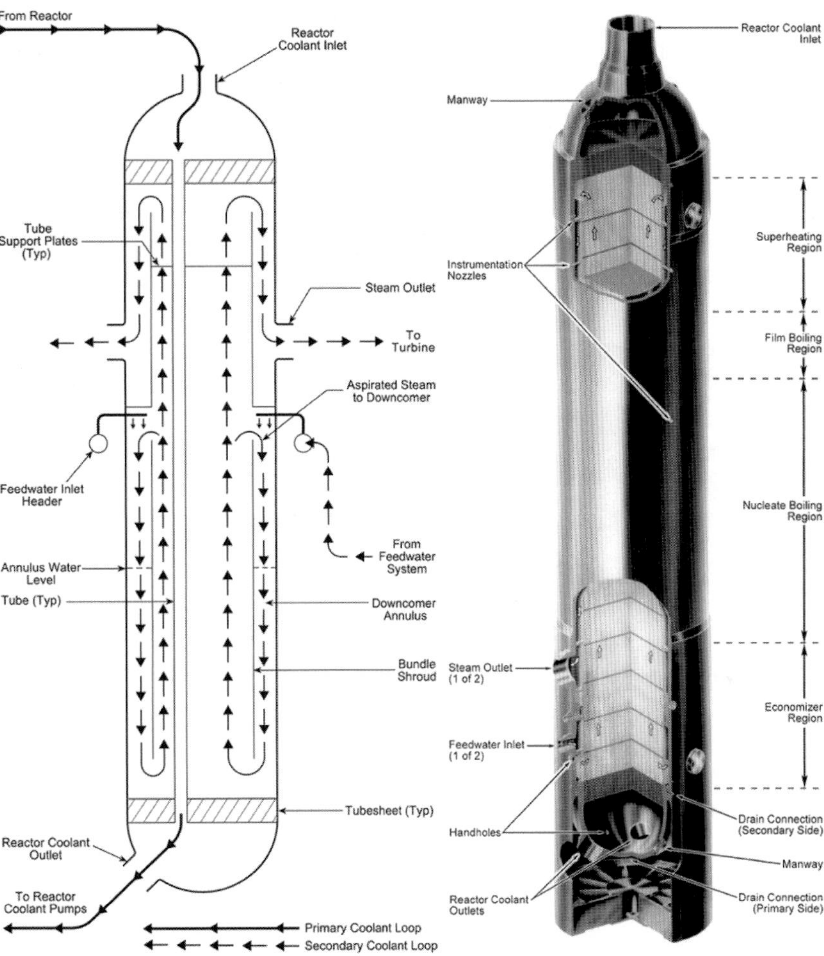

FIGURE 9.26 Simplified schematic of the flow circuits in a once-through nuclear steam generator. Reprinted from Ref. [2] with permission (courtesy the Babcock & Wilcox Company).

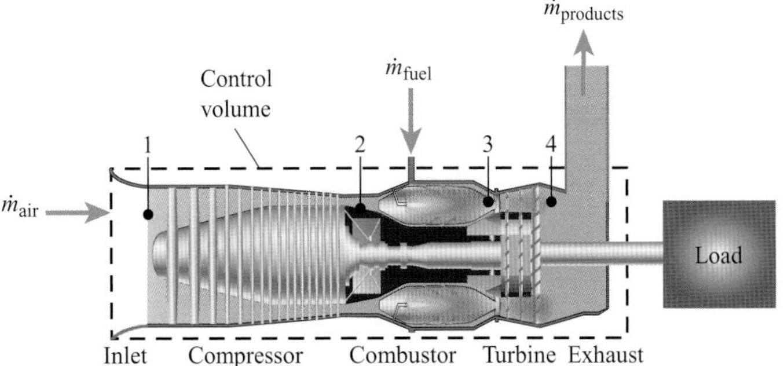

FIGURE 9.27 Schematic diagram of a gas-turbine engine. The nature of the load depends on the application. For example, the load could be an electrical generator for stationary power generation, whereas for marine propulsion, the load would be a gearbox that drives a propeller.

the air before it enters the combustor), multistage compression with intercooling, and multistage expansion with reheat can be employed to improve performance and/or efficiency. Our focus, however, is the simple system of Fig. 9.27. For more information on complex systems, see for example Refs. [7, 8].

Gas-turbine engines are frequently used in stationary electrical power generation units and range in size from tens of kilowatts to hundreds of megawatts. Such engines are also common choices for ship propulsion and pipeline pumping systems. Figures 9.28 and 9.29 show several applications of gas-turbine engines. Their application to trucks and automobiles has been explored for many years; however, the engine characteristics and the highly transient power demands of these applications are generally not well matched.

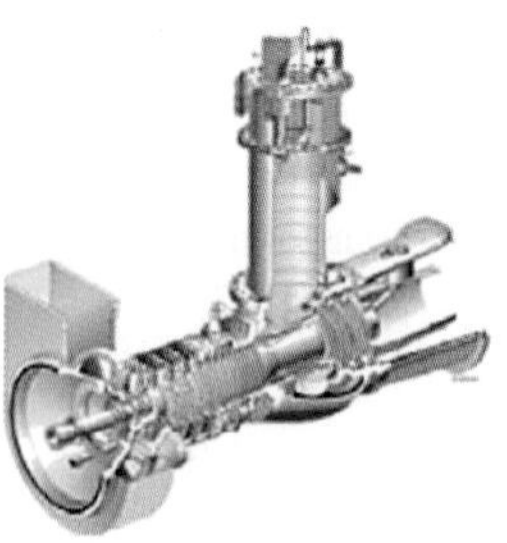

FIGURE 9.28 Gas-turbine engines are available in a wide range of sizes. The GE 9H engine (left) has a power output of 480 MW, whereas the much smaller GE10 engine (right) produces 11.25 MW. Note the array of combustors utilized in the GE 9H engine and the single vertical can combustor used in the GE 10 engine. Photographs courtesy of General Electric Company.

FIGURE 9.29 Gas- and liquid-fired gas-turbine engines drive the pumps for the Trans-Alaska Pipeline. This above-ground section of the pipeline uses a support system designed to prevent thawing of the permafrost beneath. Photograph courtesy of DOE/NREL.

In the following sections, we apply our knowledge of the basic conservation principles and analyze the thermodynamic "cycle" associated with the gas-turbine engine to gain a firm understanding of its operation.

9.2a Basic Operation of a Gas-Turbine Engine

A schematic of a generic, single-shaft, gas-turbine engine is shown in Fig. 9.27. The primary objective of the gas-turbine engine is to drive an electrical generator. This is accomplished by the following sequence of processes:

1–2: Air from the atmosphere enters the engine open system (control volume) at low speed. A multistage compressor compresses the air to a high pressure before entering the combustor. Typical gas-turbine compressors operate with pressure ratios of 12–22 or greater. Numerous compressor stages (10–15) are required to achieve the overall pressure ratio since each stage typically operates with a pressure ratio of only 1.1 or 1.2 [3].

2–3: Fuel is injected and burned in the combustor. The high-temperature strength of the turbine blades is the major factor limiting the turbine inlet temperature. The blades in the first stage of a modern turbine can withstand temperatures exceeding 1430 °C [10] if they are made from a single crystal alloy, using internal blade cooling, or if they are coated with ceramic material.

3–4: A multistage gas turbine expands the high-temperature, high-pressure gas. The shaft power from the turbine drives the compressor and the electrical generator. Gas-turbine engines typically employ one to four turbine stages.

4–1: The hot gas from the turbine output is exhausted to the atmosphere or used as a high-temperature source in a cogeneration (Section 9.3a) or combined-cycle power plant (Section 9.3b). We can analyze the gas-turbine engine as a "cycle"; however, the exhaust flow does not return to the inlet to the compressors. We will consider the atmosphere as a large heat sink to complete the cycle.

Before investigating the details of the gas-turbine engine "cycle," we perform simple, integral, open-system (control volume) analyses to help understand the basic operation of this engine.

9.2b Integral Open System (Control Volume) Analysis

As indicated by the dashed line in Fig. 9.27, we consider an open system that crosses the air inlet plane, cuts through the exhaust duct, cuts through the turbine output shaft, and otherwise surrounds the entire engine. This open system is reproduced in Fig. 9.30, which shows the mass flows (top) and energy flows (bottom).

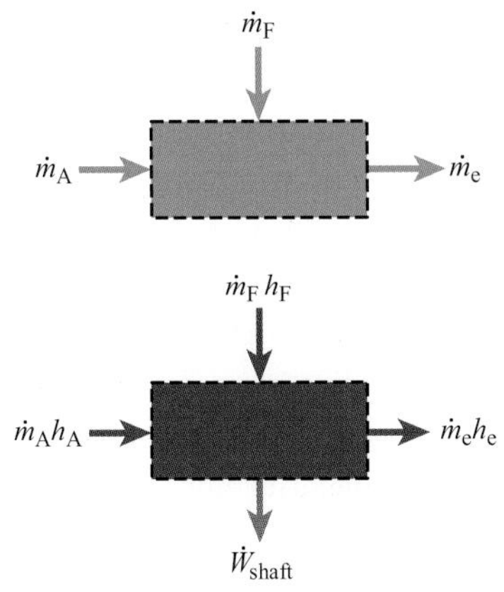

FIGURE 9.30 Integral open systems (control volumes) for application of mass conservation (top) and energy conservation (bottom) to the analysis of a gas-turbine engine.

ASSUMPTIONS

The assumptions invoked in the application of mass and energy conservation to this open system are the following:

i. The flow is steady and at steady state.

ii. There are no heat interactions between the engine and the surroundings (i.e., $\dot{Q}_{cv} = 0$).

iii. The kinetic and potential energies of the air, fuel, and exhaust streams are negligible.

MASS CONSERVATION

Air and fuel enter the open system as separate streams, and a single stream of combustion products exits as shown in Fig. 9.30 (top). Conservation of mass is thus expressed by Eq. 3.14b as follows:

$$\dot{m}_A + \dot{m}_F = \dot{m}_e. \tag{9.15a}$$

This can also be recast by introducing the fuel–air ratio F/A

$$\dot{m}_A(1 + F/A) = \dot{m}_e. \tag{9.15b}$$

Similarly to the operation of turbojet engines, fuel–air ratios for gas-turbine engines are small and of the order of 1:100.

ENERGY CONSERVATION

The starting point for energy conservation is Eq. 5.16:

$$\dot{Q}_{cv,\,net\,in} - \dot{W}_{cv,\,net\,out} = \sum_{k=1}^{M\,outlets} \dot{m}_{out,k}\left[h_k + \tfrac{1}{2}V_{avg,k}^2 + g(z_k - z_{ref})\right] - \sum_{j=1}^{N\,inlets} \dot{m}_{in,j}\left[h_j + \tfrac{1}{2}V_{avg,j}^2 + g(z_j - z_{ref})\right].$$

As shown in Fig. 9.30 (bottom), the only energy flows across the control surface are the enthalpy flows in (air and fuel) and out (combustion products) and the shaft power out. The energy conservation equation thus simplifies to

$$0 - \dot{W}_{shaft} = \dot{m}_e h_e - \dot{m}_F h_F - \dot{m}_A h_A,$$

or

$$\dot{W}_{shaft} = \dot{m}_F h_F + \dot{m}_A h_A - \dot{m}_e h_e. \tag{9.16a}$$

Combining this with mass conservation (Eq. 9.15b) results in

$$\dot{W}_{shaft} = \dot{m}_F h_F + \dot{m}_A h_A - \dot{m}_A(1 + F/A)h_e. \tag{9.16b}$$

Note that all the enthalpies in this expression are standardized values to account for chemical transformations.

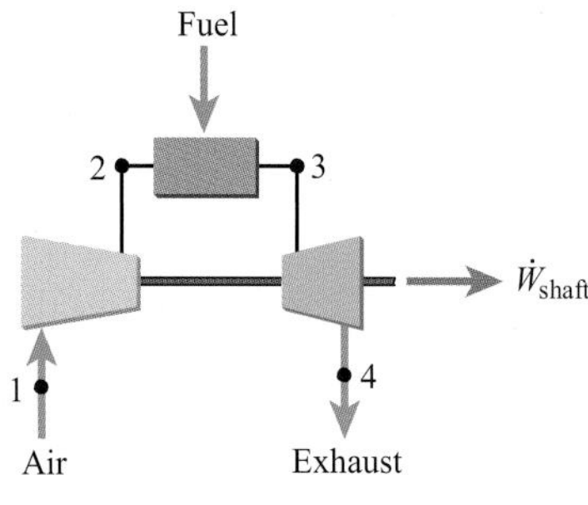

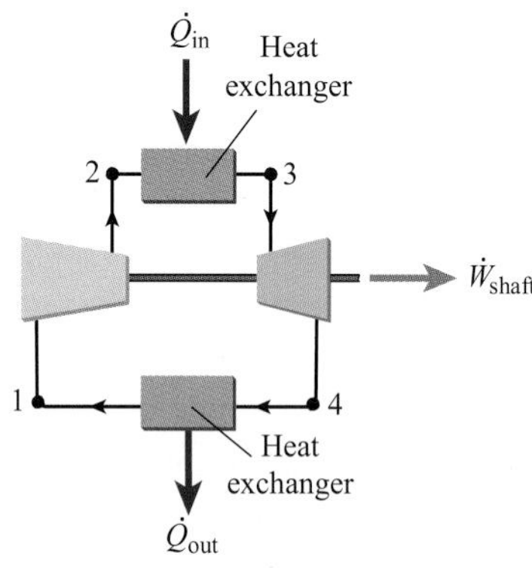

FIGURE 9.31 A gas-turbine engine (top schematic) can be idealized using an air-standard cycle (bottom schematic) in which heat-transfer processes replace the combustion process (2–3) and exhaust process (4–1).

9.2c Cycle Analysis and Performance Measures

Like the turbojet, the gas-turbine engine does not execute a true thermodynamic cycle. In both these engines, the working fluid undergoes a transformation from air and fuel to combustion products without a return to the original state as is required for a thermodynamic cycle. Nevertheless, we can define state points (see Fig. 9.31, top) and analyze the engine component by component (i.e., state to state) to define a "cycle." To do so, however, requires a treatment of the combustion process, a fairly lengthy procedure, as is demonstrated in Chapter 12. To gain some insight into the operation of gas-turbine engines, while avoiding the complexities imposed by the combustion process, we again define and analyze an air-standard cycle for this engine.

AIR-STANDARD BRAYTON CYCLE

Figure 9.32 illustrates how the air-standard Brayton cycle models the actual gas-turbine cycle. Using air as the working fluid, the combustion process is replaced by a heat-addition process, and the exhaust and intake processes are replaced with a loop-closing heat-rejection process. The ideal (reversible) air-standard Brayton cycle is thus defined by the following steady-flow processes and is illustrated using P–v and T–s coordinates in Fig. 9.32.

Ideal Air-Standard Brayton Cycle

State Change	Process
1–2:	Adiabatic and reversible (isentropic) compression,
2–3:	Constant-pressure heat addition,
3–4:	Adiabatic and reversible (isentropic) expansion, and
4–1:	Constant-pressure heat rejection.

AIR-STANDARD THERMAL EFFICIENCY

Applying the definition of the thermal efficiency of a work-producing cycle (Eq. 9.1) to the air-standard Brayton cycle yields

$$\eta_{\text{th, Brayton}} = \frac{\dot{W}_{\text{shaft}}}{\dot{Q}_{\text{in}}} = \frac{\dot{W}_{\text{turbine, out}} - \dot{W}_{\text{compressor, in}}}{\dot{Q}_{\text{in}}}.$$

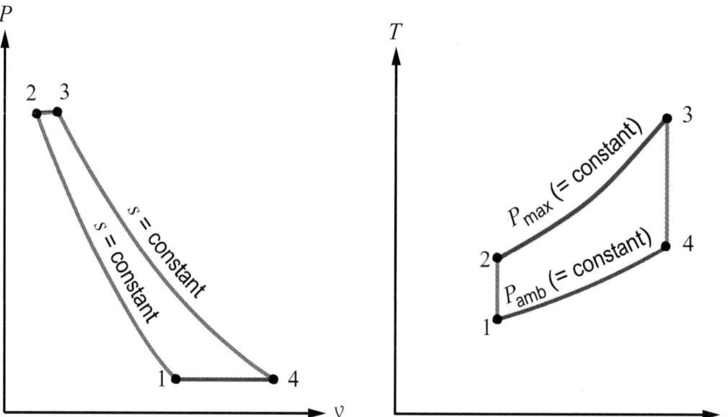

FIGURE 9.32 The ideal air-standard cycle for a gas-turbine engine, or ideal Brayton cycle, consists of an isentropic compression (1–2), a constant-pressure heat addition (2–3), an isentropic expansion (3–4), and a constant-pressure heat rejection (4–1).

Recognizing from an overall energy balance on the engine that $\dot{W}_{\text{shaft}} = \dot{Q}_{\text{in}} - \dot{Q}_{\text{out}}$ (see Fig. 9.31, bottom), we rewrite this as

$$\eta_{\text{th, Brayton}} = \frac{\dot{Q}_{\text{in}} - \dot{Q}_{\text{out}}}{\dot{Q}_{\text{in}}} = 1 - \frac{\dot{Q}_{\text{out}}}{\dot{Q}_{\text{in}}}.$$

Neglecting changes in kinetic and potential energies across the heat-exchange devices, we can relate the $\dot{Q}$s to enthalpy changes, that is,

$$\dot{Q}_{\text{in}} = \dot{m}(h_3 - h_2)$$

and

$$\dot{Q}_{\text{out}} = \dot{m}(h_4 - h_1).$$

Thus,

$$\eta_{\text{th, Brayton}} = 1 - \frac{h_4 - h_1}{h_3 - h_2}. \tag{9.17a}$$

This result applies to a cycle with either ideal (reversible) or nonideal (irreversible) components.

We can also relate Eq. 9.17a to the various state-point temperatures using the approximate ideal-gas calorific equation of state (Eq. 2.31e),

$$h_4 - h_1 = c_{p,\text{avg},1-4}(T_4 - T_1),$$
$$h_3 - h_2 = c_{p,\text{avg},2-3}(T_3 - T_2),$$

and so

$$\eta_{\text{th, Brayton}} = 1 - \frac{c_{p,\text{avg},1-4}(T_4 - T_1)}{c_{p,\text{avg},2-3}(T_3 - T_2)}. \tag{9.17b}$$

If we approximate the analysis by assuming by assuming that c_p is not a function of temperature,[3] the specific heat terms cancel from the equation to give

$$\eta_{\text{th, Brayton}} = 1 - \frac{T_4 - T_1}{T_3 - T_2}. \tag{9.17c}$$

For the ideal cycle, both the compression and expansion processes are isentropic, and the temperatures can be related to the pressure ratio, defined as

$$R_{\text{P}} \equiv P_2/P_1 = P_3/P_4,$$

using the second-law state relationship, Eq. 7.13a, as follows:

$$T_2 = T_1\left(\frac{P_2}{P_1}\right)^{(\gamma-1)/\gamma} = T_1 R_{\text{P}}^{(\gamma-1)/\gamma}$$

and

$$T_4 = T_3\left(\frac{P_4}{P_3}\right)^{(\gamma-1)/\gamma} = T_3\left(R_{\text{P}}^{-1}\right)^{(\gamma-1)/\gamma},$$

[3] This is technically an invalid assumption since we know indeed that $c_p = c_p(T)$; however, c_p for air does not vary greatly over the temperature range of interest, so this is a useful approximation.

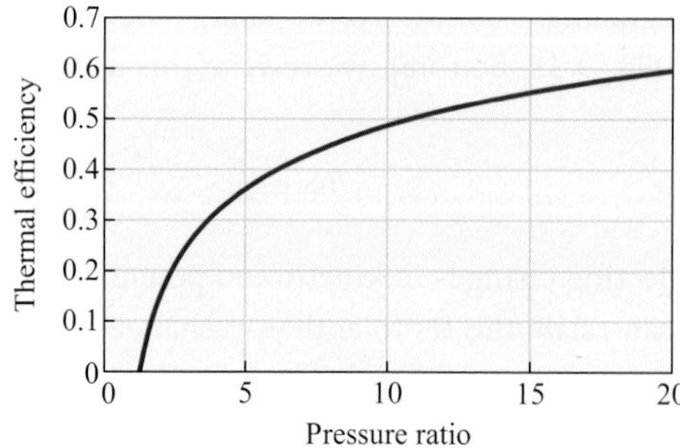

FIGURE 9.33 The ideal air-standard Brayton cycle efficiency depends only on the pressure ratio $R_P = P_2/P_1$ and the specific heat ratio γ of the working fluid. Plot values are based on $\gamma = 1.4$.

or

$$T_4 = T_3(R_P)^{-(\gamma-1)/\gamma}.$$

Substituting these relationships back into Eq. 9.17c yields

$$\eta_{\text{th, Brayton}} = 1 - \frac{T_3(R_P)^{-(\gamma-1)/\gamma} - T_1}{T_3 - T_1(R_P)^{(\gamma-1)/\gamma}},$$

which simplifies to

$$\eta_{\text{th, Brayton}} = 1 - (R_P)^{-(\gamma-1)/\gamma}. \tag{9.18}$$

This interesting result shows that the ideal Brayton cycle thermal efficiency is a function only of the pressure ratio (or temperature ratio); it is independent of both the air inlet temperature T_1 and the turbine inlet temperature T_3. Figure 9.33 illustrates the dependence of thermal efficiency on pressure ratio for a specific-heat ratio γ of 1.4. Here we see that at low pressure ratios, say 2–7, the thermal efficiency is strongly dependent on the pressure ratio, whereas the dependence is much weaker at large values of pressure ratio, say, >15.

POWER AND SIZE

A useful measure of engine performance is the ratio of the net power produced to the mass flow rate through the engine, $\dot{W}_{\text{shaft}}/\dot{m}$. With a value for this parameter, one can get a feel for how large a device needs to be in order to perform a particular task, where size enters through the cross-sectional flow area (i.e., $\dot{m} = \rho VA$). Using the air-standard Brayton cycle (Fig. 9.32) we can express $\dot{W}_{\text{shaft}}/\dot{m}$ as follows:

$$\begin{aligned}\frac{\dot{W}_{\text{shaft}}}{\dot{m}} &= \frac{\dot{W}_{\text{turbine, out}} - \dot{W}_{\text{compressor, in}}}{\dot{m}} \\ &= (h_3 - h_4) - (h_2 - h_1),\end{aligned}$$

or, furthermore, if we assume a constant specific heat,

$$\frac{\dot{W}_{\text{shaft}}}{\dot{m}} = c_{p,\text{avg}}[(T_3 - T_4) - (T_2 - T_1)].$$

We now treat T_1 and T_3 as given parameters and apply the isentropic property relationships to eliminate T_2 and T_4; this yields

$$\frac{\dot{W}_{\text{shaft}}}{\dot{m}} = c_{p,\text{avg}}\left[T_3\left(1 - R_P{}^{(1-\gamma)/\gamma}\right) + T_1\left(1 - R_P{}^{(\gamma-1)/\gamma}\right)\right]. \tag{9.19}$$

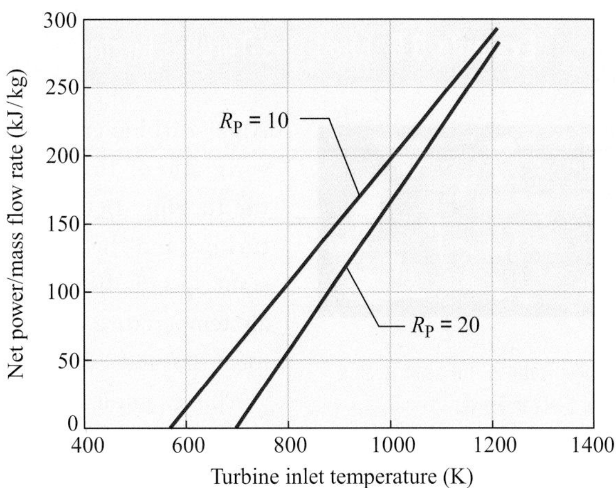

FIGURE 9.34 The net shaft power delivered per unit of mass flow for an ideal air-standard Brayton cycle is linearly related to the turbine inlet temperature T_3. The turbine inlet temperature, in turn, is essentially proportional to the fuel flow rate. Plot values are based on $T_1 = 300$ K, $\gamma = 1.4$, and $c_{p,\text{avg}} = 1.007$ kJ/kg·K.

Figure 9.34 illustrates this relationship for two values of the pressure ratio R_P. Here we see that a minimum turbine inlet temperature T_3 is required to produce any net power. At the zero value of $\dot{W}_{\text{shaft}}/\dot{m}$, all the turbine power is used to drive the compressor. Above this minimum value of the turbine inlet temperature, $\dot{W}_{\text{shaft}}/\dot{m}$ increases linearly with T_3, where the slope depends on the pressure ratio R_P. Although this analysis is based on the air-standard cycle, we can connect it with the real engine by realizing that the temperature difference across the combustor, $T_3 - T_2$, is, to first order, proportional to the fuel flow rate. Thus we can construe Fig. 9.34 as a plot of $\dot{W}_{\text{shaft}}/\dot{m}$ versus $\dot{m}_F$ (or F/A).

PROCESS THERMAL EFFICIENCY AND SPECIFIC FUEL CONSUMPTION

For a real gas-turbine engine, we can define a process thermal efficiency with reference to Fig. 9.30. Starting with Eq. 9.1,

$$\eta_{\text{process, Brayton}} = \frac{\text{useful work produced}}{\text{energy supplied}},$$

and recognizing that the energy supplied is associated with the chemical energy in the fuel, we can write

$$\eta_{\text{process, Brayton}} = \frac{\dot{W}_{\text{shaft}}}{\dot{m}_F HV_F}, \tag{9.20}$$

See Example 12.7 for a formal definition of HV_F and Appendix H for HV_F values for various fuels.

where HV_F is the heating value of the fuel.

A frequently used measure of engine efficiency is the **specific fuel consumption**, which we define as

$$sfc \equiv \frac{\text{rate fuel consumed}}{\text{rate work delivered}} = \frac{\dot{m}_F}{\dot{W}_{\text{shaft}}}. \tag{9.21}$$

Comparing this definition with Eq. 9.20, we see that the specific fuel consumption is inversely proportional to the process efficiency; thus, the smaller the value of *sfc*, the higher the efficiency. The specific fuel consumption is also frequently used to quantify spark-ignition and diesel engine fuel efficiency.

Example 9.6 Ideal Gas-Turbine Engine

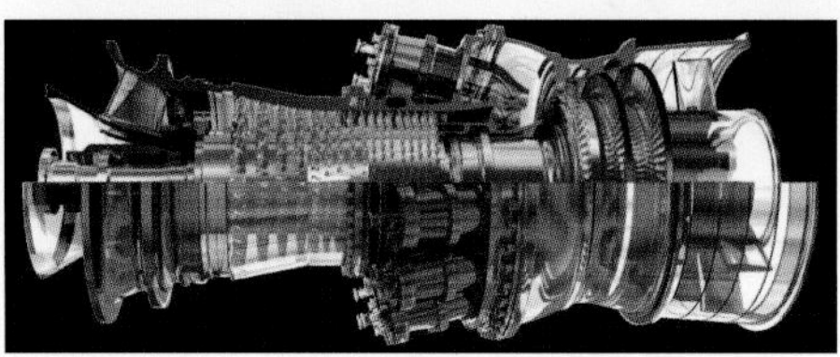

GE Gas Turbine 7F.04 specifications are used in this example. (Credit: General Electric Power.)

A gas-turbine engine is measured to have a compressor pressure ratio of 16.2 and an air temperature of 1260 °C entering the turbine. Determine the temperature at the outflow of the turbine and the gas-turbine engine efficiency. Assume a constant specific heat ratio of $\gamma = 1.4$ and an inflow atmospheric air temperature of 50 °C. Approximate the process using the ideal Brayton cycle.

This example uses actual experimental data.

Solution

Known R_P, T_1, T_3, and P_1

Find $\eta_{\text{th, Brayton}}$, T_2, and T_4

Sketch

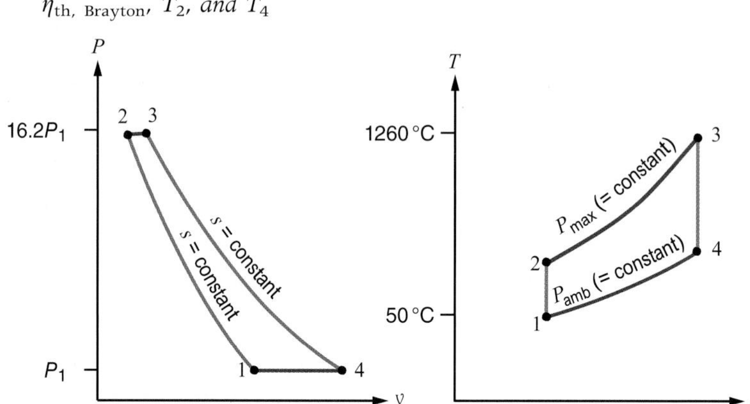

Modeling, Premises and Assumptions

i. Approximate the gas-turbine engine using the Brayton cycle for an ideal gas.
ii. The specific heats of the air components are constant and equal (i.e., $c_p = 1200\,\text{J/kg}\cdot\text{K}$, $\gamma = 1.4$).
iii. Air enters at an atmospheric pressure of 101 kPa and temperature 50°C.

Analysis The efficiency of the Brayton cycle $\eta_{\text{th, Brayton}} = 1 - (R_P)^{(1-\gamma)/\gamma}$.

$$\eta_{th,\text{Brayton}} = 1 - (R_P)^{(1-\gamma)/\gamma} = 1 - (16.2)^{(-0.4/14)} = 0.5487;$$

i.e., 54.87% of the heat transfer to the air is converted to shaft work that can be used to drive an electrical generator.

We can find the high pressure in the Brayton cycle knowing the pressure ratio and the inlet pressure:

$$P_2 = P_3 = R_P P_1 = 16.2(101\,\text{kPa}) = 1632\,\text{kPa} = 1.632\ \text{MPa}.$$

The temperatures at the outflow of the compressor and turbine can be found using the isentropic relationship between pressure and temperature:

$$T_2 = T_1\left(\frac{P_2}{P_1}\right)^{(\gamma-1)/\gamma} = T_1 R_P{}^{(\gamma-1)/\gamma}$$

$$T_2 = (323\,\text{K})(16.2)^{0.4/14} = 715.8\,\text{K}$$

and

$$T_4 = T_3\left(\frac{P_4}{P_3}\right)^{(\gamma-1)/\gamma} = T_3 R_P{}^{(\gamma-1)/\gamma},$$

$$T_4 = (1533\text{ K})\left(\frac{1}{16.2}\right)^{0.4/1.4} = 691.8\text{ K}.$$

Comments Note that the absolute temperature must be used in the isentropic equations. The temperature difference across the turbine is greater than the temperature difference across the compressor. This gives a net power output from the Brayton cycle.

Example 9.7 Actual Gas-Turbine Engine

Find the thermal efficiency for the gas-turbine engine having the conditions stated in Example 9.6 but with the actual compressor and turbine having an isentropic efficiency of 85%.

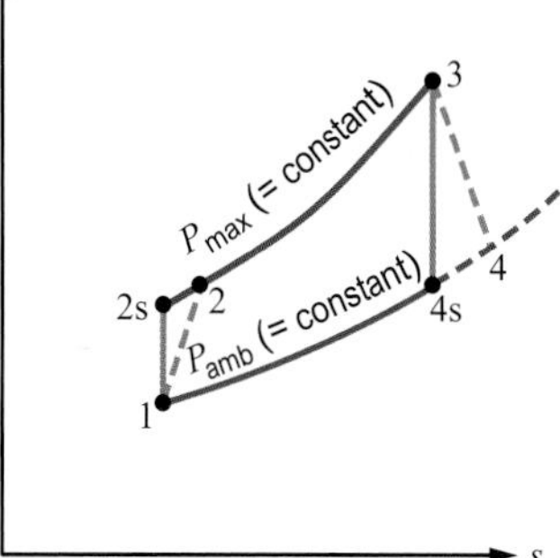

Solution

Using the definition of isentropic efficiency, we can find the temperature at the outflow of the compressor and turbine:

$$\eta_{\text{s, compressor}} = \frac{h_{2s} - h_1}{h_2 - h_1} = \frac{c_p(T_{2s} - T_1)}{c_p(T_2 - T_1)}.$$

Solving for T_2,

$$T_2 = T_1 + (T_{2s} - T_1)/\eta_{\text{s,compressor}} = 323 + (715.8 - 323)/0.85 = 785.1\text{ K}$$

$$\eta_{\text{s,turbine}} = \frac{h - h_4}{h - h_{4s}} = \frac{c_p(T_3 - T_4)}{c_p(T_3 - T_{4s})}.$$

Solving for T_4,

$$T_4 = T_3 - \eta_{\text{s, turbine}}(T_3 - T_{4s}) = 1533\text{K} - 0.85(1533 - 691.8)\text{K} = 818.0\text{ K}.$$

The thermal efficiency of the actual cycle can be found using Eq. 9.17c:

$$\eta_{\text{th, Brayton}} = 1 - \frac{h_4 - h_1}{h_3 - h_2}.$$

For an ideal gas,

$$\eta_{\text{th, actual}} = 1 - \frac{h_4 - h_1}{h_3 - h_2} = 1 - \frac{c_p(T_4 - T_1)}{c_p(T_3 - T_2)} = 1 - \frac{(818 - 323)\text{K}}{(1533 - 785.1)\text{K}} = 0.3381$$
$$= 33.81\%.$$

Comments The thermal efficiency for the actual gas-turbine cycle between the minimum and maximum temperatures is less than that for the ideal gas-turbine.

9.3 Modified Power Cycles

The Rankine cycle is often used for a modified purpose. When the waste heat from the Rankine cycle is used as process heat, the cycle is referred to as a *cogeneration cycle*: that is, it produces two forms of useful energy. The *combined power cycle* extracts additional power output by using the waste heat from a high-temperature Brayton cycle as the heat inflow to a steam Rankine cycle.

9.3a Cogeneration Cycles

The Rankine cycles in Section 9.1 and the Brayton cycles in Section 9.2 both have waste heat that is removed from the working fluid. The cogeneration cycle uses this waste heat for a secondary use such as heating a building or providing a source of process heat in a manufacturing or chemical process. Some applications using process heat include: pulp and paper production, vulcanization of rubber (170 °C), food processing, ethanol production, oil refining, plastics, and metals manufacturing. In some cases, process heat is needed at a higher temperature or pressure. In these applications, the process heat is extracted from the turbine stages, as shown in Figure 9.35.

A cogeneration power cycle has two objectives: to produce electrical power from the generator connected to the turbine and to produce process heat. The efficiency of a cogeneration cycle, sometimes called the utilization factor, includes both these desired outputs:

$$\eta_{\text{cogen}} = \frac{\text{useful work produced} + \text{process heat}}{\text{energy supplied}} = \frac{\dot{W}_{\text{turbine, out}} - \dot{W}_{\text{pump, in}} + \dot{Q}_{\text{process}}}{\dot{Q}_{\text{in}}} = \frac{\dot{Q}_{\text{out}}}{\dot{Q}_{\text{in}}}. \tag{9.22}$$

If the outflow from the turbine is used for process heat, the utilization factor can be over 90%. For the cogeneration cycle in Fig. 9.35, we have:

$$\eta_{\text{cogen}} = \frac{(1-y)(h_5 - h_6)}{(h_3 - h_2)}. \tag{9.23}$$

FIGURE 9.35 Rankine cycle with cogeneration.

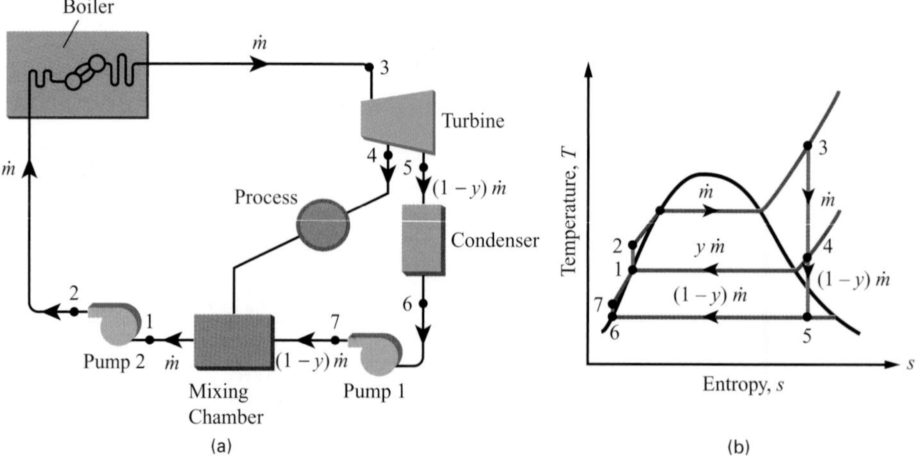

9.3b Combined Cycle Power Plants

As we found in Section 9.2, the exhaust gases from a gas turbine engine are at a high temperature, often over 600 °C. A combined cycle power plant uses these high-temperature exhaust gases as an available source of heat for a steam power cycle. This combined cycle system is often used in power plants operating with natural gas.

Figure 9.36 shows a schematic of a combined cycle power plant. The hot exhaust gases from the gas turbine pass through the heat recovery steam generator (HRSG) and boil the water in the Rankine cycle.

The thermal efficiency of the combined cycle is

$$\begin{aligned}\eta_{\text{combined}} &= \frac{\text{net work gas cycle} + \text{net work steam cycle}}{\text{energy supplied}}\\ &= \frac{\dot{W}_{\text{gas, turbine, out}} - \dot{W}_{\text{gas compressor, in}} + \dot{W}_{\text{steam turbine, out}} - \dot{W}_{\text{steam pump, in}}}{\dot{Q}_{\text{in}}}\\ &= 1 - \frac{\dot{Q}_{\text{out}}}{\dot{Q}_{\text{in}}}.\end{aligned} \tag{9.24}$$

The combined cycle efficiency is

$$\eta_{\text{combined}} = 1 - \frac{\dot{m}_{\text{water}}(h_9 - h_6)}{\dot{m}_{\text{gas}}(h_3 - h_2)}. \tag{9.25}$$

The heat recovery steam generator, is a counterflow heat exchanger that removes heat from the turbine exhaust gas and heats the feedwater in the steam cycle. The counterflow heat exchanger was described in Section 8.6 and is shown for the combined cycle in Fig. 9.37. Notice that the water temperature does not increase continuously during the heat transfer. Starting at the water inflow, location 7, the

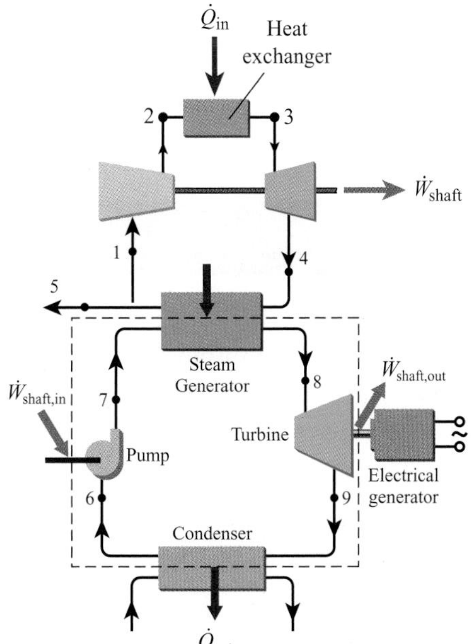

FIGURE 9.36 Combined cycle power plant.

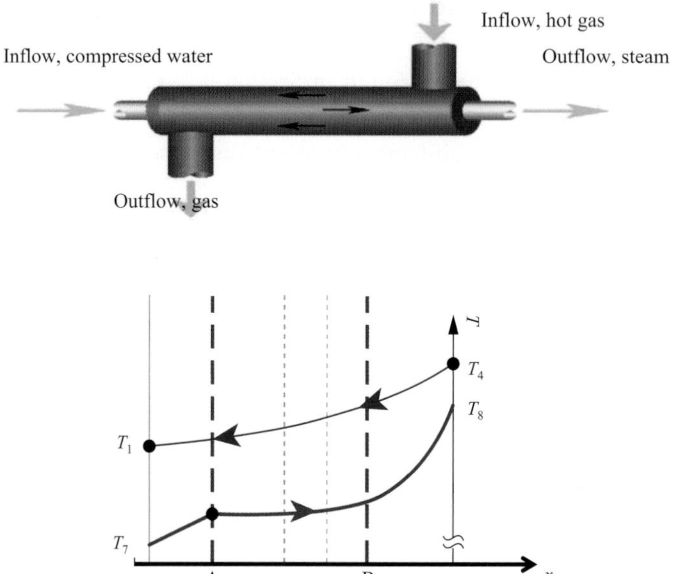

FIGURE 9.37 Schematic diagram of the temperature distributions through a heat recovery steam generator (HRSG).

compressed liquid is heated to saturated conditions. Approximating the water pressure as uniform through the heat exchanger, boiling from saturated liquid to vapor occurs at constant temperature. If the steam is then superheated, the temperature will increase to state 8. For heat transfer to occur from the hot gases to the cooler water, the temperature of the gas must be less than the temperature of the water at every x-position along the heat exchanger. In addition, the temperature difference between the gas and water must be at least 5 to 10 °C for the heat transfer to occur at a practical rate. A smaller temperature difference between the gas and water will reduce the rate of heat transfer and will require a larger, more expensive, HRSG.

We see from Fig. 9.37 that the temperature difference between the gas and water is smallest at point A, called the "pinch point." The pinch point conditions must be calculated to ensure that the temperature of the water remains at least 10 °C lower than the temperature of the gas. The analysis of the heat recovery steam generator is considered in Example 9.8 below.

Example 9.8 Heat Recovery Steam Generator

The Rankine superheat cycle in Example 9.3 is used in a combined cycle power plant. The exhaust from the gas turbine is 877 K with a flow rate of 436 kg/s. The outflow gas from the HRSG is 423 K. Determine the water flow rate for the Rankine cycle and the pinch point temperature difference.

Solution

Known T_1, T_4, $\dot{m}_{gas}$, *and* P_7

Find $\dot{m}_{water}$, ΔT_A

Sketch

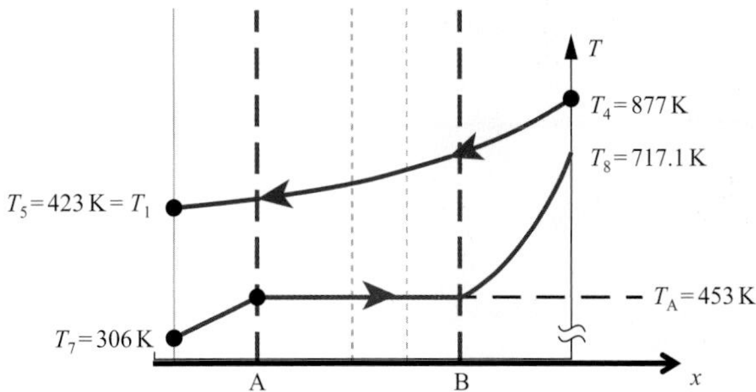

Modeling, Premises and Assumptions

i. Steady, adiabatic heat recovery steam generator with counterflow.
ii. Approximate the gas enthalpy using the specific heat of air at the average temperature (i.e., $c_p(650\text{ K}) = 1.0632\text{ kJ/kg·K}$).

Analysis From Example 9.3, the steam properties are:

Property	State 7	State A	State 8
P (MPa)	1.00	1.00	1.00
T (K)	306.05	453.03	717.1
h (kJ/kg)	138.75	762.52	3358.3

We can find the mass flow rate of the water by equating the rate of heat transfer from the gas to the rate of heat transfer to the water:

$$\dot{m}_{\text{gas}} c_p (T_4 - T_5) = \dot{m}_{\text{water}} (h_8 - h_7)$$

$$\dot{m}_{\text{water}} = \frac{\dot{m}_{\text{gas}} c_p (T_4 - T_5)}{h_8 - h_7} = \frac{(436\text{ kg/s})(1.0632\text{ kJ/kg·K})(877 - 423)\text{K}}{(3358.3 - 138.75)\text{kJ/kg}} = 65.37\text{ kg/s}.$$

We will now do a similar analysis for just the left part of the heat exchanger, from the water inflow to location A where the water is saturated liquid at 1 MPa. This will allow us to find the temperature of the gas at location A.

$$\dot{m}_{\text{gas}} c_p \left(T_{\text{A, gas}} - T_5\right) = \dot{m}_{\text{water}} \left(h_{\text{A, water}} - h_7\right)$$

$$T_{\text{A, gas}} = T_5 + \frac{\dot{m}_{\text{water}} \left(h_{\text{A, water}} - h_7\right)}{\dot{m}_{\text{gas}} c_p} = 423\text{K} +$$

$$= \frac{(65.37\text{ kg/s})(3358.3 - 138.75)\text{kJ/kg}}{(436\text{kg/s})(1.0632\text{kJ/kg·K})} = 511.0\text{ K}.$$

The temperature difference between the gas and water at the pinch point is

$$\Delta T_{\text{A}} = T_{\text{A, gas}} - T_{\text{A, water}} = 510.95 - 453.03 = 57.62\text{ K}.$$

Comments The temperature difference at the pinch point is greater than is needed for efficient heat transfer. The steam cycle can be operated at a higher HRSG pressure and temperature. This will decrease ΔT_{A} and increase the efficiency of the combined cycle.

FIGURE 9.38 The PW4084 turbofan nominally develops 385.9 kN (86,760 lb_f) of thrust at takeoff. The diameter of the fan is 2.84 m (112 in). Photograph courtesy of Pratt and Whitney.

9.4 Turbojet Engines

Modern jet engines are amazing machines. The PW4084 turbofan engine shown in Fig. 9.38 weighs 66.7 kN (15,000 lb_f) and produces more than 385 kN (86,760 lb_f) of thrust. Two of these, or similar, engines power the Boeing 777 aircraft, which weighs more than 2446 kN (550,000 lb_f) and can carry more than 300 passengers.

The two types of jet engines used in aircraft applications, the turbojet and the turbofan, were introduced in Chapter 1. In the following sections, we apply our knowledge of the basic conservation principles and analyze the thermodynamic "cycle" associated with the turbojet engine in order to gain a firm understanding of its operation. We will use the same methods as those previously applied to the gas-turbine engine.

9.4a Basic Operation of a Turbojet Engine

A schematic of a generic single-shaft turbojet engine is shown in Fig. 9.39; an actual engine and its application in a military aircraft are illustrated in Fig. 9.40. The propulsive thrust of the turbojet engine results from the production of a high-velocity exit jet. This is accomplished by the following sequence of processes:

1–2: Air enters the engine open system (control volume) at flight speed at station 1 and is slowed to a lower velocity at station 2. This velocity decrease from station 1 to station 2 results from the flow pattern outside the engine (1–1a) and a physical diffuser (1a–2). A pressure rise (the ram effect) results from this slowing of the air.

2–3: A multistage compressor (Figure 9.41a) compresses the air to a high pressure before entering the combustor. Typical turbojet compressors operate with pressure ratios of 10–15 or greater (see Fig. 9.42). Numerous compressor stages are required to achieve the overall pressure ratio since each stage typically

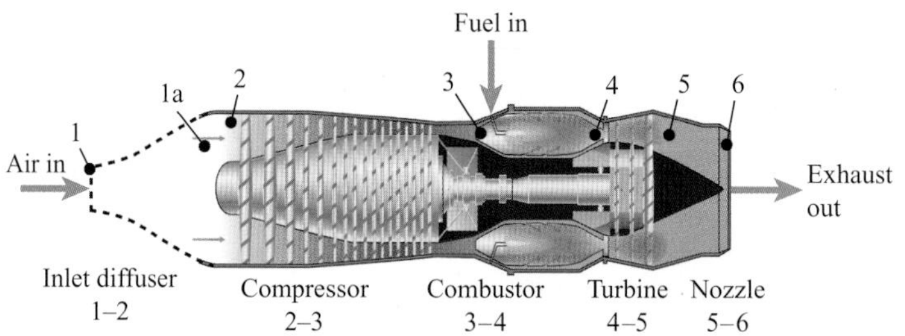

FIGURE 9.39 Schematic diagram of a turbojet engine. The gases flow through the annular space between the turbomachinery components and the engine housing.

FIGURE 9.40 Two Pratt and Whitney J52-P-408A turbojet engines power the EA-6B Prowler military aircraft. A single engine produces an intermediate thrust of 49.8 kN (11,200 lb_f), weighs 10.3 kN (2318 lb_f), has a length of 3 m (119 in), and has a maximum diameter of 0.79 m (31.1 in). Photographs courtesy of Pratt and Whitney.

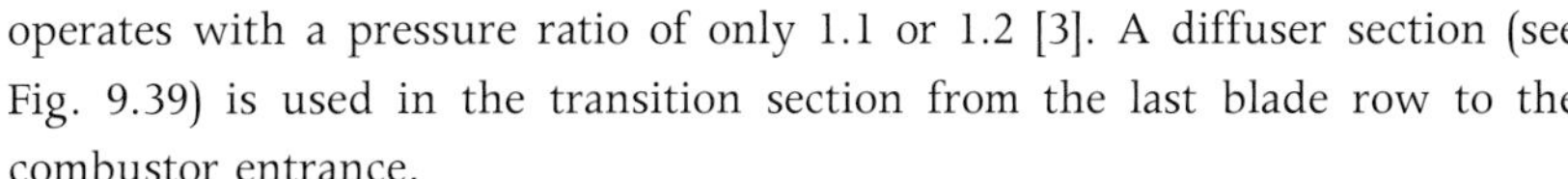

operates with a pressure ratio of only 1.1 or 1.2 [3]. A diffuser section (see Fig. 9.39) is used in the transition section from the last blade row to the combustor entrance.

3–4: Fuel is injected and burned in the combustor. A portion of the high-pressure air enters the combustor, while the remaining air cools the combustor liner and dilutes the combustion products to a sufficiently low temperature to allow proper operation of the turbine. The high-temperature strength of the turbine blades is the major factor limiting the turbine inlet temperature.

4–5: A multistage gas turbine expands the high-temperature, high-pressure gas. The shaft power from the turbine drives the compressor and any auxiliary equipment. Turbojet engines typically employ one to four turbine stages.

5–6: The partially expanded hot gases complete their expansion to ambient pressure (state 6) by passing through a nozzle. For converging nozzles, the exit velocity is limited by the choked conditions at the exit. High-speed aircraft employ converging–diverging nozzles to produce supersonic exit velocities, as shown in Figure 9.41b.

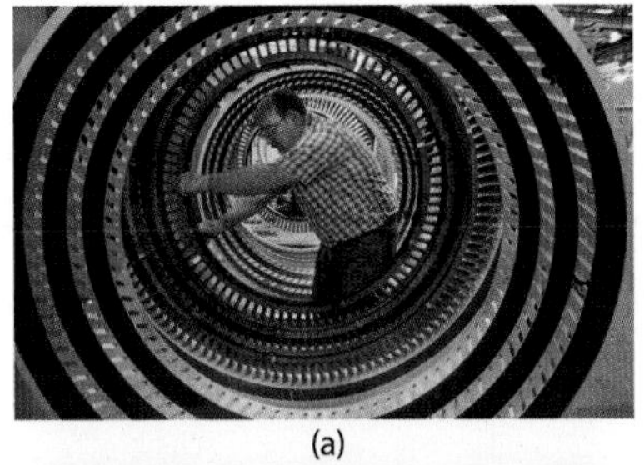

(a)

(b)

FIGURE 9.41 (a) Jet engine compressor shroud (AGEfotostock). **(b)** This F100AB jet engine utilizes a variable-geometry, converging–diverging exhaust nozzle (photograph courtesy of US Air Force).

Before investigating the details of the turbojet "cycle," we perform simple, integral open system analyses to help understand the basic operation of this engine.

9.4b Integral Open System Analysis of a Turbojet

Consider an open system (control volume) that cuts across the inlet plane (station 1 in Fig. 9.39), follows the external surface of the engine housing, and cuts across the exhaust plane (station 6 in Fig. 9.39). For simplicity, we model the engine as a cylindrical tube. The control surface also cuts through the fuel feedline and through the engine mounts. Figure 9.43 shows the flows of mass, energy, and linear momentum, respectively, for this integral open system.

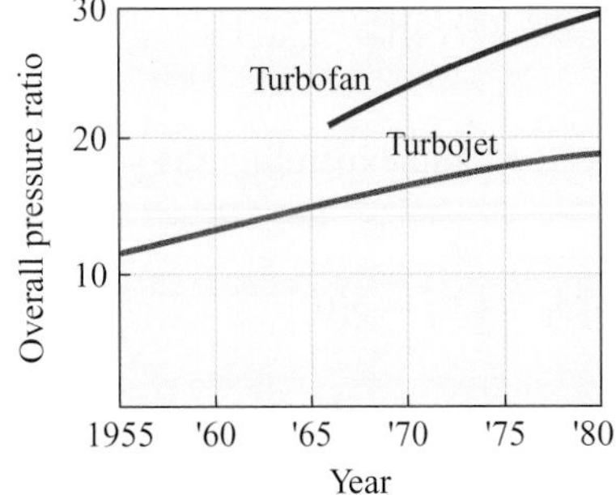

FIGURE 9.42 Overall pressure ratios increase as jet engine technology improves. Reprinted from Ref. [4] (public domain).

ASSUMPTIONS

Before applying the basic conservation principles, we list the assumptions used to simplify our analysis:

i. Steady-state and steady-flow conditions prevail.

ii. There are no heat interactions between the engine and the surroundings (i.e., $\dot{Q}_{cv} = 0$).

iii. Auxiliary power systems are neglected; thus, all the turbine power is used by the compressor and $\dot{W}_{cv} = 0$.

iv. Potential energy changes are negligible (i.e., $z_1 = z_6$).

v. The kinetic and potential energies of the fuel stream are negligible.

vi. All the axial forces acting on the open system are lumped into a single force F_t, which we designate as the total thrust.

vii. The pressure is uniform over the control surface.

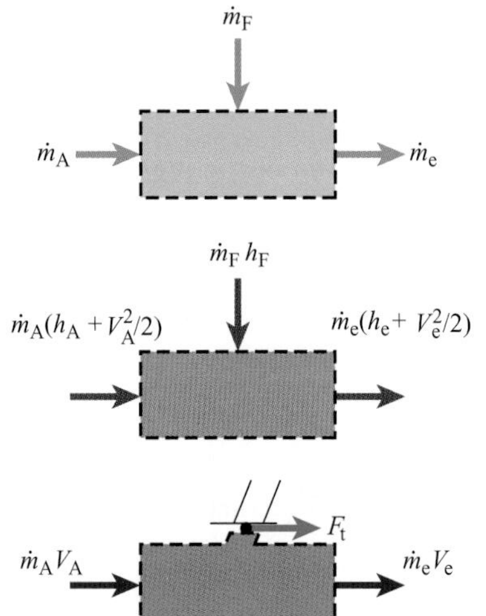

FIGURE 9.43 Integral open systems for turbojet engine for mass conservation (top), energy conservation (middle), and axial momentum conservation (bottom).

MASS CONSERVATION

Air and fuel enter the open system as separate streams, whereas a single stream of diluted combustion products exits (Fig. 9.43, top). Conservation of mass is thus expressed by Eq. 3.14b and expanded as follows:

$$\sum_{j=1}^{N\ \text{inlets}} \dot{m}_{\text{in},j} = \sum_{k=1}^{M\ \text{outlets}} \dot{m}_{\text{out},k},$$

or

$$\dot{m}_A + \dot{m}_F = \dot{m}_e \quad \textbf{(9.26)}$$

where $\dot{m}_A, \dot{m}_F, \dot{m}_e$ are the mass flow rates of the air, fuel, and exhaust products, respectively. Rearranging Eq. 9.26 to introduce the fuel–air ratio,

$$F/A \equiv \dot{m}_F/\dot{m}_A, \quad \textbf{(9.27)}$$

we obtain

$$\dot{m}_A(1 + F/A) = \dot{m}_e. \quad \textbf{(9.28)}$$

Since overall fuel–air ratios of turbojet engines are of the order of 1:100, the approximation that $\dot{m}_e \approx \dot{m}_A$ is sometimes useful, depending on the issue.

ENERGY CONSERVATION

The starting point for energy conservation is Eq. 5.16:

$$\dot{Q}_{\text{cv, net in}} - \dot{W}_{\text{cv, net out}} = \sum_{k=1}^{M\ \text{outlets}} \dot{m}_{\text{out},k}\left[h_k + \tfrac{1}{2}V_k^2 + g(z_k - z_{\text{ref}})\right] - \sum_{j=1}^{N\ \text{inlets}} \dot{m}_{\text{in},j}\left[h_j + \tfrac{1}{2}V_j^2 + g(z_j - z_{\text{ref}})\right].$$

We simplify this statement by applying assumptions ii–vi and expanding the summations, which gives

$$0 - 0 = \dot{m}_e\left[h_e + \tfrac{1}{2}V_e^2 + g(0)\right] - \dot{m}_F\left[h_F + \tfrac{1}{2}V_F^2 + g(0)\right] - \dot{m}_A\left[h_A + \tfrac{1}{2}V_A^2 + g(0)\right].$$

Note that this analysis is similar to that for a gas-turbine cycles except that now the kinetic energies of the air and exhaust streams are not negligible. The primary objective of a jet engine is to create thrust from the high-speed engine exhaust.

Upon rearrangement and substitution of Eq. 9.28 for the exhaust mass flow rate, this equation becomes

$$\dot{m}_A(1 + F/A)\left[h_e + \tfrac{1}{2}V_e^2\right] = \dot{m}_F h_F + \dot{m}_A\left[h_A + \tfrac{1}{2}V_A^2\right]. \quad \textbf{(9.29)}$$

Anticipating that the magnitude of the exhaust jet velocity is key to the performance of the engine, we isolate the kinetic energy terms in Eq. 9.29 after dividing through by $\dot{m}_A$:

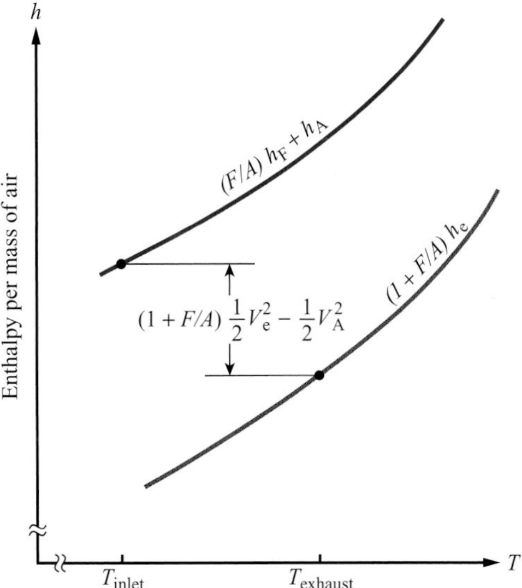

FIGURE 9.44 An h–T diagram for a turbojet engine with air and fuel entering at T_{inlet} and products exiting at $T_{exhaust}$.

$$\underbrace{(1+F/A)\tfrac{1}{2}V_e^2-\tfrac{1}{2}V_A^2}_{\substack{\text{Kinetic energy change}\\ \text{per mass of air}}} = \underbrace{[(F/A)h_F+h_A]}_{\substack{\text{Enthalpy of}\\ \text{reactants per mass}\\ \text{of air}}} - \underbrace{[1+(F/A)]h_e}_{\substack{\text{Enthalpy of}\\ \text{products per}\\ \text{mass of air}}}. \quad (9.30)$$

Assuming that the air, fuel, and products all behave as ideal gases, we will interpret Eq. 9.30 using the enthalpy–temperature diagram shown in Fig. 9.44. Fixing the inlet temperature of the air (and fuel) at T_{inlet} establishes the enthalpy value of the reactants (per unit mass of air), which is shown as a point on the reactant line in Fig. 9.44. Because the chemical energy contained in the products is less than that in the reactants, the product enthalpy (per unit mass of air) lies below that of the reactants. As we see in Fig. 9.44, the value of the exhaust temperature thus controls the kinetic energy change: the higher the exhaust temperature, the smaller the kinetic energy change.

Example 9.9 Turbojet Engine

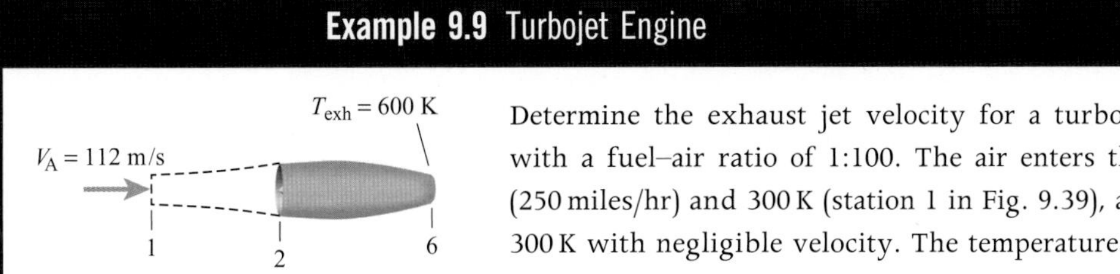

Determine the exhaust jet velocity for a turbojet engine operating with a fuel–air ratio of 1:100. The air enters the engine at 112 m/s (250 miles/hr) and 300 K (station 1 in Fig. 9.39), and the fuel enters at 300 K with negligible velocity. The temperature at the exhaust plane is 600 K. Assume the following simplified thermodynamic properties[4] for the air, fuel, and products:

i. The specific heats of the air, and products are constants and equal (i.e., $c_{p,A} = c_{p,e} = 1200\,J/kg{\cdot}K$).
ii. The enthalpy of the inflow fuel is 4×10^7 J/kg. The reference state temperature is 300 K.

Solution

Known (F/A), V_A, T_A, T_F, T_e

Find V_e

Sketch

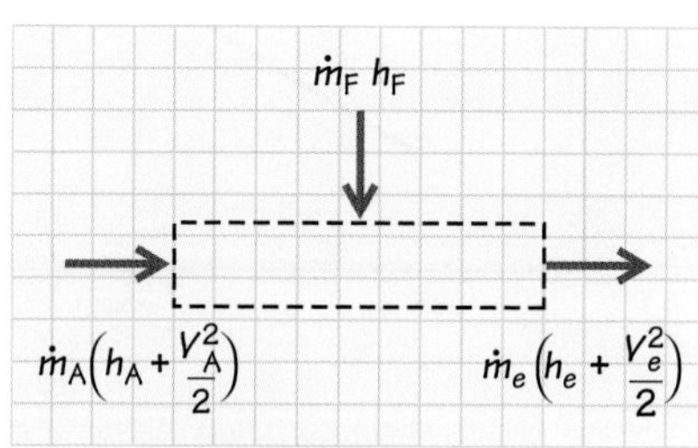

Modeling, Premises and Assumptions

i. Simplified thermodynamic properties as given
ii. Ideal-gas behavior
iii. $(Ke)_F$ is negligible
iv. Changes in potential energy can be neglected
v. Adiabatic process ($\dot{Q}_{cv} = 0$)

Analysis Assumptions ii, iii, and iv allow us to use conservation of energy as expressed by Eq. 9.30 directly. Solving Eq. 9.30 for the unknown exhaust velocity yields

$$V_e = \left[\frac{(F/A)h_F + h_A + \frac{1}{2}V_A^2 - (1 + F/A)h_e}{\frac{1}{2}(1 + F/A)}\right]^{1/2}.$$

Using the simplified thermodynamic properties given and the ideal-gas calorific equation of state for the sensible enthalpy change (Eq. 2.31e), we can write the standardized enthalpies for the three constituents using the reference temperature, 300 K. (A further description of standardized enthalpies will be presented in Chapter 12, Eq. 12.12.) Thus we have

$$\begin{aligned} h_A(T) &= c_p(T - T_{ref}) \\ &= (1200\,kJ/kg{\cdot}K)(300 - 300)K = 0\ kJ/kg \end{aligned}$$

and

$$\begin{aligned} h_e(T) &= c_p(T_e - T_{ref}) \\ &= (1200\,kJ/kg{\cdot}K)(600 - 300)K = 360.0\ kJ/kg. \end{aligned}$$

Substituting these values into our expression for V_e yields

$$V_e = \left[\frac{(0.01)(4\times10^7 J/kg)\left[\frac{kg{\cdot}m^2/s^2}{J}\right] + 0 + 0.5(112 m/s)^2 - (1.01)(3.6\times10^5 J/kg)\left[\frac{kg{\cdot}m^2/s^2}{J}\right]}{0.5(1.01)}\right]^{1/2}$$

$V_e = 291 m/s.$

[4] These assumptions are invoked for pedagogical, not engineering, reasons. With these assumptions, computations are greatly simplified by the avoidance of the need to specify the detailed composition of the products in order to calculate h_e. The simplified properties, however, are chosen to yield reasonable results [5].

Comments As expected, we find the exhaust velocity to be much greater than the air inlet velocity. Note how the simplified thermodynamics trivializes the computations of h_A, h_F, and h_e, while retaining the concept of standardized enthalpies necessary to deal with reacting flows.

MOMENTUM CONSERVATION

Momentum conservation is an important concept in fluid mechanics [10,11] in the same way that energy conservation is a key concept in thermodynamics. Because of the importance of this principle in understanding jet engines, we briefly present this concept here. In Chapter 3, we introduced the generic conservation principle expressed by Eq. 3.2:

$$\dot{X}_{\text{in}} - \dot{X}_{\text{out}} + \dot{X}_{\text{generated}} = \dot{X}_{\text{stored}}.$$

For linear momentum conservation, we identify $\dot{X}_{\text{in}}$ and $\dot{X}_{\text{out}}$ as the momentum flows $\dot{m}V_{\text{in}}$ and $\dot{m}V_{\text{out}}$, while $\dot{X}_{\text{generated}}$ is the sum of all the forces acting on the open system (control volume), $\sum F_{\text{CV}}$. For a steady state, $\dot{X}_{\text{stored}}$, i.e. $d(MV)_{\text{CV}}/dt$, is zero.

We now apply these ideas to the simple open system shown at the bottom of Fig. 9.43. The arrows represent the pertinent axial momentum flows and the axial force. With the assumption of uniform pressure at the control surface, the net pressure force is zero and hence is not shown. Cutting through the engine mount exposes the thrust force F_t. The reaction to this thrust force on the aircraft side of the motor mount drives the aircraft forward (to the left).

For an open system having a single inlet and a single outlet, both with uniform velocities, Eq. 3.2 expresses the conservation of momentum:

$$\dot{m}V_{\text{in}} - \dot{m}V_{\text{out}} + \sum F_{\text{CV}} = 0.$$

The axial-direction scalar component of this equation results from the substitution of the axial momentum flows and thrust force from Fig. 9.43, as follows:

$$\dot{m}_A V_A - \dot{m}_e V_e + F_t = 0. \tag{9.31}$$

Applying mass conservation (Eq. 9.28) and solving for the thrust force yields

$$F_t = \dot{m}_A[(1 + F/A)V_e - V_A], \tag{9.32a}$$

which can be approximated by the relationship

$$F_t \cong \dot{m}_A[V_e - V_A]. \tag{9.32b}$$

From Eqs. 9.32a or 9.32b we see that the thrust is directly proportional to the product of the air mass flow rate and the exhaust jet velocity. If we were to substitute our final energy conservation expression, Eq. 9.30, into Eq. 9.32, we could then identify the combustion process as the ultimate source of the thrust.

Self-Test 9.6

For an air mass flow rate of 50 kg/s, determine the thrust force of the engine in Example 9.9.

(Answer: 9.1 kN)

9.4c Turbojet Cycle Analysis

Analyzing the turbojet "cycle" allows us to determine the various thermodynamic states needed to calculate the thrust from Eqs. 9.32. An ideal cycle is shown in Fig. 9.45. We can also determine various measures of propulsive efficiency from this cycle analysis.

GIVEN CONDITIONS

We begin by defining parameters that are typically treated as known quantities:

- The inlet (atmospheric) pressure P_{amb} ($= P_1$) is given. For an aircraft at altitude, this pressure is significantly less than the standard sea-level value. The inlet (atmospheric) temperature T_1 is also known.
- The inlet velocity V_1, which is typically the flight speed of the aircraft, is given.
- The compressor pressure ratio P_3/P_2 is known.
- The turbine inlet temperature T_4 is given. This temperature is usually determined by metallurgical considerations and must be below values that lead to turbine blade failure.[5]
- The pressure at the exhaust nozzle outlet matches the ambient pressure (i.e., $P_6 = P_{amb}$).
- The isentropic efficiencies of each component are given.

ASSUMPTIONS

To the given conditions, we add the following assumptions:

i. The kinetic energies of the flow can be neglected at all stations *except* at the inlet (station 1) and the outlet (station 6).

ii. Potential energy changes are negligible throughout the engine.

iii. Accessory loads to the engine are negligible; thus, the power produced by the turbine equals the power input to the compressor (i.e., $\dot{W}_{t,out} = \dot{W}_{c,in}$).

APPROACH

With the given conditions and assumptions, we proceed to perform a state-to-state (1–2–3–etc.) device-by-device analysis (including the diffuser, compressor,

[5] An alternative to specifying the turbine inlet temperature is to specify the fuel–air ratio supplied to the engine, as this quantity allows T_4 to be determined.

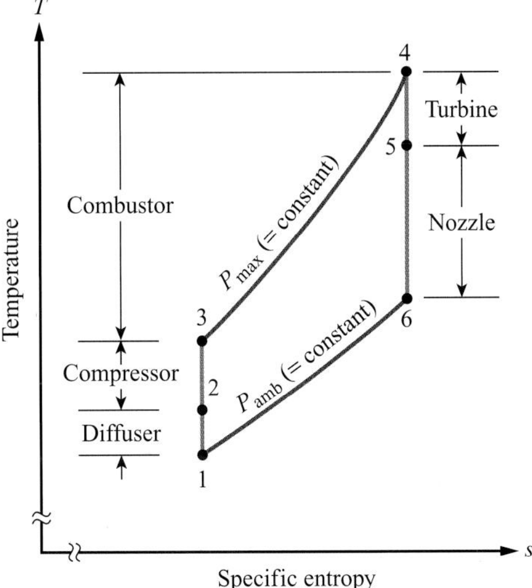

FIGURE 9.45 *T–s* diagram for an ideal turbojet cycle. See Fig. 9.39 for the definition of the diffuser (1–2).

combustor, etc.) that ultimately concludes at state 6 at the jet exit. Defining state 6, in particular determining the exit velocity V_6, is one of our major objectives. Knowing the value of this velocity allows us to determine the thrust produced by the engine (see Eqs. 9.32), that is, the useful output for which the engine was designed. Complicating factors in performing this step-by-step analysis are the addition of fuel and the combustion process. In particular, the presence of combustion means we have to deal with mixtures of combustion products (CO, H_2O, CO, O_2, etc.), beginning somewhere between states 3 and 4 and continuing from state 4 through state 6. An analysis of the combustion process will be applied in Chapter 12. We now explore what is known as the **air-standard turbojet cycle.** This model provides great simplification, yet it retains the essential features of the operation of a real turbojet engine:

- The working fluid is air.
- The combustion process is replaced by a constant-pressure heat-addition process, where the heat-addition rate is equivalent to the product of the fuel flow rate and the fuel heating value (i.e., $\dot{Q}_{\text{added}} = \dot{m}_{\text{F}} = HV_{\text{F}}$).
- The exhaust and intake processes are replaced by a constant-pressure heat-rejection process that closes the cycle.

See Table 8.1 for a first-law analysis of each component.

We now work through an ideal cycle (see Fig. 9.45) in which all the isentropic efficiencies are assumed to be unity. The introduction of non-unity isentropic efficiencies into the analysis is left as an exercise for the reader. The open systems used for the analysis of individual components of the turbojet are shown in Fig. 9.46.

Diffuser (1–2) (Figure 9.46a) The outlet state 2 is determined by application of energy conservation (Eq. 8.5 or Table 8.1) and the recognition that the process 1–2 is isentropic; thus,

$$h_2 = h_1 + \tfrac{1}{2}V_1^2 \tag{9.33a}$$

and

$$s_2 = s_1. \tag{9.33b}$$

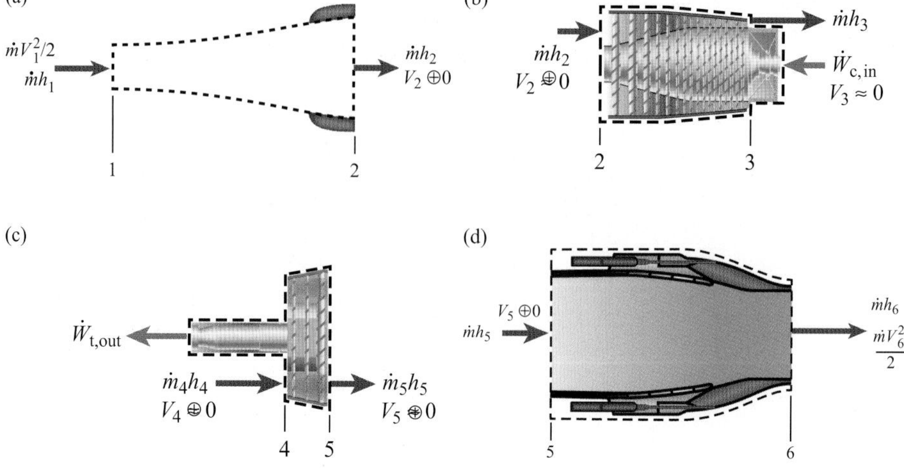

FIGURE 9.46 Open systems used for component analysis of a turbojet: **(a)** diffuser control volume (CV), **(b)** compressor CV, **(c)** turbine CV, **(d)** nozzle CV.

Knowledge of h_2 and s_2 allows the determination of all other state-2 properties. Specifically, we desire P_2:

$$(h_2, s_2) \Rightarrow P_2, T_2, \text{etc.} \tag{9.33c}$$

Compressor (2–3) (Figure 9.46b) Since the pressure ratio P_3/P_2 is given and the process is isentropic, state 3 is easily defined:

$$P_3 = P_2(P_3/P_2) \tag{9.34a}$$

and

$$s_3 = s_2. \tag{9.34b}$$

Knowing two independent properties at state 3 defines the state and allows us to find other desired properties. Thus,

$$(P_3, s_3) \Rightarrow h_2, T_3, \text{ etc.} \tag{9.34c}$$

Now knowing the enthalpy at state 3, the compressor work can be found from the conservation of energy (Eq. 8.11 or Table 8.1):

$$\dot{W}_{\text{c, in}} = \dot{m}(h_3 - h_2), \tag{9.34d}$$

or

$$\frac{\dot{W}_{\text{c, in}}}{\dot{m}} = h_3 - h_2. \tag{9.34e}$$

We will use this information later to help define state 5.

Combustor (3–4) With the assumptions that the combustor pressure is constant ($P_4 = P_3$) and that the turbine inlet temperature T_4 is given, the analysis of the combustor, or air-standard heat-addition process, is trivial since state 4 is defined by P_4 and T_4. Thus,

$$(P_4, T_4) \Rightarrow h_4, s_2, \text{etc.} \tag{9.35}$$

Turbine (4–5) (Figure 9.46c) The enthalpy at the turbine exit can be determined by equating the power produced by the turbine (Eq. 8.25 or Table 8.1) with the previously determined power input to the compressor, that is,

$$\dot{W}_{t,out} = \dot{W}_{c,in} \tag{9.36a}$$

or

$$\dot{m}(h_4 - h_5) = \dot{m}(h_3 - h_2). \tag{9.36b}$$

Solving for h_5 yields

$$h_5 = h_4 + h_2 - h_3, \tag{9.36c}$$

where h_2, h_3, and h_4 are all known from our previous definitions of states 2, 3, and 4. Assuming an isentropic efficiency of unity, we also know that the turbine is isentropic, so that

$$s_5 = s_4. \tag{9.36d}$$

State 5 is thus fully defined since two independent properties are known. The other desired properties can be found, that is,

$$(h_5, s_5) \Rightarrow P_5, T_5, \text{etc.} \tag{9.36e}$$

Nozzle (5–6) (Figure 9.46d) State 6 is easy to determine since the pressure is given and the expansion is assumed to be isentropic, so

$$P_6 = P_{amb} \tag{9.37a}$$

and

$$s_6 = s_5. \tag{9.37b}$$

Thus two independent properties are known and state 6 is defined. We can now find any other desired property at state 6,

$$(P_6, s_2) \Rightarrow h_6, T_6, \text{etc.} \tag{9.37c}$$

Knowing the enthalpy at state 6, the exhaust velocity is calculated from energy conservation (Eq. 8.5 or Table 8.1) as follows:

$$h_5 + 0 = h_6 + \tfrac{1}{2}V_6^2 \tag{9.37d}$$

or

$$V_6 = [2(h_5 - h_6)]^{1/2}. \tag{9.37e}$$

This completes the thermodynamic analysis of the cycle and allows the thrust to be computed from Eq. 9.32. A detailed implementation of this approach is presented below in Example 9.10.

9.4d Propulsive Efficiency

One measure of propulsive efficiency is the ratio of the power produced by the thrust, $\dot{W}_{thrust}$, to the rate at which energy is released in the combustor, that is,

$$\eta_{prop} \equiv \frac{\dot{W}_{thrust}}{\dot{m}(h_4 - h_3)}, \tag{9.38}$$

FIGURE 9.47 Thrust measurements are conducted in this jet engine test cell. Photograph courtesy of Mitsubishi Heavy Industries, Ltd.

where the actual chemical energy release has been replaced with the air-standard-cycle heat addition, $\dot{m}(h_4 - h_3)$. The power produced by the thrust is the product of the thrust F_t and the aircraft flight speed V_{flight}:

$$\dot{W}_{thrust} = F_t V_{flight}. \quad (9.39a)$$

For conditions where V_{flight} equals the inlet velocity V_1, the thrust power can be related through Eq. 9.32b to the inlet and exit velocities as follows:

$$\dot{W}_{thrust} = \dot{m}(V_6 - V_1)V_1. \quad (9.39b)$$

Substituting this expression back into the definition of propulsive efficiency (Eq. 9.38) yields

$$\eta_{prop} = \frac{(V_6 - V_1)V_1}{h_4 - h_3}. \quad (9.40)$$

Since all the quantities on the right-hand side are known from our cycle analysis, the propulsive efficiency is readily calculated.

9.4e Other Performance Measures

Two other commonly used performance measures relate engine flow rates to the production of thrust (Fig. 9.47). The **specific thrust** is the ratio of the thrust produced to the mass flow rate of air through the engine:

$$\text{specific thrust} \equiv F_t/\dot{m}_A, \quad (9.41)$$

with SI units of N/(kg/s) or m/s. Units typically used in industry are lb/(lb/s) or kg/(kg/s). The **specific fuel consumption** is the ratio of the fuel mass flow rate to the thrust produced:

$$\text{specific fuel consumption}\,(SFC) \equiv \frac{\dot{m}_F}{F_t}, \quad (9.42)$$

with SI units of (kg/s)/N or s/m. Units typically used in industry are lb/hr/lb or kg/ hr/kg.

Example 9.10 Turbojet Engine Cycle

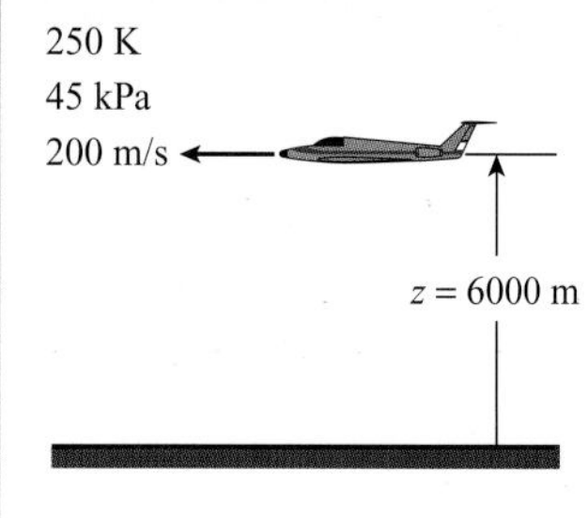

Consider a turbojet-powered aircraft flying at 200 m/s at 6000-m altitude where the ambient temperature and pressure are 250 K and 45 kPa, respectively. The pressure ratio of the compressor is 6 and the turbine inlet temperature is 970 K. At these conditions, the fuel flow rate is 0.68 kg/s and the overall air–fuel ratio is 75:1. Estimate the thrust developed by the engine. Also determine the specific thrust, the specific fuel consumption, and the propulsive efficiency. Employ an air-standard-cycle analysis with the following constant-pressure specific heats c_p and specific-heat ratios γ for

	c_p (kJ/kg·K)	γ
Diffuser, 1–2	1.04	1.40
Compressor, 2–3	1.06	1.40
Combustor, 3–4	1.10	1.37
Turbine, 4–5	1.145	1.35
Nozzle, 5–6	1.098	1.37

Solution

Known P_1, T_1, V_1, T_4, $\dot{m}_F$, A/F

Find F_t, *specific thrust*, *SFC*

Sketch The process stations are defined in Fig. 9.39, and the various processes are shown on a T–s diagram in Fig. 9.45 (repeated below with temperatures labeled).

Modeling, Premises and Assumptions

i. All processes are ideal (reversible). The diffuser, compressor, turbine, and nozzle all operate adiabatically.
ii. The working fluid is air (ideal gas) with constant specific heats given separately for each process. This use of quasi-constant specific heats approximates the effects of variable specific heats.
iii. The mass flow rate is identical at each station; that is, the effect of fuel addition on the total flow is small.
iv. The inlet and exhaust pressures are equal to the ambient pressure (i.e., $P_1 = P_6 = P_{\text{amb}}$).
v. All the turbine output power is used to drive the compressor.
vi. The velocities at states 2, 3, 4, and 5 are negligible.

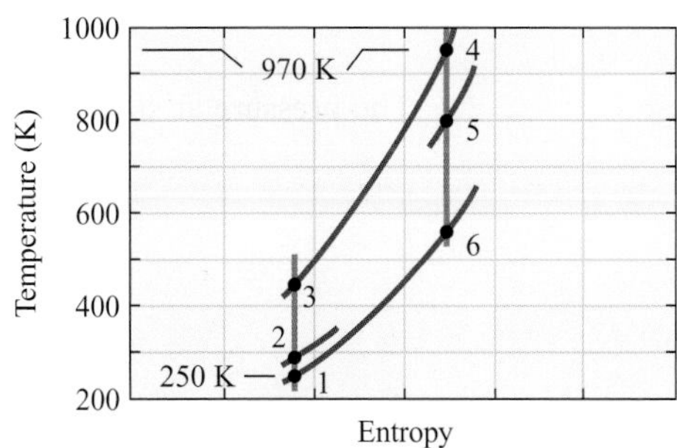

State	1	2	3	4	5	6
P (kPa)	**45.0**	58.3	349.8	349.8	169.1	**45.0**
T (K)	**250**	269.2	449.2	**970**	803.4	561.9
V (m/s)	**200**	**~0**	**~0**	**~0**	**~0**	728

Analysis Since we assume reversible and adiabatic processes for both the compression and expansion, the process sequences 1–2–3 and 4–5–6 are isentropic. With the further assumption of ideal-gas behavior, the pressures and temperatures in these sequences are related by the isentropic property relationship (Eq. 7.13a)

$$\frac{P_a}{P_b} = \left(\frac{T_a}{T_b}\right)^{\gamma/(\gamma-1)}. \tag{A}$$

Furthermore, we can apply the simplified ideal-gas calorific equation of state (Eq. 2.31e)

$$\Delta h = c_p \Delta T \tag{B}$$

for any of the processes (1–2–3–4–5–6). These two relationships, together with appropriate statements of conservation of energy, are sufficient to define all the states (1–6) and to calculate the exhaust velocity. Knowing this velocity, the thrust is calculated from momentum conservation (Eqs. 9.32).

We begin by organizing the results of our calculations in a table. Known or assumed values are shown as bold entries in the table. All other values need to be calculated.

The temperature at the diffuser outlet is determined by combining energy conservation (Eq. 9.33a) with our simplified calorific equation of state:

$$h_2 - h_1 = c_p(T_2 - T_1) = \tfrac{1}{2}\mathrm{V}_1^2,$$

or

$$T_2 = \frac{V_1^2}{2c_p} + T_1$$

$$T_2 = \frac{(200\,\mathrm{m/s})^2}{2(1040\,\mathrm{J/kg{\cdot}K})} + 250\ \mathrm{K} = 269.2\ \mathrm{K}$$

$$[=]\frac{(\mathrm{m/s})^2}{(\mathrm{J/kg{\cdot}K})}\left[\frac{1\,\mathrm{J}}{\mathrm{N{\cdot}m}}\right]\left[\frac{1\,\mathrm{N}}{\mathrm{kg{\cdot}m/s^2}}\right] = \mathrm{K}.$$

The pressure at state 2, calculated from Eq. A, is

$$P_2 = P_1\left(\frac{T_2}{T_1}\right)^{\gamma/(\gamma-1)}$$

$$= 45\,\mathrm{kPa}\left(\frac{269.2\ \mathrm{K}}{250\,\mathrm{K}}\right)^{1.4/0.4} = 58.3\ \mathrm{kPa}.$$

The compressor pressure ratio is given, so the pressure at the compressor outlet is

$$P_3 = P_2\left(\frac{P_3}{P_2}\right)$$

$$= 58.3(6)\ \mathrm{kPa} = 349.8\ \mathrm{kPa}.$$

The compressor outlet temperature is therefore

$$T_3 = T_2\left(\frac{P_3}{P_2}\right)^{\gamma/(\gamma-1)}$$

$$= 269.2\ \mathrm{K}\ (6)^{0.4/1.4} = 449.2\ \mathrm{K}.$$

From energy conservation, the compressor input power per unit mass flow can be determined. Combining this with Eq. B yields

$$\begin{aligned}\frac{\dot{W}_{\mathrm{c,in}}}{\dot{m}} &= h_3 - h_2 = c_p(T_3 - T_2)\\ &= \left(1.06\frac{\mathrm{kJ}}{\mathrm{kg{\cdot}K}}\right)(449.2 - 269.2)\,\mathrm{K} = 190.8\ \mathrm{kJ/kg}.\end{aligned}$$

The heat-addition (combustion) process occurs at constant pressure ($P_4 = P_3$), and the combustor outlet temperature (i.e., the turbine inlet temperature) is given ($T_4 = 970$ K). No calculation is required, and the P_4 and T_4 values are entered in the table.

The turbine output power equals the compressor input power. Combining this knowledge with energy conservation allows us to calculate T_5 as follows:

$$\frac{\dot{W}_{\mathrm{c,in}}}{\dot{m}} = \frac{\dot{W}_{\mathrm{t,out}}}{\dot{m}} = h_4 - h_2 = c_p(T_4 - T_5),$$

or

$$\begin{aligned}T_5 &= T_4 - \frac{\dot{W}_{\mathrm{c,in}}/\dot{m}}{c_p}\\ &= 970\ \mathrm{K} - \frac{190.8\ \mathrm{kJ/kg}}{1.145\ \mathrm{kJ/kg{\cdot}K}} = 803.4\ \mathrm{K}.\end{aligned}$$

Using this turbine outlet temperature, we can calculate the outlet pressure using the isentropic process relationship (Eq. A) as follows:

$$\begin{aligned}P_5 &= P_4\left(\frac{T_5}{T_4}\right)^{\gamma/(\gamma-1)}\\ &= 349.8\left(\frac{803.4}{970}\right)^{1.35/0.35}\ \mathrm{kPa} = 169.1\ \mathrm{kPa}.\end{aligned}$$

The final process in the cycle is the expansion of the hot, high-pressure gases through the exhaust nozzle. For this isentropic expansion,

$$\begin{aligned}T_6 &= T_5\left(\frac{P_6}{P_5}\right)^{\gamma/(\gamma-1)},\\ &= 803.4\left(\frac{45}{169.1}\right)^{0.37/1.37}\ \mathrm{K} = 561.9\ \mathrm{K}.\end{aligned}$$

where P_6 is the ambient pressure. The exhaust velocity is calculated from energy conservation (Eq. 9.37d):

$$h_5 + 0 = h_6 + \tfrac{1}{2}V_6^2.$$

Solving for V_6 then gives

$$V_6 = [2(h_5 - h_6)]^{1/2}.$$

Substituting Eq. B, we relate the velocity to the temperature difference:

$$V_6 = \left[2c_p(T_5 - T_6)\right]^{1/2}$$
$$= \left[2(1098)(803.4 - 561.9)\right]^{1/2} = 728$$
$$[=]\left(\frac{\mathrm{J}}{\mathrm{kg{\cdot}K}}\mathrm{K}\left[\frac{\mathrm{N{\cdot}m}}{1\,\mathrm{J}}\right]\left[\frac{\mathrm{kg{\cdot}m/s^2}}{1\,\mathrm{N}}\right]\right)^{1/2} = \mathrm{m/s}.$$

We can now calculate the thrust. Given the air–fuel ratio and the fuel flow rate, we determine the air mass flow rate (Eq. 9.27) as

$$\dot{m}_{\mathrm{A}} = (A/F)\dot{m}_{\mathrm{F}},$$
$$= 75(0.68\ \mathrm{kg/s}) = 51\ \mathrm{kg/s}.$$

Using the combined overall mass and momentum conservation relationship for the thrust (Eq. 9.32a), we calculate

$$F_t = \dot{m}_{\mathrm{A}}[(1 + (F/A))V_6 - V_1],$$
$$= (51)\left[\left(1 + \frac{1}{75}\right)728 - 200\right] = 27,400$$
$$[=]\frac{\mathrm{kg}}{\mathrm{s}}\frac{\mathrm{m}}{\mathrm{s}}\left[\frac{1\mathrm{N}}{\mathrm{kg{\cdot}m/s^2}}\right] = \mathrm{N}.$$

The specific thrust is (Eq. 9.41)

$$F_t/\dot{m}_{\mathrm{A}} = \frac{27{,}400}{51} = 537\ \mathrm{m/s},$$

and the specific fuel consumption (Eq. 9.42) is

$$SFC = \dot{m}_{\mathrm{F}}/F_t = \frac{0.68}{27{,}400}\,\mathrm{s/m} = 2.48 \times 10^{-5}\ \mathrm{s/m}.$$

The reader should verify the units associated with these two calculations. Our final desired quantity is the propulsive efficiency, which we can calculate directly from Eq. 9.40 as follows:

$$\eta_{\mathrm{prop}} = \frac{V_1(V_6 - V_1)}{h_4 - h_3} = \frac{V_1(V_6 - V_1)}{c_p(T_4 - T_3)}$$
$$= \frac{200(728 - 200)}{1100(970 - 449.2)} = 0.184,$$

or, expressed as a percentage,

$$\eta_{\mathrm{prop}} = 18.4\%.$$

Comments In this example, constant specific heats were assumed, to make the evaluation of thermodynamic properties nearly trivial so that attention could be focused on the cycle analysis proper. To capture some of the temperature dependence of the specific heats, different values were assigned to each process. An alternative to adopting these simplified properties is to use air tables (Appendix C).

Although we use SI units throughout this book, it is useful to digress here to illustrate the use of non-SI units for specific fuel consumption. Using the awkward kilogram-force unit, which is numerically equal to a kilogram-mass in standard gravity, we convert the *SFC* to the industry-standard kg/hr/kg or lb/hr/lb units as follows:

$$F_t = 27{,}400\,\text{N}\frac{1\,\text{kg}_f}{9807\,\text{N}} = 2794\,\text{kg}_f$$

and

$$\dot{m}_F = 0.68\frac{\text{kg}}{\text{s}}\left[\frac{3600\ \text{s}}{\text{hr}}\right] = 2448\,\text{kg/hr};$$

therefore,

$$SFC\frac{\dot{m}_F}{F_t} = \frac{2448}{2794} = 0.876\,\text{kg/hr/kg}\ \text{or}\ \text{lb/hr/lb}.$$

How would you modify this example to account for the mass flow rate through the turbine being greater than that through the compressor?

This value is consistent with the *SFC* values tabulated for various engines in Ref. [6].

Note the relatively low value of 18.4% for the ideal propulsive efficiency for this particular example. Considering irreversibilities in each component would lower this value even further. We once again see how relatively inefficient some devices are in producing a useful effect from a heat input. For comparison, see Examples 9.2 and 9.3 for Rankine cycle efficiencies.

9.5 Other Gas Power Cycles

Previously we have studied three different gas power cycles: the Carnot cycle, the Stirling cycle, and the Brayton cycle. In this section, we briefly summarize these three cycles and then introduce two other gas power cycles.

The Carnot cycle, introduced in Example 7.4 and studied further in Section 7.4e, is the idealized power cycle that gives the maximum power output from a device operating between given high- and low-temperature reservoirs. The ideal Carnot cycle, shown in Figures 7.20, 7.21, and 7.22, includes four processes:

Ideal Carnot Cycle

State change	Process
1–2:	Reversible heat addition at constant temperature,
2–3:	Reversible adiabatic expansion (isentropic),
3–4:	Reversible heat rejection at constant temperature, and
4–1:	Reversible adiabatic compression (isentropic)

The Stirling cycle in Example 6.2, Figure 7.25, and Section 7.4e is another ideal power cycle between given high- and low-temperature reservoirs that includes four processes:

Ideal Stirling Cycle

State change	Process
1–2:	Reversible heat addition at constant temperature,
2–3:	Reversible heat rejection at constant volume,
3–4:	Reversible heat rejection at constant temperature, and
4–1:	Reversible heat addition at constant volume

We have just studied the Brayton air power cycle, which can be used for power plant applications and to create thrust from a jet engine. The T–s diagram for the Brayton cycle is shown in Figures 9.32 and 9.45.

Ideal Air-Standard Brayton Cycle

State change	Process
1–2:	Adiabatic and reversible (isentropic) compression,
2–3:	Constant-pressure heat addition,
3–4:	Adiabatic and reversible (isentropic) expansion, and
4–1:	Constant-pressure heat rejection.

Self-Test 9.7

Using the cycle descriptions above, sketch the air power cycles on T–s and P–v diagrams.

(Answers: See Figs. 7.20, 7.25, and 9.32)

Two other gas power cycles commonly studied are the Otto cycle and Diesel cycle.

9.5a Otto Cycle

The Otto cycle is an idealized air cycle that is often used as a crude model for the power cycle in the four-stroke gas engine found in automobiles.

The Otto cycle was already analyzed in Examples 4.2 and 7.5. Here we will study the Otto cycle in more detail and determine the cycle efficiency.

IDEAL AIR-STANDARD OTTO CYCLE

The four-stroke engine cycle, initially described in Section 1.2b, includes four strokes as shown in Figure 9.48. The actual processes are shown on the P–v plot in Figure 9.49a. A cycle approximation used for analysis is shown in Figure 9.50b.

A. **Intake stroke** where a fresh fuel–air mixture enters the cylinder as the piston is drawn downward, increasing the volume at constant pressure. (Process 5–1)

B. **Compression stroke** where the piston rises in the closed cylinder causing a decrease in the volume and an increase in pressure. (Process 1–2) Before the piston reaches the top of its travel, the mixture is ignited. Since combustion of the fuel–air mixture occurs very quickly, there is little change in the volume as the pressure increases. (Process 2–3)

C. **Expansion stroke** (Power stroke) where the high-pressure ignited fuel–air mixture expands and the pressure decreases. (Process 3–4)

D. **Exhaust stroke** where the exhaust valve opens and the combustion gases leave the cylinder as first the pressure drops and then the piston rises. (Process 4–1–5)

See Appendix 3A at the end of Chapter 3 for more details.

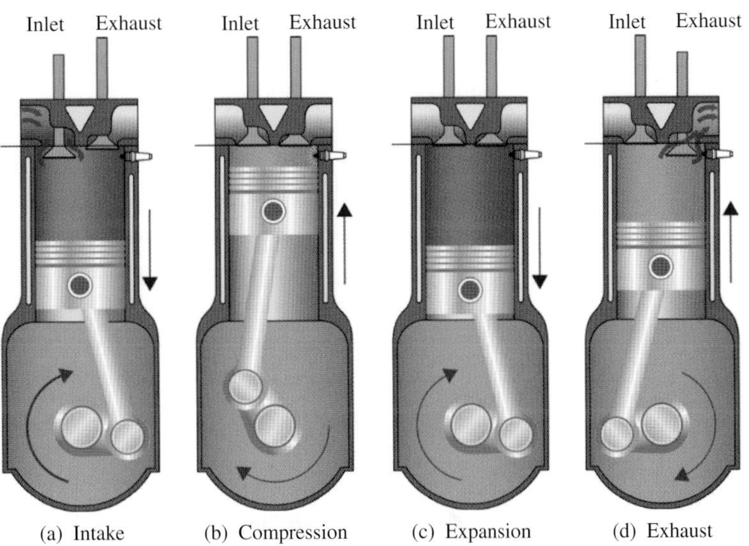

FIGURE 9.48 The mechanical cycle of the four-stroke spark-ignition engine consists of the intake stroke **(a)**, the compression stroke **(b)**, the expansion stroke **(c)**, and the exhaust stroke **(d)**. This sequence of events, however, does not execute a thermodynamic cycle since the intake air and exhaust air are not at the same state. Adapted with permission from Internal Combustion Engine Fundamentals, John Heywood. © McGraw-Hill Education.

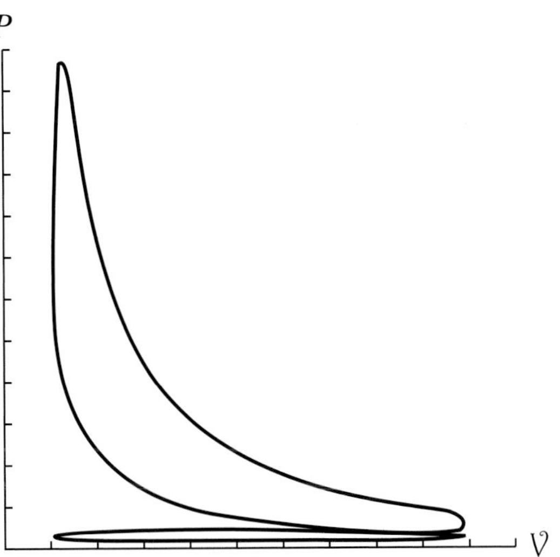

FIGURE 9.49 Actual P–$\mathcal{V}$ curve of a four-stroke internal combustion engine.

The air-standard Otto cycle is often used as a crude model for the four-stroke internal combustion engine. The air-standard Otto cycle defines a closed cycle with heat transfer as shown in Figure 9.50.

In the four-stroke engine, the mixture composition changes as the fuel-air mixture is combusted. Combustion of the fuel-air mixture in the four-stroke engine begins near the end of the compression stroke and has the effect of raising the temperature and pressure in the cylinder very quickly with little change in volume. (Example 12.6 analyzes this combustion process in the four-stroke engine.) The air-standard Otto cycle does not include the changes in chemical composition of the gas but approximates this change in pressure by heat addition at constant volume from state 2–3.

Figure 9.49 shows the pressure and volume changes in an actual internal combustion engine. The exhaust (decreasing $\mathcal{V}$) and intake strokes (increasing $\mathcal{V}$) can be located in the figure as the change in specific volume at relatively low pressure. At high loads, the net power requirement for the exhaust and intake strokes is small compared to the power output from the compression and expansion strokes. (When the engine is idling, the net power required for the exhaust and intake strokes is similar to the net power output from the compression and expansion strokes.) The air-standard Otto cycle neglects the exhaust and intake strokes and only approximates the pressure, temperature, and volume changes in the compression and expansion strokes.

In the four-stroke engine, the piston is at the bottom of travel at the beginning of the exhaust stroke and at the end of the intake stroke so the volume is the same at these states. The mass of the gas mixture is also the same at these two states since the valves are closed during the compression and expansion strokes. The temperature and pressure of the gas mixture, however, is higher at the beginning of the exhaust stroke. In the air-standard Otto cycle, the beginning of the exhaust stroke is defined as state 4 and the end of the intake stroke is defined as state 1. The air-standard Otto cycle describes the replacement of hot exhaust gases with cool intake mixture as a heat removal process at constant volume.

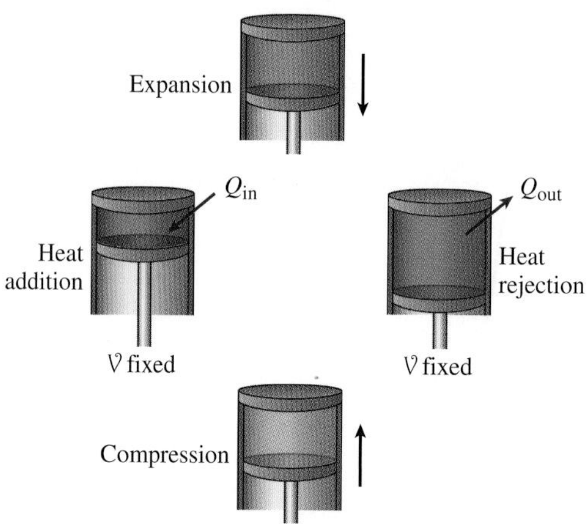

FIGURE 9.50 Air-standard Otto cycle.

The four processes in the air-standard Otto cycle are:

Ideal Air-Standard Otto Cycle

State change	Process
1–2:	Adiabatic and reversible (isentropic) compression,
2–3:	Constant-volume heat addition,
3–4:	Adiabatic and reversible (isentropic) expansion, and
4–1:	Constant-volume heat rejection.

Figure 9.51 shows the Otto cycle on P–$\forall$ and T–s diagrams. The adiabatic, reversible compression (1–2) and adiabatic, reversible expansion (3–4) are isentropic

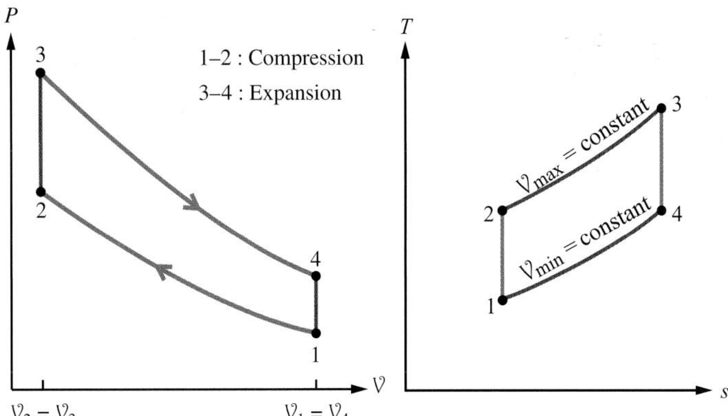

FIGURE 9.51 *P*–V and *T*–s diagrams for the Otto cycle.

processes. For an isentropic process of an ideal gas, the pressure and volume variations follow a polytropic process with $n = \gamma$.

In general, the cycle efficiency is defined as the ratio of the desired work output to the heat impact. Note that there is heat transfer in two processes of the Otto cycle, so that $W_{net} = Q_H - Q_L$:

$$\eta_{th,\,Otto} = \frac{W_{net}}{Q_H} = \frac{Q_H - Q_L}{Q_H} = 1 - \frac{Q_L}{Q_H}.$$

Heat transfer to the system occurs during the constant-volume processes. Since there is no *P*–V work done in a constant-volume process, the heat transfer can be found using the energy equation: $Q_H = {}_2Q_3 = U_3 - U_2 = mc_v(T_3 - T_2)$ and $Q_L = -{}_4Q_1 = U_4 - U_1 = mc_v(T_4 - T_1)$.

The cycle efficiency can be written in terms of the cycle temperatures:

$$\eta_{th,\,Otto} = 1 - \frac{Q_L}{Q_H} = 1 - \left(\frac{T_4 - T_1}{T_3 - T_2}\right) = 1 - \frac{T_1\left(\frac{T_4}{T_1} - 1\right)}{T_2\left(\frac{T_3}{T_2} - 1\right)}.$$

Processes 1–2 and 3–4 are isentropic processes of an ideal gas. For an isentropic process, the ratios of temperatures and volumes can be related using Equation 7.15:

$$\frac{T_2}{T_1} = \left(\frac{V_1}{V_2}\right)^{\gamma-1} \quad \text{and} \quad \frac{T_3}{T_4} = \left(\frac{V_4}{V_3}\right)^{\gamma-1}.$$

Since the volumes at states 2 and 3 are the same and the volumes at states 1 and 4 are the same, $V_1/V_2 = V_4/V_3$; therefore $T_2/T_1 = T_3/T_4$ or $T_4/T_1 = T_3/T_2$. Substituting this into the efficiency equation above gives

$$\eta_{th,\,Otto} = \frac{W_{net}}{Q_H} = 1 - \frac{T_1}{T_2}.$$

Now using the above isentropic relationship from Eq. 7.15, we can find the cycle efficiency in terms of the compression ratio, $R = V_1/V_2$:

$$\eta_{th,\,Otto} = \frac{W_{net}}{Q_H} = 1 - \frac{T_1}{T_2} = 1 - \left(\frac{V_2}{V_1}\right)^{\gamma-1} = 1 - \frac{1}{R^{\gamma-1}} = 1 - R^{1-\gamma}. \tag{9.43}$$

Example 9.11 Otto Cycle

Consider an air-standard Otto cycle with a compression ratio of 7:1 and a heat input of 2100 kJ/kg. Determine the efficiency and the net work produced by the cycle.

Solution

Known $R = \forall_1/\forall_2 = 7$, $Q_H = 2100$ kJ/kg

Find $\eta_{th,\,Otto}$ and W_{net}

Sketch The Otto cycle is described by the P–$\forall$ and T–s diagrams in Fig. 9.51.

Modeling, Premises and Assumptions

i. The working fluid is air (ideal gas) with constant specific heats.
ii. The compression and expansion processes are adiabatic and reversible (isentropic).

Analysis Using Eq. 9.43, we can determine the efficiency of the Otto cycle:

$$\eta_{th,\,Otto} = \frac{W_{net}}{Q_H} = 1 - R^{1-\gamma} = 1 - 7^{(1-1.4)} = 0.541.$$

Given Q_H, the net power output can be found.

$$\eta_{th,\,Otto} = \frac{W_{net}}{Q_H} = 1 - R^{1-\gamma} = 1 - 7^{(1-1.4)} = 0.541$$
$$W_{net} = \eta_{th,\,Otto}\, Q_H = 0.541\ (2100\ \text{kJ/kg}) = 1\ 135.8\ \text{kJ/kg}$$

Self-Test 9.8

Repeat Example 9.11 for a compression ratio of 10:1. Is a higher compression ratio desirable? Why?

(Answers: $\eta_{th,\,Otto} = 0.602$, $W_{net} = 1264.0\,kJ/kg$. A higher compression ratio is desirable since the cycle has a greater efficiency and there will be a higher net work output for a given heat source.)

9.5b Diesel Cycle

The diesel cycle is an idealized air cycle that is often used to approximate the power cycle in the four-cycle diesel engine found in trucks and some automobiles. Figure 9.52 shows the P–v and T–s diagrams for the diesel cycle. Instead of using a spark plug, the diesel engine ignites the air–fuel mixture by sufficiently increasing the pressure in cylinder. The intake air is first compressed in the cylinder without any fuel. The fuel is then injected at a high pressure and atomized to spray throughout the chamber for efficient combustion. After combustion starts, the gas expands at nearly constant

Ideal Air-Standard Diesel Cycle

State change	Process
1–2:	Adiabatic and reversible (isentropic) compression,
2–3:	Constant-pressure heat addition,
3–4:	Adiabatic and reversible (isentropic) expansion, and
4–1:	Constant-volume heat rejection.

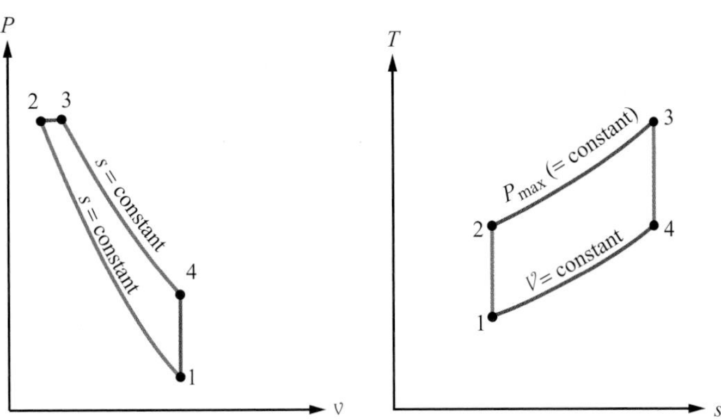

FIGURE 9.52 The ideal air-standard diesel cycle: (a) P–v diagram; (b) T–s diagram.

pressure from states 2 to 3. The ideal air-standard diesel cycle approximates the combustion process as heat added to the cycle. The combusted exhaust gases exit the diesel engine at the local atmospheric pressure at state 4.

We desire to find the cycle efficiency for the diesel cycle in terms of state properties. We start with the basic definition of efficiency in terms of the desired work output from the required heat transfer into the cycle. Note that there is heat transfer in two processes of the diesel cycle, so that $W_{\text{net}} = Q_{\text{H}} - Q_{\text{L}}$. Thus,

$$\eta_{\text{th, diesel}} = \frac{W_{\text{net}}}{Q_{\text{H}}} = \frac{Q_{\text{H}} - Q_{\text{L}}}{Q_{\text{H}}} = 1 - \frac{Q_{\text{L}}}{Q_{\text{H}}}.$$

Heat transfer to the system occurs during the constant-pressure process from states 2 to 3. During the heat addition, there is an increase in the internal energy and P–$\mathcal{V}$ work done by the system. The P–$\mathcal{V}$ work at constant pressure is found as follows:

$${}_2W_3 = P_2(\mathcal{V}_3 - \mathcal{V}_2) = MP_2(v_3 - v_2) = MR(T_3 - T_2).$$

The heat addition is then found using conservation of energy for the system:

$$Q_{\text{H}} = {}_2Q_3 = U_3 - U_2 + {}_2W_3 = Mc_v(T_3 - T_2) + MR(T_3 - T_2) = Mc_p(T_3 - T_2).$$

Heat transfer from the system occurs during the constant-volume process from states 4 to 1. Since there is no P–v work done in the constant-volume process, the heat transfer can be found using the energy equation with ${}_4W_1 = 0$:

$$Q_L = -{}_4Q_1 = U_4 - U_1 = Mc_v(T_4 - T_1).$$

The cycle efficiency can be written in terms of the cycle temperatures:

$$\eta_{\text{th, diesel}} = 1 - \frac{Q_{\text{L}}}{Q_{\text{H}}} = 1 - \frac{c_v(T_4 - T_1)}{c_p(T_3 - T_2)} = 1 - \frac{c_v T_1\left(\dfrac{T_4}{T_1} - 1\right)}{c_p T_2\left(\dfrac{T_3}{T_2} - 1\right)},$$

Substituting for the ratio of specific heats, $\gamma = c_p/c_v$, gives:

$$\eta_{th,\text{ diesel}} = 1 - \frac{T_1\left(\dfrac{T_4}{T_1} - 1\right)}{\gamma T_2\left(\dfrac{T_3}{T_2} - 1\right)}. \tag{9.44}$$

Self-Test 9.9

Using the cycle descriptions above, sketch the Otto and diesel air power cycles on T–s and P–v diagrams.

(Answers: See Figs. 9.37 and 9.38)

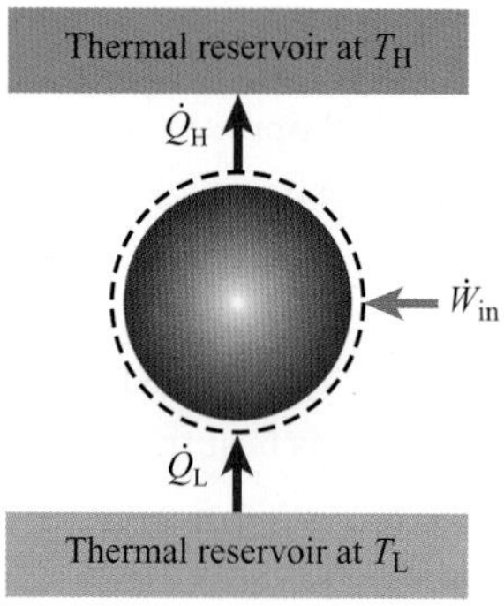

9.6 Refrigerators and Heat Pumps

FIGURE 9.53 A reversed cycle uses an input of work to transfer energy from a low-temperature reservoir to a high-temperature reservoir. See Fig. 6.4.

In this section, we explore so-called reversed cycles, in which a work or power input moves energy from a low-temperature thermal reservoir to a high-temperature thermal reservoir (Fig. 9.53). In a refrigerator, one seeks to maintain the cold space at a desired low temperature, whereas for a heat pump, the high-temperature space is the focus. Refrigerators are common in food processing and storage – there may even be a refrigerator similar to the one pictured in Fig. 9.54 in the room in which you are reading this book. The refrigeration cycle is also used to cool a room or building (Fig. 9.55). Heat pumps are becoming increasingly popular for residential and other space-heating applications. We introduced these devices in our discussion of the second law of thermodynamics in Chapter 6 without an explanation of how they might accomplish their desired tasks. The objective now is to fill this gap. Here we present and discuss refrigerators and heat pumps that operate on a **vapor-compression cycle.** Other types of reversed cycles exist; however, discussion of these goes beyond the scope of this book. For more information, we refer the interested reader to Refs. [7, 8].

9.6a Energy Conservation for a Reversed Cycle

Consider a steady-flow device that operates on an arbitrary reversed cycle, as shown in Fig. 9.53. As indicated by the dashed line, the control surface surrounds the device. The only energy flows crossing the control surface are a power input and two heat

FIGURE 9.54 Rear view of a small refrigerator suitable for use in a student dormitory room.

FIGURE 9.55 Refrigerated test cells or "cold rooms" are used for vehicle testing. Photograph courtesy of General Motors.

interactions, one with the low-temperature reservoir and one with the high-temperature reservoir. We thus express the conservation of energy simply as

$$\sum \dot{E}_{\text{in}} = \sum \dot{E}_{\text{out}},$$

or

$$\dot{W}_{\text{in}} + \dot{Q}_{\text{L}} = \dot{Q}_{\text{H}}. \tag{9.45}$$

We again define $\dot{W}_{\text{in}}$, $\dot{Q}_{\text{L}}$, and $\dot{Q}_{\text{H}}$ as positive values showing the magnitude and not the direction of energy transfer. The direction of energy transfer is accounted for when writing the conservation of energy equation. Although the cyclic device contained within our control surface may be quite complex, the overall energy conservation expression describing its operation is quite simple.

9.6b Performance Measures

As we saw in Chapter 6, the coefficient of performance is used to quantify how well a reversed-cycle device performs its job, in the same way that the thermal efficiency is used to characterize the performance of a power-producing cycle; that is,

$$COP \equiv \beta \equiv \frac{\text{desired energy}}{\text{energy that costs}}. \tag{9.46}$$

For a refrigerator, the energy to be removed from the low-temperature space is the desired energy; thus,

$$\beta_{\text{refrig}} = \frac{\dot{Q}_{\text{L}}}{\dot{W}_{\text{in}}}, \tag{9.47a}$$

where the power input comprises the energy that costs. Applying conservation of energy (Eq. 9.45), we can also express this definition as

$$\beta_{\text{refrig}} = \frac{\dot{Q}_L}{\dot{Q}_{\text{H}} - \dot{Q}_{\text{L}}} = \frac{1}{\dot{Q}_{\text{H}}/\dot{Q}_{\text{L}} - 1}. \tag{9.47b}$$

In refrigeration applications, the unit *ton* is frequently used to quantify the energy removal rate from the cold space. *One ton of refrigeration* equals 3.517 kW or 200 Btu/min.

Similarly, the desired energy for a heat pump is that delivered to the high-temperature space; thus,

$$\beta_{\text{heat pump}} = \frac{\dot{Q}_{\text{H}}}{\dot{W}_{\text{in}}}, \tag{9.48a}$$

or

$$\beta_{\text{heat pump}} = \frac{\dot{Q}_{\text{H}}/\dot{Q}_{\text{L}}}{\dot{Q}_{\text{H}}/\dot{Q}_{\text{L}} - 1}. \tag{9.48b}$$

See Example 6.3 in Chapter 6 to consolidate these definitions.

The definitions expressed in Eqs. 9.47 and 9.48 apply to both ideal and real devices.

To get a handle on the upper limits of the coefficient of performance for a device operating between two fixed temperatures, we resurrect the ideal of the Carnot cycle,

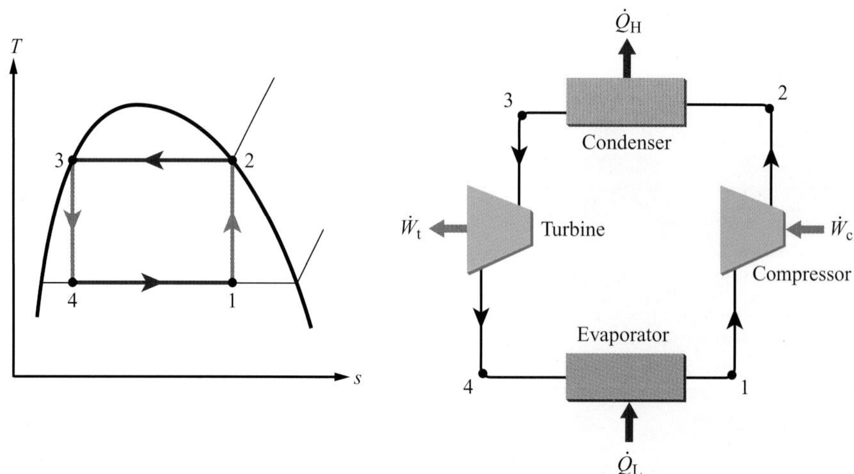

FIGURE 9.56 The reversed Carnot cycle establishes thermodynamic limits for the performance of refrigerators and heat pumps operating between two fixed temperatures.

but now operating in reverse. As you may recall from Chapter 6, the Carnot cycle is an ideal cycle in which all processes are performed reversibly (i.e., there is no friction or other source of irreversibility.) Figure 9.56 illustrates a steady-flow, reversed Carnot cycle. The processes involved and the devices that accomplish them are as follows:

State points	Process	Device
1–2;	Adiabatic, reversible (isentropic) compression	Ideal compressor
2–3;	Reversible heat rejection at constant temperature	Ideal condenser
3–4;	Adiabatic, reversible (isentropic) expansion	Ideal turbine
4–1;	Reversible heat addition at constant temperature	Ideal evaporator

Recognizing that the net power supplied is $\dot{W}_{in} = \dot{W}_c - \dot{W}_t$, we can express the coefficients of performance for the reversed-Carnot cycle refrigerator and heat pumps by Eqs. 9.47 and 9.48, respectively. We can relate the heat-transfer rates to the temperatures in the condenser and evaporator by applying the definition of entropy for a reversible, constant-temperature process. Expressing the heat-transfer rates on a per-unit-mass basis, we write

$$q_L = \dot{Q}_L/\dot{m},$$
$$q_L = T_L(s_1 - s_4)$$

and

$$q_H = \dot{Q}_H/\dot{m},$$
$$q_H = T_H(s_2 - s_3).$$

Substituting these expressions for q_L and q_H into Eq. 9.47b yields

$$\beta_{refrig,\,Carnot} = \frac{1}{T_H/T_L - 1}, \tag{9.49a}$$

where we recognize from Fig. 9.56 that $s_1 - s_4 \equiv s_2 - s_3$. Similarly, Eq. 9.48b is transformed to

$$\beta_{\text{heat pump, Carnot}} = \frac{T_H/T_L}{T_H/T_L - 1}. \qquad (9.49b)$$

We need to point out that, although theoretically possible, the reversed Carnot cycle illustrated in Fig. 9.56 is impractical for several reasons. First, compressors do not function well with wet, liquid–vapor mixtures. In practice, the refrigerant is frequently slightly superheated upon entering the compressor. Second, the power produced by a turbine in a reversed Carnot cycle would be quite small, and the complexity and expense required to recover this small amount of energy preclude the use of a turbine. In real refrigerators and heat pumps, the working fluid expands *irreversibly* through a simple throttling device or expansion valve. The refrigerator pictured in Fig. 9.54 uses a length of capillary tube to achieve the desired expansion.

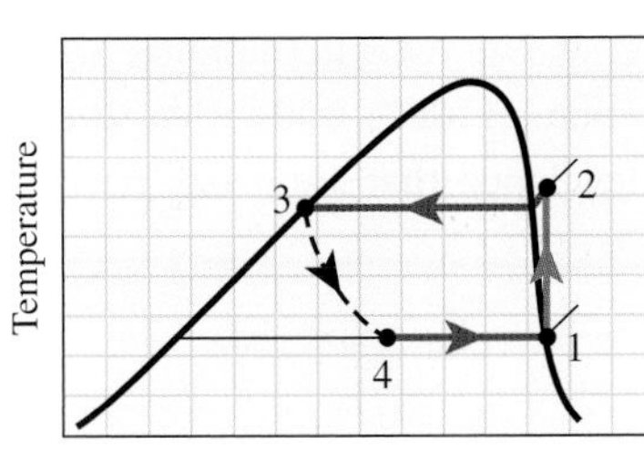

FIGURE 9.57 The ideal vapor-compression refrigeration cycle comprises an isentropic compression (1–2), a constant-pressure heat rejection in which the working fluid condenses (2–3), a constant-enthalpy expansion (3–4), and a constant-pressure heat addition in which the working fluid evaporates (4–1).

9.6c Vapor-Compression Refrigeration Cycle

The ideal vapor-compression refrigeration cycle is illustrated in T–s coordinates in Fig. 9.57, and a schematic diagram of the components used to effect this cycle is shown in Fig. 9.58. In the *ideal* vapor-compression refrigeration cycle, the compression process takes place adiabatically and reversibly, and hence isentropically, and the heat-transfer processes in the condenser and evaporator are assumed to be reversible. The throttling process, however, is by definition highly irreversible; thus, the ideal cycle contains an irreversible process.

A variety of working fluids are used in refrigerators and heat pumps. The NIST online database provides thermodynamic and transport properties for eight modern refrigerants. Among these, tetrafluoroethane (CH_3CH_2F), Refrigerant 134a (R-134a), and chlorodifluoromethane ($CHClF_2$), Refrigerant 22 (R-22), are commonly used in residential, commercial, and automotive systems. These refrigerants are formulated to minimize ozone destruction in the upper atmosphere. Older refrigerants (R-11 and R-12) contain chlorine, which reacts catalytically to destroy ozone. Refrigerant leakage from refrigeration systems and the improper disposal of refrigerants, combined with other sources of these chlorofluorohydrocarbons (CFCs), has over several decades

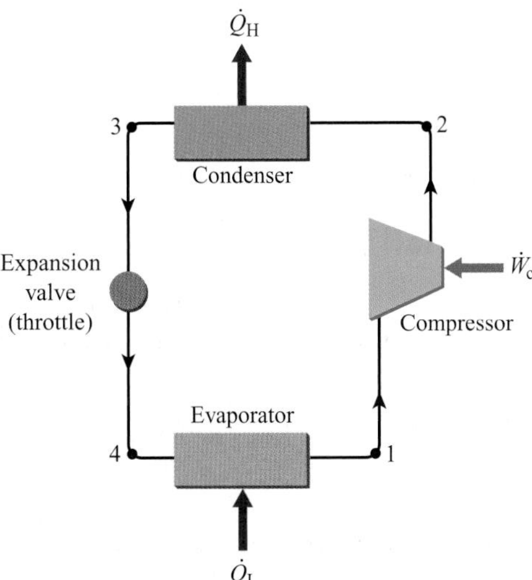

FIGURE 9.58 Schematic of components used to implement the vapor-compression refrigeration cycle.

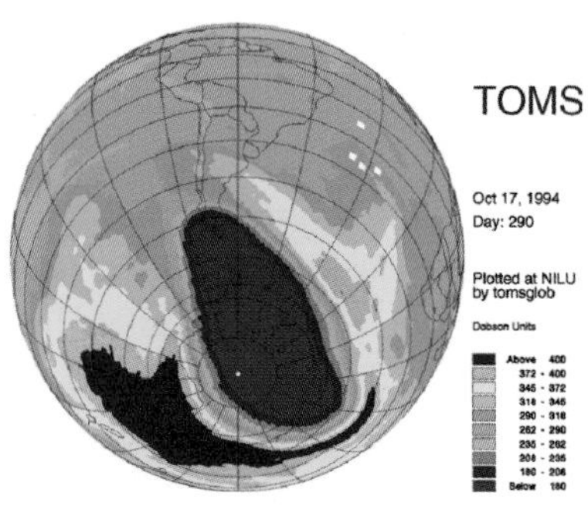

FIGURE 9.59 Meteor-3 TOMS (Total Ozone Monitoring System) image shows the ozone hole (purple) over Antarctica on October 17, 1994. TOMS data (NASA) plotted by Norwegian Institute for Air Research.

resulted in the "ozone hole" over Antarctica (Fig. 9.59). An international treaty (Montreal Protocol, 1987) phased out the production of CFCs and other ozone-depleting substances. New refrigerants are being developed that are ozone friendly and are not heat-trapping greenhouse gases.

CYCLE ANALYSIS

We now analyze the vapor-compression cycle (Fig. 9.57) and develop relationships to evaluate refrigerator and heat pump coefficients of performance.

Compressor (1–2) Assuming the compressor to be adiabatic and neglecting potential and kinetic energy changes of the entering and exiting fluid, the steady-flow conservation of energy expression (Eq. 5.15e) simplifies to

$$\dot{W}_c = \dot{m}(h_2 - h_1), \tag{9.50}$$

where $\dot{W}_c$ is the power input to the compressor. Note that Eq. 9.50 applies to both reversible and irreversible compression, provided the process is adiabatic.

Condenser (2–3) Again neglecting changes in kinetic and potential energies, conservation of energy applied to the working fluid in the condenser yields

$$\dot{Q}_H = \dot{m}(h_2 - h_3), \tag{9.51}$$

where $\dot{Q}_H$ is the heat-transfer rate to the surroundings.

Expansion valve (3–4) Assuming adiabatic operation and neglecting potential and kinetic energy changes, conservation of energy applied across the expansion valve yields

$$h_3 = h_4. \tag{9.52}$$

As discussed in Chapter 8, throttling is characterized as a constant-enthalpy process. Note that the downstream state for the expansion valve is in the liquid–vapor mixture region (Fig. 9.57). The mixture quality x_4 is readily determined from h_4 since P_4 is usually known.

Evaporator (4–1) With the usual assumptions, conservation of energy applied to the working fluid flowing through the evaporator yields

$$\dot{Q}_L = \dot{m}(h_1 - h_4), \tag{9.53}$$

where $\dot{Q}_L$ is the heat-transfer rate supplied to the refrigerant.

COEFFICIENTS OF PERFORMANCE

Substituting Eqs. 9.50–9.53 into the coefficient of performance definitions (Eqs. 9.47 and 9.48) yields

$$\beta_{\text{refrig}} = \frac{\dot{Q}_L}{\dot{W}_{\text{in}}} = \frac{h_1 - h_4}{h_2 - h_1} \tag{9.54a}$$

and

$$\beta_{\text{heat pump}} = \frac{\dot{Q}_H}{\dot{W}_{\text{in}}} = \frac{h_2 - h_3}{h_2 - h_1}. \tag{9.54b}$$

Other than the restrictions that the compression and throttling processes are adiabatic, these relationships apply to both ideal (reversible) and real (irreversible) systems. Furthermore, we note that, although the ideal cycle presented in Fig. 9.57 shows the working fluid entering the compressor as a saturated vapor, the entering fluid (state point 1) in a real compressor is likely to have been slightly superheated in order to avoid any possibility of moisture. Similarly, state point 3 may lie in the subcooled liquid region in a real device. Because no specific designation of the state points is assumed in their derivation, Eqns. 9.54a and 9.54b still apply.

In the air conditioning and heat pump industries, the efficiency of a unit is usually described using the seasonal energy efficiency ratio (SEER). This is an averaged performance of the unit over the cooling season or heating season. It is defined as

$$\mathrm{SEER} = \frac{\text{rate of cooling/heating in BTU/hr}}{\text{power input in W}}. \tag{9.55a}$$

The SEER value has inconsistent units BTU/W·hr. The units of the SEER value are often not included. Using the conversion of 3.41214 BTU/hr = 1 W to achieve consistent units gives us the relationship between SEER values and the coefficient of performance:

$$\mathrm{SEER} = 3.41214\,\beta. \tag{9.55b}$$

The following examples illustrate the application of the vapor-compression cycle to refrigerators and heat pumps.

Example 9.12 Freezer Refrigeration Cycle

It is desired to maintain the cold space in a freezer at 0 F. The freezer utilizes a vapor-compression cycle and operates in a room in which the air temperature is 70 F. Determine the requirements for the pressures in the evaporator and condenser of this freezer (i.e., the minimum or maximum pressures required). The refrigerant is R-134a.

Image courtesy of Haier Group America.

Solution

Known Maximum T_L, minimum T_H, R-134a

Find Saturation pressures corresponding to $T_{L,max}$ and $T_{H,min}$

Sketch

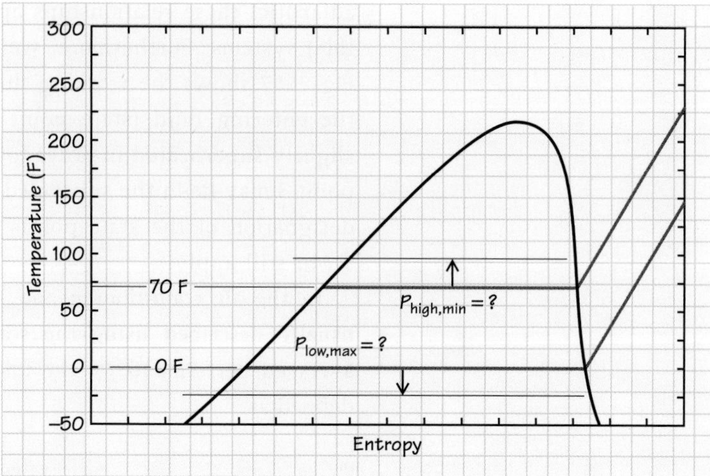

Analysis The bulk of the analysis required to solve this problem is done in the creation of the sketch. To keep the freezer cold space at 0 F requires that energy be removed from this space. This is accomplished in the evaporator of the vapor-compression cycle. The maximum possible low temperature of the working fluid is identical to the temperature of the cold space, 0 F. This represents a reversible heat-transfer process since both the system (the working fluid) and the surroundings (the cold space) are at the same temperature. The corresponding pressure in the R-134a is the saturation pressure at 0 F. From the NIST WebBook we find this to be

$$P_{\text{sat}}(T = T_{\text{sat}} = 0\text{ F}) = 1.4406\text{ atm}.$$

That this pressure is the *maximum possible* low pressure can be seen on the sketch. Since the real heat-transfer process in the evaporator will require that the temperature of the refrigerant be lower than the temperature of the cold space, the actual low pressure will correspond to P_{sat} ($T = T_{\text{sat}} < 0$ F). This lower pressure is indicated by the downward-pointing arrow originating at the $P_{\text{low,max}}$ line.

Similar logic can be applied to the heat-rejection process in the condenser. Note that the real heat-transfer process will now involve a temperature difference in which the refrigerant is *hotter* than the surroundings. This is indicated by the upward-pointing arrow on the sketch. From the NIST WebBook,

$$P_{\text{high, min}} = P_{\text{sat}}(T_{\text{sat}} = 70\text{ F}) = 5.8384\text{ atm}.$$

Table 6.1 in Chapter 6 lists many irreversibilities. Second on the list is heat transfer across a finite temperature difference.

Comments This example can also be used to illustrate the effect of irreversible heat transfer (for $\Delta T > 0$) on the performance of a vapor-compression refrigeration device. Using $T_{\text{L}} = 0$ F (255 K) and $T_{\text{H}} = 70$ F (294 K) as the reservoir temperatures for the operation of a reversed Carnot cycle in which all processes are both *internally* and *externally* reversible, we obtain the coefficient of performance for the refrigerator:

$$\beta_{\text{refng, Carnot}} = \frac{1}{T_{\text{H}}/T_{\text{L}} - 1}$$

$$= \frac{1}{\dfrac{294}{255} - 1} = 6.54.$$

For the case in which the heat transfer is accomplished irreversibly (for $\Delta T > 0$), we assume for purposes of illustration that the temperature difference is 20 F for both the evaporator and condenser; thus,

$$T_L = 0 - 20\,\text{F} = -20\ \text{F}(244\ \text{K})$$
$$T_H = 70 + 20\,\text{F} = 90\ \text{F}(350\ \text{K}).$$

For the internally reversible Carnot cycle operating between these two temperatures, the coefficient of performance is

$$\beta_{\text{refrig, Carnot}} = \frac{1}{\dfrac{305}{244} - 1} = 4.0.$$

From this, we see that irreversible (real) heat-transfer processes significantly decrease the otherwise ideal performance of a vapor-compression refrigerator.

Example 9.13 Refrigeration Cycle

Consider an ideal vapor-compression refrigeration cycle operating between pressures of 0.14 and 0.80 MPa. The flow rate of the refrigerant (R-134a) is 0.04 kg/s. Determine the rate at which energy is removed from the cold space, the rate at which energy is rejected to the surroundings, the power input to the compressor, and the cycle coefficient of performance. Also compare the cycle *COP* with that of a reversed Carnot cycle operating between the same two pressures.

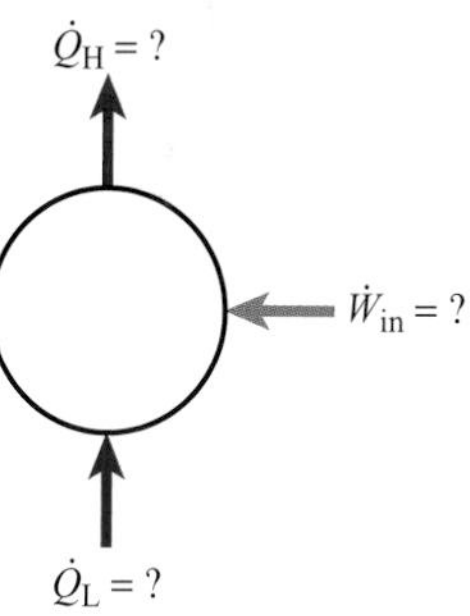

Solution

Known P_H, P_L, $\dot{m}$, R-134a

Find $\dot{Q}_L$, $\dot{Q}_H$, $\dot{W}_c$, β_{refrig}, β_{Carnot}

Sketch

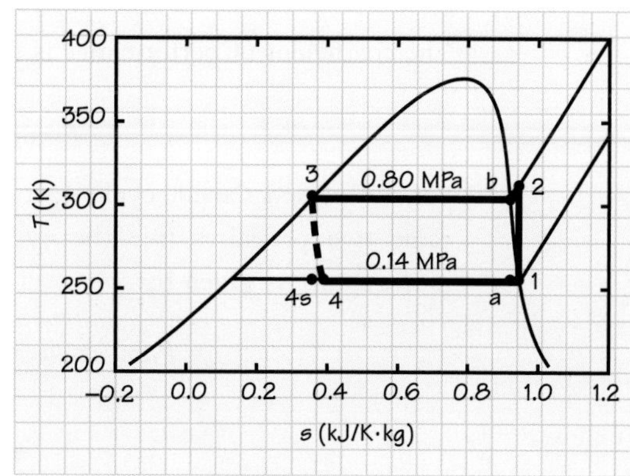

Modeling, Premises and Assumptions

i. Steady flow
ii. Ideal vapor-compression cycle
iii. Changes in kinetic and potential energy negligible for all components

Analysis The key state points (1–2–3–4–1) are identified on the sketch for the ideal cycle. Determining the mass-specific enthalpy at each of the points allows us to find the first four desired quantities. Since state 1 lies on the saturated vapor line at 0.14 MPa, all other desired state-1 properties can be obtained from the NIST resources or by a different table look-up. The properties for the saturated liquid at state 3 are similarly obtained. Using the NIST database with property values based on the ASHRAE reference state,[6] we construct the following table:

State	1	2	3	4
Region	Saturated vapor	Superheated vapor	Saturated liquid	Liquid–vapor mixture
P (MPa)	0.14	0.80	0.80	0.14
T (K)	254.39	?	304.48	254.39
h (kJ/kg)	239.18	?	95.501	$h_4 = h_3$
s (kJ/kg·K)	0.9446	$s_2 = s_1$	0.3541	?

To find h_4, we recognize that the adiabatic, reversible, compression process is isentropic (i.e., $s_2 = s_1$); thus, $h_2 = h_2(P_2, s_2)$. Again using the NIST database, we obtain

$$\begin{aligned} h_2 &= h_2(0.80\ \text{MPa},\ 0.9946\ \text{kJ/kg·K}) \\ &= 275.39\ \text{kJ/kg}, \end{aligned}$$

and the corresponding temperature is

$$T_2 = 312.13\ \text{K}$$

Recognizing that the enthalpy change is zero across the expansion valve, we have

$$h_4 = h_3 = 95.501\ \text{kJ/kg}.$$

First-law analyses of the individual components (Eqs. 9.53, 9.51, and 9.50) yield

$$\begin{aligned} \dot{Q}_L &= \dot{m}(h_1 - h_4) \\ &= 0.04(239.18 - 95.501)\ \text{kW} = 5.747\ \text{kW}, \\ \dot{Q}_H &= \dot{m}(h_2 - h_3) \\ &= 0.04(275.39 - 95.501)\text{kW} = 7.196\ \text{kW}, \end{aligned}$$

and

$$\begin{aligned} \dot{W}_c &= \dot{m}(h_2 - h_1) \\ &= 0.04(275.39 - 239.18)\text{kW} = 1.448\ \text{kW}. \end{aligned}$$

The coefficient of performance is then (Eq. 9.47a)

$$\beta_{refrig} = \frac{\dot{Q}_L}{\dot{W}_c} = \frac{5.747}{1.448} = 3.97.$$

For a reversed Carnot cycle operating within the liquid–vapor region between the same two pressures (see the cycle a–b–3–4s–a on the sketch), the coefficient of performance is given by

$$\begin{aligned} \beta_{refrig,\,Carnot} &= \frac{1}{\dfrac{T_H}{T_L} - 1} \\ &= \frac{1}{\dfrac{304.48}{254.39} - 1} = 5.08. \end{aligned}$$

Comments We note that all the energy rates are dictated by the refrigerant flow rate and the high- and low-pressure set points. As we saw in the previous example, the pressure set points are determined by the temperature requirements of the application.

[6] Various reference states are available as options. The ASHRAE standard sets $u = s = 0$ at −40 °C for saturated liquid. Other examples in this chapter use the NIST default rather than the ASHRAE reference state. Care must be exercised in using property data from different sources, as different reference states may be employed.

Thus, more refrigerant flow provides more cooling capacity for a given application. We also note that the coefficient of performance is much greater than unity. We also observe that the Carnot *COP* is approximately 28% higher than that of the ideal vapor-compression cycle.

Self-Test 9.10

A home air conditioner operating on an ideal vapor-compression refrigeration cycle (using R134-a) steadily removes heat at a rate of 500 kJ/min. If the minimum and maximum operating pressures are 0.1 and 0.9 MPa, respectively, determine the coefficient of performance and refrigerant mass flow rate of the air conditioner.

(Answer: $\beta_{refrig} = 2.89, \dot{m} = 0.063\ kg/s$*)*

Example 9.14 Heat Pump

As shown in the diagram below, a heat pump is used to heat a swimming pool by extracting energy from the ambient air at 295.3 K (72 F) and transferring it to the water in the pool at 283.1 K (50 F). The heat pump operates with R-22 between 0.653 MPa absolute (~80 psig) and 1.342 MPa (~180 psig) and delivers 24.62 kW (84,000 Btu/hr) to the water. The R-22 vapor is superheated to 291.44 K (65 F) before it enters the compressor, and the R-22 liquid is subcooled to 299.78 (80 F) at the condenser outlet. The scroll compressor has an isentropic efficiency of 65%. Assuming no losses other than those associated with the compressor, plot the vapor-compression cycle in *T–s* coordinates and determine the cycle coefficient of performance and the mass flow rate of the refrigerant.

(Credit: Mario Tama / Staff / Getty Images News.)

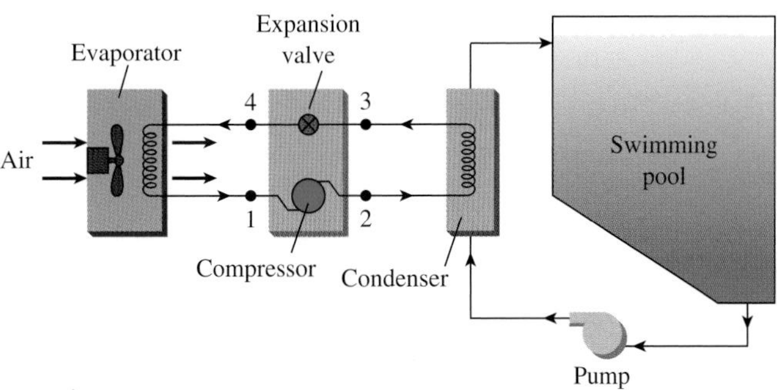

Solution

Known P_{high}, P_{low}, T_1, T_3, $\dot{Q}_H$, T_{H_2O}, T_{air}, R-22

Find T–s diagram, $\beta_{heat\ pump}$, $\dot{m}_{R\text{-}22}$

Modeling, Premises and Assumptions

i. Steady flow
ii. Adiabatic (but nonideal) compression
iii. Negligible changes in potential and kinetic energy for all components
iv. No pressure losses through coils and interconnecting plumbing

Analysis To create a T–s diagram, we first determine the properties at the points designated 1, 2, 3, and 4 in the sketch. From the assumption that there are no pressure losses through the coils, we can write

$$P_1 = P_4 = P_{low} = 0.653 \text{ MPa}$$

and

$$P_2 = P_3 = P_{high} = 1.342 \text{ MPa}.$$

The temperature at the compressor inlet, T_1, is given. With T_1 and P_1 known, we use the NIST database to find s_1 and h_1, anticipating the use of the latter to calculate the *COP*:

For state 1,

$$s_1(291.44\,\text{K}, 0.053 \text{ MPa}) = 1.7646\,\text{kJ/kg·K},$$
$$h_1(291.44\,\text{K}, 0.653\,\text{MPa}) = 415.53\,\text{kJ/kg}.$$

To define state 2, we use the definition of the compressor isentropic efficiency (Eq. 8.18) as follows:

$$\eta_{isen,c} = \frac{h_{2s} - h_1}{h_2 - h_1},$$

where h_{2s} is defined as follows and evaluated using the NIST database:

$$h_{2s}(s_{2s} = s_1, P_2) = h_{2s}(1.7646\,\text{kJ/kg·K}, 1.342 \text{ MPa})$$
$$= 434.09\,\text{kJ/kg}.$$

Solving the defining relationship for $\eta_{isen,c}$ for h_2 yields

$$h_2 = \frac{h_{2s} - h_1}{\eta_{isen,c}} + h_1$$
$$= \frac{434.09 - 415.53}{0.65} + 415.53 \text{ kJ/kg}$$
$$= 444.08 \text{ kJ/kg}.$$

With P_2 and h_2 known, all other state-2 properties can be determined. We summarize those for state 2:

$$h_2(340.70\,\text{K}, 1.342\,\text{MPa}) = 444.08 \text{ kJ/kg},$$
$$s_2(340.70\,\text{K}, 1.342\,\text{MPa}) = 1.7943\,\text{kJ/kg·K}.$$

Since state 3 lies on the high-pressure isobar, $P_3 = 1.342$ MPa, and the temperature $T_3 = 299.78$ K is given, we can obtain the subcooled liquid properties directly from the NIST database:

$$h_3(299.78\,\mathrm{K}, 1.342\,\mathrm{MPa}) = 232.33\ \mathrm{kJ/kg},$$

$$s_3(299.78\,\mathrm{K}, 1.342\,\mathrm{MPa}) = 1.1105\ \mathrm{kJ/kg{\cdot}K}.$$

The enthalpy across the expansion valve is constant (as it is a throttling process), so $h_4 = h_3$. Since $P_4 = P_{\mathrm{low}}$, h_4 and P_4 define the state. To determine s_4, however, requires that we first determine the quality x_4, which we accomplish as follows (Eq. 2.37d):

$$x_4 = \frac{h_4 - h_{\mathrm{f},4}}{h_{\mathrm{g},4} - h_{\mathrm{f},4}} = \frac{232.33 - 210.21}{408.09 - 210.21} = 0.1118,$$

and (Eq. 2.37c)

$$\begin{aligned} s_4 &= (1 - x_4)s_{\mathrm{f},4} + x_4 s_{\mathrm{g},4} \\ &= (1 - 0.1118)1.0363 + 0.1118(1.7387)\ \mathrm{kJ/kg{\cdot}K} = 1.115\ \mathrm{kJ/kg{\cdot}K}. \end{aligned}$$

The temperature $T_4 = T_{\mathrm{sat}}\ (P_{\mathrm{sat}} = P_4 = 0.653\ \mathrm{MPa})$; thus, the properties at state 4 are fully defined as follows:

$$T_4 = 281.76\,\mathrm{K},$$

$$\mathrm{h}_4(x_4 = 0.1118, 0.653\ \ \mathrm{MPa}) = 232.33\,\mathrm{kJ/kg}.$$

$$s_4 = (x_4 = 0.1118, 0.653\,\mathrm{MPa}) = 1.115\,\mathrm{kJ/kg{\cdot}K}$$

Using values of T and s at each state point, we can create the following plot:

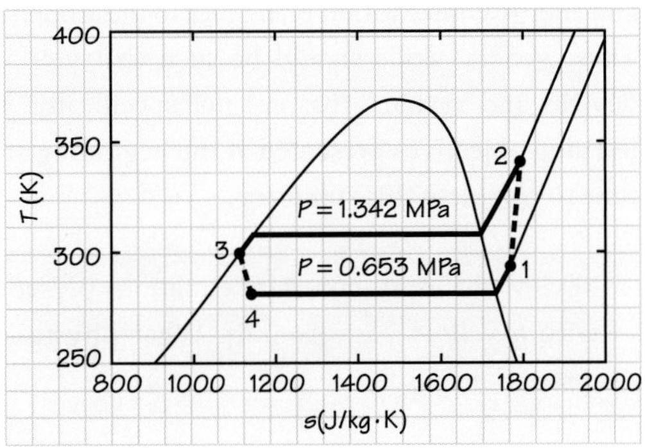

To evaluate the refrigerant flow rate, we apply the following steady-flow energy balance for the condenser:

$$\dot{Q}_{\mathrm{H}} = \dot{m}_{\mathrm{R\text{-}22}}(h_2 - h_3).$$

Solving for $\dot{m}_{\mathrm{R\text{-}22}}$, we get

$$\dot{m}_{\mathrm{R\text{-}22}} = \frac{\dot{Q}_{\mathrm{H}}}{h_2 - h_3} = \frac{24.62}{(444.08 - 232.33)} = 0.1163$$

$$[=]\frac{\mathrm{kW}}{\mathrm{kJ/kg}}\left[\frac{1\,\mathrm{kJ/s}}{\mathrm{kW}}\right] = \mathrm{kg/s}.$$

For a heat pump, the coefficient of performance is given by Eq. 9.54b:

$$\begin{aligned} \beta_{\mathrm{heat,\,pump}} &= \frac{\dot{Q}_{\mathrm{H}}}{\dot{W}_{\mathrm{c}}} = \frac{h_2 - h_3}{h_2 - h_1} \\ &= \frac{444.08 - 232.33}{444.08 - 415.33} = 7.4. \end{aligned}$$

Comment Coefficients of performance for most heat pump applications are much smaller than the value obtained in this example (they are equal to say, 3–5). The reason for this difference is the relatively high temperature used in the evaporator for this application. (For a fixed temperature in the condenser, the *COP* falls with decreasing evaporator temperature, which you can easily show to be true with some simple, reversed-Carnot-cycle calculations.) In this application, air is available at a high temperature (72 F), unlike in a typical heating application where the air temperature would be much less. This higher air temperature allows a higher temperature (and pressure) to be utilized in the evaporator while maintaining a temperature difference adequate to achieve the desired heat transfer.

SUMMARY

After studying this chapter, you should have a basic understanding of the operation of fossil-fueled steam power plants, turbojet engines, gas-turbine engines, refrigerators and heat pumps, and air conditioners. Specifically, you should be able to represent the various thermodynamic cycles associated with each of these systems using appropriate thermodynamic coordinates (e.g., T–s and P–v diagrams) and be able to model these systems to estimate measures of efficiency and performance. As a result of your study of this chapter, your ability to analyze complex thermal-fluid systems should be greatly enhanced. Specifically, your skills in selecting open systems and in applying mass, momentum, and energy conservation principles to these open systems should be quite well developed. These general skills should be such that you are comfortable analyzing thermal-fluid devices and systems that are not specifically discussed in this chapter. The detailed learning objectives presented at the beginning of this chapter should be reviewed at this time to complete this summary.

KEY EQUATIONS

Review the most important equations presented in this chapter (i.e., those boxed with a yellow background). What physical principles do they express? What restrictions apply?

CHAPTER 9 KEY CONCEPTS AND DEFINITIONS CHECKLIST

Numbers following arrows refer to Questions and Problems at the end of the chapter.

9.1 Fossil-Fueled Steam Power Plants

- ☐ Basic Rankine cycle ➔ Questions 9.1, 9.2, Problem 9.1
- ☐ Thermal efficiency ➔ Problems 9.1, 9.3
- ☐ Isentropic efficiency of pump and turbine ➔ Problems 9.2, 9.4
- ☐ Rankine cycle with superheat ➔ Questions 9.2, 9.3, Problem 9.3
- ☐ Rankine cycle with reheat ➔ Question 9.2, Problem 9.31
- ☐ Rankine cycle with regeneration ➔ Question 9.2, Problem 9.45
- ☐ Rankine cycle and variants: T–s diagrams ➔ Questions 9.2, 9.3
- ☐ Steam extraction ➔ Question 9.5
- ☐ Feedwater heater ➔ Question 9.6
- ☐ Overall efficiency ➔ Question 9.7

9.2 Gas-Turbine Engines

- ☐ Integral open system (control volume) analyses for mass and energy ➔ Problem 9.63

- ☐ Air-standard Brayton cycle ➔ Question 9.8
- ☐ Process thermal efficiency ➔ Problem 9.63

9.3 Modified Power Cycles

- ☐ Cogeneration ➔ Questions 9.9, 9.10, Problem 9.83
- ☐ Combined cycles ➔ Questions 9.11, 9.12, Problem 9.86

9.4 Jet Engines

- ☐ Two major types ➔ Question 9.13
- ☐ Turbojet "cycle" processes ➔ Question 9.14
- ☐ Integral open system (control volume) analyses for mass, energy, and momentum ➔ Problem 9.89
- ☐ Air-standard cycle analysis ➔ Problem 9.89
- ☐ Propulsive efficiency ➔ Problem 9.91
- ☐ Specific thrust ➔ Problem 9.91
- ☐ Specific fuel consumption ➔ Problem 9.91

9.5 Other Gas Power Cycles

- ☐ Otto Cycle ➔ Question 9.15, Problem 9.94
- ☐ Diesel Cycle ➔ Question 9.16, Problem 9.95

9.6 Refrigerators and Heat Pumps

- ☐ Reversed cycles ➔ Question 9.17, Problem 9.111
- ☐ Coefficients of performance ➔ Problems 9.111, 9.112
- ☐ Vapor-compression-cycle processes ➔ Question 9.19, Problem 9.115
- ☐ Vapor-compression-cycle T–s diagram ➔ Question 9.18, Problem 9.115
- ☐ Cycle analysis ➔ Problem 9.117

REFERENCES

1. *Thermophysical Properties of Fluid Systems*, National Institute of Standards and Technology, Gaithersburg, MD, 2000, http://webbook.nist.gov/chemistry/fluid.
2. *Steam: Its Generation and Use*, 41st edn, S. C. Stulz and J. B. Kitto, Eds., Babcock & Wilcox, Barberton, OH, 2005.
3. Mattingly, J. D., *Elements of Gas Turbine Propulsion*, McGraw-Hill, New York, 1996.
4. Hopkins, K. N., "Turbopropulsion Combustion – Trends and Challenges," AIAA Paper 80–1199, 1980.
5. Turns, S. R., *An Introduction to Combustion: Concepts and Applications*, 3rd edn, McGraw-Hill, New York, 2012.
6. St. Peter, J., *The History of Aircraft Turbine Engine Development in the United States*, International Gas Turbine Institute of the American Society of Mechanical Engineers, Atlanta, 1999.
7. Moran, M. J., and Shapiro, H. N., *Fundamentals of Engineering Thermodynamics*, 3rd edn, Wiley, New York, 1995.
8. Çengel, Y. A., and Boles, M. A., *Thermodynamics: An Engineering Approach*, 4th edn, McGraw-Hill, New York, 2002.
9. Drbal, L. F., *Power Plant Engineering*, Chapman & Hall, New York, NY, 1996.
10. GE Gas-Turbine Engine website, https://powergen.gepower.com.

Some end-of-chapter problems were adapted with permission from the following:

Look, D. C., Jr., and Sauer, H. J., Jr., *Engineering Thermodynamics*, PWS, Boston, 1986.
Myers, G. E., *Engineering Thermodynamics*, Prentice Hall, Englewood Cliffs, NJ, 1989.

QUESTIONS

9.1 List the components of a simple Rankine cycle. Sketch the components and show how they interconnect.

9.2 Draw T–s diagrams for (a) the basic Rankine cycle without superheat, (b) the basic Rankine cycle with superheat, (c) the Rankine cycle with reheat, and (d) the Rankine cycle with regeneration.

9.3 Sketch a T–s diagram of the Rankine cycle with superheat and include the vapor dome. Now modify the diagram for a turbine and pump that are not isentropic.

9.4 Explain how a cooling tower works.

9.5 Define *steam extraction*. Why is steam extracted from a turbine in a steam power plant?

9.6 What is the purpose of a feedwater heater in a steam power plant? Draw sketches of closed and open feedwater heaters, labeling the inflows and outflows. For each inflow, identify the device where the flow originated. For each outflow, identify the device that the flow will enter.

9.7 What factors are important in determining the overall efficiency of the conversion of fuel energy to electricity in a fossil-fueled power plant?

9.8 List the components of a stationary gas-turbine engine used for electrical power generation. Sketch the components and show how they interconnect.

9.9 What are the objectives (outputs) of a cogeneration cycle?

9.10 Sketch a simple cogeneration cycle. How is a Rankine cycle modified to provide cogeneration?

9.11 Sketch a combined cycle power plant. What two standard cycles are combined?

9.12 How does a combined cycle plant increase the thermal efficiency?

9.13 Distinguish between turbojet and turbofan jet engines.

9.14 List the components of a turbojet engine. Sketch the components and show how they interconnect.

9.15 Sketch an Otto cycle and list the conditions across each process.

9.16 Sketch a Diesel cycle and list the conditions across each process.

9.17 List the components of a vapor-compression-cycle refrigerator. Sketch the components and show how they interconnect.

9.18 Draw a T–s diagram for an ideal vapor-compression refrigeration cycle.

9.19 List the components of an air conditioner that operates on a vapor-compression cycle. Sketch the components and show how they interconnect.

Chapter 9 Problem Subject Areas

9.1–9.30	Steam power plants with simple Rankine cycle, including superheat
9.31–9.44	Rankine cycle with reheat
9.45–9.61	Rankine cycle with regeneration
9.62–9.82	Gas-turbine engines
9.83–9.88	Modified power cycles
9.89–9.92	Turbojet engines

9.93–9.110	Other gas power cycles
9.111–9.141	Refrigerators and heat pumps
9.142–9.154	FE exam problems

PROBLEMS

9.1–9.30 Steam power plants with simple Rankine cycle, including superheat

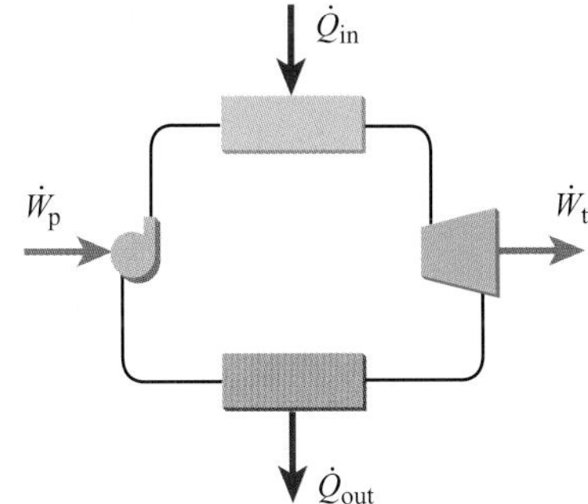

9.1 Consider an ideal Rankine cycle operating between 5 kPa and 10 MPa and using H_2O as the working fluid. Steam enters the turbine as saturated vapor at a flow rate of 8 kg/s.

A. Plot the ideal cycle on T–s coordinates.
B. Determine the power produced by the turbine.
C. Determine the power required to drive the pump.
D. Determine the thermal efficiency of the cycle.

9.2 Repeat Problem 9.1 but assume now that the turbine isentropic efficiency is 95% and the pump isentropic efficiency is 85%. Also compare the steam quality at the turbine exit with the value from Problem 9.1.

9.3 Consider an ideal Rankine cycle operating between 5 kPa and 10 MPa with a maximum cycle temperature of 750 K. The working fluid is H_2O.

A. Plot the cycle in T–s coordinates.
B. Determine the fraction of the total energy imparted to the H_2O in the boiler that is associated with the superheater section (i.e., $\dot{Q}_{3-3'}/\dot{Q}_{2-3'}$) in Fig. 9.12.
C. Determine the cycle efficiency.
D. Compare the result from part C to the cycle efficiency for the same cycle, but without superheat (see Problem 9.1). Discuss.

9.4 Consider a Rankine cycle operating between 5 kPa and 10 MPa with a maximum cycle temperature of 750 K. The working fluid is H_2O. Determine both the cycle thermal efficiency and the isentropic efficiency of the turbine if the steam quality at the turbine exit is 92%. Assume ideal operation of the pump.

9.5 Consider the ideal Rankine cycle and operating conditions specified in Problem 9.1. River water is to be used as the cooling fluid in the steam condenser. The river water is available at 295 K, and the discharge temperature is to be no greater than 5 K above the inlet temperature. The overall heat-transfer coefficient associated with the condenser is 2500 W/m^2·K. Assuming primarily counterflow operation, estimate the heat exchange area required for this condenser.

(Credit: MichaelUtech / Vetta / Getty Images.)

9.6 Conduct a design sensitivity study of an ideal Rankine cycle (with superheat) by performing the following tasks. The fluid is H_2O.

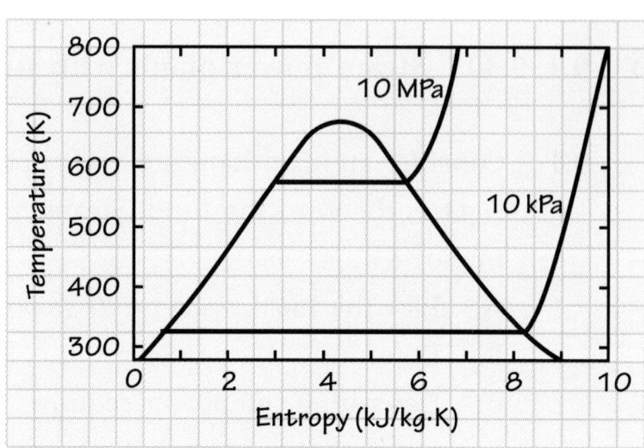

A. **Base case:** For the base case (turbine inlet temperature and pressure of 700 K and 10 MPa, respectively, and a condenser pressure of 10 kPa), calculate the following quantities:

i. The quality x of the steam exiting the turbine,

ii. The work delivered by the turbine per unit mass of working fluid (i.e., $\dot{W}_{turb}/\dot{m}$),

iii. The heat rejected in the condenser per unit mass of working fluid (i.e., $\dot{Q}_{out,\,cond}/\dot{m}$),

iv. The pump work per unit mass of working fluid, $\dot{W}_{pump}/\dot{m}$,

v. The heat added to the working fluid in the boiler per unit mass of working fluid (i.e., $\dot{Q}_{in,\,boiler}/\dot{m}$), and

vi. The thermal efficiency of the cycle.

Perform all calculations without the aid of a computer; the NIST online database, however, may be useful to obtain any required thermodynamic properties.

B. **Turbine inlet temperature:** Determine the quality of the steam exiting the turbine and the cycle efficiency as a function of the turbine inlet temperature. Use the following values: 600 K, 700 K (base case), 800 K, and 900 K. All other parameters are as in the base case. Use spreadsheet software to perform your calculations (of x and η) and to plot the results of your computations (of x and η) as functions of turbine inlet temperature.

C. **Turbine inlet pressure:** Determine the turbine exit steam quality and the cycle thermal efficiency as a function of turbine inlet pressure. Use the following values: 5 MPa, 10 MPa (base case), 15 MPa, and 20 MPa. All other parameters are as in the base case. Use spreadsheet software and plot as in Part B.

D. **Turbine exhaust pressure:** Perform calculations and plot in the same manner as in parts B and C. Now vary the turbine outlet pressure (the condenser pressure) using the following values: 5 kPa, 10 kPa (base case), 50 kPa, and 100 kPa.

E. **Summary:** Write a few brief paragraphs discussing your results from Parts B, C, and D.

9.7 Solve Problem 9.6 using EES.

9.8 Determine the quality of the steam exiting the turbine and the cycle thermodynamic efficiency of the base-case cycle in Problem 9.6 if the turbine has an isentropic efficiency of 96% and the pump has an isentropic efficiency of 92%. Compare these results to those of the ideal cycle calculated in part A of Problem 9.6.

9.9 In a Rankine cycle, steam leaves the boiler at 400 °C and 5 MPa. The condenser pressure is 0.01 MPa. The water entering the pump is saturated. Determine the first-law thermal efficiency and compare it to that of a Carnot cycle operating between the same temperature extremes.

9.10 An engineer proposes to eliminate the condenser in an ideal Rankine cycle in order to form an open cycle. Lake water is available at 16 °C. The lake water is pumped to 5 MPa and then heated at constant pressure to 400 °C. The "spent" steam is rejected back into the lake. Determine the first-law thermal efficiency of this process and compare it with that of Problem 9.9. What are some of the practical problems with this proposed cycle?

9.11 A Rankine-cycle steam power plant operates with a turbine inlet temperature 1000 F and an exhaust pressure of 1 psia. If the turbine isentropic efficiency is 0.85, determine the maximum turbine inlet pressure that can be used if the exit quality is to be 0.90.

9.12 Consider a 50-MW Rankine-cycle steam power plant. The steam enters the turbine at 1000 F and 1200 psia and leaves at 1 psia with a quality of 0.90. Water leaves the condenser at 80 F and 1 psia. The power supplied to the pump may be neglected. Determine the thermal efficiency for the power plant and the mass flow rate ($\mathrm{Mlb_m/hr}$) of steam in the boiler.

9.13 A solar-driven Rankine-cycle power plant uses refrigerant R-134a as the working fluid. During operation (i.e., daytime), the turbine inlet state is 220 F and 200 psia. The air-cooled condenser operates at a pressure of 80 psia. The pump and turbine isentropic efficiencies are 0.70 and 0.80, respectively. The work output is 100 hp when the solar-collector input is 200 $\mathrm{Btu/hr\cdot ft^2}$. The R-134a leaves the condenser as saturated liquid. Determine (a) the thermal efficiency, (b) the refrigerant mass flow rate ($\mathrm{klb_m/hr}$), and (c) the required collector surface area ($\mathrm{ft^2}$).

9.14 Consider a 600-MW Rankine-cycle steam power plant. The boiler pressure is 1000 psia and the condenser pressure is 1 psia. The turbine inlet temperature is 1000 F and the pump inlet state is saturated liquid. The turbine and pump isentropic efficiencies are 100%. Determine the thermal efficiency of the power plant.

9.15 Consider a 600-MW Rankine-cycle steam power plant. The boiler pressure is 1000 psia and the condenser pressure is 1 psia. The turbine inlet temperature is 1000 F. The condenser pressure is 1 psia and the pump inlet state is saturated liquid. The turbine isentropic efficiency is 0.90 and the pump efficiency is 0.85. Determine the thermal efficiency of the power plant.

9.16 A Rankine-cycle steam power plant operates at a boiler pressure of 5 MPa and a condenser pressure of 0.01 MPa. The turbine inlet temperature is 500 °C. The water entering the pump is saturated liquid. The mass flow rate through the boiler is 60 kg/s. The first-law thermal efficiency of the power plant is 0.30. Determine the thermal net power (MW) produced by the power plant.

9.17 Consider a 100-MW Rankine-cycle steam power plant. The peak cycle temperature is 1000 F. The condenser pressure is 1 psia. Plot the first-law thermal efficiency and mass flow rate (Mlb_m/hr) as functions of the boiler pressure (from 1 to 1000 psia using a semilogarithmic scale). Perform all calculations without the aid of EES or a spreadsheet on a computer. The NIST online database, however, may be useful to obtain any required thermo dynamic properties.

9.18 Use EES to solve Problem 9.17. Find the cycle efficiency for at least six values of the boiler pressure, so as to create a smooth plot.

9.19 Consider a 100-MW Rankine-cycle steam power plant. The boiler pressure is 600 psia and the condenser pressure is 1 psia. Plot the first-law thermal efficiency and the mass flow rate (lb_m/hr) as functions of the turbine inlet temperature (up to 1200 F). Perform all calculations without the aid of EES or a spreadsheet on a computer. The NIST online database, however, may be useful to obtain any required thermo dynamic properties.

9.20 Use EES to solve Problem 9.19. Find the cycle efficiency for at least six values of the boiler pressure, so as to create a smooth plot.

9.21 Consider a Rankine-cycle steam power plant operating between a heat source at 500 °C and a heat sink at 20 °C. For safety reasons the boiler pressure cannot be more than 10 MPa. The mass flow rate of the steam is 50 kg/s. The pumping process occurs in the liquid region. Determine the maximum thermal efficiency and the electrical generating capacity (MW) of this power plant.

9.22 Steam leaves the boiler of a 100-MW Rankine-cycle power plant at 700 F and 500 psia. The steam enters the turbine at 650 F and 475 psia. The turbine has an efficiency of 0.85 and exhausts at 2 psia. In the condenser, the water is subcooled to 100 F at 2 psia by lake water at 55 F. The pump efficiency is 0.75 and its discharge pressure is 550 psia. The water enters the boiler at 95 F and 530 psia. Determine the first-law thermal efficiency for this cycle and the mass flow rate (lb_m/hr) in the boiler.

9.23 Consider a 100-MW Rankine-cycle steam power plant. Temperatures and pressures at the inlet states of the four basic components are given in the table.

The turbine and the pump are both adiabatic. Determine (a) the mass flow rate (lb_m) in the boiler, (b) the heat-transfer rates across the power plant boundary, (c) the work rates across the power plant boundary, and (d) the power plant thermal efficiency.

State	Location	T (F)	P (psia)
1	Turbine inlet	1600	800
2	Condenser inlet	200	1
3	Pump inlet	100	1
4	Boiler inlet	102	800

9.24 A simple Rankine-cycle steam power plant operates with a steam flow rate of 500,000 lb_m/hr. Steam enters the turbine at 500 psia and 1000 F. The turbine exhaust (condenser inlet) conditions are 1.0 psia with a steam quality of 0.90. Determine the following:

A. The turbine output (kW)
B. The condenser heat rejection rate (Btu/hr)
C. The turbine isentropic efficiency (%)
D. Each term in the second-law entropy balance for the condensing process
E. The approximate pump work (kW)
F. The thermal efficiency of the plant (%)
G. The ideal thermal efficiency of the plant (%)

9.25 In a power plant operating on a Rankine cycle, steam at 400 kPa and quality 100% enters the turbine; the pressure is 3.5 kPa in the condenser. Determine the cycle efficiency. How is the efficiency affected if the turbine inlet conditions are changed to 4.0 MPa and 350 °C?

9.26 Solve Problem 9.25 using EES.

9.27 In a power plant operating on a simple Rankine cycle (Fig. 9.7), steam at 400 kPa and quality 100% enters the turbine; the pressure is 3.5 kPa in the condenser as described in Problem 9.25. Modifying your EES program from Problem 9.26, vary the boiler pressure across the range from 400 kPa to 4.0 MPa for a simple Rankine cycle with saturated vapor entering the turbine. Plot the cycle efficiency as a function of the boiler pressure. Use at least five values of the boiler pressure in order to generate a smooth plot.

9.28 In a power plant operating on a Rankine cycle with superheat (Fig. 9.12), steam at 4.0 MPa and 700 K enters the turbine; the pressure is 3.5 kPa in the condenser. Modifying your EES program from Problem 9.26, vary the boiler pressure across the range from 400 kPa to 4.0 MPa for a Rankine cycle with superheat, with steam at 700 K entering the turbine. Plot the cycle efficiency as a function of the boiler pressure. Use at least five values of the boiler pressure to generate a smooth plot.

9.29 Consider a Rankine cycle using R-134a as the working fluid. Saturated vapor leaves the boiler at 85 °C and the condenser temperature is 40 °C. What is the cycle efficiency?

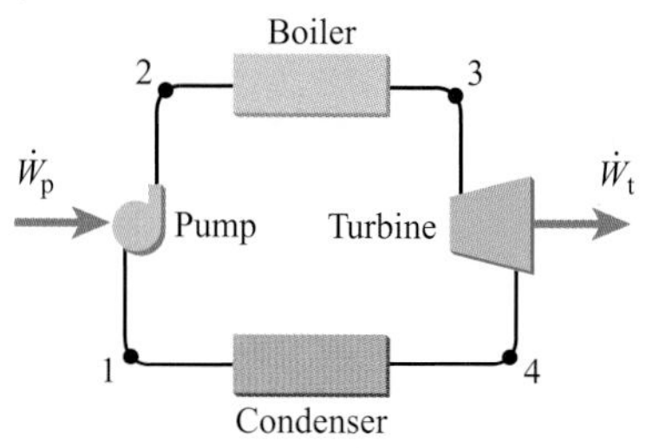

9.30 The following data are for a simple steam power plant as shown in the sketch:

$P_1 = 10$ psia, $T_1 = 160$ F,
$P_2 = 500$ psia,
$P_3 = 480$ psia, $T_3 = 800$ F,
$P_4 = 10$ psia,
steam flow rate = 250,000 lb_m/hr,
turbine isentropic efficiency = 87%.

Determine the following:

A. Pipe size between condenser and pump if the velocity is not to exceed 20 ft/s
B. Minimum pump work (hp)
C. Net power output of plant (kW)
D. Condenser heat rejection rate (Btu/hr)
E. Heat input at boiler (Btu/hr)
F. Thermal efficiency (%)
G. Maximum possible thermal efficiency (%)

9.31–9.44 Rankine cycle with reheat

9.31 Add reheat to the base-case ideal Rankine cycle described in Problem 9.6. The exit pressure of the high-pressure turbine is 2.2 MPa. The steam exiting the high-pressure turbine is reheated back to 700 K before it enters the low-pressure turbine. Calculate the thermal efficiency for this ideal cycle with reheat and compare it with that of the base cycle without reheat. Discuss your results.

9.32 Solve Problems 9.6 and 9.31 using EES.

9.33 Use your EES program to solve for the thermal efficiency in Problem 9.31 when the condenser pressure is varied from 5 kPa to 50 kPa. Plot the thermal efficiency as a function of pressure in this range. Use at least five values for the compressor pressure, so as to generate a smooth curve. Discuss your results.

9.34 Consider an ideal Rankine cycle with superheat and reheat. The maximum temperature of the superheated and reheated steam is 750 K. The maximum pressure in the cycle is 10 MPa and the minimum pressure is 5 kPa. Steam enters the high-pressure turbine at 750 K and exits as saturated vapor. The steam is then reheated to 750 K before expanding in the low-pressure turbine. The steam flow rate is 90 kg/s.

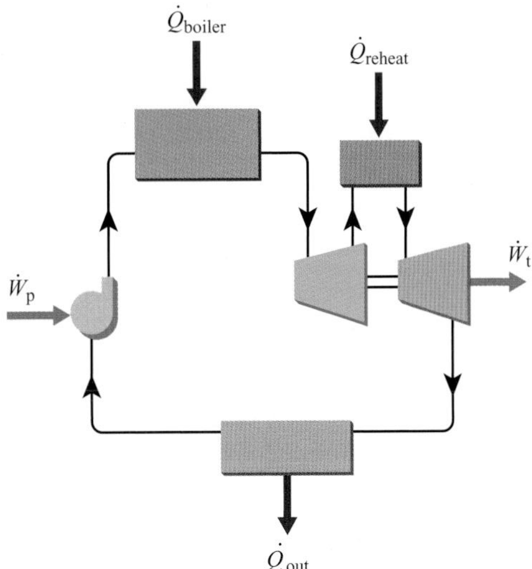

A. Plot the cycle in T–s coordinates.

B. Determine the total heat-addition rate in the boiler (economizer, steam generator, and superheater combined), the heat-addition rate in just the superheater section, and the heat-addition rate in the reheater.

C. Determine the net power produced by the cycle.

D. Determine the cycle thermal efficiency and compare with the thermal efficiency of an ideal cycle without reheat that operates between the same high and low pressures (see Problem 9.3 C).

9.35 Consider a 100-MW reheat-Rankine-cycle steam power plant. Superheated steam enters the high-pressure turbine at 800 °C and 10 MPa. The steam enters the reheater at 700 °C and 4 MPa and is reheated at constant pressure to 800 °C. The steam expands through the low-pressure turbine and enters the condenser as saturated vapor at 0.01 MPa. The water enters the pump as a saturated liquid at 0.01 MPa. The turbines and pump are adiabatic. The pump is reversible. Determine the thermal efficiency of the power plant and mass flow rate (kg/s) of the steam.

9.36 Consider a 200-MW reheat-Rankine-cycle steam power plant. Superheated steam enters the high-pressure turbine at 1000 K and 12 MPa. The steam enters the reheater at 800 K and 3 MPa and is reheated at constant pressure to 1000 K. It expands through the low-pressure turbine and enters the condenser as saturated vapor at 50 kPa. The water enters the pump as a saturated liquid at 50 kPa. The turbines and pump are adiabatic. The pump is reversible. Determine the thermal efficiency of the power plant and mass flow rate (kg/s) of the steam.

9.37 Consider a 500-MW reheat-Rankine-cycle steam power plant. The boiler pressure is 8 MPa, the reheater pressure is 1 MPa, and the condenser pressure is 6 kPa. Both turbine inlet temperatures are 800 K. The water leaving the condenser is saturated liquid. Determine the thermal efficiency of the power plant and the boiler mass flow rate (kg/s). Assume isentropic turbines and pump.

9.38 Consider a 600-MW reheat-Rankine-cycle steam power plant. The boiler pressure is 1000 psia, the reheater pressure is 200 psia, and the condenser pressure is 1 psia. Both turbine inlet temperatures are 1000 F. The water leaving the condenser is saturated liquid. Determine the thermal efficiency of the power plant and the boiler mass flow rate (lb_m/hr). Assume isentropic turbines and pump.

9.39 Consider a 400-MW reheat-Rankine-cycle steam power plant. The boiler pressure is 6 MPa, the reheater pressure is 700 kPa, and the condenser pressure is 6 kPa. The exit temperatures of the boiler and the reheater are both 800 K. Assume that the turbines and the pump are isentropic. The water leaving the condenser is saturated liquid. Determine the thermal efficiency of the power plant.

9.40 Consider a 400-MW reheat-Rankine-cycle steam power plant. The boiler pressure is 6 MPa, the reheater pressure is 700 kPa, and the condenser pressure is 6 kPa. The exit temperatures of the boiler and the reheater are both 800 K. The turbines and the pump have isentropic efficiencies of 0.90 and 0.85, respectively. The water leaving the condenser is saturated liquid. Determine the thermal efficiency of the power plant.

9.41 Plot the cycles in Problems 9.39 and 9.40 on one T–s diagram. Discuss the differences between these cycles and which cycle will have the higher cycle efficiency.

9.42 Consider a 600-MW reheat-Rankine-cycle steam power plant. The boiler pressure is 1000 psia, the reheater pressure is 100 psia, and the condenser pressure is 1 psia. The exit temperaturess of the boiler and the reheater are both 1000 F. The turbines and the pump have isentropic efficiencies of 0.90 and 0.85, respectively. The water leaving the condenser is saturated liquid. Determine the thermal efficiency of the power plant.

9.43 Consider an ideal Rankine cycle modified with the addition of a single reheat stage. The working fluid is H_2O. The maximum pressure in the cycle is 10.7 MPa, and the temperature of the steam exiting both the superheater and the reheater is 800 K. The pressure in the condenser is 0.01 MPa. A "rule of thumb" for the reheat cycle is that the optimum pressure for the reheat process is approximately one-fourth of the maximum pressure in the cycle. This problem explores the validity of this guideline.

A. Show the T–s state points for reheat cycles employing these conditions and the following reheat pressures:

a. 4.675 MPa ($= 0.25P_{max} + 2$ MPa)

b. 2.675 MPa ($= 0.25P_{max}$)

c. 0.675 MPa ($= 0.25P_{max} - 2$ MPa).

Show all three cycles on a single T–s diagram.

B. Calculate the thermal efficiency for the three reheat cycles given in part A.

C. Create an engineering graph showing the effect of reheat pressure on cycle thermal efficiency by plotting $(\eta_{th}/\eta_{th,\,w/o\ reheat} - 1)100\%$ versus $q_{reheat}(kJ/kg)$.

D. Does the above rule of thumb hold true? Discuss your results.

Perform all calculations without the aid of a computer; the NIST online database, however, may be useful to obtain any required thermodynamic properties.

9.44 Use EES to solve Problem 9.43.

9.45–9.61 Rankine cycle with regeneration

9.45 Consider an ideal regenerative Rankine cycle utilizing an open feedwater heater, as shown in Fig. 9.17. The turbine inlet conditions are 10 MPa and 700 K, and the condenser pressure is 10 kPa. The mass flow rate of the steam entering the turbine is 12 kg/s. Steam is extracted between turbine stages at 2.2 MPa.

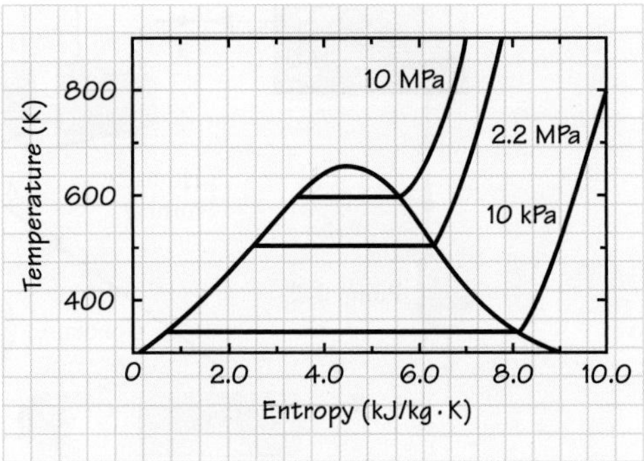

A. Determine the fraction of the total flow extracted from the turbine and the mass flow rate of the extracted steam.

B. Determine the power (kW) required to pump the condensate into the feedwater heater.

C. Determine the rate of heat addition (kW) in the steam generator section (state 2 to state 3).

D. Determine the power produced by the high-pressure portion of the turbine. Also determine the power delivered by the low-pressure turbine stages, after the extraction steam is removed.

E. Find the total power produced by the two turbine stages.

F. Determine the thermal efficiency.

9.46 Compare your results for the Rankine cycle with an open feedwater heater in Problem 9.45 with your results for the simple Rankine cycle with superheat in Problem 9.6A. Both problems consider Rankine cycles operating across the same high and low pressures and the same maximum temperature.

A. Compare the rate of heat addition in the steam generator on a per unit mass basis.

B. Compare the total power produced by the two turbine stages (per unit mass of total working fluid).

C. Compare the cycle efficiency for the two cycles.

9.47 Repeat Problem 9.45 for a closed feedwater heater where the extraction steam is pumped to the boiler pressure before being combined with the condenser feedwater, as shown in Fig. 9.18. Assume isentropic turbines and pumps.

9.48 Repeat Problem 9.45 for a closed feedwater heater where the extraction steam pressure is dropped and then combined with the condenser feedwater, as shown in Fig. 9.19. Assume isentropic turbines and pumps.

9.49 Consider the steam power plant shown in the sketch. This problem deals only with the turbines and the reheat process, with the key state points identified as follows:

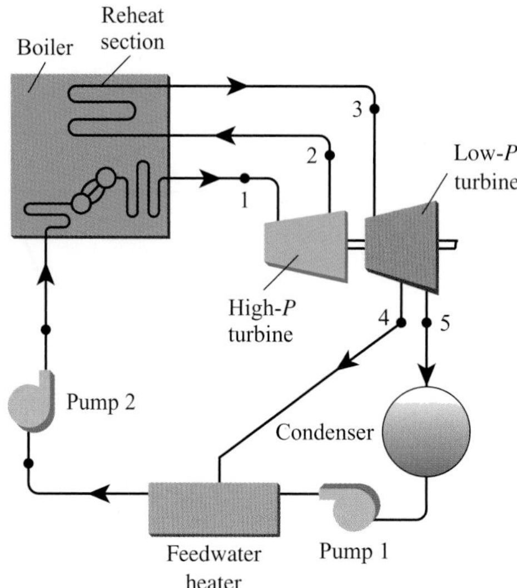

1. Superheated steam from the boiler enters the high-pressure turbine with a known flow rate $\dot{m}_1$ and a known enthalpy h_1.

2. The steam exits the high-pressure turbine and enters the reheat section of the boiler with a known enthalpy h_2.

3. The steam exits the reheater and enters the low-pressure turbine with a known enthalpy h_3.

4. A portion of the steam is extracted from the low-pressure turbine with a known enthalpy h_4. The flow rate $\dot{m}_4$ of the extracted steam is known.

5. The remaining steam exits the low-pressure turbine with a known enthalpy h_5.

A. Reproduce the above sketch and on the new sketch draw a control surface that will allow you to find the rate at which heat is added to the steam in the reheater section, $\dot{Q}_{cv,in}$. Use a dashed line to denote the control surface, and draw an arrow to represent $\dot{Q}_{cv,in}$.

B. Explicitly list your assumptions, and then simplify the general mass and energy conservation relationships as necessary to write a simplified expression that can be used to determine $\dot{Q}_{cv,in}$.

C. On your sketch also draw a control surface that will allow you to find the power output, $\dot{W}_{cv,out}$, of both turbines combined. Use a dashed line to denote the control surface, and draw an arrow to represent $\dot{W}_{cv,out}$.

D. Explicitly list your assumptions, and then simplify the general mass and energy conservation relationships as necessary to write a simplified expression that can be used to determine $\dot{W}_{cv,out}$. Treat $\dot{m}_1$ and $\dot{m}_4$ as known quantities, applying mass conservation as necessary to eliminate any other flow rate(s).

9.50 Consider a 600-MW regenerative-Rankine-cycle steam power plant with an open feedwater heater, as shown in Fig. 9.17. The boiler pressure is 8 MPa, the extraction pressure is 1.5 MPa, and the condenser pressure is 6 kPa. The high-pressure turbine inlet temperature is 820 K. The water leaving the condenser is saturated liquid. The exit state of the feedwater heater is saturated liquid. Determine the thermal efficiency of the power plant and the boiler mass flow rate (kg/s).

9.51 Repeat Problem 9.50 for a turbine having an isentropic efficiency of 90% and a pump having an isentropic efficiency of 85%.

9.52 Repeat Problem 9.50 for a closed feedwater heater where the extraction steam is pumped to the boiler pressure before being combined with the condenser feedwater, as shown in Fig. 9.18. Assume isentropic turbines and pumps.

9.53 Repeat Problem 9.50 for a closed feedwater heater where the extraction steam pressure is dropped and then the steam is combined with the condenser feedwater, as shown in Fig. 9.19. Assume isentropic turbines and pumps.

9.54 Consider a 600-MW regenerative-Rankine-cycle steam power plant with an open feedwater heater, as shown in Fig. 9.17. The boiler pressure is 1000 psia, the extraction pressure is 200 psia, and the condenser pressure is 1 psia. The high-pressure turbine inlet temperature is 1000 F. The water leaving the condenser is saturated liquid. The exit state of the feedwater heater is also saturated liquid. Determine the thermal efficiency of the power plant and the boiler mass flow rate (lb_m/hr).

9.55 Consider a 600-MW regenerative-Rankine-cycle steam power plant. The boiler pressure is 1000 psia, the extraction pressure is 200 psia, and the condenser pressure is 1 psia. The boiler exit temperature is 1000 F. The pressure in the open feedwater heater is 200 psia. The water leaving the condenser is saturated liquid. The turbines and the pumps have isentropic efficiencies of 0.90 and 0.85, respectively. Determine the thermal efficiency of the power plant and the boiler mass flow rate.

9.56 Repeat Problem 9.54 for a closed feedwater heater where the extraction steam is pumped to the boiler pressure before being combined with the condenser feedwater, as shown in Fig. 9.18. Assume isentropic turbines and pumps.

9.57 Repeat Problem 9.54 for a closed feedwater heater where the extraction steam pressure is dropped and then the steam is combined with the condenser feedwater, as shown in Fig. 9.19. Assume isentropic turbines and pumps.

9.58 A regenerative Rankine cycle operates with a first-stage turbine inlet state of 800 °C and 5 MPa and a second-stage exhaust pressure of 0.01 MPa. A single open feedwater heater, as shown in Fig. 9.17, is used with an extraction pressure of 0.7 MPa. Assume the exit states of the condenser and of the feedwater heater are both saturated liquid. The turbine and pump efficiencies are 100%. Determine the first-law thermal efficiency for the cycle and the total turbine work rate per unit boiler mass flow rate (kJ/kg).

9.59 Repeat Problem 9.58 for a turbine having an isentropic efficiency of 90% and a pump having an isentropic efficiency of 85%.

9.60 Repeat Problem 9.58 for a closed feedwater heater where the extraction steam is pumped to the boiler pressure before being combined with the condenser feedwater, as shown in Fig. 9.18. Assume isentropic turbines and pumps.

9.61 Repeat Problem 9.58 for a closed feedwater heater where the extraction steam pressure is dropped and the steam is then combined with the condenser feedwater, as shown in Fig. 9.19. Assume isentropic turbines and pumps.

9.62–9.82 Gas-turbine engines

9.62 Show that Eq. 9.18 can be obtained from a simplification of the expression preceding it in the text.

9.63 Consider a gas-turbine engine operating with 30 kg/s of air entering at 300 K and a pressure ratio of 12.5. Using an ideal air-standard cycle as a model of this engine, determine the ideal thermal efficiency and the net shaft power delivered to the load. Assume the turbine inlet temperature is 1050 K.

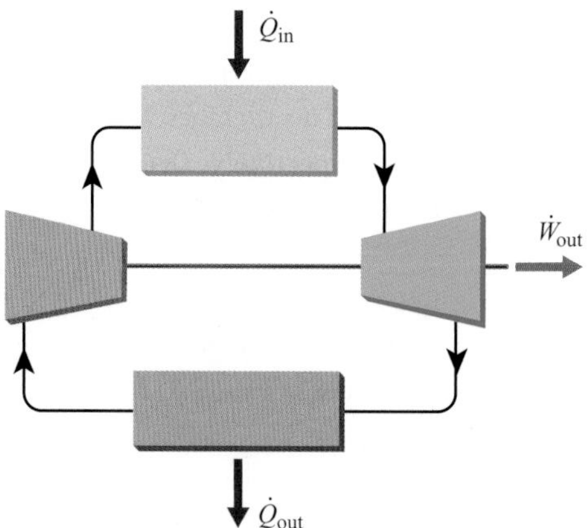

9.64 Repeat Problem 9.63, but assume a turbine inlet temperature of 1200 K.

9.65 Use EES to solve Problem 9.63 with a variable turbine inlet temperature. Plot the thermal efficiency as a function of the turbine inlet temperature from 900 K to 1200 K in increments of 50 K.

9.66 Repeat Problem 9.63, but use a pressure ratio of 15.

9.67 Use EES to solve Problem 9.63 with a variable pressure ratio. Plot the thermal efficiency as a function of pressure ratio from 10 to 15. Use six different values of the pressure ratio to generate a smooth curve.

9.68 Repeat Problem 9.63 assuming a turbine isentropic efficiency of 96% and a compressor efficiency of 95%. Compare your results with those from Problem 9.63.

9.69 Consider an air-standard Brayton cycle. The conditions at the compressor inlet are 100 kPa and 20 °C. The pressure ratio is 7:1 and the turbine inlet temperature is 800 °C. Determine (a) the compressor work (kJ/kg), (b) the heat added (kJ/kg), (c) the turbine work (kJ/kg), and (d) the cycle thermal efficiency.

9.70 The cycle shown in the sketch is used to cool the passenger cabin of a commercial aircraft. Air is the working fluid. Assuming an ideal compression process, determine the following:

A. The net work required (hp) per ton of refrigeration (1 ton = 12,000 Btu/hr)
B. The heat rejected at the heat exchanger (Btu/hr)
C. The turbine isentropic efficiency (%)
D. The coefficient of performance

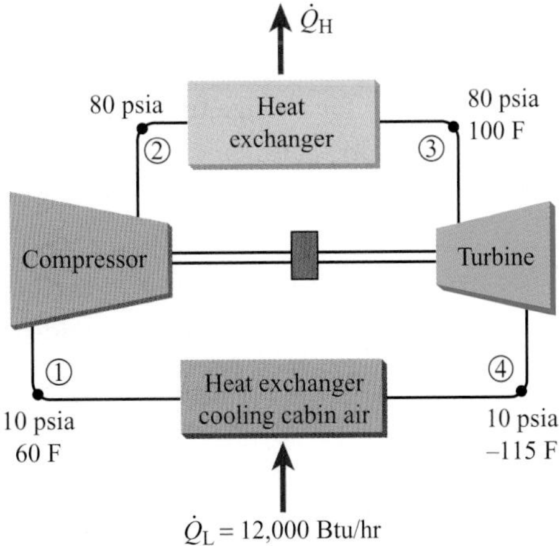

9.71 Consider a Brayton cycle using air as the working fluid. The air enters the compressor at 102 kPa and 15 °C and exits at 612 kPa. If the maximum cycle temperature is 800 °C, what is the ideal cycle efficiency?

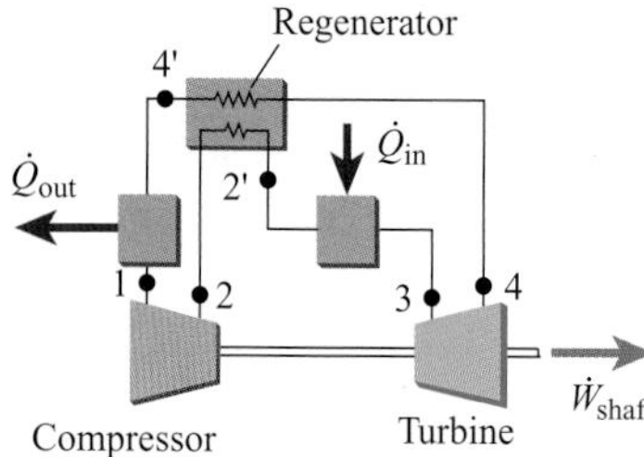

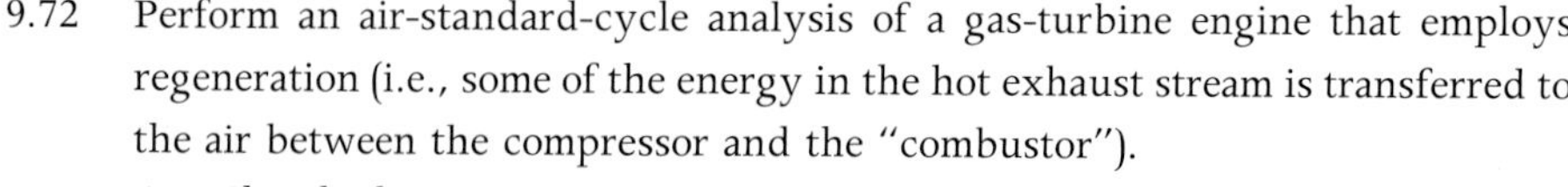

9.72 Perform an air-standard-cycle analysis of a gas-turbine engine that employs regeneration (i.e., some of the energy in the hot exhaust stream is transferred to the air between the compressor and the "combustor").

A. Sketch the regenerative cycle using T–s coordinates and label the state points.

B. From your analysis, derive a formula for the ideal cycle efficiency expressed in terms of temperatures.

C. Use your result from part B to calculate the thermal efficiency for an ideal cycle in which the air enters at 300 K, the regenerator preheats the air from 75 K, and the pressure ratio is 12.5. The turbine inlet temperature is 1050 K.

D. Compare your result from part C with the thermal efficiency calculated in Problem 9.63. Discuss.

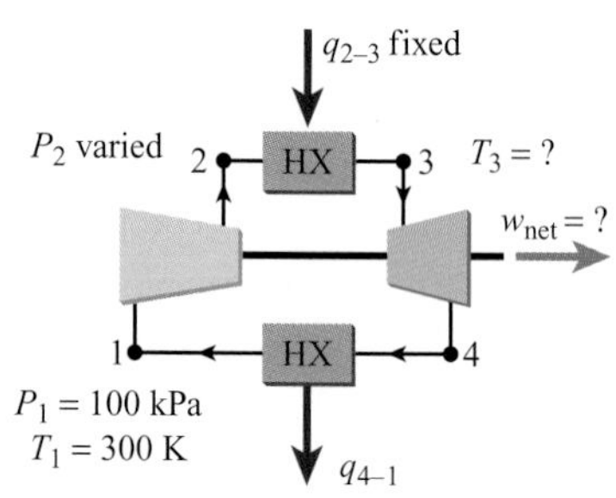

9.73 Consider an ideal air-standard Brayton cycle operating with inlet temperature (T_1) and pressure (P_1) of 300 K and 100 kPa, respectively. The heat added per unit mass, q_{2-3}, is 1050 kJ/kg. Assume the following constant values of the specific heats: $c_p = 1.005$ kJ/kg·K and $c_v = 0.718$ kJ/kg·K. Investigate the effects of the compressor pressure ratio (P_2/P_1) by performing the tasks listed. Use a spreadsheet to perform all repetitive calculations; however, make sure that you check at least one calculation by hand.

A. Sketch the cycle on T–s and P–v diagrams indicating the key state points with bold dots.

B. Calculate the turbine inlet temperature (T_3), the net work per unit mass, and the thermal efficiency as functions of the compressor pressure ratio for a range from 10 to 25. Create a single computer-generated plot of your results.

C. Discuss your results.

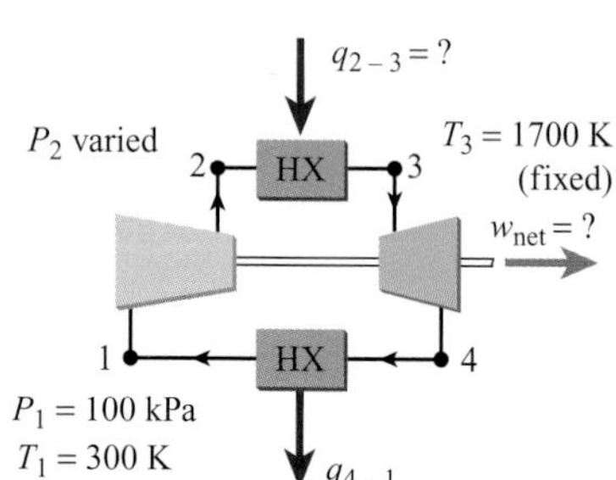

9.74 Consider the same situation described in Problem 9.73, but now the turbine inlet temperature is a given quantity ($T_3 = 1700$ K) and the heat added per unit mass, q_{2-3}, is unknown.

A. Calculate the heat added per unit mass, q_{2-3}, the net work per unit mass, w_{net}, and the thermal efficiency η_{th} as functions of the compressor pressure ratio for a range from 10 to 25. Create a single computer-generated plot of your results.

B. Discuss your results and compare them with the results of Problem 9.73.

9.75 Consider the situation described in Problem 9.74, but now neither the compressor nor the turbine is ideal. Assume that the isentropic efficiency of the compressor is 0.82 and the isentropic efficiency of the turbine is 0.87.

A. Sketch the cycle on T–s and P–v diagrams indicting the key state points with bold dots. Show both the isentropic and actual state points for conditions at the compressor and turbine outlets.

B. Calculate the heat added per unit mass, q_{2-3}, the net work per unit mass, w_{net}, and the thermal efficiency η_{th} as functions of the compressor pressure ratio for a range from 10 to 25. Also calculate the ratio of the cycle thermal efficiency to the ideal cycle thermal efficiency for the same pressure ratio. Create a single computer-generated plot of your results.

C. Discuss your results and compare them with the results of Problem 9.74. Use additional graphs as necessary.

9.76 Air enters the compressor of a 100-hp Brayton cycle at 530 R and 14.7 psia. The constant-pressure heating process occurs at 80 psia with a peak temperature of 2200 R. Heat is rejected at 530 R. Determine the thermal efficiency and the mass flow rate (lb_m/h) for the cycle.

9.77 A 75-kW Brayton cycle has a compressor inlet state at 293 K and 1 atm and a turbine inlet state at 1200 K and 4 atm. Determine the thermal efficiency and the mass flow rate (kg/hr) for both helium and air as the working fluid. Which fluid would be better? Explain.

9.78 Air enters the compressor of a Brayton cycle at 20 °C and 1 atm. The turbine inlet state is at 800 °C and 3.4 atm. Determine the net power per unit mass flow rate (kJ/kg) and the cycle thermal efficiency.

9.79 Solve Problem 9.78 using helium as the working fluid instead of air. Compare your results to the results of Problem 9.78.

9.80 Consider a 200-hp Brayton cycle using air. The compressor and turbine are both adiabatic but irreversible. There are pressure drops during heating and cooling. The following temperatures and pressures have been measured at the inlet states around the cycle:

State	T (R)	P (psia)
1	530	15.0
2	830	60.0
3	1960	58.8
4	1400	15.3

Assuming constant specific heats, determine (a) the thermal efficiency of the cycle and (b) the mass flow rate (lb_m/hr) of the air.

9.81 Air enters the compressor of a Brayton cycle at 250 K and 0.1 MPa. The pressure ratio is 4 to 1 and the maximum cycle temperature is 1300 K. The compressor and turbine efficiencies are 0.85. Using data from Appendix C.2 to account for variable specific heats, determine (a) the net specific work (kJ/kg), (b) the thermal efficiency, and (c) the ratio of compressor to turbine work.

9.82 Air enters the compressor of a 100-hp Brayton cycle at 530 R and 14.7 psia. The air is heated at 80 psia to 2200 R. The compressor and turbine efficiencies are each 0.90. After leaving the turbine, the air is cooled at 14.7 psia to 530 R. Determine the cycle thermal efficiency assuming constant specific heats.

9.83–9.88 Modified power cycles

9.83 Consider a cogeneration system using an ideal Rankine cycle, as shown in Fig. 9.35. The turbine inlet conditions are 10 MPa and 700 K, and the condenser pressure is 10 kPa. The mass flow rate of the steam entering the turbine is 12 kg/s. Steam is extracted between turbine stages at 3 kg/s and 2.2 MPa to be used for process heat in a paper mill. The condenser water is pumped to 2.2 MPa and is mixed with the process steam in the mixing chamber. The outflow of the mixing chamber is saturated water at 2.2 MPa.

A. Determine the power (kW) produced by the high- and low-pressure turbines.

B. Determine the process heat (kW) produced.

C. Determine the power (kW) required to pump the condensate into the mixing chamber (pump 1) and then to pump the feedwater to the boiler pressure (pump 2).

D. Determine the rate of heat addition (kW) in the steam generator section (state 2 to state 3).

E. Determine the utilization factor (the thermal efficiency of the cogeneration cycle).

F. (Optional; to be used if Problem 9.45 is also assigned.) Compare the utilization factor with the thermal efficiency in Problem 9.45. Why is the utilization factor for the cogeneration cycle greater?

9.84 A cogeneration plant as shown in Fig. 9.35 is designed to produce 600-MW power and 50 MW in process heat. The boiler pressure is 8 MPa, the extraction pressure is 1.5 MPa, and the condenser pressure is 6 kPa. The high-pressure turbine inlet temperature is 820 K. The condenser water is pumped to a pressure of 1.5 MPa before being mixed with the process water. The outflow of the mixing chamber is saturated liquid at 1.5 MPa.

A. Determine the energy per mass of steam (kJ/kg) passing through the high- and low-pressure turbines.

B. Determine the energy per mass of water (kJ/kg) required in the low- and high-pressure feedwater pumps.

C. Determine the flow rate (kg/hr) required through the boiler and in the process steam to produce 600 MW of power and 50 MW of process heat.

D. Determine the utilization factor for this cogeneration cycle.

9.85 A cogeneration plant as shown in Fig. 9.35 is designed to produce 600 MW power and 50 MW in process heat. The boiler pressure is 1000 psia, the extraction pressure is 200 psia, and the condenser pressure is 1 psia. The high-pressure turbine inlet temperature is 1000 F. The condenser water is pumped to a pressure of 200 psia before being mixed with the process water. The outflow of the mixing chamber is saturated liquid at 200 psia.

A. Determine the energy per mass of steam (Btu/lb_m) passing through the high- and low-pressure turbines.

B. Determine the energy per mass of water (Btu/lb_m) required in the low- and high-pressure feedwater pumps.

C. Determine the flow rate (lb_m/hr) required through the boiler and in the process steam to produce 600 MW of power and 50 MW of process heat.

D. Determine the utilization factor for this cogeneration cycle.

9.86 The combined cycle plant shown in Fig. 9.36 uses a gas-turbine engine operating with 250 kg/s of air entering at 300 K and a pressure ratio of 12.5. Use an ideal air-standard cycle as a model of this engine, and assume the turbine inlet temperature is 1400 K. The steam cycle operates in a simple Rankine cycle between 10 MPa and 70 kPa.

A. Determine the net power output from the gas-turbine cycle.

B. Determine the maximum mass flow rate in the steam cycle. Use a counter-flow heat exchanger between the gas-turbine cycle and the steam cycle. Assume that the maximum temperature of the steam is the temperature of the exhaust gas from the gas turbine. Be sure to check that the temperature of the steam is never greater than the local temperature of the gas in the heat exchanger.

C. Determine the net power output from the steam cycle.

D. What is the thermal efficiency for the combined cycle? Compare this to the thermal efficiency for the gas-turbine cycle alone.

9.87 The combined cycle plant shown in Fig. 9.36 uses a gas-turbine engine operating with 450 lb_m/s of air entering at 80 F and a pressure ratio of 12.5. Use an ideal air-standard cycle as a model of this engine, and assume the turbine inlet temperature is 2060 F. The steam cycle operates in a simple Rankine cycle between 1400 psi and 10 psi.

A. Determine the net power output from the gas-turbine cycle.

B. Determine the maximum mass flow rate in the steam cycle. Use a counter-flow heat exchanger between the gas-turbine cycle and the steam cycle. Assume that the maximum temperature of the steam is the temperature of the exhaust gas from the gas turbine. Be sure to check that the temperature of the steam is never greater than the local temperature of the gas in the heat exchanger.

C. Determine the net power output from the steam cycle.

D. What is the thermal efficiency for the combined cycle? Compare this to the thermal efficiency for the gas-turbine cycle alone.

9.88 The combined cycle plant shown in Fig. 9.36 uses a gas-turbine engine operating a Brayton cycle with air as the working fluid. The air enters the compressor at 102 kPa and 15 °C and exits at 612 kPa. The maximum cycle temperature is 800 °C and the air mass flow rate is 180 kg/s. The steam cycle operates in a simple Rankine cycle between 10 MPa and 70 kPa.

A. Determine the net power output from the gas-turbine cycle.

B. Determine the maximum mass flow rate in the steam cycle. Use a counter-flow heat exchanger between the gas-turbine cycle and the steam cycle. Assume that the maximum temperature of the steam is the temperature of the exhaust gas from the gas turbine. Be sure to check that the temperature of the steam is never greater than the local temperature of the gas in the heat exchanger.

C. Determine the net power output from the steam cycle.

D. What is the thermal efficiency for the combined cycle? Compare this to the thermal efficiency for the gas-turbine cycle alone.

9.89–9.92 Turbojet engines

9.89 A turbojet-powered aircraft is flying at 220 m/s and 8000-m altitude, where the ambient temperature and pressure are 236 K and 35 kPa, respectively.

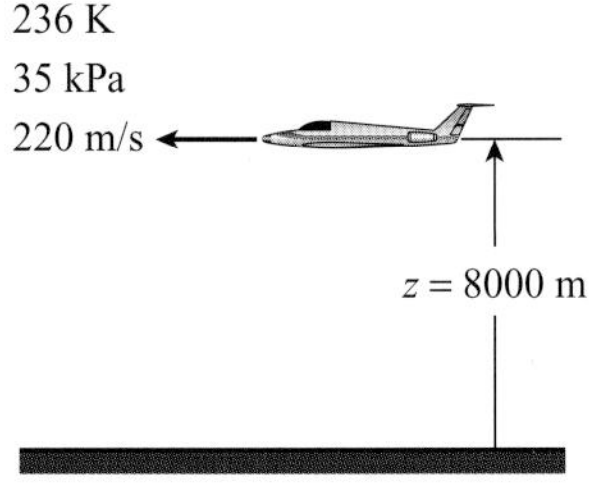

The pressure ratio of the compressor is 6 and the turbine inlet temperature is 970 K. Estimate the thrust developed by the engine when the air flow rate is 55 kg/s. Employ an air-standard-cycle analysis using the constant-pressure specific heat c_p and specific-heat ratio γ for air and neglecting the fuel properties.

9.90 A turbojet-powered aircraft is flying at 530 mph and 34,000-ft altitude where the ambient temperature and pressure are −62 F and 3.63 psia, respectively. The pressure ratio of the compressor is 6 and the turbine inlet temperature is 1200 F. Estimate the thrust developed by the engine when the air flow rate is 120 lb_m/s. Employ an air-standard-cycle analysis using the constant-pressure specific heat c_p and specific-heat ratio γ for air and neglecting the fuel properties.

9.91 A turbojet-powered aircraft is flying at 240 m/s and 7000-m altitude, where the ambient temperature and pressure are 240 K and 41 kPa, respectively. The pressure ratio of the compressor is 6 and the turbine inlet temperature is 970 K. At these conditions, the fuel flow rate is 0.68 kg/s and the overall air–fuel ratio is 75:1. Estimate the thrust developed by the engine. Also determine the specific thrust, the specific fuel consumption, and the propulsive efficiency. Employ an air-standard-cycle analysis using the constant-pressure specific heat c_p and specific-heat ratio γ for air and neglecting the fuel properties.

9.92 Consider the turbojet engine application presented in Example 9.10, with all conditions and properties the same except that now the compressor and turbine have isentropic efficiencies of 94% and 96%, respectively. Repeat the cycle analysis and performance calculations, creating a table as shown in Example 9.10. Also sketch the cycle on T–s coordinates. Use the actual values of temperature, but the values for specific entropy need only be qualitative. Discuss your results and compare them with those of the original example.

9.93–9.110 Other gas power cycles

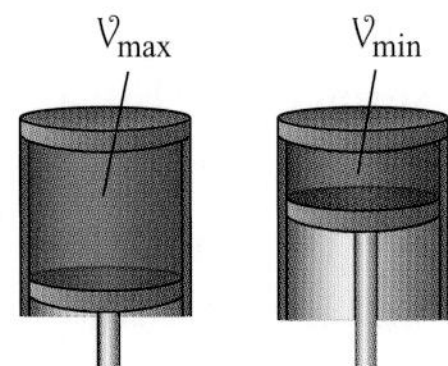

9.93 Consider a power cycle in which air is the working fluid. The air is contained in a piston–cylinder assembly and undergoes the following processes:

1–2: Internally reversible, adiabatic compression from the maximum volume V_{max} to the minimum volume V_{min}, where $V_{max} = 10V_{max}$.

2–3: Constant-volume heat addition to a specified value of $T_3 = 1000$ K.

3–4: Constant-pressure heat addition until the volume equals $3V_{min}$.

4–5: Internally reversible, adiabatic expansion from V_4 $(= 3V_{min})$ to the maximum volume V_{max}.

5–1: Constant-volume heat rejection to return to the initial state.

A. Carefully sketch the cycle using P–V and T–s coordinates. Label the state points 1, 2, 3, etc.

B. Given the following properties for air, complete the table below,

$c_p = 1.008$ kJ/kg·K (a constant value),

$c_v = 0.721$ kJ/kg·K (a constant value),

$R_{air} = 0.287$ kJ/kg·K,

State	1	2	3	4
P (kPa)	100		3333	
$\mathcal{V}$ (m^3)	0.001	0.0001	0.0001	0.0003
T (K)	300		1000	
m (kg)				

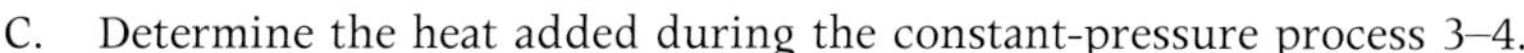

C. Determine the heat added during the constant-pressure process 3–4.

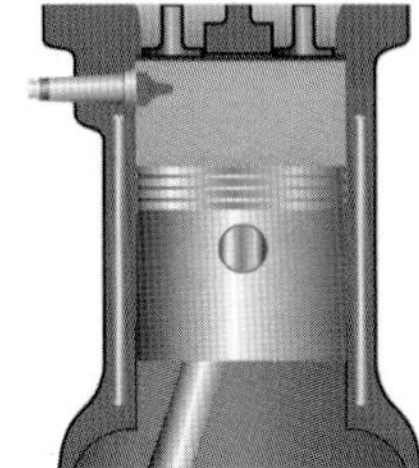

9.94 An air-standard Otto cycle is often used as a very simplified model of a spark-ignition engine. The following sequence of processes applied to a fixed mass of air in a piston–cylinder assembly constitutes the Otto cycle:

1–2: Internally reversible, adiabatic compression from the maximum volume $\mathcal{V}_{max}$ to the minimum volume $\mathcal{V}_{min}$.

2–3: Constant-volume heat addition to a peak cycle temperature T_3.

3–4: Internally reversible, adiabatic expansion from $\mathcal{V}_{min}$ to $\mathcal{V}_{max}$.

4–1: Constant-volume heat rejection to return to the initial state.

Sketch this cycle using P–$\mathcal{V}$ and T–S coordinates, labeling each state point (1, 2, 3, and 4).

9.95 Consider a fixed mass of air trapped in a piston–cylinder assembly. The air-standard diesel cycle is defined by the following sequences of processes applied to this system:

1–2: Internally reversible, adiabatic compression from the maximum volume $\mathcal{V}_{max}$ to the minimum volume $\mathcal{V}_{min}$.

2–3: Constant-pressure heat addition to a peak cycle temperature T_3. Note that $\mathcal{V}_{min} < \mathcal{V}_3 < \mathcal{V}_{max}$.

3–4: Internally reversible, adiabatic expansion to $\mathcal{V}_{max}$.

4–1: Constant-volume heat rejection to return to the initial state. Sketch this cycle using P–$\mathcal{V}$ and T–S coordinates, labeling each state point (1, 2, 3, and 4).

9.96 Consider the air-standard Otto cycle described in Problem 9.94 operating with a compression ratio of 8 $(= \mathcal{V}_{max}/\mathcal{V}_{min})$. The air just prior to compression is at 293 K and 1 atm. The maximum cycle temperature is 2500 K. Assuming constant specific heats, determine the temperature (K) and pressure (atm) at the start of each process and the thermal efficiency for the cycle.

9.97 Consider the air-standard Otto cycle described in Problem 9.94 operating with a compression ratio of 10 $(= \mathcal{V}_{max}/\mathcal{V}_{min})$. The air just prior to compression is at 70 F and 14.7 psia. The maximum cycle temperature is 4500 F. Assuming constant specific heats, determine the temperature and pressure at the beginning of each process and the thermal efficiency for the cycle.

9.98 An old V-8 spark-ignition engine has a 4-in-diameter cylinder and a 4-inch-long stroke. The intake conditions are atmospheric (70 F and 14.7 psia). The engine has a compression ratio of 9. (See Appendix 3A to review the geometrical relationships.) The peak cycle temperature is 4000 F. Assuming constant specific heats and using the Otto cycle as a model, determine the following quantities:

A. The net work (Btu) per cylinder per cycle

B. The pressure (psia) after compression

C. The peak pressure (psia) after combustion

D. The mean effective pressure (psia) (i.e., the net work divided by displacement volume)

E. The first-law thermal efficiency of the engine

F. The horsepower output of the engine running at 2000 rev/min, taking into account that, in such a four-stroke engine, two revolutions of the crankshaft are required for each power stroke

G. The specific fuel consumption [i.e., the mass flow rate (lb_m/hr) of fuel used per horsepower], assuming that the heating value of the fuel in this case is 19 kBtu/lb_m of fuel

9.99 Consider the air-standard diesel cycle described in Problem 9.95 operating with a compression ratio of 15 $(= \mathcal{V}_{max}/\mathcal{V}_{min})$. The air just prior to compression is at 20 °C and 0.1014 MPa. The maximum cycle temperature is 2500 K. Assuming constant specific heats, determine the temperature and pressure at the beginning of each process and the thermal efficiency for the cycle.

9.100 Determine the efficiency for the air-standard cycle indicated in the sketch. Assume the pressures and temperatures are known quantities.

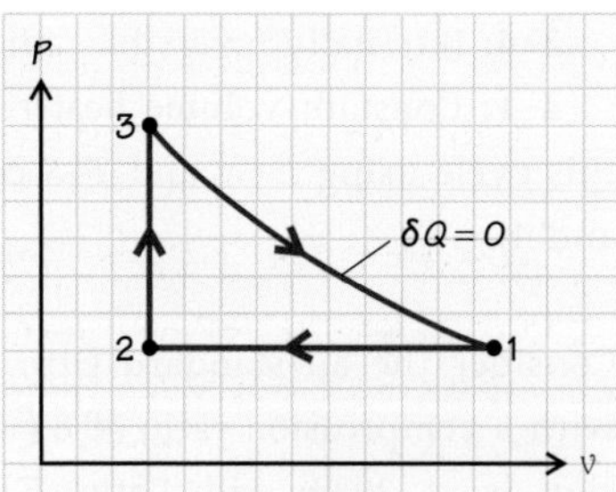

9.101 Rework Problem 9.100 for the following cycle.

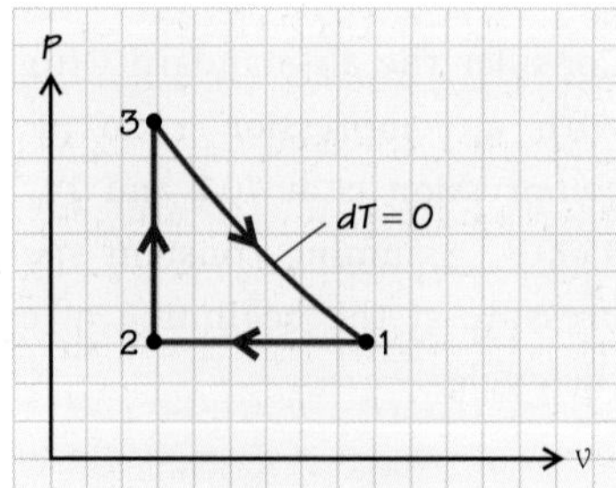

9.102 Rework Problem 9.100 for the following cycle.

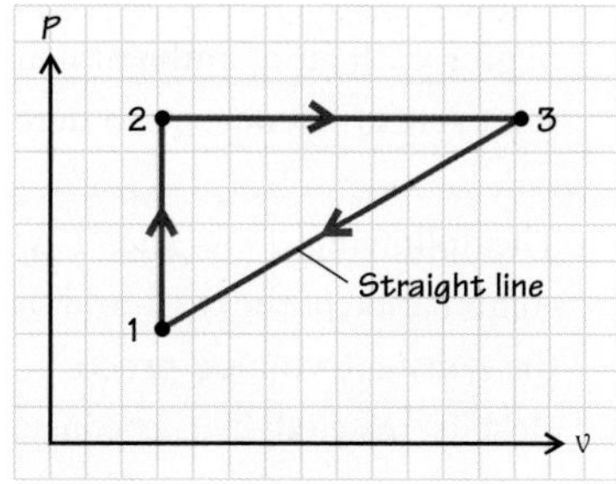

9.103 Express the thermal efficiency of a Carnot-cycle heat engine in terms of the isentropic compression ratio $\left(r_{\text{isen}} \equiv \mathcal{V}_{\text{large}}/\mathcal{V}_{\text{small}}\right)$.

9.104 An inventor proposes a reversible nonflow cycle using air. The cycle consists of the following three processes:

1–2: Constant-volume compression from 101 kPa and 15 °C to 700 kPa.

2–3: Constant-pressure heat addition during which the volume is tripled.

3–1: A process that appears as a straight line on a P–$\mathcal{V}$ diagram.

Draw P–$\mathcal{V}$ and T–S diagrams for the cycle. Compute the net work of the cycle in kJ/kg and Btu/lb$_m$.

9.105 In a compression-ignition engine, air originally at 120 F is compressed to a temperature of 980 F. Compression obeys the relationship $P\mathcal{V}^{1.34}$ = constant. Determine (a) the compression ratio required (i.e., the ratio of the volume before compression to the volume after compression), (b) the work of compression (Btu/lb$_m$), and (c) the heat transfer (Btu/lb$_m$).

9.106 In a compression-ignition engine, air originally at 50°C is compressed to a temperature of 550 °C. The compression obeys the relationship $P\mathcal{V}^{1.34}$ = constant. Determine (a) the compression ratio required (i.e., the ratio of the volume before compression to the volume after compression), (b) the work of compression (kJ/kg), and (c) the heat transfer (kJ/kg).

9.107 The compression stroke for a four-stroke-cycle spark-ignition engine is approximated as a reversible adiabatic process. Assume that the cylinder volume at the bottom center position is 400 in^3, the compression ratio $(= \mathcal{V}_1/\mathcal{V}_2)$ is 9, and the cylinder is initially charged with air at 15 psia and 90 F. Determine (a) the temperature and pressure of the air after compression and (b) the horsepower required for this compression process if the engine speed is 2000 rev/min. (Note: In a four-stroke cycle, there is one compression stroke for every two crankshaft revolutions.)

9.108 Consider the air-standard Otto cycle described in Problem 9.94. A particular cycle operates with a compression ratio of 7:1 and has a heat input of 2100 kJ/kg. The pressure and temperature at maximum volume before compression (state 1) are 100 kPa and 15 °C, respectively. Determine the net work produced by the cycle.

9.109 Consider the air-standard Otto cycle described in Problem 9.94, operating with a compression ratio of 8:1. In the constant-volume heat-addition process (state 2–state 3), 1800 kJ/kg of energy is transferred to the air. If the cycle begins with air at ambient conditions of 100 kPa and 15 °C (state 1), determine the cycle efficiency. Also determine the heat rejected to the atmosphere.

9.110 The following P–v and T–s diagrams illustrate the so-called dual cycle, a combination of the Otto- and diesel-cycle processes. Heat addition occurs in both the constant-volume process, 2–3, and in the constant-pressure process, 3–4. Heat is rejected in the constant-volume process, 5–1. The compression and expansion processes, 1–2 and 4–5, respectively, are isentropic. Assuming that air is the working fluid, determine the cycle thermal efficiency η_{dual} as a function of the following three parameters: $r_v(\equiv v_1/v_2)$, $R_p(\equiv P_3/P_2)$, and $\beta(\equiv v_4/v_3)$.

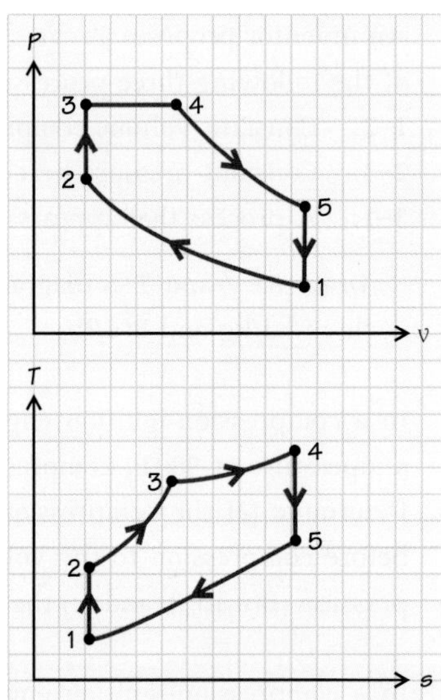

9.111–9.141 Refrigerators and heat pumps

9.111 A reversible refrigerator removes energy from one thermal reservoir at 4 °C and delivers energy to another thermal reservoir at 26 °C. Determine the coefficient of performance for this device. Also determine the coefficient of performance for a reversible heat pump operating between the same two reservoir temperatures.

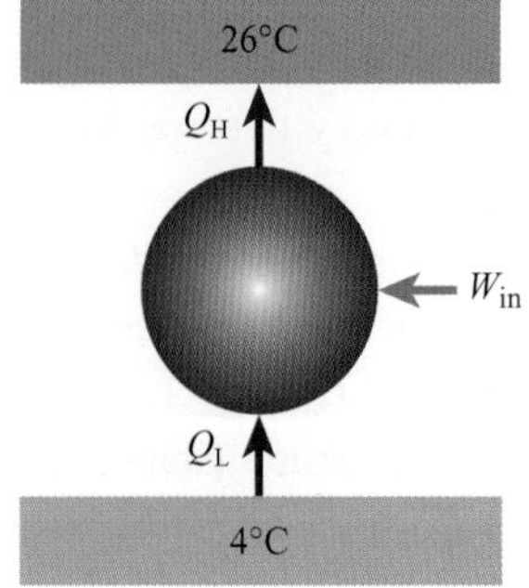

9.112 Plot the Carnot coefficient of performance for a heat pump delivering energy into a thermal reservoir at 294.2 K (70 F) as a function of the temperature of the cold reservoir. Use a range of cold reservoir temperatures from 255 K (~0 F) to 290 K (~62 F).

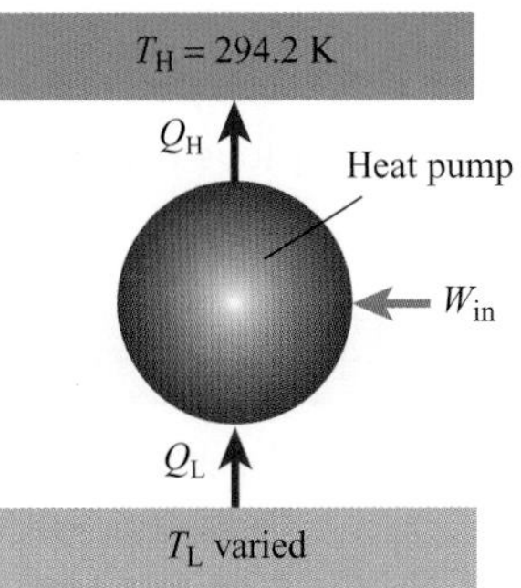

9.113 A heat pump uses the ground as a source to deliver 20,000 Btu/hr to heat a home. The heat pump coefficient of performance is 3.23. Determine (a) the electrical power required to operate the heat pump in steady state, (b) the operating cost per hour if electricity is purchased at 6.2 cents/kW·hr, and (c) the rate at which energy is removed from the ground expressed in both kW and Btu/hr.

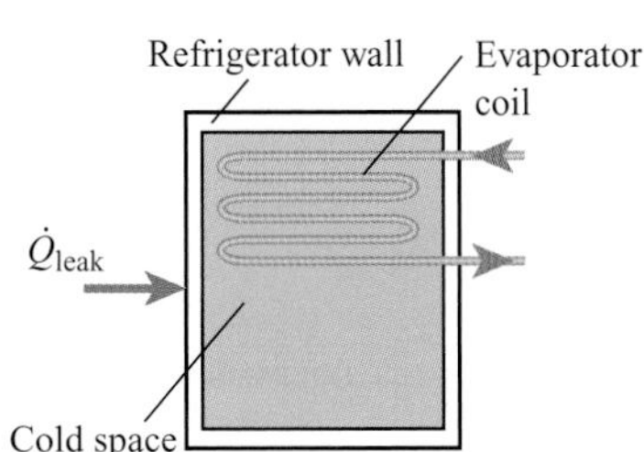

9.114 Heat leaks from the air in a kitchen through the walls of a refrigerator into the refrigerated cold space at a rate of 1.43 kW. Determine the electrical power required to maintain a steady temperature in the cold space if the refrigerator coefficient of performance is 2.6. Also determine the rate at which heat is transferred to the kitchen air from the refrigerator condenser coil and the net rate at which energy is transferred from the entire refrigerator to the air in the kitchen.

9.115 Consider an ideal vapor-compression cycle using R-134a and operating between pressures of 0.30 and 1.68 MPa. The refrigerant flow rate is 0.035 kg/s. Plot the cycle using T–s coordinates with the aid of the NIST plotting software and calculate the following quantities: $\dot{Q}_H$, $\dot{Q}_L$, $\dot{W}_{in}$, β_{refrig}, and $\beta_{heat\ pump}$.

9.116 Repeat Problem 9.115 using R-22. Compare your results with those of Problem 9.115.

9.117 Consider an ideal vapor-compression cycle using R-134a and operating between pressures of 0.28 and 1.75 MPa. Determine the coefficient of performance for the cycle for application (a) in a refrigerator and (b) in a heat pump.

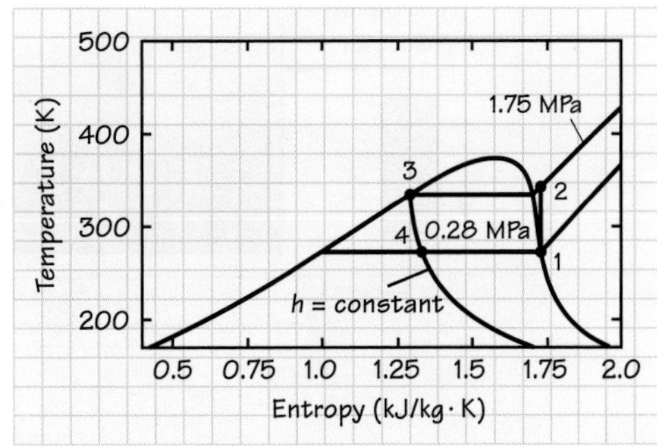

9.118 Repeat Problem 9.117 using EES. Modify your EES program to find the coefficient of performance between pressures of 0.28 and 1.5 MPa. Plot T–s diagrams for the original cycle conditions and for the new conditions. Use the two T–s diagrams to explain the change in the coefficient of performance.

9.119 Consider a vapor-compression-cycle heat pump that uses R-134a as the working fluid. The flowrate of the R-134a is 0.08 kg/s. The temperatures and pressures at various points in the cycle are as follows:

State	Location	T(K)	P(MPa)
1	Compressor inlet/evaporator outlet	278	0.29
2	Compressor outlet/condenser inlet	332	1.10
3	Condenser outlet/expansion value inlet	314	1.10
4	Expansion value outlet/evaporator inlet	—	0.29

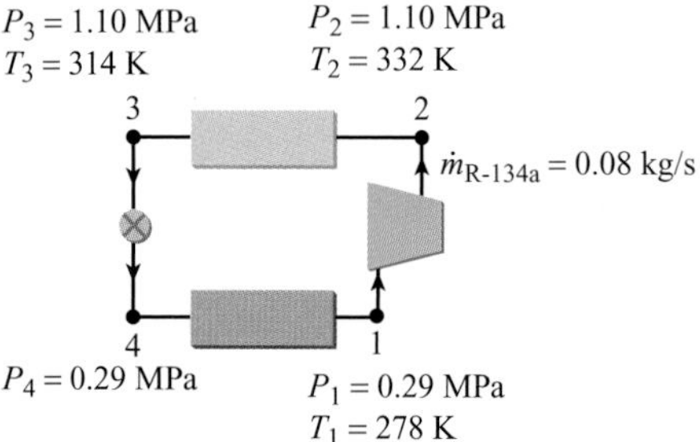

A. Sketch the cycle on a T–s diagram, exaggerating as necessary to show how the cycle deviates from the ideal cycle.
B. Determine the power input to the compressor.
C. Determine the isentropic efficiency of the compressor.
D. Determine the heat pump coefficient of performance.

9.120 Consider the heat-pump-heated swimming pool discussed in Example 9.14. The rectangular (7.6 m × 12 m) pool is filled to an average depth of 2.3 m. Estimate the rate at which the temperature of the pool water increases for steady operation of the heat pump. Express your result in K/s and in F/hr. Explicitly list your assumptions.

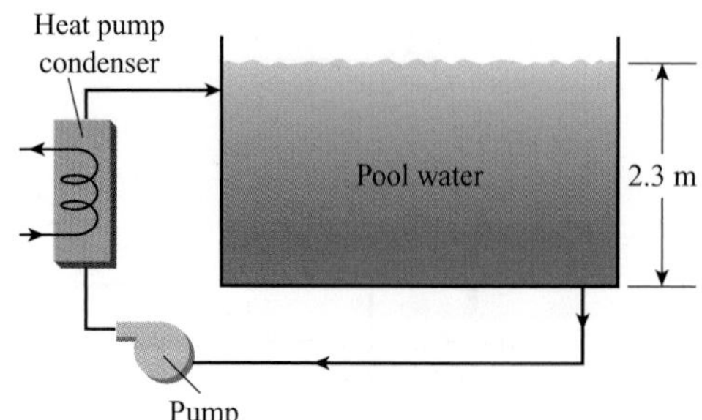

9.121 Consider the heat pump described in Example 9.14. The heat pump now operates between 0.60 MPa and 1.4 MPa. Plot the vapor-compression cycle in T–s coordinates (use NIST) and determine the cycle coefficient of performance and the mass flow rate of the R-22.

9.122 Determine the heat pump coefficient of performance for an ideal vapor-compression refrigeration cycle operating between 0.653 and 1.342 MPa. The working fluid is R-22. Compare your result with that for the real cycle described in Example 9.14.

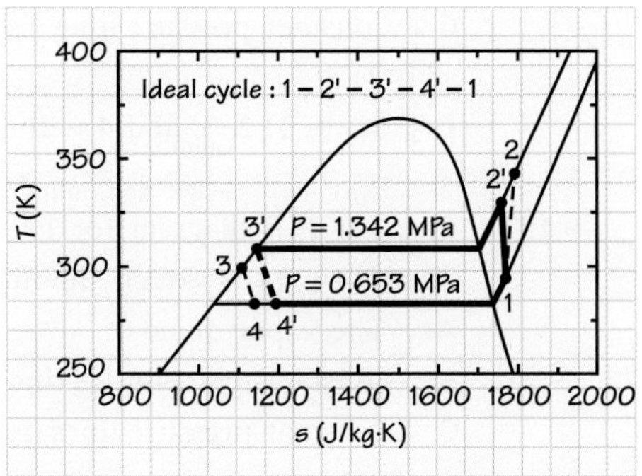

9.123 Determine the influence of the evaporator temperature on the heat pump coefficient of performance for an ideal vapor-compression cycle. The working fluid is R-22. The pressure is fixed in the condenser at 1.342 MPa. Vary the evaporator temperature over a reasonable range that includes $T_{sat}(P_{sat} = 0.653\,\text{MPa})$, the condition specified in Problem 9.122. Discuss your results.

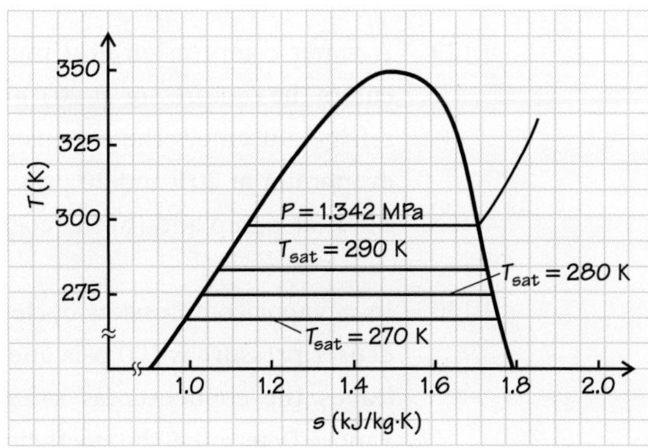

9.124 An architect wants to use an electrical work input to keep the temperature inside a building at 20 °C when the outside temperature is –20 °C. Determine the maximum heat-transfer rate (kW) that can be supplied to the building (per kW of electrical input) by (a) a resistance heater; (b) a vapor-compression cycle using R-134a with T_{evap} = –20 °C and T_{cond} = 20 °C, saturated vapor leaving the evaporator, and saturated liquid leaving the condenser; and (c) the best possible method.

9.125 A 1-ton (200 Btu/min) vapor-compression refrigerator uses R-134a as the refrigerant. Evaporation occurs at 10 F and condensation occurs at 100 F. The compressor is reversible and adiabatic. The refrigerant leaves the evaporator at 20 F and leaves the condenser at 95 F. The heating and cooling processes are both at constant pressure. Determine the steady-state power (hp) into the compressor.

9.126 Create a *T–s* diagram for Problem 9.125 using the NIST software plotting capability. Your sketch should show the following:

A. The vapor dome

B. Constant-temperature lines (isotherms) for 10, 20, 95, and 100 F

C. Constant-pressure lines (isobars) for evaporation at 10 F and condensation at 100 F

D. States 1, 2, 3, and 4 with state 1 as that at the compressor inlet

9.127 Create a *P–h* diagram for Problem 9.125 using the NIST software plotting capability. Your sketch should show the following:

A. The vapor dome

B. Constant-temperature lines (isotherms) for 10, 20, 95, and 100 F

C. Constant-pressure lines (isobars) for evaporation at 10 F and condensation at 100 F

D. States 1, 2, 3, and 4 with state 1 at the compressor inlet

9.128 A vapor-compression refrigerator using R-134a has a maximum pressure of 120 psia and a minimum pressure of 15 psia. The R-134a leaves the condenser as a saturated liquid and leaves the evaporator as a saturated vapor. The compressor is reversible and adiabatic. Determine (a) the maximum room temperature, (b) the minimum food-compartment temperature, and (c) the coefficient of performance.

9.129 A vapor-compression heat pump circulates R-134a at 300 lb_m/hr. The R-134a enters the compressor at 20 F and 20 psia and leaves at 180 F and 180 psia. The refrigerant enters the expansion valve at 100 °F and 160 psia and leaves the evaporator at 20 F and 20 psia. Determine (a) the heating capacity (kBtu/hr) of the heat pump and (b) the coefficient of performance.

9.130 In a vapor-compression cycle using R-134a as the working fluid, the condenser pressure is 100 psia and the evaporator pressure is 20 psia. The R-134a leaves the condenser as saturated liquid and the evaporator as saturated vapor. Determine the coefficient of performance both as an air conditioner and as a heat pump.

9.131 In a vapor-compression refrigeration cycle that circulates R-134a at a rate of 5 kg/min, the refrigerant enters the compressor at −10 °C and 0.18 MPa and leaves at 70 °C and 0.7 MPa. The R-134a leaves the condenser as saturated liquid at the compressor outlet pressure. Determine the refrigeration capacity (kW) and the coefficient of performance. Also suggest two modifications to this refrigeration cycle to increase its coefficient of performance. Use *T–s* diagrams or other means to illustrate how your suggestions will help.

9.132 Repeat Problem 9.131 using EES. Find the coefficient of performance and generate T–s diagrams for the original cycle conditions and your two modified cycles having an improved coefficient of performance.

9.133 The temperatures and pressures shown in the table were measured for a vapor-compression refrigerator that uses R-134a as the refrigerant.

State	T (°C)	P (MPa)
1	−30	0.08
2	60	0.60
3	15	0.60
4		0.08

The refrigerator operates between a cold region at T_C and a hot region at T_H. The positive-displacement compressor (piston–cylinder device) has a cylinder volume and operating speed such that it processes 1 m^3/min of R-134a at the inlet of the compressor.

A. Determine the minimum value of T_C and the maximum value of T_H.
B. Determine the coefficient of performance.
C. Determine the heat-transfer rate from the region at T_C.

9.134 A vapor-compression refrigerator using R-134a is to provide chilled water at 40 F and 14.7 psia. Originally the water is at 70 F and 14.7 psia. The pressure in the evaporator is 40 psia and that in the condenser is 100 psia. The R-134a enters the compressor at 40 F and leaves at 120 F. The power supplied to the compressor is 1 hp. The R-134a leaves the condenser as saturated liquid. Determine the coefficient of performance for the refrigerator and the mass flow rate (lb_m/hr) of chilled water that can be produced.

9.135 In a vapor-compression cycle, evaporation occurs at −10 °C and condensation occurs at 40 °C. The compressor is reversible and adiabatic. R-134a exits the evaporator at −5 °C and exits the condenser at 38 °C. Determine the work into the compressor per unit mass of R-134a (kJ/kg).

9.136 The capacity and compressor-work characteristics for a 3-ton (600-Btu/min) vapor-compression air conditioner are given in the table, along with the seasonal house-cooling needs. If electricity costs 5 cents/kW·hr, determine the seasonal operating cost and the seasonal coefficient of performance.

Ambient temperature (F)	Number of hours	Cooling load (Btu/hr)	Compressor power at capacity (kW)	Air-conditioner capacity (Btu/hr)
100	100	36,000	5.8	37,000
90	500	24,000	5.5	39,000
80	1200	12,000	5.2	41,000

9.137 A computer facility in the Sahara Desert is to be maintained at 15 °C by a vapor-compression refrigeration system that uses water as the refrigerant. The water leaves the evaporator as a saturated vapor at 10 °C. The compressor is reversible and adiabatic. The pressure in the condenser is 0.01 MPa and the water is saturated liquid as it leaves the condenser. Determine (a) the coefficient of performance for the cycle and (b) the power input per unit mass flow rate of water (kJ/kg).

9.138 A 3-ton (600-Btu/min) vapor-compression refrigeration cycle using R-134a as the refrigerant maintains a freezer compartment at 0 F. The room air is at 70 F. Satisfactory operation requires a minimum temperature difference between the room air and the condensing R-134a of 10 F. Similarly, a 10-F minimum temperature difference between the evaporating R-134a and the freezing compartment is required. The R-134a enters the compressor ($\eta_{isen} = 0.85$) as a saturated vapor. The R-134a leaves the condenser as a saturated liquid.

A. Perform a complete thermal analysis of the cycle.

B. Determine the horsepower required per ton of refrigeration (hp/ton). That is, determine $\dot{W}/\dot{Q}_e$, where $\dot{W}$ is the power supplied to the compressor (hp) and $\dot{Q}_e$ is the heat-transfer rate to the evaporator (tons). (Note: 1 ton of refrigeration = 200 Btu/min.)

9.139 A 3-ton (600-Btu/min) vapor-compression refrigeration cycle uses R-134a to maintain a freezing compartment at 0 F. The room air is at 70 F. The R-134a condenses at 80 F and evaporates at −10 F. It enters the compressor with 10 F of superheat. The compressor efficiency is 0.85. It leaves the condenser with 10 F of subcooling.

A. Perform a complete thermal analysis of the cycle.

B. Determine the horsepower required per ton of refrigeration (hp/ton). That is, determine W/Q_e, where W is the power supplied to the compressor (hp) and Q_e is the heat-transfer rate to the evaporator (tons). (Note: 1 ton of refrigeration = 200 Btu/min.)

9.140 An old 10-kW vapor-compression heat pump using R-12 maintains a building at 20 °C. The inlet state to the compressor is saturated vapor at −30 °C. The compressor exit state is at 50 °C and 0.7 MPa. The inlet state to the expansion valve is saturated liquid at 25 °C. Perform a complete thermal analysis of the heat pump.

9.141 A Carnot refrigerator removes energy at a rate of 300 Btu/hr from a low-temperature thermal reservoir at −160 F. The high-temperature reservoir for the refrigerator is the atmosphere at 40 F. A Carnot engine operating between a reservoir at 1140 F and the atmosphere (40 F) drives this Carnot refrigerator.

A. At what rate must heat be supplied (Btu/hr) to the Carnot engine from the 1140 F reservoir?

B. What is the power input to the Carnot refrigerator (Btu/hr)?

C. What are the coefficients of performance of the Carnot refrigerator and the thermal efficiency of the Carnot engine?

9.142–9.154 FE exam questions

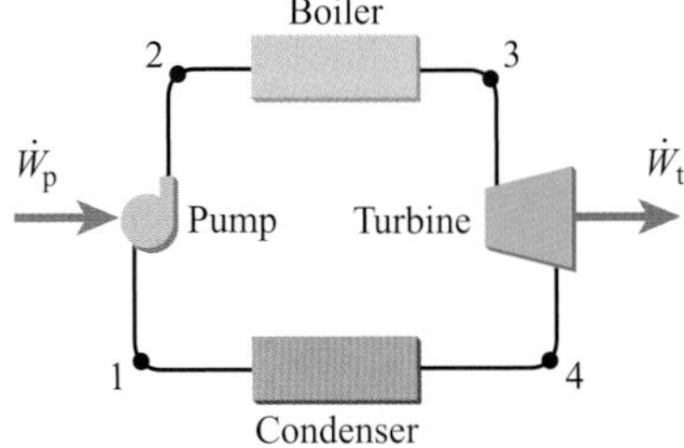

Rankine cycle with superheat Consider a Rankine-cycle steam power plant with superheat having a flow rate of 5 kg/s. The temperatures and pressures at the inlet states of the four basic components are given in the table. The turbine and the pump are both adiabatic.

State	Location	T (K)	P (kPa)	h (kJ/kg)
1	Pump inlet	333.21	20	251.42
2	Boiler inlet	333.41	5000	256.48
3	Turbine inlet	700	5000	3261.9
4	Condenser inlet	333.21	20	2221.1

9.142 For the conditions above, determine the rate of heat transfer in the boiler. a. 3 MW, b. 15 MW, c. 1.5 MW, d. 30 MW.

9.143 For the conditions above, determine the net power output. a. 5.20 MW, b. 1.04 MW, c. 5.18 MW, d. 25 kW.

9.144 For the conditions above, determine the thermal efficiency of the cycle. a. 34.5%, b. 15.3%, c. 36.7%, d. 16.8%.

Rankine cycle with reheat Consider an ideal Rankine cycle with superheat and reheat having a flow rate of 8 kg/s. The maximum temperature of the superheated and reheated steam is 720 K. The maximum pressure in the cycle is 8 MPa and the minimum pressure is 5 kPa. Steam enters the high-pressure turbine at 720 K and exits as saturated vapor. The steam is then reheated to 720 K before expanding in the low-pressure turbine. The steam flow rate is 90 kg/s.

State	Location	T (K)	P (kPa)	h (kJ/kg)
1	Pump inlet	306.02	5	137.75
2	Boiler inlet	306.22	8000	145.77
3	High-pressure turbine inlet	720	8000	3265.2
4	Reheat to boiler	457.93	1118	2781.2
5	Low-pressure turbine inlet	720	1118	3363
6	Condenser inlet	306.02	5	2304.8

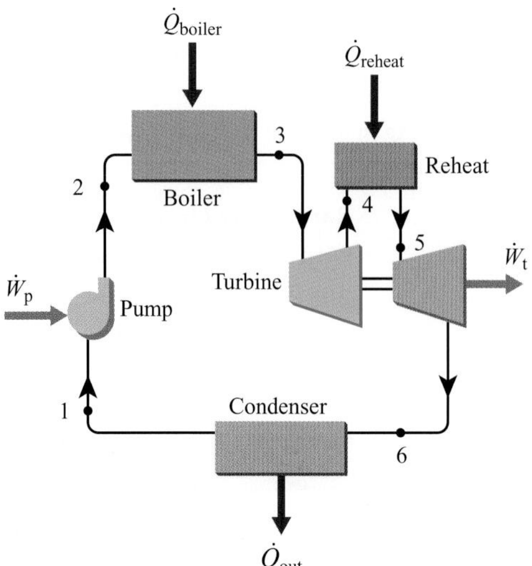

9.145 For the conditions above, determine the rate of heat transfer during reheat, $\dot{Q}_{reheat}$. a. 3.12 MW, b. 25.0 MW, c. 582 kW, d. 4.65 MW.

9.146 For the conditions above, determine the net power output. a. 16.1 MW, b. 7.7 MW, c. 2.01 MW, d. 8.5 MW.

9.147 For the conditions above, determine the thermal efficiency of the cycle. a. 55.5%, b. 54.3%, c. 34.7%, d. 64.5%.

Brayton cycle Consider a gas-turbine engine operating with 20 kg/s of air entering at 300 K and a pressure ratio of 11. Using an ideal air-standard cycle as a model of this engine, determine the ideal thermal efficiency and the net shaft power delivered to the load. Assume the turbine inlet temperature is 1050 K.

9.148 For this Brayton cycle, find the temperature of the air entering the combustor. a. 750 K, b. 529 K, c. 595 K, d. 1050 K.

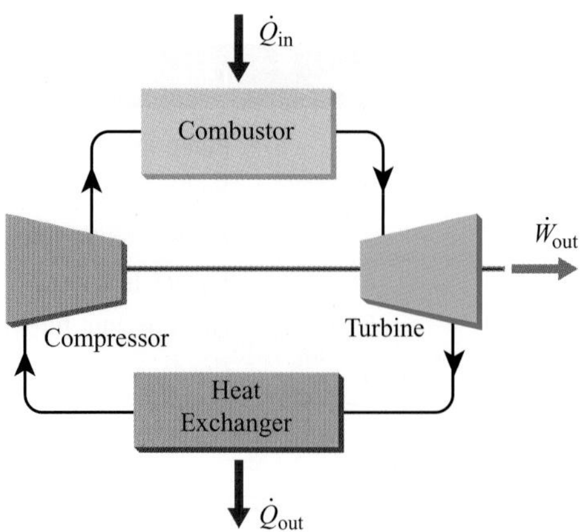

9.149 For the above Brayton cycle, find the net power output from the gas-turbine engine. a. 5.9 MW, b. 1.5 MW, 4.9 MW, 4.5 MW.

9.150 For the above Brayton cycle, determine the ideal thermal efficiency. a. 49.6%, b. 60.1%, c. 42.2%, d. 35.4%.

Refrigeration cycle Consider an ideal vapor-compression cycle using R-134a and operating between pressures of 0.28 and 1.75 MPa. The flow rate in the refrigeration cycle is 0.5 kg/s. Determine the coefficient of performance for the cycle for application (a) in a refrigerator and (b) in a heat pump.

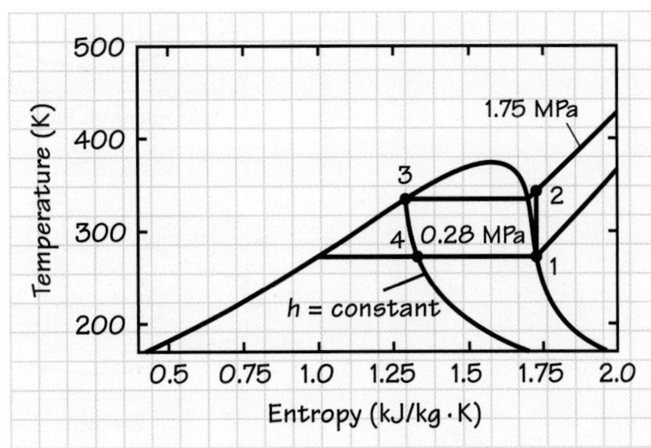

State	T (K)	P (MPa)	h (kJ/kg)
1	271.92	0.28	397.89
2	341.42	1.75	435.98
3	334.84	1.75	290.27
4	271.92	0.28	290.27

9.151 For this refrigeration cycle, find the rate of heat transfer in the evaporator. a. 72.9 kW, b. 108 kW, 53.8 kW, d. 19.0 kW.

9.152 For the above refrigeration cycle, find the power required to operate the compressor. a. 72.9 kW, b. 108 kW, 53.8 kW, d. 19.0 kW.

9.153 For the above refrigeration cycle, what is the coefficient of performance when the cycle is operating as a refrigerator? a. 3.83, b. 2.83, c. 0.354, d. 0.261.

9.154 For the above refrigeration cycle, what is the coefficient of performance when the cycle is operating as a heat pump? a. 3.83, b. 2.83, c. 0.354, d. 0.261.

Appendix 9A Turbojet Engine Analysis Revisited

This appendix provides a more rigorous analysis of momentum conservation for a turbojet engine. Earlier in the chapter we modeled the engine as a simple cylindrical

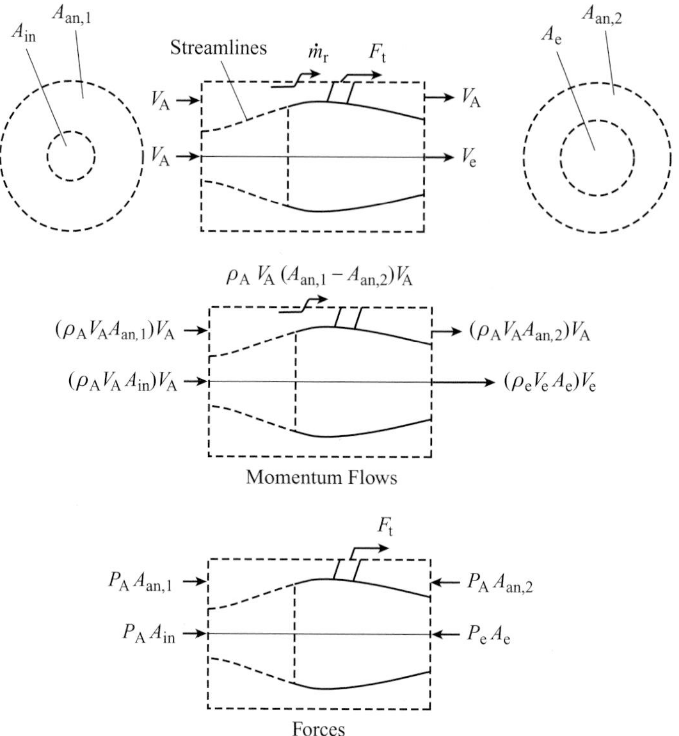

FIGURE 9A.1 To analyze a turbojet engine, we employ a control volume that extends beyond the physical boundaries of the engine. The air that enters the engine proper flows between the streamlines.

tube with equal inlet and exit areas. In this appendix, we consider a more realistic geometry. Figure 9A.1 shows an engine propelling an aircraft at a steady speed V_A. The reference frame is attached to the engine. The open system (control volume) shown in Fig. 9A.1 extends upstream from the physical inlet of the engine such that the air enters this open system at a uniform pressure equal to the ambient condition and at the flight velocity. Air crossing the area A_{in} enters the engine proper, flowing between the streamlines, while air crossing the annular area $A_{an,1}$ flows around the outside of the engine. A portion of this air exits the open system through the cylindrical side, whereas the remainder flows out through the annular area $A_{an,2}$ at the exit plane. Mass conservation requires that the mass flow rate through the cylindrical side $\dot{m}_r$ equals the difference between that entering $A_{an,1}$ and that exiting at $A_{an,2}$; that is,

$$\dot{m}_r = \rho_A V_A (A_{an,1} - A_{an,2}). \tag{9A.1}$$

The center panel of Fig. 9A.1 shows all the x-direction momentum flows associated with our large open system. Similarly, the lowest panel shows all the forces acting in the x-direction. Here we allow the pressure at the engine exit to be different from the ambient pressure. Assuming uniform velocities at the inlets and outlets, we apply the conservation of momentum to yield

$$F_t = (\dot{m}_e V_e - \dot{m}_A V_A) + (P_e - P_A) A_e. \tag{9A.2}$$

We note that, if the pressure in the exit plane of the engine P_e equals the ambient pressure P_A, Eq. 9A.2 yields the same result as that given by Eq. 9.31 in our simple analysis.

CHAPTER 10
Ideal-Gas Mixtures

LEARNING OBJECTIVES

After studying Chapter 10, you should:

- Be able to express the composition of a gas mixture using both mole and mass fractions.
- Be able to calculate the thermodynamic properties of an ideal-gas mixture knowing the mixture composition and the properties of the constituent gases.
- Understand the concept of reference states and the role of the standard state in determining values for entropies and related properties.

CHAPTER 10 OVERVIEW

CHAPTER 10 CONSIDERS IDEAL gas mixtures. We will apply the ideal-gas properties from Chapter 2 to calculate the thermodynamic properties of nonreacting ideal-gas mixtures. These mixture properties will then be used in the conservation equations from Chapter 5 and entropy calculations from Chapter 7. With this analysis, we can study the mixing of two or more gases, the heating/cooling or compression/expansion of a mixture, and the operation of steady-state devices that use a mixture of ideal gases. Until this point, we have considered air as a simple ideal gas. In this chapter, we will look at the air properties considering the composition of the air.

DESIGN AND ANALYSIS OF SYSTEMS		
ANALYSIS OF PRACTICAL DEVICES		
CONSERVATION OF MASS	CONSERVATION OF ENERGY (First Law of Thermodynamics)	SECOND LAW OF THERMODYNAMICS (Entropy)
PROPERTIES OF MATTER		
FRAMEWORKS FOR ANALYSIS, KEY CONCEPTS, AND DEFINITIONS		

FIGURE 10.1 Hierarchical arrangements of the topics in our study of mixtures.

10.1 Ideal-Gas Mixtures

So far we have confined our thermodynamics analysis to pure substances. In Chapter 2, a pure substance was defined as follows:

A pure substance ***is a substance that has a homogeneous and unchanging chemical composition.***

TABLE 10.1 Approximate Composition, Apparent Molecular Weight, and Gas Constant for Dry Air

Constituent	Mole %
N_2	78.08
O_2	20.95
Ar	0.93
CO_2	0.036
Ne, He, CH_4, others	0.003

$$\mathcal{M}_{air} = 28.97 \text{ kg/kmol}$$
$$R_{air} = 287.0 \text{ J/kg·K}$$

The chemical composition of a pure substance is homogeneous. That is, in the region being analyzed, the chemical composition does not vary spatially. The elements of the periodic table are pure substances. Compounds, such as CO_2 and H_2O, are also pure substances. Note that we have previously considered mixtures of the liquid and vapor states within a system being studied. When we studied a mixture of phases, we used the average properties (e.g., specific volume, specific enthalpy) of the mixture for our analysis. We keep in mind, however, that the liquid phase and vapor phase do not mix homogeneously; the liquid has a higher density and will settle to the bottom of the vessel or container. Although the phase of the fluid is not homogeneous, the chemical composition is homogeneous. All water has the chemical composition of H_2O regardless of its state as solid, liquid, or vapor. A liquid–vapor mixture of water, therefore, can be analyzed as a pure substance.

We have analyzed air as a pure substance. However, we know that air is a mixture of different molecules. The standard composition of air can be found in Table C.1 and is repeated here as Table 10.1. We have considered this mixture of molecules in air to be a pure substance because the ratio of molecules defining the chemical composition does not change with temperature or pressure during the processes that we have studied. If the temperature of the air is very low (−194 °C), the air will condense to the liquid state. Since air is a mixture of different molecules, some molecules will turn to liquid first. From the chemical composition of air, the nitrogen will condense at the highest temperature. If the nitrogen molecules have condensed to the liquid state, the vapor mixture has changed in chemical composition and cannot be analyzed as a pure substance during the condensation process.

Many practical devices involve mixtures of pure substances. Air conditioning and combustion systems are common examples of such. In the former application, water vapor becomes an important component of air, which already is a mixture of several components, whereas combustion systems deal with reactant mixtures of fuel and oxidizer and with product mixtures of various components. In both these examples, we can treat the various gas streams as ideal-gas mixtures with reasonable accuracy for a wide range of conditions. For air conditioning and humidification systems, this results from the small amounts of water involved in the mixtures. For combustion, the high temperatures typically involved result in mixtures of low density. Applications

FIGURE 10.2 Condensation of water vapor from combustion products (milehightraveler / E+ / Getty Images).

of the concepts developed in this section are found in Chapters 3, 5, and 11 for combustion systems in Chapter 12.

10.2 Specifying Mixture Composition

Some specific examples in Chapter 11 include the operation of evaporative coolers (Example 11.3) and household dehumidifiers (Example 11.4).

To characterize the composition of a mixture, we define two important and useful quantities: the constituent mole fractions and mass fractions. Consider a multicomponent mixture of gases composed of N_1 moles of species 1, N_2 moles of species 2, etc. The **mole fraction of species *i***, X_i is defined as the fraction of the total number of moles in the system that are species i:

$$X_i \equiv \frac{N_i}{N_1 + N_2 + \cdots} = \frac{N_i}{N_{tot}}. \tag{10.1a}$$

Similarly, the **mass fraction of species *i*, Y_i**, is the fraction of the total mixture mass that is associated with species i:

$$Y_i \equiv \frac{M_i}{M_1 + M_2 + \cdots + M_i + \cdots} = \frac{M_i}{M_{tot}}. \tag{10.1b}$$

Conservation of mass for reacting systems uses mole and mass fractions. See Examples 12.1 and 12.2 in Chapter 12.

Note that, by definition, the sum of all the constituent mole (or mass) fractions must be unity; that is,

$$\sum_{i=1}^{J} X_i = 1, \tag{10.2a}$$

$$\sum_{i=1}^{J} Y_i = 1, \tag{10.2b}$$

where J is the total number of species in the mixture.

The molecular weight, $\mathcal{M}$, gives the weight in kg of one kmol of molecules. For example, one kmol of carbon weighs 12 kg. Knowing the molecular weight and mole fraction of each component in the mixture, we can find the apparent molecular weight of the mixture:

$$M_{mix} = \sum_{i=1}^{J} X_i \mathcal{M}_i. \tag{10.3}$$

We can readily convert mole fractions and mass fractions from one to another using the molecular weights of the species of interest and the apparent molecular weight of the mixture. We will derive this conversion by finding the mass of species i in one mole of the mixture in two ways. The mass of species i is the total mass of a mole of mixture, M_{mix}, times the mass fraction Y_i of species i. The mass of species i can also be found using the mole fraction X_i times the molecular weight of species i:

$$M_i = Y_i M_{mix} = N_{tot} X_i \mathcal{M}_i.$$

Equating these expressions gives

$$Y_i = X_i \mathcal{M}_i / \mathcal{M}_{mix}, \tag{10.4a}$$

Solving for the mole fraction gives

$$X_i = Y_i \mathcal{M}_{mix} / \mathcal{M}_i. \tag{10.4b}$$

The apparent mixture molecular weight $\mathcal{M}_{mix}$ can also be found using the mass fractions of the species in the mixture. Starting with Equation 10.2a, we substitute for each mole fraction using Equation 10.4b. Solving for $\mathcal{M}_{mix}$,

$$\mathcal{M}_{mix} = \frac{1}{\sum_{i=1}^{J} (Y_i / \mathcal{M}_i)}. \tag{10.5}$$

Example 10.1 Molecular Weight of Air

Using the mole fractions in Table 10.1, find the apparent molecular weight of dry air.

Solution

For each component of dry air, we need to know the molecular weight and mole fraction. This information is found in Table 10.1 (also found in Appendix C.1), and the molecular weights are found in Table E.1 or the NIST tables. This information is put into spreadsheet form:

Constituent	X_i	$\mathcal{M}_i$ (kg/kmol)
N_2	0.7808	28.01
O_2	0.2095	32
Ar	0.0093	39.948
CO_2	0.00036	44.01
Ne, He, CH_4, others	Neglected here	
TOTAL	0.99996	28.963

The apparent molecular weight is calculated using Eq. 10.3:

$$\mathcal{M}_{mix} = \sum_{i=1}^{J} X_i \mathcal{M}_i$$

$$\mathcal{M}_{mix} = 0.7808(28.01) + 0.2095(32) + 0.0093(39.948) + 0.00036(44.01)$$

$$\mathcal{M}_{mix} = 28.963.\ \text{kg/kMol}$$

This is very close to the tabulated value of 28.97, the difference being due to round-off errors.

10.3 State (P–v–T) Relationships for Mixtures

In our treatment of mixtures, we assume a double "ideality": first, that the pure constituent gases obey the ideal-gas equation of state (Eq. 2.26a), and, second, that when these pure components mix, an ideal solution results. In an **ideal solution,** the behavior of any one component is uninfluenced by the presence of any other component.

Let us explore the characteristics of an ideal solution, considering, first, the thermodynamic property pressure. Consider a fixed volume $\mathcal{V}$ containing two or more different species. If the mixture (solution) is ideal, gas molecules of species A are free to roam through the entire volume $\mathcal{V}$, as if no other species were present. The pressure that molecules A exert on the wall is less than the total pressure, however, since other non-A molecules also collide with the wall. We can thus define the partial pressure of species A, P_A, by applying the ideal-gas equation of state just to the A molecules:

$$P_A \mathcal{V} = N_A R_u T. \tag{10.6}$$

Since each species behaves independently, similar expressions can be written for each:

$$P_B \mathcal{V} = N_B R_u T,$$
$$\vdots$$
$$P_i \mathcal{V} = N_i R_u T,$$

etc. Summing this set of equations for a mixture containing J species yields

$$\begin{aligned}(P_A + P_B + \cdots + P_i + \cdots + P_J)\mathcal{V} \\ = (N_A + N_B + \cdots + N_i + \cdots + N_J)R_u T.\end{aligned} \tag{10.7}$$

Because the sum of the number of moles of each constituent is the total number of moles in the system, N_{tot}, this equation becomes

$$\sum_{i=1}^{J} P_i \mathcal{V} = N_{tot} R_u T, \tag{10.8}$$

where P_i, is the partial pressure of the ith species. Because the mixture as a whole obeys the ideal-gas law,

$$P\mathcal{V} = N_{tot} R_u T, \tag{10.9}$$

the sum of the partial pressures must be identical to the total pressure, that is,

$$\sum_{i=1}^{J} P_i = P. \tag{10.10}$$

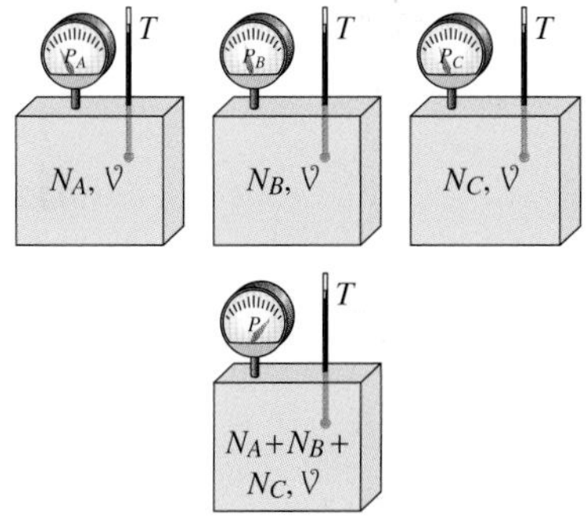

FIGURE 10.3 In an ideal-gas mixture, the sum of the pressures associated with each component isolated in the same volume at the same temperature is identical to the total pressure observed when all species are confined together in the same volume at the same temperature.

This statement is known as **Dalton's law of partial pressures** and is schematically illustrated in Fig. 10.3. Here we see four containers, each having the same volume $\mathcal{V}$. Gas A fills one container, gas B another, and gas C the third. Each gas has the same temperature T. The absolute pressure of the gas in each container is measured to be P_A, P_B, and P_C, respectively. We now conduct a thought experiment in which gases A, B, and C are transferred to the fourth container, again at temperature T. The absolute pressure of the mixture in the fourth container is found to be the sum of the absolute pressures measured when the gases were segregated. Thus the contribution of a single species to the total pressure in a gas mixture is the same as that of the pure species occupying the same total volume at the same temperature. This is a defining characteristic of an ideal-gas mixture (i.e., an ideal solution). In nonideal mixtures (solutions), the total pressure is not necessarily equal to the sum of the pure component pressures as in the thought experiment.

The partial pressure can be related to the mixture composition and total pressure by dividing Eq. 10.6 by Eq. 10.9,

In Chapter 11, the partial pressure of water vapor is used to define the relative humidity of moist air. See Eq. 11.12.

$$\frac{P_i \mathcal{V}}{P \mathcal{V}} = \frac{N_i R_u T}{N_{tot} R_u T},$$

which simplifies to

$$\frac{P_i}{P} = \frac{N_i}{N_{tot}} \equiv X_i, \tag{10.11a}$$

or

$$P_i = X_i P. \tag{10.11b}$$

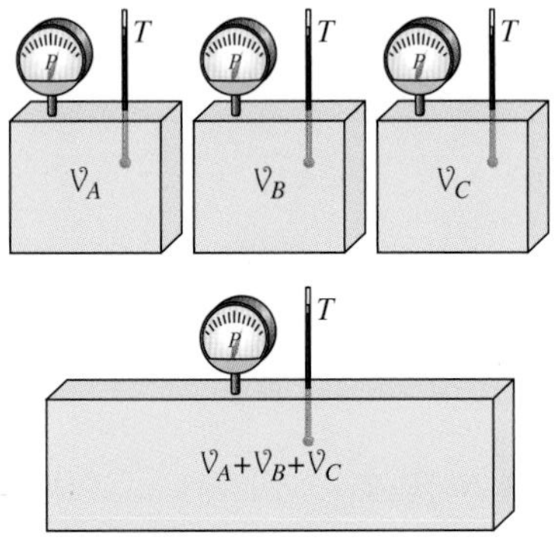

FIGURE 10.4 In an ideal-gas mixture, the sum of the volumes associated with each component isolated at the same temperature and the same pressure is identical to the total volume when all species are confined together at the same temperature and the same pressure.

Ideal-gas mixtures also exhibit the property that when component gases having different volumes but identical pressures and temperatures are brought together, the mixture volume is the sum of the pure component volumes (Fig. 10.4). Mathematically, we express this idea by writing the ideal-gas equation of state for each pure constituent,

$$P\mathcal{V}_A = N_A R_u T,$$
$$P\mathcal{V}_B = N_B R_u T,$$
$$P\mathcal{V}_C = N_C R_u T,$$

and summing to yield

$$P(\mathcal{V}_A + \mathcal{V}_B + \mathcal{V}_C) = (N_A + N_B + N_C) R_u T.$$

Since

$$P\mathcal{V}_{tot} = N_{tot} R_u T,$$

we conclude that the individual volumes, or partial volumes, must equal the total volume when combined, that is,

$$\mathcal{V}_{tot} = \mathcal{V}_A + \mathcal{V}_B + \mathcal{V}_C. \tag{10.12}$$

This view of an ideal-gas mixture is frequently referred to as *Amagat's model.*

The partial volumes also can be related to the mixture composition by defining a volume fraction,

$$\frac{\mathcal{V}_i}{\mathcal{V}} = \frac{N_i}{N_{tot}}. \tag{10.13}$$

Comparing Eqs. 10.11a and 10.13, we see that the three measures of mixture composition – the ratio of the partial pressure to the total pressure, the volume fraction, and the mole fraction – are all equivalent for an ideal-gas mixture:

$$\frac{P_i}{P} = \frac{\mathcal{V}_i}{\mathcal{V}} = \frac{N_i}{N} \ (= X_i). \tag{10.14}$$

For insight into the behavior of nonideal mixtures, the reader is referred to Refs. [4, 5].

10.4 Calorific Relationships for Mixtures

From our earlier discussion of the calorific equation of state for ideal gases in Chapter 2, we see that a reference temperature is required to evaluate the specific internal energy and enthalpy (see Eqs. 2.29c and 2.31c). If one is concerned only with a single pure substance, as opposed to a mixture, then Eqs. 2.29c and 2.31c would suffice to describe the internal energy and enthalpy changes for all thermodynamic processes, as only differences in state are of importance. Moreover, any choice of reference temperature would yield the same results. For example, the change in the specific internal energy associated with a change of temperature from T_1 to T_2 can be found from Eq. 2.29c

$$u(T_2) - u(T_2) = \int_{T_{\text{ref}}}^{T_2} c_v dT - \int_{T_{\text{ref}}}^{T_1} c_v dT = \int_{T_1}^{T_2} c_v dT,$$

and the change in the specific enthalpy is calculated from Eq. 2.31c as

$$h(T_2) - h(T_2) = \int_{T_{\text{ref}}}^{T_2} c_p dT - \int_{T_{\text{ref}}}^{T_1} c_p dT = \int_{T_1}^{T_2} c_p dT.$$

For nonreacting gases, we note that the reference temperature drops out and is not needed. In reacting flows, however, the reference temperature is needed to calculate changes in the specific internal energy and enthalpy. The changes in properties due to chemical reactions, such as combustion, must also be included. A detailed discussion of the standard reference state and calculation of properties for combusting gases will be presented in Chapter 12.

In Chapter 12, the sensible enthalpy change, $\Delta\bar{h}_{s,i}(T)$, will be defined as the enthalpy change associated with only a temperature change. The sensible enthalpy change is found in the tables in Appendix D labeled as $\bar{h}^{\circ}(T) - \bar{h}_f^{\circ}(298\ \text{K})$. If there is no chemical reaction from state 1 to state 2, the difference in the sensible enthalpies will give the change in enthalpy during the process, $\bar{h}_i(T_2) - \bar{h}_i(T_1)$. We see in these tables that the specific heat at constant volume is not given. Knowing the specific heat at constant pressure, we can easily find the specific heat at constant volume using $\bar{c}_\text{v} = \bar{c}_p - R_\text{u}$ and $c_v = c_p - R$. Similarly, the specific internal energy can be found by recalling that $h = u + P\text{v}$. Solving for the specific internal energy and substituting for $P\text{v} = RT$ we find that:

$$u = h - RT \quad \text{and} \quad \bar{u} = \bar{h} - R_\text{u}T.$$

In this chapter, we consider mixtures without chemical reactions. The calorific relationships for ideal-gas mixtures are straightforward mass-fraction or mole-fraction weightings of the pure-species individual specific calorific properties:

$$u_{\text{mix}} = \sum_{i=1}^{J} Y_i u_i, \tag{10.15a}$$

$$h_{\text{mix}} = \sum_{i=1}^{J} Y_i h_i, \tag{10.15b}$$

$$c_{v,\text{mix}} = \sum_{i=1}^{J} Y_i c_{v,i}, \tag{10.15c}$$

$$c_{p,\text{mix}} = \sum_{i=1}^{J} Y_i c_{p,i}, \tag{10.15d}$$

and, for quantities expressed per mole,

$$\bar{u}_{\text{mix}} = \sum_{i=1}^{J} X_i \bar{u}_i, \tag{10.15e}$$

$$\bar{h}_{\text{mix}} = \sum_{i=1}^{J} X_i \bar{h}_i, \tag{10.15f}$$

$$\bar{c}_{v,\text{mix}} = \sum_{i=1}^{J} X_i \bar{c}_{v,i}, \tag{10.15g}$$

$$\bar{c}_{p,\text{mix}} = \sum_{i=1}^{J} X_i \bar{c}_{p,i}, \tag{10.15h}$$

The analysis of a jet engine combustor in Chapter 12 (Example 12.8) uses concepts developed here: mole and mass fractions and standardized properties of ideal-gas mixtures.

where the subscript i represents the ith species and J is the total number of species in the mixture. Since the specific heats, internal energies, and enthalpies of the constituent ideal-gas species depend only on temperature, the same is true for the mixture calorific properties; for example, $u_{\text{mix}} = u_{\text{mix}}$ (T only), etc. Molar-specific enthalpies for a number of species are tabulated in Appendix D.

Example 10.2 Specific Heat of Air

Using the mole fractions in Table 10.1, find the apparent specific heats of dry air.

Solution

For each component of dry air, we need to know the molecular weight and mole fraction. This information is found in Table 10.1 (it is also found in Appendix C.1), and the molecular weights are found in Table E.1 or the NIST resources. This information is put into spreadsheet form:

Constituent	X_i	$\mathcal{M}_i$ (kg/kmol)	Y_i
N_2	0.7808	28.01	0.75511
O_2	0.2095	32	0.23147
Ar	0.0093	39.948	0.001283
CO_2	0.00036	44.01	0.00055
Ne, He, CH_4, others	Mixture neglected here		
TOTAL	0.99996	28.962	0.99996

For each constituent, we can find the mass fraction of the species using Eq. 10.4a. For example, the mass fraction for nitrogen is

$$Y_{N_2} = X_{N_2}\mathcal{M}_{N_2}/\mathcal{M}_{mix} = 78.08\%\left(\frac{28.01}{28.962}\right) = 75.51\%.$$

These results are tabulated above in the last column. Using these mass fractions and the specific heats from Table E.1 or the NIST resources, we can calculate the apparent specific heats for air using Eqs. 10.15c and 10.15d.

Constituent	Y_i	c_v (kJ/kg · K)	c_p (kJ/kg · K)
N_2	0.75511	0.743	1.04
O_2	0.23147	0.658	0.918
Ar	0.001283	0.312	0.522
CO_2	0.00055	0.657	0.846
Ne, He, CH_4, others		Neglected here	
TOTAL	0.99996	0.718	1.005

The apparent specific heat at constant volume is

$$c_{v,\text{air}} = \sum_{i=1}^{J} Y_i c_{v,i} = 0.75511(0.743\,\text{kJ/kg·K}) + 0.23147(0.658\ \text{kJ/kg·K}) + 0.01283(0.312\ \text{kJ/kg·K}) = 0.718\ \text{kJ/kg·K}.$$

The apparent specific heat at constant volume is:

$$c_{p,\text{air}} = \sum_{i=1}^{J} Y_i c_{p,i} = 0.75511(1.04\,\text{kJ/kg·K}) + 0.23147(0.918\ \text{kJ/kg·K}) + 0.01283(0.522\ \text{kJ/kg·K}) = 1.005\ \text{kJ/kg·K}.$$

These are the apparent specific heats for air listed in Table E.1.

10.5 Second-Law Relationships for Mixtures

To find the entropy of a mixture, we add the entropies of the components. The entropy at a particular state is found relative to a **standard reference state**. The standard-state temperature $T^\circ = 298.15$ K, and the standard-state pressure, $P^\circ = 1$ atm (101,325 Pa), consistent with the JANAF, Chemkin, and NASA thermodynamic databases [1–3]. In Tables D.1–D.13, the third law of thermodynamics is used to set zero values; thus, the standard-state entropy values are always larger than zero.

Although not needed to describe enthalpies, the standard-state pressure is required to calculate entropies at pressures other than one atmosphere (see Eq. 7.11). The values of entropy at standard pressure are

$$s^\circ(T_1) = s(T_1, P^\circ) = \int_{T_{ref}}^{T} c_p \frac{dT}{T}, \tag{10.16}$$

where the superscript $^\circ$ is used to denote that the entropy is at the standard-state pressure.[1]

The entropy for a gas that is not at the standard-state pressure can be found using Eq. 7.11:

$$s_1(T_1, P_1) = s^\circ(T_1) - R \ln \frac{P_1}{P^\circ} \tag{10.17a}$$

The entropy can also be found on a molar basis as:

$$\bar{s}_1(T_1, P_1) = \bar{s}^\circ(T_1) - R_u \ln \frac{P_1}{P^\circ}. \tag{10.17b}$$

If a constant specific heat is assumed, these calculations for entropy become

$$s_1(T_1, P_1) = c_p \ln \frac{T_1}{T^\circ} - R \ln \frac{P_1}{P^\circ}. \tag{10.18a}$$

and

$$\bar{s}_1(T_1, P_1) = \bar{c}_p \ln \frac{T_1}{T^\circ} - R_u \ln \frac{P_1}{P^\circ}. \tag{10.18b}$$

The entropy of a mixture is calculated as a weighted sum of the constituents:

$$s_{mix}(T, P) = \sum_{i=1}^{J} Y_i s_i(T, P_i), \tag{10.19a}$$

$$\bar{s}_{mix}(T, P) = \sum_{i=1}^{J} X_i \bar{s}_i(T, P_i). \tag{10.19b}$$

Unlike the ideal-gas calorific relationships, pressure is now required as a second independent variable. Here the pure-species entropies (s_i and $\bar{s}_i$) depend on the species partial pressures, as explicitly indicated in Eqs. 10.19. Equations 10.18 can be applied to evaluate the constituent entropies in Eqs. 10.19 from standard-state ($P^\circ = 1$ atm) values as

$$s_i(T, P_i) = s_i^\circ(T) - R \ln \frac{P_i}{P^\circ}, \tag{10.20a}$$

[1] The use of this superscript is redundant for ideal-gas enthalpies, which exhibit no temperature dependence; for entropies, however, the superscript is important.

$$\bar{s}_i(T,P) = \bar{s}_i^{\circ}(T) - R_u \ln \frac{P_i}{P^{\circ}}. \quad (10.20b)$$

where $P_i = X_i P$. Ideal-gas standard-state molar-specific entropies are tabulated in Appendix D for several species.

When using a second-law analysis, entropy differences for each species are usually found. Taking a difference in entropies for a species using Eq. 10.20a gives

$$\Delta s_i(T,P_i) = s_{i,2}(T_2, P_{i,2}) - s_{i,1}(T_1, P_{i,1}) = s_{i,2}^{\circ}(T_2) - R \ln \frac{P_{i,2}}{P^{\circ}} - s_{i,1}^{\circ}(T_1) - R \ln \frac{P_{i,1}}{P^{\circ}}.$$

The natural log terms can be combined to give

$$\Delta s_i(T,P_i) = s_{i,2}(T_2,P_{i,2}) - s_{i,1}(T_1,P_{i,1}) = s_{i,2}^{\circ}(T_2) - s_{i,1}^{\circ}(T_1) - R \ln \frac{P_{i,2}}{P_{i,1}}. \quad (10.21a)$$

This can also be written on a molar basis:

$$\Delta \bar{s}_i(T,P_i) = \bar{s}_{i,2}(T_2,P_{i,2}) - \bar{s}_{i,1}(T_1,P_{i,1}) = \bar{s}_{i,2}^{\circ}(T_2) - \bar{s}_{i,1}^{\circ}(T_1) - R_u \ln \frac{P_{i,2}}{P_{i,1}}. \quad (10.21b)$$

Example 10.3 Entropy of Air

Using the mole fractions in Table 10.1, find the apparent entropy of dry air at 300 K and 1 atmosphere.

Solution

For each component of dry air, we need to know the standard-state entropy at 300 K. We use the entropy values from the Appendix B tables for nitrogen, oxygen, and carbon dioxide. The NIST tables were used for argon. This information is put into spreadsheet form:

Constituent	X_i	$\bar{s}^{\circ}$ (kJ/kmol·K)	$\bar{s}_i$ (kJ/kmol·K)
N_2	0.7808	191.691	193.748
O_2	0.2095	205.224	218.220
Ar	0.0093	154.87	193.763
CO_2	0.00036	213.966	279.895
Ne, He, CH_4, others	Neglected here		
TOTAL	0.99996		

We derived Eq. 10.11a by showing that the ratio of the partial pressure of each component of a mixture to the total pressure is the mole fraction of the species:

$$\frac{P_i}{P} = \frac{N_i}{N_{\text{tot}}} \equiv X_i.$$

We can now find the entropy of each species in the air mixture using Eq. 10.20b. For example, for nitrogen we find

$$\begin{aligned}\bar{s}_i(T, P) &= \bar{s}_i(T, P^\circ) - R_u \ln\left(\frac{P_i}{P^\circ}\right) = \bar{s}_i(T, P^\circ) - R_u \ln(X_i)\\ &= 191.691\,\text{kJ/kmol·K} - (8.314472\,\text{kJ/kmol·K}) \ln(0.7808)\\ &= 193.748\,\text{kJ/kmol·K}.\end{aligned}$$

The calculated entropy for each species at its partial pressure is listed in the rightmost column of the table above. Note that the specific entropy is greater than the value at standard pressure since the entropy (randomness) of the species will increase as the pressure (the constraining effect) decreases.

We can now apply Eq. 10.19b to find the specific entropy of the air mixture:

$$\begin{aligned}\bar{s}_{\text{mix}}(T, P) &= \sum_{i=1}^{J} X_i \bar{s}_i(T, P_i)\\ &= 0.7808(193.748\,\text{kJ/kmol·K}) + 0.2095(218.220\,\text{kJ/kmol·K})\\ &\quad + 0.0093(193.763\,\text{kJ/kmol·K}) + 0.00036(279.895\,\text{kJ/kmol·K})\\ &= 198898\,\text{kJ/kmol·K}\end{aligned}$$

$$s_{\text{mix}}(T, P) = \bar{s}_{\text{mix}}(T, P)/\mathcal{M}_{\text{mix}} = \frac{198.898\,\text{kJ/kmol·K}}{28.97\,\text{kg/kg·K}} = 6.8657\,\text{kg/kg·K}.$$

Comments Note that this calculated entropy is different from that listed in Table C.2. The reason is that Table C.2 defines zero entropy for air at a temperature of 78.903 K whereas Tables D define the entropy of each gas from 0 K. Care must be taken when finding the entropy at a particular state. A calculation of the change in entropy during a process will give the same solution regardless of the standard-reference state temperature selected.

Example 10.4 Mixing of Ideal Gases

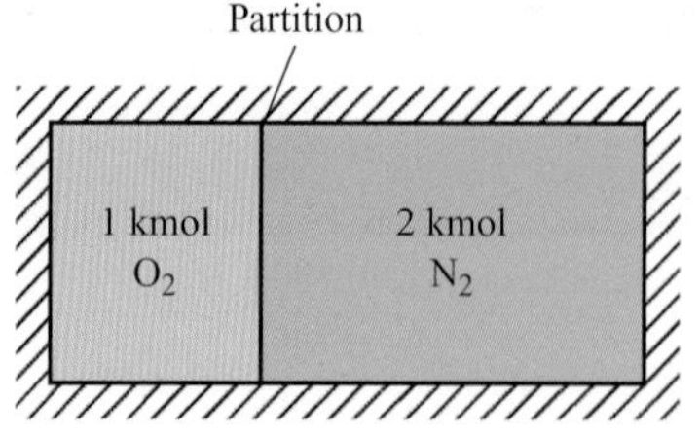

Consider 1 kmol of O_2 and 2 kmol of N_2, both at 298 K and 1 atm, separated by a partition. The partition is removed and the O_2 and N_2 mix. Determine the entropy change associated with this mixing process.

Solution

Known $N_{O_2}, N_{N_2}, T_1, P_1$

Find $S_{\text{final}} - S_{\text{init}}$

Sketch

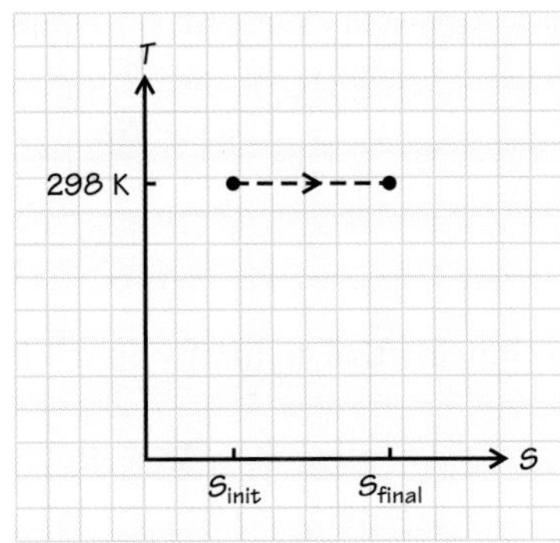

Modeling, Premises and Assumptions

i. Ideal-gas behavior
ii. Adiabatic process

Analysis We now have an understanding of mixtures with which to revisit Example 7.9 and calculate the change in entropy during the mixing process. For the mixing of ideal gases with no heat or work interactions, conservation of energy tells us that the initial and final temperatures are equal. To calculate the entropy change, we therefore only need to be concerned with the changing partial pressures. The entropy change is expressed as follows:

$$\begin{aligned}\Delta S &= S_{\text{final}} - S_{\text{init}} \\ &= \left[N_{O_2}\bar{s}_{O_2,2}(T, P_{O_2,2}) + N_{N_2}\bar{s}_{N_2,2}(T, P_{N_2,2})\right] \\ &\quad -\left[N_{O_2}\bar{s}_{O_2,1}(T, P_{O_2,1}) + N_{N_2}\bar{s}_{N_2,1}(T, P_{N_2,1})\right].\end{aligned}$$

For an ideal gas, the entropy of an individual species is given by Eq. 10.17b, that is,

$$\bar{s}_i(T, P_i) = \bar{s}_i(T, P_{ref}) - R_u \ln \frac{P_i}{P_{\text{ref}}}.$$

At the initial unmixed state, the pressures of both the N_2 and O_2 are the given value (i.e., $P_{O_2} = P_{N_2} = 1\,\text{atm}$). At the final mixed state, the total pressure is still 1 atm, whereas the partial pressures of each constituent are given by (Eq. 10.11b):

$$\begin{aligned}P_{O_2,2} &= X_{O_2}P_{\text{tot}} = \frac{N_{O_2}}{N_{\text{mix}}}P_{\text{tot}} \\ &= \left(\frac{1}{1+2}\right)1\,\text{atm} = 0.333\,\text{atm}\end{aligned}$$

and

$$\begin{aligned}P_{N_2,2} &= X_{N_2}P_{\text{tot}} = \frac{N_{N_2}}{N_{\text{mix}}}P_{\text{tot}} \\ &= \left(\frac{2}{1+2}\right)1\,\text{atm} = 0.667\,\text{atm}.\end{aligned}$$

Explicitly substituting the ideal-gas expressions for $\bar{s}_i$ into our expanded relationship for ΔS gives

$$\begin{aligned}\Delta S = &\left\{N_{O_2}\left[\bar{s}_{O_2,2}(T, P_{\text{ref}}) - R_u \ln \frac{P_{O_2,2}}{P_{\text{ref}}}\right] + N_{N_2}\left[\bar{s}_{N_2,2}(T, P_{\text{ref}}) - R_u \ln \frac{P_{N_2,2}}{P_{\text{ref}}}\right]\right\} \\ &-\left\{N_{O_2}\left[\bar{s}_{O_2,1}(T, P_{\text{ref}}) - R_u \ln \frac{P_{O_2,1}}{P_{\text{ref}}}\right] + N_{N_2}\left[\bar{s}_{N_2,1}(T, P_{\text{ref}}) - R_u \ln \frac{P_{N_2,1}}{P_{\text{ref}}}\right]\right\}.\end{aligned}$$

The $\bar{s}_i$ terms all cancel in this expression to yield

$$\Delta S = N_{O_2}\left[-R_u \ln \frac{P_{O_2,2}}{P_{ref}} - \left(-R_u \ln \frac{P_{O_2,1}}{P_{ref}}\right)\right] + N_{N_2}\left[-R_u \ln \frac{P_{N_2,2}}{P_{ref}} - \left(-R_u \ln \frac{P_{N_2,1}}{P_{ref}}\right)\right].$$

Recognizing that ln (a/b) = ln a − ln b, we reduce this to

$$\Delta S = R_u\left[N_{O_2} \ln \frac{P_{O_2,1}}{P_{O_2,2}} + N_{N_2} \ln \frac{P_{N_2,1}}{P_{N_2,2}}\right].$$

Substituting numerical values, we obtain our final result:

$$\Delta S = \left(8.314 \frac{\text{kJ}}{\text{kmol·K}}\right)\left[(1 \text{ kmol}) \ln \left(\frac{1 \text{ atm}}{0.333 \text{ atm}}\right) + (2 \text{ kmol}) \ln \left(\frac{1 \text{ atm}}{0.666 \text{ atm}}\right)\right]$$
$$= 15.88 \text{ kJ/K}.$$

Comment This example confirms what was stated in Chapter 7: the mixing of two gases increases the entropy of the system and is an irreversible process. What if we were to perform the same mixing process, but using a single constituent, say, O_2? In this scenario, the partial pressure of the O_2 at the end of the mixing is the same as at the initial state (i.e., P_{init} = 1 atm = P_{final} = $P_{O_2,final}$). With no change in pressure, ΔS is then zero. We can also view this process as a reversible one since we could put the partition back in place after the mixing takes place and retrieve the initial state, treating the O_2 molecules as indistinguishable.

Self-Test 10.1

Consider two solid metals (copper and iron) separated by a partition in a container, as depicted in Fig. 7.13b. The copper is originally at 400 K and the iron is at 300 K. Determine the total change in entropy when the partition is removed, the two metals are brought into contact, and the temperature is allowed to equilibrate.

(Answer: $\Delta S_{total} = 19 J/K$)

Example 10.5 Mixing of Ideal Gases at Different Temperatures

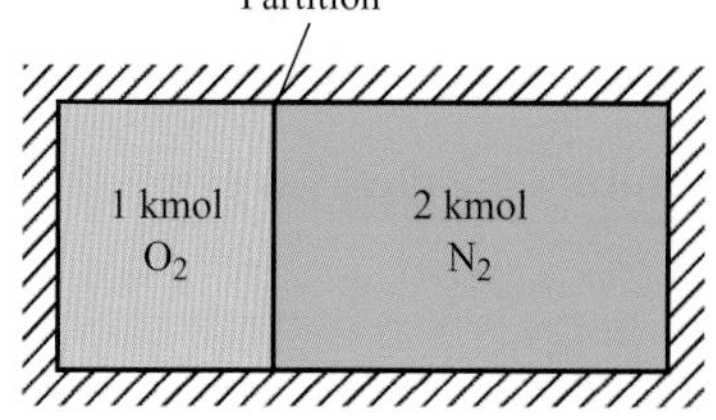

Consider 1 kmol of O_2 and 2 kmol of N_2 separated by a partition. The oxygen is at 295 K and 100 kPa. The nitrogen is at 320 K and 150 kPa. The partition is removed and the O_2 and N_2 mix. Determine the final temperature and pressure of the mixture and the entropy change associated with this mixing process.

Solution

Known $N_{O_2}, T_{O_2,1}, P_{O_2,1}, N_{N_2}, T_{N_2,1}, P_{N_2,1}$

Find $T_2, P_2, S_{final} - S_{init}$

Modeling, Premises and Assumptions

i. Ideal-gas behavior

ii. Adiabatic process

Analysis This problem is similar to the previous example but we now have a more general case where the two gases are initially at different temperatures and pressures. For the mixing of ideal gases with no heat or work interactions, conservation of energy tells us that the initial and final energies of the system are equal. At steady state, the temperature of the oxygen and nitrogen will be the same. We will first use the conservation of energy equation to find the temperature at the final state:

$$N_{O_2}\bar{u}_{O_2,1}(T_{O_2,1}) + N_{N_2}\bar{u}_{N_2,1}(T_{N_2,1}) = N_{O_2}\bar{u}_{O_2,2}(T_2) + N_{N_2}\bar{u}_{N_2,2}(T_2).$$

If exact gas properties are used from NIST resources, solving for the final temperature will require some iteration: guess a final temperature and verify that the equation above is satisfied; if not, the guess of the final temperature is updated. Instead, we will assume whether constant specific heats as listed in Table E.1:

$$\begin{aligned} N_{O_2}\bar{c}_{v,O_2}\,T_{O_2,1} + N_{N_2}\bar{c}_{v,N_2}T_{N_2,1} &= N_{O_2}\bar{c}_{v,O_2}T_2 + N_{N_2}\bar{c}_{v,N_2}T_2 N_{O_2} \\ \mathcal{M}_{O_2}c_{v,O_2}T_{O_2,1} + N_{N_2}\mathcal{M}_{N_2}c_{v,N_2}T_{N_2,1} &= N_{O_2}\mathcal{M}_{O_2}c_{v,O_2}T_2 + N_{N_2}\mathcal{M}_{N_2}c_{v,N_2}T_2 \\ &= (N_{O_2}\mathcal{M}_{O_2}c_{v,O_2} + N_{N_2}\mathcal{M}_{N_2}c_{v,N_2})T_2; \end{aligned}$$

thus

$$(1\,\text{kmol})\left(32.00\frac{\text{kg}}{\text{kmol}}\right)\left(0.658\frac{\text{kJ}}{\text{kg·K}}\right)(295\,\text{K}) + (2\,\text{kmol})\left(28.01\frac{\text{kg}}{\text{kmol}}\right)\left(0.743\frac{\text{kJ}}{\text{kg·K}}\right)(320\,\text{K})$$

$$= \left[(1\,\text{kmol})\left(32.00\frac{\text{kg}}{\text{kmol}}\right)\left(0.658\frac{\text{kJ}}{\text{kg·K}}\right) + (2\,\text{kmol})\left(28.01\frac{\text{kg}}{\text{kmol}}\right)\left(0.743\frac{\text{kJ}}{\text{kg·K}}\right)\right]T_2.$$

Solving, we find that the final (steady-state) temperature is $T_2 = 311.6$ K

We also need to find the final pressure of the mixture. In this problem, we know that the total system volume does not change during the mixing process. We will use the ideal gas law to find the final pressure of the mixture.

$$\begin{gathered} \mathcal{V}_{O_2,1} + \mathcal{V}_{N_2,1} = \mathcal{V}_2 \\ \frac{N_{O_2}R_u T_{O_2,1}}{P_{O_2,1}} + \frac{N_{N_2}R_u T_{N_2,1}}{P_{N_2,1}} = \frac{(N_{O_2} + N_{N_2})R_u T_2}{P_2}. \end{gathered}$$

We note that R_u appears in every term and can be cancelled from the equation:

$$\frac{(1\text{ kmol})(295\,\text{K})}{100\,\text{kPa}} + \frac{(2\text{ kmol})(320\,\text{K})}{150\,\text{kPa}} = \frac{(1+2)\text{kmol}(311.6\,\text{K})}{P_2}$$

$$P_2 = \frac{(1+2)\,\text{kmol}(311.6\,\text{K})}{\dfrac{(1\,\text{kmol})(295\,\text{K})}{100\,\text{kPa}} + \dfrac{(2\,\text{kmol})(320\,\text{K})}{150\,\text{kPa}}} = 129.53\text{ kPa}.$$

The entropy change is expressed as follows:

$$\begin{aligned} \Delta S = S_{\text{final}} - S_{\text{init}} &= \Delta S_{O_2} + \Delta S_{N_2} \\ &= N_{O_2}\left[\bar{s}_{O_2}(T_{O_2,2}, P_{O_2,2}) - \bar{s}_{O_2}(T_{O_2,1}, P_{O_2,1})\right] \\ &\quad + N_{N_2}\left[\bar{s}_{N_2}(T_{N_2,2}, P_{N_2,2}) - \bar{s}_{N_2}(T_{N_2,1}, P_{N_2,1})\right]. \end{aligned}$$

For an ideal gas, the entropy of an individual species is given by Eq. 10.17b, that is,

$$\bar{s}_i(T, P_i) = \bar{s}(T, P_{\text{ref}}) - R_u \ln\frac{P_i}{P_{\text{ref}}}.$$

For each species, we can find the change in the specific entropy:

$$\Delta\bar{s}_i = \left[\bar{s}(T_{i,2}, P_{\text{ref}}) - R_{\text{u}} \ln \frac{P_{i,2}}{P_{\text{ref}}}\right] - \left[\bar{s}(T_{i,1}, P_{\text{ref}}) - R_{\text{u}} \ln \frac{P_{i,1}}{P_{\text{ref}}}\right].$$

The two natural log terms can be combined to give

$$\Delta\bar{s}_i = [\bar{s}(T_{i,2}, P_{\text{ref}}) - \bar{s}(T_{i,1}, P_{\text{ref}})] + R_{\text{u}} \ln \frac{P_{i,1}}{P_{i,2}}.$$

For each species, the partial pressure at the final state is used in the equation above. The partial pressure of each species is given by (Eq. 10.11b):

$$P_{O_2} = X_{O_2} P_{\text{tot}} = \frac{N_{O_2}}{N_{\text{mix}}} P_{\text{tot}}$$

$$= \left(\frac{1}{1+2}\right)(129.53\text{ kPa}) = 43.18\text{ kPa}$$

and

$$P_{N_2} = X_{N_2} P_{\text{tot}} = \frac{N_{N_2}}{N_{\text{mix}}} P_{\text{tot}}$$

$$= \left(\frac{2}{1+2}\right)(129.53\text{ kPa}) = 86.36\text{ kPa}.$$

Explicitly substituting the ideal-gas expressions for $\bar{s}_i$ into our expanded relationship for ΔS gives

$$\Delta S = N_{O_2}\left[\bar{s}_{O_2,2}(T_2, P_{\text{ref}}) - \bar{s}_{O_2,1}(T_{O_2,1}, P_{\text{ref}}) + R_{\text{u}} \ln \frac{P_{O_2,1}}{P_{O_2,2}}\right]$$

$$+ N_{N_2}\left[\bar{s}_{N_2,2}(T_2, P_{\text{ref}}) - \bar{s}_{N_2,1}(T_{N_2,1}, P_{\text{ref}}) + R_{\text{u}} \ln \frac{P_{N_2,1}}{P_{N_2,2}}\right].$$

Substituting numerical values, we obtain our final result:

$$\Delta S = (1\,\text{kmol})\left[206.32\frac{\text{kJ}}{\text{kmol·K}} - 204.71\frac{\text{kJ}}{\text{kmol·K}} + \left(8.3145\frac{\text{kJ}}{\text{kmol·K}}\right) \ln\left(\frac{100\,\text{kPa}}{43.18\,\text{kPa}}\right)\right]$$

$$+ (2\,\text{kmol})\left[192.77\frac{\text{kJ}}{\text{kmol·K}} - 193.54\frac{\text{kJ}}{\text{kmol·K}} + \left(8.3145\frac{\text{kJ}}{\text{kmol·K}}\right) \ln \frac{150}{86.36}\right]$$

$$= 8.593\,\text{kJ/K} + 7.642\,\text{kJ/K} = 16.23\,\text{kJ/K}.$$

Comment Since the temperature of each species changes during the process, the entropies at the reference pressure do not cancel in the calculations. During the process, the nitrogen decreases in temperature; however, the entropy of nitrogen increases owing to the decrease in the partial pressure of nitrogen during the process.

Self-Test 10.2

 Repeat Example 10.5 with 1 kmol of oxygen and 4 kmol of nitrogen.

(Answer: $\Delta S_{total} = 21.25\,J/K$)

10.6 Gibbs Free Energy

The Gibbs free energy or Gibbs function, G, is a composite property involving enthalpy and entropy and is defined as

$$G \equiv H - TS, \tag{10.22a}$$

and, per unit mass,

$$g \equiv h - Ts. \tag{10.22b}$$

Molar-specific quantities can also be obtained:

$$\bar{g} \equiv \bar{h} - T\bar{s}. \tag{10.22c}$$

The Gibbs function is an important second-law property, which has many uses in dealing with ideal-gas mixtures. For example, the Gibbs function is used to determine the equilibrium composition of reacting gas mixtures (see Chapter 13).

For an ideal-gas mixture, the mass- or molar-specific Gibbs function is a weighted sum of the pure-species mass- or molar-specific Gibbs functions:

$$g_{\text{mix}}(T, P) = \sum_{i=1}^{J} Y_i\, g_i(T, P_i), \tag{10.23a}$$

$$\bar{g}_{\text{mix}}(T, P) = \sum_{i=1}^{J} X_i\, \bar{g}_i(T, P_i), \tag{10.23b}$$

where

$$g_i(T, P_i) = h_i(T) - Ts_i(T, P_i) = h_i(T) - T\left[s_i^\circ(T) - R\ln\frac{P_i}{P^\circ}\right] \tag{10.24a}$$

and

$$\bar{g}_i(T, P_i) = \bar{h}_i - T\left[\bar{s}_i^\circ(T) - R_{\text{u}}\ln\frac{P_i}{P^\circ}\right]. \tag{10.24b}$$

The connection to the mixture composition is through the ideal-gas relationship $P_i = X_i P$ (Eq. 10.11b).

The **Helmholtz free energy**, A, is also a composite property, defined similarly to the Gibbs free energy but with the internal energy replacing the enthalpy; that is,

$$A \equiv U - TS, \tag{10.25a}$$

or, per unit mass,

$$a \equiv u - Ts. \tag{10.25b}$$

Molar-specific quantities relate in the same manner as the mass-specific quantities in Eq. 10.25b. The Helmholtz free energy is useful in defining equilibrium conditions for reacting systems at constant volume and temperature. Although we make no use of the Helmholtz free energy in this book, you should be aware of its existence.

SUMMARY

This chapter introduced the reader to mixtures of ideal gases, and the myriad thermodynamic and thermophysical properties that are used throughout this book. The chapter also showed how these properties relate to one another through equations of state, calorific equations of state, and second-law (or Gibbs) relationships. As a further summary of this chapter, reviewing the learning objectives presented at the outset of this chapter is recommended.

KEY EQUATIONS

Review the most important equations presented in this chapter (i.e., those boxed with a light red background). What physical principles do they express? What restrictions apply?

CHAPTER 10 KEY CONCEPTS AND DEFINITIONS CHECKLIST

Numbers following arrows refer to the Questions and Problems at the end of the chapter.

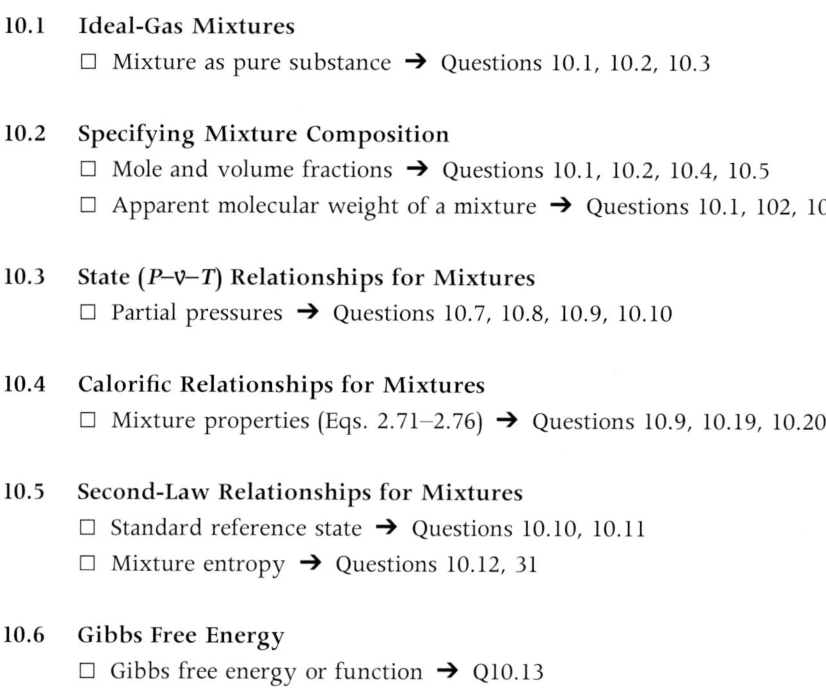

10.1 Ideal-Gas Mixtures

- ☐ Mixture as pure substance ➔ Questions 10.1, 10.2, 10.3

10.2 Specifying Mixture Composition

- ☐ Mole and volume fractions ➔ Questions 10.1, 10.2, 10.4, 10.5
- ☐ Apparent molecular weight of a mixture ➔ Questions 10.1, 102, 10.6

10.3 State (P–v–T) Relationships for Mixtures

- ☐ Partial pressures ➔ Questions 10.7, 10.8, 10.9, 10.10

10.4 Calorific Relationships for Mixtures

- ☐ Mixture properties (Eqs. 2.71–2.76) ➔ Questions 10.9, 10.19, 10.20

10.5 Second-Law Relationships for Mixtures

- ☐ Standard reference state ➔ Questions 10.10, 10.11
- ☐ Mixture entropy ➔ Questions 10.12, 31

10.6 Gibbs Free Energy

- ☐ Gibbs free energy or function ➔ Q10.13

REFERENCES

1. Kee, R. J., Rupley, F. M., and Miller, J. A., "The Chemkin Thermodynamic Data Base," Sandia National Laboratories Report SAND 87–8215 B, March 1991.
2. Stull, D. R., and Prophet, H., *JANAF Thermochemical Tables*, 2nd edn, NSRDS-NBS 37, National Bureau of Standards, June 1971. (The 3rd edn is available from NIST.)
3. Gordon, S., and McBride, B. J., "Computer Program for Calculation of Complex Chemical Equilibrium Compositions, Rocket Performance, Incident and Reflected Shocks, and Chapman–Jouguet Detonations," NASA SP-273, 1976.
4. Moran, M. J., and Shapiro, H. N., *Fundamentals of Engineering Thermodynamics,* 3rd edn, Wiley, New York, 1995.
5. Van Wylen, G. J., Sonntag, R. E., and Borgnakke, C., *Fundamentals of Classical Thermodynamics,* 4th edn, Wiley, New York, 1994.

QUESTIONS

10.1 Define a "pure substance".

10.2 Give an example of a mixture that can be analyzed as a pure substance.

10.3 Give an example of a mixture that cannot be analyzed as a pure substance.

10.4 Consider an ideal-gas mixture. Explain the differences and similarities among the following measures of composition: mole fraction, volume fraction, and mass fraction.

10.5 Derive an expression relating the mole fraction and mass fraction of species i in a mixture.

10.6 How is the apparent molecular weight used in thermodynamics calculations?

10.7 How does the partial pressure of a constituent in an ideal-gas mixture relate to the mole fraction of that constituent?

10.8 Compare and contrast Dalton's and Amagat's views of an ideal-gas mixture.

10.9 Explain in words how the mass-specific and molar-specific enthalpies of an ideal-gas mixture relate to the corresponding properties of the constituent species.

10.10 In Tables D, at what reference temperature and pressure is the entropy zero?

10.11 In Table C.2, at what reference temperature and pressure is the entropy zero? (**HINT**: refer to the table title and footnote.)

10.12 Refer to Eqs. 10.20 to determine how the specific entropy of a species changes when the pressure is reduced below P_{ref}. Explain in words why this is expected.

10.13 Using symbols, define the Gibbs free energy (of function) in terms of other thermodynamic properties.

Chapter 10	**Problem Subject Areas**
10.1–10.8	Specifying mixture composition
10.9–10.18	Ideal-gas mixtures: P–v–T relationships
10.19–10.27	Calorific relationships for mixtures
10.28–10.38	Second-law relationships for mixtures

PROBLEMS

10.1–10.8 Specifying mixture composition

10.1 A mixture of ideal gases contains 1 kmol of CO_2, 2 kmol of H_2O, 0.1 kmol of O_2, and 7.896 kmol of N_2. Determine the mole fractions and the mass fractions of each constituent. Also determine the apparent molecular weight of the mixture.

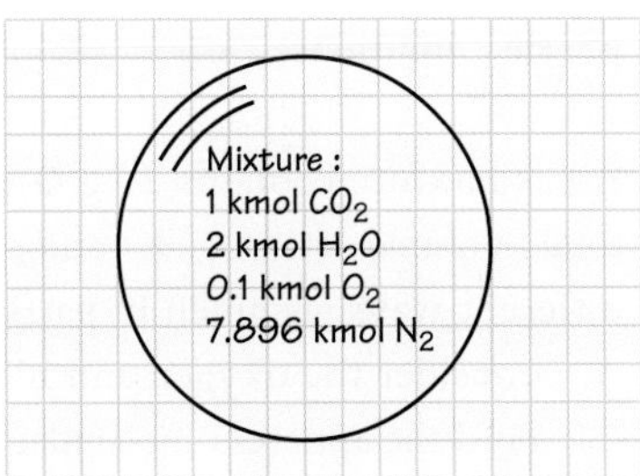

10.2 Determine the apparent molecular weight of synthetic air created by mixing 1 kmol of O_2 with 3.76 kmol of N_2. Also determine the mole and mass fractions of the O_2 in the mixture.

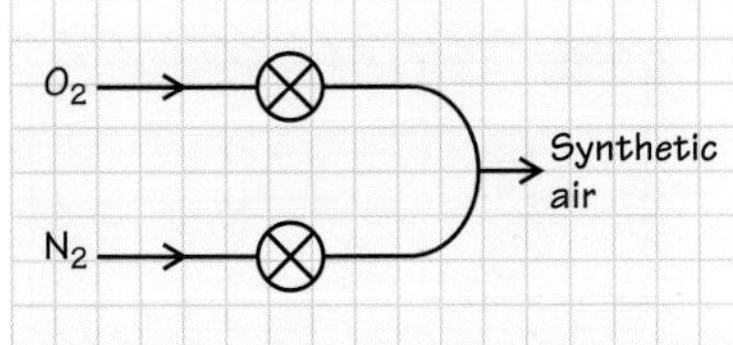

10.3 On a hot (95 F) summer day the relative humidity reaches an uncomfortable 85%. At this condition, the water vapor mole fraction is 0.056. Treating the moist air as an ideal-gas mixture of water vapor and dry air, determine the mass fraction of water vapor and the apparent molecular weight of the *moist* air. Treat the *dry* air as a simple substance with a molecular weight of 28.97 kg/kmol.

10.4 A gas mixture of O_2, N_2, and CO_2 contains 5.5, 3, and 1.5 kmol of each species, respectively. Determine the volume fractions of each component. Also determine the mass (kg) and molecular weight (kg/kmol) of the mixture.

10.5 For air containing 75.53% N_2, 23.14% O_2, 1.28% Ar, and 0.05% CO_2, by mass, determine the gas constant and its molecular weight. How do these values compare for a mass-based composition of 76.7% N_2 and 23.3% O_2?

10.6 Natural gas (methane) is mixed with air in a molar ratio of 1:6 (1 kmol of methane with 6 kmol of air). Determine the volume fractions of each component and the apparent molecular weight (kg/kmol) of the mixture. For this mixture, determine the volume of 3 kg at 300 K and 101 kPa (atmospheric conditions).

10.7 The NIST software approximates the molar content of air as 78.12% N_2, 20.96% O_2 and 0.92% Ar. Find the apparent molecular weight for the mixture. For this mixture, determine the volume of 3 kg at 300 K and 101 kPa (atmospheric conditions).

10.8 A gas mixture contains 2 kg O_2 and 5 kg N_2. For this mixture, determine the total volume at 300 K and 101 kPa (atmospheric conditions) in two different ways, as given in parts A and B below:

A. Consider the oxygen and nitrogen as being in separate chambers.

B. Find the apparent molecular weight, $\mathcal{M}_{mix}$, and gas constant, R_{mix}, for the mixture. Use R_{mix} in the ideal gas law to find the volume of the mixture.

C. Compare the total volumes that you found in parts a and b. Why do both methods give the same answer?

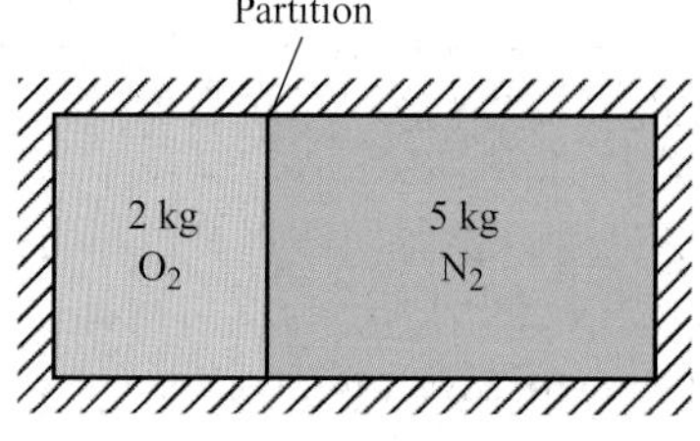

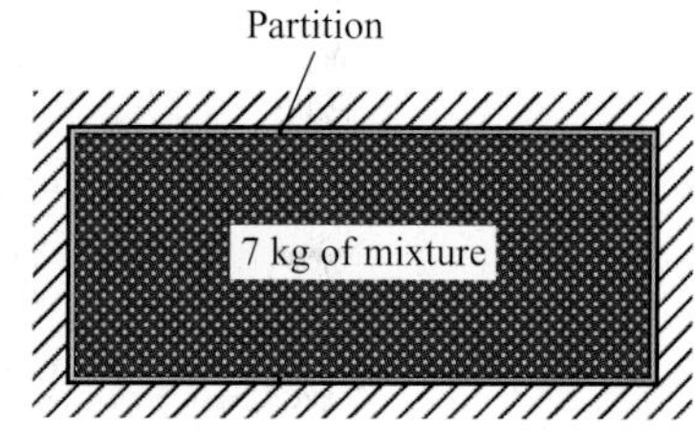

10.9–10.18 Ideal-gas mixtures: (P–$\mathcal{V}$–T) relationships

10.9 A 0.5-m^3 rigid vessel contains 1 kg of carbon monoxide and 1.5 kg of air at 15 °C. The composition of the air on a mass basis is 23.3% O_2 and 76.7% N_2. What are the partial pressures (kPa) of each component?

10.10 The gravimetric (mass) analysis of a gaseous mixture yields CO_2 = 32%, O_2 = 56.5%, and N_2 = 11.5%. The mixture is at a pressure of 3 psia. Determine (a) the volumetric composition and (b) the partial pressure of each component.

10.11 A 17.3-liter tank contains a mixture of argon, helium, and nitrogen at 298 K. The argon and helium mole fractions are 0.12 and 0.35, respectively. If the partial pressure of the nitrogen is 0.8 atm, determine (a) the total pressure in the tank, (b) the total number of moles (kmol) in the tank, and (c) the mass of the mixture in the tank.

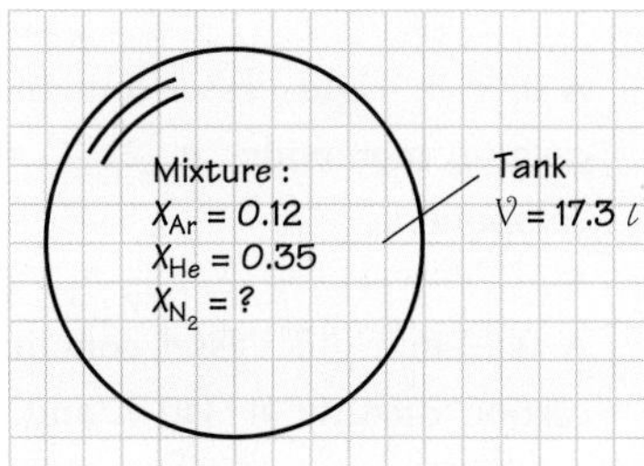

10.12 An instrument for the analysis of trace hydrocarbons in air, or in the products of combustion, uses a flame ionization detector. The flame in this device is fueled by a mixture of 40% (vol.) hydrogen and 60% (vol.) helium. The fuel mixture is contained in a 42.8-liter tank at 1500 psig and 298 K. The atmospheric pressure is 100 kPa. Determine (a) the partial pressure of each constituent in the tank and (b) the total mass of the mixture. Assume ideal-gas behavior for the mixture.

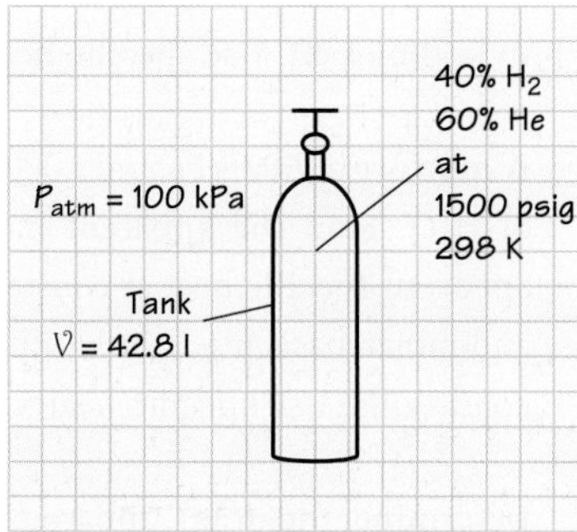

10.13 A rigid tank contains 5 kg of O_2, 8 kg of N_2, and 10 kg of CO_2 at 100 kPa and 1000 K. Assume that the mixture behaves as an ideal gas. Determine (a) the mass fraction of each component, (b) the mole fraction of each component, (c) the partial pressure of each component, (d) the average molar mass (apparent molecular weight) and the gas constant of the mixture, and (e) the volume of the mixture in m^3.

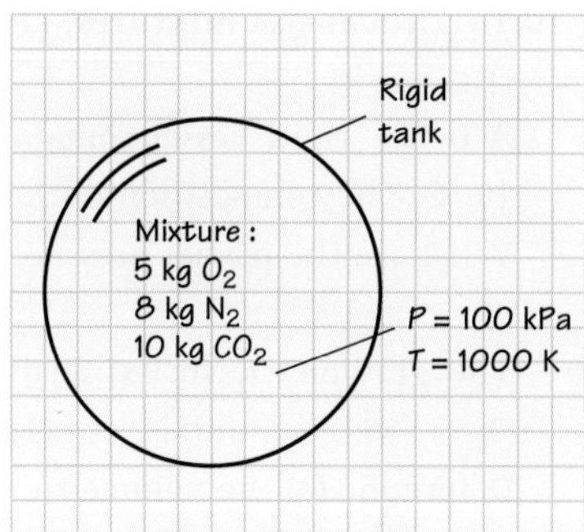

10.14 A 3-ft^3 rigid vessel contains a 50–50 mixture of N_2 and CO (by volume). Determine the mass of each component for $T = 65$ F and $P = 30$ psia.

10.15 A 1-m^3 tank contains nitrogen at 30 °C and 500 kPa. In an isothermal process, CO_2 is forced into this tank until the pressure is 1000 kPa. What is the mass (kg) of each gas present at the end of this process?

10.16 A 0.08-m^3 rigid vessel contains a 50–50 (by volume) mixture of nitrogen and carbon monoxide at 21 °C and 2.75 MPa. Determine the mass of each component.

10.17 A 0.12-m^3 rigid vessel contains a 50–50 (by mass) mixture of nitrogen and carbon dioxide at 340 K and 1.5 MPa. Determine the number of moles and partial pressure of each component.

10.18 A 3-m^3 rigid vessel contains a 50–50 mixture of N_2 and O_2 (by mass). Determine the number of moles of each component for $T = 350$ K and $P = 500$ kPa.

10.19–10.27 Calorific relationships for mixtures

10.19 A mixture at 400 kPa and 500 K contains 2 kmol O_2 and 8 kmol N_2. Find the apparent molecular weight of the mixture. Determine the apparent specific heat at constant volume using a mass-basis ($c_{v,\text{mix}}$ in kJ/kg · K) and molar-basis ($\bar{c}_{v,\text{mix}}$ in kJ/kmol · K). Use specific heat values for each species from Tables D.

10.20 A mixture at 500 kPa and 1000 K contains 2 kmol O_2, 8 kmol N_2, and 0.2 kmol of H_2O. Find the apparent molecular weight of the mixture. Determine the apparent specific heat at constant volume using a mass-basis ($c_{v,\text{mix}}$ in kJ/kg · K) and molar-basis ($\bar{c}_{v,\text{mix}}$ in kJ/kmol·K). Assume that water is an ideal gas in these conditions. Use specific heat values for each species from Tables D.

10.21 Determine the total apparent specific heat at constant pressure ($c_{p,\text{mix}}$ in kJ/kg · K) for a fuel–air reactant mixture containing 1 kmol CH_4, 2.5 kmol O_2, and 9.4 kmol N_2 at 500 K and 1 atm. Use specific heat values for each species from Table E.1.

10.22 A mixture of products of combustion contains the following constituents at 2000 K: 3 kmol of CO_2, 4 kmol of H_2O, and 18.8 kmol of N_2. Determine the following quantities:

A. The mole fraction of each constituent in the mixture

B. The apparent molecular weight of the mixture

C. The apparent constant-volume specific heat using a mass-basis ($c_{v,mix}$ in kJ/kg·K) and molar-basis ($\bar{c}_{v,mix}$ in kJ/kmol·K). Use specific heat values for each species from Appendix E.

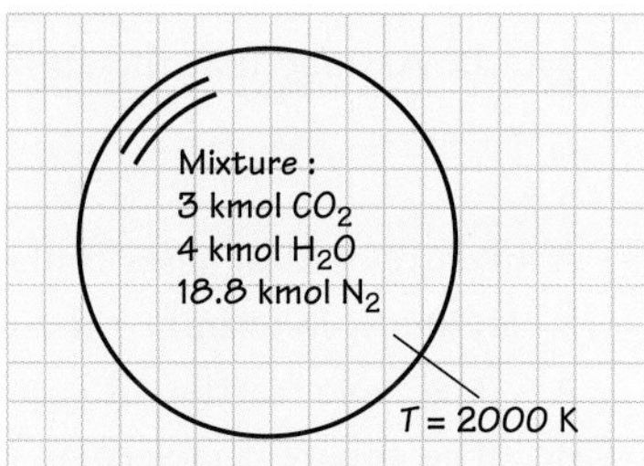

10.23 A 5-m^3 tank contains 2 kmol of oxygen and 5 kmol of nitrogen at 300 K.

a. Find the total pressure of the mixture.

b. Find the apparent constant-volume specific heat using a mass-basis ($c_{v,mix}$ in kJ/kg·K) and molar-basis ($\bar{c}_{v,mix}$ in kJ/kmol·K). Use specific heat values for each species from Table E.1.

c. Find the heat transfer required to raise the temperature of the mixture to 500 K using the apparent specific heat from part b. Assume that the apparent specific heat remains constant during this process.

d. Find the heat transfer required to raise the temperature of the mixture to 500 K using sensible enthalpy values, $\bar{h}^{\circ}(T) - \bar{h}_f^{\circ}(298\ \text{K})$, from Table D.

10.24 Determine the total apparent specific heat at constant volume ($c_{v,mix}$ in kJ/kg·K) for a fuel–air reactant mixture containing 1 kmol CH_4, 2.5 kmol O_2, and 9.4 kmol N_2 at 500 K and 1 atm. What heat transfer is required to cool the mixture from 500 K to 300 K? Use specific heat values for each species from Table E.1.

10.25 A mixture of products of combustion contains the following constituents at 1500 K: 3 kg of CO_2, 1 kg of H_2O, and 14 kg of N_2. Determine the following quantities:

A. The mole fraction of each constituent in the mixture

B. The apparent molecular weight of the mixture

C. The volume of the mixture when the pressure is 800 kPa

D. The apparent specific heat at constant volume using a mass basis ($c_{v,mix}$ in kJ/kg·K) and molar basis ($\bar{c}_{v,mix}$ in kJ/kmol·K). Use specific heat values for each species from Tables D.

d. Find the heat transfer required to cool the temperature of the mixture from 1500 K to 700 K at constant volume using the apparent specific heat from part D. Assume that the apparent specific heat remains constant during this process.

10.26 A mixture of ideal gases contains 1 kmol of CO_2, 0.2 kmol of H_2O, 2 kmol of O_2, and 6 kmol of N_2 at 600 K. Determine the following quantities:

A. The mole fraction of each constituent in the mixture

B. The apparent molecular weight of the mixture

C. The volume of the mixture when the pressure is 1.2 Mpa

D. The apparent specific heat at constant volume using a mass basis ($c_{v,\text{mix}}$ in kJ/kg·K) and molar-basis ($\bar{c}_{v,\text{mix}}$ in kJ/kmol·K). Use specific heat values for each species from Table E.1.

E. Find the heat transfer required to raise the temperature of the mixture from 600 K to 900 K at constant volume using the apparent specific heat from part D. Assume that the apparent specific heat remains constant during this process.

10.27 A mixture of ideal gases contains 0.5 kmol of CO_2, 2 kmol of O_2, and 7 kmol of N_2 at 700 K. Determine the following quantities:

A. The mole fraction of each constituent in the mixture

B. The apparent molecular weight of the mixture

C. The volume of the mixture when the pressure is 600 kPa

D. The apparent specific heat at constant pressure using a mass basis ($c_{p,\text{mix}}$ in kJ/kg·K) and molar basis ($\bar{c}_{p,\text{mix}}$ in kJ/kmol·K). Use specific heat values for each species from Table E.1.

E. Find the heat transfer required to decrease the temperature of the mixture from 900 K to 500 K at constant pressure using the apparent specific heat from part D. Assume that the apparent specific heat remains constant during this process.

10.28–10.38 Second-law relationships for mixtures

10.28 Determine the change in specific entropy (kJ/kg·K) of CO for a process from 4 atm and 2500 K to 1 atm and 1400 K.

10.29 Determine the change in specific entropy (kJ/kg·K) of CO_2 for a process from 1 atm and 500 K to 3 atm and 1400K.

10.30 Calculate the change in entropy for mixing 2 kmol of O_2 with 6 kmol of N_2. Both species are initially at 1 atm and 300 K, as is the final mixture.

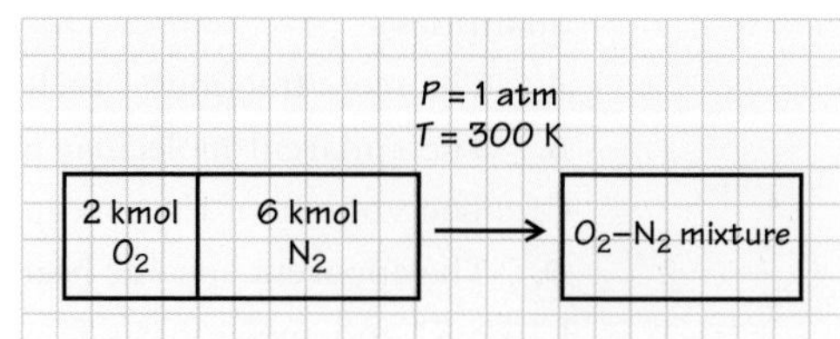

10.31 Calculate the change in entropy on mixing 4 kmol of O_2 with 1 kmol of CO_2. Both species are initially at 2 atm and 500 K, as is the final mixture.

10.32 Consider 2 kmol of O_2 and 5 kmol of N_2 separated by a partition. The oxygen is at 400 K and 200 kPa. The nitrogen is at 300 K and 450 kPa. The partition is removed and the O_2 and N_2 mix. Determine the final temperature and pressure of the mixture and the entropy change associated with this mixing process.

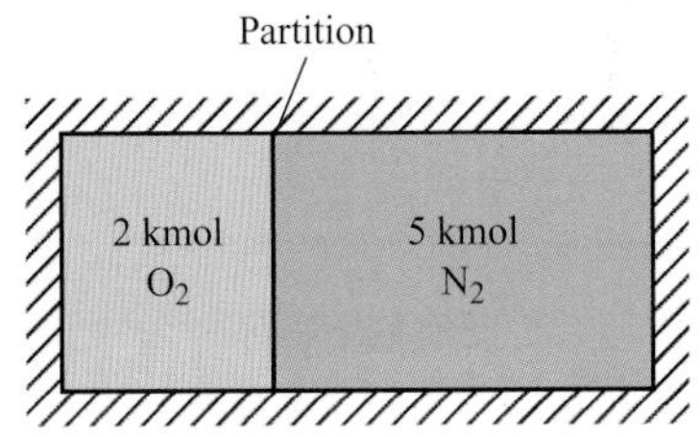

10.33 Consider 2 kg of CO_2 and 5 kg of N_2 separated by a partition. The carbon dioxide is at 300 K and 200 kPa. The nitrogen is at 500 K and 350 kPa. The partition is removed and the CO_2 and N_2 mix. Determine the final temperature and pressure of the mixture and the entropy change associated with this mixing process.

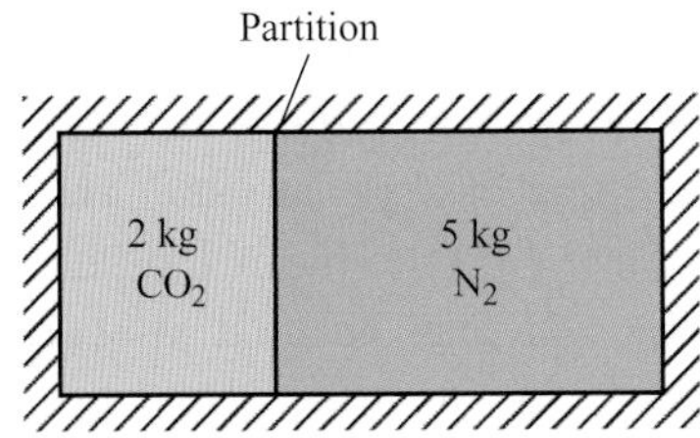

10.34 A mixture of 15% CO_2, 12% O_2, and 73% N_2, (by volume) expands to a final volume six times greater than its initial volume. The corresponding temperature change is 1000 °C to 750 °C. Determine the entropy change (kJ/kg·K).

10.35 Determine the change in entropy (kJ/kg·K) of a mixture of 60% N_2 and 40% CO_2 by volume for a reversible adiabatic increase in volume by a factor of 5. The initial temperature is 540 °C.

10.36 Determine the change in entropy (kJ/kg·K) of a mixture of 80% N_2 and 20% O_2 by volume for a reversible adiabatic decrease in volume by a factor of 4. The initial temperature is 400 K.

10.37 A partition separating a chamber into two compartments is removed. The first compartment initially contains oxygen at 600 kPa and 100 °C; the second compartment initially contains nitrogen at the same pressure and temperature. The oxygen compartment volume is twice that of the compartment containing nitrogen. The chamber is isolated from the surroundings. Determine the change of entropy associated with the mixing of the O_2 and N_2. **HINT**: Assume that the process is isothermal.

10.38 Consider two compartments of the same chamber separated by a partition. Both compartments contain nitrogen at 600 kPa and 100 °C; however, the volume of one compartment is twice that of the other. The partition is removed. Assuming the chamber is isolated from the surroundings, determine the entropy change of the N_2 associated with this process.

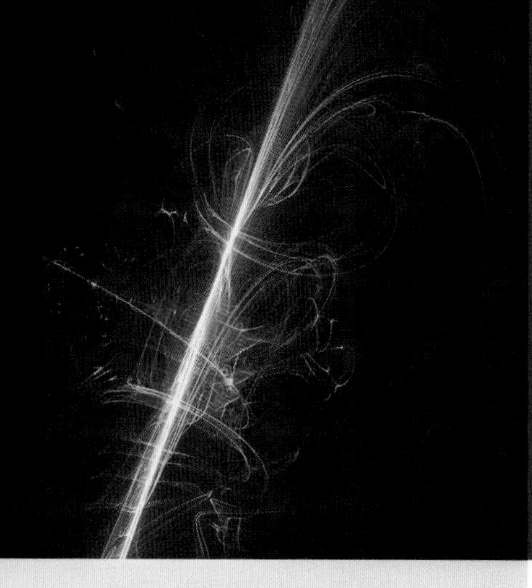

CHAPTER 11
Air–Vapor Mixtures

LEARNING OBJECTIVES

After studying Chapter 11, you should be able to:

- List several devices that process moist air.
- Understand the concepts of specific humidity, relative humidity, and dew point.
- Apply these concepts in the analysis of evaporative coolers, humidifiers, air conditioners, dehumidifiers, cooling towers, or other systems involving moist air.
- Apply the conservation of mass and conservation of energy equations to find the specific or relative humidity of the outflow of a moist air device.
- Use dry-bulb temperature and wet-bulb temperatures to find the relative humidity.

CHAPTER 11 OVERVIEW

CHAPTER 10 CONSIDERED MIXTURES OF ideal gases. In this chapter, we will apply a mixture analysis to investigate air–water mixtures, referred to as moist air. Since the water content in the air is relatively low, the partial pressure of the water is low. At low partial pressures, the water vapor can be approximated as an ideal gas and the moist air is an ideal-gas mixture. This chapter will first define some terms commonly used for moist air: specific humidity, relative humidity, and dew point. The analysis of moist air will then be used in several common applications: evaporative coolers, humidifiers, air conditioners, dehumidifiers, and cooling towers.

DESIGN AND ANALYSIS OF SYSTEMS		
ANALYSIS OF PRACTICAL DEVICES		
CONSERVATION OF MASS	CONSERVATION OF ENERGY (First Law of Thermodynamics)	SECOND LAW OF THERMODYNAMICS (Entropy)
PROPERTIES OF MATTER		
FRAMEWORKS FOR ANALYSIS, KEY CONCEPTS, AND DEFINITIONS		

FIGURE 11.1 Hierarchical arrangements of the topics in our study of mixtures.

11.1 Air Conditioning, Humidification, and Related Systems

This section focuses primarily on devices that provide comfortable indoor climates: humidifiers, dehumidifiers, and air conditioners. We also consider cooling towers. What distinguishes these devices from others we have studied is the important role that the moisture in the air now plays. The commonplace statement, "It's not the heat, but the humidity," implies a strong connection between the amount of moisture in the air and human comfort. In this section, we apply mass and energy conservation principles together with thermodynamic property relationships for mixtures to develop an understanding of these devices. We also introduce and apply the specialized concepts and nomenclature that are normally associated with these devices: specific humidity, relative humidity, and dew point.

11.2 Physical Systems

Figures 11.2–11.6 schematically illustrate the following devices:

- evaporative coolers (Fig. 11.2),
- humidifiers (Fig. 11.2),
- air conditioners (Figs. 11.3 and 11.4),
- dehumidifiers (Figs. 11.3 and 11.5), and
- cooling towers (Fig. 11.6).

Evaporative coolers (Fig. 11.2) function by blowing warm dry air through a water spray or a porous medium saturated with liquid water. Because the warm air supplies the energy to evaporate some of the liquid water, the air is cooled. This mechanical method of cooling resembles the natural evaporation of perspiration that regulates the body temperatures of humans and other animals that perspire. Evaporative coolers, also known as "swamp coolers," work best when the moisture content of the incoming air is relatively low, as the addition of moisture can detract from the comfort associated with the cooling. Evaporative coolers are inexpensive because of their simplicity and use less energy than refrigeration-cycle-based air conditioners.

Humidifiers function in the same manner as evaporative coolers; the desired effect, however, is now the increased moisture content of the air rather than a lower temperature.

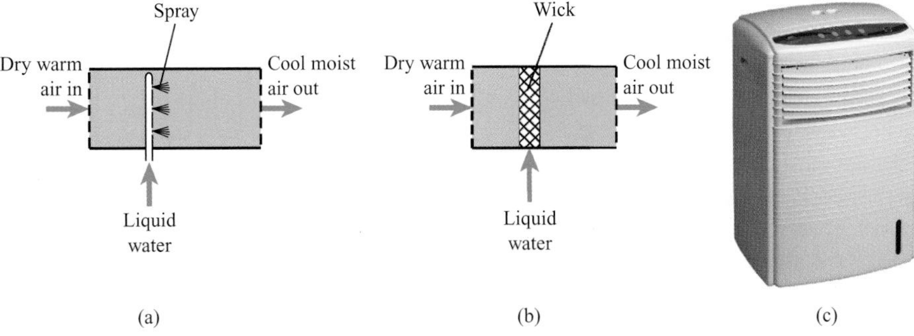

FIGURE 11.2 Evaporative coolers and humidifiers both work on the same principles. The air to be cooled or moistened is drawn through a duct in which water is **(a)** sprayed or **(b)** contained in a wick or other porous material. Internal mass and energy transfers result in cool, moist air exiting the duct. **(c)** Evaporative cooler photo. Image courtesy of Beaumark Ltd.

Air-conditioning systems (Figs. 11.3 and 11.4) rely on active cooling (and, sometimes, heating). Most residential and other relatively small-scale systems use a vapor-compression refrigeration cycle to provide the cooling. In Fig. 11.4, we see that the air-conditioner cooling coil is the evaporator of the core refrigeration cycle. The condenser is located outdoors, adjacent to the evaporator, in the interior in a window unit or at a distant location in central systems. The condensate from the cold side is usually led to the outside. On particularly hot and humid days, you can easily observe dripping air conditioners. Large-scale air-conditioning systems, such as those used in office buildings, shopping malls, and other large buildings, or a complex of buildings, utilize systems that chill water rather than air, as chilled water is more easily circulated through a large building or complex of buildings than is air. A "chiller" is an integral part of such systems. Cycles other than the simple vapor-compression refrigeration cycle are also employed when efficiency is particularly important, or when particular circumstances dictate. Among these are cascade refrigeration cycles and gas-absorption systems. Discussion of these systems is beyond the scope of this book and we refer the interested reader to Refs. [1, 2].

Dehumidifiers (Fig. 11.5) operate in a manner very similar to air conditioners. The condenser (heating coil), however, is located in the air stream following the evaporator (cooling coil). The moisture condensed from the incoming air is caught in a container or led through a tube to a drain. The figure also shows photographs of a home dehumidifier.

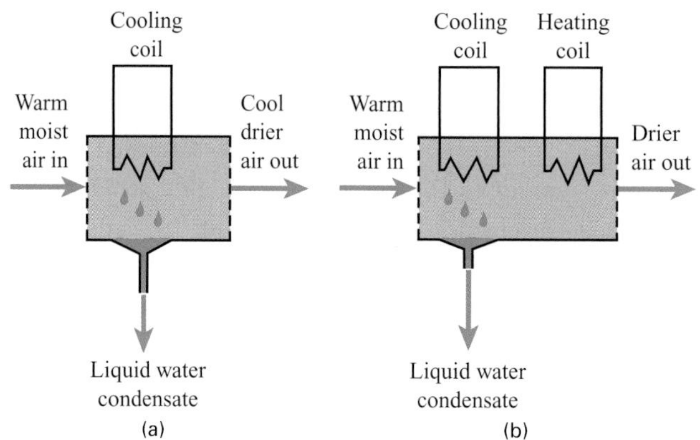

FIGURE 11.3 Air is cooled and moisture removed at the cooling coil in an air conditioner **(a** and **b)** and a dehumidifier **(b)**. In some air conditioners the cold air may also be heated to provide a desired exit temperature **(b)**.

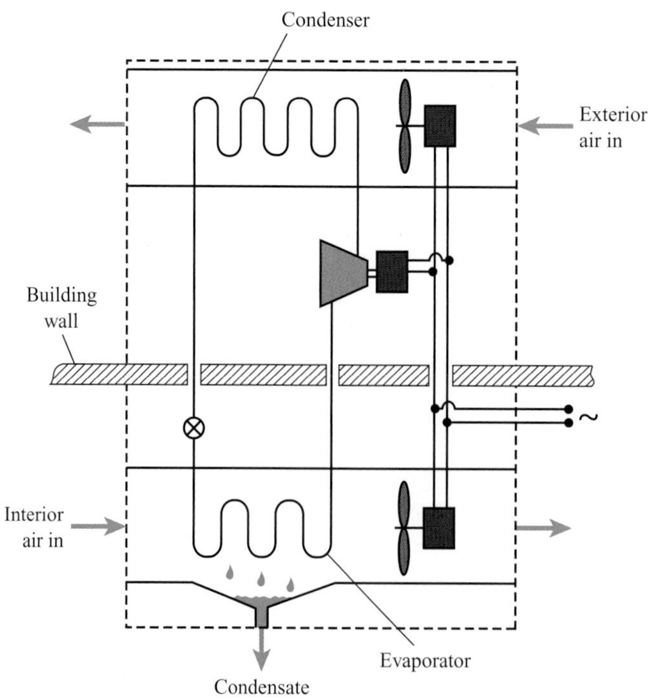

FIGURE 11.4 Schematic diagram of a residential air-conditioning system. At the heart of the air conditioner is the vapor-compression refrigeration cycle comprising an evaporator, compressor, condenser, and expansion valve. For window or wall units all components are housed in a single cabinet, whereas for central systems the evaporator is usually part of the indoor furnace assembly and the condenser is a stand-alone unit located outdoors.

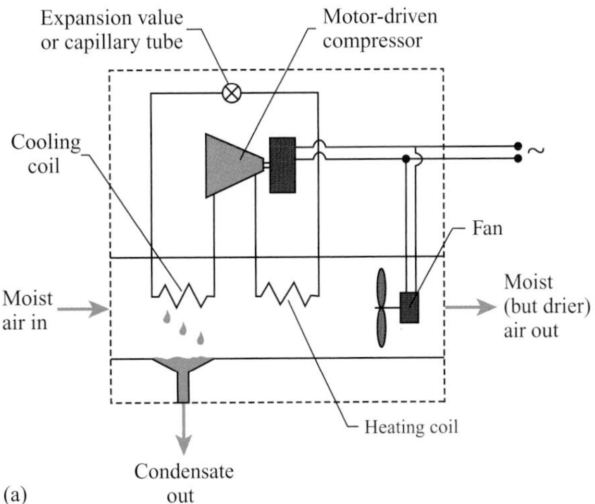

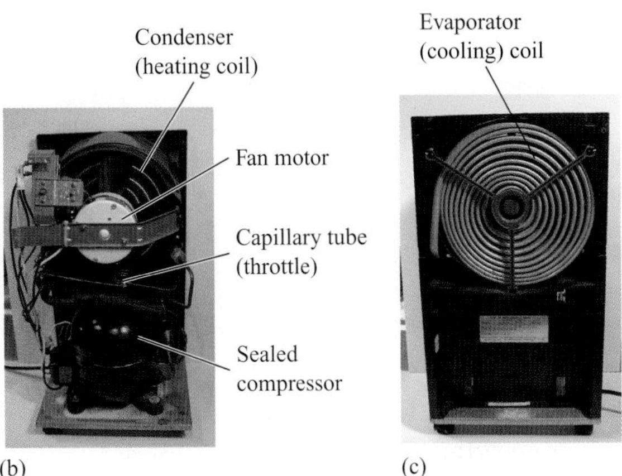

FIGURE 11.5 (a) Schematic diagram of home dehumidifier. Note the similarity to the air-conditioner system illustrated in Fig. 11.4, with the primary difference that the condenser (heating coil) now heats the dehumidified air. Photograph of residential dehumidifier. Front view **(b)** and rear view **(c)**. (Images courtesy S. R. Turns.)

The final system we present, a cooling tower, differs significantly in both scale and application from those already discussed. Cooling towers are huge structures and frequently dominate the view of steam power generating plants. Heights range from less than 15 m (50 ft) to more than 175 m (575 ft). Cooling towers provide a semiclosed loop to the cooling side of steam condensers, as shown in Fig. 11.6. The cooling water in a steam condenser exits at temperatures around 40°C (100 F). This hot water is sprayed into the upflowing air stream in the tower. The water droplets in the spray cool as they partially evaporate in the air. This process is similar to that used by the evaporative cooler previously discussed, except the desired effect is to cool the water, not the air. The cooled water is collected at the bottom of the tower and pumped back to the condenser. Make-up water is added to compensate for the evaporation. Depending on the tower design, the air stream can be pulled through the tower by fans or can flow as a result of natural convection since the density of the warm, moist air is less than that of the ambient air. So-called drift eliminators are used to prevent a large carryover of the spray out of the tower with the air. One can frequently observe a row of small clouds above a tower formed by the condensation of the moisture from the tower air.

11.3 General Analysis

To develop a basic understanding of all these devices or systems, we apply basic mass and energy conservation principles to a generic open system (control volume) in which

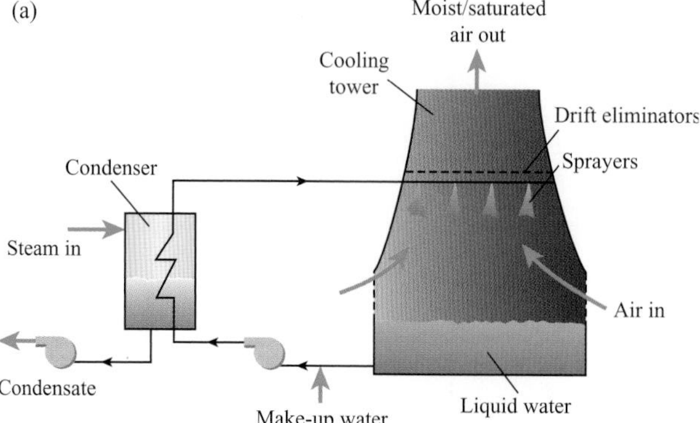

(b)

FIGURE 11.6 (a) Schematic diagram of a cooling tower used in steam power plants. Hot water is sprayed in the cooling tower and cooled by the air that enters at the base of the tower. The cold water is then pumped to the steam condenser. **(b)** Photo of cooling towers from Ratcliffe power station.
(Image by Bill Brooks / Moment / Getty Images.)

any or all the important processes associated with these devices occur. Figure 11.7a shows this open system. Here moist air enters at station 1 and is then cooled or heated, or both, while liquid water is removed (station 2) or added (station 3), or both. The air stream then exits at station 4 with more or less moisture than it had upon entering.

11.3a Assumptions

To perform our analysis, we assume the following:

- The flow is steady.
- The flow is at steady state with no accumulation or loss of mass within the open system.
- No air is dissolved in the liquid water.
- The kinetic and potential energies of all streams are negligible.
- Both the dry air and water vapor behave as ideal gases.

11.3b Mass Conservation

We can write three different mass conservation expressions for the situation shown in Fig. 11.7a: one for the dry air[1] alone, one for the H_2O alone, and an overall or combined relationship. The individual flows of dry air and H_2O are shown in Fig. 11.7b.

[1] Note that *dry air* has the composition defined in Appendix C and should not be confused with the "simple air" defined to aid the analysis of combustion processes. The molecular weight of dry air is 28.97 kg/kmol.

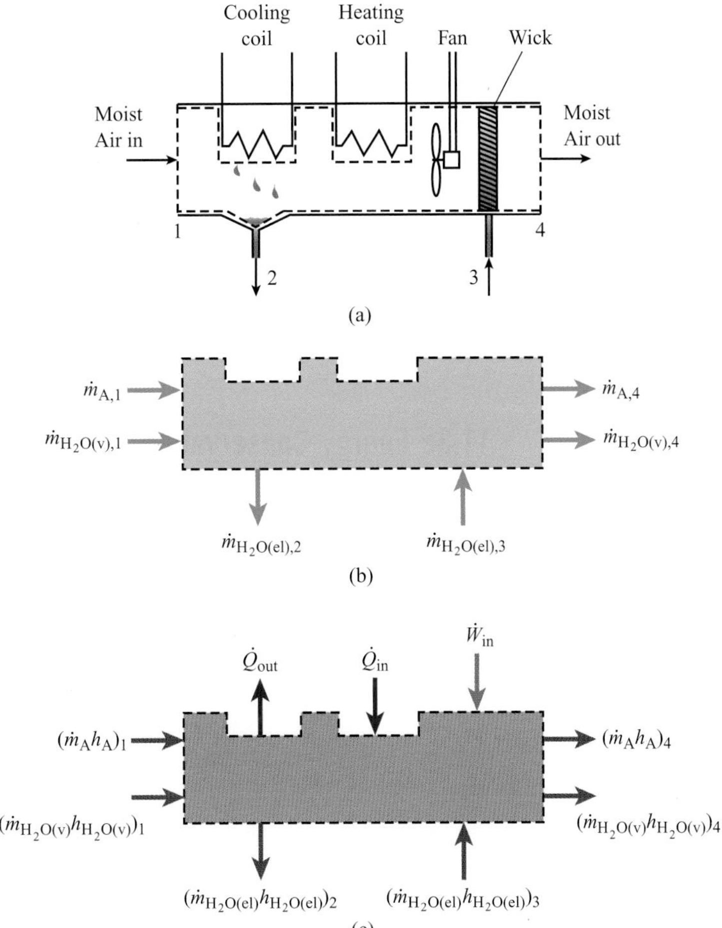

FIGURE 11.7 (a) General open system for analysis of moisture-laden air streams. **(b)** Open system illustrating mass flows. **(c)** Open system illustrating energy flows. In **(b)** and **(c)** the subscript $H_2O(l)$ refers to liquid water and the subscript $H_2O(v)$ refers to water vapor.

Dry Air. With the assumptions just listed, mass conservation for the dry air is written

$$\dot{m}_{A,1} = \dot{m}_{A,4} \equiv \dot{m}_A = \text{constant}. \tag{11.1}$$

Because no dry air enters or exits at any station other than 1 or 4 (see Fig. 11.7b), we obtain the obvious result that the mass flow of dry air in at station 1 equals the flow out at station 4. Because there is only a single value for the dry-air flow rate, we drop the subscripts 1 and 4 and designate this flow rate simply as $\dot{m}_A$.

H_2O. We apply mass conservation to the chemical compound H_2O, taking into account that the H_2O exists in two phases, liquid and vapor. Simply, the sum of all H_2O mass flows entering must equal the sum of all H_2O flows exiting, that is,

$$\sum_i^{\text{inlets}} \dot{m}_{H_2O,i} = \sum_i^{\text{outlets}} \dot{m}_{H_2O,i}, \tag{11.2a}$$

or

$$\dot{m}_{H_2O(v),1} + \dot{m}_{H_2O(l),3} = \dot{m}_{H_2O(v),4} + \dot{m}_{H_2O(l),2}. \tag{11.2b}$$

Overall. Overall mass conservation is formally the sum of Eqs. 11.1 and 11.2, or, physically, the sum of all mass flows in must equal the sum of all mass flows out:

$$\dot{m}_{A} + \dot{m}_{H_2O(v),1} + \dot{m}_{H_2O(l),3} = \dot{m}_{A} + \dot{m}_{H_2O(v),4} + \dot{m}_{H_2O(l),2}. \qquad (11.3)$$

The mass flow rate of moist air is thus the sum of the dry-air mass flow and the corresponding mass flow of water vapor (e.g., $\dot{m}_{\text{moist air},1} = \dot{m}_{A} + \dot{m}_{H_2O(v),1}$).

From Eqs. 11.2–11.3, we see that, conceptually, mass conservation is straightforward and intuitive. Later, this result will be rewritten by expressing the amount of water vapor using the specific humidity.

11.3c Energy Conservation

With our assumptions, the general steady-flow expression of energy conservation,

$$\sum \dot{E}_{in} = \sum \dot{E}_{out},$$

simplifies to

$$\begin{aligned} &\dot{Q}_{in} + \dot{W}_{in} + \dot{m}_{A}h_{A,1} + \left(\dot{m}_{H_2O(v)}h_{H_2O(v)}\right) + \left(\dot{m}_{H_2O(l)}h_{H_2O(l)}\right)_3 \\ &= \dot{Q}_{out} + \left(\dot{m}_{H_2O(l)}h_{H_2O(l)}\right)_2 + \dot{m}_{A}h_{A,4} + \left(\dot{m}_{H_2O(v)}h_{H_2O(v)}\right)_4. \end{aligned} \qquad (11.4)$$

In the analysis of the devices and systems just discussed here, the combined application of mass and energy conservation allows the determination of unknown moisture fractions or temperatures; for example, one may desire to determine the outlet temperature and relative humidity for an air-conditioner stream given the inlet conditions and cooling rate.

11.4 Some New Concepts and Definitions

In this section, we define and discuss commonly used terms related to mixtures of air and water vapor.

11.4a Psychrometry

The term **psychrometry** refers generally to the art and science of measuring the moisture content in dry air; thus, the concepts discussed in the following are topics within the field of psychrometry. The water content of air is important indoors and outside. Indoors we often control the humidity of the air to be within a comfortable range. In the summer, weather reports often refer to the "heat index" to describe the "heaviness" of the humid air as causing a feeling of hotter air temperature. (We have probably heard or made comments about the "dry heat" in desert climates being less uncomfortable.) In the summer, we can reduce the humidity of the inside air using an air conditioner or dehumidifier. Reducing the water content of warm air increases the comfort level. In the winter, the air indoors can become too dry and cause the contraction of wood furniture and the panels in wood doors. A humidifier might be used in the winter to increase the moisture of the inside air and improve comfort.

The concepts of wet-bulb and dry-bulb temperatures and the use of the psychrometric chart, although generally a part of psychrometrics, are included in a later section. References [1, 2], among many others, provide information on these topics.

11.4b Thermodynamic Treatment of Water Vapor in Dry Air

With the previously given assumption that moist air is a mixture of two ideal gases – dry air and water vapor – the total pressure of the mixture is expressed (Eq. 10.10) as

$$P = P_{\mathrm{A}} + P_{\mathrm{H_2O(v)}}. \tag{11.5}$$

The temperature, however, is uniform throughout the gas-phase mixture, and so we write

$$T = T_{\mathrm{A}} = T_{\mathrm{H_2O(v)}}. \tag{11.6}$$

Usually the total pressure (e.g., the local barometric pressure) is known, whereas the partial pressure of the water vapor depends on how much moisture is present within the mixture. For an ideal-gas mixture, the mole fraction and mass fraction of the water vapor in the dry air–water vapor mixture are defined, respectively, as

$$X_{\mathrm{H_2O(v)}} = \frac{N_{\mathrm{H_2O(v)}}}{N_{\mathrm{A}} + N_{\mathrm{H_2O(v)}}} = \frac{P_{\mathrm{H_2O(v)}}}{P}, \tag{11.7}$$

and

$$Y_{\mathrm{H_2O(v)}} = \frac{M_{\mathrm{H_2O(v)}}}{M_{\mathrm{A}} + M_{\mathrm{H_2O(v)}}}. \tag{11.8}$$

In psychrometry, however, the composition is usually defined by the humidity ratio or relative humidity rather than by using mole or mass fractions. We now define and discuss these composition variables.

11.4c Humidity Ratio

The **humidity ratio** ω (also known as the **specific humidity** or the **absolute humidity**) is defined as the mass of water vapor per unit mass of dry air:

$$\omega \equiv \frac{M_{\mathrm{H_2O(v)}}}{M_{\mathrm{A}}}. \tag{11.9}$$

We can easily relate the humidity ratio to the water-vapor mass fraction as follows:

$$Y_{\mathrm{H_2O(v)}} = \frac{M_{\mathrm{H_2O(v)}}}{M_{\mathrm{H_2O(v)}} + M_{\mathrm{A}}} = \frac{M_{\mathrm{H_2O(v)}}/M_{\mathrm{A}}}{M_{\mathrm{H_2O(v)}}/M_{\mathrm{A}} + 1} = \frac{\omega}{\omega + 1}. \tag{11.10a}$$

Solving for ω gives

$$\omega = \frac{Y_{H_2O(v)}}{1 - Y_{H_2O(v)}}. \tag{11.10b}$$

Applying the relationship between mass and mole fractions (Eqs. 10.4), we also obtain

$$X_{H_2O} = \frac{N_{H_2O}}{N_A + N_{H_2O}} = \frac{M_{H_2O}/\mathcal{M}_{H_2O}}{M_A/\mathcal{M}_A + M_{H_2O}/\mathcal{M}_{H_2O}} = \frac{\omega \mathcal{M}_A}{\mathcal{M}_{H_2O} + \omega \mathcal{M}_A}. \tag{11.10c}$$

Solving for ω gives

$$\omega = \frac{X_{H_2O(v)} \mathcal{M}_{H_2O}}{\left(1 - X_{H_2O(v)}\right) \mathcal{M}_A}. \tag{11.10d}$$

The humidity ratio can also be related to the water-vapor partial pressure. In Example 11.2, we will see that water vapor at its low partial pressure is accurately approximated as an ideal gas. Applying the ideal-gas equation of state independently to both the water vapor and the dry air (see Eq. 10.6), we have

$$M_{H_2O(v)} = \frac{P_{H_2O(v)} \mathcal{V}_{mix}}{(R_u/\mathcal{M}_{H_2O})T}$$

and

$$M_A = \frac{P_A \mathcal{V}_{mix}}{(R_u/\mathcal{M}_A)T} = \frac{\left(P - P_{H_2O(v)}\right) \mathcal{V}_{mix}}{(R_u/\mathcal{M}_A)T}.$$

Substituting these expressions into our definition of the humidity ratio (Eq. 11.9) yields

$$\omega = \frac{\mathcal{M}_{H_2O}}{\mathcal{M}_A} \frac{P_{H_2O(v)}}{P - P_{H_2O(v)}} \tag{11.11a}$$

or

$$\omega = 0.622 \frac{P_{H_2O(v)}}{P - P_{H_2O(v)}}, \tag{11.11b}$$

where numerical values for $\mathcal{M}_{H_2O}(= 18.016)$ and $\mathcal{M}_A(= 28.97)$ have been substituted. The inverse of this relationship is sometimes useful:

$$P_{H_2O(v)} = \frac{\omega P}{0.622 + \omega}. \tag{11.11c}$$

11.4d Relative Humidity

The relative humidity is the ratio of the actual amount of water vapor in the air at a given temperature and the maximum possible amount that the air can hold at that same temperature. Before expressing this definition symbolically, we explore its meaning using a concrete example.

Consider a mixture of dry air and water vapor at 25 °C with a total pressure of 1 atm (101.325 kPa) and a water-vapor partial pressure of 2.3393 kPa. The state of the water vapor is indicated as point A in Fig. 11.8, and the water-vapor mole fraction can be found by applying Eq. 11.7:

$$X_{H_2O(v)} = \frac{P_{H_2O(v)}}{P} = \frac{2.3393 \text{ kPa}}{101.325 \text{ kPa}} = 0.02309.$$

We now conduct a thought experiment in which we increase the amount of moisture present while maintaining the total pressure fixed and the temperature at 25 °C. This process continues until the vapor pressure of the water in the air equals the saturation pressure at 25 °C, point B in Fig. 11.8. Further addition of moisture results in a two-phase system with liquid water condensing out of the air–water vapor mixture. The amount of moisture in the air at point B is thus the maximum quantity that the air can hold at 25 °C, and the vapor pressure is given by $P_v = P_{sat}$ (25 °C) = 3.1699 kPa. The corresponding mole fraction is

$$X_{H_2O(v)\max @25°C} = \frac{P_{sat}(25°C)}{P} = \frac{3.1699 \text{ kPa}}{101.325 \text{ kPa}} = 0.03128.$$

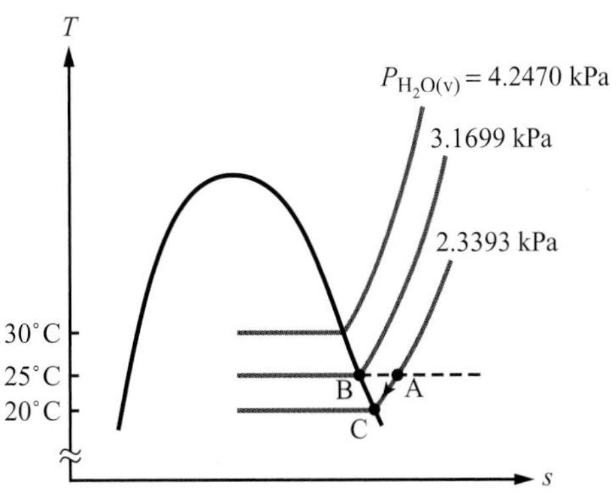

FIGURE 11.8 A T–s diagram for H_2O (not to scale) shows isobars corresponding to saturation temperatures of 20 °C, 25 °C, and 30 °C. For air at 25 °C, the maximum possible water vapor partial pressure is 3.1699 kPa. At this condition, the air is saturated with water (point B) and the relative humidity is 100%. Partial pressures less than this result in relative humidities less than 100%; for example, at point A the relative humidity is 73.8% [= (2.3393 / 3.1699) × 100%] and the corresponding dew-point temperature is T_{DP} = 20 °C (point C).

The relative humidity ϕ for this experiment is thus the ratio of the actual water vapor content $X_{H_2O(v)actual}$ to the maximum possible, $X_{H_2O(v)\max @25°C}$, that is,

$$\frac{X_{H_2O(v)actual}}{X_{H_2O(v)\max @25°C}} \frac{0.02309}{0.03128} = 0.738 \quad \text{or} \quad 73.8\%.$$

We now generalize and define the **relative humidity** as

$$\phi \equiv \frac{X_{H_2O(v)actual@T}}{X_{H_2O(v)\max @T}} = \frac{P_{H_2O(v)}(T)}{P_{sat}(T)}, \tag{11.12}$$

which is frequently expressed on a percentage basis by multiplying by 100%. Note that using a mass basis to define relative humidity also yields the result that $\phi = P_{H_2O(v)}(T)/P_{sat}(T)$. Note also that this definition of relative humidity assumes that the H_2O liquid–vapor equilibrium is unaffected by the presence of the air. Although the air does alter this equilibrium, the effect is very small.

11.4e Dew Point

The **dew point,** or **dew-point temperature** T_{DP}, is the temperature at which moisture begins to condense out of an air–water vapor mixture upon cooling. You are most likely familiar with the appearance of liquid water on the outside of a can or glass containing a cold beverage on a hot, humid day. Here the temperature at the outside surface of the container is at or below the dew point. Point C on Fig. 11.8 indicates the dew point (20 °C) associated with a 25 °C mixture having a relative humidity of 73.8%. Formally, the dew point is defined as

$$T_{DP} \equiv T_{sat}\left(P_{sat} = P_{H_2O(v)}\right), \tag{11.13}$$

where $P_{H_2O(v)}$ is the partial pressure of water vapor in the moist air under consideration.

The dew point is easily related to both the humidity ratio and the relative humidity. Knowing the dew point, we can find the humidity ratio by using Eq. 11.11a, where $P_{H_2O(v)} = P_{sat}(T_{DP})$. With the additional knowledge of the mixture temperature, the relative humidity can also be determined from its definition in Eq. 11.12. We illustrate these relationships with the following example.

Example 11.1 Conditions of Moist Air

(Credit: Image Source / Getty Images.)

On August 6, 2002, the local weather station at State College, Pennsylvania, reported an ambient temperature of 18 °C (64 F) and a dew point of 8 °C (46 F). The local barometric pressure (not corrected to sea level) was 29.16 in Hg. Determine the relative humidity, the humidity ratio, the water-vapor mole fraction, and the water-vapor mass fraction associated with the reported measurements.

Solution

Known $T,\ T_{DP},\ P$

Find $\phi,\ \omega,\ X_{H_2O(v)},\ Y_{H_2O(v)}$

Sketch

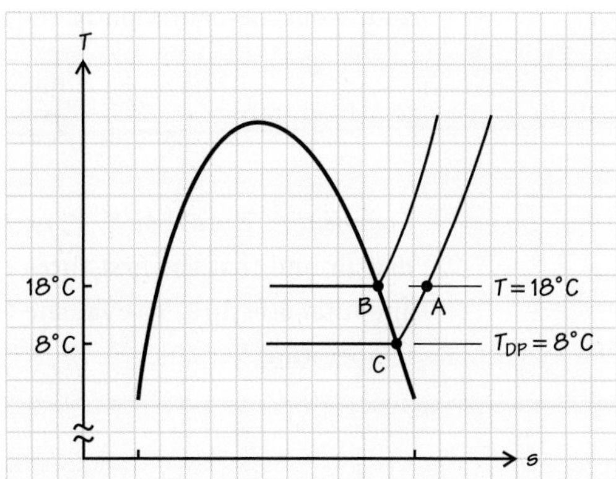

Modeling, Premises and Assumptions

i. Ideal-gas behavior of air and $H_2O(v)$
ii. H_2O liquid–vapor equilibrium unaffected by air

Analysis Given the dew-point temperature, we can determine the partial pressure of the water vapor in the moist air by recognizing that $P_A = P_C = P_{sat}(T_{DP})$ (see the sketch). From the NIST database,

$$P_{sat}(T_{DP}) = 1.073 \text{ kPa} = P_{H_2O(v)}.$$

With this value and the total pressure (P = 29.16 in -Hg = 98.75 kPa), we can find three of the four requested quantities. For an ideal-gas mixture, the mole fraction is (Eq. 11.7)

$$X_{H_2O(v)} = \frac{P_{H_2O(v)}}{P} = \frac{1.073 \text{ kPa}}{98.75 \text{ kPa}} = 0.01087,$$

the mass fraction is (see Eqs. 10.4a and 10.5)

$$\begin{aligned} Y_{H_2O(v)} &= X_{H_2O(v)} \frac{\mathcal{M}_{H_2O}}{\mathcal{M}_{mix}} \\ &= X_{H_2O(v)} \frac{\mathcal{M}_{H_2O}}{X_{H_2O(v)}\mathcal{M}_{H_2O} + \left(1 - X_{H_2O(v)}\right)\mathcal{M}_A} \\ &= 0.01087 \frac{18.016}{0.01087(18.016) + (1 - 0.01087)28.97} \text{ kg}_{H_2O}/\text{kg}_{mix} \\ &= 0.00679 \text{ kg}_{H_2O}/\text{kg}_{mix}, \end{aligned}$$

and the humidity ratio is (Eq. 11.11a)

$$\begin{aligned} \omega &= \frac{\mathcal{M}_{H_2O}}{\mathcal{M}_A} \frac{P_{H_2O(v)}}{P - P_{H_2O(v)}} \\ &= \frac{18.016}{28.97} \frac{1.073}{(98.75 - 1.073)} \text{kg}_{H_2O}/\text{kg}_A = 0.00683 \text{ kg}_{H_2O}/\text{kg}_A. \end{aligned}$$

To determine the relative humidity requires finding the saturation pressure associated with the ambient temperature of 18 °C (i.e., point B on the sketch). Using the NIST WebBook, we find

$$P_B = P_{sat}(T = 18\,°\text{C}) = 2.0647 \text{ kPa},$$

and, from the definition of the relative humidity ϕ (Eq. 11.12),

$$\phi = \frac{P_{H_2O(v)}(T)}{P_{sat}(T)} = \frac{1.073\ \text{kPa}}{2.0647\ \text{kPa}} = 0.52 \text{ or } 52\%.$$

Comment Note that X, Y, ω, and ϕ are all measures of the amount of water vapor in dry air and that the first three can be found from a knowledge of only the dew point and total pressure. To evaluate the relative humidity requires, in addition, the ambient temperature.

Self-Test 11.1

Air in a room has the properties as described in Example 11.1. The air is then heated to 75 F. Determine the absolute humidity and relative humidity at this new temperature. Comment on your results.

(Answer: $\omega = 0.00683$, $\phi = 36.3\%$. The absolute humidity is unchanged, whereas the relative humidity decreases.)

Example 11.2 Water Vapor as an Ideal Gas

Show that, for the conditions of Example 11.1, the water vapor can be treated as an ideal gas.

Solution

We follow the method used in Example 2.11 to test the ideal-gas assumption. For an ideal gas,

$$\frac{P\mathrm{v}}{RT} = 1,$$

but more generally,

$$\frac{P\mathrm{v}}{RT} = Z,$$

where Z is the compressibility factor. For a real gas, the closer Z is to unity, the closer its P–v–T behavior is to that of an ideal gas. Using the NIST online WebBook we determine the real-gas specific volume for the water vapor at points A, B, and C on the sketch in Example 11.1 and then evaluate Z. For example, at point C,

$$P = 1.073\ \text{kPa},$$
$$T = (8 + 273.15)\ \text{K} = 281.15\ \text{K},$$
$$\mathrm{v} = 120.83\ \text{m}^3/\text{kg},$$

and the gas constant is

$$R = Ru/\mathcal{M}_{H_2O} = 8314.47\ \text{J/kmol·K}/18.016\ \text{J/kg·K} = 461.505\ \text{J/kg·K}.$$

Thus,

$$Z = \frac{(1.073 \times 10^3\ \mathrm{N/m^2})(120.83\ \mathrm{m^3/kg})}{(461.505\ \mathrm{J/kg{\cdot}K})(281.15\ \mathrm{K})} = 0.9992.$$

Clearly, 0.9992 is close to unity. Similar calculations at points A and B yield, respectively, $Z = 0.9994$ and 0.9988. Again, we see that ideal-gas behavior is an excellent approximation.

Comment This example shows that the ideal-gas law can be used to analyze the water content in moist air.

11.5 Recast Conservation Equations

To facilitate the use of the mass and energy conservation equations previously developed, specifically Eqs. 11.2–11.4, we employ the definition of the humidity ratio (i.e., $\dot{m}_{\mathrm{H_2O(v)},i} = \dot{m}_{\mathrm{A}}\omega_i$) to recast these expressions as follows:

H_2O mass conservation (11.2b):

$$\dot{m}_{\mathrm{A}}\omega_1 + \dot{m}_{\mathrm{H_2O(l)},3} = \dot{m}_{\mathrm{A}}\omega_4 + \dot{m}_{\mathrm{H_2O(l)},2}, \tag{11.14}$$

overall mass conservation (11.3):

$$\dot{m}_{\mathrm{A}}(1+\omega_1) + \dot{m}_{\mathrm{H_2O(l)},3} = \dot{m}_{\mathrm{A}}(1+\omega_4) + \dot{m}_{\mathrm{H_2O(l)},2}, \tag{11.15}$$

overall energy conservation (11.4):

$$\begin{aligned} &\dot{Q}_{\mathrm{in}} + \dot{W}_{\mathrm{in}} + \dot{m}_{\mathrm{A}}\left(h_{\mathrm{A},1} + \omega_1 h_{\mathrm{H_2O(v)},1}\right) + \dot{m}_{\mathrm{H_2O(l)},3}h_{\mathrm{H_2O(l)},3} \\ &= \dot{Q}_{\mathrm{out}} + \dot{m}_{\mathrm{A}}\left(h_{\mathrm{A},4} + \omega_4 h_{\mathrm{H_2O(v)},4}\right) + \dot{m}_{\mathrm{H_2O(l)},2}h_{\mathrm{H_2O(l)},2}. \end{aligned} \tag{11.16}$$

We illustrate the use of these equations in the following examples.

Example 11.3 Evaporative Cooler

An evaporative cooler is used to cool a student apartment in Tempe, Arizona. Hot, moist air from outside enters the cooler at 106 F (314.2 K) with a relative humidity of 16% at a volumetric flow rate of 3000 ft^3/min (1.416 m^3/s). As shown in the sketch, a fan driven by a 1/8-hp (93.2-W) electric motor pulls the air through the cooler. The air exits the cooler into the apartment at 78 F (298.7 K) with an unknown humidity. The barometric pressure is 100 kPa. Determine the mass flow rate of liquid water that must be supplied to the cooler and the relative humidity of the cool air stream.

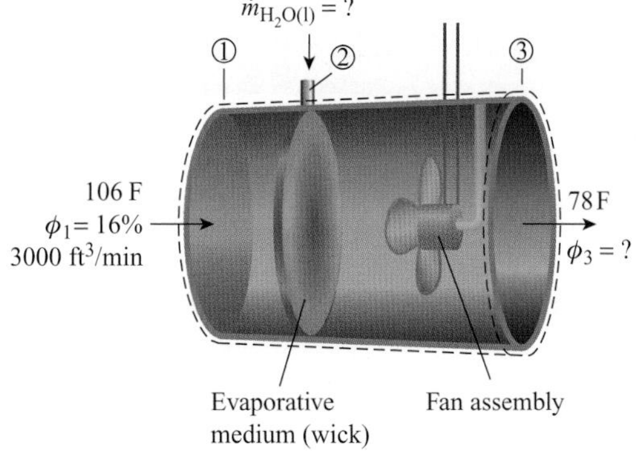

Solution

Known $P_{atm}, T_1, T_3, \phi_1, \dot{V}_1, \dot{W}_{elec}$

Find $\dot{m}_{H_2O(l),2}, \phi_3$

Sketch

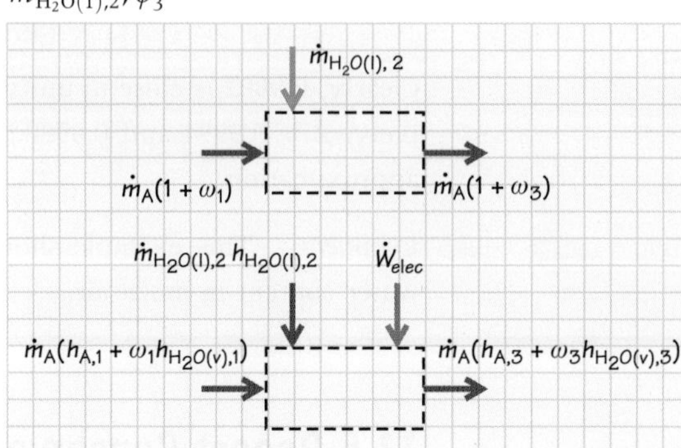

Modeling, Premises and Assumptions

i. The flow is steady and at steady state.
ii. Kinetic and potential energies are negligible.
iii. The process is adiabatic ($\dot{Q}_{cv} = 0$).
iv. The pressure is essentially uniform ($P_1 = P_2 = P_3 = P_{atm}$).
v. Air and $H_2O(v)$ behave as ideal gases.
vi. Liquid water at station 2 enters at T_3 and P_{atm}.

Analysis Our overall strategy is to write mass and energy conservation expressions using the corresponding open system sketches shown. This results in two equations with two unknowns, ω_3 and $\dot{m}_{H_2O(l),2}$, as all other quantities are easily determined or approximated. Before writing these conservation expressions, we determine the properties of the air, water vapor, and liquid water at stations 1, 2, and 3 and the mass flow rate of the entering dry air, $\dot{m}_A$. Using the definition of relative humidity (Eq. 11.12) and the NIST database, we can find $P_{H_2O(v),1}$:

$$\phi_1 = \frac{P_{H_2O(v),1}(314.2\text{ K})}{P_{sat}(314.2\text{ K})},$$

or

$$\begin{aligned} P_{H_2O(v),1} &= \phi_1 P_{sat}(314.2\text{ K}) \\ &= 0.16(7.8085\text{ kPa}) = 1.2494\text{ kPa}. \end{aligned}$$

Using the NIST WebBook for H_2O properties and Tables C.2 and C.3 for those of air, we create the following table to organize our calculations:

Station	1	2	3
T (K)	314.2	298.7	298.7
P_{H_2O} (kPa)	1.2494	100	?
h_{H_2O} (kJ/kg)	2577.5	107.22	2547.5 (est.)
$c_{p,A}$(kJ/kg·K)	1.0072	—	1.007
h_A(kJ/kg)	440.34	—	424.73

The enthalpy of the water vapor at station 3 is estimated by assuming $h_{H_2O(v),3} \approx h_{sat,3}$. Since the water vapor acts essentially as an ideal gas, the enthalpy of the vapor will not be affected significantly by the pressure.

To determine $\dot{m}_A$, we combine the definitions of the mass and volume flow rates (Eq. 3.10) and the humidity ratio (Eq. 11.9) as follows:

$$\dot{m}_1 = \rho_1 \dot{\mathcal{V}}_1 = \dot{m}_A(1 + \omega_1),$$

where (Eq. 11.11b)

$$\omega_1 = 0.622 \frac{P_{H_2O(v),1}}{P_1 - P_{H_2O(v),1}}$$
$$= 0.622 \frac{1.2494}{(100 - 1.2494)} = 0.00787.$$

We obtain the mixture density at station 1 by applying the ideal-gas equation of state, where the mixture molecular weight is determined using the water-vapor mole fraction from Eq. 11.7, that is,

$$X_{H_2O(v),1} = \frac{P_{H_2O(v),1}}{P_1} = \frac{1.2494}{100} = 0.01249,$$

$$\mathcal{M}_1 = X_{H_2O(v),1} \mathcal{M}_{H_2O} + \left(1 - X_{H_2O(v),1}\right) \mathcal{M}_A$$
$$= 0.01249(18.016 \text{ kg/kmol}) + (1 - 0.01249)(28.97 \text{ kg/kmol})$$
$$= 28.83 \text{ kg/kmol},$$

and the apparent density of the air/water mixture is

$$\rho_1 = \frac{P_1 \mathcal{M}_1}{R_u T_1} = \frac{(100 \times 10^3 \text{ N/m}^2)(28.83 \text{ kg/kmol})}{(8314.47 \text{ J/kmol·K})(314.2 \text{ K})} \left[\frac{1\text{J}}{\text{N·m}}\right] = 1.104 \text{ kg/m}^3.$$

Therefore,

$$\dot{m}_A = \frac{\rho_1 \dot{\mathcal{V}}_1}{1 + \omega_1} = \frac{(1.104 \text{ kg/m}^3)(1.416 \text{ m}^3/\text{s})}{1 + 0.00787}$$
$$= 1.551 \text{ kg/s}.$$

With these preliminary calculations out of the way, we now write mass and energy conservation using our sketches as guides. For H_2O conservation,

$$\sum^{\text{inlets}} \dot{m}_{H_2O,i} = \sum^{\text{outlets}} \dot{m}_{H_2O,i},$$

so

$$\dot{m}_{H_2O(v),1} + \dot{m}_{H_2O(v),2} = \dot{m}_{H_2O(v),3},$$

using Eq. 11.9 we can write the conservation of mass using the mass flow rate of air,

$$\omega_1 \dot{m}_A + \dot{m}_{H_2O(l),2} = \omega_3 \dot{m}_A.$$

This can be rearranged as

$$\dot{m}_{H_2O(l),2} = \dot{m}_A(\omega_3 - \omega_1) \quad \textbf{(A)}$$

For energy conservation,

$$\sum \dot{E}_{in} = \sum \dot{E}_{out};$$

we include the energy inflow of the moist air at location 1 and the liquid water at location 2. The only outflow is that of the moist air. Thus,

$$\dot{m}_A h_{A,1} + \dot{m}_{H_2O(v),1} h_{H_2O(v),1} + \dot{m}_{H_2O(l),2} h_{H_2O(l),2} + \dot{W}_{elec} = \dot{m}_A h_{A,3} + \dot{m}_{H_2O(v),3} h_{H_2O(v),3}.$$

Substituting for $\dot{m}_{H_2O}$ using Eq. 11.9 gives

$$\dot{m}_A\left(h_{A,1} - \omega_1 h_{H_2O(v),1}\right) + \dot{m}_{H_2O(l),2} h_{H_2O(l),2} + \dot{W}_{elec} = \dot{m}_A\left(h_{A,2} - \omega_3 h_{H_2O(v),3}\right). \qquad \textbf{(B)}$$

Substituting Eq. A into Eq. B and solving for the unknown ω_3 yields

$$\omega_3 = \frac{\left(h_{A,1} - h_{A,3}\right) + \omega_1\left(h_{H_2O(v),1} - h_{H_2O(l),2}\right)}{h_{H_2O(v),3} - h_{H_2O(l),2}} + \frac{\dot{W}_{elec}}{\dot{m}_A\left(h_{H_2O(v),3} - h_{H_2O(l),2}\right)},$$

which is evaluated as follows:

$$\omega_3 = \frac{(440.34 - 424.73)\ \text{kJ/kg} + 0.00787(2577.5 - 107.22)\ \text{kJ/kg}}{(2547.5 - 107.22)\ \text{kJ/kg}}$$

$$+ \frac{0.0932\ \text{kJ/s}}{(1.551\ \text{kg/s})(2547.5 - 107.22)\ \text{kJ/kg}} = 0.01436 + 0.0000246 = 0.01438.$$

Returning to the H_2O mass conservation (Eq. A), we can now find the liquid-water flow rate:

$$\dot{m}_{H_2O(l),2} = \dot{m}_A(\omega_3 - \omega_1) = 1.551(0.01438 - 0.00787)\text{kg/s} = 0.0101\ \text{kg/s}.$$

To find the exit relative humidity ϕ_3, we apply Eqs. 11.11b and 11.12 in succession, as follows:

$$\omega_3 = 0.622\frac{P_{H_2O(v),3}}{P_3 - P_{H_2O(v),3}},$$

or

$$P_{H_2O(v),3} = \frac{P_3}{\dfrac{0.622}{\omega_3} + 1} = \frac{100\ \text{kPa}}{\dfrac{0.622}{0.01438} + 1} = 2.260\ \text{kPa}.$$

Thus,

$$\phi_3 = \frac{P_{H_2O(v),3}}{P_{sat}(T_3)} = \frac{2.260\ \text{kPa}}{3.2754\ \text{kPa}} = 0.69 \quad \text{or} \quad 69\%.$$

Comments Using the value for $h_{H_2O(v),3}$ as the actual state of the water vapor at station 3 (h = 2548.0 kJ/kg at 2.260 kPa, 298.7 K) rather than our original estimate (2547.5 kJ/kg) has an insignificant effect on the determination of ω_3 (0.01437 versus 0.01438), as anticipated. We note that the contribution of $\dot{W}_{elec}$ to the final result for ω_3 is quite small (0.0000246, as revealed in our calculations).

Note that evaporative coolers work best in hot, dry climates where the addition of moisture does not significantly decrease comfort. In this example, a substantial quantity of water (0.0101 kg/s ≈ 9.6 gal/hr) is used to achieve the desired cooling.

Self-Test 11.2

Repeat Example 11.3 for an outside relative humidity of 25%.

(Answer: ϕ_3 = 90.2%)

Example 11.4 Dehumidifier

(Credit: DonNichols / E+ / Getty Images.)

Consider a household dehumidifier system as pictured in Fig. 11.5 b, c with the component arrangement as shown in Fig. 11.5a. The following information is known about this system:

inlet mass flow rate (dry air + water vapor) = 0.15 kg/s,
inlet pressure = 98 kPa,
inlet temperature = 295 K (71.3 F),
inlet relative humidity = 85%,
outlet pressure = 100 kPa,
condensate flow rate = 3.284×10^{-4} kg/s (~0.3 gal/hr),
condensate temperature = 281 K (46.2 F),
total electrical power supplied (compressor and fan motor drives combined) = 688.7 W, and
heat loss from the dehumidifier system to the surroundings = 15 W.

Determine the following quantities:

A. Total mass flow rate (dry air + water vapor) at the outlet
B. Specific humidity at the outlet
C. Temperature and relative humidity at the outlet

Solution

Known $\dot{m}_1, \dot{m}_2, P_1, P_3, T_1, T_2, \phi_1$

Find $\dot{m}_3, \omega_3, T_3, \phi_3$

Sketch

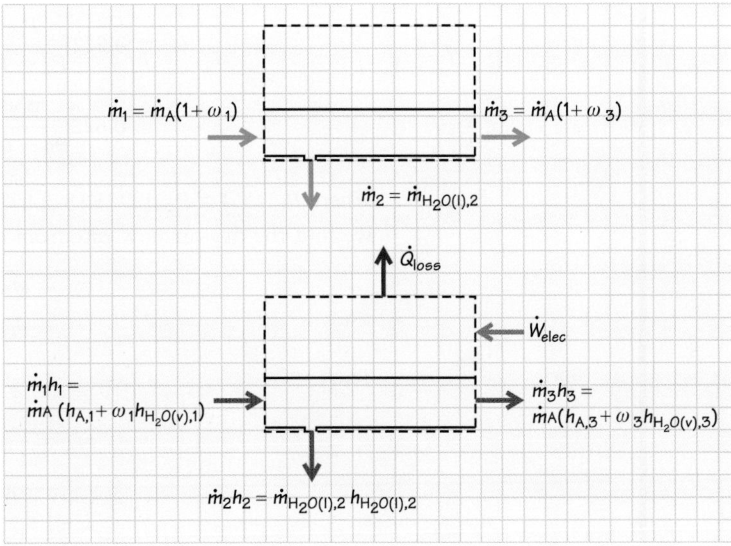

Modeling, Premises and Assumptions

i. The flow is steady and at steady state.
ii. Kinetic and potential energies are negligible.
iii. Air and $H_2O(v)$ behave as ideal gases.
iv. $P_{H_2O(l),2} = P_3$.

Analysis To find the outlet mass flow rate $\dot{m}_3$, we write the following statement of overall mass conservation:

$$\sum^{\text{inlets}} \dot{m}_i = \sum^{\text{outlets}} \dot{m}_i.$$

or

$$\dot{m}_1 = \dot{m}_2 + \dot{m}_3.$$

Thus,

$$\begin{aligned}\dot{m}_3 &= \dot{m}_1 - \dot{m}_2\\ &= 0.15 \text{ kg/s} - 3.284 \times 10^{-4} \text{ kg/s} = 0.1497 \text{ kg/s}.\end{aligned}$$

To find the specific humidity at the outlet (station 3), we apply mass conservation to the H_2O alone as follows:

$$\dot{m}_{H_2O(v),1} = \dot{m}_{H_2O(l),2} + \dot{m}_{H_2O(v),3}.$$

Using Eq. 11.9, we can write the water vapor flow rates in terms of the air flow rates:

$$\dot{m}_A\omega_1 = \dot{m}_{H_2O(l),2} + \dot{m}_A\omega_3.$$

Solving for ω_3 yields

$$\omega_3 = \omega_1 - \frac{\dot{m}_{H_2O(l),2}}{\dot{m}_A}.$$

The specific humidity at station 1 can be found from Eq. 11.11b combined with the definition of relative humidity (Eq. 11.12):

$$\begin{aligned}\omega_1 &= 0.622\frac{P_{H_2O(v),1}}{P_1 - P_{H_2O(v),1}} = 0.622\frac{\phi_1 P_{sat}(T_1)}{P_1 - \phi_1 P_{sat}(T_1)}\\ &= 0.622\frac{0.85(2.6212)}{98 - 0.85(2.6212)} = 0.01447,\end{aligned}$$

where $P_{sat}(T_1) = P_{sat}(295 \text{ K}) = 2.6212$ kPa from the NIST WebBook. Now knowing ω_1, we find the dry-air flow rate from the definition of ω:

$$\begin{aligned}\dot{m}_1 &= \dot{m}_A + \dot{m}_{H_2O(v),1}\\ &= \dot{m}_A + \omega_1\dot{m}_A = \dot{m}_A(1 + \omega_1).\end{aligned}$$

Thus,

$$\begin{aligned}\dot{m}_A &= \frac{\dot{m}_1}{1 + \omega_1}\\ &= \frac{0.15 \text{ kg/s}}{1 + 0.01447} = 0.1479 \text{ kg/s},\end{aligned}$$

and ω_3 is then evaluated as

$$\omega_3 = 0.01447 - \frac{3.284 \times 10^{-4}\ \text{kg/s}}{0.1479\ \text{kg/s}} = 0.01225.$$

To find the outlet temperature T_3, we apply overall conservation of energy to our open system using the second of our two sketches as a guide; i.e.,

$$\sum \dot{E}_{\text{in}} = \sum \dot{E}_{\text{out}}.$$

There is an inflow of moist air and an outflow of liquid water and humidified air/vapor mixture. An electric motor drives the fan and the heat transfer out of the open system is given by:

$$\dot{m}_1 h_1 + \dot{W}_{\text{elec}} = \dot{m}_3 h_3 + \dot{m}_3 h_3 + \dot{Q}_{\text{loss}}.$$

Expanding the $\dot{m}h$ terms into air and water vapor components yields

$$\dot{m}_{\text{A}}\left(h_{\text{A},1} + \omega_1 h_{\text{H}_2\text{O(v)},1}\right) + \dot{W}_{\text{elec}} = \dot{m}_{\text{H}_2\text{O(l)},2}\, h_{\text{H}_2\text{O(l)},2} + \dot{m}_{\text{A}}\left(h_{\text{A},3} + \omega_3 h_{\text{H}_2\text{O(v)},3}\right) + \dot{Q}_{\text{loss}}.$$

Known quantities in this expression are $\dot{m}_{\text{A}}$, ω_1, $\dot{W}_{\text{elec}}$, $\dot{m}_{\text{H}_2\text{O(l)},2}$, ω_3, and $\dot{Q}_{\text{loss}}$, whereas the enthalpies, $h_{\text{A},1}$, $h_{\text{H}_2\text{O(v)},1}$, and $h_{\text{H}_2\text{O(l)},2}$, are all easily evaluated from known temperatures and pressures. The remaining unknown quantities, $h_{\text{A},3}$ and $h_{\text{H}_2\text{O(v)},3}$, are functions only of the unknown temperature, if we assume negligible pressure dependence for $h_{\text{H}_2\text{O(v)},3}$ and use the approximation

$$h_{\text{H}_2\text{O(v)},3} \approx h_{\text{H}_2\text{O(v)}}(T_{\text{sat}} = T_3).$$

To proceed, we substitute $c_{p,\text{A,avg}}(T_3 - T_1)$ for $h_{\text{A},3} - h_{\text{A},1}$ to simplify our iterative calculation and rearrange our energy balance to separate the unknown and known quantities on the left- and right-hand sides, respectively, that is,

$$\dot{m}_{\text{A}} c_{p,\text{A,avg}}(T_3 - T_1) + \dot{m}_{\text{A}}\omega_3 h_{\text{H}_2\text{O(v)},3}(T_3) = \dot{m}_{\text{A}}\omega_1 h_{\text{H}_2\text{O(v)},1} - \dot{m}_{\text{H}_2\text{O(l)},2} h_{\text{H}_2\text{O(l)},2} + \dot{W}_{\text{elec}} - \dot{Q}_{\text{loss}}.$$

We obtain the following enthalpies for H_2O from the NIST WebBook:

	Station 1	Station 2
T (K)	295	281
$P_{\text{H}_2\text{O}}$ (kPa)	2.2280*	100
h (kJ/kg)	2541.0	33.094

$^*P_{\text{H}_2\text{O(v)},1} = \phi_1 P_{\text{sat}}(T_1) = 0.85(2.6212) = 2.2280$ kPa.

From Table C.3, we see that for the temperature range of interest an appropriate value for $c_{p,\text{A,avg}}$ is 1.007 kJ/kg·K. Inserting numerical values into the energy balance equation yields

$$(0.1479\ \text{kg/s})(1.007\ \text{kJ/kg·K})(T_3 - 295\ \text{K}) + (0.1479\ \text{kg/s})(0.01225)h_{\text{sat vap}}(T_3) = (0.1479\ \text{kg/s})(0.01447)(2541.0\ \text{kJ/kg}) - (3.284 \times 10^{-4}\ \text{kg/s})(33.094\ \text{kJ/kg}) + 0.6887\ \text{kJ/s} - 0.015\ \text{kJ/s},$$

where we recognize that each term has units of kW. Performing the indicated multiplications and rearranging and combining terms results in

$$(0.1489\ \mathrm{kJ/s{\cdot}K})T_3 + (0.001812\ \mathrm{kg/s})h_{\mathrm{sat\ vap}}(T_3) = 50.0368\ \mathrm{kJ/s},$$

or

$$T_3 = 336.04\ \mathrm{K} - (0.01217\ \mathrm{kg{\cdot}K/kJ})h_{\mathrm{sat\ vap}}(T_3).$$

The form of this equation is quite amenable to iteration: we guess a value for T_3 and evaluate our expression for an improved value of T_3. The process can be repeated using this improved value to reevaluate the right-hand side. We begin by guessing $T_3 = 305$ K to obtain our first iterate:

$$T_3 = 336.04\ \mathrm{K} - (0.01217\ \mathrm{kg{\cdot}K/kJ})(2558.9)\ \mathrm{kJ/kg} = 304.898\ \mathrm{K}.$$

Further iteration yields

$$T_3 = 336.04\ \mathrm{K} - (0.01217\ \mathrm{kg{\cdot}K/kJ})(2558.7)\ \mathrm{kJ/kg} = 304.901\ \mathrm{K},$$

which is quite close to the input value; thus, we conclude that, with proper rounding,

$$T_3 = 304.9\ \mathrm{K}.$$

With knowledge now of T_3, we can determine the relative humidity ϕ_3, that is,

$$\phi_3 = \frac{P_{\mathrm{H_2O(v)},3}}{P_{\mathrm{sat}}(T_3)},$$

where $P_{\mathrm{H_2O(v)},3}$ is obtained from the relationship between ω_3 and $P_{\mathrm{H_2O(v)},3}$ (Eq. 11.11c):

$$P_{\mathrm{H_2O(v)},3} = \frac{\omega_3 P_3}{0.622 + \omega_3}$$

$$= \frac{0.01225(100\ \mathrm{kPa})}{0.622 + 0.01225} + 1.931\ \mathrm{kPa}.$$

Thus,

$$\phi_3 = \frac{1.931\ \mathrm{kPa}}{P_{\mathrm{sat}}(304.9\ \mathrm{K})} = \frac{1.931\ \mathrm{kPa}}{4.693\ \mathrm{kPa}} = 0.411 \text{ or } 41.1\%.$$

Comments We note that the relative humidity of the outlet stream (41.1%) is much less than that of the entering stream (85%) and well within the generally accepted comfort zone of 40%–60% relative humidity.

Although the calculations here are in places quite lengthy, the principles involved – mass and energy conservation – are straightforward and easy to apply once all the constituent terms are evaluated. Drawing and labeling open systems showing all the mass flows and energy flows is essential to proper application of conservation principles.

11.6 Humidity Measurement

Modern instruments used to measure humidity operate on a variety of principles. Some devices depend upon sensing elements whose electrical properties (resistivity, impedance, or capacitance) vary with moisture content. Strain-gage-based humidity measurement instruments sense the expansion of a sensor as it absorbs moisture from the air. Chilled mirror devices are used to determine the dew point and, hence, the humidity. All these devices, and others employing different instrumental techniques, are based on electrical or electronic principles. Simple psychrometers that measure wet-bulb and dry-bulb temperatures are also used to measure humidity. We consider the principles involved in the use of psychrometers in the following subsections.

11.6a Adiabatic Saturation

Consider the adiabatic saturation device illustrated in Fig. 11.9. Here moist air of unknown relative humidity, ϕ_1, enters a long duct. As the air passes through the duct, additional moisture is added to the stream from evaporation of the liquid water contained in the duct floor. By insulating the duct, the energy needed to evaporate the water comes at the expense of the sensible energy of the air; hence, as the air passes through the duct, not only does its moisture content increase but its temperature drops as well. If the duct is sufficiently long, the air becomes saturated, i.e., its relative humidity is 100%. The outlet temperature is then the **adiabatic saturation temperature**. To complete our description of the adiabatic saturation process, we note that make-up water is supplied to the duct at the outlet temperature of the saturated air, T_3. In the analysis that follows, we show how measurements of the two temperatures, T_1 and T_3, can be used to determine the humidity (ϕ_1 or ω_1) of the entering moist air.

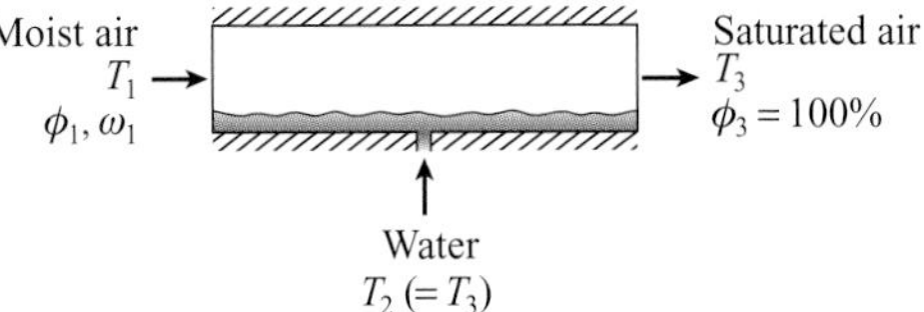

FIGURE 11.9 Moist air enters the adiabatic saturation device at T_1 with relative humidity ϕ_1 and exits with a relative humidity of 100% (saturated) at a lower temperature T_3. Make-up water is added at the outlet temperature.

We begin by writing mass conservation for the water as

$$\dot{m}_A \omega_1 + \dot{m}_{\ell,2} = \dot{m}_A \omega_3$$

or

$$\dot{m}_{\ell,2} = (\omega_3 - \omega_1)\dot{m}_A,$$

where we denote this liquid phase of the water using the subscript ℓ. Assuming adiabatic operation with no work interactions, energy conservation for the adiabatic saturator is given by

$$\dot{m}_A(h_{A,1} + \omega_1 h_{v,1}) + \dot{m}_{\ell,2} h_{\ell,2} = \dot{m}_A(h_{A,3} + \omega_3 h_{v,3}),$$

where we denote the vapor phase of the water using the subscript v. Using the water mass conservation expression to eliminate $\dot{m}_{\ell,2}$ in the above, then simplifying and rearranging yield

$$h_{A,1} - h_{A,3} - \omega_3(h_{v,3} - h_{\ell,2}) + \omega_1(h_{v,1} - h_{\ell,2}) = 0.$$

We simplify further by noting that since $T_2 = T_3$ and $h_{\ell,2} = h_{f,3}$,

$$h_{v,3} - h_{\ell,2} = h_{v,3} - h_{f,3} = h_{fg}(T_3).$$

Furthermore, we can express the enthalpy difference of the air and the approximate enthalpy of the water vapor at 1 as

$$h_{A,1} - h_{A,3} = c_{p,\text{avg}}(T_1 - T_3)$$

and

$$h_{v,1} \approx h_g(T_1).$$

We make these substitutions into the combined energy and mass conservation expression and solve for ω_1 to yield our final result:

$$\omega_1 = \frac{c_{p,\text{avg}}(T_1 - T_3) - \omega_3 h_{fg}(T_3)}{h_f(T_3) - h_g(T_1)}. \tag{11.17}$$

Note that the right-hand side of Eq. 11.17 can be evaluated from knowledge of T_1, T_3, and the total pressure. We illustrate this in the following example.

Example 11.5 Adiabatic Saturation Device

An adiabatic saturation device operates at 1 atm. Moist air enters at 25 °C and exits at 20 °C with 100% relative humidity. Determine the relative humidity of the air at the inlet.

Solution

Known: T_1, T_3, P

Find: ϕ_1

Sketch:

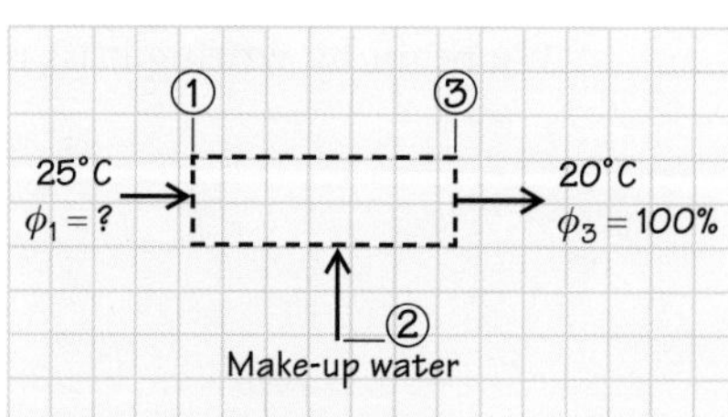

Modeling, Premises and Assumptions

i. Ideal-gas behavior of air and water vapor
ii. Constant pressure
iii. Adiabatic process (given)

Analysis We first employ Eq. 11.11b to find ϕ_1, use this value to find $P_{v,1}$ from Eq. 11.11c, and then use $P_{v,1}$ to obtain the relative humidity from its defining relationship (Eq. 11.12). To perform these operations requires the following properties, which are obtained from the NIST WebBook:

$$c_{p,\mathrm{A}}\,(22.5\ ^\circ\mathrm{C}) = 1.0065\ \mathrm{kJ/kg{\cdot}K},$$
$$h_\mathrm{g}\,(20\ ^\circ\mathrm{C}) = 2537.4\ \mathrm{kJ/kg},$$
$$h_\mathrm{f}\,(20\ ^\circ\mathrm{C}) = 83.914\ \mathrm{kJ/kg},$$
$$P_\mathrm{sat}\,(20\ ^\circ\mathrm{C}) = 2.3393\ \mathrm{kPa},$$
$$h_\mathrm{g}\,(25\ ^\circ\mathrm{C}) = 2546.5\ \mathrm{kJ/kg},$$

and

$$h_\mathrm{fg}\,(20\,^\circ\mathrm{C}) = h_\mathrm{g}\,(20\,^\circ\mathrm{C}) - h_\mathrm{f}\,(20\,^\circ\mathrm{C}) = 2453.5\ \mathrm{kJ/kg}.$$

To apply Eq. 11.17, we must first determine ω_3. At the exit the relative humidity is 100%; hence, $P_\mathrm{v}(T_3) = P_\mathrm{sat}(T_3) = 2.3393$ kPa. We can now calculate ω_3 using Eqn. 11.11b, i.e.,

$$\omega_3 = 0.622\frac{P_\mathrm{v}(T_3)}{P - P_\mathrm{v}(T_3)} = 0.622\left(\frac{2.3393\ \mathrm{kPa}}{101.325\ \mathrm{kPa} - 2.3393\ \mathrm{kPa}}\right)$$
$$= 0.0146995\ (\text{dimensionless}).$$

We now determine ω_1 (Eq. 11.17):

$$\omega_1 = \frac{c_{p,\mathrm{A}}(T_1 - T_3) - \omega_3 h_\mathrm{fg}(T_3)}{h_\mathrm{f}(T_3) - h_\mathrm{g}(T_1)}$$
$$= \frac{(1.0065\,\mathrm{kJ/kg{\cdot}K})(25 - 20)^\circ\mathrm{C} - 0.0146995(2453.5\ \mathrm{kJ/kg})}{83.914\ \mathrm{kJ/kg} - 2546.5\ \mathrm{kJ/kg}}$$
$$= 0.012603\ (\text{dimensionless}).$$

Using Eqns. 11.11c and 11.12, we obtain the relative humidity as follows:

$$p_{v,1} = \frac{\omega_1 P}{0.622 + \omega_1} = \frac{0.012603(101.325\ \mathrm{kPa})}{0.622 + 0.012603}$$
$$= 2.0122\ \mathrm{kPa}$$

and so

$$\phi_1 = \frac{P_{\mathrm{v},1}}{P_\mathrm{sat}(T_1)} = \frac{2.0122\ \mathrm{kPa}}{3.1699\ \mathrm{kPa}} = 0.635 \quad \text{or} \quad 63.5\%.$$

Comment Note the importance of the various relationships among ω, ϕ, and P_v in solving this otherwise straightforward problem.

11.6b Wet- and Dry-Bulb Temperatures

As mentioned previously, simple psychrometers can be used to determine the humidity of air. Figure 11.10 shows a sling psychrometer and a slightly more sophisticated psychrometer that uses a fan to blow air over the wick. Both of these devices comprise two thermometers: one thermometer bulb is covered by a wick moistened with distilled water, while the bulb of the second thermometer is bare. The sling psychrometer is whirled overhead to promote the evaporation of water from the wick. The flow from the fan performs the same function in the fan-driven psychrometer. As the water evaporates, the temperature of the "wet bulb" falls reaching a steady value known as the **wet-bulb temperature**. The other thermometer records the **dry-bulb temperature**. Although psychrometers are not adiabatic saturation devices, the processes involved at normal temperatures and pressures result in the wet-bulb temperature being approximately equal to the adiabatic saturation temperature. This allows Eq. 11.17 to be used to determine the humidity, where

$$T_1 = T_{\text{dry-bulb}} \equiv T_{\text{DB}} \tag{11.18a}$$

and

$$T_3 = T_{\text{wet-bulb}} \equiv T_{\text{WB}}. \tag{11.18b}$$

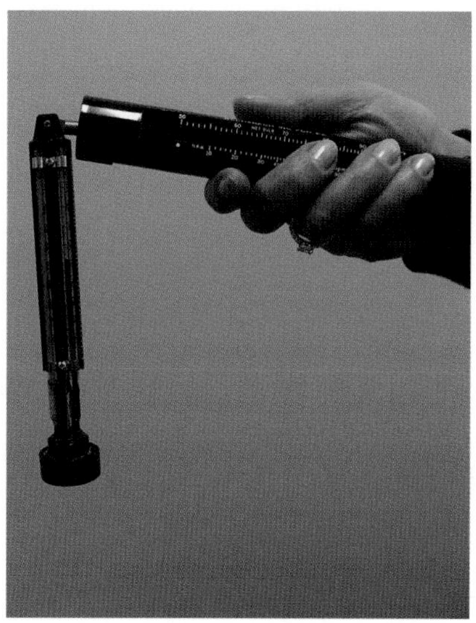

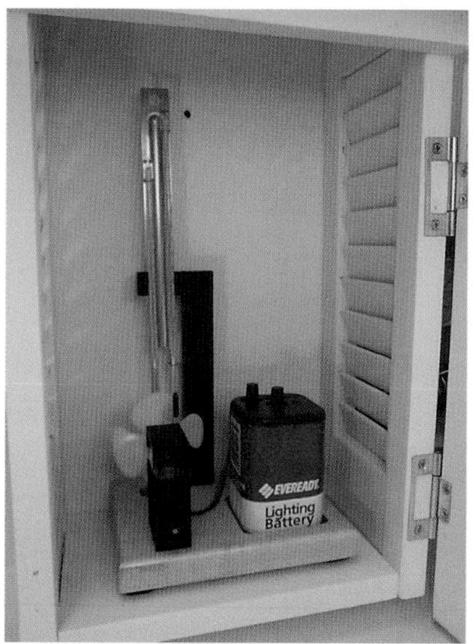

FIGURE 11.10 Sling psychrometers (top) and fan-driven psychrometers (bottom) are simple devices used to measure the humidity of ambient air. Photograph (right) courtesy of NovaLynx Corporation, www.novalynx.com.

11.6c The Psychrometric Chart

For a fixed value of the total pressure (e.g., 1 atm or 1 bar), the relationships among the wet- and dry-bulb temperatures T_{WB} and T_{DB} and the specific and relative humidities ω and ϕ can be related graphically. This graphical representation is known as a psychrometric chart. Appendix I presents a psychrometric chart based on a total pressure of 1 atm. Also indicated on the psychrometric chart are the specific enthalpy of moist air, $h_{\text{A}} + \omega h_{\text{v}}$, and the specific volume of dry air. Since ambient pressures do not deviate significantly from 1 atm, psychrometric charts can be used with reasonable accuracy for most ambient pressures.

Example 11.6 Sling Psychrometer

A sling psychrometer records a dry-bulb temperature of 25 °C and a wet-bulb temperature of 20 °C in a student-filled classroom. Use a psychrometric chart to estimate the specific humidity, ω, and relative humidity, ϕ, of the air in the classroom.

Solution

Known T_{DB}, T_{WB}

Find ω, ϕ

Sketch

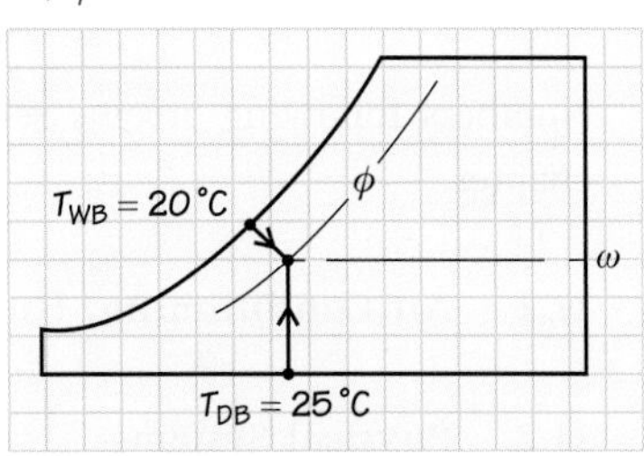

Modeling, Premises and Assumptions $P \approx 1$ atm

Analysis: Using the sketch as a guide, we enter Fig. I.1 following lines of constant T_{DB} and T_{WB} and mark their point of intersection. This intersection point lies between lines of constant relative humidity of 60% and 70%. Visual interpolation estimates ϕ to be 64%. A horizontal line extended from the intersection point to the axis on the right yields a specific humidity (or humidity ratio) estimate of 0.0127 kg/kg.

Comment Since the conditions given in this example are consistent with those of Example 11.5 (i.e., $T_{DB} = T_1 = 25$ °C and $T_{WB} = T_3 = 20$ °C), we can compare the results of the two examples. The relative humidities found were 63.5% and 64%, respectively. The corresponding specific humidities are 0.0126 and 0.0127. The agreement for both quantities is quite good.

SUMMARY

This chapter has applied ideal-mixture analysis to moist air. In Example 11.2 we verified that water vapor can be accurately analyzed as an ideal gas since the partial pressure of water vapor is very low. The water content of moist air is often described using the specific humidity or relative humidity. The dew point, wet-bulb temperature, and dry-bulb temperature are also used to describe the water content in moist air. With these new definitions for moist air, several applications have been studied including evaporative coolers, humidifiers, air conditioners, dehumidifiers, and cooling towers. As a further summary of this chapter, reviewing the learning objectives presented at the outset of the chapter is recommended.

KEY EQUATIONS

Review the most important equations presented in this chapter (i.e., those boxed with a yellow background). What physical principles do they express? What restrictions apply?

REFERENCES

1. Moran, M. J., and Shapiro, H. N., *Fundamentals of Engineering Thermodynamics*, 3rd edn, Wiley, New York, 1995.
2. Çengel, Y. A., and Boles, M. A., *Thermodynamics: An Engineering Approach*, 4th edn, McGraw-Hill, New York, 2002.

Some end-of-chapter problems were adapted with permission from the following:

Look, D. C., Jr., and Sauer, H. J., Jr., *Engineering Thermodynamics*, PWS, Boston, 1986.
Myers, G. E., *Engineering Thermodynamics*, Prentice Hall, Englewood Cliffs, NJ, 1989.

CHAPTER 11 KEY CONCEPTS AND DEFINITIONS CHECKLIST

Numbers following arrows refer to the Questions and Problems at the end of the chapter.

11.1 Air Conditioning, Humidification, and Related Systems

11.2 Physical Systems

- ☐ Evaporative cooler, humidifier, air conditioner, dehumidifier, and cooling tower ➔ Question 11.1, Problems 11.2, 11.3, 11.4, 11.5

11.3 General Analysis

- ☐ Integral open system mass and energy conservation ➔ Question 11.6

11.4 Some New Concepts and Definitions

- ☐ Psychrometry ➔ Questions 11.7, 11.14
- ☐ Humidity ratio (or specific humidity or absolute humidity) ➔ Question 11.8
- ☐ Relative humidity ➔ Question 11.9
- ☐ Dew point ➔ Question 11.10

11.5 Recast Conservation Equations

- ☐ System performance analysis ➔ Questions 11.11, 11.17

11.6 Humidity Measurement

- ☐ Adiabatic saturation ➔ Questions 11.12, 11.62
- ☐ Wet bulb and dry bulb measurements ➔ Questions 11.13, 11.65

QUESTIONS

11.1 Sketch an evaporative cooler. Discuss its operation.

11.2 Sketch a humidifier. Discuss its operation.

11.3 Sketch an air conditioner. Discuss its operation.

11.4 Sketch a dehumidifier. Discuss its operation.

11.5 Sketch a cooling tower. Discuss its operation.

11.6 For the devices in Questions 11.1 to 11.5, write the conservation of mass equation.

11.7 Define the term "psychrometry." How is psychrometry included in a weather report?

11.8 Define the humidity ratio, ω.

11.9 Define the relative humidity, ϕ.

11.10 Define the dew point, T_{DP}.

11.11 Use the humidity ratio ω to rewrite the conservation of mass equations in Question 11.6 in terms of the air mass flow rate

11.12 Define the adiabatic saturation temperature.

11.13 Define the wet-bulb temperature and the dry-bulb temperature. How are these temperatures used in psychrometry?

Chapter 11 Problem Subject Areas

11.1–11.11	Some new concepts and definitions
11.12–11.61	Recast the conservation equations
11.62–11.65	Humidity measurement

PROBLEMS

11.1–11.11 Some new concepts and definitions

11.1 For the purpose of testing air conditioners, the following indoor and outside air temperatures and relative humidities are defined for a so-called moderate climate:

Indoors: $T_A = 27\ ^\circ C\ (80.6\ F)$, $\phi = 0.48$,
Outside: $T_A = 35\ ^\circ C\ (95\ F)$, $\phi = 0.41$.

Determine the specific humidity and dew point associated with these two conditions.

11.2 Air at 100 F and 20 psia has a dew-point temperature of 70 F. Determine the relative humidity and the humidity ratio. Also determine the mixture volume (ft^3) containing 1 lb_m of dry air.

11.3 Combustion products (10% carbon dioxide, 20% water vapor, and 70% nitrogen by volume) flow in a chimney at 94 °C and 1 atm. Determine (a) the dew-point temperature (°C), (b) the humidity ratio, and (c) the relative humidity.

11.4 Combustion products (10% carbon dioxide, 20% water vapor, and 70% nitrogen by volume) flow in a pipe at 94 °C and 1.36 atm. Determine (a) the dew-point temperature (°C), (b) the humidity ratio, and (c) the relative humidity.

11.5 A mixture [10 mol CO_2, 70 mol N_2, and 20 mol H_2O (liquid plus vapor)] is at 40 °C and 1 atm. This temperature is below the dew point and the mixture is saturated with water vapor. Determine the number of moles of water that are vapor and the number of moles of water that are liquid.

11.6 A 10-ft by 20-ft by 8-ft room contains moist air at 90 F and 14.7 psia. The partial pressure of the water vapor is 0.3 psia. Determine the following quantities:

A. The dew-point temperature (F), relative humidity, and humidity ratio
B. The total mass (lb_m) of the water vapor in the room
C. The volume (ft^3) that would be occupied by this amount of water if it were a saturated liquid at 90 F.

11.7 Moist air at 100 F, 20 psia, and 40% relative humidity is flowing in a pipe. Determine (a) the dew-point temperature (F) and (b) the humidity ratio.

11.8 A 20-ft by 12-ft by 8-ft room contains an air–water vapor mixture at 80 F. The barometric pressure is 14.7 psia and the measured partial pressure of the water vapor is 0.2 psia. Calculate (a) the relative humidity, (b) the humidity ratio, (c) the dew-point temperature, and (d) the mass (lb_m) of water vapor contained in the room.

11.9 A 4-m by 6-m by 2.4-m room contains an air–water vapor mixture at a total pressure of 100 kPa and a temperature of 25 °C. The partial pressure of the water vapor is 1.4 kPa. Determine (a) the humidity ratio, (b) the dew point, and (c) the total mass of water vapor in the room.

11.10 Consider a room containing moist air at 24 °C, 60% relative humidity, and 1 atm. Determine the following:

A. The humidity ratio
B. The mixture enthalpy (kJ/kg)
C. The dew-point temperature (°C)
D. The mixture specific volume (m^3/kg)

11.11 An air–water vapor mixture at 100 F (37.8 °C) temperature contains 0.02 lb_m water vapor per pound mass of dry air. The barometric pressure is 28.561 in Hg (96.7 kPa). Calculate the relative humidity and dew-point temperature.

11.12–11.61 Recast conservation equations

11.12 On an extremely cold winter day, the air temperature and relative humidity in the living space of a home are 295.3 K (72 F) and 18%, respectively. After operating a humidifier for 8 hr, the relative humidity in the house has increased to 40%, the low end of the comfort zone. The air temperature remains at 295.3 K (72 F) and the atmospheric pressure is 100 kPa. Determine the specific humidity before and after humidifying the air. Also determine the average rate at which moisture was added to the air if the volume of the humidified space is 238 m^3. Express your result in both kg/s and gal/hr.

(Credit: Hoxton / Tom Merton / Getty Images.)

11.13 Determine the water-vapor mole and mass fractions before and after the humidifying of the air as described in Problem 11.12.

11.14 At state 1, a piston–cylinder device contains 0.004 m^3 of moist air initially at 20 °C and 0.10135 MPa with a 70% relative humidity. The volume is then decreased to 0.002 m^3 at state 2 as the piston is pushed inward in a reversible isothermal process.

A. Complete the following table:

Property	State 1	State 2
Relative humidity		
Humidity ratio		
Dew-point temperature (°C)		
Mass of dry air (g)		
Mass of water vapor (g)		
Mass of liquid water (g)		

B. Does the total internal energy of the contents of the cylinder increase, decrease, or remain constant during this process? Explain.

11.15 A dry gas mixture (60% oxygen, 40% helium, by volume) is at 94 °C and 3.4 atm. Moisture is then added. The properties at the final state are 94 °C, 3.4 atm, and 20% relative humidity. Determine (a) the dew-point temperature (°C), (b) the molar composition of the moist gas, and (c) the humidity ratio.

11.16 Moist air initially at 150 F, 20 psia, and 60% relative humidity is compressed reversibly and isothermally in a piston–cylinder device to a pressure of 40 psia. Determine the following:

A. The dew-point temperature (F) at the initial state

B. The humidity ratio at the initial state

C. The mass fraction of water that is liquid at the final state [i.e., $M_{\text{liquid}}/(M_{\text{liquid}} + M_{\text{vapor}})$ at the final state]

11.17 A dry mixture of carbon dioxide and nitrogen (30% carbon dioxide and 70% nitrogen by volume) is initially at 90 F and 14.7 psia. Moisture is then added to the mixture until it becomes saturated at 90 F and 14.7 psia. Determine the following quantities:

A. The molar mass of the original, dry mixture

B. The composition (by volume) of the saturated mixture

C. The ratio of the mixture volume when saturated to the mixture volume when dry

11.18 A 1-kg mixture of air and water vapor at 20 °C, 1 atm, and 75% relative humidity is confined in a cylinder by a frictionless piston. This mixture is compressed isothermally until the pressure is 2 atm. Determine (a) the final relative humidity and humidity ratio and (b) the mass (kg) of water vapor condensed.

11.19 Combustion products (10% carbon dioxide, 20% water vapor, and 70% nitrogen, by volume) enter a pipe at 200 F and 14.7 psia. The mass flow rate of the products is 6 lb_m/min. The products are cooled as they flow through the pipe and leave at 100 F and 14.7 psia. Determine (a) the mass flow rate (lb_m/min) of water condensed and (b) the heat-transfer rate (Btu/min).

11.20 Moist natural gas enters a valve at 40 °C, 0.2 MPa, and 70% relative humidity with a volumetric flow rate of 20 m^3/min. An analysis of the dry gases in the natural gas shows 80% methane and 20% hydrogen by volume as the gas enters. The moist natural gas leaves the valve at 0.15 MPa. Determine the moisture condensation rate (kg/min).

11.21 A tank contains moist air at 37.7 °C, 14.7 psia, and 27% relative humidity. Determine the following:

A. The dew-point temperature (°C) of the mixture

B. The temperature (°C) at which condensation will begin if the mixture is cooled down

C. The heat transfer (kJ/m^3) required to cool the contents of the tank to 10 °C

11.22 A 10-ft^3 tank initially contains a moist gas (30% carbon dioxide, 50% nitrogen, and 20% water vapor by volume) at 300 F and 50 psia. Heat transfer to the atmosphere reduces the mixture temperature to 150 F. Determine the following:

A. The final pressure (psia) in the tank

B. The mass (lb_m) of water condensed

C. The volume (ft^3) of the condensed liquid water

D. The molar composition of the moist gas at the final state

E. The heat transfer (Btu)

11.23 One of the many methods used for drying air is to cool it below the dew-point temperature so that condensation or freezing of the moisture takes place. To what temperature must atmospheric air be cooled to have a humidity ratio of 0.00433? To what temperature must this air be cooled if the pressure is 10 atm?

11.24 One method of removing moisture from atmospheric air is to cool the air so that the moisture condenses or freezes out. A laboratory experiment requires a humidity ratio of 0.00620. To what temperature must the air be cooled at a pressure of 0.1 MPa to achieve this humidity?

11.25 Air enters an air compressor at 70 F (21.1 °C) and 14.7 psia (101.3 kPa) with a 50% relative humidity. The air is compressed to 50 psia (344.7 kPa) and sent to an intercooler. If condensation of water vapor from the air is to be prevented, what is the lowest temperature to which the air can be cooled in the intercooler?

11.26 A 4-lb_m mass of air at 80 F (26.7 °C) and 50% relative humidity mixes with a 1-lb_m mass of air at 60 F (15.6 °C) and 50% relative humidity. The pressure is 1 atm. Determine (a) the relative humidity of the mixture and (b) the dew-point temperature of the mixture.

11.27 Air is compressed in a compressor from 30 °C, 60% relative humidity, and 101 kPa to 414 kPa and then cooled in an intercooler before entering a second stage of compression. What is the minimum temperature (°C) to which the air can be cooled so that condensation does not take place?

11.28 A 0.5-m^3 tank contains an air–water vapor mixture at 100 kPa and 35 °C with a 70% relative humidity. The tank is cooled until the water vapor begins to condense. Determine the temperature at which condensation begins and the heat transfer for the process.

11.29 An evaporative cooler inducts outside air at 46 °C (114.8 F) and 100 kPa having a relative humidity of 16%. The air is cooled to 29 °C (84.2 F).

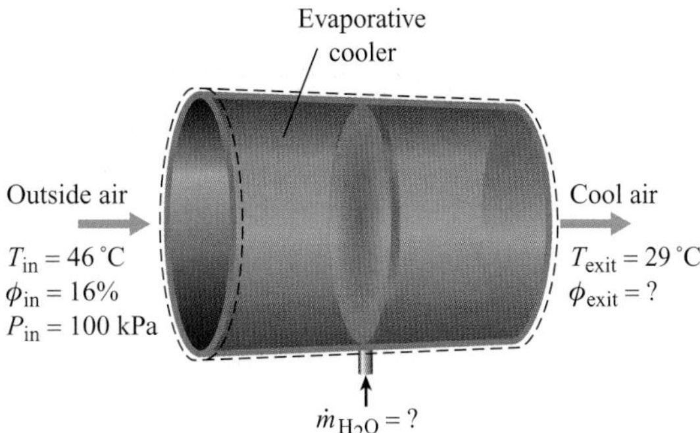

Determine the following quantities:

A. The mass of water added by the cooler per unit mass of dry air

B. The relative humidity of the outlet stream

C. The mass and volumetric flow rates of the added water when the volumetric flow rate of the entering moist air is 2500 ft^3/min

11.30 An air-conditioning unit inducts outside air at 46 °C (114.8 F) and 100 kPa having a relative humidity of 16%. The air is cooled and dehumidified to provide an outlet temperature of 29 °C (84.2 F) and relative humidity of 39%. The coefficient of performance for the vapor-compression refrigeration system contained in the air-conditioning unit is 2.92.

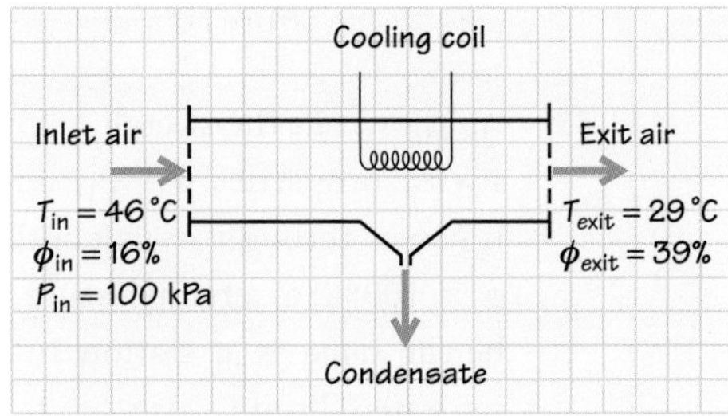

Determine the following quantities:

A. The mass of water removed per unit mass of dry air

B. The cooling rate per unit mass of dry air

C. The input power to the vapor-compression refrigeration system per unit mass of dry air

11.31 Consider a home dehumidifier, as shown in Fig. 11.5. Moist air at 78 F and relative humidity 85% enters at a volumetric flow rate of 330 ft^3/min. The air exits at 92 F with a relative humidity of 50%, and the condensate exits at 52 F. Determine the electrical power supplied to the dehumidifier in kW, and determine the condensate flow rate in kg/s and gal/min. Assume that the dehumidifier is overall adiabatic and that the pressure is uniform at 100 kPa.

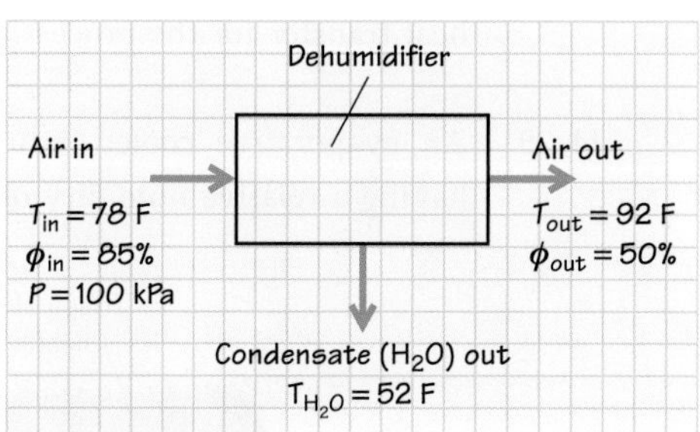

11.32 Cooling water from a power plant steam condenser enters a cooling tower at 307 K (93 F) with a flow rate of 3500 kg/s. The water exits the tower at 295 K (71.4 F). Air enters at the base of the tower at 290 K (62.4 F) with a relative humidity of 32% and exits at the top, saturated, at 304 K (87.6 F). Make-up water is provided at 295 K (71.4 F). The pressure in the tower is 96 kPa. Determine the flow rate of the make-up water and the flow rate of the air (on a dry air basis) through the tower.

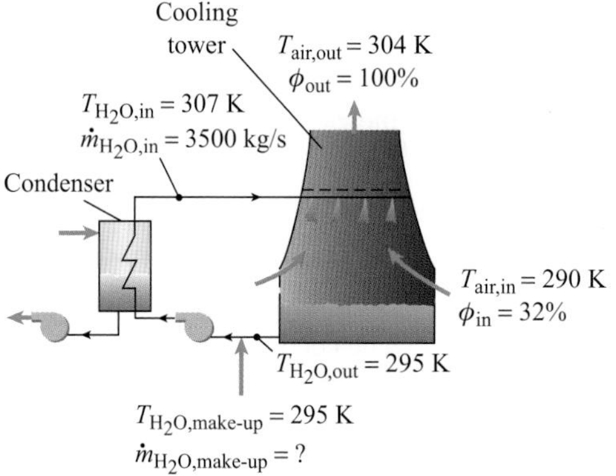

11.33 As shown in the sketch below, moist air enters an insulated tube at station 1 with properties T_1, P_1, and ω_1. Moisture is added to the air stream as it passes through a wick and exits at station 2 saturated with moisture (i.e., $\phi_2 = 100\%$) at a temperature T_2. The pressure at station 2 can be assumed to be the same as at station 1. The make-up water is supplied to the wick as saturated liquid, also at T_2. This process is referred to as an adiabatic saturation process and the device that accomplishes this is an adiabatic saturator.

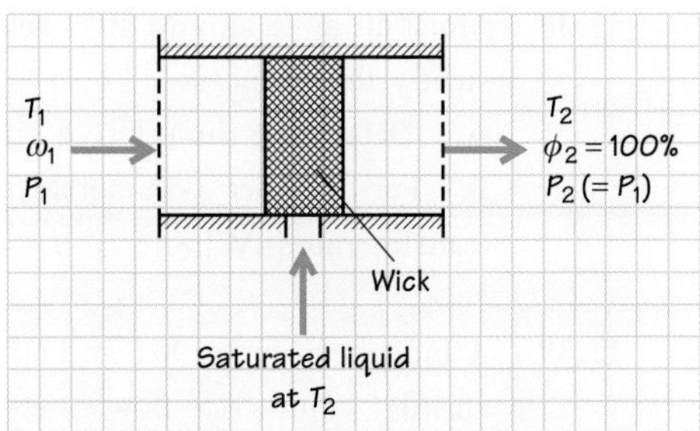

A. Derive an expression that allows you to calculate ω_1 from a knowledge of T_1, P_1, T_2, P_2 $(= P_1)$, and P_{sat} (T_2).

B. Apply your result from part A to obtain a numerical result for the specific humidity ω_1 and relative humidity ϕ_1 for the following conditions: $T_1 =$ 299.8 K (80 F), $T_2 = 293.1$ K (68 F), and $P = 1$ atm.

(Note: Sling psychrometers, and similar devices used to measure humidity, provide conditions quite similar to an adiabatic saturator. In these devices, T_1 is defined as the dry-bulb temperature and T_2 is defined as the wet-bulb temperature.)

11.34 Moist air enters an air conditioner at 305.3 K, 137.9 kPa, and 80% relative humidity. The mass flow rate of dry air entering is 45.4 kg/min. The moist air leaves the air conditioner at 285.3 K, 124.1 kPa, and 100% relative humidity. The condensed moisture leaves at 285.3 K and 124.1 kPa. Determine the heat-transfer rate (kW) from the moist air.

11.35 A moist gas (10% carbon dioxide, 70% nitrogen, 20% water vapor by volume) enters a reversible adiabatic turbine at 250 °C and 0.4 MPa. The turbine exit pressure is 0.10135 MPa. The inlet volumetric flow rate is 3 m^3/min. Determine the following:

A. The mass flow rate (kg/min) of dry gas mixture

B. The exit temperature (°C)

C. The mass flow rate (kg/min) of liquid water at the exit

D. The power produced (kW)

11.36 An adiabatic compressor receives air from the atmosphere at 60 F, 14.7 psia, and 75% relative humidity and discharges the air at 100 psia. The compressor isentropic efficiency is 0.80. Determine the relative humidity and the humidity ratio of the discharged air.

11.37 A moist gas (10% carbon dioxide, 70% nitrogen, 20% water vapor by volume) enters a reversible adiabatic turbine at 150°C and 0.4 MPa. The turbine exit pressure is 0.10135 MPa. The inlet volumetric flow rate is 3 m^3/min. Determine the following quantities:

A. The mass flow rate (kg/min) of dry gas mixture

B. The exit temperature (°C)

C. The mass flow rate (kg/min) of liquid water at the exit

D. The power output (kW)

11.38 Moist air at 283 K and 40% relative humidity is heated at 1 atm to 305.3 K in a steady-flow process.

A. Determine the relative humidity at the exit.

B. For an inlet volumetric flow rate of 1000 ft^3/min, determine the heat-transfer rate (kW).

11.39 Atmospheric air at 90 F and 60% relative humidity flows over a cooling coil at an inlet volumetric flow rate of 2000 ft^3/min. The cooling coil temperature is 40 F. The condensed liquid is removed from the system at 60 F. The air is then electrically heated until a final state of 80 F and 40% relative humidity is reached.

A. Determine the mass flow rate (lb_m/min) of the liquid water leaving the cooling section.

B. Determine the heat-transfer rate (Btu/min) in the cooling section.

C. Determine the electrical work-transfer rate (Btu/min)

11.40 A home heating system in Madison, Wisconsin, is designed to heat the outside air at 4.4 °C (40 F) and 100% relative humidity to 26.7 °C (80 F). Assume the pressure to be 1 atm everywhere. Determine the relative humidity at 26.7 °C and the heat transfer to the air (kJ/kg).

11.41 During mild exercise (e.g., walking) a person breathes in air at the rate of 20 liters/min and has a metabolism rate of 300 W. Atmospheric air is at 20 °C and 30% relative humidity, and the "air" leaving the mouth is at 35 °C and 80% relative humidity. Determine the following:

A. The energy loss from a person due to breathing (W)

B. The energy-transfer rate (W) to the dry air that is breathed

C. The water loss (kg/min) from a person due to breathing

11.42 Moist air at 90 F, 14.7 psia, and 80% relative humidity is removed from the top of a large room at a rate of 500 ft^3/min. The air is then cooled in an air conditioner at constant pressure to 60 F and returned to the bottom of the room. Condensed moisture leaves the air conditioner at 60 F and 14.7 psia. Determine (a) the rate of moisture removal (gal/hr) by the air conditioner and (b) the heat-transfer rate (Btu/min) rejected by the air conditioner.

11.43 Moist air enters an adiabatic humidifier at state 1 [21.1 °C (70 F), 1 atm, 10% relative humidity] with a volumetric flow rate of 500 ft^3/min. The air leaves the humidifier at state 2 [23.9 °C (75 F), 1 atm, 70% relative humidity]. The change in the condition of the moist air is brought about by the injection of steam at state 3 from a boiler, as shown in the diagram. The boiler is supplied with 23.3 °C (74 F) water at state 4. An adiabatic valve between the boiler and the humidifier causes the steam pressure to drop from boiler pressure to the humidifier pressure of 1 atm.

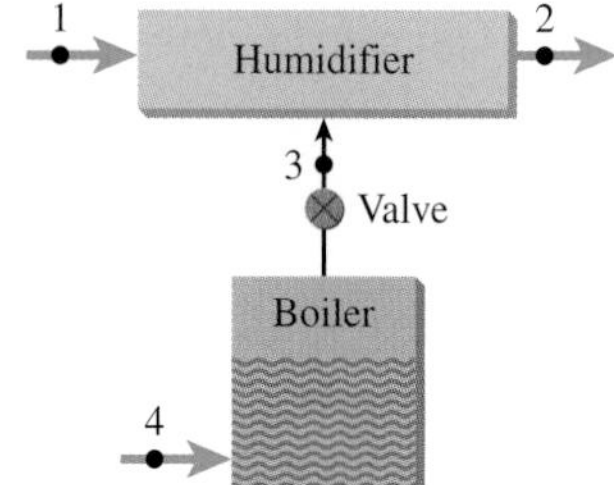

A. Determine the mass flow rate (kg/min) of the water at state 4.

B. Determine the heat-transfer rate to the boiler (kW).

C. Determine the pressure (kPa) in the boiler.

11.44 Moist air at 30 °C, 1 atm, and 20% relative humidity flows steadily into an adiabatic mixing chamber. Steam at 1 atm is sprayed into the chamber to humidify the air. The air leaves the chamber completely saturated at 30 °C and 1 atm.

A. Determine the temperature (°C) and/or quality of the added steam.

B. Determine the mass of steam required per mass of dry air (kg/kg).

11.45 Cold air enters an air conditioner at 4.4 °C (40 F) with 50% relative humidity at a volumetric flow rate of 1000 ft^3/min. The air is then heated. Next, moisture at 21.1 °C (70 F) and 1 atm is added adiabatically to bring the air to 26.7 °C (80 F) with a 40% relative humidity. The entire process occurs at 1 atm. Determine (a) the rate of moisture addition to the air (kg/min), (b) the heat-transfer rate (W), and (c) the air temperature (°C) after heating.

11.46 Moist air steadily enters a room at 20 °C and 1 atm with a dew-point temperature of 7 °C at a rate of 15 m^3/min. As the moist air passes through the room it is heated at a rate of 7 kW by a source at 65 °C. Liquid water at 20 °C is also supplied to the room at a rate of 5 kg/hr and is evaporated into the moist air. Determine the temperature (°C) and relative humidity of the moist air as it leaves the room.

11.47 A stream of moist air is obtained by mixing a 1500 ft^3/min flow of air at 26.7 °C (80 F) and 80% relative humidity with a 500 ft^3/min flow of air at 15.6 °C (60 F) and 50% relative humidity. The mixing is adiabatic and occurs at constant pressure (1 atm). Determine the temperature (°C), humidity ratio, and relative humidity of the exit stream.

11.48 An evaporative cooler for an automobile consists of an 18-in-long by 8-in-diameter cylinder that hangs on the outside of the car. At 55 mph the cooler processes 4500 lb_m of dry air per hour. During desert driving, the air is at 100 F with 10% relative humidity. Inside the cooler, the air passes through a moistened burlap cloth and leaves the cooler at 80 F. This cooled air flows through a duct to the passenger compartment. The cooler can be considered adiabatic, and all processes occur at 14.7 psia. The burlap moisture is supplied from a 5-gal container outside the car at 100 F. Kinetic energy changes are negligible.

A. Determine the humidity ratio and the dew point (F) of the desert air.

B. Determine the relative humidity of the processed air entering the car.

C. If the container is full at the start, determine the time (hr) for which the car can be driven at these conditions before the 5-gal container runs dry.

11.49 Cooling water from an internal combustion engine is to be cooled from 150 to 110 F. The pressure is 14.7 psia. Compare the following three methods with respect to the air and/or city water required (lb_m/1000 gallons of engine cooling water):

A. Heat transfer to city water. The city water temperature increases from 70 to 130 F.

B. Heat transfer to dry atmospheric air. The air temperature increases from 90 to 120 F.

C. A cooling tower that receives moist air at 90 F with a 50% relative humidity and discharges it at 95 F with 90% relative humidity. The evaporated water is made up by city water at 70 F.

11.50 Humid air enters a dehumidifier with an enthalpy of 21.6 Btu/lb_m for the dry air and 1100 Btu/lb_m for the water vapor. There is 0.02 lb_m of vapor per pound of dry air at entrance and 0.009 lb_m of vapor per pound of dry air at exit. The dry air at exit has an enthalpy of 13.2 Btu/lb_m, and the vapor at exit has an enthalpy of 1085 Btu/lb_m. Condensate with an enthalpy of 22 Btu/lb_m leaves the dehumidifier. The dry air flow rate is 287 lb_m/min. Determine (a) the amount of moisture removed from the air (lb_m/min) and (b) the rate at which heat must be removed.

11.51 Air is supplied to a room from outside, where the temperature is 20 F (−6.7 °C) and the relative humidity is 60%. The room is to be maintained at 70 F (21.1 °C) and 50% relative humidity. What mass of water must be supplied per mass of air supplied to the room?

11.52 Saturated air at 40 F (4.4 °C) is first preheated and then saturated adiabatically. This saturated air is then heated to a final condition of 105 F (40.6 °C) and 28% relative humidity. To what temperature must the air initially be heated in the preheat coil?

11.53 An air–water vapor mixture enters an air-conditioning unit at a pressure of 150 kPa, a temperature of 30 °C, and a relative humidity of 80%. The mass flow rate of dry air entering is 1 kg/s. The air–vapor mixture leaves the air-conditioning unit at 125 kPa, 10 °C, and 100% relative humidity. The condensed moisture leaves at 10 °C. Determine the heat-transfer rate (kW) for this process.

11.54 Air at 40 °C and 300 kPa with a relative humidity of 35% expands in a reversible adiabatic nozzle. To how low a pressure (kPa) can the gas be expanded if no condensation is to take place? What is the exit velocity (m/s) at this condition?

11.55 In an air-conditioning unit, air enters at 80 F, 60% relative humidity, and standard atmospheric pressure. The volumetric flow rate of the entering air (dry basis) is 71,000 ft^3/min. The moist air exits at 57 F and 90% humidity. Calculate the following:

A. The cooling capacity of the air-conditioning unit (Btu/hr)

B. The rate of water removal from the unit (lb_m/hr)

C. The dew point of the air leaving the conditioner

11.56 An air–water vapor mixture enters a heater–humidifier unit at 5 °C, 100 kPa, and 50% relative humidity. The flow rate of dry air is 0.1 kg/s. Liquid water at 10 °C is sprayed into the mixture at the rate of 0.0022 kg/s. The mixture leaves the unit at 30 °C and 100 kPa. Calculate (a) the relative humidity at the outlet and (b) the rate of heat transfer to the unit.

11.57 A room is maintained at 75 F and 50% relative humidity. The outside air is available at 40 F and 50% relative humidity. The return air from the room is cooled and dehumidified by mixing it with fresh air from the outside. The air flowing into the room is 60% outdoor air and 40% return air by mass. Determine the temperature, relative humidity, and specific humidity of the mixed air going into the room.

11.58 An air–water vapor mixture at 14.7 psia, 85 F, and 50% relative humidity is contained in a 15-ft^3 tank. At what temperature will condensation begin? If the tank and mixture are cooled an additional 15 F, how much water will condense from the mixture?

11.59 Consider 1000 ft^3/min (dry basis) of air at 14.7 psia, 90 F, and 60% relative humidity passing over a cooling coil with a mean surface temperature of 40 F. A water spray on the coil ensures that the exiting air is saturated at the coil temperature. What is the required cooling capacity of the coil in tons? (Note: 1 ton of refrigeration = 12,000 Btu/hr.)

11.60 A flow of 2832 liters/s (6000 ft^3/min, dry basis) of air at 26.7 °C (80 F) dry-bulb temperature, 50% relative humidity, and standard atmospheric pressure enters an air-conditioning unit. The air exits at temperature 13.9 °C (57 F) with 90% relative humidity. Determine the following:

A. The cooling capacity of the air-conditioning unit (kW and tons)

B. The rate of water removal from the unit (kg/hr and lb_m/hr)

C. The dew point of the air leaving the conditioner (°C and F)

11.61 A gas-turbine engine burns liquid octane with 300% "theoretical" air (i.e., the octane to air ratio is 1:3) in a steady-flow constant-pressure process. The air and octane both enter the engine at 25 °C. The exhaust gases leave the engine at 725 °C. The overall engine is well insulated. Determine the following:

A. The power output for a fuel flow rate of 39.5 kg/hr

B. The composition of the exhaust products on a volumetric basis (dry)

C. The dew-point temperature of the exhaust gases

11.62–11.65 Humidity measurement

11.62 Moist air at 100 F, 20 psia, and 40% relative humidity is flowing in a pipe. Determine (a) the dew-point temperature, (b) the humidity ratio, and (c) the adiabatic saturation temperature.

11.63 The temperature and adiabatic saturation temperature for a room of moist air are measured to be 30 °C and 20 °C, respectively.

A. Determine the humidity ratio and relative humidity if the pressure is 1 atm.

B. Determine the humidity ratio if the pressure is 0.95 atm with the same temperature and adiabatic saturation temperature as in Part A.

11.64 In a classroom the temperature is 78 F, the adiabatic saturation temperature is 64 F, and the barometric pressure is 29.175 in Hg.

A. Determine the humidity ratio.

B. Determine the partial pressures (in Hg) of the dry air and of the water vapor.

C. Determine the relative humidity.

D. Determine the density ($lb_{m,mix}/ft^3$).

E. Determine the dew-point temperature.

11.65 Without using the psychrometric chart, determine the humidity ratio and the relative humidity of an air–water-vapor mixture with a dry-bulb temperature of 32 °C and a thermodynamics wet-bulb temperature of 25 °C. The barometric pressure is 101 kPa. Check your results using the psychrometric chart.

CHAPTER 12
Reacting Systems

OVERALL LEARNING OBJECTIVES

After studying Chapter 12, you should:

- Be able to apply atom (element) conservation principles to reacting systems or flows.
- Understand the various ways that stoichiometry is expressed for combustion systems and be able to use this information to formulate mass conservation expressions.
- Understand the concept of standardized properties, in particular, standardized enthalpies, and their application to ideal-gas systems involving chemical reactions.
- Be able to apply standardized enthalpies (or internal energies) to solve first-law problems for reacting systems.

CHAPTER 12 OVERVIEW

IN THIS CHAPTER, we extend the overarching mass and energy conservation principles to reacting systems. In addition to conserving mass in an overall sense, the mass of individual elements must also be conserved in the transformation of reactants to products. Complicating energy conservation (the first law of thermodynamics) is the need to account for the potential energies associated with the making and breaking of chemical bonds. To accomplish this, we introduce the concept of standardized enthalpies. We introduce other new concepts related to mass conservation (e.g., various measures of stoichiometry) and energy conservation (e.g., adiabatic flame temperatures and fuel heating values). The chapter also applies the theoretical developments to practical steady-flow systems (e.g., furnaces, boilers, and combustors). Although the chapter focuses on combustion, the ideas developed here apply to other reacting systems as well.

<table>
<tr><td colspan="3">DESIGN AND ANALYSIS OF SYSTEMS</td></tr>
<tr><td colspan="3">ANALYSIS OF PRACTICAL DEVICES</td></tr>
<tr><td>CONSERVATION OF MASS</td><td>CONSERVATION OF ENERGY (First Law of Thermodynamics)</td><td>SECOND LAW OF THERMODYNAMICS (Entropy)</td></tr>
<tr><td colspan="3">PROPERTIES OF MATTER</td></tr>
<tr><td colspan="3">FRAMEWORKS FOR ANALYSIS, KEY CONCEPTS, AND DEFINITIONS</td></tr>
</table>

FIGURE 12.1 Chapter 12 encompasses and expands upon nearly all the concepts and major principles introduced in previous chapters (see the lower levels of the figure). Chapter 12 also expands upon many applications (the top levels) and includes combustion devices.

Historical Context

The modern understanding of combustion has its roots in the careful experiments conducted by **Antoine Laurent Lavoisier** (1743–1794) and subsequently published in 1798 [1]. Lavoisier observed that metals gained mass when heated in the presence of oxygen, whereas at higher temperatures the metal oxides decomposed back to the original metal and oxygen. These experiments were conducted at a time when the idea of phlogiston was in competition with the newer caloric theory. (Neither theory has stood the test of time.) Phlogiston was thought to be an imponderable (massless) fluid associated with combustion. For example, it was thought that when charcoal burned, its phlogiston escaped and combined with the air. Combustion would then be complete, or terminate, when all the phlogiston had escaped or when the air was saturated with phlogiston, as would occur for combustion in a closed vessel. Lavoisier's experiments made these ideas untenable.

(Credit: AGEfotostock.)

(Credit: Panopticon Lavoisier/Institute & Museum of History of Science, Florence.)

FIGURE 12.2 Combustion is important in many industrial processes. Cast iron is melted in this cupola furnace in which coke (carbon) is oxidized to form CO, which in turn further oxidizes to produce CO_2 (photllurg / iStock / Getty Images Plus).

12.1 Mass Conservation for Reacting Systems

In many systems of engineering interest, chemical reactions can be important. In particular, combustion reactions liberate thermal energy that is converted into useful work. Three examples of this were highlighted in Chapter 1: fossil-fueled steam power plants, spark-ignition engines, and jet engines. Other examples in which combustion plays an important role are engines of many types (e.g., rocket engines, diesel engines, and gas-turbine engines), space heating with gas- or oil-fired furnaces, a wide variety of industrial heating and processing operations, and metal smelting, to name just a few (Fig. 12.2). This section of the book provides the mass conservation framework to analyze both closed and open systems in which combustion occurs.

12.1a Atom Balances

In dealing with a system, the mass of each chemical element remains constant during any process, assuming, of course, that no nuclear transformations occur. For a process in which the system goes from state 1 to state 2 (see Fig. 12.3),

A recommended digression at this juncture is to return to Chapter 10 to read or review the section on specifying mixture composition. See specifically Eqs. 10.1–10.5.

$$M_i(\text{state 1}) = M_i(\text{state 2}) \qquad \text{for } i = 1, 2, \ldots, I, \tag{12.1}$$

where M_i is the mass of element i and I is the number of elements involved. For example, consider a pressure vessel filled with a mixture of fuel and air. If we ignore the small amount of CO_2 in the air, all the carbon is associated with the fuel molecules (e.g., C_xH_y). If the mixture is ignited with a spark, a flame will travel through the pressure vessel, transforming the fuel and air to combustion products. After the passage of the flame, all the carbon that was initially present in the fuel is now found in the CO_2, CO, and soot (Fig. 12.4), if any, of the combustion products. To keep track of the carbon, and all other elements involved, we write an expression that describes the overall chemical reaction as follows:

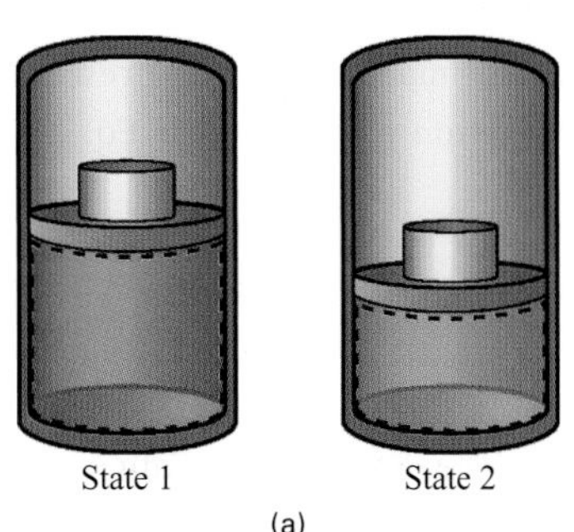

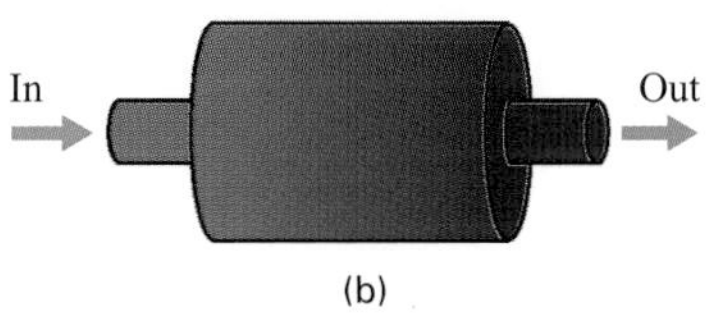

FIGURE 12.3 (a) The mass of any element is conserved in a system in which a chemical reaction takes place in going from state 1 to state 2. **(b)** For a steady flow reactor, the mass flow rate of any element has the same value at the inlet and the outlet.

$$C_xH_y + aO_2 + bN_2 \rightarrow cCO_2 + dCO + eH_2O + fH_2 + gO_2 + hN_2, \tag{12.2}$$

where a, b, c, ... are the number of moles of each respective species. In Eq. 12.2, we assume for simplicity that the only product species are those shown; however, other minor species, such as OH, O, NO, and H, can exist, depending on the specific conditions. These species could easily be included in Eq. 12.2. (We note, however, that in the developments and examples that follow, we restrict our combustion products to be those of complete combustion, i.e., CO_2, H_2O, N_2, and any excess O_2 that may be present. The dissociation of the combustion products, i.e., incomplete or nonideal combustion, is treated in the next chapter. For more information, we refer the reader to Ref. [2] or similar resources.) Since there are four elements involved in Eq. 12.2 (C, H, O, and N), we are able to write four specific element conservation relations:

Reminder: Equations with yellow backgrounds express key concepts and are the most important relationships in the chapter.

$$N_{C_xH_y}\left(\frac{N_C}{N_{C_xH_y}}\right) = N_{CO_2}\left(\frac{N_C}{N_{CO_2}}\right) + N_{CO}\left(\frac{N_C}{N_{CO}}\right), \tag{12.3a}$$

$$N_{C_xH_y}\left(\frac{N_H}{N_{C_xH_y}}\right) = N_{H_2O}\left(\frac{N_H}{N_{H_2O}}\right) + N_{H_2}\left(\frac{N_H}{N_{H_2}}\right), \tag{12.3b}$$

$$N_{O_2}(\text{Reac})\left(\frac{N_O}{N_{O_2}(\text{Reac})}\right) = N_{CO_2}\left(\frac{N_O}{N_{CO_2}}\right) + N_{CO}\left(\frac{N_O}{N_{CO}}\right) + N_{H_2O}\left(\frac{N_O}{N_{H_2O}}\right) + N_{O_2}\left(\frac{N_O}{N_{O_2}}\right), \tag{12.3c}$$

and

FIGURE 12.4 Flaring of refinery gases. Note the solid carbon (soot) escaping from the flare downstream of the yellow flame (Daniel Cogliano / EyeEm / Getty Images Plus).

$$N_{N_2}(\text{Reac})\left(\frac{N_N}{N_{N_2}(\text{Reac})}\right) = N_{N_2}\left(\frac{N_N}{N_{N_2}}\right). \tag{12.3d}$$

Each of these expressions states that the number of atoms of a particular element in the reactants (i.e., the left-hand sides of Eqs. 12.3a–12.3d) equals the number of atoms of that element in the products (i.e., the right-hand sides of Eqs. 12.3a–12.3d). For example, the term on the left-hand side of Eq. 12.3a is the number of moles of carbon in the reactants, which can be expressed by multiplying the number of moles of fuel ($N_{C_xH_y}$) in the reactant mixture by the number of moles of carbon present in each mole of fuel [i.e., $(N_c/N_{C_xH_y})(\equiv x)$]. For the reaction as written in Eq. 12.2, $N_{C_xH_y}$ is unity; thus,

$$N_{C_xH_y}\left(\frac{N_c}{N_{C_xH_y}}\right) = 1 \times x.$$

This amount of reactant carbon is equal to the amount of carbon in the products (i.e., the element carbon is conserved). The amount of carbon in the products is then the number of moles of each carbon-containing species multiplied by the respective fraction of carbon in each species. For our example, carbon is present in two product species; thus, the right-hand terms in Eq. 12.3a become

$$N_{CO_2}\left(\frac{N_C}{N_{CO_2}}\right) = c \times 1$$

and

$$N_{CO}\left(\frac{N_C}{N_{CO}}\right) = d \times 1$$

We interpret Eqs. 12.3b–12.3d similarly using the combustion reaction expressed in Eq. 12.2. Equations 12.3a–12.3d are then rewritten as follows:

$$x = c + d \quad (\text{C balance}), \tag{12.4a}$$

$$y = 2e + 2f \quad (\text{H balance}) \tag{12.4b}$$

$$2a = 2c + d + e + 2g \quad (\text{O balance}), \tag{12.4c}$$

and

$$2b = 2h \quad (\text{N balance}). \tag{12.4d}$$

Equations 12.4a–12.4d can be generalized as

$$N_i(\text{state 1}) = N_i(\text{state 2}) \quad \text{for } i = 1, 2, \ldots, I, \tag{12.5}$$

where N_i is the number of moles of element i and I is the number of elements involved. Note that Eq. 12.5 and Eq. 12.1 are equivalent since multiplying the number of moles of element I by its atomic weight yields the mass of element i (i.e., $M_i = N_i \mathcal{M}_i$).

We can easily extend the element conservation principle to open systems (control volumes). Consider a steady-state, steady flow through an open system having a single inlet and a single outlet, as shown in Fig. 3.7a. For this situation, we write

$$\underset{\substack{\text{Mass flow of element } i\\ \text{into the open system}}}{\dot{m}_{i,\text{in}}} = \underset{\substack{\text{Mass flow of element } i\\ \text{out of the open system}}}{\dot{m}_{i,\text{out}}} \quad \text{for } i = 1, 2, \ldots, I. \tag{12.6}$$

Equation 12.6 can also be expressed in terms of the individual element mass fractions $Y_{e,i}$, and the total mass flow rate; that is,

$$(\dot{m}, Y_{e,i})_{\text{in}} = (\dot{m}, Y_{e,i})_{\text{out}}.$$

because $\dot{m}_{\text{in}} = \dot{m}_{\text{out}}$ for steady flow (overall mass conservation), this equation becomes

$$(Y_{e,i})_{\text{in}} = (Y_{e,i})_{\text{out}} \quad \text{for} \quad i = 1, 2, \ldots, I. \tag{12.7}$$

Note that the element mass fraction $Y_{e,i}$ comprises contributions from every species in the mixture that contains the ith element. Equation 12.7 is the steady-flow open-system (control-volume) equivalent to Eq. 12.1 for a closed system.

Example 12.1 Combustion of Propane with Air

One mole of propane (C_3H_8) is burned with 47.6 kmol of air to produce a mixture of CO_2, H_2O, O_2, and N_2. Assume that the air consists only of O_2 and N_2 and that there are 3.76 kmol of N_2 for each kmol of O_2 (i.e., the air is 21% O_2 and 79% N_2 by volume).

A. Determine the number of moles of each of the product species in the product mixture.

B. Determine the mole fractions of each of the product species in the product mixture.

C. Determine the mass fractions of each of the product species in the product mixture.

Solution

Known $N_{C_3H_8}$, N_{air}

Find N_j, X_j, Y_j for $j = CO_2, H_2O, O_2, N_2$

Sketch

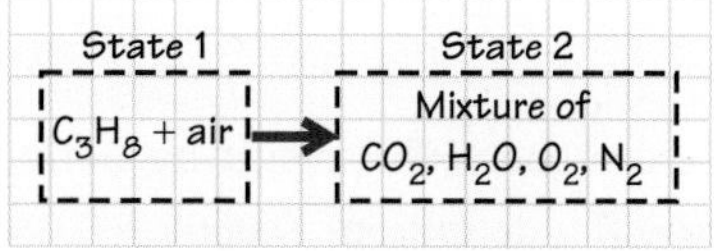

Modeling, Premises and Assumptions

i. The only products are CO_2, H_2O, O_2, and N_2.

ii. No nuclear transformations occur.

iii. Air is 21% O_2 and 79% N_2 (given).

Analysis Although the problem statement does not indicate whether we are considering a closed system or a steady-flow reactor (open system/control volume), such knowledge is not required to solve the problem. We begin by writing a reaction equation, following Eq. 12.2:

(1) $C_3H_8 + 47.6(0.21\ O_2 + 0.79\ N_2) \rightarrow a\,CO_2 + b\,H_2O + c\,O_2 + d\,N_2.$

We now write atom conservation expressions (Eqs. 12.4) for C, H, O, and N, respectively:

C: $1\ (3) = a\ (1)$
H: $1\ (8) = b\ (2)$, or $b = 4$
O: $47.6 \times 0.21\ (2) = a\ (2) + b\ (1) + c\ (2)$;

substituting a and b and solving for c yields $c = 5$.

N: $47.6 \times 0.79\ (2) = d\ (2)$, or $d = 37.6$.

Since the coefficients a, b, c, and d represent the number of moles of CO_2, H_2O, O_2, and N_2, respectively, we have completed part A:

$$N_{CO_2} = 3, \qquad N_{O_2} = 5,$$
$$N_{H_2O} = 4, \qquad N_{N_2} = 37.6.$$

To determine the mole fractions of each of the product species, we apply the definition of a mole fraction (Eq. 10.1a):

$$X_j = \frac{N_j}{N_{tot}}.$$

For our product mixture,

$$\begin{aligned} N_{tot} &= N_{CO_2} + N_{H_2O} + N_{O_2} + N_{N_2} \\ &= 3 + 4 + 5 + 37.6 \text{ kmol} \\ &= 49.6 \text{ kmol}. \end{aligned}$$

Thus,

$$\begin{aligned} X_{CO_2} &= \frac{3}{49.6} = 0.0605\,(\text{dimensionless}), \\ X_{H_2O} &= \frac{4}{49.6} = 0.0806, \\ X_{O_2} &= \frac{5}{49.6} = 0.1008, \\ X_{N_2} &= \frac{37.6}{49.6} = 0.7581. \end{aligned}$$

We can check this result by noting that the sum of the mole fractions should equal unity (by definition): $0.0605 + 0.0806 + 0.1008 + 0.7581 = 1.000$, as expected. To find the mass fractions of the product species, we apply Eq. 10.4a, which relates the mass fractions to the now-known mole fractions:

$$Y_j = X_j \frac{\mathcal{M}_j}{\mathcal{M}_{mix}},$$

where the apparent molecular weight of the mixture is given by

$$\mathcal{M}_{mix} = \sum_{j=1}^{J} X_j \mathcal{M}_j. \tag{10.3}$$

Using values from Appendix D for the molecular weights $\mathcal{M}_j$ for the species in the mixture, we can calculate $\mathcal{M}_{mix}$ as follows:

$$\begin{aligned}\mathcal{M}_{\text{mix}} &= 0.0605(44.011) + 0.0806(18.016) + 0.1008(31.999) \\ &\quad + 0.7581(28.013)\ \text{kg/kmol} \\ &= 28.577\ \text{kg/kmol}.\end{aligned}$$

We now calculate the mass fractions:

$$Y_{CO_2} = X_{CO_2}\frac{\mathcal{M}_{CO_2}}{\mathcal{M}_{\text{mix}}} = 0.0605\left(\frac{44.011}{28.577}\right) = 0.0932,$$

$$Y_{H_2O} = X_{H_2O}\frac{\mathcal{M}_{H_2O}}{\mathcal{M}_{\text{mix}}} = 0.0806\left(\frac{18.016}{28.577}\right) = 0.0508,$$

$$Y_{O_2} = X_{O_2}\frac{\mathcal{M}_{O_2}}{\mathcal{M}_{\text{mix}}} = 0.1008\left(\frac{31.999}{28.577}\right) = 0.1129,$$

$$Y_{N_2} = X_{N_2}\frac{\mathcal{M}_{N_2}}{\mathcal{M}_{\text{mix}}} = 0.7581\left(\frac{28.013}{28.577}\right) = 0.7431.$$

The mass fractions, like the mole fractions, should also sum to unity (i.e., 0.0932 + 0.0508 + 0.1129 + 0.7431 = 1.0000).

Comments Although care must be exercised in writing the combustion reaction and the element conservation equations, the procedure is quite straightforward. Note that species with molecular weights larger than the mean mixture molecular weight (CO_2, O_2) have mass fractions larger than their corresponding mole fractions, whereas the converse is true for the lighter species (H_2O, N_2). We note also the large amount of N_2 present in both the reactants and products. Although in this example the N_2 was not involved in the reaction, small quantities do react at high temperatures to form the pollutant NO.

Self-Test 12.1

Write a balanced expression for the combustion of methane (CH_4) with 4.76*a* moles of air, using the assumptions of Example 12.1.

(Answer: $CH_4 + a(O_2 + 3.76N_2) \rightarrow CO_2 + 2H_2O + (a - 2)O_2 + 3.76aN_2$)

12.1b Stoichiometry

The **stoichiometric** quantity of air (or other oxidizer) is just that amount needed to completely burn a given quantity of fuel. For example, the stoichiometric amount of air for the combustion of a hydrocarbon fuel is that amount, and no more, that allows all the carbon in the fuel to oxidize to CO_2 and all the hydrogen to oxidize to water (H_2O). If more than a stoichiometric quantity of air is supplied, the mixture is said to be fuel lean, or just **lean;** supplying less than the stoichiometric air results in a fuel-rich, or **rich,** mixture. Spark-ignition engines operate with essentially stoichiometric mixtures when used in automobiles that employ an exhaust catalyst to reduce pollutant emissions (CO, NO, and unburned hydrocarbons) (Fig. 12.5). Fossil-fueled power plants operate with slightly lean mixtures for optimum efficiency.

FIGURE 12.5 A feedback control system maintains a stoichiometric fuel–air mixture to allow a three-way catalytic converter to simultaneously reduce emissions of carbon monoxide, unburned hydrocarbons, and nitric oxide (Einar Muoni / Alamy Stock Photo).

The stoichiometric air–fuel ratio (by mass) is determined by writing element conservation expressions, as has been discussed here, assuming that the fuel reacts to form an ideal set of products. For a hydrocarbon fuel given by C_xH_y, the stoichiometric reaction is expressed as

$$C_xH_y + a(O_2 + 3.76N_2) \rightarrow xCO_2 + (y/2)H_2O + 3.76aN_2, \tag{12.8}$$

where

$$a = x + y/4. \tag{12.9}$$

For simplicity, we assume that the air consists of 21% O_2 and 79% N_2 (by volume); that is, for each mole of O_2 in air, there are 3.76 moles of N_2. From Eq. 12.8, we see that the stoichiometric air–fuel ratio is given by

$$(A/F)_{stoic} = \left(\frac{M_{air}}{M_{fuel}} \text{ or } \frac{\dot{m}_{air}}{\dot{m}_{fuel}}\right)_{stoic} = \frac{4.76a}{1}\frac{\mathcal{M}_{air}}{\mathcal{M}_{fuel}}, \tag{12.10}$$

where $\mathcal{M}_{air}$ and $\mathcal{M}_{fuel}$ are the molecular weights of the air and fuel, respectively.

In many combustion applications, the stoichiometry is the single most important factor in determining a system's performance. The following interrelated measures are frequently used to indicate quantitatively the stoichiometry of a fuel–oxidizer mixture: the equivalence ratio, percent stoichiometric air, and percent excess air.

The **equivalence ratio** is defined as

$$\Phi = \frac{(A/F)_{stoic}}{(A/F)} = \frac{(F/A)}{(F/A)_{stoic}}. \tag{12.11a}$$

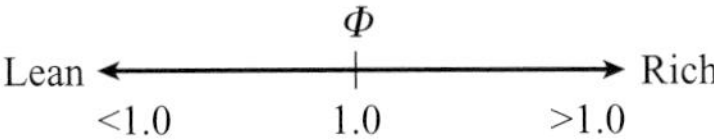

FIGURE 12.6 Equivalence ratios less than unity designate lean mixtures, and equivalence ratios greater than unity designate rich mixtures.

From this definition, we see that for fuel-rich mixtures $\Phi > 1$ and for fuel-lean mixtures $\Phi < 1$. For a stoichiometric mixture, $\Phi = 1$ (Fig. 12.6). The **percent stoichiometric air**, which relates to the equivalence ratio, is defined as

$$\% \text{ stoichiometric air} = \frac{100\%}{\Phi}, \tag{12.11b}$$

and the **percent excess air** is given by

$$\% \text{ excess air} = \frac{(1-\Phi)}{\Phi} \times 100\%. \tag{12.11c}$$

All three measures of stoichiometry are dimensionless or equivalently are expressed as a percentage.

Example 12.2 Fossil-Fuel Combustor

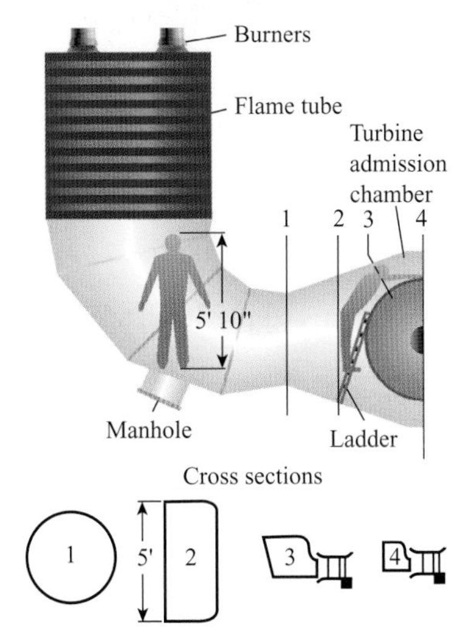

FIGURE 12.7 Schematic of silo-type gas-turbine combustor with ceramic-tile-lined chamber and access for inspection of the combustion chamber and turbine inlet. Reprinted from Ref. [3] by permission of the American Society of Mechanical Engineers.

A 35-MW gas-turbine engine is used to generate electricity and burns natural gas in a silo-type combustor, as shown in Fig. 12.7. The natural gas consists primarily of CH_4 with small quantities of several other light hydrocarbons and can be represented as $C_{1.16}H_{4.32}$. To achieve low emissions of pollutant oxides of nitrogen ($NO_x \equiv NO + NO_2$), the burners operate lean at an equivalence ratio of 0.4. Determine the operating air–fuel ratio (A/F), the percent excess air, and the oxygen (O_2) mole fraction in the combustion products, assuming complete combustion.

Solution

Known Φ, $C_{1.16}H_{4.32}$

Find A/F, % excess air, X_{O_2}

Sketch

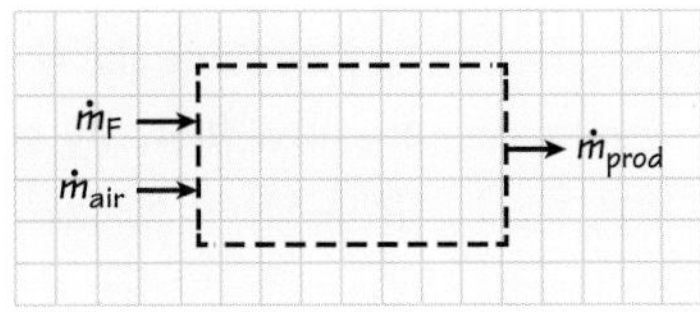

Modeling, Premises and Assumptions

i. Air is 21% O_2 and 79% N_2.
ii. Complete combustion occurs with no dissociation; that is, the products consist only of CO_2, H_2O, O_2, and N_2.

Analysis To find A/F, we start by applying the definition of the equivalence ratio (Eq. 12.11a):

$$A/F = \frac{(A/F)_{\text{stoic}}}{\Phi};$$

we can find $(A/F)_{stoic}$ from Eq. 12.10:

$$(A/F)_{stoic} = 4.76a\frac{\mathcal{M}_{air}}{\mathcal{M}_{fuel}}.$$

We calculate a (Eq. 12.9) as

$$a = x + y/4 = 1.16 + 4.32/4$$
$$= 2.24.$$

The apparent molecular weight of the air is (Eq. 10.3)

$$\mathcal{M}_{air} = 0.21\mathcal{M}_{O_2} + 0.79\mathcal{M}_{N_2}$$
$$= [0.21(31.999) + 0.79(28.013)]\text{kg/kmol}$$
$$= 28.85 \text{ kg/kmol}.$$

The molecular weight of the fuel is found directly from its chemical formula:

$$\mathcal{M}_{fuel} = 1.16\mathcal{M}_C + 4.32\mathcal{M}_H$$
$$= [1.16(12.011) + 4.32(1.00794)]\text{ kg/kmol}$$
$$= 18.287 \text{ kg/kmol}.$$

Substituting these results into Eq. 12.10 yields

$$(A/F)_{stoic} = 4.76(2.24)\frac{28.85}{18.287}$$
$$= 16.82 \text{ kg/kg}.$$

Applying the definition of the equivalence ratio (Eq. 12.11a), we obtain an operating air–fuel ratio of

$$A/F = \frac{16.82}{0.4} = 42.05\,\text{kg}_{air}/\text{kg}_{fuel}.$$

To determine the percent excess air, we directly apply its definition (Eq. 12.11c):

$$\%\text{ excess air} = \frac{(1-\Phi)}{\Phi} \times 100\%.$$

Thus,

$$\%\text{ excess air} = \frac{(1-0.4)}{0.4} \times 100\% = 150\%.$$

To determine the O_2 mole fraction, we need to determine the composition of the product mixture, first, by writing the combustion reaction (Eq. 12.2) for our assumed set of species, and, second, by performing C, H, O, and N atom balances. From Eq. 12.2 we have

$$C_{1.16}H_{4.32} + \frac{a}{\Phi}(O_2 + 3.76N_2) \rightarrow bCO_2 + cH_2O + dO_2 + eN_2,$$

where we have divided the stoichiometric coefficient a $(\equiv x + y/4)$ by Φ to account for the excess air supplied. The atom balances are as follows:

C: $1.16 = b.$

H: $4.32 = 2c$, or $c = 2.16.$

O: $\frac{a}{\Phi}(2) = 2b + c + 2d;$

$$\frac{2.24}{0.4}(2) = 2(1.16) + 2.16 + 2\text{d}.$$

Solving for d gives

$$d = 3.36.$$

$$\text{N: } \frac{a}{\Phi}(37.6)2 = 2e;$$

$$e = \frac{a(3.76)}{\Phi} = \frac{2.24(.3.76)}{0.4} = 21.06.$$

Thus, $a = 2.24$, $b = 1.16$, $c = 2.16$, $d = 3.36$, and $e = 21.06$. Since we now know the number of moles of each product species, the O_2 mole fraction is found from its definition (Eq. 10.1a):

$$X_{O_2} = \frac{N_{O_2}}{N_{mix}} = \frac{N_{O_2}}{N_{CO_2} + N_{H_2O} + N_{O_2} + N_{N_2}}$$
$$= \frac{3.36}{1.16 + 2.16 + 3.36 + 21.06}$$
$$= 0.121.$$

Comments Note that a large quantity of excess air is supplied to the combustor. This excess air results in the combustion temperatures being relatively low, which in turn prevents large quantities of the pollutants NO and NO_2 from being formed. Apart from pollutant emission concerns, the product gases entering the turbine must be sufficiently cool to prevent the turbine blades from overheating. In fact, the hot products may be diluted with additional air before entering the turbine to achieve the proper low temperature.

Self-Test 12.2

The gas turbine of Example 12.2 is operated with a natural gas flow rate of 12 kg/s. Determine the required air mass flow rates to operate the engine at equivalence ratios of 0.4 and 1.0.

(Answer: 201.84 kg/s, 504.6 kg/s)

12.2 Energy Conservation for Reacting Systems

Although a detailed study of reacting systems resides in the realm of chemical engineering, combustion systems are of wide engineering interest. In Chapter 1, we discussed the importance of combustion in the production of electricity in the United States. Table 1.1 shows that approximately 63% of the electricity produced in 2018 had its origin in the combustion of some fossil or renewable fuel [4]. The widespread use of combustion to provide useful forms of energy makes understanding the application of the energy conservation principle to reacting systems particularly important.

Our treatment here is quite basic. More information can be found in books dedicated to combustion (e.g., Refs. [2, 5]). Interestingly, our previous statements of energy conservation, Eqs. 5.6b and 5.16, already apply to reacting systems. For closed systems, energy conservation is expressed as (Eq. 5.6b)

$$_1(Q_{in,net})_2 - {}_1(Q_{out,net})_2 = \Delta E_{sys},$$

where (Eq. 5.7)

$$\Delta E_{sys} \equiv E_{sys,2} - E_{sys,1}.$$

For open systems, we express energy conservation as (Eq. 5.16)

$$\dot{Q}_{cv,\,net\,in} - \dot{W}_{cv,\,net\,out} = \sum_{k=1}^{M\,outlets} \dot{m}_{out,k}\left[h_k + \tfrac{1}{2}V^2_{avg,k} + g(z_k - z_{ref})\right] - \sum_{j=1}^{N\,inlets} \dot{m}_{in,j}\left[h_j + \tfrac{1}{2}V^2_{avg,j} + g(z_j - z_{ref})\right].$$

To apply these expressions for energy conservation to reacting systems, some elaboration is helpful. First, we revisit standardized properties and then consider two special cases: constant-pressure combustion and constant-volume combustion.

Standardized properties were mentioned in Chapter 10, where we discussed how to deal with ideal-gas mixtures.

12.2a Standardized Properties

We now develop the concept of standardized properties, which accounts for the potential energy changes associated with the breaking and making of chemical bonds. These standardized properties are particularly important in dealing with energy conservation for combustion systems, where it is this rearrangement of chemical bonds that results in the creation of hot products of combustion. The thermodynamic enthalpy property, h, is the property most frequently used in expressions of energy conservation for reacting systems.

From our earlier discussion of the calorific equation of state for ideal gases (Chapter 2), we see that a reference temperature is required to evaluate the specific internal energy and enthalpy (see Eqs. 2.29c and 2.31c). If one is concerned only with a single pure substance, as opposed to a reacting system where both reactant and product species are present, then Eqs. 2.29d and 2.31d would suffice to describe the internal energy and enthalpy changes for all thermodynamic processes, as only differences in states are of importance. Moreover, any choice of reference temperature would yield the same results. For example, the enthalpy change associated with a change of temperature from T_1 to T_2 is calculated from Eq. 2.31d as

$$h(T_2) - h(T_2) = \int_{T_{ref}}^{T_2} c_p dT - \int_{T_{ref}}^{T_1} c_p dT = \int_{T_1}^{T_2} c_p dT,$$

a straightforward calculation in which the reference temperature drops out. In reacting systems, however, we must include the energy stored in chemical bonds in our accounting. To accomplish this, the concept of standardized enthalpies is extremely valuable. For any species, we define a **standardized enthalpy** that is the sum of an enthalpy that takes into account the energy associated with chemical bonds (or lack thereof), the **enthalpy of formation**, h_f, and an enthalpy associated only with a temperature change, the **sensible enthalpy change**, Δh_s. Thus, the molar standardized enthalpy for species i is expressed as

$$\underset{\substack{\text{Standardized}\\ \text{enthalpy at}\\ \text{temperature } T}}{\bar{h}_i(T)} = \underset{\substack{\text{Enthalpy of}\\ \text{formation at}\\ \text{standard}\\ \text{reference state}\\ (T_{\text{ref}},\, P^\circ)}}{\bar{h}^\circ_{\text{f},i}(T_{\text{ref}})} + \underset{\substack{\text{Sensible}\\ \text{enthalpy change}\\ \text{in going from } T_{\text{ref}}\\ \text{to } T}}{\Delta\bar{h}_{\text{s},i}(T)}, \tag{12.12}$$

where

$$\Delta\bar{h}_{\text{s},i} \equiv \bar{h}_i(T) - \bar{h}^\circ_{\text{f},i}(T_{\text{ref}}).$$

To make practical use of Eq. 12.12, it is necessary to define a **standard reference state** (Table 12.1). We will employ a standard-state temperature, $T_{\text{ref}} = 25\,^\circ\text{C}$ (298.15 K), and standard-state pressure, $P_{\text{ref}} = P^\circ = 1$ atm (101,325 Pa), to be consistent with the JANAF, Chemkin, and NASA thermodynamic databases (see Appendix D). Although not needed to describe enthalpies, the standard-state pressure is required to calculate entropies at pressures other than one atmosphere (see Eq. 7.10a). In addition to defining T_{ref} and P_{ref}, we adopt the convention that enthalpies of formation are zero for the elements in their naturally occurring state at the reference-state temperature and pressure. For example, hydrogen exists as diatomic molecules at 25 °C and 1 atm; hence,

$$\left(\bar{h}^\circ_{\text{f},\text{H}_2}\right)_{298} = 0,$$

where the superscript $^\circ$ is used to denote that the value is for the standard-state pressure. (The use of the superscript is redundant for ideal-gases enthalpies, which exhibit no pressure dependence; for entropies, however, the superscript is important.)

TABLE 12.1 Standard Reference State

$T_{\text{ref}} \equiv 298.15$ K
$P_{\text{ref}} = 101,325$ Pa
For elements in their naturally occurring state (solid, liquid, or gas) at T_{ref} and P_{ref}*,
$h_i(T_{\text{ref}}) \equiv 0$

* For example, the standard-state enthalpies for C(s), N_2(g), and O_2(g) are all zero.

To form hydrogen atoms at the standard state requires the breaking of a rather strong chemical bond (Fig. 12.8). The bond dissociation energy for H_2 at 298 K is

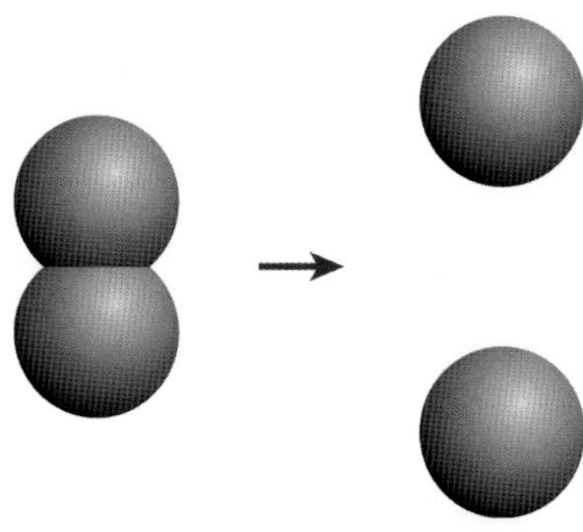

FIGURE 12.8 Shared electrons form a covalent bond in molecular hydrogen. An energy input is required to break this bond to form two hydrogen atoms.

$436\,\text{MJ/kmol}_{\text{H}_2}$. Breaking this bond creates two H atoms; thus, the enthalpy of formation for atomic hydrogen is half the value of the H_2 bond dissociation energy:

$$\left(\bar{h}^{\circ}_{\text{f},\text{H}_2}\right) = 218\,\text{MJ/kmol}_{\text{H}}.$$

Thus, enthalpies of formation have a clear physical interpretation as the net change in enthalpy associated with breaking the chemical bonds of the standard-state elements and forming new bonds to create the compounds of interest.

Representing the standardized enthalpy graphically provides a useful way to understand and use this concept. In Fig. 12.9, the standardized enthalpies for atomic hydrogen (H) and diatomic hydrogen (H_2) are plotted versus temperature starting from 200 K. At 298.15 K, we see that $\bar{h}_{\text{H}_2}$ is zero (by definition of the standard-state reference condition), and the standardized enthalpy of atomic hydrogen equals its enthalpy of formation, since the sensible enthalpy at 298.15 K is zero. At the temperature indicated (4000 K), we see the contribution of the additional sensible enthalpy to the standardized enthalpy. In Appendix D, enthalpies of formation at the reference state are given, and sensible enthalpies are tabulated as a function of temperature for a number of species of importance in combustion. Enthalpies of formation for reference temperatures other than the standard state 298.15 K are also tabulated.

Although the choice of a zero datum for the enthalpy of elements at the reference state is arbitrary (being merely a pragmatic choice), the third law of thermodynamics is used to set zero values for entropies; thus, the standard-state entropy values are always larger than zero.

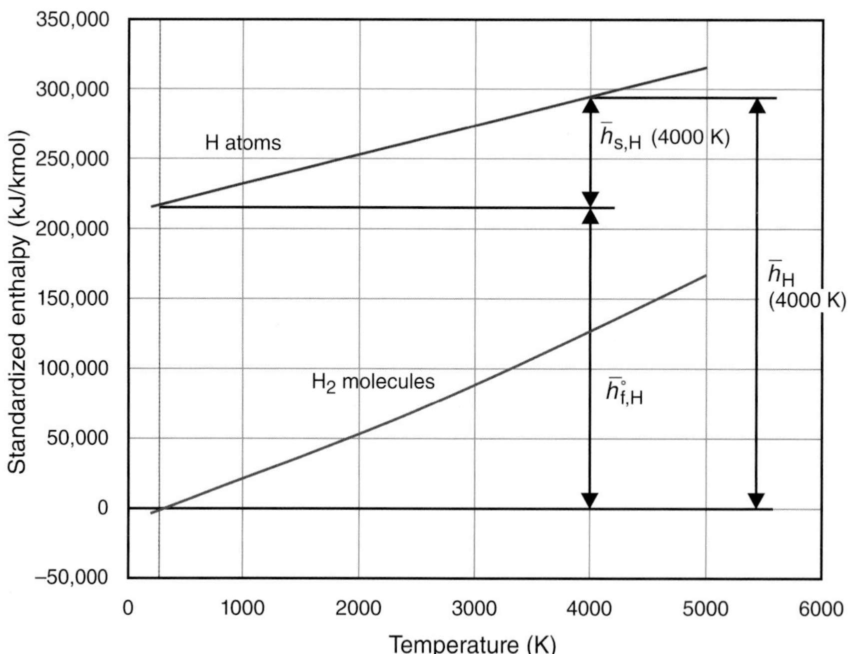

FIGURE 12.9 Graphical interpretation of standardized enthalpy, enthalpy of formation, and sensible enthalpy.

Example 12.3 Molar- and Mass-Specific Properties

Determine the standardized molar- and mass-specific enthalpies and entropies of N_2 and N at 3000 K and 2.5 atm.

Solution

Known N_2, N, T, P

Find $\bar{h}$, h, $\bar{s}$, s

Modeling, Assumptions and Premises Ideal-gas behavior

Analysis Tables D.7 and D.8 can be used to determine the molar-specific standard-state enthalpies and entropies for N_2 and N, respectively. The standard-state (1-atm) entropy can be converted to the value at 2.5 atm using the molar-specific form of Eq. 7.10a.

The sum of the enthalpies of formation and sensible enthalpies is the standardized enthalpies we seek (Eq. 12.12). At 3000 K, these values are

$$\begin{aligned}\bar{h}_{N_2} &= \bar{h}^{\circ}_{f,N_2}(298.15\text{ K}) + \Delta\bar{h}_{s,N_2}(3000\text{ K}) = 0\text{ kJ/kmol} + 92{,}730\text{ kJ/kmol}\\ &= 92{,}730\text{ kJ/kmol}\end{aligned}$$

and

$$\begin{aligned}\bar{h}_{N} &= \bar{h}^{\circ}_{f,N}(298.15\text{ K}) + \Delta\bar{h}_{s,N}(3000\text{ K}) = 472{,}628\text{ kJ/kmol} + 56{,}213\text{ kJ/kmol}\\ &= 528{,}841\text{ kJ/kmol}.\end{aligned}$$

Because the standard-state pressure is 1 atm, application of Eq. 7.10a yields

$$\bar{s}(T,P) - \bar{s}(T, 1\text{ atm}) = -R_u \ln \frac{P\text{ (atm)}}{1\text{ atm}}$$

or

$$\bar{s}(T,P) - \bar{s}^{\circ}(T) = -R_u \ln \frac{P\text{ (atm)}}{1\text{ atm}}.$$

Thus,

$$\bar{s}(3000\text{ K},\ 2.5\text{ atm}) = \bar{s}^{\circ}(3000\text{ K}) - R_u \ln 2.5.$$

Using values from Tables D.7 and D.8, we evaluate this equation as follows:

$$\bar{s}_{N_2} = 266.810 - 8.31447\ \ln 2.5 = 259.192\text{ kJ/kmol·K}$$

and

$$\bar{s}_{N} = 201.199 - 8.31447 \ln 2.5 = 193.581\text{ kJ/kmol·K}.$$

To compute the mass-specific enthalpies and entropies, we divide the molar-specific values by the molecular weight of the respective species; for example,

$$h_{N_2} = \frac{\bar{h}}{\mathcal{M}_{N_2}} = \frac{92{,}730\text{ kJ/kmol}}{28.013\text{ kg/kmol}} = 3310.25\text{ kJ/kg}.$$

The following table summarizes these calculations:

	$\mathcal{M}$ (kg/kmol)	$\bar{h}$ (kJ/kmol)	h (kJ/kg)	$\bar{s}$ (kJ/kmol·K)	s (kJ/kg·K)
N_2	28.013	92,730	3310.25	259.192	9.2526
N	14.007	528,841	37,755	193.581	13.8203

Comments This example illustrates the use of the tables in Appendix D to calculate standardized properties at temperatures and pressure not equal to the reference-state values. Because we assume ideal-gas behavior, the pressure has no effect on the enthalpy, and, furthermore, the entropy is a simple function of pressure (see of Eq. 7.10a). The conversion from the reference-state pressure (1 atm) to the pressure at the desired state (2.5 atm) is thus straightforward.

Self-Test 12.3

Determine the standardized mass-specific enthalpies and entropies of O and O_2 at 2000 K and atmospheric pressure.

(Answer: $h_{O_2} = 1849.09\ kJ/kg,\ s_{O_2} = 8.396\ kJ/kg{\cdot}K,\ h_O = 17.806.8\ kJ/kg,\ s_O = 12.571\ kJ/kg{\cdot}K$*)*

12.2b Constant-Pressure Combustion

Consider a piston–cylinder arrangement (Fig. 12.10) that contains a gaseous closed system consisting of a mixture of fuel and oxidizer at the initial state and of a mixture of combustion products (i.e., CO_2, H_2O, etc.) at the final state. The freely moving piston ensures that the initial and final states are at the same pressure.

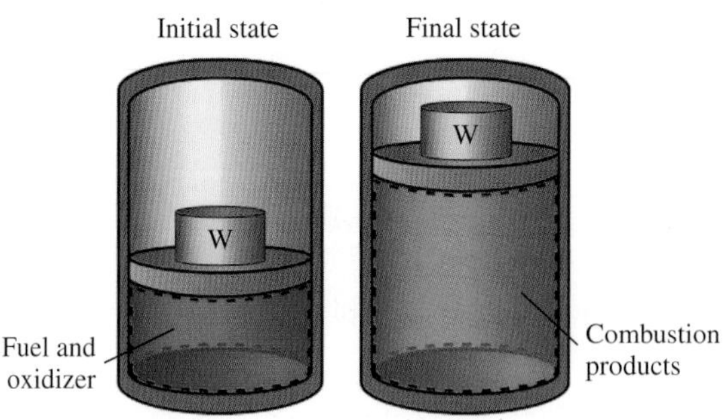

FIGURE 12.10 At the initial state, the piston–cylinder assembly contains an unreacted mixture of fuel and oxidizer. After the combustion process occurs, the cylinder contains the products of combustion.

We simplify Eq. 5.6 by neglecting the system kinetic energy – we assume the system is at rest – and by assuming that the small change in the system potential energy associated with any change in the system center of mass is insignificant. This is easily shown to be true. With these assumptions, the only contribution to the system energy is the internal energy U; thus, Eq. 5.6 becomes

$$_1(Q_{\text{in, net}})_2 - {}_1(W_{\text{out, net}})_2 = U_2 - U_1, \quad \textbf{(12.13a)}$$

where the subscripts 1 and 2 refer to the unburned and burned mixture states, respectively. We now further assume that the piston is frictionless and that the combustion process occurs sufficiently slowly that the pressure in the cylinder is uniform and constant. With these assumptions, the work ${}_1W_2$ is simply evaluated as $P(\mathcal{V}_2 - \mathcal{V}_1)$, and Eq. 12.13a becomes

$${}_1(Q_{\text{in, net}})_2 - P(\mathcal{V}_2 - \mathcal{V}_1) = U_2 - U_1.$$

The $P\mathcal{V}$ terms are moved to the right-hand side of the equation, and we recognize the enthalpy:

$${}_1(Q_{\text{in, net}})_2 = U_2 + P\mathcal{V}_2 - U_1 - P\mathcal{V}_1 = H_2 - H_1, \qquad \textbf{(12.13b)}$$

or

$${}_1(Q_{\text{in, net}})_2 = M(h_2 - h_1), \qquad \textbf{(12.13c)}$$

where h_1 and h_2 are the specific enthalpies of the reactant and product mixtures, respectively. Since the heat transfer for a combustion process is usually from the hot system to the cooler surroundings, we rewrite Eq. 12.13c and explicitly denote the reactant and product states as

Recall that standardized enthalpies account for the energy changes resulting from the rearrangement of chemical bonds. See Eq. 12.12.

$$Q_{\text{out}} = M(h_{\text{reac}} - h_{\text{prod}}). \qquad \textbf{(12.13d)}$$

The key to using Eqs. 12.13b–d is the recognition that the enthalpies therein are standardized enthalpies. With knowledge of the initial and final mixture compositions and temperatures, Eq. 12.13d could easily be evaluated to find the heat transfer Q_{out} using tabulated values of enthalpies (Appendix D).

The maximum temperature for a constant-pressure combustion process occurs when the process is adiabatic (i.e., when there is no heat loss from the system). This temperature is referred to as the **constant-pressure adiabatic flame temperature** and is denoted T_{ad}. For this situation, ${}_1Q_2$ in Eqs. 12.13a and 12.13b is zero. Thus,

$$H_2 - H_1 = 0,$$

or

$$h_2 - h_1 = 0.$$

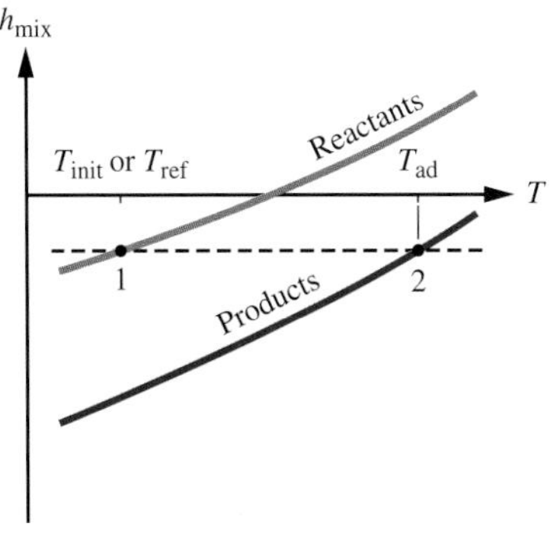

FIGURE 12.11 Standardized enthalpy versus temperature for reactant and product mixtures, illustrating the adiabatic flame temperature T_{ad}.

The subscript 1 refers to reactants at an initial temperature T_{init}, and the subscript 2 refers to products at the adiabatic flame temperature T_{ad}, so we can write

$$H_{\text{prod}}(T_{\text{ad}}) = H_{\text{reac}}(T_{\text{init}}), \qquad \textbf{(12.14a)}$$

or

Using h–T plots is invaluable in analyzing constant-pressure combustion problems.

$$h_{\text{prod}}(T_{\text{ad}}) = h_{\text{reac}}(T_{\text{init}}), \qquad \textbf{(12.14b)}$$

This adiabatic constant-pressure combustion process is shown as a horizontal line on a graph of H (or h) versus T, as illustrated in Fig. 12.11.

Example 12.4 Constant-Pressure Adiabatic Flame Temperature

Methane is the primary constituent of natural gas. Estimate the constant-pressure adiabatic flame temperature for a stoichiometric mixture of methane and air using the following conditions and assumptions or approximations. The pressure is 1 atm and the initial reactant temperature is 298 K. Assume the products consist only of undissociated CO_2, H_2O, and N_2 and that the composition of air is 3.76 kmol of N_2 for each

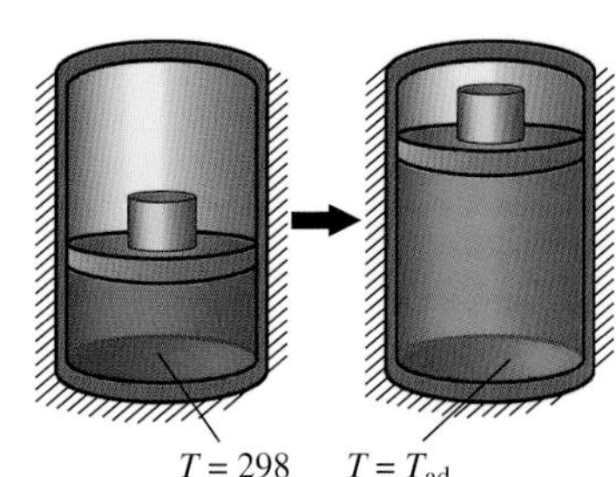

kmol of O_2. Neglect any moisture in the air. Note that the adiabatic flame temperature can be calculated directly, without iteration or the use of tabulated enthalpies, if we use a constant approximate value for the product mixture specific heat to determine the sensible enthalpies. Implement this approach by using constant specific heats for the product species evaluated at 1200 K ($\approx [T_{\text{init}} + T_{\text{ad}}]/2$, where T_{ad} is guessed to be about 2100 K).

Solution

Known CH_4–air mixture is stoichiometric ($\Phi = 1$), T_{init}

Find T_{ad}

Sketch See Fig. 12.11.

Modeling, Premises and Assumptions

i. $N_2/O_2 = 3.76$
ii. Ideal gases
iii. $c_{p,i} = c_{p,i}$ (1200 K)
iv. No dissociation of product species

Analysis For a stoichiometric mixture, all the carbon and hydrogen in the fuel appears in the CO_2 and H_2O in the products, respectively (see Eqs. 12.8 and 12.9):

$$CH_4 + 2(O_2 + 3.76N_2) \rightarrow CO_2 + 2H_2O + 2(3.76)N_2;$$

thus, $N_{CO_2} = 1$, $N_{H_2O} = 2$, and $N_{N_2} = 7.52$. We obtain the needed properties from Appendices D and H as follows:

Species	Enthalpy of formation at 298 K, $\bar{h}^o_{f,i}$ (kJ/kmol)	Specific heat at 1200 K, $\bar{c}_{p,i}$ (kJ/kmol·K)
CH_4	−74,831	—
CO_2	−393,546	56.21
H_2O	−241,845	43.87
N_2	0	33.71
O_2	0	—

We apply the first law (Eq. 12.14a),

$$H_{\text{prod}} = \sum_{\text{prod}} N_i \bar{h}_i = H_{\text{reac}} = \sum_{\text{reac}} N_i \bar{h}_i,$$

where the species enthalpies are evaluated using the assumed (simplified) calorific equation of state:

$$\bar{h}_i(T) = \bar{h}^o_{f,i}(298) + \bar{c}_{p,i}(T - 298).$$

We now evaluate the reactant and product enthalpies, which are:

$$\begin{aligned} H_{\text{reac}} &= (1)(-74{,}831) + 2(0) + 7.52(0)\,\text{kJ} \\ &= -74{,}831\,\text{kJ} \\ H_{\text{prod}} &= \sum N_i\left[\bar{h}^o_{f,i} + \bar{c}_{p,i}(T_{\text{ad}} - 298)\right] \\ &= (1)[-393{,}546 + 56.21(T_{\text{ad}} - 298)] \\ &\quad + (2)[-241{,}845 + 43.87(T_{\text{ad}} - 298)] \\ &\quad + (7.52)[0 + 33.71(T_{\text{ad}} - 298)]. \end{aligned}$$

Equating H_{prod} to H_{reac} and solving for T_{ad} yields

$$T_{ad} = 2318\ K.$$

Comment The value of the adiabatic flame temperature for a stoichiometric mixture given in Table H.l is 2226 K, which is approximately 100 K lower than our estimate. Our neglect of dissociation and the simplified evaluation of product enthalpies are responsible for this difference.

Self-Test 12.4

Using the assumptions of Example 12.4, determine the heat transferred per kilogram of fuel for the stoichiometric combustion of the CH_4–air mixture if the combustion products are at a temperature of 500 K.

(Answer: 45,011 kJ/kg$_{CH_4}$. Note that the use of c_ps at 1200 K is a poor choice.)

12.2c Constant-Volume Combustion

We now consider another idealized case: combustion at a fixed volume. In reciprocating internal combustion engines, the combustion process occurs with the piston moving relatively slowly near the top of its stroke (Fig. 12.12). Although the volume is not fixed in this case, the volume change throughout the combustion process typically is not large. For this reason, constant-volume combustion is frequently used as a primitive model for the combustion event in reciprocating internal combustion engines.

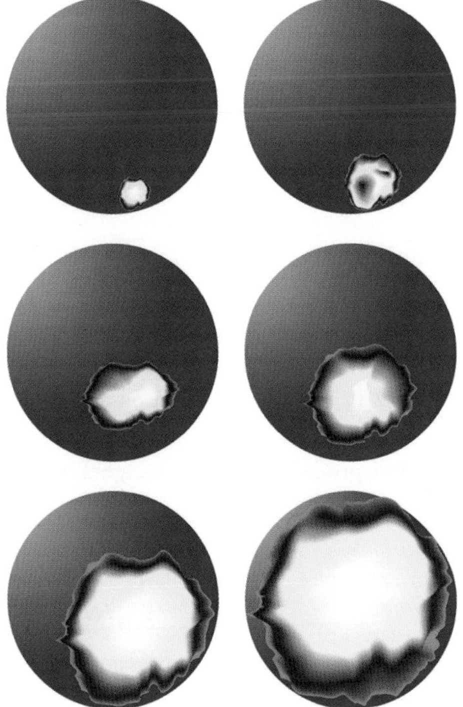

FIGURE 12.12 In a spark-ignition engine, a flame initiated at the spark plug propagates across the combustion chamber. Combustion begins before the piston reaches top center and ends after top center. Here the blue regions represent the unburned fuel–air mixture, and the red regions represent the burned gases.

We start our analysis with the same conservation of energy expression as that presented at the outset of our constant-pressure combustion analysis, Eq. 12.13a:

$$_1(Q_{in,\,net})_2 - {}_1(W_{out,\,net})_2 = U_2 - U_1.$$

With the volume fixed, there is no compression or expansion work; thus, this expression simplifies to

$$_1(Q_{in,\,net})_2 = U_2 - U_1. \tag{12.15a}$$

Recognizing again that the heat transfer is most likely to be from the system to the surroundings and that state 1 is identified with the reactants and state 2 with the products, we rewrite Eq. 12.15a as

$$Q_{out} = U_{reac} - U_{prod}, \tag{12.15b}$$

or

$$Q_{out} = M(u_{Reac} - u_{prod}). \tag{12.15c}$$

To evaluate the constant-volume adiabatic flame temperature, we set $Q_{out} = 0$; thus,

$$U_{reac}(T_{init}, P_{init}) = U_{prod}(T_{ad}, P_f), \tag{12.16a}$$

where U is the standardized internal energy of the mixture and the subscripts init and f refer to the initial and final states, respectively. Equation 12.16a can be represented graphically as a horizontal line in U–T coordinates, similarly to the H–T graph used to illustrate the solution of Eq. 12.14 for the constant-pressure adiabatic flame temperature (cf. Fig. 12.11). Because Appendix D provides values for H (or h), rather than U (or u), we employ the definition of enthalpy ($H \equiv U + P\mathcal{V}$) to express Eq. 12.16a using enthalpies:

$$H_{reac} - H_{prod} - \mathcal{V}(P_{init} - P_f) = 0. \tag{12.16b}$$

Applying the ideal-gas law to eliminate the $P\mathcal{V}$ terms,

$$P_{init}\mathcal{V} = \sum_{reac} N_j R_u T_{init} = N_{reac} R_u T_{init},$$

$$P_f\mathcal{V} = \sum_{prod} N_j R_u T_{ad} = N_{prod} R_u T_{ad},$$

we obtain

$$H_{reac} - H_{prod} - R_u(N_{reac}T_{init} - N_{prod}T_{ad}) = 0. \tag{12.17}$$

Equation 12.17 can be expressed on a per-mass-of-mixture basis by dividing by the mass of the mixture, M_{mix}, and further simplifying, using the definitions of the reactant and product molecular weights, i.e.,

$$M_{mix}/N_{reac} \equiv \mathcal{M}_{reac}$$

and

$$M_{mix}/N_{prod} \equiv \mathcal{M}_{prod}.$$

Our energy conservation expression on a per-mass-of-mixture basis is then expressed as

$$h_{reac} - h_{prod} - R_u\left(\frac{T_{init}}{\mathcal{M}_{reac}} - \frac{T_{ad}}{\mathcal{M}_{prod}}\right) = 0. \tag{12.18}$$

To evaluate the adiabatic flame temperature from either Eq. 12.17 or Eq. 12.18 requires knowledge of the mixture composition. Ignoring any species dissociation, we can use simple atom balances (see Eqs. 12.3 and 12.4) to define the composition. At high temperatures, however, dissociation can be quite important, resulting in adiabatic flame temperatures as much as several hundred kelvins lower than when dissociation is ignored. To solve this problem, our energy conservation expressions must be coupled to relationships defining the equilibrium composition of the products. These same caveats also apply to the problem of evaluating constant-pressure adiabatic flame temperatures. For more information, the reader is referred to Ref. [2].

Example 12.5 Constant-Volume Adiabatic Flame Temperature

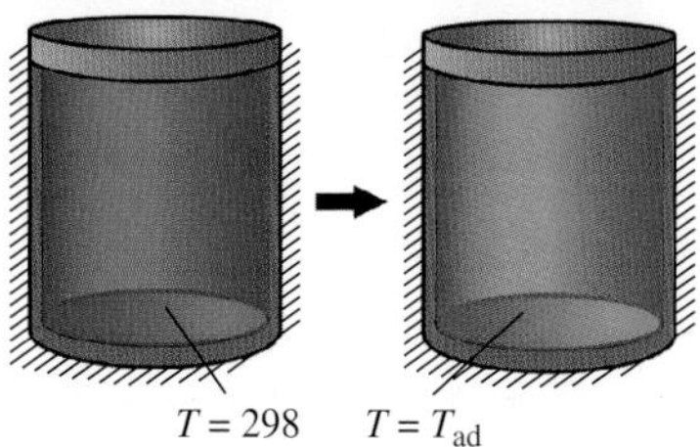

Compare the *constant-volume* adiabatic flame temperature to the *constant-pressure* adiabatic flame temperature for a stoichiometric CH_4–air mixture using the same conditions and assumptions as in Example 12.4. Also determine the final pressure.

Solution

Known CH_4–air mixture is stoichiometric ($\Phi = 1$), T_{init}, P_{init}

Find T_{ad}, P_f

Sketch

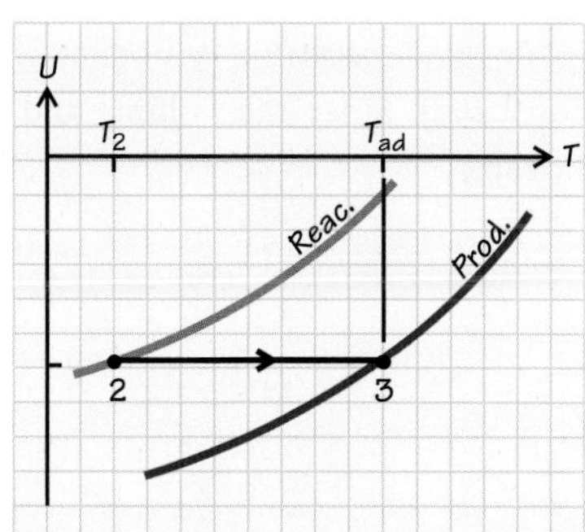

Modeling, Premises and Assumptions See Example 12.4.

Analysis The reactant and product mixtures have the same compositions as in Example 12.4. Furthermore, the reactant mixture properties are the same as well.

For constant-volume combustion, conservation of energy is expressed by Eq. 12.17:

$$H_{reac} - H_{prod} - R_u\left(N_{reac}T_{init} - N_{prod}T_{ad}\right) = 0.$$

Substituting the definitions of ideal-gas mixture enthalpies (Eq. 10.15f) for H_{reac} and H_{prod}, we obtain

$$\sum_{reac} N_j\bar{h}_j - \sum_{prod} N_j\bar{h}_j - R_u\left(N_{reac}T_{init} - N_{prod}T_{ad}\right) = 0.$$

The first term in this expression has the same numerical value as in Example 12.4. Similarly, the expression for the second term is also identical to that in Example 12.4; thus

$$\mathrm{H}_{\mathrm{react}} = [(1)(-74{,}831) + 2(0) + 7.52(0)]\,\mathrm{kJ}$$
$$= -74{,}831\,\mathrm{kJ},$$

$$\begin{aligned} H_{\mathrm{prod}} &= \{(1)[-393{,}546 + 56.21(T_{\mathrm{ad}} - 298)] \\ &\quad + (2)[-241{,}845 + 43.87\,(T_{\mathrm{ad}} - 298)] \\ &\quad + (7.52)[0 + 33.71(T_{\mathrm{ad}} - 298)]\}\ \mathrm{kJ} \\ &= -877{,}236 + 397.45(T_{\mathrm{ad}} - 298)\,\mathrm{kJ}, \end{aligned}$$

and

$$R_{\mathrm{u}}\left(N_{\mathrm{reac}} T_{\mathrm{init}} - N_{\mathrm{prod}} T_{\mathrm{ad}}\right) = 8.315(10.52)(298 - T_{\mathrm{ad}}),$$

where $N_{\mathrm{reac}} = N_{\mathrm{prod}} = 10.52\,\mathrm{kmol}$. Substituting these expressions into our conservation of energy equation and solving for the constant-volume adiabatic flame temperature $T_{\mathrm{ad,\ vol.\ constant}}$ yields

For other fuels, N_{reac} does not equal N_{prod}.

$$T_{\mathrm{ad,\,vol.\,constant}} = 2889\ \mathrm{K}.$$

We see that this value is about 570 K higher than that for the constant-pressure case:

$$T_{\mathrm{ad,\,press.\,constant}} = 2318\ \mathrm{K}.$$

Without reading the next sentence, can you reason why the constant-volume case should result in a higher adiabatic flame temperature than that for constant-pressure combustion? The answer is that no work is done on the surroundings during the constant-volume process; hence, the sensible internal energy of the products (and, therefore, the temperature) will be greater than for the constant-pressure case.

The final pressure is obtained by application of the ideal-gas equation of state (Eq. 2.26d). Since the volume and the number of moles are constant (see the sidebar above),

$$\frac{P_{\mathrm{init}}}{T_{\mathrm{init}}} = \frac{P_{\mathrm{f}}}{T_{\mathrm{ad}}},$$

or

$$P_{\mathrm{f}} = P_{\mathrm{init}} \frac{T_{\mathrm{ad}}}{T_{\mathrm{init}}}.$$

Thus,

$$P_{\mathrm{f}} = (1\ \mathrm{atm}) \frac{2889\,\mathrm{K}}{298\,\mathrm{K}} = 9.69\ \mathrm{atm}.$$

Comments We note that, for the same initial conditions, the constant-volume adiabatic flame temperature is much higher than that for constant-pressure combustion. We also note the large increase in pressure when the volume is fixed.

Self-Test 12.5

Redo Example 12.5 for a stoichiometric butane (C_4H_{10}) and air mixture.

(Answer: 2987 K, 10 atm)

Example 12.6 Compression and Combustion Processes in an SI Engine

A crude model of the compression and combustion processes of a spark-ignition (SI) engine consists of (i) a polytropic compression from bottom center (state 1) to top center (state 2) and (ii) adiabatic constant-volume combustion (state 2 to state 3), respectively. These two idealized processes are illustrated in the sketch. Determine the temperature and pressure at states 2 and 3. The engine compression ratio ($CR \equiv \mathcal{V}_1/\mathcal{V}_2$) is 8, the polytropic exponent is 1.3, and the initial temperature and pressure (state 1) are 298 K and 0.5 atm, respectively. The fuel is isooctane (C_8H_{18}) and the air–fuel ratio is the stoichiometric value.

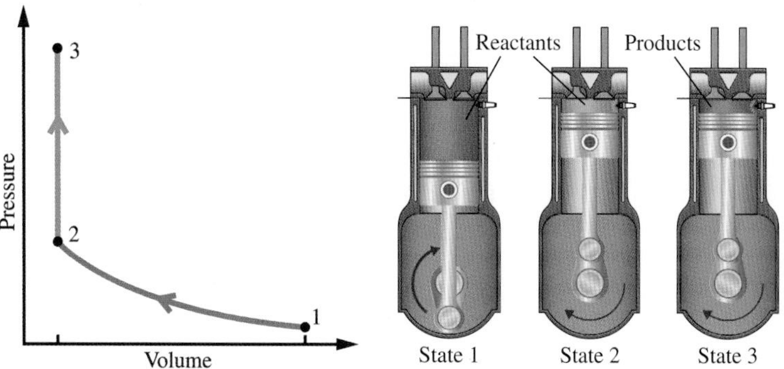

Solution

Known Stoichiometric C_8H_{18}–air mixture, T_1, P_1, $\mathcal{V}_1$, $\mathcal{V}_2$, n

Find P_2, T_2, P_3, T_3

Sketch

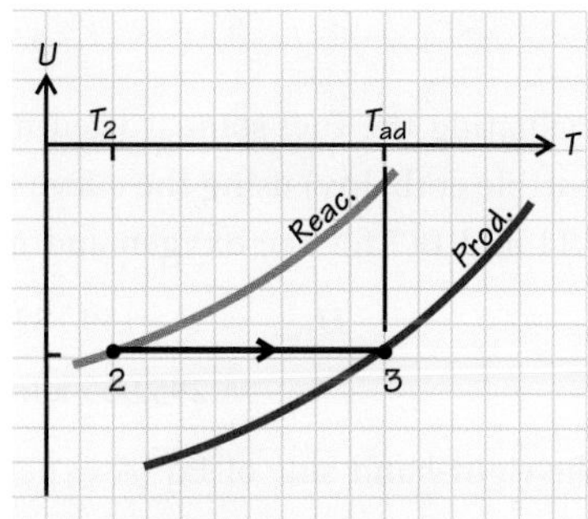

Modeling, Premises and Assumptions

i. Ideal-gas behavior
ii. Polytropic process (1–2)
iii. Adiabatic constant-volume process (2–3)
iv. No dissociation of combustion products
v. Simple air

Analysis To determine the properties at state 2, we apply the polytropic process relationship, $P\mathcal{V}^n$ = constant (Eq. 4.9b), and the ideal-gas equation of state (Eq. 2.26c) as follows:

$$P_2 = P_1\left(\frac{\mathcal{V}_1}{\mathcal{V}_2}\right)^n = 0.5\text{ atm }(8)^{1.3}$$

$$= 7.46\text{ atm or } 756\text{ kPa}$$

and

$$T_2 = T_1\left(\frac{P_2}{P_1}\right)\left(\frac{\mathcal{V}_2}{\mathcal{V}_1}\right) = 298\,\text{K}\left(\frac{7.46\text{ atm}}{0.5\text{ atm}}\right)\left(\frac{1}{8}\right)$$
$$= 556\text{ K}.$$

For process 2–3, we apply the stoichiometric combustion equations (Eqs. 12.8 and 12.9) to determine the composition of the reactant and product mixtures:

$$C_xH_y + a(O_2 + 3.76\,N_2) \rightarrow x CO_2 + \left(\frac{y}{2}\right)H_2O + 3.76a\,N_2,$$

where

$$a = x + \frac{y}{4} = 8 + \frac{18}{4} = 12.5,$$

or

$$C_8H_{18} + 12.5(O_2 + 3.76N_2) \rightarrow 8\,CO_2 + 9\,H_2O + 47\,N_2.$$

To obtain the temperature at state 3, we evaluate the conservation of energy statement (Eq. 12.16a) that

$$U_{\text{reac}}(T_2) = U_{\text{prod}}(T_3),$$

or, equivalently from a rearrangement of Eq. 12.18,

$$H_{\text{reac}}(T_2) = H_{\text{prod}}(T_3) + R_u\left(N_{\text{reac}}T_2 - N_{\text{prod}}T_3\right). \quad \textbf{(A)}$$

Expanding the left-hand side of Eq. A yields

$$H_{\text{reac}}(T_2) = \sum_{\text{reac}} N_j\bar{h}_j(556\text{ K})$$
$$= N_{C_8H_{18}}\bar{h}_{C_8H_{18}}(556\text{ K}) + N_{O_2}\bar{h}_{O_2}(556\text{ K})$$
$$+ N_{N_2}\bar{h}_{N_2}(556\,\text{K}).$$

Evaluating the fuel standardized molar enthalpy (the enthalpy of formation plus the sensible enthalpy) using the curve-fit coefficients given in Table H.2, and using Tables D.11 and D.7 for the oxygen and nitrogen enthalpies, we obtain

$$H_{\text{reac}} = 1(-159{,}182) + 12.5(0 + 7865) + 47(0 + 7592)\,\text{kJ}$$
$$= 295{,}955\text{ kJ}.$$

The right-hand side of Eq. A expands as follows:

$$H_{\text{prod}}(T_3) + R_u\left(N_{\text{reac}}T_2 - N_{\text{prod}}T_3\right)$$
$$= N_{CO_2}\bar{h}_{CO_2}(T_3) + N_{H_2O}\bar{h}_{H_2O}(T_3) + N_{N_2}\bar{h}_{N_2}(T_3)$$
$$+ R_u\left(N_{\text{reac}}T_2 - N_{\text{prod}}T_3\right).$$

Substituting numerical values for the various numbers of moles, we get

$$H_{\text{prod}}(T_3) + R_u\left(N_{\text{reac}}T_2 - N_{\text{prod}}T_3\right)$$
$$= 8\,\bar{h}_{CO_2}(T_3) + 9\,\bar{h}_{H_2O}(T_3) + 47\,\bar{h}_{N_2}(T_3)$$
$$+ R_u(60.5T_2 - 64T_3),$$

where we have made use of the fact that

$$N_{\text{reac}} = [1 + 12.5(1 + 3.76)] = 60.5\text{ kmol}$$

and

$$N_{\text{prod}} = (8 + 9 + 47) = 64\text{ kmol}.$$

We now solve Eq. A iteratively. Guessing $T_3 = 3000\,\mathrm{K}$ and evaluating the product species enthalpies from Tables D.2 (CO_2), D.6 (H_2O), and D.7 (N_2), we obtain

$$\begin{aligned} &H_{\text{prod}}(T_3) + R_{\text{u}}\left(N_{\text{reac}} T_2 - N_{\text{prod}} T_3\right) \\ &\quad = 8(-393{,}546 + 152{,}3891) + 9(-241{,}845 + 126{,}563) \\ &\qquad + 47(0 + 97{,}730) + 83145[60.5(298) - 64(3000)]\ \mathrm{kJ} \\ &\quad = -50{,}950\,\mathrm{kJ}. \end{aligned}$$

Comparing the value for the left-hand side of Eq. A (+295,955 kJ) with this estimated value for the right-hand side (–50,950 kJ), we conclude that our guess for T_3 is too low. Repeating our calculation for $T_3 = 3500\,\mathrm{K}$ yields

$$\begin{aligned} &8(-393{,}546 + 184{,}120) + 9(-241{,}845 + 154{,}795) \\ &+47(0 + 111{,}315) + 8.3145\,[60.5(298) - 64(3500)]\ \mathrm{kJ} \\ &\quad = 1{,}060{,}401\,\mathrm{kJ}. \end{aligned}$$

We thus conclude that the value of T_3 must lie between our two guesses, as shown in the following table:

T_3(K)	$H_{\text{prod}} + R_{\text{u}}(N_{\text{reac}}T_2 - N_{\text{prod}}T_3)$(kJ)
3000	–50,950
?	295,955
3500	1,060,401

Applying linear interpolation, we estimate that

$$T_3 = 3156\,\mathrm{K}.$$

Applying the ideal-gas equation of state (Eq. 2.26d) for a fixed volume, we can obtain the pressure at state 3 as follows:

$$\mathcal{V} = \frac{N_{\text{prod}} R_{\text{u}} T_3}{P_3} = \frac{N_{\text{reac}} R_{\text{u}} T_2}{P_2}.$$

Solving for P_3 yields

$$\begin{aligned} P_3 &= P_2 \frac{N_{\text{prod}} T_3}{N_{\text{reac}} T_2} \\ &= 7.46\ \mathrm{atm} \frac{64\ \mathrm{kmol}(3156\,\mathrm{K})}{60.5\ \mathrm{kmol}(556\,\mathrm{K})} \\ &= 44.8\ \mathrm{atm} \quad \text{or} \quad 4540\ \mathrm{kPa}. \end{aligned}$$

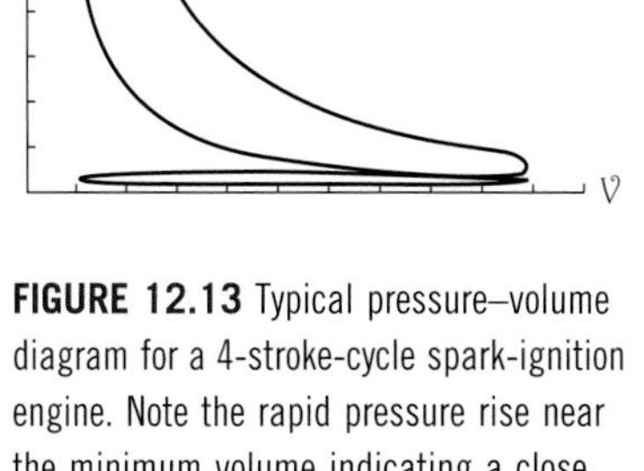

FIGURE 12.13 Typical pressure–volume diagram for a 4-stroke-cycle spark-ignition engine. Note the rapid pressure rise near the minimum volume indicating a close approximation to the constant-volume process shown in the sketch at the start of this example.

Comments Note that the final temperature and pressure are quite high. Although these values are only estimates, they are typical of real engines. See Fig. 12.13. We also note that our estimation of T_3 neglects dissociation of the combustion products, which results in an overestimation of both T_3 and P_3. Taking dissociation into account yields a final temperature of 2804 K and a pressure of 40.5 atm.

12.2d Enthalpy of Combustion

Knowing how to express the enthalpy of mixtures of reactants and mixtures of products allows us to define the enthalpy of reaction. When dealing specifically with combustion reactions, the enthalpy of reaction is usually called the enthalpy of combustion. The definition of the **enthalpy of reaction**, or the **enthalpy of combustion**, ΔH_R, is

$$\Delta H_R(T) = H_{prod}(T) - H_{reac}(T), \tag{12.19a}$$

where T may be any temperature, although a reference-state value of 298.15 K is frequently used. The enthalpy of combustion is illustrated graphically in Fig. 12.14. Note that the standardized enthalpy of the products lies below that of the reactants. For example, at 25 °C and 1 atm, the reactants' enthalpy of a stoichiometric mixture of CH_4 and air is −74,831 kJ per kmol of fuel. At the same conditions (25 °C, 1 atm), the combustion products have a standardized enthalpy of −877,236 kJ for the combustion of 1 kmol of fuel. Thus,

$$\Delta H_R = -877{,}236 - (-74{,}831) = -802{,}405\,\text{kJ}\,(\text{per kmol}\,CH_4)$$

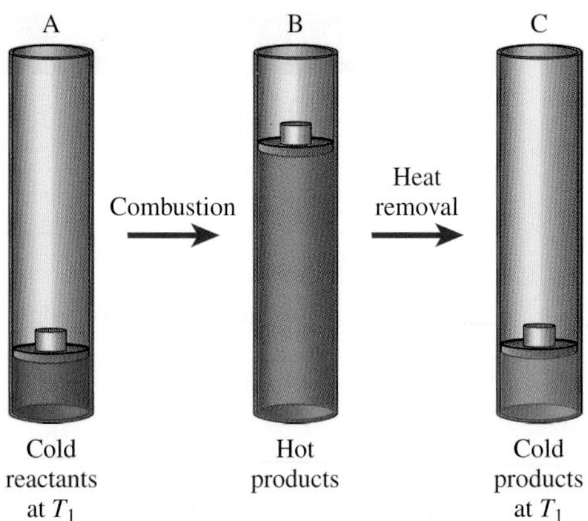

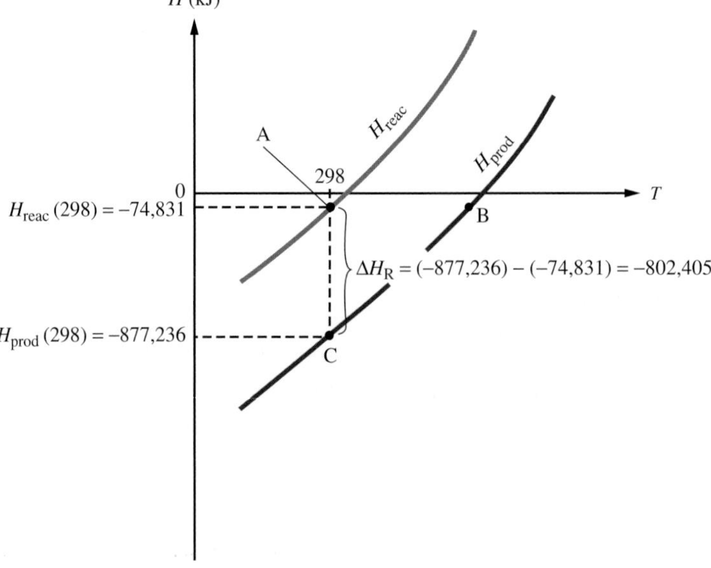

FIGURE 12.14 The enthalpy of reaction is illustrated using values for the reaction of one mole of methane with a stoichiometric quantity of air. The water in the products is assumed to be in the vapor state.

This value is usually expressed on a per-mass-of-fuel basis:

$$\Delta h_R\left(\text{kJ/kg}_{\text{fuel}}\right) = \Delta H_R/\mathcal{M}_{\text{fuel}}, \tag{12.19b}$$

or

$$\Delta h_R = (-802{,}405/16.043) = -50{,}016\,\text{kJ/kg}_{\text{fuel}}.$$

Note that the value of the enthalpy of combustion depends on the temperature chosen for its evaluation. Because the enthalpies of both the reactants and products vary with temperature, the vertical distance between the H_{prod} and H_{reac} lines in Fig. 12.14 is not constant.

12.2e Heating Values

Fuel heating values were used to define the efficiency of gas-turbine engines in Chapter 9. See Eq. 9.20.

The **heat of combustion**, Δh_c (known also as the **heating value**), is numerically equal to the enthalpy of reaction, but has opposite sign. The **upper** or **higher heating value**, *HHV*, is calculated assuming that all the water in the products has condensed to liquid. In contrast, the **lower heating value**, *LHV*, is calculated assuming none of the water condenses. The combustion reaction liberates the greatest amount of energy when the water condenses; thus, the adjectives "higher" and "lower" describe the two cases (condensation, no condensation), respectively. For CH_4, the higher heating value is approximately 11% larger than the lower one. Heating values are provided in Appendix H for a variety of hydrocarbon fuels, for stoichiometric combustion at 298.15 K and 1 atm.

Example 12.7 Heating Values of *n*-Decane

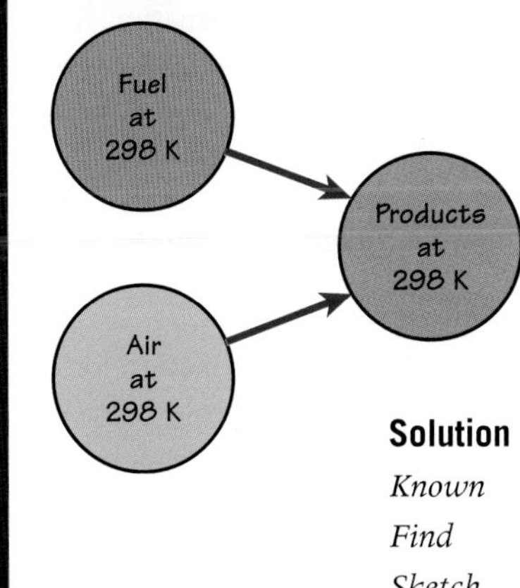

Determine the higher and lower heating values of gaseous *n*-decane ($C_{10}H_{22}$) for stoichiometric combustion with air at 298.15 K. For this condition, 15.5 kmol of O_2 reacts with each kmol of $C_{10}H_{22}$ to produce 10 kmol of CO_2 and 11 kmol of H_2O. Assume that air can be represented as a mixture of O_2 and N_2 in which there are 3.76 kmol of N_2 for each kmol of O_2. Express the results per unit mass of fuel. The molecular weight of *n*-decane is 142.284.

Solution

Known T, compositions of reactant and product mixtures, $\mathcal{M}_{C_{10}H_{22}}$

Find Δh_c (higher and lower values)

Sketch

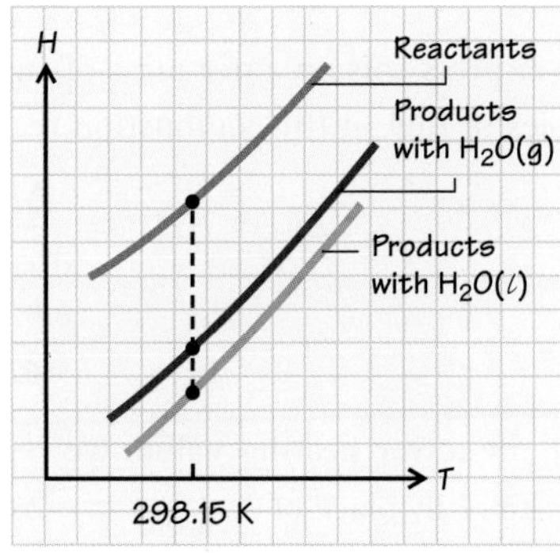

Modeling, Premises and Assumptions

i. Ideal-gas behavior
ii. Simple air
iii. Ideal combustion products (CO_2, H_2O, and N_2 only)

Analysis For 1 kmol of $C_{10}H_{22}$, stoichiometric combustion can be expressed from the given information as (see also Eqs. 12.4 and 12.5):

$$\begin{aligned} &C_{10}H_{22}(g) + 15.5(O_2 + 3.76\,N_2) \\ &\quad \rightarrow 10\,CO_2 + 11\,H_2O(\ell\,\text{or}\,g) + 15.5\,(3.76)\,N_2. \end{aligned}$$

For either the higher or lower heating value,

$$\Delta H_c = -\Delta H_R = H_{reac} - H_{prod},$$

where the numerical value of H_{prod} depends on whether the H_2O in the products is liquid (defining the higher heating value) or gaseous (defining the lower heating value). The sensible enthalpy changes for all species involved are zero because both the reactants and products are at the same temperature of 298.15 K, as indicated in the sketch. Our calculation is further simplified because the enthalpies of formation of the O_2 and N_2 are also zero at 298.15 K. We calculate the reactant and product mixture enthalpies as

$$H_{reac} = \sum_{reac} N_i \bar{h}_i \quad \text{and} \quad H_{prod} = \sum_{prod} N_i \bar{h}_i.$$

Expanding and substituting yields

$$\Delta H_{c,H_2O(\ell)} = \overline{HHV}(\text{molar basis}) = (1)\,\bar{h}^\circ_{f,C_{10}H_{22}} - \left(10\,\bar{h}^\circ_{f,CO_2} + 11\,\bar{h}^\circ_{f,H_2O(\ell)}\right).$$

Note that the N_2 contribution as a reactant cancels with the N_2 contribution as a product. Table D.6 (Appendix D) gives the enthalpy of formation for gaseous water; the enthalpy of vaporization, h_{fg}, is obtained from Table B.1 or the NIST resources. We thus calculate the enthalpy of formation of the liquid water as follows:

$$\begin{aligned} \bar{h}^\circ_{f,H_2O(\ell)} &= \bar{h}^\circ_{f,H_2O(g)} - \bar{h}_{fg} \\ &= -241{,}847\ \text{kJ/mol} - (45{,}876 - 1889)\ \text{kJ/mol} \\ &= -285{,}834\ \text{kJ/mol}. \end{aligned}$$

To complete our calculation, we obtain the enthalpies of formation for CO_2 and the fuel from Appendices D.2 and H, respectively. The higher heating value is thus

$$\begin{aligned} \Delta H_{c,H_2O(\ell)} &= \overline{HHV}\,(\text{molar basis}) = (1\ \text{kmol})\,(-249{,}659\,\text{kJ/kmol}) \\ &\quad - [10\,\text{kmol}(-393{,}546\ \text{kJ/kmol}) + 11\,\text{kmol}\,(-285{,}834\ \text{kJ/kmol})] \\ &= 6{,}829{,}975\ \text{kJ/kmol}_{fuel}. \end{aligned}$$

To express this on a per-mass-of-fuel basis, we need only to divide by the number of moles of fuel in the combustion reaction and the fuel molecular weight, that is,

$$\begin{aligned} \Delta h_c = HHV &= \frac{\Delta H_c}{M_{C_{10}H_{22}}} = \frac{\Delta H_c}{N_{C_{10}H_{22}}\mathcal{M}_{C_{10}H_{22}}} \\ &= \frac{6{,}829{,}975\ \text{kJ/kmol}}{142.284\ \text{kg/kmol}} = 48{,}002\ \text{kJ/kg}_{fuel}. \end{aligned}$$

For the lower heating value, we repeat these calculations using $\bar{h}^\circ_{f,H_2O(g)} = -241{,}847$ kJ/kmol in place of $\bar{h}^\circ_{f,H_2O(\ell)} = -285{,}834\,\text{kJ/kmol}$. The result is

$$\Delta H_c = \overline{LHV} = 6{,}345{,}986 \, \text{kJ/kmol}_{\text{fuel}}$$

and

$$\Delta h_c = LHV = 44{,}601 \, \text{kJ/kg}_{\text{fuel}}.$$

Comment We note that the difference between the higher and lower heating values is approximately 7%. What practical implications does this have, say, for a home heating furnace? Note also that heating values are typically expressed on a per-mass-of-fuel basis (see Appendix H).

Self-Test 12.6

Determine the enthalpy of reaction per unit mass of fuel for the conditions given in Example 12.6 when the products are at a temperature of 1800 K.

(Answer: $\Delta h_R = -14{,}114$ kJ/kg fuel)

12.3 Steady-Flow Applications

In this section, we consider various steady-flow combustion devices.[1] Furnaces, boilers, and like devices can be simple stand-alone units used to heat a space or to supply hot water (see Fig. 12.15), or they can be part of complex systems such as those used for electric power generation (see Fig. 12.16 and also Fig. 1.3). For residential and commercial applications, fuels are typically natural gas or fuel oil, whereas for large-scale power generation, coal and natural gas are the most commonly used fuels. For

FIGURE 12.15 Gas-fired furnaces (left) and hot-water heaters (right) are common combustion appliances found in homes (Jupiterimages / Stockbyte / Getty Images).

[1] Our requirement for steady flow thus rules out spark-ignition and diesel engines.

FIGURE 12.16 This heat-recovery steam generator is a huge heat-exchange device in which the hot products of combustion from a gas-turbine engine react with supplemental fuel to heat water and produce steam. Expansion of this steam through a steam turbine provides additional useful power (Thossaphol Somsri / Alamy Stock Photo).

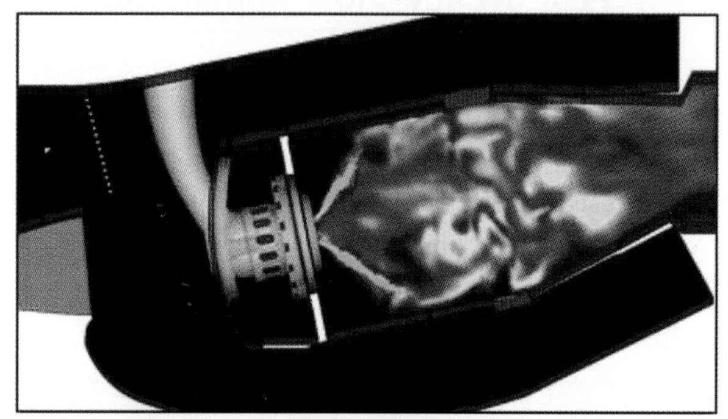

FIGURE 12.17 Modern numerical techniques are applied to simulate the complex phenomena occurring in a realistic jet engine combustor. Image courtesy of Parviz Moin and Center for Integrated Turbulence Simulations, Stanford University.

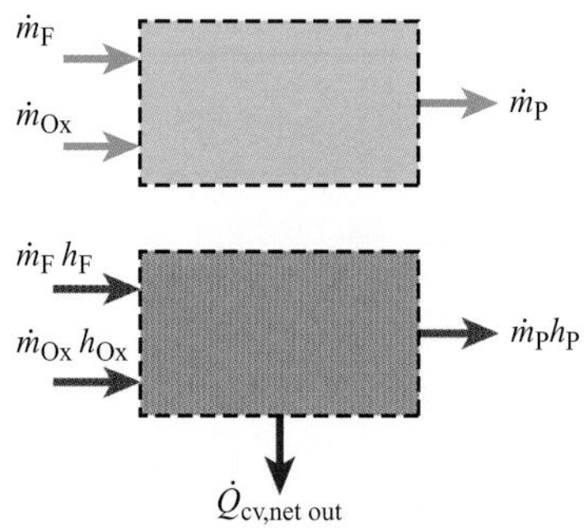

FIGURE 12.18 Open system for analysis of steady-flow, constant-pressure combustion devices. Mass flows are shown at the top, and energy flows shown at the bottom. The subscripts F, Ox, and P refer to fuel, oxidizer, and products, respectively.

these devices, the useful transfer of energy occurs by heat transfer from the hot combustion products. Combustors, however, provide a stream of hot combustion products that is used directly to provide power or thrust, rather than relying on heat transfer to extract energy from the combustion products. For example, hot products expand through turbines to produce power in stationary electric power generation systems; in rocket engines, the hot products expand through a converging–diverging nozzle to produce thrust. Turbojet engines utilize both turbines and nozzles to expand a combustion product stream to useful effect. A jet engine combustor is illustrated in Fig. 12.17, and subsequently in Figs. 12.19 and 12.20 also.

The objective of this section is to present a brief and general analysis of these devices; additional discussions, analyses, and examples were presented in Chapter 9 where these devices were integrated into larger systems. We begin by defining an appropriate open system (control volume) and listing useful assumptions. Figure 12.18 shows a simple open system with entering flows of fuel and oxidizer.[2] A single stream of combustion products exits the open system.

[2] The oxidizer is usually air; however, other oxidizers, such as pure O_2, are employed in some applications. For example the Space Shuttle main engines burned H_2 with pure O_2.

12.3a Assumptions

We employ the following assumptions to simplify our analysis:

- Steady-state, steady flow
- Constant pressure
- Uniform inlet and exit conditions
- Negligible kinetic and potential energy changes
- No work interactions other than flow work

12.3b Mass Conservation

For steady flow, the general integral expression of mass conservation (Eq. 3.14b) simplifies to

$$\dot{m}_{\mathrm{F}} + \dot{m}_{\mathrm{Ox}} = \dot{m}_{\mathrm{P}}, \tag{12.20}$$

which states that the mass flow of products out of the open system (control volume) simply equals the sum of the two incoming mass flows. Note that any moisture, or other diluent, is implicitly included in the oxidizer flow rate.

12.3c Energy Conservation

The steady-flow energy equation that applies to our multi-stream open system is Eq. 5.16:

$$\begin{aligned}
&\dot{Q}_{\mathrm{cv,in}} - \dot{Q}_{\mathrm{cv,out}} + \dot{W}_{\mathrm{cv,in}} - \dot{W}_{\mathrm{cv,out}} \\
&= \sum_{k=1}^{M\text{ outlets}} \dot{m}_{\mathrm{out},k}\left[h_k + \tfrac{1}{2}V^2_{\mathrm{avg},k} + g(z_k - z_{\mathrm{ref}})\right] \\
&\quad - \sum_{j=1}^{N\text{ inlets}} \dot{m}_{\mathrm{in},j}\left[h_j + \tfrac{1}{2}V^2_{\mathrm{avg},j} + g(z_j - z_{\mathrm{ref}})\right].
\end{aligned}$$

Applying this to our specific system with the aforementioned assumptions yields

$$0 - \dot{Q}_{\mathrm{cv,out}} + 0 + 0 = \dot{m}_{\mathrm{P}}h_{\mathrm{P}} - [\dot{m}_{\mathrm{F}}h_{\mathrm{F}} + \dot{m}_{\mathrm{Ox}}h_{\mathrm{Ox}}],$$

which can be rearranged as

$$\dot{m}_{\mathrm{F}}h_{\mathrm{F}} + \dot{m}_{\mathrm{Ox}}h_{\mathrm{Ox}} - \dot{m}_{\mathrm{P}}h_{\mathrm{P}} = \dot{Q}_{\mathrm{cv,out}}. \tag{12.21}$$

It is extremely important to note that the enthalpies appearing in Eq. 12.21 are *standardized enthalpies*, that is, enthalpies that account for the making and breaking of chemical bonds in the conversion of reactants to products. The heat-transfer term in Eq. 12.21 is the energy transferred to the surroundings. For furnaces, boilers, and similar devices, this is the energy used to heat air or water or to generate steam. This term also includes any residual heat losses, which frequently can be neglected for well-insulated systems. For engine combustors, this term represents only residual losses, which again are usually negligible.

The following sections illustrate the application of the principles developed above to a turbojet engine combustor and to a steam generator. The developments below can be easily modified to apply to any constant-pressure, steady-flow combustion system.

12.3d Engine Combustor Analysis

In our simplified analysis of the turbojet engine cycle in Chapter 9, we replaced the actual combustion process with a constant-pressure heat-addition process. We now go beyond this simplification and explore how the combustion process can be treated more realistically, yet still simply.

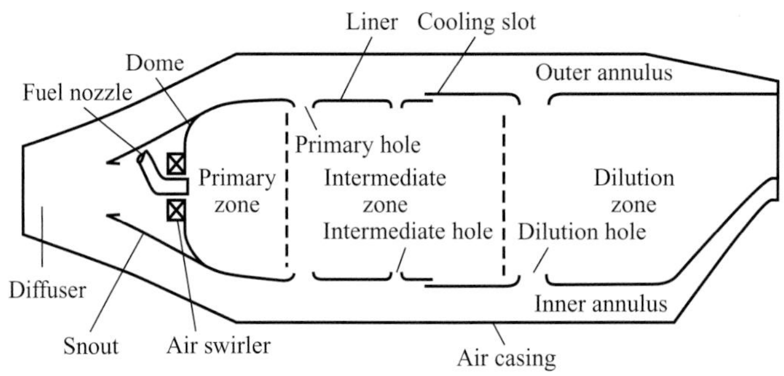

FIGURE 12.19 Schematic of annular jet-engine combustor showing various air passageways and combustion zones. Reprinted from Ref. [6] by permission of A. H. Lefebvre and Taylor & Francis.

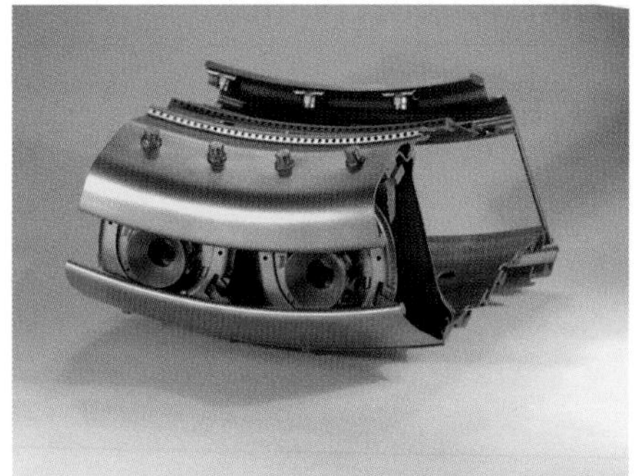

FIGURE 12.20 Segment of CFM56–7 turbofan jet-engine combustor produced by CFM International (a joint company of Snecma, France, and General Electric, USA). Fuel nozzles (not shown) fit into the large holes at the back (left). Note the cooling holes in the louvers of the combustor liner. The CFM56–7 engine is used in the Boeing 737 aircraft.

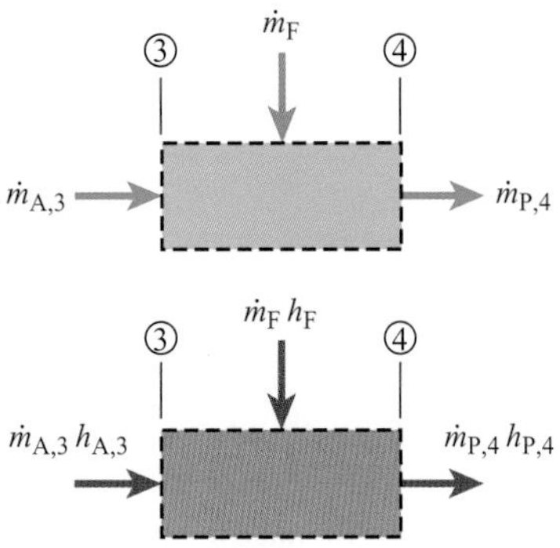

FIGURE 12.21 Open systems (control volumes) used for the jet-engine combustor. Mass flows for mass conservation are shown at the top, and energy flows for energy conservation at the bottom.

A modern jet-engine combustor is a complex device. Multiple fuel injectors and intricate airflow passages provide the proper air–fuel mixture ratios at various locations. Air pathways also provide the necessary cooling of the combustor itself and the dilution of the hot products to temperatures safe for the turbine. Figure 12.19 schematically illustrates the various components and flow passages, and Fig. 12.20 shows a photograph of a modern turbofan combustor. Our treatment here ignores this complexity, treating the combustor as a constant-pressure steady-flow reactor with two inlet streams, one for the air and one for the fuel, and a single outlet stream of dilute combustion products. Figure 12.21 shows the open system (control volume) for this situation.

ASSUMPTIONS

We add an additional assumption to the five assumptions above (Section 12.3a): The combustion process occurs without any heat transfer, i.e., the process is adiabatic ($\dot{Q}_{cv} = 0$).

MASS CONSERVATION

Conservation of mass for the combustor follows that of the overall engine in that the mass flow rate of the product stream is the sum of the total air and fuel flow rates, that is,

$$\dot{m}_{A,3} + \dot{m}_F = \dot{m}_{P,4}, \tag{12.22}$$

where the subscripts A, F, and P refer to air, fuel, and products, respectively, and the numerical subscripts correspond to both the station designations shown in Fig. 12.22 and the state points of the cycle shown in Fig. 12.23. The three flow rates are also interrelated through the specification of the operating air–fuel ratio:

$$\dot{m}_{A,3} = (A/F)\dot{m}_F \tag{12.23a}$$

and

$$\dot{m}_{P,4} = [1 + (A/F)]\dot{m}_F. \tag{12.23b}$$

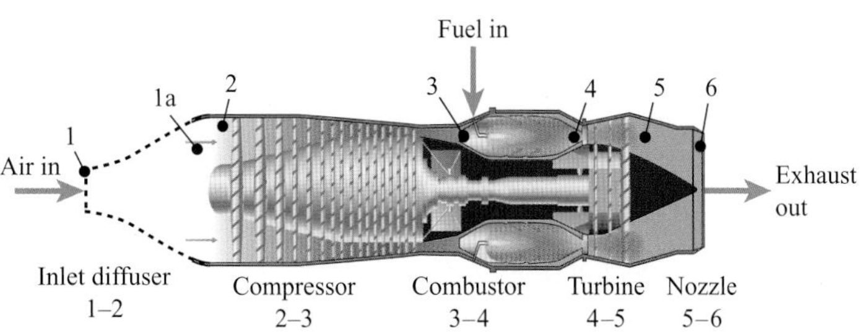

FIGURE 12.22 Schematic diagram of a turbojet engine. The gases flow through the annular space between the turbomachinery components and the engine housing.

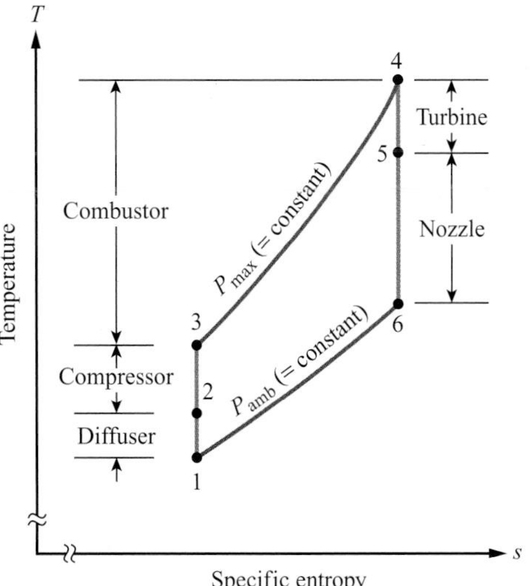

FIGURE 12.23 T–s diagram for ideal turbojet cycle. See Fig. 12.22 for the definition of the diffuser (1–2).

ENERGY CONSERVATION

Applying the assumptions above to the general steady-flow energy conservation equation (Eq. 5.16),

$$\dot{Q}_{\text{cv, net, in}} - \dot{W}_{\text{cv, net out}} = \sum_{k=1}^{M \text{ outlets}} \dot{m}_{\text{out},k}\left[h_k + \tfrac{1}{2}V^2_{\text{avg},k} + g(z_k - z_{\text{ref}})\right] - \sum_{j=1}^{N \text{ inlets}} \dot{m}_{\text{in},j}\left[h_j + \tfrac{1}{2}V^2_{\text{avg},j} + g(z_j - z_{\text{ref}})\right],$$

yields

$$0 - 0 = \dot{m}_{\text{P},4}\, h_{\text{P},4} - \dot{m}_{\text{F}}\, h_{\text{F},3} - \dot{m}_{\text{A},3}\, h_{\text{A},3}. \quad \textbf{(12.24a)}$$

The conservation of energy expressed by Eq. 12.24a is quite simple: the rate of flow of enthalpy *in* equals the rate of flow of enthalpy *out*; that is,

$$\underbrace{\dot{m}_{\text{F}}\, h_{\text{F},3} + \dot{m}_{\text{A},3}\, h_{\text{A},3}}_{\text{Enthalpy flow in}} = \underbrace{\dot{m}_{\text{P},4}\, h_{\text{P},4}}_{\text{Enthalpy flow out}}. \quad \textbf{(12.24b)}$$

Dividing this expression by the air mass flow rate yields

$$\frac{\dot{m}_{\text{F}}\, h_{\text{F},3} + \dot{m}_{\text{A},3}\, h_{\text{A},3}}{\dot{m}_{\text{A},3}} = \frac{\dot{m}_{\text{P},4}\, h_{\text{P},4}}{\dot{m}_{\text{A},3}},$$

or

$$(F/A)h_{\text{F}} + h_{\text{A},3} = (1 + F/A)h_{\text{P},4}. \quad \textbf{(12.24c)}$$

If the fuel enters at the same temperature as the air, this equation can be interpreted as follows:

$$\underbrace{h_{\text{reactants}}(T_3)}_{\substack{\text{Specific enthalpy of reactant} \\ \text{mixture at } T_3 \text{ per unit mass of air}}} = \underbrace{h_{\text{products}}(T_4)}_{\substack{\text{Specific enthalpy of product} \\ \text{mixture at } T_4 \text{ per unit mass of air}}}. \quad \textbf{(12.25)}$$

From this point, any complexity in the analysis results from determining the specific enthalpies, particularly that of the products, $h_{\text{P},4}$. The solution of Eq. 12.25 amounts to finding the temperature of the exiting products. This is shown schematically on the h–T diagram in Fig. 12.24.

Determining $h_{\text{P},4}$ requires that the product composition be known or calculated. The easiest way to do this is to assume that the combustion process is "complete," that is, that all the carbon in a hydrocarbon fuel goes to carbon dioxide, and all the fuel

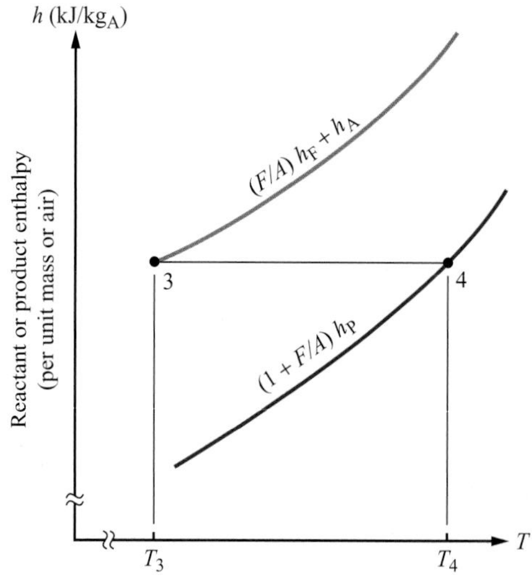

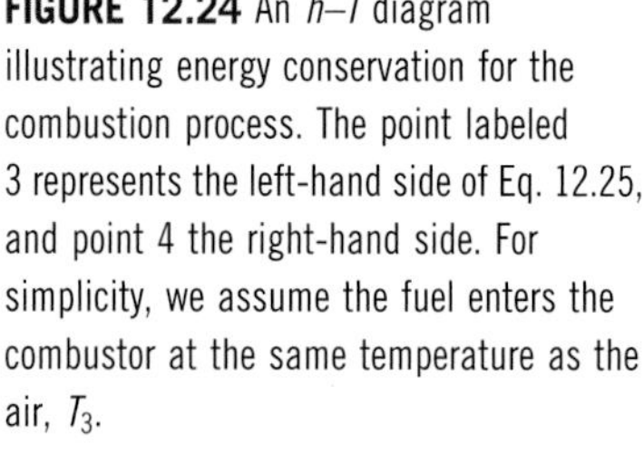

FIGURE 12.24 An h–T diagram illustrating energy conservation for the combustion process. The point labeled 3 represents the left-hand side of Eq. 12.25, and point 4 the right-hand side. For simplicity, we assume the fuel enters the combustor at the same temperature as the air, T_3.

hydrogen to water vapor. A second, and generally more realistic, assumption is that chemical equilibrium prevails at the combustor outlet temperature T_4 and pressure P_4 ($= P_3$).

Example 12.8 Turbo-Jet Combustor

Consider a turbo-jet powered aircraft flying at 200 m/s at 6000-m altitude. At these conditions, air enters the combustor at 349.8 kPa and 449.2 K. The fuel, liquid n-dodecane ($C_{12}H_{26}$), is sprayed into the combustor at 298 K to provide an overall air–fuel ratio of 75:1. The corresponding equivalence ratio is $\Phi = 0.1988$. Determine the temperature of the combustion products at the combustor exit. Also determine the constant-pressure specific heat of the product mixture.

Solution

Known P, T_A, T_F, A/F

Find T_P, $c_{p,P}$

Sketch

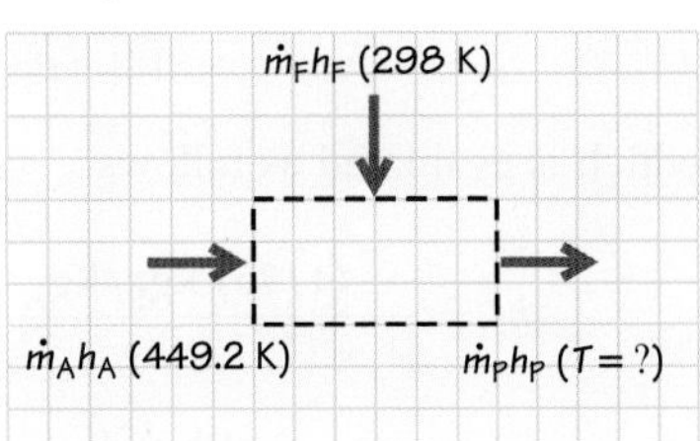

Modeling, Premises and Assumptions

i. Steady flow
ii. Constant pressure
iii. Adiabatic process ($\dot{Q}_{cv} = 0$)
iv. Negligible kinetic and potential energies
v. Ideal-gas behavior
vi. No dissociation
vii. Simple air (3.76 kmol of N_2 per kmol of O_2)

Analysis With these assumptions, we can apply the conservation of energy expressed by Eq. 12.24c to find the unknown outlet temperature T_P as follows. Because T_A and T_F are known, the left-hand side of Eq. 12.24c is easily evaluated using data from Table D.11 (O_2), Table D.7 (N_2), and Table H.1 ($C_{12}H_{26}$):

$$\begin{aligned}\bar{h}_{O_2}(449.2\text{K}) &= \bar{h}^\circ_{f,O_2}(298\,\text{K}) + \Delta h_{s,O_2}(449.2\text{K}) = 0 + 4539\ \text{kJ/kmol}\\ &= 4539\ \text{kJ/kmol},\\ \bar{h}_{N_2}(449.2\,\text{K}) &= \bar{h}^\circ_{f,N_2}(298\,\text{K}) + \Delta h_{s,N_2}(449.2\text{K}) = 0 + 4423\ \text{kJ/kmol}\\ &= 4423\,\text{kJ/kmol},\end{aligned}$$

and

$$\begin{aligned}h_A(449.2) &= \frac{\bar{h}_A(449.2)}{\mathcal{M}_A} = \frac{X_{O_2}\bar{h}_{O_2} + X_{N_2}\bar{h}_{N_2}}{\mathcal{M}_A}\\ &= \frac{0.21(4539\ \text{kJ/kmol}) + 0.79(4423\ \text{kJ/kmol})}{28.85\text{kg/kmol}}\\ &= 154.2\ \text{kJ/kmol}.\end{aligned}$$

Because the fuel enters as a liquid at the reference-state temperature 298 K, there is no sensible contribution to the standardized enthalpy [$h_{s,F}$ (298 K) = 0]. Furthermore, the enthalpy of formation for the gaseous fuel (Table H.1) needs to be decreased by the enthalpy of vaporization as follows:

$$\mathcal{M}_F = 170.337 \text{ kg/kmol},$$
$$\bar{h}^\circ_{f,F(vapor)} = -292{,}162 \text{ kJ/kmol},$$
$$h_{fg} = 256 \text{ kJ/kg}.$$

where the values are taken from Table H.1. Thus,

$$h_{F(liquid)} = \frac{\bar{h}^\circ_{f,F(vapor)}}{\mathcal{M}_F} - h_{fg} = \frac{-292{,}162 \text{ kJ/kmol}}{170.337 \text{ kg/kmol}} - 256 \text{ kJ/kg} = -1971.2 \text{ kJ/kg}.$$

Rearranging Eq. 12.24c to isolate the enthalpy of the products yields

$$\frac{(F/A)}{1+(F/A)} h_F + \frac{1}{1+(F/A)} h_A = h_P,$$

which is evaluated as follows:

$$h_P = \frac{(1/75) \text{ kg}_F/\text{kg}_A}{[1+(1/75)]\text{kg}_{mix}/\text{kg}_A}(-1971.2 \text{ kJ/kg}_F) + \frac{1 \text{ kg}_A/\text{kg}_A}{[1+(1/75)]\text{kg}_{mix}/\text{kg}_A}(154.2 \text{ kJ/kg}_A) = 126.2 \text{ kJ/kg}_{mix}.$$

Also see Example 12.2.

Our problem now is to determine what temperature of the products yields this value of the mass-specific enthalpy. We first determine the products' mixture composition using element balances (C, H, O, N) for the given stoichiometry and then combine Eqs. 10.15f and 10.3:

$$h_P = \frac{\bar{h}_P}{\mathcal{M}_P} = \frac{\sum X_i \bar{h}_i(T)}{\sum X_i \mathcal{M}_i},$$

where we guess a value for T to evaluate the individual species $\bar{h}_i$, using the tables in Appendix D. Iteration then provides a final result. We begin this procedure by writing a combustion reaction equation by combining Eqs. 12.8, 12.9, and 12.11b:

$$C_xH_y + \frac{x+y/4}{\Phi}(O_2 + 3.76\, N_2) \rightarrow b\, CO_2 + c\, H_2O + d\, O_2 + e\, N_2.$$

For $C_{12}H_{26}$, this becomes

$$C_{12}H_{26} + \frac{18.5}{\Phi}(O_2 + 3.76\, N_2) \rightarrow b\, CO_2 + c\, H_2O + d\, O_2 + e\, N_2.$$

Element conservation yields

C : $b = 12$,

H : $26 = 2c$ or $c = 13$,

O : $2(18.5)/\Phi = 2b + c + 2d$ or $d = 18.5((1-\Phi)/\Phi)$, and

N : $[2(18.5)3.76]/\Phi = 2e$ or $e = 69.56/\Phi$.

The mole fraction associated with each product constituent can be determined from

$$X_i = \frac{N_i}{N_{\text{tot}}},$$

where $N_{\text{tot}} = b + c + d + e$. Using the given value of $\Phi = 0.1988$ (the reader should verify that this value is correct, for $C_{12}H_{26}$ for which $A/F = 75.1$), we evaluate the X_is:

$$X_{CO_2} = \frac{b}{N_{\text{tot}}} = \frac{12}{449.458} = 0.0267,$$

$$X_{H_2O} = \frac{c}{N_{\text{tot}}} = \frac{13}{449.458} = 0.0289,$$

$$X_{O_2} = \frac{d}{N_{\text{tot}}} = \frac{74.558}{449.458} = 0.1659,$$

and

$$X_{N_2} = \frac{349.899}{449.458} = 0.7785.$$

We begin our iteration process by guessing a product temperature of 1000 K. The following table summarizes the calculation of $\bar{h}_P$ using the Appendix D tabulations for the various $\bar{h}_i$:

	X_i	$\bar{h}_i(1000\text{K}) = \bar{h}^\circ_{f,i}(298\text{ K}) + \Delta\bar{h}_{s,i}(1000\text{ K})$
CO_2	0.0267	−393,546 + 33,425 = −360,121 kmol/kg
H_2O	0.0289	−241,845 + 25,993 = −215,852 kmol/kg
O_2	0.1654	0 + 22,721 = 22,721 kmol/kg
N_2	0.7785	0 + 21,468 = 21,468 kmol/kg

Thus,

$$\sum X_i\bar{h}_i = \bar{h}_P(1000\text{ K}) = 4628.9\text{ kJ/kmol}$$
$$\mathcal{M}_P = \sum X_i\mathcal{M}_i = 28.81\text{ kg/kmol},$$

and

$$h_P(1000\text{ K}) = \frac{4628.9\text{ kJ/kmol}}{28.813\text{ kg/kmol}} = 160.65\text{ kJ/kg}.$$

Since $h_P(1000\text{ K}) > h_P = 126.2$ kJ/kg, we need to try a lower temperature. Using $T =$ 900 K yields $h_P(900\text{ K}) = 65.76$ kJ/kg. From this, we conclude that

$$900\text{ K} < T < 1000\text{ K}.$$

A simple linear interpolation using these two values yields $T = 964$ K, further iteration homes in on $T = 970$ K.

The product mixture specific heat is now evaluated from Eq. 10.15h using $\bar{c}_{p,i}$ (970 K) values from Appendix D:

$$\begin{aligned}\bar{c}_{p,P} &= \sum X_i\bar{c}_{p,i}\\ &= (0.0267\text{ kmol}_{CO_2}/\text{kmol}_P)(53.993\text{ kJ/kmol}_{CO_2}\cdot\text{K})\\ &\quad + (0.0289\text{ kmol}_{H_2O}/\text{kmol}_P)(40.890\text{ kJ/kmol}_{CO_2}\cdot\text{K})\\ &\quad + (0.1659\text{ kmol}_{O_2}/\text{kmol}_P)(34.791\text{ kJ/kmol}_{O_2}\cdot\text{K})\\ &\quad + (0.7785\text{ kmol}_{N_2}/\text{kmol}_P)(32.573\text{ kJ/kmol}_{N_2}\cdot\text{K})\\ &= 33.753\text{ kJ/kmol}_P\cdot\text{K}\end{aligned}$$

and

$$\bar{c}_{p,\mathrm{P}} = \frac{\bar{c}_{p,\mathrm{P}}}{\mathcal{M}_\mathrm{P}} = \frac{33.753\,\mathrm{kJ/kmol{\cdot}K}}{28.813\,\mathrm{kg/kmol}} = 1.171\,\mathrm{kJ/kg{\cdot}K}.$$

Comments The need to deal with the multicomponent product mixture complicates computations, although the principles involved (element conservation) are straightforward. This example also illustrates the importance of being able to work with both mass-specific and molar-specific properties.

Using our final result that $T = 970\,\mathrm{K}$, we can test the assumption that dissociation is negligible using software from Ref. [2]. Calculation of the equilibrium mixture composition allowing for dissociation shows that the mole fractions of all minor species (CO, H_2, OH, etc.) are less than 10^{-7}. These small values justify our original assumption of negligible dissociation.

12.3e Boiler Analysis

So far in our analyses of the Rankine cycle and its variants we have treated the energy input to the system as a heat-addition process. We now examine this process in greater detail. Figure 12.25 schematically shows how the hot products of combustion from the furnace section of the boiler[3] are used to heat water and to produce and to superheat steam. The economizer, steam generator, and superheater are, in effect, specialized heat exchangers (see Chapter 8). Although not shown in Fig. 12.25, reheater components and their associated inlet and outlet streams may also be part of the overall system. Furthermore, most power plants preheat the combustion air to improve efficiency (see Fig. 12.26). Our purpose here is not to provide a detailed analysis of any particular furnace and boiler system but, rather, to show how the fundamental principles discussed in previous chapters can be applied to obtain a more complete understanding of the steam production part of a power plant. The following example illustrates this synthesis.

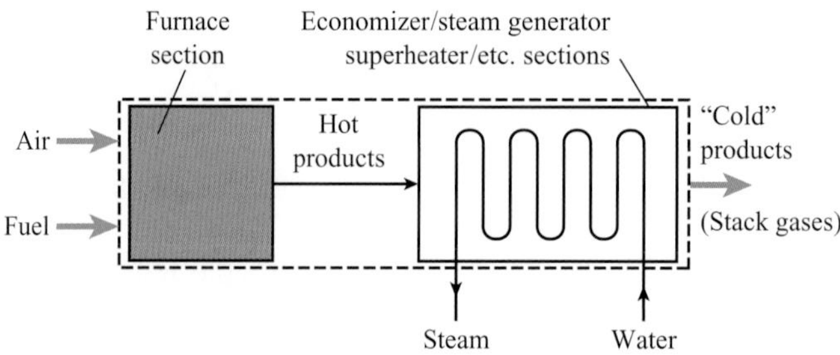

FIGURE 12.25 Schematic diagram of a steam power plant boiler. In an actual boiler, the furnace section is integrated with the boiler components.

[3] We use the generic term *boiler* to refer to the package of components used to produce steam in a power plant.

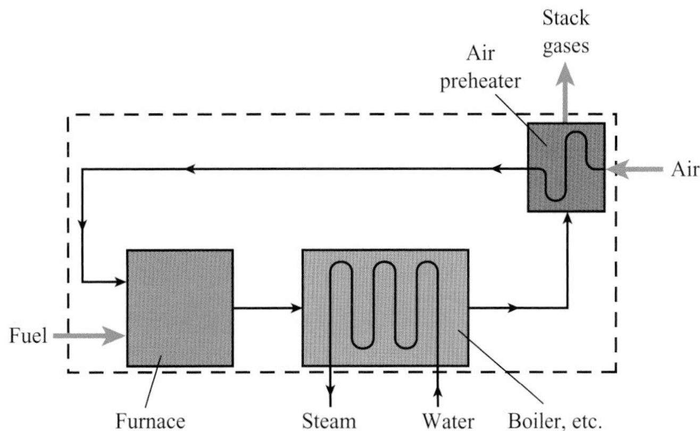

FIGURE 12.26 Schematic diagram of a steam power plant boiler with preheating of the combustion air using hot stack gases. The heat exchangers used for this purpose can be either recuperators (steady-flow devices) or regenerators (energy-storage devices). See Chapter 8.

Example 12.9 Steam Generator

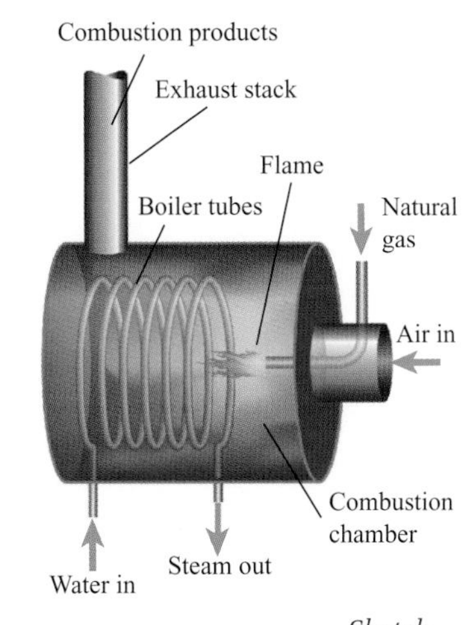

Consider a simple, natural-gas-fired steam generator as shown in the sketch. Air and fuel (assumed to be CH_4) both enter at 298 K and 1 atm, and the products of combustion exit the stack at 500 K. The fuel mass flow rate is 3 kg/s, and the air is supplied at 105% of the theoretical (stoichiometric) rate. The air can be treated as a simple mixture of 21% O_2 and 79% N_2 (i.e., 3.76 kmol of N_2 for each kmol of O_2, with a molecular weight 28.85 kg/kmol). Water enters the boiler at 400 K and 12 MPa; superheated steam exits with negligible pressure drop at 900 K. Determine the mass flow rate of the steam.

Solution

Known $\dot{m}_F$, 105% theoretical air, T_F, T_A, P_F, P_A, T_P, $T_{H_2O,in}$, $T_{H_2O,out}$, P_{H_2O}

Find $\dot{m}_{H_2O}$

Sketch

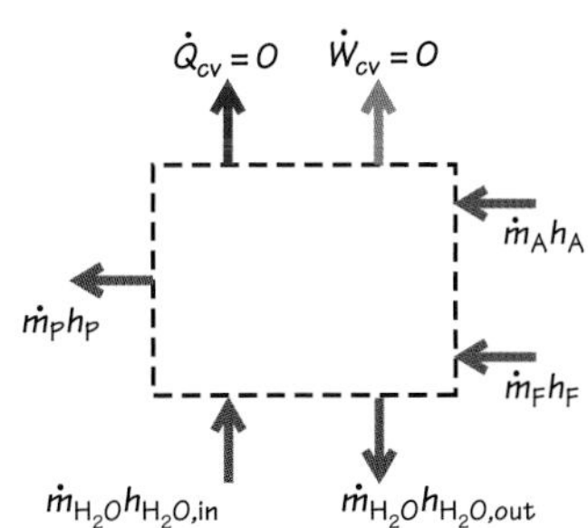

Modeling, Premises and Assumptions

i. The flow is steady.
ii. Air, fuel, and products behave as ideal gases.
iii. Natural gas consists only of CH_4.
iv. The control volume is adiabatic and produces no work (i.e., $\dot{Q}_{cv} = \dot{W}_{cv} = 0$).
v. The air is 21% O_2 and 79% N_2.
vi. No pressure losses occur due to friction (i.e., $P_{H_2O,in} = P_{H_2O,out}$).
vii. Complete combustion occurs with negligible dissociation.
viii. Changes in kinetic and potential energy are negligible for all streams.

Analysis We begin by applying mass conservation, recognizing that the water/steam side of the system is isolated from the combustion side; thus,

$$\dot{m}_{H_2O,\,in} = \dot{m}_{H_2O,\,out}$$

and

$$\dot{m}_A + \dot{m}_F = \dot{m}_P,$$

where A, F, and P indicate air, fuel, and products, respectively. The air flow rate can be found using the given stoichiometry, and the product flow rate can be found from our mass conservation expression.

Before proceeding with these calculations, we apply energy conservation to our control volume to illustrate the overall approach to finding the unknown steam (water) flow rate. With our assumptions, the general steady-flow energy equation for a multistream control volume (Eq. 5.16),

$$\dot{Q}_{cv,\,net\,in} - \dot{W}_{cv,\,net\,out} = \sum_{k=1}^{M\,outlets} \dot{m}_{out,k}\left[h_k + \tfrac{1}{2}V_{avg,k}^2 + g(z_k - z_{ref})\right] - \sum_{j=1}^{N\,inlets} \dot{m}_{in,j}\left[h_j + \tfrac{1}{2}V_{avg,j}^2 + g(z_j - z_{ref})\right],$$

simplifies as follows:

$$0 - 0 = \dot{m}_P(h_P + 0 + 0) + \dot{m}_{H_2O,\,out}(h_{H_2O,\,out} + 0 + 0) - \dot{m}_A(h_A + 0 + 0) - \dot{m}_F(h_F + 0 + 0) - \dot{m}_{H_2O,\,in}(h_{H_2O,\,in} + 0 + 0),$$

or

$$\dot{m}_A h_A + \dot{m}_F h_F - \dot{m}_P h_P = \dot{m}_{H_2O}(h_{H_2O.out} - h_{H_2O.in}).$$

The unknown water flow rate is then

$$\dot{m}_{H_2O} = \frac{\dot{m}_A h_A + \dot{m}_F h_F - \dot{m}_P h_P}{h_{H_2O.out} - h_{H_2O.in}}.$$

With the big picture complete, we now focus on the details. To find $\dot{m}_A$, we combine the definitions of the air–fuel ratio (Eq. 12.10) and percent stoichiometric air (Eqs. 12.11):

$$\dot{m}_A = \dot{m}_F(A/F)_{stoic}\frac{\%\ \text{stoichiometric air}}{100\%}.$$

For the fuel composition expressed as C_xH_y, we can find the stoichiometric air–fuel ratio using Eqs. 12.8–12.10:

$$(A/F)_{stoic} = \left(\frac{\dot{m}_A}{\dot{m}_F}\right)_{stoic} = \frac{4.76(x + y/4)\mathcal{M}_{air}}{\mathcal{M}_{fuel}}.$$

For methane (CH_4),

$$(A/F)_{stoic} = \frac{[4.76(1+1)\ \text{kmol}_A/\text{kmol}_F](28.85\ \text{kg}_A/\text{kmol}_A)}{(16.04\ \text{kg}_F/\text{kmol}_F)}$$

$$= 17.1\,\text{kmol}_{A,\,stoic}/\text{kmol}_{F,\,stoic}.$$

The air flow rate is then

$$\dot{m}_A = 3\text{kg}_F/\text{s}\left(17.1\,\text{kg}_{A,\,stoic}/\text{kg}_{F,\,stoic}\right)\frac{1.05\,\text{kg}_{A,\,stoic}}{1\,\text{kg}_{F,\,stoic}} = 53.9\ \text{kg}_A/\text{s},$$

and the products flow rate is

$$\dot{m}_P = 53.9\ \text{kg}_A + 3\ \text{kg}_F = 56.9\ \text{kg}_P.$$

Next we find the five enthalpies involved in our energy conservation expression. The water and steam enthalpies are found using the NIST resources:

$$h_{H_2O,\,in} = h_{H_2O}(400\,\text{K},12\ \text{MPa}) = 541.05\ \text{kJ/kg},$$
$$h_{H_2O,\,out} = h_{H_2O}(900\,\text{K},12\ \text{MPa}) = 3676.2\ \text{kJ/kg}.$$

Because we are dealing with a combustion process, we must employ standardized enthalpies for the fuel, air, and products. (The water/steam system is non-reacting, so simple sensible enthalpies suffice.) The enthalpy of the air is expressed as

$$\bar{h}_A = X_{O_2}\bar{h}_{O_2} + X_{N_2}\bar{h}_{N_2},$$

where the standardized enthalpy of each species i is the sum of the enthalpy of formation and the sensible enthalpy change (Eq. 12.12):

$$\bar{h}_i(T) = \bar{h}^{\circ}_{f,i} + \Delta\bar{h}_{s,i}(T).$$

For O_2 and N_2 at 298 K and 1 atm, both contributions to the standardized enthalpy are zero; that,

$$\bar{h}_A(298\,\text{K}) = 0.21(0+0) + 0.79(0+0) = 0.$$

The mass-specific enthalpy is also zero:

$$h_A = \bar{h}_A/\mathcal{M}_A = (0\ \text{kJ/kmol})/(28.85\ \text{kg/kmol}) = 0\ \text{kJ/kg}.$$

The enthalpy of the entering fuel, found using the enthalpy of formation of methane from Table H.1, is

$$\bar{h}_F\,(298\,\text{K}) = \bar{h}_{CH_4}(298\text{K}) = \bar{h}^{\circ}_{f,CH_4} + \Delta h_{s,CH_4} = (-74,831 + 0)\,\text{kJ/kmol},$$

and (Eq. 2.5)

$$h_T = \overline{h_F}/\mathcal{M}_F = (-74{,}831\,\text{kJ/kmol})/(16.04\,\text{kg/kmol}) = -4665.3\,\text{kJ/kg}.$$

We now need to determine the composition of the combustion products stream so that we can apply Eq. 10.15f to determine the product enthalpy:

$$\bar{h}_p = \sum X_j\bar{h}_i.$$

See Examples 12.1 and 12.2 to review stoichiometry concepts.

The reaction of CH_4 with 105% stoichiometric air can be written (Eq. 12.2)

$$CH_4 + 1.05(2)(O_2 + 3.76N_2) \rightarrow b\,CO_2 + c\,H_2O + d\,O_2 + e\,N_2.$$

The unknown coefficients are found from atom balances as follows:

C: $b = 1$,
H: $4 = 2c$ or $c = 2$,
O: $1.05\,(2)\,2 = 2b + c + 2d$ or $d = (4.2 - 2b - c)/2 = 0.1$, and
N: $1.05\,(2)\,3.76\,(2) = 2e$ or $e = 7.896$.

The total number of moles of products (per mole of CH_4 burned) is

$$N_P = b + c + d + e$$
$$= 1 + 2 + 0.1 + 7.896 = 10.996,$$

and the mole fraction of each constituent is

$$X_{CO_2} = N_{CO_2}/N_P = 1/10.996 = 0.0909,$$

$$X_{H_2O} = N_{H_2O}/N_P = 2/10.996 = 0.1819,$$

$$X_{O_2} = N_{O_2}/N_P = 0.1/10.996 = 0.0091,$$

and

$$X_{N_2} = N_{N_2}/N_P = 7.896/10.996 = 0.7181.$$

For each constituent, we calculate its molar-specific standardized enthalpy using data from Appendix D at the stack temperature, 500 K:

$$\begin{aligned}
\bar{h}_{CO_2}(500\,\text{K}) &= (-393{,}546 + 8301)\,\text{kJ/kmol} = -385{,}245\ \text{kJ/kmol},\\
\bar{h}_{H_2O}(500\,\text{K}) &= (-241{,}845 + 6947)\ \text{kJ/kmol} = -234{,}898\ \text{kJ/kmol},\\
\bar{h}_{O_2}(500\,\text{K}) &= (0 + 6097)\ \text{kJ/kmol} = 6097\ \text{kJ/kmol},\\
\bar{h}_{N_2}(500\,\text{K}) &= (0 + 5920)\,\text{kJ/kmol} = 5920\ \text{kJ/kmol}.
\end{aligned}$$

Using Eq. 10.15f, we find the mixture molar-specific standardized enthalpy:

$$\begin{aligned}
\bar{h}_P = &\ (0.0909\ \text{kmol}_{CO_2}/\text{kmol}_P(-385{,}245\ \text{kJ/kmol}_{CO_2})\\
&+(0.1819\ \text{kmol}_{H_2O}/\text{kmol}_P)(-234{,}898\ \text{kJ/kmol}_{H_2O})\\
&+(0.0091\ \text{kmol}_{O_2}/\text{kmol}_P)(6097\ \text{kJ/kmol}_{O_2})\\
&+(0.7181\ \text{kmol}_{N_2}/\text{kmol}_P)(5920\ \text{kJ/kmol}_{N_2})\\
= &\ -73{,}440\ \text{kJ/kmol}_P.
\end{aligned}$$

Converting this to a mass-specific basis for our energy balance requires the product mixture molecular weight, which is calculated from Eq. 10.3:

$$\begin{aligned}
\mathcal{M}_P &= \sum X_i \mathcal{M}_i\\
&= 0.0909(44.011) + 0.1819(18.016)\\
&\quad + 0.0091(31.999) + 0.7181(28.013)\\
&= 27.685\ \text{kg/kmol}_P.
\end{aligned}$$

Thus,

$$h_P = \bar{h}_P/\mathcal{M}_P = (-73{,}440\ \text{kJ/kmol}_P)/(27.685\ \text{kg/kmol}_P) = -2652.7\ \text{kJ/kg}.$$

Before proceeding, we summarize our intermediate results:

	$\dot{m}$(kg/s)	h(kJ/kg)
Air	53.9	0
Fuel	3.0	–4665.3
Products	56.9	–2652.7
Water	?	541.05
Steam	?	3676.2

With this information, we can accomplish our original objective of determining the boiler water flow rate. Using the result previously derived from energy conservation, we calculate

$$\dot{m}_{H_2O} = \frac{\dot{m}_A h_A + \dot{m}_F h_F - \dot{m}_P h_P}{h_{H_2O,\,out} - h_{H_2O,\,in}}$$

$$= \frac{59.3\,kg/s(0\,kJ/kg) + 3.0\,kg/s(-4665.3\,kJ/kg) - 56.9\,kg/s(-2652.7\,kJ/kg)}{(3676.2 - 541.05)\,kJ/kg}$$

$$= 43.68\ kg/s.$$

Comments This example integrates many concepts and topics. Among these are: element, mass, and energy conservation for control volumes; stoichiometry; properties of ideal-gas mixtures; and standardized enthalpies. Application of these principles provides a quantitative and realistic understanding of how a boiler works.

This example also affords an opportunity to evaluate the cost of raising steam. Using the US national average residential price of natural gas for 2014 of \$10.97 per thousand cubic feet, we can estimate the price per kJ of steam energy, using the following formula:

$$\frac{\$}{E_{steam}} = \frac{\dot{E}_F}{\dot{E}_{steam}}\frac{\$}{E_F} = \frac{\dot{m}_F\,HHV}{\dot{m}_{steam}h_{l-v}}\frac{\$}{E_F},$$

where HHV is the higher heating value of the fuel and

$$\frac{\$}{E_F} = \frac{\$}{\mathcal{V}_F}\frac{1}{\rho_F}\frac{1}{HHV}.$$

Thus,

$$\frac{\$}{E_{steam}} = \frac{\$}{\mathcal{V}_F}\frac{\dot{m}_F}{\dot{m}_{steam}h_{l-v}\rho_F}.$$

Using the numbers from our example and converting to SI units yields

$$\frac{\$}{E_{steam}} = \frac{\$10.97}{1000\,ft^3}\left[\frac{1\,ft^3}{(0.3048)^3 m^3}\right]\frac{3\,kg/s}{43.68\ kg/s(3135.15\,kJ/kg)\,0.6559\ kg/m^3}$$

$$= 1.30 \times 10^{-5}\ \$/kJ_{steam}.$$

We can also estimate the operating cost associated with supplying fuel, which is

$$\frac{\$}{time} = \frac{\$}{\mathcal{V}_F}\frac{\dot{m}_F}{\rho_F}$$

$$= \frac{\$10.97}{1000\ ft^3}\left[\frac{1\,ft^3}{(0.3048)^3 m^3}\right]\frac{3\ kg/s}{0.6559\,kg/m^3}\left[\frac{3600\ s}{hr}\right]$$

$$= \$6380/hr.$$

Clearly this is an important cost in the operation of this boiler.

Self-Test 12.7

☑ Using the same temperatures, redo Example 12.9 for the stoichiometric combustion of methane and air.

(Answer: $\dot{m}$ = 43.84 kg/s)

12.3f Power Plant Overall Energy Utilization

Figure 12.27 shows a control volume containing an entire power plant. Energy enters this control volume with the fuel ($\dot{m}_F h_F$) and the air ($\dot{m}_A h_A$); it exits with the stack gases ($\dot{m}_P h_P$), the heat rejected in the condenser(s) and any residual heat losses, combined as $\dot{Q}_{out}$, and the useful electrical power out, $\dot{W}_{elec}$. Using the general idea that an efficiency is the ratio of a desired energy output to an input energy that costs, we define an overall power-plant efficiency as

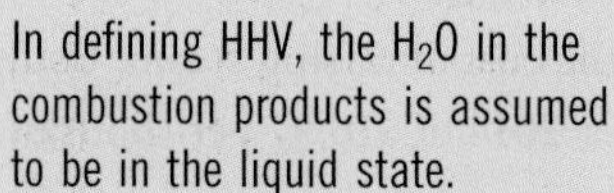

$$\begin{aligned}\eta_{OA} &\equiv \frac{\text{useful electrical power produced}}{\text{supplied fuel energy rate}}\\ &= \frac{\dot{W}_{elec}}{\dot{m}_F\, HHV},\end{aligned} \tag{12.26a}$$

where HHV is the higher heating value of the fuel. Recall that the HHV is the energy removed from the products of adiabatic combustion in order to cool them back to the initial temperature of the reactants, usually assumed to be 298 K. Consider, for example, the control volume shown in Fig. 12.27. If the stack gases exited at the same temperature as the entering fuel and air, all the chemical energy released during combustion would go into $\dot{W}_{elec}$ and $\dot{Q}_{out}$; there would be no loss of energy associated with the stack gases. In reality, a stack loss always exists because of the limitations of heat-exchange equipment. Another way of looking at the overall efficiency of a steam power plant is to consider that the overall efficiency is the product of: the efficiency of the boiler in transferring energy from the combustion products to the working fluid, η_{boiler}; the thermal efficiency of the closed steam cycle, $\eta_{th,Rankine}$; and the efficiency of the electrical generator, η_{gen}. That is,

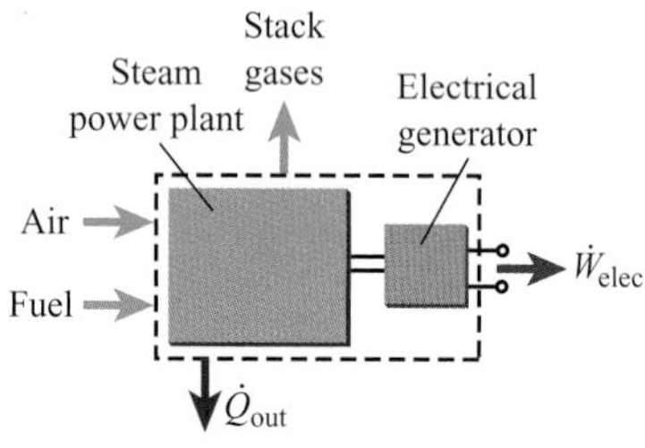

FIGURE 12.27 System boundary for evaluation of overall power-plant efficiency.

$$\eta_{OA} = \eta_{boiler}\eta_{th,\,Rankine}\,\eta_{gen}. \tag{12.26b}$$

Typical values of boiler efficiencies range from 85% to 90%, whereas electrical generator efficiencies for large power plants range from 97% to 98%. Thermal efficiencies for Rankine cycles with reheat and regeneration are of the order of 30% to 40% in many power plants. Advanced cycles yield higher efficiencies.

Example 12.10 Boiler Efficiency

Determine the efficiency of the boiler operating as discussed in Example 12.9.

Solution

Known $\dot{m}_A, \dot{m}_F, \dot{m}_P, h_A, h_F, h_P$

Find η_{boiler}

Sketch

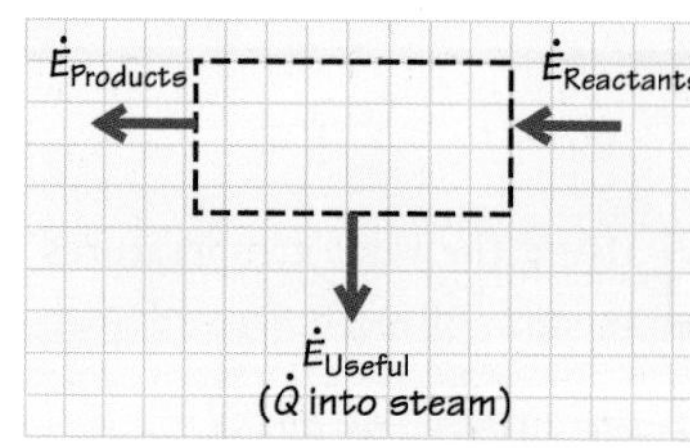

Modeling, Premises and Assumptions These remain as in Example 12.9.

Analysis We first define the boiler efficiency to be the ratio of the energy delivered to the water to the chemical energy released by the fuel, that is,

$$\eta_{\text{boiler}} = \frac{\dot{Q}_{\text{steam}}}{\dot{m}_{\text{F}}\text{HHV}},$$

where *HHV* is the higher heating value of the fuel (CH_4). From Example 12.9,

$$\begin{aligned}\dot{Q}_{\text{steam}} &= \dot{m}_{\text{H}_2\text{O}}(h_{\text{H}_2\text{O,out}} - h_{\text{H}_2\text{O,in}}) \\ &= \dot{m}_{\text{A}} h_{\text{A}} + \dot{m}_{\text{F}} h_{\text{F}} - \dot{m}_{\text{P}} h_{\text{P}} \\ &= 59.3\ \text{kg/s}(0\ \text{kJ/kg}) + 3.0\ \text{kg/s}(-4665.3\ \text{kJ/kg}) - 56.9\ \text{kg/s}(-2652.7\ \text{kJ/kg}) \\ &= 136{,}900\ \text{kJ/s or kW}.\end{aligned}$$

The maximum possible thermal energy that can be extracted from the combustion process is that associated with cooling the products to the reference temperature, 298 K and condensing the water vapor in the products; thus

$$\begin{aligned}\dot{Q}_{\text{max possible}} &= \dot{m}_{\text{F}} HHV \\ &= 3.0\,\text{kg/s}(55{,}528\,\text{kJ/kg}) = 166{,}600\,\text{kJ/s or kW},\end{aligned}$$

where the higher heating value for CH_4 is from Table H.1. Applying our definition of boiler efficiency results in

$$\eta_{\text{boiler}} = \frac{136{,}900\ \text{kW}}{166{,}600\ \text{kW}} = 0.822 \text{ or } 82.2\%.$$

Comment Approximately 17.8% (= 100% – 82.2%) of the fuel energy goes up the stack, serving no useful purpose. Note, however, that the boiler efficiency is much greater than the efficiencies associated with the steam cycle. What physical factors determine the boiler efficiency?

Self-Test 12.8

Determine the efficiency of the boiler operating as discussed in Self Test 12.7.

(Answer: 82.5%)

SUMMARY

This chapter has shown how the overarching principles of mass and energy conservation apply to reacting systems, with a focus on combustion systems. Mass conservation was expanded to the idea that the mass of each chemical element is conserved during reactions. The concept of element balances was developed, and important measures of stoichiometry were presented. We saw the importance of using standardized enthalpies (and internal energies) in our energy conservation expressions to account for the energies associated with the breaking and making of chemical bonds. The chapter illustrated the application of mass and energy conservation to both closed and open combustion systems and presented a number of concepts important to the study of combustion; these include adiabatic flame temperatures, enthalpy of reaction, and higher and lower heating values.

KEY EQUATIONS

Review the most important equations presented in this chapter (i.e., those boxed with a yellow background). What physical principles do they express? What restrictions apply?

CHAPTER 12 KEY CONCEPTS AND DEFINITIONS CHECKLIST

Answer the Questions and solve the Problems following the arrows to demonstrate mastery of the listed concepts and definitions.

12.1 Mass Conservation for Reacting Systems

- ☐ Atom balances ➔ Problem 12.1
- ☐ Combustion equation ➔ Problem 12.15
- ☐ Stoichiometry ➔ Question 12.2, Problems 12.7, 12.8
- ☐ Equivalence ratio ➔ Question 12.4, Problem 12.23A
- ☐ Percent stoichiometric air ➔ Problems 12.20, 12.21
- ☐ Percent excess air ➔ Problems 12.20, 12.21

12.2 Energy Conservation for Reacting Systems

- ☐ Standardized enthalpies ➔ Question 12.5, Problem 12.37
- ☐ Constant-pressure adiabatic flame temperature ➔ Problem 12.42
- ☐ *H–T* diagrams for constant-pressure combustion ➔ Question 12.6, Problem 12.41
- ☐ Enthalpy of reaction ➔ Question 12.7
- ☐ Heating values (*HHV* and *LHV*) ➔ Question 12.8, Problems 12.50, 12.52
- ☐ Constant-volume adiabatic flame temperature ➔ Problems 12.46, 12.47
- ☐ *U–T* diagrams for constant-volume combustion ➔ Questions 12.9, 12.10

12.3 Steady-Flow Applications ➔ Problems 12.59, 12.72, 12.68

REFERENCES

1. Lavoisier, A. L., *Elements of Chemistry, in a New Systematic Order, Containing all of the Modern Discoveries*, translated by R. Kerr, Dover, New York, 1965.
2. Turns, S. R., *An Introduction to Combustion: Concepts and Applications*, 3rd edn, McGraw-Hill, New York, 2012.
3. Maghon, H., Berenbrink, P., Termeulen, H., and Gartner, G., "Progress in NO_x and CO Emission Reduction of Gas Turbines," ASME 90-JPGC/GT-4, ASME/IEEE Power Generation Conference, Boston, Oct. 21–25, 1990.
4. Energy Information Agency, U.S. Department of Energy, "Monthly Energy Review Apirl 2019," http://www.eia.gov/totalenergy/data/monthly/index.cfm#electricity, Release Date: Apirl 25, 2019.
5. Warnatz, J., Maas, U., and Dibble, R. W., *Combustion*, Springer-Verlag, Berlin, 1996.
6. Lefebvre, A. H., *Gas Turbine Combustion*, Taylor & Francis, Hemisphere, Washington, DC, 1983.

Some end-of-chapter problems were adapted with permission from the following:

Myers, G. E., *Engineering Thermodynamics*, Prentice Hall, Englewood Cliffs, NJ, 1989.

Look, D. C., Jr., and Sauer, H. J., Jr. *Engineering Thermodynamics*, PWS, Boston, 1986.

QUESTIONS

12.1 What quantities are conserved when a fuel and air burn to form combustion products?

12.2 What does it mean that a fuel burns with air in stoichiometric proportions?

12.3 Distinguish between the terms "rich" and "lean."

12.4 What is the physical meaning of the equivalence ratio Φ? What is implied when Φ is greater than unity? Less than unity?

12.5 Explain the concept of standardized properties applied to chemically reacting systems.

12.6 Sketch an adiabatic constant-pressure combustion process in H–T coordinates. Show lines representing $H_{reac}(T)$ and $H_{prod}(T)$.

12.7 What is the sign (positive or negative) associated with the enthalpy of reaction for exothermic reactions? For endothermic reactions?

12.8 Explain what is meant by the heating value of a fuel. What distinguishes the "higher" heating value from the "lower" heating value?

12.9 Sketch an adiabatic constant-volume combustion process in U–T coordinates. Show lines representing $U_{reac}(T)$ and $U_{prod}(T)$.

12.10 Using the sketches created in Questions 12.6 and 12.9 show how decreasing the temperature of the initial reactants results in a decreased temperature for the products.

Chapter 12 Problem Subject Areas

12.1–12.6	Combustion equation and element conservation
12.7–12.17	Stoichiometric conditions
12.18–12.36	Nonstoichiometric conditions
12.37–12.40	Standardized enthalpies
12.41–12.45	Constant-pressure adiabatic flame temperatures
12.46–12.49	Constant-volume adiabatic flame temperatures
12.50–12.55	Heating values
12.56–12.75	Steady-flow combustion
12.76–12.78	Combustion products dew-point temperature

PROBLEMS

12.1–12.6 Combustion equation and element conservation

12.1 Balance the chemical equations and find the mass fuel–oxidizer ratio for the following equations:

A. $C_5H_{12} + a\,O_2 = b\,CO_2 + c\,H_2O$

B. $C_3H_8 + a\,O_2 = b\,CO_2 + c\,H_2O$

12.2 Determine whether the following mixtures are stoichiometric:

A. $C_2H_4 + 4\,O_2$

B. $C_8H_{18} + 12\,O_2$

12.3 Determine whether the following reactions are stoichiometric:

A. $2C_3H_8 + 7O_2 = 6CO + 8H_2O$

B. $CH_4 + 2O_2 = CO_2 + 2H_2O$

12.4 A mass of particular solid fuel (70% carbon, 20% hydrogen, and 10% water by mass) reacts stoichiometrically with air. Assume that the air is 79% N_2 and 21% O_2 by volume. Determine the balanced chemical equation. Your final equation should be for an amount of fuel containing 1 mol of carbon.

12.5 A certain amount of Hexane (C_6H_{14}) burns with air (21% O_2, 79% N_2) in stoichiometric proportions. Write the overall chemical reaction for this situation and determine the mole fractions for the C_6H_{14}, O_2, and N_2 in the reactant mixture. Also determine the mole fraction of each species in the product mixture.

12.6 Set up the necessary combustion equations and determine the mass of air required to burn 1 lb_m of pure carbon to equal masses of CO and CO_2. Assume that the air is 79% N_2 and 21% O_2 by volume.

12.7–12.17 Stoichiometric conditions

12.7 Derive an expression for the stoichiometric air–fuel ratio (by mass) of an arbitrary hydrocarbon fuel C_xH_y.

12.8 Write the stoichiometric combustion reaction for methanol (CH_3OH) and air (21% O_2, 79% N_2) and determine the stoichiometric air–fuel ratio (by mass).

12.9 Propane reacts completely with a stoichiometric amount of hydrogen peroxide (H_2O_2) to form carbon dioxide and water. Determine the mass of water formed per mass of propane.

12.10 Liquid hydrogen peroxide (H_2O_2) is used as the oxidizer to burn 1 kmol of propane in a stoichiometric reaction. Initially, the hydrogen peroxide and propane are each at 40 °C and 2 atm. The final state of the exhaust product mixture (carbon dioxide and water) is 800 K and 10 atm. Determine the water (kmol) produced.

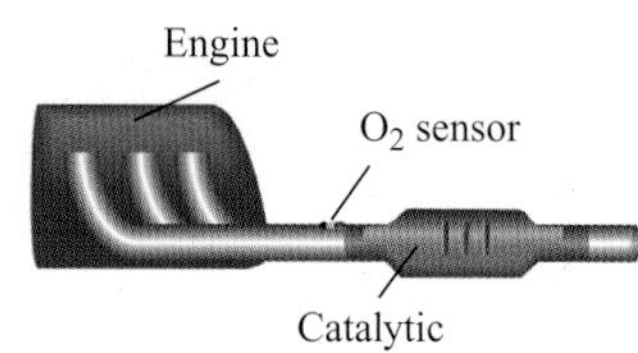

12.11 For correct operation of the catalytic converter in an automobile, the air–fuel ratio of the engine must be precisely controlled to near the stoichiometric value. This control is achieved using feedback from an O_2 sensor located in the exhaust stream.

A. If the equivalent composition of gasoline is given by C_8H_{15}, determine the value of the stoichiometric air–fuel ratio (by mass). Assume that air is a simple mixture of O_2 and N_2 in molar proportions of 1:3.76.

B. Assuming complete combustion at stoichiometric conditions, determine the mole fractions of each constituent in the exhaust stream.

C. Determine the molar mass (apparent molecular weight) of the exhaust gas mixture.

12.12 Ethanol (C_2H_5OH) is burned at stoichiometric conditions in a space heater at atmospheric pressure. Assume the air is 79% N_2 and 21% O_2 by volume.

A. Determine the mass air–fuel ratio and the mass of water formed per mass of fuel.

B. Determine the dry analysis of the exhaust gases in percentage by volume. (Note: For a dry analysis, all the water is removed from the products.)

12.13 Carbon is burned with exactly the right amount of air (79% N_2 and 21% O_2 by volume) to form carbon dioxide.

A. Determine the air–fuel ratio (by mass).

B. Determine the mass of carbon dioxide per mass of fuel.

C. Determine the mass of carbon dioxide formed per mass of air.

12.14 Calculate and compare the stoichiometric air–fuel mass ratios for the following common fuels. Assume that air is a simple mixture of O_2 and N_2 in molar proportions of 1:3.76.

A. Natural gas (assume it is essentially all methane, CH_4)

B. Liquified petroleum gas (assume it is essentially all propane, C_3H_8)

C. Typical gasoline blend, a $C_{7.9}H_{14.8}$ equivalent

D. Typical light diesel blend, a $C_{12.3}H_{22.2}$ equivalent

E. Methanol, CH_3OH

12.15 Assuming complete combustion, determine the total CO_2 production (kg/day) for a mid-sized city on a typical day from the following sources:

A. Automotive: Each one of 30,000 cars goes 15 miles at an average 10 miles/gal of octane gasoline. See Appendix H or any needed properties.

B. Human: Each of 200,000 people uses 1 lb_m of glucose ($C_6H_{12}O_6$) per day.

C. Heating systems: Each of 40,000 home-heating systems burns natural gas (assumed to be methane) at a rate of 1000 ft^3/day at 70 F and 14.7 psia.

D. Steam power plant: The local power plant produces 180 MW_e using coal (treat as 100% carbon) at a rate of 1.5 kW·hr/$lb_{m,coal}$.

12.16 Solving this problem provides information needed to solve Problem 12.17. An automobile is traveling at 55 mph on a highway. The car's engine is operating at 2000 revolutions per minute. The engine fills its cylinders with air each time two revolutions are completed, i.e., it operates on a four-stroke cycle. The volume of air ingested during each cycle is 95% of the engine's displacement of 1.5 liters. The air enters the cylinders at 0.7 atm and 305 K. Determine the following:

A. The volumetric flow rate of the air at the entering conditions. Calculate in units of m^3/s and ft^3/min (CFM).

B. The mass flow rate of the air (kg/s).

C. The volumetric flow rate of the air at standard temperature and pressure (STP $\equiv$ 101.325 kPa and 298.15 K). Calculate in units of m^3/s and ft^3/min (SCFM).

12.17 The car in Problem 12.16 operates on liquefied petroleum gas (LPG). Assume the LPG can be approximated as C_3H_8. Assume the air composition is that of "simple air," i.e., 1 kmol of O_2 for every 3.76 kmol of N_2. The engine operates at stoichiometric conditions, i.e., the mass air-to-fuel ratio is 15.6 kg_{air}/kg_{fuel}. The temperature and pressure of the exhaust gases are 950 °C and 98 kPa, respectively. Determine the following:

A. The molecular weight of the exhaust gases (kg/kmol)

B. The mole fraction of CO_2 in the exhaust stream

C. The mass fraction of CO_2 in the exhaust stream

D. The mass flow rate of CO_2 in the exhaust stream (kg/s)

E. The volumetric flow rate of CO_2 in the exhaust stream at the exhaust conditions (m^3/s)

F. The volumetric flow rate of CO_2 in the exhaust stream at STP (m^3/s)

12.18–12.36 Nonstoichiometric conditions

12.18 A hydrocarbon fuel is burned with air. Assume the air to be a simple mixture of 21% O_2 and 79% N_2. On a dry basis (all the water vapor has been removed from the products), a volumetric analysis of the products of combustion yields the following:

CO_2: 7.8%,
CO: 1.1%,
O_2: 8.3%,
N_2: 82.8%.

Determine the following:

A. The composition of the fuel on a mass basis
B. The percent theoretical air
C. The air–fuel ratio by mass

12.19 Air (21% O_2, 79% N_2) and natural gas (CH_4) are supplied to an industrial furnace at an equivalence ratio Φ of 0.95. The air mass flow rate is 4 kg/s. Determine the mass flow rate of the CO_2 exiting the furnace.

(Credit: AGEfotostock.)

12.20 Propane (C_3H_8) and air (21% O_2 and 79% N_2) burn at an air–fuel mass ratio of 20:1. Determine (a) the equivalence ratio Φ, (b) the percent stoichiometric air, and (c) the percent excess air.

(Credit: Gado Images / Photodisc / Getty Images.)

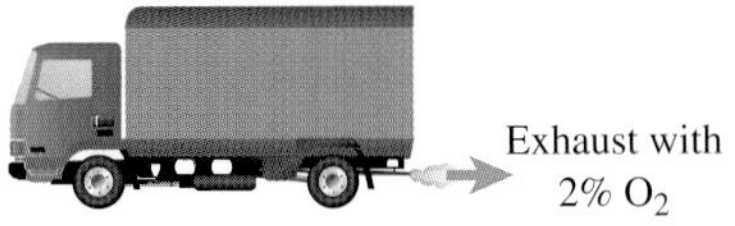

12.21 In a propane-fueled truck, 2% (by volume) oxygen is measured in the exhaust stream of the running engine. Assuming complete combustion without dissociation, determine the air–fuel ratio (mass) supplied to the engine. Also determine the percent theoretical air, the percent excess air, and the equivalence ratio.

12.22 Natural gas is burned to produce hot water to heat a clothing store. Assuming that the natural gas can be approximated as methane (CH_4) and that the air is a simple mixture of O_2 and N_2 in molar proportions of 1:3.76, determine the following:

A. The molar air–fuel ratio for stoichiometric combustion

B. The mass air–fuel ratio for stoichiometric conditions

C. The operating air–fuel ratio (mass) of the heater when 4% (molar or volume) oxygen, O_2, is present in the flue gases

D. The percent excess air, the percent theoretical air, and the equivalence ratio associated with the conditions given in part C

12.23 The Dassault Falcon aircraft is powered by two TF37 turbofan jet engines. At a cruise condition, each engine consumes fuel at a rate of 0.232 kg/s with a concomitant air consumption of 16.4 kg/s. The fuel blend can be approximated as $C_{12}H_{22}$, and the air composition can be assumed to be 21% O_2 and 79% N_2.

(Credit: Peter Brogden / Alamy Stock Photo.)

A. Determine the equivalence ratio for the combustion process.

B. Determine the composition of the combustion products exiting the combustor assuming complete combustion with no dissociation. Express your results as mole fractions.

12.24 Ethane burns with 150% stoichiometric air. Assume the air is 79% N_2 and 21% O_2 by volume. Combustion goes to completion. Determine (a) the air–fuel ratio by mass and (b) the mole fraction (percentage) of each product.

12.25 Rework Problem 12.24 but use propane as the fuel.

12.26 Ethanol (C_2H_5OH) is burned in a space heater at atmospheric pressure. Assume the air is 79% N_2 and 21% O_2 by volume.

A. For combustion with 20% excess air, determine the mass air–fuel ratio and the mass of water formed per mass of fuel.

B. For combustion with 180% stoichiometric air, determine the dry analysis of the exhaust gases in percentage by volume. (Note: For a dry analysis, all the water is removed from the products.)

12.27 Methane (CH_4) is burned with air (79% N_2 and 21% O_2 by volume) at atmospheric pressure. The **molar** analysis of the flue gas yields CO_2 = 10.00%, O_2 = 2.41%, CO = 0.52%, and N_2 = 87.07%. Balance the combustion equation and determine the mass air–fuel ratio, the percentage of stoichiometric air, and the percentage of excess air.

12.28 Determine the air–fuel ratio by mass when a liquid fuel with a composition of 16% hydrogen and 84% carbon by mass is burned with 15% excess air.

12.29 Compute the composition of the flue gases (percentage by volume on a dry basis) resulting from the combustion of C_8H_{18} with 114% stoichiometric air. (Note: For a dry analysis, all the water is removed from the products.)

12.30 A liquid petroleum fuel having a hydrogen/carbon ratio by weight of 0.169 is burned in a heater with an air–fuel ratio of 17 by mass. Determine the volumetric analysis of the exhaust gas on both wet and dry bases. (Note: For a dry analysis, all the water is removed from the products.)

12.31 A gas mixture (60% methane, 30% ethane, and 10% nitrogen by volume) undergoes a complete reaction with 120% theoretical air (79% N_2 and 21% O_2 by volume). Determine the composition (in mole fractions) of the dry products. (Note: For a dry analysis, as is requested here, all the water is removed from the products.)

12.32 One mole of a hydrocarbon fuel (CH_x) is burned with excess air. The volumetric analysis of the dry products (with H_2O removed) yields:

N_2: 83.6%, O_2: 5.0%
CO_2: 10.4%, CO: 1.0%

A. Determine the approximate composition of the fuel on a mass basis.
B. Determine the percent theoretical air.
C. Determine the equivalence ratio.

12.33 Determine the equivalence ratio of a mixture of 1 mol of methane and 7 mol of air. Is the mixture lean or rich?

12.34 Derive that $(x + y/4)/\Phi$ is the oxidizer coefficient a in the combustion reaction

$$C_xH_y + a(O_2 + 3.76\ N_2) \longrightarrow \text{Products},$$

where Φ is the air–fuel equivalence ratio. Start by considering the case where $\Phi = 1$; i.e., by doing C, H, O, and N balances you get $a = x + y/4$ when the mixture is in stoichiometric proportions. Then show how Φ enters in when the mixture is not stoichiometric.

12.35 Compare the mass of CO_2 produced per mass of fuel burned, i.e., the emission index of CO_2 (kg_{CO2}/kg_{fuel}), for the following fuels for operation at stoichiometric ($\Phi = 1.0$) and lean ($\Phi = 0.7$) conditions. Assume that all the fuel carbon appears in the products as CO_2. Treat the air as a simple mixture of N_2 (79%) and O_2 (21%).

i. Methane (CH_4)
ii. Propane (C_3H_8)
iii. Gasoline (C_8H_{15}-equivalent)
iv. Diesel fuel ($C_{15}H_{22}$-equivalent)

12.36 Coal is a physical mixture of many compounds. However, from an elemental (ultimate) analysis of dry ash-free coal, we can artificially represent the fuel as $C_vH_wN_xS_yO_z$.

Open-pit coal mining (agnormark / iStock / Getty Images Plus).

Assuming that air is a simple mixture of O_2 and N_2 in molar proportions of 1:3.76, answer the following:

A. Determine the coefficient a in the general combustion equation $C_vH_wN_xS_yO_z + (a/\Phi)(O_2 + 3.76N_2) \rightarrow$ Products. Assume that the fuel nitrogen appears as molecular N_2 in the products and that the fuel sulfur is oxidized to SO_2.

B. Determine the mass air–fuel ratio for the stoichiometric combustion of Pittsburgh #8 coal, defined as the following equivalent fuel: $C_{65}H_{52}NSO_3$.

C. A 500-MW power plant burns Pittsburgh #8 coal with an equivalence ratio Φ of 0.9. The mass flow rate of the combustion products exiting the combustor/boiler is 383.3 kg/s. Determine the air and fuel mass flow rates supplied to the combustor.

D. Determine the mole fraction (expressed as parts per million) of the SO_2 in the product stream from part C.

E. Determine the mass flow rate of SO_2 (see parts C and D) entering the pollution control device (scrubber) for the power plant.

12.37–12.40 Standardized enthalpies

12.37 Determine the standardized enthalpies of the following pure species at 4 atm and 2500 K: H_2, H_2O, and OH.

12.38 Use the curve-fit coefficients for the standardized enthalpy from Table H.2 to verify the enthalpies of formation at 298.15 K in Table H.1 for methane, propane, and hexane.

12.39 Determine the total standardized enthalpy H (kJ) for a fuel–air reactant mixture containing 1 kmol CH_4, 2.5 kmol O_2, and 9.4 kmol N_2 at 500 K and 1 atm. Also determine the mass-specific standardized enthalpy h (kJ/kg) for this mixture.

12.40 A mixture of products of combustion contains the following constituents at 2000 K: 3 kmol of CO_2, 4 kmol of H_2O, and 18.8 kmol of N_2. Determine the following quantities:

A. The mole fraction of each constituent in the mixture
B. The total standardized enthalpy H of the mixture
C. The mass-specific standardized enthalpy h of the mixture

12.41–12.45 Constant-pressure adiabatic flame temperatures

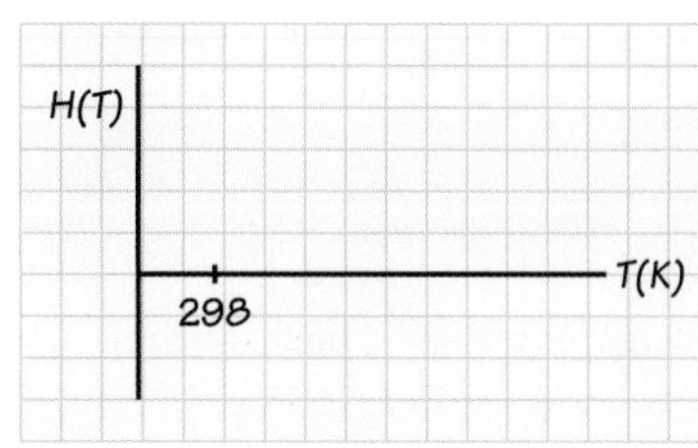

12.41 Consider the combustion of a fuel at constant pressure. Reproduce the coordinate system in the sketch and indicate and appropriately label the following items:

A. H_{reac}
B. H_{prod}
C. The adiabatic flame temperature (T_{ad}) for the reactants at 298 K

12.42 A piston–cylinder arrangement initially contains 0.002 kmol of H_2 and 0.01 kmol of O_2 at 298 K and 1 atm. The mixture is ignited and burns adiabatically at constant pressure. Determine the final temperature assuming the products contain only H_2O and the excess reactant. Also determine the work done during the process. Sketch the process on H–T and P–$\mathcal{V}$ coordinates.

12.43 Hydrogen burns with 200% stoichiometric air in a piston–cylinder arrangement. The process is carried out adiabatically and at constant pressure. The initial temperature and pressure are 298 K and 1 atm, respectively. Determine the final temperature and the work performed per mass of mixture. Assume the air is a simple mixture of 21% O_2 and 79% N_2 by volume. Also sketch the process in H–T and P–$\mathcal{V}$ coordinates.

12.44 A mixture of 5 g of ethane and a stoichiometric amount of oxygen is contained in a piston–cylinder device at 25 °C and 1 atm. The piston moves freely without friction. A spark ignites the mixture and a complete reaction occurs at a pressure of 1 atm. The products are cooled to 25 °C by the end of the reaction. Determine the following:

A. The initial volume (cm^3) of the reactants
B. The masses of liquid water and water vapor in grams
C. The work transfer (kJ)
D. The increase in internal energy of the gases in the cylinder
E. The heat interaction (MJ)

12.45 Initially, 10 cm^3 of dry air and a stoichiometric amount of gunpowder (carbon) are contained behind a 20-g bullet in a long barrel at 25 °C and 1 atm. The gunpowder is then ignited and completely burned. As the pressure builds up

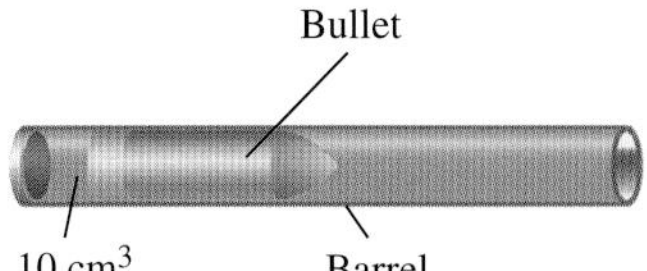

behind the bullet, the bullet accelerates forward. Assume the process occurs so fast that there is no time for any heat interaction with the barrel or with the bullet. The temperature of the products at the time when the bullet leaves the barrel is estimated to be 700 K and the pressure has returned to 1 atm. Determine (i) the initial gunpowder mass and (ii) the bullet velocity leaving the barrel.

12.46–12.49 Constant-volume adiabatic flame temperatures

12.46 A rigid spherical pressure vessel initially contains 0.002 kmol of H_2 and 0.01 kmol of O_2 at 298 K and 1 atm. The mixture is ignited and burns adiabatically. Determine the final temperature and pressure assuming the products contain only H_2O and the excess reactant. Sketch the process in U–T and P–$\mathcal{V}$ coordinates.

12.47 Determine the adiabatic constant-volume flame temperature for a stoichiometric mixture of propane (C_3H_8) and air. The reactants are at 1 atm and 298 K. Assume the air to be a simple mixture of 21% O_2 and 79% N_2. Furthermore, assume complete combustion with no dissociation. Sketch the process in H–T and P–$\mathcal{V}$ coordinates.

12.48 Determine the heat transfer in the constant-volume combustion of 1 kg of carbon indicated in the following reaction:

$$C + 1.5\,O_2 \rightarrow CO_2 + 0.5\,O_2.$$

The reactants are at 298 K and the products are at 500 K.

12.49 A stoichiometric mixture of gaseous octane and oxygen reacts completely at 25 °C in a nonflow, constant-volume process. The initial pressure is 1 atm. Assume all the water in the products is liquid. Determine the heat transfer in MJ per kg of octane.

12.50–12.55 Heating values

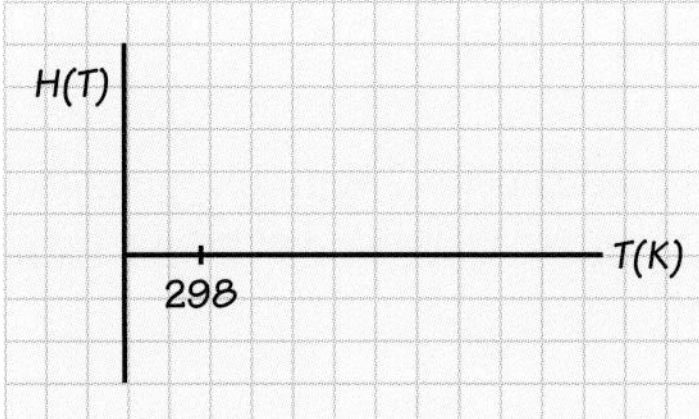

12.50 Consider the combustion of a fuel at constant pressure. Reproduce the coordinate system in the sketch and indicate and appropriately label the following items:

A. H_{reac}
B. H_{prod}
C. Heating value (HV)

12.51 Determine the lower and higher heating values (kJ/kg) of butane at 298 K. Compare your results with the values in Appendix H.

12.52 Determine the lower and higher heating values (kJ/kg) of propane at 298 K. Compare your results with the values in Appendix H.

12.53 Determine the lower and higher heating values (kJ/kg) of acetylene at 298 K. Compare your results with the values in Appendix H.

12.54 Determine the mass of CO_2 produced (kg) per unit of energy released (kJ), i.e., the CO_2 emission, for the complete combustion of the following fuels. The higher heating value (*HHV*) expresses the energy released per mass of fuel burned.

Fuel	Composition	Higher heating value (kJ/kg)
Methane	CH_4	55,528
Natural gas	$C_{1.16}H_{4.32}N_{0.11}$	50,000
Gasoline	$C_{7.9}H_{14.8}$	47,300
Diesel	$C_{12.3}H_{22.2}$	44,800
Ethanol	C_2H_6O	29,700
Coal (Pbgh. #8)	$C_{65}H_{52}NSO_3$	32,550
Hydrogen	H_2	142,000

Rank the fuels from the least CO_2 producing to the greatest. Discuss your results.

12.55 In Problem 12.54, you calculated the CO_2 emission factor for several hydrocarbon fuels. Derive an expression for the CO_2 emission factor for an arbitrary hydrocarbon C_xH_y in which the hydrogen-to-carbon ratio y/x appears explicitly. Assume that all the fuel carbon appears in the products as CO_2 and use "simple" air. The atomic weights of C and H are 12.011 and 1.008, respectively. Use your expression to evaluate four of the same fuels as before:

i. Methane (CH_4)
ii. Propane (C_3H_8)
iii. Gasoline (C_8H_{15}-equivalent)
iv. Diesel fuel ($C_{15}H_{22}$-equivalent)

Plot the CO_2 emission factor as a function of the hydrogen-to-carbon ratio y/x.

12.56–12.75 Steady-flow combustion

12.56 Consider a stoichiometric reaction involving liquid octane and oxygen in a steady-flow reactor. The reactants enter at 25 °C and 1 atm, and the products exit at the same conditions. Assuming liquid water in the products, determine the heat transfer from the reactor in MJ per kg of fuel.

12.57 For a steady-flow stoichiometric reaction of gaseous octane with oxygen, each reactant enters at 25 °C and 1 atm and the products leave at 25 °C and 1 atm. Determine the heat transfer (MJ per kg fuel).

12.58 Ethane flows into a combustion chamber at 1.5 kg/min along with zero percent excess air. The fuel, oxidizer, and products are all at 25 °C. The reaction is complete and the water is liquid. Determine the heat-transfer rate from the combustion chamber (MW).

12.59 A 50–50 blend (by volume) of methane and propane is burned with 100% excess air in a steady-flow combustion chamber. The fuel mixture and the air each enter the combustion chamber at 298 K and 1 atm. Assuming the reaction is complete and exhaust products leave the combustion chamber at 1000 K and 1 atm, determine the heat transfer per mass of fuel burned (MJ/kg).

12.60 Propane is completely burned in a steady-flow process with 100% excess air. The propane and the air each enter the control volume at 25 °C and 1 atm, and the combustion products leave at 600 K and 1 atm. Determine the heat transfer $(\mathrm{MJ/kg_{C_3H_8}})$ to the surroundings.

12.61 A stoichiometric mixture of carbon monoxide and air, initially at 25 °C and 1 atm, reacts completely in an adiabatic, constant-pressure, steady-flow process. Determine the exit temperature (K) of the products.

12.62 Determine the adiabatic flame temperature (K) for a mixture of methane and 200% theoretical air that reacts completely in a steady-flow process at 1 atm. The methane and air enter the reaction at 298 K.

12.63 A cutting torch burns a steady flow of acetylene gas at 25 °C and 1 atm with stoichiometric air at 25 °C and 1 atm. The products are at 1 atm. Assume the reaction is complete (no dissociation) and adiabatic. Determine the exit temperature.

12.64 Hydrogen gas and 100% excess air each enter a steady-flow combustion chamber at 25 °C and 1 atm. A complete reaction occurs with a heat loss of 40 MJ/kmol of fuel. Determine the temperature (K) of the product gas.

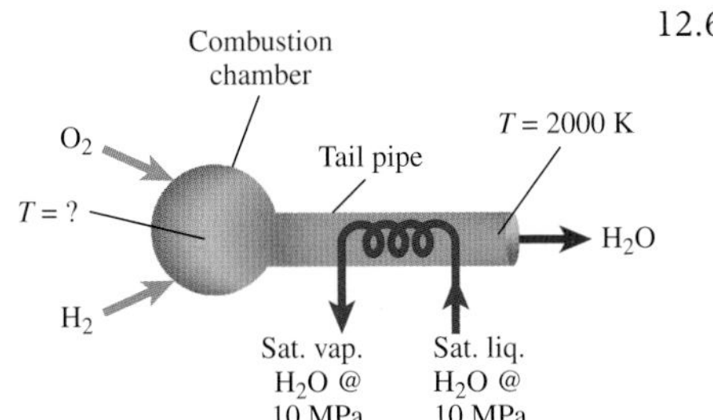

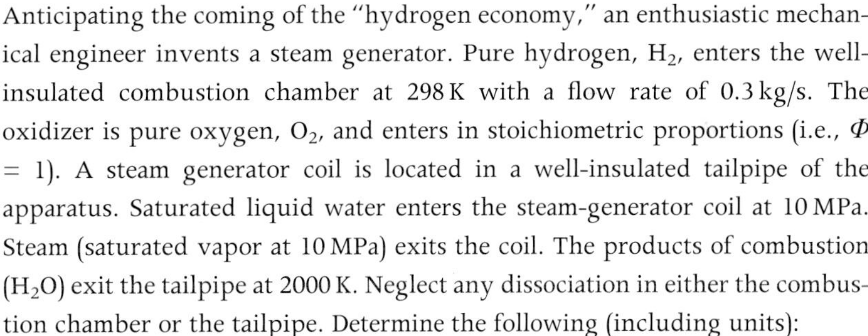

12.65 Anticipating the coming of the "hydrogen economy," an enthusiastic mechanical engineer invents a steam generator. Pure hydrogen, H_2, enters the well-insulated combustion chamber at 298 K with a flow rate of 0.3 kg/s. The oxidizer is pure oxygen, O_2, and enters in stoichiometric proportions (i.e., $\Phi = 1$). A steam generator coil is located in a well-insulated tailpipe of the apparatus. Saturated liquid water enters the steam-generator coil at 10 MPa. Steam (saturated vapor at 10 MPa) exits the coil. The products of combustion (H_2O) exit the tailpipe at 2000 K. Neglect any dissociation in either the combustion chamber or the tailpipe. Determine the following (including units):

A. The mass flow rate of the O_2

B. The maximum temperature in the combustion chamber assuming all the H_2 and O_2 react to form H_2O at constant pressure

C. The heat removed from the product stream in the tailpipe section

D. The mass flow rate of the steam through the steam-generator coil

12.66 Consider the theoretical combustion of methane in a steady-flow process at 1 atm. Determine the heat transfer per kmol and per kg of fuel from the combustion chamber for the following cases:

A. The products and the reactants are at the same temperature of 15.6 °C.

B. The air and the methane enter at 5 °C and 60 °C, respectively, and the products exit at 449 °C.

12.67 Determine the mass air–fuel ratio if gaseous propane (C_3H_8) is burned with air at 1 atm in a steady-flow reactor for the following conditions. The propane and air enter the reactor at 25 °C, and the combustion products must exit at less than 1425 °C.

12.68 A natural gas–fired boiler raises steam for a pulp and paper manufacturing power plant. The air and natural gas enter at 298 K and 1 atm. Water enters the boiler at 305 K and 10 MPa at a flow rate of 38.7 kg/s. Steam exits the boiler/superheater at 780 K and 10 MPa. The boiler efficiency (defined in Example 12.10) is 85%.

A. Determine the mass flow rate of the natural gas (CH_4).

B. Determine the temperature of the products of combustion exiting the boiler. Assume stoichiometric combustion with negligible dissociation.

12.69 Air enters a turbojet engine with a velocity of 150 m/s (at station 1) and a mass flow rate of 23.6 kg/s. Fuel is supplied to the engine at 0.3165 kg/s. The mass-specific standardized enthalpy of the air and fuel are 1.872 kJ/kg$_A$ and −3210.4 kJ/kg$_F$, respectively, and the standardized enthalpy of the combustion products exiting the engine is −129.9 kJ/kg$_e$. Estimate the thrust developed by the engine.

12.70 Redo Example 9.9 using standardized enthalpies for the air, fuel, and products based on the ideal-gas properties given in Appendices D and H. Assume that the fuel is *n*-decane ($C_{10}H_{22}$) and that it enters the combustor at 298 K. Retain the assumption of simple air (1 kmol of O_2 per 3.76 kmol of N_2) and assume complete combustion with no dissociation.

12.71 Show that Eq. 12.25 can be expressed equivalently on (a) a per unit mass of mixture basis and (b) a per unit mass of fuel basis.

12.72 Consider the turbojet combustor described in Example 12.8. Determine the temperature of the combustion products exiting the combustor if the air–fuel ratio is reduced to 50:1.

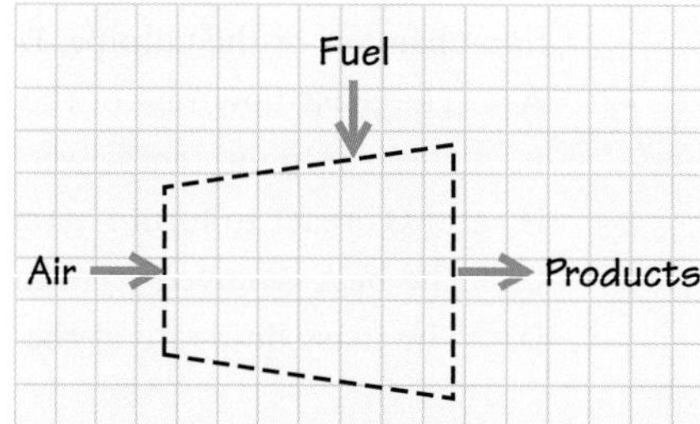

12.73 The maximum temperature that the turbine blades can withstand limits the performance of a turbojet engine. This temperature, in turn, is controlled by the temperature of the combustion products exiting the combustor. Determine the minimum air–fuel ratio that can be supplied to a turbojet engine such that the combustor outlet temperature does not exceed 1200 K (1700 F) when the air enters at 400 K. Assume the fuel is $C_{10}H_{22}$ and enters the combustor at 298 K. Also assume the usual simplified composition of air.

Aircraft turbine blades. Photograph courtesy of NASA.

12.74 Create a spreadsheet model of a turbojet combustor in which the air temperature, fuel temperature, and air–fuel ratio are input quantities, and the combustion product temperature is an output quantity. Use iso-octane (C_8H_{18}) and simple air as the reactants.

12.75 Consider a gas-turbine engine. Air enters the compressor at 298 K and 1 atm with a flow rate of 685 kg/s. The compressor has a pressure ratio of 18 and an isentropic efficiency of 0.85. The air from the compressor enters the combustion chamber where natural gas (CH_4) at 298 K is injected, mixed, and burned with the air. The mass air–fuel ratio is 50. The hot products of combustion expand through a turbine back to 1 atm. The isentropic efficiency of the turbine is 0.92.

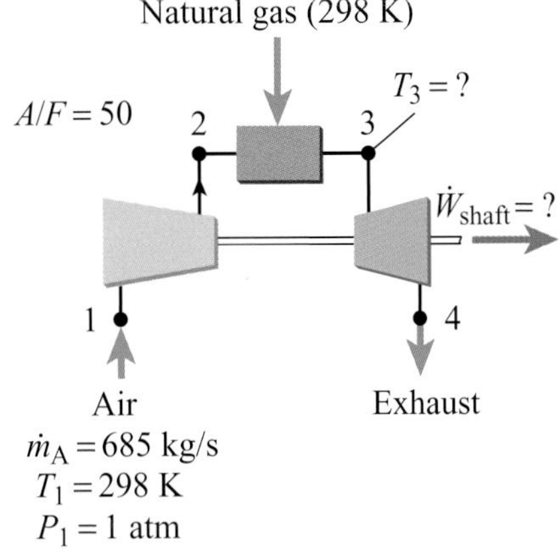

Solve parts A and B below using the following assumptions:

i. Combustion is "complete" (i.e., the only product species are CO_2, H_2O, O_2, and N_2).

ii. The air has a simplified composition with a molar O_2-to-N_2 ratio of 1:3.76 and a specific-heat ratio γ of 1.4.

iii. All species behave as ideal gases with properties obtained from Appendix D.

iv. The combustor is adiabatic and all potential and kinetic energy changes are negligible.

A. Determine the temperature of the gases at the outlet of the combustor. (Note: This is also the turbine inlet temperature.)

B. Determine the shaft power delivered by the engine.

C. Determine the engine process thermal efficiency using the product of the fuel flow rate and the fuel higher heating value as the input energy rate.

12.76–12.78 Combustion products dew-point temperature

12.76 Ethane burns with 150% stoichiometric air. Assume complete combustion with no dissociation. Determine (a) the mole percentage of each product species and (b) the dew-point temperature of the products at 1 atm.

12.77 Rework Problem 12.76 but use propane as the fuel.

12.78 Consider the combustion of benzene, C_6H_6, with air. Determine the dew-point temperature of the combustion products if the mass air–fuel ratio is 20:1. The pressure is 1 atm.

CHAPTER 13
Chemical and Phase Equilibrium

OVERALL LEARNING OBJECTIVES

After studying Chapter 13, you should:

- Be able to explain how the second law of thermodynamics relates to chemical and phase equilibrium.
- Be able to calculate equilibrium constants from tabulations of the Gibbs function of formation data.
- Be able to calculate the equilibrium composition of systems involving a single equilibrium reaction.
- Be able to calculate the equilibrium composition of systems involving two or more equilibrium reactions.

CHAPTER 13 OVERVIEW

THE CHAPTER EXAMINES the principles of chemical equilibrium and phase equilibrium as extensions of the second law. We revisit entropy and define two other second-law properties: the Gibbs function and the Helmholtz free energy. The chapter explores how equilibrium relates to these three properties. The chapter focuses on the conditions of fixed temperature and pressure to explore chemical equilibrium. The equilibrium constant is defined and used to determine the detailed composition of a system. Simple, single, equilibrium reactions (dissociations) are investigated. The equilibrium constant approach is extended to multiple equilibria. The chapter also develops how minimization of the Gibbs function establishes the conditions for liquid–vapor (nonreacting) equilibrium.

<table>
<tr><td colspan="3">DESIGN AND ANALYSIS OF SYSTEMS</td></tr>
<tr><td colspan="3">ANALYSIS OF PRACTICAL DEVICES</td></tr>
<tr><td>CONSERVATION OF MASS</td><td>CONSERVATION OF ENERGY (First Law of Thermodynamics)</td><td>SECOND LAW OF THERMODYNAMICS (Entropy)</td></tr>
<tr><td colspan="3">PROPERTIES OF MATTER</td></tr>
<tr><td colspan="3">FRAMEWORKS FOR ANALYSIS, KEY CONCEPTS, AND DEFINITIONS</td></tr>
</table>

FIGURE 13.1 Chapter 13 expands upon the second law of thermodynamics and introduces new thermodynamics properties and a few new concepts. For our purposes, these ideas allow us to determine the chemical and phase equilibrium composition of simple compressible substances. Although not highlighted in this figure, we will employ the conservation of mass (element conservation) and conservation of energy in our developments.

Historical Context

Josiah Willard Gibbs (1839-1903) physicist. Photograph undated, retouched (Bettmann / Contributor / Getty Images).

The evolution of the modern concept of chemical equilibrium has a long history. Lindauer [1] indicates that the development of the concept of chemical affinity – the tendency of substances to react – was the first step in its evolution. The concept of affinity can be traced back to **Hippocrates** (*ca.*470–*ca.*370 BCE) and later to **Albertus Magnus** (*ca.* 1200–1280) [1]. **Torbern Olof Bergman** (1735–1784) studied displacement reactions of the type AB + C → AC + B and ranked substances according to affinities [1]. For example, in this arbitrary reaction, substance C has a greater affinity than substance B. The idea that reactions are influenced by the quantity (concentration) of the reactants is generally attributed to **Claude Louis Berthollet** (1748–1822). This idea led Berthollet to the concept of chemical forces, a combination of reaction affinity and the quantity of reactants. Bertholett wrote [2] that a reaction ends when there is "equilibrium of contending forces." This is quite close to the modern idea that equilibrium exists when the forward and reverse rates of a reaction are equal. The earliest expressions of the law of mass action – that the reaction velocity is the product of an affinity constant and the concentrations of the reactants raised to powers – were first developed by **Cato Maximilian Guldberg** (1833–1902) and **Peter Waage** (1839–1900) [3]. Major contributors to understanding the thermodynamics of reacting systems were **Pierre Eugene Marcelin Berthelot** (1827–1907) and **Hans Peter Jørgen Julius Thomsen** (1826–1909), cited as the founders of thermochemistry by Lindauer [1]; and **Jacobus Hendricus van't Hoff** (1852–1911) and **Josiah Willard Gibbs** (1839–1903). Lindauer [1] writes . . . *our present concept of chemical equilibrium differs little from the form in which it was presented in van't Hoff's "Studies in Chemical Dynamics," which was published in 1884*. Gibbs, whose mathematical expression of equilibrium we present in this chapter, is noted for his many contributions to science – not only chemical thermodynamics, but also statistical mechanics, vector calculus, and optics [4].

13.1 Thermodynamic Equilibrium Revisited

In Chapter 1, we introduced the concept of thermodynamic equilibrium. In subsequent chapters, we utilized this concept in developing and applying the first and second laws of thermodynamics. Our definition of thermodynamic equilibrium requires that a system has no unbalanced potentials or drivers to promote a change of state. The state of a system in thermodynamic equilibrium remains unchanged for all time. For a system to be in thermodynamic equilibrium, we require that the system be simultaneously in thermal equilibrium, mechanical equilibrium, chemical equilibrium, and phase equilibrium.

We quickly review these four conditions: **Thermal equilibrium** exists when a system both has a uniform temperature and is at the same temperature as its surroundings. **Mechanical equilibrium** occurs when the pressure throughout the system is uniform and there are no unbalanced forces at the system boundaries. Another condition for thermodynamic equilibrium requires the system to be in **chemical equilibrium.** For chemical equilibrium to prevail, the mole fractions

Hydrogen is produced by the steam reforming of natural gas. A key step in the process involves the equilibrium reaction $CO + H_2O \leftrightharpoons CO_2 + H_2$. (Photograph courtesy of Linde AG.)

(or any other composition variable) are unchanging with time. **Phase equilibrium** relates to conditions in which a substance can exist in more than one physical state, i.e., any combination of vapor, liquid, and solid. Phase equilibrium requires that the amount of a substance in any one phase not change with time.

In this chapter, we explore the principles governing chemical and phase equilibrium and establish ways to calculate equilibrium compositions of multi-species gas mixtures and multi-phase substances. Applications of these equilibria abound (see sidebar photo for one example).

13.2 The Second Law and Equilibrium

In Chapter 7, we introduced entropy and two other second-law-based properties. In Chapter 6, we indicated that the second law could be recast as a series of statements defining conditions for equilibrium in terms of these three properties. Our development of chemical and phase equilibrium in this chapter is based on these concepts.

These second-law statements are mathematically expressed, respectively, in terms of entropy (S), the Gibbs free energy (G), and the Helmholtz free energy (A), as follows:

$$(dS)_{u,\mathrm{v}} \geq 0, \tag{13.1a}$$

Equations with yellow backgrounds express key concepts and are the most important relationships in the chapter.

$$(dG)_{T,P} \leq 0, \tag{13.1b}$$

and

$$(dA)_{T,\mathrm{v}} \leq 0. \tag{13.1c}$$

The subscripts refer to the properties that are held fixed during the establishment of equilibrium (Fig. 13.2). The **Gibbs free energy,** or **Gibbs function,** is defined in terms of other thermodynamic properties as

$$G \equiv H - TS, \tag{13.2}$$

and the **Helmoltz free energy** is defined as

See Section 2.3c in Chapter 2.

$$A \equiv U - TS. \tag{13.3}$$

These three equilibrium criteria (Eqs. 13.1a–c) state, respectively, that (i) equilibrium occurs when the entropy is a maximum for a simple system at constant internal energy and volume, (ii) equilibrium occurs when the Gibbs free energy is a minimum for a simple system at constant pressure and temperature, and (iii) equilibrium occurs when the Helmholtz free energy is a minimum for a simple system at constant temperature and volume. Furthermore, these three statements are very general statements of equilibrium and apply to pure substances, solutions, and reacting and nonreacting mixtures. The only restriction is that they apply to simple thermodynamic

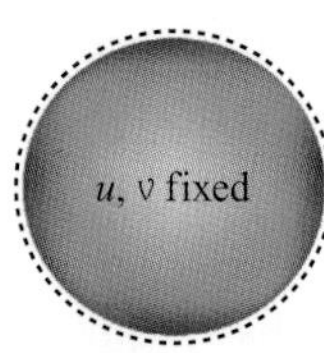

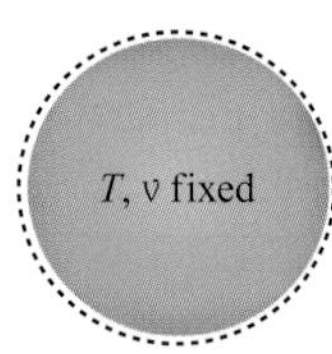

FIGURE 13.2 Criteria for equilibrium depend on which pair of properties is held constant. For many practical applications, fixed temperature and pressure is the most useful case.

systems, that is, systems unaffected by gravity, magnetic forces, electrical forces, and surface forces other than normal forces (see Chapter 2).

For many practical applications, the Gibbs function criterion (Eq. 13.1b) is the most useful for equilibria. In later sections, we focus on the application of this to chemical and phase equilibrium, with particular attention given to combustion applications and to liquid–vapor phase equilibrium. However, before working with the Gibbs function, a new property, we explore the entropy-based criterion because entropy is a familiar property from previous chapters.

13.3 Equilibrium for Conditions of Fixed Internal Energy and Volume

13.3a General Considerations

Before dealing specifically with chemical equilibrium, we want to illustrate the connections between the increase in entropy principle discussed in Chapter 7, i.e.,

$$dS_{\text{isolated}} \geq 0, \tag{13.4}$$

and the equilibrium criterion Eq. 13.1a,

$$(dS)_{u,\text{v}} \geq 0. \tag{13.1a}$$

We note the subscripts u and v denote that any changes to the system under consideration occur at both constant internal energy and constant volume. To maintain the internal energy constant requires either that any heat and work interactions cancel ($\delta Q = \delta W$) or that both are zero ($\delta Q = \delta W = 0$). Since the volume is constant, no moving boundary work is possible (i.e., $\delta W = Pd\mathcal{V} = 0$); we therefore determine that neither heat nor work is allowed. The absence of these interactions is our definition of an isolated system. Thus, we see that Eqs. 13.4 and 13.1a are equivalent. From the perspective of equilibrium, Eq. 13.1a indicates that an isolated system will proceed to an equilibrium state only by processes that involve an increase of entropy. Thus, we conclude that the equilibrium state must be a state of maximum entropy. Similar arguments can be applied to the equilibrium criteria given by Eqs. 13.1b and 13.1c.

13.3b Application to Chemical Equilibrium

Chemical equilibrium is important in any process in which chemical reactions take place. Criteria for chemical equilibrium provide no information about the rate of reaction but, rather, determine the final state given a set of initial reactants and the two intensive properties held constant for the process. Defining the final state in reacting systems requires determining the product composition (i.e., the values of the various species mole fractions).

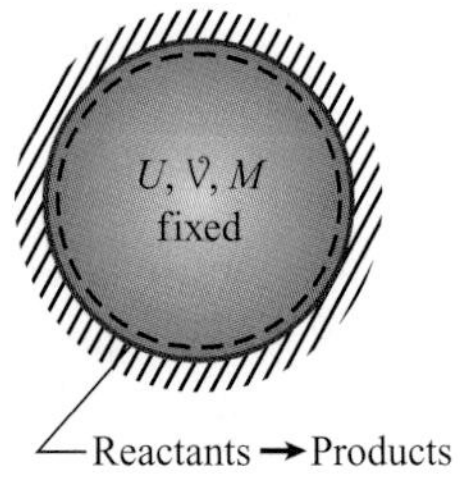

FIGURE 13.3 Chemical equilibrium occurs at a state of maximum entropy for an isolated system.

To apply Eq. 13.1a, we consider a fixed-volume adiabatic reaction vessel in which a fixed mass of reactants forms products (Fig. 13.3). As the reactions proceed, both the temperature and pressure rise until a final equilibrium condition is reached. We apply the second-law statement (Eq. 13.1a) in combination with the first law to determine this final state (i.e., its temperature, pressure, and composition). It is the addition of the composition variable that makes the present analysis different from those for non-reacting systems. Rather than work in general terms, we will illustrate this problem by considering a single specific reaction. Carbon monoxide burns in oxygen to form carbon dioxide via the reaction

$$\mathrm{CO} + \tfrac{1}{2}\mathrm{O}_2 \rightarrow \mathrm{CO}_2. \tag{13.5}$$

The CO_2 will dissociate at typical product temperatures for adiabatic reactions. We assume that the products consist only of CO_2, CO, and O_2, recognizing that the O_2 molecule may also dissociate. (We deal with that in a later section.) With this assumption, we write

$$\underset{\text{reactants}}{\left[\mathrm{CO} + \tfrac{1}{2}\mathrm{O}_2\right]_{\text{cold}}} \rightarrow \underset{\text{products}}{\left[(1-\alpha)\mathrm{CO}_2 + \alpha\mathrm{CO} + \tfrac{1}{2}\alpha\mathrm{O}_2\right]_{\text{hot}}}, \tag{13.6}$$

where α is the fraction of the CO_2 that has dissociated. We can calculate the adiabatic flame temperature as a function of the dissociated fraction α using the procedures developed in Chapter 12. For example, with $\alpha = 1$, no chemical bonds are broken or formed; thus the mixture temperature, pressure, and composition remain unchanged. With complete combustion (i.e., no dissociation and $\alpha = 0$), the maximum conversion of chemical bond energy to sensible energy occurs, and the temperature and pressure would be the highest possible allowed by the first law. The adiabatic temperature as a function of the dissociated fraction α is shown in Fig. 13.4.

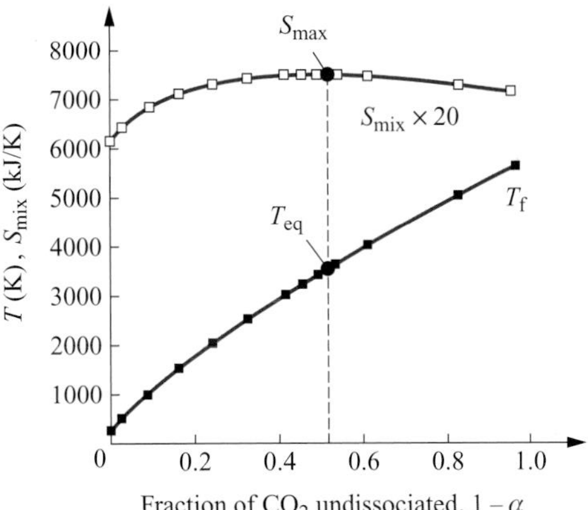

FIGURE 13.4 We determine the equilibrium composition of a chemically reacting, isolated, fixed-mass system by locating the condition of maximum entropy. Adapted from Ref. [8] © McGraw-Hill Education.

Our next step is to see how the second law constrains the dissociated fraction, and, hence, the adiabatic temperature. Our criterion (Eq. 13.1a) states that equilibrium will be achieved at the condition where the system entropy is a maximum. We can determine this maximum by calculating the system entropy as a function of the dissociated fraction. The system entropy is calculated by summing the entropies of the three components (CO_2, CO, and O_2):

$$S_{\text{mix}}(T_{\text{f}}, P) = \sum_{i=1}^{3} N_i \bar{s}_i(T_{\text{f}}, P_i) = (1-\alpha)\bar{s}_{\text{CO}_2} + \alpha\bar{s}_{\text{CO}} + \tfrac{1}{2}\alpha\bar{s}_{\text{O}_2}, \tag{13.7}$$

where N_i is the number of moles of species i in the mixture. As discussed in Chapter 7, the individual species entropies can be expressed as

See Eq. 7.10a in Chapter 7.

$$\bar{s}_i = \bar{s}_i^{\circ}(T_{\text{ref}}) + \int_{T_{\text{ref}}}^{T_f} \bar{c}_{p,i} \frac{dT}{T} - R_u \ln \frac{P_i}{P^{\circ}}, \tag{13.8}$$

where ideal-gas behavior is assumed and P_i is the partial pressure of the ith species. The results of our calculations for several values of the dissociated fraction are shown in Fig. 13.4. Here we see that the maximum entropy occurs near $1 - \alpha = 0.5$. From this we can determine the number of moles of each constituent ($N_{CO_2} = 1 - 0.5 = 0.5$; $N_{CO} = 0.5$; and $N_{CO_2} = 0.5/2 = 0.25$) and the corresponding mole fractions ($X_{CO_2} = 0.5/1.25 = 0.40$; $X_{CO} = 0.5/1.25 = 0.40$; and $X_{CO_2} = 0.25/1.25 = 0.20$). We can also see that the equilibrium adiabatic temperature lies between 3000 K and 4000 K. To determine the equilibrium state pressure, we can apply the ideal-gas law with our knowledge now of the temperature and composition.

We now review how the second law is at the heart of this example. The statement that $(dS)_{u,v} \geq 0$, or more precisely that

$$\left(\frac{dS}{d\alpha}\right)_{u,v} = 0, \tag{13.9}$$

caused us to choose the state of maximum entropy as the equilibrium state, as indicated in Fig. 13.4. The idea that spontaneous change occurs in the direction of increasing entropy requires that the composition of the system shift toward the point of maximum entropy when approaching from either side, since dS is positive. Once the maximum entropy is reached, no further change in composition is allowed, since this would require the system entropy to decrease, in violation of the second law (Eq. 13.1a).

13.4 Chemical Equilibrium for Conditions of Fixed Temperature and Pressure

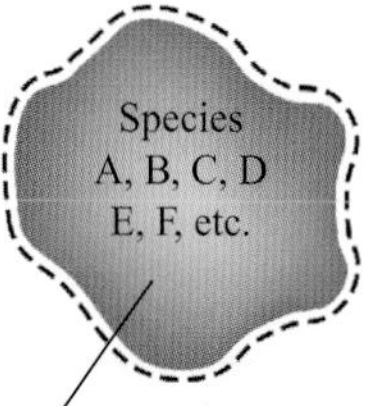

FIGURE 13.5 In a system at a fixed temperature and pressure, species mole fractions are in equilibrium when the system Gibbs free energy is a minimum.

As indicated previously, the criteria for chemical equilibrium provide no information about the rate of reaction but, rather, determine the final state given a set of initial reactants and the two intensive properties that are to be held constant for the process. In this section, we apply the criterion for equilibrium when the two intensive properties held constant are temperature and pressure. For most practical systems, determining equilibrium composition at a fixed temperature and pressure is the most useful choice of intensive properties.

13.4a Determining the Equilibrium State

APPLYING CRITERION FOR EQUILIBRIUM

For conditions of fixed temperature and pressure, we apply the second law as expressed by Eq. 13.1b, $(dG)_{T,P} \leq 0$, to determine the equilibrium composition (Fig. 13.5). The Gibbs function attains a minimum in equilibrium, in contrast to the maximum in entropy we saw for the fixed-energy and fixed-volume case (Fig. 13.4). This minimization is mathematically expressed as

$$(dG)_{T,P} = 0, \tag{13.10}$$

As an example, consider the dissociation of CO_2 discussed in the previous section. For this situation, we obtain the value of the dissociated fraction α such that the mixture Gibbs function is a minimum by the application of

$$\left(\frac{dG}{d\alpha}\right)_{T,P} = 0.$$

Figure 13.6 shows the Gibbs function versus the dissociated fraction for a temperature of 2500 K and 1 atm. The dissociated fraction at equilibrium (i.e., at the minimum G value) is approximately 0.13 and the corresponding mole fractions are $X_{CO_2} = 0.82$, $X_{CO} = 0.12$, and $X_{O_2} = 0.06$.

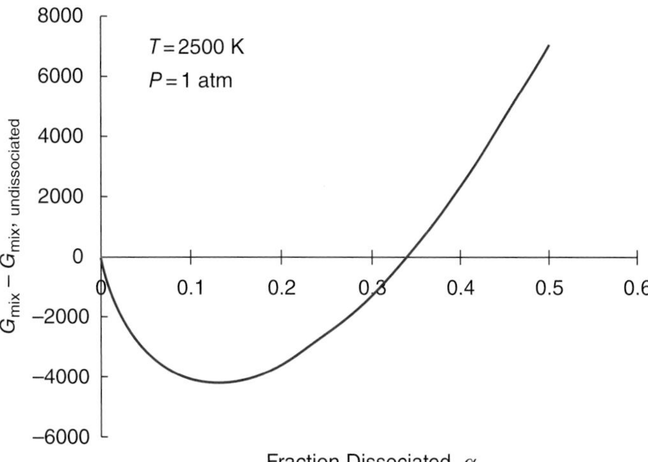

FIGURE 13.6 Chemical equilibrium for conditions of fixed T and P is determined by minimizing the Gibbs function. For the reaction $CO_2 \Leftrightarrow CO + 0.5O_2$, the minimum Gibbs function occurs at a dissociated fraction of approximately 0.13.

In the development that follows, we consider only reacting mixtures of ideal gases, which is a reasonable restriction for many engineering applications, including combustion. For more general treatments of chemical equilibrium, we refer the reader to other textbooks [5, 6].

STANDARD-STATE GIBBS FUNCTION CHANGE AND EQUILIBRIUM CONSTANT

We begin by considering the following general equilibrium reaction among species A, B, C, etc.:

$$a\text{A} + b\text{B} + \cdots \Leftrightarrow e\text{E} + f\text{F} + \cdots \tag{13.11}$$

The Gibbs function for the mixture containing the species A, B, C, etc. can be expressed as

$$G_{\text{mix}} = \sum N_i \mu_{i,T}, \tag{13.12}$$

where N_i, is the number of moles of the ith species and $\mu_{i,T}$ is the chemical potential of the ith species at the temperature of interest. For a mixture of *ideal gases*, the **chemical potential** for the ith species is simply the Gibbs function per mole of i:

$$\mu_{i,T} = \bar{g}_{i,T} = \bar{g}^{\circ}_{i,T} + R_u T \ln(P_i/P^{\circ}), \tag{13.13}$$

where $\bar{g}^{\circ}_{i,T}$ is the Gibbs function of the *pure* species at the standard-state pressure (i.e., $P_i = P^{\circ}$), and P_i is the partial pressure. We define the standard-state pressure P° to be 1 atm, which, as we will see later, greatly simplifies our calculations.

Before proceeding in our development, we introduce the **Gibbs function of formation**, $\Delta\bar{g}^{\circ}_{f,i}$, which has great utility in equilibrium calculations:

$$\Delta\bar{g}^{\circ}_{f,i}(T) \equiv \bar{g}^{\circ}_{i}(T) - \sum v'_j \bar{g}^{\circ}_{j}(T), \tag{13.14}$$

where the v'_j are the stoichiometric coefficients of the elements required to form one mole of the compound of interest. For example, in the formation of one mole of CO from naturally occurring carbon and molecular oxygen, i.e., $C + \frac{1}{2}O_2 \rightarrow CO$, the coefficients are $v'_{O_2} = \frac{1}{2}$ and $v'_C = 1$. As with enthalpies of formation, the Gibbs functions of formation of the naturally occurring elements are assigned values of zero at the reference state. Appendix D provides tabulations of Gibbs functions of formation over a range of temperatures for many species in the C–H–O–N system. Tabulations of Gibbs functions of formation as a function of temperature greatly simplify the calculation of equilibrium constants, as developed below. In addition to those in Appendix D, tabulations for over 1000 species can be found in the JANAF tables [7].

With these definitions, we proceed in our development of how to determine the equilibrium state of a reacting system at fixed temperature and pressure. Substituting Eq. 13.13 into Eq. 13.12 yields

$$G_{mix} = \sum N_i \bar{g}_{i,T} = \sum N_i \left[\bar{g}^{\circ}_{i,T} + R_u T \ln(P_i/P^{\circ})\right]. \tag{13.15}$$

We now apply the equilibrium criterion $(dG)_{T,P} = 0$ (Eq. 13.10) to Eq. 13.15; thus,

$$\sum dN_i \left[\bar{g}^{\circ}_{i,T} + R_u T \ln(P_i/P^{\circ})\right] + \sum N_i\, d\left[\bar{g}^{\circ}_{i} + R_u T \ln(P_i/P^{\circ})\right] = 0. \tag{13.16}$$

For conditions of constant temperature and pressure, the second term in Eq. 13.16 is zero. This results from noting that $d(\ln P_i) = dP_i/P_i$ and, furthermore, that $\sum dP_i = 0$, because all changes in the partial pressures must sum to zero as the total pressure is constant. Thus,

$$dG_{mix} = 0 = \sum dN_i \left[\bar{g}^{\circ}_{i,T} + R_u T \ln(P_i/P^{\circ})\right]. \tag{13.17}$$

From the chemical system described by Eq. 13.11, the change in the number of moles of each species is directly proportional to its stoichiometric coefficient, so we have

$$\begin{aligned} dN_A &= -\kappa a, & dN_B &= -\kappa b, \ \ldots \\ dN_E &= +\kappa e, & dN_F &= +\kappa f, \ \ldots \end{aligned} \tag{13.18}$$

We can demonstrate that these relationships hold by considering the following simple specific equilibrium reaction: $1O_2 \Leftrightarrow 2O$. Assume for the sake of argument that there are 10 moles of O_2 in our reacting mixture and that the only other species is the O atom. We now force a change of the number of O_2 moles from ten to seven (i.e., $dN_{O_2} = -3$). Because the total number of atoms must be conserved, the number of moles of O atoms must increase by a factor of six (i.e., $dN_O = +6$); every time one mole of O_2 disappears, two moles of O atoms are generated. This result is consistent with the requirement of Eq. 13.18 that $-dN_A/a = +dN_E/e = \kappa$ (i.e., $-dN_{O_2}/1 = +dN_O/2$ or $[-(-3)/1 = +6/2]$).

We substitute Eq. 13.18 into Eq. 13.17 and cancel the proportionality constant κ to obtain

$$\begin{aligned} &-a\left[\bar{g}^\circ_{A,T} + R_u T \ln\left(P_A/P^\circ\right)\right] - b\left[\bar{g}^\circ_{B,T} + R_u T \ln\left(P_B/P^\circ\right)\right] - \cdots \\ &+e\left[\bar{g}^\circ_{E,T} + R_u T \ln\left(P_E/P^\circ\right)\right] + f\left[\bar{g}^\circ_{F,T} + R_u T \ln\left(P_F/P^\circ\right)\right] + \cdots = 0. \end{aligned} \tag{13.19}$$

We rearrange Eq. 13.19 and group the natural log terms to yield

$$\begin{aligned} &-\left(e\bar{g}^\circ_{E,T} + f\bar{g}^\circ_{F,T} + \ldots - a\bar{g}^\circ_{A,T} - b\bar{g}^\circ_{B,T} - \cdots\right) \\ &= R_u T \ln \frac{(P_E/P^\circ)^e (P_E/P^\circ)^f \cdots}{(P_A/P^\circ)^a (P_B/P^\circ)^b \cdots}. \end{aligned} \tag{13.20}$$

We now define the **standard-state Gibbs function change** ΔG°_T

$$\Delta G^\circ_T \equiv \left(e\bar{g}^\circ_{E,T} + f\bar{g}^\circ_{F,T} + \cdots - a\bar{g}^\circ_{A,T} - b\bar{g}^\circ_{B,T} - \cdots\right), \tag{13.21a}$$

which appears on the left-hand side of Eq. 13.20. The standard-state Gibbs function change can also be conveniently expressed in terms of the Gibbs function of formation at the temperature T of interest:

$$\Delta G^\circ_T \equiv \left(e\Delta\bar{g}^\circ_{f,E}(T) + f\Delta\bar{g}^\circ_{f,F}(T) + \cdots - a\Delta\bar{g}^\circ_{f,A}(T) - b\Delta\bar{g}^\circ_{f,B}(T) - \cdots\right). \tag{13.21b}$$

The argument of the natural logarithm in Eq. 13.20 is defined as the **equilibrium constant** K_p:

$$K_P \equiv \frac{(P_E/P^\circ)^e (P_F/P^\circ)^f \cdots}{(P_A/P^\circ)^a (P_B/P^\circ)^b \cdots}. \tag{13.22}$$

Chemical equilibrium at constant temperature and pressure can now be succinctly expressed by the following:

$$\Delta G^\circ_T = -R_u T \ln K_p, \tag{13.23a}$$

or

$$K_p = \exp\left(-\Delta G^\circ_T / R_u T\right). \tag{13.23b}$$

It is important to remind the reader that Eq. 13.23, a quite simple relationship, resulted from the second-law statement that chemical equilibrium for conditions of fixed temperature and pressure is achieved when the Gibbs function for the system is a minimum. In solving problems involving chemical equilibrium, **element**

conservation is usually required in conjunction with the second-law concepts just discussed. The following simple example illustrates how to apply Eqs. 13.23a, b and element conservation to the problem of finding the equilibrium composition of a gas mixture at a fixed temperature and pressure.

Example 13.1 Dissociation of Oxygen

At high temperatures, oxygen molecules dissociate to produce O atoms via the reaction

$$O_2 \Leftrightarrow O + O.$$

Determine the equilibrium partial pressures and mole fractions of O and O_2 at 1 atm and 3000 K.

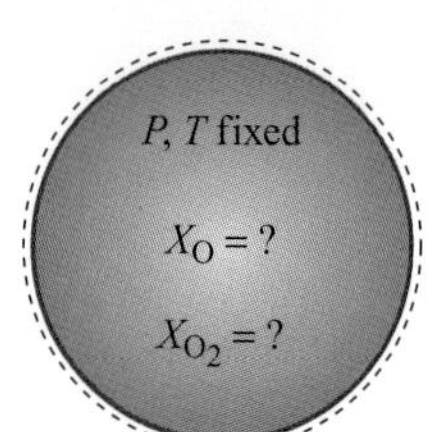

Solution

Known Equilibrium, P, T

Find $P_O, P_{O_2}, X_O, X_{O_2}$

Modeling, Premises and Assumptions Ideal-gas behavior

Analysis To solve this problem, we perform the following sequence of calculations: First, we calculate the equilibrium constant K_p defined in Eq. 13.23b using thermodynamic property data from Appendix D; second, we use this K_p value to calculate the partial pressures P_O and P_{O_2}, using Eq. 13.22; and last, we relate these partial pressures to their corresponding mole fractions (Eq. 10.14).

Using Tables D.11 and D.12 we evaluate the standard-state Gibbs function change (Eq. 13.21b) for the dissociation of O_2 at 3000 K as

$$\begin{aligned}\Delta G_T^\circ &= 2\Delta \bar{g}^\circ_{f,O}(T) - 1\Delta \bar{g}^\circ_{f,O_2}(T) \\ &= [2(54{,}554) -, 1(0)]\,\text{kJ/kmol} \\ &= 109{,}108\,\text{kJ/kmol}.\end{aligned}$$

From Eq. 13.23b, we now evaluate K_p:

$$\begin{aligned}K_p &= \exp\left[-\Delta G_T^\circ / R_u T)\right] \\ &= \exp\left[\frac{-109{,}109\,\text{J/kmol}}{8.314\,(\text{kJ/kmol·K})(3000\,\text{K})}\right] = 0.0126.\end{aligned}$$

Note that K_p must be dimensionless because the argument of an exponential cannot have dimensions. The partial pressures relate to K_p (Eq. 13.22) as follows:

$$K_p = \frac{(P_O/P^\circ)^2}{(P_{O_2}/P^\circ)} = \frac{P_O^2}{P_{O_2}}\left(\frac{1}{P^\circ}\right).$$

Recognizing that the partial pressures sum to the total pressure (Eq. 10.10), we have

$$P_{\text{tot}} = P = P_O + P_{O_2},$$

or

$$P_{O_2} = P - P_O.$$

Substituting this expression for P_{O_2} into our K_p definition (Eq. 13.22) yields

$$K_p = \frac{P_O^2}{P - P_O}\left(\frac{1}{P^\circ}\right),$$

or

$$P_O^2 + K_p P^\circ P_O - K_p P^\circ (P) = 0.$$

For a total pressure P of 1 atm and a reference pressure of 1 atm, this becomes

$$P_O^2 + 0.0126(1\,\text{atm})P_O - 0.0126(1\,\text{atm})(1\,\text{atm}) = 0.$$

We find the useful root of this quadratic equation to be

$$P_O = 0.1061 \text{ atm}.$$

Thus,

$$\begin{aligned} P_{O_2} &= P - P_O = 1 \text{ atm} - 0.1061 \text{ atm} \\ &= 0.8939 \text{ atm}. \end{aligned}$$

For ideal-gas mixtures, the component mole fractions equal the ratio of each constituent partial pressure to the total pressure (Eq. 10.11). Thus, the corresponding mole fractions are

$$X_O = P_O/P = 0.1061\,\text{atm}/1\text{atm} = 0.1061$$

and

$$X_{O_2} = P_{O_2}/\text{P} = 0.8939\,\text{atm}/1\,\text{atm} = 0.8939.$$

Comment We note how the evaluation of $\Delta G_T^\circ$was easily accomplished by using the Gibbs function of formation at 3000 K. We also note that element conservation was not explicitly treated but was implicit in our statement that the atomic oxygen (O) and molecular oxygen (O_2) partial pressures equaled the total pressure.

INTERPRETING THE EQUILIBRIUM CONSTANT

From the definition of K_p (Eq. 13.22) and its relation to ΔG_T° (Eq. 13.23), we can obtain a qualitative indication of whether a particular reaction (expressed by Eq. 13.11) favors products (goes strongly to completion) or favors reactants (undergoes very little reaction) at equilibrium. Consider the case of a positive value of the standard-state Gibbs function change, i.e., ($\Delta G_T^\circ > 0$). From Eq. 13.23b, $K_p = \exp\left(-\Delta G_T^\circ / R_u T\right)$, we see that then K_p must be less than unity because the argument of the exponential, $-\Delta G_T^\circ / R_u T$, will always be negative for this case, as ΔG_T°, R_u, and T are all positive quantities. (Recall that the exponential function is unity when its argument is zero ($\exp x = 1$, $x = 0$), larger than unity when its argument is positive ($\exp x > 1$, $x > 0$), and smaller than unity when its argument is negative ($\exp x < 1$, $x < 0$). From the definition of K_p (Eq. 13.22),

$$K_P \equiv \frac{(P_E/P^\circ)^e (P_F/P^\circ)^f \cdots}{(P_A/P^\circ)^a (P_B/P^\circ)^b \cdots},$$

we conclude that reactants will be favored when K_p is less than unity. We can similarly demonstrate that when the value of the standard-state Gibbs-function change is negative, i.e., $\Delta G_T^\circ < 0$, the equilibrium constant will be greater than unity, and hence, products will be favored. Large values of K_p indicate that a reaction will go strongly to completion, with products dominating the composition at equilibrium.

We can obtain physical insight into this behavior by appealing to the definition of ΔG in terms of the enthalpy and entropy changes associated with the reaction. From Eq. 13.2, we can write

$$\Delta G_T^\circ = \Delta H^\circ - T\Delta S^\circ,$$

which can be substituted into Eq. 13.23b to yield

$$K_p = e^{-\Delta H^\circ/R_u T} e^{\Delta S^\circ/R_u}. \quad \textbf{(13.24)}$$

For K_p to be greater than unity, the condition that favors products, the enthalpy change for the reaction, ΔH°, should be negative. In addition, the entropy change, ΔS°, should be positive. Such a reaction is exothermic; the system energy is lowered in going from reactants to products and the entropy increases. (Recall Fig. 12.11 illustrating the enthalpy change for a combustion reaction, which, of course, is highly exothermic and with a large increase in entropy.)

Consider the reverse reaction to that presented in Example 13.1 (i.e., $O + O \Leftrightarrow O_2$). Evaluate K_p through Eq. 13.23b and develop an expression for the partial pressure as in Eq. 13.22. Compare these results to those from Example 13.1 and comment on them.

(Answer: $K_{p,rev} = 79.398 = (P_{O_2}/P_O^2)/\,(1/P^\circ)$. Note that the equilibrium constant for this reverse reaction is the reciprocal of the forward reaction; for this reverse reaction, $K_{p,\text{rev}} = 1/K_{p,\text{forward}}$. As we will see below, the large value of $K_{p,rev}$ indicates that this reaction strongly favors the products, in this case, O_2.)

13.4b Factors Affecting Chemical Equilibrium

TEMPERATURE

Although we commonly refer to the equilibrium *constant*, it is anything but a constant with respect to temperature. To remind the reader of this temperature dependence, we frequently denote the equilibrium constant as $K_p\,(T)$. We can explore the origins of this temperature dependence by analyzing Eq. 13.23b, $K_p(T) = \exp(-\Delta G_T^\circ/R_u T)$. Here we see a dual temperature dependence: first, the explicit reciprocal temperature dependence within the exponential, and second, the implicit temperature dependence associated with the standard-state Gibbs function change, ΔG_T°. For endothermic reactions ($\Delta G_T^\circ > 0$), increasing the temperature has the effect of increasing $K_p\,(T)$. This in turn results in the equilibrium composition shifting toward the products side of the reaction. For a dissociation reaction (e.g., the $O_2 \Leftrightarrow O + O$ reaction in Example 13.1), we expect to find a greater proportion of O atoms as the temperature is increased at a fixed pressure. For exothermic reactions ($\Delta G_T^\circ < 0$), the converse obtains. Hence, we conclude that increasing the temperature always favors the endothermic direction of any reaction.

Van't Hoff developed a relationship between the enthalpy of reaction (see Chapter 12) and the temperature dependence of the equilibrium constant. This famous relationship bears his name (the **van't Hoff equation**) and is expressed as

$$\frac{d\ln K_p}{dT} = \frac{\Delta H_T^\circ}{R_u T^2}. \tag{13.25}$$

Here ΔH_T° is the temperature-dependent enthalpy of reaction at the standard-state pressure. This equation follows from the combination and manipulation of the definition of the equilibrium constant $K_p(T) = \exp\left(-\Delta G_T^\circ / R_u T\right)$, the definition of the Gibbs function change $\Delta G_T^\circ = \Delta H^\circ - T\Delta S^\circ$ and the TdS (or Gibbs) equation $TdS = dH - \mathcal{V}dP$. Assuming that the enthalpy of reaction does not vary much over the temperature range of interest, a good assumption for many reactions, Eq. 13.25 can be integrated and evaluated between two temperatures, T_1 and T_2:

$$\ln\frac{K_p(T_2)}{K_p(T_1)} = \frac{-\Delta H^\circ}{R_u}\left(\frac{1}{T_2} - \frac{1}{T_1}\right), \tag{13.26}$$

From this result, we see, once again, that for an exothermic reaction ($\Delta H < 0$) the equilibrium constant decreases with increasing temperature, and conversely for an endothermic reaction.

In the following example we explore the temperature dependence of the dissociation of carbon dioxide. This example also adds a bit more complexity to our equilibrium analysis as we are now dealing with three species; Example 13.1 considered the simplest case of only two species.

Example 13.2 Dissociation of Carbon Dioxide at Different Temperatures

Ideal combustion, as discussed in Chapter 12, assumes that all the fuel carbon and hydrogen are completely converted to CO_2 and H_2O, respectively. At typical combustion temperatures, however, carbon dioxide will dissociate. Consider the dissociation of CO_2 as a function of temperature via the reaction

$$CO_2 \Leftrightarrow CO + \tfrac{1}{2}O_2.$$

Find the composition of the mixture (i.e., the mole fractions of CO_2, CO, and O_2) that results from subjecting originally pure CO_2 to temperatures of 1500, 2000, 2500, and 3000 K. The pressure is 1 atm.

Solution

Known Initially all CO_2, P, T

Find X_i

Modeling, Premises and Assumptions

Ideal-gas behavior

The only species present in the mixture are CO_2, CO, and O_2

Analysis To find the three unknown mole fractions, X_{CO_2}, X_{CO} and X_{O_2}, we will need three equations. The first will be an equilibrium expression, Eq. 13.23b. The other two equations will come from element conservation expressions that state that the total amounts of C and O are constant, regardless of how they are distributed among the three species, because the original mixture was pure CO_2.

To implement Eq. 13.23b, we need to evaluate the standard-state Gibbs function change (Eq. 13.21b) at each temperature. For the reaction

$$1CO_2 \Leftrightarrow 1CO + \tfrac{1}{2}O_2,$$

we recognize that $a = 1$, $b = 1$, and $CO = \frac{1}{2}$. We use these coefficients to evaluate the standard-state Gibbs function change. For example at $T = 2500\,K$,

$$\begin{aligned}\Delta G_T^\circ &= \left[\left(\tfrac{1}{2}\right)\Delta\bar{g}^\circ_{f,O_2} + (1)\Delta\bar{g}^\circ_{f,CO} - (1)\Delta\bar{g}^\circ_{f,CO_2}\right]_{T=2500}\\ &= \left[\left(\tfrac{1}{2}\right)0 + (1)(-327{,}245) - (-396{,}152)\right]\,kJ/kmol\\ &= 68{,}907\,kJ/kmol.\end{aligned}$$

The values for the Gibbs function of formation are found in Tables D.11, D.1, and D.2. From the definition of K_p we have

$$K_p = \frac{(P_{CO}/P^\circ)^1 (P_{O_2}/P^\circ)^{0.5}}{(P_{CO_2}/P^\circ)^1}.$$

We can rewrite K_p in terms of the mole fractions by recognizing that $P_i = X_i P$. Thus,

$$K_p = \frac{X_{CO}X_{O_2}^{0.5}}{X_{CO_2}}(P/P^\circ)^{0.5}.$$

Substituting this expression into Eq. 13.23b, we have

$$\begin{aligned}\frac{X_{CO}X_{O_2}^{0.5}(P/P^\circ)^{0.5}}{X_{CO_2}} &= \exp\left(\frac{-\Delta G_T^\circ}{R_u T}\right)\\ &= \exp\left[\frac{-68{,}907\,kJ/kmol}{(8.3145\,kJ/kmol{\cdot}K)(2500\,K)}\right] \qquad \textbf{(I)}\\ &= 0.03635\ (\text{dimensionless}).\end{aligned}$$

We create a second equation to express the conservation of elements:

$$\frac{\text{number of carbon atoms}}{\text{number of oxygen atoms}} = \frac{1}{2} = \frac{X_{CO} + X_{CO_2}}{X_{CO} + 2X_{CO_2} + 2X_{O_2}}.$$

We can make the problem more general by defining the C/O ratio to be a parameter Z that can take on different values depending on the initial composition of the mixture:

$$Z \equiv \frac{X_{CO} + X_{CO_2}}{X_{CO} + 2X_{CO_2} + 2X_{O_2}}.$$

From this definition we form our second equation,

$$(Z-1)X_{CO} + (2Z-1)X_{CO_2} + 2ZX_{O_2} = 0. \qquad \textbf{(II)}$$

To obtain a third and final equation, we require that all the mole fractions sum to unity:

$$\sum X_i = 1,$$

or

$$X_{CO} + X_{CO_2} + X_{O_2} = 1. \qquad \textbf{(III)}$$

Simultaneous solution of Eqs. I, II, and III for any given values of P, T, and Z yields values for the mole fractions X_{CO}, X_{CO_2}, and X_{O_2}. By using Eqs. II and III to eliminate X_{CO_2} and X_{O_2}, Eq. I becomes

$$X_{CO}(1 - 2Z + ZX_{CO})^{0.5}(P/P^\circ)^{0.5} \\ -[2Z - (1+Z)X_{CO}]\exp\left(-\Delta G_T^\circ/R_u T\right) = 0.$$

This expression is easily solved for X_{CO} using any equation solver. We apply Newton–Raphson iteration implemented using spreadsheet software to obtain X_{CO}. The other unknowns, X_{CO_2} and X_{O_2}, are then recovered using Eqs. II and III. For $P = 1$ atm, $Z = \frac{1}{2}$, and for the four temperatures, we obtain the results shown in Table 13.1.

TABLE 13.1 Equilibrium Compositions at Various Temperatures for $CO_2 \Leftrightarrow CO + \frac{1}{2}O_2$ at 1 atm

	$T = 1500$ K	$T = 2000$ K	$T = 2500$ K	$T = 3000$ K
ΔG_T° (kJ/kmol)	1.5268×10^8	1.1046×10^8	6.8907×10^7	2.7878×10^7
K_p	4.823×10^{-6}	0.00492	0.03635	0.3270
X_{CO}	3.601×10^{-4}	0.0149	0.1210	0.3581
X_{CO_2}	0.9994	0.9777	0.8185	0.4629
X_{O_2}	1.801×10^{-4}	0.0074	0.0605	0.179

Comments From a purely mathematical point of view, we first note that the absolute value of the argument of the exponential in $K_p(T) = \exp\left(-\Delta G_T^\circ/R_u T\right)$ decreases with increasing temperature, for two reasons: first, values of ΔG_T° decrease; and second, the temperature dependence is reciprocal. These temperature dependences result in K_p becoming larger with temperature, as can be seen in Table 13.1. We also note that when the temperature is increased, the composition shifts in the endothermic direction, as expected. Since energy is absorbed when CO_2 breaks down into CO and O_2 (i.e., the reaction is endothermic), increasing the temperature produces a shift to the right, to the $CO + \frac{1}{2}O_2$ side.

We also note that the dissociation of CO_2 is one source of carbon monoxide, a regulated pollutant, from internal combustion engines.

PRESSURE

We explore the effect of pressure by substituting $X_i P$ for the partial pressures P_i in the definition of the equilibrium constant K_p:

$$K_p \equiv \frac{(P_E/P^\circ)^e (P_F/P^\circ)^f \cdots}{(P_A/P^\circ)^a (P_B/P^\circ)^b \cdots} = \frac{(X_E P/P^\circ)^e (X_F P/P^\circ)^f \cdots}{(X_A P/P^\circ)^a \cdot (X_B P/P^\circ)^b \cdots}.$$

Collecting all the P/P° terms yields

$$K_P \equiv \frac{(X_E)^e (X_F)^f \cdots}{(X_A)^a (X_B)^b \cdots}(P/P^\circ)^{e+f+\ldots-a-b-\ldots}. \tag{13.27}$$

From Eq. 13.27, we see that the P/P° factor will increase with pressure when the sum of the product coefficients $(e+f+g+\cdots)$ exceeds the sum of the reactant coefficients $(a+b+c+\cdots)$. For K_p to remain constant as the pressure increases, the product mole fractions must decrease while the reactants mole fractions must increase. The converse holds when the sum of the reactant coefficients $(a+b+c+\cdots)$ exceeds the sum of the products coefficients $(e+f+g+\cdots)$. The summary result is that when the pressure increases, the reaction will shift in the direction of the least number of moles in the reaction $a\text{A} + b\text{B} + \cdots \leftrightarrow e\text{E} + f\text{F} + \cdots$. For example, with a dissociation reaction (e.g., $N_2 \leftrightarrow N + N$), increasing the pressure decreases the amount of dissociation. There is no effect of pressure on the composition (expressed as mole fractions) when the reactant and product coefficient sums are equal (e.g., $CO + H_2O \leftrightarrow CO_2 + H_2$).

Example 13.3 Dissociation of Oxygen at Low Total Pressure

Revisit the dissociation of oxygen molecules considered in Example 13.1:

$$O_2 \Leftrightarrow O + O.$$

Determine the equilibrium partial pressures and mole fractions of O and O_2 at 3000 K, but now at a pressure of 0.1 atm, one-tenth of that considered previously.

Solution

Known Equilibrium, P, T

Find $P_O, P_{O_2}, X_O, X_{O_2}$

Modeling, Premises and Assumptions Ideal-gas behavior

Analysis We begin by noting that neither ΔG_T° nor K_p depends on pressure. Because the temperature is the same as that in Example 13.1, we simply repeat the values for these quantities here:

$$\Delta G_T^\circ = 109{,}108 \text{ kJ/kmol, and}$$
$$K_p = 0.0126 \text{ (dimensionless)}$$

The pressure effect enters into the problem through the relationship of the equilibrium constant to the partial pressures, i.e.,

$$K_p = \frac{(P_O/P^\circ)^2}{(P_{O_2}/P^\circ)} = \frac{P_O^2}{P_{O_2}}\left(\frac{1}{P^\circ}\right).$$

Again, recognizing that the partial pressures sum to the total pressure, we have

$$P_{\text{tot}} = P = P_O + P_{O_2},$$

or

$$P_{O_2} = P - P_O.$$

Eliminating P_{O_2} in our expression for K_p and simplifying yields

$$P_O^2 + K_p P^\circ P_\circ - K_p P^\circ (P) = 0.$$

Here we see the pressure dependence explicitly appearing in the third term.

Evaluating this expression with the lower pressure of 0.1 atm, we obtain

$$P_O^2(\text{atm}) + 0.0126(1\,\text{atm})P_O(\text{atm}) - 0.0126(1 \text{ atm})(0.1 \text{ atm}) = 0.$$

We find the useful root to be

$$P_O = 0.02975 \text{ atm}.$$

Thus,

$$P_{O_2} = 0.1\,\text{atm} - 0.02975\,\text{atm} = 0.07025\,\text{atm}$$

and

$$X_O = \frac{0.02975}{0.1} = 0.2975,$$

$$X_{O_2} = \frac{0.07025}{0.1} = 0.7025.$$

Comment Note how the dissociation is enhanced at the lower pressure. At 0.1 atm, the O-atom mole fraction is nearly three times its value at 1 atm.

Example 13.4 Dissociation of Carbon Dioxide at Different Pressures

We note that engine combustion occurs at pressures well above ambient. Explore the effect of pressure on the dissociation of CO_2 considered in Example 13.2:

$$CO_2 \Leftrightarrow CO + \tfrac{1}{2}O_2.$$

Find the composition of the mixture (i.e., the mole fractions of CO_2, CO, and O_2) for the dissociation of pure CO_2 for pressures of 0.1, 1, 10, and 100 atm at 2500 K.

Solution

Known Initially all CO_2, T, various pressures

Find X_i

Modeling, Premises and Assumptions Ideal-gas behavior

Analysis We begin by recognizing that, as in Example 13.2, three equations are needed to find the three mole fractions. As Example 13.3, we note again that the equilibrium constant is independent of pressure. Therefore, we can start our calculations at the point at which K_p is related to the partial pressures of the component gases (Eq. 13.22):

$$K_p = \frac{(P_{CO}/P^\circ)^1 (P_{O_2}/P^\circ)^{0.5}}{(P_{CO_2}/P^\circ)^1}.$$

We introduce the pressure by relating the partial pressures to the mole fractions and the total pressure, i.e., $P_i = X_iP$. Thus,

$$K_p = \frac{X_{CO}X_{O_2}^{0.5}}{X_{CO_2}}(P/P^\circ)^{0.5}.$$

From Example 13.2, K_p was found to be 0.03565 for 2500 K; thus, we have

$$\frac{X_{CO}\,X_{O_2}^{0.5}(P/P^\circ)^{0.5}}{X_{CO_2}} = 0.03635. \qquad \text{(I)}$$

As before, we generate a second equation that expresses the conservation of elements:

$$\frac{\text{number of carbon atoms}}{\text{number of oxygen atoms}} = Z = \frac{X_{CO} + X_{CO_2}}{X_{CO} + 2X_{CO_2} + 2X_{O_2}}.$$

Rearranging the above results in

$$(Z - 1)X_{CO} + (2Z - 1)X_{CO_2} + 2ZX_{O_2} = 0. \quad \text{(II)}$$

To obtain a third and final equation, we require that all the mole fractions sum to unity:

$$\sum X_i = 1,$$

or

$$X_{CO} = X_{CO_2} + X_{O_2} = 1. \quad \text{(III)}$$

By using Eqs. II and III to eliminate X_{CO_2} and X_{O_2}, Eq. I becomes

$$X_{CO}(1 - 2Z + ZX_{CO})^{0.5}(P/P^\circ)^{0.5} - [2Z - (1 + Z)X_{CO}]K_p = 0.$$

The pressure dependence appears explicitly in the $(P/P^\circ)^{0.5}$term. We solve this expression for X_{CO} with the following values: $Z = 0.5$, recalling that the C-to-O-atom ratio (Z) is based on the given condition that we started with pure CO_2; $K_p = 0.03635$; and $P^\circ = 1$ atm. Newton–Raphson iteration was implemented using spreadsheet software for this solution. The other unknowns, X_{CO} and X_{O_2}, are then recovered using Eqs. II and III. The results are shown in Table 13.2 for the four pressures.

TABLE 13.2 Equilibrium Compositions at Various pressures for $CO_2 \Leftrightarrow CO + \frac{1}{2}O_2$ at 2500 K

	$P = 0.1$ atm	$P = 1$ atm	$P = 10$ atm	$P = 100$ atm
X_{CO}	0.2260	0.1210	0.0602	0.0289
X_{CO_2}	0.6610	0.8185	0.9096	0.9566
X_{O_2}	0.1130	0.0605	0.0301	0.0145

Comments We see that the amount of dissociation decreases as the pressure is increased. At 0.1 atm, the oxygen mole fraction is more than 10% of the mixture, whereas at 100 atm, the oxygen mole fraction is about 1.5%. This result is consistent with our analysis that when the pressure increases, the reaction will shift in the direction of the least number of moles of dissociated products.

Figure 13.7 shows the combined results of Examples 13.2 and 13.4, along with additional results, for the temperature and pressure dependencies on the dissociation reaction $CO_2 \Leftrightarrow CO + \frac{1}{2}O_2$. Increasing the temperature and decreasing the pressure both promote the dissociation.

PRESENCE OF INERT GASES

In many practical systems, inert gases may be present. Such inert gases are frequently called just inerts or referred to as diluents. For example, in combustion systems,

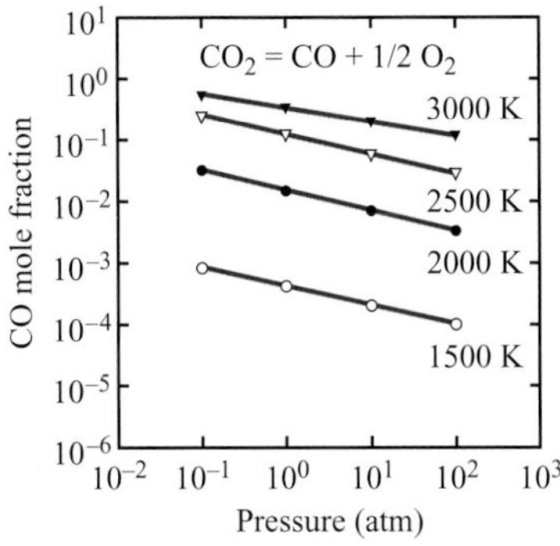

FIGURE 13.7 Increasing the temperature and/or decreasing the pressure promotes the dissociation of CO_2 to CO and O_2. Adapted from Ref. [8] with permission, © McGraw-Hill Education.

molecular nitrogen frequently can be treated as an inert because very little N_2 dissociation occurs at typical combustion temperatures. Also, in the study of reacting systems, inert gases, such as argon or helium, are sometimes added.

For a fixed total pressure, an inert gas exerts its influence on equilibrium by reducing the partial pressures of each constituent (and the mole fractions). We can see how this affects the equilibrium composition by analyzing Eq. 13.27, i.e.,

$$K_P \equiv \frac{(X_E)^e (X_F)^f \cdots}{(X_A)^a (X_B)^b \cdots} (P/P^\circ)^{e+f+\cdots-a-b-\cdots}.$$

We explicitly introduce the presence of an inert gas by expressing the mole fractions as $X_i = N_i / N_{mix}$, where N_i is the number of moles of species i participating in the equilibrium reaction, and N_{mix} is the total number of moles in the mixture, including the diluent. (Note that, with both the temperature and temperature fixed, the volume must change to accommodate the diluent.) Substituting this into Eq. 13.27 yields

$$K_P \equiv \frac{(N_E)^e (N_F)^f \cdots}{(N_A)^a (N_B)^b \cdots} N_{mix}^{a+b+\cdots-e-f-\cdots} (P/P^\circ)^{e+f+\cdots-a-b-\cdots}. \tag{13.28}$$

For the case of a reaction in which the product coefficient sum is greater than that of the reactants (e.g., $A \Leftrightarrow E + F$), the $N_{mix}^{a+b+\cdots-e-f-\cdots}$ factor will decrease with the addition of the diluent. This in turn requires the ratio $\left[(N_E)^e (N_F)^f \cdots\right] / \left[(N_A)^a (N_B)^b \cdots\right]$ to increase. In other words, the reaction will shift to the product side. Conversely, for the case of a reaction in which the reactant coefficient sum is greater than that of the products (e.g., $A + B \Leftrightarrow E$), the $N_{mix}^{a+b+\cdots-e-f-\cdots}$ factor will increase with the addition of the diluent. This in turn requires the ratio $\left[(N_E)^e (N_F)^f \cdots\right] / \left[(N_A)^a (N_B)^b \cdots\right]$ to decrease, i.e., the reaction will shift to the reactant side. For equimolar equilibria (e.g. $A + B \Leftrightarrow C + D$), the ratio $\left[(N_E)^e (N_F)^f \cdots\right] / \left[(N_A)^a (N_B)^b \cdots\right]$ is unaffected by the addition of a diluent. Thus, the equilibrium shifts associated with the addition of a diluent at fixed temperature and pressure are similar to those associated with reducing the total pressure.

Example 13.5 Dissociation of Oxygen Mixed with an Inert Gas

We revisit the dissociation reaction $O_2 \Leftrightarrow O + O$ considered in Example 13.1 to determine the effect of dilution with argon, an inert gas. The temperature and pressure remain as before, 3000 K and 1 atm, respectively. The initial mixture consists of one kmol of pure O_2 and one kmol of argon producing an equimolar mixture of the two elements.

Determine the number of moles of O and O_2, their mole fractions, and partial pressures at equilibrium. Also compare the percentage of the original O_2 that is dissociated with that for the undiluted situation of Example 13.1.

Solution

Known Equilibrium, P, T

Find $X_O, X_{O_2}, P_O, P_{O_2}$, percentage O_2 dissociated

Modeling, Premises and Assumptions

Ideal-gas behavior

Argon does not react with O_2 or O

Analysis Because the temperature is the same as in Example 13.1 (3000 K), the equilibrium constant also has the same value ($K_p = 0.0126$). We use this value with the equilibrium constant expression involving the number of moles (Eq. 13.28) to account explicitly for the addition of the inert Ar. For the equilibrium $O_2 \Leftrightarrow O + O$, Eq. 13.28 becomes

$$K_p = \frac{N_O^2}{N_{O_2} N_{mix}} \left(\frac{P}{P^\circ}\right).$$

We are given that there is initially one mole each of O_2 and Ar. Recognizing that two oxygen atoms (O) are created for each molecule of oxygen (O_2) dissociated, the number of moles of O_2 at equilibrium is the initial number of moles of O_2 molecules less those dissociated, i.e.,

$$N_{O_2} = 1 - N_O/2.$$

The total number of moles in the mixture at equilibrium is the sum of the number of moles of oxygen molecules, oxygen atoms, and argon atoms:

$$\begin{aligned} N_{mix} &= N_{O_2} + N_O + N_{Ar} \\ &= (1 - N_O/2) + N_O + 1. \end{aligned}$$

Substituting these expressions for N_{O_2} and N_{mix} into Eq. 13.28 yields

$$K_p = \frac{N_O^2}{(1 - N_O/2)([1 - N_O/2] + N_O + 1)} \left(\frac{P}{P^\circ}\right).$$

For the given pressure of 1 atm, the P/P° term is unity. With this substitution, we rearrange the above to obtain the following quadratic equation:

$$\left(0.25 + 1/K_p\right) N_O^2 + 0.5 N_O - 2 = 0.$$

Using the value 0.0126 for K_p, we solve for N_O. The physically realistic value is

$$N_O = 0.1554 \text{ kmol}.$$

The number of moles of molecular oxygen and the total number of moles in the mixture are then

$$N_{O_2} = (1 - N_O/2) = 1 - 0.1554/2 = 0.9223 \text{ kmol, and}$$

$$N_{mix} = N_{O_2} + N_O + N_{Ar} = 0.9223 + 0.1554 + 1 = 2.0777 \text{ kmol}.$$

With these values, we can determine the mole fractions:

$$\begin{aligned} X_O &= \frac{N_O}{N_{mix}} = \frac{0.1554 \text{ kmol}}{2.0777 \text{ kmol}} = 0.07478, \\ X_{O_2} &= \frac{N_{O_2}}{N_{mix}} = \frac{0.9223 \text{ kmol}}{2.0777 \text{ kmol}} = 0.4439, \text{ and} \\ X_{Ar} &= \frac{N_{Ar}}{N_{mix}} = \frac{1 \text{ kmol}}{2.0777 \text{ kmol}} = 0.4813. \end{aligned}$$

We verify that the three mole fractions sum to unity. The partial pressures are obtained as $P_i = X_i P_{tot}$. Because the total pressure is 1 atm, the partial pressures are numerically equal to the mole fractions, with units of atm:

$$P_O = 0.07478 \text{ atm},$$
$$P_{O_2} = 0.4439 \text{ atm, and}$$
$$P_{Ar} = 0.4813 \text{ atm}.$$

We define the percentage of molecular oxygen dissociated is given by

$$\begin{aligned}\% \ O_2 \text{ dissociated} &\equiv \frac{N_O}{N_{O_2} + N_O} \times 100\% \\ &= \frac{X_O}{X_{O_2} + X_O} \times 100\% = \frac{0.07478}{0.4439 + 0.07478} \times 100\% \\ &= 14.4\%.\end{aligned}$$

We can compare this to the undiluted situation of Example 13.1, where

$$\begin{aligned}\% \ O_2 \text{ dissociated (no inert gas)} &= \frac{X_O}{X_{O_2} + X_O} \times 100\% \\ &= \frac{0.1061}{0.8939 + 0.1061} \times 100\% \\ &= 10.6\%.\end{aligned}$$

Comment We see that the addition of argon results in a substantial increase in the percentage of molecular oxygen dissociated. This numerical result is consistent with our analysis of the effect of adding an inert gas when considering a dissociation reaction.

PRINCIPLE OF LE CHÂTELIER

The effects of pressure, temperature, and the addition of an inert gas discussed above illustrate the **principle of Le Châtelier,** which states that any system initially in a state of equilibrium when subjected to a change (e.g., changing the pressure, temperature, or the total number of moles by including an inert gas) will shift in composition in such a way as to minimize the change. For an increase in pressure, this translates to the equilibrium shifting in the direction to produce fewer moles. For the $CO_2 \Leftrightarrow CO + \frac{1}{2}O_2$ reaction, this means a shift to the left, to the CO_2 side. For equimolar reactions, pressure has no effect. When the temperature is increased, the composition shifts in the endothermic direction. Since energy is absorbed when CO_2 breaks down into CO and O_2, increasing the temperature produces a shift to the right, to the $CO + \frac{1}{2}O_2$ side. The effect of adding a diluent at fixed temperature and pressure is to reduce the partial pressures and, hence, to cause shift to the side of the reaction with the greater number of moles.

$$O_2 \Leftrightarrow 2O \quad \text{(R1)}$$
$$N_2 \Leftrightarrow 2N \quad \text{(R2)}$$
$$H_2 \Leftrightarrow 2H \quad \text{(R3)}$$
$$CO_2 \Leftrightarrow CO + \tfrac{1}{2}O_2 \quad \text{(R4)}$$
$$H_2O \Leftrightarrow OH + H \quad \text{(R5)}$$
$$H_2O + CO \Leftrightarrow CO_2 + H_2 \quad \text{(R6)}$$
$$O_2 + N_2 \Leftrightarrow 2NO \quad \text{(R7)}$$

FIGURE 13.8 Typical equilibrium reactions considered for the products of hydrocarbon–air combustion.

13.4c Multiple Equilibrium Reactions

In practical, high-temperature systems, multiple equilibrium reactions will be important. Figure 13.8 shows a typical situation for hydrocarbon–air combustion products.

To illustrate, in principle, how to deal with multiple equilibria, we consider the species and equilibria shown in Fig. 13.8. We define the problem as finding the mole fractions of the 11 species (O_2, O, N_2, N, H_2, H, H_2O, OH, CO_2, CO, and NO) for a given temperature and pressure. Three atom ratios (Z_1, Z_2, and Z_3 in Eqs. 13.29 and 13.31) can be determined by specifying the equivalence ratio (see Chapter 12), the fuel composition C_xH_y, and the O_2-to-N_2 ratio in the air:

$$Z_1 \equiv \frac{\text{No. of C atoms}}{\text{No. of O atoms}} = \frac{X_{CO_2} + X_{CO}}{2X_{O_2} + X_O + 2X_{CO_2} + X_{CO} + X_{H_2O} + X_{OH} + X_{NO}}, \quad (13.29)$$

$$Z_2 \equiv \frac{\text{No. of H atoms}}{\text{No. of O atoms}} = \frac{2X_{H_2O} + 2X_{H_2} + X_H + X_{OH}}{2X_{O_2} + X_O + 2X_{CO_2} + X_{CO} + X_{H_2O} + X_{OH} + X_{NO}}, \text{ and} \quad (13.30)$$

$$Z_3 \equiv \frac{\text{No. of N atoms}}{\text{No. of O atoms}} = \frac{2X_{N_2} + X_N + X_{NO}}{2X_{O_2} + X_O + 2X_{CO_2} + X_{CO} + X_{H_2O} + X_{OH} + X_{NO}}. \quad (13.31)$$

A fourth equation, expressing element (C, H, O, N) conservation, completes the set of equations.

$$\sum X_i = 1 = X_{O_2} + X_O + X_{H_2} + X_H + X_{H_2O} + X_{OH} + X_{CO_2} + X_{CO} + X_{N_2} + X_N + X_{NO}. \quad (13.32)$$

To solve for the 11 unknown mole fractions requires seven additional equations. These we obtain by writing equilibrium expressions for the seven reactions, R1–R7 shown in Fig. 13.8, as follows:

$$\left(X_O^2/X_{O_2}\right) P/P^\circ = K_{p,R1}, \quad (13.33a)$$

$$\left(X_N^2/X_{N_2}\right) P/P^\circ = K_{p,R2}, \quad (13.33b)$$

$$\left(X_H^2/X_{H_2}\right) P/P^\circ = K_{p,R3}, \quad (13.33c)$$

$$\left(X_{CO}\, X_{O_2}^{1/2}/X_{CO_2}\right) (P/P^\circ)^{1/2} = K_{p,R4}, \quad (13.33d)$$

$$(X_{OH}X_H/X_{H_2O})P/P^\circ = K_{p,R5}, \quad (13.33e)$$

$$\frac{X_{CO_2}X_{H_2}}{X_{H_2O}X_{CO}} = K_{p,R6}, \text{ and} \quad (13.33f)$$

$$\frac{X_{NO}^2}{X_{O_2}X_{N_2}} = K_{p,R7}. \quad (13.33g)$$

Once the seven equilibrium constants are evaluated from Eq. 13.23b for the given temperature, our equation set is closed. Mathematically, we have a set of 11 nonlinear algebraic equations to solve for the unknown mole fractions. The generalized Newton's method or other techniques can be applied to solve these equations using a computer.

This specific example illustrates the basic approach to solving multiple equilibrium problems: (i) apply element conservation for the specific system, (ii) relate the individual unknown species mole fractions via individual equilibrium reactions, and (iii) solve the resulting system of nonlinear algebraic equations.

The computer code developed by Olikara and Borman [9] was developed specifically to solve the problem described above. Other frequently encountered software packages capable of solving complex equilibria include the NASA chemical equilibrium code [10] and STANJAN [11]. Another easy-to-use code, GASEQ, is available as freeware [12] from the Internet.

13.5 Phase Equilibrium

We now consider conditions for equilibrium between two phases of a pure substance. A familiar example of this is the equilibrium between the liquid and vapor phases of H_2O (Fig. 13.9).

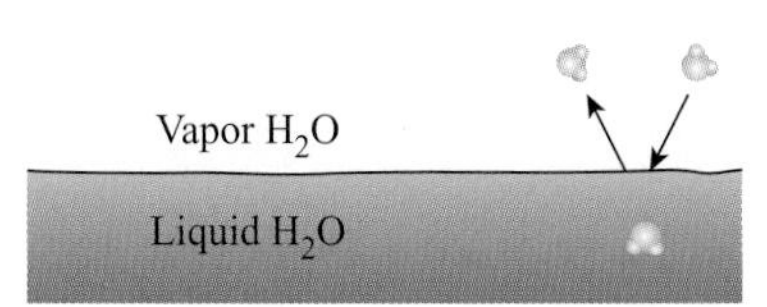

FIGURE 13.9 For a water–steam system in equilibrium at a given temperature, there is only one value of pressure that allows the system to remain in equilibrium. This is the saturation pressure discussed in Chapter 2.

The questions we wish to address are the following: For a given temperature, what pressure must exist at equilibrium? Or, alternatively, for a given pressure, what temperature must prevail for equilibrium? Like the chemical equilibrium problem, for phase equilibrium we consider a system at constant temperature and pressure. As a result, the appropriate mathematical statement of the second law of thermodynamics controlling equilibrium is Eq. 13.10,

$$\left(dG_{\text{sys}}\right)_{T,P} = 0$$

where G_{sys} is the Gibbs function for the two-phase system.

For a liquid–vapor system, the Gibbs function can be expressed as

$$G_{\text{sys}} = N_{\text{f}}\,\bar{g}_{\text{f}} + N_{\text{g}}\,\bar{g}_{\text{g}}, \tag{13.34}$$

where the subscripts f and g refer to the liquid and vapor phases, respectively. Applying the condition of Eq. 13.10 yields

$$dG_{\text{sys}} = dN_{\text{f}}\,\bar{g}_{\text{f}} + dN_{\text{g}}\,\bar{g}_{\text{g}} = 0.$$

Because any evaporating liquid appears as vapor and, conversely, any condensing vapor appears as liquid, the change in the number of moles of one phase is the negative of the change in the other phases, and so

$$dN_{\text{f}} = -dN_{\text{g}}.$$

The condition for phase equilibrium then becomes

$$dG_{\text{sys}} = 0 = dN_{\text{f}}\left(\bar{g}_{\text{f}} - \bar{g}_{\text{g}}\right),$$

or

$$\bar{g}_{\text{f}} = \bar{g}_{\text{g}}. \tag{13.35a}$$

Our conclusion is that, for liquid–vapor equilibrium, the molar Gibbs function of the liquid phase must equal the molar Gibbs function of the vapor phase. Since the molecular weight of any substance is the same in both phases, we can also express Eq. 13.35a on a mass basis,

$$g_{\text{f}} = g_{\text{g}}. \tag{13.35b}$$

The ideas developed here apply not only to liquid–vapor equilibrium but to any pair of coexisting phases (i.e., vapor–solid and liquid–solid); thus, the general expression of phase equilibrium is

$$\bar{g}_{\text{phase 1}} = \bar{g}_{\text{phase 2}}, \tag{13.36a}$$

or

$$g_{\text{phase 1}} = g_{\text{phase 2}}. \tag{13.36b}$$

The following example illustrates the criterion that $g_f = g_g$ applies to the equilibrium between the liquid and vapor phases of H_2O.

Example 13.6

Use the saturation properties for H_2O at 100°C from the NIST resources to show that the Gibbs function criterion (Eq. 13.35b) is met.

H_2O (g) $g_g = ?$

H_2O (f) $g_f = ?$

Solution

From the NIST WebBook we obtain the following property data at 100°C:

Phase	*h* (kJ/kg)	*s* (kJ/kg·K)
Liquid	419.17	1.3072
Vapor	2675.6	6.3541

The Gibbs function is defined by Eq. 13.2:

$$g \equiv h - Ts,$$

where T is the absolute temperature. We use this definition and the property data to calculate the Gibbs function for each phase as follows:

$$\begin{aligned} g_f &= h_f - Ts_f \\ &= 419.17\ \text{kJ/kg} - (100 + 273.15)(\text{K})1.3072\ \text{kJ/kg·K} \\ &= -68.612\ \text{kJ/kg} \end{aligned}$$

and

$$\begin{aligned} g_g &= h_g - Ts_g \\ &= 2675.6\ \text{kJ/kg} - (100 + 273.15)(\text{K})7.3541\ \text{kJ/kg·K} \\ &= -68.582\ \text{kJ/kg}. \end{aligned}$$

Comment Although these two values (–68.612 and –68.582 kJ/kg) are not identical, they differ by only 0.044% and, most likely, are within the accuracy of the curve fits used to generate the NIST property data.

The **Gibbs phase rule** is one of the many important contributions of Gibbs to the field of chemical thermodynamics. Consider a nonreacting thermodynamic system at equilibrium. We define $\mathcal{P}$ to be the number of phases present, C to be the number of components (distinct chemical substances), and F the number of intensive properties (the degrees of freedom) that must be specified to fix the state of *each phase* of the system. With these definitions, the Gibbs phase rule is expressed as

$$F = C - \mathcal{P} + 2. \tag{13.37}$$

Let's see how this rule relates to our existing knowledge of the properties of H_2O. Because we are dealing with a single chemical substance, H_2O, the value of C is one (unity). Consider first either the superheat region or the subcooled-liquid region, where only one phase is present (all vapor in the superheat region, or all liquid in the subcooled-liquid region); thus, $\mathcal{P}$ is also one (unity) in these regions. Applying the phase rule for this situation, we have

$$F = C - \mathcal{P} + 2 = 1 - 1 + 2 = 2.$$

The consequence of this is that two intensive properties are required to determine the state of the single-phase system; two intensive properties can be varied without changing the number of phases present. For example, we can vary the temperature and pressure independently and still remain within the single-phase region; the same hold for any two intensive properties. Any two intensive properties can be used to determine the state of the single-phase system.

Consider now states within the liquid–vapor region (wet mixture region). Here two phases (liquid and vapor) coexist; hence, the value of $\mathcal{P}$ is 2. The number of substances is still just one; thus, $C = 1$, as before. With these values, we apply the Gibbs phase rule again:

$$F = C - \mathcal{P} + 2 = 1 - 2 + 2 = 1.$$

We conclude that now only a single intensive property establishes the state of each phase. For example, knowing either the temperature (or the pressure) defines the specific volume, specific internal energy, specific entropy, etc., of both the liquid and vapor phases. If we vary the pressure, the temperature of either the liquid or vapor phase cannot be independently varied without changing the number of phases present. For example, increasing the system temperature above the saturation temperature would move the system into the single-phase superheat region. Knowing the saturation properties, however, is not sufficient to determine the overall state of a system in the liquid–vapor region. To define the overall properties of a two-phase system requires knowing the relative proportions of the phases. We have used the property *quality* many times before to express these proportions.

Last, consider three phases of H_2O in equilibrium: solid, liquid, and vapor. For this case,

$$F = C - \mathcal{P} + 2 = 1 - 3 + 2 = 0.$$

Here, at the triple point, there are zero degrees of freedom ($F = 0$). No property can be varied without changing the number of phases. There is only a single pressure and a single temperature at which all three phases coexist. Again, knowledge of the relative proportions of the three phases is necessary to establish overall mixture properties such as specific volume. At the H_2O triple point, the temperature and pressure are 0.01 °C (273.16 K) and 0.61165 kPa, respectively.

Figure 2.38 in Chapter 2 provides a phase diagram for H_2O that graphically illustrates the ideas discussed here.

The **Clapeyron equation** and the **Clausius–Clapeyron equation** are analytical expressions that relate the slope of the saturation-pressure–saturation-temperature curve to the enthalpy of vaporization, h_{fg}. Before we start our derivation of these famous equations, we state the results:

$$\text{Clapeyron equation:}\quad \frac{dP_{sat}}{dT_{sat}} = \frac{h_g - h_f}{T_{sat}\left(v_g - v_f\right)} = \frac{h_{fg}}{T_{sat}\left(v_g - v_f\right)}. \tag{13.38}$$

$$\text{Clausius–Clapeyron equation:}\quad \frac{d\ln P_{sat}}{dT_{sat}} = \frac{h_{fg}}{RT_{sat}^2}. \tag{13.39}$$

These equations have many uses. For example: measurements of the saturation pressure as a function of temperature can be used to obtain values of the enthalpy of vaporization; the equations provide a useful form for fitting actual $P_{sat}-T_{sat}$ data; they provide a basis for simplified models of evaporation processes, such as droplet evaporation in combustion processes [8]; among others.

To arrive at Eqs. 13.38 and 13.39, we apply several property relations, introduce some mathematics, and use some approximations.

In Chapter 2, we introduced the concept of state relations, i.e., the state of a simple compressible substance is defined by two independent intensive properties. This implies that knowledge of any two properties allows the determination of any other property. For the situation at hand, we want to relate the specific entropy to temperature and specific volume:

$$s = s(T, v)$$

In Chapter 4, we introduce the idea that, mathematically, all thermodynamic properties are *point functions* or *exact differentials*. The test for *exactness* of a function $x = x(y, z)$ is that the differential dx, expressed as

$$dx = Mdy + Ndz,$$

is exact if

$$\left(\frac{\partial M}{\partial z}\right)_y = \left(\frac{\partial N}{\partial y}\right)_z.$$

(The subscripts on the partial derivatives remind the reader of what variable was held constant during the differentiation, a common practice in the field of thermodynamics.) Because entropy s is an exact differential, we use the first of these mathematical relationships and write

$$ds = \left(\frac{\partial s}{\partial T}\right)_v dT + \left(\frac{\partial s}{\partial v}\right)_T dv.$$

During a phase change, the temperature is constant and $dT = 0$; hence,

$$ds = \left(\frac{\partial s}{\partial v}\right)_T dv. \tag{13.40}$$

To proceed toward our goal, the partial derivative $(\partial s/\partial v)_T$ can be related to the saturation temperature and pressure by performing the test for exactness on one of the Gibbs relations. In Chapter 7, we introduced two of these relations, the T–ds equations (Eqs. 7.6 and 7.7). We repeat them here and add two additional Gibbs relations to complete the set of four equations, each arranged in the form $dx = Mdy + Ndz$:

$$du = Tds - Pdv, \tag{13.41}$$

$$dh = Tds + vdP, \tag{13.42}$$

$$da = -Pdv - sdT, \tag{13.43}$$

$$dg = vdP - sdT. \tag{13.44}$$

The new relations (Eq. 13.43 and 13.44) are easily derived from the definitions of the Helmholtz free energy ($a \equiv u - Ts$, Eq. 13.3) and the Gibbs free energy ($g \equiv h - Ts$, Eq. 13.2) together with the first two T–ds equations (Eqs. 13.41 and 13.42). For our purposes, we focus on Eq. 13.43 and use the exactness test, which would hold for any of these equations. Noting that (see above) $x = a$, $y = v$, and $z = T$, we obtain

$$\left(\frac{\partial s}{\partial v}\right)_T = \left(\frac{\partial P}{\partial T}\right)_v.$$

Substituting this result into Eq. 13.40 yields

$$ds = \left(\frac{\partial P}{\partial T}\right)_v dv. \tag{13.45}$$

For a phase change, the saturation pressure is only a function of the saturation temperature; thus,

$$\left(\frac{\partial P}{\partial T}\right)_v = \frac{dP_{sat}}{dT_{sat}}.$$

We also note from Eq. 13.42 that

$$ds = \frac{dh}{T} - \frac{vdP}{T}$$

where we recognize that for a phase change $dP = 0$. Substituting these two results into Eq. 13.45 yields

$$\frac{dP_{sat}}{dT_{sat}} dv = \frac{dh}{T}. \tag{13.46}$$

We now integrate Eq. 13.46 over the phase change from liquid to vapor at a constant temperature and rearrange to isolate dP_{sat}/dT_{sat} on the left-hand side; these operations yield the Clapeyron equation (Eq. 13.38):

$$\frac{dP_{sat}}{dT_{sat}} = \frac{h_g - h_f}{T_{sat}(v_g - v_f)} = \frac{h_{fg}}{T_{sat}(v_g - v_f)}.$$

With additional approximations, first, that $v_g \gg v_f$ and, second, that the pressure is sufficiently low that the saturated vapor can be approximated as an ideal gas ($v_g = RT/P$), we recover the Clausius–Clapeyron equation (Eq. 13.39):

$$\frac{d \ln P_{sat}}{dT_{sat}} = \frac{h_{fg}}{RT_{sat}^2}.$$

Note that we also employed the fact that $dP_{sat} = P_{sat}\, d(\ln P_{sat})$ to obtain this final result. Furnishing the details of these last steps is left as an exercise for the reader.

SUMMARY

At the chapter outset, we indicated the usefulness of the second law in defining thermodynamic equilibrium. We revisited entropy, a second-law property, and defined two other properties: the Gibbs function and the Helmholtz free energy. Depending upon which two intensive properties are fixed, equilibrium is achieved when the system entropy is a maximum (u, v fixed), when the Gibbs function is a minimum (T, P fixed), or when the Helmholtz free energy is a minimum (T, v fixed). One focus of the chapter was on chemical equilibrium for conditions of fixed temperature and pressure. The equilibrium constant was defined and used to determine the detailed composition of a system at a fixed temperature and pressure. Single and multiple equilibria were considered. In addition to chemical (reacting) equilibrium, we considered liquid–vapor (nonreacting) equilibria. The Gibbs phase rule was introduced, and the Clapeyron and Clausius–Clapeyron equations were presented.

KEY EQUATIONS

Review the most important equations presented in this chapter (i.e., those boxed with a yellow background). What physical principles do they express? What restrictions apply?

CHAPTER 13 KEY CONCEPTS AND DEFINITIONS CHECKLIST

Answer the Questions and solve the Problems following the arrows to demonstrate mastery of the listed concepts and definitions.

13.1 Thermodynamic Equilibrium Revisited

- ☐ Four requirements for thermodynamic equilibrium → Question 13.1
- ☐ Interpreting requirements for thermodynamic equilibrium → Question 13.1

13.2 The Second Law and Equilibrium

- ☐ Maximization of entropy → Question 13.2, Problem 13.1
- ☐ Minimization of Gibbs free energy (Gibbs function) → Question 13.3, Problem 13.2

13.3 Equilibrium for Conditions of Fixed Internal Energy and Volume

- ☐ Relation of temperature and composition to maximization of entropy → Problem 13.3

13.4 Chemical Equilibrium for Conditions of Fixed Temperature and Pressure

- ☐ Gibbs free energy (Gibbs function) → Problems 13.4, 13.9
- ☐ Chemical equilibrium → Problem 13.11

- ☐ Standard-state Gibbs function change ➔ Problems 13.14, 13.18
- ☐ Equilibrium constant ➔ Question 13.4, Problems 13.19, 13.26
- ☐ Van't Hoff equation ➔ Problems 13.28, 13.29
- ☐ Equilibrium composition: single equilibrium reaction ➔ Question 13.6, Problems 13.30, 13.32
- ☐ Equilibrium composition: multiple equilibrium reactions ➔ Problem 13.52

13.5 Phase Equilibrium

- ☐ General considerations ➔ Questions 13.7, 13.8, Problem 13.57
- ☐ Gibbs phase rule ➔ Problem 13.60
- ☐ Clapeyron and Clausius–Clapeyron equations ➔ Problems 13.63, 13.65

REFERENCES

1. Lindauer, M. W., "The Evolution of the Concept of Chemical Equilibrium from 1775 to 1923," *Journal of Chemical Education*, Vol. 39, No. 8, August 1962.
2. Berthollet, C. T. (translated by M. Farrell), *Researches into the Laws of Chemical Affinity*, John Murray, London, 1804.
3. Lengyel, S., "Chemical Kinetics and Thermodynamics," *Computers Math. Applic.*, Vol. 17, No. 1–3, pp. 443–455, 1989.
4. Wheeler, L. P., *Josiah Willard Gibbs, The History of a Great Mind*, Shoe String Press, North New Haven, CT, 1970.
5. Atkins, P. W., *Physical Chemistry*, 6th edn, Oxford University Press, Oxford, 1999.
6. DeNevers, N., *Physical and Chemical Equilibrium for Chemical Engineers*, Wiley, New York, 2002.
7. Stull, D. R., and Prophet, H., *JANAF Thermochemical Tables*, 2nd edn, NSRDS-NBS 37, National Bureau of Standards, June 1971.
8. Turns, S. R., *An Introduction to Combustion*, 3rd edn, McGraw-Hill, New York, 2012.
9. Olikara, C., and Borman, G. L., "A Computer Program for Calculating Properties of Equilibrium Combustion Products with Some Applications to I. C. Engines," SAE Paper 750468, 1975.
10. Gordon, S., and McBride, B. J., "Computer Program for Calculation of Complex Chemical Equilibrium Compositions, Rocket Performance, Incident and Reflected Shocks, and Chapman–Jouguet Detonations," NASA SP-273, 1976.
11. Reynolds, W. C., "The Element Potential Method for Chemical Equilibrium Analysis: Implementation in the Interactive Program STANJAN," Department of Mechanical Engineering, Stanford University, January 1986.
12. Morley, C., "Gaseq, A Chemical Equilibrium Program for Windows," available for download at http://www.c.morley.ukgateway.net/.

Some end-of-chapter problems were adapted with permission from the following:

Look, D. C., Jr., and Sauer, H. J., Jr., *Engineering Thermodynamics*, PWS, Boston, 1986.

Myers, G. E., *Engineering Thermodynamics*, Prentice Hall, Englewood Cliffs, NJ, 1989.

QUESTIONS

13.1 List and discuss the conditions necessary for a system to be in thermodynamic equilibrium.

13.2 Explain how the second law of thermodynamics governs the equilibrium composition of a reacting system at fixed internal energy and volume.

13.3 Explain how the second law of thermodynamics governs the equilibrium composition of a reacting system at fixed temperature and pressure.

13.4 Define the standard-state Gibbs function change ΔG_T°. What is the relationship of ΔG_T° to the equilibrium constant K_p?

13.5 Consider an arbitrary chemical equilibrium. What is the physical significance of $K_p \ll 1$ and $K_p \gg 1$?

13.6 Consider the following equilibrium reaction at 1 atm and 2500 K:

$$CO_2 \leftrightarrow CO + \tfrac{1}{2}O_2.$$

A. Which is the exothermic direction for this reaction? To the left or to the right?

B. If the temperature is increased to 4000 K, but the pressure is unchanged, will the equilibrium shift to the left or to the right?

C. If the pressure is increased to 5.0 atm, but the temperature is unchanged, will the equilibrium shift to the left or to the right?

13.7 Explain how the second law of thermodynamics governs the equilibrium between the phases of a simple substance.

13.8 Consider a thermodynamic system consisting of liquid H_2O and H_2O vapor. Which of the following conditions is/are necessary for thermodynamic equilibrium to prevail in the system? Note that there may be more than one correct answer.

A. $T_{\text{liq}} = T_{\text{vap}}$

B. $\bar{g}_{\text{liq}} > \bar{g}_{\text{vap}}$

C. $\bar{g}_{\text{liq}} = \bar{g}_{\text{vap}}$

D. $\bar{g}_{\text{liq}} < \bar{g}_{\text{vap}}$

Chapter 13 Problem Subject Areas

13.1–13.3	Second law and chemical equilibrium
13.4 –13.13	Gibbs free energy (Gibbs function)
13.14 –13.18	Standard-state Gibbs function change
13.19 –13.27	Equilibrium constant
13.28 –13.29	Van't Hoff equation
13.30 –13.39	Equilibrium composition: single equilibrium reaction
13.40 –13.42	Equilibrium composition: effect of temperature
13.43 –13.46	Equilibrium composition: effect of pressure
13.47 –13.51	Equilibrium composition: effect of inert gas
13.52 –13.54	Equilibrium composition: multiple equilibrium reactions
13.55 –13.59	Phase equilibrium
13.60 –13.61	Gibbs phase rule
13.62 –13.65	Clapeyron and Clausius–Clapeyron equations

PROBLEMS

13.1–13.3 Second law and chemical equilibrium

13.1 Consider a system at fixed internal energy and volume in which the following equilibrium is maintained:

$$A + B \Leftrightarrow C.$$

The reaction is exothermic as written. Draw a sketch showing the system entropy and temperature as functions of the mole fraction of species C.

13.2 Consider a system at fixed temperature and pressure in which the following equilibrium is maintained:

$$A + B \Leftrightarrow C.$$

The reaction is exothermic as written. Draw a sketch showing the system Gibbs free energy (Gibbs function) as a function of the mole fraction of species C.

13.3 Using the property data in Appendix D, reproduce Fig. 13.4. Note that the initial temperature and pressure are 298 K and 1 atm, respectively. Spreadsheet software is recommended to facilitate your calculations. Plot the mixture temperature, entropy, and pressure as functions of the fraction of CO_2 dissociated. **HINT**: Start by calculating the adiabatic flame temperature as a function of the dissociated fraction.

13.4–13.13 Gibbs free energy (Gibbs function)

13.4 Consider one kmol of O_2 at 1 atm and 2600 K. Calculate the Gibbs function of the O_2 using the Gibbs function definition ($G = H - TS$).

13.5 Consider one kmol of O atoms at 1 atm and 2600 K. Calculate the Gibbs function of the O atoms using the Gibbs function definition ($G = H - TS$).

13.6 Consider one kmol of CO_2 at 1 atm and 2500 K. Calculate the Gibbs function of the CO_2 using the Gibbs function definition ($G = H - TS$).

13.7 Consider one kmol of CO at 1 atm and 2500 K. Calculate the Gibbs function of the CO using the Gibbs function definition ($G = H - TS$).

13.8 Use spreadsheet (or other) software to calculate and plot the Gibbs function of one kmol of O_2 for temperatures from 1000 K to 3000 K. The pressure is 1 atm. Repeat your calculations and plot on the same graph the Gibbs function for two kmol of O atoms.

13.9 Consider one kmol of H_2O at 1500 K. Calculate the Gibbs function of the H_2O for pressures of 1.0 and 100 atm.

13.10 Consider one kmol of O_2 at 2400 K. Calculate and plot the Gibbs function of the O_2 for pressures of 0.1, 1.0, 10, and 100 atm.

13.11 Consider the equilibrium reaction for the dissociation of one kmol of carbon dioxide $CO_2 \Leftrightarrow CO + \frac{1}{2}O_2$. Calculate and plot the Gibbs function versus the dissociated fraction for the CO_2–CO–O_2 mixture for conditions of 2500 K and 1 atm. Use spreadsheet or other software. **HINTS**: You will need to use the definition $\bar{g}_i(T) = \bar{h}_i(T) + T\bar{s}_i(T)$ where $\bar{s}_i(T) = \bar{s}_i^\circ(T) - R_u \ln P_i/P^\circ$. Also, to make your plot instructive, subtract the value of the undissociated CO_2 Gibbs function from your G_{mix} values, i.e., plot $G_{\text{mix}}(\alpha) - G_{\text{mix}}(\alpha = 0)$.

13.12 Perform the tasks requested in problem 13.11 but for conditions of 3000 K and 0.5 atm. Use spreadsheet or other software. **HINTS**: Same as above.

13.13 In the combustion of a hydrocarbon fuel with insufficient oxygen for complete combustion, it is frequently assumed that all the carbon in the fuel appears as CO and CO_2 in the products, and that all the hydrogen in the fuel appears as H_2 and H_2O. Furthermore, these species are assumed to be in equilibrium following the reaction $H_2O + CO \Leftrightarrow CO_2 + H_2$, which is known as the water–gas shift reaction. Consider a mixture of these four species at 1200 K and 1 atm. Initially, 2 kmol of H_2O and 1 kmol of CO are present. For the given conditions, calculate and plot the Gibbs function for the H_2O–CO–CO_2–H_2 mixture as a function of the number of kmol of CO_2 formed. Use spreadsheet or other software. **HINT**: You will need to use the definition $\bar{g}_i(T) = \bar{h}_i(T) + T\bar{s}_i(T)$ where $\bar{s}_i(T) = \bar{s}_i^{\circ}(T) - R_u \ln P_i/P^{\circ}$.

13.14–13.18 Standard-state Gibbs function change

13.14 Consider the equilibrium between molecular oxygen and oxygen atoms $O_2 \Leftrightarrow 2\,O$, at 2000 K and 1 atm. Calculate the standard-state Gibbs function change for this reaction in two ways: (1) using Eq. 13.21a and (2) using Eq. 13.21b. Compare your results and discuss. **HINT**: You will need to use the definition $\bar{g} = \bar{h} - T\bar{s}$ for the first method.

13.15 Consider the equilibrium between molecular oxygen and oxygen atoms $O_2 \Leftrightarrow 2\,O$, at 3200 K and 1 atm. Calculate the standard-state Gibbs function change for this reaction in two ways: (1) using Eq. 13.21a and (2) using Eq. 13.21b. Compare your results and discuss. **HINT**: You will need to use the definition $\bar{g} = \bar{h} - T\bar{s}$ for the first method.

13.16 Consider the equilibrium between molecular nitrogen and nitrogen atoms $N_2 \Leftrightarrow 2\,N$, at 3000 K and 1 atm. Calculate the standard-state Gibbs function change for this reaction in two ways: (1) using Eq. 13.21a and (2) using Eq. 13.21b. Compare your results and discuss. **HINT**: You will need to use the definition $\bar{g} = \bar{h} - T\bar{s}$ for the first method.

13.17 Consider the equilibrium between molecular nitrogen and nitrogen atoms $N_2 \Leftrightarrow 2\,N$, at 4500 K and 1 atm. Calculate the standard-state Gibbs function change for this reaction in two ways: (1) using Eq. 13.21a and (2) using Eq. 13.21b. Compare your results and discuss. **HINT**: You will need to use the definition $\bar{g} = \bar{h} - T\bar{s}$ for the first method.

13.18 In a steam reforming process, natural gas (or other hydrocarbon or coal) is used to produce hydrogen. In this process, the water–gas shift equilibrium reaction $H_2O + CO \Leftrightarrow CO_2 + H_2$ is important. Calculate the standard-state Gibbs function change for this reaction for temperatures of 1000 K, 1500 K, and 2000 K. Discuss the implications of your calculations.

13.19–13.27 Equilibrium constant

13.19 Consider the reaction $O_2 \Leftrightarrow 2\,O$. Determine a numerical value for the equilibrium constant K_p for the reaction as written for temperatures of 500 K and 5000 K.

13.20 Consider the reaction $N_2 \Leftrightarrow 2\,N$. Determine a numerical value for the equilibrium constant K_p for the reaction as written for temperatures of 500 K and 5000 K. Compare your results with Problem 13.19 and discuss.

13.21 Consider the reaction $CO_2 \Leftrightarrow CO + \frac{1}{2}O_2$. Determine a numerical value for the equilibrium constant K_p for the reaction as written for a temperature of 1800 K.

13.22 Consider the reaction $H_2O + CO \Leftrightarrow CO_2 + H_2$. Determine a numerical value for the equilibrium constant K_p for the reaction as written for a temperature of 1200 K.

13.23 Consider the reaction $N_2 + O_2 \Leftrightarrow 2\,NO$. Determine a numerical value for the equilibrium constant K_p for the reaction as written for temperatures of 1000 K and 3500 K. Discuss.

13.24 Consider the reaction $NO_2 \Leftrightarrow NO + \frac{1}{2}O_2$. Determine a numerical value for the equilibrium constant K_p for the reaction as written for temperatures of 300 K and 2500 K. Discuss.

13.25 Consider the reaction $2\,H_2 + O_2 \Leftrightarrow 2\,H_2O$. Express the equilibrium constant K_p for the reaction using appropriate partial pressures. Also write an expression for the equilibrium constant when the reaction is written from right to left.

13.26 Express the equilibrium constant K_p for the following reactions using appropriate partial pressures: $2\,H_2 + O_2 \Leftrightarrow 2\,H_2O$ and $H_2 + \frac{1}{2}O_2 \Leftrightarrow H_2O$. How do the two K_ps relate?

13.27 Consider a mixture of molecular nitrogen (N_2) and atomic nitrogen (N) in equilibrium at 3 atm at an unknown temperature. Determine the value of the equilibrium constant K_p for the reaction $N_2 \Leftrightarrow 2N$ if the partial pressure of the N_2 is 2.995 atm. Does K_p have units or is it dimensionless?

13.28–13.29 Van't Hoff equation

13.28 Calculate the enthalpy of reaction for $CO + \frac{1}{2}O_2 \Leftrightarrow CO_2$ at 2000 K. Use your value of ΔH at 2000 K, along with the K_p value at 2000 K of 20.325, to estimate the equilibrium constant for a temperature of 2500 K.

13.29 Consider the equilibrium dissociation of carbon dioxide $CO_2 \Leftrightarrow CO + \frac{1}{2}O_2$. At 2500 K, the equilibrium constant is 0.03635. Calculate the enthalpy of reaction for this reaction at 2500 K and use this to estimate the equilibrium constant for a temperature of 3000 K. Compare your estimated value with the exact value from Table 13.1 in Example 13.2 and discuss.

13.30–13.39 Equilibrium composition: single equilibrium reaction

13.30 Consider the dissociation of oxygen molecules $O_2 \Leftrightarrow O + O$ at 3800 K and 2 atm. Determine the equilibrium partial pressures and mole fractions of the O atoms and molecular oxygen.

13.31 Consider an initial equimolar mixture of molecular oxygen and carbon monoxide. For the equilibrium reaction $CO + \frac{1}{2}O_2 \Leftrightarrow CO_2$ determine the equilibrium partial pressures and the mole fractions of each species at 2200 K for a total pressure of 1 atm.

13.32 Consider an initially equimolar mixture of water vapor and carbon monoxide. For the equilibrium reaction $H_2O + CO \Leftrightarrow CO_2 + H_2$ determine the equilibrium partial pressures and the mole fractions of each species at 1500 K for a total pressure of 1.4 atm.

13.33 Consider the isolated equilibrium reaction $\frac{1}{2}N_2 + \frac{1}{2}O_2 \Leftrightarrow NO$. Determine the equilibrium mole fraction of NO at 1 atm and 4000 K. Assume that there are equal proportions of N and O atoms in the mixture.

13.34 Carbon monoxide and oxygen (O_2) exist in equilibrium with carbon dioxide. Determine the equilibrium composition (mole fractions) at 3200 K and 1 atm of an initial mixture of 2 kmol of carbon monoxide and 2 kmol of oxygen.

13.35 Carbon monoxide and water react to form carbon dioxide and hydrogen (H_2), the so-called water–gas shift reaction. Determine the equilibrium composition (mole fractions) of a mixture at 1100 K and 1 atm initially containing 1 kmol of carbon monoxide and 1 kmol of water.

13.36 Carbon monoxide and water react to form carbon dioxide and hydrogen (H_2), the so-called water–gas shift reaction. Determine the equilibrium composition (mole fractions) of a mixture at 1100 K and 1 atm initially containing 1 kmol of carbon monoxide and 2 kmol of water.

13.37 Hydrogen and oxygen react to form water. Determine the equilibrium composition (mole fractions) at 4000 K and 1 atm of an initial mixture of 1 kmol of hydrogen (H_2) and 1 kmol of oxygen (O_2).

13.38 Nitric oxide (NO) is a major pollutant formed in high-temperature combustion processes, but typically in less than equilibrium amounts. Equilibrium values thus can be considered to be upper limits. For the equilibrium reaction $N_2 + O_2 \Leftrightarrow 2\,NO$ determine the equilibrium partial pressures and the mole fractions (in parts per million) of the nitric oxide at combustion-like conditions of 2200 K and 10 atm. The initial mole fractions of O_2 and N_2 prior to the formation of any NO are 0.21 and 0.79, respectively, the approximate values for air.

13.39 The conversion of nitric oxide (NO) to nitrogen dioxide (NO_2) is important in atmospheric chemistry. National Ambient Air Quality Standards are set for NO_2, a major component of photochemical smog. Nitric oxide and oxygen are contained in an experimental reactor and allowed to react until the equilibrium $NO_2 \Leftrightarrow NO + \frac{1}{2}O_2$ is achieved. Determine the equilibrium partial pressures and the mole fractions of each species for conditions of 300 K and 1 atm. The ratio of O atoms to N atoms is 10.

13.40–13.42 Equilibrium composition: effect of temperature

13.40 Carbon monoxide and oxygen (O_2) react to form carbon dioxide. Determine the equilibrium composition (mole fractions) of a mixture at 298 K and 1 atm initially containing 2 kmol of carbon monoxide and 1 kmol of oxygen. Repeat for a temperature of 2000 K and discuss.

13.41 Molecular nitrogen has a strong triple bond and, thus, high temperatures are required to achieve any significant dissociation, as compared with less strongly bonded diatomic molecules. Explore the temperature dependence of the equilibrium reaction $N_2 \Leftrightarrow N + N$ by calculating the equilibrium mole fractions of N and N_2 and the fraction of N_2 dissociated at 1 atm for temperatures of 1000, 2000, 3000, and 4000 K. Plot your results on a logarithmic scale. Also compare the dissociation fraction for 3000 K with the corresponding dissociation fraction for oxygen, which you can calculate from Example 13.1.

13.42 In high-temperature combustion processes the principal oxide of nitrogen formed is nitric oxide (NO). At ambient atmospheric temperatures, however, the principal oxide of nitrogen present is nitrogen dioxide (NO_2). For the equilibrium reaction $NO_2 \Leftrightarrow NO + \frac{1}{2}O_2$determine the equilibrium partial pressures and the mole fractions of each species for a combustion-like temperature of 2200 K and an ambient temperature of 300 K, both at 1 atm. The only species present are O_2, NO, and NO_2, and the ratio of O atoms to N atoms is 10.

13.43–13.46 Equilibrium composition: effect of pressure

13.43 Consider the equilibrium reaction $N_2 \Leftrightarrow N + N$. Determine the equilibrium partial pressures and mole fractions of N and N_2 at 1 atm and 3000 K. Repeat for a pressure of 0.1 atm. Compare and contrast your results with those for $O_2 \Leftrightarrow O + O$ presented in Example 13.1. Discuss.

13.44 Consider the dissociation of oxygen molecules, $O_2 \Leftrightarrow O + O$. Determine the equilibrium mole fraction of O atoms and the fraction of the O_2 that is dissociated as functions of pressure for a fixed temperature of 2500 K. Use pressures of 0.01, 0.1, 1.0, and 10 atm. Plot your results using logarithmic scales.

13.45 Consider an initial equimolar mixture of molecular oxygen and carbon monoxide. For the equilibrium reaction $CO + \frac{1}{2}O_2 \Leftrightarrow CO_2$ determine the equilibrium partial pressures and the mole fractions of each species at 2200 K for total pressures of 1 and 5 atm.

13.46 Methanol-based fuel cells have been proposed for automotive applications. In this concept, steam reforming of the methanol is used to produce hydrogen. The hydrogen is then the fuel for the operation of the fuel cell. In this process, the water–gas shift equilibrium reaction $H_2O + CO \Leftrightarrow CO_2 + H_2$ is important. Determine the mole fraction of hydrogen in the equilibrium mixture at 1000 K for operation at both 1 and 5 atm. Initially, there are 10 kmol of H_2O for every kmol of CO present.

13.47–13.51 Equilibrium composition: effect of inert gas

13.47 Water at high temperature dissociates into hydrogen (H_2) and oxygen (O_2). A mixture of 1 kmol of water vapor and 10 kmol of nitrogen (N_2) is placed in a piston–cylinder device at an initial state of 298 K and 1 atm. The mixture is then heated with an electric heater at constant pressure to 4000 K. Assume the nitrogen does not react and that the H_2 and O_2 do not dissociate. Determine the mixture composition (mole fractions) and the percent dissociation at the final state.

13.48 Consider the dissociation of one kmol of carbon dioxide at 2500 K and 10 atm following the reaction $CO_2 \Leftrightarrow CO + \frac{1}{2}O_2$. Determine the equilibrium partial pressures and the mole fractions for the CO_2, CO, and O_2 when 9 kmol of N_2 is present in the mixture. Also, determine the dissociated fraction and compare it to the value when there is no N_2 present. Use the information in Table 13.2 for the undiluted case.

13.49 Carbon monoxide and oxygen (O_2) react to form carbon dioxide. Determine the equilibrium composition (mole fractions) at 3200 K and 1 atm of an initial mixture of 2 kmol of carbon monoxide, 1 kmol of oxygen, and 3.774 kmol of nitrogen. **HINT**: Treat the nitrogen as an inert species.

13.50 In high-temperature combustion processes the principal oxide of nitrogen formed is nitric oxide (NO). After being emitted into the atmosphere, the NO oxidizes to form nitrogen dioxide (NO_2). For the equilibrium reaction $NO_2 \Leftrightarrow NO + \frac{1}{2}O_2$ determine the equilibrium mole fractions of NO and NO_2 in the atmosphere (air) for an ambient temperature of 298 K and pressure of 1 atm. Initially, 10 ppm (1×10^{-5} mole fraction) of NO is present. There is no initial NO_2. The O_2 mole fraction is 0.21. Assume that the N_2 in the atmosphere does not participate in the reaction and can be considered as an inert gas. **HINTS**: Equation 13.27 is a good starting point. Also, because the O_2 mole fraction is much, much greater than either the NO or NO_2 mole fractions, it can be considered to be constant.

13.51 In high-temperature combustion processes the principal oxide of nitrogen formed is nitric oxide (NO). At ambient atmospheric temperatures, however, the principal oxide of nitrogen present is nitrogen dioxide (NO_2). (See Problem 13.50.) For the equilibrium reaction $NO_2 \Leftrightarrow NO + \frac{1}{2}O_2$ determine the equilibrium constant such that mole fractions of NO and NO_2 would be equal in an ambient environment of air at 1 atm. Use this K_p value to estimate the temperature at which the mole fractions are equal. Initially, 3000 ppm (3×10^{-3} mole fraction) of NO is present. There is no initial NO_2. The O_2 mole fraction is 0.21. Assume that the N_2 in the atmosphere does not participate in the reaction and can be considered as an inert gas. **HINTS**: See the hints for Problem 13.50.

13.52–13.54 Equilibrium composition: multiple equilibrium reactions

13.52 Add the $O_2 \Leftrightarrow O + O$ equilibrium reaction to the problem discussed in Example 13.4. Determine the equilibrium mole fractions of the product species (CO_2, CO, O_2, and O) at 0.1 atm and 2500 K.

13.53 Air (0.21 O_2 and 0.79 N_2 mole fractions) is heated to 3500 K at 1 atm. The O_2 and N_2 both dissociate ($O_2 \Leftrightarrow O + O$, $N_2 \Leftrightarrow N + N$) and nitric oxide is formed ($O_2 + N_2 \Leftrightarrow 2NO$).

A. Algebraically formulate the problem to determine the equilibrium mole fractions of the product species (O_2, O, N_2, N, and NO).

B. Solve the problem as formulated in part A. **HINT**: You might consider using software that can solve a system of non-linear algebraic equations.

13.54 Water vapor is heated to 3000 K at 1 atm. The H_2O partially dissociates to produce OH, H_2, H, O_2, and O species.

A. Algebraically formulate the problem to determine the equilibrium mole fractions of the H_2O, OH, H_2, H, O_2, and O in the mixture.

B. Solve the problem as formulated in part A. **HINT**: This will require software that can solve a system of non-linear algebraic equations.

13.55–13.59 Phase equilibrium

13.55 Using data for H_2O from the NIST resources, verify that the condition for phase equilibrium (Eq. 13.35) is met. Use temperatures of 300 and 600 K.

13.56 Using data for H_2O from the NIST resources, verify that the condition for phase equilibrium (Eqs. 13.35) is met for a wet mixture at 455 K having a quality of 0.90.

13.57 Ammonia is used as a refrigerant in large-scale refrigeration systems. Using data for ammonia (NH_3) from the NIST resources, verify that the condition for phase equilibrium (Eqs. 13.35) is met. Use temperatures of 300 and 400 K.

13.58 Heptane is frequently used as a single-component surrogate fuel to represent gasoline. Using data for heptane from the NIST resources, verify that the condition for phase equilibrium (Eqs. 13.35) is met. Use temperatures of 350 and 450 K.

13.59 Consider the liquid–vapor equilibrium of H_2O at 20 °C in which N_2 is added to the gas phase to obtain a total pressure of 1 atm. Assuming that the N_2 is both inert and insoluble in the liquid H_2O, how does the N_2 affect the equilibrium pressure (or partial pressure) of the H_2O vapor? To answer this, compute the partial pressure of the H_2O vapor in the $H_2O(g)$–N_2 mixture. The total pressure is fixed at 1 atm.

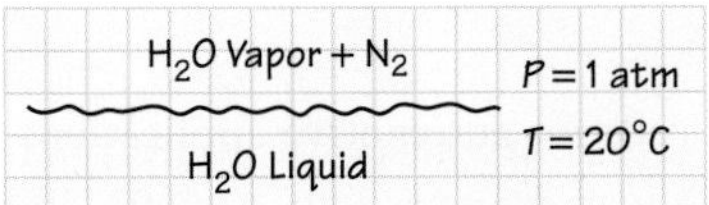

H_2O Vapor + N_2
$P = 1$ atm
$T = 20°C$
H_2O Liquid

13.60–13.61 Gibbs phase rule

13.60 Determine the number of degrees of freedom associated with the following systems. Also, use the NIST resources, and any other resources you can find, to determine numerical values associated with the pressure and temperature boundaries (lines, regions, point) associated with each system.

A. Liquid H_2O

B. Vapor H_2O

C. Liquid and solid H_2O

D. Liquid and vapor H_2O
E. Solid and vapor H_2O
F. Solid, liquid, and vapor H_2O

13.61 Determine the number of degrees of freedom associated with the following systems. Also, use the NIST resources, and any other resources you can find, to determine numerical values associated with the pressure and temperature boundaries (lines, regions, point) associated with each system.
A. Liquid CO_2
B. Vapor CO_2
C. Liquid and solid CO_2
D. Liquid and vapor CO_2
E. Solid and vapor CO_2
F. Solid, liquid, and vapor CO_2

13.62–13.65 Clapeyron and Clausius–Clapeyron equations

13.62 Derive the Clausius–Clapeyron equation from the Clapeyron equation, assuming that the gas-phase specific volume is much, much greater than that of the liquid phase, and that the pressure is sufficiently low that the vapor behaves as an ideal gas.

13.63 Use the Clausius–Clapeyron to determine saturation pressures from 373.12 K to 500 K and the percentage difference from the values from NIST or the steam tables in Appendix B. Use the known enthalpy of vaporization and saturation pressure at 373.12 K for your calculations. Plot your results (P_{sat} vs. T_{sat}). Based on your work, discuss the usefulness of the Clausius–Clapeyron equation.

13.64 Heptane is often used as a single-component fuel to model the behavior of gasoline. Volatility is important to the cold-start performance of gasoline in an automobile. Use the Clausius–Clapeyron to determine saturation pressures (kPa and psia) from 275 K (35.3 F) to 290 K (62.3 F) and the percentage difference from the values from the NIST database. Use the NIST enthalpy of vaporization at 282.5 K and the NIST vapor pressure at 275 K for your calculations. Plot your results (P_{sat} vs. T_{sat}) and discuss.

13.65 Methanol has been proposed for use in fuel cells for automotive applications. The vapor pressure for methanol (CH_3OH) was measured to be 4.02 and 352 kPa at 273 K and 373 K, repectively. Use these data to estimate the enthalpy of vaporization (kJ/kg) of methanol. Compare your result with the actual value (from the NIST resources or other source) at the average of the two temperatures.

APPENDIX A
Timeline

Lifetime	Key people	Date	Developments in thermal-fluid sciences	Date	World events
287–212 BCE	Archimedes	250 BCE	Laws of buoyancy formulated	264–241	First Punic War
1452–1519	Leonardo da Vinci	*ca.* 1506–1510	Formulated one-dimensional, steady, incompressible continuity equation	1492	Columbus first European to sail to Caribbean islands
1548–1620	Simon Stevin	1586	Solved hydrostatic paradox	1584	Sir Walter Raleigh lands expedition in Roanoke and names land Virginia
				1603	Shakespeare's *Hamlet* written
1642–1727	Isaac Newton	1686	Established principles of momentum conservation and formulated stress–strain relationship for "Newtonian" fluids; *Principia* published	1689	Peter the Great becomes Czar of Russia and attempts westernization
1663–1729	Thomas Newcomen	1712	Coal-burning steam engine/pump put into service	1707	United Kingdom of Britain formed by union of England, Scotland, and Wales
1700–1782	Daniel Bernoulli	1738	Published *Hydrodynamica*		
1707–1783	Leonhard Euler	1750	Formulated integral and differential forms of continuity equation and derived "Euler equation" for inviscid fluid (*ca.* 1750); provided modern form of Bernoulli equation	1755	Samuel Johnson's *Dictionary* published
1717–1783	Jean le Rond d'Alembert	1752	Published paper containing famous paradox		
1724–1792	John Smeaton	1759	First (?) use of scale-model experiments; conducted systematic parametric testing of water wheels; founded the British Society of Civil Engineers		
1736–1819	James Watt	1765	Improved steam engine using separate condenser		
1718–1798	Antoine Chézy	1770	Measured channel and pipe flow resistances; developed first resistance formula *ca.* 1775		Boston Massacre
1743–1794	Antoine Lavoisier	1777	Formulated concept of element conservation	1775–1783	American Revolution
1734–1809	Pierre Louis Georges Du Buat	1786	French military engineer; measured resistance in pipes and channels; helped resolved d'Alembert's paradox	1787	US Constitution
1753–1814	Benjamin Thompson (Count Rumford)	1798	Cannon boring experiments show motion and heat are related	1789–1795	French Revolution
					Malthus publishes *Essay on the Principles of Population*
1773–1829	Thomas Young	1801	Defined energy as capacity to do work		Napoleonic Wars: 1796–1815
1768–1830	Jean Baptiste Joseph Fourier	1807	First to formulate theory of dimensional analysis *ca.* 1807–1822		

1785–1836	Louis Marie Henri Navier	1822	Extends Euler's formulation of momentum conservation to viscous fluids – "Navier–Stokes equation"		Schubert's Symphony No. 8 *(The Unfinished)* composed
1796–1832	Sadi Carnot	1824	Formulated second law of thermodynamics and Carnot cycle		Beethoven composes Symphony No. 9 *(Choral)*
1797–1884	Gotthilf Heinrich Ludwig Hagen	1839	Developed empirical law that flow rate is proportional to the fourth power of the tube diameter for laminar flow		Charles Goodyear discovers vulcanization process for rubber
1799–1869	Jean Louis Poiseuille	1839	Obtained same result as Hagen in nearly coincident studies		Opium War between Britain and China 1839–1842
1814–1878	Julius Robert Mayer	1842	Conservation of energy principle stated		Ether first used as surgical anesthesia
1797–1886	Claude Barre de Saint-Venant	1843	Developed generalized form of momentum conservation for fluids		Charles Dickens' *A Christmas Carol* published
1789–1857	Augustin Louis de Cauchy	1845	Extended Euler's formulation of momentum conservation to viscous fluids		*The Raven and Other Poems* of Edgar Allen Poe published
1781–1840	Simeon Denis Poisson	1845	Extended Euler's formulation of momentum conservation to viscous fluids		Annexation of Texas to the United States
1819–1903	George Gabriel Stokes	1845	Extended Euler's formulation of momentum conservation to viscous fluids– "Navier–Stokes equation"		
1821–1894	Hermann Helmholtz	1847	Extended conservation of energy principle		Liberia established as independent republic
1824–1907	William Thomson (Lord Kelvin)	1848	Establishes concept of absolute temperature and defines "Kelvin" scale		Marks and Engels publish *Communist Manifesto*
1818–1889	James Prescott Joule	1849	Measured mechanical equivalent of heat		British annex Punjab
1822–1888	Rudolf Clausius	1850	Stated that the energy of the universe is constant and created the word *entropy*		Invention of the Bunsen gas burner
1798–1895	Franz Neumann	1856	Developed theory for Hagen–Poiseuille flow (independently from Hagenbach)		Treaty of Paris ends Crimean War
1833–1910	Eduard Hagenbach	1856	Developed theory for Hagen–Poiseuille flow (independently from Neumann)		Neanderthal skull discovered
1803–1858	Henry Philibert Gaspard Darcy	1857	Performed experimental study that conclusively showed the importance of wall roughness		Financial panic in United States and Europe
				1861–1865	US Civil War
1810–1879	William Froude	1868	Proposed drag laws for ship hulls from model testing		Louisa Alcott's *Little Women* and Dostoyevsky's *The Idiot* published

(cont.)

Lifetime	Key people	Date	Developments in thermal-fluid sciences	Date	World events
1842–1912	Osborne Reynolds	1874	Developed analogy among mass, heat, and momentum transfer (i.e., the Reynolds analogy)	1870–1871	Franco-Prussian War
1839–1903	Josiah Willard Gibbs	1876	First to develop generalized thermodynamic relationships		Invention of telephone (Bell) and internal combustion engine (Otto)
1844–1906	Ludwig Boltzmann	1877	Pioneer of statistical-mechanics-derived microscopic interpretation of entropy		Invention of phonograph (Edison)
1842–1919	John William Strutt (Lord Rayleigh)	1877	Published "Method of Dimensions"		Edison's invention of electric light bulb
1842–1912	Osborne Reynolds	1883	Defined critical parameter (i.e., Reynolds number) for transition from laminar to turbulent flow		Eruption of Krakatoa volcano in Indonesia
1857–1899	Aimeé Vaschy	1892	Developed concepts of dimensional analysis and similitude; published "Sur les lois similitude en physique"		Tchaikowsky's *Nutcracker Suite;* diesel engine patented
1842–1912	Osborne Reynolds	1895	Analyzed turbulent flows by decomposing variables into mean and fluctuating components (i.e., the Reynolds decomposition)	1894–1895	Sino-Japanese War
1858–1947	Max Planck	1900	Advanced understanding of second law and entropy from a macroscopic viewpoint; won Nobel Prize for quantum theory of energy	1903	Wright brothers first powered flight at Kittyhawk, NC
1875–1953	Ludwig Prandtl	1904	"Inventor" of boundary layers and separation; "Father of Modern Fluid Mechanics"	1904–1905	Russo-Japanese War
1882–1962	Dimitri Riabouchinsky	1911	Engineering applications of dimensional analysis and similitude		Chinese Republic replaces Manchu Dynasty
1881–1963	Theodor von Karman	1911	Numerous contributions to fluid mechanics; published studies of vortex street that bears his name		Roald Amundsen reaches South Pole
1883–1970	Paul Heinrich Blasius	1913	First to plot friction coefficient as a function of Reynolds number and relative roughness; student of Prandtl		Niels Bohr formulates theory of atomic structure
1867–1940	Edgar Buckingham	1914	Engineering applications of dimensional analysis and similitude	1914–1918	World War One

1871–1951	Moritz Weber	1919	Gave modern form to principles of similitude; gave names to Reynolds number and Froude number	1917–1920	Russian Revolution
1882–1961	Percy Bridgeman	1922	Author of classic monograph *Dimensional Analysis* (1922) and Nobel Prize winner for work and invention in high-pressure physics (1946)	1927– 1929	Economic collapse in Germany “Black Friday” world economic crisis begins Spain becomes a republic
				1932	Aldous Huxley's *Brave New World* published
1894–1979	Johann Nikuradse	1933	Performed meticulous experiments to quantify wall roughness effects in pipe flows using various-sized sand grains; student of Prandtl		Famine in USSR and Roosevelt's “New Deal” in USA; Hitler appointed chancellor in Germany
1859–1958	William Frederick Durand	1934	Developed dimensionless analysis for aeronautics; published six-volume reference, *Aerodynamic Theory*		Mao Zedong begins “Long March” north; German plebiscite elects Hitler as Führer
1886–1975	Geoffrey Ingram Taylor	1935	Developed statistical theories of turbulence; pioneer in the study of hydrodynamic stability	1938	Persia renamed Iran Orson Welles radio broadcast of *War of the Worlds*
				1939	Germany invades Poland; World War Two begins
1900–1977	Joseph Keenan	1941	Modern codifier of thermodynamics and author of classical textbook		Japan attacks Pearl Harbor; Manhattan Project begins
1880–1953	Lewis Ferry Moody	1944	Presents paper containing useful summary of friction data for internal flows (Moody chart)		Tennessee Williams's *The Glass Menagerie* presented
				1945	World War Two ends; United Nations created
1886–1984	Jerome Clarke Hunsaker	1947	First formal control volume analyses presented in *Engineering Applications of Fluid Mechanics* with B. G. Rightmire; developed modern wind tunnel (1914)	1947	Jackie Robinson joins Brooklyn Dodgers and breaks color barrier; Chuck Yeager breaks the sound barrier
				1947–1951	Marshall Plan for European recovery

APPENDIX B

Thermodynamic Properties of H_2O

TABLE B.1 Saturation Properties of Water and Steam – Temperature Increments

T (K)	*P* (kPa)	v (m^3/kg) sat. liquid	v (m^3/kg) sat. vapor	*u* (kJ/kg) sat. liquid	*u* (kJ/kg) sat. vapor	*h* (kJ/kg) sat. liquid	*h* (kJ/kg) sat. vapor	*s* (kJ/kg·K) sat. liquid	*s* (kJ/kg·K) sat. vapor
273.16	0.611650	0.0010002	205.99	0	2374.9	0.00061	2500.9	0	9.1555
274	0.650030	0.0010002	194.43	3.5435	2376.1	3.5442	2502.5	0.012952	9.1331
275	0.698460	0.0010001	181.60	7.7590	2377.5	7.7597	2504.3	0.028309	9.1066
276	0.750070	0.0010001	169.71	11.971	2378.8	11.972	2506.1	0.043600	9.0804
277	0.805020	0.0010001	158.70	16.181	2380.2	16.182	2508.0	0.058825	9.0544
278	0.863500	0.0010001	148.48	20.388	2381.6	20.389	2509.8	0.073985	9.0287
279	0.925700	0.0010001	139.00	24.593	2383.0	24.594	2511.6	0.089083	9.0031
280	0.991830	0.0010001	130.19	28.795	2384.3	28.796	2513.4	0.10412	8.9779
281	1.062200	0.0010002	122.01	32.996	2385.7	32.997	2515.3	0.11909	8.9528
282	1.136800	0.0010003	114.40	37.194	2387.1	37.195	2517.1	0.13401	8.9280
283	1.216000	0.0010003	107.32	41.391	2388.4	41.392	2518.9	0.14886	8.9034
284	1.300000	0.0010004	100.74	45.586	2389.8	45.587	2520.8	0.16366	8.8791
285	1.389100	0.0010005	94.602	49.779	2391.2	49.780	2522.6	0.17840	8.8549
286	1.483600	0.0010006	88.887	53.971	2392.6	53.973	2524.4	0.19308	8.8310
287	1.583600	0.0010008	83.560	58.162	2393.9	58.163	2526.2	0.20771	8.8073
288	1.689500	0.0010009	78.592	62.351	2395.3	62.353	2528.1	0.22228	8.7838
289	1.801600	0.0010011	73.955	66.540	2396.7	66.542	2529.9	0.23680	8.7605
290	1.920100	0.0010012	69.625	70.727	2398.0	70.729	2531.7	0.25126	8.7374
291	2.045400	0.0010014	65.581	74.914	2399.4	74.916	2533.5	0.26568	8.7145
292	2.177900	0.0010016	61.801	79.099	2400.8	79.101	2535.3	0.28003	8.6918
293	2.317800	0.0010018	58.267	83.284	2402.1	83.286	2537.2	0.29434	8.6693
294	2.465500	0.0010020	54.960	87.468	2403.5	87.471	2539.0	0.30860	8.6471
295	2.621300	0.0010022	51.865	91.652	2404.8	91.654	2540.8	0.32280	8.6250
296	2.785700	0.0010025	48.966	95.835	2406.2	95.837	2542.6	0.33696	8.6031
297	2.959100	0.0010027	46.251	100.02	2407.6	100.02	2544.4	0.35106	8.5814
298	3.141800	0.0010030	43.705	104.20	2408.9	104.20	2546.2	0.36512	8.5599
299	3.334300	0.0010032	41.318	108.38	2410.3	108.38	2548.0	0.37913	8.5385
300	3.536900	0.0010035	39.078	112.56	2411.6	112.56	2549.9	0.39309	8.5174
301	3.750200	0.0010038	36.976	116.74	2413.0	116.75	2551.7	0.40700	8.4964
302	3.974600	0.0010041	35.002	120.92	2414.4	120.93	2553.5	0.42087	8.4756
303	4.210600	0.0010044	33.147	125.10	2415.7	125.11	2555.3	0.43469	8.4550
304	4.458700	0.0010047	31.403	129.28	2417.1	129.29	2557.1	0.44846	8.4346
305	4.719400	0.0010050	29.764	133.46	2418.4	133.47	2558.9	0.46219	8.4144
306	4.993299	0.0010053	28.222	137.64	2419.8	137.65	2560.7	0.47587	8.3943
307	5.280799	0.0010057	26.770	141.82	2421.1	141.83	2562.5	0.48950	8.3744

TABLE B.1 (cont.)

T (K)	*P* (kPa)	v (m³/kg) sat. liquid	v (m³/kg) sat. vapor	*u* (kJ/kg) sat. liquid	*u* (kJ/kg) sat. vapor	*h* (kJ/kg) sat. liquid	*h* (kJ/kg) sat. vapor	*s* (kJ/kg·K) sat. liquid	*s* (kJ/kg·K) sat. vapor
308	5.582599	0.0010060	25.403	146.00	2422.5	146.01	2564.3	0.50310	8.3546
309	5.899199	0.0010063	24.116	150.18	2423.8	150.19	2566.1	0.51664	8.3351
310	6.231199	0.0010067	22.903	154.36	2425.2	154.37	2567.9	0.53015	8.3156
311	6.579299	0.0010071	21.759	158.54	2426.5	158.55	2569.7	0.54361	8.2964
312	6.944099	0.0010074	20.680	162.72	2427.8	162.73	2571.5	0.55702	8.2773
313	7.326199	0.0010078	19.663	166.90	2429.2	166.91	2573.2	0.57040	8.2584
314	7.726299	0.0010082	18.702	171.08	2430.5	171.09	2575.0	0.58373	8.2396
315	8.145199	0.0010086	17.795	175.26	2431.9	175.27	2576.8	0.59702	8.2210
316	8.583499	0.0010090	16.938	179.44	2433.2	179.45	2578.6	0.61027	8.2025
317	9.041899	0.0010094	16.129	183.62	2434.5	183.63	2580.4	0.62348	8.1842
318	9.521299	0.0010099	15.363	187.80	2435.9	187.81	2582.2	0.63664	8.1660
319	10.022989	0.0010103	14.639	191.98	2437.2	191.99	2583.9	0.64977	8.1480
320	10.546989	0.0010107	13.954	196.16	2438.5	196.17	2585.7	0.66285	8.1302

T (K)	*P* (MPa)	v (m³/kg) sat. liquid	v (m³/kg) sat. vapor	*u* (kJ/kg) sat. liquid	*u* (kJ/kg) sat. vapor	*h* (kJ/kg) sat. liquid	*h* (kJ/kg) sat. vapor	*s* (kJ/kg·K) sat. liquid	*s* (kJ/kg·K) sat. vapor
320	0.010547	0.0010107	13.954	196.16	2438.5	196.17	2585.7	0.66285	8.1302
325	0.013532	0.0010130	11.039	217.07	2445.2	217.08	2594.6	0.72768	8.0430
330	0.017214	0.0010155	8.8050	237.98	2451.8	238.00	2603.3	0.79154	7.9592
335	0.021719	0.0010181	7.0788	258.9	2458.3	258.93	2612.1	0.85447	7.8787
340	0.027189	0.0010209	5.7339	279.84	2464.8	279.87	2620.7	0.91650	7.8013
345	0.033784	0.0010239	4.6776	300.79	2471.2	300.82	2629.3	0.97766	7.7267
350	0.041683	0.0010270	3.8419	321.75	2477.6	321.79	2637.7	1.0380	7.6549
355	0.051081	0.0010303	3.1759	342.73	2483.9	342.78	2646.1	1.0975	7.5857
360	0.062195	0.0010337	2.6414	363.73	2490.1	363.79	2654.4	1.1562	7.5190
365	0.075261	0.0010373	2.2098	384.75	2496.2	384.82	2662.5	1.2142	7.4545
370	0.090536	0.0010410	1.8590	405.79	2502.3	405.88	2670.6	1.2715	7.3923
375	0.108310	0.0010449	1.5722	426.86	2508.2	426.97	2678.5	1.3281	7.3321
380	0.128860	0.0010490	1.3364	447.96	2514.1	448.09	2686.2	1.3839	7.2738
385	0.152529	0.0010532	1.1414	469.09	2519.8	469.25	2693.9	1.4392	7.2174
390	0.179649	0.0010575	0.97928	490.25	2525.4	490.44	2701.3	1.4938	7.1627
395	0.210609	0.0010620	0.84389	511.45	2530.9	511.67	2708.6	1.5478	7.1097
400	0.245779	0.0010667	0.73024	532.69	2536.2	532.95	2715.7	1.6013	7.0581
405	0.285589	0.0010715	0.63441	553.98	2541.4	554.28	2722.6	1.6541	7.0081
410	0.330459	0.0010765	0.55323	575.31	2546.5	575.66	2729.3	1.7065	6.9593
415	0.380879	0.0010817	0.48418	596.69	2551.4	597.10	2735.8	1.7583	6.9119
420	0.437309	0.0010870	0.42520	618.13	2556.2	618.60	2742.1	1.8097	6.8656
425	0.500259	0.0010926	0.37463	639.62	2560.7	640.17	2748.1	1.8606	6.8205
430	0.570269	0.0010983	0.33110	661.18	2565.1	661.80	2753.9	1.9110	6.7764
435	0.647879	0.0011042	0.29350	682.8	2569.3	683.52	2759.5	1.9610	6.7333
440	0.733679	0.0011103	0.26090	704.5	2573.3	705.31	2764.7	2.0106	6.6911
445	0.828249	0.0011166	0.23255	726.26	2577.1	727.19	2769.7	2.0598	6.6498
450	0.932209	0.0011232	0.20781	748.11	2580.7	749.16	2774.4	2.1087	6.6092

TABLE B.1 (cont.)

T (K)	*P* (MPa)	v (m^3/kg) sat. liquid	v (m^3/kg) sat. vapor	*u* (kJ/kg) sat. liquid	*u* (kJ/kg) sat. vapor	*h* (kJ/kg) sat. liquid	*h* (kJ/kg) sat. vapor	*s* (kJ/kg·K) sat. liquid	*s* (kJ/kg·K) sat. vapor
455	1.046289	0.0011299	0.18616	770.05	2584.0	771.23	2778.8	2.1571	6.5694
460	1.170988	0.0011369	0.16715	792.07	2587.2	793.41	2782.9	2.2053	6.5303
465	1.306987	0.0011442	0.15041	814.2	2590.1	815.69	2786.6	2.2532	6.4917
470	1.455187	0.0011517	0.13564	836.42	2592.7	838.09	2790.0	2.3007	6.4538
475	1.616086	0.0011594	0.12255	858.75	2595.0	860.62	2793.1	2.3480	6.4164
480	1.790586	0.0011675	0.11094	881.19	2597.1	883.28	2795.8	2.3950	6.3794
485	1.979285	0.0011758	0.10061	903.76	2598.9	906.09	2798.1	2.4418	6.3428
490	2.183184	0.0011845	0.091390	926.45	2600.5	929.04	2800.0	2.4884	6.3066
495	2.402884	0.0011935	0.083149	949.28	2601.7	952.15	2801.4	2.5348	6.2708
500	2.639283	0.0012029	0.075764	972.26	2602.5	975.43	2802.5	2.5810	6.2351
505	2.893182	0.0012127	0.069131	995.38	2603.0	998.89	2803.1	2.6271	6.1997
510	3.165582	0.0012228	0.063161	1018.7	2603.2	1022.5	2803.2	2.6731	6.1645
515	3.457181	0.0012334	0.057776	1042.1	2603.0	1046.4	2802.7	2.7189	6.1293
520	3.769080	0.0012445	0.052910	1065.8	2602.4	1070.5	2801.8	2.7647	6.0942
525	4.101980	0.0012561	0.048503	1089.6	2601.4	1094.8	2800.3	2.8104	6.0591
530	4.456979	0.0012682	0.044503	1113.7	2599.9	1119.3	2798.2	2.8561	6.0239
535	4.834978	0.0012809	0.040868	1138	2597.9	1144.2	2795.5	2.9019	5.9885
540	5.236977	0.0012942	0.037556	1162.5	2595.5	1169.3	2792.2	2.9476	5.9530
545	5.664076	0.0013083	0.034535	1187.3	2592.5	1194.7	2788.1	2.9935	5.9171
550	6.117276	0.0013231	0.031772	1212.4	2588.9	1220.5	2783.3	3.0394	5.8809
555	6.597675	0.0013387	0.029242	1237.8	2584.8	1246.6	2777.7	3.0855	5.8443
560	7.106274	0.0013553	0.026920	1263.5	2579.9	1273.1	2771.2	3.1319	5.8071
565	7.644473	0.0013729	0.024786	1289.6	2574.4	1300.1	2763.9	3.1785	5.7693
570	8.213272	0.0013917	0.022820	1316	2568.0	1327.5	2755.5	3.2254	5.7307
575	8.814071	0.0014118	0.021005	1343	2560.9	1355.4	2746.0	3.2727	5.6912
580	9.448070	0.0014334	0.019328	1370.4	2552.7	1383.9	2735.3	3.3205	5.6506
585	10.117686	0.0014567	0.017773	1398.4	2543.5	1413.1	2723.3	3.3690	5.6087
590	10.821674	0.0014820	0.016329	1426.9	2533.2	1443.0	2709.9	3.4181	5.5654
595	11.563662	0.0015095	0.014984	1456.2	2521.5	1473.7	2694.8	3.4680	5.5203
600	12.345649	0.0015399	0.013728	1486.4	2508.3	1505.4	2677.8	3.5190	5.4731
605	13.167635	0.0015735	0.012552	1517.4	2493.4	1538.1	2658.7	3.5713	5.4234
610	14.033620	0.0016112	0.011446	1549.6	2476.4	1572.2	2637.0	3.6252	5.3707
615	14.943604	0.0016541	0.010400	1583.2	2456.9	1608.0	2612.3	3.6811	5.3142
620	15.901586	0.0017039	0.0094067	1618.6	2434.3	1645.7	2583.9	3.7396	5.2528
625	16.908567	0.0017634	0.0084538	1656.5	2407.8	1686.3	2550.7	3.8019	5.1851
630	17.969544	0.0018374	0.0075279	1697.7	2375.8	1730.7	2511.1	3.8698	5.1084
635	19.086516	0.0019353	0.0066074	1744.3	2335.7	1781.2	2461.8	3.9463	5.0181
640	20.265481	0.0020767	0.0056451	1799.7	2281.1	1841.8	2395.5	4.0375	4.9027
645	21.515425	0.0023527	0.0044553	1880.5	2185.2	1931.1	2281.0	4.1722	4.7147
647.096	22.064000	0.0031056	0.0031056	2015.7	2015.7	2084.3	2084.3	4.4070	4.4070

TABLE B.2 Saturation Properties of Water and Steam – Pressure Increments

P (kPa)	*T* (K)	v (m³/kg) sat. liquid	v (m³/kg) sat. vapor	*u* (kJ/kg) sat. liquid	*u* (kJ/kg) sat. vapor	*h* (kJ/kg) sat. liquid	*h* (kJ/kg) sat. vapor	*s* (kJ/kg·K) sat. liquid	*s* (kJ/kg·K) sat. vapor
1.0	280.12	0.0010001	129.18	29.298	2384.5	29.299	2513.7	0.10591	8.9749
2.0	290.64	0.0010014	66.987	73.426	2398.9	73.428	2532.9	0.26056	8.7226
3.0	297.23	0.0010028	45.653	100.97	2407.9	100.98	2544.8	0.35429	8.5764
4.0	302.11	0.0010041	34.791	121.38	2414.5	121.39	2553.7	0.42239	8.4734
5.0	306.02	0.0010053	28.185	137.74	2419.8	137.75	2560.7	0.47620	8.3938
6.0	309.31	0.0010065	23.733	151.47	2424.2	151.48	2566.6	0.52082	8.3290
7.0	312.15	0.0010075	20.524	163.34	2428.0	163.35	2571.7	0.55903	8.2745
8.0	314.66	0.0010085	18.099	173.83	2431.4	173.84	2576.2	0.59249	8.2273
9.0	316.91	0.0010094	16.199	183.24	2434.4	183.25	2580.2	0.62230	8.1858
10.0	318.96	0.0010103	14.670	191.80	2437.2	191.81	2583.9	0.6492	8.1488
11.0	320.83	0.0010111	13.412	199.64	2439.7	199.65	2587.2	0.67372	8.1154
12.0	322.57	0.0010119	12.358	206.90	2442.0	206.91	2590.3	0.69628	8.0849
13.0	324.18	0.0010126	11.462	213.65	2444.1	213.67	2593.1	0.71717	8.0570
14.0	325.70	0.0010134	10.691	219.98	2446.1	219.99	2595.8	0.73664	8.0311
15.0	327.12	0.0010140	10.020	225.93	2448.0	225.94	2598.3	0.75486	8.0071
16.0	328.46	0.0010147	9.4306	231.55	2449.8	231.57	2600.6	0.77201	7.9846
17.0	329.74	0.0010154	8.9087	236.88	2451.4	236.90	2602.9	0.78820	7.9636
18.0	330.95	0.0010160	8.4431	241.95	2453.0	241.96	2605.0	0.80355	7.9437
19.0	332.10	0.0010166	8.0252	246.78	2454.5	246.80	2607.0	0.81813	7.9250
20.0	333.21	0.0010172	7.6480	251.40	2456.0	251.42	2608.9	0.83202	7.9072
21.0	334.27	0.0010177	7.3056	255.83	2457.4	255.85	2610.8	0.84530	7.8903
22.0	335.28	0.0010183	6.9936	260.09	2458.7	260.11	2612.5	0.85800	7.8743
23.0	336.26	0.0010188	6.7079	264.18	2460.0	264.20	2614.2	0.87020	7.8589
24.0	337.20	0.0010193	6.4453	268.13	2461.2	268.15	2615.9	0.88191	7.8442
25.0	338.11	0.0010198	6.2032	271.93	2462.4	271.96	2617.4	0.89319	7.8302
26.0	338.99	0.0010203	5.9792	275.62	2463.5	275.64	2619.0	0.90407	7.8167
27.0	339.84	0.0010208	5.7713	279.18	2464.6	279.21	2620.4	0.91457	7.8037
28.0	340.67	0.0010213	5.5778	282.64	2465.7	282.66	2621.8	0.92472	7.7912
29.0	341.47	0.0010218	5.3972	285.99	2466.7	286.02	2623.2	0.93455	7.7791
30.0	342.25	0.0010222	5.2284	289.24	2467.7	289.27	2624.5	0.94407	7.7675
31.0	343.00	0.0010227	5.0702	292.41	2468.7	292.44	2625.8	0.95331	7.7562
32.0	343.74	0.0010231	4.9215	295.49	2469.6	295.52	2627.1	0.96228	7.7453
33.0	344.45	0.0010236	4.7816	298.49	2470.5	298.52	2628.3	0.97100	7.7348
34.0	345.15	0.0010240	4.6497	301.41	2471.4	301.45	2629.5	0.97948	7.7246
35.0	345.83	0.0010244	4.5251	304.27	2472.3	304.30	2630.7	0.98774	7.7146
36.0	346.50	0.0010248	4.4072	307.05	2473.1	307.09	2631.8	0.99579	7.7050
37.0	347.14	0.0010252	4.2955	309.77	2474.0	309.81	2632.9	1.0036	7.6956
38.0	347.78	0.0010256	4.1895	312.43	2474.8	312.47	2634.0	1.0113	7.6865
39.0	348.40	0.0010260	4.0888	315.04	2475.6	315.08	2635.0	1.0188	7.6776
40.0	349.01	0.0010264	3.9930	317.58	2476.3	317.62	2636.1	1.0261	7.6690
42.0	350.18	0.0010271	3.8146	322.52	2477.8	322.56	2638.0	1.0402	7.6524
44.0	351.32	0.0010279	3.6520	327.27	2479.3	327.31	2639.9	1.0537	7.6365
46.0	352.40	0.0010286	3.5031	331.83	2480.6	331.88	2641.8	1.0667	7.6214
48.0	353.45	0.0010293	3.3663	336.24	2481.9	336.29	2643.5	1.0792	7.6069

TABLE B.2 (cont.)

P (kPa)	T (K)	ν (m^3/kg) sat. liquid	ν (m^3/kg) sat. vapor	u (kJ/kg) sat. liquid	u (kJ/kg) sat. vapor	h (kJ/kg) sat. liquid	h (kJ/kg) sat. vapor	s (kJ/kg·K) sat. liquid	s (kJ/kg·K) sat. vapor
50.0	354.47	0.0010299	3.2400	340.49	2483.2	340.54	2645.2	1.0912	7.5930
52.0	355.45	0.0010306	3.1232	344.60	2484.4	344.66	2646.8	1.1028	7.5797
54.0	356.40	0.0010312	3.0148	348.59	2485.6	348.64	2648.4	1.1140	7.5669
56.0	357.32	0.0010319	2.9139	352.45	2486.8	352.51	2649.9	1.1248	7.5545
58.0	358.21	0.0010325	2.8198	356.20	2487.9	356.26	2651.4	1.1353	7.5426
60.0	359.08	0.0010331	2.7317	359.84	2489.0	359.91	2652.9	1.1454	7.5311
62.0	359.92	0.0010337	2.6492	363.39	2490.0	363.45	2654.2	1.1553	7.5200
64.0	360.74	0.0010342	2.5716	366.84	2491.0	366.91	2655.6	1.1649	7.5093
66.0	361.54	0.0010348	2.4986	370.20	2492.0	370.27	2656.9	1.1742	7.4989
68.0	362.32	0.0010354	2.4298	373.48	2493.0	373.55	2658.2	1.1833	7.4888
70.0	363.08	0.0010359	2.3648	376.68	2493.9	376.75	2659.4	1.1921	7.4790
72.0	363.82	0.0010364	2.3033	379.80	2494.8	379.88	2660.6	1.2007	7.4695
74.0	364.55	0.0010370	2.2450	382.86	2495.7	382.94	2661.8	1.2091	7.4602
76.0	365.26	0.0010375	2.1897	385.84	2496.5	385.92	2663.0	1.2172	7.4512
78.0	365.96	0.0010380	2.1371	388.77	2497.4	388.85	2664.1	1.2252	7.4425
80.0	366.64	0.0010385	2.0871	391.63	2498.2	391.71	2665.2	1.2330	7.4339
82.0	367.30	0.0010390	2.0394	394.43	2499.0	394.51	2666.3	1.2407	7.4256
84.0	367.95	0.0010395	1.9940	397.18	2499.8	397.26	2667.3	1.2482	7.4175
86.0	368.59	0.001040	1.9506	399.87	2500.6	399.96	2668.3	1.2555	7.4096
88.0	369.22	0.0010404	1.9091	402.51	2501.3	402.60	2669.3	1.2626	7.4018
90.0	369.84	0.0010409	1.8694	405.10	2502.1	405.20	2670.3	1.2696	7.3943
92.0	370.44	0.0010414	1.8313	407.65	2502.8	407.75	2671.3	1.2765	7.3869
94.0	371.04	0.0010418	1.7949	410.15	2503.5	410.25	2672.2	1.2833	7.3796
96.0	371.62	0.0010423	1.7599	412.61	2504.2	412.71	2673.1	1.2899	7.3726
98.0	372.19	0.0010427	1.7262	415.02	2504.9	415.13	2674.1	1.2964	7.3656
100.0	372.76	0.0010432	1.6939	417.40	2505.6	417.50	2674.9	1.3028	7.3588
101.325 (atmos)	373.12	0.0010434	1.6732	418.95	2506.0	419.06	2675.5	1.3069	7.3544

P (MPa)	T (K)	ν (m^3/kg) sat. liquid	ν (m^3/kg) sat. vapor	u (kJ/kg) sat. liquid	u (kJ/kg) sat. vapor	h (kJ/kg) sat. liquid	h (kJ/kg) sat. vapor	s (kJ/kg·K) sat. liquid	s (kJ/kg·K) sat. vapor
0.15	384.50	0.0010527	1.1593	466.97	2519.2	467.13	2693.1	1.4337	7.2230
0.20	393.36	0.0010605	0.88568	504.49	2529.1	504.70	2706.2	1.5302	7.1269
0.25	400.56	0.0010672	0.71866	535.08	2536.8	535.34	2716.5	1.6072	7.0524
0.30	406.67	0.0010732	0.60576	561.10	2543.2	561.43	2724.9	1.6717	6.9916
0.35	412.01	0.0010786	0.52418	583.88	2548.5	584.26	2732.0	1.7274	6.9401
0.40	416.76	0.0010836	0.46238	604.22	2553.1	604.65	2738.1	1.7765	6.8955
0.45	421.05	0.0010882	0.41390	622.65	2557.1	623.14	2743.4	1.8205	6.8560
0.50	424.98	0.0010925	0.37481	639.54	2560.7	640.09	2748.1	1.8604	6.8207
0.55	428.61	0.0010967	0.34260	655.16	2563.9	655.76	2752.3	1.8970	6.7886
0.60	431.98	0.0011006	0.31558	669.72	2566.8	670.38	2756.1	1.9308	6.7592
0.65	435.13	0.0011044	0.29259	683.36	2569.4	684.08	2759.6	1.9623	6.7322
0.70	438.10	0.0011080	0.27277	696.23	2571.8	697.00	2762.8	1.9918	6.7071
0.75	440.90	0.0011114	0.25551	708.40	2574.0	709.24	2765.6	2.0195	6.6836

TABLE B.2 (cont.)

P (MPa)	*T* (K)	v (m^3/kg) sat. liquid	v (m^3/kg) sat. vapor	*u* (kJ/kg) sat. liquid	*u* (kJ/kg) sat. vapor	*h* (kJ/kg) sat. liquid	*h* (kJ/kg) sat. vapor	*s* (kJ/kg·K) sat. liquid	*s* (kJ/kg·K) sat. vapor
0.80	443.56	0.0011148	0.24034	719.97	2576.0	720.86	2768.3	2.0457	6.6616
0.85	446.09	0.0011180	0.22689	731.00	2577.9	731.95	2770.8	2.0705	6.6409
0.90	448.50	0.0011212	0.21489	741.55	2579.6	742.56	2773.0	2.0940	6.6213
0.95	450.81	0.0011242	0.20410	751.67	2581.2	752.74	2775.1	2.1165	6.6027
1.00	453.03	0.0011272	0.19436	761.39	2582.7	762.52	2777.1	2.1381	6.5850
1.10	457.21	0.0011330	0.17745	779.78	2585.5	781.03	2780.6	2.1785	6.5520
1.20	461.11	0.0011385	0.16326	796.96	2587.8	798.33	2783.7	2.2159	6.5217
1.30	464.75	0.0011438	0.15119	813.11	2589.9	814.60	2786.5	2.2508	6.4936
1.40	468.19	0.0011489	0.14078	828.36	2591.8	829.97	2788.8	2.2835	6.4675
1.50	471.44	0.0011539	0.13171	842.83	2593.4	844.56	2791.0	2.3143	6.4430
1.60	474.52	0.0011587	0.12374	856.60	2594.8	858.46	2792.8	2.3435	6.4199
1.70	477.46	0.0011634	0.11667	869.76	2596.1	871.74	2794.5	2.3711	6.3981
1.80	480.26	0.0011679	0.11037	882.37	2597.2	884.47	2795.9	2.3975	6.3775
1.90	482.95	0.0011724	0.10470	894.48	2598.2	896.71	2797.2	2.4227	6.3578
2.00	485.53	0.0011767	0.099585	906.14	2599.1	908.50	2798.3	2.4468	6.3390
2.10	488.01	0.0011810	0.094938	917.39	2599.9	919.87	2799.3	2.4699	6.3210
2.20	490.4	0.0011852	0.090698	928.27	2600.6	930.87	2800.1	2.4921	6.3038
2.30	492.71	0.0011894	0.086815	938.79	2601.1	941.53	2800.8	2.5136	6.2872
2.40	494.94	0.0011934	0.083244	949.00	2601.6	951.87	2801.4	2.5343	6.2712
2.50	497.10	0.0011974	0.079949	958.91	2602.1	961.91	2801.9	2.5543	6.2558
2.60	499.20	0.0012014	0.076899	968.55	2602.4	971.67	2802.3	2.5736	6.2409
2.70	501.23	0.0012053	0.074066	977.93	2602.7	981.18	2802.7	2.5924	6.2264
2.80	503.21	0.0012091	0.071429	987.07	2602.9	990.46	2802.9	2.6106	6.2124
2.90	505.13	0.0012129	0.068968	995.99	2603.1	999.51	2803.1	2.6283	6.1988
3.00	507.00	0.0012167	0.066664	1004.7	2603.2	1008.3	2803.2	2.6455	6.1856
3.10	508.83	0.0012204	0.064504	1013.2	2603.2	1017.0	2803.2	2.6623	6.1727
3.20	510.61	0.0012241	0.062475	1021.5	2603.2	1025.4	2803.1	2.6787	6.1602
3.30	512.35	0.0012278	0.060564	1029.7	2603.2	1033.7	2803.0	2.6946	6.1479
3.40	514.05	0.0012314	0.058761	1037.7	2603.1	1041.8	2802.9	2.7102	6.1360
3.50	515.71	0.0012350	0.057058	1045.5	2602.9	1049.8	2802.6	2.7254	6.1243
3.60	517.33	0.0012385	0.055446	1053.1	2602.8	1057.6	2802.4	2.7403	6.1129
3.70	518.92	0.0012421	0.053918	1060.7	2602.6	1065.3	2802.1	2.7549	6.1018
3.80	520.48	0.0012456	0.052467	1068.1	2602.3	1072.8	2801.7	2.7691	6.0908
3.90	522.01	0.0012491	0.051089	1075.3	2602.0	1080.2	2801.3	2.7831	6.0801
4.00	523.50	0.0012526	0.049776	1082.5	2601.7	1087.5	2800.8	2.7968	6.0696
4.10	524.97	0.0012560	0.048525	1089.5	2601.4	1094.7	2800.3	2.8102	6.0592
4.20	526.41	0.0012594	0.047332	1096.4	2601.0	1101.7	2799.8	2.8234	6.0491
4.30	527.83	0.0012629	0.046192	1103.2	2600.6	1108.7	2799.2	2.8363	6.0391
4.40	529.22	0.0012663	0.045102	1109.9	2600.1	1115.5	2798.6	2.8490	6.0293
4.50	530.59	0.0012696	0.044059	1116.5	2599.7	1122.2	2797.9	2.8615	6.0197
4.60	531.93	0.0012730	0.043059	1123.0	2599.2	1128.9	2797.3	2.8738	6.0102
4.70	533.25	0.0012764	0.042100	1129.5	2598.7	1135.5	2796.5	2.8859	6.0009
4.80	534.55	0.0012797	0.041180	1135.8	2598.1	1141.9	2795.8	2.8978	5.9917
4.90	535.83	0.0012831	0.040296	1142.0	2597.6	1148.3	2795.0	2.9095	5.9826
5.00	537.09	0.0012864	0.039446	1148.2	2597.0	1154.6	2794.2	2.9210	5.9737

TABLE B.2 (cont.)

P (MPa)	T (K)	v (m³/kg) sat. liquid	v (m³/kg) sat. vapor	u (kJ/kg) sat. liquid	u (kJ/kg) sat. vapor	h (kJ/kg) sat. liquid	h (kJ/kg) sat. vapor	s (kJ/kg·K) sat. liquid	s (kJ/kg·K) sat. vapor
5.0	537.09	0.0012864	0.039446	1148.2	2597.0	1154.6	2794.2	2.9210	5.9737
5.5	543.12	0.0013029	0.035642	1177.9	2593.7	1185.1	2789.7	2.9762	5.9307
6.0	548.73	0.0013193	0.032448	1206.0	2589.9	1213.9	2784.6	3.0278	5.8901
6.5	554.01	0.0013356	0.029727	1232.7	2585.7	1241.4	2778.9	3.0764	5.8516
7.0	558.98	0.0013519	0.027378	1258.2	2581	1267.7	2772.6	3.1224	5.8148
7.5	563.69	0.0013682	0.025330	1282.7	2575.9	1292.9	2765.9	3.1662	5.7793
8.0	568.16	0.0013847	0.023526	1306.2	2570.5	1317.3	2758.7	3.2081	5.7450
8.5	572.42	0.0014013	0.021923	1329.0	2564.7	1340.9	2751.0	3.2483	5.7117
9.0	576.49	0.0014181	0.020490	1351.1	2558.5	1363.9	2742.9	3.2870	5.6791
9.5	580.40	0.0014352	0.019199	1372.6	2552.0	1386.2	2734.4	3.3244	5.6473
10.0	584.15	0.0014526	0.018030	1393.5	2545.2	1408.1	2725.5	3.3606	5.6160
10.5	587.75	0.0014703	0.016965	1414.0	2538.0	1429.4	2716.1	3.3959	5.5851
11.0	591.23	0.0014885	0.015990	1434.1	2530.5	1450.4	2706.3	3.4303	5.5545
11.5	594.58	0.0015071	0.015093	1453.8	2522.6	1471.1	2696.1	3.4638	5.5241
12.0	597.83	0.0015263	0.014264	1473.1	2514.3	1491.5	2685.4	3.4967	5.4939
12.5	600.96	0.0015461	0.013496	1492.3	2505.6	1511.6	2674.3	3.5290	5.4638
13.0	604.00	0.0015665	0.012780	1511.1	2496.5	1531.5	2662.7	3.5608	5.4336
13.5	606.95	0.0015877	0.012112	1529.9	2487.0	1551.3	2650.5	3.5921	5.4032
14.0	609.82	0.0016097	0.011485	1548.4	2477.1	1571.0	2637.9	3.6232	5.3727
14.5	612.60	0.0016328	0.010895	1566.9	2466.6	1590.6	2624.6	3.6539	5.3418
15.0	615.31	0.0016570	0.010338	1585.3	2455.6	1610.2	2610.7	3.6846	5.3106
15.5	617.94	0.0016824	0.0098106	1603.8	2444.1	1629.9	2596.1	3.7151	5.2788
16.0	620.50	0.0017094	0.0093088	1622.3	2431.8	1649.7	2580.8	3.7457	5.2463
16.5	623.00	0.0017383	0.0088299	1641.0	2418.9	1669.7	2564.6	3.7765	5.2130
17.0	625.44	0.0017693	0.0083709	1659.9	2405.2	1690.0	2547.5	3.8077	5.1787
17.5	627.82	0.0018029	0.0079292	1679.2	2390.5	1710.8	2529.3	3.8394	5.1431
18.0	630.14	0.0018398	0.0075017	1699.0	2374.8	1732.1	2509.8	3.8718	5.1061
18.5	632.41	0.0018807	0.0070856	1719.3	2357.8	1754.1	2488.8	3.9053	5.0670
19.0	634.62	0.0019268	0.0066773	1740.5	2339.1	1777.2	2466.0	3.9401	5.0256
19.5	636.79	0.0019792	0.0062725	1762.8	2318.5	1801.4	2440.8	3.9767	4.9808
20.0	638.90	0.002040	0.0058652	1786.4	2295.0	1827.2	2412.3	4.0156	4.9314
20.5	640.96	0.0021126	0.0054457	1812.0	2267.6	1855.3	2379.2	4.0579	4.8753
21.0	642.98	0.0022055	0.0049961	1841.2	2233.7	1887.6	2338.6	4.1064	4.8079
21.5	644.94	0.0023468	0.0044734	1879.1	2186.9	1929.5	2283.1	4.1698	4.7181
22.0	646.86	0.0027044	0.0036475	1951.8	2092.8	2011.3	2173.1	4.2945	4.5446
22.064	647.096	0.0031056	0.0031056	2015.7	2015.7	2084.3	2084.3	4.4070	4.4070

TABLE B.3 Superheated Vapor (Steam)*

TABLE B.3a Isobaric Data for $P = 0.006$ MPa

T (K)	P (MPa)	ρ (kg/m^3)	v (m^3/kg)	u (kJ/kg)	h (kJ/kg)	s (kJ/kg·K)
309.31	0.006	0.042135	23.733	2424.2	2566.6	8.3290
320	0.006	0.040708	24.565	2439.7	2587.1	8.3940
340	0.006	0.038291	26.116	2468.4	2625.1	8.5092
360	0.006	0.036151	27.662	2497.0	2663.0	8.6176
380	0.006	0.034239	29.206	2525.7	2701.0	8.7202
400	0.006	0.032522	30.748	2554.6	2739.0	8.8179
420	0.006	0.030969	32.290	2583.5	2777.3	8.9112
440	0.006	0.029559	33.831	2612.7	2815.7	9.0005
460	0.006	0.028272	35.371	2642.1	2854.3	9.0863
480	0.006	0.027092	36.911	2671.6	2893.1	9.1689
500	0.006	0.026008	38.450	2701.4	2932.1	9.2485
520	0.006	0.025006	39.990	2731.4	2971.4	9.3255
540	0.006	0.024079	41.529	2761.7	3010.9	9.4000
560	0.006	0.023219	43.068	2792.2	3050.6	9.4722
580	0.006	0.022418	44.607	2822.9	3090.6	9.5424
600	0.006	0.02167	46.146	2853.9	3130.8	9.6105
620	0.006	0.020971	47.685	2885.1	3171.2	9.6769
640	0.006	0.020315	49.224	2916.6	3211.9	9.7415
660	0.006	0.019699	50.763	2948.3	3252.9	9.8046
680	0.006	0.01912	52.301	2980.4	3294.2	9.8661
700	0.006	0.018573	53.840	3012.6	3335.7	9.9263
720	0.006	0.018057	55.379	3045.2	3377.4	9.9851
740	0.006	0.017569	56.917	3078.0	3419.5	10.043
760	0.006	0.017107	58.456	3111.0	3461.8	10.099
780	0.006	0.016668	59.995	3144.4	3504.3	10.154
800	0.006	0.016251	61.533	3178.0	3547.2	10.209

* Property values generated from NIST Database 23: *REFPROP* Version 7.0 (August 2002).

TABLE B.3b Isobaric Data for $P = 0.035$ MPa

T (K)	P (MPa)	ρ (kg/m^3)	v (m^3/kg)	u (kJ/kg)	h (kJ/kg)	s (kJ/kg·K)
345.83	0.035	0.22099	4.5251	2472.3	2630.7	7.7146
360	0.035	0.21197	4.7176	2493.5	2658.6	7.7939
380	0.035	0.20052	4.9871	2523.1	2697.7	7.8994
400	0.035	0.1903	5.2549	2552.5	2736.5	7.9989
420	0.035	0.1811	5.5218	2581.9	2775.2	8.0934
440	0.035	0.17278	5.7879	2611.4	2813.9	8.1835
460	0.035	0.16519	6.0535	2640.9	2852.8	8.2699
480	0.035	0.15826	6.3187	2670.7	2891.8	8.353
500	0.035	0.15189	6.5837	2700.6	2931.0	8.433
520	0.035	0.14602	6.8484	2730.7	2970.4	8.5102
540	0.035	0.14059	7.113	2761.1	3010.0	8.5849
560	0.035	0.13555	7.3775	2791.6	3049.8	8.6573
580	0.035	0.13086	7.6419	2822.4	3089.9	8.7276
600	0.035	0.12648	7.9062	2853.4	3130.1	8.7958
620	0.035	0.12239	8.1704	2884.7	3170.7	8.8623
640	0.035	0.11856	8.4346	2916.2	3211.4	8.927
660	0.035	0.11496	8.6987	2948.0	3252.5	8.9901
680	0.035	0.11157	8.9627	2980.0	3293.7	9.0517
700	0.035	0.10838	9.2268	3012.3	3335.3	9.1119
720	0.035	0.10537	9.4908	3044.9	3377.1	9.1708
740	0.035	0.10251	9.7547	3077.7	3419.1	9.2284
760	0.035	0.099813	10.019	3110.8	3461.4	9.2848
780	0.035	0.097251	10.283	3144.2	3504.0	9.3402
800	0.035	0.094818	10.547	3177.8	3546.9	9.3944
820	0.035	0.092503	10.81	3211.7	3590.1	9.4477
840	0.035	0.090299	11.074	3245.9	3633.5	9.5
363.08	0.07	0.42287	2.3648	2493.9	2659.4	7.479

TABLE B.3c Isobaric Data for $P = 0.07$ MPa

T (K)	P (MPa)	ρ (kg/m³)	v (m³/kg)	u (kJ/kg)	h (kJ/kg)	s (kJ/kg·K)
380	0.07	0.40301	2.4813	2519.9	2693.5	7.5709
400	0.07	0.38205	2.6175	2550.0	2733.2	7.6727
420	0.07	0.3633	2.7525	2579.9	2772.6	7.7687
440	0.07	0.3464	2.8868	2609.7	2811.8	7.8599
460	0.07	0.33106	3.0206	2639.6	2851.0	7.9471
480	0.07	0.31706	3.154	2669.5	2890.3	8.0307
500	0.07	0.30422	3.2871	2699.6	2929.7	8.1111
520	0.07	0.2924	3.42	2729.9	2969.3	8.1886
540	0.07	0.28147	3.5528	2760.3	3009.0	8.2636
560	0.07	0.27134	3.6854	2790.9	3048.9	8.3362
580	0.07	0.26193	3.8179	2821.8	3089.0	8.4066
600	0.07	0.25314	3.9503	2852.9	3129.4	8.475
620	0.07	0.24494	4.0827	2884.2	3170.0	8.5416
640	0.07	0.23725	4.215	2915.8	3210.8	8.6064
660	0.07	0.23003	4.3472	2947.6	3251.9	8.6696
680	0.07	0.22324	4.4794	2979.6	3293.2	8.7312
700	0.07	0.21685	4.6116	3012.0	3334.8	8.7915
720	0.07	0.2108	4.7437	3044.5	3376.6	8.8504
740	0.07	0.20509	4.8758	3077.4	3418.7	8.9081
760	0.07	0.19968	5.0079	3110.5	3461.1	8.9645
780	0.07	0.19455	5.14	3143.9	3503.7	9.0199
800	0.07	0.18968	5.2721	3177.5	3546.6	9.0742
820	0.07	0.18504	5.4041	3211.5	3589.7	9.1275
840	0.07	0.18063	5.5361	3245.7	3633.2	9.1798
860	0.07	0.17643	5.6681	3280.1	3676.9	9.2313

TABLE B.3d Isobaric Data for $P = 0.1$ MPa

T (K)	P (MPa)	ρ (kg/m³)	v (m³/kg)	u (kJ/kg)	h (kJ/kg)	s (kJ/kg·K)
372.76	0.1	0.59034	1.6939	2505.6	2674.9	7.3588
380	0.1	0.57824	1.7294	2517.0	2689.9	7.3986
400	0.1	0.54761	1.8261	2547.8	2730.4	7.5025
420	0.1	0.52038	1.9217	2578.2	2770.3	7.5999
440	0.1	0.49592	2.0165	2608.3	2810.0	7.6921
460	0.1	0.47378	2.1107	2638.4	2849.5	7.7799
480	0.1	0.45361	2.2045	2668.5	2889.0	7.864
500	0.1	0.43514	2.2981	2698.7	2928.6	7.9447
520	0.1	0.41815	2.3915	2729.1	2968.2	8.0226
540	0.1	0.40247	2.4847	2759.6	3008.1	8.0978
560	0.1	0.38794	2.5777	2790.3	3048.1	8.1705
580	0.1	0.37444	2.6707	2821.3	3088.3	8.2411
600	0.1	0.36185	2.7635	2852.4	3128.8	8.3096
620	0.1	0.3501	2.8563	2883.8	3169.4	8.3763
640	0.1	0.33909	2.9491	2915.4	3210.3	8.4411
660	0.1	0.32876	3.0418	2947.2	3251.4	8.5044
680	0.1	0.31904	3.1344	2979.3	3292.8	8.5661
700	0.1	0.30988	3.227	3011.7	3334.4	8.6264
720	0.1	0.30124	3.3196	3044.3	3376.2	8.6854
740	0.1	0.29307	3.4122	3077.1	3418.3	8.7431
760	0.1	0.28533	3.5047	3110.3	3460.7	8.7996
780	0.1	0.27799	3.5972	3143.6	3503.4	8.855
800	0.1	0.27102	3.6897	3177.3	3546.3	8.9093
820	0.1	0.2644	3.7822	3211.3	3589.5	8.9626
840	0.1	0.25809	3.8747	3245.5	3632.9	9.015
860	0.1	0.25207	3.9671	3280.0	3676.7	9.0665
880	0.1	0.24633	4.0595	3314.7	3720.7	9.1171

TABLE B.3e Isobaric Data for $P = 0.15$ MPa

T (K)	P (MPa)	ρ (kg/m^3)	v (m^3/kg)	u (kJ/kg)	h (kJ/kg)	s (kJ/kg·K)
384.5	0.15	0.8626	1.1593	2519.2	2693.1	7.223
400	0.15	0.82612	1.2105	2544.0	2725.6	7.3058
420	0.15	0.78408	1.2754	2575.2	2766.5	7.4057
440	0.15	0.74658	1.3394	2605.9	2806.8	7.4995
460	0.15	0.71278	1.403	2636.4	2846.9	7.5884
480	0.15	0.6821	1.4661	2666.9	2886.8	7.6733
500	0.15	0.65407	1.5289	2697.3	2926.6	7.7547
520	0.15	0.62835	1.5915	2727.8	2966.6	7.833
540	0.15	0.60463	1.6539	2758.5	3006.6	7.9086
560	0.15	0.58268	1.7162	2789.4	3046.8	7.9816
580	0.15	0.5623	1.7784	2820.4	3087.1	8.0524
600	0.15	0.54333	1.8405	2851.6	3127.7	8.1212
620	0.15	0.52562	1.9025	2883.1	3168.4	8.188
640	0.15	0.50903	1.9645	2914.7	3209.4	8.253
660	0.15	0.49348	2.0264	2946.6	3250.6	8.3164
680	0.15	0.47886	2.0883	2978.8	3292.0	8.3782
700	0.15	0.46508	2.1502	3011.1	3333.7	8.4386
720	0.15	0.45209	2.212	3043.8	3375.6	8.4976
740	0.15	0.4398	2.2738	3076.7	3417.7	8.5554
760	0.15	0.42817	2.3355	3109.8	3460.2	8.6119
780	0.15	0.41714	2.3973	3143.3	3502.9	8.6674
800	0.15	0.40667	2.459	3177.0	3545.8	8.7217
820	0.15	0.39671	2.5207	3210.9	3589.0	8.7751
840	0.15	0.38724	2.5824	3245.1	3632.5	8.8275
860	0.15	0.3782	2.6441	3279.7	3676.3	8.879
880	0.15	0.36958	2.7058	3314.4	3720.3	8.9296
900	0.15	0.36135	2.7674	3349.5	3764.6	8.9794

TABLE B.3f Isobaric Data for $P = 0.3$ MPa

T (K)	P (MPa)	ρ (kg/m^3)	v (m^3/kg)	u (kJ/kg)	h (kJ/kg)	s (kJ/kg·K)
406.67	0.3	1.6508	0.60576	2543.2	2724.9	6.9916
420	0.3	1.5906	0.62868	2565.8	2754.4	7.063
440	0.3	1.5101	0.6622	2598.4	2797.1	7.1623
460	0.3	1.4387	0.69507	2630.3	2838.8	7.255
480	0.3	1.3746	0.72749	2661.7	2879.9	7.3426
500	0.3	1.3165	0.75958	2692.9	2920.8	7.426
520	0.3	1.2635	0.79144	2724.0	2961.5	7.5057
540	0.3	1.2149	0.82312	2755.2	3002.1	7.5824
560	0.3	1.1700	0.85467	2786.4	3042.8	7.6564
580	0.3	1.1285	0.8861	2817.7	3083.6	7.7279
600	0.3	1.0900	0.91744	2849.2	3124.4	7.7973
620	0.3	1.0541	0.94871	2880.9	3165.5	7.8646
640	0.3	1.0205	0.97992	2912.7	3206.7	7.93
660	0.3	0.98904	1.0111	2944.8	3248.1	7.9937
680	0.3	0.95951	1.0422	2977.1	3289.7	8.0558
700	0.3	0.93173	1.0733	3009.6	3331.6	8.1165
720	0.3	0.90553	1.1043	3042.4	3373.6	8.1757
740	0.3	0.88079	1.1353	3075.3	3416.0	8.2337
760	0.3	0.85738	1.1663	3108.6	3458.5	8.2904
780	0.3	0.8352	1.1973	3142.1	3501.3	8.346
800	0.3	0.81415	1.2283	3175.9	3544.3	8.4005
820	0.3	0.79414	1.2592	3209.9	3587.6	8.4539
840	0.3	0.7751	1.2901	3244.2	3631.2	8.5064
860	0.3	0.75696	1.3211	3278.7	3675.1	8.558
880	0.3	0.73966	1.352	3313.6	3719.2	8.6087
900	0.3	0.72314	1.3829	3348.7	3763.5	8.6586
920	0.3	0.70734	1.4138	3384.1	3808.2	8.7076

TABLE B.3g Isobaric Data for $P = 0.5$ MPa

T (K)	P (MPa)	ρ (kg/m^3)	v (m^3/kg)	u (kJ/kg)	h (kJ/kg)	s (kJ/kg·K)
424.98	0.5	2.668	0.37481	2560.7	2748.1	6.8207
440	0.5	2.5579	0.39095	2587.6	2783.1	6.9015
460	0.5	2.429	0.41169	2621.6	2827.4	7.0001
480	0.5	2.3153	0.43191	2654.5	2870.5	7.0917
500	0.5	2.2135	0.45176	2686.8	2912.7	7.1779
520	0.5	2.1215	0.47136	2718.8	2954.5	7.2599
540	0.5	2.0376	0.49076	2750.6	2996.0	7.3382
560	0.5	1.9607	0.51002	2782.4	3037.4	7.4134
580	0.5	1.8898	0.52915	2814.1	3078.7	7.486
600	0.5	1.8242	0.54820	2846.0	3120.1	7.5561
620	0.5	1.7631	0.56717	2878.0	3161.5	7.6241
640	0.5	1.7063	0.58608	2910.1	3203.1	7.6901
660	0.5	1.6531	0.60493	2942.4	3244.8	7.7542
680	0.5	1.6032	0.62374	2974.8	3286.7	7.8168
700	0.5	1.5564	0.64252	3007.5	3328.8	7.8777
720	0.5	1.5123	0.66126	3040.4	3371.1	7.9373
740	0.5	1.4706	0.67997	3073.6	3413.5	7.9955
760	0.5	1.4313	0.69867	3106.9	3456.3	8.0524
780	0.5	1.394	0.71734	3140.5	3499.2	8.1082
800	0.5	1.3587	0.73599	3174.4	3542.4	8.1629
820	0.5	1.3252	0.75463	3208.5	3585.8	8.2165
840	0.5	1.2932	0.77325	3242.9	3629.5	8.2691
860	0.5	1.2629	0.79186	3277.5	3673.4	8.3208
880	0.5	1.2339	0.81045	3312.4	3717.6	8.3716
900	0.5	1.2062	0.82904	3347.6	3762.1	8.4216
920	0.5	1.1798	0.84762	3383.0	3806.8	8.4708
940	0.5	1.1545	0.86619	3418.7	3851.8	8.5191

TABLE B.3h Isobaric Data for $P = 0.7$ MPa

T (K)	P (MPa)	ρ (kg/m^3)	v (m^3/kg)	u (kJ/kg)	h (kJ/kg)	s (kJ/kg·K)
438.1	0.7	3.666	0.27277	2571.8	2762.8	6.7071
440	0.7	3.6453	0.27433	2575.5	2767.6	6.718
460	0.7	3.4477	0.29005	2612.3	2815.3	6.8242
480	0.7	3.2775	0.30511	2647.0	2860.5	6.9204
500	0.7	3.1274	0.31976	2680.5	2904.3	7.0099
520	0.7	2.9929	0.33413	2713.4	2947.3	7.0941
540	0.7	2.8712	0.34828	2745.9	2989.7	7.1742
560	0.7	2.7603	0.36228	2778.3	3031.9	7.2508
580	0.7	2.6585	0.37616	2810.5	3073.8	7.3244
600	0.7	2.5645	0.38994	2842.7	3115.7	7.3953
620	0.7	2.4775	0.40364	2875.0	3157.6	7.464
640	0.7	2.3965	0.41728	2907.4	3199.5	7.5306
660	0.7	2.3209	0.43086	2939.9	3241.5	7.5952
680	0.7	2.2502	0.4444	2972.6	3283.7	7.6582
700	0.7	2.1839	0.4579	3005.4	3326.0	7.7195
720	0.7	2.1215	0.47137	3038.5	3368.5	7.7793
740	0.7	2.0626	0.48481	3071.8	3411.1	7.8378
760	0.7	2.0071	0.49823	3105.3	3454.0	7.895
780	0.7	1.9545	0.51163	3139.0	3497.1	7.9509
800	0.7	1.9047	0.52501	3172.9	3540.4	8.0058
820	0.7	1.8575	0.53837	3207.1	3584.0	8.0595
840	0.7	1.8125	0.55172	3241.6	3627.8	8.1123
860	0.7	1.7697	0.56505	3276.3	3671.8	8.1641
880	0.7	1.729	0.57838	3311.3	3716.1	8.215
900	0.7	1.6901	0.59169	3346.5	3760.7	8.2651
920	0.7	1.6529	0.60499	3382.0	3805.5	8.3143
940	0.7	1.6174	0.61829	3417.7	3850.5	8.3628

TABLE B.3i Isobaric Data for $P = 1$ MPa

T (K)	P (MPa)	ρ (kg/m³)	v (m³/kg)	u (kJ/kg)	h (kJ/kg)	s (kJ/kg·K)
453.03	1	5.145	0.19436	2582.7	2777.1	6.585
460	1	5.0376	0.19851	2597.0	2795.5	6.6253
480	1	4.7658	0.20983	2634.9	2844.7	6.7301
500	1	4.5323	0.22064	2670.6	2891.2	6.825
520	1	4.3268	0.23112	2705.0	2936.1	6.9131
540	1	4.1431	0.24137	2738.7	2980.1	6.996
560	1	3.9771	0.25144	2772.0	3023.4	7.0748
580	1	3.8258	0.26138	2804.9	3066.3	7.1501
600	1	3.6871	0.27122	2837.7	3109.0	7.2224
620	1	3.5591	0.28097	2870.5	3151.5	7.2921
640	1	3.4404	0.29066	2903.3	3194.0	7.3596
660	1	3.33	0.3003	2936.2	3236.5	7.425
680	1	3.227	0.30988	2969.2	3279.1	7.4885
700	1	3.1305	0.31943	3002.3	3321.7	7.5504
720	1	3.04	0.32895	3035.6	3364.5	7.6107
740	1	2.9547	0.33844	3069.1	3407.5	7.6695
760	1	2.8744	0.3479	3102.7	3450.6	7.727
780	1	2.7984	0.35734	3136.6	3494.0	7.7833
800	1	2.7265	0.36677	3170.7	3537.5	7.8384
820	1	2.6583	0.37618	3205.1	3581.2	7.8924
840	1	2.5936	0.38557	3239.6	3625.2	7.9454
860	1	2.532	0.39495	3274.4	3669.4	7.9974
880	1	2.4733	0.40432	3309.5	3713.8	8.0484
900	1	2.4174	0.41367	3344.8	3758.5	8.0986
920	1	2.3639	0.42302	3380.4	3803.4	8.148
940	1	2.3129	0.43236	3416.3	3848.6	8.1966
960	1	2.264	0.4417	3452.4	3894.1	8.2444

TABLE B.3j Isobaric Data for $P = 1.5$ MPa

T (K)	P (MPa)	ρ (kg/m³)	v (m³/kg)	u (kJ/kg)	h (kJ/kg)	s (kJ/kg·K)
471.44	1.5	7.5924	0.13171	2593.4	2791.0	6.443
480	1.5	7.3885	0.13534	2612.3	2815.3	6.4942
500	1.5	6.9775	0.14332	2652.6	2867.6	6.6009
520	1.5	6.629	0.15085	2690.1	2916.4	6.6967
540	1.5	6.3249	0.1581	2726.1	2963.2	6.7851
560	1.5	6.0549	0.16515	2761.0	3008.8	6.8678
580	1.5	5.8121	0.17206	2795.3	3053.4	6.9462
600	1.5	5.5915	0.17884	2829.2	3097.5	7.0209
620	1.5	5.3897	0.18554	2862.9	3141.2	7.0925
640	1.5	5.2039	0.19216	2896.4	3184.7	7.1616
660	1.5	5.0319	0.19873	2929.9	3228.0	7.2282
680	1.5	4.8721	0.20525	2963.4	3271.3	7.2929
700	1.5	4.7231	0.21173	2997.0	3314.6	7.3556
720	1.5	4.5836	0.21817	3030.7	3358.0	7.4167
740	1.5	4.4527	0.22458	3064.5	3401.4	7.4762
760	1.5	4.3295	0.23097	3098.5	3445.0	7.5343
780	1.5	4.2133	0.23734	3132.7	3488.7	7.5911
800	1.5	4.1036	0.24369	3167.0	3532.6	7.6466
820	1.5	3.9997	0.25002	3201.6	3576.6	7.701
840	1.5	3.9011	0.25634	3236.4	3620.9	7.7543
860	1.5	3.8075	0.26264	3271.4	3665.3	7.8066
880	1.5	3.7184	0.26893	3306.6	3710.0	7.858
900	1.5	3.6335	0.27522	3342.1	3754.9	7.9084
920	1.5	3.5525	0.28149	3377.8	3800.0	7.958
940	1.5	3.4752	0.28775	3413.8	3845.4	8.0068
960	1.5	3.4013	0.29401	3450.0	3891.0	8.0548
980	1.5	3.3305	0.30026	3486.5	3936.9	8.1021

TABLE B.3k Isobaric Data for $P = 2$ MPa

T (K)	P (MPa)	ρ (kg/m^3)	v (m^3/kg)	u (kJ/kg)	h (kJ/kg)	s (kJ/kg·K)
485.53	2	10.042	0.099585	2599.1	2798.3	6.339
500	2	9.5781	0.10441	2632.6	2841.4	6.4265
520	2	9.0447	0.11056	2674.0	2895.1	6.5319
540	2	8.5934	0.11637	2712.6	2945.4	6.6267
560	2	8.2008	0.12194	2749.5	2993.4	6.7141
580	2	7.853	0.12734	2785.3	3040.0	6.7959
600	2	7.5406	0.13262	2820.4	3085.6	6.8732
620	2	7.2573	0.13779	2855.0	3130.6	6.947
640	2	6.9983	0.14289	2889.4	3175.1	7.0176
660	2	6.7599	0.14793	2923.5	3219.4	7.0857
680	2	6.5395	0.15292	2957.6	3263.4	7.1514
700	2	6.3346	0.15786	2991.6	3307.4	7.2151
720	2	6.1436	0.16277	3025.8	3351.3	7.277
740	2	5.9648	0.16765	3059.9	3395.2	7.3372
760	2	5.7969	0.1725	3094.2	3439.3	7.3959
780	2	5.639	0.17734	3128.7	3483.4	7.4532
800	2	5.4901	0.18215	3163.3	3527.6	7.5092
820	2	5.3493	0.18694	3198.1	3572.0	7.564
840	2	5.2159	0.19172	3233.1	3616.5	7.6176
860	2	5.0894	0.19649	3268.3	3661.2	7.6702
880	2	4.9691	0.20124	3303.7	3706.1	7.7219
900	2	4.8547	0.20599	3339.3	3751.3	7.7726
920	2	4.7456	0.21072	3375.2	3796.6	7.8224
940	2	4.6415	0.21545	3411.3	3842.2	7.8714
960	2	4.5421	0.22016	3447.6	3887.9	7.9196
980	2	4.4469	0.22488	3484.2	3934.0	7.967
1000	2	4.3558	0.22958	3521.1	3980.2	8.0137

TABLE B.3l Isobaric Data for $P = 3$ MPa

T (K)	P (MPa)	ρ (kg/m^3)	v (m^3/kg)	u (kJ/kg)	h (kJ/kg)	s (kJ/kg·K)
507	3	15.001	0.066664	2603.2	2803.2	6.1856
520	3	14.309	0.069888	2637.1	2846.7	6.2704
540	3	13.442	0.074391	2682.9	2906.1	6.3825
560	3	12.729	0.078561	2724.7	2960.4	6.4812
580	3	12.119	0.082512	2764.1	3011.6	6.5712
600	3	11.587	0.086307	2801.9	3060.8	6.6546
620	3	11.113	0.089986	2838.7	3108.6	6.733
640	3	10.687	0.093576	2874.7	3155.5	6.8073
660	3	10.299	0.097096	2910.3	3201.6	6.8783
680	3	9.9443	0.10056	2945.6	3247.3	6.9465
700	3	9.6174	0.10398	2980.7	3292.6	7.0122
720	3	9.3147	0.10736	3015.7	3337.7	7.0758
740	3	9.033	0.1107	3050.6	3382.7	7.1374
760	3	8.77	0.11403	3085.6	3427.7	7.1973
780	3	8.5236	0.11732	3120.6	3472.6	7.2557
800	3	8.292	0.1206	3155.8	3517.6	7.3126
820	3	8.0738	0.12386	3191.0	3562.6	7.3682
840	3	7.8678	0.1271	3226.4	3607.7	7.4226
860	3	7.6729	0.13033	3262.0	3653.0	7.4758
880	3	7.488	0.13355	3297.8	3698.4	7.528
900	3	7.3124	0.13675	3333.7	3744.0	7.5792
920	3	7.1454	0.13995	3369.9	3789.7	7.6295
940	3	6.9863	0.14314	3406.2	3835.7	7.6789
960	3	6.8344	0.14632	3442.8	3881.8	7.7275
980	3	6.6893	0.14949	3479.7	3928.1	7.7752
1000	3	6.5506	0.15266	3516.7	3974.7	7.8223
1020	3	6.4177	0.15582	3554.0	4021.5	7.8686

TABLE B.3m Isobaric Data for $P = 4$ MPa

T (K)	P (MPa)	ρ (kg/m^3)	v (m^3/kg)	u (kJ/kg)	h (kJ/kg)	s (kJ/kg·K)
523.5	4	20.09	0.049776	2601.7	2800.8	6.0696
540	4	18.831	0.053103	2648.4	2860.8	6.1824
560	4	17.642	0.056683	2696.9	2923.7	6.2968
580	4	16.675	0.05997	2740.9	2980.8	6.397
600	4	15.857	0.063063	2782.1	3034.3	6.4878
620	4	15.147	0.066018	2821.4	3085.4	6.5716
640	4	14.52	0.068870	2859.4	3134.9	6.6501
660	4	13.958	0.071643	2896.6	3183.2	6.7244
680	4	13.449	0.074354	2933.2	3230.6	6.7953
700	4	12.985	0.077014	2969.4	3277.5	6.8632
720	4	12.558	0.079634	3005.4	3323.9	6.9285
740	4	12.163	0.082219	3041.1	3370.0	6.9917
760	4	11.796	0.084775	3076.8	3415.9	7.0529
780	4	11.454	0.087306	3112.5	3461.7	7.1124
800	4	11.134	0.089817	3148.2	3507.4	7.1702
820	4	10.833	0.092309	3183.9	3553.1	7.2267
840	4	10.55	0.094785	3219.7	3598.9	7.2818
860	4	10.283	0.097247	3255.7	3644.7	7.3357
880	4	10.03	0.099696	3291.8	3690.6	7.3885
900	4	9.791	0.10213	3328.1	3736.6	7.4402
920	4	9.5636	0.10456	3364.5	3782.8	7.4909
940	4	9.3473	0.10698	3401.2	3829.1	7.5407
960	4	9.1412	0.10939	3438.0	3875.6	7.5897
980	4	8.9446	0.1118	3475.1	3922.3	7.6378
1000	4	8.7568	0.1142	3512.4	3969.1	7.6851
1020	4	8.5771	0.11659	3549.9	4016.2	7.7317
1040	4	8.405	0.11898	3587.6	4063.5	7.7776

TABLE B.3n Isobaric Data for $P = 6$ MPa

T (K)	P (MPa)	ρ (kg/m^3)	v (m^3/kg)	u (kJ/kg)	h (kJ/kg)	s (kJ/kg·K)
548.73	6	30.818	0.032448	2589.9	2784.6	5.8901
560	6	29.166	0.034287	2629.1	2834.8	5.9807
580	6	26.947	0.03711	2687.2	2909.8	6.1125
600	6	25.247	0.039608	2737.6	2975.2	6.2233
620	6	23.865	0.041903	2783.5	3034.9	6.3211
640	6	22.697	0.044058	2826.5	3090.8	6.4099
660	6	21.686	0.046112	2867.5	3144.2	6.4921
680	6	20.795	0.048089	2907.2	3195.7	6.569
700	6	19.998	0.050006	2945.9	3246.0	6.6418
720	6	19.277	0.051875	2984.0	3295.2	6.7112
740	6	18.62	0.053706	3021.5	3343.8	6.7777
760	6	18.017	0.055504	3058.7	3391.8	6.8417
780	6	17.46	0.057274	3095.7	3439.4	6.9036
800	6	16.943	0.059022	3132.6	3486.7	6.9635
820	6	16.461	0.06075	3169.4	3533.9	7.0217
840	6	16.01	0.062461	3206.1	3580.9	7.0784
860	6	15.587	0.064157	3242.9	3627.9	7.1336
880	6	15.188	0.06584	3279.8	3674.8	7.1876
900	6	14.812	0.067512	3316.7	3721.8	7.2404
920	6	14.457	0.069173	3353.8	3768.8	7.2921
940	6	14.119	0.070825	3391.0	3815.9	7.3427
960	6	13.799	0.072469	3428.4	3863.2	7.3924
980	6	13.494	0.074105	3465.9	3910.5	7.4413
1000	6	13.204	0.075735	3503.6	3958.0	7.4892
1020	6	12.927	0.077358	3541.5	4005.6	7.5364
1040	6	12.662	0.078977	3579.6	4053.4	7.5828
1060	6	12.409	0.08059	3617.9	4101.4	7.6285

TABLE B.3o Isobaric Data for $P = 8$ MPa

T (K)	P (MPa)	ρ (kg/m^3)	v (m^3/kg)	u (kJ/kg)	h (kJ/kg)	s (kJ/kg·K)
568.16	8	42.507	0.023526	2570.5	2758.7	5.745
580	8	39.647	0.025223	2619.0	2820.7	5.8531
600	8	36.218	0.027611	2684.7	2905.6	5.9971
620	8	33.704	0.02967	2740.2	2977.6	6.1151
640	8	31.713	0.031532	2789.9	3042.1	6.2176
660	8	30.065	0.033261	2835.8	3101.9	6.3096
680	8	28.658	0.034894	2879.3	3158.4	6.394
700	8	27.431	0.036455	2921.0	3212.6	6.4726
720	8	26.343	0.037961	2961.5	3265.2	6.5466
740	8	25.367	0.039422	3001.1	3316.5	6.6168
760	8	24.482	0.040847	3040.0	3366.8	6.6839
780	8	23.673	0.042242	3078.5	3416.4	6.7484
800	8	22.929	0.043612	3116.6	3465.5	6.8105
820	8	22.242	0.044961	3154.5	3514.2	6.8706
840	8	21.602	0.046292	3192.3	3562.6	6.9289
860	8	21.006	0.047606	3229.9	3610.8	6.9856
880	8	20.447	0.048907	3267.6	3658.8	7.0409
900	8	19.922	0.050196	3305.2	3706.8	7.0948
920	8	19.427	0.051475	3342.9	3754.7	7.1474
940	8	18.96	0.052744	3380.7	3802.6	7.199
960	8	18.517	0.054004	3418.6	3850.6	7.2495
980	8	18.097	0.055257	3456.6	3898.6	7.299
1000	8	17.698	0.056503	3494.7	3946.8	7.3476
1020	8	17.318	0.057742	3533.1	3995.0	7.3953
1040	8	16.956	0.058976	3571.5	4043.3	7.4423
1060	8	16.61	0.060205	3610.2	4091.8	7.4885
1080	8	16.279	0.06143	3649.0	4140.5	7.5339

TABLE B.3p Isobaric Data for $P = 10$ MPa

T (K)	P (MPa)	ρ (kg/m^3)	v (m^3/kg)	u (kJ/kg)	h (kJ/kg)	s (kJ/kg·K
584.15	10	55.463	0.01803	2545.2	2725.5	5.616
600	10	49.773	0.020091	2619.1	2820.0	5.7756
620	10	45.151	0.022148	2689.7	2911.2	5.9253
640	10	41.84	0.0239	2748.7	2987.7	6.0468
660	10	39.259	0.025472	2801.1	3055.8	6.1516
680	10	37.145	0.026922	2849.2	3118.4	6.2451
700	10	35.355	0.028285	2894.5	3177.4	6.3305
720	10	33.804	0.029582	2937.8	3233.7	6.4098
740	10	32.437	0.030829	2979.7	3288.0	6.4843
760	10	31.215	0.032036	3020.6	3341.0	6.5549
780	10	30.112	0.033209	3060.7	3392.8	6.6222
800	10	29.107	0.034356	3100.2	3443.7	6.6867
820	10	28.185	0.035479	3139.3	3494.1	6.7489
840	10	27.335	0.036584	3178.1	3543.9	6.8089
860	10	26.546	0.037671	3216.7	3593.4	6.8671
880	10	25.81	0.038744	3255.1	3642.6	6.9237
900	10	25.123	0.039804	3293.5	3691.6	6.9787
920	10	24.478	0.040854	3331.9	3740.4	7.0324
940	10	23.87	0.041893	3370.3	3789.2	7.0849
960	10	23.297	0.042924	3408.7	3837.9	7.1362
980	10	22.755	0.043947	3447.2	3886.7	7.1864
1000	10	22.241	0.044963	3485.8	3935.5	7.2357
1020	10	21.752	0.045972	3524.6	3984.3	7.284
1040	10	21.287	0.046976	3563.4	4033.2	7.3315
1060	10	20.844	0.047975	3602.5	4082.2	7.3782
1080	10	20.421	0.048969	3641.6	4131.3	7.4241
1100	10	20.017	0.049959	3681.0	4180.6	7.4693

TABLE B.3q Isobaric Data for $P = 12$ MPa

T (K)	P (MPa)	ρ (kg/m^3)	v (m^3/kg)	u (kJ/kg)	h (kJ/kg)	s (kJ/kg·K)
597.83	12	70.106	0.014264	2514.3	2685.4	5.4939
600	12	68.549	0.014588	2528.8	2703.8	5.5247
620	12	59.113	0.016917	2628.8	2831.8	5.7346
640	12	53.502	0.018691	2701.7	2926.0	5.8843
660	12	49.499	0.020202	2762.6	3005.0	6.0059
680	12	46.392	0.021555	2816.6	3075.3	6.1108
700	12	43.857	0.022801	2866.3	3139.9	6.2045
720	12	41.718	0.023971	2912.9	3200.5	6.2899
740	12	39.87	0.025082	2957.4	3258.4	6.3692
760	12	38.245	0.026147	3000.4	3314.2	6.4436
780	12	36.796	0.027177	3042.3	3368.4	6.514
800	12	35.49	0.028177	3083.3	3421.4	6.5811
820	12	34.303	0.029152	3123.7	3473.5	6.6454
840	12	33.215	0.030107	3163.6	3524.9	6.7073
860	12	32.213	0.031044	3203.2	3575.7	6.7671
880	12	31.284	0.031966	3242.5	3626.1	6.8251
900	12	30.419	0.032874	3281.7	3676.2	6.8813
920	12	29.611	0.033771	3320.7	3726.0	6.9361
940	12	28.853	0.034658	3359.7	3775.6	6.9895
960	12	28.14	0.035536	3398.7	3825.2	7.0416
980	12	27.468	0.036406	3437.8	3874.6	7.0926
1000	12	26.832	0.037269	3476.9	3924.1	7.1425
1020	12	26.229	0.038126	3516.0	3973.5	7.1915
1040	12	25.657	0.038976	3555.3	4023.0	7.2395
1060	12	25.112	0.039821	3594.7	4072.5	7.2867
1080	12	24.593	0.040662	3634.2	4122.1	7.3331
1100	12	24.098	0.041498	3673.9	4171.8	7.3787

TABLE B.3r Isobaric Data for $P = 14$ MPa

T (K)	P (MPa)	ρ (kg/m^3)	v (m^3/kg)	u (kJ/kg)	h (kJ/kg)	s (kJ/kg·K)
609.82	14	87.069	0.011485	2477.1	2637.9	5.3727
620	14	77.66	0.012877	2549.9	2730.1	5.5229
640	14	67.43	0.01483	2646.6	2854.2	5.72
660	14	61.128	0.016359	2719.6	2948.6	5.8653
680	14	56.586	0.017672	2781.1	3028.5	5.9847
700	14	53.047	0.018851	2836.0	3099.9	6.0882
720	14	50.153	0.019939	2886.5	3165.7	6.1808
740	14	47.711	0.020959	2934.1	3227.5	6.2655
760	14	45.602	0.021929	2979.5	3286.5	6.3441
780	14	43.747	0.022859	3023.3	3343.3	6.418
800	14	42.094	0.023756	3066.0	3398.5	6.4879
820	14	40.605	0.024628	3107.7	3452.5	6.5545
840	14	39.252	0.025477	3148.8	3505.5	6.6184
860	14	38.012	0.026307	3189.5	3557.8	6.6799
880	14	36.871	0.027122	3229.7	3609.4	6.7392
900	14	35.813	0.027923	3269.7	3660.6	6.7967
920	14	34.829	0.028712	3309.5	3711.4	6.8526
940	14	33.91	0.02949	3349.1	3762.0	6.9069
960	14	33.048	0.030259	3388.7	3812.3	6.96
980	14	32.237	0.03102	3428.2	3862.5	7.0117
1000	14	31.472	0.031774	3467.8	3912.6	7.0623
1020	14	30.749	0.032521	3507.4	3962.7	7.1119
1040	14	30.064	0.033262	3547.1	4012.8	7.1605
1060	14	29.414	0.033998	3586.8	4062.8	7.2082
1080	14	28.795	0.034729	3626.7	4112.9	7.255
1100	14	28.205	0.035455	3666.7	4163.1	7.301
1120	14	27.641	0.036178	3706.8	4213.3	7.3463

TABLE B.3s Isobaric Data for $P = 16$ MPa

T (K)	P (MPa)	ρ (kg/m^3)	v (m^3/kg)	u (kJ/kg)	h (kJ/kg)	s (kJ/kg·K)
620.5	16	107.42	0.009309	2431.8	2580.8	5.2463
640	16	85.058	0.011757	2579.1	2767.2	5.5427
660	16	74.683	0.01339	2670.5	2884.8	5.7237
680	16	67.984	0.014709	2742.1	2977.5	5.8622
700	16	63.065	0.015857	2803.5	3057.2	5.9778
720	16	59.195	0.016893	2858.6	3128.9	6.0788
740	16	56.014	0.017853	2909.6	3195.3	6.1697
760	16	53.32	0.018755	2957.7	3257.8	6.253
780	16	50.988	0.019613	3003.7	3317.5	6.3306
800	16	48.935	0.020435	3048.1	3375.1	6.4035
820	16	47.103	0.021230	3091.4	3431.1	6.4726
840	16	45.452	0.022001	3133.8	3485.8	6.5386
860	16	43.951	0.022753	3175.5	3539.5	6.6018
880	16	42.576	0.023488	3216.7	3592.5	6.6627
900	16	41.308	0.024208	3257.5	3644.8	6.7215
920	16	40.134	0.024916	3298.0	3696.7	6.7785
940	16	39.042	0.025614	3338.3	3748.2	6.8338
960	16	38.021	0.026301	3378.5	3799.4	6.8877
980	16	37.063	0.026981	3418.6	3850.3	6.9403
1000	16	36.163	0.027653	3458.7	3901.1	6.9916
1020	16	35.313	0.028318	3498.8	3951.8	7.0418
1040	16	34.51	0.028977	3538.8	4002.5	7.091
1060	16	33.749	0.029631	3579.0	4053.1	7.1391
1080	16	33.026	0.030279	3619.2	4103.7	7.1864
1100	16	32.338	0.030924	3659.5	4154.3	7.2329
1120	16	31.682	0.031564	3700.0	4205.0	7.2785
1140	16	31.056	0.0322	3740.5	4255.7	7.3235

TABLE B.3t Isobaric Data for $P = 18$ MPa

T (K)	P (MPa)	ρ (kg/m^3)	v (m^3/kg)	u (kJ/kg)	h (kJ/kg)	s (kJ/kg·K)
630.14	18	133.3	0.0075017	2374.8	2509.8	5.1061
640	18	110	0.0090911	2489.2	2652.8	5.3314
660	18	91.08	0.010979	2613.3	2810.9	5.5749
680	18	80.955	0.012353	2698.9	2921.2	5.7397
700	18	74.095	0.013496	2768.4	3011.3	5.8704
720	18	68.943	0.014505	2829.0	3090.1	5.9813
740	18	64.839	0.015423	2883.9	3161.6	6.0793
760	18	61.439	0.016276	2935.0	3228.0	6.1679
780	18	58.544	0.017081	2983.4	3290.9	6.2495
800	18	56.029	0.017848	3029.8	3351.0	6.3257
820	18	53.809	0.018584	3074.7	3409.2	6.3975
840	18	51.825	0.019296	3118.4	3465.7	6.4656
860	18	50.034	0.019986	3161.3	3521.0	6.5307
880	18	48.403	0.02066	3203.5	3575.3	6.5931
900	18	46.908	0.021318	3245.2	3628.9	6.6533
920	18	45.529	0.021964	3286.5	3681.8	6.7115
940	18	44.251	0.022598	3327.5	3734.3	6.7679
960	18	43.06	0.023223	3368.3	3786.3	6.8227
980	18	41.948	0.023839	3409.0	3838.1	6.876
1000	18	40.904	0.024448	3449.5	3889.6	6.9281
1020	18	39.922	0.025049	3490.0	3940.9	6.9789
1040	18	38.995	0.025645	3530.6	3992.2	7.0286
1060	18	38.118	0.026234	3571.1	4043.3	7.0774
1080	18	37.287	0.026819	3611.7	4094.4	7.1251
1100	18	36.497	0.0274	3652.3	4145.5	7.172
1120	18	35.745	0.027976	3693.1	4196.6	7.2181
1140	18	35.029	0.028548	3733.9	4247.8	7.2633

TABLE B.3u Isobaric Data for $P = 20$ MPa

T (K)	P (MPa)	ρ (kg/m^3)	v (m^3/kg)	u (kJ/kg)	h (kJ/kg)	s (kJ/kg·K)
638.9	20	170.5	0.005865	2295.0	2412.3	4.9314
640	20	160.5	0.006231	2328.4	2453.0	4.995
660	20	112.09	0.008921	2543.7	2722.1	5.4104
680	20	96.059	0.01041	2650.1	2858.3	5.6139
700	20	86.38	0.011577	2730.2	2961.8	5.7639
720	20	79.525	0.012575	2797.4	3048.9	5.8867
740	20	74.257	0.013467	2857.0	3126.3	5.9927
760	20	70.002	0.014285	2911.5	3197.2	6.0872
780	20	66.444	0.01505	2962.5	3263.5	6.1733
800	20	63.396	0.015774	3010.9	3326.4	6.253
820	20	60.736	0.016465	3057.5	3386.8	6.3276
840	20	58.379	0.017129	3102.7	3445.3	6.398
860	20	56.268	0.017772	3146.8	3502.2	6.465
880	20	54.357	0.018397	3190.1	3558.0	6.5291
900	20	52.615	0.019006	3232.7	3612.8	6.5907
920	20	51.015	0.019602	3274.8	3666.8	6.6501
940	20	49.538	0.020186	3316.5	3720.2	6.7075
960	20	48.168	0.020761	3358.0	3773.2	6.7633
980	20	46.891	0.021326	3399.2	3825.7	6.8174
1000	20	45.696	0.021884	3440.3	3878.0	6.8702
1020	20	44.574	0.022435	3481.3	3930.0	6.9217
1040	20	43.518	0.022979	3522.2	3981.8	6.972
1060	20	42.521	0.023518	3563.2	4033.5	7.0213
1080	20	41.577	0.024052	3604.1	4085.1	7.0695
1100	20	40.682	0.024581	3645.1	4136.7	7.1168
1120	20	39.831	0.025106	3686.1	4188.3	7.1633
1140	20	39.021	0.025627	3727.3	4239.8	7.2089

TABLE B.3v Isobaric Data for $P = 26$ MPa (Supercritical)

T (K)	P (MPa)	ρ (kg/m^3)	v (m^3/kg)	u (kJ/kg)	h (kJ/kg)	s (kJ/kg·K)
660	26	365.88	0.002733	2010.9	2082.0	4.386
680	26	167.08	0.005985	2448.8	2604.4	5.1692
700	26	134.79	0.007419	2591.2	2784.1	5.43
720	26	118.02	0.008473	2688.8	2909.1	5.6062
740	26	106.98	0.009348	2767.2	3010.2	5.7447
760	26	98.863	0.010115	2834.6	3097.6	5.8613
780	26	92.512	0.010809	2895.3	3176.3	5.9635
800	26	87.325	0.011451	2951.2	3248.9	6.0554
820	26	82.963	0.012054	3003.7	3317.1	6.1396
840	26	79.211	0.012625	3053.7	3382.0	6.2178
860	26	75.927	0.01317	3101.9	3444.3	6.2912
880	26	73.015	0.013696	3148.6	3504.7	6.3606
900	26	70.403	0.014204	3194.2	3563.5	6.4267
920	26	68.039	0.014697	3239.0	3621.1	6.4899
940	26	65.882	0.015179	3283.0	3677.6	6.5507
960	26	63.902	0.015649	3326.5	3733.3	6.6094
980	26	62.074	0.01611	3369.5	3788.4	6.6661
1000	26	60.378	0.016562	3412.2	3842.9	6.7211
1020	26	58.797	0.017008	3454.7	3896.9	6.7747
1040	26	57.318	0.017447	3497.0	3950.6	6.8268
1060	26	55.93	0.01788	3539.2	4004.0	6.8777
1080	26	54.623	0.018307	3581.2	4057.2	6.9274
1100	26	53.389	0.01873	3623.3	4110.2	6.976
1120	26	52.222	0.019149	3665.3	4163.1	7.0237
1140	26	51.115	0.019564	3707.3	4216.0	7.0704
1160	26	50.062	0.019975	3749.4	4268.7	7.1163
1180	26	49.06	0.020383	3791.5	4321.4	7.1614
1200	26	48.104	0.020788	3833.7	4374.2	7.2057

TABLE B.3w Isobaric Data for $P = 28$ MPa (Supercritical)

T (K)	P (MPa)	ρ (kg/m^3)	v (m^3/kg)	u (kJ/kg)	h (kJ/kg)	s (kJ/kg·K)
660	28	455.47	0.0021955	1897.4	1958.9	4.1923
680	28	210.73	0.0047453	2347.1	2480.0	4.9705
700	28	157.04	0.0063676	2533.7	2712.0	5.3072
720	28	133.92	0.0074672	2647.0	2856.1	5.5104
740	28	119.74	0.0083511	2733.9	2967.7	5.6634
760	28	109.74	0.0091127	2806.9	3062.0	5.7892
780	28	102.1	0.0097942	2871.3	3145.6	5.8977
800	28	95.978	0.010419	2930.1	3221.9	5.9943
820	28	90.896	0.011002	2984.9	3293.0	6.0821
840	28	86.571	0.011551	3036.8	3360.2	6.1632
860	28	82.818	0.012075	3086.5	3424.6	6.2389
880	28	79.511	0.012577	3134.5	3486.6	6.3102
900	28	76.563	0.013061	3181.1	3546.8	6.3779
920	28	73.906	0.013531	3226.8	3605.6	6.4425
940	28	71.494	0.013987	3271.6	3663.2	6.5044
960	28	69.286	0.014433	3315.8	3719.9	6.5641
980	28	67.254	0.014869	3359.5	3775.8	6.6217
1000	28	65.374	0.015297	3402.8	3831.1	6.6775
1020	28	63.626	0.015717	3445.8	3885.8	6.7318
1040	28	61.994	0.016131	3488.5	3940.2	6.7845
1060	28	60.465	0.016538	3531.1	3994.2	6.8359
1080	28	59.029	0.016941	3573.6	4047.9	6.8862
1100	28	57.675	0.017339	3615.9	4101.4	6.9353
1120	28	56.395	0.017732	3658.3	4154.8	6.9833
1140	28	55.183	0.018121	3700.6	4208.0	7.0304
1160	28	54.033	0.018507	3743.0	4261.2	7.0767
1180	28	52.938	0.01889	3785.3	4314.3	7.122
1200	28	51.895	0.01927	3827.8	4367.3	7.1666

TABLE B.3x Isobaric Data for $P = 32$ MPa (Supercritical)

T (K)	P (MPa)	ρ (kg/m^3)	v (m^3/kg)	u (kJ/kg)	h (kJ/kg)	s (kJ/kg·K)
660	32	516.64	0.001936	1823.6	1885.6	4.0689
680	32	351.66	0.002844	2100.2	2191.2	4.5242
700	32	217.87	0.00459	2395.5	2542.3	5.0338
720	32	172.38	0.005801	2553.3	2739.0	5.3111
740	32	148.9	0.006716	2661.8	2876.7	5.4999
760	32	133.76	0.007476	2747.9	2987.1	5.6471
780	32	122.84	0.00814	2821.2	3081.7	5.77
800	32	114.42	0.00874	2886.5	3166.2	5.877
820	32	107.62	0.009292	2946.2	3243.6	5.9726
840	32	101.96	0.009808	3002.0	3315.9	6.0597
860	32	97.13	0.010295	3054.9	3384.4	6.1403
880	32	92.934	0.01076	3105.6	3449.9	6.2156
900	32	89.235	0.011206	3154.5	3513.1	6.2866
920	32	85.935	0.011637	3202.1	3574.4	6.354
940	32	82.962	0.012054	3248.6	3634.3	6.4184
960	32	80.262	0.012459	3294.3	3692.9	6.4802
980	32	77.791	0.012855	3339.3	3750.6	6.5396
1000	32	75.517	0.013242	3383.7	3807.5	6.597
1020	32	73.413	0.013622	3427.7	3863.6	6.6527
1040	32	71.458	0.013994	3471.5	3919.3	6.7067
1060	32	69.633	0.014361	3514.9	3974.4	6.7592
1080	32	67.924	0.014722	3558.1	4029.3	6.8105
1100	32	66.318	0.015079	3601.2	4083.8	6.8605
1120	32	64.804	0.015431	3644.3	4138.0	6.9094
1140	32	63.374	0.015779	3687.2	4192.1	6.9572
1160	32	62.019	0.016124	3730.1	4246.1	7.0041
1180	32	60.734	0.016465	3773.0	4299.9	7.0502
1200	32	59.511	0.016804	3816.0	4353.7	7.0953

TABLE B.4 Compressed Liquid (Water)*

TABLE B.4a Isobaric Data for $P = 5$ MPa

T (K)	P (MPa)	ρ (kg/m^3)	v (m^3/kg)	u (kJ/kg)	h (kJ/kg)	s (kJ/kg·K)
273.15	5	1002.3	0.000998	0.044068	5.0	0.0001
280	5	1002.3	0.000998	28.731	33.7	0.1039
300	5	998.74	0.001001	112.15	117.2	0.3917
320	5	991.56	0.001009	195.5	200.5	0.66066
340	5	981.68	0.001019	278.9	284.0	0.91364
360	5	969.62	0.001031	362.5	367.7	1.1528
380	5	955.65	0.001046	446.5	451.7	1.38
400	5	939.91	0.001064	530.9	536.2	1.5968
420	5	922.46	0.001084	616.1	621.5	1.8047
440	5	903.27	0.001107	702.2	707.7	2.0053
460	5	882.22	0.001134	789.6	795.2	2.1998
480	5	859.08	0.001164	878.6	884.5	2.3897
500	5	833.51	0.001200	970.0	976.0	2.5764
520	5	804.92	0.001242	1064.3	1070.5	2.7618
537.09	5	777.37	0.001286	1148.2	1154.6	2.921

* Property values generated from NIST Database 23: REFPROP Version 7.0 (August 2002).

TABLE B.4b Isobaric Data for $P = 10$ MPa

T (K)	P (MPa)	ρ (kg/m^3)	v (m^3/kg)	u (kJ/kg)	h (kJ/kg)	s (kJ/kg·K)
273.15	10	1004.8	0.0009952	0.1171	10.069	0.0003376
280	10	1004.7	0.0009954	28.659	38.613	0.10355
300	10	1001.0	0.0009991	111.74	121.73	0.39029
320	10	993.7	0.0010063	194.78	204.84	0.65846
340	10	983.84	0.0010164	277.92	288.08	0.91079
360	10	971.85	0.001029	361.27	371.56	1.1493
380	10	957.99	0.0010439	444.93	455.37	1.3759
400	10	942.42	0.0010611	529.06	539.67	1.5921
420	10	925.19	0.0010809	613.84	624.65	1.7994
440	10	906.28	0.0011034	699.52	710.55	1.9992
460	10	885.59	0.0011292	786.38	797.67	2.1928
480	10	862.94	0.0011588	874.8	886.39	2.3816
500	10	838.02	0.0011933	965.25	977.18	2.5669
520	10	810.36	0.001234	1058.4	1070.7	2.7504
540	10	779.15	0.0012835	1155.3	1168.1	2.9341
560	10	743.0	0.0013459	1257.6	1271.1	3.1213
580	10	699.05	0.0014305	1368.8	1383.1	3.3177
584.15	10	688.42	0.0014526	1393.5	1408.1	3.3606

TABLE B.4c Isobaric Data for $P = 15$ MPa

T (K)	P (MPa)	ρ (kg/m^3)	v (m^3/kg)	u (kJ/kg)	h (kJ/kg)	s (kJ/kg·K)
273.15	15	1007.3	0.000993	0.17746	15.069	0.0004468
280	15	1007.0	0.000993	28.581	43.476	0.10316
300	15	1003.1	0.000997	111.34	126.29	0.38886
320	15	995.83	0.001004	194.1	209.16	0.65627
340	15	985.98	0.001014	276.99	292.2	0.90797
360	15	974.05	0.001027	360.07	375.47	1.1459
380	15	960.3	0.001041	443.45	459.07	1.3719
400	15	944.88	0.001058	527.26	543.13	1.5875
420	15	927.86	0.001078	611.68	627.85	1.7942
440	15	909.22	0.0011	696.94	713.44	1.9933
460	15	888.87	0.001125	783.3	800.18	2.186
480	15	866.68	0.001154	871.1	888.4	2.3738
500	15	842.36	0.001187	960.75	978.56	2.5578
520	15	815.52	0.001226	1052.8	1071.2	2.7395
540	15	785.52	0.001273	1148.2	1167.3	2.9208
560	15	751.28	0.001331	1248.3	1268.2	3.1042
580	15	710.78	0.001407	1355.3	1376.4	3.294
600	15	659.41	0.001517	1474.9	1497.7	3.4994
615.31	15	603.52	0.001657	1585.3	1610.2	3.6846

TABLE B.4d Isobaric Data for $P = 20$ MPa

T (K)	P (MPa)	ρ (kg/m^3)	v (m^3/kg)	u (kJ/kg)	h (kJ/kg)	s (kJ/kg·K)
273.15	20	1009.7	0.00099	0.22569	20.033	0.00046962
280	20	1009.4	0.000991	28.496	48.31	0.10271
300	20	1005.3	0.000995	110.94	130.84	0.38741
320	20	997.93	0.001002	193.44	213.48	0.65408
340	20	988.09	0.001012	276.07	296.31	0.90516
360	20	976.22	0.001024	358.89	379.38	1.1426
380	20	962.58	0.001039	442.0	462.77	1.368
400	20	947.31	0.001056	525.5	546.62	1.583
420	20	930.48	0.001075	609.58	631.08	1.7891
440	20	912.09	0.001096	694.44	716.37	1.9874
460	20	892.08	0.001121	780.32	802.74	2.1794
480	20	870.3	0.001149	867.53	890.51	2.3662
500	20	846.53	0.001181	956.44	980.07	2.5489
520	20	820.44	0.001219	1047.6	1071.9	2.7291
540	20	791.5	0.001263	1141.6	1166.9	2.9082
560	20	758.86	0.001318	1239.7	1266.0	3.0885
580	20	721.04	0.001387	1343.5	1371.3	3.2731
600	20	675.11	0.001481	1456.8	1486.4	3.4682
620	20	613.23	0.001631	1588.6	1621.2	3.6891
638.9	20	490.19	0.00204	1786.4	1827.2	4.0156

TABLE B.4e Isobaric Data for $P = 30$ MPa

T (K)	P (MPa)	ρ (kg/m^3)	v (m^3/kg)	u (kJ/kg)	h (kJ/kg)	s (kJ/kg·K)
273.15	30	1014.5	0.0009857	0.28791	29.858	0.0002688
280	30	1014.0	0.0009862	28.307	57.893	0.10164
300	30	1009.6	0.0009905	110.16	139.87	0.38444
320	30	1002.1	0.0009979	192.15	222.08	0.64972
340	30	992.24	0.0010078	274.29	304.52	0.89961
360	30	980.48	0.0010199	356.62	387.21	1.1359
380	30	967.04	0.0010341	439.19	470.21	1.3603
400	30	952.05	0.0010504	522.11	553.62	1.5742
420	30	935.59	0.0010688	605.53	637.6	1.7791
440	30	917.67	0.0010897	689.63	722.33	1.9761
460	30	898.25	0.0011133	774.62	808.02	2.1666
480	30	877.23	0.00114	860.75	894.95	2.3516
500	30	854.45	0.0011703	948.32	983.43	2.5322
520	30	829.66	0.0012053	1037.7	1073.9	2.7095
540	30	802.5	0.0012461	1129.5	1166.9	2.885
560	30	772.42	0.0012946	1224.4	1263.2	3.0601
580	30	738.56	0.001354	1323.4	1364.0	3.237
600	30	699.47	0.0014296	1428.5	1471.4	3.4189
620	30	652.41	0.0015328	1542.9	1588.9	3.6116
640	30	591.01	0.001692	1674.8	1725.5	3.8283

TABLE B.4f Isobaric Data for $P = 50$ MPa

T (K)	P (MPa)	ρ (kg/m^3)	v (m^3/kg)	u (kJ/kg)	h (kJ/kg)	s (kJ/kg·K)
273.15	50	1023.8	0.000977	0.28922	49.126	−0.0010315
280	50	1022.9	0.000978	27.862	76.742	0.098824
300	50	1017.8	0.000982	108.63	157.76	0.37828
320	50	1010.1	0.00099	189.68	239.18	0.64102
340	50	1000.3	0.001	270.92	320.9	0.88875
360	50	988.72	0.001011	352.32	402.89	1.123
380	50	975.62	0.001025	433.9	485.15	1.3454
400	50	961.12	0.00104	515.76	567.78	1.5573
420	50	945.31	0.001058	597.99	650.88	1.7601
440	50	928.2	0.001077	680.73	734.6	1.9548
460	50	909.8	0.001099	764.15	819.1	2.1426
480	50	890.05	0.001124	848.42	904.59	2.3245
500	50	868.88	0.001151	933.75	991.29	2.5015
520	50	846.15	0.001182	1020.4	1079.5	2.6744
540	50	821.66	0.001217	1108.6	1169.5	2.8442
560	50	795.16	0.001258	1198.9	1261.8	3.012
580	50	766.3	0.001305	1291.7	1356.9	3.1789
600	50	734.55	0.001361	1387.6	1455.7	3.3463
620	50	699.21	0.00143	1487.7	1559.2	3.516
640	50	659.19	0.001517	1593.4	1669.3	3.6907

TABLE B.5 Vapor Properties: Saturated Solid–Vapor (Sublimation Line: 200–273.16 K)*

T (K)	P (kPa)	ρ_g (kg/m^3)	v_g (m^3/kg)	u_g (kJ/kg)	h_g (kJ/kg)	s_g (kJ/kg·K)
200	0.000162	1.758E-06	568840	2273.5	2365.8	12.3800
205	0.000343	3.628E-06	275640	2280.4	2375.1	12.08
210	0.000701	7.231E-06	138290	2287.4	2384.3	11.795
215	0.001385	1.395E-05	71663	2294.3	2393.6	11.524
220	0.002653	2.613E-05	38277	2301.3	2402.8	11.267
225	0.004938	4.755E-05	21031.000	2308.2	2412.1	11.021
230	0.008947	8.429E-05	11864.000	2315.2	2421.3	10.788
235	0.015806	0.0001457	6861.300	2322.1	2430.6	10.565
240	0.027271	0.0002462	4061.300	2329.1	2439.8	10.352
245	0.046015	0.000407	2457.100	2336.0	2449.1	10.149
250	0.076029	0.000659	1517.400	2343.0	2458.3	9.9545
255	0.12316	0.0010467	955.360	2349.9	2467.6	9.7685
260	0.19583	0.0016324	612.590	2356.8	2476.8	9.5903
265	0.30594	0.0025024	399.610	2363.7	2486.0	9.4195
270	0.47008	0.0037742	264.960	2370.6	2495.1	9.2556
273.15	0.61115	0.0048508	206.15	2374.9	2500.9	9.1558
273.16	0.61166	0.0048546	205.990	2374.9	2500.9	9.1555

* Property values generated from NIST Database 23: REFPROP Version 7.0 (August 2002).

APPENDIX C
Thermodynamic and Thermo-Physical Properties of Air

TABLE C.1 Approximate Composition, Apparent Molecular Weight, and Gas Constant for Dry Air

Constituent	Mole%
N_2	78.08
O_2	20.95
Ar	0.93
CO_2	0.036
Ne, He, CH_4, others	0.003

$$\mathcal{M}_{air} = 28.97 \text{ kg/kmol}$$

$$R_{air} = 287.0 \text{ J/kg·K}$$

TABLE C.2 Thermodynamic Properties of Air at 1 atm*

T (K)	h (kJ/kg)	u (kJ/kg)	s° (kJ/kg·K)	c_p (kJ/kg·K)	c_v (kJ/kg·K)	$\gamma\ (= c_p/c_v)$
200	325.42	268.14	3.4764	1.007	0.716	1.406
210	335.49	275.32	3.5255	1.007	0.716	1.405
220	345.55	282.49	3.5723	1.006	0.716	1.405
230	355.62	289.67	3.6171	1.006	0.716	1.404
240	365.68	296.85	3.6599	1.006	0.716	1.404
250	375.73	304.02	3.7009	1.006	0.717	1.404
260	385.79	311.20	3.7404	1.006	0.717	1.403
270	395.85	318.38	3.7783	1.006	0.717	1.403
280	405.91	325.56	3.8149	1.006	0.717	1.403
290	415.97	332.74	3.8502	1.006	0.718	1.402
300	426.04	339.93	3.8844	1.007	0.718	1.402
310	436.11	347.12	3.9174	1.007	0.719	1.401
320	446.18	354.31	3.9494	1.008	0.719	1.401
330	456.26	361.51	3.9804	1.008	0.720	1.400
340	466.34	368.72	4.0105	1.009	0.721	1.400
350	476.43	375.94	4.0397	1.009	0.721	1.399
360	486.53	383.16	4.0682	1.010	0.722	1.399
370	496.64	390.39	4.0959	1.011	0.723	1.398
380	506.75	397.63	4.1228	1.012	0.724	1.398
390	516.88	404.89	4.1491	1.013	0.725	1.397
400	527.02	412.15	4.1748	1.014	0.727	1.396
410	537.17	419.42	4.1999	1.016	0.728	1.395
420	547.33	426.71	4.2244	1.017	0.729	1.395
430	557.51	434.01	4.2483	1.018	0.731	1.394
440	567.70	441.33	4.2717	1.020	0.732	1.393

TABLE C.2 (cont.)

T (K)	h (kJ/kg)	u (kJ/kg)	s° (kJ/kg·K)	c_p (kJ/kg·K)	c_v (kJ/kg·K)	$\gamma\ (=c_p/c_v)$
450	577.90	448.66	4.2947	1.021	0.734	1.392
460	588.13	456.01	4.3171	1.023	0.735	1.391
470	598.36	463.38	4.3392	1.025	0.737	1.390
480	608.62	470.76	4.3608	1.026	0.739	1.389
490	618.89	478.16	4.3819	1.028	0.741	1.388
500	629.18	485.58	4.4027	1.030	0.743	1.387
510	639.50	493.01	4.4231	1.032	0.745	1.386
520	649.83	500.47	4.4432	1.034	0.747	1.385
530	660.18	507.95	4.4629	1.036	0.749	1.384
540	670.55	515.45	4.4823	1.038	0.751	1.383
550	680.94	522.97	4.5014	1.040	0.753	1.382
560	691.35	530.51	4.5201	1.042	0.755	1.381
570	701.79	538.07	4.5386	1.045	0.757	1.380
580	712.25	545.65	4.5568	1.047	0.759	1.378
590	722.73	553.26	4.5747	1.049	0.762	1.377
600	733.23	560.89	4.5924	1.051	0.764	1.376
610	743.76	568.55	4.6098	1.054	0.766	1.375
620	754.30	576.22	4.6269	1.056	0.769	1.374
630	764.88	583.92	4.6438	1.058	0.771	1.373
640	775.47	591.65	4.6605	1.061	0.773	1.371
650	786.09	599.40	4.6770	1.063	0.776	1.370
660	796.74	607.17	4.6932	1.066	0.778	1.369
670	807.41	614.96	4.7093	1.068	0.781	1.368
680	818.10	622.78	4.7251	1.070	0.783	1.367
690	828.81	630.63	4.7408	1.073	0.786	1.366
700	839.55	638.50	4.7562	1.075	0.788	1.365
710	850.32	646.39	4.7715	1.078	0.790	1.364
720	861.11	654.30	4.7866	1.080	0.793	1.362
730	871.92	662.24	4.8015	1.082	0.795	1.361
740	882.76	670.21	4.8162	1.085	0.798	1.360
750	893.62	678.20	4.8308	1.087	0.800	1.359
760	904.50	686.21	4.8452	1.090	0.802	1.358
770	915.41	694.24	4.8595	1.092	0.805	1.357
780	926.34	702.30	4.8736	1.094	0.807	1.356
790	937.29	710.39	4.8875	1.097	0.809	1.355
800	948.27	718.49	4.9014	1.099	0.812	1.354
820	970.30	734.77	4.9285	1.104	0.816	1.352
840	992.41	751.15	4.9552	1.108	0.821	1.350
860	1014.62	767.61	4.9813	1.113	0.825	1.348
880	1036.91	784.16	5.0069	1.117	0.830	1.346
900	1059.29	800.80	5.0321	1.121	0.834	1.344
920	1081.76	817.52	5.0568	1.125	0.838	1.343
940	1104.30	834.32	5.0810	1.129	0.842	1.341
960	1126.93	851.21	5.1048	1.133	0.846	1.339
980	1149.64	868.18	5.1283	1.137	0.850	1.338
1000	1172.43	885.22	5.1513	1.141	0.854	1.336
1020	1195.29	902.34	5.1739	1.145	0.858	1.335
1040	1218.23	919.53	5.1962	1.149	0.861	1.333

TABLE C.2 (cont.)

T (K)	h (kJ/kg)	u (kJ/kg)	s° (kJ/kg·K)	c_p (kJ/kg·K)	c_v (kJ/kg·K)	$\gamma\ (=c_p/c_v)$
1060	1241.23	936.80	5.2181	1.152	0.865	1.332
1080	1264.31	954.13	5.2397	1.156	0.868	1.331
1100	1287.46	971.53	5.2609	1.159	0.872	1.329
1120	1310.67	989.00	5.2818	1.162	0.875	1.328
1140	1333.95	1006.54	5.3024	1.165	0.878	1.327
1160	1357.29	1024.13	5.3227	1.169	0.881	1.326
1180	1380.69	1041.79	5.3427	1.172	0.884	1.325
1200	1404.15	1059.51	5.3624	1.175	0.887	1.324
1220	1427.67	1077.29	5.3819	1.177	0.890	1.323
1240	1451.25	1095.12	5.4010	1.180	0.893	1.322
1260	1474.88	1113.01	5.4199	1.183	0.896	1.321
1280	1498.56	1130.95	5.4386	1.186	0.898	1.320
1300	1522.30	1148.95	5.4570	1.188	0.901	1.319
1320	1546.09	1166.99	5.4751	1.191	0.903	1.318
1340	1569.92	1185.08	5.4931	1.193	0.906	1.317
1360	1593.81	1203.23	5.5108	1.195	0.908	1.316
1380	1617.74	1221.42	5.5282	1.198	0.911	1.315
1400	1641.71	1239.65	5.5455	1.200	0.913	1.315
1420	1665.74	1257.93	5.5625	1.202	0.915	1.314
1440	1689.80	1276.25	5.5793	1.204	0.917	1.313
1460	1713.91	1294.61	5.5960	1.206	0.919	1.312
1480	1738.05	1313.02	5.6124	1.208	0.921	1.312
1500	1762.24	1331.46	5.6286	1.210	0.923	1.311
1520	1786.46	1349.94	5.6447	1.212	0.925	1.310
1540	1810.73	1368.46	5.6605	1.214	0.927	1.310
1560	1835.03	1387.02	5.6762	1.216	0.929	1.309
1580	1859.36	1405.61	5.6917	1.218	0.931	1.309
1600	1883.73	1424.24	5.7070	1.219	0.932	1.308
1620	1908.14	1442.90	5.7222	1.221	0.934	1.307
1640	1932.58	1461.60	5.7372	1.223	0.936	1.307
1660	1957.05	1480.33	5.7520	1.224	0.937	1.306
1680	1981.55	1499.09	5.7667	1.226	0.939	1.306
1700	2006.08	1517.88	5.7812	1.227	0.940	1.305
1750	2067.54	1564.99	5.8168	1.231	0.944	1.304
1800	2129.19	1612.27	5.8516	1.235	0.947	1.303
1850	2190.99	1659.72	5.8854	1.238	0.951	1.302
1900	2252.96	1707.33	5.9185	1.241	0.954	1.301
1950	2315.08	1755.09	5.9508	1.244	0.957	1.300
2000	2377.34	1803.00	5.9823	1.247	0.959	1.299
2050	2439.73	1851.04	6.0131	1.249	0.962	1.298
2100	2502.26	1899.21	6.0432	1.252	0.965	1.298
2150	2564.90	1947.50	6.0727	1.254	0.967	1.297
2200	2627.67	1995.91	6.1016	1.256	0.969	1.296
2250	2690.54	2044.43	6.1298	1.259	0.971	1.296
2300	2753.53	2093.05	6.1575	1.261	0.974	1.295
2350	2816.61	2141.78	6.1846	1.263	0.976	1.294

* Property values generated from NIST Database 23: *REFPROP* Version 7.0 (August 2002). Reference state: h (78.903 K) = 0.0; s (78.903 K) = 0.0.

TABLE C.3a Thermo-Physical Properties of Air (100–1000 K at 1 atm)*

T (K)	ρ (kg/m^3)	c_p (kJ/kg·K)	μ (μPa·s)	ν (m^2/s)	k (W/m·K)	α (m^2/s)	Pr
100	3.6043	1.0356	7.1551	1.985E-06	0.010116	2.710E-06	0.73245
120	2.9772	1.0211	8.4995	2.855E-06	0.011996	3.946E-06	0.72349
140	2.5403	1.0142	9.7899	3.854E-06	0.013802	5.357E-06	0.71940
160	2.2169	1.0105	11.029	4.975E-06	0.015540	6.936E-06	0.71723
180	1.9674	1.0084	12.221	6.212E-06	0.017213	8.677E-06	0.71593
200	1.7688	1.0071	13.370	7.559E-06	0.018829	1.057E-05	0.71508
220	1.6068	1.0063	14.479	9.011E-06	0.020392	1.261E-05	0.71449
240	1.4721	1.0059	15.552	1.056E-05	0.021908	1.480E-05	0.71407
260	1.3584	1.0058	16.592	1.221E-05	0.023381	1.711E-05	0.71376
280	1.2610	1.0061	17.601	1.396E-05	0.024817	1.956E-05	0.71354
300	1.1767	1.0066	18.582	1.579E-05	0.026220	2.214E-05	0.71339
320	1.1030	1.0075	19.536	1.771E-05	0.027594	2.483E-05	0.71330
340	1.0380	1.0087	20.465	1.972E-05	0.028944	2.764E-05	0.71324
360	0.98022	1.0103	21.372	2.180E-05	0.030272	3.057E-05	0.71323
380	0.92856	1.0122	22.256	2.397E-05	0.031583	3.361E-05	0.71326
400	0.88208	1.0144	23.121	2.621E-05	0.032880	3.675E-05	0.71331
420	0.84004	1.0169	23.966	2.853E-05	0.034164	3.999E-05	0.71340
440	0.80183	1.0198	24.794	3.092E-05	0.035437	4.334E-05	0.71352
460	0.76695	1.0230	25.605	3.339E-05	0.036702	4.678E-05	0.71366
480	0.73497	1.0264	26.400	3.592E-05	0.037960	5.032E-05	0.71384
500	0.70556	1.0301	27.180	3.852E-05	0.039212	5.395E-05	0.71403
520	0.67842	1.0340	27.946	4.119E-05	0.040458	5.767E-05	0.71425
540	0.65329	1.0382	28.698	4.393E-05	0.041699	6.148E-05	0.71449
560	0.62995	1.0424	29.438	4.673E-05	0.042935	6.538E-05	0.71475
580	0.60823	1.0469	30.166	4.960E-05	0.044167	6.936E-05	0.71503
600	0.58795	1.0514	30.883	5.253E-05	0.045395	7.343E-05	0.71532
620	0.56898	1.0561	31.589	5.552E-05	0.046618	7.758E-05	0.71562
640	0.55120	1.0608	32.284	5.857E-05	0.047837	8.181E-05	0.71593
660	0.53450	1.0656	32.970	6.168E-05	0.049050	8.612E-05	0.71626
680	0.51878	1.0704	33.646	6.486E-05	0.050259	9.051E-05	0.71659
700	0.50396	1.0752	34.313	6.809E-05	0.051462	9.497E-05	0.71693
720	0.48996	1.0800	34.972	7.138E-05	0.052659	9.951E-05	0.71727
740	0.47672	1.0848	35.623	7.473E-05	0.053851	1.041E-04	0.71761
760	0.46417	1.0896	36.266	7.813E-05	0.055036	1.088E-04	0.71796
780	0.45227	1.0943	36.901	8.159E-05	0.056215	1.136E-04	0.71831
800	0.44097	1.0989	37.529	8.511E-05	0.057388	1.184E-04	0.71866
820	0.43022	1.1035	38.151	8.868E-05	0.058553	1.233E-04	0.71901
840	0.41997	1.1081	38.765	9.230E-05	0.059712	1.283E-04	0.71936
860	0.41021	1.1125	39.374	9.599E-05	0.060863	1.334E-04	0.71971
880	0.40089	1.1169	39.976	9.972E-05	0.062007	1.385E-04	0.72006
900	0.39198	1.1212	40.573	1.035E-04	0.063143	1.437E-04	0.72041
920	0.38346	1.1254	41.164	1.074E-04	0.064271	1.489E-04	0.72075
940	0.37530	1.1295	41.749	1.112E-04	0.065392	1.543E-04	0.72109
960	0.36749	1.1335	42.329	1.152E-04	0.066505	1.597E-04	0.72143
980	0.35999	1.1374	42.904	1.192E-04	0.067610	1.651E-04	0.72176
1000	0.35279	1.1412	43.474	1.232E-04	0.068708	1.707E-04	0.72210

* Values generated from NIST Database 23: *REFPROP* Version 7.0 (August 2002).

TABLE C.3b Thermo-Physical Properties of Air (1000–2300 K at 1 atm)*

T (K)	ρ (kg/m^3)	c_p (kJ/kg·K)	μ (μPa·s)	ν (m^2/s)	k (W/m·K)	α (m^2/s)	Pr
1000	0.35281	1.1412	43.474	1.23E-04	0.068708	6.871E-06	0.72210
1100	0.32074	1.1590	46.258	1.44E-04	0.074077	7.408E-06	0.72372
1200	0.29402	1.1745	48.941	1.66E-04	0.079254	7.925E-06	0.72530
1300	0.27141	1.1881	51.539	1.90E-04	0.084248	8.425E-06	0.72683
1400	0.25202	1.1999	54.063	2.15E-04	0.089070	8.907E-06	0.72835
1500	0.23523	1.2103	56.524	2.40E-04	0.093733	9.373E-06	0.72986
1600	0.22053	1.2194	58.930	2.67E-04	0.098251	9.825E-06	0.73138
1700	0.20756	1.2274	61.286	2.95E-04	0.10263	1.026E-05	0.73293
1800	0.19603	1.2345	63.600	3.24E-04	0.10690	1.069E-05	0.73451
1900	0.18571	1.2409	65.877	3.55E-04	0.11105	1.111E-05	0.73614
2000	0.17643	1.2466	68.119	3.86E-04	0.11509	1.151E-05	0.73781
2100	0.16803	1.2517	70.332	4.19E-04	0.11904	1.190E-05	0.73954
2200	0.16039	1.2564	72.518	4.52E-04	0.12290	1.229E-05	0.74134
2300	0.15342	1.2607	74.681	4.87E-04	0.12668	1.267E-05	0.74320

* Values generated from NIST Database 23: *REFPROP* Version 7.0 (August 2002).

APPENDIX D

Thermodynamic Properties of Ideal Gases and Carbon

Tables D.1–D.13 present values for $\bar{c}_p(T), \bar{h}^\circ(T) - \bar{h}^\circ_{\mathrm{f,ref}}, \bar{h}^\circ_{\mathrm{f}}(T), \bar{s}^\circ(T)$, and $\Delta\bar{g}^\circ_{\mathrm{f}}(T)$ at standard reference state ($T = 298.15$ K, P = 1 atm) for various species of the C–H–O–N system (with ideal-gas values for gaseous species).

Note that the enthalpy of formation and the Gibbs function of formation for compounds are calculated from the elements as follows:

$$\begin{aligned}\bar{h}^\circ_{\mathrm{f},i}(T) &= \bar{h}^\circ_i(T) - \sum_{j\,\text{elements}} v'_j \bar{h}^\circ_j(T), \bar{g}^\circ_{\mathrm{f},i}(T) \\ &= \bar{g}^\circ_i(T) - \sum_{j\,\text{elements}} v'_j \bar{g}^\circ_j(T) \\ &= \bar{h}^\circ_{\mathrm{f},i}(T) - T\bar{s}^\circ_i(T) - \sum_{j\,\text{elements}} v'_j\left[-T\bar{s}^\circ_j(T)\right].\end{aligned}$$

Sources: Tables D.1–D.12 were generated from Key, R. J., Rupley, F. M., and Miller, J. A., "The Chemkin Thermodynamic Data Base," Sandia Report, SAND87-8215B, March 1991. Table D.13 is from Myers, G. E., *Engineering Thermodynamics,* Prentice-Hall, Englewood Cliffs, NJ, 1989. Table D.14 is from Key, R. J., *et al.*, ibid.

TABLE D.1 CO (Molecular Weight = 28.010, Enthalpy of Formation at 298 K = −110,541 kJ/kmol)

T (K)	$\bar{c}_p$ (kJ/kmol · K)	$\bar{h}^\circ(T) - \bar{h}^\circ_f(298)$ (kJ/kmol)	$\bar{h}^\circ_f(T)$ (kJ/kmol)	$\bar{s}^\circ(T)$ (kJ/kmol · K)	$\Delta \bar{g}^\circ_f(T)$ (kJ/kmol)
200	28.687	−2835	−111,308	186.018	−128,532
298	29.072	0	−110,541	197.548	−137,163
300	29.078	54	−110,530	197.728	−137,328
400	29.433	2979	−110,121	206.141	−146,332
500	29.857	5943	−110,017	212.752	−155,403
600	30.407	8955	−110,156	218.242	−164,470
700	31.089	12,029	−110,477	222.979	−173,499
800	31.860	15,176	−110,924	227.180	−182,473
900	32.629	18,401	−111,450	230.978	−191,386
1000	33.255	21,697	−112,022	234.450	−200,238
1100	33.725	25,046	−112,619	237.642	−209,030
1200	34.148	28,440	−113,240	240.595	−217,768
1300	34.530	31,874	−113,881	243.344	−226,453
1400	34.872	35,345	−114,543	245.915	−235,087
1500	35.178	38,847	−115,225	248.332	−243,674
1600	35.451	42,379	−115,925	250.611	−252,214
1700	35.694	45,937	−116,644	252.768	−260,711
1800	35.910	49,517	−117,380	254.814	−269,164
1900	36.101	53,118	−118,132	256.761	−277,576
2000	36.271	56,737	−118,902	258.617	−285948
2100	36.421	60,371	−119,687	260.391	−294,281
2200	36.553	64,020	−120,488	262.088	−302,576
2300	36.670	67,682	−121,305	263.715	−310,835
2400	36.774	71,354	−122,137	265.278	−319,057
2500	36.867	75,036	−122,984	266.781	−327,245
2600	36.950	78,727	−123,847	268.229	−335,399
2700	37.025	82,426	−124,724	269.625	−343,519
2800	37.093	86,132	−125,616	270.973	−351,606
2900	37.155	89,844	−126,523	272.275	−359,661
3000	37.213	93,562	−127,446	273.536	−367,684
3100	37.268	97,287	−128,383	274.757	−375,677
3200	37.321	101,016	−129,335	275.941	−383,639
3300	37.372	104,751	−130,303	277.090	−391,571
3400	37.422	108,490	−131,285	278.207	−399,474
3500	37.471	112,235	−132,283	279.292	−407,347
3600	37.521	115,985	−133,295	280.349	−415,192
3700	37.570	119,739	−134,323	281.377	−423,008
3800	37.619	123,499	−135,366	282.380	−430,796
3900	37.667	127,263	−136,424	283.358	−438,557
4000	37.716	131,032	−137,497	284.312	−446,291
4100	37.764	134,806	−138,585	285.244	−453,997
4200	37.810	138,585	−139,687	286.154	−461,677
4300	37.855	142,368	−140,804	287.045	−469,330
4400	37.897	146,156	−141,935	287.915	−476,957
4500	37.936	149,948	−143,079	288.768	−484,558
4600	37.970	153,743	−144,236	289.602	−492,134
4700	37.998	157,541	−145,407	290.419	−499,684
4800	38.019	161,342	−146,589	291.219	−507,210
4900	38.031	165,145	−147,783	292.003	−514,710
5000	38.033	168,948	−148,987	292.771	−522,186

TABLE D.2 CO_2 (Molecular Weight = 44.011, Enthalpy of Formation at 298 K = –393, 546 kJ/kmol)

T (K)	$\bar{c}_p$ (kJ/kmol · K)	$\bar{h}^\circ(T) - \bar{h}^\circ_f(298)$ (kJ/kmol)	$\bar{h}^\circ_f(T)$ (kJ/kmol)	$\bar{s}^\circ(T)$ (kJ/kmol · K)	$\Delta\,\bar{g}^\circ_f(T)$ (kJ/kmol)
200	32.387	–3423	–393,483	199.876	–394,126
298	37.198	0	–393,546	213.736	–394,428
300	37.280	69	–393,547	213.966	–394,433
400	41.276	4003	–393,617	225.257	–394,718
500	44.569	8301	–393,712	234.833	–394,983
600	47.313	12,899	–393,844	243.209	–395,226
700	49.617	17,749	–394,013	250.680	–395,443
800	51.550	22,810	–394,213	257.436	–395,635
900	53.136	28,047	–394433	263.603	–395,799
1000	54.360	33,425	–394,659	269.268	–395,939
1100	55.333	38,911	–394,875	274.495	–396,056
1200	56.205	44,488	–395,083	279.348	–396,155
1300	56.984	50,149	–395,287	283.878	–396,236
1400	57.677	55,882	–395,88	288.127	–396,301
1500	58.292	61,681	–395,691	292.128	–396,352
1600	58.836	67,538	–395,897	295.908	–396,389
1700	59.316	73,446	–396,110	299.489	–396,414
1800	59.738	79,399	–396,332	302.892	–396,425
1900	60.108	85,392	–396,564	306.132	–396,424
2000	60.433	91,420	–396,808	309.223	–396,410
2100	60.717	97,477	–397,065	312.179	–396,384
2200	60.966	103,562	–397,338	315.009	–396,346
2300	61.185	109,670	–397,626	317.724	–396,294
2400	61.378	115,798	–397,931	320.333	–396,230
2500	61.548	121,944	–398,253	322.842	–396,152
2600	61.701	128,107	–398,594	325.259	–396,061
2700	61.839	134,284	–398,952	327.590	–395,957
2800	61.965	140,474	–399,329	329.841	–395,840
2900	62.083	146,677	–399,725	332.018	–395,708
3000	62.194	152,891	–400,140	334.124	–395,562
3100	62.301	159,116	–400,573	336.165	–395,403
3200	62.406	165,351	–401,025	338.145	–395,229
3300	62.510	171,597	–401,495	340.067	–395,041
3400	62.614	177,853	–401,983	341.935	–394,838
3500	62.718	184,120	–402,489	343.751	–394,620
3600	62.825	190,397	–403,013	345.519	–394,388
3700	62.932	196,685	–403,553	347.242	–394,141
3800	63.041	202,983	–404,110	348.922	–393,879
3900	63.151	209,293	–404,684	350.561	–393,602
4000	63.261	215,613	–405,273	352.161	–393,311
4100	63.369	221,945	–405,878	353.725	–393,004
4200	63.474	228,287	–406,499	355.253	–392,683
4300	63.575	234,640	–407,135	356.748	–392,346
4400	63.669	241,002	–407,785	358.210	–391,995
4500	63.753	247,373	–408,451	359.642	–391,629
4600	63.825	253,752	–409,132	361.044	–391,247
4700	63.881	260,138	–409,828	362.417	–390,851
4800	63.918	266,528	–410,539	363.763	–390,440
4900	63.932	272,920	–411,267	365.081	–390,014
5000	63.919	279,313	–412,010	366.372	–389,572

TABLE D.3 H_2 (Molecular Weight = 2.016, Enthalpy of Formation at 298 K = 0 kJ/kmol)

T (K)	$\bar{c}_p$ (kJ/kmol · K)	$\bar{h}^\circ(T) - \bar{h}^\circ_f(298)$ (kJ/kmol)	$\bar{h}^\circ_f(T)$ (kJ/kmol)	$\bar{s}^\circ(T)$ (kJ/kmol · K)	$\Delta\,\bar{g}^\circ_f(T)$ (kJ/kmol)
200	28.522	−2818	0	119.137	0
298	28.871	0	0	130.595	0
300	28.877	53	0	130.773	0
400	29.120	2954	0	139.116	0
500	29.275	5874	0	145.632	0
600	29.375	8807	0	150.979	0
700	29.461	11,749	0	155.514	0
800	29.581	14,701	0	159.455	0
900	29.792	17,668	0	162.950	0
1000	30.160	20,664	0	166.106	0
1100	30.625	23,704	0	169.003	0
1200	31.077	26,789	0	171.687	0
1300	31.516	29,919	0	174.192	0
1400	31.943	33,092	0	176.543	0
1500	32.356	36,307	0	178.761	0
1600	32.758	39,562	0	180.862	0
1700	33.146	42,858	0	182.860	0
1800	33.522	46,191	0	184.765	0
1900	33.885	49,562	0	186.587	0
2000	34.236	52,968	0	188.334	0
2100	34.575	56,408	0	190.013	0
2200	34.901	59,882	0	191.629	0
2300	35.216	63,388	0	193.187	0
2400	35.519	66,925	0	194.692	0
2500	35.811	70,492	0	196.148	0
2600	36.091	74,087	0	197.558	0
2700	36.361	77,710	0	198.926	0
2800	36.621	81,359	0	200.253	0
2900	36.871	85,033	0	201.542	0
3000	37.112	88,733	0	202.796	0
3100	37.343	92,455	0	204.017	0
3200	37.566	96,201	0	205.206	0
3300	37.781	99,968	0	206.365	0
3400	37.989	103,757	0	207.496	0
3500	38.190	107,566	0	208.600	0
3600	38.385	111,395	0	209.679	0
3700	38.574	115,243	0	210.733	0
3800	38.759	119,109	0	211.764	0
3900	38.939	122,994	0	212.774	0
4000	39.116	126,897	0	213.762	0
4100	39.291	130,817	0	214.730	0
4200	39.464	134,755	0	215.679	0
4300	39.636	138,710	0	216.609	0
4400	39.808	142,682	0	217.522	0
4500	39.981	146,672	0	218.419	0
4600	40.156	150,679	0	219.300	0
4700	40.334	154,703	0	220.165	0
4800	40.516	158,746	0	221.016	0
4900	40.702	162,806	0	221.853	0
5000	40.895	166,886	0	222.678	0

TABLE D.4 H (Molecular Weight = 1.008, Enthalpy of Formation at 298 K = 217,979 kJ/kmol)

T (K)	$\bar{c}_p$ (kJ/kmol · K)	$\bar{h}^\circ(T) - \bar{h}^\circ_f(298)$ (kJ/kmol)	$\bar{h}^\circ_f(T)$ (kJ/kmol)	$\bar{s}^\circ(T)$ (kJ/kmol · K)	$\Delta \bar{g}^\circ_f(T)$ (kJ/kmol)
200	20.786	−2040	217,346	106.305	207,999
298	20.786	0	217,977	114.605	203,276
300	20.786	38	217,989	114.733	203,185
400	20.786	2117	218,617	120.713	198,155
500	20.786	4196	219,236	125.351	192,968
600	20.786	6274	219,848	129.141	187,657
700	20.786	8353	220,456	132.345	182,244
800	20.786	10,431	221,059	135.121	176,744
900	20.786	12,510	221,653	137.569	171,169
1000	20.786	14,589	222,234	139.759	165,528
1100	20.786	16,667	222,793	141.740	159,830
1200	20.786	18,746	223,329	143.549	154,082
1300	20.786	20,824	223,843	145.213	148,291
1400	20.786	22,903	224,335	146.753	142,461
1500	20.786	24,982	224,806	148.187	136,596
1600	20.786	27,060	225,256	149.528	130,700
1700	20.786	29,139	225,687	150.789	124,777
1800	20.786	31,217	226,099	151.977	118,830
1900	20.786	33,296	226,493	153.101	112,859
2000	20.786	35,375	226,868	154.167	106,869
2100	20.786	37,453	227,226	155.181	100,860
2200	20.786	39,532	227,568	156.148	94,834
2300	20.786	41,610	227,894	157.072	88,794
2400	20.786	43,689	228,204	157.956	82,739
2500	20.786	45,768	228,499	158.805	76,672
2600	20.786	47,846	228,780	159.620	70,593
2700	20.786	49,925	229,047	160.405	64,504
2800	20.786	52,003	229,301	161.161	58,405
2900	20.786	54,082	229,543	161.890	52,298
3000	20.786	56,161	229,772	162.595	46,182
3100	20.786	58,239	229,989	163.276	40,058
3200	20.786	60,318	230,195	163.936	33,928
3300	20.786	62,396	230,390	164.576	27,792
3400	20.786	64,475	230,574	165.196	21,650
3500	20.786	66,554	230,748	165.799	15,502
3600	20.786	68,632	230,912	166.384	9350
3700	20.786	70,711	231,067	166.954	3194
3800	20.786	72,789	231,212	167.508	−2967
3900	20.786	74,868	231,348	168.048	−9132
4000	20.786	76,947	231,475	168.575	−15,299
4100	20.786	79,025	231,594	169.088	−21,470
4200	20.786	81,104	231,704	169.589	−27,644
4300	20.786	83,182	231,805	170.078	−33,820
4400	20.786	85,261	231,897	170.556	−39,998
4500	20.786	87,340	231,981	171.023	−46,179
4600	20.786	89,418	232,056	171.480	−52,361
4700	20.786	91,497	232,123	171.927	−58,545
4800	20.786	93,575	232,180	172.364	−64,730
4900	20.786	95,654	232,228	172.793	−70,916
5000	20.786	97,733	232,267	173.213	−77,103

TABLE D.5 OH (Molecular Weight = 17.007, Enthalpy of Formation at 298 K = 38,986 kJ/kmol)

T (K)	$\bar{c}_p$ (kJ/kmol · K)	$\bar{h}^\circ(T) - \bar{h}^\circ_f(298)$ (kJ/kmol)	$\bar{h}^\circ_f(T)$ (kJ/kmol)	$\bar{s}^\circ(T)$ (kJ/kmol · K)	$\Delta \bar{g}^\circ_f(T)$ (kJ/kmol)
200	30.140	−2948	38,864	171.607	35,808
298	29.932	0	38,985	183.604	34,279
300	29.928	55	38,987	183.789	34,250
400	29.718	3037	39,030	192.369	32,662
500	29.570	6001	39,000	198.983	31,072
600	29.527	8955	38,909	204.369	29,494
700	29.615	11,911	38,770	208.925	27,935
800	29.844	14,883	38,599	212.893	26,399
900	30.208	17,884	38,410	216.428	24,885
1000	30.682	20,928	38,220	219.635	23,392
1100	31.186	24,022	38,039	222.583	21,918
1200	31.662	27,164	37,867	225.317	20,460
1300	32.114	30,353	37,704	227.869	19,017
1400	32.540	33,586	37,548	230.265	17,585
1500	32.943	36,860	37,397	232.524	16,164
1600	33.323	40,174	37,252	234.662	14,753
1700	33.682	43,524	37,109	236.693	13,352
1800	34.019	46,910	36,969	238.628	11,958
1900	34.337	50,328	36,831	240.476	10,573
2000	34.635	53,776	36,693	242.245	9194
2100	34.915	57,254	36,555	243.942	7823
2200	35.178	60,759	36,416	245.572	6458
2300	35.425	64,289	36,276	247.141	5099
2400	35.656	67,843	36,133	248.654	3746
2500	35.872	71,420	35,986	250.114	2400
2600	36.074	75,017	35,836	251.525	1060
2700	36.263	78,634	35,682	252.890	−275
2800	36.439	82,269	35,524	254.212	−1604
2900	36.604	85,922	35,360	255.493	−2927
3000	36.759	89,590	35,191	256.737	−4245
3100	36.903	93,273	35,016	257.945	−5556
3200	37.039	96,970	34,835	259.118	−6862
3300	37.166	100,681	34,648	260.260	−8162
3400	37.285	104,403	34,454	261.371	−9457
3500	37.398	108,137	34,253	262.454	−10,745
3600	37.504	111,882	34,046	263.509	−12,028
3700	37.605	115,638	33,831	264.538	−13,305
3800	37.701	119,403	33,610	265.542	−14,576
3900	37.793	123,178	33,381	266.522	−15,841
4000	37.882	126,962	33,146	267.480	−17,100
4100	37.968	130,754	32,903	268.417	−18,353
4200	38.052	134,555	32,654	269.333	−19,600
4300	38.135	138,365	32,397	270.229	−20,841
4400	38.217	142,182	32,134	271.107	−22,076
4500	38.300	146,008	31,864	271.967	−23,306
4600	38.382	149,842	31,588	272.809	−24,528
4700	38.466	153,685	31,305	273.636	−25,745
4800	38.552	157,536	31,017	274.446	−26,956
4900	38.640	161,395	30,722	275.242	−28,161
5000	38.732	165,264	30,422	276.024	−29,360

TABLE D.6 H_2O (Molecular Weight = 18.016, Enthalpy of Formation at 298 K = −241,845 kJ/kmol, Enthalpy of Vaporization = 44,010 kJ/kmol)

T (K)	$\bar{c}_p$ (kJ/kmol · K)	$\bar{h}^\circ(T) - \bar{h}^\circ_f(298)$ (kJ/kmol)	$\bar{h}^\circ_f(T)$ (kJ/kmol)	$\bar{s}^\circ(T)$ (kJ/kmol · K)	$\Delta\bar{g}^\circ_f(T)$ (kJ/kmol)
200	32.255	−3227	−240,838	175.602	−232,779
298	33.448	0	−241,847	188.715	−228,608
300	33.468	62	−241,865	188.922	−228,526
400	34.437	3458	−242,858	198.686	−223,929
500	35.337	6947	−243,822	206.467	−219,085
600	36.288	10,528	−244,753	212.992	−214,049
700	37.364	14,209	−245,638	218.665	−208,861
800	38.587	18,005	−246,461	223.733	−203,550
900	39.930	21,930	−247,209	228.354	−198,141
1000	41.315	25,993	−247,879	232.633	−192,652
1100	42.638	30,191	−248,475	236.634	−187,100
1200	43.874	34,518	−249,005	240.397	−181,497
1300	45.027	38,963	−249,477	243.955	−175,852
1400	46.102	43,520	−249,895	247.332	−170,172
1500	47.103	48,181	−250,267	250.547	−164,464
1600	48.035	52,939	−250,597	253.617	−158,733
1700	48.901	57,786	−250,890	256.556	−152,983
1800	49.705	62,717	−251,151	259.374	−147,216
1900	50.451	67,725	−251,384	262.081	−141,435
2000	51.143	72,805	−251,594	264.687	−135,643
2100	51.784	77,952	−251,783	267.198	−129,841
2200	52.378	83,160	−251,955	269.621	−124,030
2300	52.927	88,426	−252,113	271.961	−118,211
2400	53.435	93,744	−252,261	274.225	−112,386
2500	53.905	99,112	−252,399	276.416	−106,555
2600	54.340	104,524	−252,532	278.539	−100,719
2700	54.742	109,979	−252,659	280.597	−94,878
2800	55.115	115,472	−252,785	282.595	−89,031
2900	55.459	121,001	−252,909	284.535	−83,181
3000	55.779	126,563	−253,034	286.420	−77,326
3100	56.076	132,156	−253,161	288.254	−71,467
3200	56.353	137,777	−253,290	290.039	−65,604
3300	56.610	143,426	−253,423	291.777	−59,737
3400	56.851	149,099	−253,561	293.471	−53,865
3500	57.076	154,795	−253,704	295.122	−47,990
3600	57.288	160,514	−253,852	296.733	−42,110
3700	57.488	166,252	−254,007	298.305	−36,226
3800	57.676	172,011	−254,169	299.841	−30,338
3900	57.856	177,787	−254,338	301.341	−24,446
4000	58.026	183,582	−254,515	302.808	−18,549
4100	58.190	189,392	−254,699	304.243	−12,648
4200	58.346	195,219	−254,892	305.647	−6742
4300	58.496	201,061	−255,093	307.022	−831
4400	58.641	206,918	−255,303	308.368	5085
4500	58.781	212,790	−255,522	309.688	11,005
4600	58.916	218,674	−255,751	310.981	16,930
4700	59.047	224,573	−255,990	312.250	22,861
4800	59.173	230,484	−256,239	313.494	28,796
4900	59.295	236,407	−256,501	314.716	34,737
5000	59.412	242,343	−256,774	315.915	40,684

TABLE D.7 N_2 (Molecular Weight = 28.013, Enthalpy of Formation at 298 K = 0 kJ/kmol)

T (K)	$\bar{c}_p$ (kJ/kmol · K)	$\bar{h}^\circ(T) - \bar{h}^\circ_f(298)$ (kJ/kmol)	$\bar{h}^\circ_f(T)$ (kJ/kmol)	$\bar{s}^\circ(T)$ (kJ/kmol · K)	$\Delta\bar{g}^\circ_f(T)$ (kJ/kmol)
200	28.793	−2841	0	179.959	0
298	29.071	0	0	191.511	0
300	29.075	54	0	191.691	0
400	29.319	2973	0	200.088	0
500	29.636	5920	0	206.662	0
600	30.086	8905	0	212.103	0
700	30.684	11,942	0	216.784	0
800	31.394	15,046	0	220.927	0
900	32.131	18,222	0	224.667	0
1000	32.762	21,468	0	228.087	0
1100	33.258	24,770	0	231.233	0
1200	33.707	28,118	0	234.146	0
1300	34.113	31,510	0	236.861	0
1400	34.477	34,939	0	239.402	0
1500	34.805	38,404	0	241.792	0
1600	35.099	41,899	0	244.048	0
1700	35.361	45,423	0	246.184	0
1800	35.595	48,971	0	248.212	0
1900	35.803	52,541	0	250.142	0
2000	35.988	56,130	0	251.983	0
2100	36.152	59,738	0	253.743	0
2200	36.298	63,360	0	255.429	0
2300	36.428	66,997	0	257.045	0
2400	36.543	70,645	0	258.598	0
2500	36.645	74,305	0	260.092	0
2600	36.737	77,974	0	261.531	0
2700	36.820	81,652	0	262.919	0
2800	36.895	85,338	0	264.259	0
2900	36.964	89,031	0	265.555	0
3000	37.028	92,730	0	266.810	0
3100	37.088	96,436	0	268.025	0
3200	37.144	100,148	0	269.203	0
3300	37.198	103,865	0	270.347	0
3400	37.251	107,587	0	271.458	0
3500	37.302	111,315	0	272.539	0
3600	37.352	115,048	0	273.590	0
3700	37.402	118,786	0	274.614	0
3800	37.452	122,528	0	275.612	0
3900	37.501	126,276	0	276.586	0
4000	37.549	130,028	0	277.536	0
4100	37.597	133,786	0	278.464	0
4200	37.643	137,548	0	279.370	0
4300	37.688	141,314	0	280.257	0
4400	37.730	145,085	0	281.123	0
4500	37.768	148,860	0	281.972	0
4600	37.803	152,639	0	282.802	0
4700	37.832	156,420	0	283.616	0
4800	37.854	160,205	0	284.412	0
4900	37.868	163,991	0	285.193	0
5000	37.873	167,778	0	285.958	0

TABLE D.8 N (Molecular Weight = 14.007, Enthalpy of Formation at 298 K = 472,629 kJ/kmol)

T (K)	$\bar{c}_p$ (kJ/kmol · K)	$\bar{h}^\circ(T) - \bar{h}^\circ_f(298)$ (kJ/kmol)	$\bar{h}^\circ_f(T)$ (kJ/kmol)	$\bar{s}^\circ(T)$ (kJ/kmol · K)	$\Delta\,\bar{g}^\circ_f(T)$ (kJ/kmol)
200	20.790	−2040	472,008	144.889	461,026
298	20.786	0	472,628	153.189	455,504
300	20.786	38	472,640	153.317	455,398
400	20.786	2117	473,258	159.297	449,557
500	20.786	4196	473,864	163.935	443,562
600	20.786	6274	474,450	167.725	437,446
700	20.786	8353	475,010	170.929	431,234
800	20.786	10,431	475,537	173.705	424,944
900	20.786	12,510	476,027	176.153	418,590
1000	20.786	14,589	476,483	178.343	412,183
1100	20.792	16,668	476,911	180.325	405,732
1200	20.795	18,747	477,316	182.134	399,243
1300	20.795	20,826	477,700	183.798	392,721
1400	20.793	22,906	478,064	185.339	386,171
1500	20.790	24,985	478,411	186.774	379,595
1600	20.786	27,064	478,742	188.115	372,996
1700	20.782	29,142	479,059	189.375	366,377
1800	20.779	31,220	479,363	190.563	359,740
1900	20.777	33,298	479,656	191.687	353,086
2000	20.776	35,376	479,939	192.752	346,417
2100	20.778	37,453	480,213	193.766	339,735
2200	20.783	39,531	480,479	194.733	333,039
2300	20.791	41,610	480,740	195.657	326,331
2400	20.802	43,690	480,995	196.542	319,612
2500	20.818	45,771	481,246	197.391	312,883
2600	20.838	47,853	481,494	198.208	306,143
2700	20.864	49,938	481,740	198.995	299,394
2800	20.895	52,026	481,985	199.754	292,636
2900	20.931	54,118	482,230	200.488	285,870
3000	20.974	56,213	482,476	201.199	279,094
3100	21.024	58,313	482,723	201.887	272,311
3200	21.080	60,418	482,972	202.555	265,519
3300	21.143	62,529	483,224	203.205	258,720
3400	21.214	64,647	483,481	203.837	251,913
3500	21.292	66,772	483,742	204.453	245,099
3600	21.378	68,905	484,009	205.054	238,276
3700	21.472	71,048	484,283	205.641	231,447
3800	21.575	73,200	484,564	206.215	224,610
3900	21.686	75,363	484,853	206.777	217,765
4000	21.805	77,537	485,151	207.328	210,913
4100	21.934	79,724	485,459	207.868	204,053
4200	22.071	81,924	485,779	208.398	197,186
4300	22.217	84,139	486,110	208.919	190,310
4400	22.372	86,368	486,453	209.431	183,427
4500	22.536	88,613	486,811	209.936	176,536
4600	22.709	90,875	487,184	210.433	169,637
4700	22.891	93,155	487,573	210.923	162,730
4800	23.082	95,454	487,979	211.407	155,814
4900	23.282	97,772	488,405	211.885	148,890
5000	23.491	100,111	488,850	212.358	141,956

TABLE D.9 NO (Molecular Weight = 30.006, Enthalpy of Formation at 298 K = 90,297 kJ/kmol)

T (K)	$\bar{c}_p$ (kJ/kmol · K)	$\bar{h}^\circ(T) - \bar{h}^\circ_f(298)$ (kJ/kmol)	$\bar{h}^\circ_f(T)$ (kJ/kmol)	$\bar{s}^\circ(T)$ (kJ/kmol · K)	$\Delta\bar{g}^\circ_f(T)$ (kJ/kmol)
200	29.374	−2901	90,234	198.856	87,811
298	29.728	0	90,297	210.652	86,607
300	29.735	55	90,298	210.836	86,584
400	30.103	3046	90,341	219.439	85,340
500	30.570	6079	90,367	226.204	84,086
600	31.174	9165	90,382	231.829	82,828
700	31.908	12,318	90,393	236.688	81,568
800	32.715	15,549	90,405	241.001	80,307
900	33.489	18,860	90,421	244.900	79,043
1000	34.076	22,241	90,443	248.462	77,778
1100	34.483	25,669	90,465	251.729	76,510
1200	34.850	29,136	90,486	254.745	75,241
1300	35.180	32,638	90,505	257.548	73,970
1400	35.474	36,171	90,520	260.166	72,697
1500	35.737	39,732	90,532	262.623	71,423
1600	35.972	43,317	90,538	264.937	70,149
1700	36.180	46,925	90,539	267.124	68,875
1800	36.364	50,552	90,534	269.197	67,601
1900	36.527	54,197	90,523	271.168	66,327
2000	36.671	57,857	90,505	273.045	65,054
2100	36.797	61,531	90,479	274.838	63,782
2200	36.909	65,216	90,447	276.552	62,511
2300	37.008	68,912	90,406	278.195	61,243
2400	37.095	72,617	90,358	279.772	59,976
2500	37.173	76,331	90,303	281.288	58,711
2600	37.242	80,052	90,239	282.747	57,448
2700	37.305	83,779	90,168	284.154	56,188
2800	37.362	87,513	90,089	285.512	54,931
2900	37.415	91,251	90,003	286.824	53,677
3000	37.464	94,995	89,909	288.093	52,426
3100	37.511	98,744	89,809	289.322	51,178
3200	37.556	102,498	89,701	290.514	49,934
3300	37.600	106,255	89,586	291.670	48,693
3400	37.643	110,018	89,465	292.793	47,456
3500	37.686	113,784	89,337	293.885	46,222
3600	37.729	117,555	89,203	294.947	44,992
3700	37.771	121,330	89,063	295.981	43,766
3800	37.815	125,109	88,918	296.989	42,543
3900	37.858	128,893	88,767	297.972	41,325
4000	37.900	132,680	88,611	298.931	40,110
4100	37.943	136,473	88,449	299.867	38,900
4200	37.984	140,269	88,283	300.782	37,693
4300	38.023	144,069	88,112	301.677	36,491
4400	38.060	147,873	87,936	302.551	35,292
4500	38.093	151,681	87,755	303.407	34,098
4600	38.122	155,492	87,569	304.244	32,908
4700	38.146	159,305	87,379	305.064	31,721
4800	38.162	163,121	87,184	305.868	30,539
4900	38.171	166,938	86,984	306.655	29,361
5000	38.170	170,755	86,779	307.426	28,187

TABLE D.10 NO_2 (Molecular Weight = 46.006, Enthalpy of Formation at 298 K = 33,098 kJ/kmol)

T (K)	$\bar{c}_p$ (kJ/kmol · K)	$\bar{h}^\circ(T) - \bar{h}^\circ_f(298)$ (kJ/kmol)	$\bar{h}^\circ_f(T)$ (kJ/kmol)	$\bar{s}^\circ(T)$ (kJ/kmol · K)	$\Delta \bar{g}^\circ_f(T)$ (kJ/kmol)
200	32.936	−3432	33,961	226.016	45,453
298	36.881	0	33,098	239.925	51,291
300	36.949	68	33,085	240.153	51,403
400	40.331	3937	32,521	251.259	57,602
500	43.227	8118	32,173	260.578	63,916
600	45.737	12,569	31,974	268.686	70,285
700	47.913	17,255	31,885	275.904	76,679
800	49.762	22,141	31,880	282.427	83,079
900	51.243	27,195	31,938	288.377	89,476
1000	52.271	32,375	32,035	293.834	95,864
1100	52.989	37,638	32,146	298.850	102,242
1200	53.625	42,970	32,267	303.489	108,609
1300	54.186	48,361	32,392	307.804	114,966
1400	54.679	53,805	32,519	311.838	121,313
1500	55.109	59,295	32,643	315.625	127,651
1600	55.483	64,825	32,762	319.194	133,981
1700	55.805	70,390	32,873	322.568	140,303
1800	56.082	75,984	32,973	325.765	146,620
1900	56.318	81,605	33,061	328.804	152,931
2000	56.517	87,247	33,134	331.698	159,238
2100	56.685	92,907	33,192	334.460	165,542
2200	56.826	98,583	33,233	337.100	171,843
2300	56.943	104,271	33,256	339.629	178,143
2400	57.040	109,971	33,262	342.054	184,442
2500	57.121	115,679	33,248	344.384	190,742
2600	57.188	121,394	33,216	346.626	197,042
2700	57.244	127,116	33,165	348.785	203,344
2800	57.291	132,843	33,095	350.868	209,648
2900	57.333	138,574	33,007	352.879	215,955
3000	57.371	144,309	32,900	354.824	222,265
3100	57.406	150,048	32,776	356.705	228,579
3200	57.440	155,791	32,634	358.529	234,898
3300	57.474	161,536	32,476	360.297	241,221
3400	57.509	167,285	32,302	362.013	247,549
3500	57.546	173,038	32,113	363.680	253,883
3600	57.584	178,795	31,908	365.302	260,222
3700	57.624	184,555	31,689	366.880	266,567
3800	57.665	190,319	31,456	368.418	272,918
3900	57.708	196,088	31,210	369.916	279,276
4000	57.750	201,861	30,951	371.378	285,639
4100	57.792	207,638	30,678	372.804	292,010
4200	57.831	213,419	30,393	374.197	298,387
4300	57.866	219,204	30,095	375.559	304,772
4400	57.895	224,992	29,783	376.889	311,163
4500	57.915	230,783	29,457	378.190	317,562
4600	57.925	236,575	29,117	379.464	323,968
4700	57.922	242,367	28,761	380.709	330,381
4800	57.902	248,159	28,389	381.929	336,803
4900	57.862	253,947	27,998	383.122	343,232
5000	57.798	259,730	27,586	384.290	349,670

TABLE D.11 O_2 (Molecular Weight = 31.999, Enthalpy of Formation at 298 K = 0 kJ/kmol)

T (K)	$\bar{c}_p$ (kJ/kmol · K)	$\bar{h}^\circ(T) - \bar{h}^\circ_f(298)$ (kJ/kmol)	$\bar{h}^\circ_f(T)$ (kJ/kmol)	$\bar{s}^\circ(T)$ (kJ/kmol · K)	$\Delta\,\bar{g}^\circ_f(T)$ (kJ/kmol)
200	28.473	−2836	0	193.518	0
298	29.315	0	0	205.043	0
300	29.331	54	0	205.224	0
400	30.210	3031	0	213.782	0
500	31.114	6097	0	220.620	0
600	32.030	9254	0	226.374	0
700	32.927	12,503	0	231.379	0
800	33.757	15,838	0	235.831	0
900	34.454	19,250	0	239.849	0
1000	34.936	22,721	0	243.507	0
1100	35.270	26,232	0	246.852	0
1200	35.593	29,775	0	249.935	0
1300	35.903	33,350	0	252.796	0
1400	36.202	36,955	0	255.468	0
1500	36.490	40,590	0	257.976	0
1600	36.768	44,253	0	260.339	0
1700	37.036	47,943	0	262.577	0
1800	37.296	51,660	0	264.701	0
1900	37.546	55,402	0	266.724	0
2000	37.788	59,169	0	268.656	0
2100	38.023	62,959	0	270.506	0
2200	38.250	66,773	0	272.280	0
2300	38.470	70,609	0	273.985	0
2400	38.684	74,467	0	275.627	0
2500	38.891	78,346	0	277.210	0
2600	39.093	82,245	0	278.739	0
2700	39.289	86,164	0	280.218	0
2800	39.480	90,103	0	281.651	0
2900	39.665	94,060	0	283.039	0
3000	39.846	98,036	0	284.387	0
3100	40.023	102,029	0	285.697	0
3200	40.195	106,040	0	286.970	0
3300	40.362	110,068	0	288.209	0
3400	40.526	114,112	0	289.417	0
3500	40.686	118,173	0	290.594	0
3600	40.842	122,249	0	291.742	0
3700	40.994	126,341	0	292.863	0
3800	41.143	130,448	0	293.959	0
3900	41.287	134,570	0	295.029	0
4000	41.429	138,705	0	296.076	0
4100	41.566	142,855	0	297.101	0
4200	41.700	147,019	0	298.104	0
4300	41.830	151,195	0	299.087	0
4400	41.957	155,384	0	300.050	0
4500	42.079	159,586	0	300.994	0
4600	42.197	163,800	0	301.921	0
4700	42.312	168,026	0	302.829	0
4800	42.421	172,262	0	303.721	0
4900	42.527	176,510	0	304.597	0
5000	42.627	180,767	0	305.457	0

TABLE D.12 O (Molecular Weight = 16.000, Enthalpy of Formation at 298 K = 249,197 kJ/kmol)

T (K)	$\bar{c}_p$ (kJ/kmol · K)	$\bar{h}^\circ(T) - \bar{h}^\circ_f(298)$ (kJ/kmol)	$\bar{h}^\circ_f(T)$ (kJ/kmol)	$\bar{s}^\circ(T)$ (kJ/kmol · K)	$\Delta\,\bar{g}^\circ_f(T)$ (kJ/kmol)
200	22.477	–2176	248,439	152.085	237,374
298	21.899	0	249,197	160.945	231,778
300	21.890	41	249,211	161.080	231,670
400	21.500	2209	249,890	167.320	225,719
500	21.256	4345	250,494	172.089	219,605
600	21.113	6463	251,033	175.951	213,375
700	21.033	8570	251,516	179.199	207,060
800	20.986	10,671	251,949	182.004	200,679
900	20.952	12,768	252,340	184.474	194,246
1000	20.915	14,861	252,698	186.679	187,772
1100	20.898	16,952	253,033	188.672	181,263
1200	20.882	19,041	253,350	190.490	174,724
1300	20.867	21,128	253,650	192.160	168,159
1400	20.854	23,214	253,934	193.706	161,572
1500	20.843	25,299	254,201	195.145	154,966
1600	20.834	27,383	254,454	196.490	148,342
1700	20.827	29,466	254,692	197.753	141,702
1800	20.822	31,548	254,916	198.943	135,049
1900	20.820	33,630	255,127	200.069	128,384
2000	20.819	35,712	255,325	201.136	121,709
2100	20.821	37,794	255,512	202.152	115,023
2200	20.825	39,877	255,687	203.121	108,329
2300	20.831	41,959	255,852	204.047	101,627
2400	20.840	44,043	256,007	204.933	94,918
2500	20.851	46,127	256,152	205.784	88,203
2600	20.865	48,213	256,288	206.602	81,483
2700	20.881	50,300	256,416	207.390	74,757
2800	20.899	52,389	256,535	208.150	68,027
2900	20.920	54,480	256,648	208.884	61,292
3000	20.944	56,574	256,753	209.593	54,554
3100	20.970	58,669	256,852	210.280	47,812
3200	20.998	60,768	256,945	210.947	41,068
3300	21.028	62,869	257,032	211.593	34,320
3400	21.061	64,973	257,114	212.221	27,570
3500	21.095	67,081	257,192	212.832	20,818
3600	21.132	69,192	257,265	213.427	14,063
3700	21.171	71,308	257,334	214.007	7307
3800	21.212	73,427	257,400	214.572	548
3900	21.254	75,550	257,462	215.123	–6212
4000	21.299	77,678	257,522	215.662	–12,974
4100	21.345	79,810	257,579	216.189	19,737
4200	21.392	81,947	257,635	216.703	–26,501
4300	21.441	84,088	257,688	217.207	–33,267
4400	21.490	86,235	257,740	217.701	–40,034
4500	21.541	88,386	257,790	218.184	–46,802
4600	21.593	90,543	257,840	218.658	–53,571
4700	21.646	92,705	257,889	219.123	–60,342
4800	21.699	94,872	257,938	219.580	–67,113
4900	21.752	97,045	257,987	220.028	–73,886
5000	21.805	99,223	258,036	220.468	–80,659

TABLE D.13 C(s) (Graphite, Molecular Weight = 12.011, Enthalpy of Formation at 298 K = 0 kJ/kmol)

T (K)	$\bar{c}_p$ (kJ/kmol · K)	$\bar{h}^\circ(T) - \bar{h}^\circ_f(298)$ (kJ/kmol)	$\bar{h}^\circ_f(T)$ (kJ/kmol)	$\bar{s}^\circ(T)$ (kJ/kmol · K)	$\Delta \bar{g}^\circ_f(T)$ (kJ/kmol)
100	1.65	−1000	0	0.88	0
200	5.03	−670	0	3.01	0
298	8.53	0	0	5.69	0
400	11.93	1050	0	8.68	0
500	14.63	2380	0	11.65	0
600	16.89	3960	0	14.52	0
700	18.58	5740	0	17.26	0
800	19.83	7660	0	19.83	0
900	20.79	9700	0	22.22	0
1000	21.54	11,820	0	24.45	0
1100	22.19	14,000	0	26.53	0
1200	22.72	16,250	0	28.49	0
1300	23.12	18,540	0	30.33	0
1400	23.45	20,870	0	32.05	0
1500	23.72	23,230	0	33.68	0
1600	23.94	25,610	0	35.22	0
1700	24.12	28,020	0	36.67	0
1800	24.28	30,440	0	38.06	0
1900	24.42	32,870	0	39.38	0
2000	24.54	35,320	0	40.63	0
2100	24.65	37,780	0	41.83	0
2200	24.74	40,250	0	42.98	0
2300	24.84	42,730	0	44.08	0
2400	24.92	45,220	0	45.14	0
2500	25.00	47,710	0	46.16	0
2600	25.07	50,220	0	47.14	0
2700	25.14	52,730	0	48.09	0
2800	25.21	55,240	0	49.00	0
2900	25.28	57,770	0	49.89	0
3000	25.34	60,300	0	50.75	0
3100	25.41	62,840	0	51.58	0
3200	25.47	65,380	0	52.39	0
3300	25.53	66,880	0	53.17	0
3400	25.60	70,490	0	53.94	0
3500	25.66	73,050	0	54.68	0
3600	25.73	75,620	0	55.40	0
3700	25.79	78,200	0	56.11	0
3800	25.86	80,780	0	56.80	0
3900	25.93	83,370	0	57.47	0
4000	26.00	85,960	0	58.13	0
4100	26.07	88,570	0	58.77	0
4200	26.14	91,180	0	59.40	0
4300	26.21	93,800	0	60.02	0
4400	26.28	96,420	0	60.62	0
4500	26.36	99,050	0	61.21	0
4600	26.43	101,690	0	61.79	0
4700	26.51	104,340	0	62.36	0
4800	26.59	106,990	0	62.92	0
4900	26.66	109,650	0	63.47	0
5000	26.74	112,320	0	64.01	0

TABLE D.14 Curve-Fit Coefficients for Thermodynamic Properties (C–H–O–N System):

$$\frac{\bar{c}_p}{R_u} = a_1 + a_2T + a_3T^2 + a_4T^3 + a_5T^4,$$

$$\frac{\bar{h}^\circ}{R_uT} = a_1 + \frac{a_2}{2}T + \frac{a_3}{3}T^2 + \frac{a_4}{4}T^3 + \frac{a_5}{5}T^4 + \frac{a_6}{T},$$

$$\frac{\bar{s}^\circ}{R_u} = a_1 \ln T + a_2T + \frac{a_3}{2}T^2 + \frac{a_4}{3}T^3 + \frac{a_5}{4}T^4 + a_7$$

Species	T (K)	a_1	a_2	a_3	a_4	a_5	a_6	a_7
CO	1000–5000	0.03025078E+02	0.14426885E–02	–0.05630827E–05	0.10185813E–09	–0.06910951E–13	–0.14268350E+05	0.06108217E+02
	300–1000	0.03262451E+02	0.15119409E–02	–0.03881755E–04	0.05581944E–07	–0.02474951E–10	–0.14310539E+05	0.04848897E+02
CO_2	1000–5000	0.04453623E+02	0.03140168E–01	–0.12784105E–05	0.02393996E–08	–0.16690333E–13	–0.04896696E+06	–0.09553959E+01
	300–1000	0.02275724E+02	0.09922072E–01	–0.10409113E–04	0.06866686E–07	–0.02117280E–10	–0.04837314E+06	0.10188488E+02
H_2	1000–5000	0.02991423E+02	0.07000644E–02	–0.05633828E–06	–0.09231578E–10	0.15827519E–14	–0.08350340E+04	–0.13551101E+01
	300–1000	0.03298124E+02	0.08249441E–02	–0.08143015E–05	0.09475434E–09	0.04134872E–11	–0.10125209E+04	–0.03294094E+02
H	1000–5000	0.02500000E+02	0.00000000E+00	0.00000000E+00	0.00000000E+00	0.00000000E+00	0.02547162E+06	–0.04601176E+01
	300–1000	0.02500000E+02	0.00000000E+00	0.00000000E+00	0.00000000E+00	0.00000000E+00	0.02547162E+06	–0.04601176E+01
OH	1000–5000	0.02882730E+02	0.10139743E–02	–0.02276877E–05	0.02174683E–09	–0.05126305E–14	0.03886888E+05	0.05595712E+02
	300–1000	0.03637266E+02	0.01850910E–02	–0.16761646E–05	0.02387202E–07	–0.08431442E–11	0.03606781E+05	0.13588605E+01
H_2O	1000–5000	0.02672145E+02	0.03056293E–01	–0.08730260E–05	0.12009964E–09	–0.06391618E–13	–0.02989921E+06	0.06862817E+02
	300–1000	0.03386842E+02	0.03474982E–01	–0.06354696E–04	0.06968581E–07	–0.02506588E–10	–0.03020811E+06	0.02590232E+02
N_2	1000–5000	0.02926640E+02	0.14879768E–02	–0.05684760E–05	0.10097038E–09	–0.06753351E–13	–0.09227977E+04	0.05980528E+02
	300–1000	0.03298677E+02	0.14082404E–02	–0.03963222E–04	0.05641515E–07	–0.02444854E–10	–0.10208999E+04	0.03950372E+02
N	1000–5000	0.02450268E+02	0.10661458E–03	–0.07465337E–06	0.01879652E–09	–0.10259839E–14	0.05611604E+06	0.04448758E+02
	300–1000	0.02503071E+02	–0.02180018E–03	0.05420529E–06	–0.05647560E–09	0.02099904E–12	0.05609890E+06	0.04167566E+02
NO	1000–5000	0.03245435E+02	0.12691383E–02	–0.05015890E–05	0.09169283E–09	–0.06275419E–13	0.09800840E+05	0.06417293E+02
	300–1000	0.03376541E+02	0.12530634E–02	–0.03302750E–04	0.05217810E–07	–0.02446262E–10	0.09817961E+05	0.05829590E+02
NO_2	1000–5000	0.04682859E+02	0.02462429E–01	–0.10422585E–05	0.01976902E–08	–0.13917168E–13	0.02261292E+05	0.09885985E+01
	300–1000	0.02670600E+02	0.07838500E–01	–0.08063864E–04	0.06161714E–07	–0.02320150E–10	0.02896290E+05	0.11612071E+02
O_2	1000–5000	0.03697578E+02	0.06135197E–02	–0.12588420E–06	0.01775281E–09	–0.11364354E–14	–0.12339301E+04	0.03189165E+02
	300–1000	0.03212936E+02	0.11274864E–02	–0.05756150E–05	0.13138773E–08	–0.08768554E–11	–0.10052490E+04	0.06034737E+02
O	1000–5000	0.02542059E+02	–0.02755061E–03	–0.03102803E–07	0.04551067E–10	–0.04368051E–14	0.02923080E+06	0.04920308E+02
	300–1000	0.02946428E+02	–0.16381665E–02	0.02421031E–04	–0.16028431E–08	0.03890696E–11	0.02914764E+06	0.02963995E+02

APPENDIX E
Various Thermodynamic Data

TABLE E.1 Critical Constants and Specific Heats for Selected Gases*

Substance	$\mathcal{M}$ (kg/kmol)	T_c (K)	P_c (10^5 Pa)	v_c (m^3/kmol)	Z_c	c_v (kJ/kg·K)	c_p (kJ/kg·K)
Acetylene (C_2H_2)	26.04	309	62.4	0.112	0.272	1.37	1.69
Air (equivalent)	28.97	133	37.7	0.0829	0.284	0.718	1.005
Ammonia (NH_3)	17.04	406	112.8	0.0723	0.242	1.66	2.15
Benzene (C_6H_6)	78.11	562	48.3	0.256	0.274	0.67	0.775
n-Butane (C_4H_{10})	58.12	425.2	37.9	0.257	0.274	1.56	1.71
Carbon dioxide (CO_2)	44.01	304.2	73.9	0.0941	0.276	0.657	0.846
Carbon monoxide (CO)	28.01	133	35.0	0.0928	0.294	0.744	1.04
Refrigerant 134a ($C_2F_4H_2$)	102.03	374.3	40.6	0.200	0.262	0.76	0.85
Ethane (C_2H_6)	30.07	305.4	48.8	0.148	0.285	1.48	1.75
Ethylene (C_2H_4)	28.05	283	51.2	0.128	0.279	1.23	1.53
Helium (He)	4.003	5.2	2.3	0.0579	0.300	3.12	5.19
Hydrogen (H_2)	2.016	33.2	13.0	0.0648	0.304	10.2	14.3
Methane (CH_4)	16.04	190.7	46.4	0.0991	0.290	1.70	2.22
Nitrogen (N_2)	28.01	126.2	33.9	0.0897	0.291	0.743	1.04
Oxygen (O_2)	32.00	154.4	50.5	0.0741	0.290	0.658	0.918
Propane (C_3H_8)	44.09	370	42.5	0.200	0.278	1.48	1.67
Sulfur dioxide (SO_2)	64.06	431	78.7	0.124	0.268	0.471	0.601
Water (H_2O)	18.02	647.1	220.6	0.0558	0.230	1.40	1.86

* Adapted from Wark, K., Jr., and Richards, D. E., *Thermodynamics,* 6th edn, McGraw-Hill, New York, 1999.

TABLE E.2 Van der Waals Constants for Selected Gases*

Substance	a [10^5 Pa· (m^3/kmol)^2]	b (m^3/kmol)	Substance	a [10^5 Pa· (m^3/kmol)^2]	b (m^3/kmol)
Acetylene (C_2H_2)	4.410	0.0510	Ethylene (C_2H_4)	4.563	0.0574
Air (equivalent)	1.358	0.0364	Helium (He)	0.0341	0.0234
Ammonia (NH_3)	4.223	0.0373	Hydrogen (H_2)	0.247	0.0265
Benzene (C_6H_6)	18.63	0.1181	Methane (CH_4)	2.285	0.0427
n-Butane (C_4H_{10})	13.80	0.1196	Nitrogen (N_2)	1.361	0.0385
Carbon dioxide (CO_2)	3.643	0.0427	Oxygen (O_2)	1.369	0.0315
Carbon monoxide (CO)	1.463	0.0394	Propane (C_3H_8)	9.315	0.0900
Refrigerant 134a ($C_2F_4H_2$)	10.05	0.0957	Sulfur dioxide (SO_2)	6.837	0.0568
Ethane (C2H6)	5.575	0.0650	Water (H_2O)	5.507	0.0304

* Adapted from Wark, K., Jr., and Richards, D. E., *Thermodynamics,* 6th edn, McGraw-Hill, New York, 1999.

APPENDIX F

Thermo-Physical Properties of Selected Gases at 1 atm

TABLE F.1a Ammonia (NH_3)*

T (K)	ρ (kg/m^3)	c_p (kJ/kg K)	μ (μPa·s)	ν (m^2/s)	k (W/m·K)	α (m^2/s)	Pr
239.824	0.8895	2.297	8.054	9.054E–06	0.0210	1.026E–05	0.8822
240	0.8888	2.296	8.059	9.068E–06	0.0210	1.028E–05	0.8820
260	0.8135	2.207	8.734	1.074E–05	0.0220	1.228E–05	0.8744
280	0.7515	2.172	9.436	1.256E–05	0.0234	1.435E–05	0.8748
300	0.6990	2.165	10.160	1.454E–05	0.0251	1.659E–05	0.8762
320	0.6538	2.174	10.902	1.668E–05	0.0271	1.904E–05	0.8759
340	0.6143	2.193	11.657	1.898E–05	0.0293	2.173E–05	0.8734
360	0.5795	2.219	12.422	2.144E–05	0.0317	2.467E–05	0.8691
380	0.5485	2.249	13.195	2.406E–05	0.0344	2.786E–05	0.8634
400	0.5207	2.283	13.971	2.683E–05	0.0372	3.130E–05	0.8572
420	0.4956	2.320	14.751	2.976E–05	0.0402	3.498E–05	0.8510
440	0.4728	2.358	15.531	3.285E–05	0.0433	3.886E–05	0.8453
460	0.4521	2.399	16.310	3.607E–05	0.0465	4.292E–05	0.8405
480	0.4331	2.440	17.088	3.945E–05	0.0498	4.714E–05	0.8369
500	0.4157	2.483	17.863	4.297E–05	0.0531	5.147E–05	0.8349
520	0.3996	2.526	18.635	4.663E–05	0.0564	5.587E–05	0.8346
540	0.3848	2.570	19.403	5.043E–05	0.0596	6.030E–05	0.8362
560	0.3710	2.615	20.167	5.436E–05	0.0628	6.471E–05	0.8401
580	0.3581	2.660	20.927	5.843E–05	0.0658	6.904E–05	0.8463
600	0.3462	2.706	21.682	6.264E–05	0.0686	7.324E–05	0.8552

* Property values generated from NIST Database 23: *REFPROP* Version 7.0 (August 2002).

TABLE F.1b Carbon Dioxide (CO_2)*

T (K)	ρ (kg/m^3)	c_p (kJ/kg·K)	μ (μPa·s)	ν (m^2/s)	k (W/m·K)	α (m^2/s)	Pr
220	2.472	0.781	11.06	4.475E–06	0.0109	5.647E–06	0.792
240	2.258	0.796	12.07	5.344E–06	0.0122	6.808E–06	0.785
260	2.079	0.814	13.06	6.282E–06	0.0137	8.079E–06	0.778
280	1.927	0.833	14.05	7.288E–06	0.0152	9.465E–06	0.770
300	1.797	0.853	15.02	8.361E–06	0.0168	1.096E–05	0.763
320	1.683	0.872	15.98	9.499E–06	0.0184	1.256E–05	0.756
340	1.583	0.890	16.93	1.070E–05	0.0201	1.426E–05	0.750
360	1.494	0.908	17.87	1.196E–05	0.0218	1.605E–05	0.745
380	1.415	0.925	18.79	1.328E–05	0.0235	1.792E–05	0.741
400	1.343	0.942	19.70	1.466E–05	0.0251	1.988E–05	0.738
420	1.279	0.958	20.59	1.610E–05	0.0268	2.190E–05	0.735
440	1.220	0.973	21.47	1.759E–05	0.0285	2.401E–05	0.733
460	1.167	0.988	22.33	1.913E–05	0.0302	2.618E–05	0.731
480	1.118	1.002	23.18	2.073E–05	0.0318	2.842E–05	0.729
500	1.073	1.015	24.02	2.237E–05	0.0335	3.072E–05	0.728
520	1.032	1.029	24.84	2.407E–05	0.0351	3.309E–05	0.727
540	0.994	1.041	25.65	2.581E–05	0.0368	3.553E–05	0.727
560	0.958	1.053	26.44	2.760E–05	0.0384	3.802E–05	0.726
580	0.925	1.065	27.23	2.944E–05	0.0400	4.057E–05	0.726
600	0.894	1.076	28.00	3.131E–05	0.0416	4.318E–05	0.725
620	0.865	1.087	28.76	3.324E–05	0.0431	4.585E–05	0.725
640	0.838	1.098	29.50	3.520E–05	0.0447	4.858E–05	0.725
660	0.813	1.108	30.24	3.721E–05	0.0462	5.136E–05	0.724
680	0.789	1.117	30.96	3.925E–05	0.0478	5.420E–05	0.724
700	0.766	1.127	31.68	4.134E–05	0.0493	5.709E–05	0.724
720	0.745	1.136	32.38	4.347E–05	0.0508	6.004E–05	0.724
740	0.725	1.145	33.07	4.563E–05	0.0523	6.304E–05	0.724
760	0.706	1.153	33.75	4.783E–05	0.0538	6.609E–05	0.724
780	0.688	1.161	34.43	5.007E–05	0.0553	6.920E–05	0.724
800	0.670	1.169	35.09	5.234E–05	0.0567	7.236E–05	0.723

* Property values generated from NIST Database 23: *REFPROP* Version 7.0 (August 2002).

TABLE F.1c Carbon Monoxide (CO)*

T (K)	ρ (kg/m^3)	c_p (kJ/kg·K)	μ (μPa·s)	ν (m^2/s)	k (W/m·K)	α (m^2/s)	Pr
200	1.7112	1.0443	12.8977	7.537E–06	0.01923	1.076E–05	0.701
220	1.5544	1.0430	13.9377	8.967E–06	0.02080	1.283E–05	0.699
240	1.4241	1.0420	14.9369	1.049E–05	0.02232	1.504E–05	0.697
260	1.3140	1.0412	15.8998	1.210E–05	0.02378	1.738E–05	0.696
280	1.2198	1.0407	16.8304	1.380E–05	0.02520	1.985E–05	0.695
300	1.1382	1.0402	17.7315	1.558E–05	0.02656	2.243E–05	0.694
320	1.0669	1.0399	18.6057	1.744E–05	0.02789	2.514E–05	0.694
340	1.0040	1.0396	19.4549	1.938E–05	0.02918	2.795E–05	0.693
360	0.9481	1.0394	20.2811	2.139E–05	0.03043	3.088E–05	0.693
380	0.8982	1.0393	21.0859	2.348E–05	0.03165	3.391E–05	0.692
400	0.8532	1.0392	21.8707	2.563E–05	0.03285	3.705E–05	0.692
450	0.7583	1.0392	23.7543	3.133E–05	0.03571	4.532E–05	0.691
500	0.6824	1.0395	25.5404	3.743E–05	0.03844	5.419E–05	0.691
550	0.6204	1.0400	27.2444	4.392E–05	0.04105	6.363E–05	0.690
600	0.5687	1.0408	28.8791	5.078E–05	0.04357	7.362E–05	0.690
650	0.5249	1.0418	30.4548	5.802E–05	0.04601	8.414E–05	0.690
700	0.4874	1.0429	31.9798	6.561E–05	0.04838	9.518E–05	0.689
750	0.4549	1.0443	33.4606	7.355E–05	0.05070	1.067E–04	0.689
800	0.4265	1.0457	34.9026	8.184E–05	0.05297	1.188E–04	0.689

* Property values generated from NIST Database 12: NIST Pure Fluids Version 5.0 (September 2002).

TABLE F.1d Helium (He)*

T (K)	ρ (kg/m^3)	c_p (kJ/kg·K)	μ (μPa·s)	ν (m^2/s)	k (W/m·K)	α (m^2/s)	Pr
100	0.487	5.149	9.78	2.008E–05	0.0737	2.914E–05	0.689
120	0.4060	5.1938	10.79	2.659E–05	0.0833	3.952E–05	0.673
140	0.3481	5.1935	11.94	3.432E–05	0.0925	5.117E–05	0.671
160	0.3046	5.1933	13.05	4.284E–05	0.1013	6.404E–05	0.669
180	0.2708	5.1932	14.11	5.212E–05	0.1098	7.806E–05	0.668
200	0.2437	5.1931	15.14	6.213E–05	0.1180	9.322E–05	0.667
220	0.2216	5.1931	16.14	7.286E–05	0.1260	1.095E–04	0.666
240	0.2031	5.1930	17.12	8.429E–05	0.1337	1.268E–04	0.665
260	0.1875	5.1930	18.08	9.641E–05	0.1413	1.451E–04	0.664
280	0.1741	5.1930	19.01	1.092E–04	0.1487	1.645E–04	0.664
300	0.1625	5.1930	19.93	1.226E–04	0.1560	1.848E–04	0.664
320	0.1524	5.1930	20.83	1.367E–04	0.1631	2.061E–04	0.663
340	0.1434	5.1930	21.72	1.514E–04	0.1701	2.284E–04	0.663
360	0.1354	5.1930	22.59	1.668E–04	0.1770	2.516E–04	0.663
380	0.1283	5.1930	23.45	1.827E–04	0.1837	2.757E–04	0.663
400	0.1219	5.1930	24.29	1.993E–04	0.1904	3.007E–04	0.663
420	0.1161	5.1930	25.13	2.164E–04	0.1969	3.266E–04	0.663
440	0.1108	5.1930	25.95	2.342E–04	0.2034	3.534E–04	0.663
460	0.1060	5.1930	26.76	2.525E–04	0.2098	3.811E–04	0.663
480	0.1016	5.1930	27.57	2.714E–04	0.2161	4.096E–04	0.663
500	0.0975	5.1930	28.36	2.908E–04	0.2223	4.389E–04	0.663
550	0.0887	5.1930	30.31	3.419E–04	0.2376	5.159E–04	0.663
600	0.0813	5.1930	32.22	3.963E–04	0.2524	5.980E–04	0.663
650	0.0750	5.1930	34.07	4.541E–04	0.2669	6.850E–04	0.663
700	0.0697	5.1930	35.89	5.152E–04	0.2811	7.768E–04	0.663
750	0.0650	5.1930	37.68	5.794E–04	0.2949	8.733E–04	0.663
800	0.0610	5.1930	39.43	6.468E–04	0.3085	9.745E–04	0.664
850	0.0574	5.1930	41.15	7.172E–04	0.3219	1.080E–03	0.664
900	0.0542	5.1930	42.85	7.907E–04	0.3350	1.190E–03	0.664
950	0.0513	5.1930	44.52	8.671E–04	0.3479	1.305E–03	0.664
1000	0.0488	5.1930	46.16	9.464E–04	0.3606	1.424E–03	0.665

* Property values generated from NIST Database 12: NIST Pure Fluids Version 5.0 (September 2002).

TABLE F.1e Hydrogen (H_2)*

T (K)	ρ (kg/m^3)	c_p (kJ/kg·K)	μ (μPa·s)	ν (m^2/s)	k (W/m·K)	α (m^2/s)	Pr
100	0.2457	11.23	4.190	1.705E–05	0.0683	2.477E–05	0.688
150	0.1637	12.61	5.561	3.397E–05	0.1010	4.894E–05	0.694
200	0.1228	13.54	6.780	5.523E–05	0.1324	7.970E–05	0.693
250	0.09820	14.05	7.903	8.047E–05	0.1606	1.164E–04	0.691
300	0.08184	14.31	8.953	1.094E–04	0.1858	1.586E–04	0.690
350	0.07016	14.43	9.946	1.418E–04	0.2103	2.077E–04	0.682
400	0.06139	14.47	10.89	1.774E–04	0.2341	2.634E–04	0.674
450	0.05457	14.49	11.80	2.162E–04	0.2570	3.249E–04	0.665
500	0.04912	14.51	12.67	2.580E–04	0.2805	3.936E–04	0.656
550	0.04465	14.53	13.52	3.027E–04	0.3042	4.689E–04	0.646
600	0.04093	14.54	14.34	3.503E–04	0.3281	5.514E–04	0.635

* Property values generated from NIST Database 12: NIST Pure Fluids Version 5.0 (September 2000).

TABLE F.1f Nitrogen (N_2)*

T (K)	ρ (kg/m^3)	c_p (kJ/kg·K)	μ (μPa s)	ν (m^2/s)	k (W/m K)	α (m^2/s)	Pr
100	3.4831	1.0718	6.97	2.000E–06	0.0099	2.644E–06	0.756
150	2.2893	1.0486	10.10	4.410E–06	0.0145	6.061E–06	0.728
200	1.7107	1.0435	12.92	7.555E–06	0.0187	1.045E–05	0.723
250	1.3666	1.0418	15.51	1.135E–05	0.0224	1.573E–05	0.721
300	1.1382	1.0414	17.90	1.572E–05	0.0259	2.182E–05	0.721
350	0.9753	1.0423	20.12	2.063E–05	0.0291	2.863E–05	0.721
400	0.8532	1.0450	22.22	2.604E–05	0.0322	3.612E–05	0.721
450	0.7584	1.0497	24.20	3.191E–05	0.0352	4.423E–05	0.721
500	0.6825	1.0564	26.08	3.821E–05	0.0381	5.290E–05	0.722
550	0.6204	1.0650	27.88	4.493E–05	0.0410	6.211E–05	0.723
600	0.5687	1.0751	29.60	5.205E–05	0.0439	7.182E–05	0.725
700	0.4875	1.0981	32.87	6.742E–05	0.0496	9.267E–05	0.728
800	0.4266	1.1223	35.93	8.424E–05	0.0552	1.153E–04	0.731
900	0.3792	1.1457	38.83	1.024E–04	0.0607	1.396E–04	0.733
1000	0.3413	1.1674	41.60	1.219E–04	0.0660	1.656E–04	0.736
1100	0.3103	1.1868	44.25	1.426E–04	0.0712	1.933E–04	0.738
1200	0.2844	1.2040	46.81	1.646E–04	0.0762	2.224E–04	0.740
1300	0.2625	1.2191	49.29	1.878E–04	0.0810	2.532E–04	0.742
1400	0.2438	1.2324	51.70	2.121E–04	0.0858	2.855E–04	0.743
1500	0.2275	1.2439	54.06	2.376E–04	0.0904	3.193E–04	0.744
1600	0.2133	1.2541	56.36	2.642E–04	0.0948	3.545E–04	0.745
1700	0.2008	1.2630	58.61	2.919E–04	0.0992	3.913E–04	0.746
1800	0.1896	1.2708	60.83	3.208E–04	0.1035	4.295E–04	0.747
1900	0.1796	1.2778	63.01	3.508E–04	0.1077	4.692E–04	0.748
2000	0.1707	1.2841	65.16	3.818E–04	0.1118	5.104E–04	0.748

* Property values generated from NIST Database 23: *REFPROP* Version 7.0 (August 2002).

TABLE F.1g Oxygen (O_2)*

T (K)	ρ (kg/m^3)	c_p (kJ/kg·K)	μ (μPa·s)	ν (m^2/s)	k (W/m·K)	α (m^2/s)	Pr
100	3.995	0.9356	7.74835	1.940E–06	0.00931	2.491E–06	0.779
150	2.619	0.9198	11.34	4.329E–06	0.01399	5.807E–06	0.745
200	1.956	0.9146	14.65	7.491E–06	0.01840	1.029E–05	0.728
250	1.562	0.9150	17.71	1.134E–05	0.02260	1.581E–05	0.717
300	1.301	0.9199	20.56	1.581E–05	0.02666	2.228E–05	0.710
350	1.114	0.9291	23.23	2.085E–05	0.03067	2.962E–05	0.704
400	0.9749	0.9417	25.75	2.641E–05	0.03469	3.779E–05	0.699
450	0.8665	0.9564	28.14	3.247E–05	0.03872	4.672E–05	0.695
500	0.7798	0.9722	30.41	3.900E–05	0.04275	5.639E–05	0.692
550	0.7089	0.9880	32.58	4.597E–05	0.04676	6.677E–05	0.688
600	0.6498	1.003	34.67	5.336E–05	0.05074	7.783E–05	0.686
700	0.5569	1.031	38.62	6.935E–05	0.05850	1.019E–04	0.681
800	0.4873	1.054	42.32	8.685E–05	0.06593	1.283E–04	0.677
900	0.4332	1.074	45.82	1.058E–04	0.07300	1.569E–04	0.674
1000	0.3899	1.090	49.15	1.261E–04	0.07968	1.875E–04	0.672
1100	0.3544	1.103	52.33	1.477E–04	0.08601	2.200E–04	0.671
1200	0.3249	1.114	55.40	1.705E–04	0.09198	2.541E–04	0.671
1300	0.2999	1.123	58.36	1.946E–04	0.09762	2.897E–04	0.672
1400	0.2785	1.131	61.23	2.199E–04	0.1030	3.267E–04	0.673
1500	0.2599	1.138	64.02	2.463E–04	0.1080	3.650E–04	0.675

* Property values generated from NIST Database 23: *REFPROP* Version 7.0 (August 2002).

TABLE F.1h Water Vapor (H_2O)*

T (K)	ρ (kg/m^3)	c_p (kJ/kg·K)	μ (μPa·s)	ν (m^2/s)	k (W/m·K)	α (m^2/s)	Pr
373.124	0.5977	2.080	12.27	2.053E–05	0.02509	2.019E–05	1.017
400	0.5549	2.009	13.28	2.394E–05	0.02702	2.423E–05	0.988
450	0.4910	1.976	15.25	3.105E–05	0.03117	3.213E–05	0.966
500	0.4409	1.982	17.27	3.917E–05	0.03586	4.105E–05	0.954
550	0.4003	2.001	19.33	4.828E–05	0.04096	5.112E–05	0.944
600	0.3667	2.027	21.41	5.838E–05	0.04637	6.239E–05	0.936
650	0.3383	2.056	23.49	6.944E–05	0.05205	7.485E–05	0.928
700	0.3140	2.087	25.56	8.142E–05	0.05796	8.847E–05	0.920
750	0.2930	2.119	27.63	9.429E–05	0.06408	1.032E–04	0.914
800	0.2746	2.153	29.67	1.080E–04	0.07039	1.191E–04	0.907
850	0.2584	2.187	31.69	1.226E–04	0.07685	1.360E–04	0.902
900	0.2440	2.222	33.69	1.380E–04	0.08347	1.539E–04	0.897
950	0.2312	2.257	35.65	1.542E–04	0.09022	1.729E–04	0.892
1000	0.2196	2.292	37.59	1.712E–04	0.09709	1.929E–04	0.888

* Property values generated from NIST Database 23: *REFPROP* Version 7.0 (August 2002).

APPENDIX G

Thermo-Physical Properties of Selected Liquids

TABLE G.1 Thermo-Physical Properties of Saturated Water*

Temperature (K)	Pressure (MPa)	Liquid density (kg/m^3)	Vapor density (kg/m^3)	Liquid c_p (kJ/kg·K)	Vapor c_p (kJ/kg·K)	Liquid viscosity (μPa·s)	Vapor viscosity (μPa·s)	Liquid therm. cond. (W/m·K)	Vapor therm. cond. (W/m·K)	Liquid Prandtl	Vapor Prandtl	Surface tension (N/m)	Liquid expansion coef. β (1/K)	Vapor expansion coef. β (1/K)	T (K)
273.16	0.000612	999.79	0.0048546	4.2199	1.8844	1791.2	9.2163	0.56104	0.017071	13.472	1.0173	0.075646	−0.000067965	0.0036807	273.16
280	0.000992	999.86	0.0076812	4.2014	1.8913	1433.7	9.3815	0.57404	0.017442	10.493	1.0173	0.074677	0.000043569	0.0035962	280
285	0.001389	999.47	0.010571	4.1927	1.8967	1239.3	9.509	0.58348	0.017729	8.9052	1.0173	0.073951	0.00011191	0.0035375	285
290	0.00192	998.76	0.014363	4.1869	1.9023	1084	9.6414	0.59273	0.018031	7.6573	1.0172	0.07321	0.0001721	0.0034814	290
295	0.002621	997.76	0.019281	4.1832	1.9081	957.87	9.7784	0.60169	0.018345	6.6594	1.017	0.072455	0.00022593	0.0034277	295
300	0.003537	996.51	0.02559	4.1809	1.9141	853.84	9.9195	0.61028	0.018673	5.8495	1.0168	0.071686	0.00027471	0.0033765	300
305	0.004719	995.03	0.033598	4.1798	1.9204	766.95	10.064	0.61841	0.019014	5.1837	1.0165	0.070903	0.00031942	0.0033276	305
310	0.006231	993.34	0.043663	4.1795	1.927	693.54	10.213	0.62605	0.019369	4.6301	1.0161	0.070106	0.00036081	0.0032811	310
315	0.008145	991.46	0.056195	4.1798	1.9341	630.91	10.364	0.63315	0.019736	4.1651	1.0156	0.069295	0.00039947	0.0032369	315
320	0.010546	989.39	0.071662	4.1807	1.9417	577.02	10.518	0.63971	0.020117	3.7711	1.0152	0.06847	0.00043586	0.0031951	320
325	0.013531	987.15	0.09059	4.1821	1.9499	530.29	10.675	0.64571	0.020512	3.4346	1.0147	0.067632	0.00047035	0.0031557	325
330	0.017213	984.75	0.11357	4.1838	1.9587	489.49	10.833	0.65118	0.020922	3.145	1.0143	0.066781	0.00050326	0.0031186	330
335	0.021718	982.2	0.14127	4.186	1.9684	453.64	10.994	0.65611	0.021345	2.8942	1.0139	0.065917	0.00053484	0.0030839	335
340	0.027188	979.5	0.1744	4.1885	1.979	421.97	11.157	0.66055	0.021784	2.6757	1.0136	0.06504	0.00056531	0.0030516	340
345	0.033783	976.67	0.21378	4.1913	1.9906	393.85	11.321	0.6645	0.022238	2.4842	1.0135	0.06415	0.00059485	0.0030218	345
350	0.041682	973.7	0.26029	4.1946	2.0033	368.77	11.487	0.668	0.022707	2.3156	1.0135	0.063248	0.00062362	0.0029945	350
355	0.05108	970.61	0.31487	4.1983	2.0173	346.3	11.654	0.67108	0.023193	2.1665	1.0137	0.062333	0.00065178	0.0029697	355
360	0.062194	967.39	0.37858	4.2024	2.0326	326.1	11.823	0.67376	0.023695	2.034	1.0142	0.061406	0.00067944	0.0029476	360
365	0.07526	964.05	0.45253	4.207	2.0493	307.87	11.992	0.67606	0.024213	1.9158	1.0149	0.060467	0.00070671	0.0029281	365
370	0.090535	960.59	0.53792	4.2122	2.0676	291.36	12.162	0.67802	0.02475	1.81	1.016	0.059517	0.00073371	0.0029114	370
373.15	0.10142	958.35	0.59817	4.2157	2.08	281.74	12.269	0.67909	0.025096	1.749	1.0169	0.058912	0.00075062	0.0029023	373.15
375	0.1083	957.01	0.63605	4.2178	2.0877	276.36	12.332	0.67966	0.025303	1.715	1.0175	0.058555	0.00076053	0.0028975	375
380	0.12885	953.33	0.7483	4.2241	2.1096	262.69	12.504	0.681	0.025875	1.6294	1.0194	0.057581	0.00078726	0.0028865	380
385	0.15252	949.53	0.87615	4.2309	2.1334	250.21	12.675	0.68205	0.026465	1.5521	1.0218	0.056596	0.00081398	0.0028786	385
390	0.17964	945.62	1.0212	4.2384	2.1594	238.77	12.848	0.68283	0.027074	1.4821	1.0247	0.055601	0.00084079	0.0028737	390
395	0.2106	941.61	1.185	4.2466	2.1877	228.27	13.02	0.68335	0.027701	1.4185	1.0282	0.054595	0.00086777	0.0028719	395
400	0.24577	937.49	1.3694	4.2555	2.2183	218.8	13.192	0.68364	0.028347	1.3607	1.0324	0.053578	0.00089499	0.0028735	400
405	0.28558	933.26	1.5763	4.2652	2.2514	209.68	13.365	0.68369	0.029013	1.308	1.0371	0.052551	0.00092254	0.0028784	405

TABLE G.1 (cont.)

Temperature (K)	Pressure (MPa)	Liquid density (kg/m^3)	Vapor density (kg/m^3)	Liquid c_p (kJ/kg·K)	Vapor c_p (kJ/kg·K)	Liquid viscosity (μPa·s)	Vapor viscosity (μPa·s)	Liquid therm. cond. (W/m·K)	Vapor therm. cond. (W/m·K)	Liquid Prandtl	Vapor Prandtl	Surface tension (N/m)	Liquid expansion coef. β (1/K)	Vapor expansion coef. β (1/K)	T (K)
410	0.33045	928.92	1.8076	4.2756	2.2871	201.43	13.538	0.68352	0.029699	1.26	1.0425	0.051514	0.0009505	0.0028868	410
415	0.38087	924.48	2.0654	4.2868	2.3254	193.78	13.711	0.68313	0.030404	1.216	1.0487	0.050468	0.00097895	0.0028987	415
420	0.4373	919.93	2.3518	4.299	2.3666	186.68	13.883	0.68253	0.031128	1.1758	1.0555	0.049411	0.001008	0.0029142	420
425	0.50025	915.27	2.6693	4.312	2.4105	180.07	14.056	0.68172	0.031873	1.139	1.063	0.048346	0.0010377	0.0029334	425
430	0.57026	910.51	3.0202	4.326	2.4573	173.91	14.228	0.6807	0.032638	1.1052	1.0712	0.047272	0.0010681	0.0029564	430
435	0.64787	905.63	3.4072	4.341	2.507	168.16	14.401	0.67948	0.033424	1.0743	1.0801	0.046189	0.0010994	0.0029834	435
440	0.73367	900.65	3.8329	4.3571	2.5597	162.77	14.573	0.67805	0.03423	1.0459	1.0898	0.045098	0.0011317	0.0030143	440
445	0.82824	895.55	4.3001	4.3743	2.6154	157.72	14.745	0.67642	0.035056	1.02	1.1001	0.043999	0.001165	0.0030495	445
450	0.9322	890.34	4.812	4.3927	2.6742	152.98	14.917	0.67459	0.035904	0.99615	1.1111	0.042891	0.0011995	0.0030889	450
455	1.0462	885.01	5.3717	4.4124	2.7362	148.52	15.089	0.67254	0.036773	0.97438	1.1228	0.041777	0.0012354	0.0031328	455
460	1.1709	879.57	5.9826	4.4334	2.8014	144.31	15.261	0.67028	0.037663	0.9545	1.1352	0.040655	0.0012727	0.0031814	460
465	1.3069	874	6.6484	4.4559	2.8701	140.34	15.434	0.66781	0.038576	0.93639	1.1483	0.039527	0.0013116	0.003235	465
470	1.4551	868.31	7.3727	4.4799	2.9422	136.58	15.606	0.66512	0.039512	0.91994	1.1621	0.038392	0.0013522	0.0032937	470
475	1.616	862.49	8.1598	4.5055	3.0181	133.02	15.779	0.66221	0.040471	0.90506	1.1767	0.037252	0.0013948	0.003358	475
480	1.7905	856.54	9.0139	4.533	3.0979	129.64	15.952	0.65907	0.041455	0.89167	1.1921	0.036105	0.0014396	0.0034282	480
485	1.9792	850.45	9.9397	4.5623	3.1819	126.43	16.126	0.65569	0.042464	0.87972	1.2083	0.034954	0.0014867	0.0035048	485
490	2.1831	844.22	10.942	4.5937	3.2705	123.37	16.3	0.65206	0.043502	0.86914	1.2255	0.033797	0.0015365	0.0035882	490
495	2.4028	837.84	12.027	4.6274	3.3641	120.45	16.476	0.64819	0.044568	0.85989	1.2436	0.032637	0.0015891	0.0036791	495
500	2.6392	831.31	13.199	4.6635	3.4631	117.66	16.653	0.64405	0.045666	0.85195	1.2628	0.031472	0.001645	0.0037781	500
505	2.8931	824.63	14.465	4.7022	3.568	114.98	16.831	0.63964	0.046799	0.84529	1.2832	0.030304	0.0017045	0.0038861	505
510	3.1655	817.77	15.833	4.744	3.6796	112.42	17.011	0.63495	0.047969	0.83991	1.3049	0.029133	0.001768	0.004004	510
515	3.4571	810.74	17.308	4.7889	3.7986	109.95	17.193	0.62997	0.049182	0.83581	1.3279	0.027959	0.001836	0.0041328	515
520	3.769	803.53	18.9	4.8375	3.9257	107.57	17.377	0.62468	0.050442	0.83301	1.3524	0.026784	0.001909	0.0042738	520
525	4.1019	796.13	20.617	4.8901	4.0622	105.27	17.564	0.61908	0.051756	0.83154	1.3786	0.025608	0.0019876	0.0044285	525
530	4.4569	788.53	22.47	4.9471	4.209	103.05	17.755	0.61315	0.05313	0.83143	1.4066	0.02443	0.0020727	0.0045986	530
535	4.8349	780.71	24.469	5.0092	4.3677	100.89	17.949	0.60688	0.054575	0.83275	1.4365	0.023253	0.0021652	0.0047862	535
540	5.2369	772.66	26.627	5.077	4.54	98.792	18.149	0.60026	0.056102	0.83558	1.4688	0.022077	0.0022659	0.0049936	540
545	5.664	764.36	28.956	5.1513	4.7277	96.746	18.353	0.59329	0.057723	0.84001	1.5031	0.020902	0.0023764	0.0052239	545

550	6.1172	755.81	31.474	5.2331	4.9332	94.746	18.563	0.58595	0.059456	0.84616	1.5402	0.01973	0.002498	0.0054805	550
555	6.5976	746.97	34.198	5.3235	5.1594	92.785	18.781	0.57826	0.061321	0.85418	1.5802	0.018561	0.0026326	0.0057678	555
560	7.1062	737.83	37.147	5.4239	5.4099	90.857	19.007	0.57021	0.063341	0.86425	1.6234	0.017396	0.0027826	0.0060909	560
565	7.6444	728.36	40.346	5.5361	5.6889	88.956	19.242	0.56181	0.065549	0.87658	1.67	0.016236	0.0029508	0.0064566	565
570	8.2132	718.53	43.822	5.6624	6.002	87.074	19.489	0.55308	0.067981	0.89146	1.7207	0.015082	0.0031409	0.0068731	570
575	8.814	708.3	47.607	5.8055	6.356	85.206	19.749	0.54405	0.070685	0.90923	1.7758	0.013937	0.0033575	0.0073509	575
580	9.448	697.64	51.739	5.9691	6.7598	83.342	20.024	0.53474	0.073721	0.93033	1.8361	0.0128	0.0036067	0.007904	580
585	10.117	686.48	56.265	6.1579	7.2252	81.477	20.318	0.52519	0.077163	0.95534	1.9025	0.011673	0.0038965	0.0085503	585
590	10.821	674.78	61.242	6.3784	7.7679	79.6	20.634	0.51543	0.081108	0.98504	1.9762	0.010559	0.0042377	0.0093144	590
595	11.563	662.45	66.738	6.6393	8.4096	77.703	20.976	0.50551	0.085682	1.0205	2.0588	0.0094591	0.0046454	0.01023	595
600	12.345	649.41	72.842	6.9532	9.1809	75.773	21.35	0.49546	0.091052	1.0634	2.1528	0.0083756	0.0051415	0.011345	600
605	13.167	635.53	79.669	7.3391	10.127	73.798	21.765	0.4853	0.097442	1.116	2.2619	0.0073112	0.0057587	0.012731	605
610	14.033	620.65	87.369	7.8268	11.315	71.759	22.229	0.47503	0.10517	1.1823	2.3917	0.006269	0.0065497	0.014496	610
615	14.943	604.55	96.15	8.4674	12.857	69.632	22.759	0.46465	0.11468	1.2689	2.5517	0.0052528	0.0076048	0.016815	615
620	15.901	586.88	106.31	9.3541	14.945	67.382	23.374	0.4541	0.12666	1.388	2.758	0.0042676	0.0090924	0.019997	620
625	16.908	567.09	118.29	10.673	17.944	64.951	24.109	0.44338	0.14223	1.5635	3.0417	0.0033194	0.011355	0.024628	625
630	17.969	544.25	132.84	12.827	22.658	62.244	25.018	0.43251	0.16344	1.8459	3.4683	0.0024169	0.015162	0.032006	630
635	19.086	516.71	151.35	16.795	31.271	59.101	26.208	0.42189	0.19479	2.3528	4.2073	0.0015728	0.022437	0.045668	635
640	20.265	481.53	177.15	25.942	52.586	55.247	27.938	0.41493	0.25001	3.4542	5.8764	0.00080882	0.039706	0.07995	640
645	21.515	425.05	224.45	93.35	191.32	49.357	31.348	0.46136	0.42459	9.9868	14.125	0.00017569	0.171	0.30797	645
647.1	22.064	322	322	—	—	39.43	39.43	0.19748	0.19748	—	—	0	—	—	647.1

* Property values generated from NIST Database 23: *REFPROP* Version 7.0 (August 2002).

TABLE G.2a R-134a (1,1,1,2-Tetrafluoroethane) – Saturated*

T (K)	ρ (kg/m^3)	c_p (kJ/kg · K)	μ (μPa · s)	ν (m^2/s)	k (W/m · K)	α (m^2/s)	Pr	β (1/K)
170	1590. 7	1.1838	2139.7	1.345E–06	0.14515	7.708E–08	17.45	0.001658
180	1564.2	1.1871	1479.1	9.456E–07	0.1391	7.492E–08	12.62	0.0017
190	1537.5	1.1950	1106.2	7.195E–07	0.1333	7.257E–08	9.91	0.001748
200	1510.5	1.2058	867.3	5.742E–07	0.1277	7.014E–08	8.19	0.001802
210	1483.1	1.2186	702.3	4.735E–07	0.1224	6.771E–08	6.99	0.001864
220	1455.2	1.2332	582.2	4.001E–07	0.1172	6.529E–08	6.13	0.001934
230	1426.8	1.2492	491.2	3.443E–07	0.1121	6.292E–08	5.47	0.002017
240	1397.7	1.2669	420.2	3.006E–07	0.1073	6.058E–08	4.96	0.002113
250	1367.9	1.2865	363.3	2.656E–07	0.1025	5.827E–08	4.56	0.002226
260	1337.1	1.3082	316.6	2.368E–07	0.0979	5.598E–08	4.23	0.002361
270	1305.1	1.3326	277.5	2.127E–07	0.0934	5.371E–08	3.96	0.002525
280	1271.8	1.3606	244.3	1.927E–07	0.0890	5.143E–08	3.74	0.002726
290	1236.8	1.3933	215.6	1.744E–07	0.0846	4.912E–08	3.55	0.002979
300	1199.7	1.4324	190.5	1.588E–07	0.0803	4.675E–08	3.40	0.003303
310	1159.9	1.4807	168.0	1.449E–07	0.0761	4.429E–08	3.27	0.003733
320	1116.8	1.5426	147.8	1.323E–07	0.0718	4.167E–08	3.18	0.004326
330	1069.1	1.6267	129.2	1.209E–07	0.0675	3.879E–08	3.12	0.005191
340	1015.0	1.7507	111.8	1.102E–07	0.0631	3.549E–08	3.10	0.006567
350	951.32	1.9614	95.1	9.996E–08	0.0586	3.14E–08	3.18	0.009102
360	870.11	2.4368	78.1	8.981E–08	0.0541	2.55E–08	3.52	0.015393
370	740.32	5.1048	58.0	7.829E–08	0.0518	1.37E–08	5.72	0.055237

* Property values generated from NIST Database 23: *REFPROP* Version 7.0 (August 2002).

TABLE G.2b Engine Oil (Unused) – Saturated*

T (K)	ρ (kg/m^3)	c_p (kJ/kg · K)	μ (μPa · s)	ν (m^2/s)	k (W/m · K)	α (m^2/s)	Pr	β (1/K)
273	899.1	1.796	3.85	0.00428	0.147	9.10E–08	47,000	0.0007
280	895.3	1.827	2.17	0.00243	0.144	8.80E–08	27,500	0.0007
290	890	1.868	0.999	0.00112	0.145	8.72E–08	12,900	0.0007
300	884.1	1.909	0.486	0.00055	0.145	8.59E–08	6,400	0.0007
310	877.9	1.951	0.253	0.000288	0.145	8.47E–08	3,400	0.0007
320	871.8	1.993	0.141	0.000161	0.143	8.23E–08	1,965	0.0007
330	865.8	2.035	0.0836	0.0000966	0.141	8.00E–08	1,205	0.0007
340	859.9	2.076	0.0531	0.0000617	0.139	7.79E–08	793	0.0007
350	853.9	2.118	0.0356	0.0000417	0.138	7.63E–08	546	0.0007
360	847.8	2.161	0.0252	0.0000297	0.138	7.53E–08	395	0.0007
370	841.8	2.206	0.0186	0.000022	0.137	7.38E–08	300	0.0007
380	836.0	2.250	0.0141	0.0000169	0.136	7.23E–08	233	0.0007
390	830.6	2.294	0.0110	0.0000133	0.135	7.09E–08	187	0.0007
400	825.1	2.337	0.00874	0.0000106	0.134	6.95E–08	152	0.0007
410	818.9	2.381	0.00698	0.00000852	0.133	6.82E–08	125	0.0007
420	812.1	2.427	0.00564	0.00000694	0.133	6.75E–08	103	0.0007
430	806.5	2.471	0.0047	0.00000583	0.132	6.62E–08	88	0.0007

* Property values from Incropera, F. P., and DeWitt, D. P., *Fundamentals of Heat and Mass Transfer*, 3rd edn, Wiley, New York, 1990.

TABLE G.2c Ethylene Glycol ($C_2H_4(OH)_2$) – Saturated*

T (K)	ρ (kg/m^3)	c_p (kJ/kg·K)	μ (Pa·s)	ν (m^2/s)	k (W/m·K)	α (m^2/s)	Pr	β (1/K)
273	1130.8	2.294	0.0651	0.0000576	0.242	9.33E–08	617	0.00065
280	1125.8	2.323	0.0420	0.0000373	0.244	9.33E–08	400	0.00065
290	1118.8	2.368	0.0247	0.0000221	0.248	9.36E–08	236	0.00065
300	1114.4	2.415	0.0157	0.0000141	0.252	9.39E–08	151	0.00065
310	1103.7	2.460	0.0107	0.00000965	0.255	9.39E–08	103	0.00065
320	1096.2	2.505	0.00757	0.00000691	0.258	9.40E–08	73.5	0.00065
330	1089.5	2.549	0.00561	0.00000515	0.260	9.36E–08	55.0	0.00065
340	1083.8	2.592	0.00431	0.00000398	0.261	9.29E–08	42.8	0.00065
350	1079.0	2.637	0.00342	0.00000317	0.261	9.17E–08	34.6	0.00065
360	1074.0	2.682	0.00278	0.00000259	0.261	9.06E–08	28.6	0.00065
370	1066.7	2.728	0.00228	0.00000214	0.262	9.00E–08	23.7	0.00065
373	1058.5	2.742	0.00215	0.00000203	0.263	9.06E–08	22.4	0.00065

* Property values from Incropera, F. P., and DeWitt, D. P., *Fundamentals of Heat and Mass Transfer*, 3rd edn, Wiley, New York, 1990.

TABLE G.2d Glycerin ($C_3H_5(OH)_3$) – Saturated*

T (K)	ρ (kg/m^3)	c_p (kJ/kg·K)	μ (μPa·s)	ν (m^2/s)	k (W/m·K)	α (m^2/s)	Pr	β (1/K)
273	1276.0	2.261	10.6	0.00831	0.282	9.77E–08	85,000	0.00047
280	1271.9	2.298	5.34	0.00420	0.284	9.72E–08	43,200	0.00047
290	1265.8	2.367	1.85	0.00146	0.286	9.55E–08	15,300	0.00048
300	1259.9	2.427	0.799	0.000634	0.286	9.35E–08	6,780	0.00048
310	1253.9	2.490	0.352	0.000281	0.286	9.16E–08	3,060	0.00049
320	1247.2	2.564	0.210	0.000168	0.287	8.97E–08	1,870	0.00050

* Property values from Incropera, F. P., and DeWitt, D. P., *Fundamentals of Heat and Mass Transfer*, 3rd edn, Wiley, New York, 1990.

TABLE G.2e Mercury (Hg) – Saturated*

T (K)	ρ (kg/m^3)	c_p (kJ/kg·K)	μ (μPa·s)	ν (m^2/s)	k (W/m·K)	α (m^2/s)	Pr	β (1/K)
273	13,595	0.1404	0.001688	1.240E–07	8.18	4.285E–06	0.0290	0.000181
300	13,529	0.1393	0.001523	1.125E–07	8.54	4.530E–06	0.0248	0.000181
350	13,407	0.1377	0.001309	9.76E–08	9.18	4.975E–06	0.0196	0.000181
400	13,287	0.1365	0.001171	8.82E–08	9.80	5.405E–06	0.0163	0.000181
450	13,167	0.1357	0.001075	8.16E–08	10.40	5.810E–06	0.0140	0.000181
500	13,048	0.1353	0.001007	7.71E–08	10.95	6.190E–06	0.0125	0.000182
550	12,929	0.1352	0.000953	7.37E–08	11.45	6.555E–06	0.0112	0.000184
600	12,809	0.1355	0.000911	7.11E–08	11.95	6.880E–06	0.0103	0.000187

* Property values from Incropera, F. P., and DeWitt, D. P., *Fundamentals of Heat and Mass Transfer*, 3rd edn, Wiley, New York, 1990.

APPENDIX H

Thermo-Physical Properties of Hydrocarbon Fuels

TABLE H.1 Selected Properties of Hydrocarbon Fuels: Enthalpy of Formation,[a] Gibbs Function of Formation,[a] Entropy,[a] and Higher and Lower Heating Values All at 298.15 K and 1 atm; Boiling Points[b] and Latent Heat of Vaporization[c] at 1 atm; Constant-Pressure Adiabatic Flame Temperature at 1 atm;[d] Liquid Density[c]

Formula	Fuel	Molecular weight (kg/kmol)	$\bar{h}_f^\circ$ (kJ/kmol)	$\Delta\bar{g}_f^\circ$ (kJ/kmol)	$\bar{s}^\circ$ (kJ/kmol·K)	*HHV** (kJ/kg)	*LHV** (kJ/kg)	Boiling pt. (°C)	h_{fg} (kJ/kg)	$T_{ad}^\dagger$ (K)	$\rho_{liq}^\ddagger$ (kg/m³)
CH_4	Methane	16.043	−74,831	−50,794	186.188	55,528	50,016	−164	509	2226	300
C_2H_2	Acetylene	26.038	226,748	209,200	200.819	49,923	48,225	−84	—	2539	—
C_2H_4	Ethene	28.054	52,283	68,124	219.827	50,313	47,161	−103.7	—	2369	—
C_2H_6	Ethane	30.069	−84,667	−32,886	229.492	51,901	47,489	−88.6	488	2259	370
C_3H_6	Propene	42.080	20,414	62,718	266.939	48,936	45,784	−47.4	437	2334	514
C_3H_8	Propane	44.096	−103,847	−23,489	269.910	50,368	46,357	−42.1	425	2267	500
C_4H_8	1-Butene	56.107	1172	72,036	307.440	48,471	45,319	−63	391	2322	595
C_4H_{10}	*n*-Butane	58.123	−124,733	−15,707	310.034	49,546	45,742	−0.5	386	2270	579
C_5H_{10}	1-Pentene	70.134	−20,920	78,605	347.607	48,152	45,000	30	358	2314	641
C_5H_{12}	*n*-Pentane	72.150	−146,440	−8201	348.402	49,032	45,355	36.1	358	2272	626
C_6H_6	Benzene	78.113	82,927	129,658	269.199	42,277	40,579	80.1	393	2342	879
C_6H_{12}	1-Hexene	84.161	−41,673	87,027	385.974	47,955	44,803	63.4	335	2308	673
C_6H_{14}	*n*-Hexane	86.177	−167,193	209	386.811	48,696	45,105	69	335	2273	659
C_7H_{14}	1-Heptene	98.188	−62,132	95,563	424.383	47,817	44,665	93.6	—	2305	—
C_7H_{16}	*n*-Heptane	100.203	−187,820	8745	425.262	48,456	44,926	98.4	316	2274	684
C_8H_{16}	1-Octene	112.214	−82,927	104,140	462.792	47,712	44,560	121.3	—	2302	—
C_8H_{18}	*n*-Octane	114.230	−208,447	17,322	463.671	48,275	44,791	125.7	300	2275	703
C_9H_{18}	1-Nonene	126.241	−103,512	112,717	501.243	47,631	44,478	—	—	2300	—
C_9H_{20}	*n*-Nonane	128.257	−229,032	25,857	502.080	48,134	44,686	150.8	295	2276	718
$C_{10}H_{20}$	1-Decene	140.268	−124,139	121,294	539.652	47,565	44,413	170.6	—	2298	—
$C_{10}H_{22}$	*n*-Decane	142.284	−249,659	34,434	540.531	48,020	44,602	174.1	277	2277	730
$C_{11}H_{22}$	1-Undecene	154.295	−144,766	129,830	578.061	47,512	44,360	—	—	2296	—
$C_{11}H_{24}$	*n*-Undecane	156.311	−270,286	43,012	578.940	47,926	44,532	195.9	265	2277	740
$C_{12}H_{24}$	1-Dodecene	168.322	−165,352	138,407	616.471	47,468	44,316	213.4	—	2295	—
$C_{12}H_{26}$	*n*-Dodecane	170.337	−292,162	—	—	47,841	44,467	216.3	256	2277	749

*Based on gaseous fuel.

† For stoichiometric combustion with air (79% N_2, 21% O_2).

‡ For liquids at 20 °C or for gases at the boiling point of the liquefied gas.

Sources:

[a] Rossini, F. D., *et al.*, *Selected Values of Physical and Thermodynamic Properties of Hydrocarbons and Related Compounds*, Carnegie Press, Pittsburgh, PA, 1953.

[b] Weast, R. C. (Ed.), *Handbook of Chemistry and Physics,* 56th edn., CRC Press, Cleveland, OH, 1976.

[c] Obert, E. F., *Internal Combustion Engines and Air Pollution*, Harper & Row, New York, 1973.

[d] Turns, S. R., *An Introduction to Combustion,* 3rd edn, McGraw-Hill, New York, 2012.

TABLE H.2 Curve-Fit Coefficients for Fuel Specific Heat and Enthalpy[a] for Reference State of Zero Enthalpy of the Elements at 298.15 K and 1 atm:

$$\bar{c}_p\,(\text{kJ/kmol·K}) = 4.181\left(a_1 + a_2\theta + a_3\theta^2 + a_4\theta^3 + a_5\theta^{-2}\right),$$
$$\bar{h}^\circ\,(\text{kJ/kmol·K}) = 4181\left(a_1\theta + a_2\theta^2/2 + a_3\theta^3/3 + a_4\theta^4/4 - a_5\theta^{-1} + a_6\right),$$
$$\text{where } \theta \equiv T(\text{K})/1000.$$

Formula	Fuel	Molecular Weight	a_1	a_2	a_3	a_4	a_5	a_6	a_8^*
CH_4	Methane	16.043	−0.29149	26.327	−10.610	1.5656	0.16573	−18.331	4.300
C_3H_8	Propane	44.096	−1.4867	74.339	−39.065	8.0543	0.01219	−27.313	8.852
$C_6H_{!4}$	Hexane	86.177	−20.777	210.48	−164.125	52.832	0.56635	−39.836	15.611
C_8H_{18}	Isooctane	114.230	−0.55313	181.62	−97.787	20.402	−0.03095	−60.751	20.232
CH_3OH	Methanol	32.040	−2.7059	44.168	−27.501	7.2193	0.20299	−48.288	5.3375
C_2H_5OH	Ethanol	46.07	6.990	39.741	−11.926	0	0	−60.214	7.6135
$C_{8.26}H_{15.5}$	Gasoline	114.8	−24.078	256.63	−201.68	64.750	0.5808	−27.562	17.792
$C_{7.76}H_{13.1}$		106.4	−22.501	227.99	−177.26	56.048	0.4845	−17.578	15.232
$C_{10.8}H_{18.7}$	Diesel	148.6	−9.1063	246.97	−143.74	32.329	0.0518	−50.128	23.514

* To obtain 0 K reference state for enthalpy, add a_8 to a_6.

[a] *Source:* From Heywood, J. B., *Internal Combustion Engine Fundamentals*, McGraw-Hill, New York, 1988, by permission of McGraw-Hill, Inc.

TABLE H.3 Curve-Fit Coefficients for Fuel Vapor Thermal Conductivity, Viscosity, and Specific Heat:[a]

$$\left.\begin{array}{l} k(\mathrm{W/m\cdot K}) \\ \mu\left(\mathrm{N\cdot s/m^2}\right)\times 10^6 \\ c_p(\mathrm{J/kg\cdot K}) \end{array}\right\} = a_1 + a_2T + a_3T^2 + a_4T^3 + a_5T^4 + a_6T^5 + a_7T^6$$

Formula	Fuel	T range (K)	Property	a_1	a_2	a_3	a_4	a_5	a_6	a_7
CH_4	Methane	100–1000	k	−1.34014990E−2	3.66307060E−4	−1.82248608E−6	5.93987998E−9	−9.14055050E−12	−6.78968890E−15	−1.95048736E−18
		70–1000	μ	2.96826700E−1	3.71120100E−2	1.21829800E−5	−7.02426000E−8	7.54326900E−11	−2.72371660E−14	0
			c_p	See Table D.2						
C_3H_8	Propane	200–500	k	−1.07682209E−2	8.38590325E−5	4.22059864E−8	0	0	0	0
		270–600	μ	−3.54371100E−1	3.08009600E−2	−6.99723000E−6	0	0	0	0
			c_p	See Table D.2						
C_6H_{14}	*n*-Hexane	150–1000	k	1.28775700E−3	−2.00499443E−5	2.37858831E−7	−1.60944555E−10	7.71027290E−14	0	0
		270–900	μ	1.54541200E+0	1.15080900E−2	2.72216500E−5	−3.26900000E−8	1.24545900E−11	0	0
			c_p	See Table D.2						
C_7H_{16}	*n*-Heptane	250–1000	k	−4.60614700E−2	5.95652224E−4	−2.98893153E−6	8.44612876E−9	−1.22927E−11	9.0127E−15	−2.62961E−18
		270–580	μ	1.54009700E+0	1.09515700E−2	1.80066400E−5	−1.36379000E−8	0	0	0
		300–755	c_p	9.46260000E+1	5.86099700E+0	−1.98231320E−3	−6.88699300E−8	−1.93795260E−10	0	0
		755–1365	c_p	−7.40308000E+2	1.08935370E+1	−1.26512400E−2	9.84376300E−6	−4.32282960E−9	7.86366500E−13	0
C_8H_{18}	*n*-Octane	250–500	k	−4.01391940E−3	3.38796092E−5	8.19291819E−8	0	0	0	0
		300–650	μ	8.32435400E−1	1.40045000E−2	8.79376500E−6	−6.84030000E−9	0	0	0
		275–755	c_p	2.14419800E+2	5.35690500E+0	−1.17497000E−3	−6.99115500E−7	0	0	0
		755–1365	c_p	2.43596860E+3	−4.46819470E+0	−1.66843290E−2	−1.78856050E−5	8.64282020E−9	−1.61426500E−12	0
$C_{10}H_{22}$	*n*-Decane	250–500	k	−5.88274000E−3	3.72449646E−5	7.55109624E−8	0	0	0	0
			μ	Not available						
		300–700	c_p	2.40717800E+2	5.09965000E+0	−6.29026000E−4	−1.07155000E−6	0	0	0
		700–1365	c_p	−1.35345890E+4	9.14879000E+1	−2.20700000E−1	2.91406000E−4	−2.15307400E−7	8.38600000E−11	−1.34404000E−14
CH_3OH	Methanol	300–550	k	−2.02986750E−2	1.21910927E−4	−2.23748473E−8	0	0	0	0
		250–650	μ	1.19790000E+0	2.45028000E−2	1.86162740E−5	−1.30674820E−8	0	0	0
			c_p	See Table D.2						
C_2H_5OH	Ethanol	250–550	k	−2.46663000E−2	1.55892550E−4	−8.22954822E−8	0	0	0	0
		270–600	μ	−6.33595000E−2	3.20713470E−2	−6.25079576E−6	0	0	0	0
			c_p	See Table D.2						

[a] *Source*: Andrews, J. R., and Biblarz, O., "Temperature Dependence of Gas .Properties in Polynomial Form," Naval Postgraduate School, NPS67–81–001, January 1981.

APPENDIX I
Psychrometric Charts

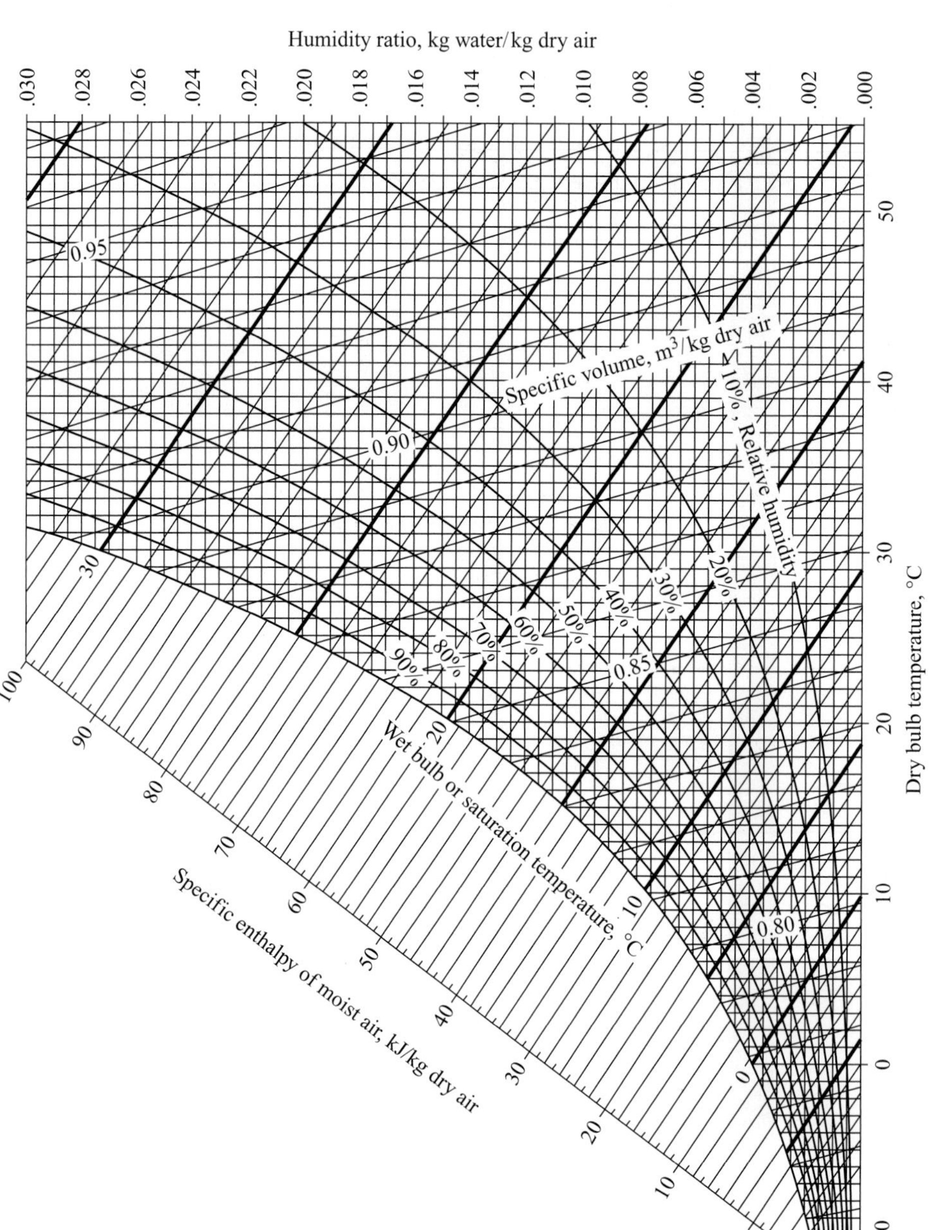

FIGURE I.1 Adapted with permission from Z. Zhang and M.B. Pate, "A Methodology for Implementing a Psychrometric Chart in a Computer Graphics System," ASHRAE Transactions, Vol. 94, Pt. 1, 1988.

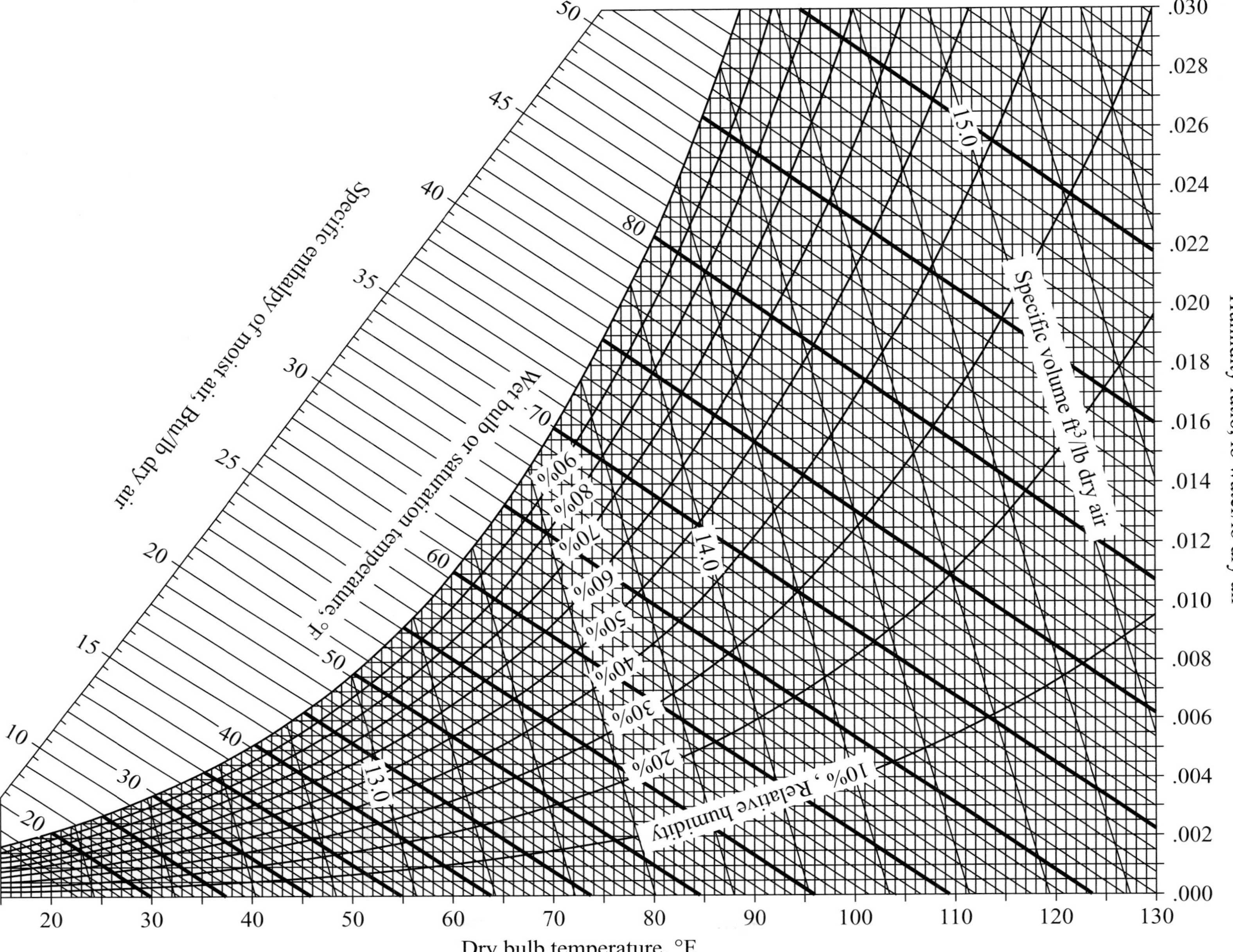

FIGURE I.2 Adapted with permission from Z. Zhang and M.B. Pate, "A Methodology for Implementing a Psychrometric Chart in a Computer Graphics System," ASHRAE Transactions, Vol. 94, Pt. 1, 1988.

Index